CHEMISTRY
Of The ALKALOIDS

Betalains account for red-violet and yellow flower colors in plants belonging to ten closely related plant families, one of which is the Cactaceae. Typical betalains are present in the yellow and red flowers of the two *Opuntia* cacti shown.
(Photographs by Dr. H. Rösler)

CHEMISTRY Of The ALKALOIDS

Edited by

S. W. Pelletier

Alumni Foundation Distinguished Professor, Department of Chemistry, The University of Georgia, Athens, Georgia

VNR

VAN NOSTRAND REINHOLD COMPANY

New York/Cincinnati/Toronto/London/Melbourne

Van Nostrand Reinhold Company Regional Offices:
New York Cincinnati Chicago Millbrae Dallas

Van Nostrand Reinhold Company Foreign Offices:
London Toronto Melbourne

Library of Congress Catalog Card Number: 71-108649

Manufactured in the United States of America

Published by Van Nostrand Reinhold Company
450 West 33rd Street, New York, N.Y. 10001

Published simultaneously in Canada by
D. Van Nostrand Company (Canada), Ltd.

15 14 13 12 11 10 9 8 7 6 5 4 3 2 1

Preface

The aim of this book is to provide the reader with a modern one-volume survey of the chemistry of the most important classes of alkaloids. The book is not encyclopedic in scope, but rather provides a selection of reactions and syntheses representative of each class of alkaloids. We sought to meet the needs of the advanced undergraduate, graduate student, postdoctorate and specialist for an abbreviated yet authoritative treatment of the major topics of alkaloid chemistry. Twenty years ago T. A. Henry's fourth edition of "Plant Alkaloids" served these needs admirably. Today when the alkaloid chemist relies so heavily on physical tools such as infrared, nmr and mass spectroscopy, optical rotatory dispersion, circular dichroism and x-ray crystallography for structural elucidation, a modern book on alkaloids must reflect this change in approach and methodology. Yet the "spectroscopic approach" can be overdone in a survey-type book. We have tried to achieve a balance by a judicious mix of the "classical" with the "modern." Chapters treating alkaloids most of whose structures were elucidated many years ago will naturally be heavy on "classical" content, while those chapters treating structural developments of recent years will reflect "modern" methods. We are aware that what we believe is a reasonable balance will not please every reader.

When this project was initiated several years ago there was no single-volume in English covering this field. Since then a fine, though rather brief, text has appeared. Our book provides considerably more detailed treatment and in addition devotes a chapter each to several topics omitted or only briefly mentioned in the aforementioned book, e.g., to our knowledge ours is the first book on alkaloids to include a chapter (13) on the very interesting betalains, the red-violet and yellow alkaloids, which occur abundantly in plants and which escaped structure elucidation until the 1960's. Also included are topics on: the Lycopodium alkaloids (Chapter 16), the C_{20}-diterpene alkaloids (Chapter 17), the C_{19}-diterpene alkaloids (Chapter 18) and the steroid alkaloids (Chapters 19 and 20). In addition an unusual chapter on alkaloids and plant systematics (Chapter 22) is provided. We hope this chapter will give the reader not only some insights into the goals of the systematist, but also an awareness of the potential and limitations of chemical data as an aid in the study of botanical systematics.

The organization of this book deserves comment. The alkaloids are organized in four distinct groups:

I. Alkaloids Containing Phenylethylamine, Isoquinolines or Related Structures (Chapters 2–8).

II. Alkaloids Containing an Indole or a Quinoline System (Chapters 9–13).

III. Alkaloids Containing Pyrrolidine, Piperidine, Pyridine Rings and Related Structures (Chapters 14–16).
IV. Terpene and Steroidal Alkaloids (Chapters 17–20).

Two final chapters treat biosynthesis (Chapter 21) and plant systematics (Chapter 22). Some readers may be surprised to see the chapter on Lycopodium alkaloids under Section III rather than II. Because the Lycopodium alkaloids are now known to be piperidine dimers, derived from lysine, it was felt that their biogenesis could be emphasized by including them among the piperidine class. With regard to biogenesis, the editor would especially like to draw attention to the lucid and unusual account of this subject by Prof. Ian D. Spenser. A reviewer in commenting on this subject stated, "... the quality of the chapter is such that many chemists and biochemists will buy the book for this chapter alone." We agree.

Three indexes have been prepared to enhance the usefulness of this book: *Author; Reaction and Reagent; Subject.* In the *Author Index* an author's name is cited not only on the page or pages where his name appears, but also wherever a reference to his work is cited. In the *Reaction and Reagent Index* we have listed most important synthetic, degradative and name reactions as well as important reagents and catalysts. The alkaloid literature is a rich source of new reactions and applications of old reactions to new problems. A rather complete listing of major reaction types and reagents should, therefore, prove of benefit to those who wish to use the book as a source of synthetic reactions for either research or teaching purposes. This index is also illustrative of the variety of reactions which can be effected by a given reagent or catalyst. The *Subject Index* lists all important subjects and compounds mentioned in the book as well as plant specie, genera and family names.

Twenty-two different authors contributed to this book. In any multi-authored book one can expect a certain degree of disunity. This book is no exception. Yet we believe that the advantages of authoritativeness and accuracy brought to the work by having experts in each field prepare the chapters, far outweigh the disunity which inevitably results. We hope the reader agrees. Worth mentioning is the fact that each chapter was independently reviewed by an expert in the field.

Special thanks are due the following chemists who reviewed and made very helpful suggestions regarding the chapters: Drs. W. Ayer, K. Brown, A. Dreiding, O. E. Edwards, H. Fales, R. Hill, J. Coke, E. Leete, G. Lyle, H. T. Openshaw, M. Shamma, Y. Sato, W. I. Taylor, K. Wiesner and W. Wildman.

S. W. Pelletier

Athens, Georgia
December 21, 1969

Contents

1

Introduction

S. W. PELLETIER
Department of Chemistry,
University of Georgia,
Athens, Georgia

THE PLANT ALKALOIDS—GENERAL INFORMATION

Definition

From ancient times man has utilized alkaloids as medicines, poisons, and magical potions. Only recently has he gained precise knowledge about the chemical structures of many of these interesting compounds. The term *alkaloid*, or "alkali-like," was first proposed by the pharmacist, W. Meissner, in 1819. It is usually applied to basic, nitrogen-containing compounds of plant origin. Two further qualifications are usually added to this definition—alkaloids have complex molecular structures, and they manifest significant pharmacological activity. Such compounds occur only in certain genera and families, rarely being universally distributed in large groups of plants. Chemical, pharmacological, and botanical properties must all be considered when classifying a compound as an alkaloid. Examples of well-known alkaloids are *morphine* (opium poppy), *nicotine* (tobacco), *quinine* (cinchona bark), *reserpine* (rauwolfia), and *strychnine* (strychnos nux-vomica) (Figure 1-1).

It should be emphasized that many widely distributed bases of plant origin, such as methyl, trimethyl and other open-chain simple alkylamines, the cholines, and the betaines are not classed as alkaloids. These are designated by some authorities as "biological amines" or "protoalkaloids." Certain authorities also class the phenylalkylamines among the "biological amines" though in this book they will be treated as genuine alkaloids. In certain cases, the distinction drawn between the "biological amines" and

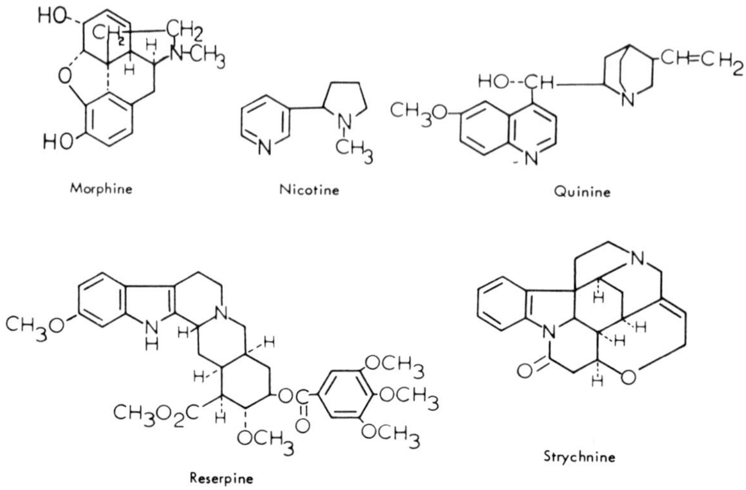

Figure 1-1

alkaloids is rather arbitrary. Alkaloids *usually* have a rather complex structure with the nitrogen atom involved in a heterocyclic ring. Yet *thiamine*, a heterocyclic nitrogeneous base, is not regarded as an alkaloid mainly because of its almost universal distribution in living matter. Other heterocyclic nitrogenous bases not classed as alkaloids by some authorities include the purines, of which caffeine, xanthine, and theobromine are examples. Interestingly, *colchicine* (Chapter 8) is classed as an alkaloid even though it is not basic, and its nitrogen atom is not incorporated in a heterocyclic ring, because of its particular pharmacological activity and limited distribution in the plant world.

Occurrence and Distribution

Today a little over two thousand alkaloids are known, and it is estimated that they are present in only 10–15% of all vascular plants. They are rarely found in cryptogamia (exception—ergot alkaloids), gymnosperms, or monocotyledons. They occur abundantly in certain dicotyledons and particularly in the families: APOCYNACEAE (dogbane, quebracho, pereiro bark), PAPAVERACEAE (poppies, chelidonium), PAPILIONACEAE (lupins, butterfly-shaped flowers), RANUNCULACEAE (aconitum, delphinium), RUBIACEAE (cinchona bark, ipecacuanha), RUTACEAE (citrus, fagara) and SOLANACEAE (tobacco, deadly nightshade, tomato, potato, thorn apple). Well-characterized alkaloids have been isolated from the roots, seeds, leaves or bark of some forty plant families, and it is probable that the remaining families will provide only an occasional alkaloid-bearing plant. PAPAVERA-

CEAE is an unusual family in that all of its species contain alkaloids. The majority of plant families occupy an intermediate position in which most species within a genus or closely related genera either do or do not contain alkaloids. Thus, all *Aconitum* and *Delphinium* species elaborate alkaloids whereas most of the other genera (*Anemone, Ranunculus, Trocleus*) in the family RANUNCULACEAE do not. It is generally true that a given genus or related genera yield the same or structurally related alkaloids; for example, seven different genera of SOLANACEAE contain hyoscyamine. It is also true that simple alkaloids often occur in numerous, botanically unrelated plants while the more complicated ones such as quinine, nicotine, and colchicine are usually limited to one specie or genus of plant and form a distinguishing characteristic of it.

Nomenclature and Classification

Because of the complexity of the compounds involved and for historical reasons, the nomenclature of alkaloids has not been systematized. The two commonly used systems classify alkaloids either according to the plant genera in which they occur or on the basis of similarity of molecular structure. Important classes of alkaloids containing generically related members are the aconitum, cinchona, ephedra, lupin, opium, rauwolfia, senecio, solanum, and strychnos alkaloids. Chemically derived alkaloid names are

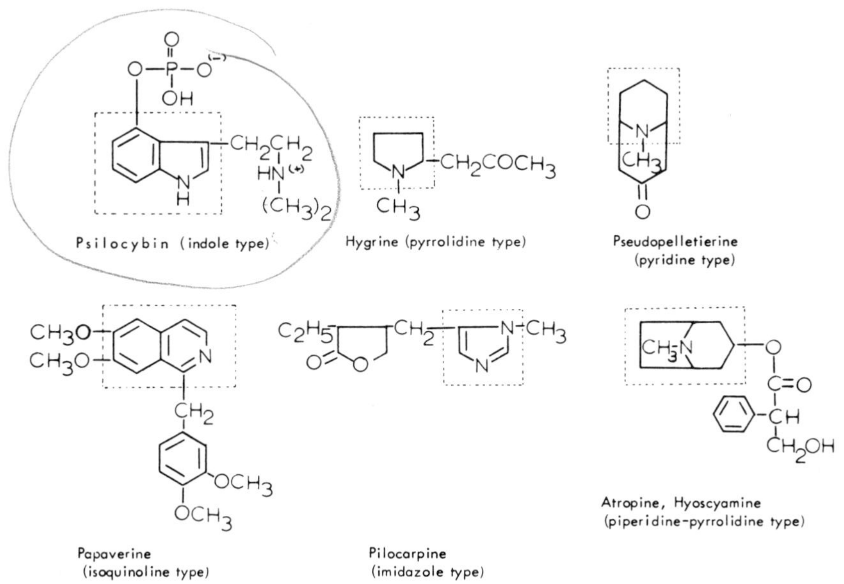

Note: The dashed lines enclose the parent chemical nucleus upon which the type name is based.

Figure 1-2

based on the skeletal feature which members of a group possess in common. Thus, indole alkaloids (e.g., psilocybin, the active principle of Mexican hallucinogenic mushrooms) contain an indole or modified indole nucleus and pyrrolidine alkaloids (e.g., hygrine) contain the pyrrolidine ring system. Other examples (Figure 1.1 and 1.2) of this classification include the pyridine, quinoline (cf. quinine), isoquinoline, imidazole, pyridine–pyrrolidine (cf. nicotine), and piperidine–pyrrolidine type alkaloids. A large number of important alkaloids have received names derived directly from those of plants—as papaverine, hydrastine, berberine. A few have been named for their physiological action, e.g., morphine, (Ger. *morphin*, God of dreams), narcotin (Gk. *narkoō*, to benumb), and emetine (Gk. *emetikos*, to vomit). Only one, the pelletierine group, has been named after a chemist—Pierre Joseph Pelletier, the discoverer of emetine (1817), colchicine (1819), quinine (1820), cinchonine (1820), strychnine (1820), brucine (1820), caffeine (1820), piperine (1821), thebaine (1835) and incidentally, also chlorophyll.

History

The beginning of alkaloid chemistry is usually dated back about 170 years when F. W. Sertürner announced the isolation of morphine in 1805. He prepared several salts of morphine and demonstrated that it was the principle responsible for the physiological effect of opium. Later (1810) Gomes treated an alcoholic extract of cinchona bark with alkali and obtained a crystalline precipitate which he named "cinchonino." Subsequent studies (1820) by P. J. Pelletier and J. B. Caventou of the Faculty of Pharmacy, Paris, showed that "cinchonino" was a mixture which they separated into two new alkaloids named quinine and cinchonine. Subsequently various investigators isolated more than two dozen additional bases from species of *Cinchona* and *Remijia*. Between 1820–1850 investigations were intensified and a large number of alkaloids of new and varied types were isolated and characterized. Among the important representatives discovered during this period were aconitine, one of the most toxic materials of plant origin known to man; atropine, a powerful mydriatic agent (4.3×10^{-6} g will cause dilation of the pupil of the eye); colchicine, the alkaloid of the meadow saffron, which is used extensively in the treatment of gout; coniine, the principle responsible for the death of Socrates when he drank the cup of poison hemlock; codeine, a close relative of morphine, and a valuable pain killer and cough repressant; hyoscyamine, the optically active form of atropine; piperine, an alkaloid of pepper plants; berberine of barberry root; strychnine, a highly poisonous alkaloid used in certain cardiac disorders and for exterminating rodents; and emetine, a powerful emetic sometimes used for treating amoebic dysentery.

Determination of Molecular Structure

Because of the complex molecular architecture of many of these compounds, little progress was made in the elucidation of their structures during the 19th century. Only within the past few years have certain of the more complex alkaloids yielded the secrets of their molecular structure to the chemist.

The first step in determining the structure of a pure alkaloid consists in ascertaining the molecular formula. This information can be obtained by a combustion analysis for the elements present (always carbon, hydrogen, nitrogen, and often oxygen) together with a molecular weight determination, or where high resolution mass spectrometry is available, by this modern technique. The chemist next proceeds to ascertain the function of oxygen and nitrogen atoms in the molecule. Oxygen is most frequently present in the form of hydroxyl or phenol ($-OH$), methoxyl ($-OCH_3$), acetoxyl ($-OCOCH_3$), benzoxyl ($-OCOC_6H_5$), carboxyl ($-CO_2H$), or ketone ($C{=}O$) groups. Oxygen in the form of a phenolic group can be recognized by alkali solubility, a color reaction with ferric chloride, acylation to an ester or alkylation to an ether; in the form of an alcohol by acetylation, dehydration or oxidation, and by characteristic absorption in the infrared in the 3.0μ region. Carboxyl groups are suggested by solubility in weak base, by esterification, and by specific absorption in the infrared. Methoxyl groups can be determined quantitatively by the method of Zeisel which involves boiling the alkaloid with hydriodic acid and determining the quantity of methyl iodide formed. The estimation of the methylenedioxy group is accomplished by reactions in which formaldehyde is split out by means of sulfuric acid. Carbonyl groups such as ketones or aldehydes may be detected by spectroscopic means (infrared, ultraviolet, nuclear magnetic resonance), as well as by standard chemical tests.

The determination of alkyl groups on nitrogen may be carried out by heating the alkaloid hydriodide salt at 200–300° and measuring the amount of alkyl iodide produced. Thus, an alkaloid bearing an $N-CH_3$ group (e.g., nicotine) will produce methyl iodide and one containing an $N-CH_2CH_3$ group (e.g., aconitine) will liberate ethyl iodide. NMR spectroscopy is also utilized for the rapid detection of $N-CH_3$ and $N-CH_2CH_3$ groups in alkaloids. In most cases the nitrogen in an alkaloid is involved in a ring structure and will usually be secondary or tertiary. Though several tests are available, it is sometimes difficult to distinguish between these forms. The most widely used method for ascertaining the environment about the nitrogen is exhaustive methylation, also known as the Hofmann degradation. This method is based on the property of quaternary ammonium hydroxides, when heated, of decomposing with loss of water and cleavage of a carbon–nitrogen linkage to give an olefin. Thus

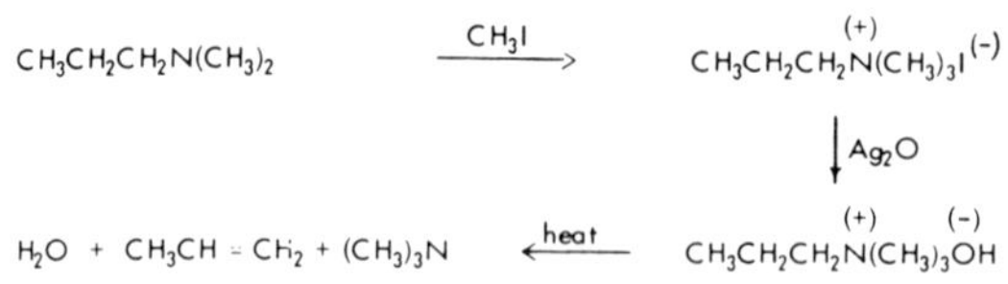

Figure 1-3

with an acyclic tertiary amine (Figure 1.3), a single methylation and decomposition eliminates the nitrogen as trimethylamine. If the nitrogen atom is involved in a cyclic structure, two or three such cycles of methylation and decomposition will be necessary to liberate the nitrogen and expose the carbon skeleton. Thus in the classic degradation of pseudopelletierine by Willstätter, the alkaloid was converted to the methohydroxide via methyl granatinine and this decomposed to yield the monoolefin, (Figure 1.4). Repetition of the methylation and decomposition yielded 1,5-

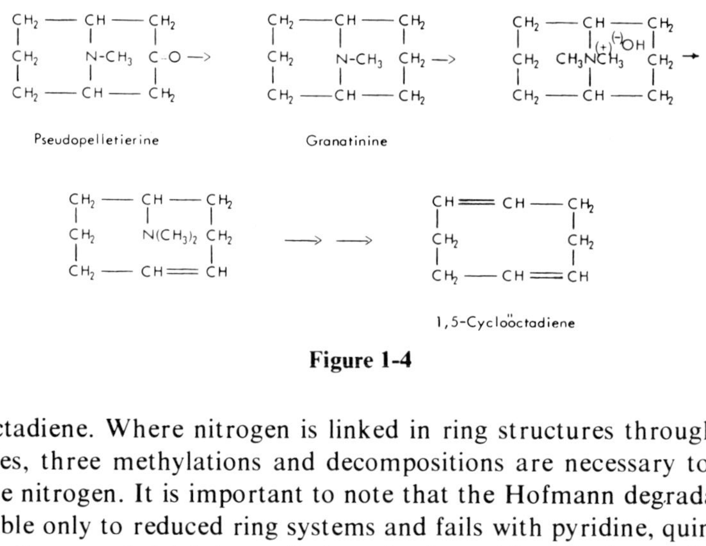

Figure 1-4

cyclooctadiene. Where nitrogen is linked in ring structures through three valencies, three methylations and decompositions are necessary to eliminate the nitrogen. It is important to note that the Hofmann degradation is applicable only to reduced ring systems and fails with pyridine, quinolines, isoquinolines, and 1,2,3,4-tetrahydroquinolines. The Emde modification of the Hofmann degradation involves treating an aqueous or alcoholic solution of the quaternary ammonium halide with sodium amalgam. Application of the Emde procedure to N,N-dimethyltetrahydroquinolinium chloride furnishes o-propyldimethylaniline. The use of the Hofmann degradation in this case does not result in ring opening, but leads to elimination of methanol and formation of N-methyltetrahydroquinoline. Another method of rupturing nitrogen-containing rings, due to von Braun, involves treating a tertiary amine with cyanogen bromide, CNBr. The bromine becomes attached to the carbon atom and the cyano group to the nitrogen. With unsymmetrical tertiary amines the product composition is determined by the nature of the substituents adjacent to the nitrogen atom.

Numerous other degradative methods, which can transform the alkaloid into stable, easily recognized compounds, are available. Thus, distillation over hot zinc dust will sometimes degrade an alkaloid to a stable aromatic derivative, e.g., morphine gives phenanthrene, and cinchonine yields quinoline and picoline. Another valuable method involves dehydrogenating the alkaloid with a catalyst such as sulfur, selenium, or palladium. During dehydrogenation peripheral groups such as hydroxyls and C-methyls are eliminated. From the dehydrogenation reaction relatively simple, easily recognized products can sometimes be isolated. These may provide a ready clue to the gross skeleton of the alkaloid. A good example of the value of this method is seen in the dehydrogenation of atisine, an alkaloid of *Aconitum heterophyllum*. The products (Figure 1.5), 1-methyl-6-ethylphenan-

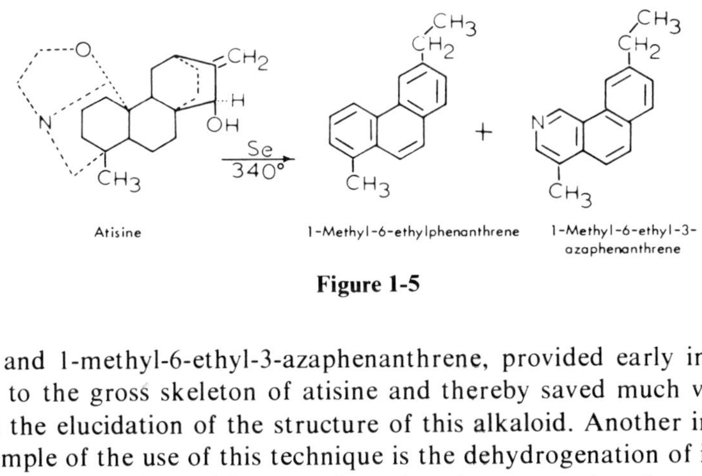

Figure 1-5

threne and 1-methyl-6-ethyl-3-azaphenanthrene, provided early information as to the gross skeleton of atisine and thereby saved much valuable time in the elucidation of the structure of this alkaloid. Another interesting example of the use of this technique is the dehydrogenation of isorubijervine, an alkaloid found in a number of *Veratrum* species. A crystalline hydrocarbon, cyclopentanophenanthrene, and a liquid aromatic amine, 2-ethyl-5-methylpyridine are the main products of the reaction and provide immediate insight into the gross structure of isorubijervine (Figure 1.6).

Other methods of degradation, which allow the removal of particular groups or the cleavage of the molecule at specific points, are available. The aim is to break the molecule into smaller identifiable fragments which will assist the chemist in visualizing the complete structure. The final part of the elucidation of an unknown alkaloid is usually concerned with the stereochemistry of the molecule, viz. the determination of the complete three-dimensional representation of all the atoms in the molecule. In the case of complex alkaloids, such degradation and detective work can often take years to complete. Fortunately, in recent times the techniques of single crystal x-ray crystallography have been refined to such a point that where a suitable heavy-atom crystalline derivative of an unknown alkaloid is available, the complete structure and stereochemistry may often be elucidated in a matter of a few weeks' time, particularly if large electronic computers are available to solve the required Fourier equations. However, even in situations where the structure of an alkaloid has been deduced by physical methods such as spectroscopy, mass spectrometry and/or x-ray crystallography, an organic chemist may still wish to explore the detailed chemistry of the compound by degradative procedures and even to develop

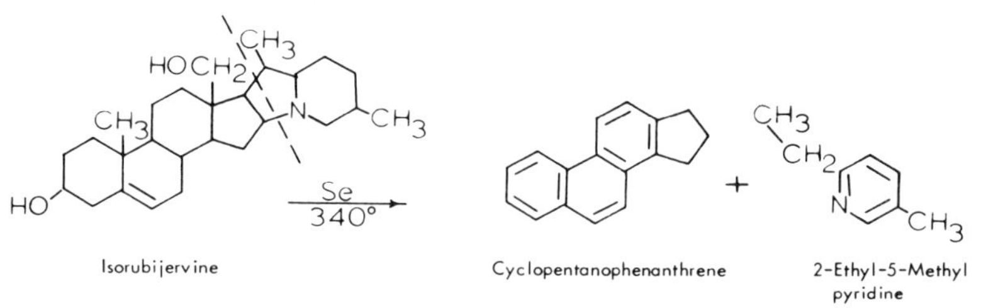

Figure 1-6

a total synthesis of the molecule. Much of our current knowledge of organic chemical reactions has been derived from detailed degradative studies of complex natural products such as alkaloids.

Biogenesis of Alkaloids

One of the most exciting and fascinating subjects pertaining to alkaloids is the mode of their synthesis in plants. Over the past few decades, chemists have proposed many biogenetic schemes for the synthesis of individual alkaloids. Most of these schemes have been based upon the idea that alkaloids are derived from relatively simple precursors such as phenylalanine, tryptophan, "acetate" units, terpene units, methionine and a few other amino acids such as ornithine. On paper at least, the structures of most alkaloids may be derived from such simple precursors using a few well-known chemical reactions. As a matter of fact several simple alkaloids have been synthesized in the laboratory from amino-acid derivatives under physiological conditions using just such biogenetic concepts. With the advent of isotopically labeled compounds, these theories have been subjected to experimental test.

The modern approach to biosynthetic studies of alkaloids involves administration of labeled precursors to selected plants and, after a suitable period of growth, isolation of the alkaloids. These are then degraded in a systematic fashion to determine the position of the labeled atoms. Using this technique, many alkaloids (e.g., morphine, nicotine, hyoscyamine, pellotine, papaverine, colchicine, gramine) have been shown to be synthesized from amino acids. The present state of research on alkaloid biogenesis represents a dramatic breakthrough. Not only are plants known to incorporate amino acids, acetate and mevalonolactone, but in specific cases large intermediates have been successfully introduced into the plant's biosynthetic system. A number of the most important of these developments are discussed in Chapter 21.

Function of Alkaloids

Unlike biogenesis, the function of alkaloids in plants remains a subject for speculation. Many authorities regard them as by-products of plant metabolism. Still others conceive of alkaloids as reservoirs for protein synthesis; as protective materials discouraging animal or insect attacks; as plant stimulants or regulators in such activities as growth, metabolism, and reproduction; or as detoxicating agents, which render harmless, by processes such as methylation, condensation, and ring closure, substances whose accumulation might otherwise cause damage to the plant. While a particular explanation may have application to a given plant, it should be remembered that 85–90% of all plants manage very well without elaborating any alkaloids. In conclusion, one might say that while much has been learned about the biogenesis and metabolism of alkaloids, their functions in the plant, if any, are still largely unknown.

GENERAL REFERENCES

1. L. F. Small, "Alkaloids," in H. Gilman, "Organic Chemistry. An Advanced Treatise," 2nd ed., Vol 2, pp. 1166–1258, John Wiley & Sons, Inc., New York, N.Y., 1943.
2. T. A. Henry, "The Plant Alkaloids," 4th ed., Blakeston, Philadelphia, Pa., 1949.
3. R. H. F. Manske and H. L. Holmes, *Alkaloids* **1–5** (1950–1955).
4. Ref. 3, **6–12** (1956–1969).
5. K. W. Bentley, "The Chemistry of the Morphine Alkaloids," Oxford Univ. Press, London, England, 1954.
6. R. Robinson, "The Structural Relations of Natural Products," Clarendon Press, Oxford, England, 1955.
7. K. W. Bentley, "The Alkaloids," Interscience Publishers, Inc., New York, N.Y., 1957.
8. E. H. Rodd, Ed., "Chemistry of Carbon Compounds," Vol. IV-C, Chaps. 23–29, pp. 1799–2130, Elsevier Publ. Co., Amsterdam, 1960.
9. A. R. Battersby, "Alkaloid Biogenesis," *Quart. Rev.* **15,** 259–286 (1961).

10. H. G. Boit, "Ergebnisse der Alkaloid—Chemie bis 1960," Akademie-Verlag, Berlin, Germany, 1961.
11. D. Ginsburg, "The Opium Alkaloids," Interscience Publishers, Inc., New York, N.Y., 1962.
12. E. Leete, "Alkaloid Biogenesis," in "Biogenesis of Natural Compounds," Chap. 17, pp. 739–796 (P. Bernfeld, Ed.), Macmillan, New York, N.Y., 1963.
13. G. L. Closs, "Alkaloids," in "Encyclopedia Britannica," pp. 637–639 (H. S. Ashmore, Ed.), W. Benton, Chicago, Ill., 1963.
14. J. B. Hendrickson, "The Alkaloids," in "The Molecules of Nature," Chap. 5, pp. 143–179, W. A. Benjamin, New York, N.Y., 1965.
15. W. I. Taylor, "Indole Alkaloids," Pergamon Press, Oxford, England, 1966.
16. S. W. Pelletier, "Alkaloids," in "The Encyclopedia of Chemistry," pp. 43–46 (G. L. Clark and G. G. Hawley, Eds.), Reinhold, New York, N.Y., 1966.
17. G. A. Swan, "An Introduction to the Alkaloids," John Wiley & Sons, Inc., New York, N.Y., 1967.
18. E. Leete, "Alkaloid Biosynthesis," *Ann. Rev. Plant Physiol.* **18,** 179–196 (1967).

2

Alkaloids Containing a Simple Aromatic Moiety

A. BROSSI AND B. PECHERER
Chemical Research Department, Hoffmann–La Roche Inc., Nutley, New Jersey

1. INTRODUCTION

With few exceptions, the alkaloids under discussion in this chapter are phenylalkylamines or derivatives thereof. The expression "simple aromatic moiety," therefore, refers almost exclusively to a phenyl or a substituted phenyl residue. Included are substances such as cinnamic acid amides or quaternary compounds, which are not alkaloids in the classical sense, but are closely related to them. No consideration is given to the antibiotic chloramphenicol, the hormone adrenaline, or to certain amino acids that would belong chemically to this class of compounds.

The manifold pharmacological properties of these relatively simple plant constituents, their biogenesis, or their botanical aspects are not discussed here, since these points receive adequate treatment in the standard texts of Winterstein and Trier,[1] Henry,[2] Guggenheim,[3] Stanek,[4] Robinson,[5] or the recent progress reports of Barton and Cohen,[6] Battersby,[7] and Fodor.[8] While this literature is mentioned occasionally, frequent references to detailed matters will be made to the classical fundamental treatises of alkaloid chemistry by Manske and Holmes[9] and Boit.[10]

2. ANILINE DERIVATIVES

2.1 Ismine

The alkaloid ismine, isolated by Highet[11] from *Amaryllidaceae* species, is the amino alcohol (3). The structure of ismine has been established by the following synthesis:[12,13]

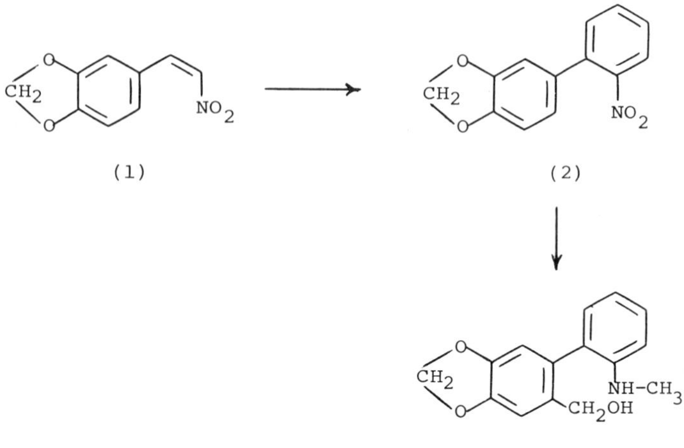

(1) (2) (3)

The Diels–Alder reaction of the nitrostyrene (1) with 1,4-diacetoxybutadiene gave the substituted biphenyl derivative (2) that was converted into ismine (3) by a sequence of standard reactions.

2.2 Anthranilic acid derivatives[10]

Anthranilic acid and N-methylanthranilic acid methyl esters are constituents of essential oils and are of importance in perfumery. Several diterpene alkaloids from *Ranunculaceae* species are amino alcohols esterified with anthranilic acid or N-acylanthranilic acids. Another naturally occurring substance of this group is damascenine (4). Ewins established its structure by synthesis from 3-methoxy-2-nitrobenzoic acid.[14] An improved synthesis of (4) and homologs was reported recently.[15]

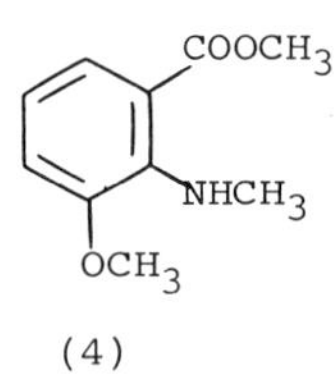

(4)

3. PHENYLALKYLAMINES

This large group can be subdivided into the following three classes: (a) benzylamines, (b) phenethylamines, and (c) α-methylphenethylamines. Most of the known phenylalkylamine alkaloids are members of class b.

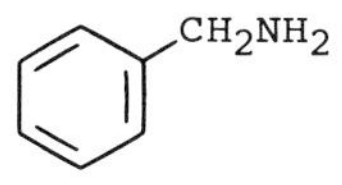

a

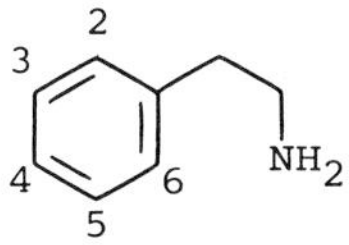

b

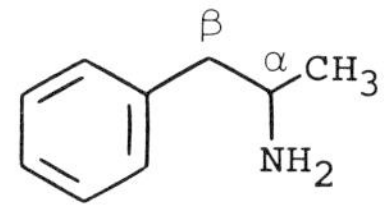

c

3.1 Benzylamines[10]

Benzylamine, which is identical with the alkaloid moringine, and N-methylbenzylamine have been isolated from various plant species. Two other representatives of this class are the nonbasic N-benzylphthalimide, which has recently been isolated from *Chondodendron* species,[16] and *p*-hydroxybenzylamine occurring in white mustard.[17]

3.1.1 Capsaicin The structure (8) for capsaicin, a pungent constituent of *Capsicum* species, was proposed by Nelson and Dawson.[18] The *trans* configuration was established by a stereospecific synthesis by Crombie *et al.*[19] starting with dihydropyran (5) that was converted via three steps into *trans*-6-methyl-4-hepten-1-ol (6). The alcohol (6) yielded the *trans*-8-methyl-6-nonenoic acid (7) on bromination and homologation. The 4-hydroxy-3-methoxy-benzylamide (8) is identical with capsaicin. The dihydro derivative of (8) has recently been isolated from paprika species.[20]*

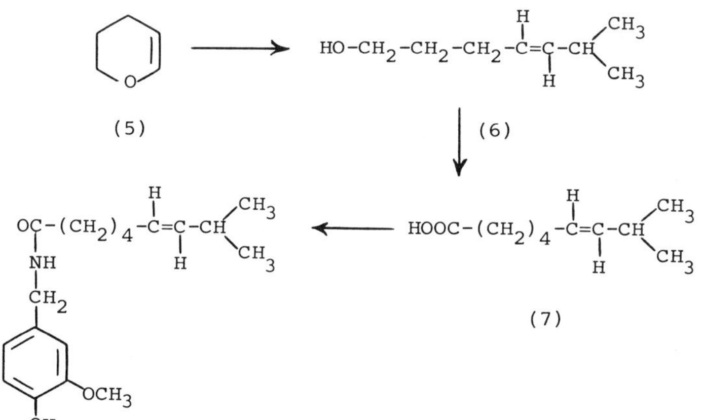

(5) (6) (7) (8)

3.1.2 The Belladines Several groups of workers, but mainly Battersby[7] and Wildman,[21] have shown that benzylamine derivatives of the norbelladine type (9) are incorporated *in vivo* into various types of *Amaryllida-*

*Recent studies on the constitution and biosynthesis of natural capsaicin have shown it to be a mixture of closely related amides, D. J. Bennett and G. W. Kirby, *J. Chem. Soc.* 442 (1968).

ceae alkaloids. Norbelladine (9) has not yet been isolated from natural sources.

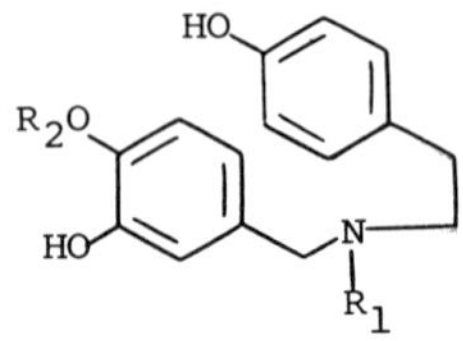

(9) $R_1 = R_2 = -H$

(10) $R_1 = -H,\ R_2 = -CH_3$

(11) $R_1 = R_2 = -CH_3$

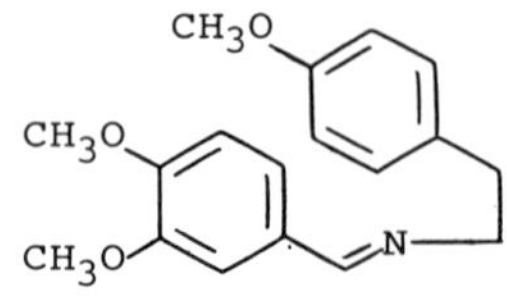

(12)

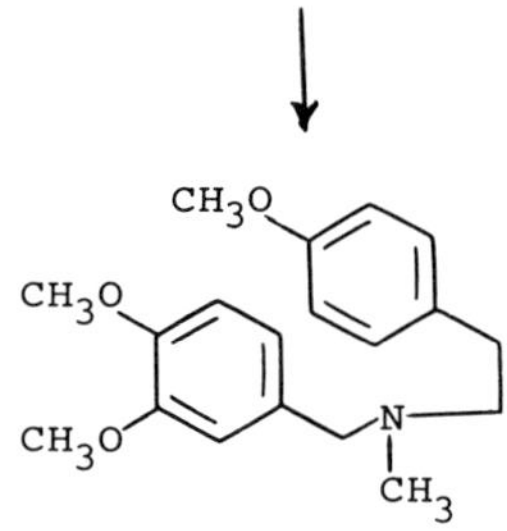

(13)

Its monomethyl ether (10) and the corresponding tertiary amine (11), however, seem to be naturally occurring alkaloids, but they have not yet been isolated in pure form.[7] It was correctly predicted that N,O-methylation of norbelladine (9) would stabilize this molecule and that the fully methylated derivative (13), which would not be susceptible to direct oxidation, might be a naturally occurring alkaloid.[22] This substance named belladine (13) is indeed an alkaloid and has been isolated from the bulbs of an *Amaryllis belladonna* hybrid.[22] It is intersting to note that belladine was synthesized in the laboratory before it was found in nature.[23] For the synthesis of (13), the Schiff base (12) was reduced and the secondary amine methylated. If one starts with isovanillin O-benzylether and tyramine O-benzylether, this scheme can be applied to the preparation of the norbelladine derivatives (10) and (11), and also to norbelladine (9) by ether cleavage of the above-mentioned secondary amines with 48% hydrobromic acid.

3.2 Phenethylamines[10,24]

The simplest naturally occurring substances of this group are the primary amine (14) and the secondary amine (15) which are derived biogenetically from the amino acid phenylalanine (16). The optically active alcohol (17), which has recently been isolated from an *Apocyanaceae* species,[25] is another interesting representative. The finding of neutral substances such as (17) among natural products is of particular interest in connection with biogenetic considerations.

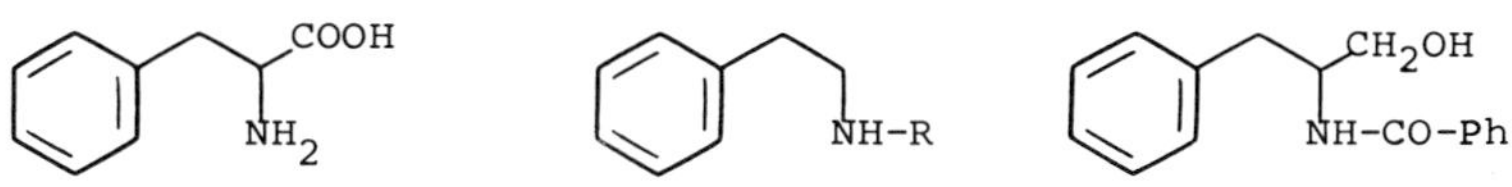

(16) (14) R = -H (17)

(15) R = $-CH_3$

The following two synthetic routes to phenethylamines, which can also be applied to the preparation of substituted analogs, are of interest. The first method can be recommended whenever the appropriate phenylacetonitrile is readily accessible. Reduction to the amine is carried out conveniently by catalytic hydrogenation in the presence of ammonia to minimize the formation of secondary amines.[26] The nitromethane approach can be applied with equal facility to the synthesis of the α-methyl substituted homologs merely by condensing the appropriate aromatic aldehyde with nitroethane instead of nitromethane. The intermediate nitrostyrenes are reduced with lithium aluminum hydride or catalytically in the presence of a palladium catalyst.[27]

The coexistence of primary amines with the corresponding secondary and tertiary amines in the same or related plant species is frequently observed in the phenylalkylamine group of alkaloids. Therefore, the chemical transformation of the simple primary amine (14) into the corresponding secondary amine (15) and tertiary amine (22) is discussed in greater detail. Condensation of phenethylamine (14) with benzaldehyde yields the Schiff base (18). Quaternization with dimethyl sulfate or methyl bromide gives the quaternary salt (19) which, upon hydrolysis, is converted directly into the secondary amine (15).[28–30] It is important to monitor the two reaction steps either by physical methods or by thin-layer chromatography, since the hydrolysis of quaternized material contaminated with the unreacted primary amine or the Schiff base leads to complex mixtures which are very difficult to separate. The following reaction sequence can also be used: Reduction of the Schiff base (18) catalytically or with sodium borohydride[31] yields the N-benzylphenethylamine (20), which can be converted into (21) either directly[32] or via the urethane (23).[33] The tertiary amine (21) can be debenzylated catalytically to (15).[34] Methylation of the secondary amine (15) obtained by either of these routes yields the tertiary amine (22).

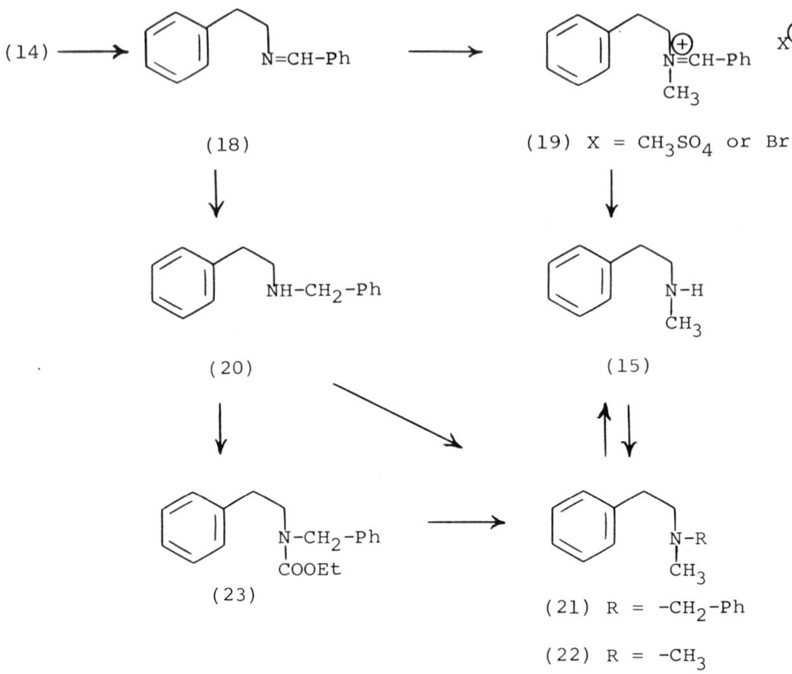

It must be mentioned that the usually reliable Eschweiler–Clarke reductive methylation procedure[35] can be applied successfully only with unsubstituted or 4-substituted phenethylamines. It has been reported that primary and secondary phenethylamines substituted with methoxy groups at the 3,4-positions[34] or 3,4,5-positions[36] are converted directly into 1,2,3,4-tetrahydroisoquinoline derivatives by this method.

3.2.1 4-Hydroxyphenethylamines[10] The following compounds originate biogenetically from the amino acid, tyrosine, and have been isolated from plant materials: the primary amine tyramine (24), the N-acetyl and N-benzoyl derivatives (25) and (26), N-methyltyramine (27), hordenine (28), and the O-methyl ether of hordenine.[37]

The most suitable starting material for the synthesis of tyramine (24) is anisaldehyde, which can be converted into tyramine O-methyl ether by either of the two procedures already mentioned. Tyramine O-methyl ether can be monomethylated or dimethylated by the reductive methylation procedure.[38] Demethylation of the various O-methyl ethers to tyramine (24), N-methyltyramine (27), and hordenine (28), respectively, can be easily accomplished by treatment with 48% hydrobromic acid. Hordenine (28) can also be prepared from tyrosine by decarboxylation and methylation or vice versa.[39]

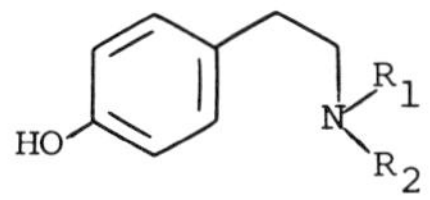

(24) $R_1 = R_2 = -H$

(25) $R_1 = -H$, $R_2 = -COCH_3$

(26) $R_1 = -H$, $R_2 = -COPh$

(27) $R_1 = -H$, $R_2 = -CH_3$

(28) $R_1 = R_2 = -CH_3$

3.2.2 3,4-Dihydroxyphenethylamines (Catecholamines)[10] The primary amine (31), dopamine, and the secondary amine (32), epinine, have been isolated from *Cytisus* species. Dopamine (31) has recently also been detected in bananas.[40] It seems plausible that these amines originate from the amino acids β-(3,4-dihydroxyphenyl)-alanine (DOPA) or tyrosine.[41] Dopamine (31) might be an intermediate in the biosynthesis of certain 1,2,3,4-tetrahydroisoquinoline alkaloids such as salsoline or salsolidine, since its condensation with acetaldehyde or its equivalents takes place under physiological conditions.[42,43] A convenient intermediate for the synthesis of the catecholamines (31) and (32) is 3,4-dimethoxyphenethylamine (30), commonly called homoveratrylamine, which can be obtained from veratrole (29) by chloromethylation, followed by reaction with cyanide and reduction. Cleavage of (30) with 48% hydrobromic acid yields (31). N-monomethylation of (30) and ether cleavage yields epinine (32).

(29) (30) (31) $R = -H$

(32) $R = -CH_3$

3.2.3 3,4,5-Trimethoxyphenethylamines[10,44] The alkaloid mescaline (33), a constituent of the cactus species *Lophophora williamsii* or *Anhalonium lewinii*, is the active hallucinatory principle of "mescal buttons." Recent investigations on some tetrahydroisoquinoline alkaloids, which coexist with mescaline (33), have shown that the physiological effects of these cactus extracts are mainly due to their mescaline content.[45,46] Mescaline (33) originates biogenetically from a hydroxylated phenylalanine equivalent which is converted by yet unknown mechanisms involving decarboxylation, hydroxylation and O-methylation.[47]

The extensive literature on the chemistry and pharmacology of mescaline has been summarized by Downing.[48] The most direct syntheses of this alkaloid are the reduction of either the nitrostyrene (36)[49] or the oxonitrile (34).[50]

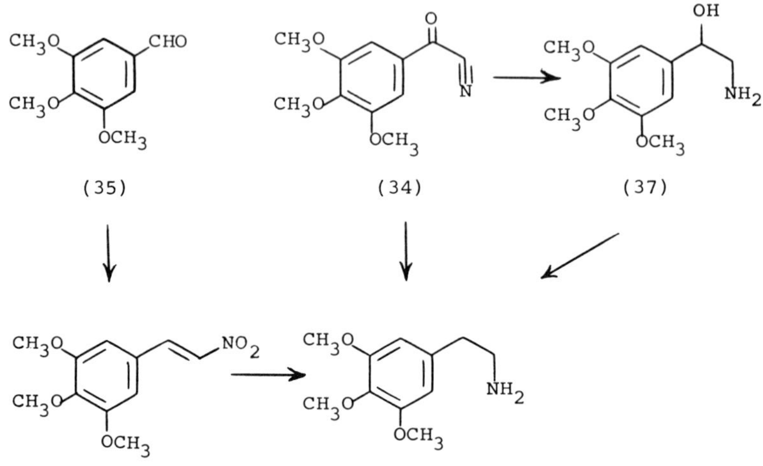

(35) (34) (37) (36) (33)

The catalytic hydrogenation of the oxonitrile (34) to either mescaline (33) or the amino alcohol (37) depends on the reaction conditions. The alcohol (37) is probably an intermediate in the reduction of (34) to (33).

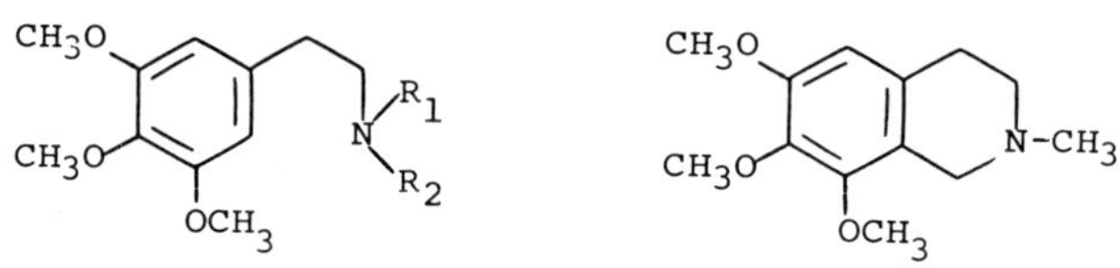

(38) $R_1 = -H$, $R_2 = -CO-CH_3$

(39) $R_1 = -H$, $R_2 = -CH_3$

(40) $R_1 = R_2 = -CH_3$

(41)

Three related cactus alkaloids are N-acetylmescaline (38), N-methylmescaline (39), and N,N-dimethylmescaline (trichocercine) (40). These three compounds can be prepared readily from mescaline (33) by methods already discussed. Eschweiler–Clarke methylation of mescaline gives the tertiary tetrahydroisoquinoline (41) directly and in high yield.[36]

3.2.4 Quaternary alkaloids[10] The two quaternary phenethyl alkaloids, candicine (42) and coryneine (43), have been isolated from cactus species. Leptodactyline (44) and salicifoline (45) are also representatives of this group. Hesperaline (46), recently isolated from a *Crucifera* species,[51] is the choline ester of veratric acid. This quaternary alkaloid, though not a phenethyl alkaloid, should be included here.

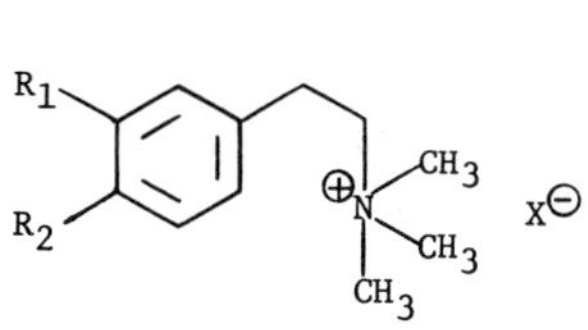

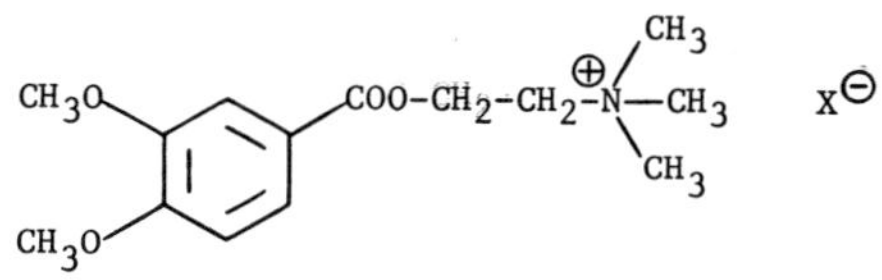

(46)

(42) R_1 = -H, R_2 = -OH

(43) R_1 = R_2 = -OH

(44) R_1 = -OH, R_2 = -H

(45) R_1 = -OH, R_2 = $-OCH_3$

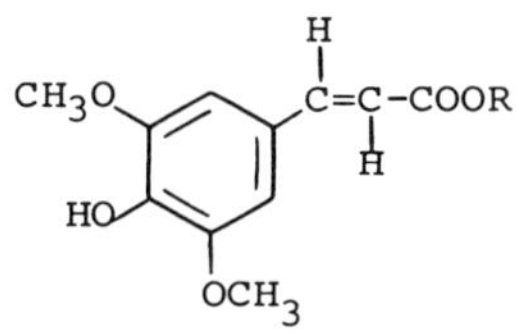

(47) R = $-CH_2-CH_2-N^{\oplus}(CH_3)_3\ X^{\ominus}$

(48) R = -H

The four quaternary compounds (42)–(45) have been isolated as iodides, which have also been prepared synthetically from the corresponding tertiary amines and methyl iodide. Hesperaline iodide (46) has been synthesized from veratroyl chloride and dimethylaminoethanol followed by quaternization with methyl iodide.[51] The alkaloids with free phenolic groups are obtained from the quaternary salts of the corresponding benzyloxy derivatives by mild acid hydrolysis or by catalytic hydrogenation.

Another quaternary alkaloid, sinapine (47), has been shown to be the choline ester of sinapic acid (48). The latter acid is identical with *trans*-4-hydroxy-3,5-dimethoxycinnamic acid* that could be converted into sinapine.[52]

*Commercially available sinapic acid has been shown to have the *trans* configuration by IR and nmr spectroscopy.

3.2.5 β-Hydroxyphenethylamines[10] Four naturally occurring substances belonging to this group have been isolated in optically active form. Two of these are (−)-octopamine (49) and (−)-norepinephrine (51), the former having been detected in citrus fruits,[53] and the latter in bananas.[40] The third is (−)-halostachine (52), and the fourth is (−)-macromerine (53), recently isolated from a *Cactus* species.[54] Another representative, aegelin, is the N-cinnamoyl derivative of the optically inactive amino alcohol (50)*:

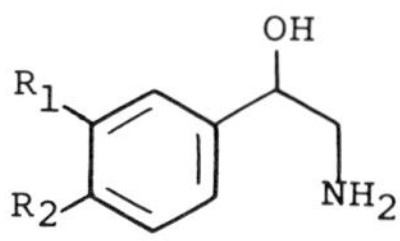

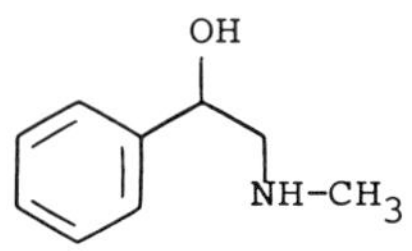

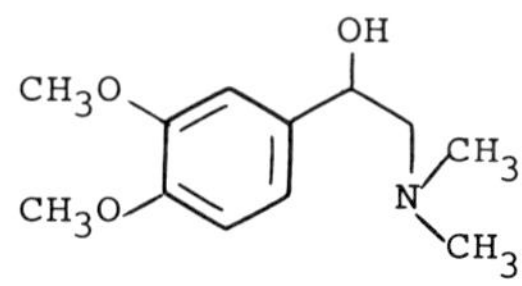

(49) R_1 = -H, R_2 = -OH

(52)

(53)

(50) R_1 = -H, R_2 = $-OCH_3$

(51) R_1 = R_2 = -OH

These amino alcohols may be synthesized from the readily available intermediates (54)–(57). (±)-Octopamine (49),[55,56] (±)-halostachine (52),[57] and (±)-macromerine (53)[54] have been obtained by reduction of the corresponding 2′-aminoacetophenone of type (54). The amino alcohol (50) has been synthesized from a cyanohydrin (56),[58] but it could also be obtained from an oxonitrile of type (55) by reduction. Lithium aluminum hydride reduction of optically active N-methylmandelamide [(57), R = -H] has been used to prepare (−)-halostachine (52),[59] thus establishing the absolute configuration of this alkaloid. Both optically active enantiomers of octopamine (49)[60] and halostachine (52)[57] have been obtained by optical resolution of their corresponding racemates. The preferred intermediates for preparing the phenolic representatives are the corresponding benzyloxy compounds.

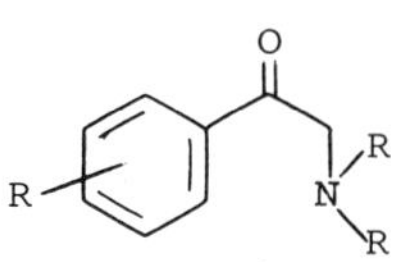

(54)

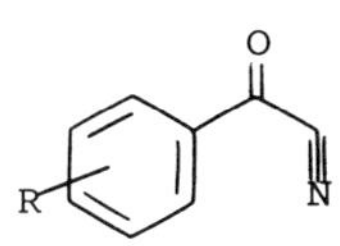

(55)

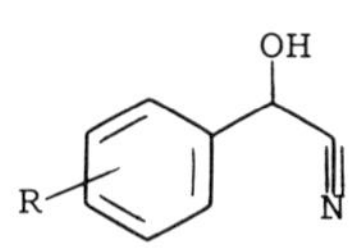

(56)

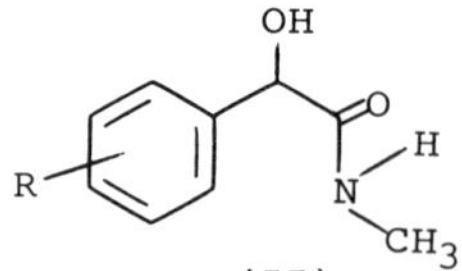

(57)

*(±)-N-Benzoyl-2-hydroxy-2-(4′-methoxyphenyl)ethylamine has recently been isolated by S. R. Johns, J. A. Lamberton, and J. R. Price, *Australian J. Chem.* **20**, 2795 (1967).

4. CINNAMIC ACID DERIVATIVES AND RELATED ALKALOIDS

Most of the nitrogenous substances treated in this section, are nonbasic, and consequently are not usually considered as alkaloids. However, they are primary plant constituents and warrant inclusion in this discussion.

4.1 Cinnamic acid amides[10]

Fagaramide (58), isolated from the root of *Fagara xanthoxyloides*, is structurally the simplest compound of this type. Its structure has been established as the *trans* N-isobutyl-3,4-methylenedioxycinnamide (58) by synthesis. Herclavin (59) has been shown to have the indicated structure by synthesis from *trans*-cinnamoyl chloride and N,O-dimethyltyramine. Aegelin (60) is the optically inactive hydroxyamide (60) as shown by a series of chemical transformations and by direct synthesis from *trans*-cinnamoyl chloride and α-aminomethyl-p-methoxybenzyl alcohol.[61,62] Another compound of this group, isolated from the stems of *Piper longum*, is piplartine (61),[63,64] identical with piperlongumine, to which originally structure (62) has been assigned.[63]

Subaphylline (63) has been synthesized by causing N-trityl-putrescine to react with the mixed anhydride of O-acetylferulic acid followed by acid hydrolysis.[65]

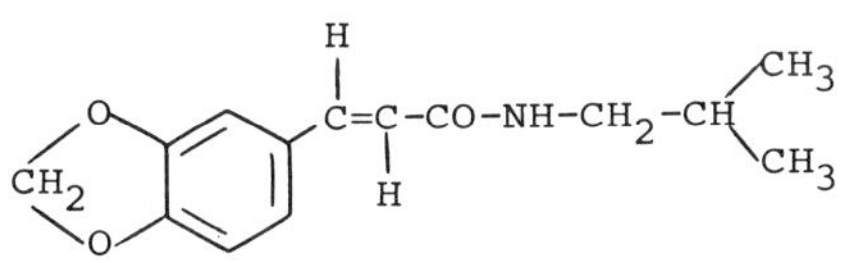

(58)

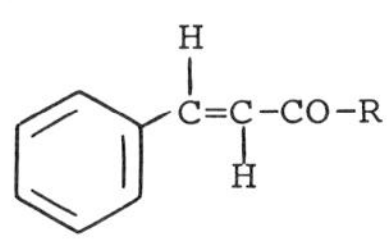

(59) R = 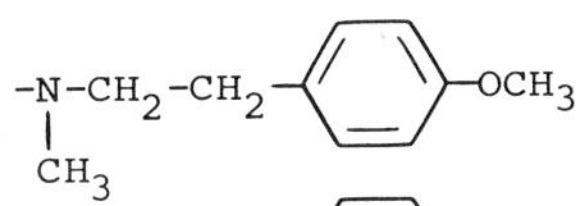

(60) R = 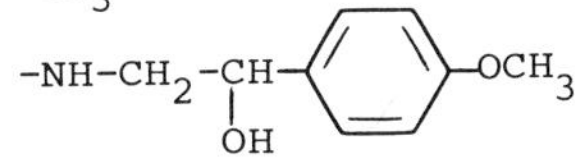

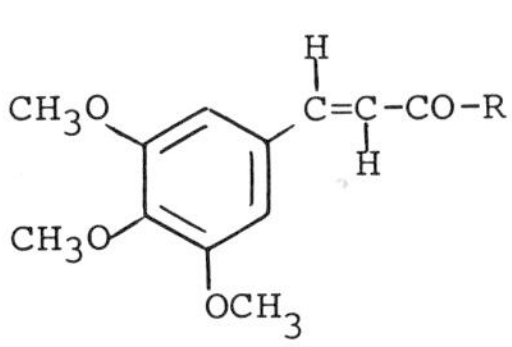

(61) R = 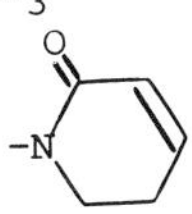

(62) R = 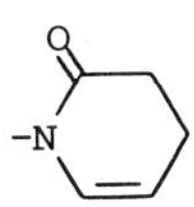

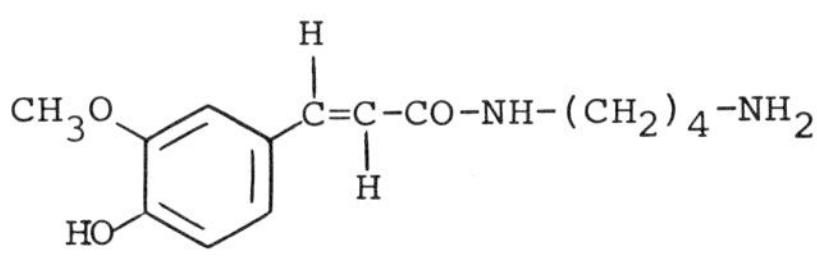

(63)

Other alkaloids containing a cinnamoyl residue are the piperidine alkaloids astrophylline,[66] orensine,[67] and iso-orensine.[67] It is interesting to note that in astrophylline and iso-orensine, the double bond of the cinnamoyl residue has the *cis* configuration, whereas in all the cinnamic acid derivatives mentioned above, the double bond is *trans* oriented.

4.2 Pepper alkaloids

This group contains some pungent principles of various pepper species which have found practical application as synergists for insecticides of the pyrethrin type. These conjugated 3,4-methylenedioxyphenyl substituted polyeneoic acids have very characteristic UV chromophores that are easily detectable.[68] Piperine (64), a major constituent of black pepper, was isolated and characterized long ago.[69] Its structure as the all *trans* 2,4-pentadieneoic acid piperidide (64) has recently been confirmed by a stereospecific synthesis.[70] A second constituent of black pepper, chavicine (65), presumably has the indicated all *cis* structure.[71] A third pepper constituent, piperettine (66), is the 3,4-methylenedioxyphenyl substituted all *trans* 2,4,6-heptatrieneoic acid piperidide (66) that has been synthesized from piperettic acid and piperidine.[72] Another constituent, which together with piperlongumine (62) occurs in *Piper longum*, is piperlongumine (67). Identification of (67) as the isobutylamide of piperic acid has been established by synthesis.[64]

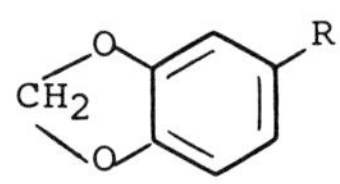

(64) R = 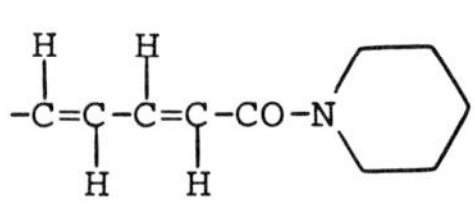

(65) R = 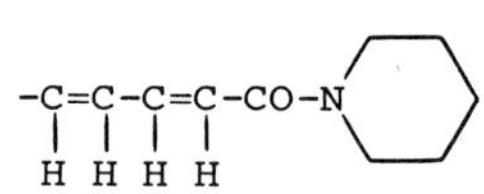

(66) R = 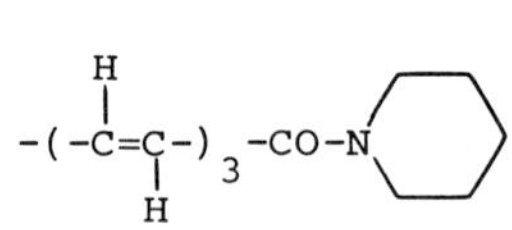

(67) R = 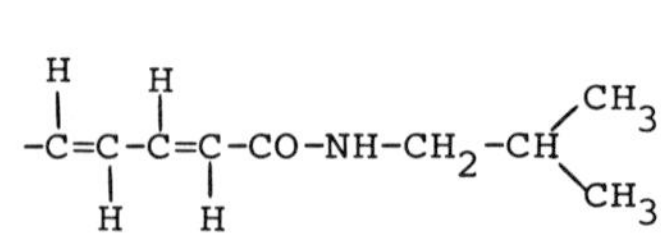

4.3 Oxazole alkaloids

The close structural relationship of aegelin (60) with the recently isolated alkaloid annuloline (68) warrants inclusion of the two currently known oxazole alkaloids in this book. Annuloline (68), which is strongly

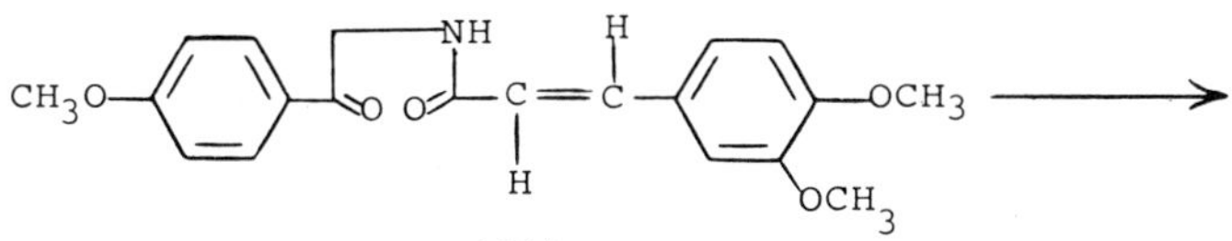

(69)

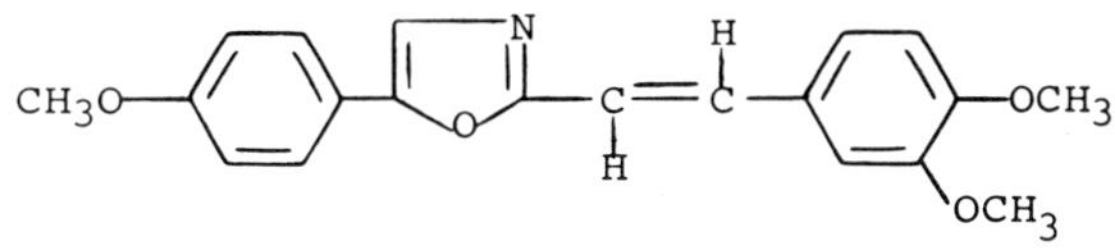

(68)

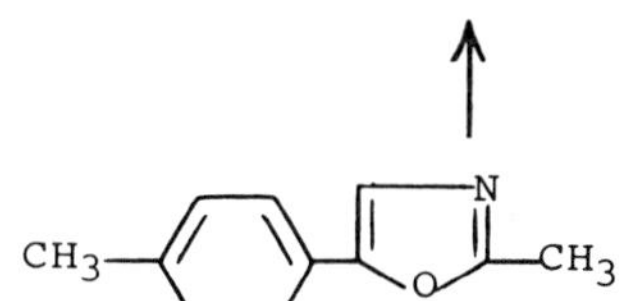

(70)

fluorescent, has been isolated from the root of *Lolium multiflorum*.[73] It has been synthesized by cyclizing *trans*-3,4-dimethoxy-N-(4-methoxyphenacyl)-cinnamide (69) with phosphorus oxychloride.[74] Another synthesis of annuloline (68) involves condensation of the readily available oxazole derivative (70) with veratraldehyde.[75] The natural pigment is probably the *trans* isomer. A photochemical conversion of the unnatural *cis* isomer into *trans* annuloline has been reported.[76]

The quaternary alkaloid N-methylhalfordine (71), isolated in the form of various salts from *Halfordia scleroxyla*,[77] is the second known oxazole alkaloid. During the degradation of (71), a phenol, halfordinol (72), along with a variety of volatile fragments, was obtained. The structure of (72) has been confirmed by mass spectroscopy[78] and an unequivocal synthesis.[75]

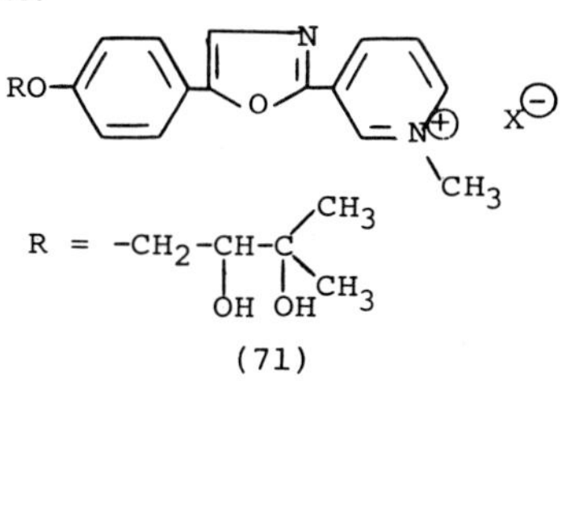

(71)

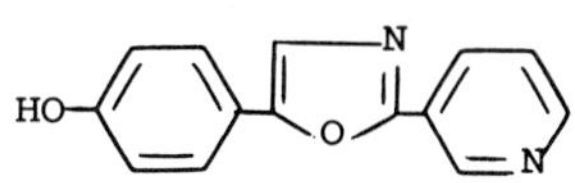

(72)

5. EPHEDRA ALKALOIDS[8,10,79]

Plants of genus *Ephedra* contain six optically active alkaloids, some of which are useful in medicine. The two major alkaloids, (−)-ephedrine and (+)-pseudoephedrine, shown in the Fischer projections (74) and (75), are diastereoisomers. The myriad of problems associated with the relative and absolute configurations of these alkaloids have long been the concern of many research groups. It is now generally accepted that the weaker of the two bases, (−)-ephedrine, has the *erythro* configuration and, therefore, the *threo* configuration is assigned to (+)-pseudoephedrine. Both alkaloids, which are reversibly interconvertible, can be reduced to the same optically active (+)-desoxyephedrine (73).[80–82] (−)-Ephedrine (74) and (+)-pseudoephedrine (75) differ, therefore, only with respect to the configuration of the carbinol function. The absolute configuration of this asymmetric center is identical with that of D-(−)-mandelic acid,[83,84] and the absolute configuration of the carbon atom bearing the methylamino group is identical with that of natural L-(+)-alanine.[84,85] An x-ray analysis of crystalline (−)-ephedrine hydrochloride has provided confirmation of these findings.[86]

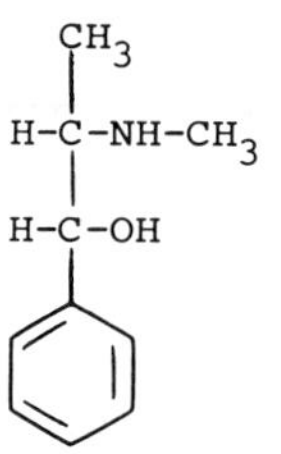

(74)

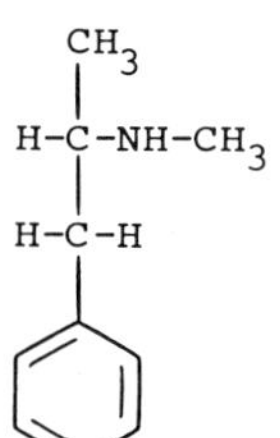

(73)

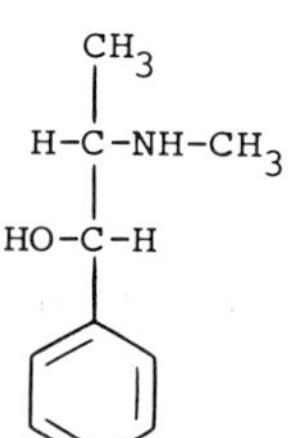

(75)

It has been shown that the other naturally occurring ephedra alkaloids, (−)-norephedrine (76) and (−)-N-methylephedrine (78), are derived from (−)-ephedrine (74), whereas the two dextro rotatory alkaloids, (+)-

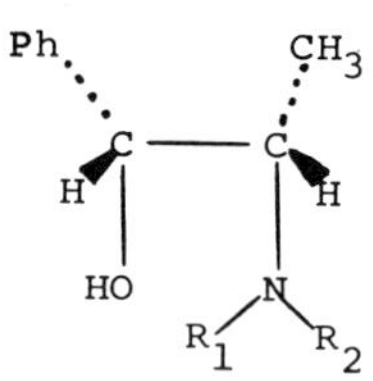

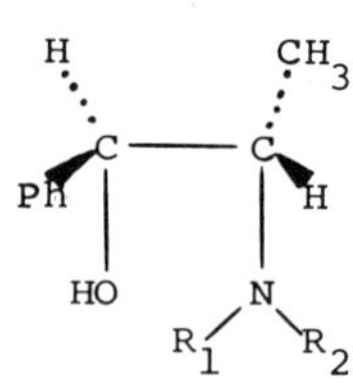

(74) $R_1 = -H,\ R_2 = -CH_3$ (75)

(76) $R_1 = R_2 = -H$ (77)

(78) $R_1 = R_2 = -CH_3$ (79)

norpseudoephedrine (77) and (+)-N-methylpseudoephedrine (79), are similarly related to (+)-pseudoephedrine (75). All six optically active alkaloids have been synthesized, characterized, and correlated.[87]

Of the many syntheses of (±) and (−)-ephedrine (74), the one shown below is of commercial interest[87]:

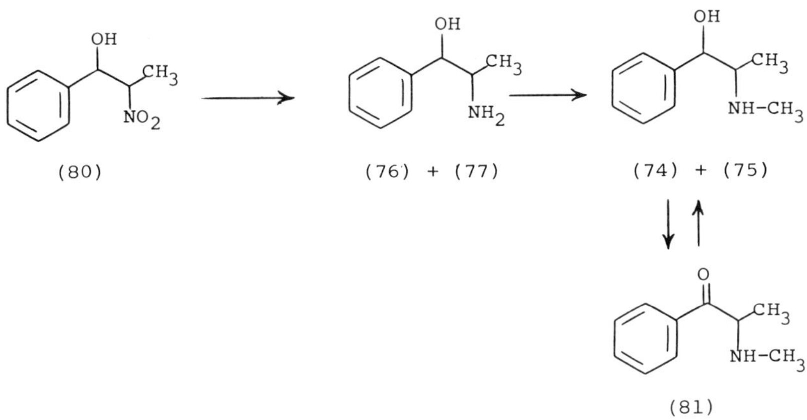

Condensation of benzaldehyde with nitroethane in the presence of potassium carbonate yields a diastero mixture of nitro alcohols (80). Reduction gives a mixture of (±)-norephedrine (76) and (±-norpseudoephedrine (77) which can be separated. Methylation of (±)-norephedrine (76) yields (±)-ephedrine (74) that can be resolved into the optical antipodes without difficulty. (±)-Norpseudoephedrine (77) can be methylated to (±-pseudoephedrine (75) that can either be isomerized to (±)-ephedrine (74) or reconverted to a diastereo mixture of (±)-(74) and (±)-(75) via the methylamino ketone (81).

6. ALKALOIDS FROM *CARDUUS* SPECIES

The isolation of two optically active alkaloids from *Carduus acanthoides L.*, a plant widely distributed in Argentina, has recently been reported.[88] The biologically active alkaloid acanthoidine (83) is the tetrahydro derivative of a second alkaloid called acanthoine (82) that is biologically inactive. Alkaline hydrolysis of the two alkaloids yielded *inter alia* the corresponding optically active diamines (84) and (85), reconvertible to (82) and (83), respectively, by treatment with ethyl formimidate. For the synthesis of (±)-acanthoidine (83), the nitropropane derivative (86) was condensed with ethyl acrylate to form the nitro ester (87). Reduction of the amide derived from ester (87) with lithium aluminum hydride and treatment of the resulting (±)-diamine (85) with ethyl formimidate gave (±)-(83).

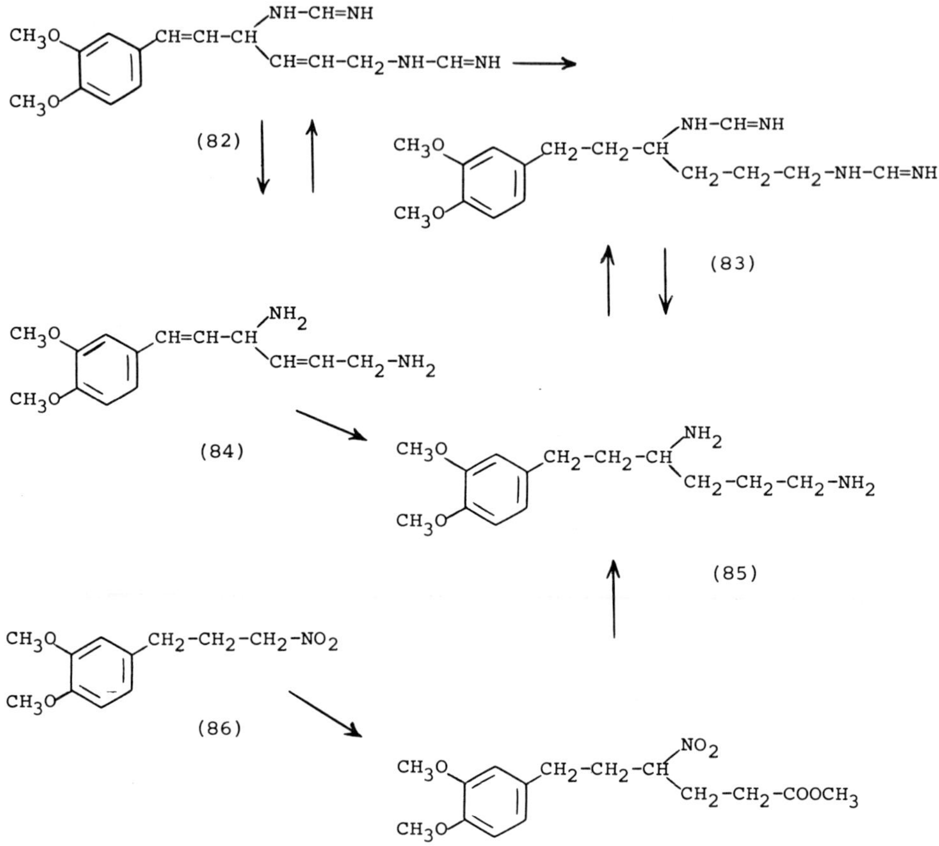

(82) (83) (84) (85) (86) (87)

7. MISCELLANEOUS

Even in a chapter on so restricted a subject as "Alkaloids Containing a Simple Aromatic Moiety," there are exceptions that do not fit into the prescribed pattern. Three such examples are given in the following paragraphs.

7.1 Leonticine (Petaline methine)

The quaternary isoquinoline structure (88) has been suggested for petaline that has been isolated from *Leontice leontopetalum L.*[89] Evidence has been presented that an accompanying alkaloid, leonticine, is identical with the primary Hofmann degradation product of petaline (88) and, therefore, has the structure (89). Since ring opening of petaline occurs so easily, it has been postulated that the alkaloid leonticine (89) is an artifact and should, henceforth, be referred to as petaline methine. Recently, these relationships have been confirmed by the total syntheses of racemic, optically active petaline, and petaline methine.[90,91]

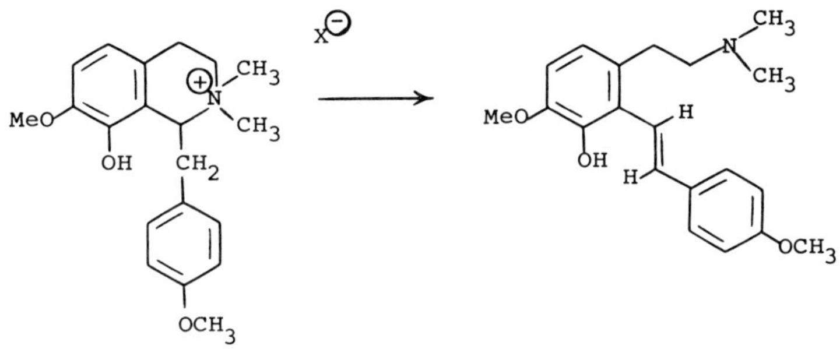

(88) (89)

7.2 Girgensohnine

This alkaloid is present in *Girgensohnia oppositiflora* along with N-methyl-piperidine.[92] Its structure (90) was established by alkaline hydrolysis which yielded piperidine, hydrogen cyanide, and *p*-hydroxybenzaldehyde. The synthesis of girgensohnine (90) has been achieved by condensation of *p*-hydroxybenzaldehyde cyanohydrin with piperidine.[92]

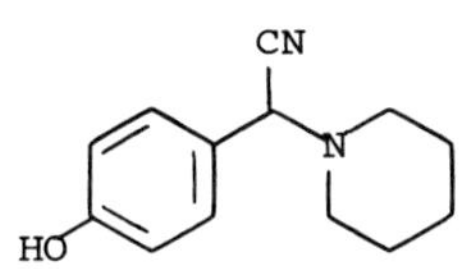

(90)

7.3 Leonurine

Structure (91) has been proposed for leonurine,[93] an alkaloid isolated from the leaves of *Leonurus sibiricus L.*,[94] based on the results of acid hydrolysis that yielded syringic acid (92) and the guanidino alcohol (93). The latter, on alkaline hydrolysis, was converted into 4-aminobutanol (94).* Leonurine is not identical with the synthetic compound (95)[93] or

*The validity of the structures (93) and (94) has been confirmed by A. Brossi and S. Teitel, 1965 (unpublished work).

with the ester (96)** that has been excluded on the basis of pK_a measurements.[93]

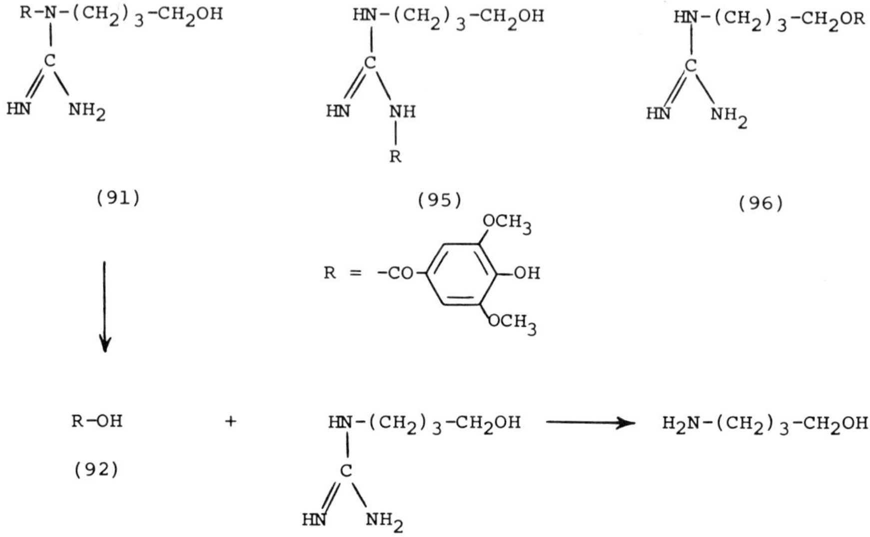

(91) (95) (96) (92) (93) (94)

REFERENCES

1. E. H. Winterstein and G. Trier, "Die Alkaloide," 2nd ed., Verlag Gebrüder Borntraeger, Berlin, Germany, 1931.
2. T. A. Henry, "The Plant Alkaloids," 4th ed., The Blakiston Co., Philadelphia, Pa., 1949.
3. M. Guggenheim, "Die biogenen Amine," 4th ed., Karger Verlag, Basel, Switzerland, 1951.
4. J. Stanek, "Alkaloidy," Československé Akademie Věd, Prague, Czechoslovakia, 1957.
5. R. Robinson, "The Structural Relations of Natural Products," Clarendon Press, Oxford, England. 1955.
6. D. H. R. Barton and T. Cohen, "Festschrift Arthur Stoll," pp. 117–143, Birkhauser-Verlag, Basel, Switzerland, 1957.
7. A. R. Battersby, "Tilden Lecture," *Proc. Chem. Soc.* 189 (1963).
8. G. Fodor, K. Nador, and I. V. Torgov, "Recent Developments in the Chemistry of Natural Carbon Compounds," Vol. 1, Hungarian Academy of Sciences, Budapest, Hungary. 1965.
9. R. H. F. Manske and H. L. Holmes, *Alkaloids* **1–8** (1950–1960).
10. H. G. Boit, "Ergebnisse der Alkaloid-Chemie bis 1960," Akademie-Verlag, Berlin, Germany. 1961.

The correct structure of the alkaloid leonurine has now been confirmed by total synthesis. It is the ester alkaloid (96) and not the guanidino alcohol (91) as stated above [Y. Kishi, S. Sugiura, S. Inoue, Y. Hayashi, and T. Goto, *Tetrahedron Letters* **5, 637–640 (1968)].

11. R. J. Highet, *J. Org. Chem.* **26**, 4767 (1961).
12. R. K. Hill and R. M. Carlson, *J. Org. Chem.* **30**, 1571 (1964).
13. R. K. Hill and R. M. Carlson, *Tetrahedron Letters* **19**, 1157 (1964).
14. A. J. Ewins, *J. Chem. Soc.* **101**, 544 (1912).
15. E. Mutschler, *Arch. Pharm.* **299**, 31 (1966).
16. I. R. C. Bick and P. S. Clezy, *J. Chem. Soc.* 2402 (1960).
17. P. O. Larsen, *Biochim. Biophys. Acta* **107**, 134 (1965).
18. E. K. Nelson and L. E. Dawson, *J. Amer. Chem. Soc.* **45**, 2179 (1923).
19. L. Crombie, S. H. Dandegaonker, and K. B. Simpson, *J. Chem. Soc.* 1025 (1955).
20. S. Kosuge, Y. Inagaki, and T. Ino, *Nippon Nogeikagaku Kaishi* **34**, 811 (1960); *Chem. Abstr.* **59**, 6724 (1963).
21. W. C. Wildman, H. M. Fales, R. J. Highet, S. W. Breuer, and A. R. Battersby, *Proc. Chem. Soc.* 180 (1962).
22. E. W. Warnhoff, *Chem. Ind. (London)* 1385 (1957).
23. A. R. Surrey, A. Mooradian, R. A. Cutler, C. M. Suter, and J. S. Buck, *J. Amer. Chem. Soc.* **71**, 2421 (1949).
24. L. Reti, *Alkaloids* **3**, 313–338 (1953).
25. A. R. Battersby and R. S. Kapil, *Tetrahedron Letters* **39**, 3529 (1965).
26. J. C. Robinson and H. R. Snyder, *Org. Syn.* **3**, 720–722 (1955).
27. V. V. Perekalin, "Unsaturated Nitro Compounds," Part 2, Chap. 2, pp. 163–190 (Russian transl.), Daniel Davey & Co., Inc., New York, N. Y., 1964.
28. J. Lee, A. Ziering, L. Berger, and S. D. Heineman, "Emil Barell Festschrift 1946," pp. 264–305, Hoffmann-La Roche, Basel, Switzerland. 1946.
29. S. Kirkwood and L. Marion, *J. Amer. Chem. Soc.* **72**, 2522 (1950).
30. E. Späth and J. Bruck, *Chem. Ber.* **70**, 2446 (1937).
31. J. H. Billman and A. C. Diesing, *J. Org. Chem.* **22**, 1068 (1957).
32. W. S. Emerson, *Org. Reactions*, **4**, 174–255 (1948).
33. F. V. Bruchhausen and J. Knabe, *Arch. Pharm.* **287**, 601 (1954).
34. J. S. Buck and R. Baltzly, *J. Amer. Chem. Soc.* **64**, 2263 (1942).
35. M. L. Moore, *Org. Reactions* **5**, 320 (1949).
36. J. Castrillon, *J. Amer. Chem. Soc.* **74**, 558 (1952).
37. G. M. Badger, B. J. Christie, and H. J. Rodda, *Australian J. Chem.* **16**, 734 (1963).
38. Y. Raoul, *Bull. Soc. Chim. Biol.* **19**, 675 (1937).
39. S. Kanao and T. Suyama, *J. Pharm. Soc. Japan* **87**, 99 (1967).
40. T. P. Waalkes, A. Sjoerdsma, C. R. Creveling, H. Weissbach, and S. Udenfriend, *Science* **127**, 648 (1958).
41. A. Pletscher, K. F. Gey, and P. Zeller, *Prog. Drug Res.* **2**, 417–590 (1960).
42. C. Schöpf and H. Bayerle, *Ann. Chem.* **513**, 190 (1934).
43. O. Kovacs and G. Fodor, *Chem. Ber.* **84**, 795 (1951).
44. L. Reti, *Alkaloids* **3**, 324–329 (1953).
45. A. Brossi, F. Schenker, and W. Leimgruber, *Helv. Chim. Acta* **47**, 2089 (1964).
46. A. Brossi, F. Schenker, R. Schmidt, R. Banziger, and W. Leimgruber, *Helv. Chim. Acta*, **49**, 403 (1966).
47. E. Leete, *Chem. Ind. (London)* 604 (1959); S. Agurell, J. Lundström, and F. Sandberg, *Tetrahedron Letters* **26**, 2433 (1967); J. L. McLaughlin and A. G. Paul, *Lloydia* **30**, 91 (1967); H. Rosenberg, J. L. McLaughlin, and A. G. Paul, *Lloydia* **30**, 100 (1967).
48. D. F. Downing, "Psychopharmacological Agents," Vol. 4–I, pp. 574–579, (M. Gordon, Ed.), Academic Press, Inc., New York, N. Y., 1964.
49. F. Benington and R. D. Morin, *J. Amer. Chem. Soc.* **73**, 1353 (1951).
50. O. M. Friedman, K. N. Parameswaran, and S. Burstein, *J. Med. Chem.* **6**, 227 (1963).
51. R. Gmelin and H. Mohrle, *Arch. Pharm.* **300**, 176 (1967).

52. E. Späth and G. Gibian, *Monatsh. Chem.* **41,** 271 (1920).
53. I. Stewart and T. A. Wheaton, *Science* **145,** 60 (1964).
54. J. E. Hodgkins, S. D. Brown, and J. L. Massingill, *Tetrahedron Letters* **14,** 1321 (1967).
55. G. Fodor, O. Kovacs, and T. Mecker, *Acta Chim. Acad. Sci. Hung.* **1,** 395 (1951); *Chem. Abstr.* **49,** 897 (1955).
56. A. D'Amico, L. Bertolini, and C. Monreale, *Chim. Ind. (Milan)* **38,** 93 (1956); *Chem. Abstr.* **50,** 13801 (1956).
57. G. P. Men'shikov and G. M. Borodina, *J. Gen. Chem. (U.S.S.R.)* **17,** 1569 (1947); *Chem. Abstr.* **42,** 2245 (1948).
58. N. Adityachaudhuri and A. Chatterjee, *J. Indian Chem. Soc.* **36,** 585 (1959).
59. R. Lukes, V. Dienstbierova, J. Kovar, and K. Blaha, *Collection Czech. Chem. Commun.* **26,** 466 (1961).
60. T. Kappe and M. D. Armstrong, *J. Med. Chem.* **7,** 569 (1964).
61. R. N. Chakravarti and B. Dasgupta, *J. Chem. Soc.* 1580 (1958).
62. A. Chatterjee and C. P. Dutta, *Tetrahedron Letters* **16,** 1797 (1966).
63. A. Chatterjee and C. P. Dutta, *Tetrahedron* **23,** 1769 (1967).
64. B. S. Joshi, V. N. Kamat, and A. K. Saksena, *Tetrahedron Letters* **20,** 2395 (1968).
65. A. Hillmann-Elies and G. Hillman, *Z. Naturforsch.* **8b,** 526 (1953).
66. H. A. Lloyd, *Tetrahedron Letters* **50,** 4537 (1965).
67. C. Schöpf and W. Merkel, *Ann. Chem.* **701,** 180 (1967).
68. A. W. Sangster and K. L. Stuart, *Chem. Rev.* **65,** 69 (1965).
69. L. Rügheimer, *Chem. Ber.* **15,** 1390 (1882).
70. H. Normant and C. Feugeus, *Compt. Rend.* **258,** 2846 (1964).
71. E. Ott and O. Lüdemann, *Chem. Ber.* **57,** 214 (1924).
72. F. Spring and J. Stark, *J. Chem. Soc.* 1177 (1950).
73. B. Axelrod and J. R. Belzile, *J. Org. Chem.* **23,** 919 (1958).
74. R. S. Karimoto, B. Axelrod, J. Wolinsky, and E. D. Schall, *Tetrahedron Letters* **3,** 83 (1962).
75. A. Brossi and E. Wenis, *J. Het. Chem.* **2,** 310 (1965).
76. W. Schunack and H. Rochelmeyer, *Arch. Pharm.* **298,** 572 (1965).
77. W. D. Crow and J. H. Hodgkin, *Tetrahedron Letters* **2,** 85 (1963).
78. W. D. Crow, J. H. Hodgkin, and J. S. Shannon, *Australian J. Chem.* **18,** 1433 (1965).
79. L. Reti, *Alkaloids* **3,** 339–362 (1953).
80. E. Schmidt, *Arch. Pharm.* **252,** 89 (1914).
81. E. Schmidt, *Arch. Pharm.* **253,** 52 (1915).
82. H. Emde, *Helv. Chim. Acta* **12,** 377 (1929).
83. K. Freudenberg, E. Schoeffel, and E. Braun, *J. Amer. Chem. Soc.* **54,** 234 (1932).
84. W. Leithe, *Chem. Ber.* **65,** 660 (1932).
85. K. Freudenberg and F. Nikolai, *Ann. Chem.* **510,** 223 (1934).
86. D. C. Phillips, *Acta Cryst.* **7,** 159 (1954); *Chem. Abstr.* **48,** 7386 (1954).
87. W. N. Nagai and S. Kanao, *Ann. Chem.* **470,** 157 (1929).
88. B. Frydman and V. Deulofeu, *Tetrahedron* **18,** 1063 (1962).
89. N. J. McCorkindale, D. S. Magrill, M. Martin-Smith, S. J. Smith, and J. B. Stenlake, *Tetrahedron Letters* **51,** 3841 (1964).
90. G. Grethe, M. Uskokovic, and A. Brossi, *Tetrahedron Letters* **15,** 1599 (1966).
91. G. Grethe, M. Uskokovic, and A. Brossi, *J. Org. Chem.* **33,** 2500 (1968).
92. N. K. Yurashevskii and N. L. Stepanova, *J. Gen. Chem. (U.S.S.R.)* **16,** 141 (1946); *Chem. Abstr.* **40,** 6754 (1946).
93. T. Goto, N. Kato, Y. Hirata, and Y. Hayashi, *Tetrahedron Letters* **13,** 545 (1962).
94. S. Kubota and S. Nakashima, *Nippon Yakurigaku Zasshi* **11,** 153 (1930); *Chem. Abstr.* **25,** 771, 1285 (1931).

3

The Isoquinoline Alkaloids

MAURICE SHAMMA
Department of Chemistry,
The Pennsylvania State University,
University Park, Pennsylvania

1. THE TETRAHYDROISOQUINOLINE ALKALOIDS

1.1 Structural Elucidation and Syntheses

A number of simple alkaloids, found mostly in members of the Cactaceae, are based upon a tetrahydroisoquinoline skeleton (1) where the C(1) sub-

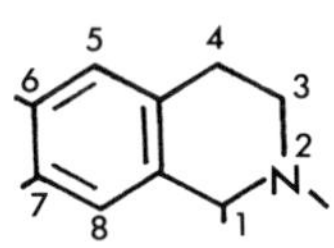

(1)

stituent can be methyl, hydroxymethyl, or isobutyl. Permanganate oxidation of these alkaloids yields substituted phthalic acids that are easily characterized. In case the C(1) side chain is isobutyl, isobutyric and isovaleric acid are also produced during the oxidation.

Two syntheses recorded for members of this group are those for calycotomine and lophocerine, and in both instances the classical Bischler-Napieralski cyclization was used.[1,2]

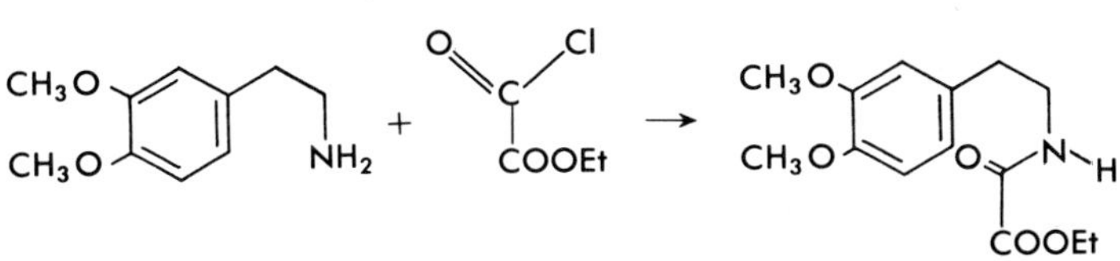

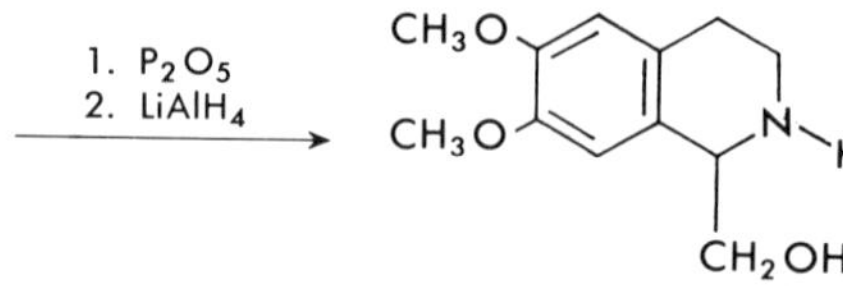

Calycotomine

And:

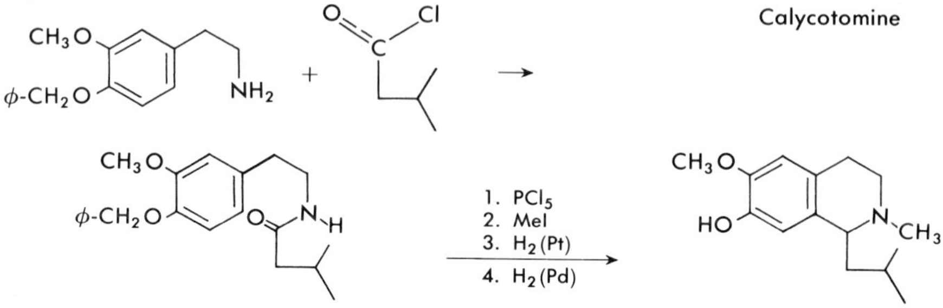

Lophocerine

Among the more complex tetrahydroisoquinoline alkaloids is pilocereine which is found in a variety of giant cacti. The original work on this unusual base was handicapped by the absence of such modern physical tools as nuclear magnetic resonance and mass spectrometry. However, permanganate oxidation yielded a mixture of isobutyric and isovaleric acid, indicating at least one isobutyl side chain in the alkaloid.

Since the presence of a diphenyl ether linkage was suspected, the potassium in liquid ammonia reduction of pilocereine and its derivatives was studied. This type of reduction cleaves diaryl ethers, but does not usually

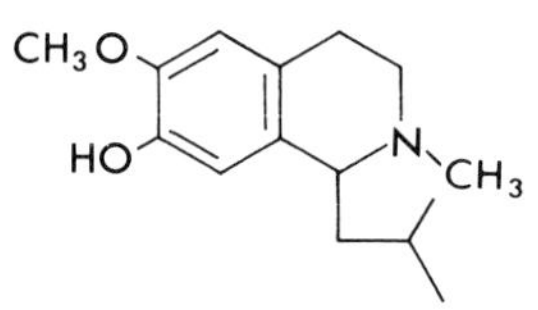

(2)

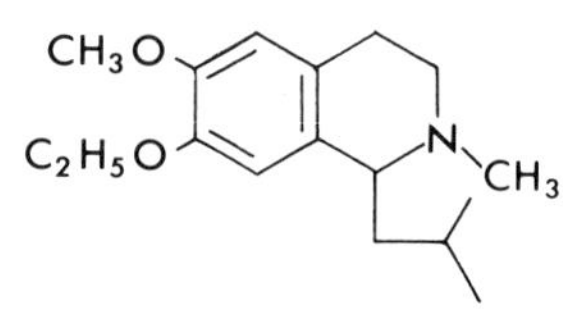

(3)

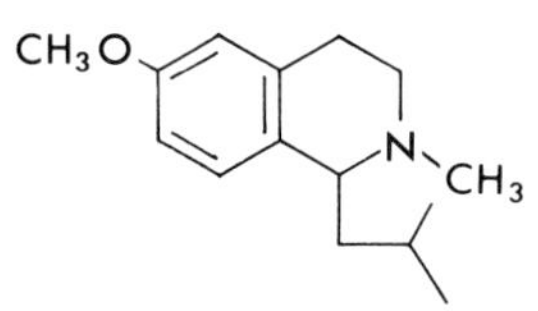

(4)

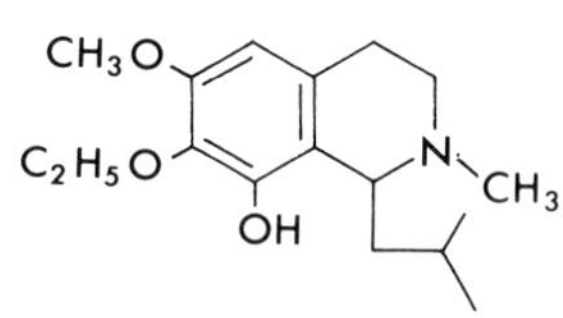

(5)

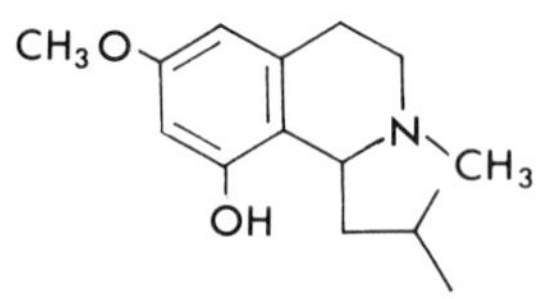

(6)

affect alkylaryl ethers. The O-ethyl ether of pilocereine was thus initially found to yield four amines that were characterized as (2)–(5).

At a later stage, it was found that the molecular ion peak for the O-methyl ether of pilocereine was situated at m/ε 757, so that pilocereine must be trimeric. Analysis of the nmr spectrum of pilocereine acetate indicated four aromatic hydrogens, three O-methyl and three N-methyl groups, one O-acetyl and three isobutyl residues. A reinvestigation of the potassium in liquid ammonia reduction of O-ethylpilocereine then showed that the tetrahydroisoquinoline 6 had also been produced, so that pilocereine must be represented by expression (7) which accommodates all of the reduction products.[3]

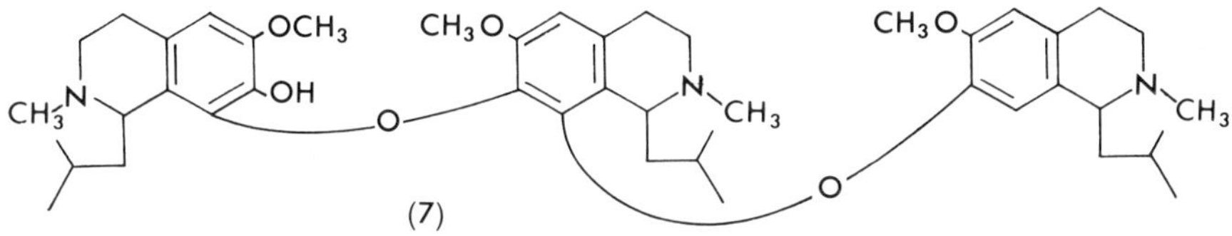

(7)

The synthesis of pilocereine has been effected by phenolic oxidative coupling of lophocerine with $K_3Fe(CN)_6$. Pilocereine (7) was isolated from the resulting mixture by thin-layer chromatography.[4]

1.2 Absolute Configuration

The absolute configuration of the tetrahydroisoquinoline alkaloids has been established by Battersby[5] who, in an experiment paralleling that of Corrodi and Hardegger (*vide infra*), oxidized the N-formyl derivative of naturally occurring (−)-salsolidine with ozone followed by peracetic acid. Subsequent acid hydrolysis to remove the protective group yielded the previously unknown aminodiacid (8) which was synthesized from L-alanine as shown.

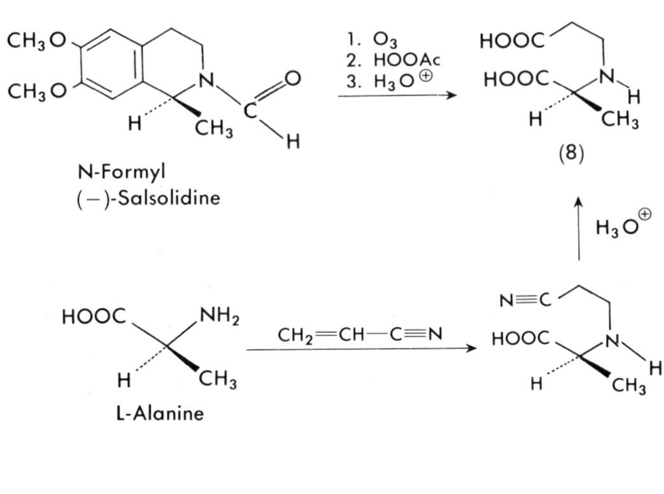

The chirality of (+)-calycotomine was determined by treating the alkaloid with *p*-toluenesulfonyl chloride in pyridine to obtain a disulfonyl derivative that was then reduced with $LiAlH_4$ to the levorotatory *p*-toluenesulfonamide 9. Since the corresponding sulfonamide from (−)-salsolidine (of known absolute configuration) was dextrorotatory, it follows that (+)-calycotomine must have a configuration opposite to that of (−)-salsolidine.

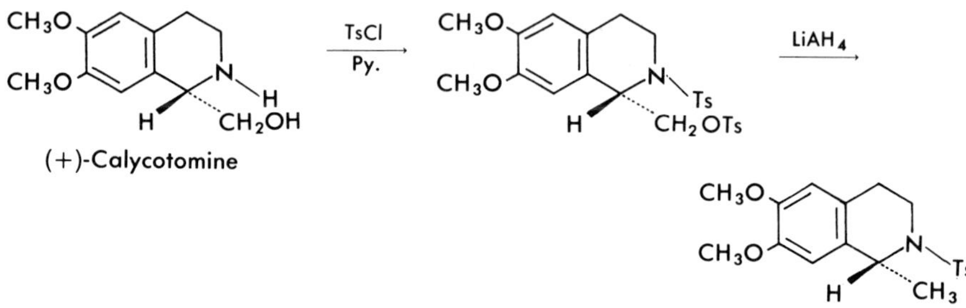

(+)-Calycotomine

(−)-(9)

2. THE BENZYLISOQUINOLINE AND PAVINE TYPE ALKALOIDS

2.1 Structural Elucidation

The benzylisoquinoline alkaloids include both the benzylisoquinoline bases of type (10) and the benzyltetrahydroisoquinolines of type (11). Quaternary alkaloids have also been isolated belonging to each of these two categories.

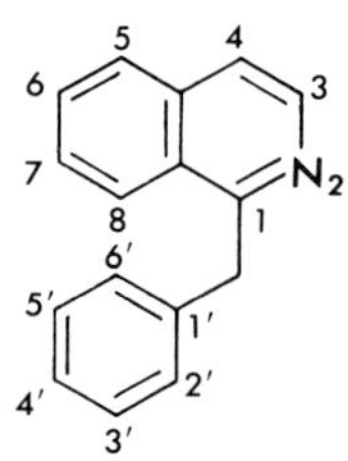

(10)

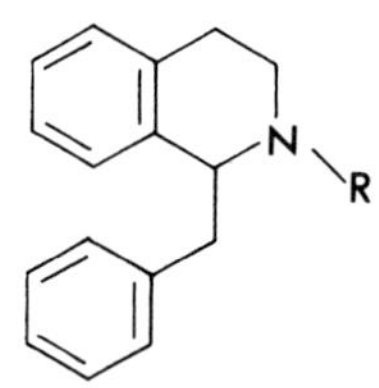

(11)

A recent structural elucidation of a benzyltetrahydroisoquinoline is that for the quaternary alkaloid tembetarine, isolated from *Fagara naranjillo*.[6] The analysis of (+)-tembetarine chloride agreed with the formula, $C_{20}H_{26}NO_4Cl$, and the usual determinations indicated the presence of two O-methyl and two phenolic hydroxyl functions. With diazomethane, the alkaloid gave N-methyl-(+)-laudanosine of known structure and absolute configuration. O,O-Diethyltembetarine iodide, obtained by treatment of the quaternary alkaloid with ethyl iodide, underwent Hofmann degradation. Subsequent oxidation of the product gave the known O-ethylisovanillic acid, thus settling the position of the two substituents in the benzylic moiety of the alkaloid. Alternatively, the O,O-diethyl quaternary salt could be converted to the free base by refluxing with ethanolamine, and oxidation of the free base gave the known lactam 11a. Finally, phenolic oxidative coupling of tembetarine according to the method of Franck gave the known aporphine (+)-laurifoline chloride.

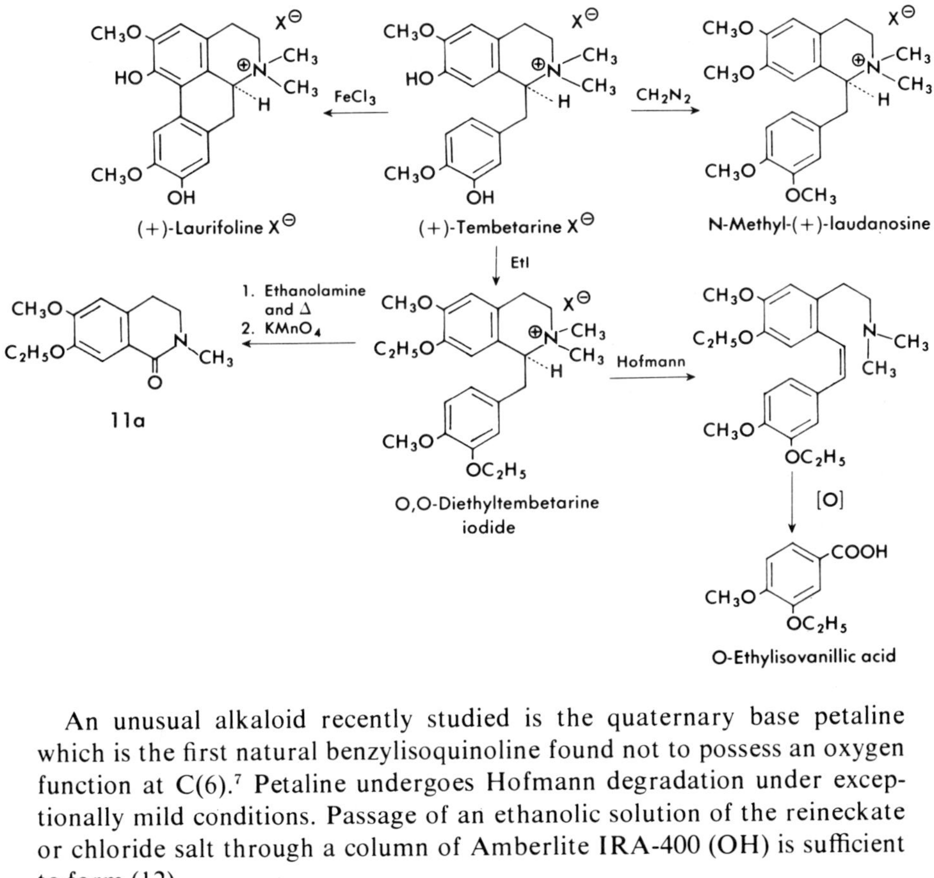

An unusual alkaloid recently studied is the quaternary base petaline which is the first natural benzylisoquinoline found not to possess an oxygen function at C(6).[7] Petaline undergoes Hofmann degradation under exceptionally mild conditions. Passage of an ethanolic solution of the reineckate or chloride salt through a column of Amberlite IRA-400 (OH) is sufficient to form (12).

Petaline methine (12) gave a positive Gibbs test indicating the absence of a substituent *para* to the phenol group. This conclusion was further confirmed by oxidizing its mesylate to *p*-methoxybenzoic acid and the mesy-

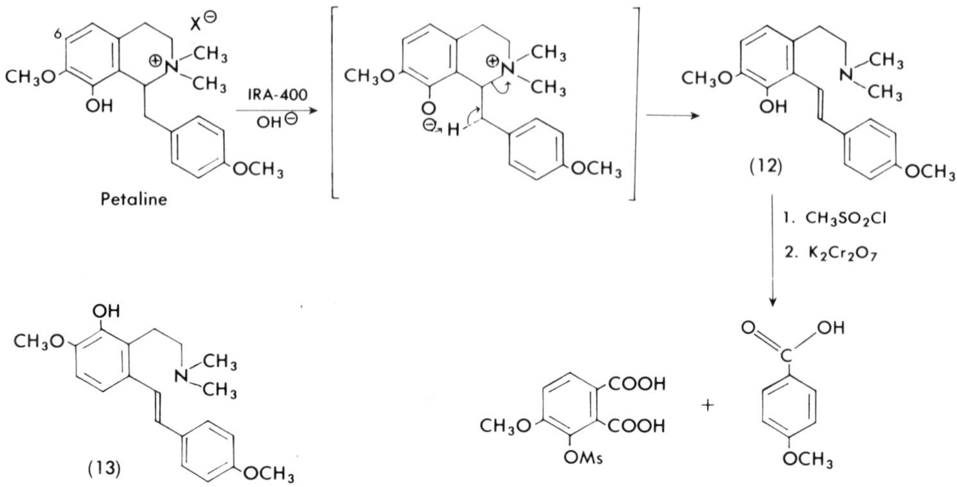

late of 3-hydroxy-4-methoxyphthalic acid. The alternate structure (13) for the methine was eliminated since it was synthesized and found to be different from (12). The structure assigned to petaline also explains the easy Hofmann degradation that the alkaloid undergoes, since the phenolic anion readily abstracts a hydrogen atom *beta* to the quaternary nitrogen as indicated.

2.2 General Methods of Synthesis

A common method of synthesis of alkaloids possessing skeletons (10) or (11) involves condensation of a properly substituted phenylethylamine and substituted phenylacetyl chloride to give an amide. The Bischler–Napieralski ring closure of the amide yields an imine that can be either dehydrogenated to a molecule of type (10), or catalytically hydrogenated, or reduced with sodium borohydride to a base of type (11). In case a substituent in the desired final product is a phenol, this function is usually protected as a benzyl ether during the course of the synthesis. A preparation of a benzyltetrahydroisoquinoline involving protective benzylation is that

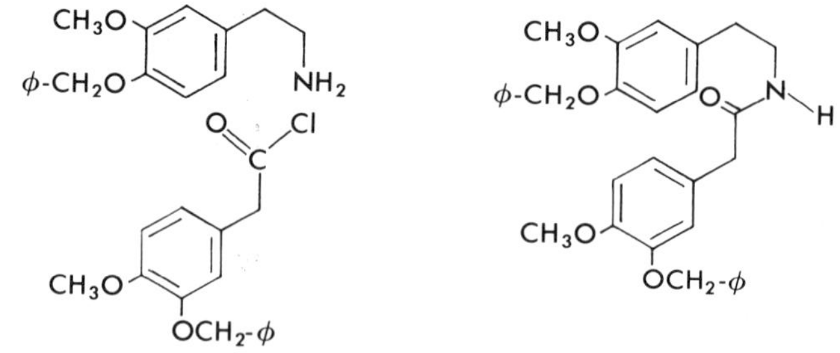

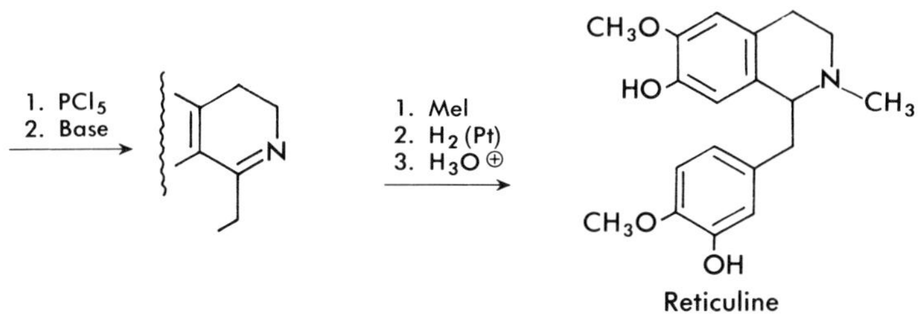

which was carried out to obtain the biogenetically important alkaloid, reticuline.[8]

One of the early biogenetic type syntheses of a benzyltetrahydroisoquinoline was carried out by Späth and Berger in 1930.[9] Condensation of homoveratraldehyde with homoveratrylamine in ether at room temperature, followed by treatment of the intermediate aldimine with hydrochloric acid, produced racemic N-norlaudanosine in 8% yield.

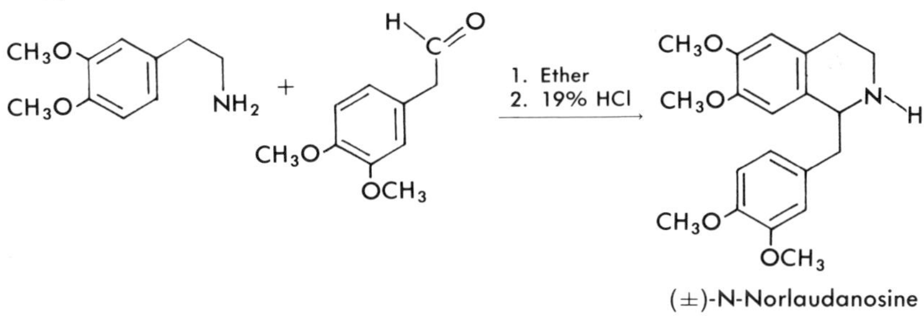

2.3 Absolute Configuration

The absolute configuration of the benzyltetrahydroisoquinoline alkaloids has been studied by Corrodi and Hardegger in a classical experiment whereby (−)-N-norlaudanosine (14) was ozonized to N-β-carboxyethyl-L-aspartic acid of known absolute configuration.[10]

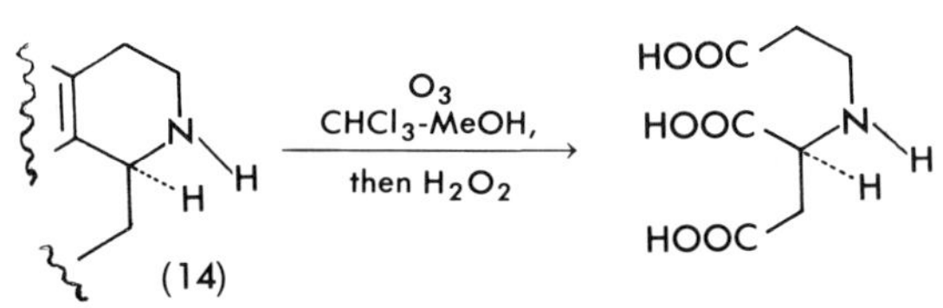

The determination of the chirality of the other benzyltetrahydroisoquinoline alkaloids ultimately revolves, therefore, around direct or indirect comparison with (−)-N-norlaudanosine or one of its derivatives.

2.4 Rearrangements and Cyclization

An interesting over-all 1,3-shift has recently been observed when N-methyl-1,2-dihydropapaverine (15) is heated with dilute acid, the product being the immonium ion (16). On the other hand, when compound (15) is treated with concentrated acid, the tetracyclic N-methylpavine (17) is formed by means of an intramolecular nucleophilic attack on the immonium ion derived from protonation of (15).[11–13]

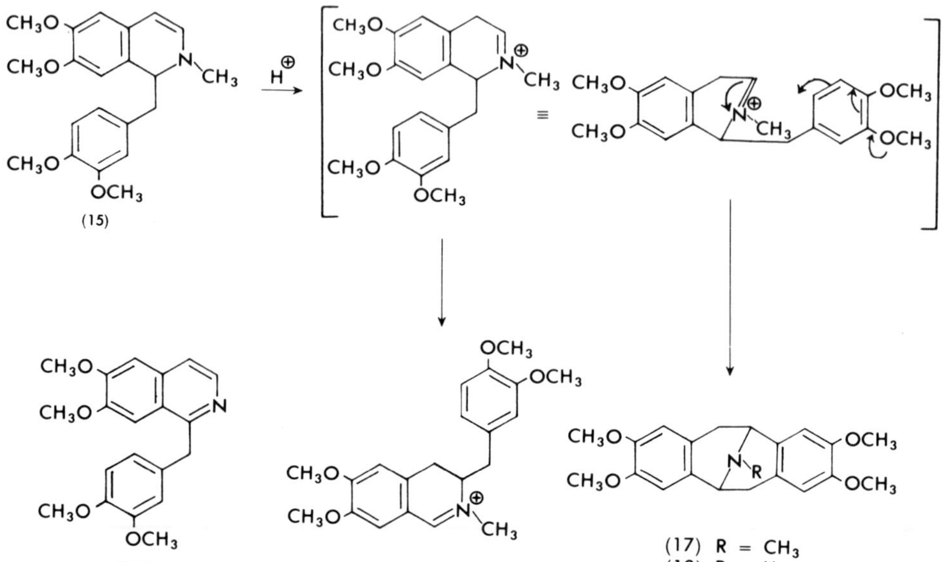

(15) (19) (16) (17) R = CH_3 (18) R = H

Pavine (18) can be obtained directly from papaverine (19) by reduction with tin and hydrochloric acid, the by-product being tetrahydropapaverine. The alkaloid argemonine, which occurs in a variety of Argemone species, was shown in 1963 to correspond to N-methylpavine (17), thus demonstrating the existence of this tetracyclic skeleton in nature.[14]

3. THE BISBENZYLISOQUINOLINE ALKALOIDS[15]

3.1 Introduction

The bisbenzylisoquinoline alkaloids, found mostly among members of the Menispermaceae and Ranunculaceae families, are a group of over sixty compounds all of which are derived biogenetically from the oxidative dimerization of simple benzylisoquinoline bases. In several cases, an eighteen or nineteen membered ring is incorporated as part of the molecule. Some of the bases possess only one diphenyl ether linkage, while others are found to have two or even three such linkages. Because of the different sites at which the oxidative dimerization can occur, several different subgroups for the bisbenzylisoquinoline alkaloids can be recognized. Thus, the bisbenzylisoquinolines represent an excellent example of the unity in diversity that is found among structurally related natural products. Some examples of bisbenzylisoquinoline subgroups are given below.

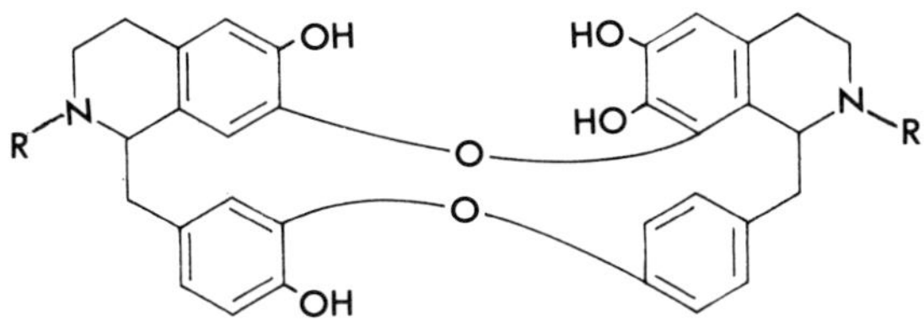

Oxyacanthine Type

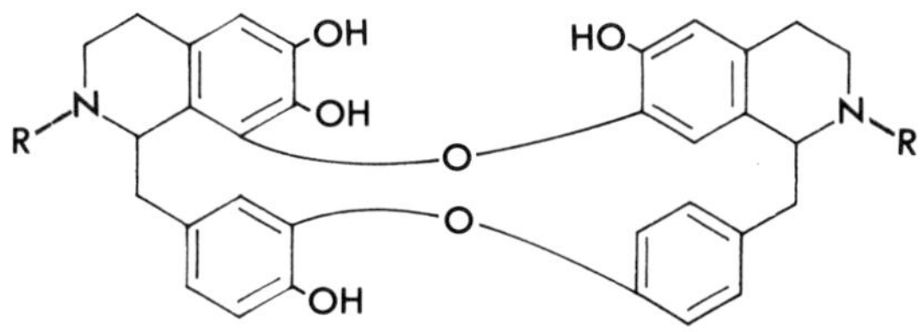

Berbamine Type

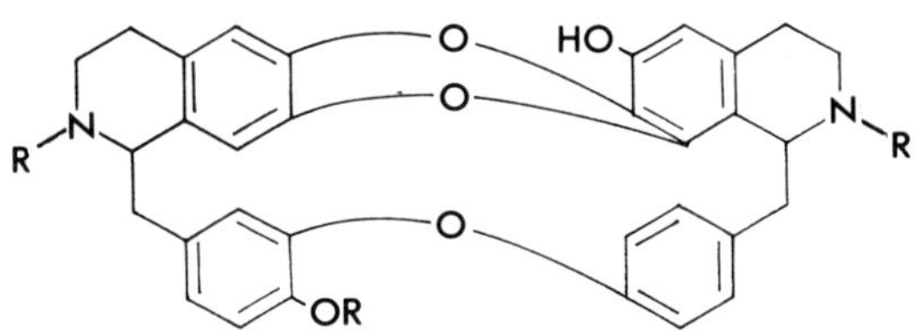

Trilobine Type

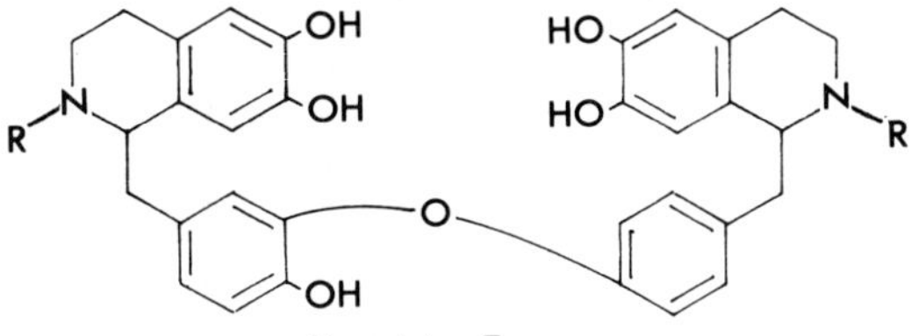

Dauricine Type

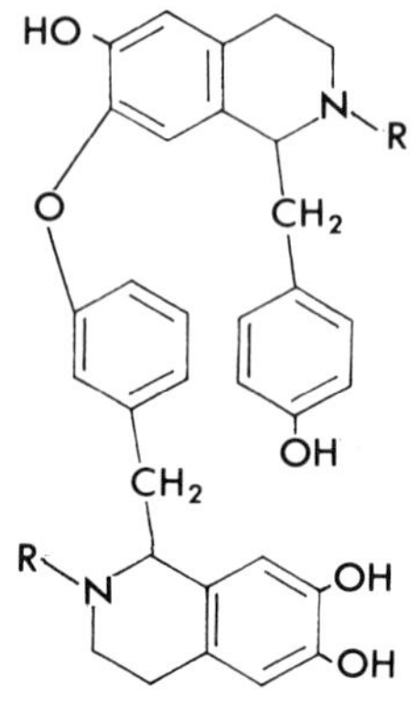

Liensinine Type

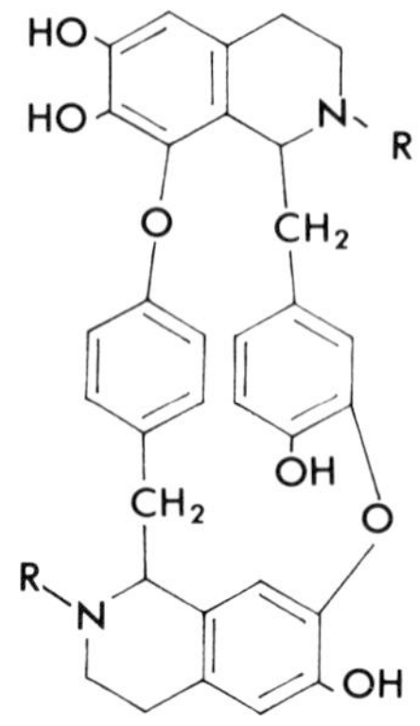

Curine Type

3.2 Structural Determination

There is such a substantial similarity in the chemistry of the different bisbenzylisoquinoline subgroups, that it would be pointless to describe the work on each individual compound in detail. The reactions of these alkaloids fall within well-defined loci, which can be summarized in the following fashion:

(1) Sodium in liquid ammonia reduction of the alkaloid, where the phenolic groups have been protected as the O-methyl or O-ethyl ethers, cleaves the bisbenzylisoquinoline system at the diphenyl ether linkage, resulting in formation of two simple benzylisoquinolines that can be easily identified.

(2) Hofmann degradation of the alkaloidal bisquaternary salt yields a diolefin that can in turn be cleaved with ozone. The resulting aldehydic fragments may then be identified as such or may be oxidized to the carboxylic acid stage prior to identification.

A degradation which exemplifies these principles is that of (−)-pycnamine (20), a bisbenzylisoquinoline alkaloid which has an analysis corresponding to $C_{37}H_{40}N_2O_6$, and possesses three O-methyl, two N-methyl groups, and one phenolic function. Sodium in liquid ammonia cleavage of O-ethylpycnamine yielded (−)-N-methylcoclaurine and (−)-O-ethylarmepavine. Hofmann degradation of the dimethiodide salt of the alkaloid gave the methyl methine (21) which upon ozonization yielded two dialdehydes, (22) and (23). The diacid (24), obtained by further oxidation of (23), was then synthesized via an Ullmann condensation. Finally, treatment of pycnamine with diazomethane gave the known alkaloid phaeanthine of the berbamine subgroup.[16]

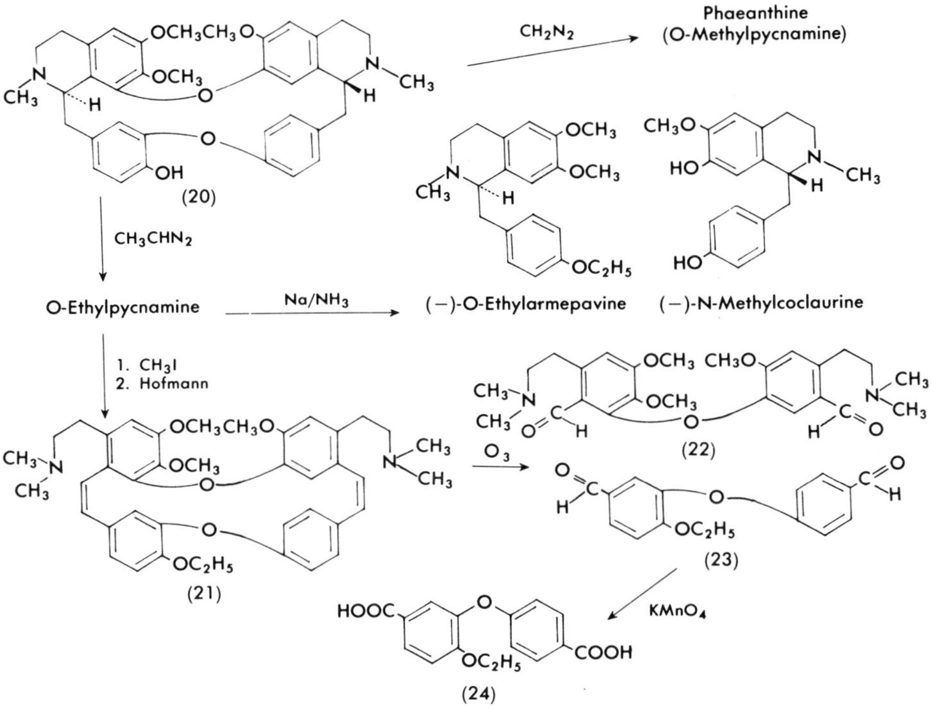

3.3 Syntheses of Bisbenzylisoquinoline Alkaloids

In most syntheses of bisbenzylisoquinoline systems, the Ullmann condensation has been used in the formation of the diphenyl ether linkage. For example, a total synthesis of racemic dauricine by Kametani and Fukumoto used this condensation in the initial step. The synthesis afforded two racemic diastereoisomers, one of which corresponded to (±)-dauricine (25).[17]

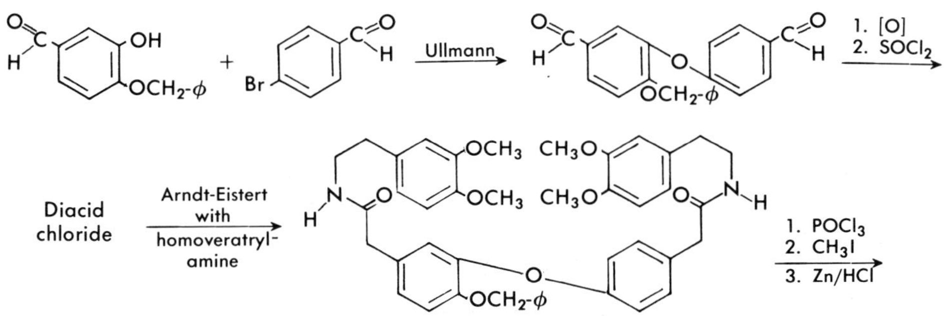

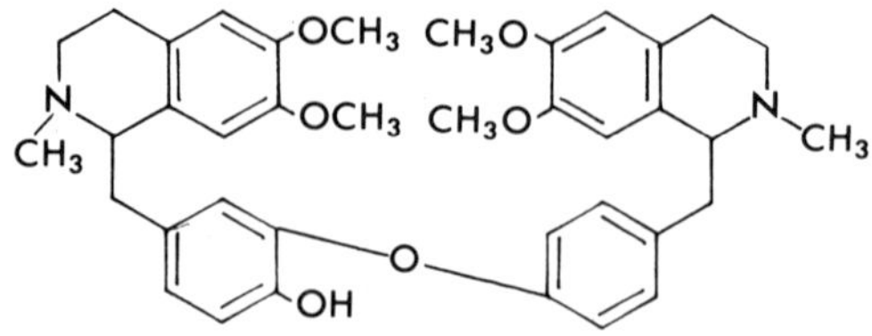

(25)

It has been found possible to obtain bisbenzylisoquinolines by phenolic oxidative coupling. Franck and his associates have demonstrated that oxidation of magnocurarine with potassium ferricyanide results in the formation of the dimer (26).[18]

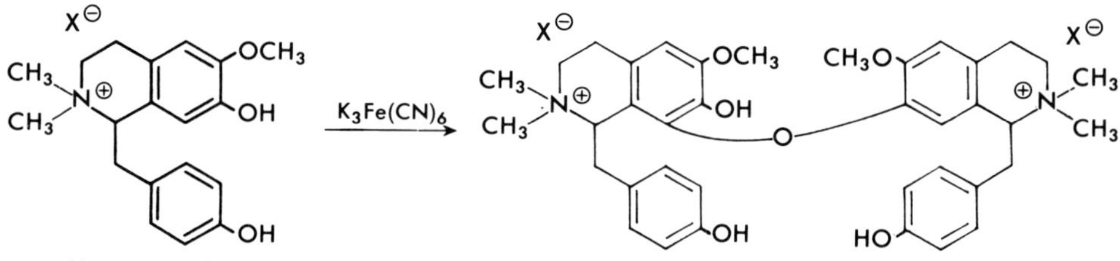

(26)

4. THE APORPHINE ALKALOIDS

4.1 Structural Elucidation

The aporphine alkaloids are an assemblage of over fifty bases, each possessing the skeleton (27). The nitrogen at position 6 may be secondary, in which case the alkaloid is called a noraporphine, or it may be tertiary, or quaternary. Positions (1) and (2) are always substituted.

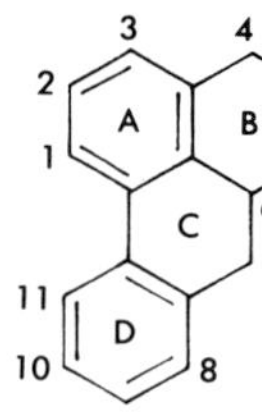

(27)

The classical methods for the structural elucidation of the aporphines follow well-established routes. If the alkaloid is phenolic, it is sometimes first O-methylated or O-ethylated. Permanganate oxidation yields a substituted phthalimide usually arising from rings A and B. When a tertiary aporphine is refluxed with an acyl halide, ring B is cleaved between (6) and (6a) with complete aromatization of ring C and formation of a substituted phenanthrene amide. Finally, two successive Hofmann degradations on a quaternary aporphine will yield a vinyl phenanthrene. This product can be oxidized to a phenanthrene carboxylic acid and then decarboxylated. The resulting substituted phenanthrene may then be compared with known compounds.

Alternatively, an aporphine of unknown constitution can be transformed to another aporphine of known structure. Such a transformation can be brought about by N-methylation, by O-methylation, or by hydrolysis of a methylenedioxy function with sulfuric acid followed by O-methylation of the resulting phenolic groups with diazomethane.

The degradation of the N-methylisocorydinium salt exemplifies some of the above principles.[19]

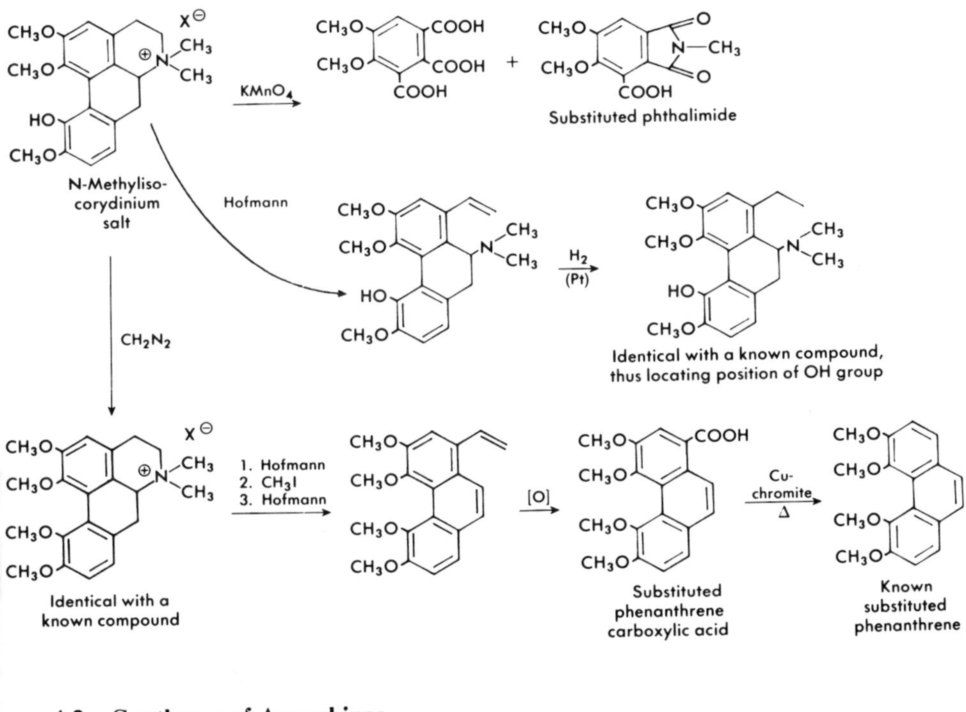

4.2 Syntheses of Aporphines

Three classical synthetic approaches to the aporphines are available. The first involves the condensation of an *o*-nitrophenylacetyl chloride with a properly substituted phenylethylamine, Bischler-Napieralski cyclization, methylation, reduction, and finally Pschorr cyclization.

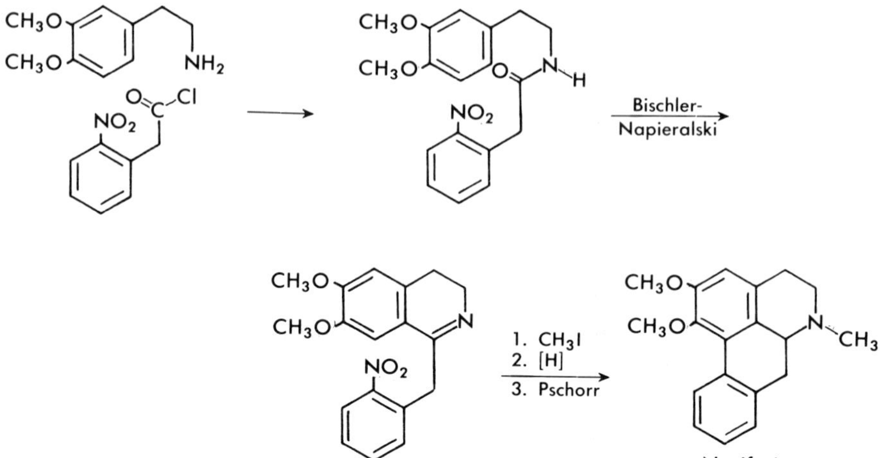

Nuciferine

The second classical route proceeds through nitration of a benzyltetrahydroisoquinoline, followed by reduction, diazotization, and cyclization.

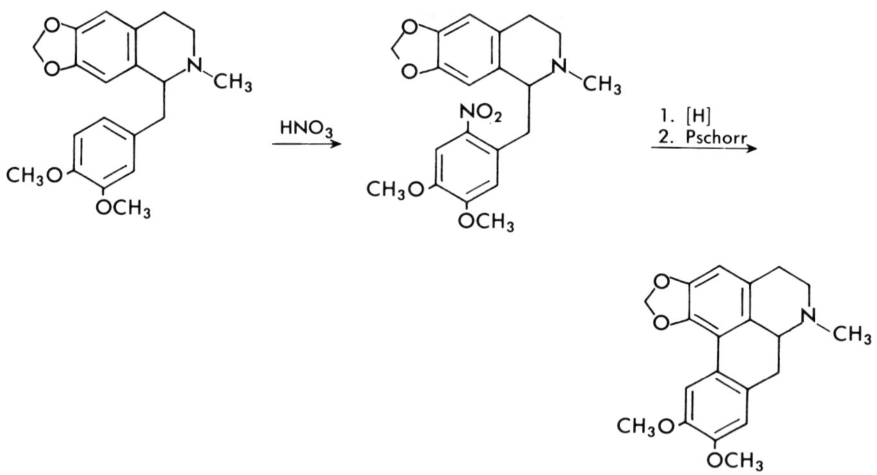

Dicentrine

The base catalyzed condensation of *o*-nitrotoluenes with isoquinolinium salts represents a third route to the aporphines. This classical approach has recently been reexamined and improved.[20]

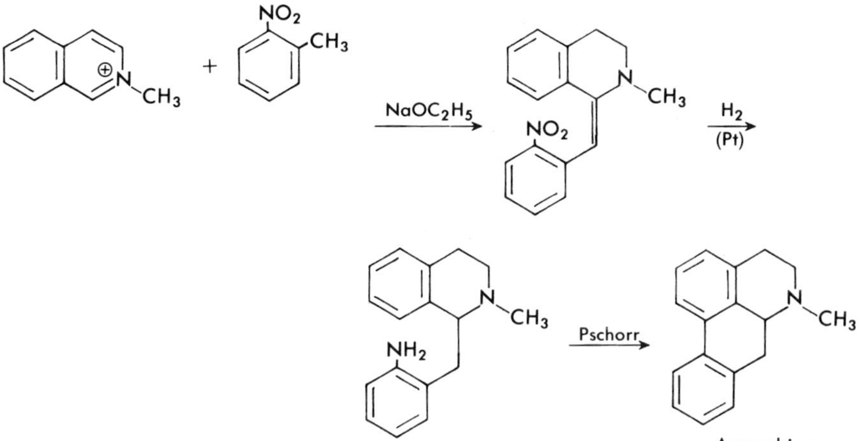

Aporphine

An interesting and novel approach to the aporphines has been described by Franck who oxidized laudanosoline methiodide with aqueous ferric chloride at room temperature to an aporphine salt in excellent yield. Deulofeu and co-workers have used this same reaction to convert the alkaloid (+)–tembetarine chloride into the aporphine (+)-laurifoline chloride. It is worth noting that in both of these oxidative cyclizations the

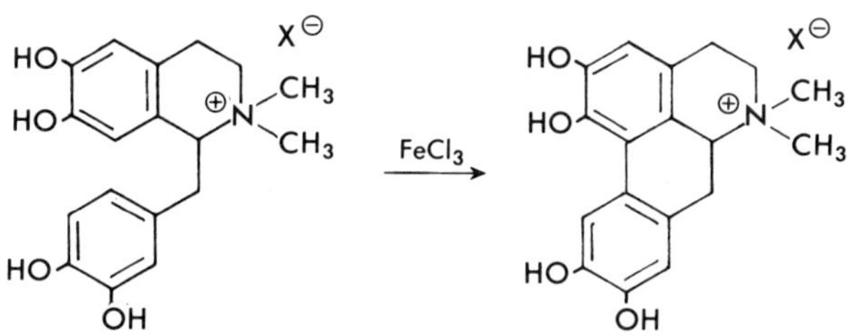

Laudanosoline methiodide

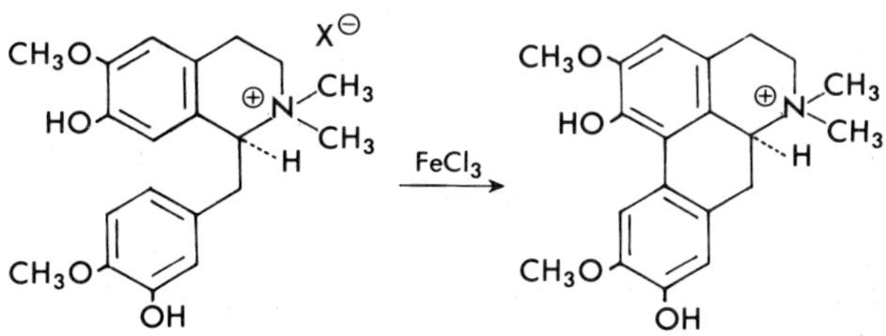

(+)-Tembetarine chloride

(+)-Laurifoline chloride

less-hindered aporphine is produced, and no 1,2,10,11-tetra-substituted aporphine is isolated.[21,22]

4.3 Stereochemistry of the Aporphines

Shamma has pointed out that the aporphines are not planar, but can exist in either of two stereochemical arrangements, (28) or (29).[23] The problem of the absolute configuration of the aporphines can be approached from several avenues. Transformation of the benzyltetrahydroisoquinoline,

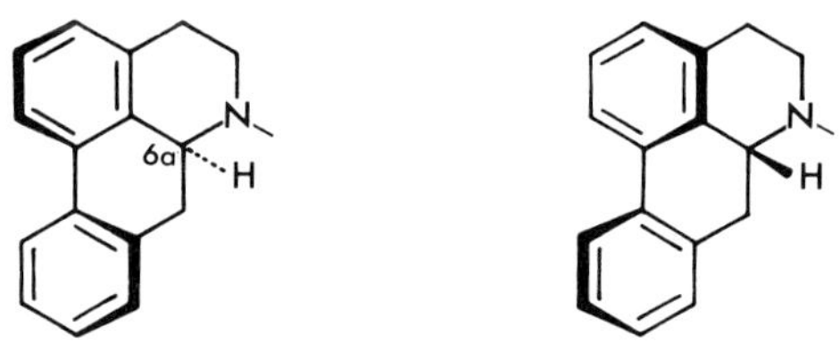

(28) (29)

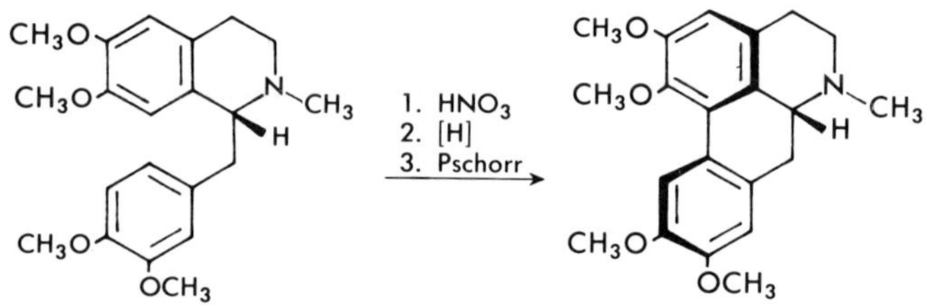

(−)-Laudanosine (−)-Glaucine (non-natural isomer)

(−)-laudanosine into (−)-glaucine indicates that in naturally occurring (+)-glaucine the C(6a) hydrogen must be alpha.

The aporphine (+)-bulbocapnine undergoes hydrogenolysis with sodium in liquid ammonia to yield (+)-morphothebaine that is the antipode of (−)-morphothebaine whose absolute configuration is known because of its relationship to codeine (see p. 47).

Optical rotatory dispersion has also been used to determine the absolute configuration of a number of aporphine alkaloids. Recently, a generalization has been made that, regardless of the positions of the substituents,[24] the high-amplitude Cotton effect centered at 235–245 mμ will be positive with aporphines of absolute configuration (28) and negative for configuration (29). Another simple rule is that aporphines, which exhibit

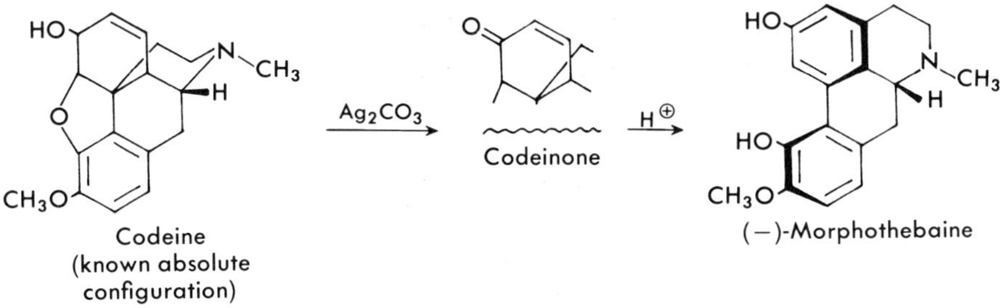

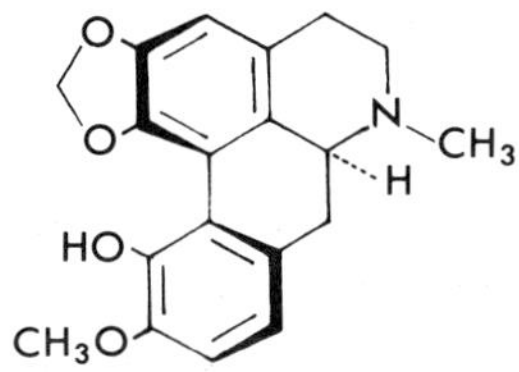

(+)-Bulbocapnine

substantial positive specific rotations at the sodium D line, will have absolute configuration (28), while those that are levoratory will have configuration (29). It has also been noted that if substituents are present at both positions (1) and (11), an aporphine will exhibit a relatively large specific rotation in the range of + 200° or − 200°. This observation can be quite useful, since almost all naturally occurring aporphines are optically active[23] The only other aporphines that may have large specific rotations are those with no substituents on ring D. Aporphines substituted at 1, 2, 10, 11 and at 1, 2, 9, 10 are usually dextrorotatory.

4.4 Ultraviolet Absorption

The positions of the maximum absorptions in the ultraviolet spectra of aporphines depend mainly upon the location of the substituents. In particular, 1,2,9,10-tetrasubstituted aporphines exhibit three absorption maxima, the central peak appearing around 280 mμ. On the other hand, 1, 2,10,11-tetrasubstituted aporphines have a central absorption maximum around 270 mμ.[23]

4.5 Aporphine-Benzylisoquinoline Dimers

Some alkaloids have been recognized as being aporphine-benzylisoquinoline dimers; two of these are thalicarpine (30) and thalmelatine (31) found in *Thalictrum* species.[25,26]

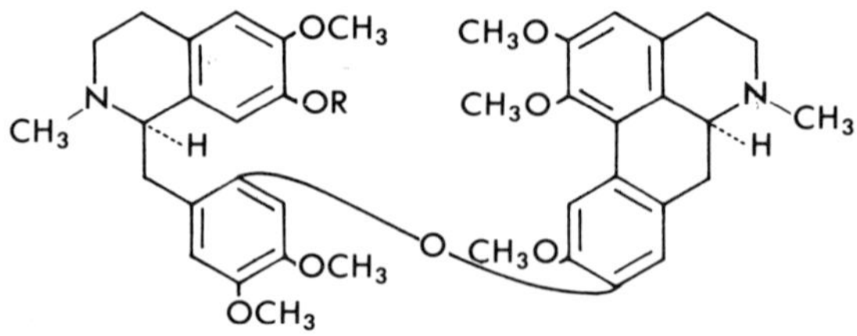

(30) R = OCH_3
(31) R = H

5. THE PROAPORPHINE ALKALOIDS

5.1. Reactions and Stereochemistry

As first pointed out by Barton,[26a] the proaporphine alkaloids can be considered to be the biogenetic precursors of the aporphines. The proaporphines possess a tetracyclic nucleus that incorporates a cyclohexadienone system in addition to the ubiquitous tetrahydroisoquinoline moiety. In some instances, one or both of the double bonds of the dienone system have been reduced.

Crotonosine, one of the first proaporphines isolated, is represented by the expression given below.[27] The alkaloid, $C_{17}H_{17}NO_3$, is dextrorotatory, has one methoxyl, one secondary amine, and one phenolic function, and when refluxed with 3*N* hydrochloric acid in methanol undergoes a dienone–phenol rearrangement to the noraporphine (32).

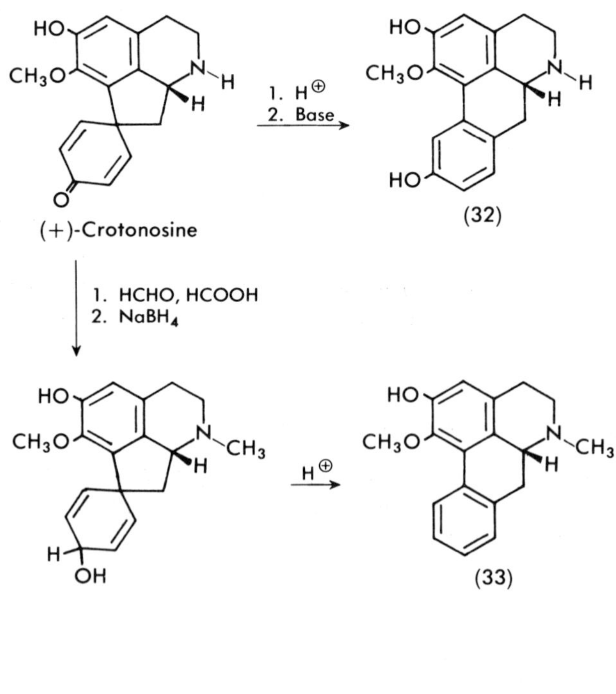

Alternatively, N-methylcrotonosine, prepared by treating crotonosine with formic acid and formaldehyde, when reduced with sodium borohydride and the product heated with mineral acid gave (−)-2-hydroxy-1-methoxyaporphine (33).

Proaporphines may be transformed to benzyltetrahydroisoquinolines by reductive cleavage with sodium in liquid ammonia, and in this fashion the alkaloid (+)-pronuciferine was reduced to (−)-armepavine. Since the absolute configuration of (−)-armepavine is known, it follows that (+)-pronuciferine must have the identical configuration at the corresponding asymmetric carbon atom. Furthermore, because (+)-pronuciferine can be converted by reduction of the carbonyl followed by acid rearrangement into the aporphine alkaloid (−)-nuciferine, the latter base must also possess the identical absolute configuration.[28]

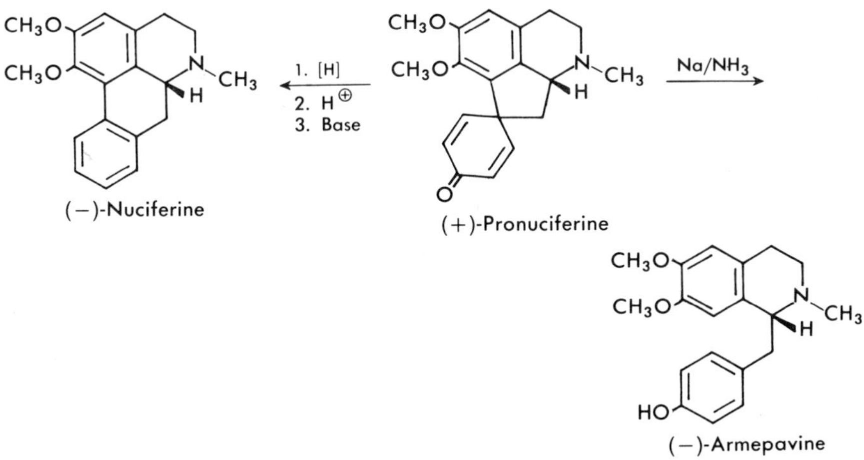

(−)-Nuciferine (+)-Pronuciferine (−)-Armepavine

5.2 Syntheses of Proaporphine Alkaloids

Bernauer has reported a synthesis of racemic pronuciferine.[29] The starting aminoester was N-methylated by the Clarke–Eschweiler procedure, and the product internally acylated to the tricyclic ketone (34). A Darzens glycidic ester synthetic sequence then gave the aldehyde (35) which was condensed with methylethinyl ketone to yield the desired product.[29]

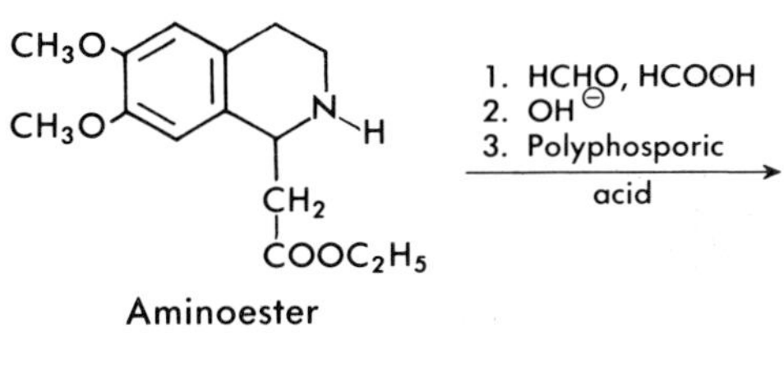

Aminoester

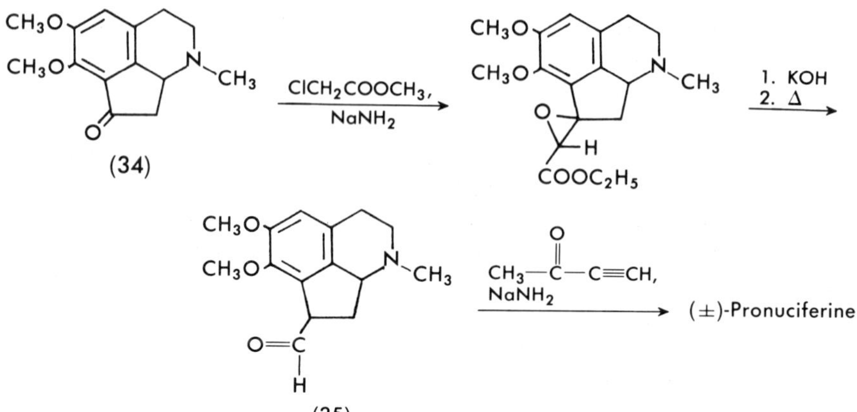

(34)

(35)

Battersby and Brown have also carried out a simple biogenetic-type synthesis of the proaporphine framework via phenolic oxidative coupling of the benzyltetrahydroisoquinoline base (36). Subsequent reduction of the carbonyl of the proaporphine followed by acid catalyzed rearrangement then gave the racemic form of the naturally occurring base isothebaine.[30]

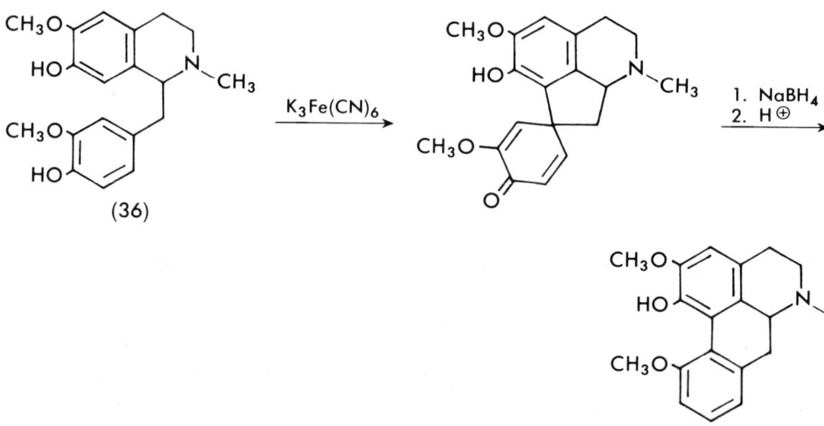

(36)

(±)-Isothebaine

6. LIRIODENINE (OR OXOAPORPHINE) TYPE ALKALOIDS

6.1 Structural Elucidation and Synthesis

The liriodenine alkaloids are a small collection of highly aromatic compounds formed in nature by the oxidation of aporphine bases. The alkaloid liriodenine itself, which can be considered the prototype for this group, was isolated from the heartwood of the yellow poplar, *Liriodendron tulipifera*. The compound is yellow, optically inactive, and has an analysis corresponding to $C_{17}H_9NO_3$. A carbonyl group absorbing at 1639 cm^{-1} could be defined as a ketone because of oxime formation. Additionally, liriodenine was shown to possess one methylenedioxy function, but no hydroxyl, O-methyl, or N-methyl groups.

Oxidation of liriodenine (37) yielded the azaanthraquinone carboxylic acid (38), which was decarboxylated to the known azaanthraquinone (39). With these data on hand, and bearing in mind the pronounced ultraviolet absorption together with the low hydrogen content of the molecule, Taylor proposed the aromatic structure (37) for liriodenine.[31]

This structural assignment was confirmed through a synthesis of the alkaloid utilizing the classical Pschorr cyclization and starting with the nitro compound (40).[31]

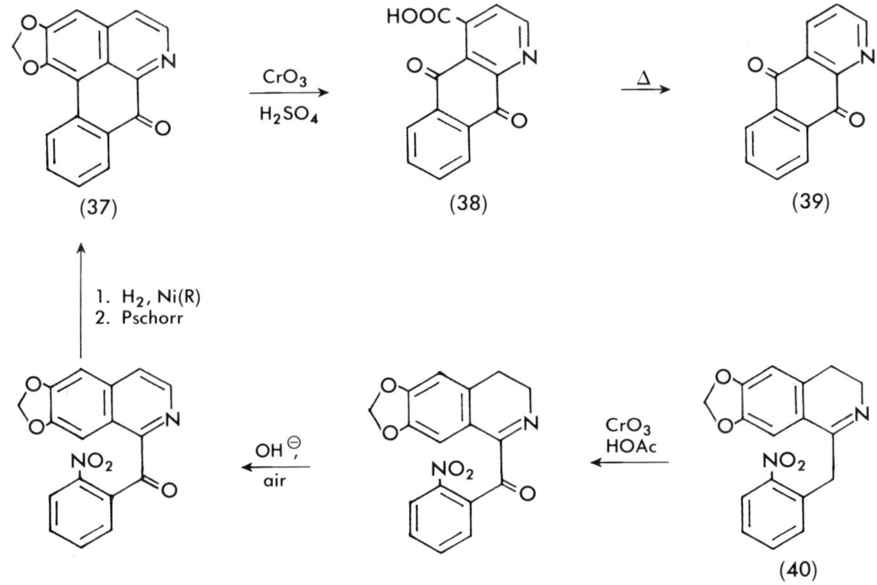

Liriodenine-type bases can also be prepared by simple chromium trioxide–pyridine oxidation of nonphenolic aporphines.[32]

7. CULARINE ALKALOIDS

7.1 Structural Elucidation and Absolute Configuration

The cularine group of alkaloids comprises less than half a dozen bases, all of which have been found in the family Papaveraceae. The alkaloids possess a unique dihydrooxepine system as indicated in expression (41) for cularine.

Cularine, $C_{20}H_{23}NO_4$, is dextrorotatory and has one N-methyl and three O-methyl functions, the fourth oxygen being part of a diphenyl ether system. The Hofmann degradations as well as the reductions and oxidations that have been carried out on cularine are summarized below. The isolation of such degradation products as 4-methoxyphthalic acid and 2, 4,5-trimethoxybenzoic acid served to define the location of the substituents on the two aromatic rings.

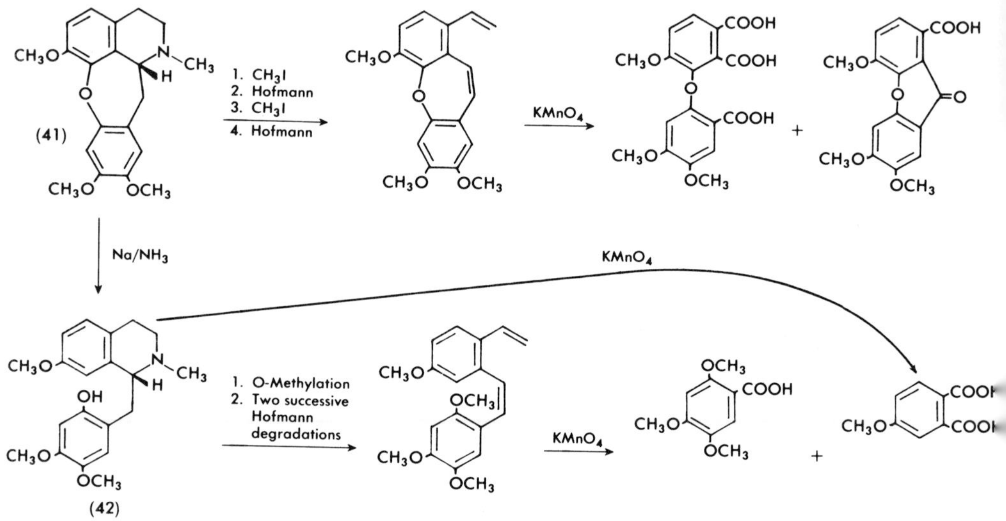

The benzylisoquinoline (42), whose absolute configuration was established by ORD, served to settle the absolute configuration of cularine.[33]

7.2 Syntheses of Cularine

Cularine has been synthesized by Kametani and Fukumoto who started with the ketone (43), originally prepared by Manske, and proceeded via a Pomeranz–Fritsch cyclization.[34]

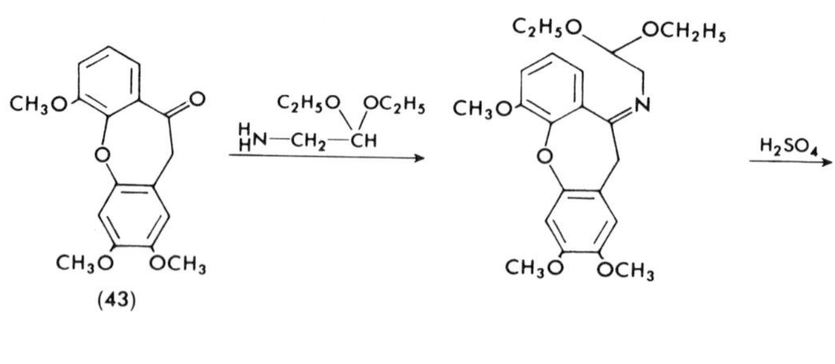

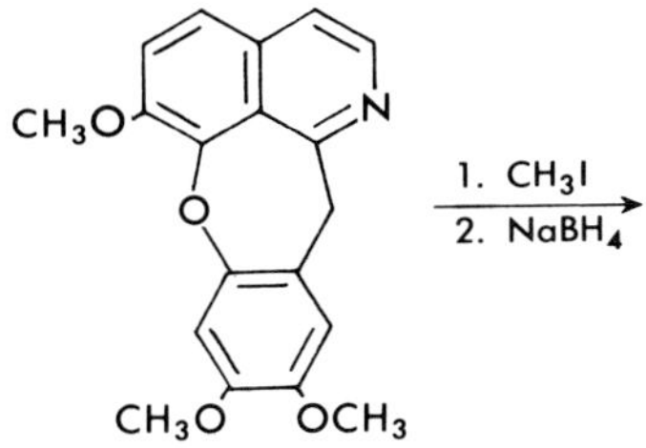

(41)

In a later synthesis, a completely different approach was used:[35]

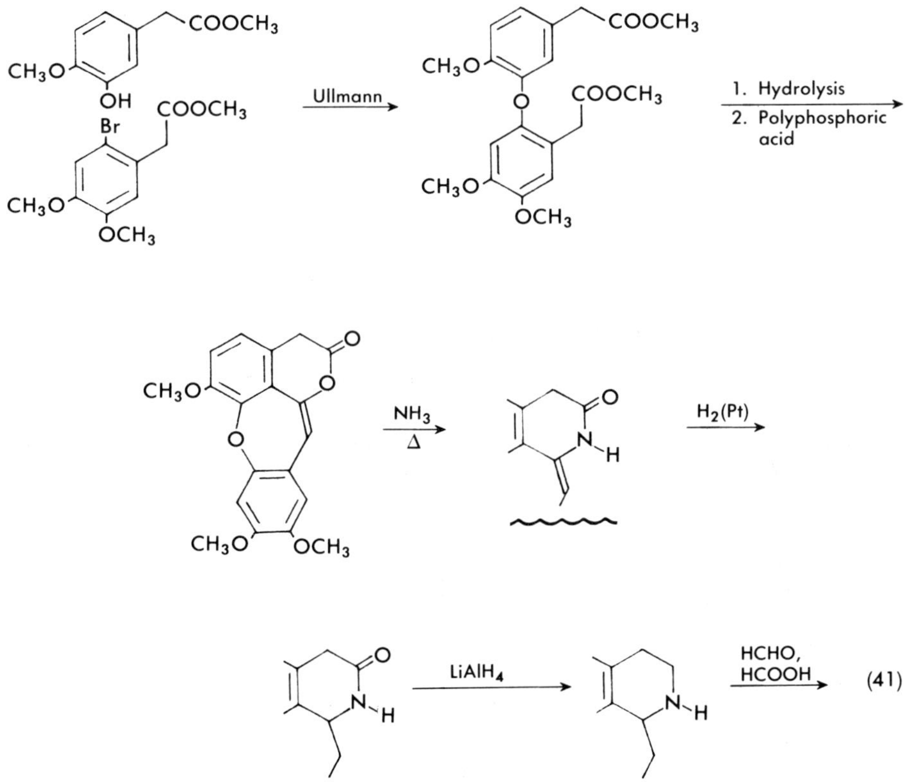

(41)

The most interesting synthetic approach in the cularine series involves the phenolic oxidative coupling of a diphenolic benzyltetrahydroisoquinoline to a dienone which then underwent acid catalyzed rearrangement to a cularine-type molecule as indicated below.[36]

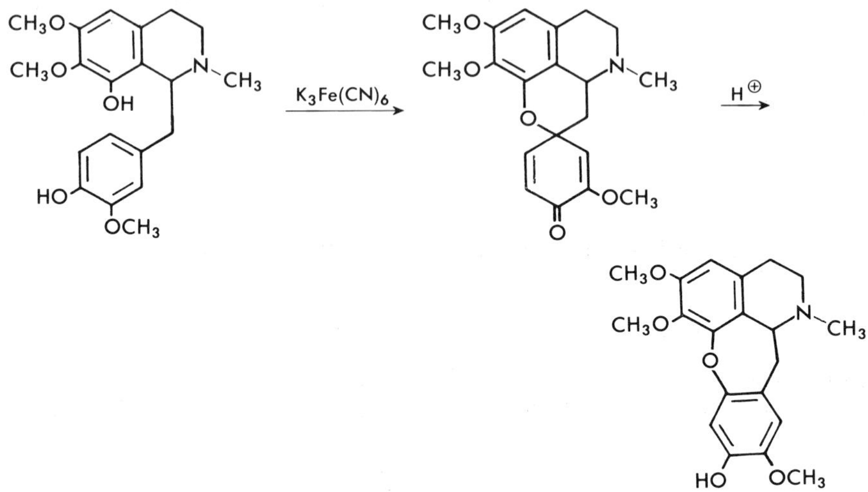

8. THE DIBENZOPYRROCOLINE ALKALOIDS

8.1 Historical Background and Structural Elucidation

In 1932, Robinson[37] and Schöpf[38] reported the results of their independent efforts to synthesize the aporphine nucleus via oxidative coupling of laudanosoline. The reaction, however, did not take the expected course, and the product was shown to be the quaternary dibenzopyrrocoline system (44) resulting from carbon-nitrogen bonding.

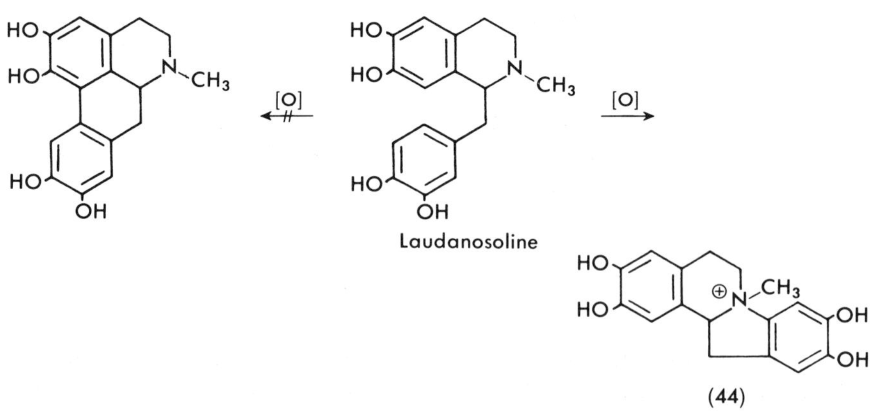

(44)

Twenty years later, Hughes and co-workers characterized two new alkaloidal salts, isolated from the bark of the Australian shrub *Cryptocaria bowiei* (Lauraceae) that were indeed shown to be based upon the dibenzopyrrocoline nucleus.

The first of these alkaloids, cryptaustoline, $C_{20}H_{24}O_4N^+X^-$, is levorotatory, and contains one phenolic hydroxyl and three O-methyl groups. It was converted to its O-methyl derivative which was then extensively degraded as shown.

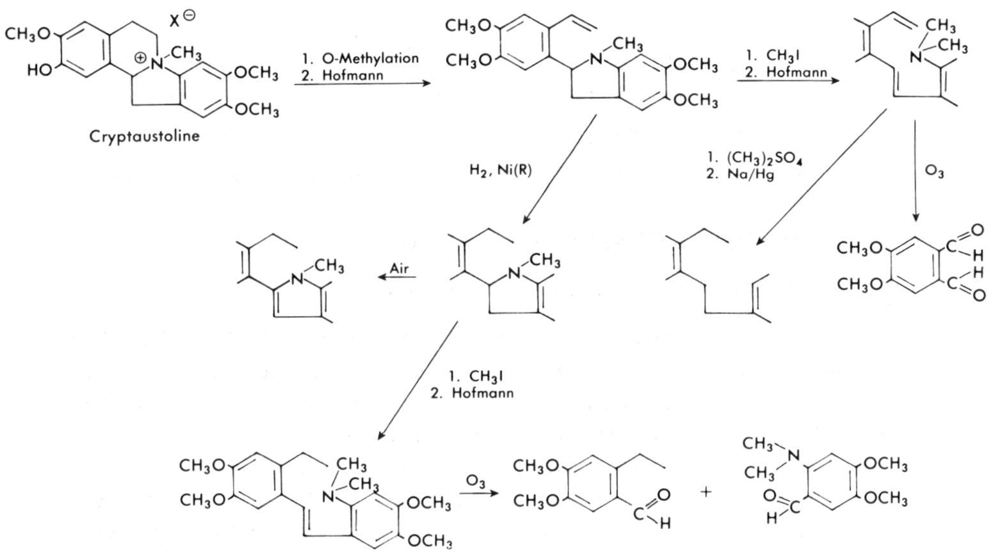

(45)

The position of the phenolic function in cryptaustoline was located by converting O-ethylcryptaustoline to 4-methoxy-5-ethoxy-2-ethylbenzaldehyde instead of to the dimethoxy aldehyde (45).[39]

Cryptowoline iodide, the second alkaloid, was subsequently shown to correspond to structure (46).

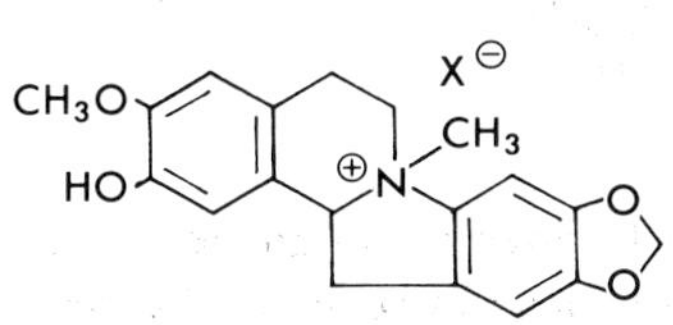

(46)

8.2 Synthesis and Absolute Configuration

The synthesis of O-methylcryptaustoline iodide was reported in the same year as the structural elucidation of the alkaloid. Following the Robinson approach, (+)-laudanosoline hydrobromide of known absolute configuration was oxidized with chloranil to the tetracyclic salt dehydrolaudanosoline (44). O-methylation of this salt followed by addition of sodium iodide yielded the desired (−)-O-methylcryptaustoline iodide identical with material derived from the natural source.[40] The preparation of synthetic (±)-cryptowoline iodide has also been described.[41]

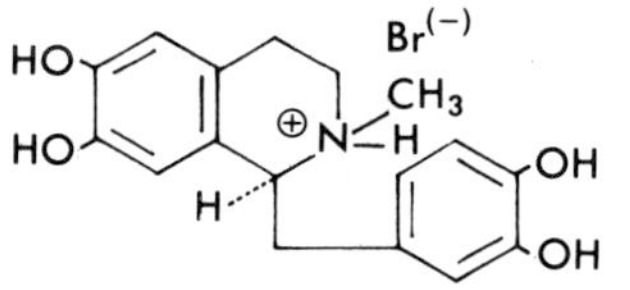

(+)-Laudanosoline hydrobromide

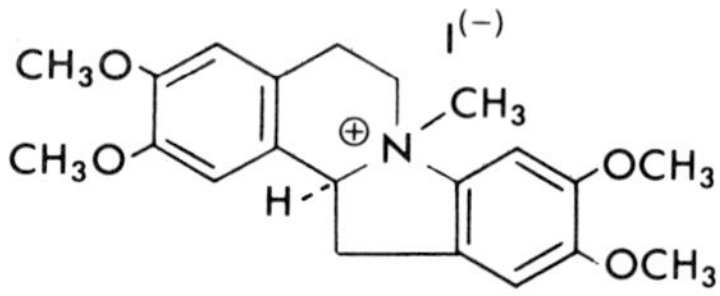

(−)-O-Methylcryptaustoline Iodide

9. THE PROTOBERBERINE ALKALOIDS

9.1 Structural Elucidation and Interconversions

The protoberberine alkaloids include over thirty-five alkaloids based on the tetracyclic structures (47) and (48). In the discussion to follow, alkaloids of type (47) will be referred to as quaternary protoberberines, while those of type (48) will be called tetrahydroprotoberberines.

Both type (47) and (48) alkaloids are almost always substituted at C(2) and (3), and usually at either C(9) and (10), or at C(10) and (11). In several cases a methyl group is located at C(13). A few of the alkaloids possess a hydroxylic function at C(1) or at C(13). Berberastine and thalidastine, two quaternary protoberberine alkaloids recently studied, were found to have an alcoholic group at C(5).

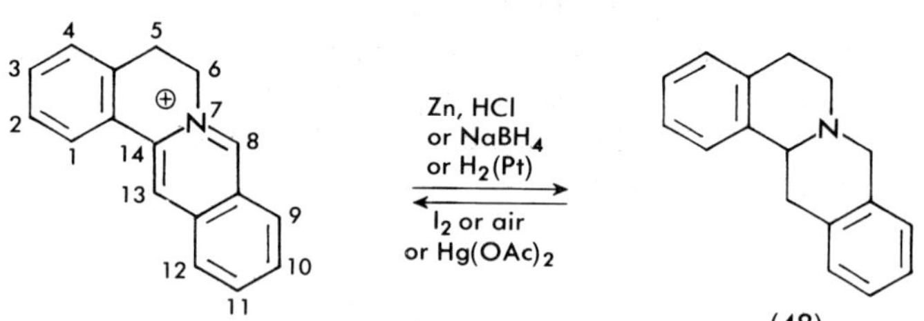

(47) (48)

Quaternary protoberberines (47) can be readily reduced to their tetrahydro analogs (48), and these in turn can be reoxidized to the quaternary stage. In the presence of excess hydroxide ion, the alkaloid berberine gives a pseudo-base, which, when heated with concentrated alkali, undergoes a Cannizzaro type oxidation-reduction with berberine itself to give dihydroberberine and oxyberberine; the latter compound corresponds to the naturally occurring berlambine.

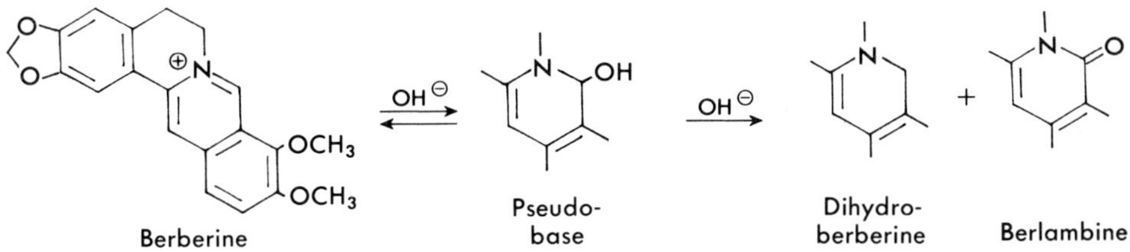

Degradative studies for purposes of structural determination are usually carried out on the tetrahydro derivatives. Oxidation of such compounds by means of potassium permanganate has been used extensively. If a free phenolic function is originally present, the alkaloid is first treated with diazomethane or diazoethane to obtain the O-alkyl derivative. This transformation usually serves the dual purpose of protecting the phenolic group and, in case of O-ethylation, labeling the phenolic hydroxyl in order to ascertain its position.

Under mild, usually neutral oxidative conditions, a substituted bicyclic lactam is formed. Stronger conditions result in the formation of substituted phthalic acids which are often characterized as their N-methyl or N-ethylimides. The Manske degradation of O,O-diethylscoulerine is a case in point.[42]

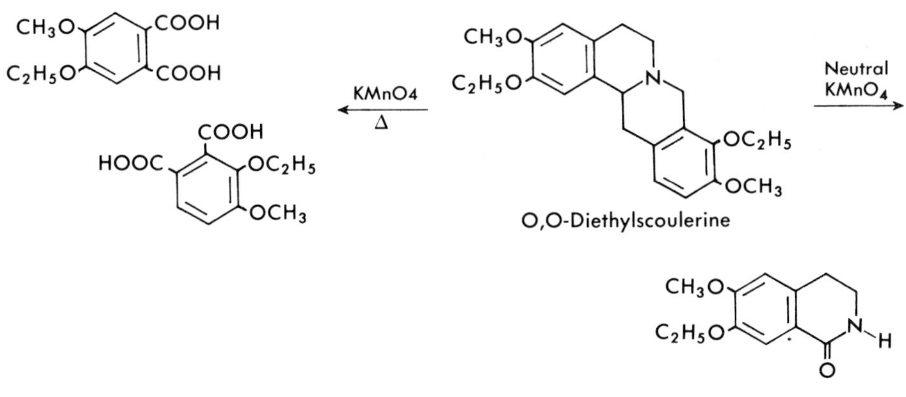

The structural elucidation of the C(13) methylated tetrahydroprotoberberines is somewhat more complex, and is based not only on the study of the products of permanganate oxidation, but also upon a scheme involving an Emde degradation. This is illustrated in the degradation of the tetrahydroprotoberberine alkaloid thalictricavine.[43]

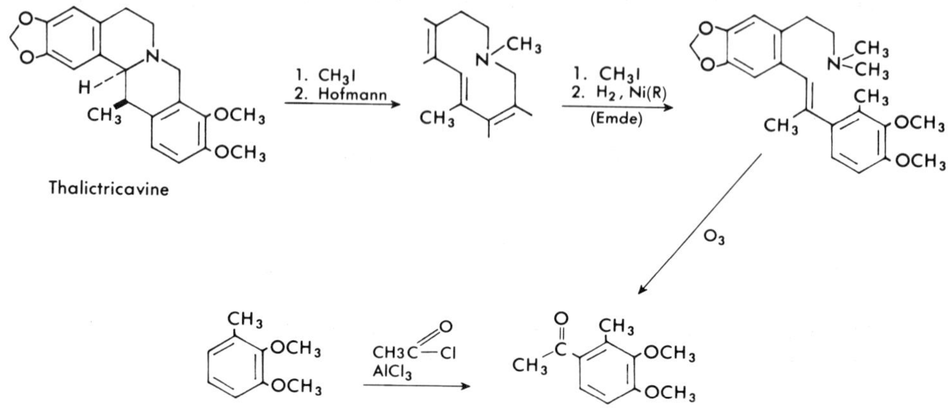

A method has also been worked out for converting simple quaternary protoberberines into C(13) methylated tetrahydroprotoberberines. This transformation, which was a harbinger of Stork's enamine alkylation method, has been utilized to interrelate the two quaternary alkaloids coptisine chloride and corysamine chloride. Treatment of acetonyl coptisine with methyl iodide results in alkylation at C(13). The addition of acetone can then be reversed, and the resulting quaternary salt is reduced with zinc in acetic acid to tetrahydrocorysamine.[44]

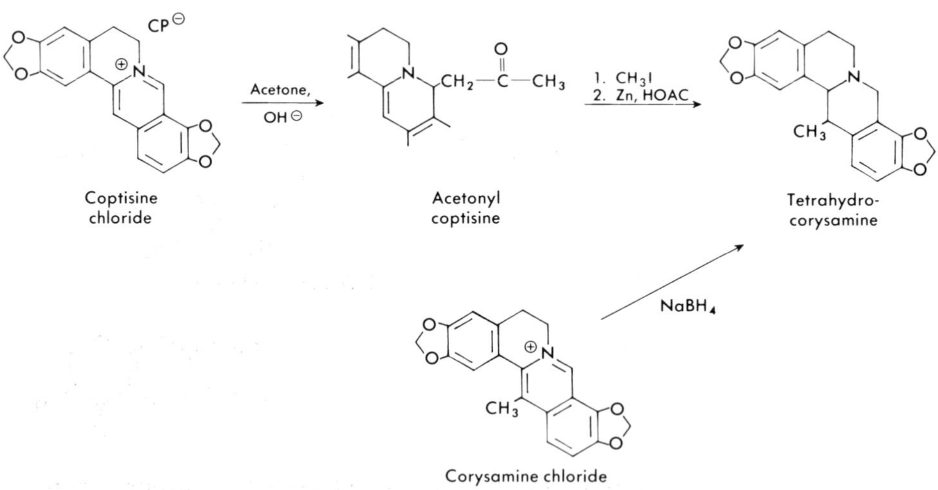

All quaternary protoberberines are colored, and the color changes involved in going from an acidic to a basic alkaloidal solution may give an insight into the nature and position of the aromatic substituents.[45]

9.2 Methods of Synthesis

The most common route to the tetrahydroprotoberberines is through the Mannich condensation of a benzyltetrahydroisoquinoline with formaldehyde. When the substituents on the benzyl moiety are hydroxyls, cyclization occurs indiscriminately to give a mixture of 9, 10, and 10, 11 substituted tetrahydroprotoberberines. But if the substituents are methoxyl groups, the main product is 10,11 substituted.

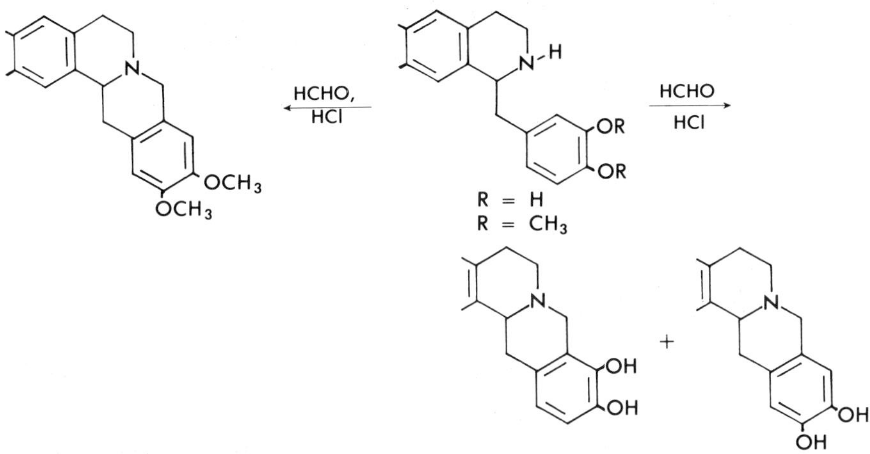

Recently, the above method of forming tetrahydroprotoberberines has been reexamined and extended.[46,47]

In 1955, Battersby and co-workers showed that tetrahydroprotoberberines can be obtained from dihydroisoquinolines, as was shown by their synthesis of norcoralydine.[48]

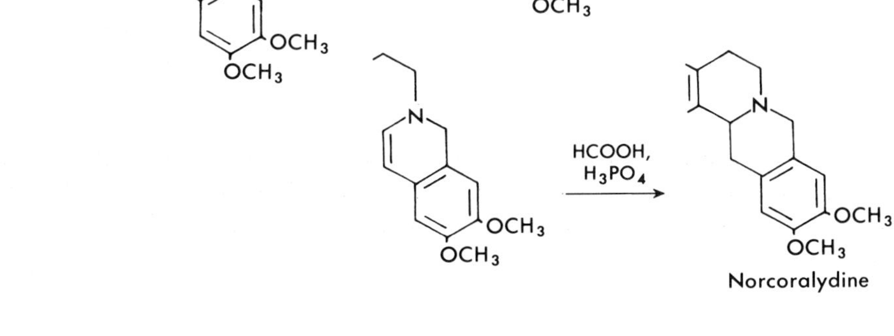

Bradsher and Dutta[49] have developed a synthesis of tetrahydroprotoberberines that involves the condensation of a substituted isoquinoline-1-carboxaldehyde (or better the corresponding oxime) with a benzyl bromide derivative. Cyclization of the resulting quaternary salt gives a dehydroberberinium salt that can be reduced to a tetrahydroprotoberberine.

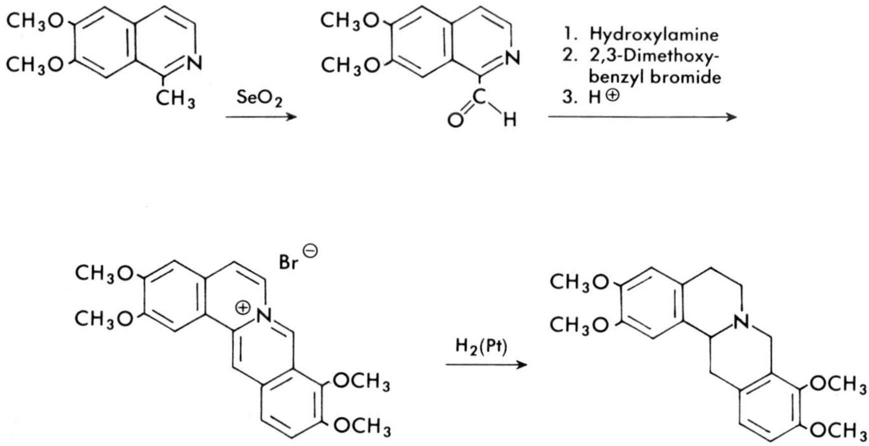

Dehydropalmatinium bromide

Tetrahydropalmatine

A related method is the acid-catalyzed cyclization of an amide such as shown below:[50]

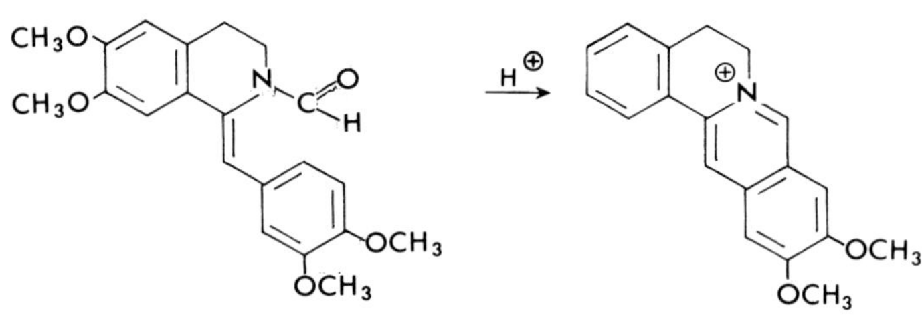

Another synthesis of a tetrahydroprotoberberine involves treatment of the quaternary halide (49) first with aqueous alkali to give compounds (50) and (51), and then with acid to achieve cyclization to (52) and (53).[51]

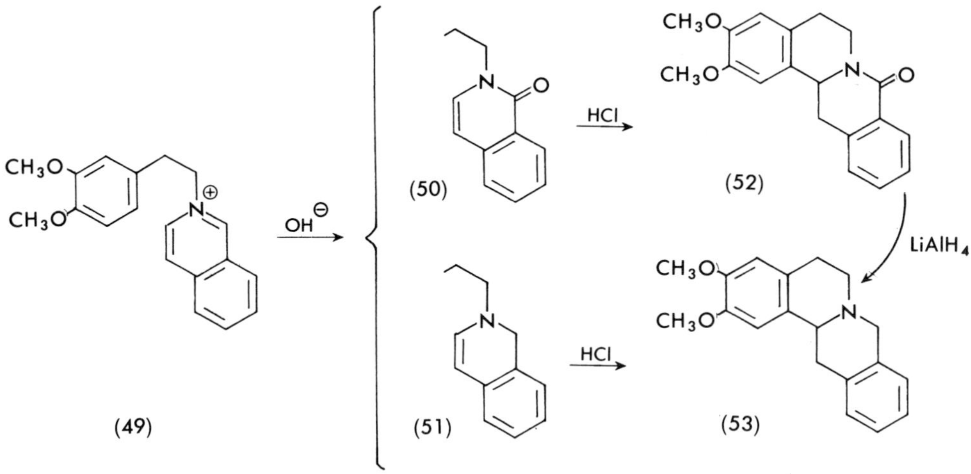

9.3 Absolute Configuration

The key compound in the study of the absolute configuration of the tetrahydroprotoberberines is (−)-N-norlaudanosine whose absolute configuration is known because of its oxidation to N-β-carboxyethyl-L-aspartic acid. Cyclization of (−)-N-norlaudanosine in the presence of formaldehyde and hydrochloric acid give (−)-norcoralydine whose absolute configuration is, therefore, elucidated.[10]

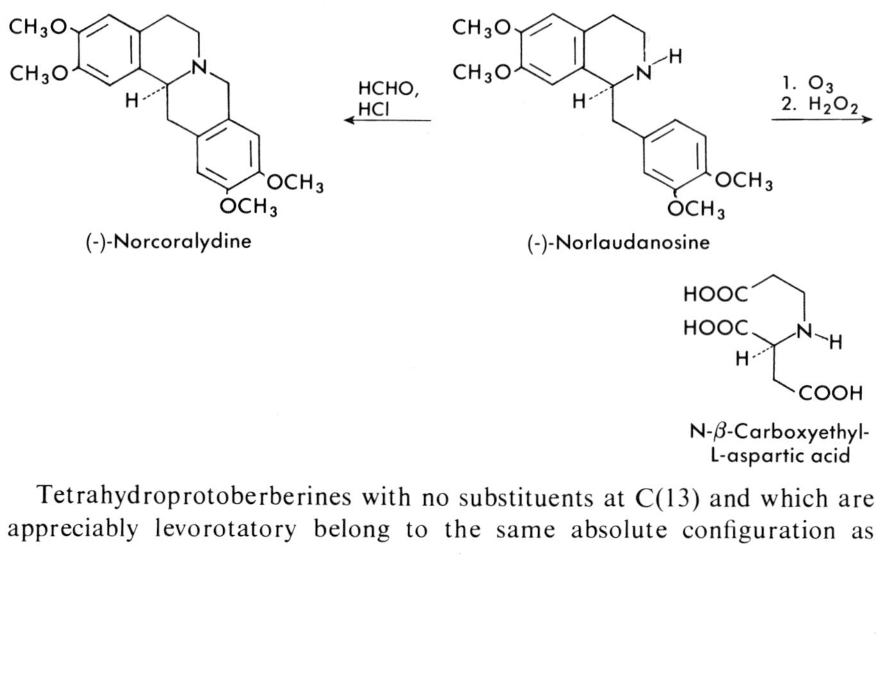

Tetrahydroprotoberberines with no substituents at C(13) and which are appreciably levorotatory belong to the same absolute configuration as

(–)-norcoralydine, while dextrorotatory bases belong to the opposite configuration. The stereochemistry of the C(13) methylated tetrahydroprotoberberines has been discussed by Jeffs.[52]

10. THE PROTOPINE ALKALOIDS

10.1 Structural Elucidation and Reactions

The protopine alkaloids incorporate the tricyclic frame (54). A few of the alkaloids are optically active due to the presence of an additional methyl group at C(13). The two main alkaloids of this group are cryptopine and protopine.

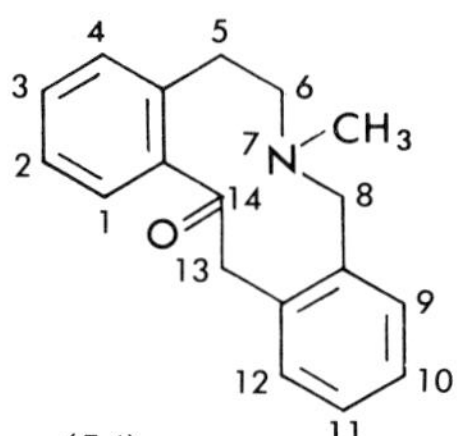

(54)

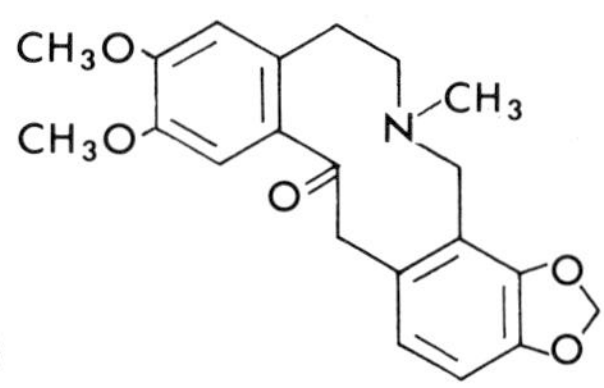

Cryptopine

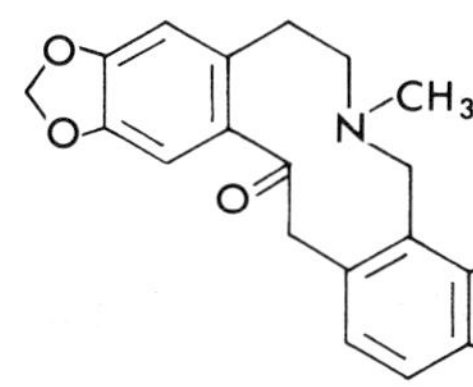

Protopine

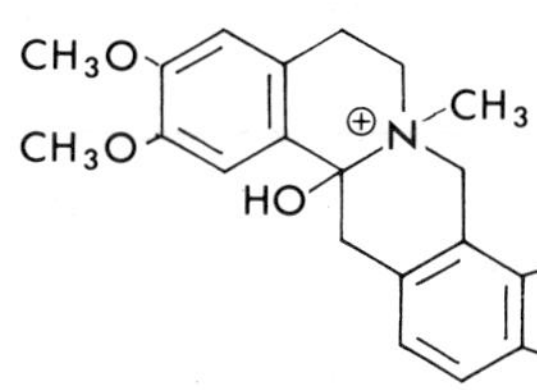

(55)

The structure of the protopine alkaloids was elucidated almost fifty years ago through the work of W. H. Perkin, jun., and his school. Some of the degradative evidence on cryptopine may be summarized as follows:

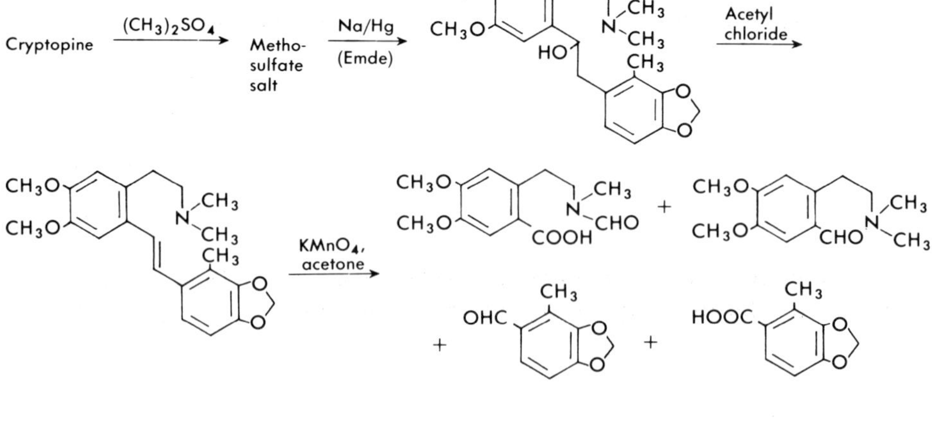

The paramount chemical fact about the protopine alkaloids is the ease of the transannular reaction between the tertiary nitrogen and the carbonyl at C(14). Owing to the amide-like interaction between the basic nitrogen and the carbonyl function, $O{=}C{<}\curvearrowleft{:}N{<}$, the infrared carbonyl band of cryptopine is at 1675 cm^{-1} rather than at the usual 1692 cm^{-1} which is to be expected of a simple conjugated ketone. However, in cryptopine methiodide the carbonyl band is at the expected wavelength of 1686 cm^{-1}, so that no transannular effect is present. Acid salts of cryptopine exhibit no infrared carbonyl bands and must, therefore, exist as the tetracyclic structure (55).

Internal cyclization of cryptopine can also be effected by treating dihydrocryptopine with acetyl chloride or phosphorus oxychloride. Two diastereoisomeric quaternary methochlorides (56) and (57), called isodihydrocryptopines, which differ in stereochemistry at C(14), can thus be obtained.

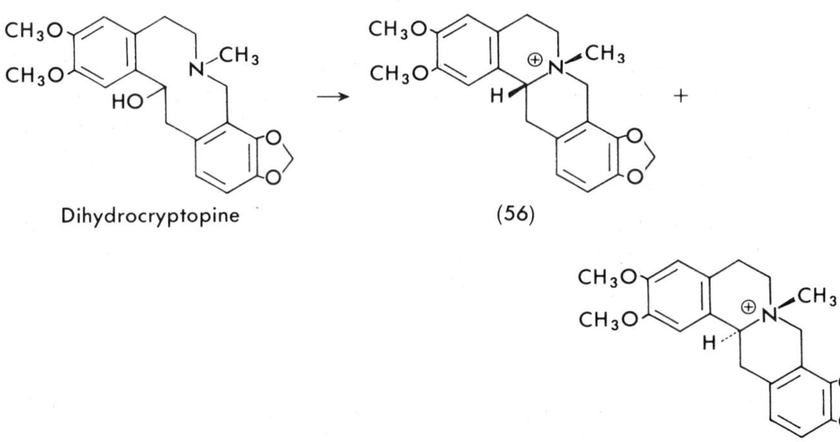

Dihydrocryptopine (56) (57)

The action of phosphorus oxychloride on cryptopine itself leads to the formation of isocryptopine chloride which has been formulated as (58).

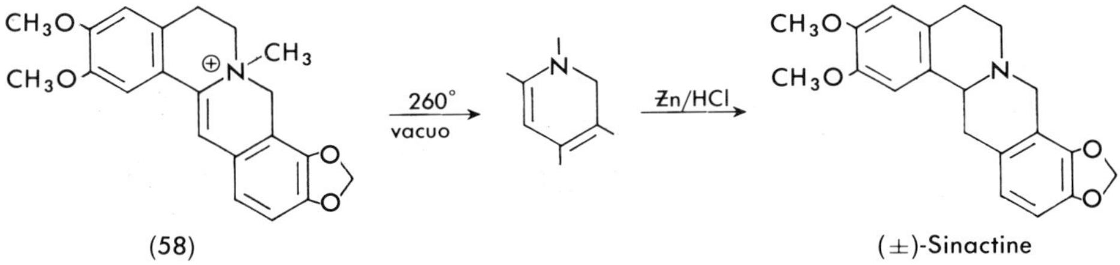

(58) (±)-Sinactine

For the purpose of relating tetrahydroprotoberberines to the protopine alkaloids, a salt such as isocryptopine chloride (58) may be heated *in vacuo* to obtain the corresponding tertiary free base which upon reduction leads to a tetrahydroprotoberberine. In this fashion, it has been possible to relate isocryptopine chloride with the alkaloid sinactine.

10.2 Syntheses of Protopine Alkaloids

The Hofmann degradation of a tetrahydroprotoberberine methohydroxide can lead to two products, (59) and (60).

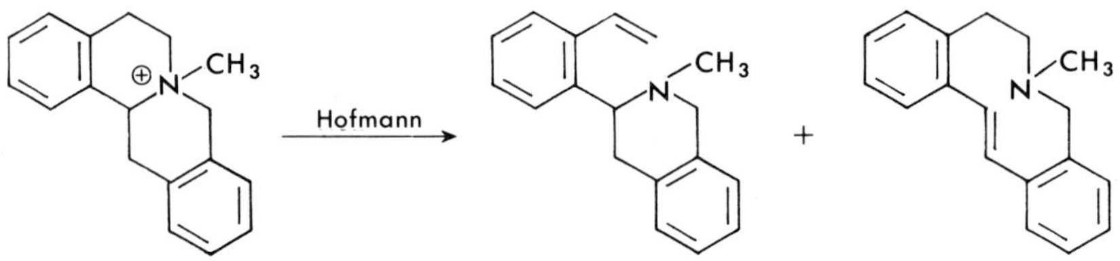

(59) (60)

Haworth and Perkin, jun., made several attempts to convert compounds of type (60) to protopine bases. This oxidation proved difficult, however, because of the great ease with which (60) cyclizes back to a tetracyclic system. It was finally found that perbenzoic acid oxidation of (60) to the N-oxide followed by treatment with hydrochloric acid in acetic acid gave the desired protopine base. The mechanism of this interesting transformation has been interpreted in terms of a transannular interaction between the N-oxide oxygen and the olefinic bond.[53]

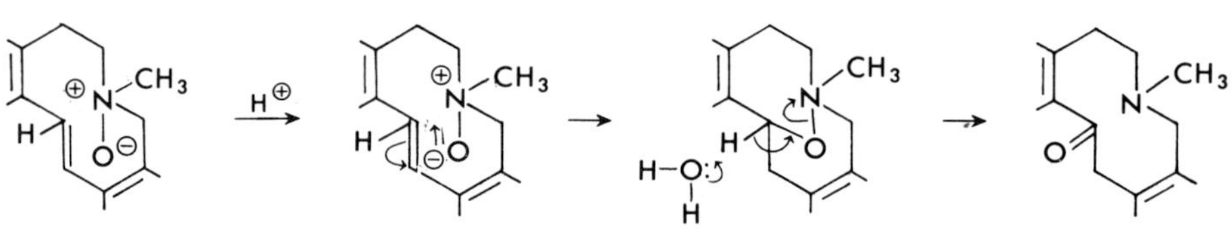

An alternate preparation of protopine salts from tetrahydroprotoberberine N-oxides has been presented by Bentley and Murray:[54]

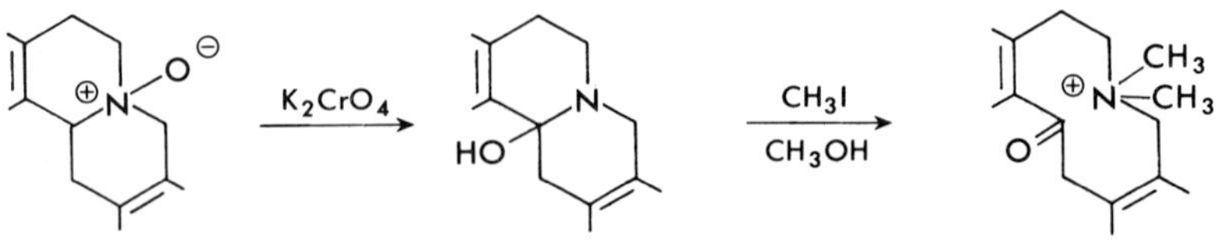

11. THE PHTHALIDEISOQUINOLINE ALKALOIDS

11.1 Structural Elucidation

The phthalideisoquinoline alkaloids are found mainly in *Papaver, Corydalis, Dicentra,* and *Hydrastis* species. Typical phthalideisoquinoline alkaloids are narcotine and hydrastine. Substituents are usually found at positions 6 and 7, and at 4′ and 5′, and in some cases a hydroxyl or methoxyl group is found at C(8).

The γ-lactone ring may not be present in all of the alkaloids of this group. For example, narceine is represented by the ketonic structure shown below. Rhoeadine has been characterized as an acetal, and this base bears a structural relationship to the phthalideisoquinolines.

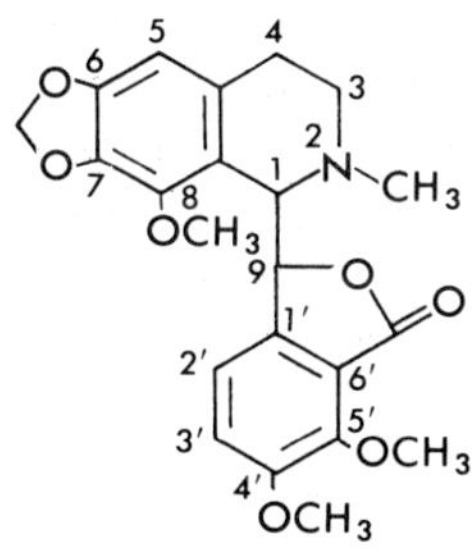

Narcotine

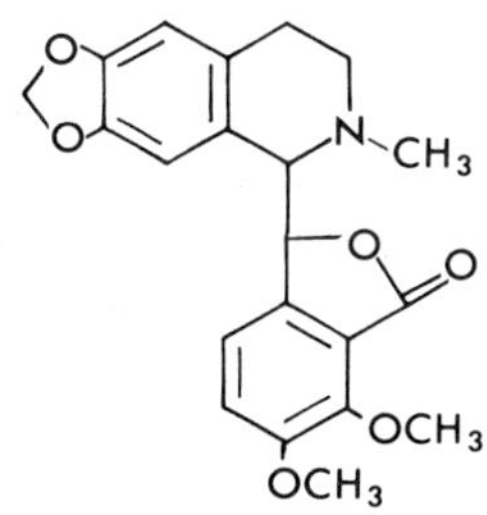

Hydrastine

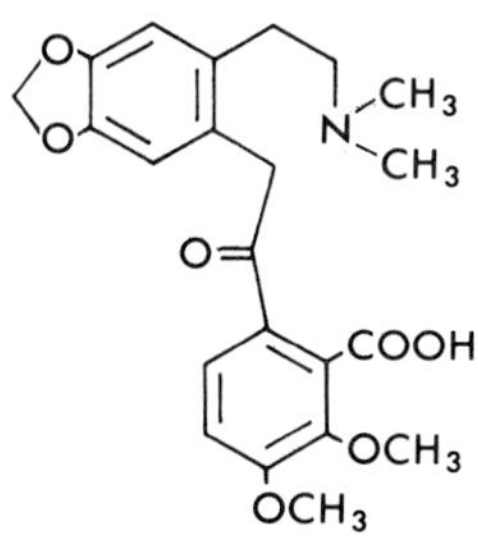

Narceine

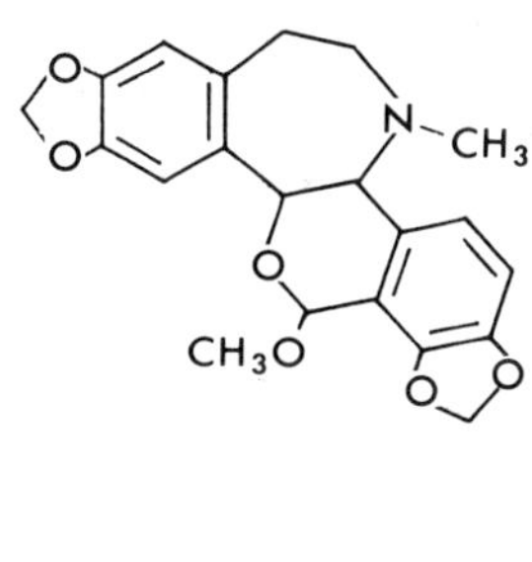

Rhoeadine

The key degradative step in the lactonic phthalideisoquinoline series is the facile cleavage of the doubly benzylic C(1)–C(9) bond. The degradation of narcotine closely parallels that of hydrastine, and the work on the latter alkaloid will be discussed in some detail.

Oxidative hydrolysis of hydrastine with dilute nitric acid results in the formation of two fragments, opianic acid and hydrastinine. Opianic acid is actually an aldehydoacid that can be reduced to the five-membered lactone meconine. The latter compound was synthesized by Perkin and his associates in 1925 starting from 2,3-dimethoxybenzaldehyde:[55]

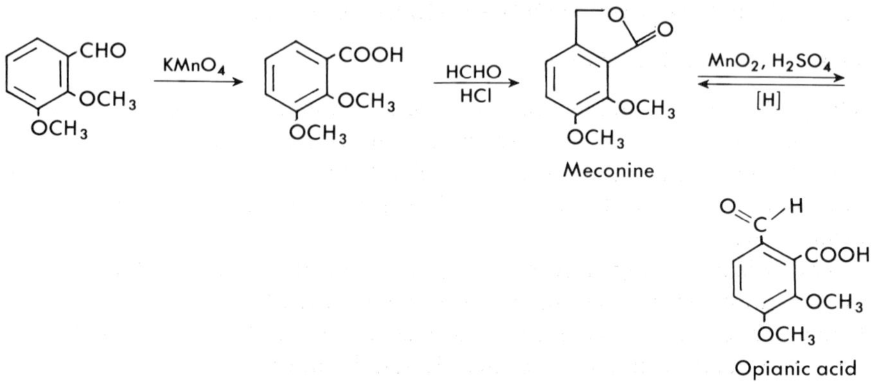

Turning now to hydrastinine, this product exists in the pseudo-base form in the solid state or in nonpolar solvents, while in aqueous or alcoholic solution it exists as an ionic quaternary form. Treatment of hydrastinine with strong alkali results in oxidation-reduction with formation of hydrohydrastinine and oxyhydrastinine. The position of the methylenedioxy substituent in hydrastinine was settled by its oxidation to hydrastic acid.

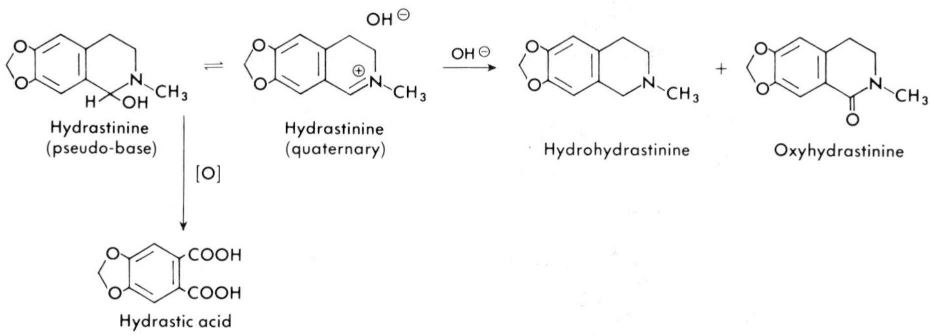

With the two cleavage products, opianic acid and hydrastinine fully characterized, and given the presence of a lactone ring in hydrastine, the structure of this alkaloid was essentially established. Robinson and Mirza have worked out a conversion of hydrastine to berberine by the series of simple transformations illustrated.

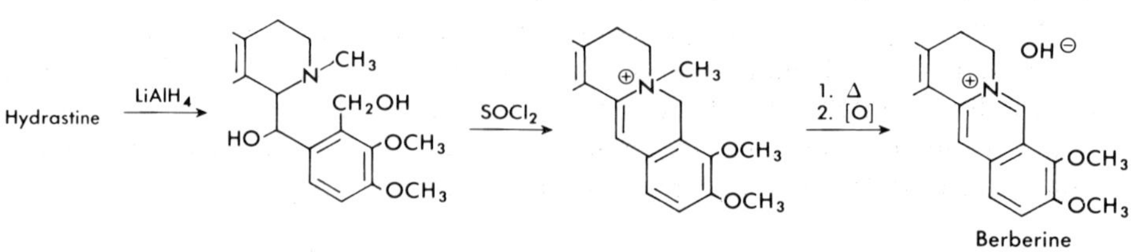

11.2 Syntheses of Phthalideisoquinolines

An early synthesis of phthalideisoquinolines is due to Hope and Robinson[56] who condensed 3,4-dimethoxy-6-nitrophthalide with hydrastinine in alcoholic solution. A mixture of two diasterisomeric nitrohydrastines resulted. This material was in turn converted to a mixture of hydrastines.

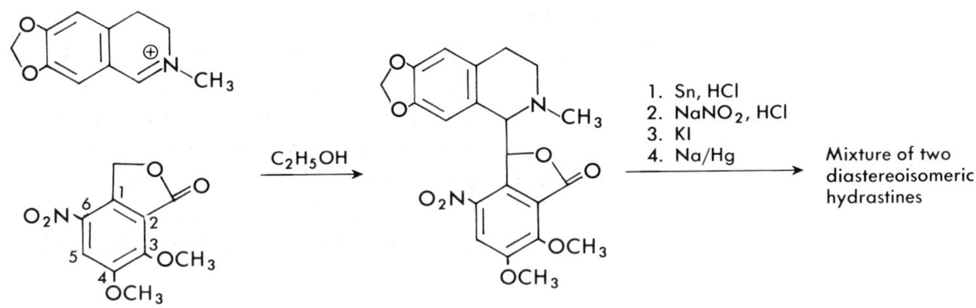

An improved synthesis, using the acid chloride of meconine-α-carboxylic acid and N-methylhomopiperonylamine as starting materials, was achieved by Haworth, Pinder, and Robinson.[57]

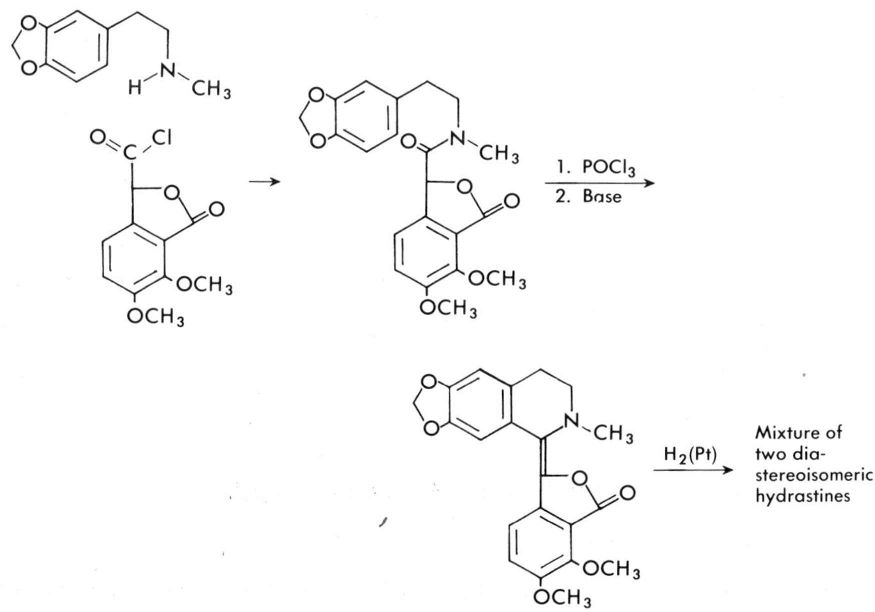

11.3 Stereochemistry

The stereochemistry and absolute configuration of the phthalideisoquinolines have been considered in detail.[58,59]

In one scheme,[59] naturally occurring (–)-α-narcotine was reduced with $LiAlH_4$, and the resulting diol was cyclized by means of methanesulfonyl chloride to a quaternary salt that on pyrolysis yielded the tetracyclic free base (61). Hydrogenolysis of the latter compound with hydrogen and palladium on carbon yielded the levorotatory tetrahydroprotoberberine (62), which falls into the *alpha* hydrogen series, so that (–)-α-narcotine must possess at C(1) the stereochemistry indicated.

The stereochemistry at C(9) was settled by a comparison of the nmr spectrum of (61), after exchange of the hydroxylic proton by deuterium, with that for the corresponding base (63) obtained from (–)-β-narcotine (which is itself formed from (–)-α-narcotine by inversion at C(9) through rigorous treatment with alkali). The splitting constants $J_{13,14}$ were found to be 9 cps for (61), and only 1.5 cps for (63), so that isomer (61) must have the *trans* arrangement of hydrogens at C(13), (14), and (63) the *cis*.[59]

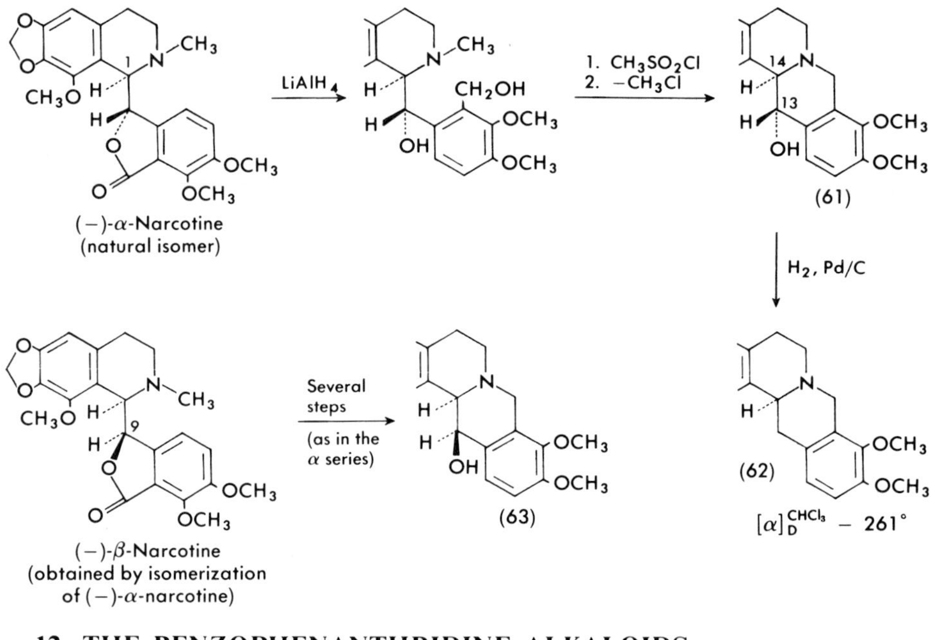

(–)-α-Narcotine (natural isomer)

(–)-β-Narcotine (obtained by isomerization of (–)-α-narcotine)

12. THE BENZOPHENANTHRIDINE ALKALOIDS

12.1 Structural Elucidation

The benzophenanthridine alkaloids can be divided into two main classes exemplified by the completely aromatic sanguinarine and chelerythrine on the one hand, and the alcoholic bases chelidonine and homochelidonine on the other. Sanguinarine and chelerythrine are found in a large variety of plants, most notably in the Papaveraceae *Sanguinaria canadensis* and *Chelidonum majus*.

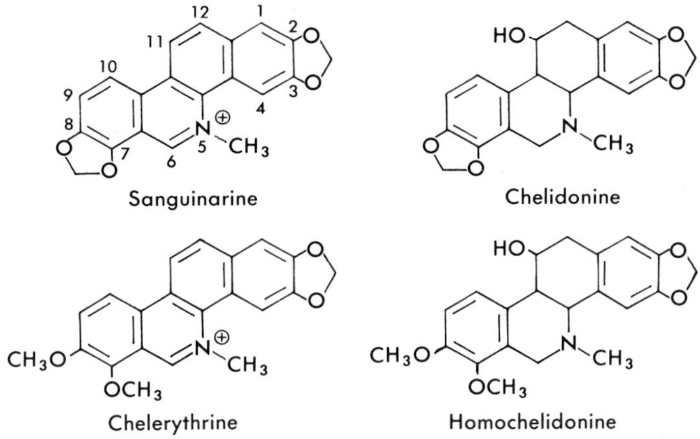

Sanguinarine salts are red, while chelerythrine salts are usually orange-red. Completely aromatic benzophenanthridine alkaloids undergo oxidation reduction when in a medium of high hydroxide ion concentration. Reduction of the alkaloids with zinc and hydrochloric acid or with sodium borohydride gives weakly basic dihydro derivatives that can be re-

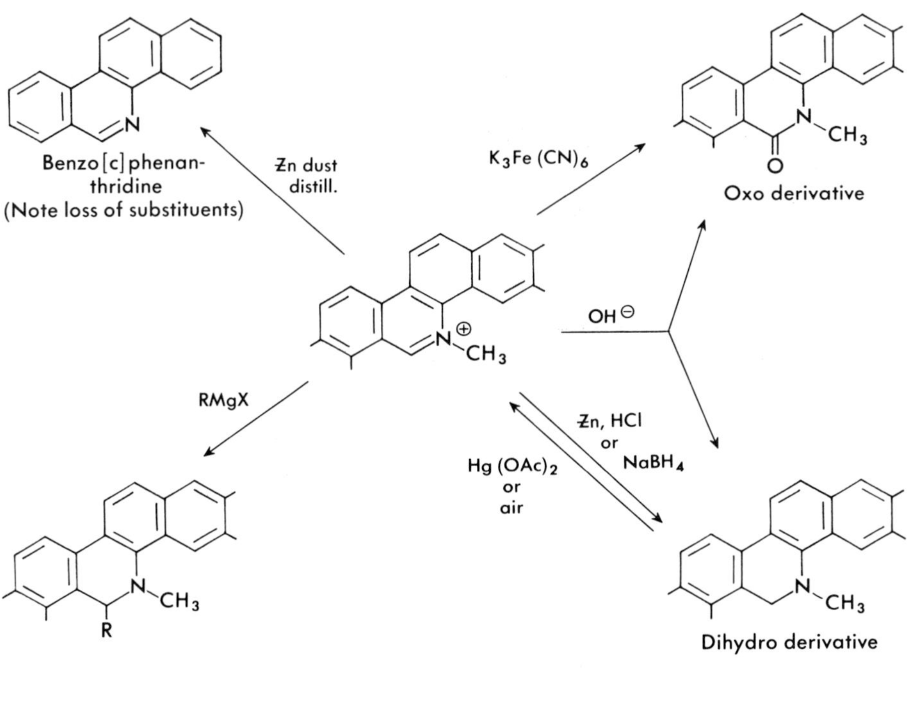

oxidized back to the starting quaternary ammonium ion. The aromatic alkaloid salts can also be alkylated by Grignard reagents to 6-alkyl-dihydro compounds. Upon zinc dust distillation, the alkaloids lead to the unsubstituted benzo[c]phenanthridine which is the nucleus from which all benzophenanthridine alkaloids are derived.

The location of the substituents in a completely aromatic benzophenanthridine alkaloid is usually determined by oxidation. Chelerythrine with mild permanganate yields the N-methylimide of hemipinic acid, while upon more energetic conditions hydrastic acid is produced.

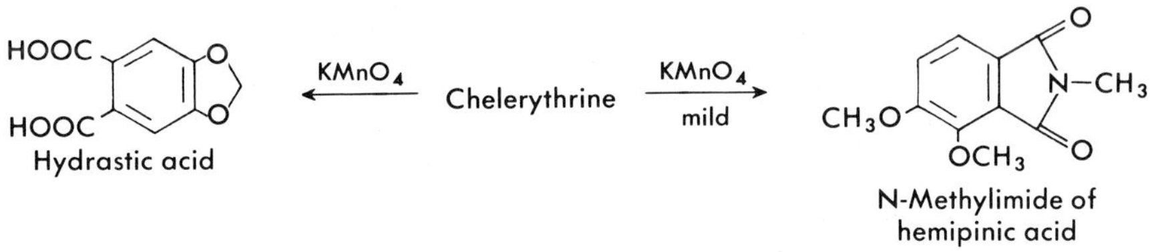

12.2 Syntheses

Bailey and Robinson[60] synthesized compound (66), an aromatic analog of the benzophenanthridine alkaloids, by starting from opianic acid and 3,4-dimethoxyacetophenone, and going through the intermediates (64) and (65).

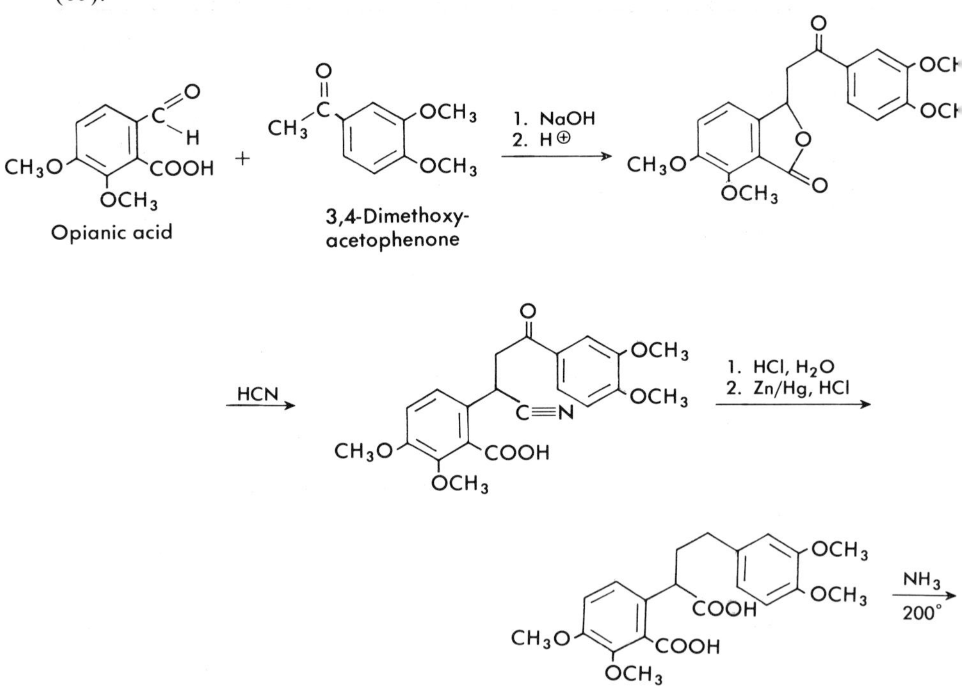

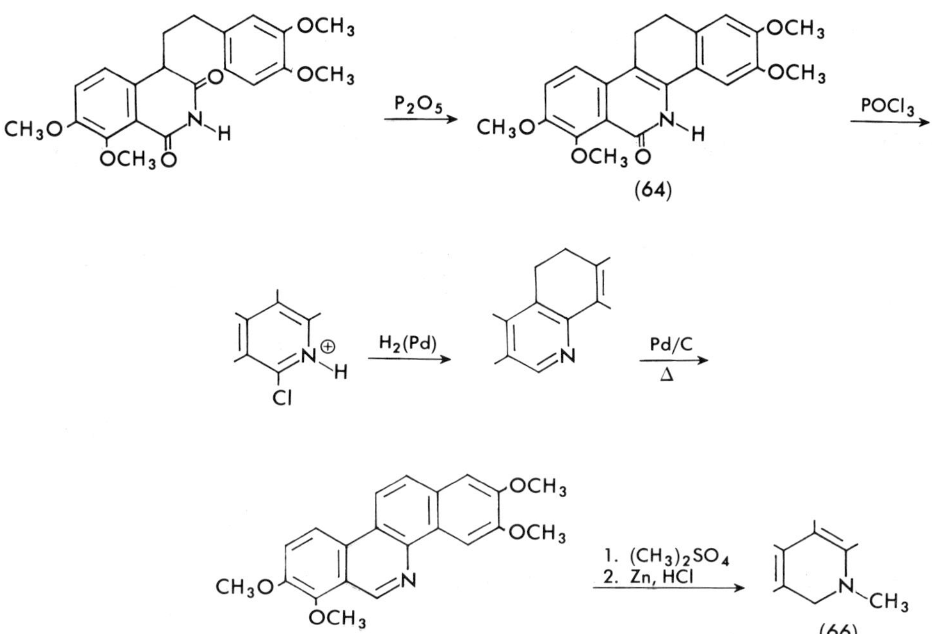

(64) (65) (66)

In an effort to extend this approach to the synthesis of chelerythrine, Bailey and Worthington[61] found that the appropriate tricyclic imide could not be cyclized to form the tetracyclic nucleus. Cyclization had to be carried out, therefore, at the diacid stage:

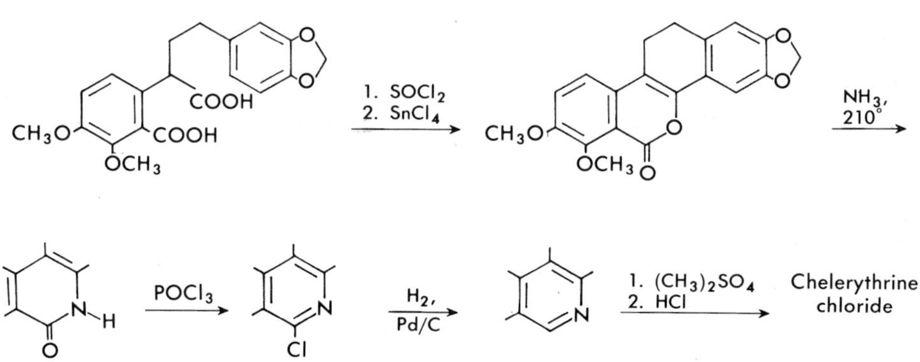

A 2,3,8,9-tetramethoxy analog of sanguinarine was also prepared by Robinson and co-workers.[62] The synthesis involves formation, via a Leuckart reaction, of the formyl derivative (67), which was subsequently converted to the desired tetracyclic structure:

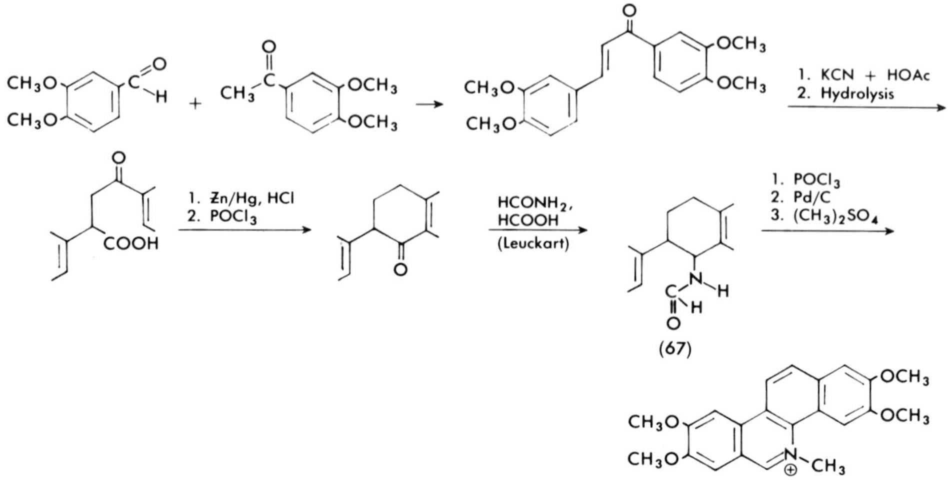

13. THE SPIROBENZYLISOQUINOLINE ALKALOIDS

13.1 Structural Elucidation

The novel and interesting structure (68a) was proposed for the Corydalis alkaloid, ochotensimine, $C_{22}H_{23}NO_4$, in 1964 on the basis of some degradative work which is outlined below, and on spectral (mostly nmr) evidence. This assignment then received final confirmation through an x-ray analysis of the methiodide salt of the companion alkaloid ochotensine (68b).[63]

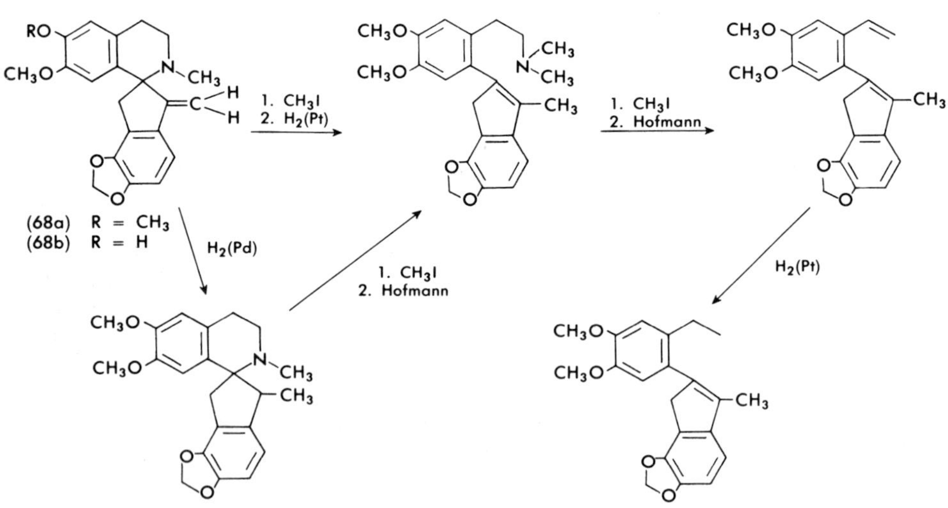

14. PROTOSTEPHANINE

14.1 Structural Elucidation

Protostephanine was isolated from *Stephania japonica*, and has been investigated by Kondo and his collaborators. The alkaloid analyzes for $C_{21}H_{27}NO_4$ and possesses one tertiary N-methyl and four O-methyl functions. Permanganate oxidation yielded *m*-hemipinic acid, while zinc dust distillation gave 1,3,6,7-tetramethoxyphenanthrene. Hofmann degradation of the alkaloid produced two methines, (69) and (70). Through the sequence—catalytic hydrogenation, Hofmann degradation, and again catalytic hydrogenation—methines (69) and (70) could be converted to the same tetramethoxydiethylbiphenyl (71). Furthermore, permanganate oxidation of the bismethine (72) resulted in formation of 1,3,6,7-tetramethoxyphenanthrenequinone (73) which could be further oxidized with hydrogen peroxide to the diphenic acid (74). Alternatively, ozonization of the bismethine (72) gave the dialdehyde (75) together with formaldehyde. This degradative sequence clearly established the structure of protostephanine, so that this alkaloid has the unique feature of incorporating a nine-membered ring.

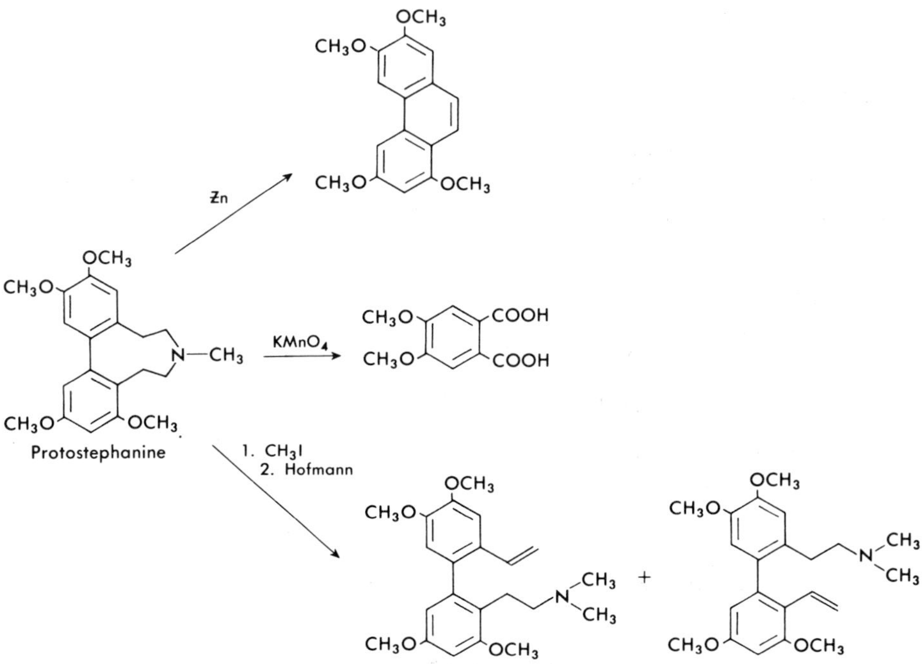

(69) (70)

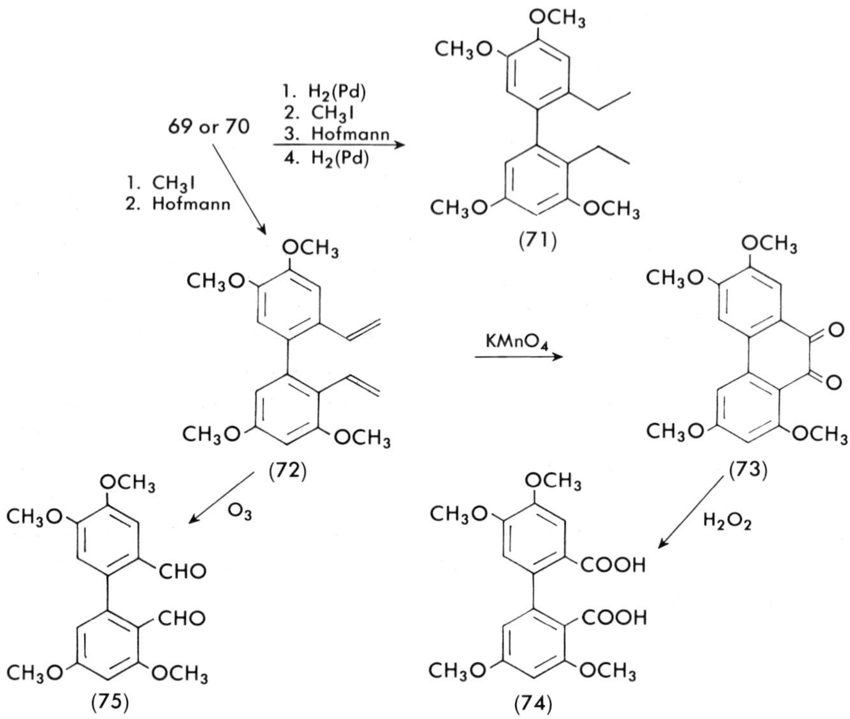

14.2 Syntheses

Takeda[64] and also Pecherer and Brossi[65] have carried out the synthesis of protostephanine, and the route used by the latter group is outlined below. The starting material was the diol (76).[65]

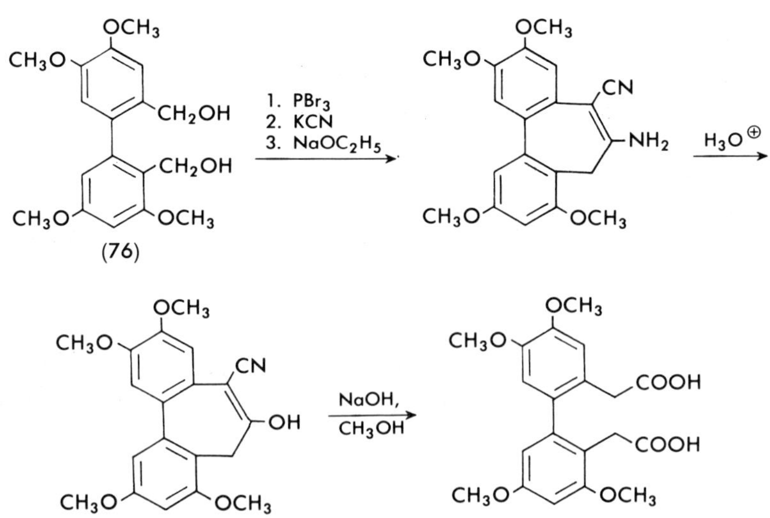

1. $LiAlH_4$, THF
2. PBr_3
3. CH_3NH_2

→ Protostephanine

15. THE PHENYLETHYLISOQUINOLINE ALKALOIDS

15.1 Androcymbine

The alkaloid androcymbine (77a), $C_{21}H_{25}NO_5$, was isolated from the leaves of *Androcymbium melanthioides*. Analytical procedures showed that one N-methyl and one phenolic hydroxyl group are present, while the NMR spectrum clearly indicated three O-methyl and one N-methyl groups, one aromatic proton, and two vinylic hydrogens. The infrared spectrum showed bands characteristic of a cross conjugated cyclohexadienone system.

Oxidation of O-methylandrocymbine (77b) gave 3,4,5-trimethoxyphthalic acid (78), while reduction with sodium in liquid ammonia afforded the phenylethyltetrahydroisoquinoline derivative (79) whose structure was confirmed by synthesis. Androcymbine must, therefore, be represented by expression (77a).[66]

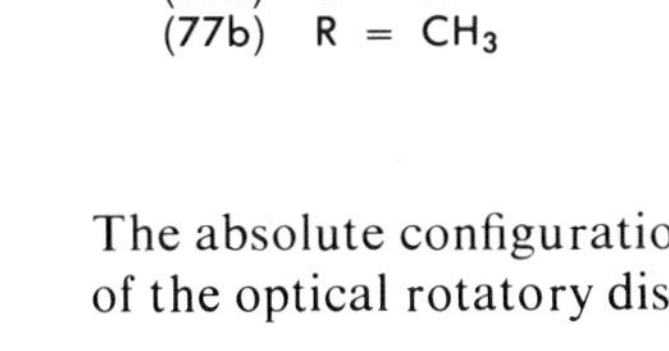

(77a) R = H
(77b) R = CH_3

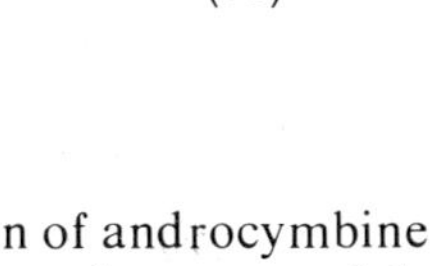

(78)

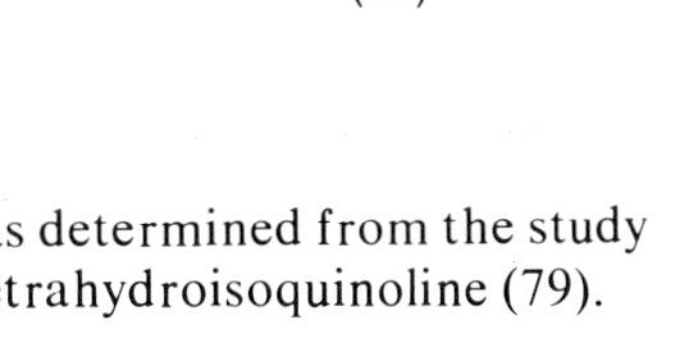

(79)

The absolute configuration of androcymbine was determined from the study of the optical rotatory dispersion curve of the tetrahydroisoquinoline (79).

15.2 Melanthioidine

A second phenylethylisoquinoline alkaloid is melanthioidine, also isolated from *A. melanthioides* and found to possess structure (80). The symmetry of the molecule is such that reductive cleavage of O,O-dimethylmelanthioidine (81) with sodium in liquid ammonia afforded the phenol (82), while the minor reduction products (83) and (84) indicated the molecule was in a head-to-tail arrangement.[67]

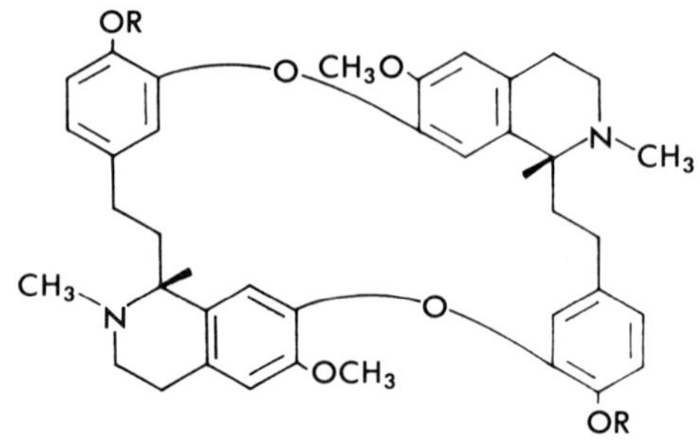

(80) R = H
(81) R = CH_3

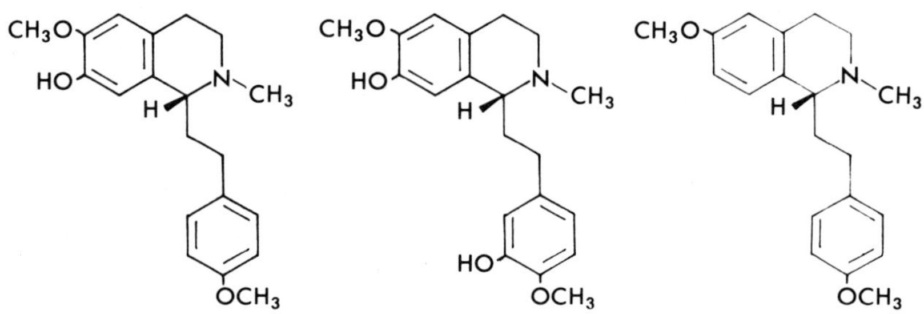

(82) (83) (84)

15.3 The Homoaporphines

Three alkaloids, namely multifloramine, floramultine, and kreysigine, are now known to be homoaporphines.[68] Their structures were originally derived from spectral and biogenetic arguments. Multifloramine was then synthesized through a phenolic oxidative coupling reaction followed by a dienone–phenol rearrangement:

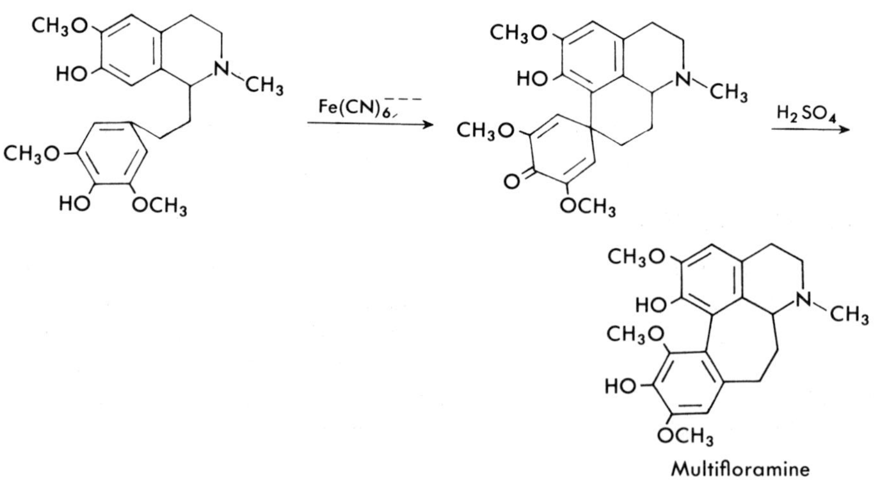

Multifloramine

16. THE ISOQUINOLONE ALKALOIDS

A minor but distinct group of alkaloids is represented by the three isoquinolones doryanine (85) from *Doryphora sassafras*,[69] and N-methylcorydaldine (86), and N-methylthalidaldine (87) from *Thalictrum fendleri*.[70]

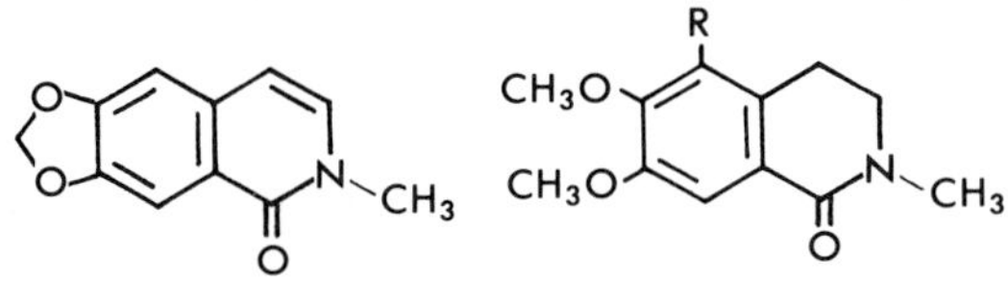

(85) (86) R = H (87) R = OCH_3

These very weakly basic alkaloids can be readily recognized because of their low molecular weight, and the presence of a lactam carbonyl peak in the infrared between 1639–1667 cm^{-1}.

N-Methylcorydaldine (86) had previously been obtained as a degradation product from the laboratory oxidation of a variety of naturally occurring benzylisoquinoline, protoberberine, and emetine type bases. Corydaldine itself was synthesized by a simple sequence.[71]

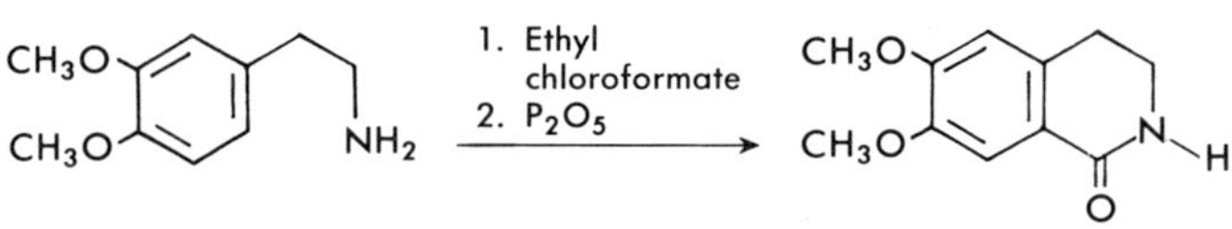

Corydaldine

The isoquinolones are probably derived in the plant from more complex precursors, such as benzylisoquinolines, by oxidative degradation.[69]

17. A PHENYLTETRAHYDROISOQUINOLINE ALKALOID

The optically active crystalline alkaloid (88) was isolated from *Cryptostylis fulva*, and the structure deduced on spectral grounds. The alkaloid was then synthesized by the route indicated. Alkaloid (88) is easily racemized in

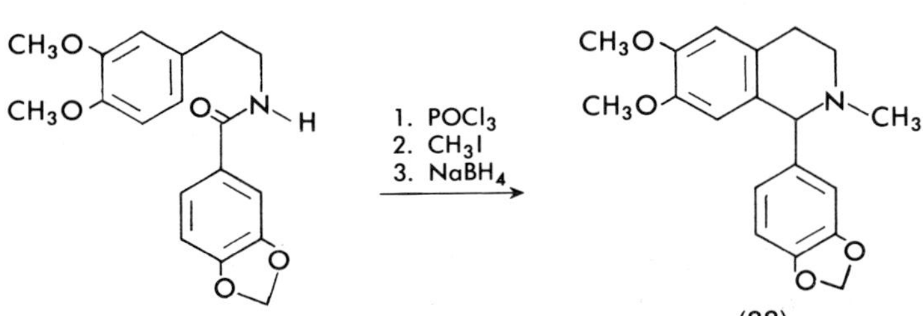

(88)

boiling ether, and is also the first phenyltetrahydroisoquinoline found in nature.[72]

18. THE PHENANTHROQUINOLIZIDINE AND THE PHENANTHROINDOLIZIDINE ALKALOIDS

The phenanthroquinolizidine and the phenanthroindolizidine alkaloids are basic pentacyclic compounds based on the systems (89) and (90), respectively, and found in members of the Lauraceae, Asclepiadaceae, and Moraceae. They are not usually considered to be isoquinoline alkaloids, but are included here simply as a matter of convenience.

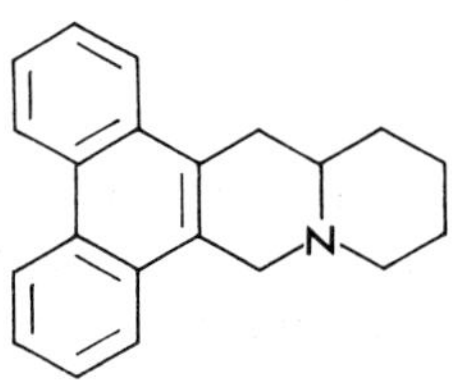

(89)

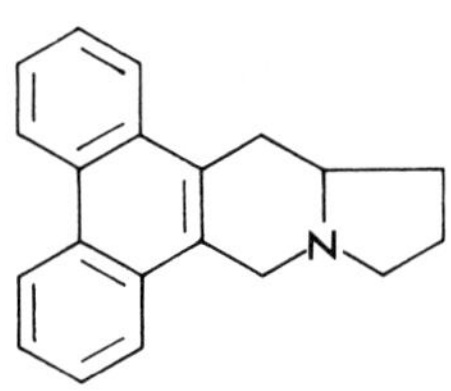

(90)

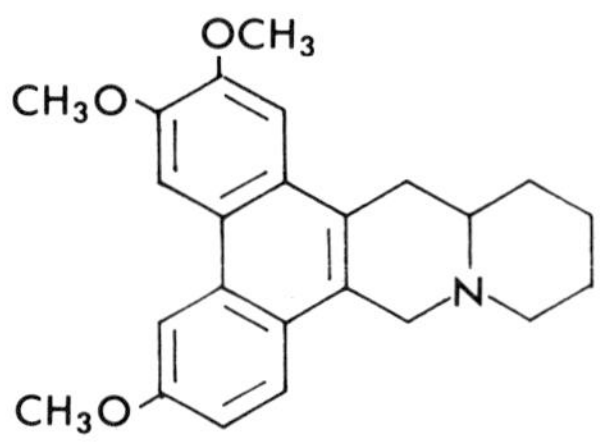

(91)

The first member of this group to be characterized was cryptopleurine (91), $C_{24}H_{27}NO_3$, which has powerful vesicant properties and is found in *Cryptocaria pleurosperma*. Gellert and Riggs in an initial study in 1954 found that (–)-cryptopleurine methiodide when refluxed in aqueous KOH easily racemized instead of undergoing a Hofmann degradation, and shortly afterwards Fridrichsons and Mathieson published a complete x-ray structural determination carried out on the racemic methiodide salt.[73]

Bradsher and Berger were first in reporting the synthesis of cryptopleurine, and their approach is outlined below.[74]

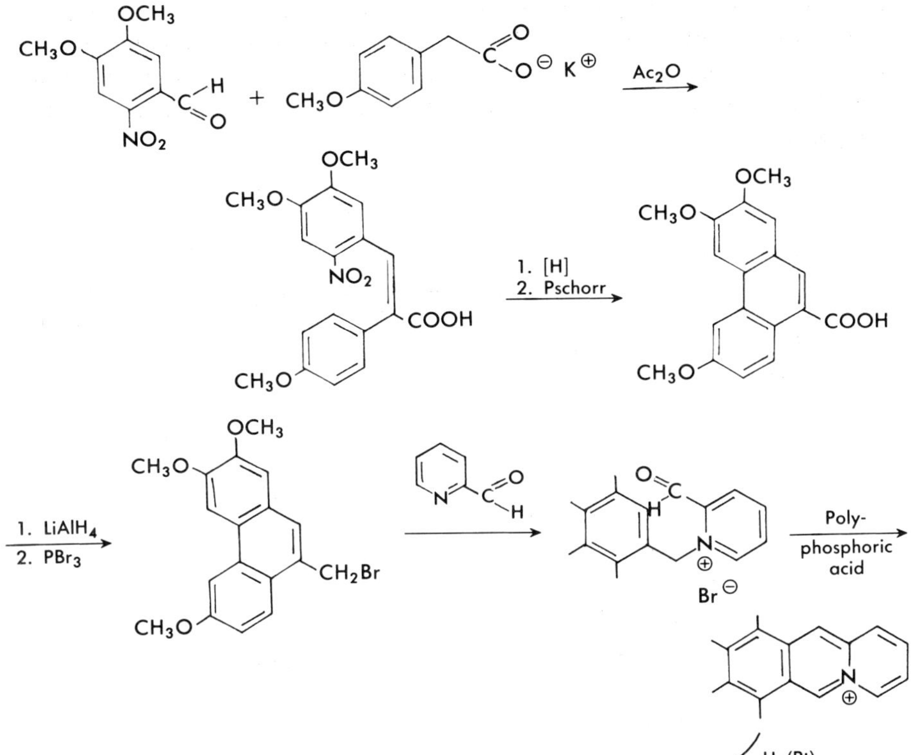

(±)-Cryptopleurine
(91)

A few months later, Marchini and Belleau offered an alternate approach to the alkaloid.[75]

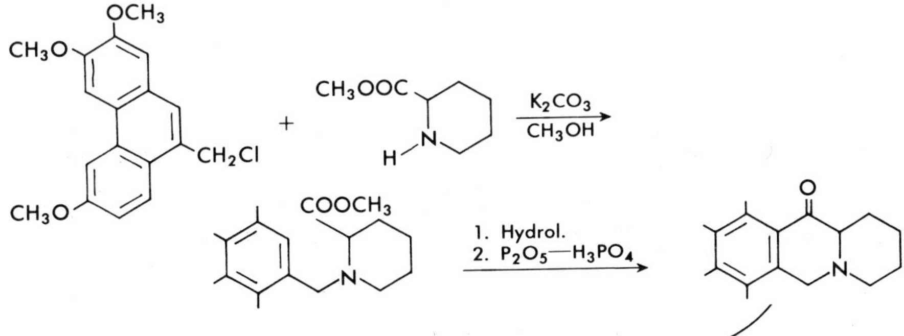

(±)-Cryptopleurine

The alkaloid (−)-tylophorine, found in *Tylophora asthmatica*, was studied by the Indian team of Govindachari, Pai, and co-workers. They observed that the compound exhibited an ultraviolet spectrum fairly close to that of cryptopleurine, and that it had an analysis corresponding to one methoxyl group more and one methylene group less than cryptopleurine. The flow sheet which follows summarizes much of the degradative work on tylophorine[76].

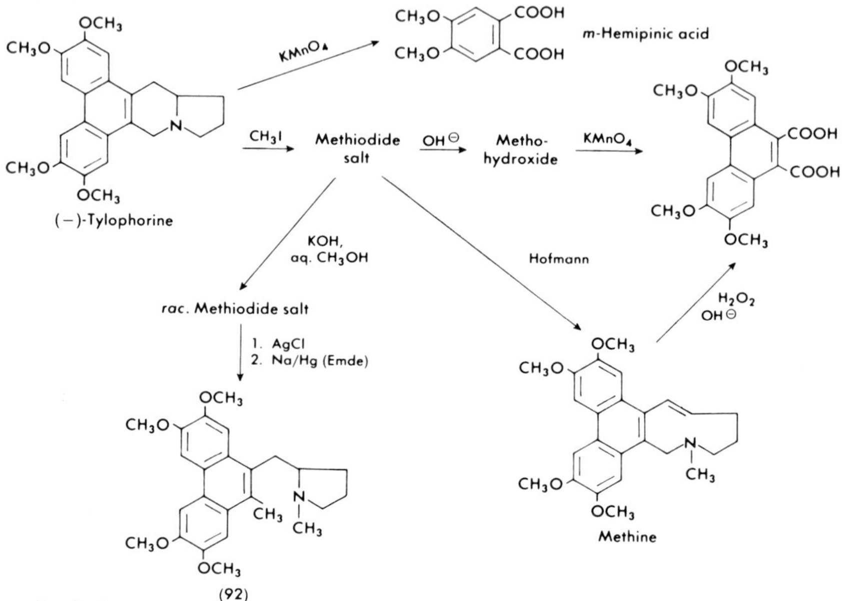

And also:

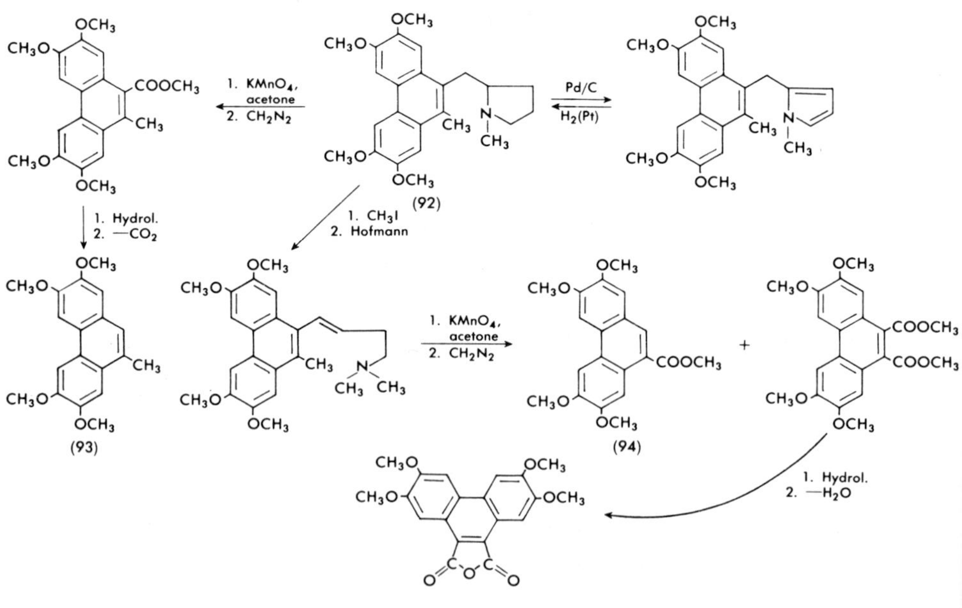

As with cryptopleurine, refluxing of tylophorine methiodide with KOH in aqueous methanol resulted in racemization, rather than in Hofmann elimination. The latter reaction could, however, be made to occur when the methohydroxide was heated without solvent under reduced pressure. The key intermediate obtained in the degradation scheme was the Emde reduction product (92), which in turn could be oxidized or made to undergo Hofmann elimination. Some of the ultimate degradation products obtained, such as *m*-hemipinic acid, the methylphenanthrene (93), and the ester (94), were identified by means of comparisons with authentic samples.

The Indian workers then followed their structural elucidation work with a total synthesis based upon the Bradsher–Belleau approach.

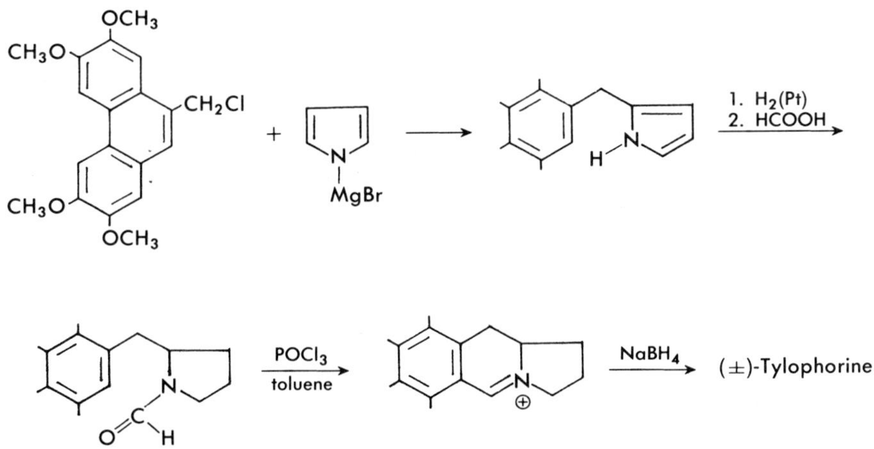

Tylophorinine, the minor alkaloid from *T. asthmatica* was shown by Govindachari, Pai, and co-workers, to be represented by the expression (95); while Gellert, Govindachari *et al.*, have found that tylocrebine is (96).

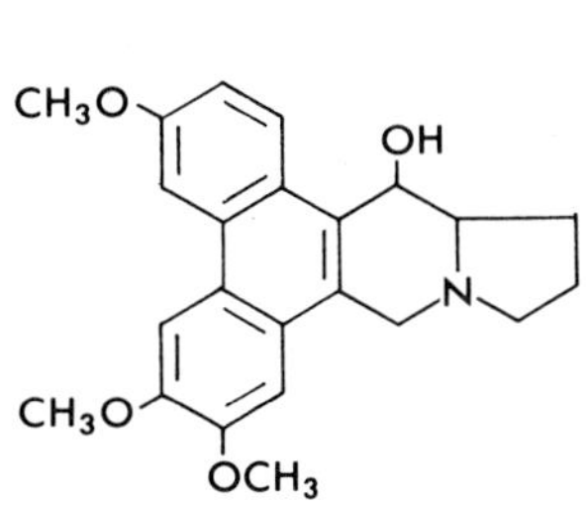

(95)

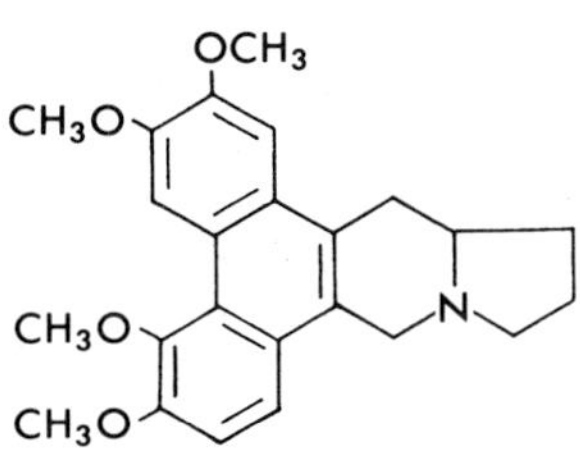

(96)

C. pleurosperma, in addition to producing cryptopleurine, also yields the alkaloid pleurospermine, which could be related to intermediates in the biogenesis of the phenanthroquinolizidine alkaloids. The structure of pleurospermine has been proven by synthesis:[77]

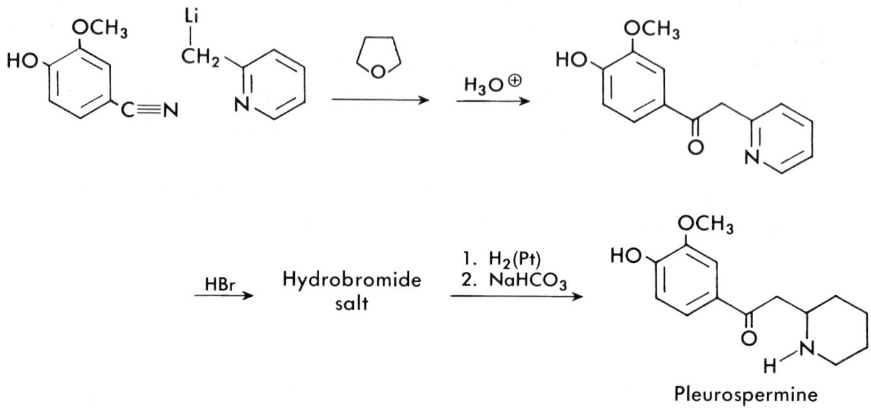

It is worth noting that in the above synthesis there was apparently no need to protect the phenolic function, and that furthermore the ketonic group was not reduced to the alcohol stage by Adams catalyst.

REFERENCES

1. A. R. Battersby and T. P. Edwards, *J. Chem. Soc.* 1909 (1959).
2. J. M. Bobbitt and T. T. Chou, *J. Org. Chem.* **24,** 1106 (1959).
3. C. Djerassi, H. W. Brewer, C. Clarke, and L. J. Durham, *J. Amer. Chem. Soc.* **84,** 3210 (1962).
4. J. M. Bobbitt, R. Ebermann, and M. Schubert, *Tetrahedron Letters* 575 (1963).
5. A. R. Battersby and T. P. Edwards, *J. Chem. Soc.* 1214 (1960).
6. S. M. Albonico, A. M. Kuck, and V. Deulofeu, *Chem. Ind.* (*London*) 1580 (1964).
7. N. J. McCorkindale, D. S. Magrill, M. Martin-Smith, S. J. Smith, and J. B. Stenlake, *Tetrahedron Letters* 3841 (1964).
8. K. W. Gopinath, T. R. Govindachari, and N. Viswanathan, *Chem. Ber.* **92,** 1657 (1959).
9. E. Spath and F. Berger, *Ber.* **63,** 2098 (1930).
10. H. Corrodi and E. Hardegger, *Helv. Chim. Acta.* **39,** 889 (1956).
11. J. Knabe and J. Kunitz, *Arch. Pharm.* **297,** 129 (1964).
12. A. R. Battersby and R. Binks, *J. Chem. Soc.* 2888 (1955).
13. S. F. Dyke and M. Sainsbury, *Tetrahedron Letters* 1545 (1964).
14. F. R. Stermitz, S. Y. Lwo, and G. Kallos, *J. Amer. Chem. Soc.* **85,** 1552 (1963).
15. For a review on the bisbenzylisoquinoline alkaloids see M. F. Grundon, "Progress in Organic Chemistry," Vol. 6, pp. 38–85, (J. W. Cook, Ed.), Academic Press, New York, N.Y., (1964).

16. F. von Bruchhausen, G. Aguilar-Santos, and C. Schafer, *Arch. Pharm.* **293,** 785 (1960).
17. T. Kametani and K. Fukumoto, *Tetrahedron Letters* 2771 (1964).
18. B. Franck, G. Blaschke, and G. Schlingloff, *Tetrahedron Letters* 439 (1962).
19. J. Comin and V. Deulofeu, *J. Org. Chem.* **19,** 1774 (1954).
20. J. A. Weisbach, C. Burns, E. Macko, and B. Douglas, *J. Med. Chem.* **6,** 91 (1963).
21. B. Franck, *Angew. Chem. Intern. Ed. Engl.* **1,** 559 (1962).
22. S. M. Albonico, A. M. Kuck, and V. Deulofeu, *Chem. Ind. (London)* 1580 (1964).
23. M. Shamma, *Experientia* **16,** 484 (1960).
24. J. Cymerman Craig and S. K. Roy, *Tetrahedron* **21,** 395 (1965).
25. M. Tomita, H. Furukawa, S. T. Lu and M. Kupchan, *Tetrahedron Letters* 4309 (1965).
26. N. M. Mollov and H. B. Dutschewska, *Tetrahedron Letters* 2219 (1964).
26a. D. H. R. Barton and T. Cohen, *Festschr. Arthur Stoll*, 117 (1957).
27. L. J. Haynes, K. L. Stuart, D. H. R. Barton, and G. W. Kirby, *Proc. Chem. Soc.* 280 (1963); 261 (1964).
28. M. P. Cava, K. Nomura, R. H. Schlessinger, K. T. Buck, B. Douglas, R. F. Raffauf, and J. A. Weisbach, *Chem. Ind. (London)* 282 (1964).
29. K. Bernauer, *Experientia* **20,** 380 (1964).
30. A. R. Battersby and T. H. Brown, *Proc. Chem. Soc.* **85,** (1964).
31. W. I. Taylor, *Tetrahedron* **14,** 42 (1961).
32. M. Tomita, T. H. Yang, H. Furukawa, and H. M. Yang, *J. Pharm. Soc. Japan* **82,** 1574 (1962); *Chem. Abstr.* **58,** 14012 (1963).
33. N. S. Bhacca, C. J. Craig, R. H. F. Manske, S. K. Roy, M. Shamma and W. A. Slusarchyk, *Tetrahedron* **22,** 1467 (1966).
34. T. Kametani and K. Fukumoto, *J. Chem. Soc.* 4289 (1963).
35. T. Kametani, S. Shibuya, S. Seino, and K. Fukumoto, *Tetrahedron Letters* 25 (1964).
36. T. Kametani, T. Kikuchi, and A. Fukumoto, *Chem. Comm.* 546 (1967).
37. R. Robinson and S. Sugasawa, *J. Chem. Soc.* 789 (1932).
38. C. Schöpf and K. Thierfelder, *Ann.* **497,** 22 (1932).
39. J. Ewing, G. K. Hughes, E. Ritchie, and W. C. Taylor, *Australian J. Chem.* **6,** 78 (1953).
40. G. H. Hughes, E. Ritchie, and W. C. Taylor, *Australian J. Chem.* **6,** 315 (1953).
41. F. Benington and R. D. Morin, *J. Org. Chem.* **32,** 1050 (1967).
42. R. H. F. Manske, *Can. J. Res.* **18B,** 414.
43. C. Tani, N. Takao, S. Takao, and K. Tagahara, *J. Pharm. Soc. Japan* **82,** 751 (1962).
44. C. Tani and T. Takao, *J. Pharm. Soc., Japan* **82,** 598 (1962).
45. M. P. Cava, T. A. Reed, and J. L. Beal, *Lloydia* **28,** 73 (1965).
46. A. R. Battersby, R. Southgate, J. Staunton, and M. Hirst, *J. Chem. Soc.* 1052 (1966).
47. T. Kametani and M. Ihara, *J. Chem. Soc.* 530 (1967).
48. A. R. Battersby, R. Binks, and P. S. Uzzell, *Chem. Ind. (London)* 1039 (1955).
49. C. K. Bradsher and N. L. Dutta, *J. Amer. Chem. Soc.* **82,** 1145 (1960).
50. N. T. L. Q. Thuan and J. Gardent, *Bull. Soc. Chim. France* 2401 (1966).
51. D. W. Brown and S. F. Dyke, *Tetrahedron Letters* 3587 (1964); *Tetrahedron* **22,** 2429 (1966).
52. P. W. Jeffs, *Experientia* **21,** 690 (1965).
53. P. R. Russell, *J. Amer. Chem. Soc.* **78,** 3115 (1956).
54. K. W. Bentley and A. W. Murray, *J. Chem. Soc.* 2497 (1963).
55. G. A. Edwards, W. H. Perkin, jun., and F. W. Stoyle, *J. Chem. Soc.* 195 (1925).
56. E. Hope and R. Robinson, *Proc. Chem. Soc.* **28,** 17 (1912); *J. Chem. Soc.* **105,** 1456, 2085 (1914).

57. R. D. Haworth, A. R. Pinder, and R. Robinson, *Nature* **165,** 529 (1950).
58. M. Ohta, H. Tani, and S. Morozumi, *Tetrahedron Letters* 859 (1963). See also: *Tetrahedron Letters* 1857 (1963); 693 (1964); *Chem. Pharm. Bull.* (Japan) **12,** 569, 1072, 1080 (1964). Additional studies on the stereochemistry of the phthalideisoquinolines are by S. Safe and R. Y. Moir, *Can. J. Chem.* **42,** 160 (1964) and by K. Blaha, J. Hrbek, Jun., J. Kovar, L. Pijewska, and F. Santavy, *Coll. Czech. Chem. Comm.* **29,** 2328 (1964).
59. A. R. Battersby and H. Spencer, *J. Chem. Soc.* 1087 (1965).
60. A. S. Bailey and R. Robinson, *J. Chem. Soc.* 1375 (1950); 2277 (1950).
61. A. S. Bailey and C. R. Worthington, *J. Chem. Soc.* 4535 (1956).
62. T. Richardson, R. Robinson, and E. Seijo, *J. Chem. Soc.* 835 (1937).
63. S. McLean, M. S. Lin, A. C. Macdonald, and J. Trotter, *Tetrahedron Letters* 185 (1966).
64. K. Takeda, *Chem. Abstr.* **60,** 5570 (1964).
65. B. Pecherer and A. Brossi, *J. Org. Chem.* **32,** 1053 (1967).
66. A. R. Battersby, R. B. Herbert, L. Pijewska, and F. Santavy, *Chem. Comm.* 228 (1965).
67. A. R. Battersby, R. B. Herbert, and F. Santavy, *Chem. Com.* 415 (1965).
68. A. R. Battersby, R. B. Bradbury, R. B. Herbert, M. H. G. Munro, and R. Ramage, *Chem. Comm.* 450 (1967).
69. S. A. Gharbo, J. L. Beal, R. H. Schlessinger, M. P. Cava, and G. Svoboda, *Lloydia* **28,** 237 (1965).
70. M. Shamma and Sr. M. A. Podczasy, unpublished results.
71. E. Spath and A. Dobrowsky, *Ber.* **58,** 1274 (1925); E. Spath and H. Epstein, *Ber.* **59,** 2791 (1926).
72. R. Leànder and B. Lüning, *Tetrahedron Letters* 1393 (1968).
73. J. Fridrichsons and A. McL. Mathieson, *Nature* **173,** 732 (1954); and *Acta Cryst.* **8,** 761 (1955).
74. C. K. Bradsher and H. Berger, *J. Amer. Chem. Soc.* **79,** 3287 (1957); **80,** 930 (1958).
75. P. Marchini and B. Belleau, *Can. J. Chem.* **36,** 581 (1958).
76. T. R. Govindachari, M. V. Lakshmikantham, K. Nagarajan, and B. R. Pai, *Chem. Ind. (London)* 1484 (1957).
77. J. W. Loder, *Australian J. Chem.* **15,** 296 (1962).

4

The Ipecacuanha Alkaloids

H. T. OPENSHAW
The Wellcome Research Laboratories
Beckenham, Kent, England

1. INTRODUCTION

The alkaloids discussed in this chapter form a structurally distinct subgroup of the very large class of isoquinoline alkaloids. They are characterised by the presence of the tricyclic 3-ethyl-hexahydrobenzo[a]quinolizine system. This system is linked through its 2-position with a second heterocyclic structure, most commonly an isoquinoline moiety, as seen in the case of emetine (1), the prototype of the group. The most important

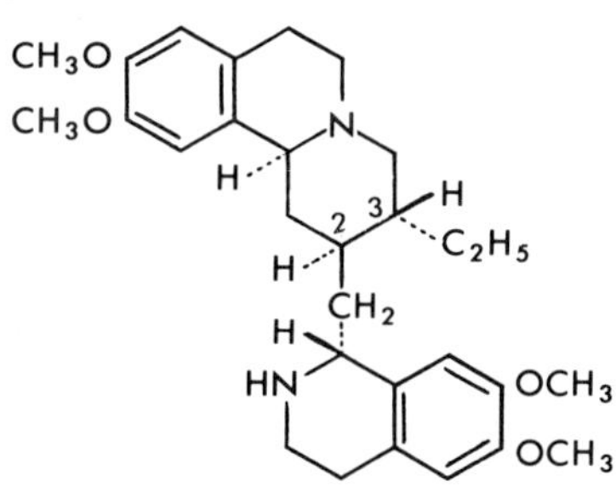

(1) Emetine

alkaloids, those of ipecac root,[1] have been known for a long time, but quite recently a number of analogs have been isolated from other plant sources, especially *Alangium lamarckii* Thwaites, and these will also be discussed briefly. The biogenesis of the alkaloids of this group is a particularly interesting problem,[2] but is only touched on here since it is discussed in greater depth in Chapter 21.

The elucidation of the structures of the ipecac alkaloids has occupied the attention of chemists intermittently over the course of more than half a century, and it was achieved almost entirely by the application of the classical methods of degradative organic chemistry. By contrast, the examination of the recently discovered analogs has relied very heavily upon modern techniques for its success. Confirmation of the correctness of the various structures has been obtained by a number of total syntheses of the alkaloids. Efficient stereospecific methods, one of which is in commercial operation, have been developed for the synthesis of emetine.

2. IPECACUANHA ROOT

Ipecacuanha is a native name for a small, shrubby, creeping plant indigenous to the tropical rain forests of Brazil and countries immediately to the north. The name means "little wayside plant that causes vomiting," and the roots had a local reputation for medicinal value in dysentery. The drug was first brought to Europe in the latter part of the 17th century. In 1688 its special value in the treatment of dysentery was confirmed by Helvetius, a Paris physician, who publicised the use of his "Brazilian root" while keeping its nature secret. The appropriate usage of this drug in treating amoebic dysentery was scientifically validated in 1912 when emetine, the principal alkaloid, was shown by E. B. Vedder to be highly toxic *in vitro* to the infective organism, *Entamoeba histolytica*, and by L. Rogers to be effective in the treatment of the disease.

Commercial ipecac consists of the root bark of either of two species of the family Rubiaceae, each of which has a confusing multiplicity of botanical synonyms: Brazilian ipecac, the more important, comes from *Psychotria ipecacuanha* Stokes (*Cephaelis ipecacuanha* Rich., *Uragoga ipecacuanha* Baill.). This species is also cultivated in India and Malaya ("Johore ipecacuanha"). Carthagena ipecac, also known as Nicaragua or Panama ipecac, is derived from *Psychotria granadensis* Benth. (*Uragoga granatensis* Baill., *Cephaelis acuminata* Karsten) which grows mainly in Colombia and Venezuela. Other related species of Rubiaceae contain varying, but usually much smaller amounts of the alkaloids, and ipecac alkaloids have also been found in small amounts in *Alangium lamarckii* (Alangiaceae) (see Section 10).

3. ISOLATION, FUNCTIONAL GROUPS, AND INTERRELATIONSHIP OF THE ALKALOIDS

In view of the long medicinal usage of ipecac, it is not surprising that it was one of the first drugs to be examined for alkaloid content. In 1817 Pelletier and Magendie isolated the crude base, "emetine." However, it was not until 1894 that Paul and Cownley showed the crude alkaloid to be a mixture of the nonphenolic emetine with two phenolic alkaloids, cephaeline and psychotrine, and obtained emetine and cephaeline in analytically pure condition. Hesse, in 1898, correctly assigned the formula $C_{28}H_{38}N_2O_4$ to cephaeline, but the correct formula for emetine, $C_{29}H_{40}N_2O_4$, was first established in 1914 by Carr and Pyman, who also showed psychotrine to be $C_{28}H_{36}N_2O_4$. In 1917, Pyman added two further alkaloids, *O*-methylpsychotrine, $C_{29}H_{38}N_2O_4$, and emetamine, $C_{29}H_{36}N_2O_4$.

The structural interrelationships of these five alkaloids were elucidated by Pyman,[3] and are shown in Chart 4.1. Emetine contains four methoxyl

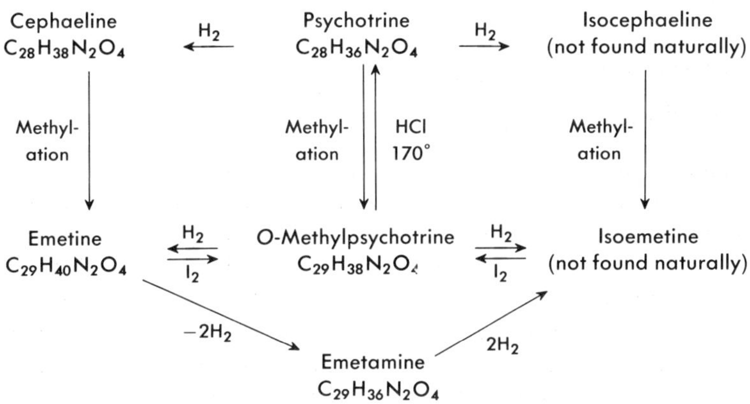

Chart 4.1

groups, cephaeline and psychotrine each contain three methoxyl and one phenolic hydroxyl group, and controlled methylation of cephaeline produces emetine. Psychotrine gives on reduction a mixture of two dihydroderivatives, one being identical with cephaeline and the other being a stereoisomer, isocephaeline. Similarly, *O*-methylpsychotrine, which can be obtained by methylation of psychotrine, gives on reduction emetine and isoemetine; the isoalkaloids do not occur in ipecac. Emetamine can also be reduced to isoemetine with the uptake of four hydrogen atoms. Mild dehydrogenation of emetine or isoemetine with iodine gives *O*-methylpsychotrine. Complete demethylation of emetine or cephaeline with hydrochloric acid at 135–140° (or hydriodic acid) affords amorphous emetoline, $C_{25}H_{32}N_2O_4$, which gives catechol color reactions.

No *N*-methyl groups are present in the alkaloids. Both emetine and cephaeline contain one tertiary and one secondary amino group; thus, emetine forms an *N*-benzoyl derivative that is a mono-acidic, tertiary base, and both alkaloids can be converted to *N*-methyl derivatives which are ditertiary bases. Psychotrine, *O*-methylpsychotrine, and emetamine are all ditertiary bases.

Emetine base is normally amorphous, but it has been obtained in a crystalline form. The base is levorotatory, but its salts are dextrorotatory, and their specific rotation varies very greatly with the concentration of the solution as well as with the nature of the solvent. The common salts of emetine are rather heavily hydrated: The hydrochloride, for example, crystallizes as a heptahydrate. The other alkaloids crystallize without difficulty, and form crystalline, hydrated salts. Cephaeline, like emetine, is levorotatory but forms dextrorotatory salts; the remaining three bases are dextrorotatory, but emetamine forms levorotatory salts. The interconversions of the alkaloids show that, in spite of their differences in rotation, they all belong to the same stereochemical series.

4. STRUCTURAL DETERMINATION

4.1. Oxidation

The main features of the structure of the ipecac alkaloids were deduced from a study of their oxidation products. The oxidation of emetine[3,4] with permanganate gives rise to *m*-hemipinic acid (2), its imide, and small amounts of 6,7-dimethoxyisoquinoline-1-carboxylic acid (3). With chromic acid, 4,5-dimethoxyphthalonimide (4) is formed. These products point to the presence of at least one 6,7-dimethoxyisoquinoline unit in emetine. In 1914 Dobbie and Fox[5] showed that the ultraviolet absorption spectrum of emetine closely resembles that of tetrahydropapaverine, and hence the molecule must contain two *o*-dimethoxybenzene units. Since the absorption spectra of fully aromatic isoquinolines such as papaverine or the acid (3) are quite different from that of emetine, the latter must be a derivative of tetrahydroisoquinoline; hence, the formation of the acid (3) in the oxidation of the alkaloid must involve a concomitant dehydrogenation.

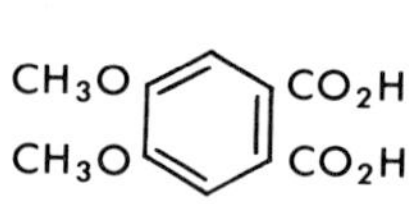

(2)

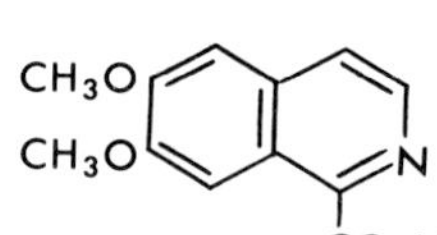

(3)

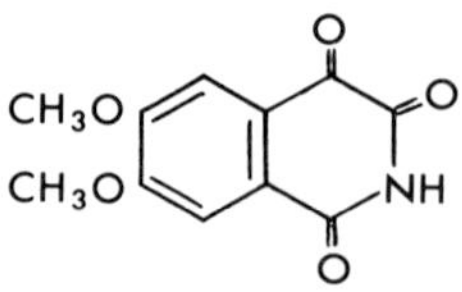

(4)

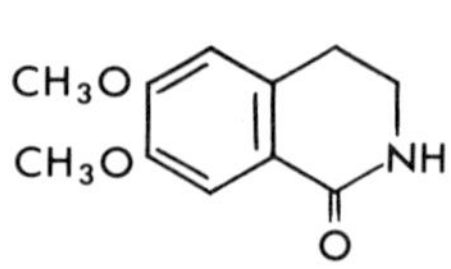

(5)

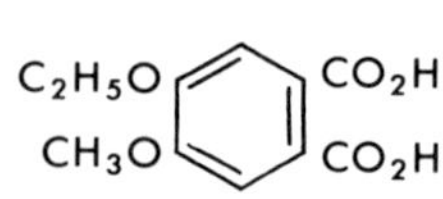

(6)

A more detailed study of the oxidation was made by Spath and Leithe.[6] They showed that gentle oxidation of emetine with alkaline permanganate gave corydaldine (5). After separation of this product, further oxidation of the residues gave *m*-hemipinic acid, and the combined yield of the two products amounted to 0.96 mole per mole of emetine. Since comparable treatment of the alkaloids papaverine and palmatine, which contain only one 6,7-dimethoxyisoquinoline unit, gave only about half this yield, the presence of two such units in emetine was suggested. Finally, the *O*-ethyl ether of cephaeline was oxidized, and a mixture of corydaldine and an ethoxymethoxytetrahydroisoquinolin-1-one was obtained. This mixture could not be separated into its components, but on further oxidation it yielded *m*-hemipinic acid (2) and 4-ethoxy-5-methoxyphthalic acid (6), which were separated and identified. Emetine therefore contains two 6,7-dimethoxytetrahydroisoquinoline units. The 1-oxo group of corydaldine represents the point of attachment of each of these units to the remainder of the molecule; since one of the nitrogen atoms is tertiary, a further linkage through this atom is also required, and Späth and Leithe were able to put forward the partial structure (7).

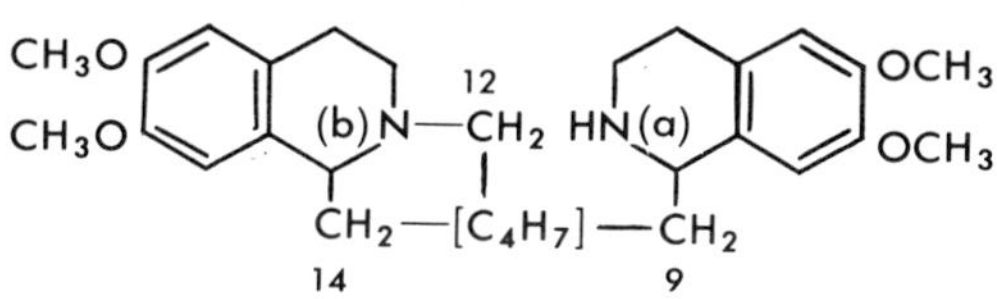

(7)

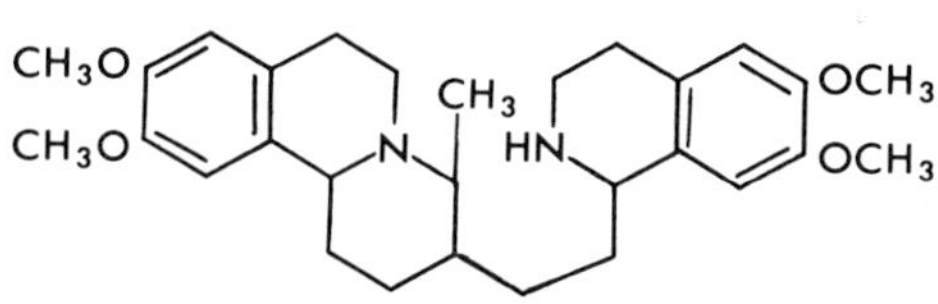

(8)

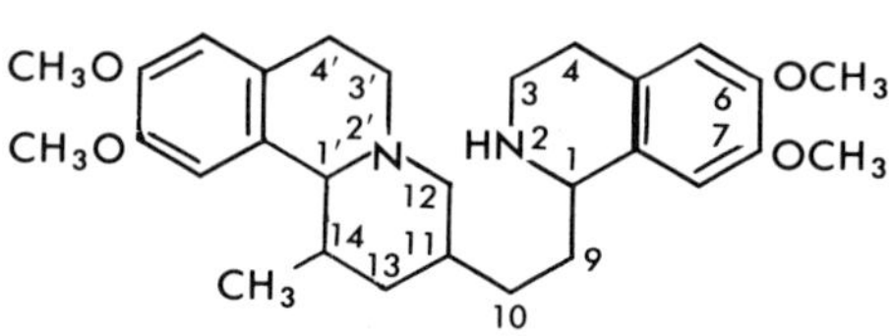

(9)

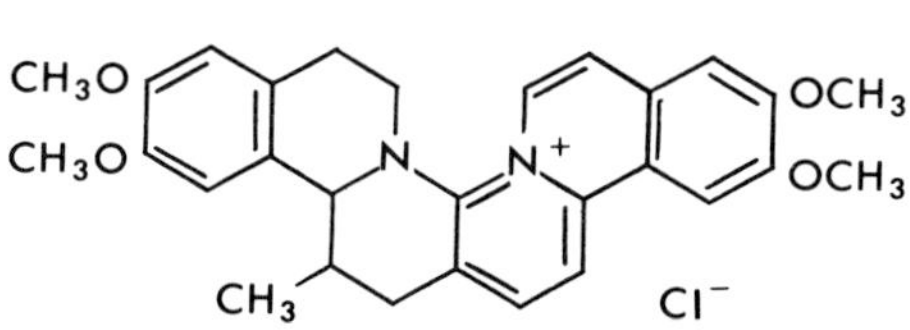

(10)

When emetine is treated with mild acidic oxidizing agents such as ferric chloride, bromine, or mercuric acetate, it does not undergo degradation, but is dehydrogenated to a bright orange-red material, which Carr and Pyman named "rubremetine." This has the properties of a quaternary, monoacidic base, and was isolated in the form of its salts, e.g., "rubremetine hydrochloride," $C_{29}H_{33}N_2O_4Cl$, which tenaciously retain water of crystallization. By the action of excess iodine on emetine, Karrer[7] obtained a product "dehydroemetine hydriodide" which he at first thought to differ from "rubremetine hydriodide." Subsequently, the identity of the two compounds was established. Because of their quaternary character, these products are perhaps better named as rubremetinium salts. The formation of the rubremetinium cation from emetine may be formulated:

$$C_{29}H_{40}N_2O_4 + H^+ - 8H \rightarrow C_{29}H_{33}N_2O_4^+$$

This highly characteristic dehydrogenation reaction is clearly of great significance in relation to the structure of the alkaloid, but it was interpreted only after the structure of emetine became known; therefore, its discussion is deferred to a later stage. Speculations concerning the course of the reaction led, however, to the propounding of various tentative structures for emetine. The most plausible of these (9), which was widely accepted for some twenty years, was put forward by Brindley and Pyman[8] in 1927. This embraced the available degradative evidence, as represented in Späth's partial structure (7), and was a modification of the structure (8) suggested by Robinson on the basis of biogenetic speculations. Brindley and Pyman pointed out that the removal of eight hydrogen atoms from emetine in the process of oxidation to rubremetine must be accompanied by some structural change that could account for the loss in basic character of one of the nitrogen atoms. This would not be the case if the change were merely the aromatisation of the two tetrahydroisoquinoline units in (8). However, if the two nitrogen atoms could be incorporated in an amidine-like structure, they would together constitute a single basic center. Accordingly, they modified Robinson's structure to (9), by the transfer of a methyl group, and represented rubremetinium chloride as (10).*

4.2 Hofmann Degradation

The partial structure (7) of Späth and Leithe, which summarizes the conclusions that may be drawn from the oxidative degradation of emetine, forms a convenient basis for the discussion of the further work leading to the complete elucidation of the structure. The numbering of the atoms in

*It is remarkable that they overlooked the fact that structure (10) requires the removal of ten hydrogen atoms from emetine!

this formula corresponds to that used by Brindley and Pyman in structure (9), and the secondary and tertiary nitrogen atoms are designated N(a) and N(b), respectively.*

Oxidation experiments have given little information as to the mode of linkage of the two isoquinoline nuclei, save that it involves the 1-position of each. The establishment of the complete carbon skeleton of emetine was achieved through studies of the Hofmann exhaustive methylation of the alkaloid. This was first examined by Hesse who obtained *N*-methylemetinemethine (11) by thermal decomposition of *N*-methylemetine dimethohydroxide. The methine was more fully characterised by Pyman, who also

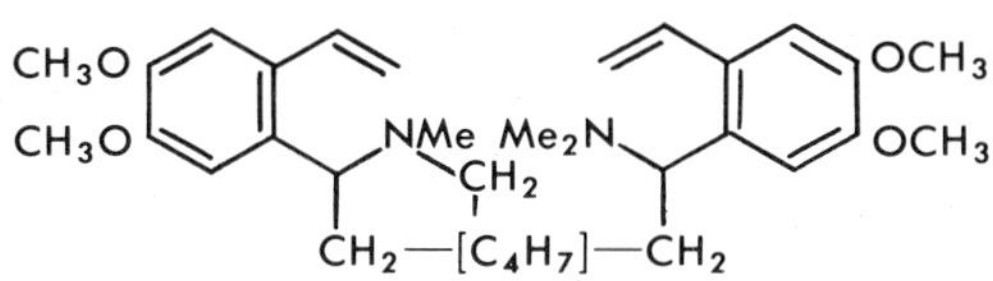

(11)

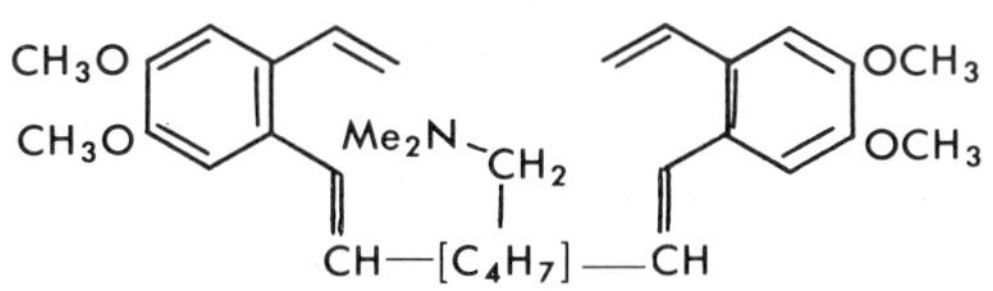

(12)

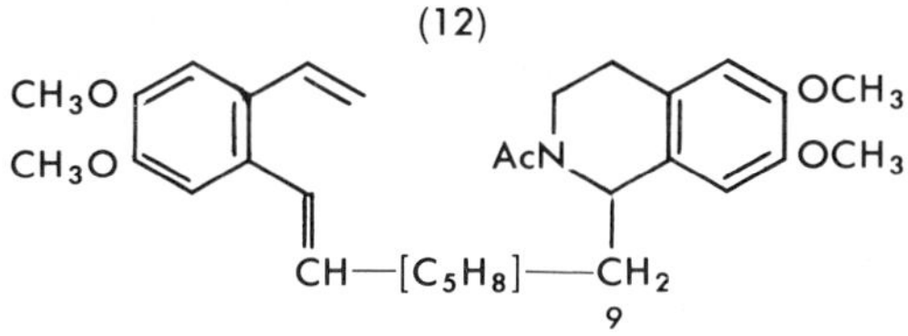

(13)

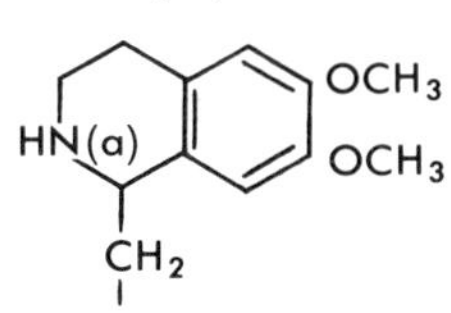

(14)

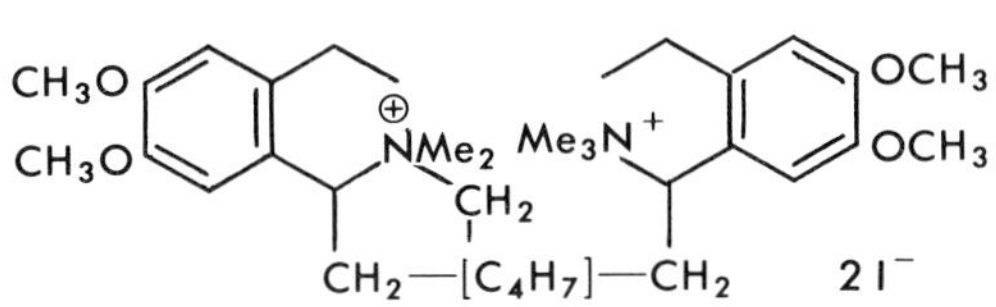

(15)

* It should be noted that Späth and Leithe had no evidence for the presence of methylene groups at positions 9, 12, and 14.

showed that, owing to the introduction of a new center of asymmetry by quaternization at N(b), methylation of emetine gives rise to two stereoisomeric *N*-methylemetine dimethiodides, both of which yielded the same methine (11). The second stage of the Hofmann degradation, carried out by Karrer,[7] resulted in the loss of N(a) as trimethylamine. The retention of N(b) shows that, in emetine, this nitrogen atom is common to two rings. The bismethine (12) gave no crystalline derivatives and its further degradation was not studied.

Ahl and Reichstein[9] protected N(a) by acetylation, and then removed N(b) by a series of three Hofmann degradations. The final product (13) gave 4,5-dimethoxyphthalonimide (4) on oxidation, and 6,7-dimethoxy-1-methylisoquinoline on zinc dust distillation,[10] showing the presence of the partial structure (14).

More extensive investigations of the Hofmann degradation of emetine were made concurrently by Späth and Pailer,[11] and by Battersby and Openshaw.[12] Because of difficulties encountered by the earlier workers, which are probably due to the accumulation of ethylenic bonds in the molecule, the intermediate methine bases were hydrogenated. Hydrogenation of *N*-methylemetinemethine (11) gave the tetrahydromethine which, with methyl iodide in the cold, gave the dimethiodide (15). This compound showed an unexpected instability, since at 100° it decomposed with the fission of one mole of trimethylamine hydriodide, leaving a monomethiodide (16) which contained one ethylenic linkage. This premature occurrence

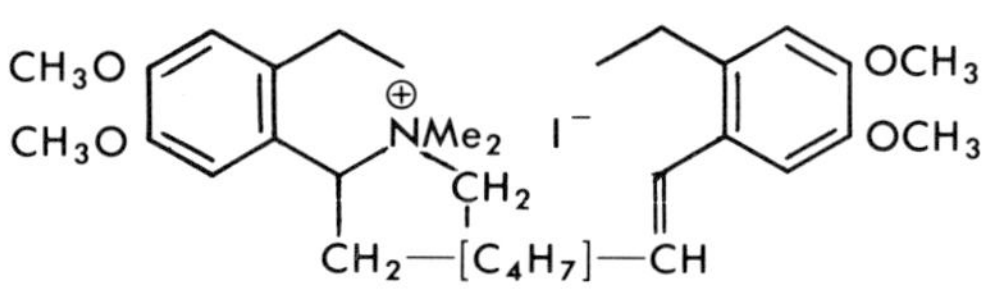

(16)

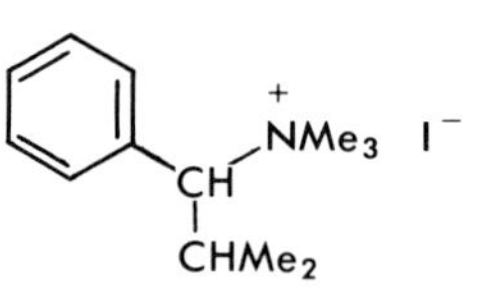

(17)

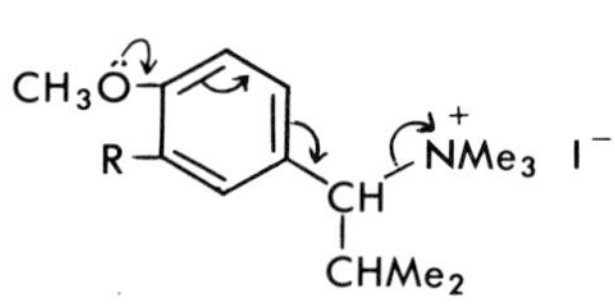

(18) R = H or CH_3O

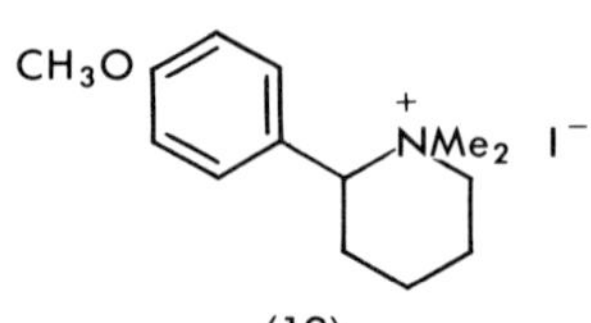

(19)

of a Hofmann-type degradation is due to the presence of an α-(*p*-methoxyphenyl)alkylamine structure, as was shown by the similar behavior of model compounds.[13,14] Thus, although the simple α-phenylalkylammonium salts (17) are stable in solution at 100°, methoxylated analogs such as (18) decompose under these conditions. The fission may be assumed to involve a carbonium ion intermediate (Ingold's[15] E1-type elimination), the separation of which will be facilitated by the influence of the *p*-methoxy group. The stability of the second quaternary center [N(b)] in (15) must be due to its presence in a cyclic structure, since the cyclic analog (19) is similarly stable. Perhaps, this may be ascribed to the reversibility of the C–N fission process. With a cyclic compound, the restraint of the connecting chain prevents the rapid removal of the carbonium ion from the proximity of the nitrogen atom, and hence, facilitates their recombination.

The further stages of the degradation are summarised in Chart 4-2. Oxidation of the tertiary base (20) corresponding to the monomethiodide (16) gave 6-ethylveratric acid ($ArCO_2H$) and an amino acid, without loss of carbon, confirming that (20) contains the group Ar—CH=CH—, and hence, that C(9) in emetine carries two hydrogen atoms. Similar oxidation of (21) showed that C(14) also originally carried two hydrogen atoms (contradicting Brindley and Pyman's structure (9) for emetine).

The final stage of the Hofmann degradation, carried out by Späth and Pailer on the unsaturated product (21), gave rise to a crystalline diene (22), ozonolysis of which gave the expected 6-ethylveratraldehyde, and an unsaturated aldehyde* (23), $C_{18}H_{26}O_3$, which could be reduced to a saturated aldehyde (24). Oxidation of (23) gave 6-ethyl-3,4-dimethoxyhydrocinnamic acid, and ozonolysis gave ethyl methyl ketone. These two products account for all but one of the carbon atoms of (23), and only one structure (25) is possible for the latter. The carbon skeleton of emetine must, therefore, be represented as (26). Pailer[16] ingeniously confirmed the symmetrical attachment of the two Ar-C-C-groups to the central 5-carbon unit by a further degradation. *N*-Acetylemetine was converted into the hexahydro derivative (27) of Ahl and Reichstein's product (13). After hydrolysis of the acetyl group, N(a) was removed from this by further Hofmann degradations to give a nitrogen-free compound (29). Ozonolysis removed the right-hand Ar-C group, and the product was identical with the aldehyde (24) arising from removal of the left-hand Ar-C moiety. The structure of the aldehyde (25) was confirmed by synthesis.

The full carbon skeleton (26) of emetine can, therefore, be depicted as (30), and in order to arrive at the structure of the alkaloid it is necessary to link the two nitrogen atoms into this skeleton, so as to produce the two iso-

*The demonstration that C(14) is a CH_2 group requires that (23) should be an aldehyde and not a ketone.

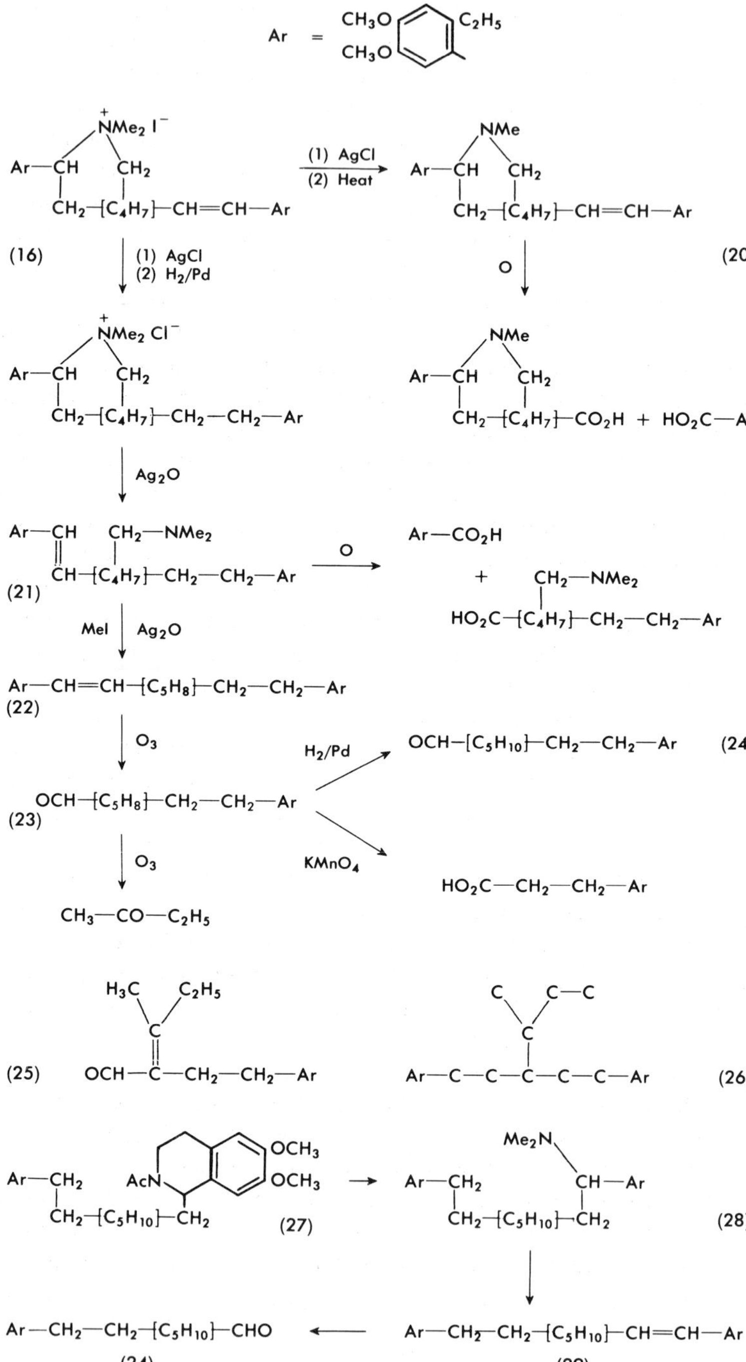

Chart 4.2

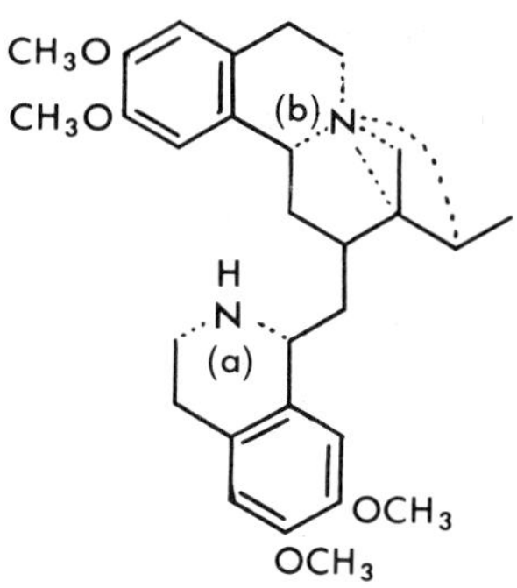

(30)

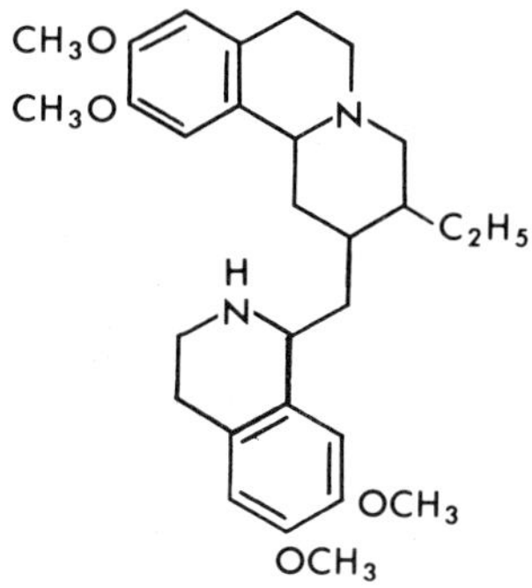

(31)

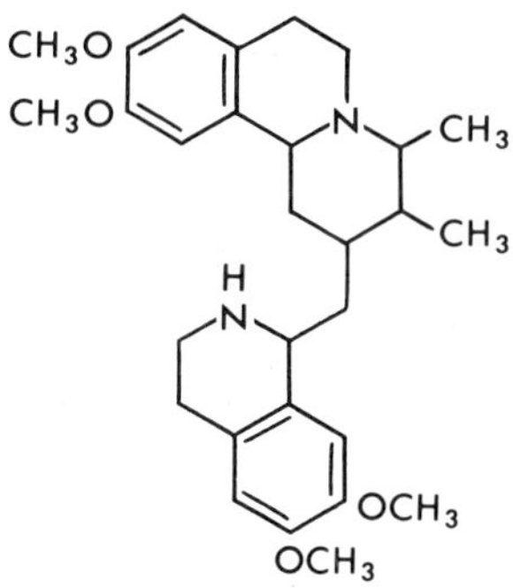

(32)

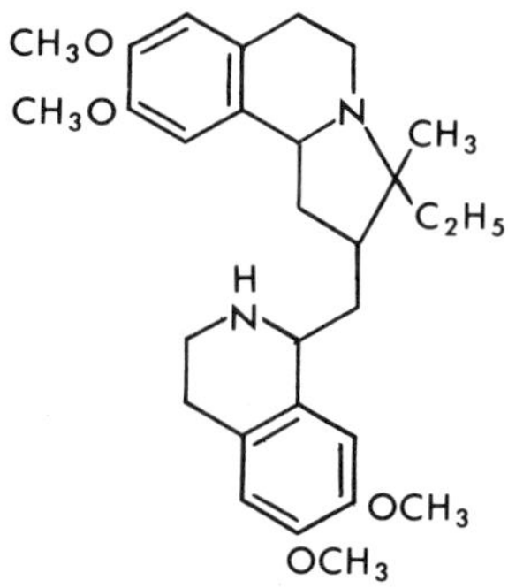

(33)

quinoline systems. Clearly, there is only one possibility for the secondary nitrogen N(a); but for the tertiary nitrogen N(b), there are three possibilities for the third linkage that are consistent with the results of the Hofmann degradation, and lead to (31), (32), and (33). R. Robinson[17] pointed out that the structure (31) could, in fact, be inferred on the basis of biogenetic considerations alone. Applying the principle first advanced by R. B. Woodward to account for the biogenesis of strychnine, he proposed that the product (34), which could be derived from two molecules of dihydroxyphenylalanine and one of formaldehyde, undergoes a Woodward-type fission of the right-hand ring, with concomitant hydrogen transfers, to give the aldehydic intermediate (35). Condensation of this with a third dihydroxyphenylalanine unit, and *O*-methylation, would then yield the structure (31).

That (31) is indeed the correct structure was finally confirmed largely through the results of dehydrogenation experiments that gave rise to identifiable pyridine derivatives. For example, Pailer and Porschinski[18] subjected *N*-methylemetine dibenzyliodide to Hofmann degradation followed by hydrogenation, to give the product (36), which on pyrolytic de-

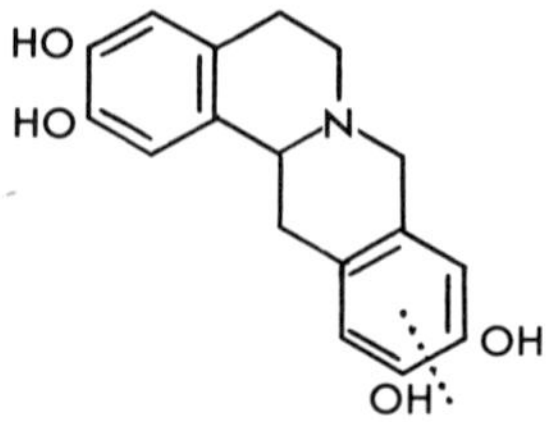

(34)

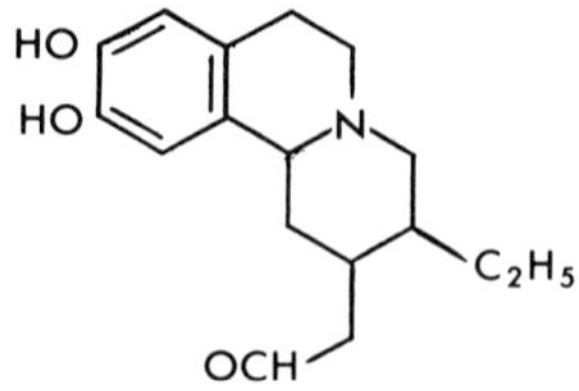

(35)

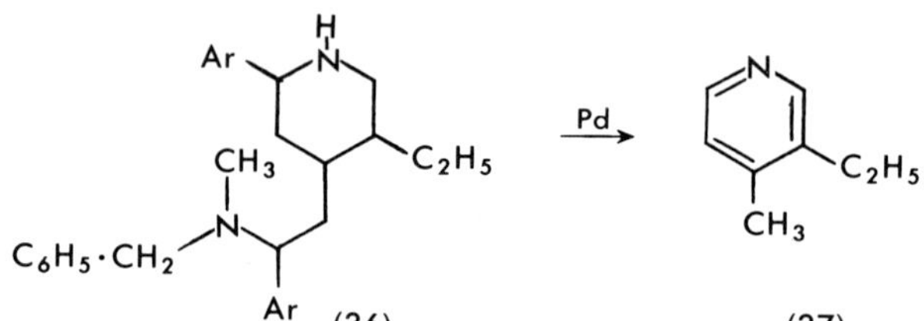

(36) (37)

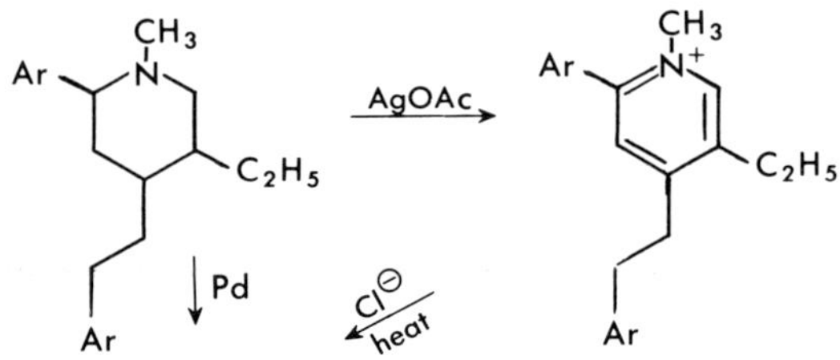

(38) (39)

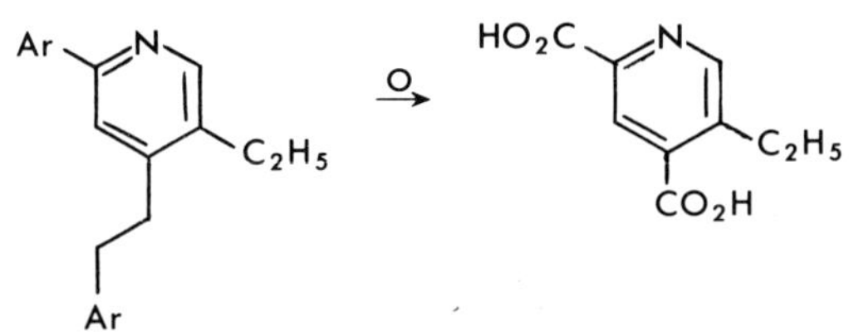

(40) (41)

(Ar = 2-ethyl-4,5-dimethoxyphenyl).

hydrogenation over palladium at 300–310° gave rise to β-collidine (37). Similarly, Battersby and Openshaw[12] hydrogenated the Hofmann degradation product (20), and dehydrogenated the resulting base (38) under relatively mild conditions (aqueous silver acetate at 180°), to give as the main product the pyridinium salt (39). Thermal decomposition of the chloride gave the related pyridine (40), which was also obtained directly by palladium dehydrogenation of (38) at 250–270°. Nitric acid oxidation converted (40) into 5-ethylpyridine-2,4-dicarboxylic acid (41), whose structure was established by its further oxidation with permanganate to pyridine-2,4,5-tricarboxylic acid and by its decarboxylation to 3-ethylpyridine.

The gross structure of emetine was thus finally confirmed, and it is shown in formula (42) with the currently accepted numbering appropriate to a derivative of benzo[a]quinolizine.

4.3 Cephaeline, psychotrine, and emetamine

The structure of emetine (42) being established, the assignment of structures to the related alkaloids involves only the location of the free phenolic group in cephaeline and psychotrine, and of the unsaturation in psychotrine and emetamine.

Pyman correctly allocated the free phenolic group to the position 6′ on the basis of various arguments, the validity of which has recently been

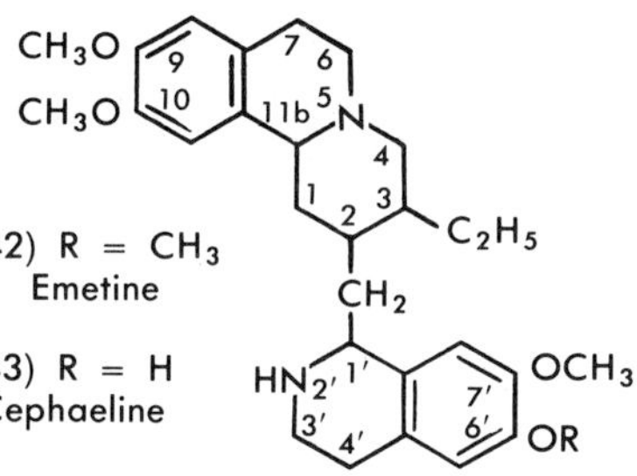

(42) R = CH_3 Emetine

(43) R = H Cephaeline

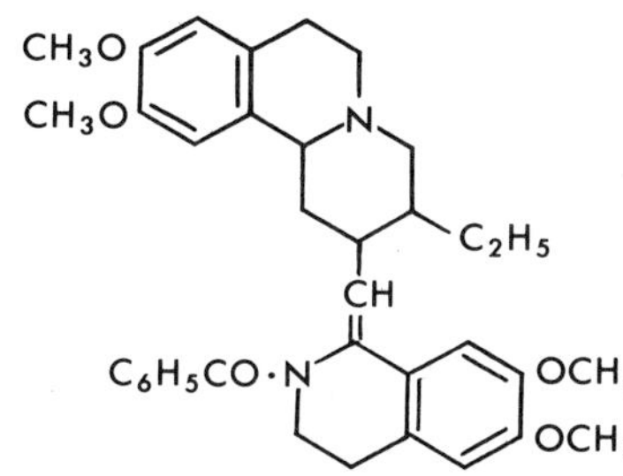

(44)

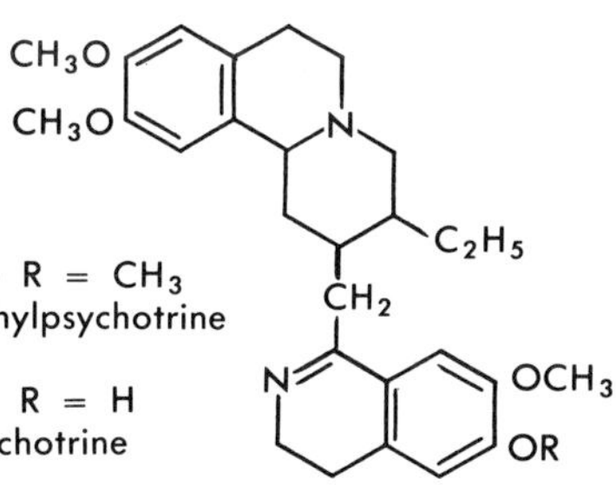

(45) R = CH_3 O-Methylpsychotrine

(46) R = H Psychotrine

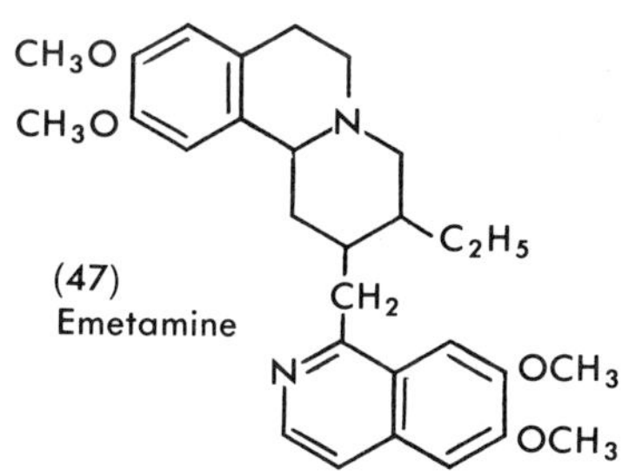

(47) Emetamine

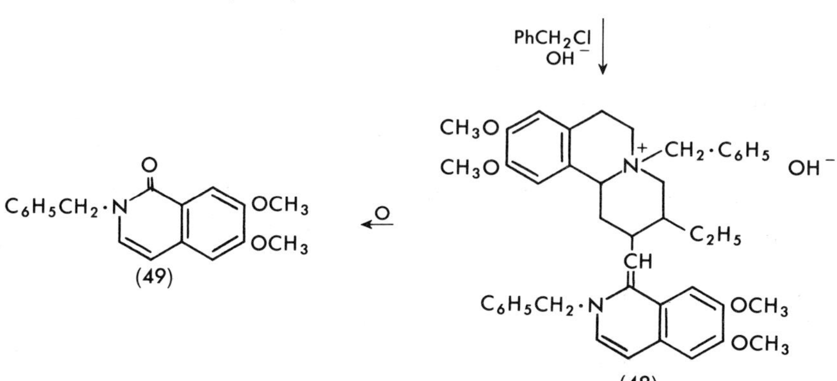

(48)

(49)

questioned.[19] Definite proof was given by Pailer and Porschinski[18] through the Hofmann degradation of *O*-ethylcephaeline to the analog of (16), which on ozonolysis gave rise to 2-ethyl-4-ethoxy-5-methoxybenzaldehyde. Cephaeline is therefore (43).

Psychotrine forms an *N*-benzoyl derivative and was, therefore, long considered to contain a secondary amino group. Pyman allocated the double bond to the position shown in (44). Karrer, Eugster, and Rüttner[10] confirmed this structure for the *N*-benzoyl derivative of *O*-methylpsychotrine by oxidation with perphthalic acid or ozone to the *N*-benzoyl derivative of corydaldine (5). However, the uv[20] and ir[21] absorption spectra of *O*-methylpsychotrine require it to be a 3,4-dihydroisoquinoline (45). The formation of the *N*-benzoyl derivative is readily accounted for by a tautomeric shift of the double bond.[22] Psychotrine has the structure (46). Reduction of psychotrine gives cephaeline and isocephaeline that differ only in their configuration at C(1′); emetine and isoemetine are similarly related.

Emetamine is a ditertiary base, and since on reduction it gives isoemetine, it is reasonable to suggest the structure (47), containing a fully aromatic isoquinoline ring. This is supported by the similarity of the uv spectrum of emetamine[23] to that of papaverine, which contains the same chromophoric groups. Confirmation of this structure was obtained by treating emetamine bisbenzylochloride with alkali, and oxidizing the resulting anhydro-base (48) with permanganate, when the known *N*-benzylisoquinolone (49) was obtained. A similar degradation of *O*-methylpsychotrine, through the anhydro-base (50), gave *N*-benzylcorydaldine (51), showing that the 3,4-double bond in (49) had not been introduced during the oxidation.

5. PROTOEMETINE

Up to 1957 only the five closely related alkaloids, already mentioned, had been isolated from ipecac and identified with certainty, and although the isolation of other bases had been claimed in early reports, it is almost certain that these were mixtures. However, by countercurrent distribution of the nonphenolic basic residues from the commercial extraction of emetine, Battersby, Davidson, and Harper[21] isolated two new alkaloids. One of these, "ipecac alkaloid A," a crystalline base, was obtained in too small an amount for complete characterisation. The other was very unstable as the free base, but formed a crystalline perchlorate, $C_{19}H_{27}NO_3 \cdot HClO_4$. Analytical and spectroscopic data indicated the presence of one 6,7-dimethoxytetrahydroisoquinoline unit and one carbonyl group, probably in the form of an aldehyde. The latter was confirmed by Tollens' test and the formation of a crystalline semicarbazone, and by its reduction to an

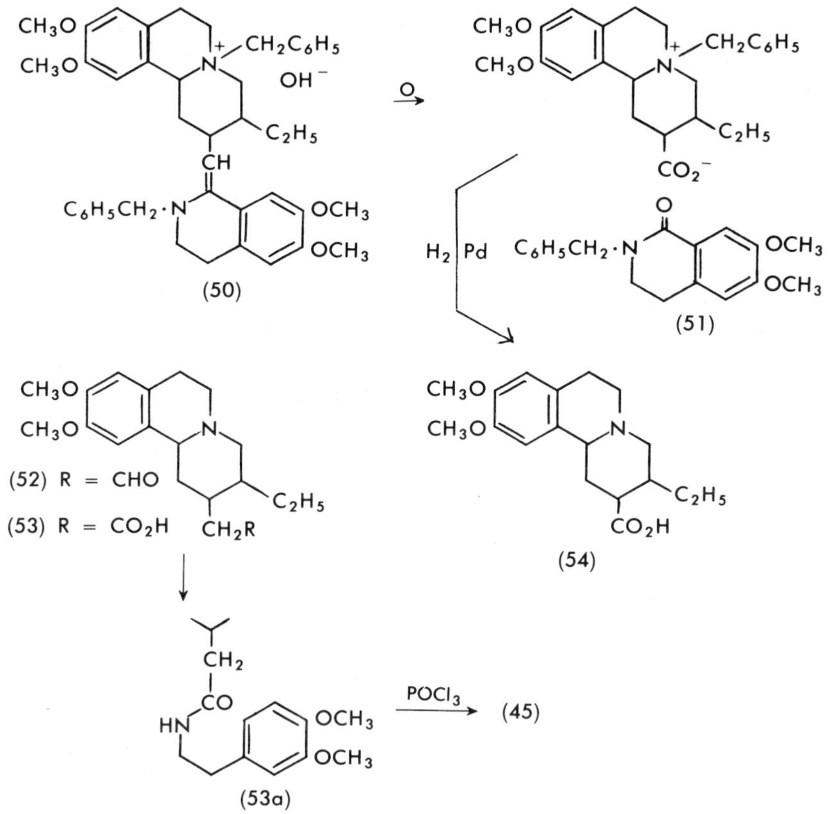

alcohol. The nitrogen atom was shown to be tertiary, and its basic strength was consistent with its occupying a position common to two rings. Thus, it seemed possible that the alkaloid might possess the structure (52), which is the *O*-methylation product of Robinson's proposed biogenetic intermediate (35). This was proven correct by converting the alkaloid, by dehydration of its oxime, into a nitrile which upon hydrolysis gave the acid (53). This acid could also be obtained by Arndt-Eistert homologation of the acid (54) obtained from *O*-methylpsychotrine as shown. Moreover, cyclization of the homoveratrylamide (53a) of (53) with phosphoryl chloride gave *O*-methylpsychotrine (45), identical with the natural alkaloid. Since emetine can be obtained from *O*-methylpsychotrine by reduction, this constituted a partial synthesis of emetine from the new alkaloid, which was appropriately named protoemetine.* The discovery of proto-

*Although protoemetine could not be converted directly into emetine by a Pictet–Spengler condensation with 3,4-dimethoxyphenethylamine,[21] it was subsequently condensed[24] with 3-hydroxy-4-methoxyphenethylamine to give cephaeline (and some isocephaeline), from which emetine can be obtained by *O*-methylation.

emetine appeared to lend powerful support to Robinson's biogenetic proposals; however, it has since transpired[2] that the "Woodward fission" of an intermediate such as (34) is not involved in the biosynthesis of the ipecac alkaloids.

6. STEREOCHEMISTRY OF EMETINE

The structure (42) contains four asymmetric centers (2,3,11b, and 1′); and, therefore, there are eight possible racemic or sixteen possible optically active forms. The complete relative and absolute configuration of these centers in emetine has been elucidated largely through the work of Battersby[25] and of van Tamelen,[26] and is depicted in Figure 4.1.

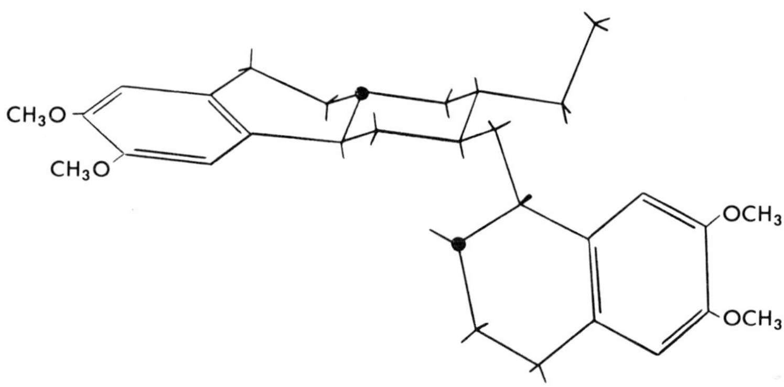

Figure 4.1 Stereo-formula of emetine. The shaded circles represent the nitrogen atoms.

Of the various conformations possible for the hexahydrobenzoquinolizine system present in emetine, the most stable is that shown in Figure 4.2, in which the hydrogen atom at C(11b) is *trans* to the unshared pair of

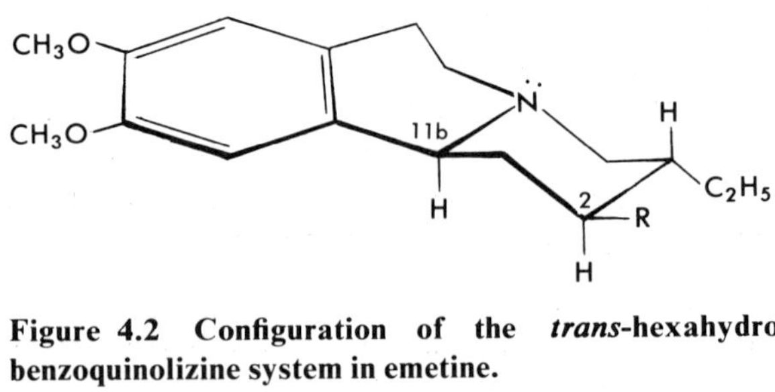

Figure 4.2 Configuration of the *trans*-hexahydrobenzoquinolizine system in emetine.

electrons on the nitrogen atom; the configuration is comparable to that of *trans*-decalin. It has been shown by Bohlmann[27] that the presence of characteristic absorption around 2750 cm^{-1} in the ir spectra of such quinolizidines is diagnostic of this *trans*- configuration. Since the "Bohlmann bands" are apparent in the spectra of emetine and of degradation products such as (54), which contain the quinolizidine structure, these must all possess the stable *trans* configuration.

The methyl ester of the amino acid (54), obtained from *O*-methylpsychotrine as described above, was found to be unchanged by heating with methanolic sodium methoxide under conditions sufficiently vigorous to allow epimerization at the position α to the methoxycarbonyl group; this group must, therefore, occupy the more stable equatorial situation (Figure 4.2, R = CO_2Me). Hence, the ester can be represented as (55) (or its mirror image). That no stereochemical inversion has occurred during the prepa-

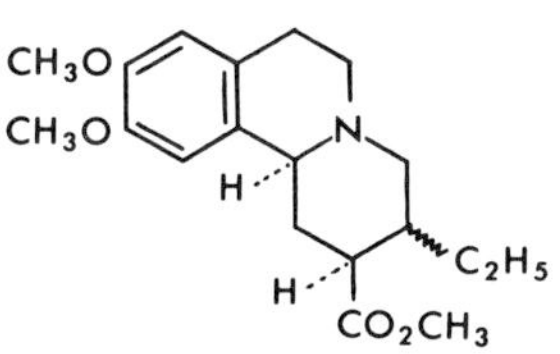

(55)

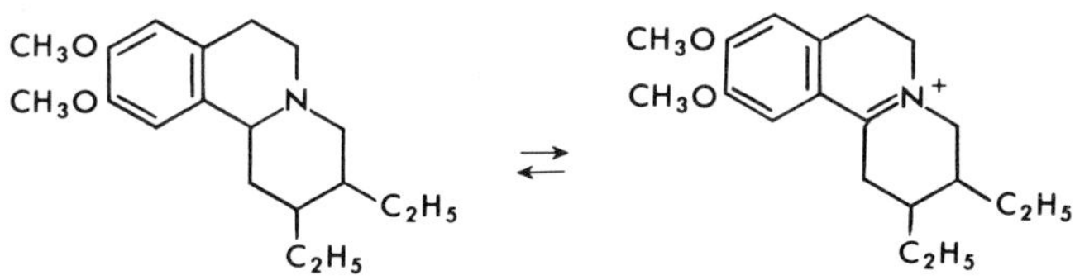

(56) (57)

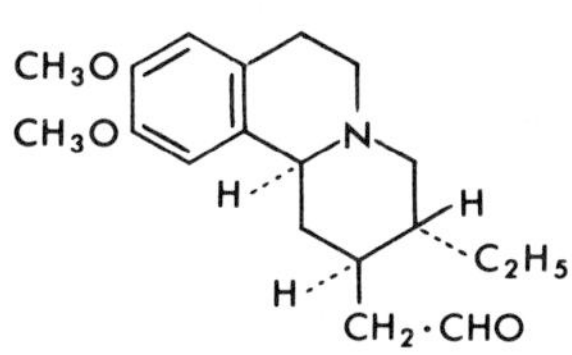

(58)
Protoemetine

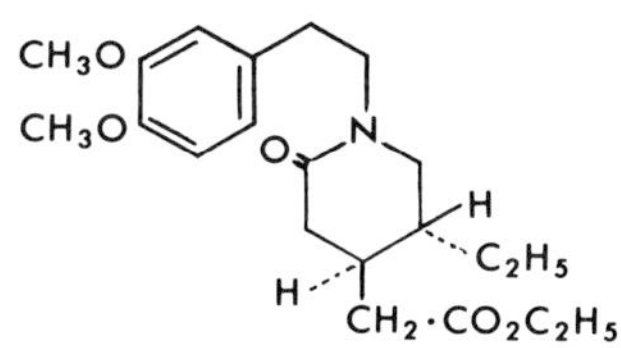

(59)

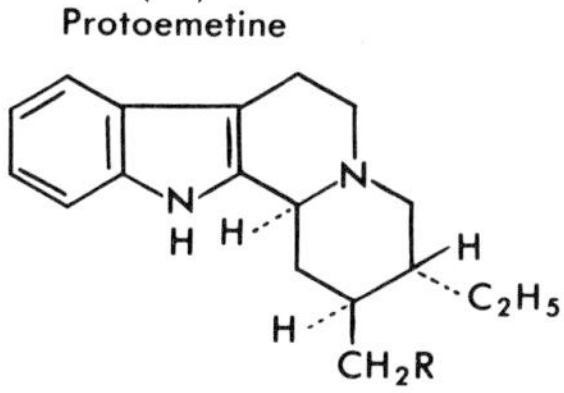

(60) R = CHO
(61) R = CH_3

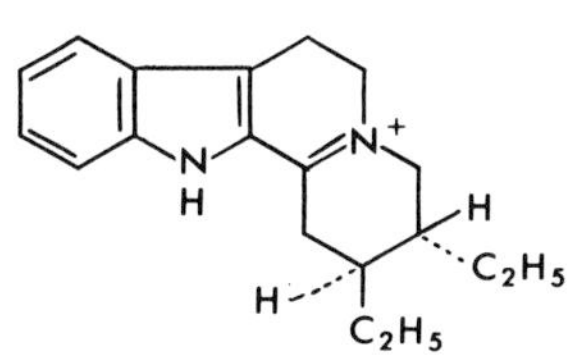

(62)

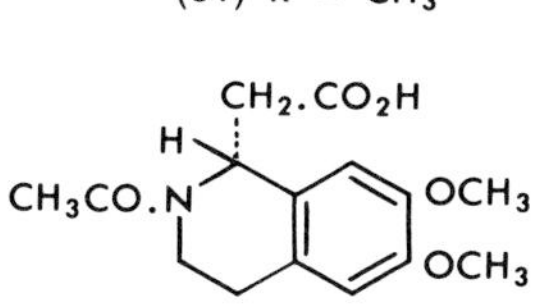

(63)

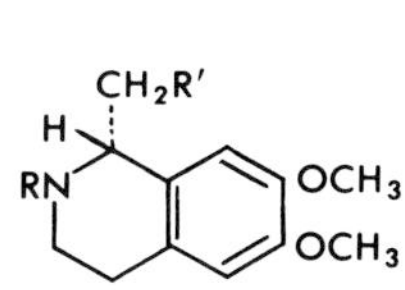

(64) R = H; R′ = $CO_2C_2H_5$
R = C_2H_5; R′ = CH_2OAc, CH_2OH, CH_3.
(65) R = H; R′ = OH

ration or isolation of this ester was shown by its correlation, through reactions not involving the asymmetric centers, with the ester of the homologous acid (53) obtained from protoemetine and shown to have the same stereochemistry as *O*-methylpsychotrine.

The orientation of the ethyl group was determined by converting protoemetine, by the Wolff–Kishner reduction, into the diethyl compound (56), which on dehydrogenation with mercuric acetate gave the quaternary salt (57). The racemic *cis*- and *trans*- forms of (56) and (57) were synthesized from *cis*- and *trans*-3,4-diethylcyclopentanone, and the synthetic *trans*- compounds had infrared absorption in solution identical with that of the corresponding optically active products from protoemetine. Hence, the ethyl group is *trans* to the adjacent aldehydic side-chain and protoemetine has the relative configuration shown in (58) (*cf.* Figure 4.2; R = CH_2CHO).

The *trans*-isomer of (56) was also correlated by van Tamelen[26] with an intermediate (59) from which Evstigneeva *et al.*[28] had previously synthesized emetine (see below) and which must, therefore, correspond in stereochemistry with the alkaloid, thus independently supporting a 2,3-*trans* configuration.

The *absolute* configuration of protoemetine was proven to be that shown in (58) by the close resemblance of the molecular rotation changes associated with the conversions (58) → (56) → (57) to those accompanying the analogous transformations of dihydrocorynantheal (60) of known absolute stereochemistry into (61) and (62). This demonstration of the stereochemical correlation of the ipecac with the indole alkaloids is of considerable biogenetic interest.

Additional evidence for the orientation assigned to the hydrogen atom at position 11b was afforded by the observation[25,26] that reduction of 5,11b-dehydro-compounds such as (57) by sodium and alcohol, sodium borohydride, or catalytic hydrogenation gives in each case only one stereoisomer (56), which must be the thermodynamically more stable isomer with the hydrogen axial.* Similarly, 5,11b-dehydro-*N*-benzoylemetine perchlorate, obtained by mercuric acetate dehydrogenation of *N*-benzoylemetine, exclusively regenerates the latter on hydrogenation.[29] This behavior on reduction parallels that known in the indole alkaloid series; for example, reduction of (62) gives (61).

Having established the stereochemical configuration of the benzoquinolizine portion of the molecule, it remained to determine the orientation of the isolated asymmetric center at position 1′ of the isoquinoline nucleus. This was achieved in two ways.

Van Tamelen[26] pointed out that, of the four asymmetric centers, only

*Only with tin and hydrochloric acid was a proportion of the isomeric reduction product obtained.

those at positions 11b and 1′ are in proximity to the ultraviolet chromophores, and these should, therefore, exert the main influence on the optical rotatory dispersion of the alkaloid. Emetine hydrobromide shows practically constant specific rotation over the wavelength range 350–700 mμ. In contrast, isoemetine hydrobromide shows an increasing positive rotation with falling wavelength over this range. This suggests that, in emetine, the two centers have opposite configurations, so that their individual dispersion effects counteract. The correctness of this assumption was established by Battersby's demonstration of the absolute configuration at C(1′). By Hofmann degradation and subsequent oxidation, *N*-acetylemetine was converted to the levorotatory acid (63), four derivatives (64) of which, obtained by methods not affecting the asymmetric center, all had rotations that became more negative as the polarity of the solvent was increased. Comparison with the rotation shifts of similar tetrahydroisoquinolines of known stereochemistry led to assignment of the absolute configuration shown, and this was confirmed by a chemical correlation with (+)-calycotomine (65) of known configuration.

Thus, emetine has the absolute stereochemistry represented in formula (66).

7. IPECOSIDE

From the nonbasic extractives of ipecac root, a nitrogenous glucoside, ipecoside, has been isolated,[30] and has been shown[31] to have the struc-

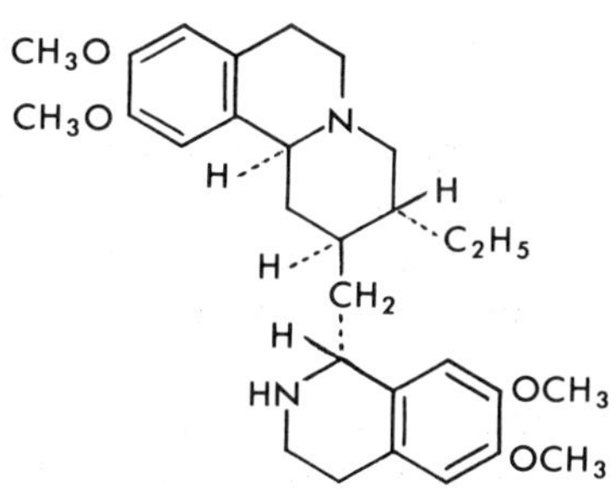

(66) Emetine

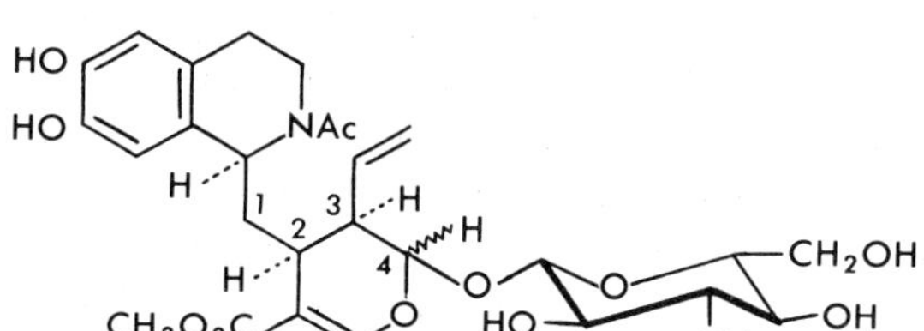

(67) Ipecoside.

ture (67). In this formula, the atoms have been numbered to show their relationship to those in the protoemetine skeleton, which is derivable from that of ipecoside by hydrolysis of the glucosyl and acetyl groups, fission of the pyran ring and closure between C(4) and the nitrogen atom. Ipecoside is thus of particular interest[2] in relation to the course of biogenesis of the ipecac alkaloids.

8. STRUCTURE OF RUBREMETINE

The dehydrogenation of solutions of emetine salts by such reagents as ferric chloride, halogens, or mercuric acetate to give the highly colored rubremetinium salts has already been mentioned in Section 4.1. The interpretation of this reaction in the light of the established structure (66) of emetine is not immediately obvious. The most straightforward suggestion, made by Karrer,[10] that the reaction involves merely the aromatization of both isoquinoline ring-systems, is untenable for several reasons. It does not account for the intense color of the salts, the loss of basicity of one nitrogen atom, or the fact that oxidation of emetamine (47) gives a similar, but not identical, product "rubremetamine." Karrer's proposal was based in part on the observation that rubremetine could be reduced by zinc and acid to a crystalline tetrahydro compound, which he suggested was a stereoisomer of emetamine and which resisted further hydrogenation.

The intense color of the rubremetinium salts requires the presence of an extended conjugated system, and their monoquaternary nature suggested that this system might involve the two nitrogen atoms in a mesomeric structure of the type found in the cyanine dyes. On this basis the structure (68) was proposed,[20,32] support being given by the demonstration that its reduction products gave pyrrole color reactions. Rubremetine is optically active, in agreement with the retention in (68) of one remaining asymmetric center at position 3.

When hydrogenated over platinum in the presence of sodium acetate, rubremetinium chloride rapidly absorbs exactly 1 mole of hydrogen, and gives a mixture of two colorless dihydrorubremetines (69), epimeric at position (11b). Both compounds show pyrrole color reactions, and both are readily reoxidised to rubremetine. Further hydrogenation occurs only extremely slowly in neutral solution, but is markedly catalyzed by acid, and both α- and β-dihydrorubremetine give rise to the same tetrahydro-compound (70), identical with that obtained by reduction with zinc and acid. This still shows pyrrole reactions, contains no NH, and must, therefore, arise by hydrogenolysis between C(11b) and the pyrrole ring. This can be rationalized in the following way.

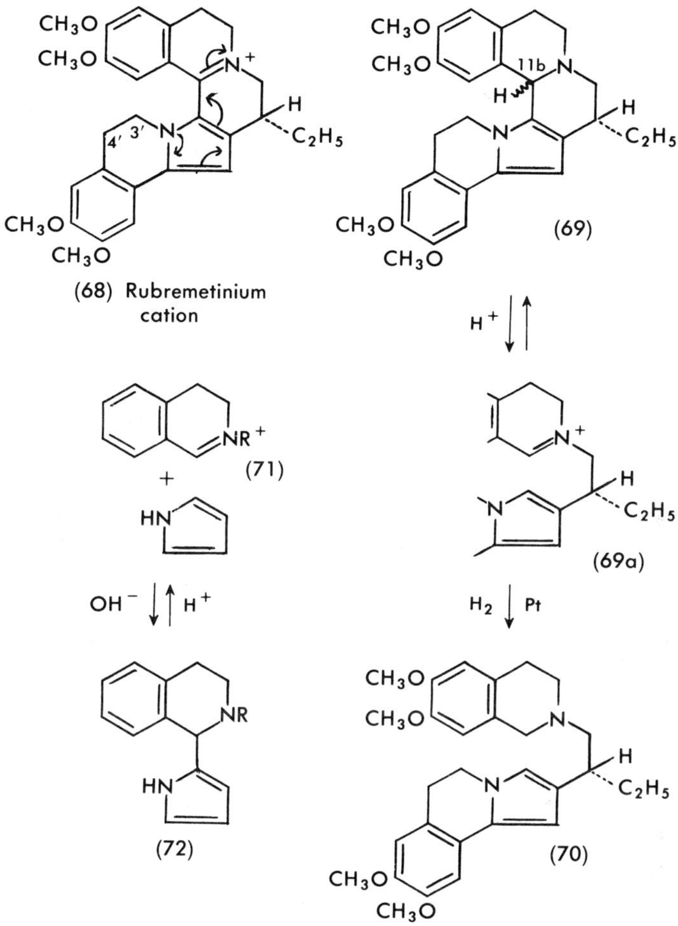

It is well known that 3,4-dihydroisoquinolinium salts such as (71) readily react at the 1-position with a variety of nucleophilic reagents to give 1-substituted tetrahydroisoquinolines, and that this reaction can be reversed by acids. Pyrrole can act as such a nucleophilic reagent to give a condensation product (72) analogous to (69). Hence in acid solution, (69) may be expected to be in equilibrium with a ring-opened form (69a) that will be susceptible to reduction at the $C{=}N^+$ linkage. In agreement with this supposition, it was found that the dihydrorubremetines undergo a rather rapid mutarotation in alcoholic hydrochloric acid solution, the two forms equilibrating through the common cation (69a). Basification of the equilibrated solution from either isomer gives a mixture of the α- and β-forms.

The behavior of rubremetamine, the oxidation product of emetamine, on reduction parallels that of rubremetine, and it therefore possesses a structure analogous to (68), but with a 3′,4′-double bond.[23] General support was given to the structure (68) for rubremetine by the synthesis and spectroscopic examination of model compounds containing the same chromophoric system,[32,33] and by the failure of *N*-methylemetine[8] and of a synthetic 1-methylemetine[34] to be oxidized to rubremetine-like products.

9. SYNTHESIS OF EMETINE

The elucidation of the gross structure (42) of emetine in 1948 initiated numerous attempts to synthesize the alkaloid, and the solution of its stereochemistry further stimulated interest in this problem. Since at least a dozen successful syntheses have been described, the following account is of necessity confined to a discussion of the more generally significant aspects.[35]

The earlier work was undertaken to obtain confirmation of the structure, and the first claim to a complete synthesis of the alkaloid was that of Evstigneeva and Preobrazhenskii,[28] although the evidence they adduced for the identity of their final product with natural *l*-emetine was inconclusive. However, their synthesis did give the desired proof of the gross structure, since oxidation of the product gave rubremetine, and a repetition[26] of their work confirmed that emetine could be obtained by their method. Other early syntheses[36,37] similarly avoided the stereochemical complications by oxidation of the final, sterically indeterminate products to *dl*-rubremetine which, since it contains only one asymmetric center, could be compared spectroscopically with the optically active product from natural emetine.

The majority of the described syntheses proceed through one of two key intermediates. In some cases, the benzoquinolizine-acetic ester (73) is first synthesized, and the derived 3,4-dimethoxyphenethylamide (homoveratrylamide) then cyclized by the Bischler–Napieralski reaction to *O*-methylpsychotrine (or stereoisomers) (45), which on reduction gives emetine and isoemetine (or stereoisomers). Alternatively, the two isoquinoline rings are formed simultaneously or successively by the cyclization of the bis-homoveratryl compound (74) and subsequent reduction. However, quite widely varying methods have been used to prepare the key intermediates. Initially these were stereochemically nonspecific, but as knowledge of the desired configurations was acquired, so the methods could be chosen to lead more specifically towards the desired isomers.

Evstigneeva and her colleagues[28] built up the central portion of the molecule in the form of 3-α-cyanopropylglutaric ester (75) as shown in Chart 4-3 (see p. 108). Hydrogenation of the cyano group in the presence of

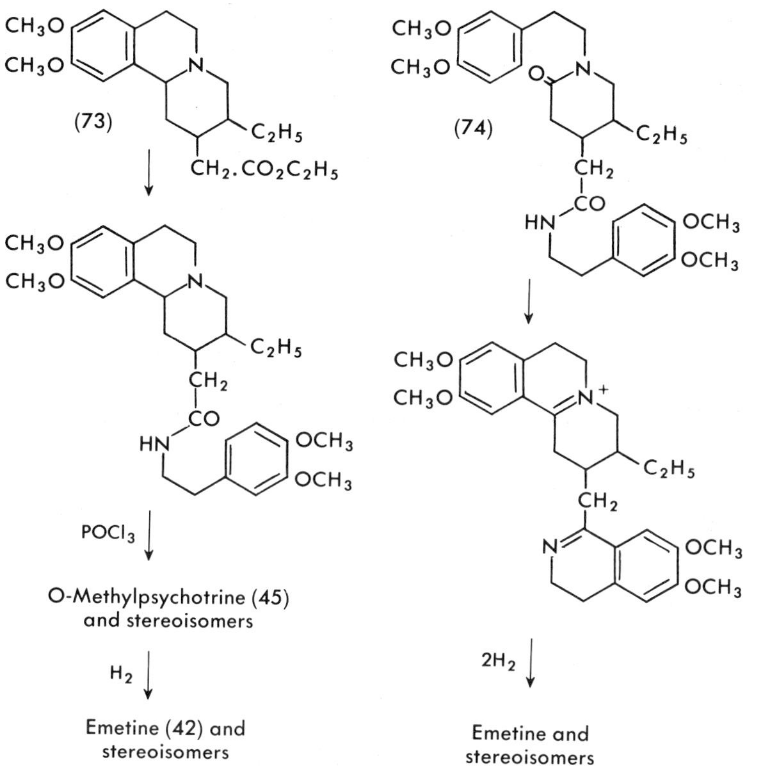

homoveratrylamine led to formation of a secondary amine that spontaneously lactamized to give (76). Cyclization of this, followed by reduction, gave the intermediate (73), whereas further condensation of (76) with homoveratrylamine produced the alternative intermediate (74). Although separation of stereoisomers was attempted at different stages, the final products were almost certainly mixtures, from which *l*-emetine was subsequently isolated through the camphorsulphonate.

Barash and Osbond[38] also synthesized the piperidone (76), starting from 3-ethyl-4-methylpyridine, and obtained two stereoisomers. Treatment of each homoveratrylamide (74) with phosphoric oxide preferentially closed the lower dihydroisoquinoline ring, and hydrogenation, which introduces a further asymmetric center at C(1′), gave two products (77) in each case. Phosphoryl chloride cyclized each of the four products to a dehydroemetinium salt (78). On reduction, a total of six different stereoisomers (out of the possible eight) was obtained, one of which was identified as *dl*-emetine and was resolved through the dibenzoyltartrate. Of the various alternative routes to the intermediate (73) which have been described, some are susceptible to stereochemical control. Thus, Bat-

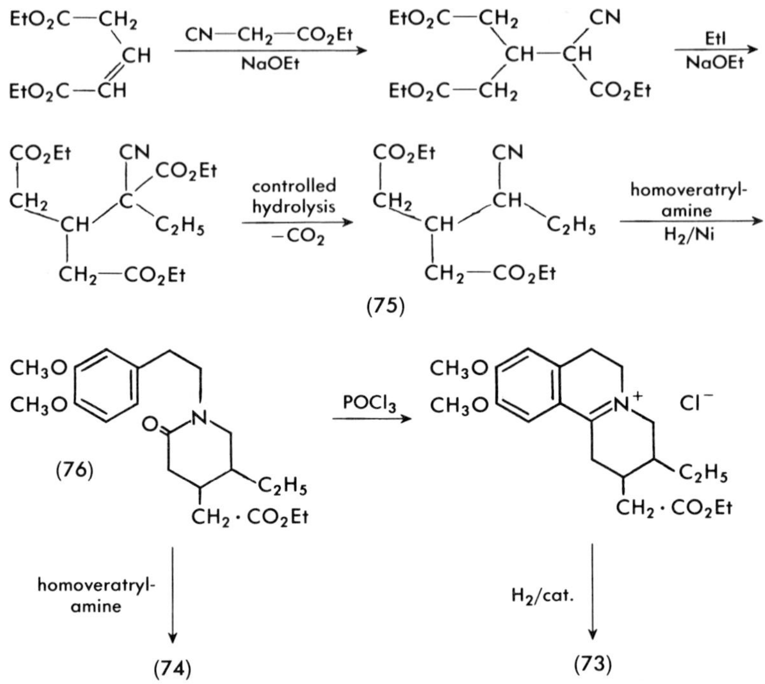

Chart 4.3

tersby and Turner[39] prepared the dihydropyridone (79) by a multistage procedure, and subjected it to Michael condensation with ethyl malonate, a reaction that can be expected to give the thermodynamically more stable adduct, in this case the *dl-trans*-isomer (80). Hydrolysis and decarboxylation gave the *trans*-form of (76), which was converted in high yield into the desired isomer (81) of (73), the hydrogenation step proceeding apparently exclusively to give the more stable *trans*-fused ring system.

One of the earliest synthetic routes[36] utilized the tricyclic ketone (82), which condenses with ethyl cyanoacetate to give the product (83), trans-

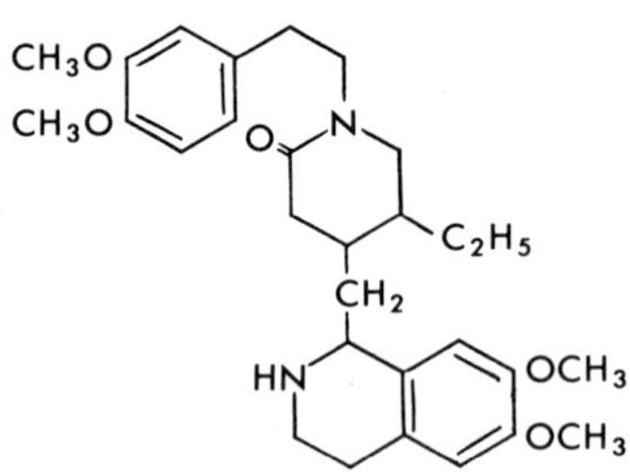

(77)

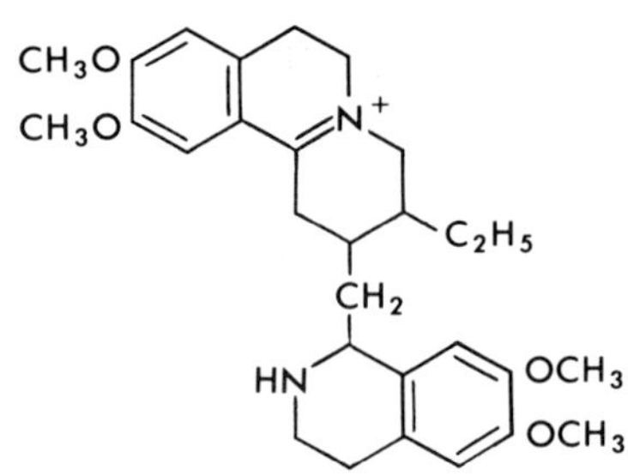

(78)

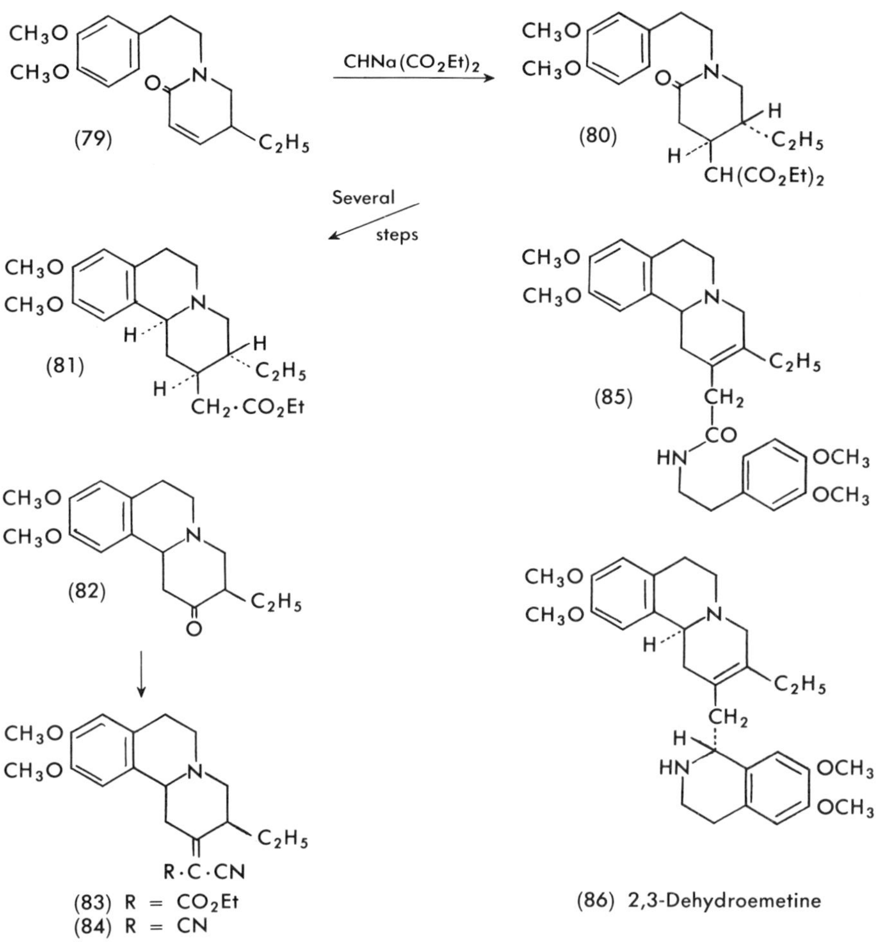

formed by conventional steps to (73).* This route was studied in more detail by Brossi,[40] who showed that hydrolysis and decarboxylation of the analogous malononitrile condensation product (84) was accompanied by a shift of the double bond into the 2,3-position. Hydrogenation of the derived homoveratrylamide (85) was difficult. It gave a mixture of isomers from which the desired 2,3-*trans*-compound could be isolated in only small yield. When the unsaturated intermediate (85) was cyclized, *dl*-2,3-dehydro-*O*-methylpsychotrine, containing only one asymmetric center [C(11b)], was obtained, and on reduction gave *dl*-2,3-dehydroemetine (86) and its C(1′) epimer, *dl*-2,3-dehydroisoemetine. The former is comparable with

*It has subsequently been shown that the product obtained by this route has the unnatural configuration at C(3).

emetine in amoebicidal properties, and has proved to be effective in the treatment of amoebic dysentery.

Although the foregoing investigations did not lead to a practically useful synthesis of emetine itself, an efficient and commercially applicable procedure has been developed[41] through the use of the intermediate (82). A simple and almost quantitative synthesis of this ketone involves the interaction of 6,7-dimethoxy-3,4-dihydroisoquinoline with the hydrochloride of the Mannich base (87). The resulting racemic ketone is converted into the required (-)-ketone (88) by refluxing with (-)-camphor-10-sulphonic acid in

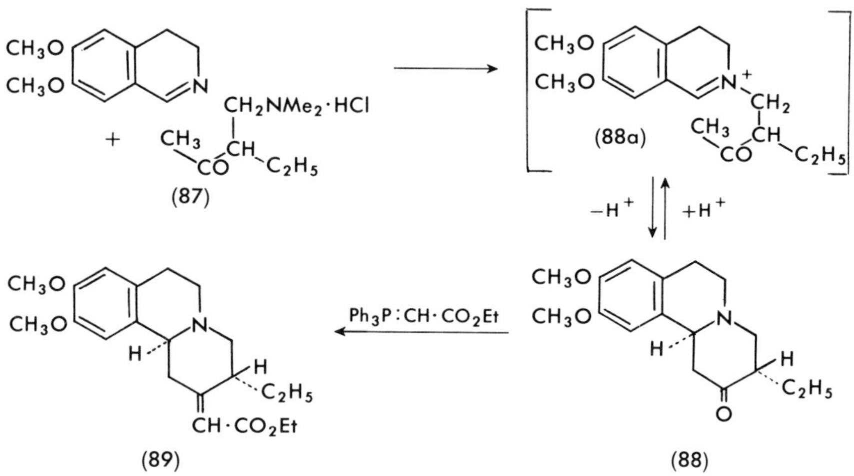

ethyl acetate solution. The salt of the (-)-base separates from the hot solution, and the (+) base remaining in solution is continuously racemized by the acid present, forming more (-)-base, until almost all has separated in the desired optically active form. The racemization proceeds through equilibration with the ring-opened dihydroisoquinolinium salt (88a), which is also an intermediate in the synthesis, combined with epimerization at C(3) through keto-enol tautomerism.

When the (-)-ketone (88) is heated with the Wittig reagent, ethoxycarbonylmethylenetriphenylphosphorane, a single, optically active, unsaturated ester (89) is formed. Hydrogenation of this ester gives over 75% of the desired optically active *trans*-isomer (81), the conversion of which to *O*-methylpsychotrine has already been described. Hydrogenation of *O*-methylpsychotrine gives a mixture of emetine and isoemetine. The latter is valueless as an amoebicide, but it need not be wasted, since it can be reconverted into *O*-methylpsychotrine in high yield by treatment with sodium hypochlorite, which forms an *N*-chloro-derivative readily dehydrochlorinated by alkali.

An entirely different synthetic approach, which avoids simultaneous production of isoemetine, was devised by Ritchie *et al.*,[42] who linked together two preformed dihydroisoquinoline units by reaction with acetone dicarboxylic acid. The resulting bis-tetrahydroisoquinolylacetone can be isolated in either the racemic or the meso form, the two forms being readily interconvertible; the latter (90) corresponds to emetine in its relative configuration at positions 11b and 1′, and may be used for its synthesis. Preferably, however, the mono-*N*-benzyl derivative (91) is used;

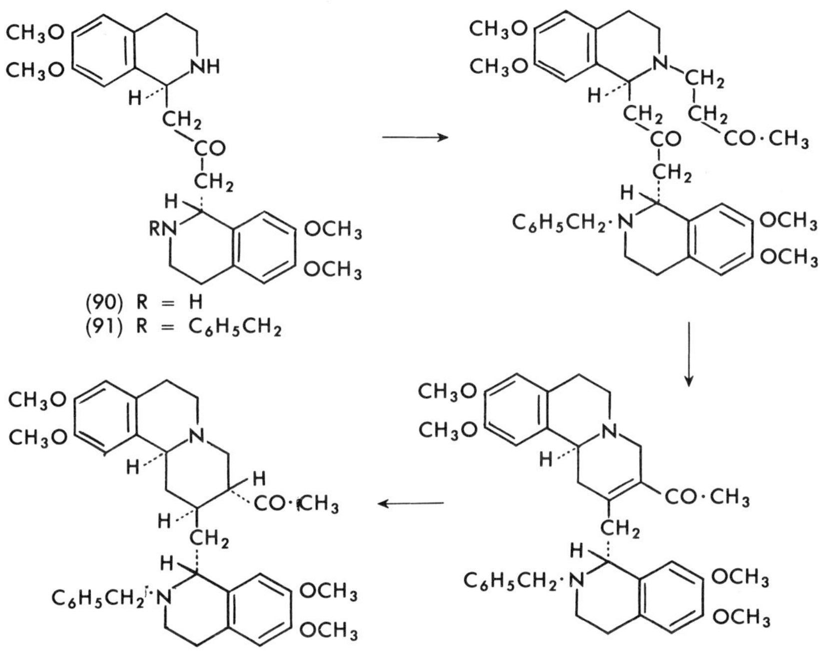

(90) R = H
(91) R = $C_6H_5CH_2$

(93) (92)

addition of methyl vinyl ketone to the secondary amino group is followed by ring closure and dehydration to an $\alpha\beta$-unsaturated ketone (92). Reduction with lithium–ammonia, followed by equilibration at C(3) with hot acid, leads to a saturated ketone (93) with the desired *trans*-configuration. Reduction of the carbonyl group by nickel desulphurization of the ethylene thioketal, and hydrogenolysis of the *N*-benzyl group, gives *dl*-emetine; similar treatment of (92) gives *dl*-2,3-dehydroemetine.

10. ALKALOIDS OF *ALANGIUM LAMARCKII* THWAITES

The Indian plant *Alangium lamarckii* Thwaites (fam. Alangiaceae) is widely used in indigenous medicine, and earlier studies showed the presence of a number of ill-defined alkaloids.[43] In 1964, Pakrashi[44] isolated emetine, cephaeline, and psychotrine from the seed kernels and root bark of *A. lamarckii*, the first reported occurrence of these alkaloids outside the Rubiaceae. Further examination[45] of the stem and root bark yielded two new alkaloids of the emetine type: alangicine, which was identified as 11-hydroxypsychotrine (94), and a desmethylpsychotrine (95). A further alkaloid from this source was identified as tubulosine (96), a novel alkaloid that had been isolated from the Argentinian plant *Pogonopus tubulosus* (DC) Schumann (fam. Rubiaceae), and also from the S. African plant *Cassinopsis ilicifolia* Kuntze (fam. Icacinaceae).[46] Later,[47] two other new alkaloids were isolated from *A. lamarckii* bark, and shown to be isotubulosine [C(1′) epimer of (96)] and a demethyltubulosine (97). The leaves of this plant[48] contain two further indolic analogs of emetine, namely deoxytubulosine (98) and its 11-hydroxy-derivative, alangimarckine (99), together with two new types, dihydroprotoemetine (100) and its 11-hydroxy-derivative, ankorine (101).

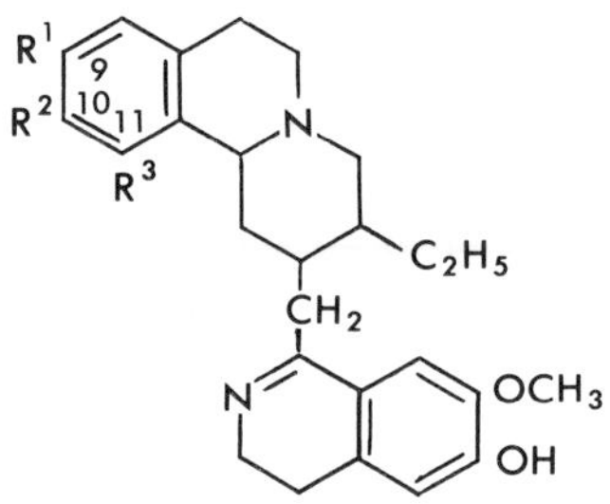

(94) $R^1 = R^2 = OCH_3$; $R^3 = OH$. Alangicine

(95) R^1 or $R^2 = OCH_3$; R^2 or $R^1 = OH$; $R^3 = H$.

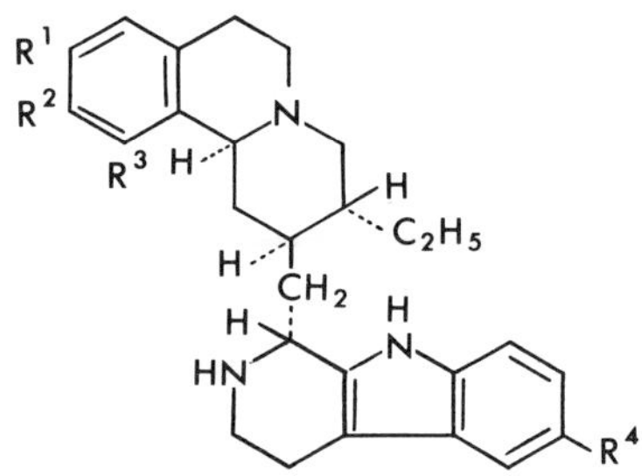

(96) $R^1 = R^2 = OCH_3$; $R^3 = H$; $R^4 = OH$. Tubulosine

(97) R^1 or $R^2 = OCH_3$; R^2 or $R^1 = OH$, $R^3 = H$, $R^4 = OH$.

(98) $R^1 = R^2 = OCH_3$; $R^3 = R^4 = H$. Deoxytubulosine

(99) $R^1 = R^2 = OCH_3$; $R^3 = OH$, $R^4 = H$. Alangimarckine (stereochemistry not determined)

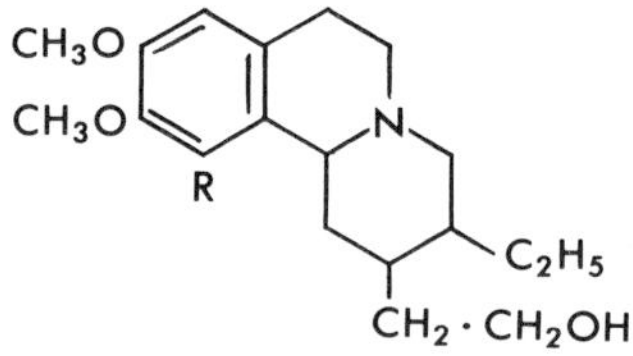

(100) R = H. Dihydroprotoemetine

(101) R = OH. Ankorine

The structures of all these alkaloids were determined largely by spectroscopic methods, and particularly by the use of mass spectrometry. The ipecac alkaloids undergo a typical fragmentation in the mass spectrometer, giving sets of peaks which are characteristic of the benzoquinolizine and isoquinoline portions.[44] It is, therefore, comparatively easy to trace the location of the structural variations occurring in the analogs; for example, in the mass spectrum of alangimarckine the peaks characteristic of the benzoquinolizine system all occur 16 m/e units higher than those in the deoxytubulosine spectrum, whereas those due to the β-carboline nucleus coincide. It is, therefore, clear that the phenolic hydroxyl group present is to be found on the benzoquinolizine nucleus, and the high field position of the NMR signal from the one isolated aromatic proton indicates that this proton occupies the 8-position, and hence that the nucleus is 9,10,11-trisubstituted. Similar comparisons between the mass spectra of ankorine and dihydroprotoemetine show the same pattern of substitution in the aromatic ring of the former, and since oxidation of *O*-ethylankorine gives 3-ethoxy-4,5-dimethoxyphthalic acid, the structure (101) can be confidently assigned to the alkaloid; hence, by analogy, alangimarckine is assumed to be (99).

Only certain of these alkaloids have been fully characterised stereochemically. Dihydroprotoemetine is identical with the reduction product of natural protoemetine of known configuration (58). Deoxytubulosine[49] and tubulosine[50] have each been synthesized from the optically active ester (81), by conversion to the tryptamide or 5-benzyloxytryptamide, cyclization with phosphoryl chloride, and reduction, followed in the case of tubulosine by hydrogenolysis of the *O*-benzyl group. The reduction step gives two epimers in each case, but deoxytubulosine is shown to correspond in configuration with emetine at C(1′) by a comparison of molecular rotation differences, and since tubulosine can be correlated with deoxytubulosine,[46] it also corresponds in configuration to emetine, as shown in (96). Demethyltubulosine (97) can be methylated to tubulosine, but the configuration of alangimarckine (99) has not been established. The natural occurrence of isotubulosine (also obtained in the foregoing synthesis) is remarkable, as no other *iso*-alkaloids of the ipecac group have been found in nature. Tubulosine and deoxytubulosine possess amoebicidal properties, but the former is more toxic than emetine, and the latter considerably less active.

REFERENCES

1. Earlier reviews: T. A. Henry, "The Plant Alkaloids," 4th ed., pp. 394–405, J. & A. Churchill, Ltd., London, England, 1949; M. M. Janot, *Alkaloids* **3**, 363–394 (1953) R. H. F. Manske, *Alkaloids* **7**, 419–442 (1960); H. G. Boit, "Ergebnisse der Alkaloid-Chemie bis 1960," pp. 370–379, Akademie-Verlag, Berlin, Germany, 1961; Cs. Szántay, "Structure and Synthesis of Ipecac Alkaloids," in "Recent Developments

in the Chemistry of Natural Carbon Compounds," Vol. 2, pp. 63–124, Akadémiai Kiadó, Budapest, Hungary, 1967.
2. A. R. Battersby, *Pure Appl. Chem.* **14,** 124 (1967); A. R. Battersby and B. Gregory, *Chem. Commun.* 134 (1968).
3. F. H. Carr and F. L. Pyman, *J. Chem. Soc.* **105,** 1591 (1914); F. L. Pyman, *J. Chem. Soc.* **111,** 419 (1917); **113,** 222 (1918).
4. A. Windaus and L. Hermanns, *Chem. Ber.* **47,** 1470 (1914).
5. J. J. Dobbie and J. J. Fox, *J. Chem. Soc.* **105,** 1639 (1914).
6. E. Späth and W. Leithe, *Chem. Ber.* **60,** 688 (1927).
7. P. Karrer, *Chem. Ber.* **49,** 2057 (1916); **50,** 582 (1917).
8. W. H. Brindley and F. L. Pyman, *J. Chem. Soc.* 1067 (1927).
9. A. Ahl and T. Reichstein, *Helv. Chim. Acta* **27,** 366 (1944).
10. P. Karrer, C. H. Eugster, and O. Rüttner, *Helv. Chim. Acta* **31,** 1219 (1948).
11. E. Späth and M. Pailer, *Monatsh. Chem.* **78,** 348 (1948).
12. A. R. Battersby and H. T. Openshaw, *J. Chem. Soc. S.* 59, 3207 (1949).
13. M. Pailer and L. Bilek, *Monatsh. Chem.* **79,** 135 (1948).
14. G. Norcross and H. T. Openshaw, *J. Chem. Soc.* 1174 (1949).
15. C. K. Ingold, "Structure and Mechanism in Organic Chemistry," p. 424, G. Bell and Sons, Ltd., London, England, 1953.
16. M. Pailer, *Monatsh. Chem.* **79,** 127, 331 (1948).
17. R. Robinson, *Nature* **162,** 524 (1948).
18. M. Pailer and K. Porschinski, *Monatsh. Chem.* **80,** 94, 101 (1949).
19. S. Teitel and A. Brossi, *J. Amer. Chem. Soc.* **88,** 4068 (1966).
20. H. T. Openshaw and H. C. S. Wood, *J. Chem. Soc.* 391 (1952).
21. A. R. Battersby, G. C. Davidson, and B. J. T. Harper, *Chem. Ind. (London)* 983 (1957); *J. Chem. Soc.* 1744, 1748 (1959).
22. A. Brossi, J. Würsch, and O. Schnider, *Chimia (Aarau)* **12,** 114 (1958).
23. A. R. Battersby, "Recent Work on Naturally Occurring Nitrogen Heterocyclic Compounds," *Chem. Soc. (London)* **3,** 36–40 (1955); A. R. Battersby, G. C. Davidson, and J. C. Turner, *J. Chem. Soc.* 3899 (1961).
24. Cs. Szántay and L. Töke, *Tetrahedron Letters* 1323 (1963); Cs. Szántay, L. Töke, and P. Kolonits, *J. Org. Chem.* **31,** 1447 (1966).
25. A. R. Battersby, R. Binks, D. Davidson, G. C. Davidson, and T. P. Edwards, *Chem. Ind. (London)* 982 (1957); *J. Chem. Soc.* 2704 (1959); 3474 (1960); A. R. Battersby and S. Garratt (née Cox), *Chem. Ind. (London)* 983 (1957); *Proc. Chem. Soc.* 86 (1959); *J. Chem. Soc.* 3512 (1959).
26. E. E. van Tamelen, P. E. Aldrich, and J. B. Hester, *J. Amer. Chem. Soc.* **79,** 4817 (1957); **81,** 507, 6214 (1959).
27. F. Bohlmann, *Angew. Chem.* **69,** 641 (1957); *Chem. Ber.* **91,** 2157 (1958).
28. R. P. Evstigneeva, R. S. Livshits, L. I. Zakharkin, M. S. Bainova, and N. A. Preobrazhenskii, *Dokl. Akad. Nauk. SSSR.* **75,** 539 (1950); R. P. Evstigneeva and N. A. Preobrazhenskii, *Tetrahedron* **4,** 223 (1958).
29. Y. Ban, M. Terashima, and O. Yonemitsu, *Chem. Ind. (London)* 568, 569 (1959).
30. P. Bellet, *Ann. Pharm. Franc.* **10,** 81 (1952); **12,** 466 (1954).
31. A. R. Battersby, B. Gregory, H. Spencer, J. C. Turner, M. M. Janot, P. Potier, P. Francois, and J. Levisalles, *Chem. Commun.* 219 (1967).
32. A. R. Battersby, H. T. Openshaw, and H. C. S. Wood, *Experientia* **5,** 114 (1949); H. T. Openshaw, "Recent Work on Naturally Occurring Nitrogen Heterocyclic Compounds," *Chem. Soc. (London)* **3,** 28–35 (1955).
33. Y. Ban and M. Terashima, *Tetrahedron Letters* 796 (1961); *Chem. Pharm. Bull.* (Tokyo) **13,** 775 (1965).

34. R. P. Evstigneeva, L. V. Lavrova, C. D. Zarankina, and N. A. Preobrazhenskii, *Zh. Obshch. Khim.* **28,** 1190 (1958).
35. Reviews: C. Viel and P. Rumpf, *Bull. Soc. Chim. France* 2335 (1962); Cs. Szántay, Ref. 1.
36. A. R. Battersby and H. T. Openshaw, *Experientia* **6,** 387 (1950); A. R. Battersby, H. T. Openshaw, and H. C. S. Wood, *J. Chem. Soc.* 2463 (1953).
37. Y. Ban, *Pharm. Bull.* (Tokyo) **3,** 53 (1955); M. Pailer and G. Beier, *Monatsh. Chem.* **88,** 830 (1957).
38. M. Barash and J. M. Osbond, *Chem. Ind. (London)* 490 (1958); J. M. Osbond, *Chem. Ind. (London)* 257 (1959); M. Barash, J. M. Osbond, and J. C. Wickens, *J. Chem. Soc.* 3530 (1959).
39. A. R. Battersby and J. C. Turner, *Chem. Ind. (London)* 1324 (1958); *J. Chem. Soc.* 717 (1960).
40. A. Brossi, M. Baumann, L. H. Chopard-dit-Jean, J. Würsch, F. Schneider, and O. Schnider, *Helv. Chim. Acta* **42,** 772 (1959); A. Brossi, M. Baumann, and O. Schnider, *Helv. Chim. Acta* **42,** 1515 (1959).
41. H. T. Openshaw and N. Whittaker, *J. Chem. Soc.* 1449, 1461 (1963).
42. J. H. Chapman, P. G. Holton, A. C. Ritchie, T. Walker, G. B. Webb, and K. D. E. Whiting, *J. Chem. Soc.* 2471 (1962); D. E. Clark, P. G. Holton, R. K. F. Meredith, A. C. Ritchie, T. Walker and K. D. E. Whiting, *J. Chem. Soc.* 2479 (1962); D. E. Clark, R. F. K. Meredith, A. C. Ritchie, and T. Walker. *J. Chem. Soc.* 2490 (1962).
43. For summary see H. G. Boit, Ref. 1, pp. 933–934.
44. S. C. Pakrashi and P. P. Ghosh-Dastidar, *Indian J. Chem.* **2,** 379 (1964); H. Budzikiewicz, S. C. Pakrashi, and H. Vorbrüggen, *Tetrahedron* **20,** 399 (1964).
45. S. C. Pakrashi and E. Ali, *Tetrahedron Letters* 2143 (1967).
46. P. Brauchli, V. Deulofeu, H. Budzikiewicz, and C. Djerassi, *J. Amer. Chem. Soc.* **86,** 1895 (1964); H. Monteiro, H. Budzikiewicz, C. Djerassi, R. R. Arndt, and W. H. Baarschers, *Chem. Commun.* 317 (1965).
47. A. Popelak, E. Haack, and H. Spingler, *Tetrahedron Letters* 1081, 5077 (1966).
48. A. R. Battersby, R. S. Kapil, D. S. Bhakuni, S. P. Popli, J. R. Merchant, and S. S. Salgar, *Tetrahedron Letters* 4965 (1966).
49. A. R. Battersby, J. R. Merchant, E. A. Ruveda, and S. S. Salgar, *Chem. Commun.* 315 (1965).
50. H. T. Openshaw and N. Whittaker, *Chem. Commun.* 131 (1966).

5

The Morphine Alkaloids

K. W. BENTLEY
Pharmaceutical Research Laboratories
Department of Organic Chemistry
Reckitt and Sons, Ltd.
Hull, England

The alkaloids of the morphine group form a subgroup of the isoquinoline alkaloids, being derived in nature from bases of the laudanosine series by oxidative processes. Their pharmacological and chemical properties are sufficiently distinct from those of other isoquinoline alkaloids to warrant their separate discussion, and it is on account of their peculiar properties that they have been subjected to more study than any other group of alkaloids of comparable size. There are nine main members of the group, and these may be divided into two enantiomorphic subgroups, namely, morphine, codeine, neopine, thebaine, and oripavine, all isolated from *Papaver* species; and sinomenine, hasubanonine, metaphenine, and protometaphenine, all isolated from Japanese plants of the *Sinomenium* and *Stephania* species. Several minor alkaloids of the two sub-groups have recently been discovered. (see 2.5 below)

1. MORPHINE, CODEINE, NEOPINE, THEBAINE, AND ORIPAVINE

These bases are all closely related. Morphine ($C_{17}H_{19}NO_3$) is a phenol and gives codeine ($C_{18}H_{21}NO_3$) upon methylation. Neopine, which is isomeric with codeine, gives dihydrocodeine upon catalytic reduction, and can only differ from codeine in the position of a double bond. Codeine is also a secondary alcohol and can be oxidized to an α,β-unsaturated ketone, codeinone ($C_{18}H_{19}NO_3$) obtainable in poor yield by the careful hydrolysis of thebaine ($C_{19}H_{21}NO_3$), which must therefore be an enol methyl ether of codeinone. Oripavine, like morphine, is a phenol, and on methylation with diazomethane adds CH_2 to give thebaine.

1.1 Elucidation of Structures

Morphine (1, R = H) and codeine (1, R = Me) have the structures shown; thebaine (2, R = Me) is one of the enol ethers of codeinone (3) and ori-

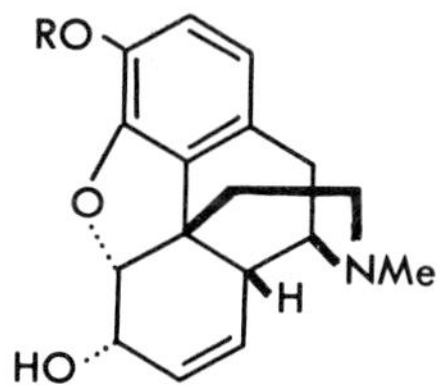

(1)

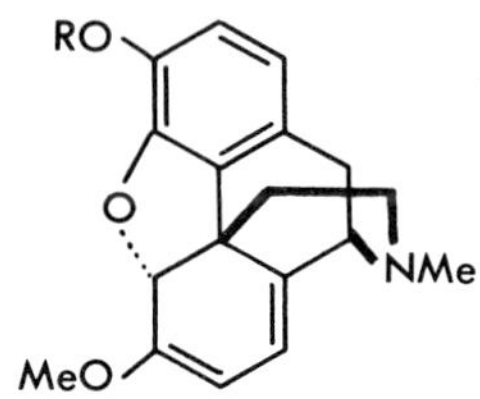

(2)

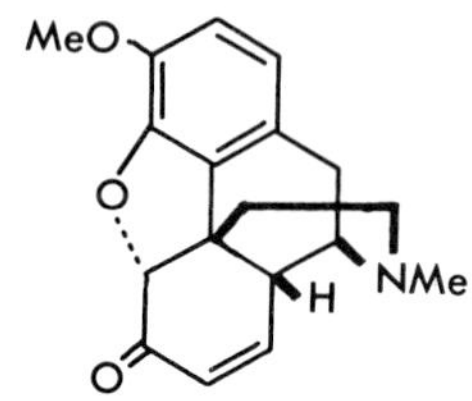

(3)

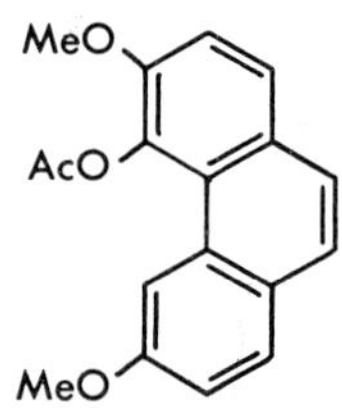

(4)

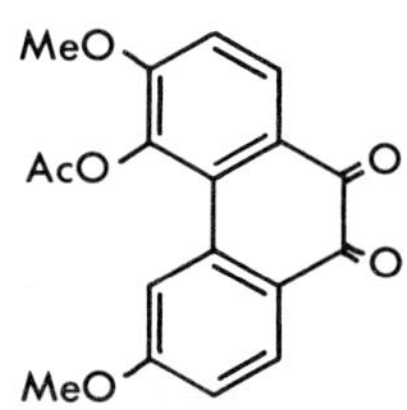

(5)

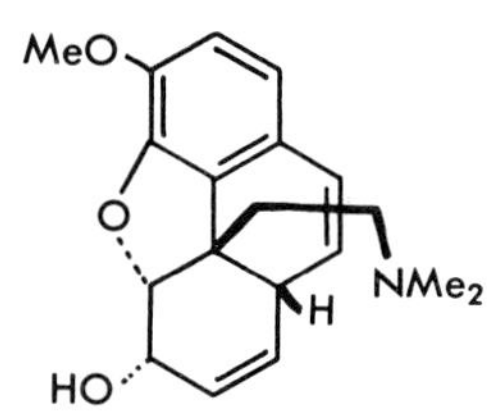

(6)

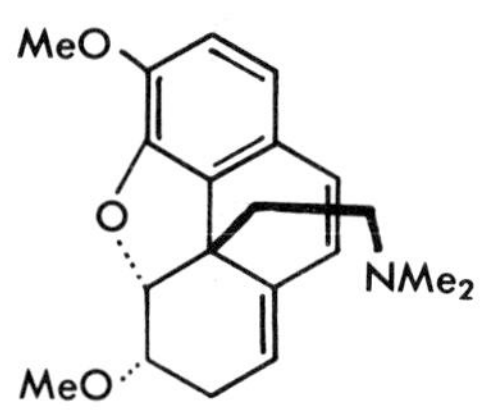

(7)

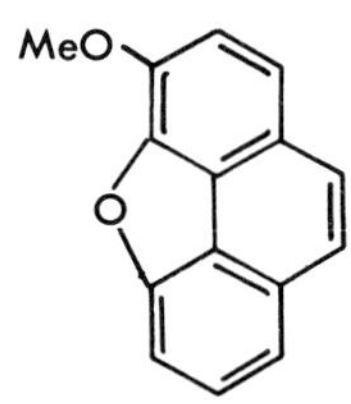

(8)

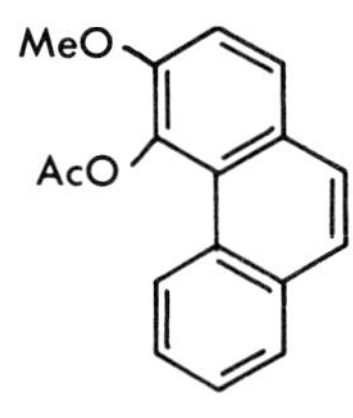

(9)

pavine is the corresponding phenol (2, R = H). The characteristic feature of the alkaloids is the ease with which they undergo degradations leading to the extrusion of a nitrogen-containing sidechain with the production of phenanthrene derivatives, or suffer aromatization to derivatives of diphenyl or phenanthrene as a result of migration of the nitrogen-containing sidechain. The key degradations and rearrangements that led Gulland and Robinson[1,2] to propose the correct structures for the alkaloids were the following:

(a) The degradation of thebaine methiodide to acetylthebaol (4, R = Me) and of codeinone methiodide to 4,6-diacetoxy-3-methoxyphenanthrene (4, R = Ac).[3]

(b) The degradation of the methiodides of α-codeimethine[4] (6), the product of Hofmann degradation of codeine methiodide, and of β-codeimethine (7) (its isomerization product) to methylmorphenol (8)[5] by bases and to acetylmethylmorphol (9)[6,7] by acetic anhydride.

(c) The dehydration and rearrangement of morphine by acids to apomorphine (10)[8–10] and the rearrangement of thebaine to morphothebaine.[11]

(d) The rearrangement of thebaine and codeinone to thebenine (11) by dilute acids.

Gulland and Robinson argued that the nitrogen-containing sidechain of the alkaloids must be so placed on a hydrogenated phenanthrene skeleton that aromatization to a phenanthrene or diphenyl system could not be accomplished without extrusion or migration of that sidechain. Since the nitrogen end of the sidechain had previously been shown to be attached at C(9)

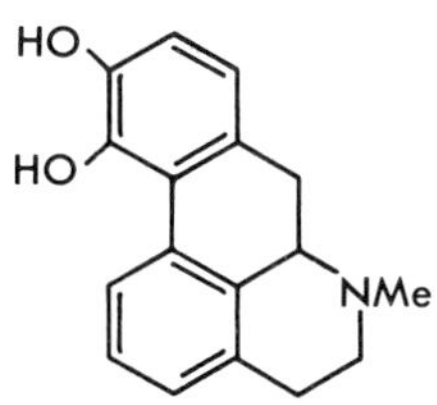

(10)

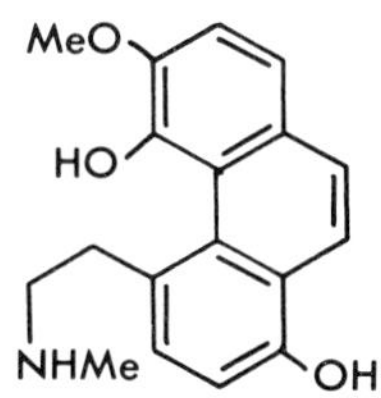

(11)

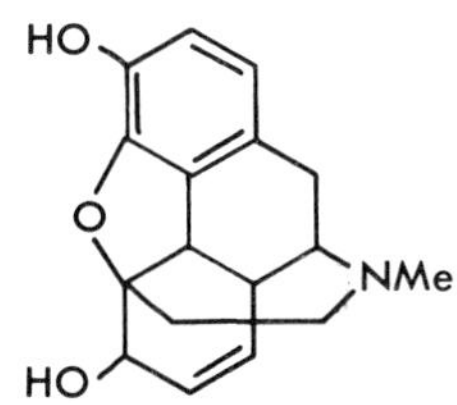

(12)

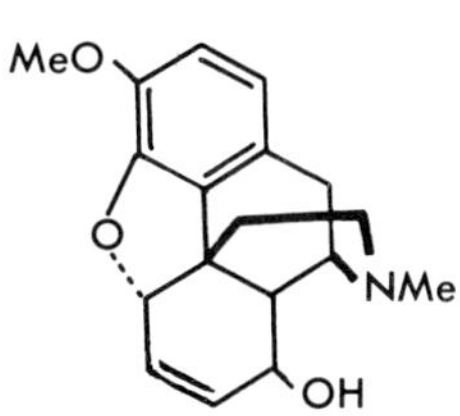

(13)

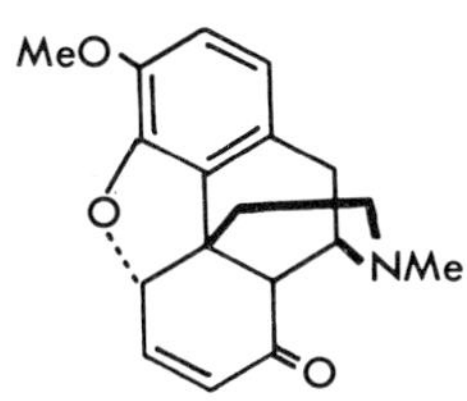

(14)

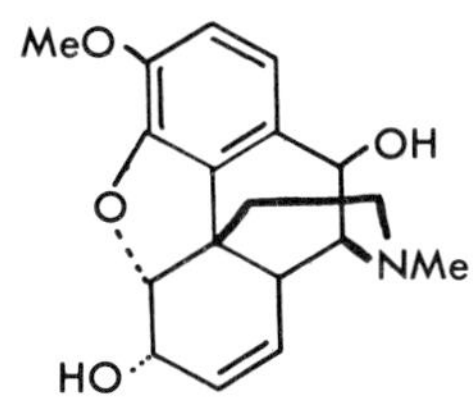

(15)

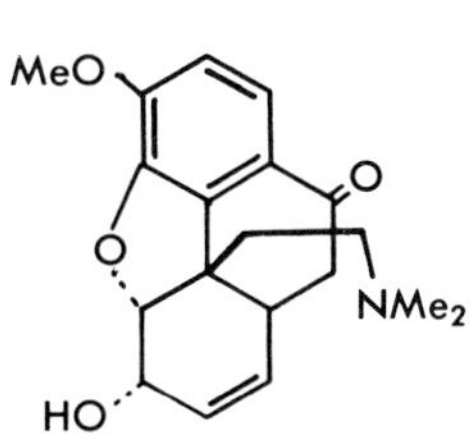

(16)

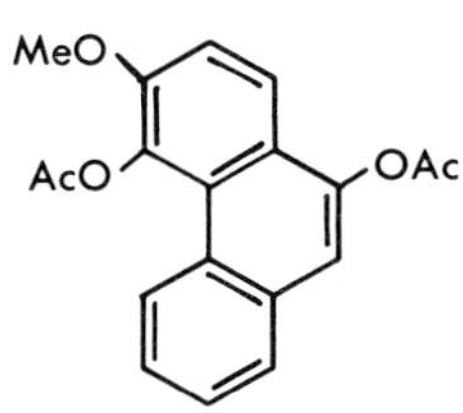

(17)

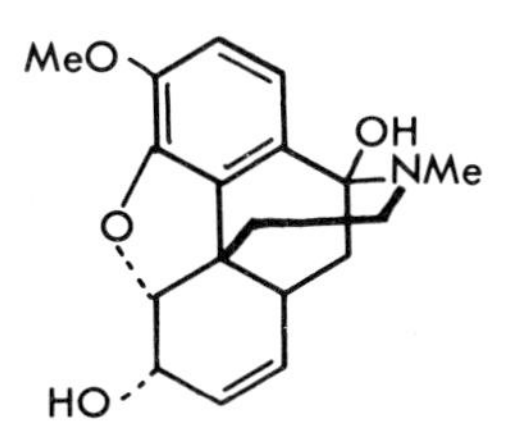

(18)

or C(10), the carbon end of the sidechain (frequently recovered as $AcOCH_2CH_2NMe_2$ during degradations effected by acetic anhydride) must be located at an angular position, so that its extrusion from that position is essential for aromatization. Only two such positions are available, namely, C(14) and C(13) and of these the latter was chosen since only in that case could a satisfactory mechanism (see formulas 75, 76, 76A, 80–82, Section 1.5.3) be devised for the rearrangement of thebaine (1, R = Me) to thebenine (11).

The earlier formulas of Knorr and Weiland (12)[12,13] in which the sidechain was attached to C(5) failed to explain universal extrusion or migration of the sidechain during aromatizations. Still earlier formulas with a C(8) attachment of the sidechain became untenable when codeine (1, R = Me) was isomerized through the chlorocodides (see Section 1.3) to ψ-codeine (13) which was oxidized to ψ-codeinone (14) and degraded to 4,8-diacetoxy-3-methoxyphenanthrene.[14]

The attachment of the nitrogen atom to C(9) or C(10) was proved by the behavior of hydroxycodeine (obtained by the oxidation of codeine) on degradation. Hofmann degradation of the methiodide of this base (15) gave a methine base (16) further degradable to acetoxyacetylmethylmorphol (17), which lost an acetoxy group on oxidation to a phenanthraquinone, unlike acetylmethylmorphol (9) obtained by a similar sequence of reactions from codeine.[15,16] The additional hydroxyl group in hydroxycodeine must, therefore, be at C(9) or C(10) [subsequently proved to be at C(10)[17]] and since the methine base (16) is a ketone the nitrogen atom must be at-

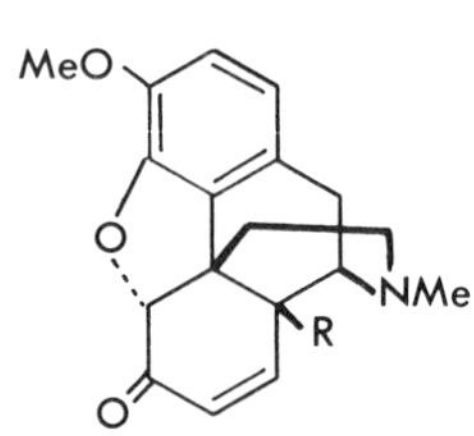

(19)

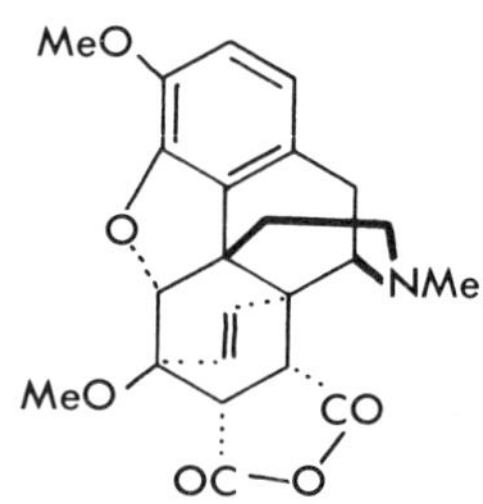

(20)

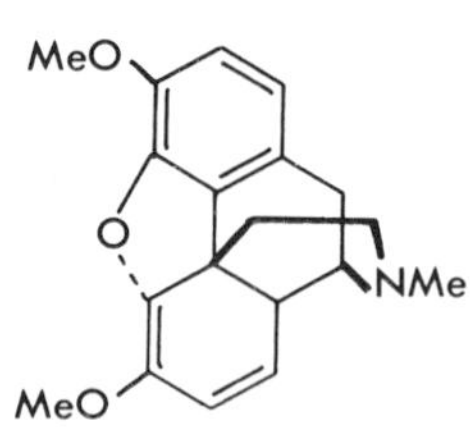

(21)

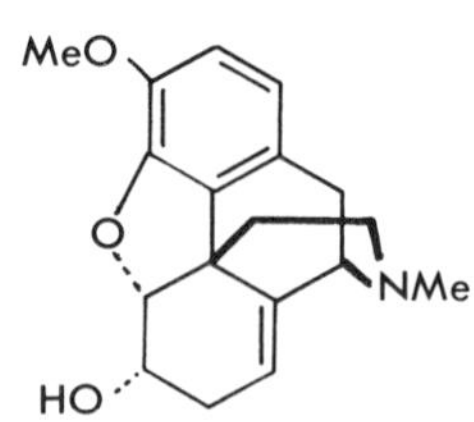

(22)

tached at C(9) or C(10); C(8) and C(14) are ruled out on other grounds. No positive proof of the attachment of the nitrogen to C(9) was obtained until morphine and codeine were synthesized.

The formulation of thebaine as the conjugated diene (2, R = Me) is in accord with its ready conversion into 14-bromocodeinone (19, R = Br) and 14-hydroxycodeinone (R = OH) by treatment with bromine or hydrogen peroxide, respectively, in acetic acid, and with its participation in the Diels–Alder reaction with dienophils to give adducts such as that (20) with maleic anhydride. The fact that these adducts are not enol ethers rules out the alternative conjugated diene structure (21) for the alkaloid. Neopine, which is isomeric with codeine, is degraded by Hofmann's method directly to β-codeimethine (7) and must, therefore, have the structure (22).

Direct evidence for the location of the sidechain at C(13) or C(5) was sought by the Beckmann rearrangement of dihydrocodeinone oxime.[18] This reaction usually gives amides, but a number of cases are known in which carbon–carbon fission occurs in oximes bearing an oxygen substituent on the α-carbon atom, the products being nitriles and aldehydes or ketones. The occurrence of such a reaction with dihydrocodeinone oxime would furnish an aldehyde (25) with the sidechain at C(13) or a ketone (26) if the sidechain is at C(5). Beckmann rearrangement did occur but it was not possible to identify positively the transformation product, or its methyl ether, as an aldehyde or ketone. The methyl ether was accordingly converted into the oxime and subjected to a further Beckmann rearrangement

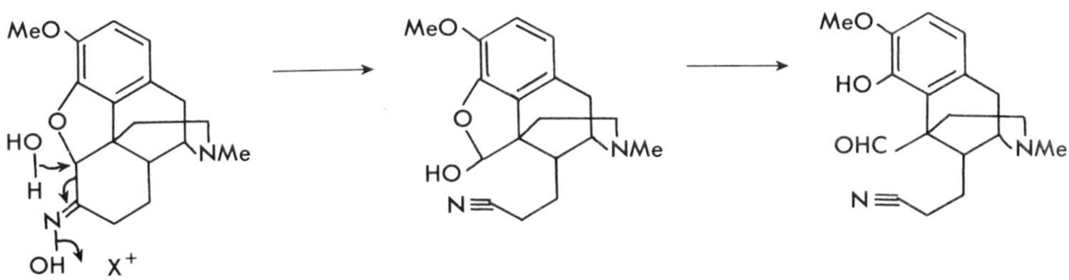

(23) (24) (25)

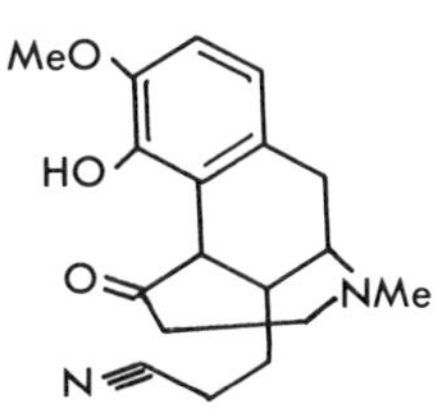

(26)

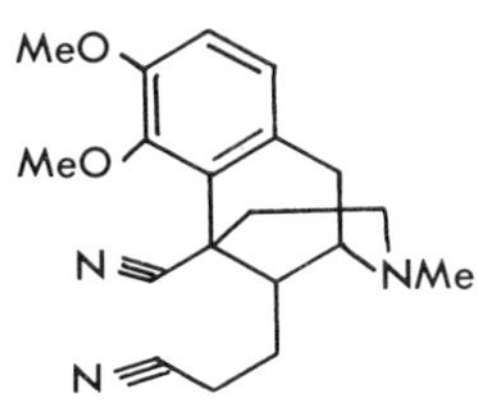

(27)

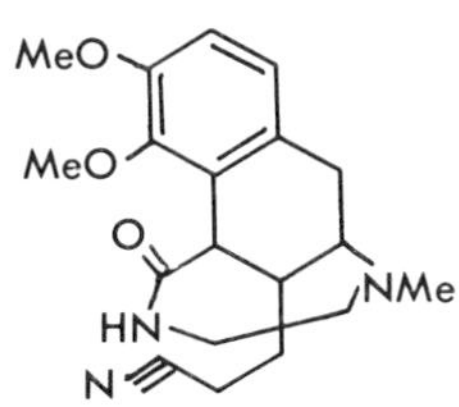

(28)

to give a product that retained all three nitrogen atoms on Hofmann degradation. Only a derivative of the aldehyde (25) would behave in this way since the derived product (27) contains the nitrogen-ring intact. The rearrangement product derived from the ketone (26) would be the amide (28), an isomer of this or an amino acid, any one of which would lose the entire nitrogen-containing sidechain on Hofmann degradation.

Accordingly, these degradations supported attachment of the sidechain at C(13), and until the total synthesis of codeine and morphine, represented the only direct evidence in favor of the Gulland and Robinson structures. The validity of these structures rested for many years entirely on their unique ability to explain all of the curious rearrangements of the bases of this group.

1.2 Total Synthesis

The first synthesis of the carbon-nitrogen skeleton of the morphine alkaloids was achieved by an application of the Bogert–Cook synthesis of hydrophenanthrenes using a suitably substituted benzyloctahydroisoquinoline. When the olefinic base (29, R = R′ = H) was heated with syrupy phosphoric acid at 150° for three days, a 50% yield of N-methylmorphinan (30, R = R′ = H) was obtained together with a small amount of the isomorphinan, differing only in stereochemistry at C(14).[19] The reaction was repeated with the bases (29, R = OMe, R′ = H) and (29, R = R′ = OMe) to give the bases (30, R = OH, R′ = H) and (30, R = OMe, R′ = OH); the last of these was shown to be identical with (±)-tetrahydrodesoxycodeine prepared by mixing the separate (+) and (−) isomers obtained from sinomenine and codeine, respectively.[20] This process, however, is not well suited to the synthesis of the full morphine structure, which was eventually achieved by the complex sequence of reactions set out in formulas (31)–(40). The reaction sequence was first carried out as far as an analog of the base (34) without the methoxyl groups, during which the correct interpretation of the unusual cyclization step (32)–(33) was elucidated.[21–24] Reduction of the base analogous to (34) obtained in this way afforded N-methylisomorphinan identical with the by-product from Grewe's synthesis of N-methylmorphinan and differing from the normal bases of the morphine group in stereochemistry at C(14). In the dimethoxy series the base (34) was identified with β-dihydrodesoxycodeine-B methyl ether obtained from β-dihydrothebainone (35) in three steps.[25] β-Dihydrothebainone differs from the normal morphine bases at C(14) and is obtained by reduction of β-thebainone-A, the major hydrolysis product of dihydrothebaine-ϕ (see p. 129). The olefin (34) was then hydrated, partially demethylated, and oxidized to β-dihydrothebainone itself (34) after resolution of the racemate.

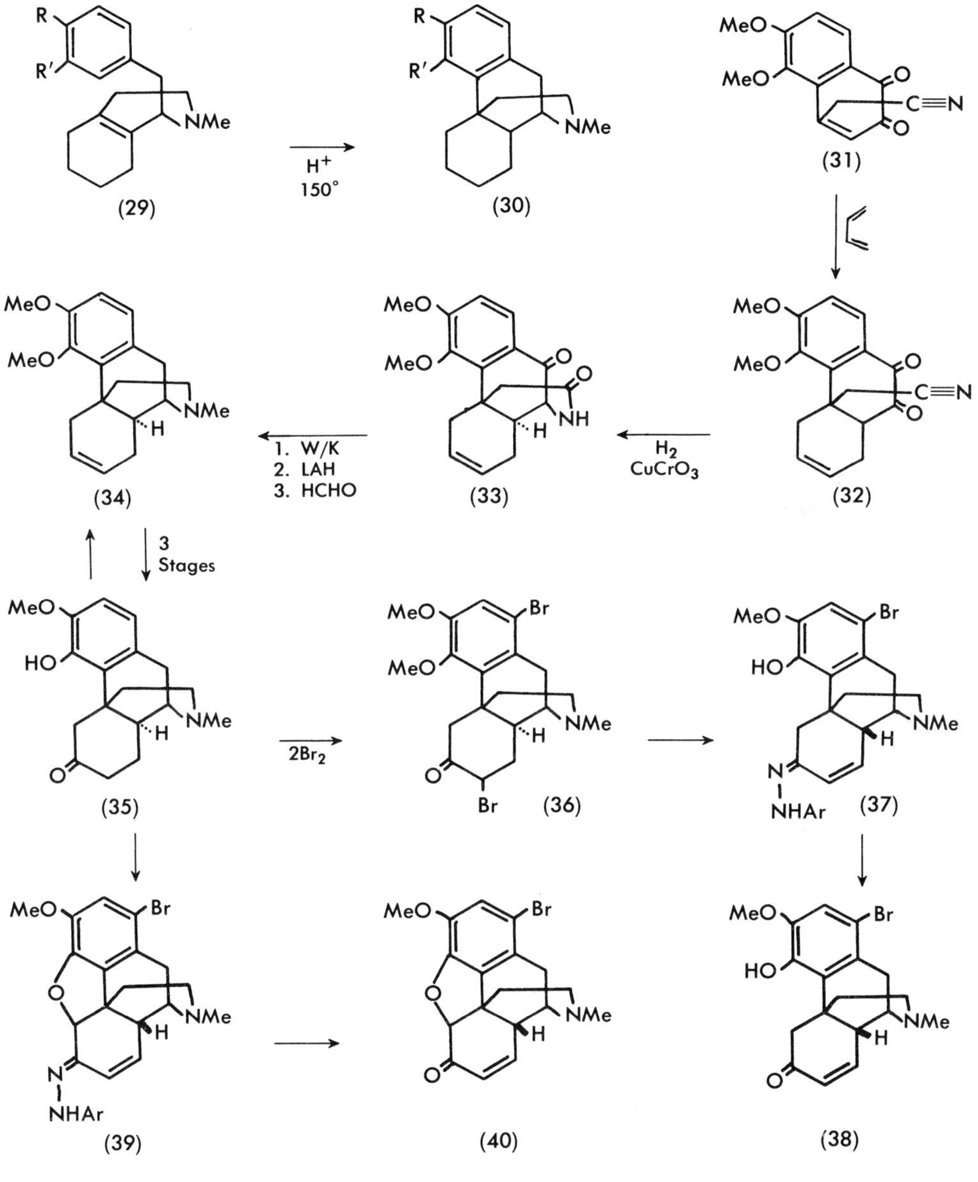

Cyclization of the oxygen-ring characteristic of morphine and epimerization of the C(14) center was then accomplished in the following way.[26,27] Earlier it had been shown[28] that bromination of dihydrothebainone [C(14) epimer of 35] yields a mono-, di-, or tribromo compound according to the amount of bromine used and that treatment of the dibromo compound with alkali affords 1-bromocodeinone with establishment of the 4,5-oxygen bridge. The bromination of the β-isomer (35) was studied and the dibromo compound (36) was found to lose hydrogen bromide when converted into the 2,4-dinitrophenylhydrazone and at the same time to undergo very rapid epimerization at C(14), the final product being 1-bromothebainone-A dinitrophenylhydrazone (37). The latter could be split by an excess of hydrochloric acid in acetone to 1-bromothebainone (38). The rapid epimerization of the hydrazone was independently verified with β-thebainone-A, itself. The tribromo-ketone with dinitrophenylhydrazine underwent hydrazone formation, dehydrobromination, epimerization at C(14), and oxygen-ring closure to give 1-bromocodeinone dinitrophenylhydrazone. Cleavage of the latter with acetone and acid gave 1-bromocodeinone (40), and reduction of this with lithium aluminum hydride yielded codeine. Demethylation of codeine with pyridine hydrochloride then gave morphine identical with the natural alkaloid.[26,27]

A second synthesis of dihydrothebainone and, hence, of morphine was subsequently achieved by Ginsburg.[29,30] The formal synthesis of thebaine was completed by the methylation of dihydrocodeinone to the enol ether (41), which is dihydrothebaine; the bromination of this to the bromide (42) with N-bromosuccinimide, followed by dehydrobromination to (43) and 1,4 loss of methanol.[31]

1.3 Stereochemistry of Morphine and Codeine

A number of attempts have been made to deduce the stereochemical arrangement at the various asymmetric centers in codeine by chemical methods, but none of these were free from ambiguities and it was not until the results of crystallographic studies were available that the problem could be regarded as finally being solved. One such attempt[32] was based on the observation that 14-hydroxycodeinone (44) on catalytic reduction and exhaustive methylation affords the cyclic ether (44A) in which the oxygen-containing ring must of necessity be *cis* fused to the phenanthrene system. Assuming that no epimerization at C(5) occurs during the alkaline degradation, 14-hydroxycodeinone must therefore have structure (44), or its mirror image. Assuming that catalytic reduction of thebaine to dihydrocodeine methyl ether and the addition of hydrogen peroxide to thebaine proceed with the establishment of the same stereochemistry at C(14) and that reduction of thebaine proceeds to a product with *cis*-hydrogens at C(6) and C(14), codeine could be represented as having the partially defined structure (44B). The assumptions made in this argument, though eminently reasonable, remain assumptions.

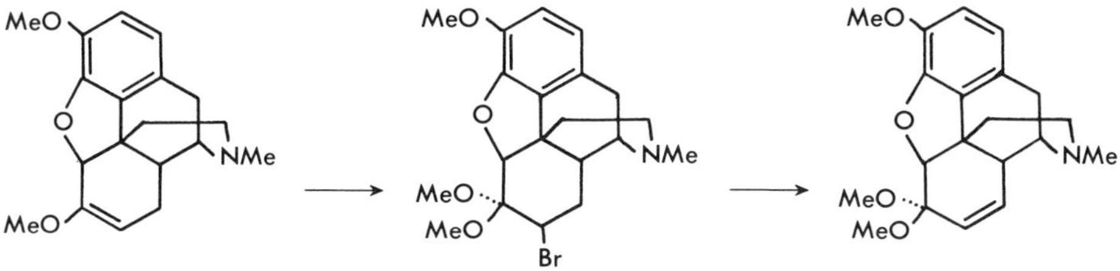

(41) (42) (43)

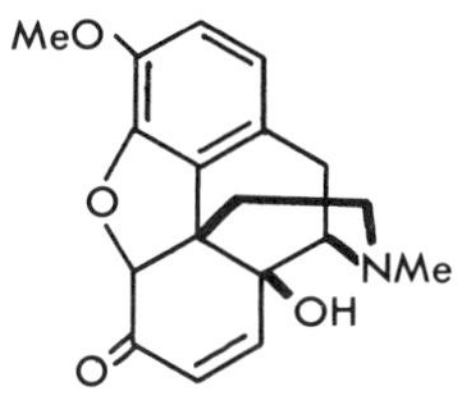

(44)

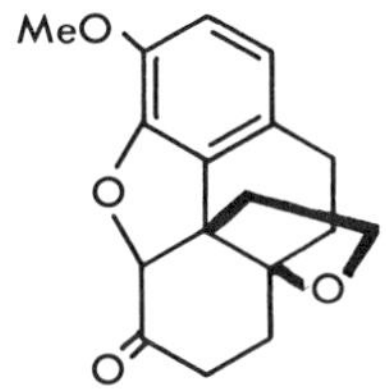

(44A)

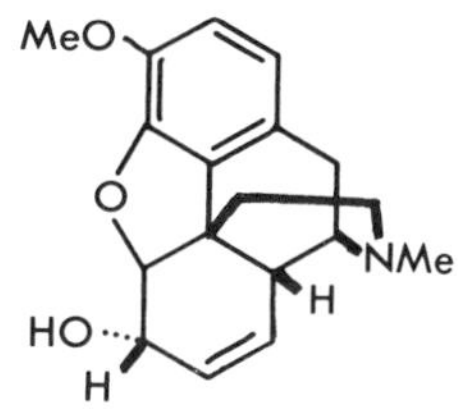

(44B)

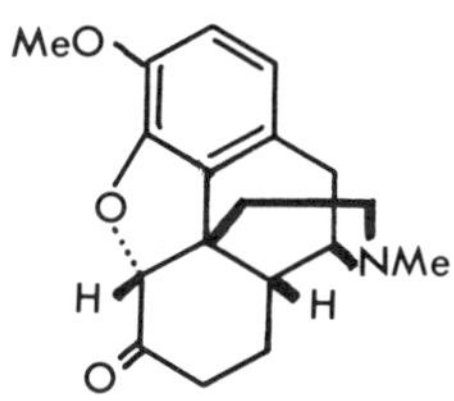

(45)

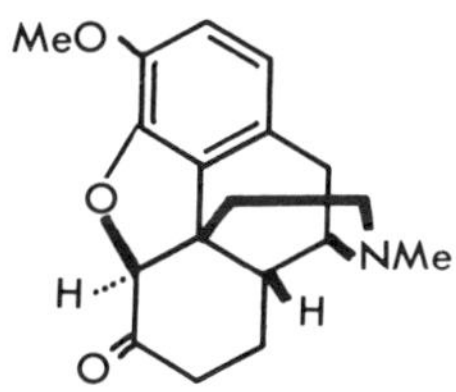

(47)

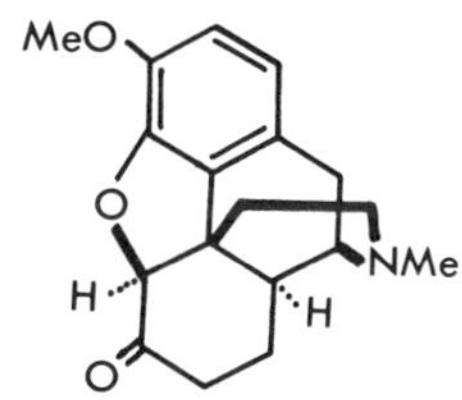

(46)

Stork[33] deduced the stereochemistry at C(5) and C(13) by regarding the oxide ring and ring C of dihydrocodeinone as analogous to α-hydrindanone in which it is known that the *trans* fused form is the more stable. He argued that, because dihydrocodeinone is not epimerized by alkali, the C(5) hydrogen and C(13) sidechain must be *trans* about the 5,13 bond. However, this argument is unsound, since the whole molecule and not just a part of it must be taken into consideration in deciding the relative stabilities of isomers. Examination of models shows that (45) with a *trans* arrangement, and (46) with a *cis* arrangement of the C(5) hydrogen and C(13) sidechain are reasonably strain free, and that epimerization of either of these at C(5) would lead to a highly strained system.[34] Epimerization of (45) at C(5) leads to (47), which is the C(14) epimer of (46), and likewise the C(5) epimer of (46) is the C(14) epimer of (45); only (45) and (46) are strain free. This criticism has recently been vindicated by the preparation of the 1-bromo derivative of (46), which is as stable to alkali as is dihydrocodeinone.[35]

Rapoport made a more direct chemical attack on the problem and endeavored to determine the relative arrangements at three pairs of asymmetric centers as follows. Ozonolysis of dihydrocodeine (48) yields the

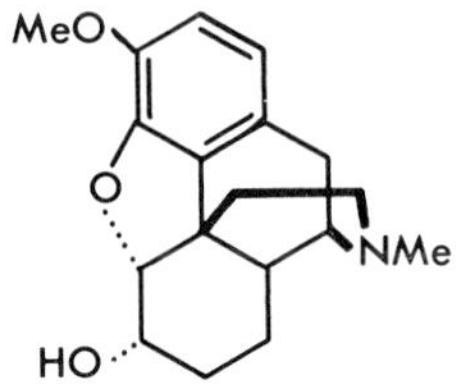

(48)

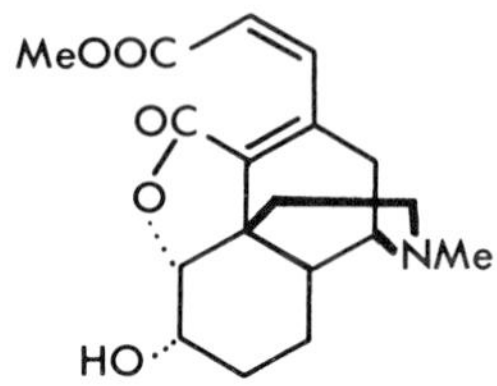

(49)

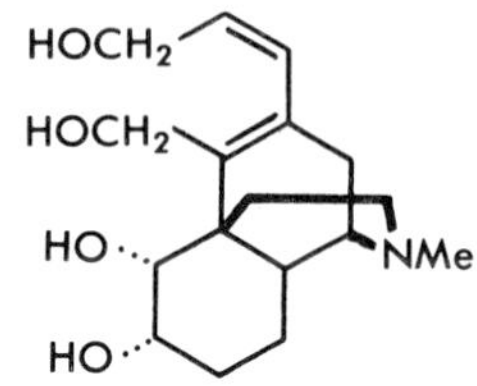

(50)

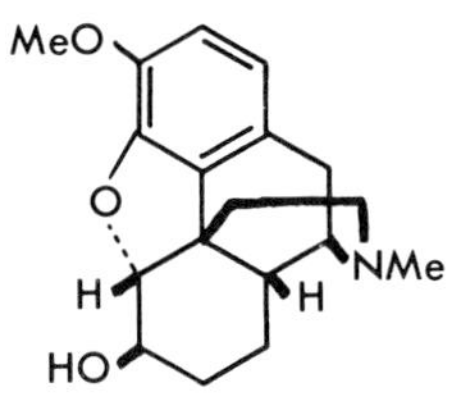

(51)

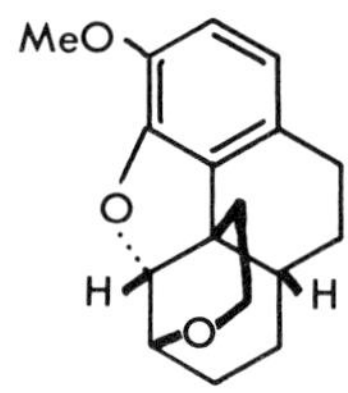

(52)

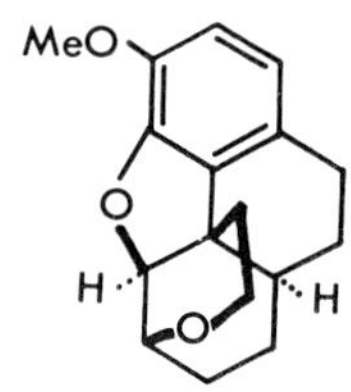

(53)

ester lactone (49), lithium aluminum hydride reduction of which yields the tetrol (50). A similar sequence of reactions with dihydroisocodeine, differing only in stereochemistry at C(6), yielded an isomer of (50). These two isomeric tetrols were oxidized by lead tetra-acetate at different rates. The one derived from codeine was oxidized three times more rapidly than its isomer, from which it was concluded that (50) is a *cis* diol and, therefore, that codeine has the oxygen functions at C(5) and C(6) in the *cis* relationship.[36]

The relationship of the C(6) hydroxyl group and C(13) sidechain was inferred from a study of the exhaustive methylation of dihydrocodeine and dihydroisocodeine with reduction after the first stage. It had been earlier shown that Hofmann degradation of tetrahydrocodeimethine was accompanied by some methylation of the C(6) hydroxyl[37] at the expense of some quaternary salt, and the degradation of tetrahydroisocodeimethine also furnished a cyclic ether 6-codiran.[38] It was argued that since the cyclic ether system must be *cis* fused to the hydrophenanthrene system, the product must have structure (52) and dihydroisocodeine must be (51) in view of the previously determined relationship of oxygen functions at C(5) and C(6). However, a strain-free model of (53) can also be constructed, and it is possible to rationalize the production of this from an alternative structure for dihydroisocodeine.[34]

Exhaustive methylation of dihydrothebainone (54), with reduction of the initial methine, affords the cyclic ether thebenone (55), the bis-isonitroso

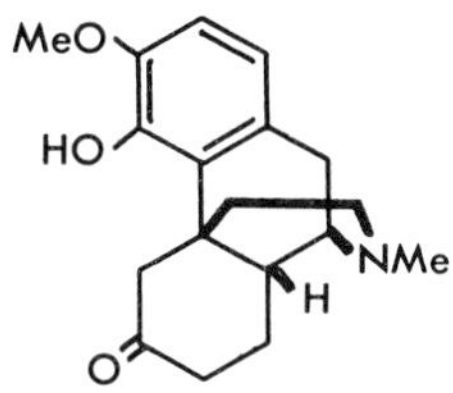

(54)

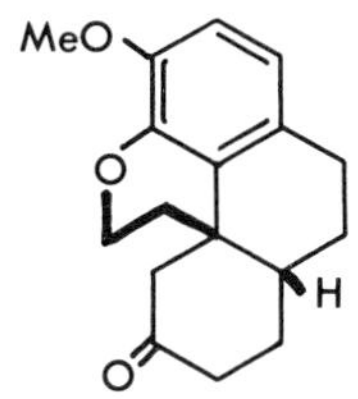

(55)

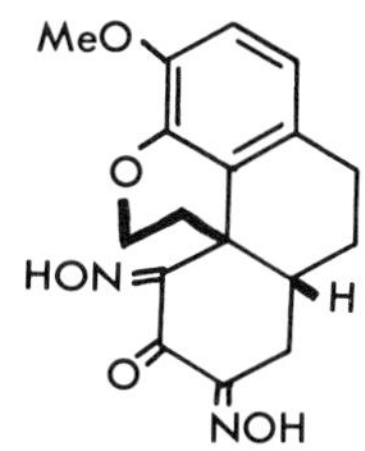

(56)

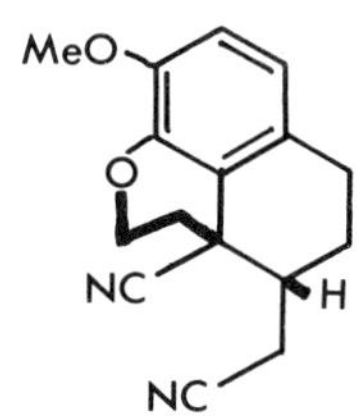

(57)

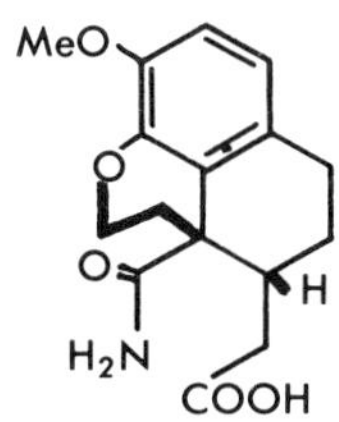

(58)

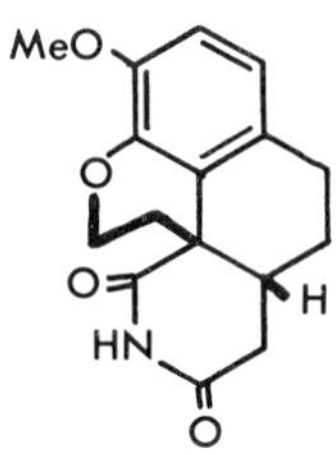

(59)

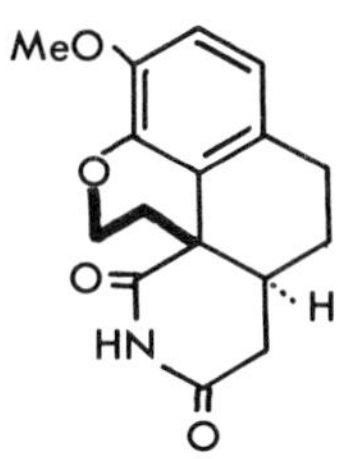

(60)

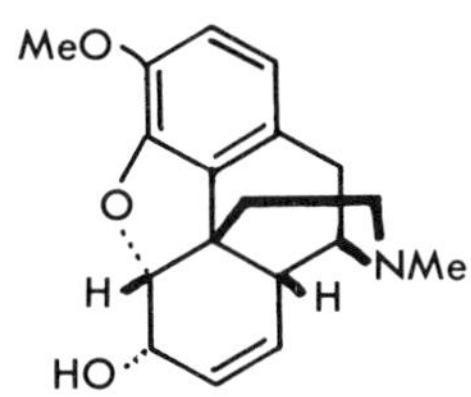

(61)

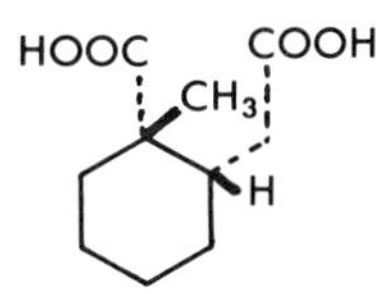

(62)

derivative of which (56) has been degraded through the dinitrile (57) and amido acid (58) to the cyclic imide (59). By contrast the C(14) epimer of thebenone, i.e., β-thebenone, can be degraded only as far as the acid isomeric with (58). It was argued that cyclization to the imide would only be possible with a *cis* arrangement of $—CONH_2$ and CH_2COOH and, hence, that thebenone and dihydrothebainone must have the defined structures (55) and (54), respectively.[39] However, a strain-free model of (60) can also be constructed, and hence, uncertainty remains.[34]

The issue was, however, put beyond doubt by x-ray studies, which revealed that codeine has the structure (61)[40] and by degradation of thebaine to the acid (62),[41] the latter process establishing not only the relative stereochemistry at C(13) and C(14), but the absolute stereochemistry as well.

On the basis of the stereochemistry for codeine indicated (61), the production and hydrolysis of the chlorocodides has been rationalized by Stork.[33,42] The conversion of codeine into α-chlorocodide (63) involves

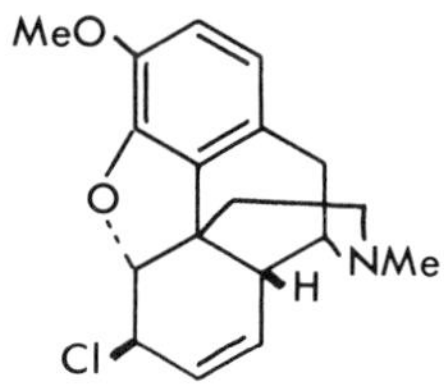
(63)

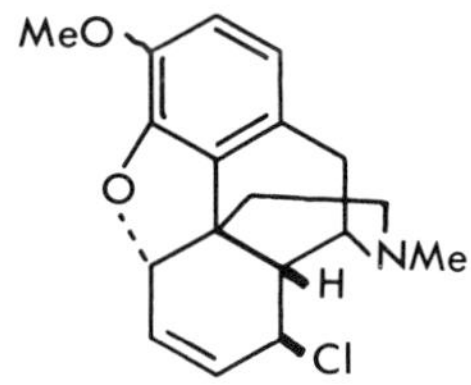
(64)

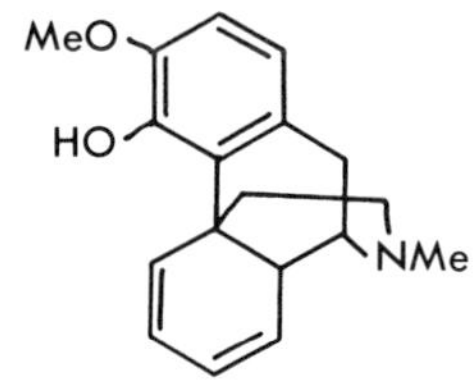
(65)

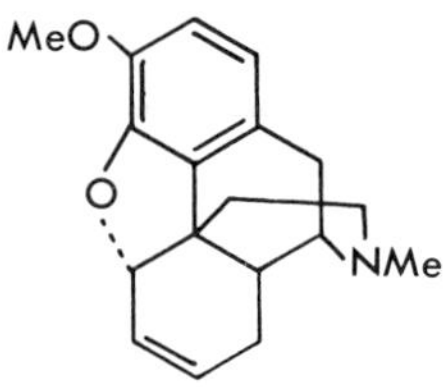
(66)

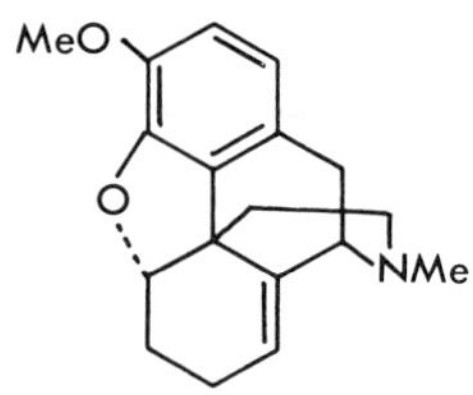
(67)

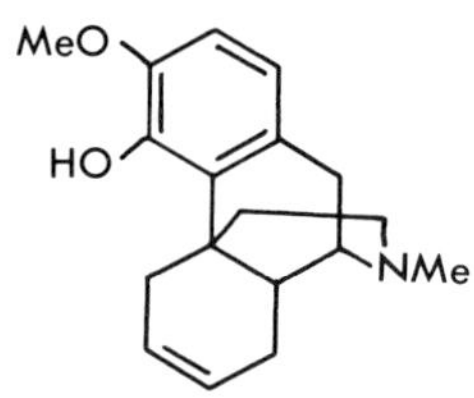
(68)

S_N2 displacement of hydroxyl by attack from the unhindered side of the molecule. Hydrolysis of the halide by the S_N2 mechanism would, however, involve attack on the hindered side, and accordingly the hydrolysis takes place by the S_N2' process to give ψ-codeine and allo-ψ-codeine with the hydroxyl at C(8). By the same token isocodeine cannot be converted into α-chlorocodide, but by an S_N2' reaction gives β-chlorocodide (64), which is obtained directly from allo-ψ-codeine. This halide on hydrolysis affords mainly isocodeine by the S_N2' process, and, since α-chlorocodide is readily transformed into the β-compound some isocodeine accompanies ψ-codeine and allo-ψ-codeine during the hydrolysis of this isomer.

Reduction of the chlorocodides and of the toluenesulfonyl esters of codeine and neopine affords a variety of desoxycodeines and dihydrodesoxycodeines, e.g., desoxycodeine-A (65), desoxycodeine-C (66), desoxycodeine-D (67), and dihydrodesoxycodeine-C (68).[43]

1.4 Reduction of Thebaine

Thebaine contains a highly reactive doubly unsaturated ether system

—O—CH—C(OMe)=CH—CH=C— (positions 1 2 3 4 5 6) that can suffer reduction in all possible ways. Lithium aluminum hydride gives 1:2 reduction to the phenolic conjugated diene β-dihydrothebaine,[44–45] whereas sodium and ammonia reduction yields the isomeric phenolic unconjugated diene dihydrothebaine-ϕ by 1:4-reduction.[46] Catalytic reduction yields the products of 3:4-reduction (neopine methyl ether[47]) and 5:6-reduction (dihydrothebaine[48]), together with the products resulting from 1:6-reduction followed by 2:5-reduction (dihydrothebainone enol methyl ether[49]), 3:6-reduction followed by 4:5-reduction (tetrahydrothebaine[50]), and 1:2, followed successively by 3:6 and 4:5-reduction (dihydrothebainol 6-methyl ether[49]). The product of 1:6-reduction, which does not survive the reducing conditions, has been independently prepared by the action of sodium ethoxide on codeine methyl ether.[49]

The hydrolysis of the two isomeric bases dihydrothebaine-ϕ (69) and β-dihydrothebaine (70) with potassium bisulfate affords β-thebainone-A (71) with the abnormal configuration at C(14),[49,44] the process being kinetically determined since thebainone-A, with the normal morphine configuration at C(14), has been shown to be the stable epimer.[27] The βγ-unsaturated ketone thebainone-B (72) has been obtained by hydrolysis with mineral acid,[46] but hydrolysis with sulfurous acid yields thebainone-C (74), which results from initial opening of the N-containing ring of (72) to (73) followed by cyclization.[51]

1.5 Molecular Rearrangements

Thebaine and its derivatives readily undergo a variety of rearrangements, almost all of which are initiated by attack of a proton or other positive ion on the cyclic ether oxygen atom. In acid solution, thebaine is readily converted into codeinone, of which it is an enol ether, and accordingly this base undergoes many of the same rearrangements.

1.5.1 Morphothebaine and apomorphine When thebaine (75) is dissolved in concentrated hydrochloric acid, an orange-red halochromic solution, from which no thebaine can be recovered, is obtained. This very probably contains the intermediate (76), which is converted by a dienone–phenol rearrangement on heating into morphothebaine (77)[52] or, by catalytic reduction of the cold solution to metathebainone (78).[53] Apomorphine (79) is presumably obtained by a similar process from morphine under the same conditions,[54] with dehydration an essential additional step for the estab-

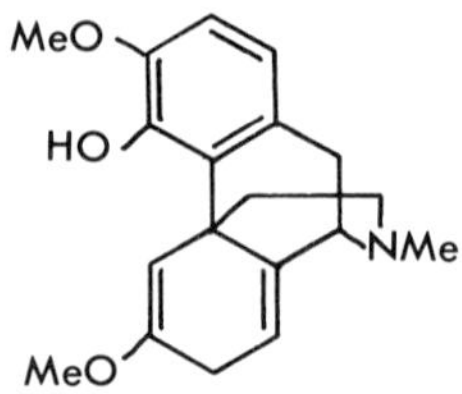
(69)

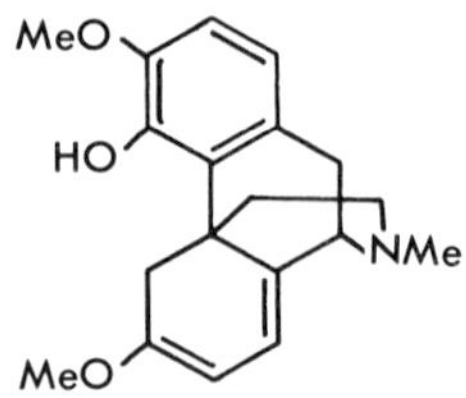
(70)

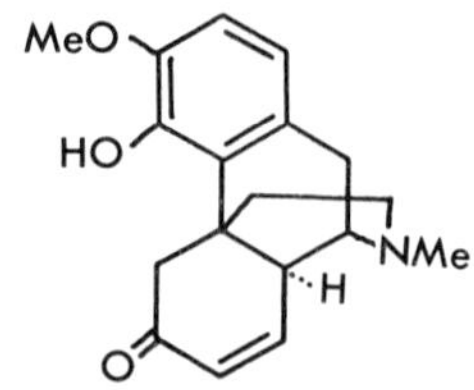
(71)

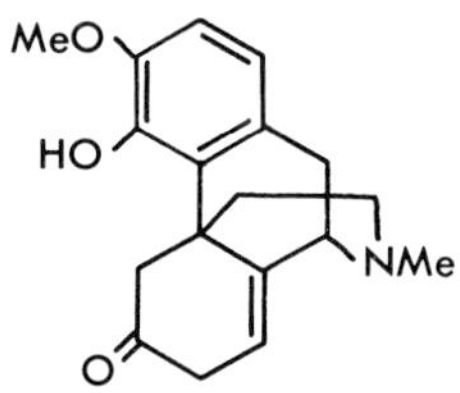
(72)

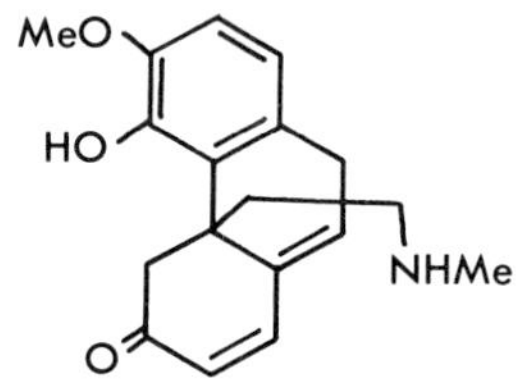
(73)

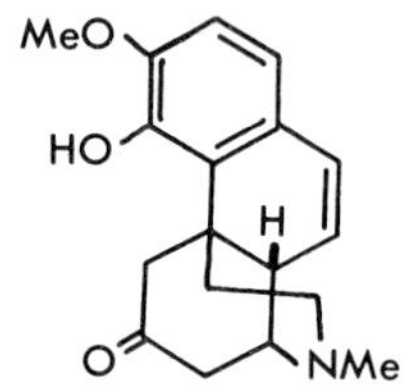
(74)

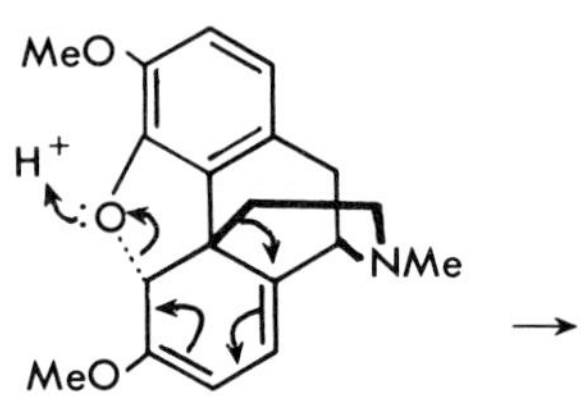
(75)

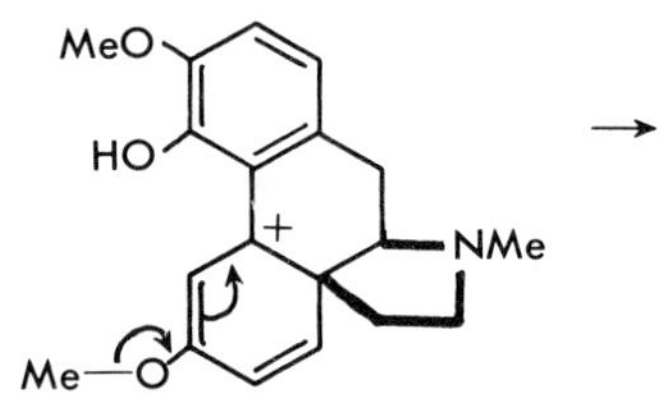

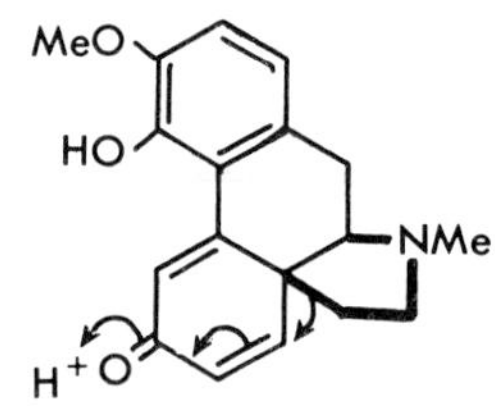
(76)

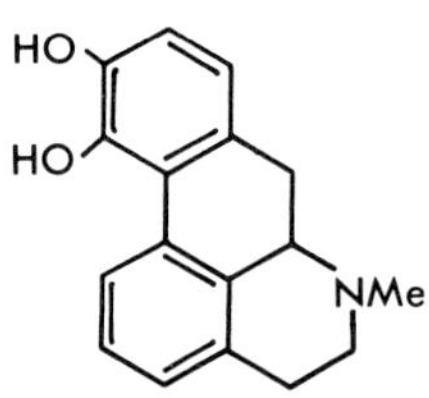
(79)

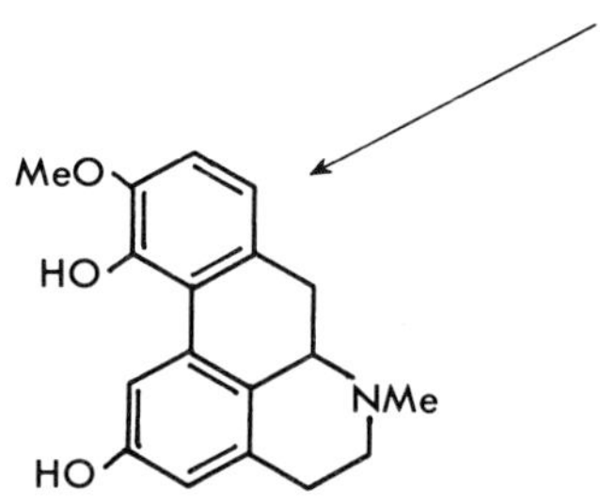
(77)

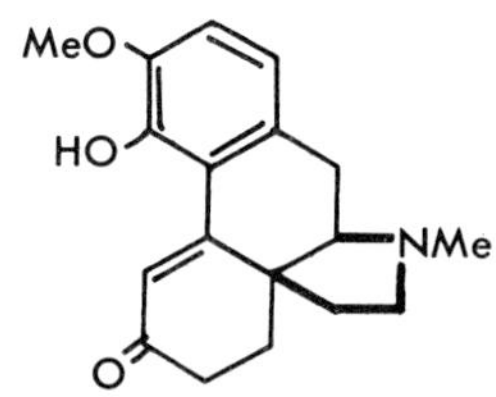
(78)

lishment of the same double-bond system as is present in (75). The structures of apomorphine and morphothebaine were established by the general methods applicable to aporphine alkaloids (see Chapter 00).

1.5.2 Metathebainone The structure (78) was assigned to this base on account of the similarity of its color reactions and those of salicylideneacetone and salicylideneacetophenone and of the elimination of the sidechain during acetolysis of the methyl ether methine to 3,4-dimethoxyphenanthrene.[55] This structure has been confirmed by the synthesis of a degradation product.[56]

1.5.3 Thebenine When thebaine is heated for a short period with dilute hydrochloric acid, rearrangement takes an entirely different course,[57] the fate of the dienone (76) being determined in this case by the availability of an unshared pair of electrons on the nitrogen atom in the free base that is present to a small extent in the dilute acid. Further attack by a proton then leads to (80), which, being the quaternary salt of a Schiff base, is readily hydrolyzed to the aminoaldehyde (81) from which thebenine (82) arises by cyclization of the phenanthrene system. The plausibility of this mechanism was one of the reasons that led Robinson to suggest the attachment of the sidechain at C(13) in the morphine alkaloids.[16,17]

The structure of thebenine was determined by Hofmann degradation of the dimethyl ether quaternary salt to the vinyl-compound (83, R = CH = CH_2) and reduction of this to 5-ethyl-3,4,8-trimethoxyphenanthrene (83, R = Et), which was shown to be identical with material prepared by synthesis.[58] Hofmann degradation of the phenol affords thebenol (84).[58,59]

1.5.4 Phenyldihydrothebaine When thebaine is treated with Grignard reagents a rearrangement occurs which follows essentially the same path as the thebenine transformation. However, since the conditions are neutral, the fate of the intermediate Schiff base quaternary salt is different in this case. With phenylmagnesium bromide the reaction path is as set out in (85)–(88). Phenyldihydrothebaine shows none of the properties usually associated with derivatives of thebaine; for example, it very strongly resists hydrolysis and hydrogenation, and these reactions when forced eventually afford a trihydric phenol (readily methylated to phenyldihydrothebaine methyl ether) and a secondary base resulting from reductive fission of the nitrogen-containing ring, respectively. Hofmann degradation affords a mixture of two methine bases, as a result of opening of the ring on either side of the nitrogen atom and further degradation of these affords an optically active nitrogen-free product containing two reducible double bonds, produced during the two stages of exhaustive methylation. The remarkable stability of the base towards acid and hydrogenation is a clear indication that ring C has become aromatic during the Grignard reaction with migration of the sidechain from C(13). Although Small[60] was unable to accept this aromatization on the grounds that the ultraviolet spectrum of the base shows no evidence of diphenyl conjugation, Robinson[61] on the basis of a plausible mechanism for the rearrangement, correctly deduced the structure (88) for phenyldihydrothebaine. He

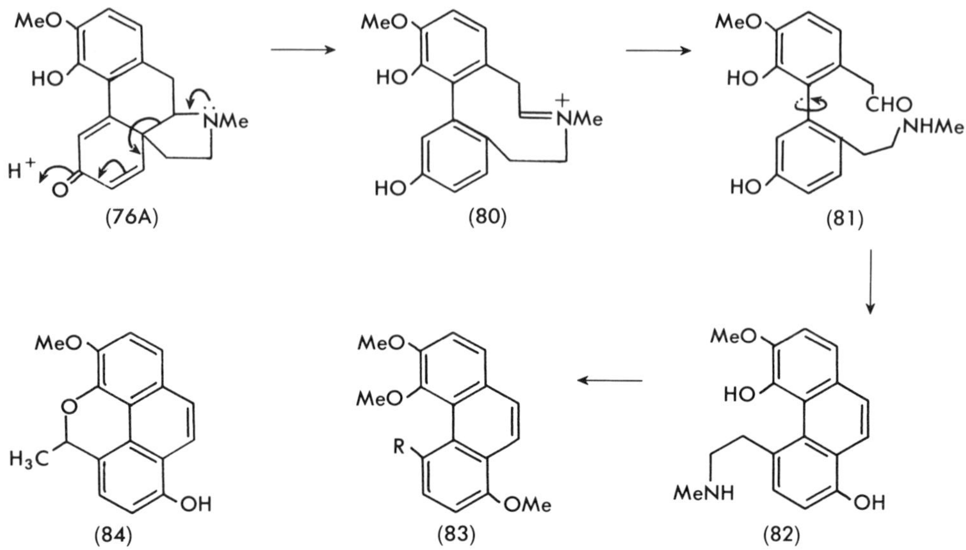

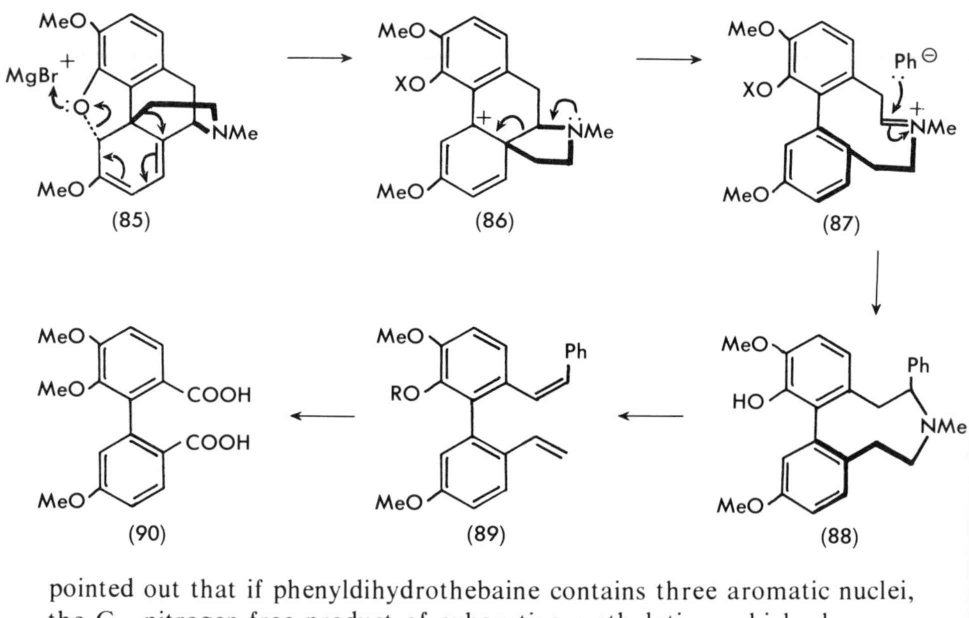

pointed out that if phenyldihydrothebaine contains three aromatic nuclei, the C_{24} nitrogen-free product of exhaustive methylation, which also contains two methoxyl groups and four carbon atoms present in two reducible double bonds, cannot contain an asymmetric carbon atom, and must owe its optical activity to molecular dissymmetry resulting from restricted rotation (89, R = H).

The structure (88) was proved correct for phenyldihydrothebaine by oxidation of the base (88) to 4-methoxyphthalic acid and of the N-free product (89, R = Me) to benzoic acid and the diphenic acid (90).[61,62] In (87) attack of the iminium salt by Ph^- can occur on either side of the molecule and isomeric bases are thus generated, differing in stereochemistry at C(9). Steric factors favor attack on one side of the molecule, however, with the result that the two isomers are not formed in equal amounts. The diphenyl system is generated during the reaction in one configuration only but it may be subsequently racemized by heat. It may be noted, also, that noncoplanar diphenyls do not exhibit diphenyl ultraviolet absorption since the planar requirements for conjugation are not met.[63,64] The parent dihydrothebaine (88, Ph=H) has been obtained by the reduction of the iminium salt (87), obtained from thebaine and magnesium iodide.[62,65]

1.5.5 Acetylthebaol The degradation of thebaine methiodide to acetylthebaol by heating with acetic anhydride and sodium acetate presumably proceeds as shown in (91–(93) and is favored by the *trans* relationship

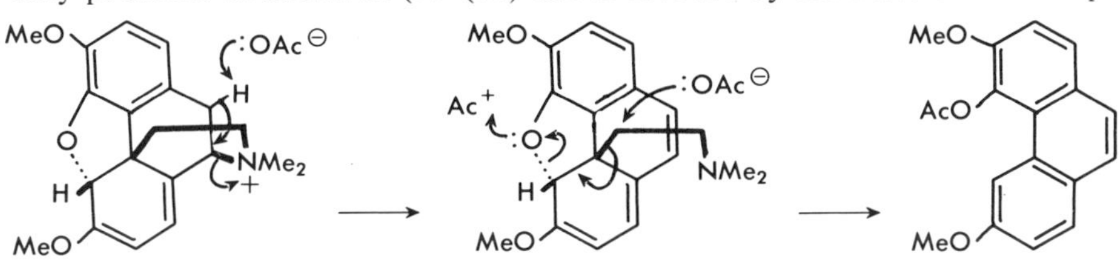

(91) (92) (93)

of the sidechain and oxygen atom at C(13) and C(5). The degradations of codeinone, ψ-codeinone and α-codeimethine to phenanthrene derivatives under the same conditions proceeds along similar lines, though in the case of the last-named base (19), dehydration also is necessary for aromatization.

1.5.6 Other Oxide-Ring Openings It will be seen that all the above rearrangements are initiated by attack of a positive ion on the oxide bridge, the resulting electron deficit at C(5) being made up by migration of the sidechain to C(14). (The degradation of thebaine methiodide to acetylthebaol can well be represented as a migration to C(14) followed by elimination of the side chain.) In the absence of an 8:14 double bond, however, this migration is not possible and rearrangement does not occur. However, if an electron pair can be supplied from elsewhere, opening of the oxide ring can still occur in such compounds. Examples of such processes are the reaction of dihydrothebaine (94, X = OMe),[66] dihydrocodeinone enol acetate (94, X = OAc),[67] and desoxycodeine-C (94, X = H)[68] with Grignard reagents, the products being the phenols (95) and (96) arising from attack by R^- at C(7) and C(5), respectively. Where X = OMe or OAc

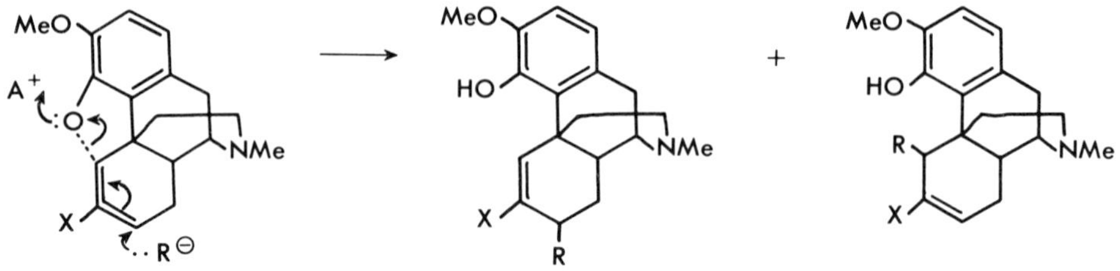

(94) (95) (96)

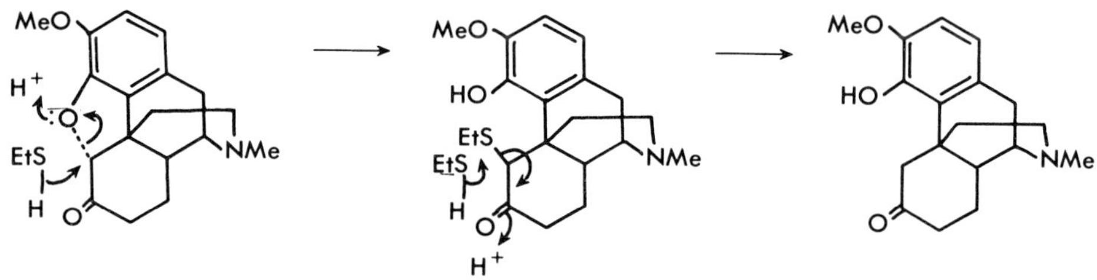

(97) (98) (99)

the products, being very sensitive to hydrolysis, were converted during isolation into the corresponding 6-ketones. Desoxycodeine-C also reacts with ethyl mercaptan and acid to give ethylthiodihydrodesoxycodeine (95, R = SEt, X = H).[69]

Dihydrocodeinone (97), which contains a carbonyl group at C(6) activating the ether bridge, reacts with ethyl mercaptan and acid to give dihydrothebainone (99) presumably by way of the thio ether (98).[70] The same process occurs on Clemmensen reduction, the dihydrothebainone being further reduced to tetrahydrodesoxycodeine (30, R = OMe, R′ = OH).[71]

1.5.7 Flavothebaone Thebaine, being a reactive conjugated diene, readily undergoes Diels–Alder condensation with dienophils. Some compounds obtained in this manner are so constructed that opening of the 4,5-oxide bridge can occur without C(13) sidechain migration and without an external supply of electrons. The electron deficit at C(5) is finally made up by the C(6) oxygen atom with migration of the added group.

Addition of benzoquinone to thebaine affords thebaine-quinone (100) which is very readily enolized to the quinol. The latter, when heated with concentrated hydrochloric acid, suffers the rearrangement (101) to flavothebaone (102).[72–74] The structure of flavothebaone has been determined in the following way. It is clearly, from its properties, a trihydric phenol and an $\alpha\beta$-unsaturated ketone. The trimethyl ether methiodide can be degraded to a methine base, the uv spectrum of which is a close approxi-

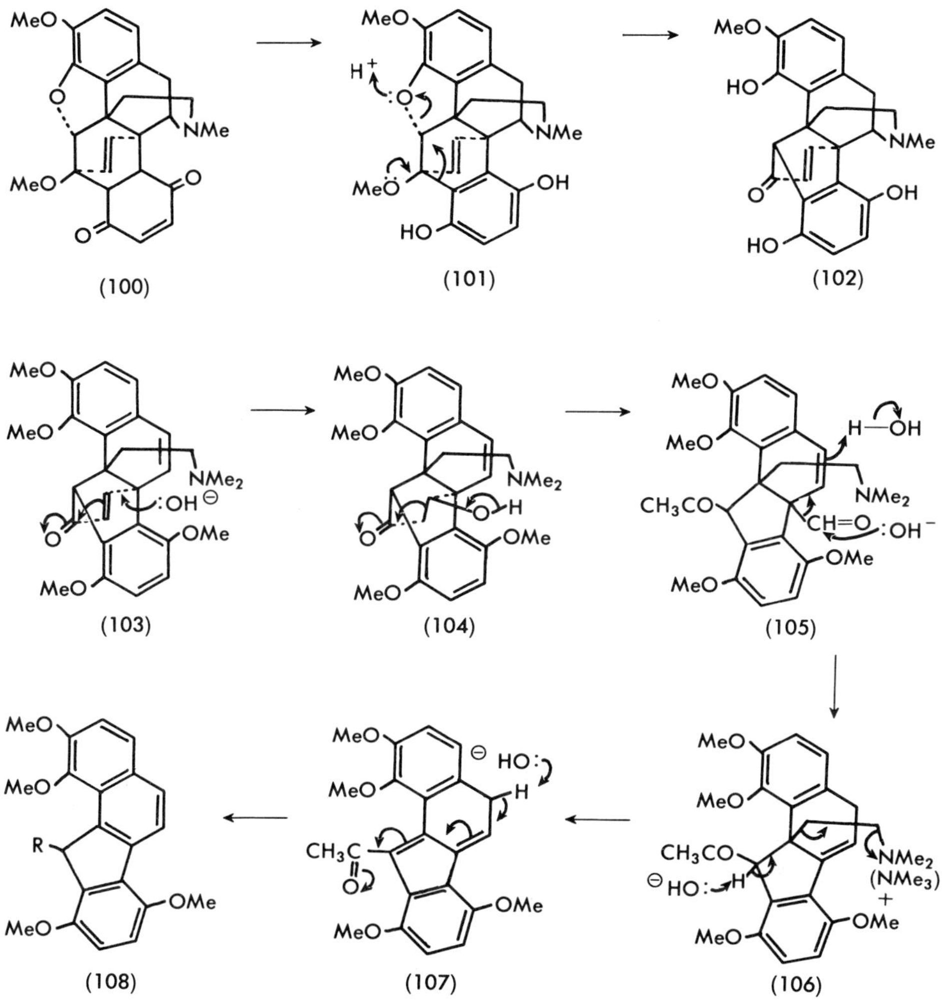

mation to a sum of those of flavothebaone and α-codeimethine (19).[73] However, this base strongly resists aromatization when heated with acetic anhydride; and it cannot be isomerized to an analog of β-codeimethine when heated with an alkali, a treatment that gives a ψ-methine with loss of a carbon atom.[75] It can be deduced from this that flavothebaone bears substituents at both angular positions C(13) and C(14). Since further reactions of the ψ-methine can only be explained if the nitrogen-containing sidechain is retained in this base at C(13) (see below), the quinol nucleus must remain attached to C(14). Since the uv spectrum of flavothebaone is quite different from that of metathebainone (78), the double bond can only be placed at C(7)–C(8). Assuming that the carbonyl group is generated from the C(6) methoxyl and that flavothebaone trimethyl ether cannot be condensed with piperonal (its dihydro-derivative can), the base must contain the system —HCR—CO—CH=CH— and flavothebaone must accordingly be represented by (101).[74,75]

1.5.8 Further Rearrangement of Flavothebaone Derivatives Flavothebaone trimethyl ether methine, which has the structure (103), upon heating with alkalies affords the ψ-methine with the gain of two hydrogen atoms and the loss of one carbon.[73] This base contains a saturated carbonyl group and a C-methyl group, which must be present together as CH_3CO— since conversion of the base into its piperonylidene derivative results in loss of the C—CH_3 group. The nitrogen-containing sidechain must occupy an angular position since it is extruded with formation of an aromatic compound of composition $C_{23}H_{22}O_5$ (optically inactive) when the methiodide is heated with sodium acetate and acetic anhydride. This neutral product is ketonic and can be hydrolyzed with production of acetic acid and a non-ketonic substance $C_{21}H_{20}O_4$, obtainable together with $C_{22}H_{22}O_4$ and the C_{23} compound by Hofmann degradation of the ψ-methine methohydroxide. The uv spectra of all three aromatic compounds are similar and closely resemble those of β-phenylnaphthalenes. This fact, together with the composition of the C_{21} compound, which contains only one carbon atom more than a tetramethoxyphenylnaphthalene ($C_{20}H_{20}O_4$), indicates that the substances are benzofluorenes.[75] The subsequent synthesis of the benzofluorene (108, R = H),[76] which is identical with the C_{21} degradation product merely confirms the deductions previously made as to the structure of the three products and the ψ-methine.[75] The C_{22} compound, which is obtainable by methylation of the C_{21} compound is clearly (108, R =CH_3) and the readily hydrolyzed C_{23} ketone is equally obviously (108, R = $COCH_3$), the optical inactivity of the latter being due to racemization under the enolizing conditions of the reactions. The C_{22} compound (108, R = CH_3) presumably arises by the methylation of (108, R = $COCH_3$) by some quaternary salt during Hofmann degradation.

Since the ψ-methine must contain the skeleton of (108, R = $COCH_3$) with the sidechain at an angular position, and since the infrared and ultraviolet spectra show that it contains a trisubstituted double bond conjugated with the quinol dimethyl ether nucleus, only structure (106) is possible for this base. It presumably arises from the methine (103) by hydration of the double bond, dealdolization of the product (104) and elimination of the formyl group (105).[75]

Two other N-free products have been isolated from the degradation of the ψ-methine. One, which accompanies the benzofluorenes on dry distillation of the methohydroxide, is undoubtably the cyclic ether (109) [compare with thebenone (55)] and the other, which is obtained together with (108, R = H) by effecting the Hofmann degradation with potassium hydroxide in cyclohexanol, has been assigned the structure (113) on spectral grounds. A vinyl group, together with a second double bond conjugated with the carbonyl group and one or both aromatic nuclei, is present in the molecule. Both types of conjugation are eliminated upon epoxidation of this double bond. For such a conjugated system to exist, the sidechain must have migrated from C(13), and the only rational mechanism leading to a satisfactory structure for this product is that set out in formulas (110)–(113).[75]

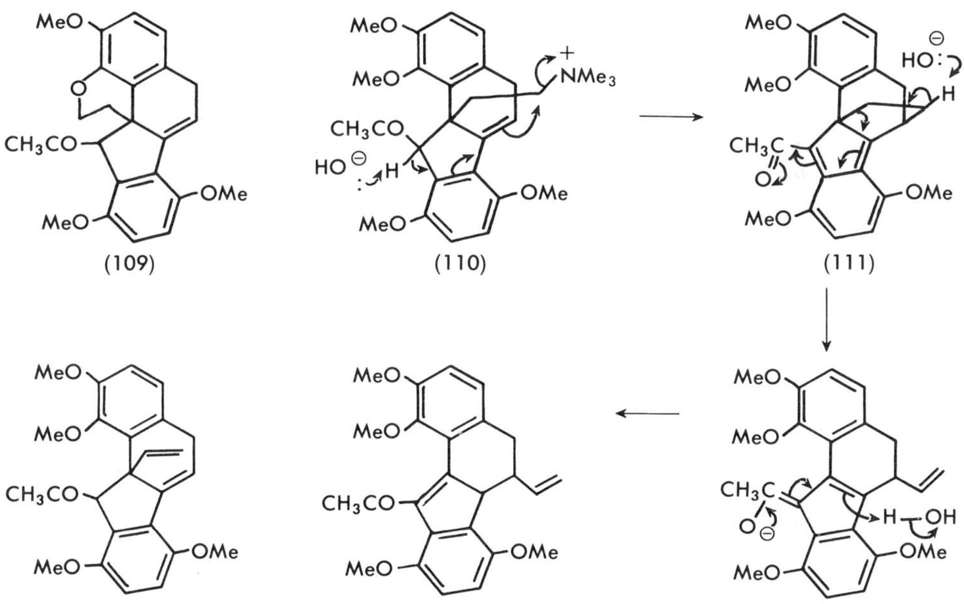

(109) (110) (111)

(114) (113) (112)

It is of interest to note that the angular vinyl compound (114), which is the simple N-free derivative of the ψ-methine, can be obtained by pyrolysis of the N-oxide, a reaction that affords the corresponding vinyl compound from the methine (103) and from bases such as α and β-codeimethine that aromatize during normal Hofmann degradation.[77–78]

Beckmann transformation of the oximes of the ψ-methine and its derivatives leads to further rearrangement.[79] The ψ-methine oxime (115) affords acetonitrile and the carbinol (116), which may be rearranged by formic acid to the benzofluorene (117, $R = CH_2CH_2NMe_2$). Compound (117) is degradable by pyrolysis of the N-oxide to the vinyl compound

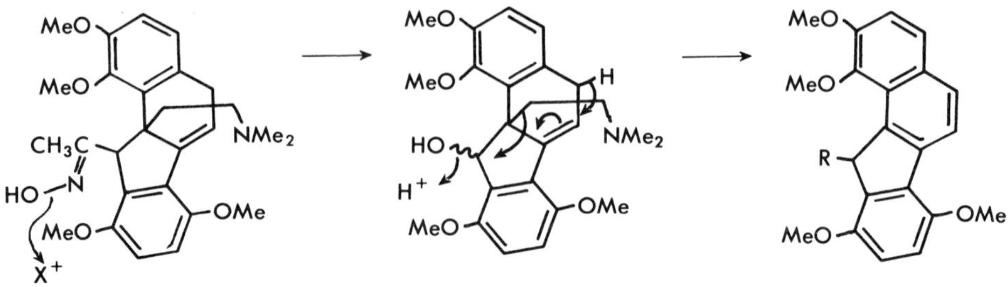

(115) (116) (117)

(117, R = CH═CH₂), which is obtained directly by the Beckmann transformation of the oxime of the N-free product (114). This difference in the behavior of the two compounds is attributable to the greater ease of migration of an unsaturated compared to a saturated group. The benzofluorenes obtained in this way are all optically active, and are racemized by alkali, indicating that the rearrangements are stereospecific.

A similar migration of the angular substituent was observed in the dehydration of the carbinol (120), obtained in stages from the product (119) of Beckmann transformation of the oxime (118), itself derived from the methine base (103).[77]

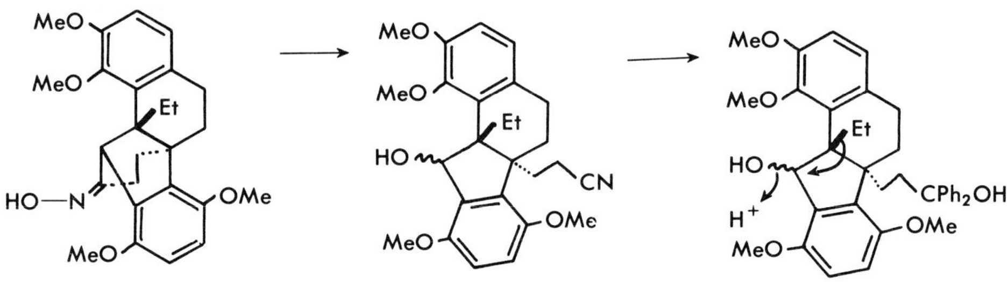

(118) (119) (120)

1.5.9 Flavonepenthone and Isonepenthone. Diels–Alder addition of phenylvinylketone to thebaine affords the ketone nepenthone (121) which is readily reduced to nepenthol (122). The latter is readily dehydrated with formic acid to the substituted codeinone nepenthene (123), which is further converted by concentrated hydrochloric acid to flavonepenthone (124).[80] Base-catalyzed rearrangement of nepenthone, however, affords isonepenthone (126) by way of the intermediate (125). Isonepenthone is an extremely easily hydrolyzed ketal, being converted by acids into the unsaturated ketone ψ-nepenthone (127). Hydrolysis of the secondary alcohol (128) derived from isonepenthone, even under mild conditions, is accompanied by dehydration, the principal product being flavonepenthone (124).[81] Further rearrangements in this series have been reported.[82–87]

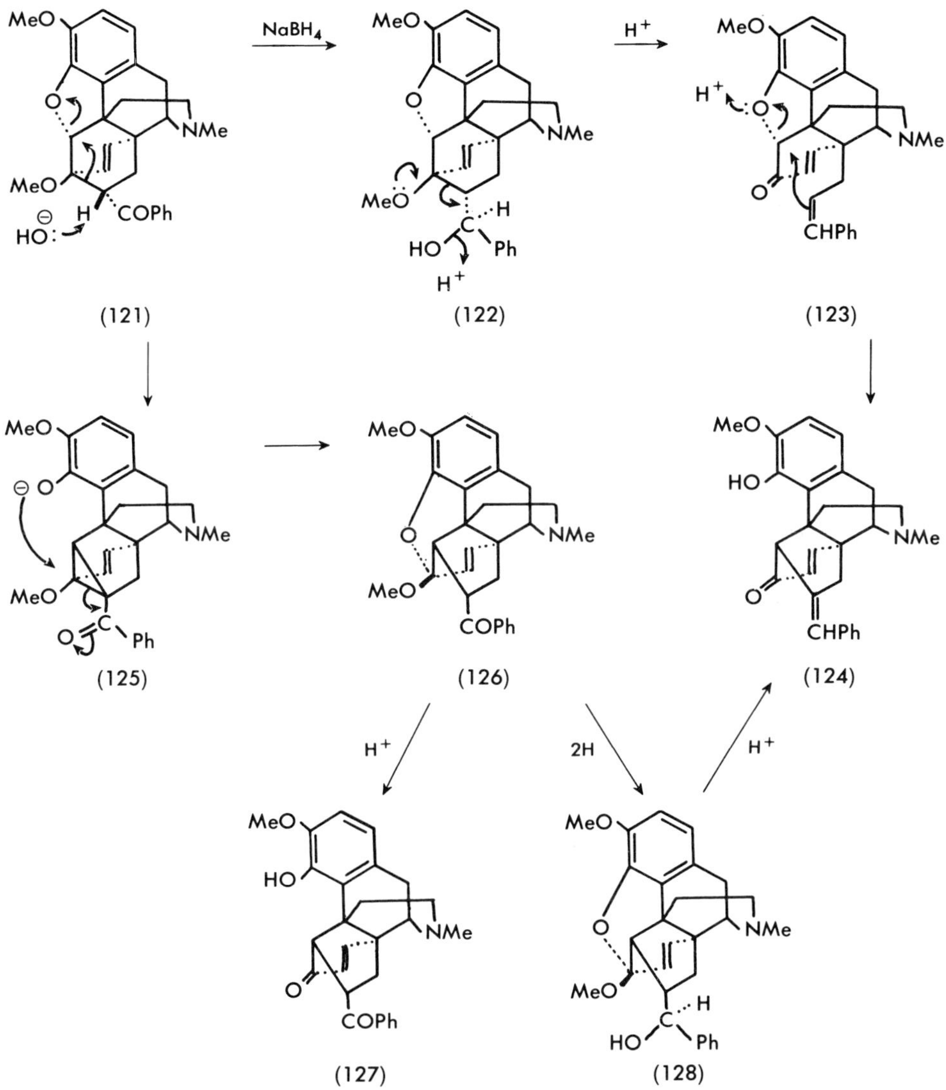

1.5.10. Thermal Rearrangement of Diels–Alder Adducts of Thebaine. The condensation of thebaine with acetylenic dienophils yields adducts susceptible to thermal rearrangement.[88] The adduct with methyl acetylenedicarboxylate (129) in this way gives the benzazocine derivative (130), which may be reduced to the diol with lithium aluminium hydride and then subjected to Hofmann degradation to give the stilbenoid base (131). A similar reaction was observed with the ethyl propiolate adduct of thebaine. The rearrangements are analogous to the aromatization of the adduct of

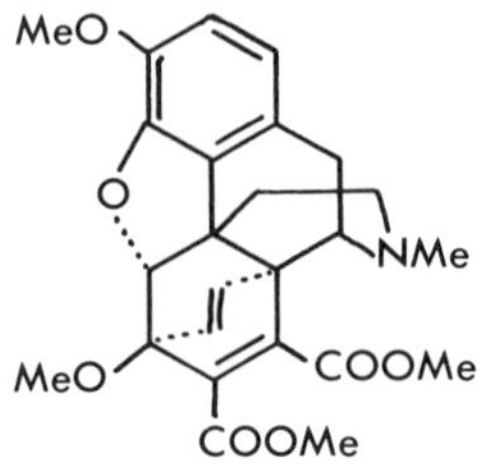

(129)

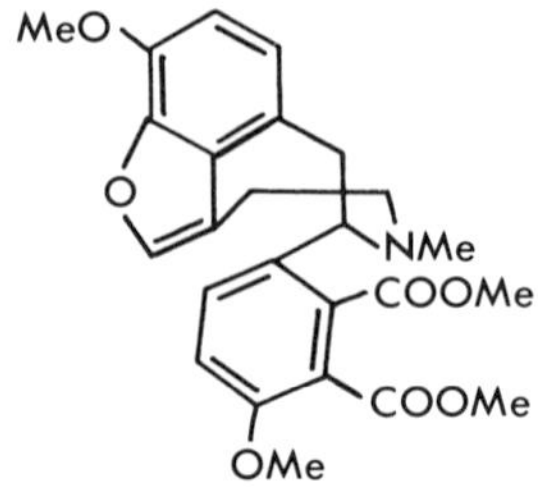

(130)

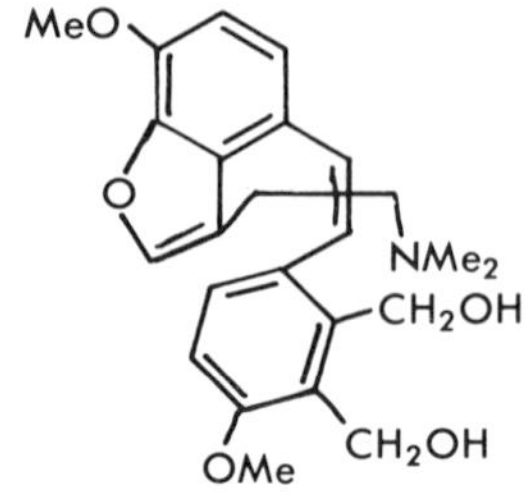

(131)

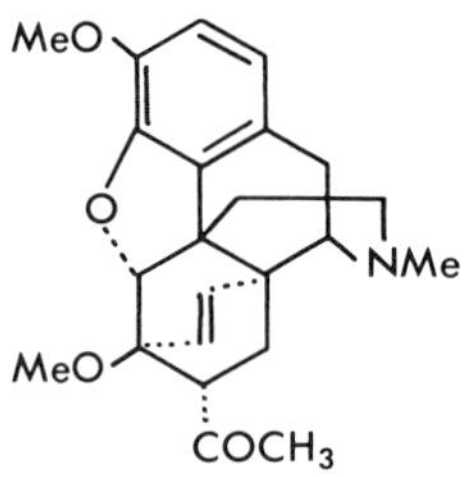

(132)

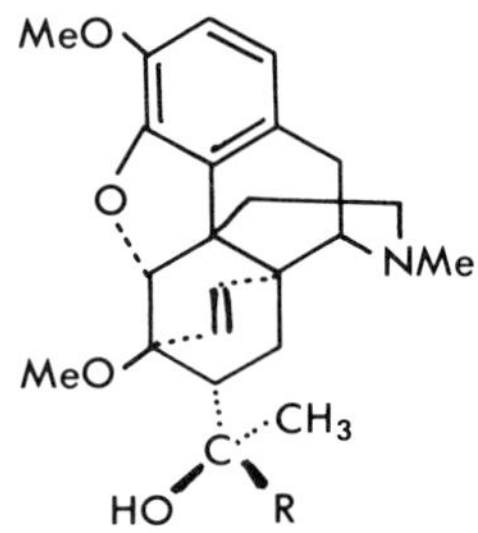

(133)

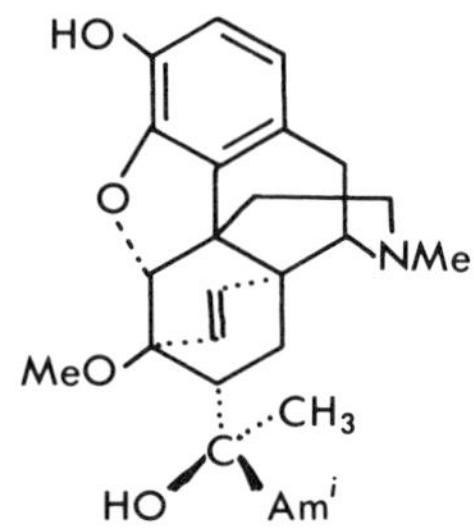

(134)

methyl acetylenedicarboxylate and cyclohexadiene, with extrusion of ethylene.

1.6. Analgesics of the Morphine Series

As an analgesic for general clinical use, morphine has some disadvantages; it causes respiratory depression, tolerance to its action rapidly develops, and it is a drug of primary addiction. Attempts have been made to eliminate some or all of these disadvantages by modification of the molecule and by the synthesis of compounds that represent only what are believed to be the active sections of the molecule. The following generalizations, to which of course there are exceptions, may be made.

(a). Activity of morphine is increased by catalytic reduction, methylation, oxidation, or elimination of the C(6) hydroxyl group, introduction of a C(14) hydroxyl or C(5) methyl group, or replacement of the N-methyl by certain groups such as N—CH_2CH_2Ph.

(b). Activity is decreased by methylation of the phenolic hydroxyl group, quaternization of the nitrogen, fission of the nitrogen-containing ring, and opening of the 4,5-oxygen bridge.

(c). Remarkably high activity is found in many derivatives of Diels–Alder adducts of thebaine. The ketone (132),[82] for example, is as active as morphine and the series of tertiary alcohols (133) derived from this contains compounds up to 500 times as potent as the alkaloid.[83] Demethylation of these alcohols affords a series of phenols of even higher activity, one of which (134) is about 10,000 times as potent as morphine.[84,89]

(d). Analgesic activity is found in bases of structures (136)–(140) and many of their derivatives; the structural resemblance of these compounds to morphine (135) is obvious.[90]

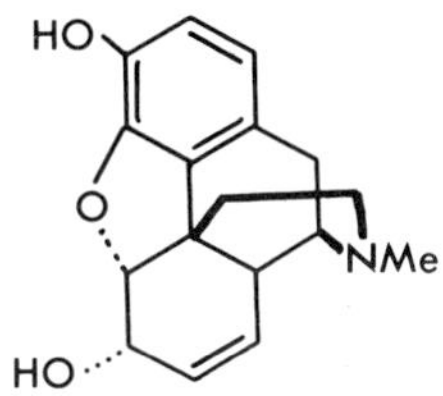

(135)

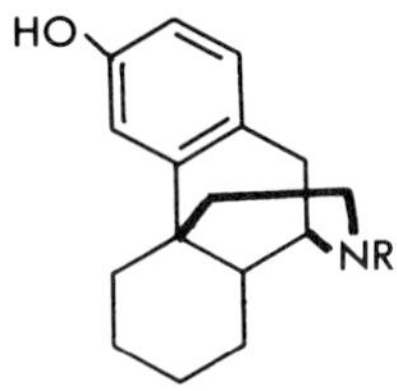

(136)

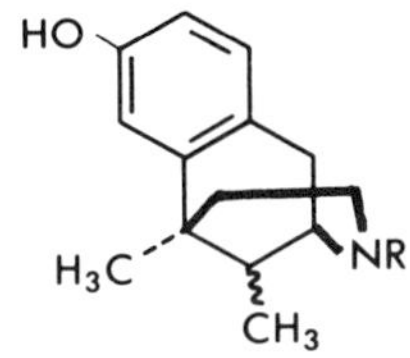

(137)

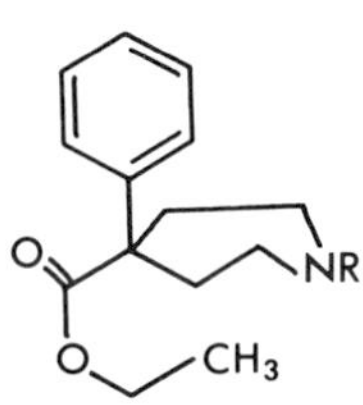

(138)

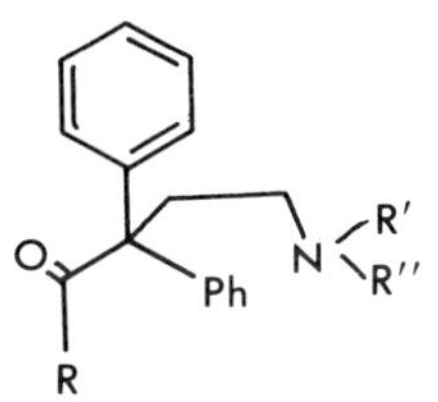

(139)

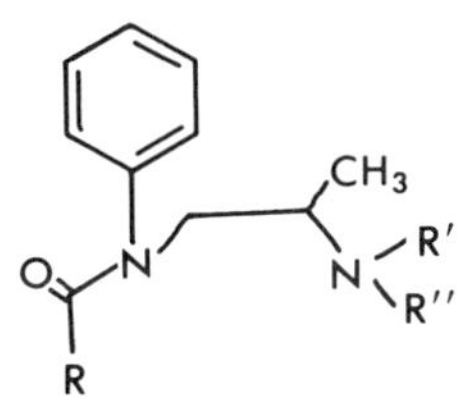

(140)

2. THE SINOMENINE GROUP

The sinomenine subgroup consists of four bases containing no 4,5-oxide bridge. Only sinomenine belongs to the morphine group, and the other three bases have been recently assigned a modified skeleton.

2.1. Sinomenine

The base, the principal alkaloid of *Sinomenium acutum*, has the composition $C_{19}H_{23}NO_4$, contains two methoxyl groups, and is a phenolic $\alpha\beta$-unsaturated ketone. It is readily degraded to a methine base, and when its methiodide is heated with benzoic anhydride, it is degraded to a phenanthrene derivative, dibenzoylsinomenol (141, R = COPh), from which diethylsinomenol (141, R = Et), identical with 3,7-dimethoxy-4,6-diethoxyphenanthrene, was prepared.[91] Catalytic reduction of sinomenine

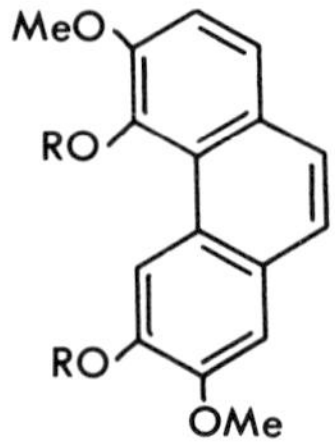
(141)

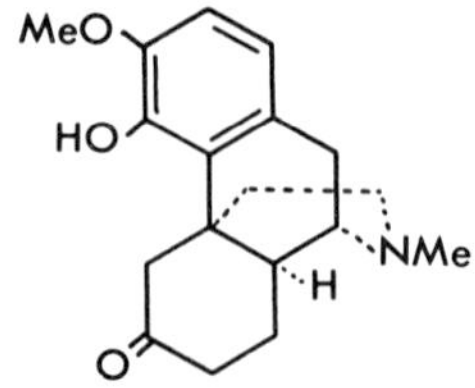
(142)

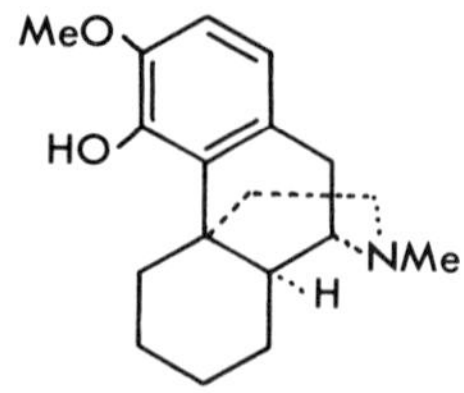
(143)

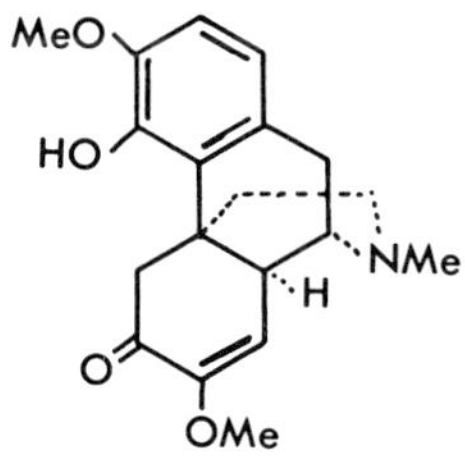
(144)

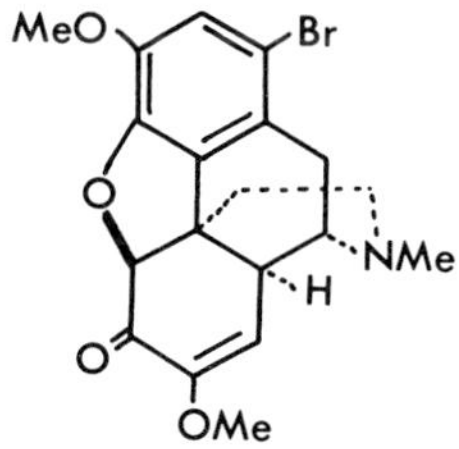
(145)

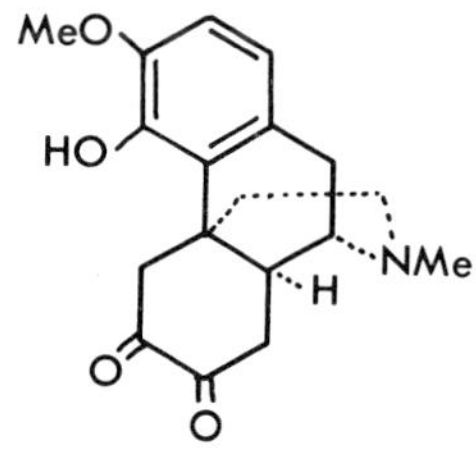
(146)

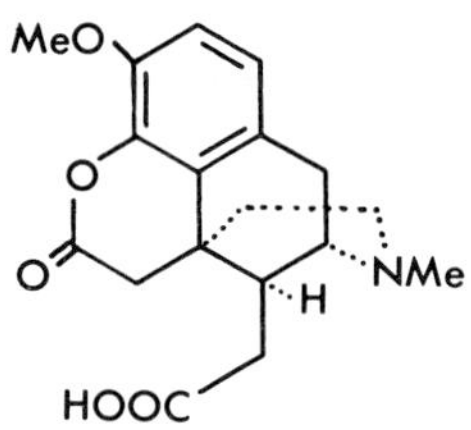
(147)

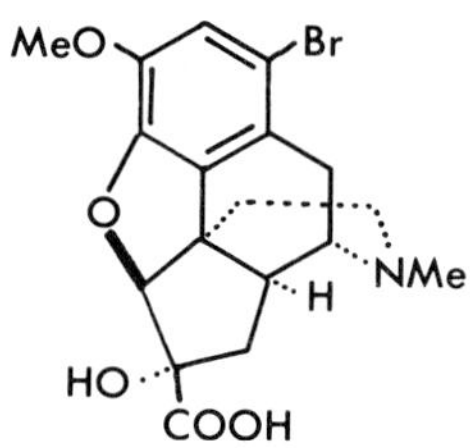
(148)

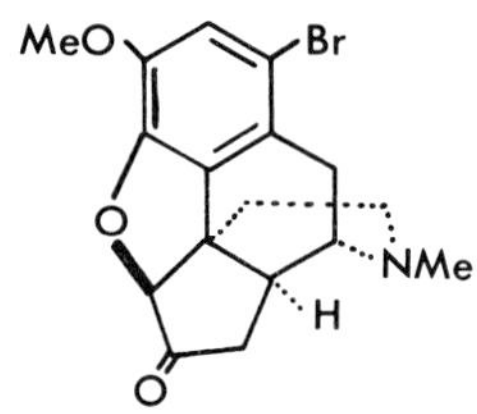
(149)

yields dihydrosinomenine, which upon further reduction with sodium amalgam is converted into the optical antipode of dihydrothebainone (142) and yields (+)-tetrahydrodesoxycodeine (143) on Clemmensen reduction. Sinomenine, therefore, belongs to a series enantiomorphic with that to which morphine and thebaine belong, and has the structure of 7-methoxy-(+)-thebainone (144).[92–94] Dihydrosinomenine has been converted into (+)-thebainone[95] and, by the method of Gates,[27] into (+)-codeine and (+)-morphine.[96] The closure of the 4,5-oxygen bridge on bromination, which is an essential step in the last of these transformations, was first observed in the bromination of sinomenine itself to 1-bromosinomeneine (1-bromo-7-methoxy-(+)-codeinone) (145).[97]

Sinomenine may be hydrolyzed to the α-diketone sinomeninone (146) and this with hydrogen peroxide is oxidized to sinomeninic acid (147).[98]

The α-diketone, 1-bromosinomeneine ketone, derived from (145) undergoes benzilic acid rearrangement to 1-bromosinomenilic acid (148),[99] which, being an α-hydroxy acid is converted by concentrated sulfuric acid into the ketone, 1-bromosinomenilone (149).[100]

2.2. Hasubanonine

This base, from *Stephania Japonica* Miers, has the composition, $C_{21}H_{29}NO_5$, and contains four methoxyl groups and an αβ-unsaturated ketone system. On Hofmann degradation it affords an alkali-soluble methine base, apparently formed by demethylation of a methoxy group. On acetolysis the methine base loses the sidechain and one of the oxygen functions to give acetylhasubanol. Hydrolysis of this yields hasubanol which affords 3,4,6,8-tetramethoxyphenanthrene (150, R = R′ = Me) on methylation. Hasubanol gives a positive test with Gibbs' reagent and would thus appear to have a free position *para* to the hydroxyl group. Only structures (150, R = H, R′ = Me) and (150, R = Me, R′ = H) fit this requirement. On the basis of this evidence, the stability of the double bond to hydrogenation, and the oxidation of hasubanonine itself to hemipinic acid, structure (151) was advanced for the alkaloid.[101–105] Recently, a new structure (152) has been advanced for the alkaloid.[106] It is claimed that nmr studies indicate the orientation of oxygen substituents shown; sodium borohydride reduction of the base of the alcohol (153) followed by acid treatment led to ketone (154), converted by stages into the base (155, R = Me). The latter is the enantiomorph of a base prepared from 14-bromocodeinone in the following way. Reduction of the bromo-ketone (156) with excess of sodium borohydride yields a base formulated as (158) on the basis of nmr studies and the following chemical evidence.[107]

(a) It affords β-codeimethine on Hofmann degradation.
(b) It may be oxidized with silver carbonate to an αβ-unsaturated ketone.
(c) Catalytic reduction affords a dihydro-base, oxidizable to a saturated ketone that on reduction with zinc yields a phenolic ketone differing from dihydrothebainone as its C(14) epimer.

Base (158) was assumed to originate via the ion (157). For this to arise, however, the bromine atom in ketone (156) would have to be on the side of the molecule opposite to that on which it is almost certainly located. The saturated ketone referred to in (c) above on Wolff–Kishner and catalytic reduction yielded a base formulated as (155, R = H) methylatable to (155, R = Me). It may be noted that as far as can be determined from the data published so far, the physical properties of the bases (155, R = H, and Me) and their derivatives are very closely similar to those of tetrahydrodesoxycodeine and its corresponding derivatives. The placing of a keto-

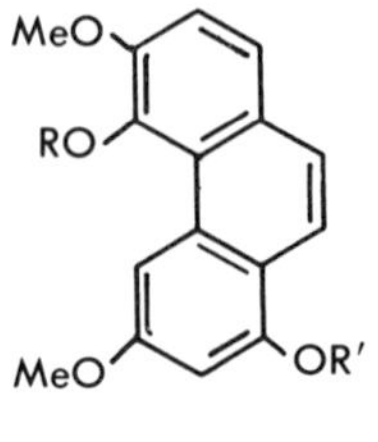

(150)

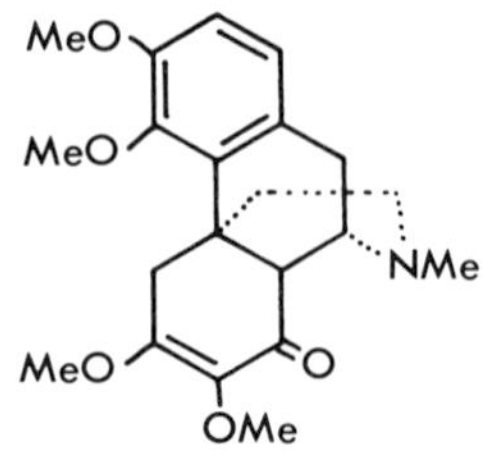

(151)

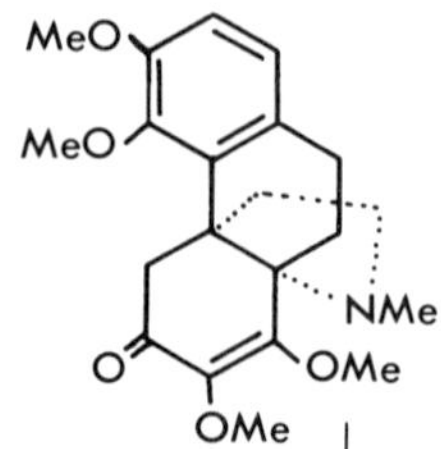

(152)

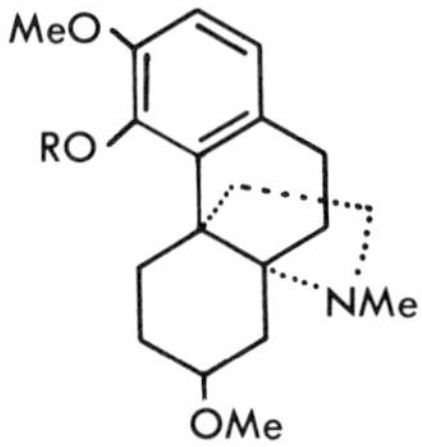

(155)

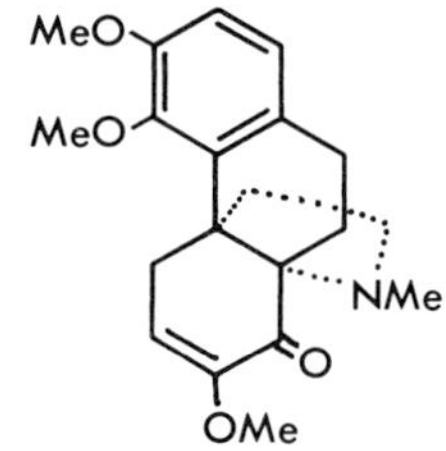

(154)

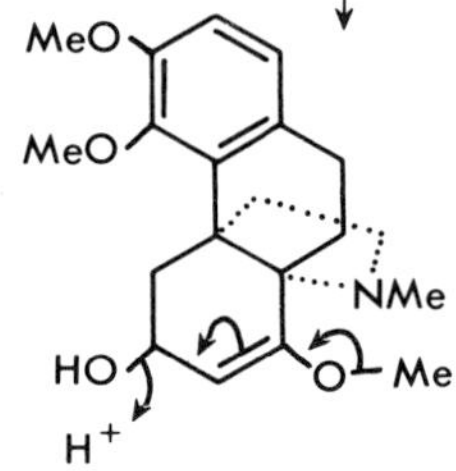

(153)

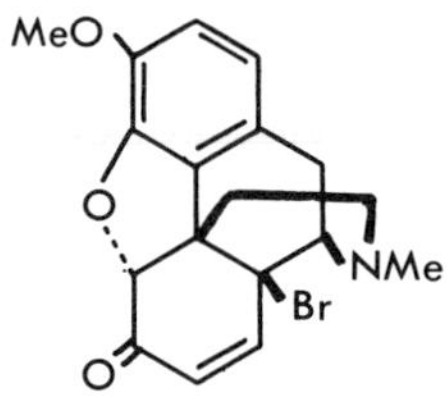

(156)

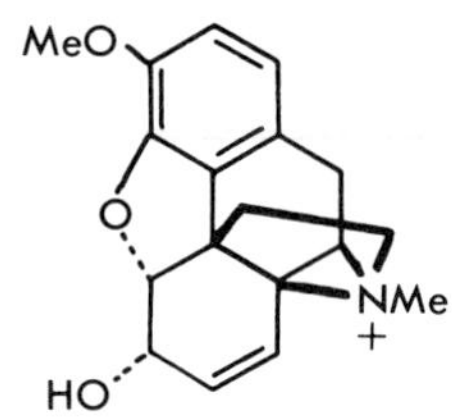

(157)

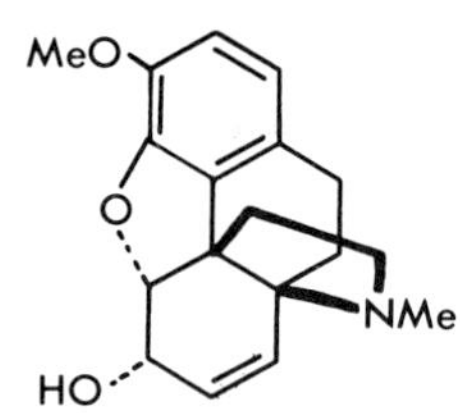

(158)

group at C(6) in (152) is at variance with the published work of Kondo, and it may be noted that if structure (151) correctly represented hasubanonine, the base presently formulated as (154) would be sinomenine methyl ether. Apparently no comparison between these two bases has, so far, been made.

2.3. Metaphenine

This alkaloid, which accompanies hasubanonine in the plant, has the composition $C_{19}H_{23}NO_5$ and contains one NMe group, two OMe groups, one OH group, and one carbonyl group. Acetolysis affords a neutral compound, $C_{20}H_{18}O_6$ that on hydrolysis and ethylation yields 3,4-dimethoxy-7,8-diethoxyphenanthrene.[108] Hydrogenation of the base gives the diol dihydrometaphenine, which yields 3,4,7,8-tetramethoxy- and 3,4,8-trimethoxyphenanthrene on acetolysis, hydrolysis, and methylation, suggesting that the carbonyl group of the alkaloid is at C(7), since alcoholic hydroxyls are generally eliminated in the acetolysis reaction.

Huang–Minlon reduction of the alkaloid afforded four desoxo-bases, one of which was formulated as (159) since on catalytic reduction it give the base assigned structure (155, R = Me) obtained from hasubanonine. A second desoxo-compound was formulated as (160) since it was oxidized by manganese dioxide to a conjugated (aromatic) ketone which on Huang–Minlon reduction again gave (155, R = Me).

Since metaphenine behaves as a masked α-diketone, and undergoes benzilic acid transformation, the alkaloid is formulated as (161). The point of attachment of the nitrogen atom in this base as well as in hasubanonine rests finally on the correct identification of the base (158) derived from 14-bromocodeinone.[109]

2.4. Prometaphenine

This base, also from *Stephania japonica* Miers, is closely related to metaphenine into which it is converted by dilute mineral acid. On spectral grounds it is assigned the structure of an equilibrium mixture of (162) and (163).[110]

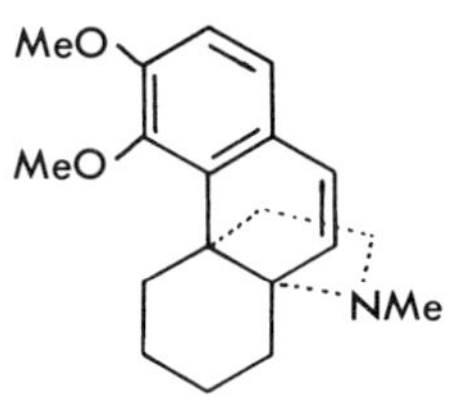

(159)

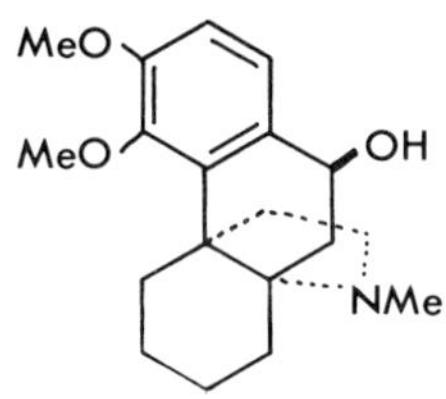

(160)

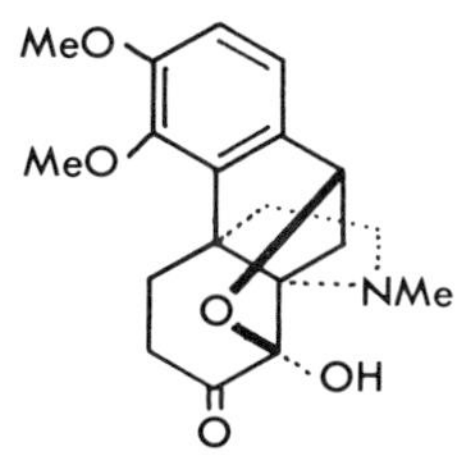

(161)

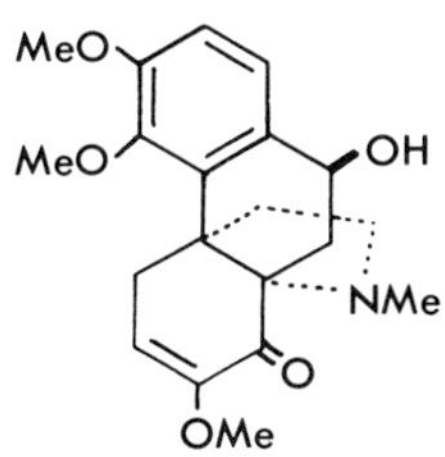

(162)

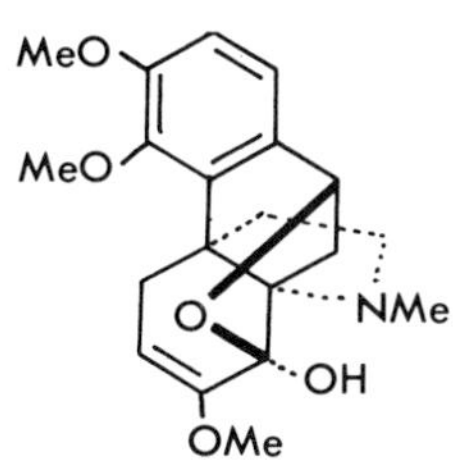

(163)

Newly discovered bases of the morphine-sinomenine group include the following: Salutaridine (164), which is an intermediate in the biogenesis of morphine,[111] and its 8,14-dihydro-compound,[112] flavinantine (165),[113–115] and the closely related amurine (166)[116] and nudaurine (167),[116] all of which belong to the same stereochemical series as morphine, and sinoacutine[117] (168, R = Me) (the antipode of salutaridine), norsinoacutine[117] (168, R = H) and cepharamine (169),[118] of the sinomenine-hasubanonine series. The alkaloids kreysiginine (170)[119] and androcymbine (171)[120] represent ring-B homologues of the main series, and are probably biogenetically related to colchicine (172).[121]

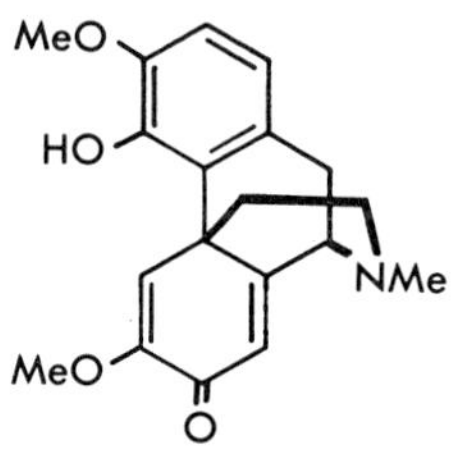

(164)

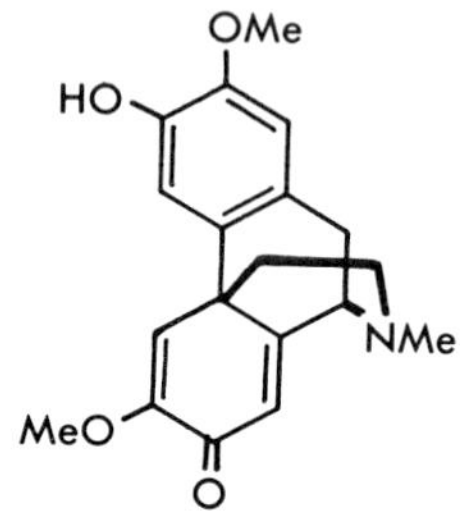

(165)

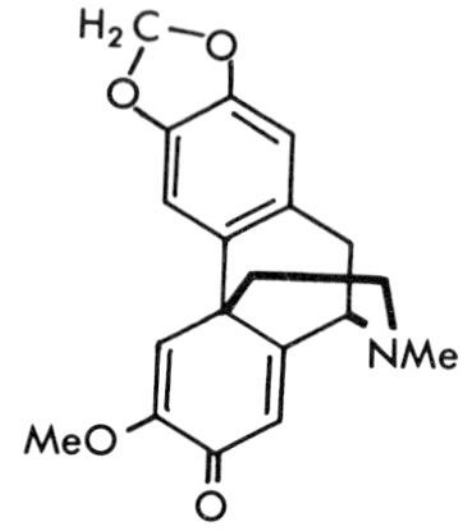

(166)

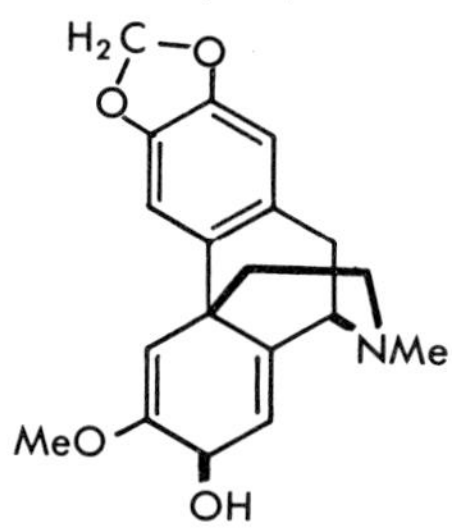

(167)

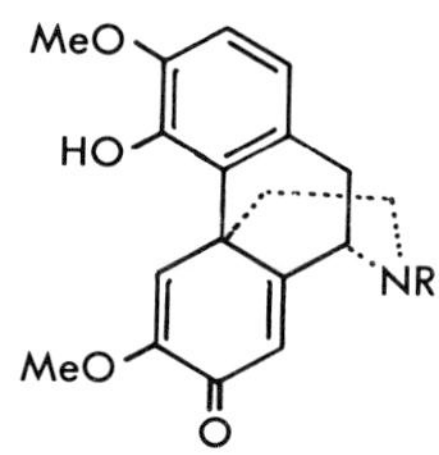

(168)

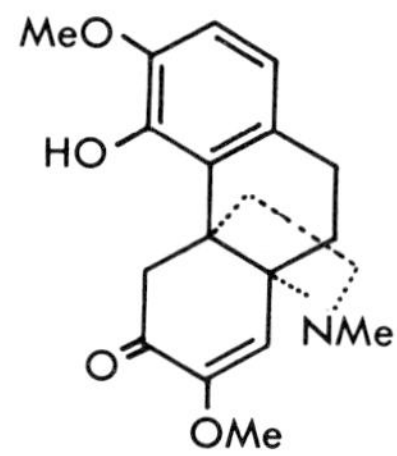

(169)

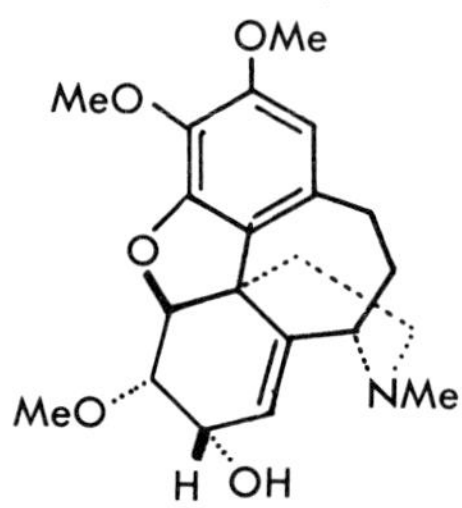

(170)

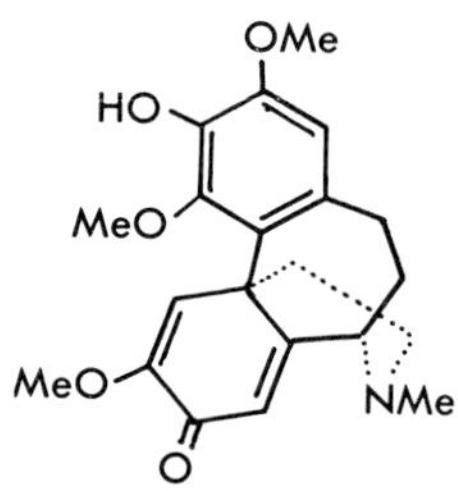

(171)

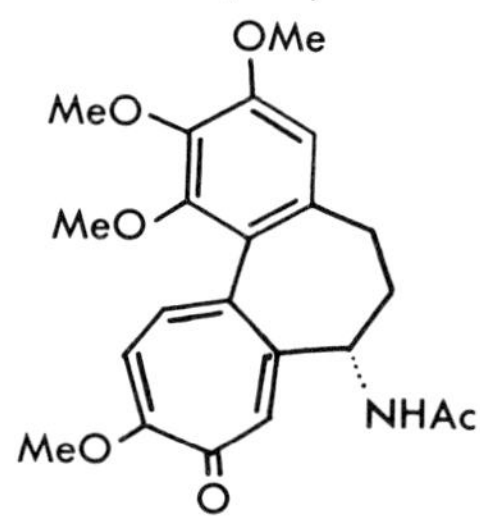

(172)

REFERENCES

1. J. M. Gulland and R. Robinson, *J. Chem. Soc.* 980 (1923).
2. J. M. Gulland and R. Robinson, *Mem. Proc. Manchester Lit. Phil. Soc.* **69,** 79 (1925).
3. R. Pschorr, C. Seydel, and W. Stöhrer, *Ber.* **35,** 4400 (1902).
4. E. Grimaux, *Compt. Rend.* **93,** 591 (1881).

5. E. Vongerichten and O. Dittmer, *Ber.* **39,** 1718 (1906).
6. O. Hesse, *Ann.* **222,** 203 (1883).
7. R. Pschorr and C. Sumuleanu, *Ber.* **33**, 1810 (1900).
8. R. Pschorr and W. Karo, *Ber.* **39**, 3124 (1906).
9. R. Pschorr, H. Einbeck, and O. Spangenberg, *Ber.* **40**, 1998 (1907).
10. R. Pschorr and W. Busch, *Ber.* **40**, 2001 (1907).
11. L. Knorr and R. Pschorr, *Ber.* **38**, 3153 (1905).
12. L. Knorr and H. Hörlein, *Ber.* **40**, 3301 (1907).
13. H. Wieland and E. Koraleck, *Ann.* **433**, 267 (1923).
14. L. Knorr and H. Hörlein, *Ber.* **40**, 2032 (1907).
15. L. Knorr and H. Hörlein, *Ber.* **40**, 2042 (1907).
16. R. Pschorr and H. Einbeck, *Ber.* **40**, 1980 (1907).
17. H. Rapoport and G. W. Stevenson, *J. Amer. Chem. Soc.* **76,** 1796 (1954).
18. C. Schöpf, *Ann.* **452,** 411 (1927).
19. R. Grewe and A. Mondon, *Ber.* **81,** 279 (1948).
20. R. Grewe, A. Mondon, and E. Nolte, *Ann.* **564,** 161 (1949).
21. M. Gates and W. F. Newhall, *J. Amer. Chem. Soc.* **70,** 2261 (1948).
22. M. Gates and W. F. Newhall, *Experientia* **5,** 285 (1949).
23. M. Gates and W. F. Newhall, *J. Amer. Chem. Soc.* **72,** 228 (1950).
24. M. Gates, R. B. Woodward, W. F. Newhall, and R. Kunzli, *J. Amer. Chem. Soc.* **72,** 114 (1950).
25. M. Gates and G. Tschudi, *J. Amer. Chem. Soc.* **72,** 4839 (1950).
26. M. Gates and G. Tschudi, *J. Amer. Chem. Soc.* **74,** 1109 (1952).
27. M. Gates and G. Tschudi, *J. Amer. Chem. Soc.* **78,** 1380 (1956).
28. C. Schöpf and T. Pfeiffer, *Ann.* **483,** 157 (1930).
29. D. Elad and D. Ginsburg, *J. Amer. Chem. Soc.* **76,** 312 (1954).
30. D. Elad and D. Ginsburg, *J. Chem. Soc.* 3052 (1954).
31. H. Rapoport, H. N. Reist, and C. H. Lowell, *J. Amer. Chem. Soc.* **78,** 5128 (1956).
32. L. F. Fieser, "Natural Products Related to Phenanthrene," 3rd ed., p. 23, Reinhold, New York, N.Y., 1949.
33. G. Stork, Alkaloids **2,** 171 (1952).
34. K. W. Bentley and H. M. E. Cardwell, *J. Chem. Soc.* 3252 (1955).
35. M. Gates and M. S. Shepard, *J. Amer. Chem. Soc.* **84,** 4125 (1962).
36. H. Rapoport and G. F. Payne, *J. Org. Chem.* **15,** 1093 (1950).
37. H. Rapoport, *J. Org. Chem.* **13,** 714 (1948).
38. H. Rapoport and G. F. Payne, *J. Amer. Chem. Soc.* **75,** 2630 (1952).
39. H. Rapoport and J. B. Lavigne, *J. Amer. Chem. Soc.* **75,** 5329 (1953).
40. D. Mackay and D. C. Hodgkin, *J. Chem. Soc.* 3261 (1955).
41. J. Kalvoda, P. Buchschacher, and O. Jeger, *Helv. Chim. Acta* **38,** 1847 (1955).
42. G. Stork and F. H. Clarke, *J. Amer. Chem. Soc.* **78,** 4619 (1956).
43. K. W. Bentley, "The Chemistry of the Morphine Alkaloids," Clarendon Press, Oxford, England, 1954.
44. H. Schmid and P. Karrer, *Helv. Chim. Acta* **33,** 863 (1950).
45. G. Stork, *J. Amer. Chem. Soc.* **74,** 768 (1952).
46. K. W. Bentley, R. Robinson, and A. E. Wain, *J. Chem. Soc.* 958 (1952).
47. L. F. Small, *J. Org. Chem.* **20,** 953 (1955).
48. M. Freund, E. Speyer, and E. Guttmann, *Ber.* **53,** 2250 (1920).
49. L. F. Small and G. Browning, *J. Org. Chem.* **3,** 618 (1939).
50. A. Skita, F. Nord, J. Reichert, and P. Stukart, *Ber.* **54,** 1560 (1921).
51. K. W. Bentley and H. M. E. Cardwell, *J. Chem. Soc.* 3245 (1955).

52. W. C. Howard, *Ber.* **17,** 527 (1884).
53. R. S. Cahn, *J. Chem. Soc.* 1038 (1933).
54. A. Matthiessen and C. R. A. Wright, *Proc. Roy. Soc. (London)* **17,** 455 (1869).
55. C. Schöpf and F. Borkowsky, *Ann.* **458,** 148 (1927).
56. K. W. Bentley, S. F. Dyke, and A. R. Marshall, *Tetrahedron* **21,** 2553 (1965).
57. O. Hesse, *Ann.* **153,** 47 (1870).
58. J. M. Gulland and C. J. Virden, *J. Chem. Soc.* 921 (1928).
59. M. Freund, *Ber.* **27,** 2961 (1894).
60. L. F. Small, L. J. Sargent, and J. A. Bralley, *J. Org. Chem.* **12,** 847 (1947).
61. R. Robinson, *Nature* **160,** 815 (1947).
62. K. W. Bentley and R. Robinson, *J. Chem. Soc.* 947 (1952).
63. L. W. Pickett, G. F. Walter, and H. France, *J. Amer. Chem. Soc.* **58,** 2296 (1936).
64. M. T. O'Shaughnessy and W. H. Rodebush, *J. Amer. Chem. Soc.* **62,** 2906 (1940).
65. K. W. Bentley, *J. Amer. Chem. Soc.* **89,** 2464 (1967).
66. L. F. Small, H. M. Fitch, and W. E. Smith, *J. Amer. Chem. Soc.* **58,** 1457 (1936).
67. L. F. Small, S. G. Turnbull, and H. M. Fitch, *J. Org. Chem.* **3,** 204 (1938).
68. L. F. Small and K. C. Yuen, *J. Amer. Chem. Soc.* **58,** 192 (1936).
69. L. F. Small and D. E. Morris, *J. Amer. Chem. Soc.* **56,** 2159 (1934).
70. T. D. Perrine and L. F. Small, *J. Org. Chem.* **17,** 1540 (1952).
71. E. Ochiai, *J. Pharm. Soc. Japan* **49,** 91 (1929).
72. C. Schöpf, K. von Gottberg, and W. Petri, *Ann.* **536,** 216 (1938).
73. K. W. Bentley and J. Dominguez, *J. Org. Chem.* **21,** 1348 (1956).
74. K. W. Bentley, J. Dominguez, and J. P. Ringe, *J. Org. Chem.* **22,** 418 (1957).
75. Ref. 74, p. 409.
76. K. W. Bentley and S. F. Dyke, *J. Chem. Soc.* 4055.
77. Ref. 74, p. 422.
78. K. W. Bentley, J. C. Ball, and J. P. Ringe, *J. Chem. Soc.* 1963 (1956).
79. Ref. 74, p. 425.
80. K. W. Bentley and J. C. Ball, *J. Org. Chem.* **23,** 1720 (1958).
81. Ref. 80, p. 1725.
82. K. W. Bentley and D. G. Hardy, *J. Amer. Chem. Soc.* **89,** 3267 (1967).
83. Ref. 82, p. 3273.
84. Ref. 82, p. 3281.
85. Ref. 82, p. 3293.
86. Ref. 82, p. 3303.
87. Ref. 82, p. 3312.
88. H. Rapoport and P. Sheldrick, *J. Amer. Chem. Soc.* **85,** 1636 (1963).
89. K. W. Bentley and D. G. Hardy, *Proc. Chem. Soc.* 220 (1963).
90. K. W. Bentley, *Endeavour* **23,** 97 (1964).
91. K. Goto and H. Sudzuki, *Bull. Chem. Soc. Japan* **4,** 163 (1929).
92. Ref. 91, p. 244.
93. H. Kondo and E. Ochiai, *Ann.* **470,** 224 (1929).
94. K. Goto and S. Mitsui, *Bull. Chem. Soc. Japan* **6,** 33 (1931).
95. K. Goto and I. Yamamoto, *Proc. Japan Acad.* **29,** 210 (1953).
96. K. Goto and I. Yamamoto, *Proc. Japan Acad.* **30,** 769 (1954).
97. K. Goto and T. Nambo, *Bull. Chem. Soc. Japan* **5,** 165 (1930).
98. K. Goto, K. Takubo, and S. Mitsui, *Ann.* **494,** 1 (1932).
99. K. Goto, H. Shishido, and K. Takubo, *Ann.* **495,** 122 (1932).
100. K. Goto and H. Shishido, *Ann.* **507,** 296 (1933).
101. H. Kondo, *J. Pharm. Soc. Japan* **58,** 46 (1938).

102. H. Kondo, *Ann. Rept. ITSUU Lab.* **1,** 50 (1950).
103. H. Kondo, M. Satomi, and T. Odera, *Ann. Rept. ITSUU Lab.* **2,** 35 (1951).
104. H. Kondo, M. Satomi, and T. Odera, *Ann. Rept. ITSUU Lab.* **4,** 45 (1953).
105. M. Satomi, *J. Pharm. Soc. Japan* **72,** 834 (1952).
106. M. Tomita, T. Ibuka, Y. Inubushi, Y. Watanabe, and M. Matsui, *Tetrahedron Letters* 2937 (1964).
107. S. Okuda, K. Tsuda, and S. Yamaguchi, *J. Org. Chem.* **27,** 4121 (1962).
108. K. Takeda, *Ann. Rept. ITSUU Lab.* **11,** 61 (1961).
109. M. Tomita, T. Ibuka, Y. Inubushi, and K. Takeda, *Tetrahedron Letters* 3605 (1964).
110. Ref. 109, p. 3617.
111. D. H. R. Barton, G. W. Kirby, W. Steglich, G. M. Thomas, A. R. Battersby, T. A. Dobson, and H. Ramuz, *J. Chem. Soc.* 2423 (1965).
112. L. J. Haynes, G. E. M. Husbands, and K. L. Stuart, *Chem. Comm.* 15 (1967).
113. C. Chambers and K. L. Stuart, *Chem. Comm.* 328 (1968).
114. K. L. Stuart, V. Teetz, and B. Franck, *Chem. Comm.* 333 (1969).
115. T. Kametani, K. Fukomoto, F. Satoh, and H. Yagi, *J. Chem. Soc.* 520 (1969).
116. W. Dopke, H. Flentje and P. W. Jeffs, *Tetrahedron* **24,** 4459 (1968).
117. J.-H. Chu, S.-Y. Lo, and Y. L. Chou, *Acta. Chim. Sinica.* **30,** 265 (1964).
118. M. Tomita and M. Kozuka, *Tetrahedron Letters* 6229 (1966).
119. A. R. Battersby, M. H. G. Munro, R. B. Bradbury, and F. Santavy, *Chem. Comm.* 695 (1968).
120. A. R. Battersby, R. B. Herbert, L. Pijewska, and F. Santavy, *Chem. Comm.* 228 (1965).
121. A. R. Battersby, R. B. Herbert, E. McDonald, R. Ramage, and J. H. Clements, *Chem. Comm.* 603 (1966).

6

Amaryllidaceae Alkaloids

WILLIAM C. WILDMAN
Department of Chemistry
Iowa State University of Science and Technology
Ames, Iowa

1. INTRODUCTION

The Amaryllidaceae alkaloids comprise a unique group of bases that have been found to occur only in the title family up to the present time. Conversely, with the exception of hordenine, alkaloids of other plant families have not been found in the Amaryllidaceae. More than one hundred alkaloids have been isolated and most of these have been reported in the past ten years. With the exception of ismine, all alkaloids can be considered to be derived from a nucleus of fifteen carbon atoms comprised of two fragments; namely, an aromatic C(6)–C(1) unit and a C(6)–C(2) hydroaromatic (or completely reduced) moiety. The large number of alkaloids found in the family may be attributed to minor variations in the degree of oxygenation or aromatic substitution within these units, to the frequent occurrence of optical antipodes within a given ring system, and to variations in the degree of hydrogenation in the hydroaromatic ring. The alkaloids can be divided conveniently into eight different ring systems that serve as the major divisions or subdivisions of this chapter. The alkaloids selected for review are either the major alkaloids of a given type or are of particular current interest. Information concerning other alkaloids containing the same ring system can be found in detailed reviews.[1,2]

Alkaloids are found in almost all genera of the subfamily Amaryllidoideae, but the high water content of the bulbs limits the total alkaloid content to a maximum of 1–2%. Isolation of alkaloids that are either major in quantity or are highly insoluble is not difficult by combined crystallization and column chromatographic techniques. But after these advantages have been utilized, there usually remains a complex mixture of minor alkaloids that are exceedingly difficult to separate. This has been illustrated dramatically in the isolation by conventional methods of four alkaloids from *Pancratium maritimum* in 1955–1956. Reinvestigation of the alkaloids of this plant by two-dimensional thin-layer chromatography showed the presence of at least thirty-seven alkaloids.[3] As another example, fifteen crystalline alkaloids were isolated from *Nerine bowdenii*, yet more than half the crude alkaloid fractions remained as a noncrystalline mixture.[4]

The Amaryllidaceae alkaloids have generated little pharmacological interest. Galanthamine is a potent anticholineesterase and analgesic, but other alkaloids or simple synthetic preparations of greater activity and reduced toxicity are available. Recent interest in the alkaloids of the family has been in the area of biosynthesis, total synthesis, and the determination of absolute configuration by methods independent of Mills' or Hudson's rules.

2. STRUCTURES OF INDIVIDUAL ALKALOIDS

2.1 Lycorine

Lycorine (1), the most common alkaloid of the family, is one of the simplest to isolate because it is almost completely insoluble in chloroform. More than twenty additional alkaloids have been assigned structures related to (1). These vary in the substitution pattern in rings A, C, and D. Interconversions of the alkaloids often may be achieved through simple hydrolytic, oxidative, and reductive methods, or by hydrogenolysis with sodium and amyl alcohol.

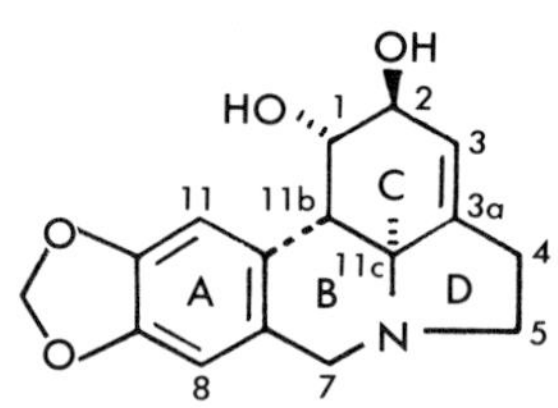

(1)

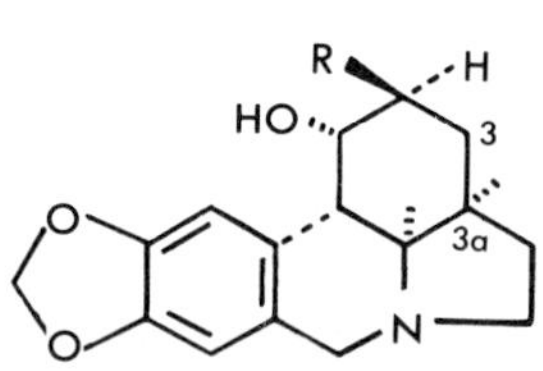

(2)

Lycorine forms a single dihydro derivative [(2) R=OH] upon catalytic hydrogenation. When treated with methyl iodide, lycorine affords two methiodides because the nitrogen atom becomes a new asymmetric center. Both quaternary methohydroxides afford lycorine anhydromethine [(3) R=CH═CH_2] under Hofmann conditions. The structure of the methine was verified by the synthesis of the dihydro derivative [(3) R=C_2H_5].[5] Hofmann degradation of lycorine α- and β-methohy-

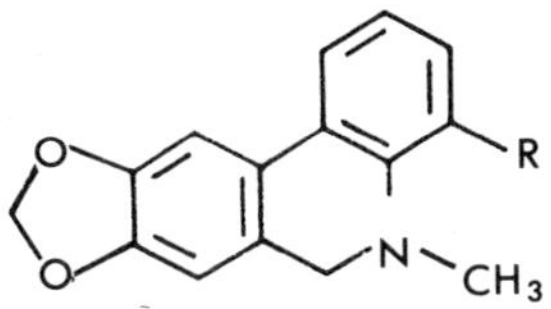

(3)

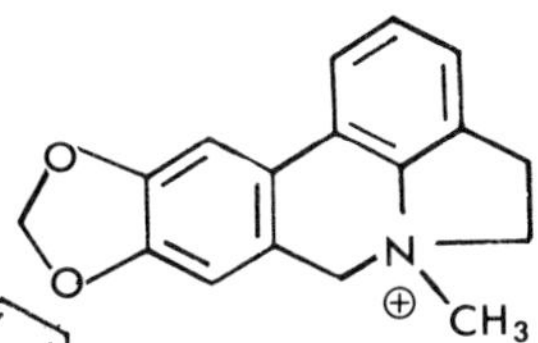

(4)

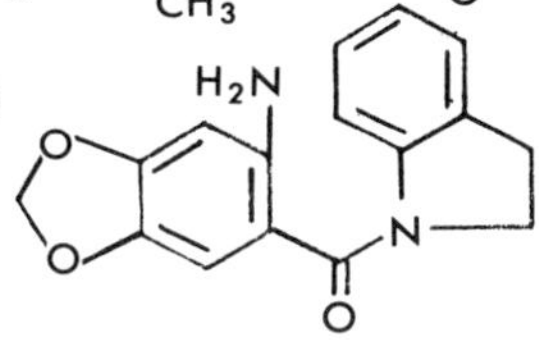

(5)

droxides in the presence of alkali gives some lycorine anhydromethine, but the major product is the anhydrolycorine salt (4). The structure of (4) has been verified by synthesis from (5) by Pschorr cyclization, lithium aluminum hydride reduction, and N-methylation.[6]

The vicinal nature of the two hydroxyl groups of lycorine was first demonstrated by the cleavage of dihydrolycorinone (6) with lead tetra-

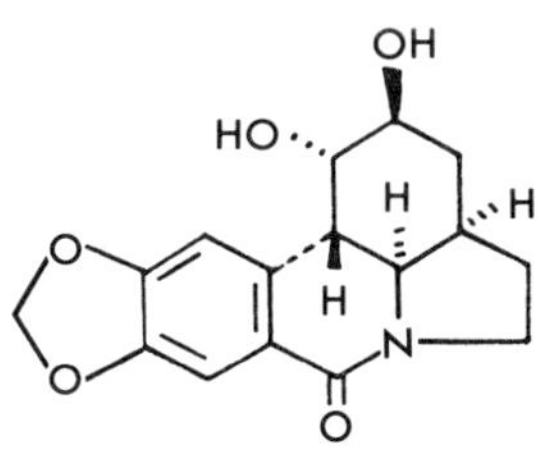

(6)

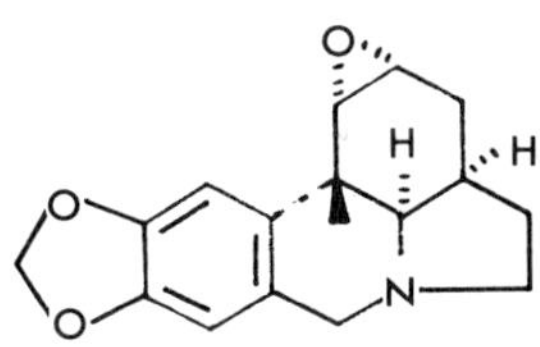

(7)

acetate. Lycorine does not form an acetonide, and the hydroxyl groups were considered to be *trans*. This assignment was substantiated by the facile conversion of 2-O-tosyldihydrolycorine to an epoxide (7) that afforded dihydrolycorine upon treatment with aqueous mineral acid and 2-O-acetyldihydrolycorine with acetic acid. Reduction of either the tosylate or the epoxide with lithium aluminum hydride produced α-dihydrocaranine [(2) R=H], thus interrelating the mono- and dihydroxy alkaloids of the group.[7] Hydroxyl substitution at C(1) can be demonstrated either by phosphorus oxychloride dehydrations of the dihydro derivatives of lycorine, caranine [(2) R = H, double bond C(3)-C(3a)], pluviine (the dimethoxy analog of caranine) to complex styrenes (8)

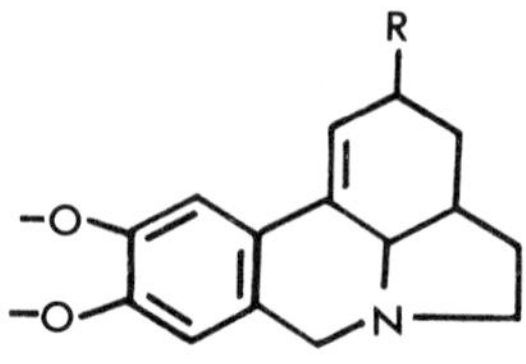
(8)

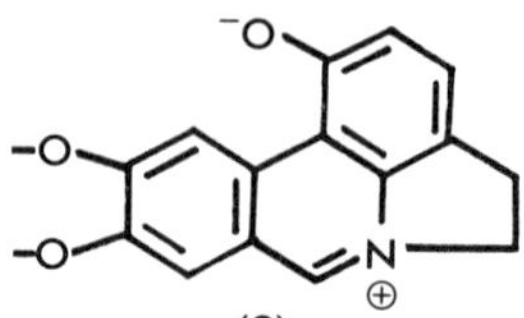
(9)

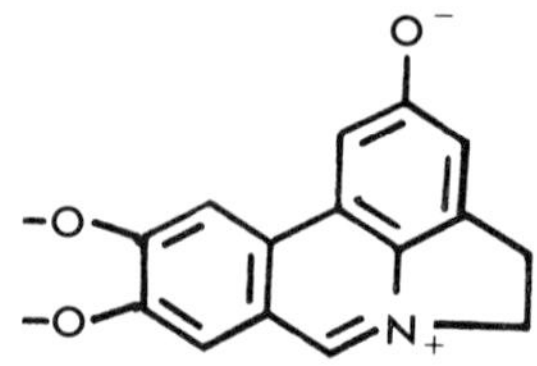
(10)

or by the conversion of the alkaloids with mild oxidizing agents to 1- and/or 2-oxyphenanthridinium betaines (9) and (10).[8] 1-O-Acetyllycorine is oxidized to 1-acetyl-2-oxolycorine by manganese dioxide, and the olefinic unsaturation of lycorine must be between C(3) and C(3a).[9] The nmr spectrum of diacetyllycorine shows one olefinic proton (δ 5.76) in agreement with the findings by oxidation.[10]

The relative stereochemistry of lycorine was assigned in the following manner. The facile dehydration of lycorine derivatives to Δ^{1-11b}-olefins suggests that the C(11b) hydrogen is *trans* to the C(1)-hydroxyl. Since a *trans* relationship between the two hydroxyl groups has been established, there remains only the configuration at C(11c) to be determined. Indirect evidence for the configuration at C(11c) shown in (1) was first obtained from dihydrolycorine [(2) R=OH]. The α-configuration of the hydrogen at C(3a) is assured because several cyclic ethers [(11) R=CN, H, or C(=NH) (OC_2H_5)] were formed when the cyanogen bromide reaction product (12) of O,O-diacetyldi-

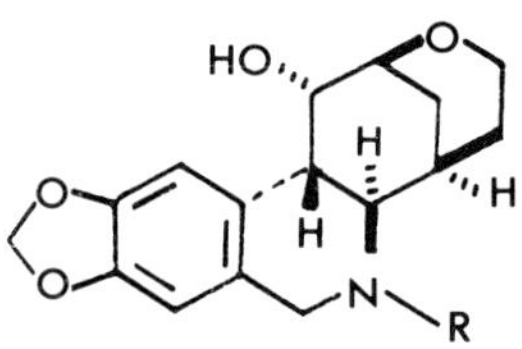
(11)

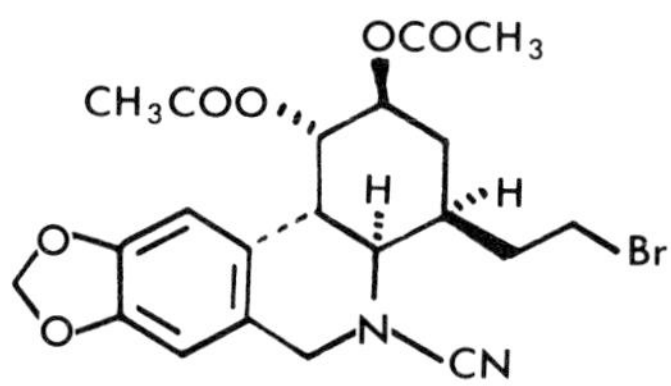
(12)

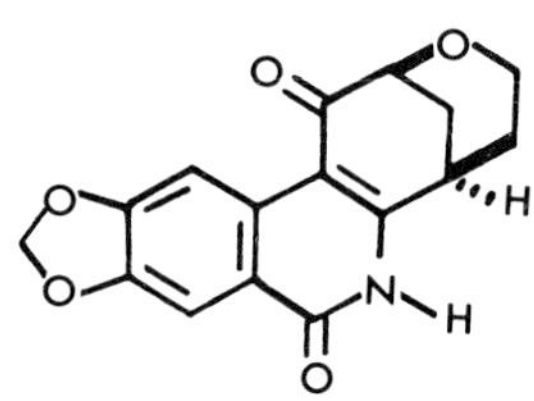
(13)

hydrolycorine was treated with ethanolic base. Ether formation of the sidechain with the C(2) rather than C(1) hydroxyl is proven through the oxidation of [(11) R = CN] to the vinylogous imide (13). Cyclic ethers such as (11) and (13) must be joined to ring C in a *cis*-1,3-diaxial manner. If the two-carbon chain at C(3a) is axial, the stereochemistry at C(11c) must be shown in (1) and (11) because a *trans*-diaxial fusion of ring D to ring C is impossible.[11] Recent syntheses of the lycoranes as well as nmr data are in accord with this assignment.[12–14]

The absolute configurations expressed in the formulas of this section were first derived from the application of Mills' rule to lycorine and 2-*epi*-lycorine and from interconversions with the lactonic alkaloids (Section 2.2), where absolute configurations may be assigned by Klyne's extension of the Hudson lactone rule. X-ray crystallographic methods have shown that α-dihydrolycorine has the absolute configuration shown in [(2) R = OH].[15] Detailed CD, ORD, and NMR studies supply supporting information.[10]

2.2 Lycorenine and Homolycorine

Although a relatively large number of alkaloids have been classified in this group, chemical degradations have been limited mainly to the title compounds since most of the alkaloids of this type are rather rare. Many of the alkaloids classified as derivatives of [2]benzopyrano[3,4*g*]indole have been assigned structures on the basis of functional group analysis and, in particular, spectral evidence for the presence of a conjugated aromatic lactone function, either in the alkaloid itself or in the oxidation products of the hemiacetal types. The recent isolation of macronine and pretazettine (Section 2.5.2) places some structures assigned only by spectral methods on a less secure basis.

Alkaloids related to [2]benzopyrano[3,4*g*]indole [*e.g.*, lycorenine, (14a)] may be oxidized by chromium trioxide or manganese dioxide to

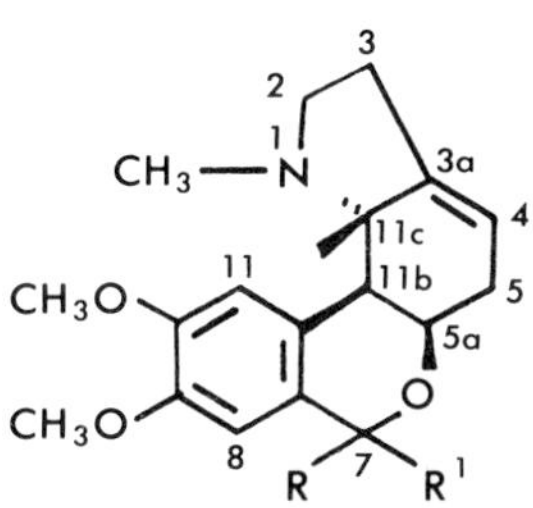

(14)

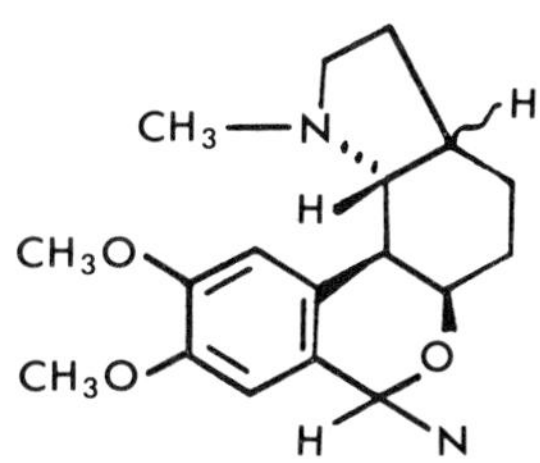

(15)

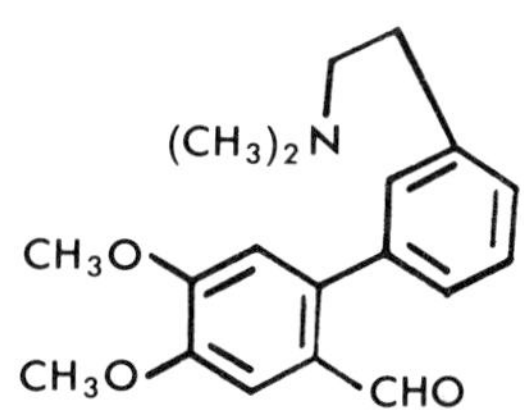

(16)

a) R = H, R′ = OH
b) R, R^1 = O

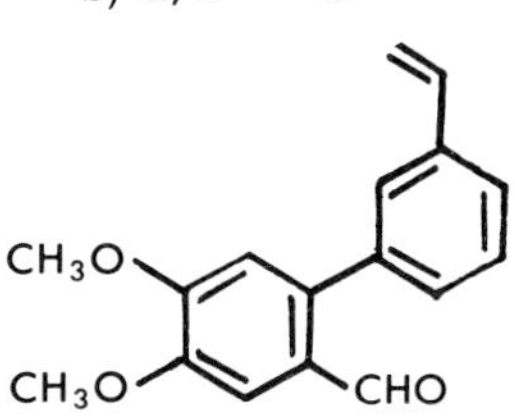

(17)

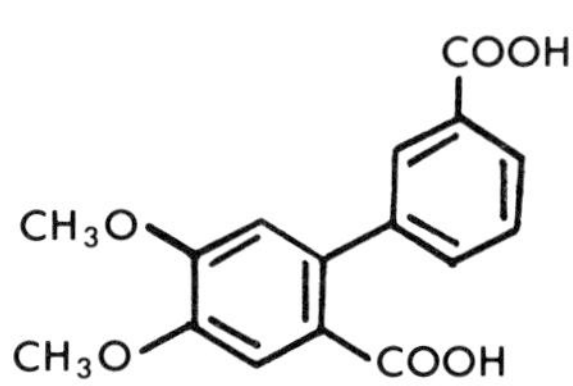

(18)

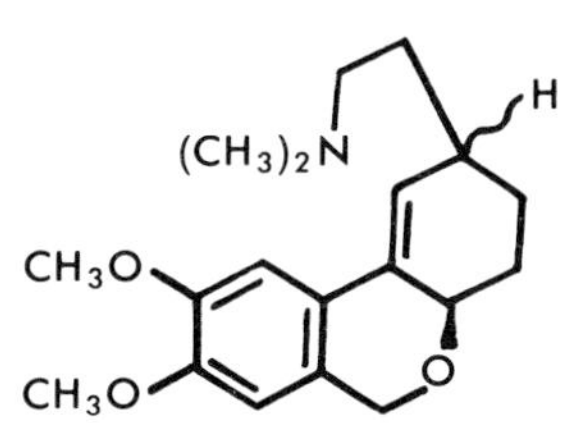

(19)

the corresponding lactones [*e.g.*, homolycorine (14b)]. Reductive methods for the reverse process are exceptionally difficult. Chromatography of the hemiacetal alkaloids on alumina often is accompanied by disproportionation to the lactone and corresponding benzyl alcohol.

Spectral and chemical functional group analysis of lycorenine (14a) shows the presence of one N-methyl, one hydroxyl, and two aromatic methoxy groups. A masked carbonyl group was suspected since lycorenine formed an oxime hydrochloride. Early hydrogenation studies designed to form dihydro derivatives were complicated both by the concommitant hydrogenolysis of the benzylic hydroxyl group and the formation of *cis* and *trans*-octahydroindoles (15).[16]

The position of attachment of $—CH_2CH_2—N—CH_3$ to the hydroaromatic ring was first demonstrated by exhaustive methylation methods. Oxidation of des-N-lycorenine [(17) formed by successive Hofmann degradations of lycorenine via 16] afforded 3,4-dimethoxybiphenyl-6,3'-dicarboxylic acid (18). Hofmann degradation of 15 ("α-isomer") gave a methine (19) in which the double bond was conjugated with the aromatic ring.

The most elegant method of structure proof involves the interconversion of lycorenine and homolycorine to alkaloids of the lycorine series. Both (14a) and (14b) are converted to tetrahydrohomolycorine (20) upon reduction with lithium aluminum hydride. Tosylation of (20) occurs at the benzylic alcohol group. Intramolecular nucleophilic attack by the basic amino function affords [(21) X = OTs] which was converted to pluviine β-methiodide [(21) X = I] by iodide ion. Pyrolysis of [(21) X = Cl] affords the parent alkaloid, pluviine.[14]

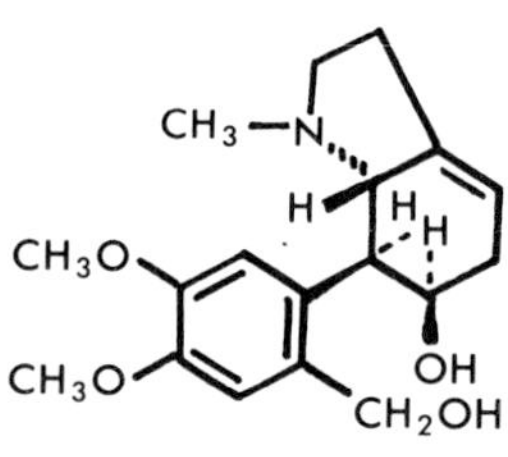

(20)

≡

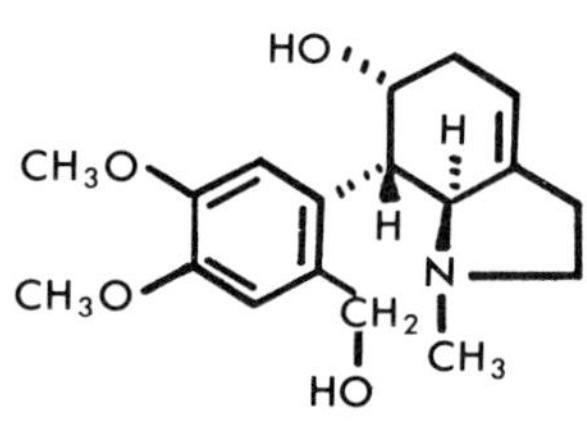

(20)

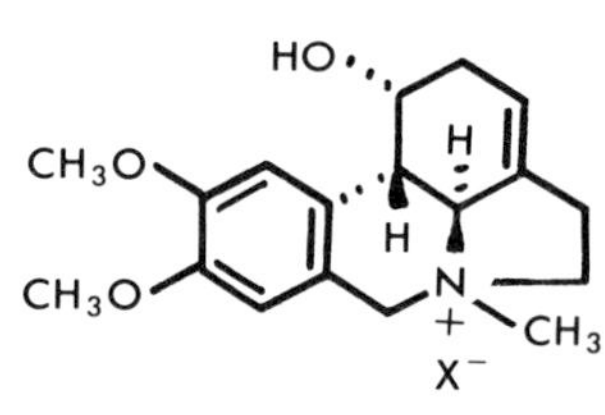

(21)

2.3 Galanthamine

This alkaloid and its dihydro-derivative (lycoramine) are common constituents of many species in the *Galanthus, Leucojum, Lycoris, Narcissus,* and *Vallota* genera. Lycoramine was studied intensively by Kondo and his co-workers in the 1930's, but correct structures for alkaloids of this group were not forthcoming until about a decade ago.

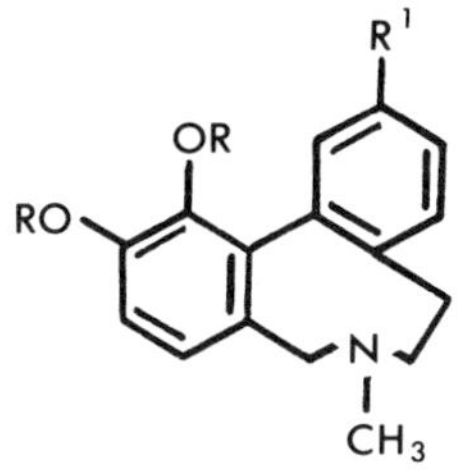

(22)

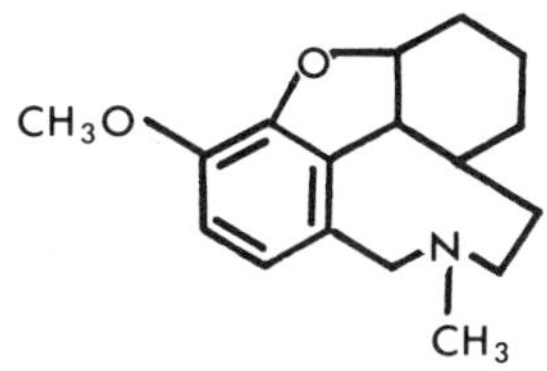

(23)

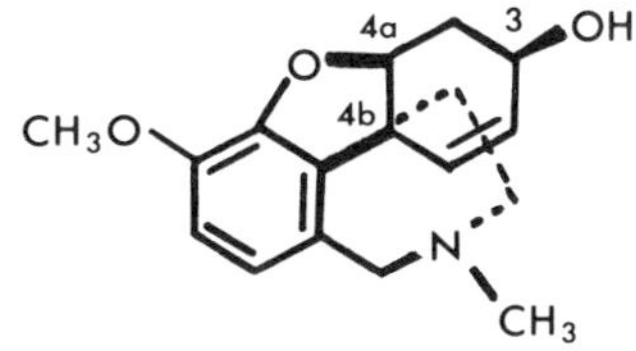

(24)

Functional group analysis of galanthamine shows the presence of one aromatic methoxyl, one hydroxyl, and one N-methyl group. The catalytic reduction of galanthamine to lycoramine provided evidence for the one isolated double bond and also established the relationship between the two alkaloids.[17] The third oxygen was considered to be associated with an ether function. Concentrated hydrobromic acid converted galanthamine to apogalanthamine [(22) R,R′ = H]. O-Methylation of [(22) R,R′ = H] provided an O,O-dimethyl derivative [(22) R = CH_3, R′ = H] whose structure was proven by synthesis.

The allylic nature of the hydroxyl group in galanthamine was demonstrated by the manganese dioxide oxidation of the alkaloid to the analogous ketone, galanthaminone.[18] Although the formation of apogalanthamine and galanthaminone could be rationalized by an allylic alcohol derivative of 23, the alternate structure (24), which was proposed originally for biogenetic reasons,[19] was found to be the correct structure for the alkaloid. Galanthaminone proved to be the key compound to show that (24) is the correct structure for the alkaloid. Galanthaminone [(24)═O instead of OH], when isolated from the oxidation reaction, is levorotatory. Chromatography of this ketone on alumina or even heating with warm ethanol is a sufficiently vigorous condition to afford racemic galanthaminone.[20] These mild conversions of optically active galanthamine to *racemic* galanthaminone were difficult to explain by any structure related to (23). However, the C(3)-ketone derived from (24) could easily undergo β-elimination of the ether bridge to form an optically inactive dienone. A racemic galanthaminone would result upon reformation of the ether linkage.[21] Conversion of (24) to [(22) R,R′ = H] would occur by dienone-phenol rearrangement through the action of strong acid. By a similar mechanism, galanthaminone affords the observed hydroxyapogalanthamine [(22) R = H, R¹ = OH]. In keeping with this formulation, galanthaminone forms a monobenzylidene derivative and lycoraminone [(24) no double bond, ═O at C(3)] affords a dibenzylidene derivative.[22]

The relative configuration depicted in (24) is derived from the observation that (−)-galanthaminone is racemized, but not epimerized at C(4a) on treatment with alkali. This would indicate the more stable *cis* ring junction at C(4a)-C(4b). Galanthamine shows a strong intramolecular hydrogen bond between the hydroxyl group at C(3) and the oxide bridge; thus, the C(3)-hydroxyl must be *cis* to it.[20] The absolute configuration of galanthamine was assigned originally by the application of Mills' rule for allylic isomers. Confirmation of (24) has been obtained by x-ray spectroscopy.[23]

2.4 5,10b-Ethanophenanthridine Alkaloids

Alkaloids derived from this ring system are very common in the Amaryllidaceae and the total number in this group isolated to date exceeds even that found for lycorine and its analogs. The large number may be attributed to the same factors that gave rise to the numerous lycorine analogs, i.e., variations in position, type, and configuration of oxygenated substituents and the presence of alkaloids with enantiomeric basic nuclei. Six bases have been selected for examination in this section both because they are representative of the entire group and because they are biosynthetic precursors of the alkaloids discussed in Sections 2.5 and 2.6.

2.4.1 Crinine and Buphanisine Crinine and buphanisine are weakly levorotatory alkaloids occurring frequently in *Crinum, Buphane,* and *Nerine* species. The structure of crinine (25a) was established in 1959 by relatively classical degradative procedures. Manganese dioxide oxidation of crinine gave an α,β-unsaturated ketone, oxocrinine[(25) R,R′ = O]. Reduction of the double bond and carbonyl group by catalytic hydrogenation and Wolff-Kishner reduction provided (−)-crinane [(25) R,R′ = H; no double bond 1-2], the structure of which was proven by synthesis.[24] Since di-

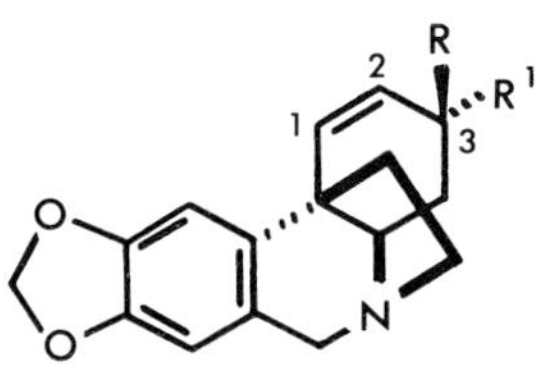

(25)

a) R = H, R′ = OH
b) R = H, R′ = OCH_3

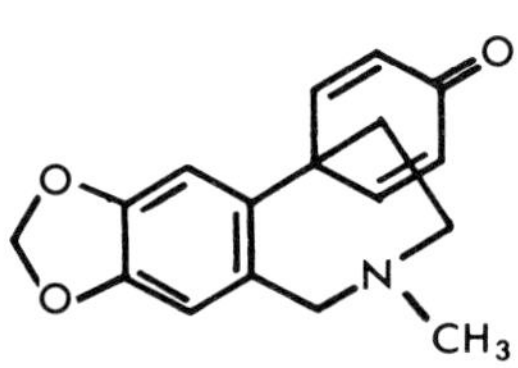

(26)

hydrooxocrinine [(25) R,R′ = O; no 1,2-unsaturation] forms a dibenzylidene derivative, only 3-hydroxy-$\Delta^{1,2}$- and 2-hydroxyl-$\Delta^{3,4}$-isomers are possible for the hydroaromatic ring substitution. The former alternative was proven correct by N-methylation of oxocrinine and treatment with base to provide an optically inactive dienone (26), the tetrahydro derivative of which has been synthesized.[25] The *cis*-methylenedioxyphenyl-hydroxyl relationship in crinine was proven by hydrogen-bonding studies and interconversions via buphanisine (25b) with the alkaloids of Sections 2.4.2 and 2.5.1. Buphanisine is hydrolyzed by acid in poor yield to (−)-crinine. The reverse conversion can be achieved by the O-methylation of crinine with sodium and methyl *p*-toluenesulfonate. Catalytic reduction of buphanisine affords a dihydro derivative, the optical antipode of which has been related to dihydrohaemanthamine (Section 2.4.2).

2.4.2 Haemanthamine and Crinamine A number of alkaloids related to buphanisine contain a hydroxyl group at C(11) and may possess an additional hydroxyl substituent at C(6) as well. The C(6),C(11)-dihydroxy alkaloids can be related to tazettine, criwelline, and macronine by simple chemical transformations. The alkaloids of primary importance are listed below as derivatives of (27).

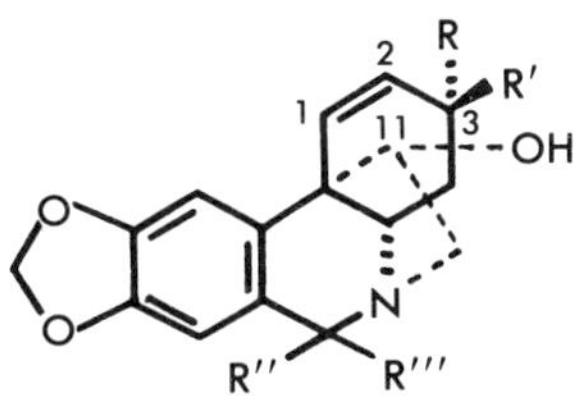

(27)

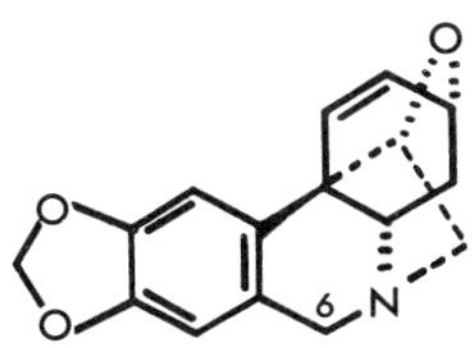

(28)

a) R, R″, R‴ = H; R′ = OCH_3 Haemanthamine
b) R′, R″, R‴ = H; R = OCH_3 Crinamine
c) R, R″ = H; R′ = OCH_3; R‴ = OH Haemanthidine
d) R′, R″ = H; R = OCH_3; R‴ = OH 6-Hydroxycrinamine

Haemanthamine (27a) forms a dihydro derivative upon catalytic reduction. Treatment of dihydrohaemanthamine with thionyl chloride followed by lithium aluminum hydride affords the optical antipode of dihydrobuphanisine [(25b) no double bond]. The C(11) location of the hydroxyl group in haemanthamine is evident from the carbonyl absorption at 5.72 μ of the ketone formed by the pyridine-chromic oxide oxidation of the alkaloid and from the product of the Hofmann degradation of the ketone, N-methyl-N-(6-phenylpiperonyl)-glycine. The configuration of the C(11)-hydroxyl group in haemanthamine is established by the presence of a hydrogen bond from the C(11)-hydroxyl group to the C(1)–C(2) double bond (3598 cm^{-1}). Dihydrohaemanthamine possesses an unbonded hydroxyl group (3625 cm^{-1}). Haemanthamine and crinamine (27b) can be interrelated through apohaemanthamine (28), the ether formed by treating either alkaloid with mineral acid. This product also establishes the *cis*-fusion of the five-membered heterocyclic ring to the hydroaromatic ring, because 3,11-oxide formation is impossible with a *trans*-fusion of these rings. The ketone obtained from the oxidation of crinamine also shows absorption at 5.72 μ and is not identical with the corresponding oxidation product of haemanthamine. The evidence cited above does not exclude the possibility that crinamine is a 1-methoxy-$\Delta^{2,3}$ allylic isomer of haemanthamine which undergoes rearrangement in acid to (28). However, the nearly identical hydrogen-bonded hydroxyl stretching frequencies of haemanthamine and crinamine and nmr evidence support structure (27b) for crinamine.[26]

2.4.3. Haemanthidine and 6-Hydroxycrinamine These alkaloids are unique among the 5,10b-ethanophenanthridine group because of the C(6)-hydroxyl substitution. The additional hydroxyl group in each can be located immediately by the facile manganese dioxide oxidation to the corresponding bridgehead lactams which have characteristic ultraviolet absorption spectra. Because Bredt's rule is valid in a 1-azabicyclo[3,2,1] octane derivative, the oxo compounds behave both spectrally and chemically as ketones rather than lactams. Either sodium borohydride or lithium aluminum hydride reduction regenerates the original alkaloid. No haemanthamine or crinamine is formed with the latter reagent.[27]

Like their 6-desoxy analogs, each reacts with mineral acid to form the same demethoxy ether, apohaemanthidine [(28) OH at C(6)]. Dihydroapohaemanthidine, upon successive treatment with thionyl chloride and lithium aluminum hydride, is converted to apohaemanthamine (28). This transformation establishes the 5,10b-ethanophenanthridine nucleus for the bases. The configurations of the C(11)-hydroxyl group in haemanthidine and 6-hydroxycrinamine were determined by the infrared hydroxyl stretching frequencies as in the case of the 6-desoxy alkaloids. The configuration of the C(3)-methoxyl group in haemanthidine and 6-hydroxycrinamine can be ascertained by several methods. Haemanthamine and haemanthidine are more levorotatory than crinamine and 6-hydroxycrinamine. Application of Mills' rule for these allylic epimers leads to the assignments shown in (27 c,d). Haemanthamine and haemanthidine afford much better yields of the corresponding *apo* derivatives [(28) OH at C(6)] than do their C(3)-methoxy epimers. Finally, haemanthidine and 6-hydroxycrinamine can be converted in good yield to tazettine and criwelline, respectively (see Section 2.5).[28]

Although stereochemical assignments have been made for the C(6)-hydroxyl group of haemanthidine and 6-hydroxycrinamine, recent nmr evidence shows that, in solution, both haemanthidine and 6-hydroxycrinamine, as well as all their derivatives having a free C(6)-hydroxyl group, exist as an equilibrating mixture of C(6)-epimers.[29] It is interesting to note that recent x-ray spectroscopic evidence shows that crystalline 6-hydroxycrinamine exists as a single C(6)-epimer in which the C(6)-hydroxyl group is *trans* to the pyrrolidine ring.[30]

2.5 [2]Benzopyrano[3,4*c*]indole Alkaloids

Tazettine (29a) is the most venerable of the alkaloids containing this nucleus. Isolation and structure studies were initiated more than thirty years ago. Although it is considered one of the more common of Amaryllidaceae alkaloids, there is evidence that, in many cases, tazettine as well as criwelline (29d) are artifacts arising from isolation techniques. Few iso-

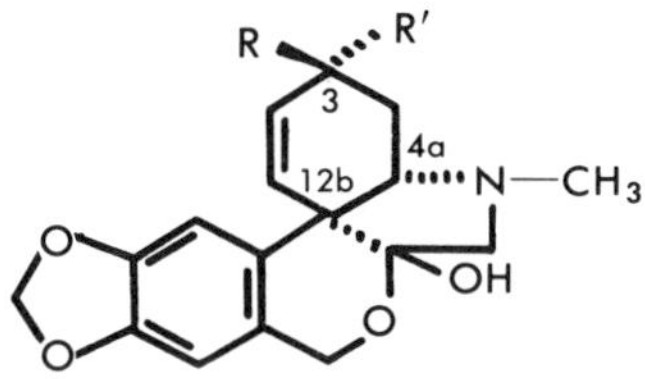

(29)

a) R = OCH_3; R′ = H
b) R = OH; R′ = H
c) R = H; R′ = OH
d) R = H; R′ = OCH_3

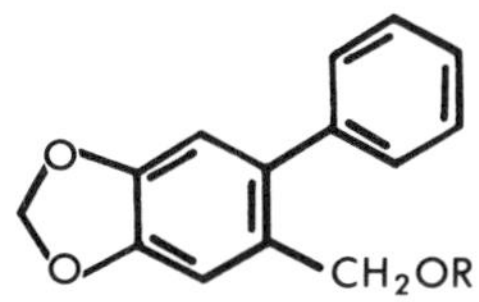

(30)

a) R = $-\underset{\underset{O}{\|}}{C}CH_2N(CH_3)_2$
b) R = H

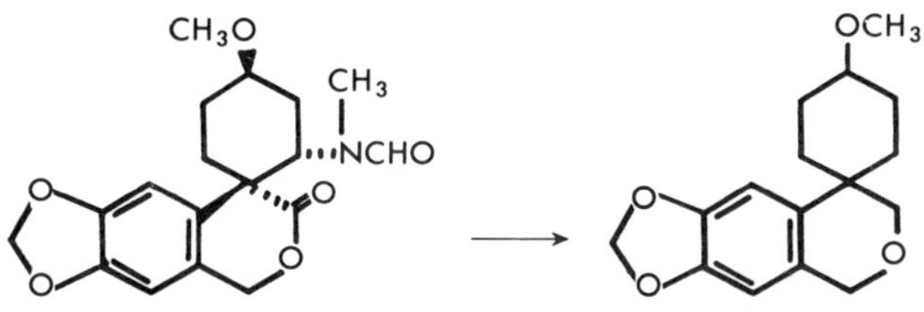

(31) (32)

lations of tazettine from plant material have avoided strong base, alumina chromatography, or prolonged standing of the crude alkaloid fraction, once isolated. A recent study of alkaloids from *Sprekelia formosissima*, a known source of tazettine, avoiding these conditions, showed no tazettine to be present in the crude alkaloid isolate. However, within a few hours, tazettine could be detected by thin-layer chromatographic techniques and upon prolonged standing, crystalline tazettine precipitates from the alkaloidal mixture.[31] The chemistry of one precursor, pretazettine, is discussed in Section 2.5.2.

2.5.1 Tazettine and Criwelline Functional group analysis of tazettine showed the presence of methylenedioxy, methoxy, and N-methyl groups. The olefinic double bond was demonstrated by catalytic hydrogenation. Earliest degradative studies provided 6-phenylpiperonyl alcohol (30b) as the major product from two successive Hofmann degradations. The structure of tazettine methine (30a) was not deduced until 1955, when several research groups became interested in the problem. The structure of (30a) was verified by synthesis.[32] While these degradation products provided meager information concerning the structure of tazettine, they gave a facile route for degradative biosynthetic studies to follow in later years.

The location of the aliphatic methoxy group was established by the isolation of 6-(4-methoxyphenyl)-piperonyl alcohol as one of the products of the Hofmann degradation of O-methyltazettine [(29a) OCH_3 instead of OH]. In a subsequent study it was shown that dihydrotazettamide (31), the manganese dioxide oxidation product of dihydrotazettine, could be converted by conventional steps to (32) which was optically inactive. Such an observation requires the methoxyl group to be in position 4 with respect to the aromatic ring.[33,34] The reduction of tazettine with lithium aluminum hydride to a diol (tazettadiol) provided the first clue to the hemiacetal function. Treatment of the diol with acid afforded a cyclic ether (33). Hofmann degradation of 33 proceeded with aromatization of the hydroaromatic ring to form (34) that was first considered the true desoxytazettine methine. The structure of (34) was proven by additional degradations and

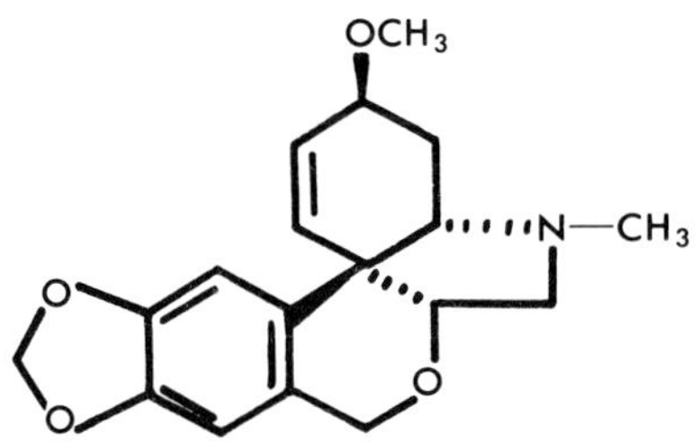

(33)

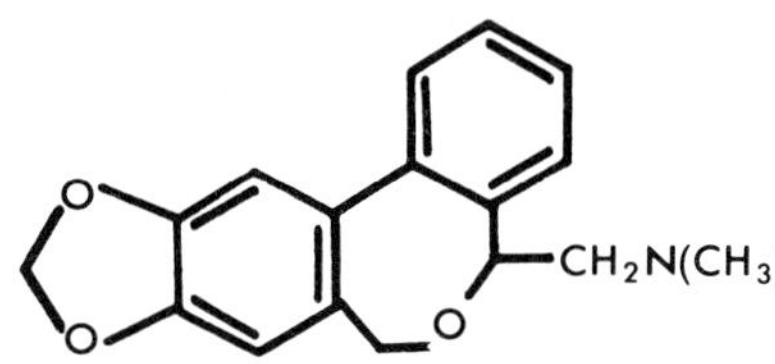

(34)

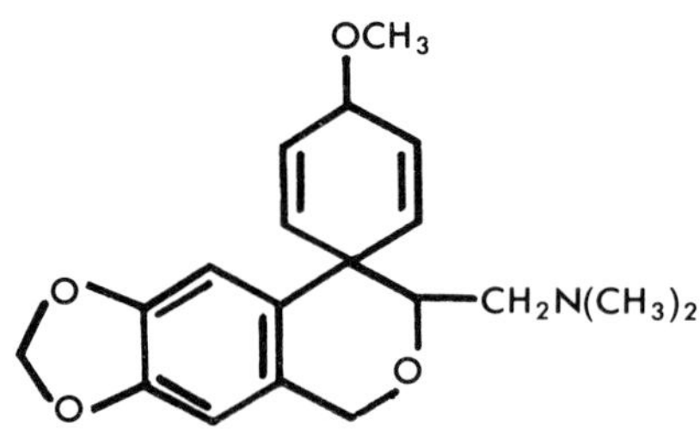

(35)

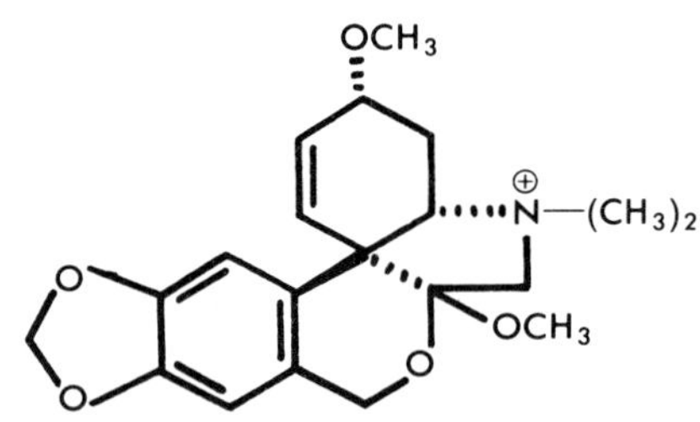

(36)

by the synthesis of degradation products. The true methine was shown to be (35), which rearranges in acid to (34).[33,35]

The methoxyl group of tazettine is cleaved by acid to provide a mixture of epimeric alcohols (29b and 29c). Both alcohols could be oxidized to the same C(3)-ketone and allylic rearrangement did not take place during the acid cleavage. O- and N-Methylation of (29b) provided the known O-methyltazettine methiodide. Under similar conditions, isotazettine (29c) gave O-methylisotazettine methiodide (36).[33]

The relative configuration of oxy-substituents at C(3) with respect to the methylamino group at C(4a) was assigned from basicity studies. Derivatives of the isotazettinol series are stronger bases than the normal series owing to the formation of a hydrogen bond between the C(3)-hydroxyl and the proton of the conjugate acid. Such bonding would be most effective in structures where the C(3)-hydroxyl and the C(4a)-methylamino groups are *pseudo* axial and axial, respectively. If this assignment is correct, the hemiketal group at C(12b) must be *cis* to the amino function at C(4a) since a *trans*-fusion is not possible. Such reasoning leads to the relative configurations depicted in (29a) for tazettine.[33]

The absolute configurations described in (29a) were assigned originally by the application of Mills' rule to various tazettine and tazettinol derivatives. Dihydrotazettine forms a methine (37) upon Hofmann degradation.

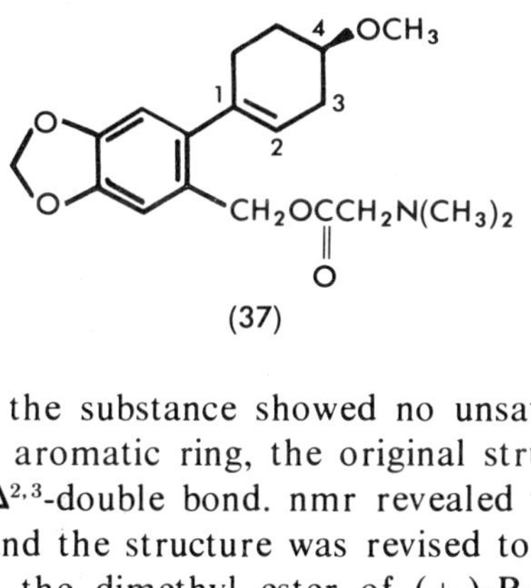

(37)

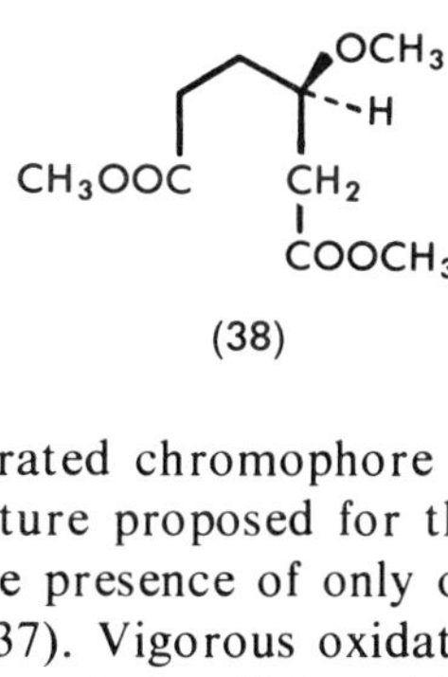

(38)

Because the substance showed no unsaturated chromophore conjugated with the aromatic ring, the original structure proposed for the methine had a $\Delta^{2,3}$-double bond. nmr revealed the presence of only one olefinic proton and the structure was revised to (37). Vigorous oxidation of (37) provided the dimethyl ester of (+)-R-β-methoxyadipic acid (38) after several steps.[36] This degradation provides chemical evidence for the absolute configuration of tazettine.

Criwelline was first reported as an alkaloid of *Crinum powellii* var. *album*. The empirical formula and functional groups were the same as those found for tazettine. The structure of the alkaloid was established by the finding that 6-hydroxycrinamine (27d) was converted to criwelline (29d) by N-methylation and treatment with base. O-Methylation of criwelline gave O-methylisotazettine methiodide (36).[28]

2.5.2 Pretazettine, Precriwelline, and Macronine Of these alkaloids, macronine was the first to be isolated. Macronine was shown to possess one N-methyl, one O-methyl, and one δ-lactone group, the carbonyl group of which is conjugated with a methylenedioxyphenyl system.[37] Although this combination of functional groups might place the alkaloid with those of Section 2.2, entirely dissociated research on the chemistry of 6-hydroxycrinamine provided a structure for the alkaloid.

6-Hydroxycrinamine is converted to criwelline by successive treatment of the alkaloid in acetone solution with methyl iodide followed by base. If the methylation is carried out in refluxing methanol, substantial quantities of the methyl ketal (39) are formed. Mild acid hydrolysis of (39), followed by treatment with base gave criwelline (29d), presumably *via* (40).

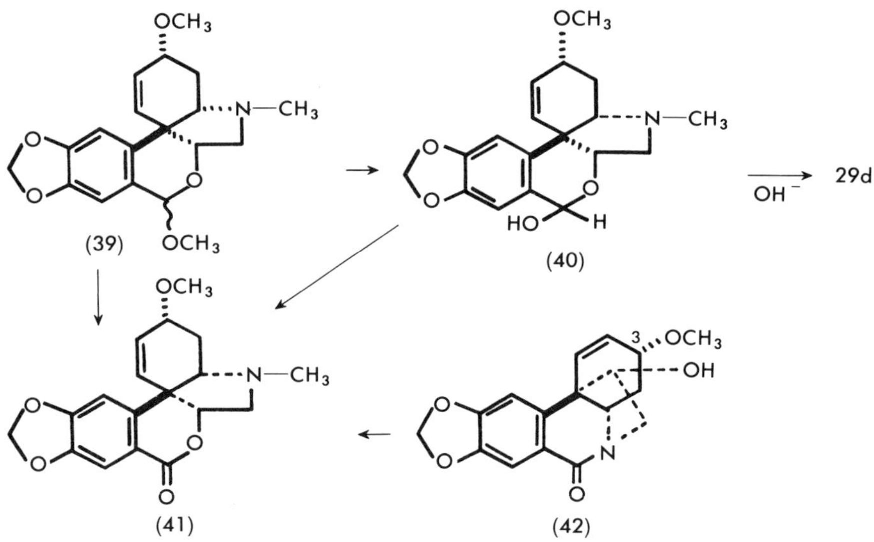

Oxidation of (39) with chromium trioxide in acetic acid gave a lactone (41), identical in all respects with macronine. It is presumed that precriwelline (40) is the intermediate in this transformation and that chromium trioxide oxidation of the hemiacetal occurs readily.[38] Precriwelline has been synthesized by the action of dilute mineral acid on the methiodide of 6-hydroxycrinamine.[39]

Manganese dioxide oxidation of 6-hydroxycrinamine yields 6-oxocrinamine (42). Hydrolysis of the strained "lactam," lactonization of the free carboxyl group with the C(11)-hydroxyl, and N-methylation yields macronine by an independent synthetic route.[39]

Pretazettine (43) is the major alkaloid of *Sprekelia formosissima.* The alkaloid is labile and is converted slowly to tazettine upon standing. Reaction conditions such as strong base, alumina chromatography, or heating

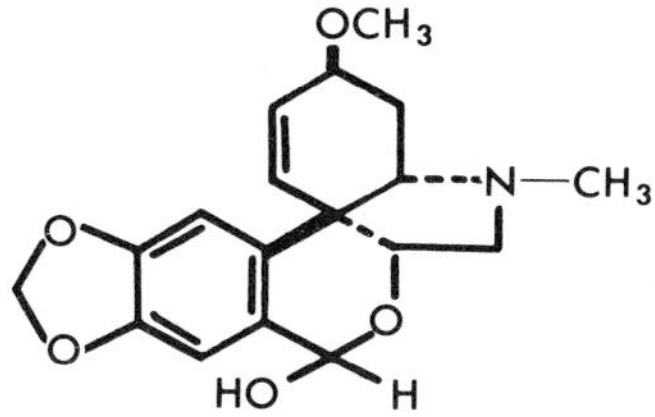

(43)

to 100° in water are sufficient to cause isomerization to tazettine. Oxidation of (43) with manganese dioxide afforded a lactone [(43) =O, instead of > CHOH] that was synthesized from haemanthidine by manganese dioxide oxidation, lactonization, and N-methylation.[31,39]

2.6 Montanine, Coccinine, and Manthine

Montanine and coccinine were named for the respective species: *Haemanthus montanus* and *H. coccineus*. The genus *Haemanthus* is native to South Africa and is cultivated as an ornamental house plant in the cooler American and European areas. Standard analytical techniques showed that both montanine and coccinine contain one methylenedioxyphenyl group, an aliphatic double bond not conjugated with the aromatic ring, one aliphatic methoxyl, and one tertiary amino group that did not contain an N-methyl function.[40] Although montanine (44a) and coccinine (44b) were not oxidized by manganese dioxide, Oppenauer oxidation of either alkaloid gave the secondary aminophenol (45a). The formation of (45a) can be envisaged as the result of oxidation of the hydroxyl group of (44a) and (44b) to the ketone, elimination of the β-amino group, and tautomerization to the phenol. O- and N-methylation of (45a), quaternization of the product with methyl iodide, and Hofmann degradation of the methohydroxide afforded an optically inactive methine (46).

N-Methylation of (45a) with formaldehyde and formic acid, followed by treatment with diazomethane, afforded a product that contained two less hydrogen atoms than (45c). This product (47) can be rationalized as a Pictet–Spengler-type cyclization from the nitrogen to the methoxy-hydroxy activated aromatic ring. O-Methylation of the cyclization product would complete the sequence. The structure of (47) was verified by synthesis. It is interesting to note that (47) is optically active only by virtue of the two types of aromatic substitution, yet it shows a significant optical rotation ($[\alpha]_D^{23}$ + 41°).

Supporting evidence for the structures and stereochemistry of these alkaloids was derived from research with the alkaloids of Section 2.4.2. Treatment of haemanthamine (27a) with mesyl chloride and pyridine,

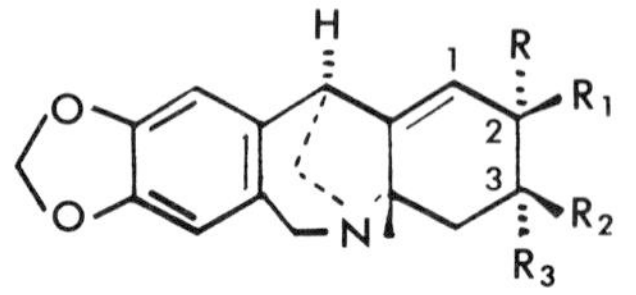

(44)

a) $R_1, R_3 = H; R = OCH_3; R_2 = OH$
b) $R, R_3 = H; R_1 = OCH_3; R_2 = OH$
c) $R_1, R_3 = H; R = OH; R_2 = OCH_3$
d) $R_1, R_3 = H; R, R_2 = OCH_3$
e) $R, R_1 = O; R_2 = OCH_3; R_3 = H$
f) $R, R_3 = H; R_1 = OH; R_2 = OCH_3$

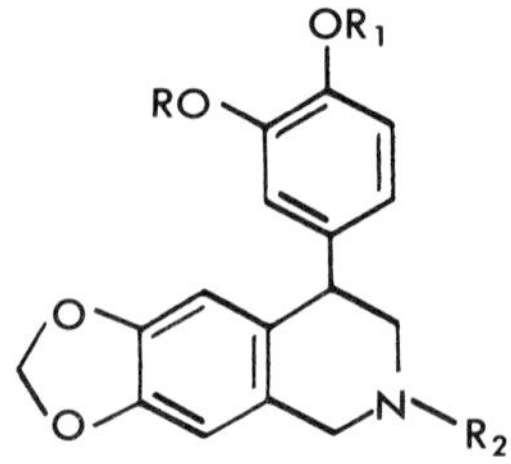

(45)

a) $R = CH_3; R_1, R_2 = H$
b) $R = H; R_1, R_2 = CH_3$
c) $R, R_1, R_2 = CH_3$

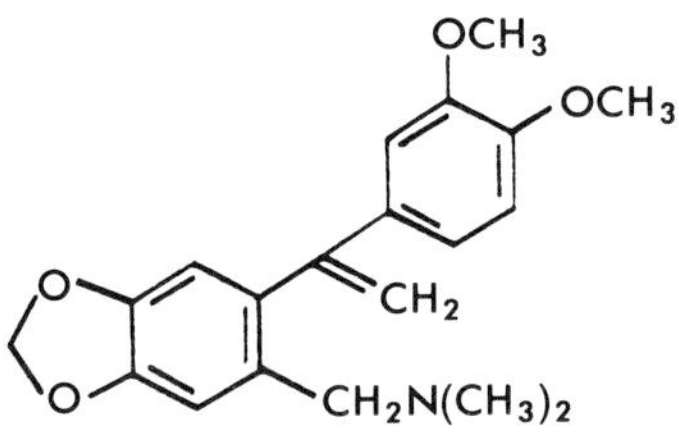

(46)

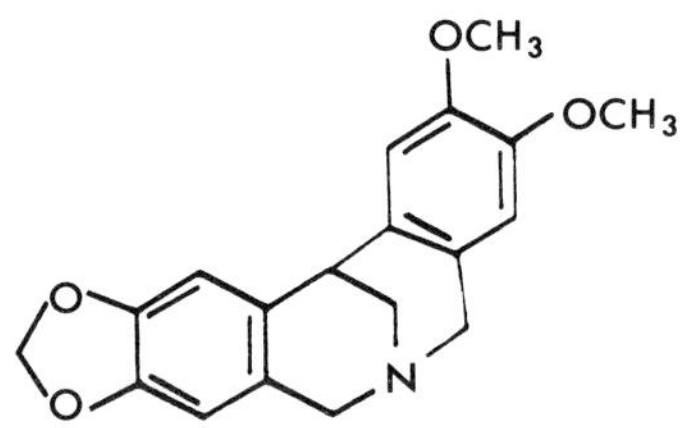

(47)

followed by hydrolysis with dilute sodium bicarbonate, afforded isohaemanthamine that was shown to be (44c). Hydrolysis of the mesylate with methanolic sodium methoxide gave manthine (44d). Proof for the gross structure of (44c) involved manganese dioxide oxidation to the corresponding ketone, N-methylation, and Hofmann degradation to an aminophenol (45b). O-Methylation of (45b) afforded (45c) that was identical with the O,N-methylation product of (45a).[41]

The stereochemistry of the rearrangement products is the foundation of the structures assigned to montanine, coccinine, and manthine, and it is important to consider the mechanism(s) and stereochemistry of the transformations. In its simplest form, the conversion of (48) to (44c) and (44d) can be described as a nucleophilic displacement by OH^- or OCH_3^- at C(2), followed by migration of the C(1)–C(2) double bond, aryl group migration, and elimination of the mesylate anion as shown. An alternate mechanism, which cannot be excluded at the present time, involves a carbonium ion formed through Ar(1–3) participation. The rearrangement requires both a C(1)–C(2) double bond and the C(11)–mesylate configuration shown in (48), since the dihydro compounds and those C(11)–epimers

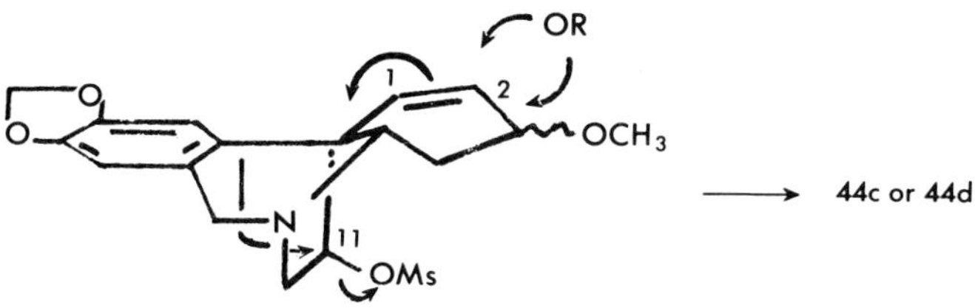

(48)

with a hydroxyl group toward the aromatic ring appear not to undergo the rearrangement. The configuration of the C(3)-methoxyl group is not altered in the course of the rearrangement. The C(3)-methoxyl of haemanthamine is *quasi*axial and becomes axial at C(3) in the rearrangement product (44c). Its axial nature is verified by the conversion of (44e) to the equatorial epimer by base. The reduction of (44e) with sodium borohydride gave *epiiso*haemanthamine (44f). The reduction did not affect the configuration at C(3), since manganese dioxide oxidation of (44f) afforded only (44e).[42]

A study of the intramolecular hydrogen bonding of the vicinal hydroxyl to methoxyl groups in the isomers of (44) (as studied by high-resolution infrared techniques) provided the relative stereochemistry of these centers. *Iso*haemanthamine (44c) showed an unbonded hydroxyl group ($\nu = 3623\ cm^{-1}$). Since the C(3)-methoxyl is axial, the C(2)-hydroxyl group of *iso*haemanthamine must be *trans* and *quasi*axial. The hydroxyl group of *epiiso*haemanthamine (44f) is completely hydrogen bonded ($\nu = 3567\ cm^{-1}$) and must represent the *cis* isomer.

The infrared spectrum of montanine in dilute solution showed that it is a *trans*-isomer (no OH-bonding). Direct comparison of *iso*haemanthamine and *epiiso*haemanthamine with montanine and coccinine was accomplished through O-methylation of each compound with potassium and methyl-*p*-toluenesulfonate. O-Methylmontanine, O-methyl*iso*haemanthamine, and manthine (44d) were found to be identical. O-Methylation of *epiiso*haemanthamine and coccinine gave the same dimethyl ether, and coccinine can be assigned structure (44b).[42]

2.7 Miscellaneous Systems

A few alkaloids cannot be grouped logically within the sections described earlier on the basis of parent ring systems. These compounds can best be considered either as trapped biosynthetic intermediates, rearrangement products related to known alkaloids, or possibly as artifacts.

2.7.1 Belladine The alkaloid has been reported to occur in an *Amaryllis belladonna* hybrid and in *Nerine bowdenii*. The structure of belladine (49) was ascertained from functional group analysis and the formation of *p*-

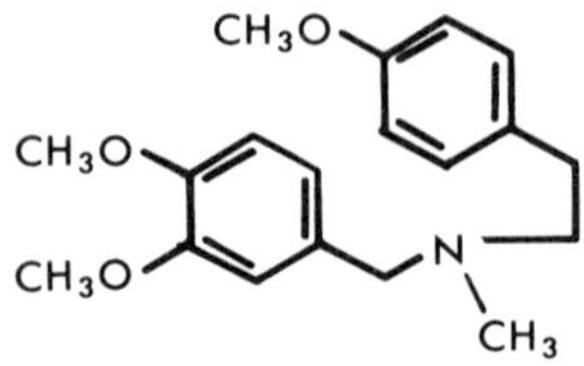

(49)

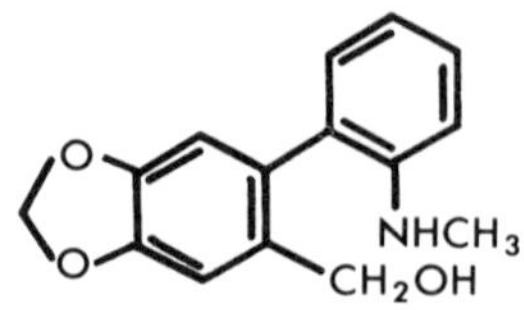

(50)

methoxystyrene and N,N-dimethyl veratrylamine on Hofmann degradation. The substance was synthesized for pharmacological studies prior to its isolation as a natural product. Belladine probably represents a completely methylated product derived from important biosynthetic intermediates.[41]

2.7.2 Ismine This substance has been isolated in low yield from *Ismine* sp., *Sprekelia formosissima,* and *Crinum powellii.* It lacks the usual fifteen-carbon nucleus of the other ring systems and may represent a natural degradation product of the alkaloids of Sections 2.4 and 2.5. It contains one methylenedioxy and one N-methyl group and forms an O,N-diacetate. The N-methyl aniline portion of the molecule was evident from the formation of a red dye with *p*-nitrobenzenediazonium chloride and the change in ultraviolet spectrum in the presence of acid. The spectrum in acid solution is very similar to 6-phenylpiperonyl alcohol. Structure (50) has been confirmed for ismine both by conversion of the base to 8,9-methylenedioxy-5-methyl-6-phenanthridone[43] and by synthesis.[44]

3. SYNTHETIC APPROACHES

Although the synthesis of Amaryllidaceae alkaloids might appear to be a modest synthetic challenge, few have been reported to date. A number of degradation and transformation products, lacking the complications of isolated unsaturation and the stereochemical aspects associated with oxygenated substituents, were the first to be prepared. (±)-Crinane,[24] the lycoranes,[12] and tetrahydrooxocrinine[25] methine were among the first to be synthesized.

Of classic simplicity is the synthesis of (±)-narwedine (51).[20] In accord

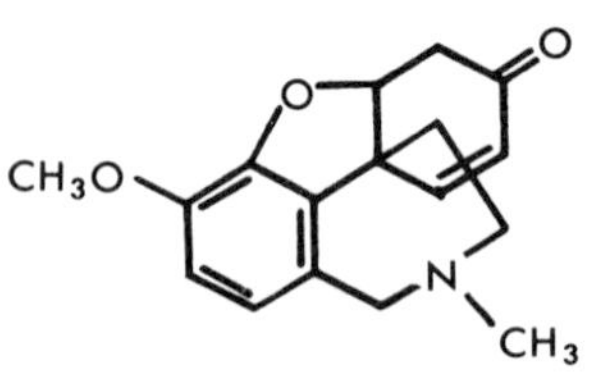

(51)

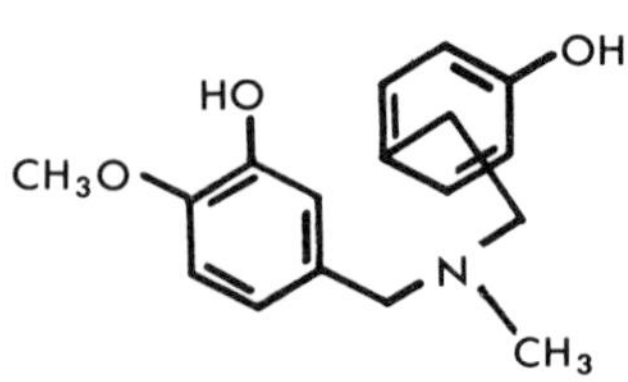

(52)

with the biogenetic theory of phenyl–phenyl oxidative coupling, (52) is converted by either manganese dioxide, lead dioxide, silver oxide, or potassium ferricyanide to (51). The maximum yield was 1.4% since polymerization of the precursor was the favored pathway. An exceptionally short synthesis of a substituted crinane nucleus has been reported.[45] The product (58) possesses oxygenated substituents at C(6) and C(11), a feature

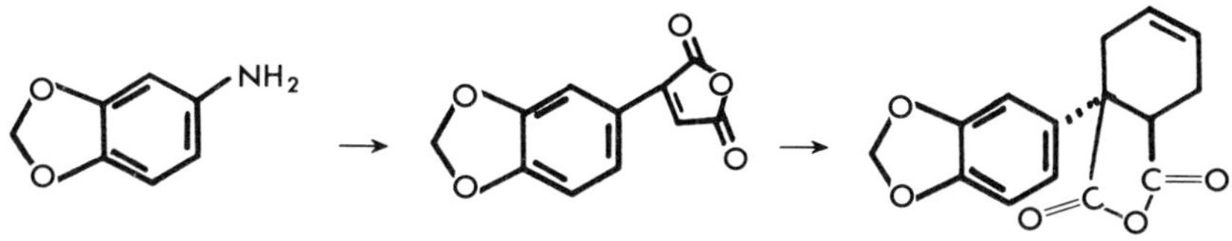

(53) (54) (55)

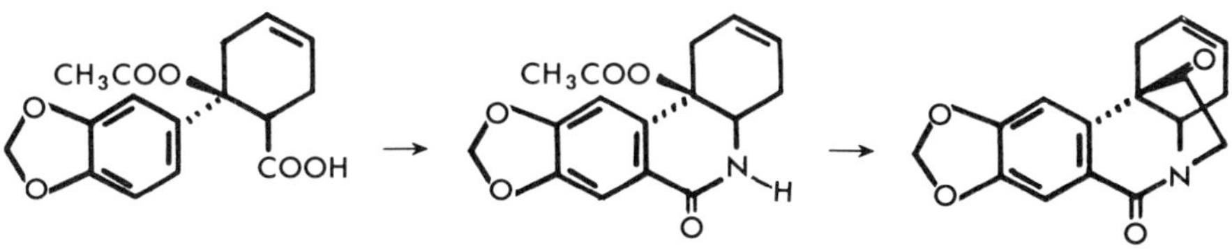

(56) (57) (58)

common to several alkaloids of Sections 2.4.2 and 2.4.3. Meerwein arylation of (53) with maleic acid provided (54) after treatment of the reaction product with acetic anhydride. Methanolysis of the butadiene adduct (55) provided a half-ester (56) that was converted to an isocyanate by the Curtius reaction and then cyclized to (57) by acid. The two-carbon bridge was elaborated by hydrolysis of the ester, conversion to the diazoketone (via the action of diazomethane on the acid chloride), and treatment with acid. Sodium borohydride reduction of (58) gave 6,11-dihydroxycrinene as a mixture of C(11)-epimers.

1-Oxocrinane has been synthesized from [(59) R = OH]. The corresponding aziridine derivative [(59) R = $\overline{CH_2CH_2N}$] was cyclized to (60) by sodium iodide. Catalytic hydrogenation of (60) and Pictet–Spengler cyclization afforded (61). Bromination of the hydrochloride of (61) fol-

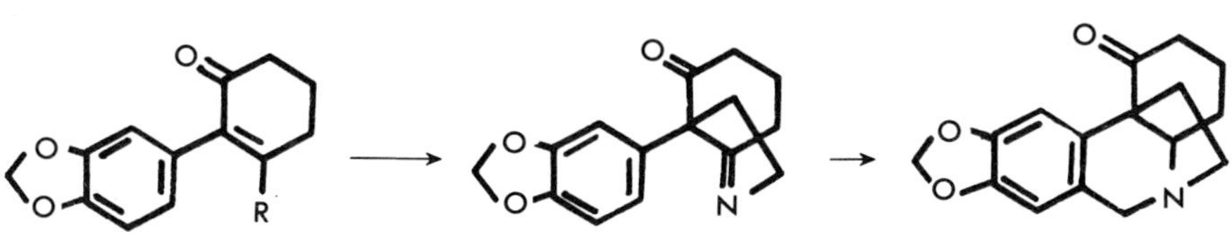

(59) (60) (61)

lowed by dehydrobromination gave an α,β-unsaturated ketone that could be reduced to the corresponding allylic alcohol and rearranged by acid to (±)-crinine.[46]

A total synthesis of (±)-crinine and (±)-*epi*crinine used (62) as a starting material. Condensation with methyl vinyl ketone provided (63). Saponification and decarboxylation gave a keto-amide that could be reduced to

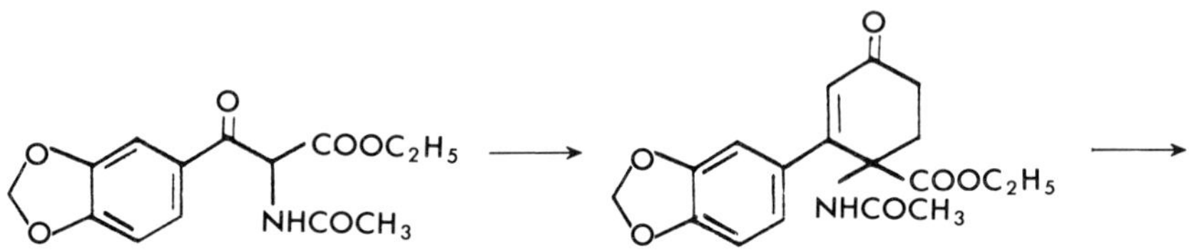

(62) (63)

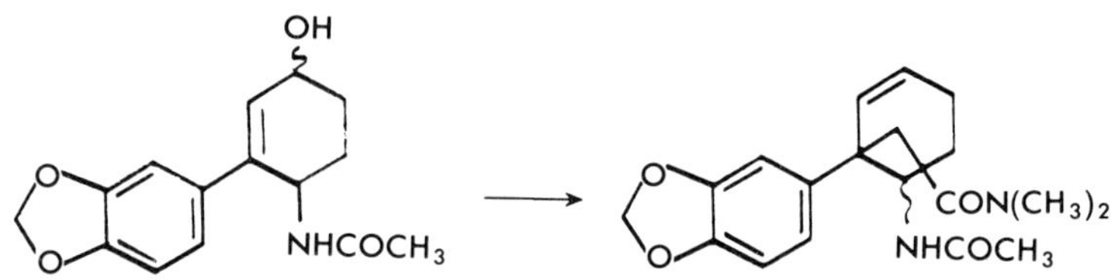

(64) (65)

the epimeric alcohols (64). The two-carbon side chain of (65) was introduced by condensation of (64) with 1,1-dimethoxy-1-dimethylaminoethane. The product (65) consisted of approximately equal parts of *cis*-and *trans*-diamides. Hydrolysis of the mixture afforded (66) that could be converted to (±)-crinane via (67). Selenium dioxide oxidation of (67) in acetic acid and acetic anhydride followed by saponification gave (±)-crinine.[47]

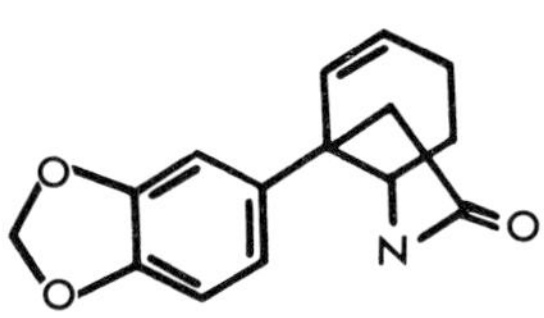

(66)

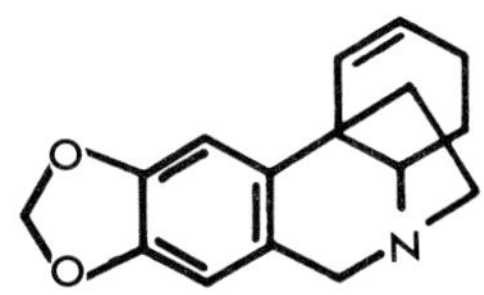

(67)

REFERENCES

1. W. C. Wildman, *Alkaloids* **6,** 289–413 (1960).
2. H. G. Boit, "Ergebnisse der Alkaloide-Chemie bis 1960," pp. 410–475, Akademie Verlag, Berlin, Germany, 1961.
3. F. Sandberg and K. H. Michel, *Lloydia* **26,** 78 (1963).
4. R. E. Lyle, E. A. Kielar, J. R. Crowder, H. M. Fales, and W. C. Wildman, *J. Amer. Chem. Soc.* **82,** 2620 (1960).
5. R. B. Kelly, W. I. Taylor, and K. Wiesner, *J. Chem. Soc.* 2094 (1953).

6. L. G. Humber, H. Kondo, K. Kotera, S. Takagi, W. I. Taylor, B. R. Thomas, Y. Tsuda, K. Tsukamoto, S. Uyeo, H. Yajima, and N. Yanaihara, *J. Chem. Soc.* 4622 (1954).
7. K. Takeda and K. Kotera, *Pharm. Bull.* (Tokyo) **5,** 234 (1957).
8. H. M. Fales and W. C. Wildman, *J. Amer. Chem. Soc.* **80,** 4395 (1958).
9. Y. Nakagawa and S. Uyeo, *J. Chem. Soc.* 3736 (1959).
10. K. Kotera, Y. Hamoda, K. Tori, A. Aono, and K. Kuriyama, *Tetrahedron Letters* 2009 (1966).
11. S. Takagi, W. I. Taylor, S. Uyeo, and H. Yajima, *J. Chem. Soc.* 4003 (1955).
12. R. K. Hill, J. A. Joule, and L. J. Loeffler, *J. Amer. Chem. Soc.* **84,** 4951 (1962).
13. N. Ueda, T. Tokuyama, and T. Sakan, *Bull. Chem. Soc. Japan* **39,** 2012 (1966).
14. T. Kitagawa, S. Uyeo, and N. Yokoyama, *J. Chem. Soc.* 3741 (1959).
15. M. Shiro, T. Sato, and H. Koyama, *Chem. Ind. (London)* 1229 (1966).
16. H. Kondo and T. Ikeda, *J. Pharm. Soc. Japan* **65,** 9–10A, 5 (1945).
17. S. Uyeo and S. Kobayashi, *Pharm. Bull. (Tokyo)* **1,** 139 (1953).
18. H. M. Fales, L. D. Guiffrida, and W. C. Wildman, *J. Amer. Chem. Soc.* **78,** 4145 (1956).
19. D. H. R. Barton and T. Cohen, "Festschrift Arthur Stoll," pp. 117–143, Birkhauser, Basel, Switzerland, 1957.
20. D. H. R. Barton and G. W. Kirby, *J. Chem. Soc.* 806 (1962).
21. Ref. 1, p. 342.
22. S. Uyeo, Handbook, XVI, IUPAC Congress, Paris, 1957.
23. D. J. Williams and D. Rogers, *Proc. Chem. Soc.* 357 (1964).
24. W. C. Wildman, *J. Amer. Chem. Soc.* **80,** 2567 (1958).
25. S. Uyeo, H. Irie, A. Yoshitake, and A. Ito, *Chem Pharm. Bull. (Tokyo)* **13,** 1084 (1956).
26. H. M. Fales and W. C. Wildman, *J. Amer. Chem. Soc.* **82,** 197 (1960).
27. S. Uyeo, H. M. Fales, R. J. Highet, and W. C. Wildman, *J. Amer. Chem. Soc.* **80,** 2590 (1958).
28. H. M. Fales, D. H. S. Horn, and W. C. Wildman, *Chem. Ind. (London)* 1415 (1959).
29. R. W. King, C. F. Murphy, and W. C. Wildman, *J. Amer. Chem. Soc.* **87,** 4912 (1965).
30. J. Karle, J. A. Estlin, and I. L. Karle, *J. Amer. Chem. Soc.* **89,** 6510 (1967).
31. W. C. Wildman and D. T. Bailey, *J. Amer. Chem. Soc.* **89,** 5514 (1967).
32. W. I. Taylor, S. Uyeo, and H. Yajima, *J. Chem. Soc.* 2962 (1955).
33. T. Ikeda, W. I. Taylor, Y. Tsuda, S. Uyeo, and H. Yajima, *J. Chem. Soc.* 4749 (1956).
34. R. J. Highet and W. C. Wildman, *Chem. Ind. (London)* 1159 (1955).
35. T. Ikeda, W. I. Taylor, Y. Tsuda, and S. Uyeo, *Chem. Ind. (London)* 411 (1956).
36. R. J. Highet and P. F. Highet, *Tetrahedron Letters* 4099 (1966).
37. H. Hauth and D. Stauffacher, *Helv. Chim. Acta* **47,** 185 (1964).
38. C. F. Murphy and W. C. Wildman, *Tetrahedron Letters* 3857 (1964).
39. D. T. Bailey and W. C. Wildman, *J. Org. Chem.* **33,** 3749 (1968).
40. W. C. Wildman and C. J. Kaufman, *J. Amer. Chem. Soc.* **77,** 1248 (1955).
41. E. W. Warnhoff, *Chem. Ind. (London)* 1385 (1957).
42. Y. Inubushi, H. M. Fales, E. W. Warnhoff, and W. C. Wildman, *J. Org. Chem.* **25,** 2153 (1960).
43. R. J. Highet, *J. Org. Chem.* **26,** 4767 (1961).
44. R. K. Hill and R. M. Carlson, *J. Org. Chem.* **30,** 1571 (1956).
45. J. B. Hendrickson, C. Foote, and N. Yoshimura, *Chem. Commun.* 165 (1965).
46. H. W. Whitlock, Jr. and G. L. Smith, *J. Amer. Chem. Soc.* **89,** 3600 (1967).
47. H. Muxfeldt, R. S. Schneider, and J. Mooberry, *J. Amer. Chem. Soc.* **88,** 3670 (1966).

7

Erythrina Alkaloids

A. MONDON
Institut für Organische Chemie
der Universität Kiel
West Germany

1. GENERAL SURVEY

1.1 History

Folkers and Koniuszy began an intense study of the erythrina alkaloids in 1937. They isolated from about fifty *Erythrina* species, almost all the alkaloids known today in this plant genus. After 1949, the major center of research shifted to the laboratories of Prelog in Zürich and Boekelheide in Rochester from which studies elucidating the structures of the alkaloids appeared. In recent years particular attention has been devoted to stereochemical problems and the absolute configuration of the alkaloids has been determined. Certain unexpected aspects of biosynthesis have been reported by Barton. In the field of synthesis, Belleau and the schools of Wiesner, Prelog, Mondon, Rapoport, and Scott have achieved significant successes. Developments in this field are surveyed by detailed reviews.[1–6]

1.2 Isolation and General Properties of the Alkaloids

Alkaloids occur in the seeds and plant parts of *Erythrina* species that grow in the tropic and subtropic zones of all parts of the world. Botanically, the genus *Erythrina* belongs to the Legume family, subfamily Papilionaceae. Notable exceptions are the occurrence of two erythrina alkaloids in the family of Menispermaceae.

The alkaloids produce paralysis of the smooth muscles similar to the effect of curare.[7–9] This effect is also obtained on oral administration and disappears when the bases are quaternized.[10] Cocculolidine has insecticidal activity.[11]

The alkaloids listed in Table 7.1 are divided into three groups according to their properties: erythroidines, erythra, and eryso alkaloids. The free bases are all crystalline and dextrorotatory. Column 4 of Table 7.1 provides information about distribution. The phenol alkaloids erysodine, erysovine, and erysopine surpass all the other alkaloids in ready availability.

The seeds of all Erythrina species contain hypaphorine, the methyl betaine of L-tryptophan.[22]

Only the erythroidines and erythra alkaloids can be extracted di-

TABLE 7.1. Natural Erythrina Alkaloids

Name	M.p.	$[\alpha]_D$	Distrib.[a]	Lit. ref.
α-Erythroidine	58–60°	+136° (water)	2	12,13,14
β-Erythroidine	99–100°	+89° (water)	2	12,13,14
14-Nordihydro-β-erythroidine[b]	144–146°	+273° (chloroform)	1[c]	11
Erythraline	106–107°	+212° (ethanol)	10	14,15,16,17
Erythramine	103–104°	+227° (ethanol)	4	14,15,18
2-Ketoerythramine	136–137°	+409°	1	19, 32a
Erythratine	170–171°	+145° (ethanol)	2	15,18
Erythratidine	120–121°	+279° (ethanol)	1	20
Erysodine	204–205°	+248° (ethanol)	24	14,21,22, 23,24
Erysovine	178–179°	+234° (ethanol)	16	21,22,23,24
Erysopine	242–243°	+225° (morpholine)	21	14,21,22, 23,24
Erysonine	241–243°	+272° (morpholine)	1	23
Dihydroerysodine	208–209°	+224° (methanol)	1[c]	25

[a]Number of Erythrina species from which the alkaloid has been isolated.
[b]Identical with Cocculolidine.
[c]Isolated from *Cocculus* species, family Menispermaceae.

rectly from aqueous plant extracts. The phenolic bases, on the other hand, are present in a water-soluble bound form from which they must first be liberated by hydrolysis with acids.[21] Their phenol groups are, for the most part, esterified with sulfoacetic acid as in erysothiovine and erysothiopine[26] and, more rarely, bound to glucose as in glucoerysodine.[14,27]

1.3 Basic Skeleton and Functional Groups

The erythrina alkaloids are derived from erythrinane, a tetracyclic spiro amine, the *cis*-[28] and *trans*-[9] forms of which are known.

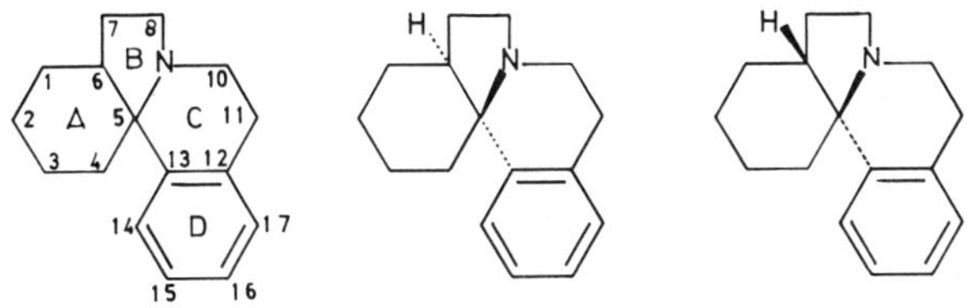

Erythrinane A/B-cis A/B-trans

The erythra and eryso alkaloids form the series of aromatic erythrina alkaloids, the major representatives of which (Table 7.2) have two conjugated double bonds in rings A and B and three free or etherified hydroxy groups at carbon atoms 3, 15, and 16.

TABLE 7.2. Main Erythrina Alkaloids

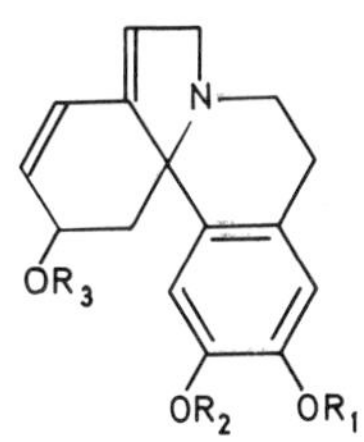

Name	Composition	R_1	R_2	R_3
Erythraline[21]	$C_{18}H_{19}NO_3$	—CH_2—		CH_3
Erysodine[21]	$C_{18}H_{21}NO_3$	H	CH_3	CH_3
Erysovine[21]	$C_{18}H_{21}NO_3$	CH_3	H	CH_3
Erysopine[21]	$C_{17}H_{19}NO_3$	H	H	CH_3
Erysonine[22]	$C_{17}H_{19}NO_3$	H	CH_3	H

The relative positions of the hydroxy and methoxy groups in the eryso alkaloids have been established through synthesis[29] or nmr spectroscopy of the deuterated phenol bases.[19]

TABLE 7.3. Minor Erythrina Alkaloids

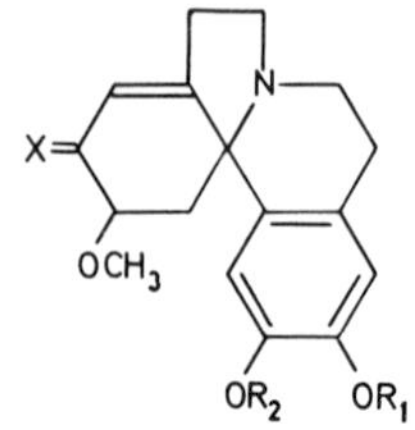

Name	Composition	R_1	R_2	X
Erythramine[18]	$C_{18}H_{21}NO_3$	—CH_2—		H_2
Dihydroerysodine[25]	$C_{18}H_{23}NO_3$	H	CH_3	H_2
Erythratine[15, 19]	$C_{18}H_{21}NO_4$	—CH_2—		H, OH
2-Ketoerythramine[19]	$C_{18}H_{19}NO_4$	—CH_2—		=O
Erythratidine[20, *]	$C_{19}H_{25}NO_4$	CH_3	CH_3	H, OH

*The structure is not rigidly proved.

The minor alkaloids listed in Table 7.3 have only one double bond in the $\Delta^{1(6)}$-position.

α- and β-Erythroidine[30, 31] together with 14-nordihydro-β-erythroidine (Cocculolidine),[11] constitute a group of lactone alkaloids, the composition of which is $C_{16}H_{19}NO_3$ and $C_{15}H_{19}NO_3$, respectively. They are derived from erythrinane by partial oxidation of the aromatic ring with loss of one or two carbon atoms. The α,β-terminology of the erythroidines was based on the assumption that they are diastereoisomers.[13] The β-form is more stable and more readily available.

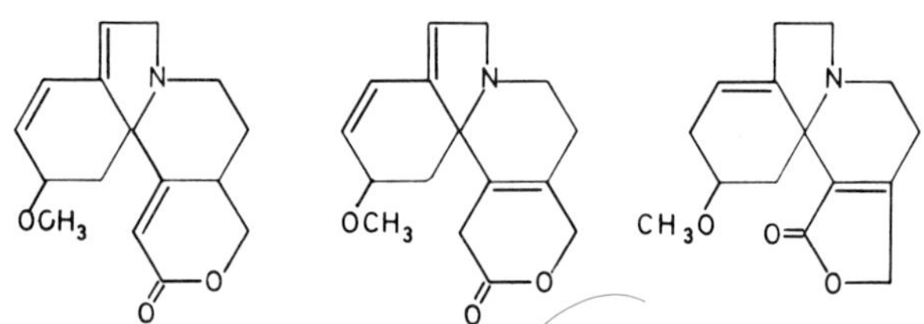

α-Erythroidine β-Erythroidine Cocculolidine

1.4 Biogenesis

The biogenesis of the aromatic erythrina alkaloids has been elucidated by Barton and his coworkers[32, 32a] according to the following scheme:

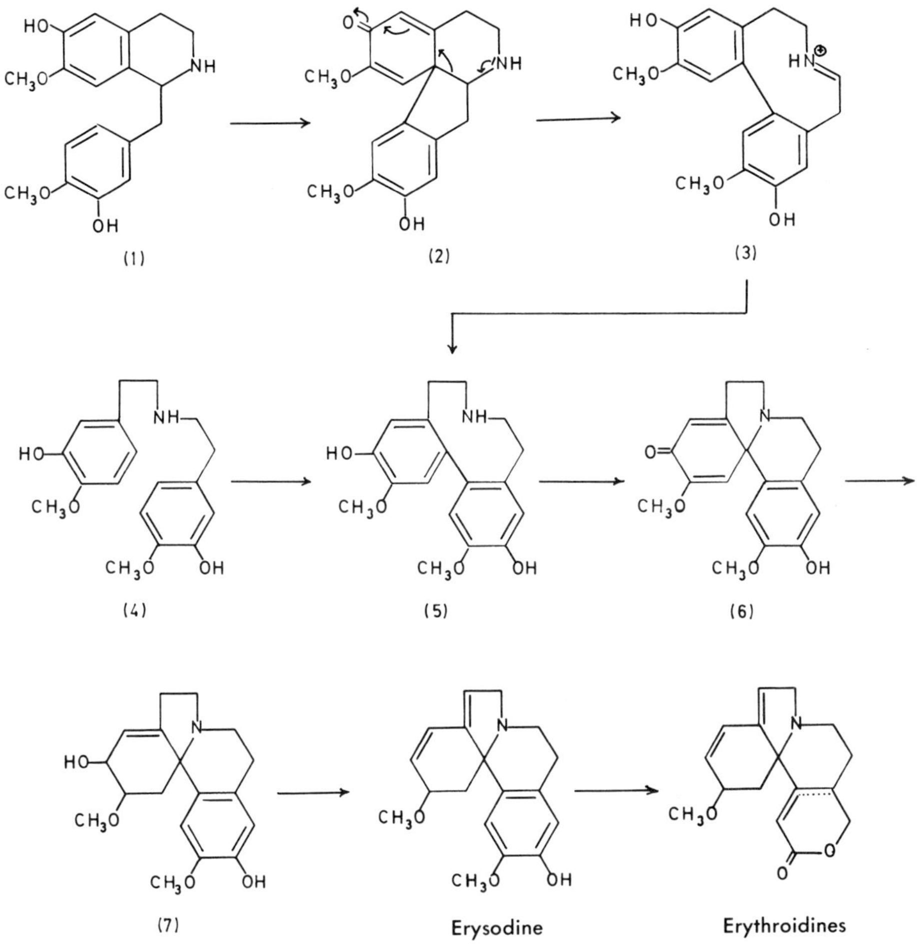

2. ELUCIDATION OF STRUCTURE OF THE AROMATIC ERYTHRINA ALKALOIDS

2.1 Basic Features

Aromatic erythrina alkaloids are tertiary bases without N-methyl groups. In addition to the methylenedioxy group, erythraline has a methoxy group and partial hydrogenation affords erythramine.[17] Erysodine and erysovine are isomers with one phenolic group and two methoxy groups, whereas erysopine possesses two ortho phenolic groups and one methoxy group. The three phenolic bases have the same basic skeleton and on methylation give erysotrine, [34,35] a trimethyl ether, which has not as yet been found in nature.

$$\left.\begin{array}{l}\text{Erysodine}\\ \text{Erysovine}\\ \text{Erysopine}\end{array}\right\} \longrightarrow \text{Erysotrine}$$

On hydrogenation the alkaloids give dihydro[17] and tetrahydro derivatives,[36,37] for example:

Erythraline → Dihydroerythraline → Tetrahydroerythraline[38]
(Erythramine)

Oxidation of erythraline gives hydrastic acid,[36] and ethylation and then oxidation of erysodine gives 4-ethoxy-5-methoxyphthalic acid[39] so that one methoxy group is bound in an aliphatic part of the molecule. Both alkaloids give indole on fusion with potassium hydroxide.[39,40] The ultraviolet absorption and Hofmann degradation of tetrahydroerythraline[41] suggest a tetrahydroisoquinoline derivative in which the nitrogen belongs to two rings.

The structure problem, however, proved to be very complex and a solution was found only when, in 1951, Carmack, McKusick, and Prelog[42] postulated a novel spiroamine system, erythrinane.

2.2 Apo-Rearrangement

Folkers, Koniuszy, and Shavel[39] rearranged erysopine and erythraline to the optically inactive saturated base apoerysopine. However, Carmack, McKusick, and Prelog,[42] obtained from erysodine and erythraline, on mild treatment with mineral acids, two different optically active bases, apoerysodine and apoerythraline. These are distinguished by high unsaturation, high negative rotation, and an ultraviolet absorption shifted to longer wavelengths with λ_{max} about 310 mμ. Both bases are rearranged further to inactive apoerysopine by heating with mineral acids.

The uv maxima at 235–240 mμ suggests that the double bonds of erysodine and erythraline are conjugated heteroannularly, and since the bases have no labile vinylamine structure, the double bonds in the postulated erythrinane ring system extend from C(1) and C(6). The aliphatic methoxy groups must then be located at C(3) or C(4) in order to permit elimination to a linearly conjugated triene.

The structural information can be summarized by the expressions (8a)–(8c) for erysopine, erysodine, and erythraline as well as (9b) and (9c) for apoerysodine and apoerythraline. The ring strain of the alkaloids disappears by rearrangement to the stable diphenyl system of apoerysopine (10) by way of (9) and (11).

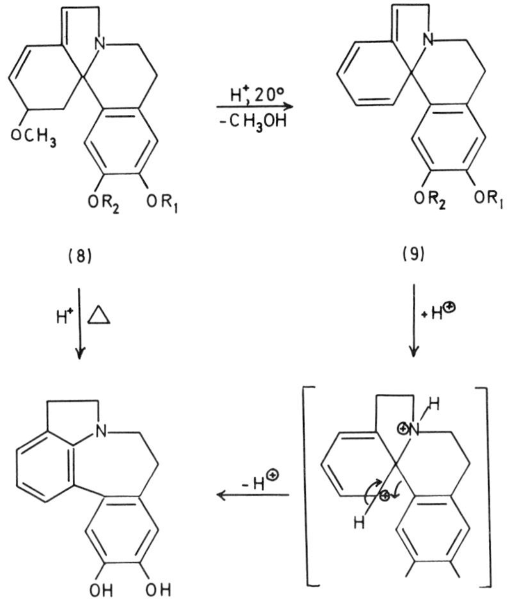

(8) (9) (10) (11)

a: $R_1 = R_2 = H$
b: $R_1 = H; R_2 = CH_3$
c: $R_1R_2 = -CH_2-$

As a by-product of the mild mineral acid treatment of erysodine there is formed, by cleavage of the aliphatic methyl ether group, the alkaloid erysonine[42,43] that is thus unequivocally related to the other eryso alkaloids.[18]

2.3 Degradation and Synthesis of Apoerysopine

Systematic Hofmann degradation of apoerysopine dimethyl ether (12),[39,42] followed by hydrogenation of the methine bases leads to (13) and (15). The ring opening sequence becomes apparent from the dehydrogenation of (13) to the indole base (14).

In 1955, the structure of apoerysopine was confirmed by Wiesner, Valenta, Manson, and Stonner[44] by synthesis of the dimethyl ether (12). The indole base (14) was also obtained synthetically.[45] Recently, Mondon, and Menz[46] obtained apoerysopine in a few steps by way of a novel ring closure.

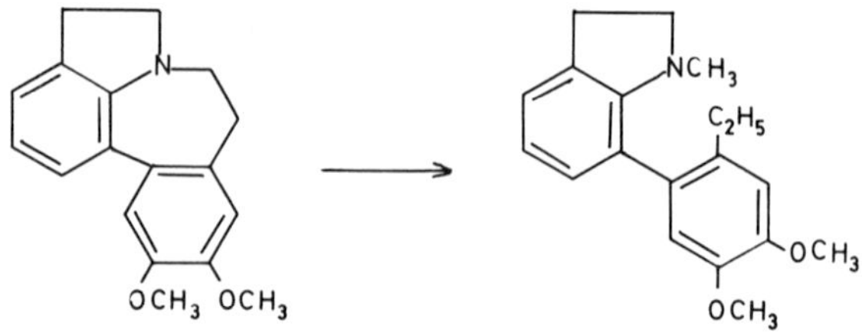

(12) (13)

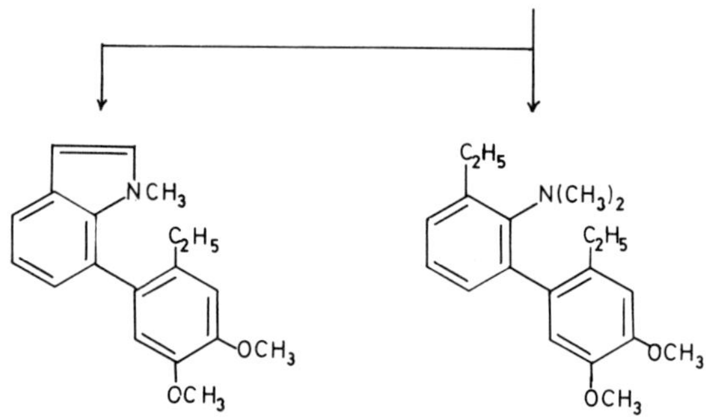

(14) (15)

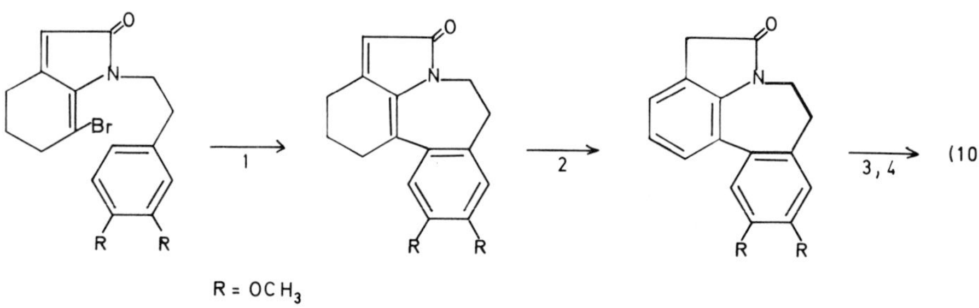

1: HCO_2H—H_3PO_4; 2: Pd—C ; 3: $LiAlH_4$; 4: HBr 48 %.

2.4 Catalytic Hydrogenation

Doubly unsaturated alkaloids such as (8) are hydrogenated to dihydro bases (16) with palladium in ethanol[17] and stereospecifically to tetrahydro bases (17) with platinum in acid solution.[17,34,36,37]

Earlier the dihydro bases were formulated with the double bond in the five-membered ring on the basis of basicity measurements.[3] Later the isomeric position for the double bond between C(1) and C(6) was demonstrated first for dihydro-β-erythroidine[47] and finally for dihydroerysodine[29] and erythramine.[32]

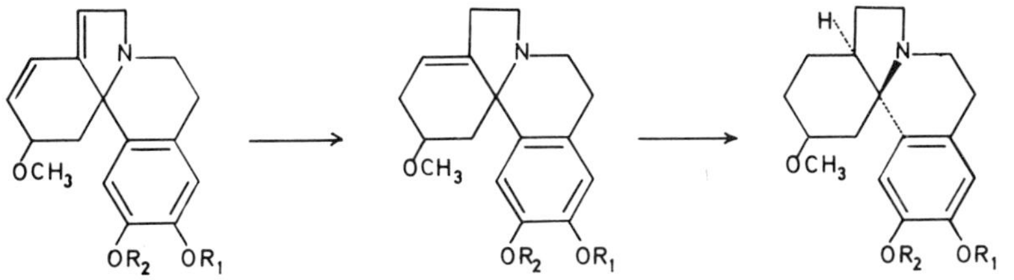

(8) (16) (17)

b: $R_1 = H$, $R_2 = CH_3$

c: $R_1R_2 = -CH_2-$

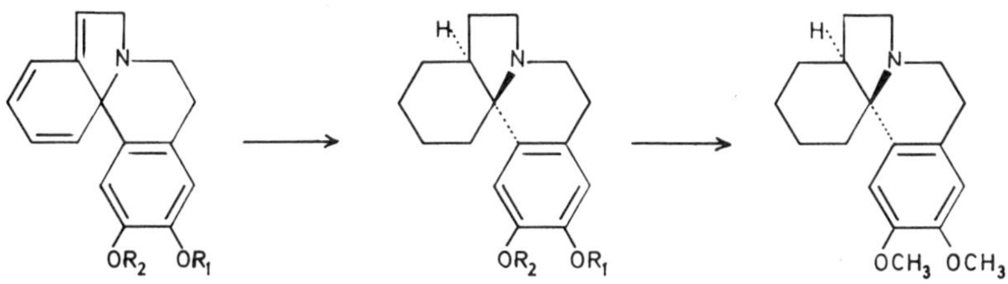

(9) (18) (19)

Hydrogenation of apoerysodine (9b) and apoerythraline (9c) gives the sterically homogeneous hexahydro bases (18b) and (18c), respectively. Each of these products can be converted to hexahydroapoerysotrine (19).[42] By way of the key compound (19), the long-favored hypothesis that erysodine and erythraline and all aromatic alkaloids related to them are derived from a common basic skeleton was confirmed.

In contrast to earlier views the tetrahydro and hexahydro bases (17) and (19) belong sterically to the *cis*-erythrinane series.[48]

2.5 Cyanogen Bromide Degradation of Dihydroerysotrine

In order to confirm the spiroamine structure of the aromatic erythrina alkaloids, dihydroerysotrine (20) was degraded with cyanogen bromide.[49] After reduction of the crude product with lithium aluminium hydride, the inactive secondary base (21) was obtained, the structure of which was proved by degradation to the dicarboxylic acid (22). The Prelog-base (21),[50] along with the $\Delta^{1,6}$-diene (24),[51] are also obtained by treatment of the unsaturated mesylate (23)[48] with strong base.

2.6 Structure and Absolute Configuration

Nowacki and Bonsma[52] confirmed the structure, which was derived chemically in 1958, by an x-ray diffraction study of erythraline hydro-

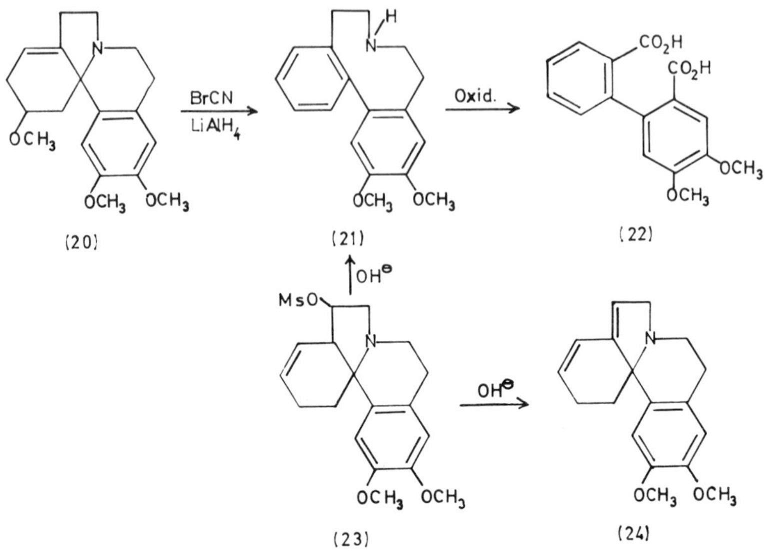

bromide. For the first time, they succeeded in proving the position of the aliphatic methyl ether group at C(3) and its *cis* arrangement with respect to nitrogen. The corresponding arrangement in the case of eryso alkaloids has since been verified by synthesis.[43]

After the (3R, 5S) configuration had been proved for α- and β-erythroidine,[47,53] the absolute configuration of the aromatic alkaloids could also be determined through comparisons.[54] From the molecular rotation differences between the triene bases (9b) and (9c) and α- and β-erythroidine or the related hexahydro bases, it may be ascertained that the base pairs have the same configuration at C(5). The rotatory dispersion curves of hexahydroapoerysotrine (19) and β-tetrahydro-β-erythroidine also coincide between 320–650 mμ.

Consideration of all the evidence leads to the 3R, 5S-absolute configuration for erysodine and erythraline also. Thus, the following general steric structures apply for the aromatic erythrina alkaloids and their di-, tetra-, and hexahydro derivatives.

Recently the absolute configuration of erythratine was established to be (2S, 3S, 5S) with the hydroxy and methoxy groups in *trans*-position.[19]

3. ELUCIDATION OF STRUCTURE OF THE LACTONE ALKALOIDS

3.1 General Properties of α- and β-Erythroidines

α- and β-erythroidine are tertiary bases with two double bonds capable of hydrogenation, an easily eliminated methyl ether group, and a lactone

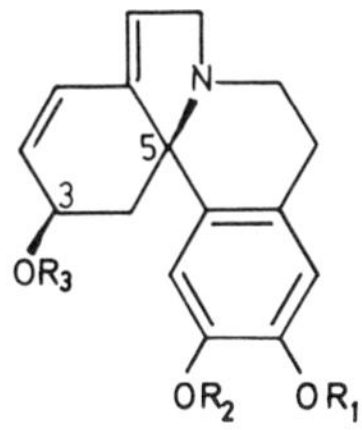

(3R,5S)-

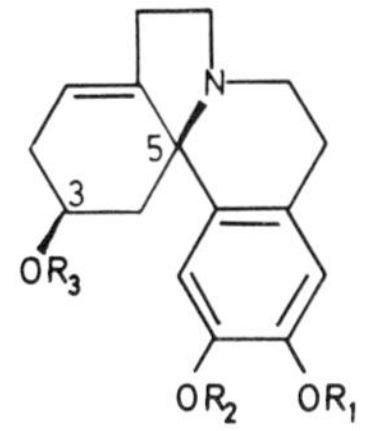

(3S,5S)-

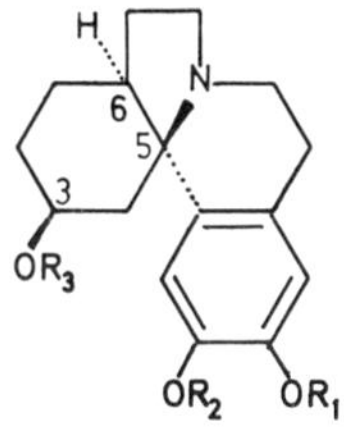

(3S,5S,6S)-

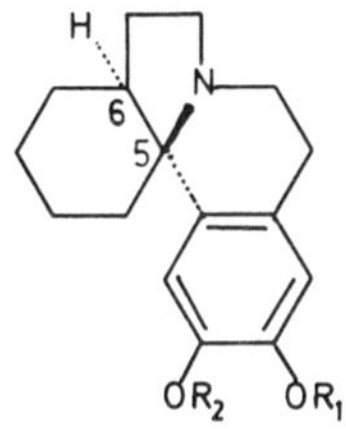

(5S,6S)-

ring.[30,38] Their infrared spectra have carbonyl bands in the same position at 5.78 μ while the ultraviolet absorption maxima of the hydrochlorides are different: 224 mμ (log ϵ, 4.5) and 238 mμ (log ϵ, 4.4), respectively. The maxima are compatible with heteroannular conjugated double bonds.[33,34] β-Erythroidine gives a characteristic red color with sulfuric acid and ferric chloride.[14,55]

The elucidation of structure of α- and β-erythroidine is credited to the comprehensive studies of Boekelheide and his co-workers.

3.2 β-Erythroidine

3.2.1 Apo and Isoapo Rearrangement β-Erythroidine (25) behaves towards mineral acids as do the alkaloids of the aromatic series. Under mild conditions optically active desmethoxy-β-erythroidine (28) is formed; on heating, inactive apo-β-erythroidine (26) results from rearrangement. By chromatography on aluminium oxide the colorless apo base (26) is converted, with double bond shift, to the yellow isoapo-β-erythroidine (29).[56–59]

The hydrogenation of (26) and (29) to (30) proves the common carbon–nitrogen skeleton and reveals the third double bond of β-erythroidine that cannot be hydrogenated in the alkaloid base itself.[60] Dehydrogenation of (26) gives (27) whose positive Ehrlich test demonstrates the transition from an indoline to an indole derivative. Permanganate oxidation of (27) leads to isatin-7-carboxylic acid and 2-aminoisophthalic acid.[60]

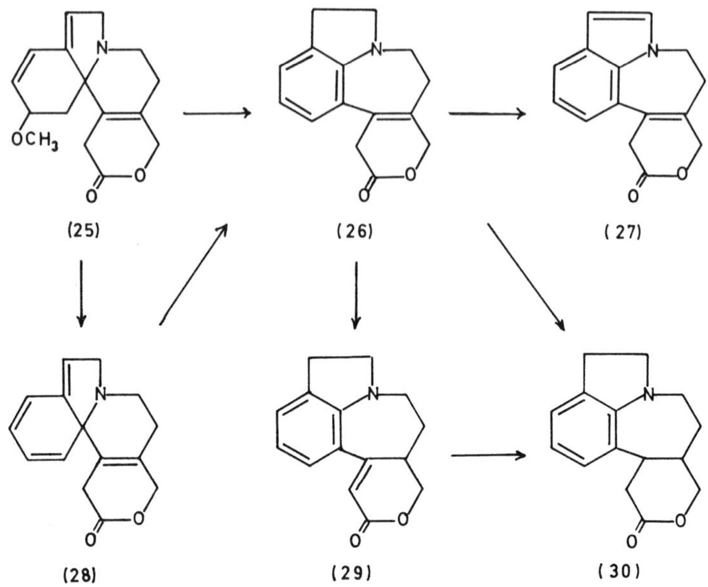

3.2.2 Hofmann Degradation of Apo-β-erythroidine Hofmann degradation of apo-β-erythroidine (26) gives further information about the ring system.[61] The degradation proceeds by way of the zwitterion (31) to the bases (32), (33), and (35) and shows the presence of the grouping $—CH_2—CH_2—\overset{|}{N}—CH_2—CH_2—$. The apo methine base (32) is rearranged by aluminium oxide to the yellow isoapo methine base (34) that gives acetaldehyde on ozonization, whereas ozonization of (32) and (35) yields formaldehyde.

3.2.3 Structure of β-Erythroidine β-Erythroidine is poorly suited for study by the Hofmann degradation, but the conditions are more favorable in the case of dihydro-β-erythroidine (36) that is obtained from 25 by hydrogenation with platinum or better with Raney nickel in alkaline solution. The 1,4-addition of hydrogen to the conjugated double bonds has been demonstrated by an x-ray determination of the structure of the dihydrobase.[47]

Hofmann degradation of (36) gives, in addition to the normal methine base, an oxygen-free base (39)[62]. This base is formed from the betaine (37) by aromatization of ring A with elimination of methanol to (38) and decarboxylation and dehydration of the cinnamic acid derivative. On oxidation (39) gives N-methylphthalimide and on cleavage with ozone, formaldehyde.

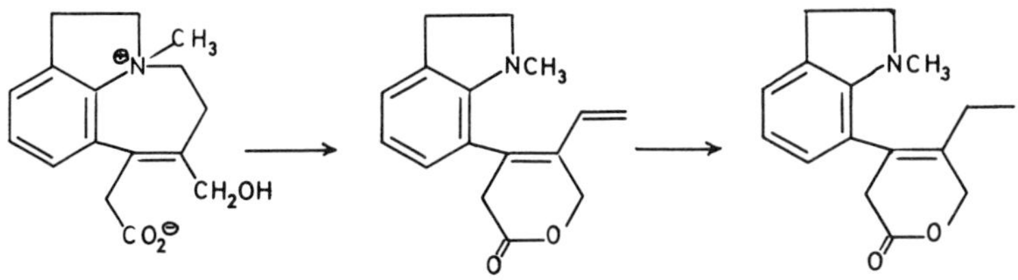

(31) (32) (33)

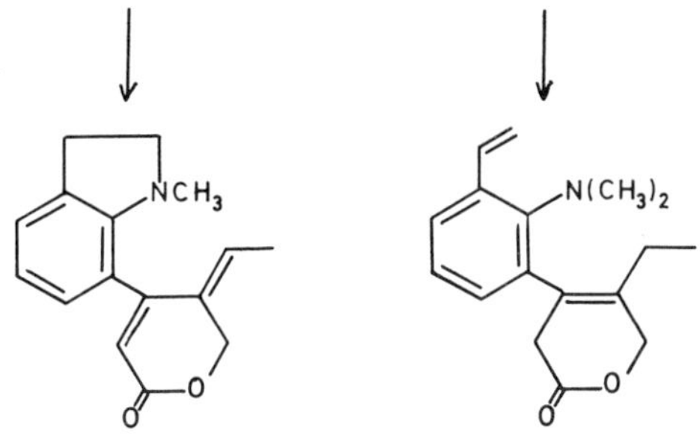

(34) (35)

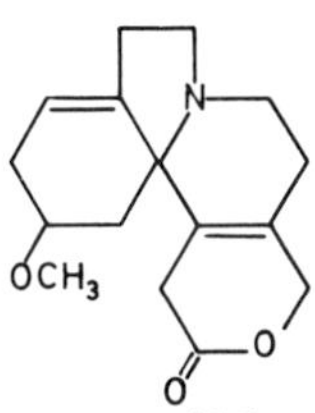

(36)

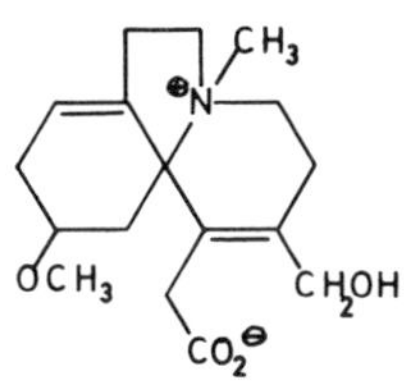

(37)

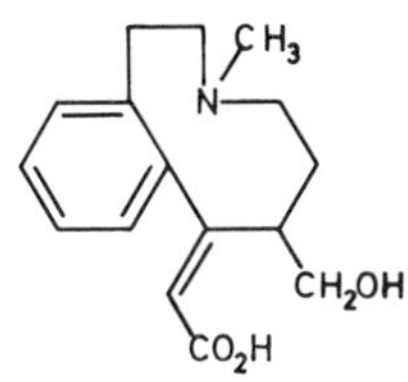

(38)

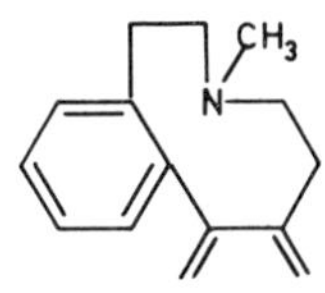

(39)

The Hofmann degradation of dehydro-β-erythroidinol (40) that is obtained by reduction of (36) with lithium aluminium hydride,[62] proceeds still better. The only degradation product suited for structure determination by further degradation is the crystalline base (41).

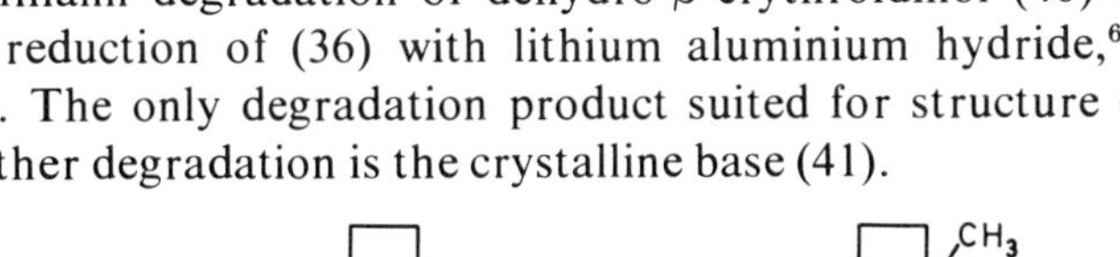

(40) (41)

The structure of the diol base (41) follows, on the one hand, from a twofold Hofmann degradation to (42), hydrogenation to (43), and oxidation to o-ethylbenzoic acid. On the other hand, an allylic hydroxyl group can be eliminated by hydrogenation of (41) with palladium on charcoal, and the resulting base (44) can be converted to (45) by Hofmann degradation and hydrogenation. Repetition of both steps gives (46) whose cleavage with ozone leads to methyl ethyl ketone.[63] In this manner, the position of both alcohol groups in (41) and the position of the lactone ring in β-erythroidine (25) are established.

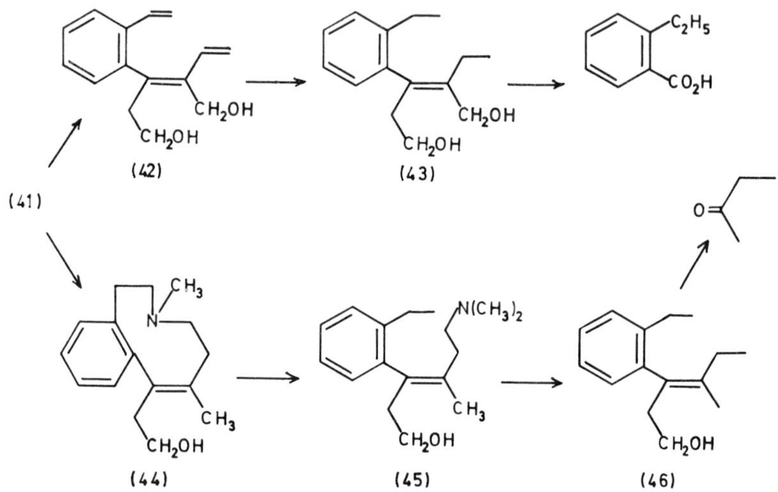

The lactone ring remains unchanged in the rearrangement of β-erythroidine to apo-β-erythroidine.[64] This may be shown simply since both compounds (25) and (26) are converted to the same anhydrobase (48) by way of their reduction products (47) and (49) through apo rearrangement and elimination of water or elimination of hydrogen chloride by means of thionylchloride.

The position of the methoxy group in β-erythroidine was determined by degradation of β-erythroidinol (47) to the diol base (50) and oxidation of the latter to 3-methoxyphthalic acid.[33] With this last proof, the structural formulas (25) and (26) for β-erythroidine and apo-β-erythroidine are completely established.

3.2.4 Confirmation of Structure Elucidation β-Erythroidine (25) is reduced with lithium aluminium hydride to the diol base (51), which is converted further by Birch reduction to (52), and by catalytic hydrogenation and elimination of water to the anhydrobase (55).[65]

Müller, Grossnickle, and Boekelheide[65] prepared compound (54) by the synthesis described by Mondon.[66] After reduction and resolution, one antipode was identified unequivocally as the natural degradation product (55).

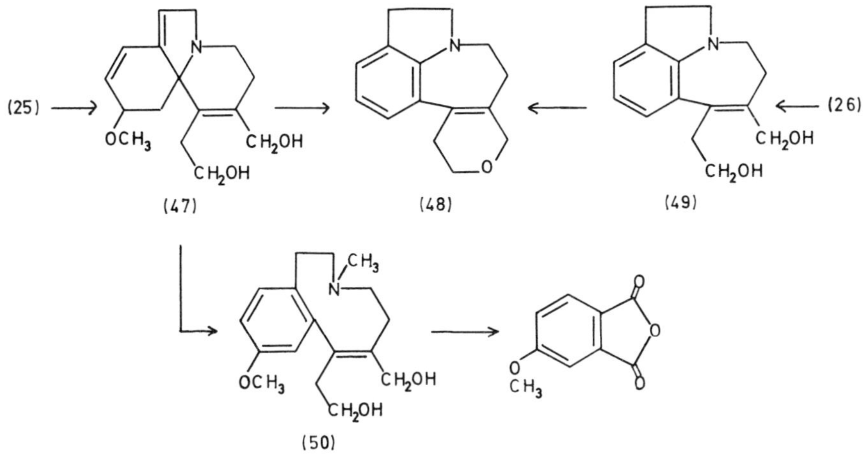

Since the steric course of the above-mentioned synthesis is known to result in *cis*-fusion of rings A and B,[48,67] the evidence for the *cis*-configuration of the anhydrobase (55) is strong. An isomeric anhydrobase (53) with *trans*-configuration is also described[65] but its stereochemistry is not rigorously proved.

Blake, Tretter, and Rapoport[68] confirmed the structure of apo-β-erythroidine by synthesis[69] in 1965.

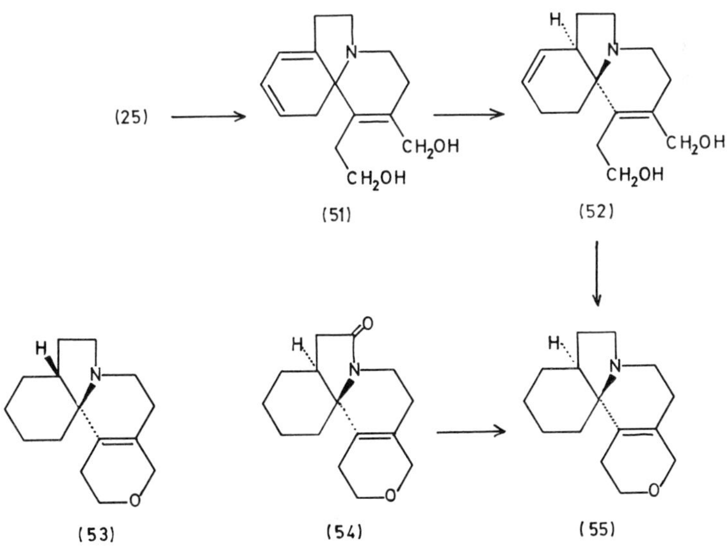

3.3 α-Erythroidine

3.3.1 Special Properties α-Erythroidine (56) is extremely sensitive to air, but it is relatively stable in the form of its salts. Under the conditions of the apo rearrangement it gives no isolable products, nor does it give any color reaction.[70] Hydrogen fluoride eliminates methanol at 20° to give desmethoxy-α-erythroidine that yields the hexahydro base on catalytic hydrogenation. Reduction of (56) with lithium aluminium hydride gives the diol base (57); with Raney nickel both of the bases are hydrogenated partially to dihydro-α-erythroidine (59) and dihydro-α-erythroidinol (58), respectively.

3.3.2 Structure of α-Erythroidine In contrast to the corresponding degradation in the β-series, the Hofmann degradation of dihydro-α-erythroidinol (58)[71] leads to an optically active base (60). α-Erythroidine has, accordingly, an additional asymmetric center and is not a diastereoisomer of β-erythroidine. The further degradation of (60) proceeds by way of several intermediate stages to an optically active ketone (62), the structure of which has been confirmed by synthesis.[72]

Of decisive importance for the elucidation of the structure was the Hofmann degradation of dihydro-α-erythroidine (59) that leads by way of the betaine (61) to the base (39) that was obtained also in the degradation of dihydro-β-erythroidine. The carbonyl group of the lactone ring has the same position in both α-and β-erythroidine; the two bases differ only in the position of the double bond, which in the case of α-erythroidine results in the additional asymmetric center at C(12).[72]

After the close relationship between the erythroidines had been recognized, it also proved possible to provide the simplest proof for their common basic skeleton. When α-erythroidine (56) is warmed with 2*N* sodium hydroxide, β-erythroidine (25) is formed by a unilateral shift of the double bond; with the dihydro bases an equilibrium, which is 95% in favor of the β-form,[72] is set up.

3.4 Absolute Configuration of the Erythroidines

It is hardly possible to determine the absolute configuration of the spiran carbon atom of the erythroidines by chemical degradation. The conditions are more favorable in the case of C(12) of α-erythroidine and C(3) of β-erythroidine.

In 1955 Godfrey, Tarbell, and Boekelheide[71] obtained, by degradation of dehydro-α-erythroidine, a dextrorotatory ketone (62) whose asymmetric center corresponds to C(12) of α-erythroidine. Hill and Schearer[73] synthesized (62) from (+)-tetrahydrofuran-3-carboxylic acid and established

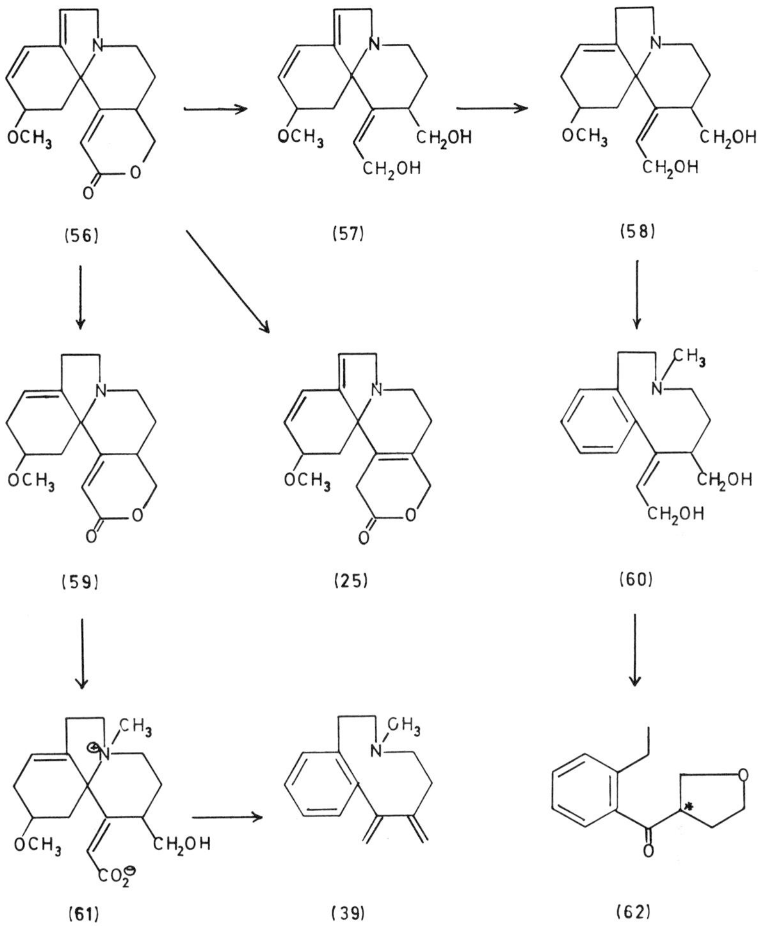

the (12S)-configuration for α-erythroidine by comparison with (+) and (−)-3-methyltetrahydrofuran.

Wenzinger and Boekelheide[53] degraded β-tetrahydro-β-erythroidine systematically to (3S)-3-methoxyadipic acid and in this manner proved the absolute (3R)-configuration of the erythroidines.

In 1963, Hanson[47] confirmed the (3R)-configuration as well as the $\Delta^{1(6)}$-position of the double bond by x-ray diffraction studies of dihydro-β-erythroidine hydrobromide. Beyond this, he succeeded in finally proving the (5S)-configuration at the spiran carbon atom. A summary of all results[74] leads to the following steric formulas for β- and α-erythroidine and dihydro-β-erythroidine, respectively.

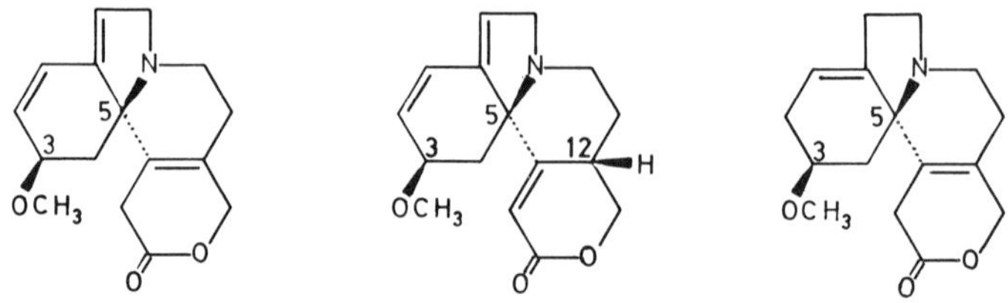

(3R,5S)- (3R,5S,12S)- (3R,5S)-

3.5 Structure of Cocculolidine

In 1966 Wada, Marumo, and Munakata[11] isolated from the fresh leaves of *Cocculus trilobus* an insecticidal alkaloid, Cocculolidine (64). Chemical and spectroscopic investigations suggested a tetracyclic tertiary base with a $\Delta^{\alpha,\beta}$-γ-lactone group, one trisubstituted double bond and a secondary methoxy group. Selective reduction with sodium borohydride afforded a base with a saturated lactone ring which was further reduced catalytically to the tetrahydro base (65).

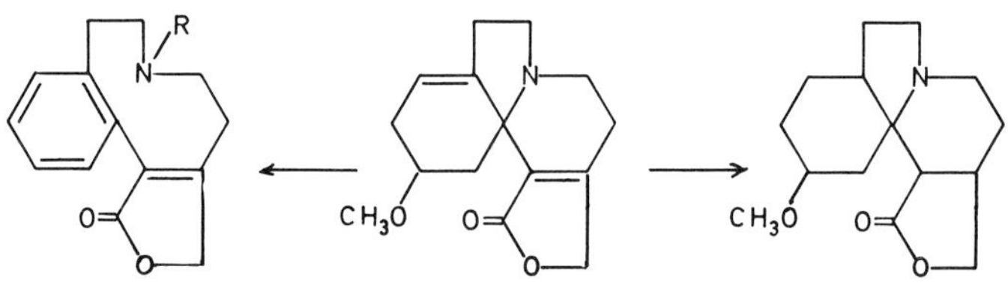

(63) (64) (65)

The main feature of structure elucidation is the degradation with cyanogen bromide to the cyanamide (63) (R = CN), corresponding to the related degradation of dihydroerysotrine.[49] Similarly the hydrochloride of (64) rearranged to the secondary amine (63) (R = H) by heating with phosphoric acid. Spectroscopic considerations confirmed the structure of the degradation products, especially the arrangement of the γ-lactone ring. On the basis of these and other data, cocculolidine was assigned the structure of 14-nordihydro-β-erythroidine (64).

4. ERYTHRINANE SYNTHESES

4.1 Synthesis of Hexahydroapoerysotrine

Soon after the postulation of the spirocyclic carbon–nitrogen skeleton for the aromatic erythrina alkaloids, Belleau[28] in 1953 succeeded in synthesizing erythrinane (R = H) (Scheme 1), which was later followed by the synthesis of 15,16-dimethoxyerythrinane (R = OCH_3).[75,76] Similar syntheses, which are distinguished by an especially smooth ring closure (Scheme 1),[77,78] were reported in 1956. After resolution of the dimethoxy base the levorotatory isomer was identified as hexahydroapoerysotrine.[76,79]

Numerous derivatives of erythrinane[80,81] have been prepared by the two last methods. 14,15,16,17-Tetrahydroerythrinane was obtained by the

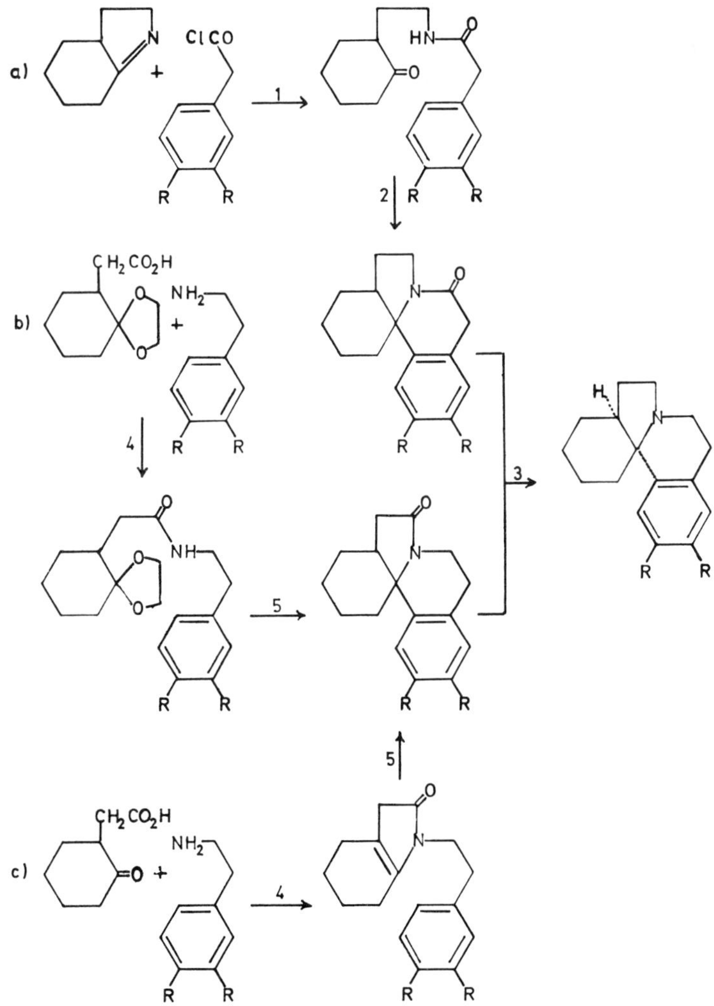

1: Schotten-Baumann; 2: Polyphosphoric acid; 3: $LiAlH_4$;
4: 180°; 5: H^+.

Scheme 1

use of cyclohexenylethylamine.[79] The erythrinane synthesis developed by Sugasawa and Yoshikawa[82] in 1960 has gained no importance as yet.

4.2 Stereochemistry of Erythrinane Ring Closure

Hexahydroapoerysotrine was earlier formulated as a *trans*-compound until it proved possible to demonstrate its relationship to the *cis*-series in the following manner. Hansen[83] obtained bifunctional erythrinanes by use of the glyoxylic ester synthesis (Scheme 2).[84]

In this procedure the precursor for the cyclization is stabilized by enamine formation, and the amine residue is eliminated again in the ring closure to the tetracyclic enol lactam. After reduction of the enol group

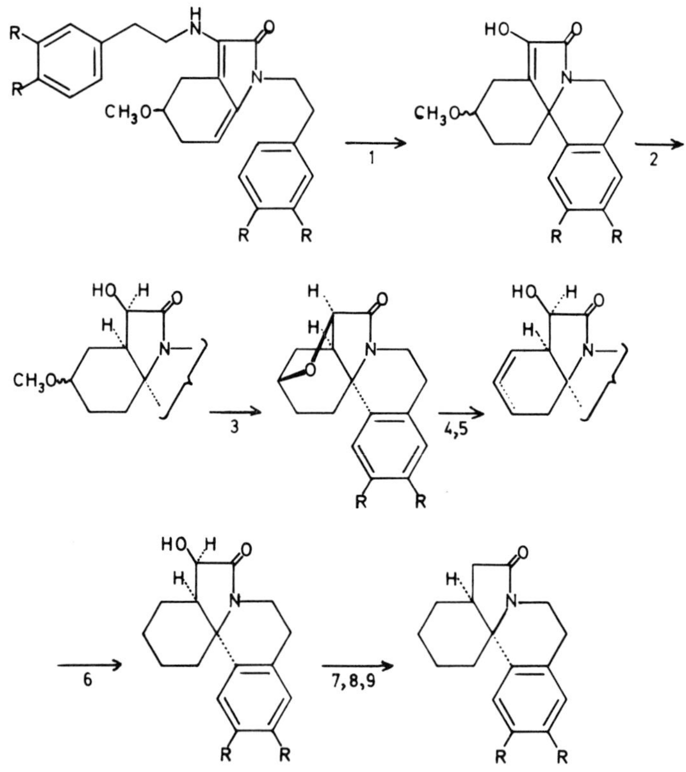

R = OCH_3

1: $H^{\oplus}$; 2: Raney-Ni, H_2; 3: H_2SO_4; 4: Ac_2O, $H^{\oplus}$;
5: $OH^{\ominus}$; 6: H_2, Pt; 7: MsCl/Py;
8: NaI, acetone; 9: Zn, AcOH.

Scheme 2

and treatment with sulfuric acid, a cyclic ether with unequivocal *cis*-fusion of rings A and B is formed. The same lactam, which was used earlier for the synthesis of hexahydroapoerysotrine, is obtained by degradation without changing the configuration. The natural base, therefore, belongs to the *cis*-erythrinane series.[48]

Reductive elimination of the phenolic groups in *cis*-15,16-dihydroxy-erythrinane leads to *cis*-erythrinane,[48] identical with the Belleau base.[28] To date, there are no known exceptions to the rule that cyclization leads exclusively to *cis*-erythrinanes.*

4.3 Syntheses of *trans*-Erythrinanes

In 1962, Faro[51] synthesized the above mentioned base (24) with the double-bond system of the natural alkaloids.

Partial hydrogenation with palladium on carbon in methanol leads to the $\Delta^{1(6)}$-base (67) by 1,4-addition of hydrogen[85]; it is completely different from the Δ^6-base (66) obtained by another route.[81,84]

Seidel[85] hydrogenated the $\Delta^{1(6)}$-base (67) with Adams catalyst in acetic acid to a mixture of *cis*- and *trans*-dimethoxyerythrinane, (69) and (70), in the ratio 3:1. The difference in stereochemistry was proved by repeated

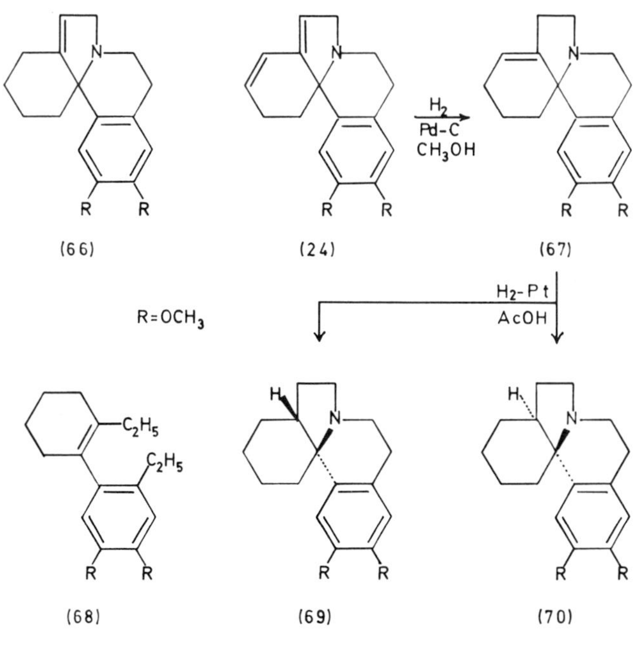

*"14,15,16,17-Tetrahydroerythrinane," described by Belleau[86] and checked by Boekelheide,[79] is actually hexahydroerythrinane.[67]

Hofmann degradation and hydrogenation of the first two methine bases; in the third degradation step the asymmetric centers disappeared and an identical product (68) is obtained from both bases.

Interestingly, *cis*- and *trans*-dimethoxyerythrinane and their optical antipodes manifest similar curariform activity.[9]

4.4 Synthesis of Tetrahydroerysonine Methyl Ether

In 1959 Prelog, Langemann, Rodig, and Ternbah[43] employed a different route for the construction of the erythrinane skeleton (Scheme 3) and succeeded for the first time in introducing a keto group at C(3).

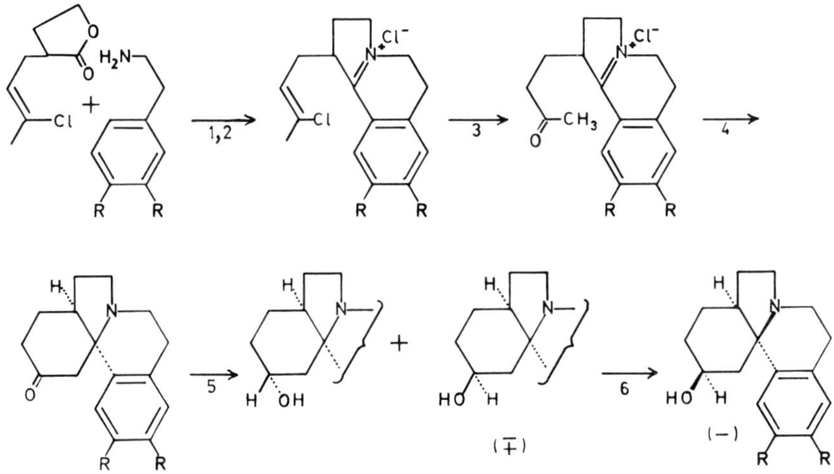

R = OCH_3

1: 200°; 2: $POCl_3$; 3: H_2SO_4, 20°; 4: $OH^{\ominus}$;
5: $NaBH_4$; 6: Resolution.

Scheme 3

After reduction of the labile keto base to a mixture of the epimeric hydroxy bases, the stereoisomer with the intramolecular hydrogen bond between the hydroxyl group and nitrogen was resolved. The levorotatory isomer proved to be identical with tetrahydroerysonine methyl ether.

4.5 Total Synthesis

Mondon and Nestler[87] succeeded in carrying out a total synthesis of erysotrine in 1964 by extending the synthesis developed by Hansen.[83] The Δ^1-7-acetoxylactam available by acetolysis of the already mentioned

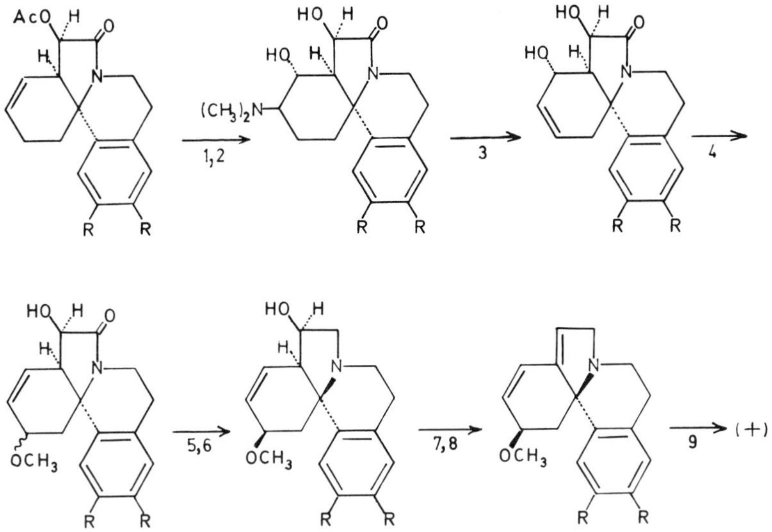

R = OCH_3

1: $C_6H_5CO_3H$; 2: $(CH_3)_2NH$; 3: Cope cleavage;
4: CH_3OH—HCl; 5: Separation by chromatography;
6: $LiAlH_4$; 7: MsCl/Py; 8: KOH; 9: Resolution.

Scheme 4

cyclic ether (Scheme 2) was converted to racemic erysotrine (Scheme 4). The dextrorotatory isomer was obtained by resolution of the racemate, and it was identical in all respects with erysotrine.[35]

Recently, dihydroerysodine has been synthesized by a biogenetic-like scheme.[29] In addition to the oxydative coupling of the phenolamine (4) to erysodienone (6),[88] another interesting feature of the synthesis is the novel rearrangement of the dienol base (71) to the enone base (72) with hydrogen chloride in ether. Reduction to a mixture of epimeric alcohols (73) and replacement of the hydroxy group by hydrogen afforded racemic dihydroerysodine that was resolved.

Methylation of the epimers (73) with diazomethane leads to racemic erythratidine and epierythratidine, each of which gives racemic erysotrine by elimination of water.* Similarly, Barton and co-workers[19] by partial synthesis obtained erythraline from natural erythratine.

*A. Mondon and Christa Fischer, unpublished. At present comparision with natural erythratidine is not possible because a sample is unavailable.

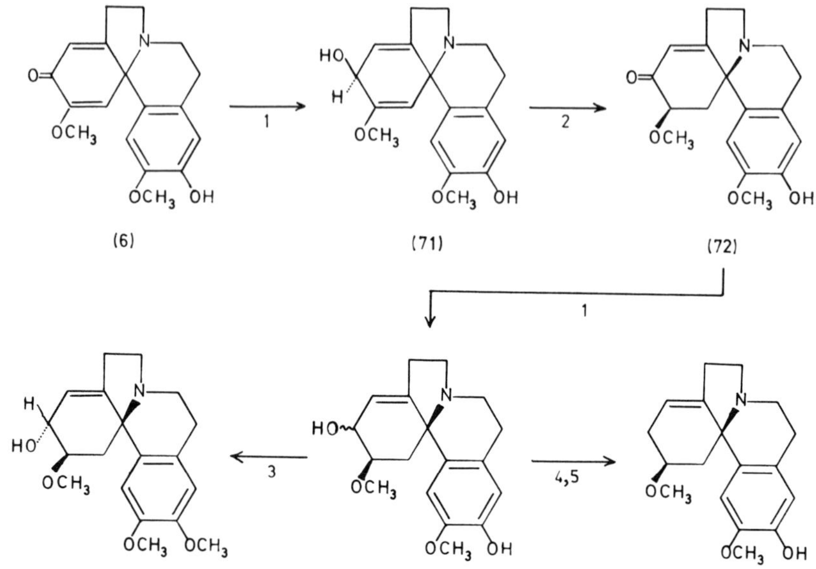

(6) (71) (72)

(73)

(±) Erythratidine (±) Dihydroerysodine

1: $NaBH_4$; 2: HCl-Ether; 3: CH_2N_2 and separation by chromatography; 4: $SOCl_2$; 5: Pd-C, H_2.

REFERENCES

1. L. Marion, *Alkaloids*, **2**, 499–511 (1952).
2. V. Boekelheide and V. Prelog, "Progress in Organic Chemistry," Vol. 3, pp. 218–266 (J. W. Cook, Ed.), Butterworths, London, England, 1955.
3. V. Prelog, *Angew. Chem.* **69**, 33 (1957); V. Prelog, "Festschrift A. Stoll," (Collection of Papers Dedicated to A. Stoll), pp. 728–745. Birkhäuser, Basel, Switzerland, 1957.
4. V. Boekelheide, *Alkaloids* **7**, 201–227 (1960).
5. H. G. Boit, "Ergebnisse der Alkaloid-Chemie bis 1960," pp. 383–401, Akademie-Verlag, Berlin, Germany, 1961.
6. R. K. Hill, *Alkaloids* **9**, 483–515 (1967).
7. K. Unna, M. Kniazuk, and J. G. Greslin, *J. Pharmacol. Exp. Therap.* **80**, 39 (1944); K. Unna and J. G. Greslin, *J. Pharmacol. Exp. Therap.*, **80**, 53 (1944).
8. D. Megirian, D. E. Leary, and I. H. Slater, *J. Pharmacol. Exp. Therap.* **113**, 212 (1955).
9. H. Lüllmann, A. Mondon, and P. R. Seidel, *Arch. Exp. Pathol. Pharmakol.* **258**, 91 (1967).
10. L. E. Craig, *Alkaloids* **5**, 281–283 (1955).
11. K. Wada, S. Marumo, and K. Munakata, *Tetrahedron Letters* 5179 (1966).
12. K. Folkers and R. T. Major, *J. Amer. Chem. Soc.* **59**, 1580 (1937).
13. K. Folkers and R. T. Major, U.S. Patent 2,385,266, Sept. 18, 1945; *Chem. Abstr.* **39**, 5408 (1945).

14. C. Lapière, thesis, University of Liège, Belgium, 1952.
15. K. Folkers and F. Koniuszy, *J. Amer. Chem. Soc.*, **62,** 436 (1940).
16. V. Deulofeu, R. Labriola, E. Hug, M. Fondovila, and A. Kauffmann, *J. Org. Chem.* **12,** 486 (1947).
17. V. Prelog, K. Wiesner, H. G. Khorana, and G. W. Kenner, *Helv. Chim. Acta* **32,** 453 (1949).
18. K. Folkers and F. Koniuszy, *J. Amer. Chem. Soc.* **61,** 1232 (1939).
19. D. H. R. Barton, R. James, G. W. Kirby, D. W. Turner, and D. A. Widdowson, *Chem. Commun.* 294 (1966).
20. V. Deulofeu, *Chem. Ber.* **85,** 620 (1952).
21. K. Folkers and F. Koniuszy, *J. Amer. Chem. Soc.* **62,** 1677 (1940).
22. K. Folkers, J. Shavel, Jr., and F. Koniuszy, *J. Amer. Chem. Soc.* **63,** 1544 (1941).
23. K. Folkers and J. Shavel, Jr., *J. Amer. Chem. Soc.* **64,** 1892 (1942).
24. R. A. Gentille and R. Labriola, *J. Org. Chem.* **7,** 136 (1942).
25. M. Tomita and H. Yamaguchi, *Pharm. Bull. Tokyo* **4,** 225 (1956).
26. K. Folkers, F. Koniuszy, and J. Shavel, Jr., *J. Amer. Chem. Soc.* **66,** 1083 (1944).
27. C. Lapière, *J. Pharm. Belg.* **6,** 71 (1951); *Chem. Abstr.* **45,** 9806 (1951).
28. B. Belleau, *J. Amer. Chem. Soc.* **75,** 5765 (1953).
29. A. Mondon and M. Ehrhardt, *Tetrahedron Letters* 2557 (1966).
30. R. T. Major and K. Folkers, U.S. Patent 2,407,714, Sept. 17, 1946; *Chem. Abstr.* **41,** 781 (1947).
31. Merck Co. and R. T. Major, British Patent 543,187, Feb. 13, 1942; *Chem. Abstr.* **37,** 232 (1943).
32. D. H. R. Barton, R. James, G. W. Kirby, and D. A. Widdowson, *Chem. Commun.* 267 (1967).
32a. D. H. R. Barton, Chem. in Britain, 330 (1967); D. H. R. Barton, R. James, G. W. Kirby, D. W. Turner and D. A. Widdowson, *J. Chem. Soc.* (London) C, 1529 (1968).
33. V. Boekelheide, J. Weinstock, M. F. Grundon, G. L. Sauvage, and E. J. Agnello, *J. Amer. Chem. Soc.* **75,** 2550 (1953).
34. F. Koniuszy, P. F. Wiley, and K. Folkers, *J. Amer. Chem. Soc.* **71,** 875 (1949).
35. R. A. Labriola, V. Deulofeu, and B. Berinzaghi, *J. Org. Chem.* **16,** 90 (1950).
36. K. Folkers and F. Koniuszy, *J. Amer. Chem. Soc.* **62,** 1673 (1940).
37. K. Folkers and F. Koniuszy, *J. Amer. Chem. Soc.* **72,** 1832 (1950).
38. K. Folkers and F. Koniuszy, *J. Amer. Chem. Soc.* **61,** 3053 (1939).
39. K. Folkers, F. Koniuszy, and J. Shavel, Jr., *J. Amer. Chem. Soc.* **73,** 589 (1951).
40. K. Folkers, F. Koniuszy, and J. Shavel, Jr., *J. Amer. Chem. Soc.* **64,** 2146 (1942).
41. G. W. Kenner, H. G. Khorana, and V. Prelog, *Helv. Chim. Acta* **34,** 1969 (1951).
42. M. Carmack, B. C. McKusick, and V. Prelog, *Helv. Chim. Acta* **34,** 1601 (1951).
43. V. Prelog, A. Langemann, O. Rodig, and M. Ternbah, *Helv. Chim. Acta* **42,** 1301 (1959).
44. K. Wiesner, Z. Valenta, A. J. Manson, and F. W. Stonner, *J. Amer. Chem. Soc.* **77,** 675 (1955).
45. J. R. Merchant, D. D. Vaghani, and C. S. Kalliaupur, *J. Indian Chem. Soc.* **40,** 315 (1963).
46. A. Mondon and H. U. Menz, *Tetrahedron* **20,** 1729 (1964).
47. A. W. Hanson, *Proc. Chem. Soc.* 52 (1963); *Acta Cryst.* **16,** 939 (1963).
48. A. Mondon and K. F. Hansen, *Tetrahedron Letters* **14,** 5 (1960).
49. V. Prelog, B. C. McKusick, J. R. Merchant, S. Julia, and M. Wilhelm, *Helv. Chim. Acta* **39,** 498 (1956).

50. A. Mondon, H. J. Nestler, H. G. Vilhuber, and M. Ehrhardt, *Chem. Ber.* **98,** 46 (1965).
51. H. P. Faro, thesis, University of Kiel, Germany, 1962.
52. W. Nowacki and G. F. Bonsma, *Z. Krist.* **110,** 89 (1958).
53. G. R. Wenzinger and V. Boekelheide, *Proc. Chem. Soc.* 53 (1963).
54. V. Boekelheide and M. Y. Chang, *J. Org. Chem.* **29,** 1303 (1964).
55. E. M. Dietz and K. Folkers, *J. Amer. Pharm. Assoc.* **35,** 48 (1946).
56. F. Koniuszy and K. Folkers, *J. Amer. Chem. Soc.* **72,** 5579 (1950).
57. F. Koniuszy and K. Folkers, *J. Amer. Chem. Soc.* **73,** 333 (1951).
58. G. L. Sauvage and V. Boekelheide, *J. Amer. Chem. Soc.* **72,** 2062 (1950).
59. G. L. Sauvage, F. M. Berger, and V. Boekelheide, *Science* **109,** 627 (1949).
60. M. F. Grundon and V. Boekelheide, *J. Amer. Chem. Soc.* **74,** 2637 (1952).
61. M. F. Grundon and V. Boekelheide, *J. Amer. Chem. Soc.* **75,** 2537 (1953).
62. V. Boekelheide and E. Agnello, *J. Amer Chem. Soc.* **73,** 2286 (1951).
63. J. Weinstock and V. Boekelheide, *J. Amer. Chem. Soc.* **75,** 2546 (1953).
64. M. F. Grundon, L. G. Sauvage, and V. Boekelheide, *J. Amer. Chem. Soc.* **75,** 2541 (1953).
65. M. Müller, T. T. Grossnickle, and V. Boekelheide, *J. Amer. Chem. Soc.* **81,** 3959 (1959).
66. A. Mondon, *Angew. Chem.* **68,** 578 (1956).
67. B. Neffgen, diplome, University of Kiel, Germany, 1967.
68. J. Blake, J. R. Tretter, and H. Rapoport, *J. Amer. Chem. Soc.* **87,** 1397 (1965).
69. J. Blake, J. R. Tretter, G. J. Juhasz, W. Bonthrone, and H. Rapoport, *J. Amer. Chem. Soc.* **88,** 4061 (1966).
70. V. Boekelheide and M. F. Grundon, *J. Amer. Chem. Soc.* **75,** 2563 (1953).
71. J. C. Godfrey, D. S. Tarbell, and V. Boekelheide, *J. Amer. Chem. Soc.* **77,** 3342 (1955).
72. V. Boekelheide and G. C. Morrison, *J. Amer. Chem. Soc.* **80,** 3905 (1958).
73. R. K. Hill and W. R. Schearer, *J. Org. Chem.* **27,** 921 (1962).
74. V. Boekelheide and G. R. Wenzinger, *J. Org. Chem.* **29,** 1307 (1964).
75. B. Belleau, *Chem. and Ind. (London)* 410 (1956).
76. B. Belleau, *Can. J. Chem.* **35,** 651 (1957).
77. A. Mondon, *Chem. Ber.* **92,** 1461 (1959).
78. Ref. 77, p. 1472.
79. V. Boekelheide, M. Müller, J. Jack, T. T. Grossnickle, and M. Y. Chang, *J. Amer. Chem. Soc.* **81,** 3955 (1959).
80. A. Mondon, G. Hasselmeier, and J. Zander, *Chem. Ber.* **92,** 2543 (1959).
81. A. Mondon, J. Zander, and H. U. Menz, *Liebigs Ann. Chem.* **667,** 126 (1963).
82. S. Sugasawa, and H. Yoshikawa, *Chem. Pharm. Bull. Tokyo* **8,** 290 (1960).
83. K. F. Hansen, thesis, University of Kiel, Germany, 1960.
84. A. Mondon, *Liebigs Ann. Chem.* **628,** 123 (1959).
85. P. R. Seidel, thesis, University of Kiel, Germany, 1965.
86. B. Belleau, *Can. J. Chem.* **35,** 663 (1957).
87. A. Mondon and H. J. Nestler, *Agnew. Chem.* **76,** 651 (1964).
88. J. E. Gervey, F. McCapra, T. Money, and A. I. Scott, *Chem. Commun.* 142 (1966).

8

Colchicine

WILLIAM C. WILDMAN
Department of Chemistry
Iowa State University of Science and Technology
Ames, Iowa

1. INTRODUCTION

Frequent medical, chemical, and botanical references have marked the long history of the active ingredient of *Colchicum autumnale* L. (Liliaceae). The most common active principle, colchicine, has been used for thousands of years in the treatment and diagnosis of gout. Although the substance is remarkably effective, the mechanism of its action is uncertain. The difference between therapeutic and toxic doses is so minute that the gout sufferer has little choice between the Scylla of the disease and the Charybdis of colchicine poisoning. A report in 1889 that colchicine could arrest mitosis at metaphase stimulated little interest until the period from 1934 to 1950 when the genetic potentialities and possible therapeutic value of the agent in combating cancer became apparent. The biological aspects of colchicine and related compounds have been reviewed in book form.[1] Other recent reviews emphasizing the chemistry of colchicine also have included work of biological interest.[2,3] This chapter summarizes the chemistry and photochemistry of colchicine and presents a brief introduction to the structure of androcymbine, an important biosynthetic precursor.

The name *Colchicum* is derived from an area near the Black Sea (Colchis). *Colchicum* species are common in most of Europe and the Middle East. Chemical studies began about 1820 and the present Zeisel methoxyl determination has its foundations in the determination of this functional group in colchicine. Possessing only one asymmetric carbon atom, the alkaloid (1) seems exceptionally simple when compared with the complex alkaloids now known, yet structural studies spanned more than 100 years. Total synthesis was achieved in 1959. Significant advances in understanding the biosynthesis of colchicine began in 1960 and progress continues to be made.

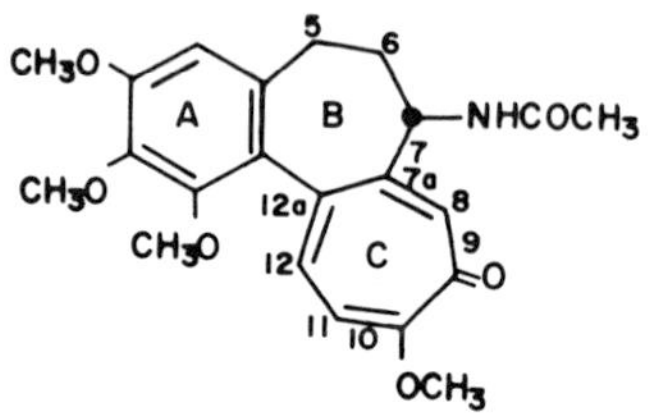

(1)

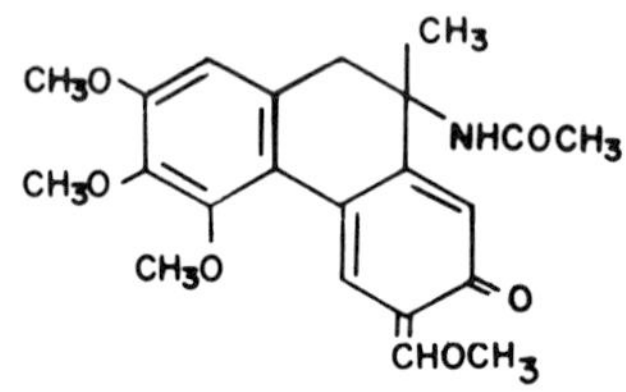

(2)

Structure (1) is known to be the correct representation for colchicine from x-ray analysis, total synthesis, and interpretation of a large amount of degradative data. The structural studies outlined below do not reflect the intense chemical interest of the 1940's in the mechanisms of the rearrangements of rings B and C (and tropolone chemistry in general) that led to the discarding of the earlier proposed Structure (2).[4] Many research groups have been active in this field, and no attempt has been made in this chapter to document individual contributions (often overlapping) in the structural work prior to 1960. Exact references to the many contributions may be found in earlier reviews.[1–3,5]

2. STRUCTURE OF COLCHICINE

Colchicine, $C_{22}H_{25}NO_6$, is not an alkaloid in the strictest sense because the nitrogen atom is not basic and it is part of an acetamido function. Four oxygen atoms are present as methoxyl groups. The remaining oxygen atom is unreactive toward reagents that effect acylation and affords no carbonyl derivatives. Acid hydrolysis of colchicine in varying degrees of stringency provides a method for the selective cleavage of the functional groups. Dilute acid affords colchiceine (3a), an acidic substance that can be methylated to colchicine and isocolchicine (3b) by diazomethane. More strenuous acid hydrolysis cleaves the acetamido group to form trimethylcolchicinic acid (4a). Under most vigorous conditions, the remaining three methoxyl groups can be removed to provide colchicinic acid (4b).

The reactions of colchicine and colchiceine in basic media have proven to be most valuable in the determination of structure, particularly for rings B and C. In these transformations, ring C has become benzenoid. Of primary importance is N-acetylcolchinol methyl ether (5a) prepared from

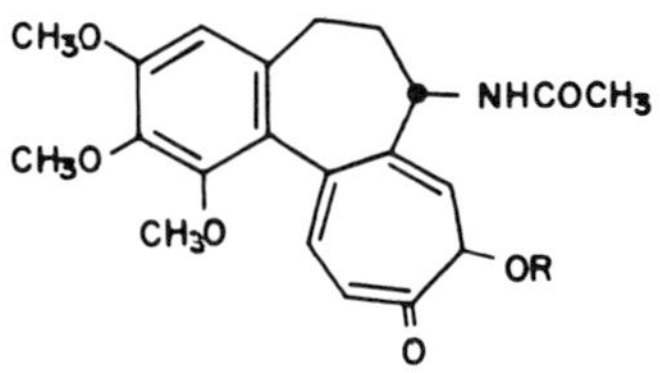

(3)

(a) R = H
(b) R = CH_3

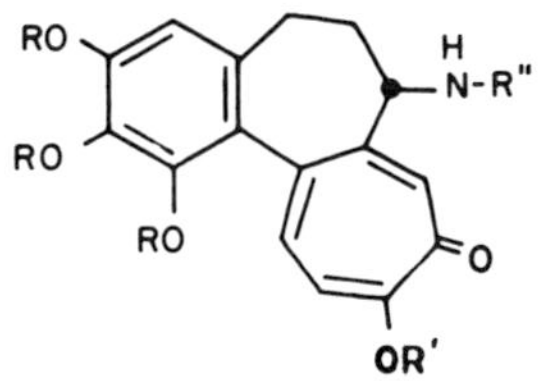

(4)

(a) R = CH_3; R′, R″ = H
(b) R, R′, R″ = H

colchiceine either by the action of alkaline hydrogen peroxide and subsequent methylation of the reaction product with diazomethane or by the O-methylation and reduction of N-acetyliodocolchinol (5b), in turn derived from colchiceine by the action of iodine and base. Acid hydrolysis of (5a) affords colchinol methyl ether (5c). Allocolchicine (6) is formed by the action of methanolic sodium methoxide on either colchicine or isocolchicine. Allocolchicine has been related to N-acetylcolchinol *via* the analogous amine.

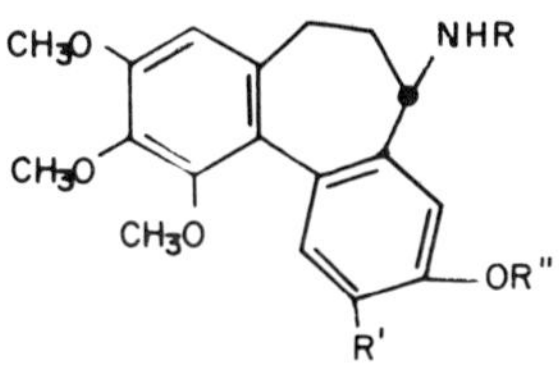

(5)

(a) R = $COCH_3$; R′ = H; R″ = CH_3
(b) R = $COCH_3$; R′ = I; R″ = CH_3
(c) R, R′ = H; R″ = CH_3

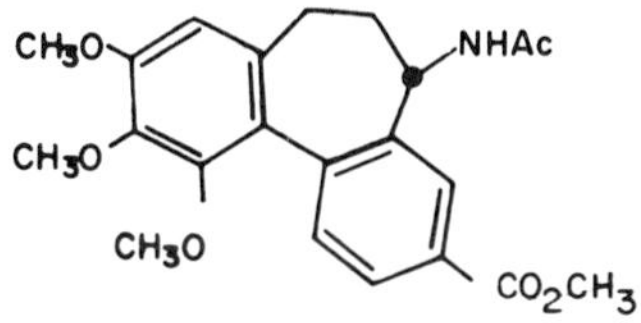

(6)

Colchicine and many of its derivatives may be oxidized to 3,4,5-trimethoxyphthalic acid. These degradations provided the first evidence for the structure of ring A and eleven of the twenty-two carbon atoms. Since one methoxyl and the acetamido group must be present in other parts of the molecule, eight carbon atoms need to be accounted for. From the molecular formula, it is clear that colchicine has eleven unsaturated centers or rings. Ring A and the acetamido group comprise five of these, and the remainder can be deduced from reductive studies. The extent of catalytic reduction of colchicine depends upon the solvent and catalyst, but hexahydrocolchicine and hexahydrocolchiceine have been prepared. These compounds show the presence of a new hydroxyl group, and it must be assumed that one mole of hydrogen is used to reduce the unreactive carbonyl group to an alcohol. The hexahydro compounds still contain one nonreducible double bond since they form epoxides with peracid. Thus, colchicine must be tricyclic and possess three double bonds in addition to those of ring A and the carbonyl groups.

The seven-membered nature of ring B was established by the unambiguous structures assigned to N-acetylcolchinol methyl ether (5a) and N-benzoylcolchinic anhydride (7). The latter compound was prepared by the oxidation of N-benzoyltrimethylcolchinic acid. Deamination of (7) with phosphorus pentoxide gave an anhydride (8) that was not naphthalenic in chemical behavior. Catalytic reduction of (8) afforded a dihydro derivative [(8), no double bond (7)–(8)], the structure of which was verified by synthesis.

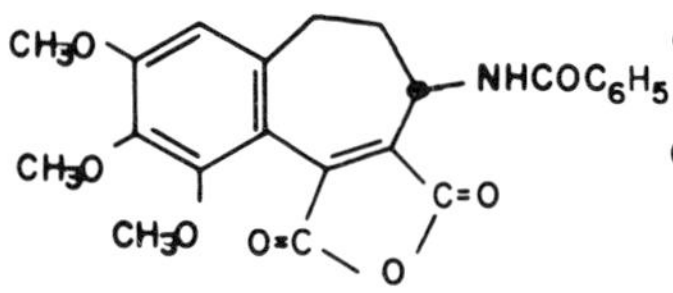

(7)

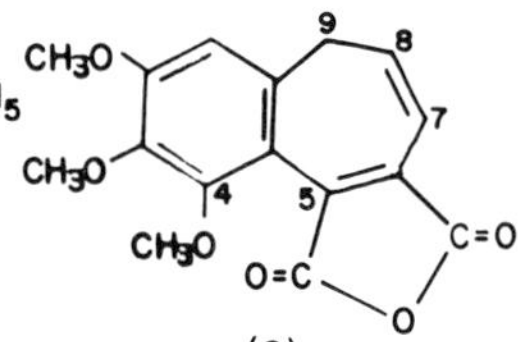

(8)

When heated with phosphorus pentoxide, N-acetylcolchinol methyl ether (5a) provides two compounds, deaminocolchinol methyl ether (9) and isodeaminocolchinol methyl ether (10). Both isomers afford the same dihydro derivative. Compounds (9) and (10) also have been prepared from (5c) by carbinol formation with nitrous acid and dehydration. This transformation is not straightforward and involves two successive Demjanow rearrangements. The intermediate carbinol has a six-membered ring B (13) that expands to seven in the dehydration step. Exhaustive methylation and Hofmann degradation of (5a) produces only (9). The methoxyl substitution pattern of the colchinol series was established by the oxidation of (9) to 2,3,4,7-tetramethoxyphenanthraquinone using sodium dichromate in acetic acid. Treatment of (9) and (10) with osmium tetroxide followed by lead tetraacetate cleavage and dilute base, afforded (11) and (12), respectively. The structures of the latter were proven by synthesis.

With these data suggesting that ring B is seven-membered, a direct synthesis of (14) [a by-product formed in the oxidation of (9) to 2,3,4,7-tetramethoxyphenanthraquinone], and two syntheses of colchinol methyl ether (5c) established the seven-membered nature of Ring B in a second manner.

The troponoid nature of ring C in colchicine was first suggested by Dewar in 1945.[6] The correctness of this nucleus has been verified by analogy to the chemical and photochemical reactions of simpler tropolones, x-ray analysis, and the total syntheses cited in Section 3. The mechanisms of the tropolone → benzene rearrangement and the photochemical transformations of tropolones have been reviewed.[7–9]

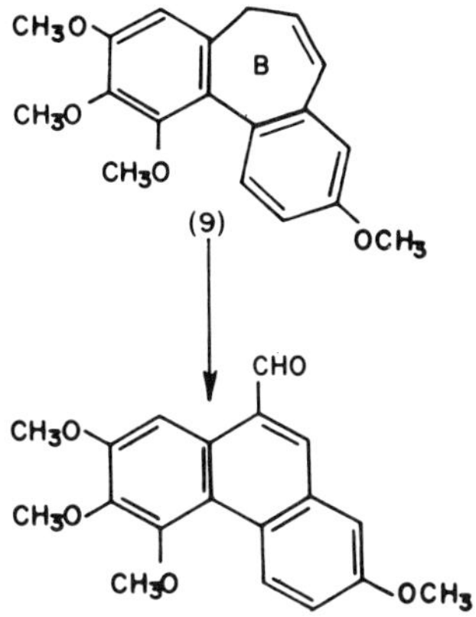

(9)

(11)

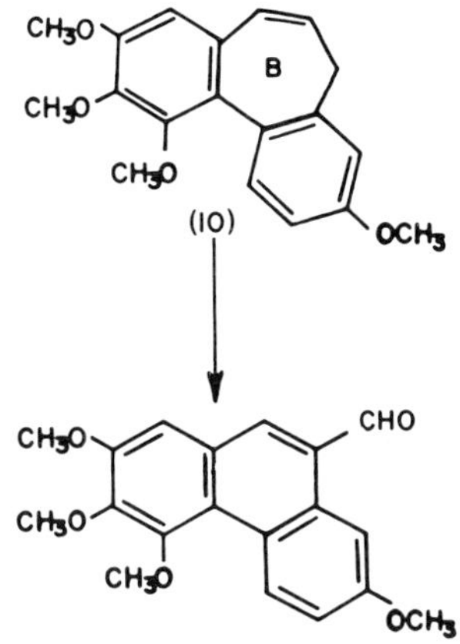

(10)

(12)

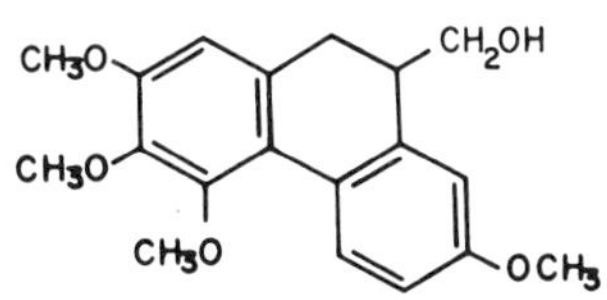

(13)

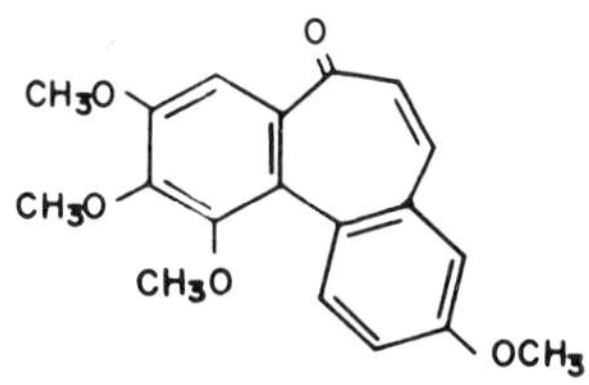

(14)

Compounds of the "*iso*" series (15) have more negative specific rotations than those of the "normal" series (16), and the former appear to be both less toxic and less physiologically active than the latter. From these generalizations and the similarity in the 7 μ region of the infrared spectra of colchiceine and compounds in the "*iso*" series, colchiceine is considered a single substance represented by (3a) and partial structure (15, R = H). Conversion to the isomeric structure (16, R = H) must be facile since colchiceine affords both (3b) and (1) upon methylation.

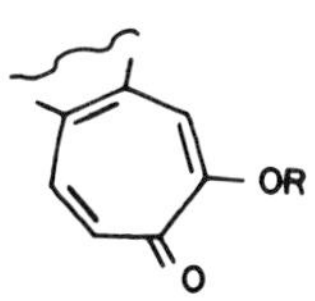

(15)

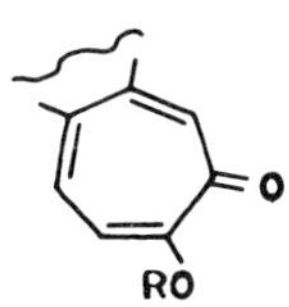

(16)

Since the degradative data could not differentiate between (1) and (3b), x-ray crystallographic studies were undertaken. Results indicated that (1) was the most likely structure for colchicine. It should be noted that the total syntheses (Section 3) involved the preparation of colchiceine derivatives, and no evidence for the relative positions of the tropolone carbonyl and methoxyl in colchicine could be obtained from these efforts.

Colchicine is optically active by virtue of the single asymmetric carbon atom at C(7). The absolute configuration at this center was established by strenuous oxidation of colchicine to N-acetyl-L-glutamic acid.[10]

3. SYNTHESES OF COLCHICINE

With conclusive degradative evidence that colchicine is represented by (1), several research groups examined routes leading to the total synthesis of the alkaloid. A key structure in two of the four syntheses reported in this section was (±)-trimethylcolchicinic acid (4a), which had been converted earlier to (1) by resolution, N-acetylation, and O-methylation. Except for the mode of construction of ring C, the syntheses of Sections 3.1 and 3.2 are similar in concept. Both construct ring C from a benzsuberone derivative and introduce nitrogen at C(7) *via* a 7-bromo compound in the late stages of the synthesis. In the third synthesis (Section 3.3), the potential amino group is present from the first step; rings A, B, and C are then formed in turn. A fourth synthesis (*dl*-colchiceine, Section 3.4) constructs rings A and an unsubstituted ring C prior to ring B.

3.1 Synthesis from 7,8,9-Trimethoxy-9-benzosuberone

A flow sheet of this synthesis is given below:

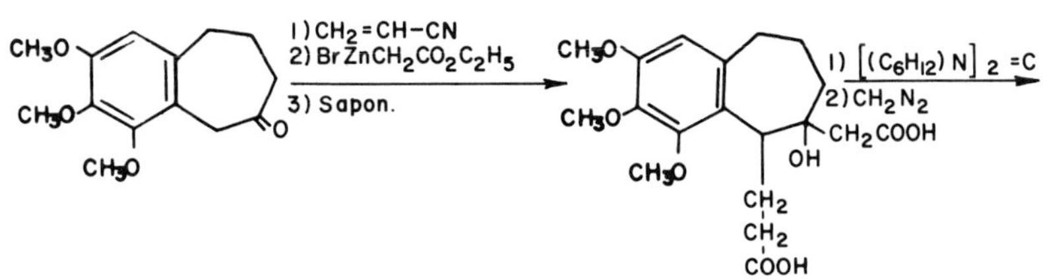

(17) (18)

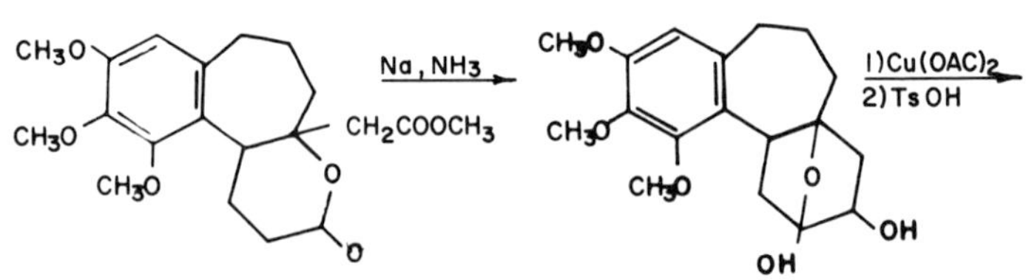

(19) (20)

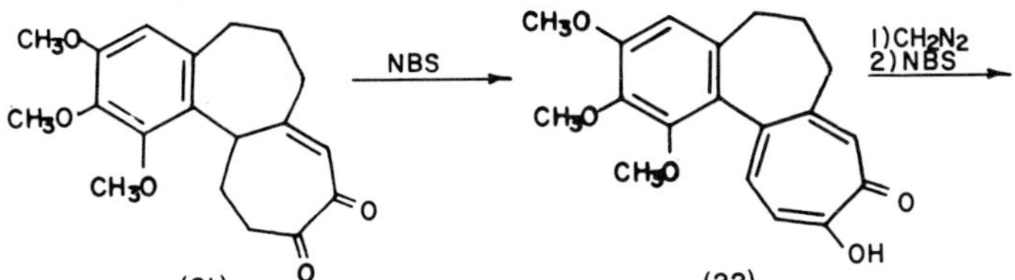

(21) (22)

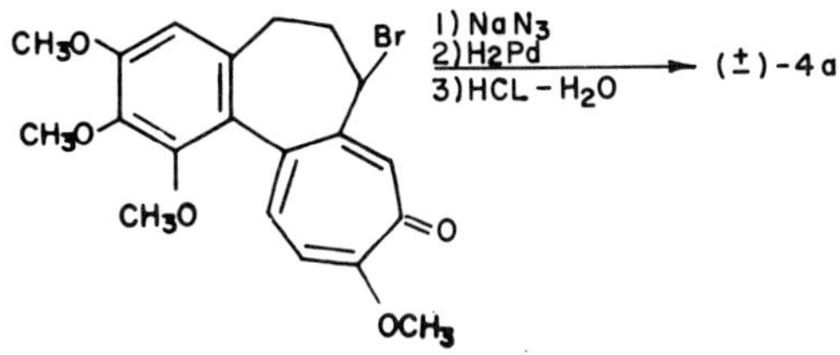

(23)

Although most of the steps in the successful synthesis outlined above appear to be straightforward, alternative procedures might be considered at many stages. The rationale behind the specific choice of intermediates and reagents at each step has been reported in detail.[11] Perhaps the key step in this synthesis is the conversion of (19) to (20) by an intramolecular acyloin condensation under carefully controlled conditions that minimized the competing reaction of the Dieckmann condensation of (19) to a six-membered ketone.

3.2 Synthesis from Purpurogallin Trimethyl Ether

A flow sheet of the synthesis is given below:

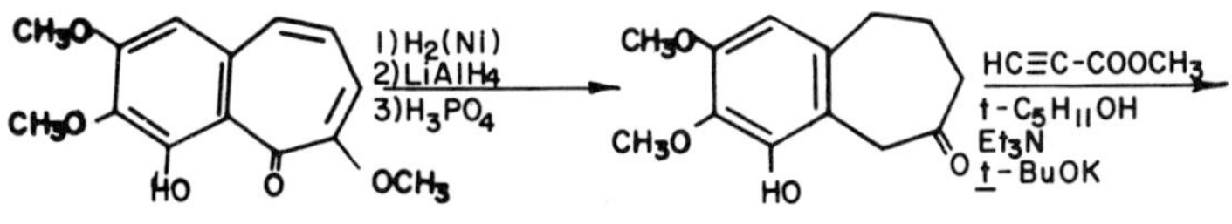

(24) (25)

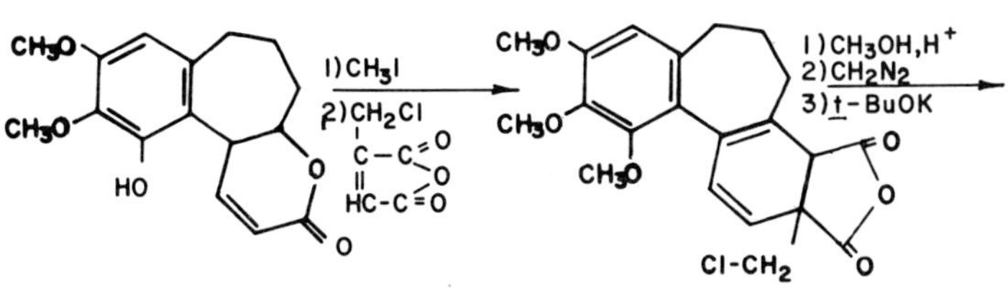

(26) (27)

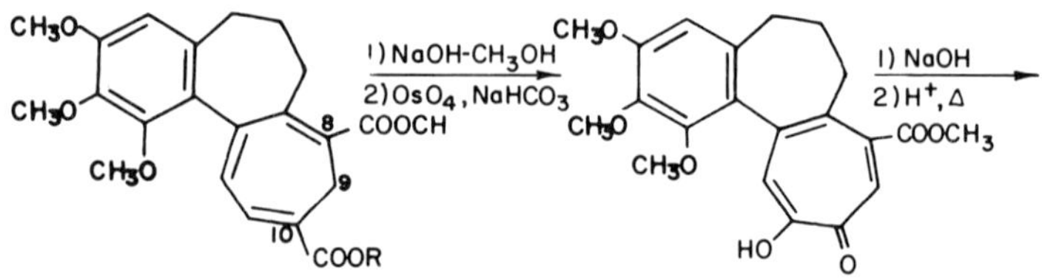

(28) (29)

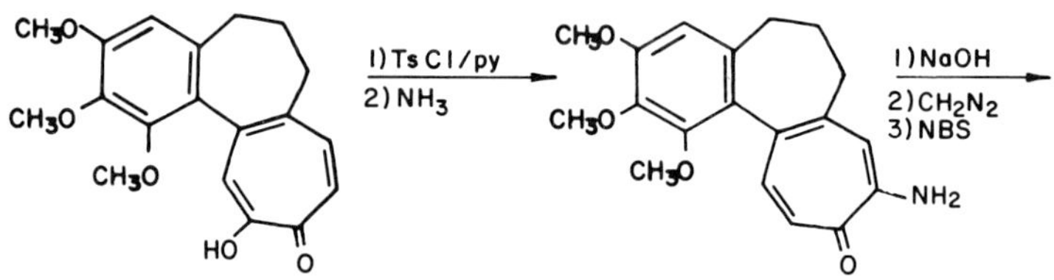

(30) (31)

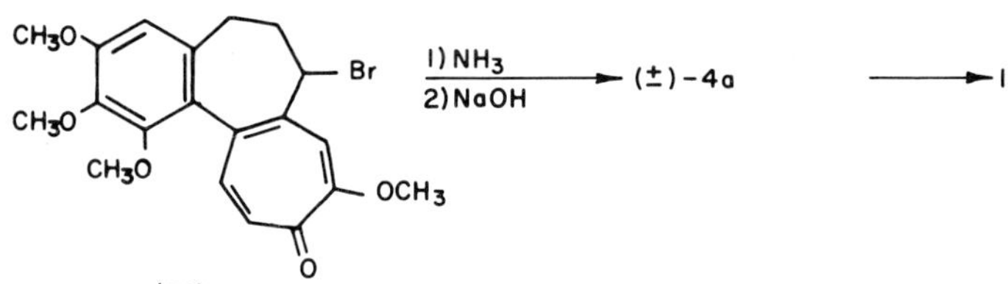

(32)

Although the synthesis of (28) is a major achievement in itself, the most crucial steps of the synthesis were in the conversion of (28) to (31). In these steps, an 8,10-disubstituted cycloheptatriene was transformed into a 9,10-tropolone intermediate. These transformations included the selective hydrolysis of (28) and oxidative decarboxylation of the intermediate by osmium tetroxide to provide (29). Hydrolysis of (29) and subsequent decarboxylation yielded. (30).[12]

3.3 Synthesis from a Substituted Isothiazole

The third synthesis of colchicine represents a significant departure from the synthetic concepts of Sections 3.1 and 3.2. Potential carbon atoms for positions 6, 7, 7a, 8, and 12a as well as the amino group in colchicine are present in the fundamental starting material (33). This compound is of particular interest because no simple derivative of isothiazole was known prior to this synthesis. Rings A, B, and C are constructed in turn. Most

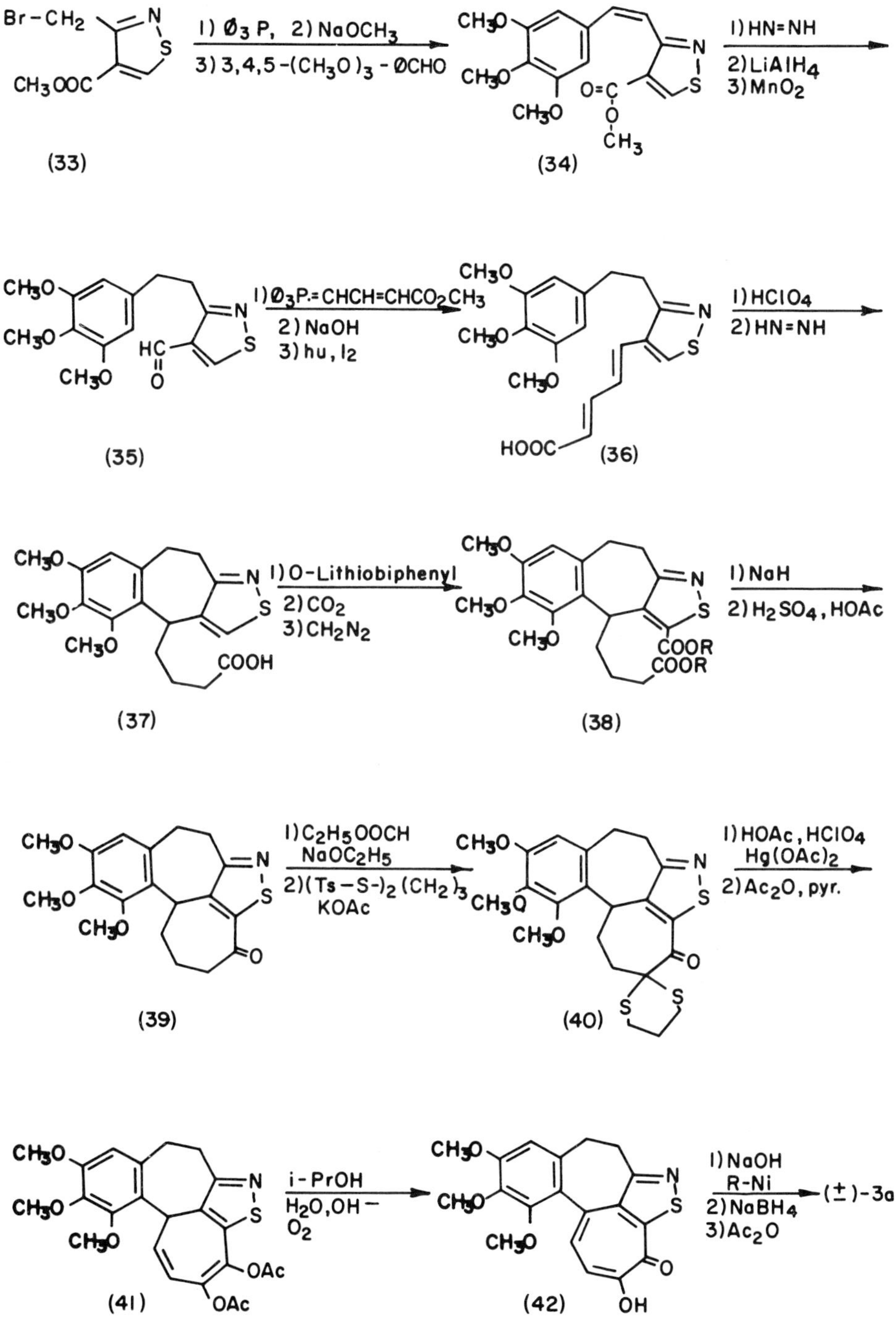

(33)
1) Ø3 P, 2) NaOCH3
3) 3,4,5-(CH3O)3 - ØCHO
(34)
1) HN=NH
2) LiAlH4
3) MnO2
(35)
1) Ø3P.=CHCH=CHCO2CH3
2) NaOH
3) hu, I2
(36)
1) HClO4
2) HN=NH
(37)
1) O-Lithiobiphenyl
2) CO2
3) CH2N2
(38)
1) NaH
2) H2SO4, HOAc
(39)
1) C2H5OOCH
NaOC2H5
2) (Ts-S-)2 (CH2)3
KOAc
(40)
1) HOAc, HClO4
Hg(OAc)2
2) Ac2O, pyr.
(41)
i-PrOH
H2O, OH−
O2
(42)
1) NaOH
R-Ni
2) NaBH4
3) Ac2O
(±)-3a

noteworthy steps include the facile cyclization of (36) to (37) and the novel sequence for tropolone synthesis involving structures (39)–(42).[13]

3.4 Synthesis of Colchiceine from *dl*-Demethoxydeoxyhexahydrocolchiceine

By a conventional series of ten steps, it was possible to synthesize the key intermediate (43) that was converted to *dl*-colchiceine by the reactions given below.[14] The elaboration of the tropolone ring system follows methods developed by Dryden, Nozoe, and Kursanov on simpler systems.

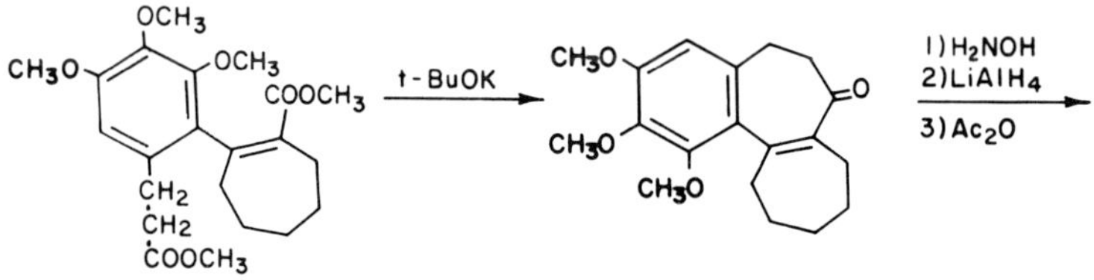

(43) (44)

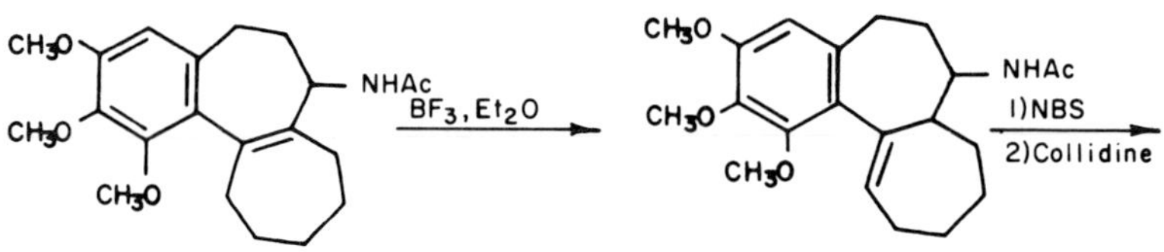

(45) (46)

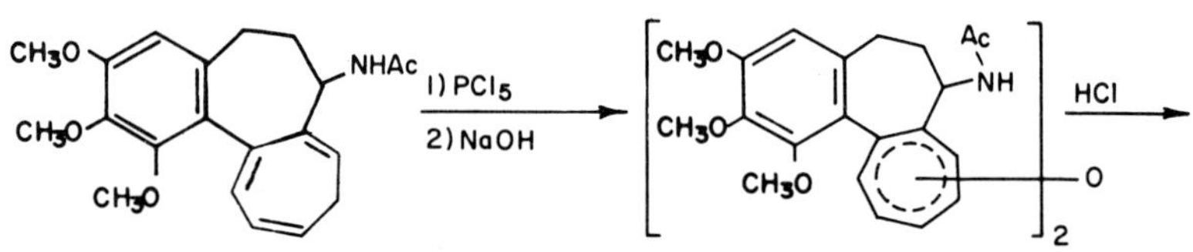

(47) (48)

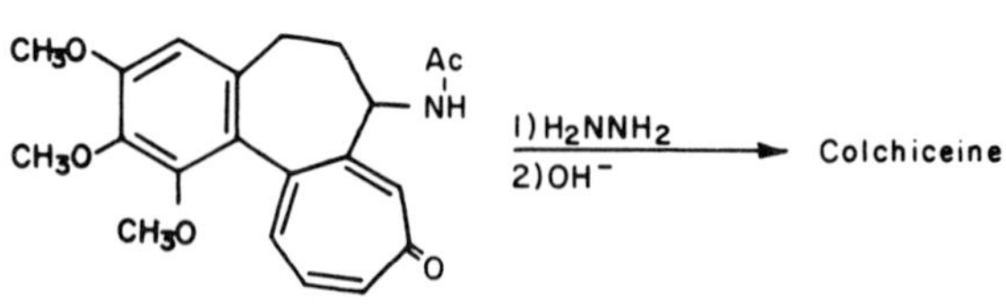

(49)

4. LUMICOLCHICINES

In the presence of ultraviolet light, colchicine is converted to a mixture of three photo isomers named α-, β-, and γ-lumicolchicine. The β- and γ-isomers occur in small amounts in various *Colchicum* and *Merendera* species.

The spectral and chemical properties of the β- and γ-lumicolchicines indicate that the troponoid ring C is no longer present in the photo products and that a tetracyclic structure has been formed.[15–18] The additional ring was postulated to arise by the rearrangement of double bonds in ring C of colchicine. The transformations may best be represented by the sequence (1) → (50) and/or (51). The presence of reactive carbonyl and enol groups was quite discernible from chemical evidence. Two *cis* isomers, (50) and (51), are possible.

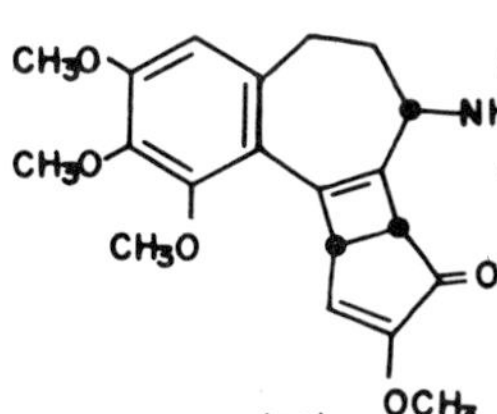

(50)

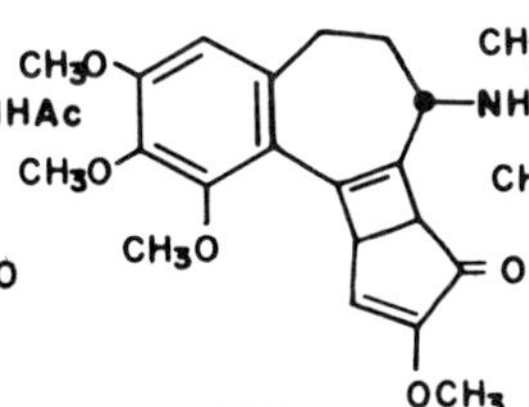

(51)

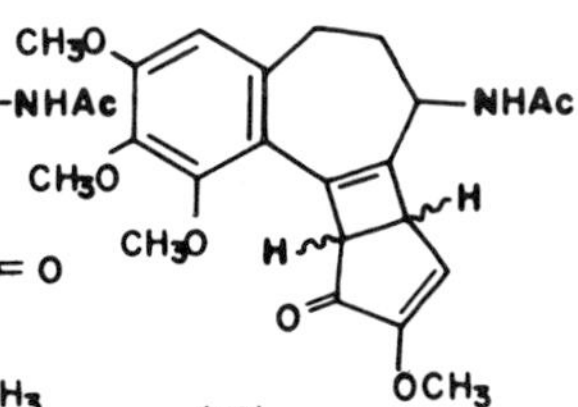

(52)

These early interpretations ignored an alternative (52) and possible transformations of ring B that might alter its size. These uncertainties have been resolved by nuclear magnetic resonance studies on β- and γ-lumicolchicine and several of their reduction products.[19] The five aliphatic protons comprising ring B of β- and γ-lumicolchicine are essentially in the same positions as those of colchicine itself, and it is clear that ring B is not modified by irradiation. Through the use of pyridine as a solvent and double resonance techniques, evidence has been provided for the presence of two bridgehead protons as well as one olefinic proton, the latter being coupled to one of the former. These data established the correctness of gross structures (50) and (51) but did not permit assignment of individual structures to β- and γ-lumicolchicine. Reduction of β-lumicolchicine with sodium borohydride gave a single product that showed hydrogen bonding between the N-H of the acetamido group and the oxygen of the newly formed hydroxyl. Such bonding requires these groups to be on the same side of the molecule and to be in close proximity. By this reasoning, the alcohol derived from β-lumicolchicine was assigned structure (53). Hydrogen bonding of this type is impossible for any alcohol derived from (52), and this structure need not be considered further for β-lumicolchicine.

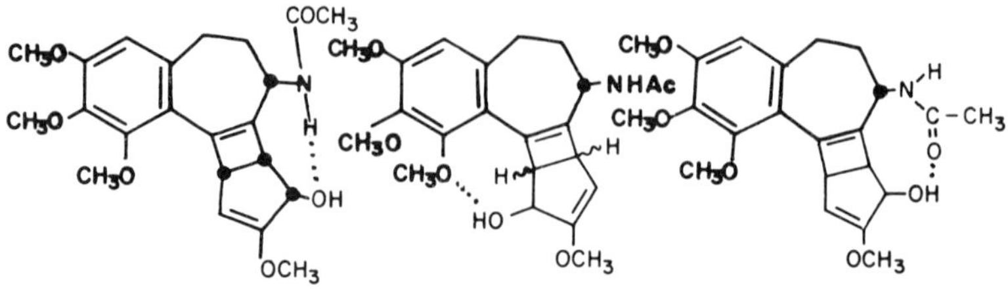

(53) (54) (55)

The alcohol derived from the sodium borohydride reduction of γ-lumicolchicine shows strong intramolecular hydrogen bonding of the hydroxyl group to oxygen. Three structures for this product have been considered [the two C:D ring fusion isomers of (54) and compound (55)]. However, the extreme similarity of β- and γ-lumicolchicine in all spectral properties suggests that (51) is the correct structure for γ-lumicolchicine.

It was recognized that α-lumicolchicine possessed different chemical properties from the β- and γ-isomers. It was not reduced under catalytic conditions and was inert to carbonyl reagents. A close relationship between α- and β-lumicolchicine was suggested from observations that (1) the two isomers showed no mixed melting point depression, (2) irradiation of the β-isomer afforded α-lumicolchicine, and (3) above 100° α-lumicolchicine is converted to the β-isomer. All spectral data led to the conclusion that the double bond of the enol ether group of β-lumicolchicine is not present in α-lumicolchicine. A photodimer of β-lumicolchicine was suspected and proven by chemical and spectral methods. Four such structures can be formulated (head-to-head, tail-to-tail, *cis*, and *trans*). The head-to-head *trans* formulation (56) was shown to be correct.[20]

Comparable chemical and spectral studies have shown that isocolchicine forms the photo product (57). A major factor in the formation of (57) [as well as (50) and (51)] may be the stability of the trimethoxystyryl conjugated system.[21,22]

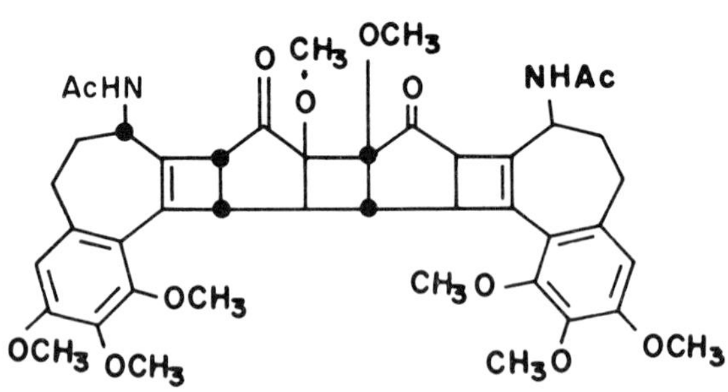

(56)

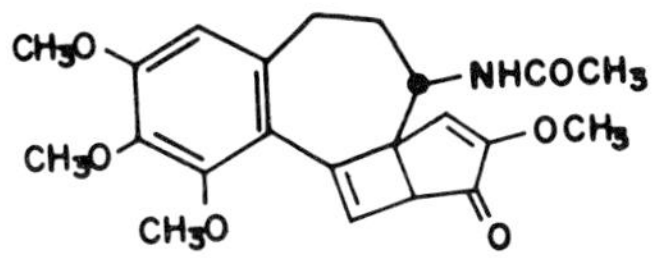

(57)

5. RELATED ALKALOIDS

Minor alkaloids accompany colchicine in many plants. These have proven to be either variations in substitution in the colchicine nucleus, photochemical transformation products, or biosynthetic precursors. A compound of major biosynthetic interest is androcymbine (58). The alkaloid was isolated from *Androcymbium melanthioides.* Uv and ir spectral data suggested a cross-conjugated cyclohexadienone system combined with

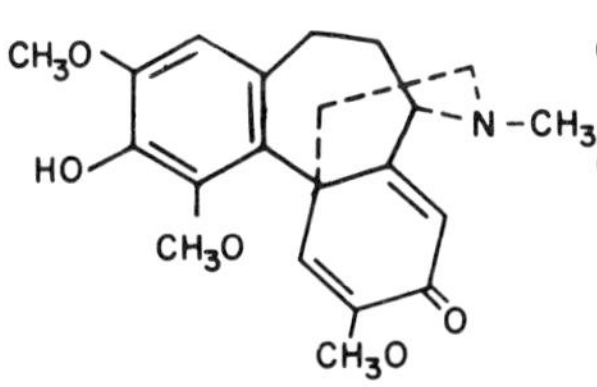

(58)

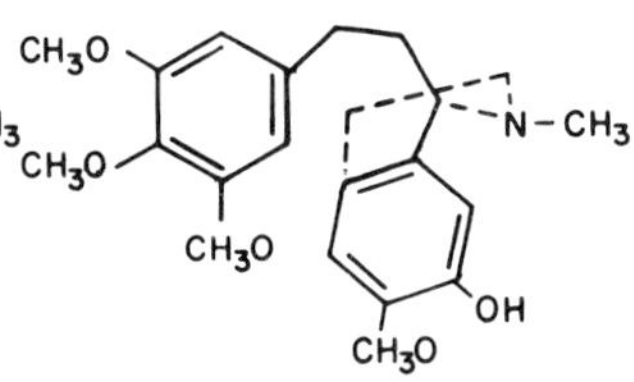

(59)

an aromatic chromophore. The nmr spectrum showed three methoxyl groups, one N-CH_3, and one aromatic proton. The olefinic protons appear as signlets (δ–6.27 and 6.83 ppm). Five protons appear as unresolved signals in the range expected for protons in benzylic positions or adjacent to nitrogen. O-Methyl-androcymbine reacts with sodium and ammonia to provide (59), a structure proven by synthesis. The phenolic hydroxyl group of (58) has not been proven by chemical evidence, but is placed at C(2) because 2-demethylcolchicine occurs in the same plant. The absolute configuration depicted in (58) rests on the observation that androcymbine and salutaridine (60) have opposite Cotton effects at 270 mμ.[23,24]

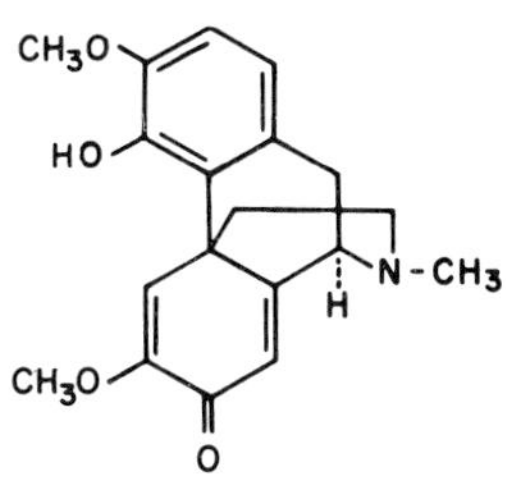

(60)

REFERENCES

1. O. J. Eigsti and P. Dustin, Jr., "Colchicine in Agriculture, Medicine, Biology and Chemistry," Iowa State College Press, Ames, Iowa, 1955.
2. J. W. Cook and J. D. Loudon, *Alkaloids* **2,** 261–329 (1952).
3. W. C. Wildman, *Alkaloids* **6,** 247–288 (1960).
4. A. Windaus, *Ann.* **439,** 59 (1924).
5. H.-G. Boit, "Ergebnisse der Alkaloid-Chemie bis 1960," pp. 28–49, Akademie-Verlag, Berlin, Germany, 1961.
6. M. J. S. Dewar, *Nature* **155,** 141 (1945).
7. W. von E. Doering and L. H. Knox, *J. Amer. Chem. Soc.* **73,** 828 (1951).
8. P. L. Pauson, *Chem. Rev.* **55,** 9 (1955).
9. O. L. Chapman, "Advances in Photochemistry," Vol. 1, pp. 323–420 (W. Albert Noyes, Jr., George S. Hammond, and J. N. Pitts, Jr., Ed.), Interscience Publishers, New York, N. Y., 1963.
10. H. Corrodi and E. Hardeggar, *Helv. Chim. Acta* **38,** 2030 (1955).
11. E. E. Van Tamelen, T. A. Spencer, Jr., D. S. Allen, Jr., and R. L. Orvis, *Tetrahedron* **14,** 8 (1961).
12. J. Schreiber, W. Leimgruber, M. Pesaro, P. Schudel, T. Threfall, and A. Eschenmoser, *Helv. Chim. Acta.* **44,** 540 (1961).
13. R. B. Woodward, "The Harvey Lectures," Series 59, pp. 31–47, Academic Press, Inc., New York, N. Y., 1965.
14. G. Sunagawa, T. Nakamura, and J. Nakazawa, *Chem. Pharm. Bull. (Tokyo),* **10,** 291 (1962).
15. F. Šantavý, *Collection Czech. Chem. Commun.* **16,** 665 (1951).
16. R. Grewe and W. Wulf, *Chem. Ber.* **84,** 621 (1951).
17. E. J. Forbes, *J. Chem. Soc.* 3864 (1955).
18. P. D. Gardner, R. L. Brandon, and G. R. Haynes, *J. Amer. Chem. Soc.* **79,** 6334 (1957).
19. O. L. Chapman, H. G. Smith, and R. W. King, *J. Amer. Chem. Soc.* **85,** 803 (1963).
20. O. L. Chapman, H. G. Smith, and R. W. King, *J. Amer. Chem. Soc.* **85,** 806 (1963).
21. O. L. Chapman, H. G. Smith, and P. A. Barks, *J. Amer. Chem. Soc.* **85,** 3171 (1963).
22. W. G. Dauben and D. A. Cox, *J. Amer. Chem. Soc.* **85,** 2130 (1963).
23. A. R. Battersby and R. B. Herbert, *Chem. Commun.* 228 (1965).
24. J. Hrbek and F. Šantavý, *Collection Czech. Chem. Commun.* **27,** 255 (1962).

9

Indole Alkaloids

NORBERT NEUSS
The Lilly Research Laboratories
Eli Lilly and Company
Indianapolis, Indiana

1. INTRODUCTION

The study of alkaloids has fascinated chemists since the beginning of the last century when in 1817 Sertürner published his now famous paper on morphine.[1]

In the last decade, a group of naturally occurring organic bases containing the indole nucleus has received a great deal of attention not only from the chemist but also from the pharmacologist, physiologist, and physician. This recent interest in indole alkaloids evolved from the discovery of the remarkable physiological properties of lysergic acid diethylamide on the one hand and of reserpine, the sedative principle of the Indian plant *Rauwolfia*, on the other. The extensive investigations of indole alkaloids as well as other constituents of the large family of flowering plants, *Apocynaceae*, were highlighted a few years ago by the introduction of two *Vinca* alkaloids, vincaleukoblastine and leurocristine, into the treatment of Hodgkin's disease and acute leukemia, respectively.

It is not surprising that this intensive search for new indole alkaloids afforded some 600-700 compounds. Of these, the structures of some 300–400 have been elucidated by the combined use of chemical methods and infrared, ultraviolet, and nuclear magnetic resonance spectroscopy, mass spectrometry, and X-ray crystallography. The importance of these methods in the structure elucidation of complicated organic molecules is the basis for the emphasis that this chapter places on the usefulness of spectral data in understanding the chemistry of various indole alkaloids. Accordingly, in the discussion of the chemistry of certain important alkaloids, a detailed interpretation of the spectral data has been included.

Since certain ring systems and substitutions have typical spectral characteristics, the chapter has been divided into sections according to structural types rather than a more conventional classification based upon the plant origin of the alkaloid. Thus, reserpine is listed as a yohimbine derivative and vincaleukoblastine as a dimeric indole–indoline alkaloid. This division appears to be logical in view of the occurrence of the same structural type in different plants.

In view of the occurrence of a large number of alkaloids of various structural types in *Rauwolfia*, *Aspidosperma*, *Strychnos*, and *Vinca* species, short sections describing these compounds as a group have also been included.

The limited scope of this chapter does not permit enumeration of all known indole alkaloids and indeed necessitates the selection of certain representatives of the various classes for inclusion. For complete coverage of phytochemical considerations, biogenetic pathways, physical data, and the chemistry of indole alkaloids, the reader is referred to monographs and review articles listed in the bibliography.

2. INDOLE

The indole system can be visualized as a fusion of a pyrrole ring and a benzene ring giving rise to two isomeric benzopyrroles: indole (1) and isoindole (2).

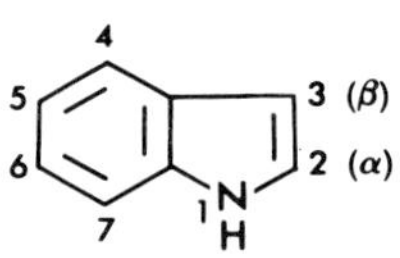

(1)

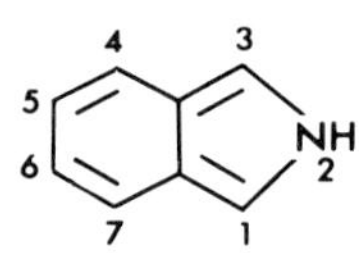

(2)

By far, the more important of the two is the indole system present in a great variety of natural products. Indole is a crystalline solid, mp 52°, soluble in most organic solvents and readily available commercially. It can be prepared by several methods; however, the most important synthesis of substituted indoles is that discovered by Emil Fischer in 1884.[2]

It involves the reaction of phenylhydrazones of a variety of aldehydes, ketones, or ketoacids in the presence of a catalyst such as zinc chloride, ethanolic hydrogen chloride, dilute sulfuric acid, glacial acetic acid, or polyphosphoric acid. Its mechanism was suggested by Robinson and Robinson[3] and is illustrated in the reaction of phenylhydrazone of acetone to give 2-methyl-indole.

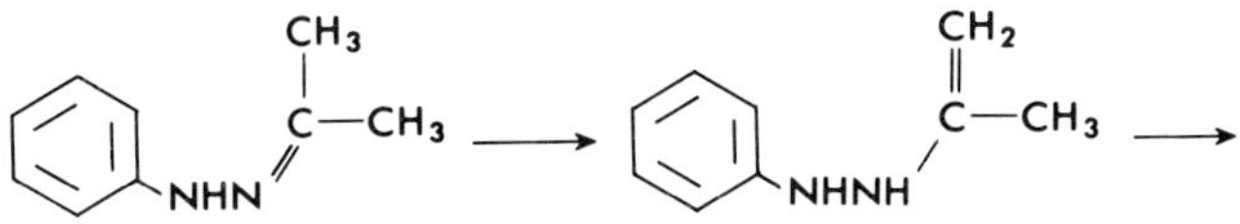

(3) (4)

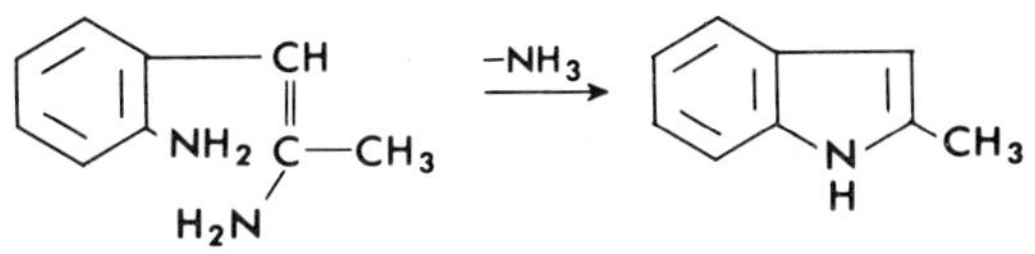

(5) (6)

An early use of ^{15}N in the study of reaction mechanism is connected with this synthesis. Thus $C_6H_5{}^{15}NHNH_2$ reacts with acetophenone to afford 2-phenylindole containing ^{15}N. Conversely, $C_6H_5NH^{15}NH_2$ with acetone yields ^{15}N free 2-methyl indole with elimination of ^{15}N containing ammonia.[4]

Indole resembles pyrrole in many respects. The nitrogen atom donates its unshared electrons to the aromatic structure. This delocalization is reduced upon salt formation with acids. Indole is a very weak base, and its true salts are unknown; however, metal derivatives are readily formed. The substitution by electrophilic reagents occurs usually in the 3-position. This can be explained on the basis of the retention of the full aromaticity in the transition state corresponding to the 3-substitution by X^+ (7).

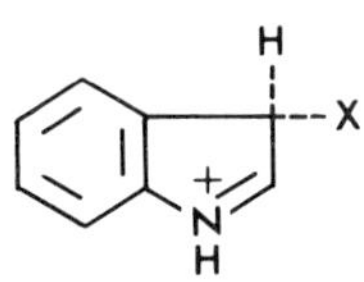

(7)

A number of alkaloids are also derived from 2,3-dihydroindole (indoline). Unlike indole, this compound is a liquid; and since it is a base, its reactions are like those of a secondary arylalkylamine.

Some indole reactions suggest the existence of the equilibrium of the indole with small amounts of a tautomeric form, indolenine (8), a structural unit also present in natural products.

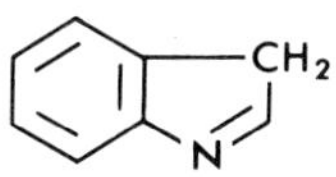

(8)

3. INDOLYL AMINES

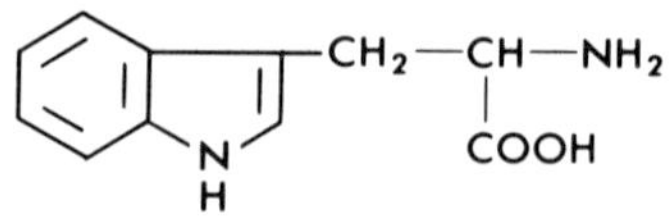

(9)

A number of substances derived from tryptophan (9), an essential amino acid, or its 5-hydroxy analog are found in nature. Of particular biological significance is 5-hydroxy-tryptamine (Serotonin) (10), which is ubiquitous in animal tissues and is also found in the plant kingdom, e.g., in *Gossypium hirsutum*[5] and *Paneolus campanulatus*.[6]

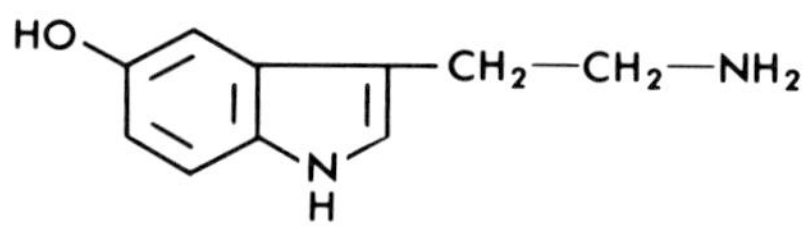

(10)

The high concentration of serotonin (10) in brain tissue suggests that it plays an important role in the modulation of neuronal activity. High levels of this biogenic amine have been observed in intestines and blood platelets. Although the precise function of serotonin (10) has not been completely defined, the enormous amount of pharmacological research concerned with its role is indicative of its importance.

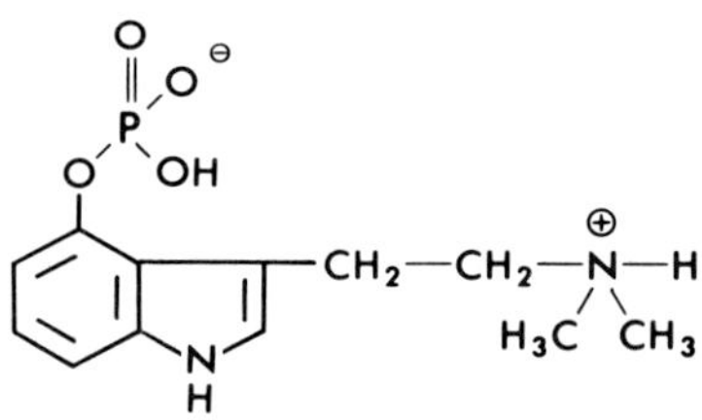

(11)

Among substances related to tryptamine one should mention here the first phosphorus-containing derivative, psilocybine (11). This compound, the hallucinogenic principle of *Psilocybe aztecorum*, the narcotic mushroom of the Aztecs, is a phosphoric acid ester of 4-hydroxy-N,N-dimethyl tryptamine.[7]

4. HARMAN TYPE (12)

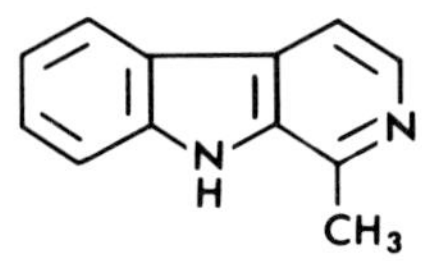

(12)

Alkaloids in this class are derived from β-carboline (13). Most of the alkaloids of this group are derivatives of 1-methyl-β-carbolines, hydrogenated at C(3) and C(4) or C(1),2,3 and C(4) and sometimes substituted at C(7) by a hydroxyl or methoxyl. Harman alkaloids have been found in various plant families, such as *Leguminoseae* (*Petalostylis*), *Malpighiaceae* (*Banisteria caapi*), *Rubiaceae* (*Leptactina*), *Zygophyllaceae* (*Peganum*), and many others.[8]

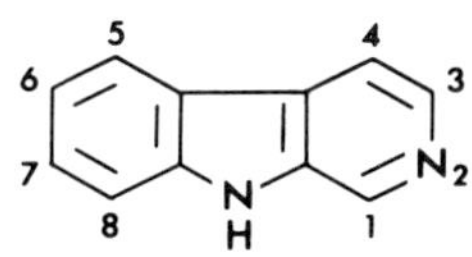

(13)

Peganum harmala alkaloids, such as harmaline (15), were used at one time in therapeutics against tremors of Parkinson's syndrome and are now recognized as potent inhibitors of monoamine oxidase.[9] The most important alkaloids to be mentioned here are harman (12), harmine (14), and harmaline (15).[8] Of the many syntheses of harman, one offers a good illustration of the utility of the Fischer indole synthesis. It consists of the ring closure of the condensation product of 2-methyl-3-hydrazinopyridine (16) and cyclohexanone.[10]

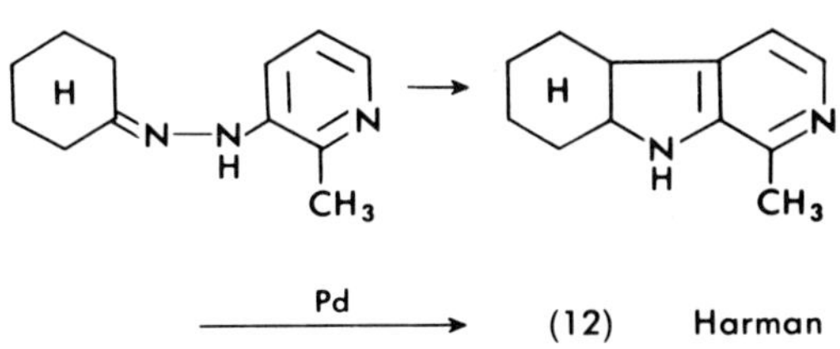

(12) Harman

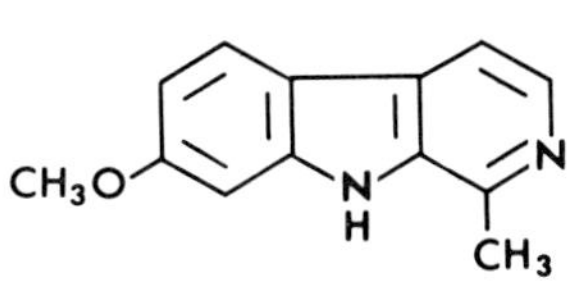

(14)

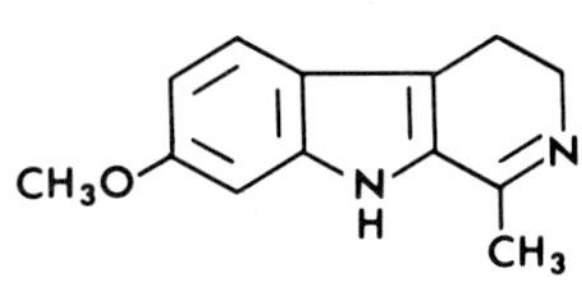

(15)

5. ELLIPTICINE-TYPE (17)

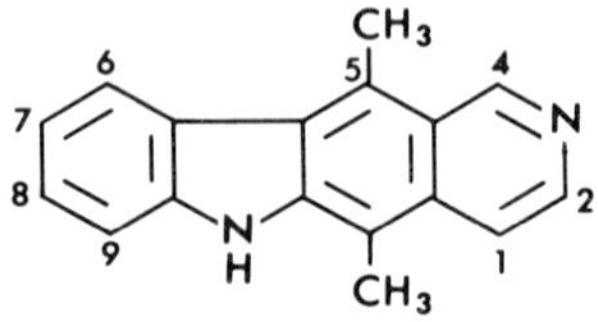

(17)

This group of aromatic compounds includes alkaloids that occur in the genera *Aspidosperma*, *Ochrosia*, and *Tabernaemontana*, members of the family *Apocynaceae*. Two typical alkaloids of this series are olivacine (18)[11] and ellipticine (17).[12]

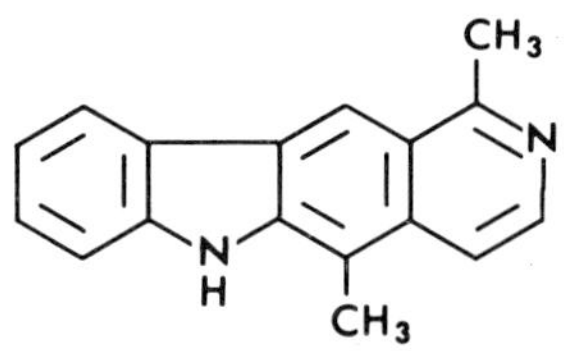

(18)

Ellipticine (17) was synthesized in an interesting sequence of only three steps.[12]

6. RAUWOLFIA ALKALOIDS

For centuries the natives of the Himalayan foothills have used the root of a flowering plant *Rauwolfia* for healing afflictions ranging from snake bite to insanity. The drug, which supposedly could control hypertension, epilepsy, insomnia, fevers, cholera, dizziness, and headaches, was obtained from the snake-like roots of *Rauwolfia*, finally found its way to chemical laboratories first in India and later in Europe and in the U.S.A. These researches culminated in the discovery of reserpine (19)[13] which was

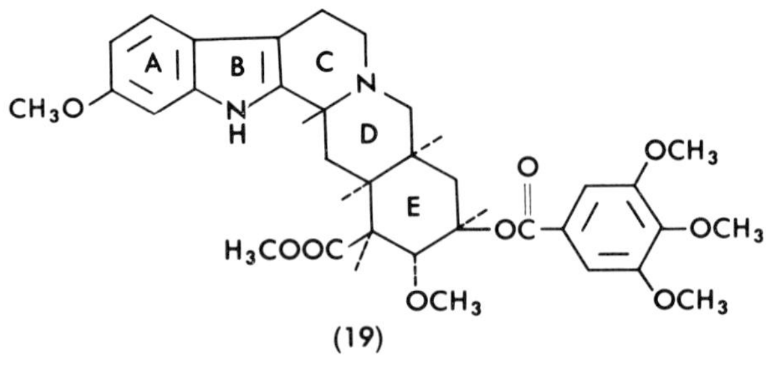

(19)

found to be the main active principle of the plant. This plant belongs to the *Apocynaceae* family. Chemical investigations of over sixty various *Rauwolfia* species[14] led to the isolation of some fifty to seventy alkaloids. Some of these are artifacts, and some have never been reisolated. The vast majority, however, have been shown to represent definite chemical entities. The basic ring systems of nearly all *Rauwolfia* alkaloids are represented in the following four compounds: yohimbine (20), ajmalicine (21), ajmaline (22), and sarpagine (23). These occur also in other plant genera.[15]

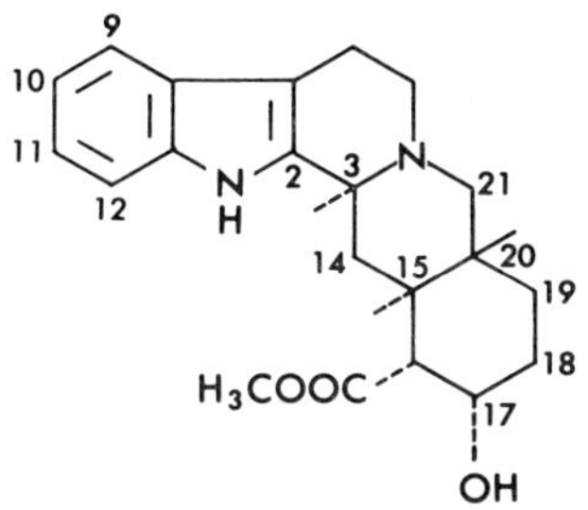

(20)

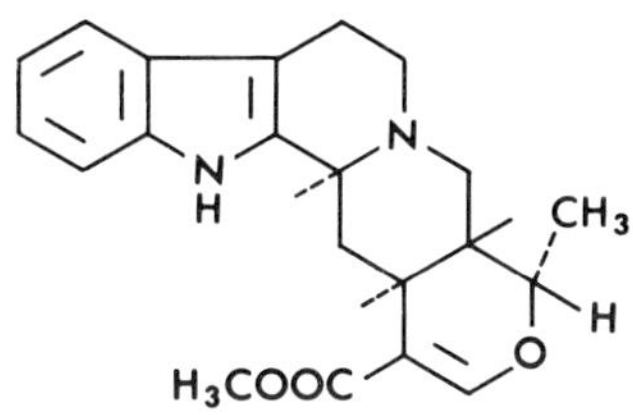

(21)

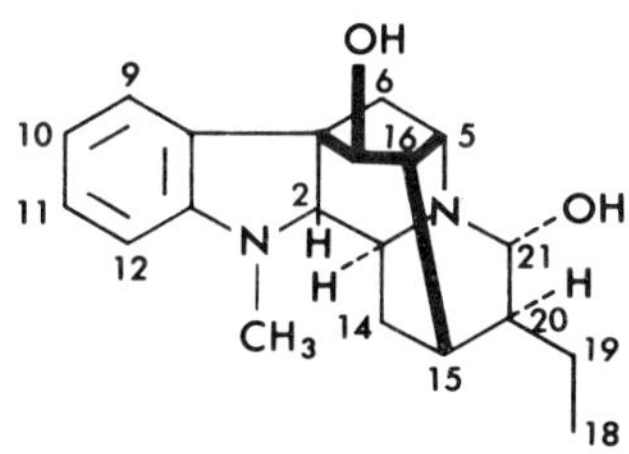

(22)

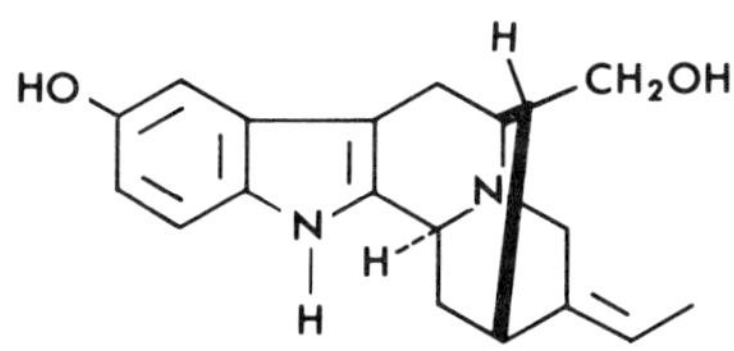

(23)

6.1 Yohimbine (20)

First found in *Corynanthe yohimbe* (*Rubiaceae*), it also occurs in several members of the *Apocynaceae* family. The structure of the compound has been confirmed by several syntheses of its derivatives and yohimbine itself. The alkaloid is an adrenolytic agent and was used in the past in veterinary medicine as an aphrodisiac.

Dehydrogenation of yohimbine (20) with selenium at 280–300° affords three products: yobyrine (24), tetrabyrine (25), and ketoyobyrine (26). The transformation leading to ketoyobyrine (26) represents a rather unique case of a rearrangement of a piperidine to an α-pyridone during dehydrogenation. It has been proposed[16] that the ring E of yohimbine is first dehydrated, then dehydrogenated (27). The benzylic C–N bond is reductively

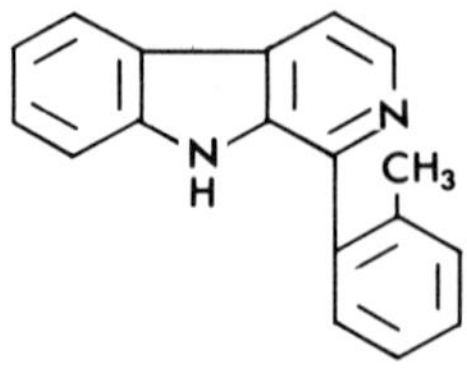

(24)

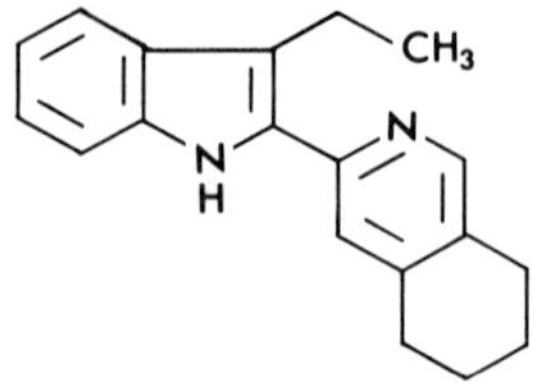

(25)

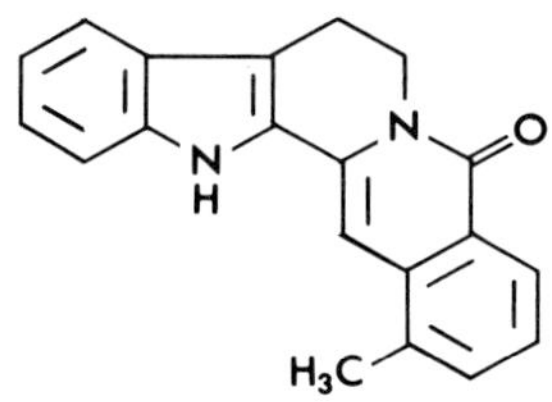

(26)

cleaved, permitting rotation about the C(14)–C(15) bond (28), lactam formation, and finally dehydrogenation to (26).

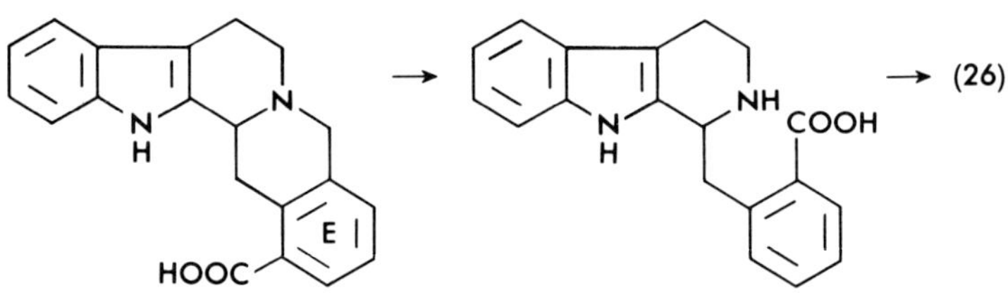

(27) (28)

The structure of yohimbine (20) was fully confirmed by the total synthesis of the alkaloid.[17]

It is interesting that all natural isomers of yohimbine possess at C(15) the α-configuration. This phenomenon must be of great biogenetic significance since this is the only nonequilibratable site among the possible six asymmetric carbon atoms in most alkaloids of the yohimbine group.[18,19]

Yohimbine and its derivatives have been isolated from various plant genera [e.g., *Rauwolfia, Amsonia, Vallesia, Aspidosperma, Vinca* (all members of the *Apocynaceae*), *Gelsemium, Strychnos* (*Loganiaceae*), *Corynanthe, Pausinystalia* (*Rubiaceae*) and *Alchornea* (*Euphorbiaceae*)].

6.1.1 Reserpine (19) This alkaloid, biologically the most important constituent of the vast group of alkaloids isolated from several species of the genus *Rauwolfia*, was largely responsible for the development of drugs acting on the central nervous system and opened new avenues of investigation in neuropharmacology. This is mainly due to the discovery of its sedative and hypnotic effects. These effects are quite different from those of other well-known drugs with central depressant action, such as phenobarbitol. This has been demonstrated dramatically by the persistance of escape behavior with loss of conditioned avoidance behavior in rats when the action of reserpine is compared with that of phenobarbitol. Single doses of reserpine result in a slowly developing fall in arterial pressure in different animal species and man. The pharmacodynamic properties of reserpine have been associated with consistant loss of serotonin, norepinephrine, and dopamine from the nervous tissues.

Reserpine (19) is a weak base that, upon alkaline hydrolysis, affords reserpic acid, $C_{22}H_{28}N_2O_5$ (29), 3,4,5-trimethoxybenzoic acid (30), and methanol. Lithium aluminum hydride reduction of the alkaloid gives ac-

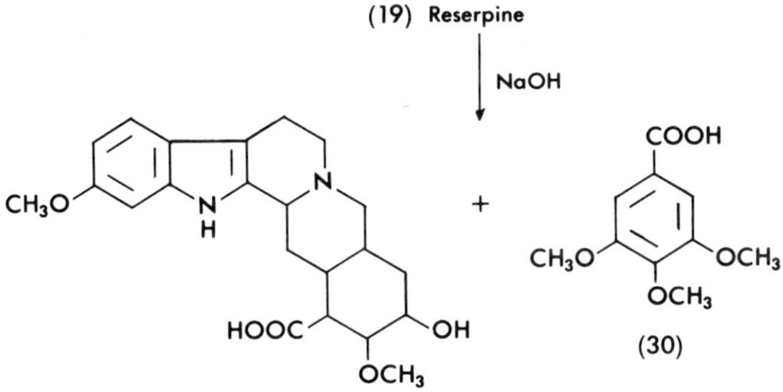

(29) (30)

cordingly 3,4,5-trimethoxybenzyl alcohol and reserpic alcohol, $C_{22}H_{30}N_2O_4$. The ultraviolet and infrared spectra of reserpine (19) and the above-mentioned derivatives are quite characteristic. By the proper use of these data and appropriate model compounds, it was possible to deduce the correct structural features of the alkaloid.[20] Chemical degradative work of another research group resulted in a more thorough substantiation of the structure.[21]

Extensive chemical work carried out by several groups resulted in the complete structure elucidation of reserpine, including the assignment of all six asymmetric centers of the alkaloid.[22] Additional support for these configurations was provided by the nmr data.[23]

The unequivocal structure proof was furnished by the brilliantly conceived total synthesis of the alkaloid.[24] If one considers the presence of six asymetric carbons in reserpine and with it the possibility of thirty-two possible racemates, the synthesis looms as a formidable task. However, each step in the synthesis has been designed to take place under sterically controlled conditions so as to yield only one racemate in the final step. This synthesis was modified and used for commercial production of the alkaloid when great demand for the drug made the cost of the totally synthetic alkaloid quite competitive with that of the alkaloid isolated from the *Rauwolfia* plant. Many chemical modifications of reserpine have been studied, but no derivative has replaced the alkaloid in its use in medicine.

Reserpine undergoes a number of facile isomerizations under a variety of conditions. Only a few are mentioned here. Under acidic conditions an epimerization of reserpine at C(3) can be easily achieved according to the following scheme, thus giving rise to isoreserpine series.

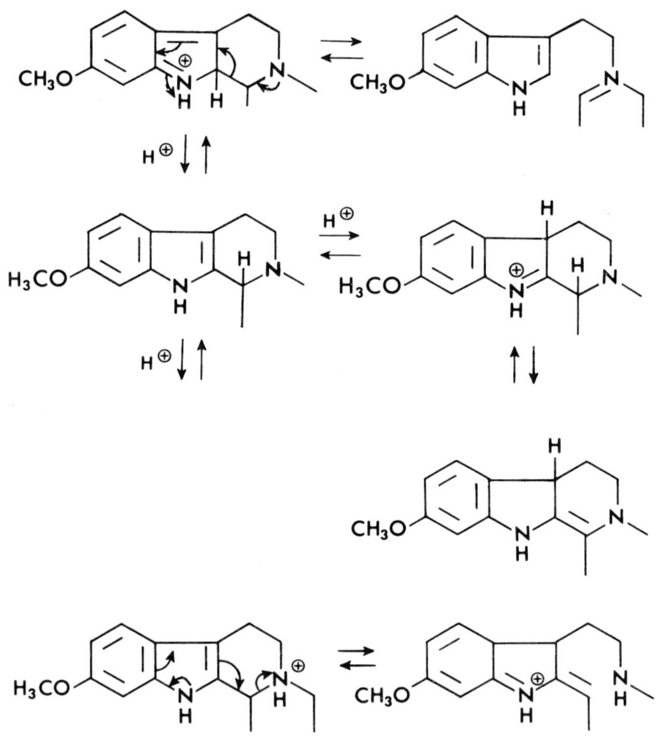

The 18-deoxycongener of this series is known to occur naturally as α-yohimbine (31). Isoreserpic acid lactone can be inverted under acidic conditions to reserpic acid lactone (32). The knowledge of this type of transformation was also used in the total synthesis of reserpine.

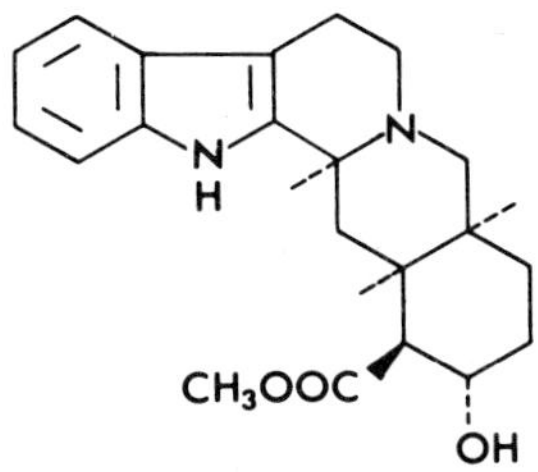

(31)

Although the cis-trans-cis conformation of the molecule is normally preserved, a prior change to the trans-trans-cis configuration is required for the formation of the lactone. This is illustrated in the following conformational structures.

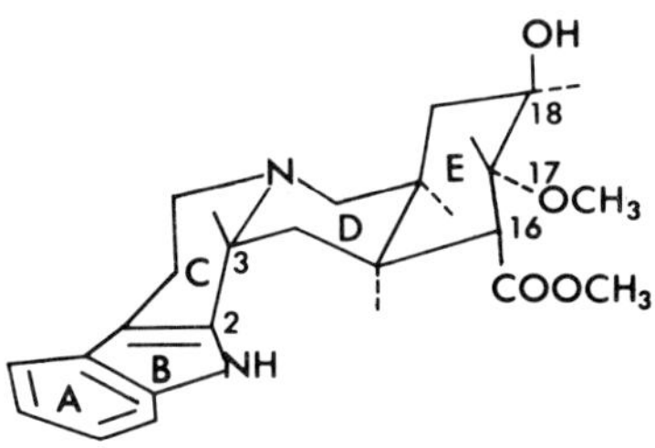

(19)

(19a) *cis, trans, cis* methyl reserpate

T° Cyclohexanone Al-Phenolate

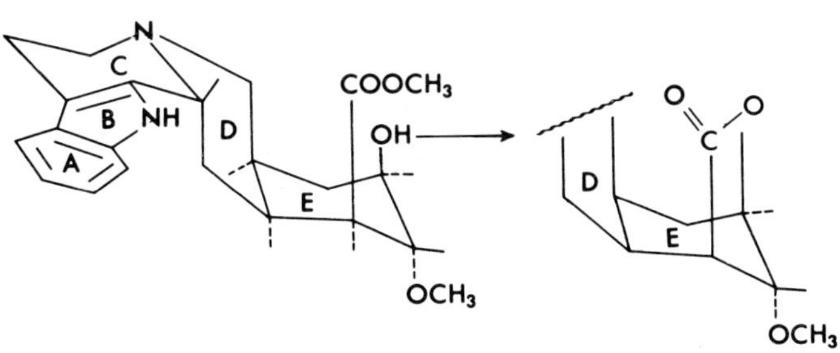

(19b) (32)

6.2 Corynantheine-Type (33)

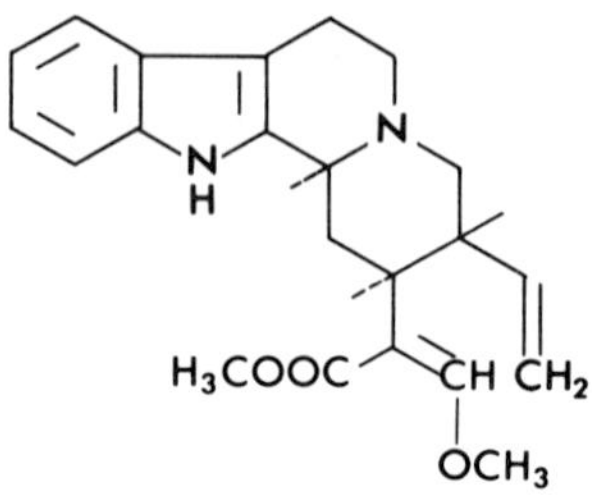

(33)

The structure and stereochemistry of corynantheine (33) [*Pseudocinchona africana* and *Corynanthe yohimbé* (*Rubiaceae*)] has been well established by the total synthesis of the alkaloid and its hydrogenation product, dihydrocorynantheine (34).[25,26]

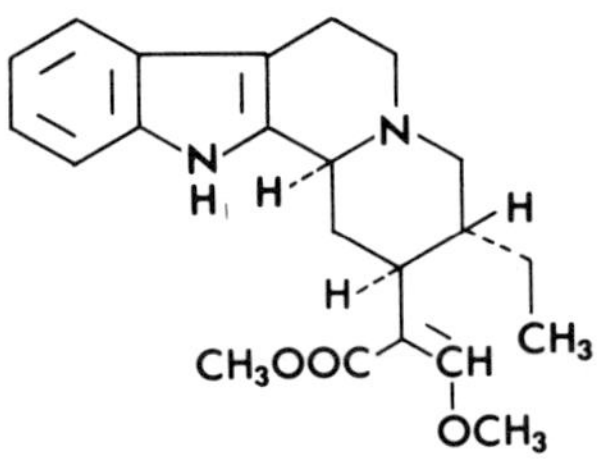

(34)

6.3 Ajmalicine (21)

This alkaloid, the weakest base among *Rauwolfia* alkaloids, occurs also in *Vinca* (*Apocynaceae*) as well as *Corynanthe yohimbé* (*Rubiaceae*). Its dehydrogenation with lead tetraacetate affords another naturally occurring alkaloid serpentine (35) that in turn can be reconverted by catalytic reduction to the starting substance.[27]

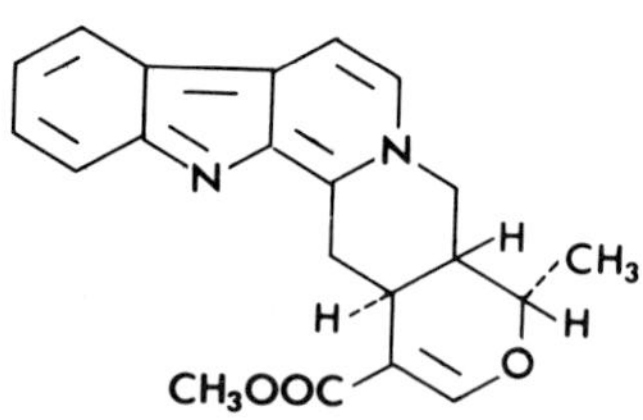

(35)

The absolute configuration of ajmalicine has been established by its conversion[28] to dihydrocorynantheane (36), a derivative of the alkaloid dihydrocorynantheine (34).

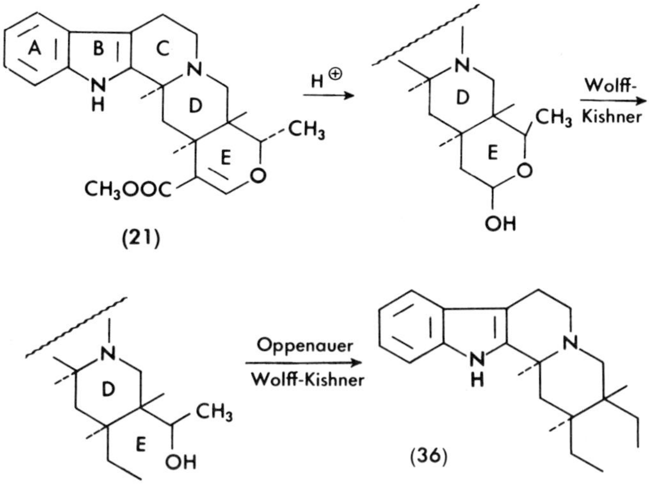

(21)

(36)

Ajmalicine (21) and its congeners can be divided into six stereochemical groups according to the chemical shifts of the C(19) methyl groups in the nmr spectra and certain bands in the infrared spectra near 3.4μ.[29] The rate of formation of methiodides and the chemical shifts of $+\,N{-}CH_3$ protons of the methiodide salts have been also used to differentiate between various stereoisomers.[30] The structure of ajmalicine itself has been proven by an elegant synthesis.[31]

6.4 Ajmaline (22)

Ajmaline (22) is one of the major alkaloids of *Rauwolfia serpentina.* It is a carbinolamine with a nitrogen at a bridgehead and, therefore, undergoes reactions of an amine and an aldehyde (formation of an oxime and reduction of ammoniacal silver nitrate solution). The reactions occurring *via* the ring-opened tautomeric aminoaldehyde lead to a series of derivatives of *chano*ajmaline (37).[32]

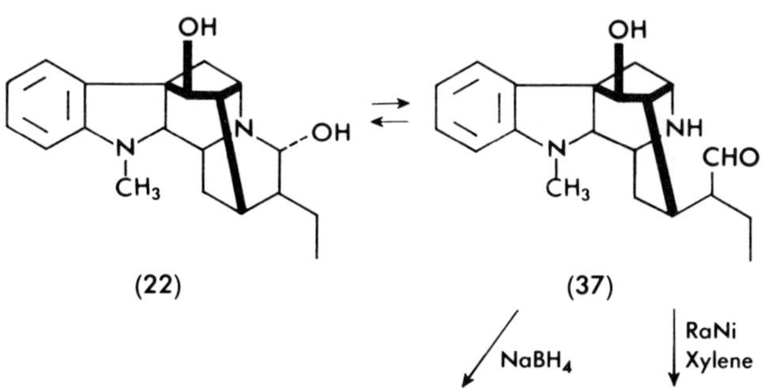

(22)

(37)

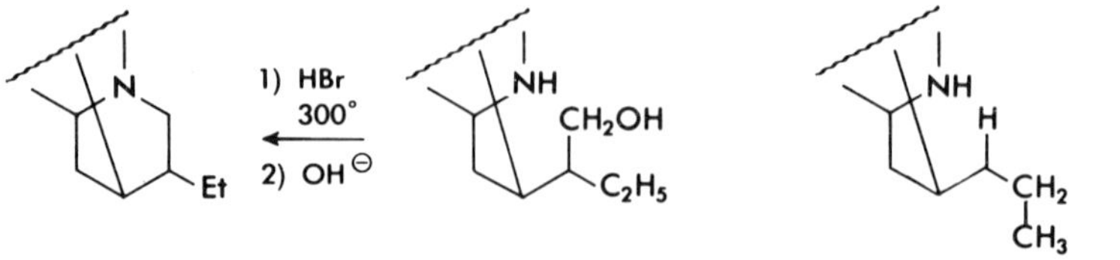

(38) Dihydrochanoajmaline Decarbonochanoajmaline

A large number of interesting transformations from the latter form has been reported. An experiment of importance in the study of the biosynthesis of the alkaloid[33] is the elimination of C(21) as carbon monoxide, resulting from the treatment of ajmaline with Raney nickel in refluxing xylene. This reaction leads to the formation of decarbonochanoajmaline.[34]

Oxidation of deoxyajmaline with lead tetraacetate in acetic acid gives rise to an indole aldehyde, deoxyajmalal B (39).

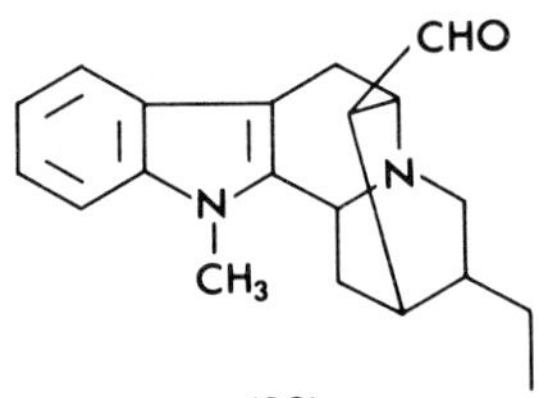

(39)

Mild dehydrogenation of deoxydihydroajmaline gives rise to two products, ajarmine (40) and ajmyrine (41). These were of great importance in the elucidation of the ring skeleton. They have also been synthesized.[35]

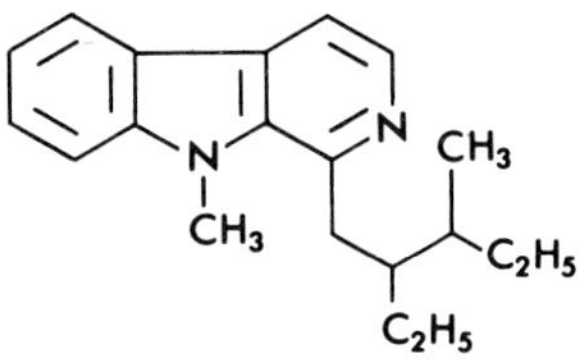

(40)

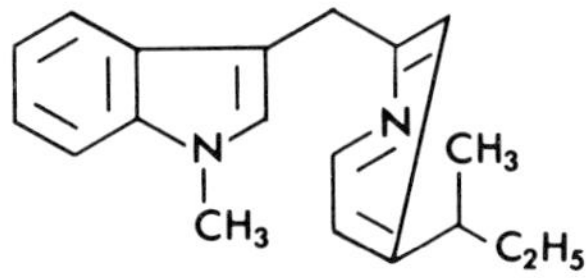

(41)

There are numerous alkaloids containing the ajmaline skeleton in several *Rauwolfia* species, as well as in other genera of the *Apocynaceae*.

7. SARPAGINE TYPE (23)

Another group of alkaloids, related to sarpagine, occurs in several *Rauwolfia* species. Sarpagine (23) is a derivative of 5-hydroxyindole. Its structure was finally established by the conversion to one of the four 21-deoxy-ajmalols (42) prepared from ajmaline.[36]

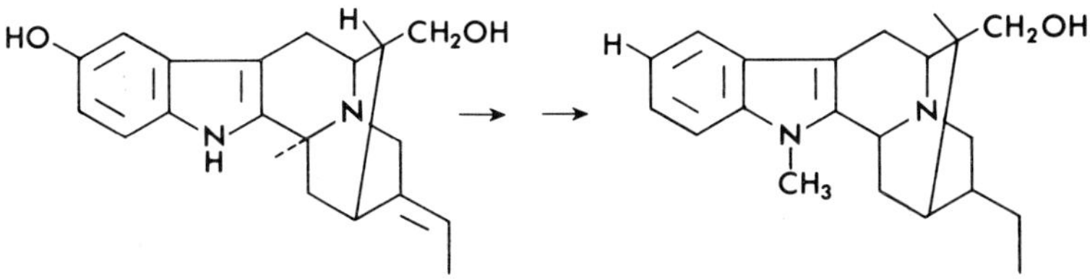

(23) (42)

8. OXINDOLE ALKALOIDS

This group of alkaloids is derived from oxindole. Oxindole itself appears to exist as a mixture of (43) and (44) as evidenced by the infrared spectra; however, the tautomerism between the lactam (45) and lactim (46) should also be considered.[37]

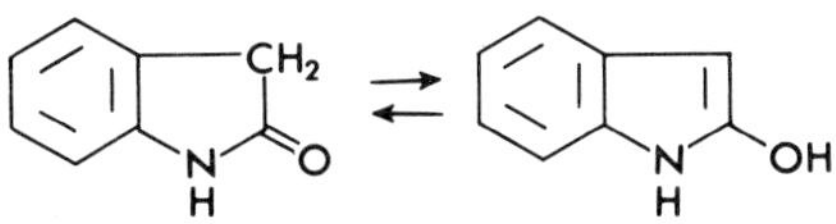

(43) A (44) B

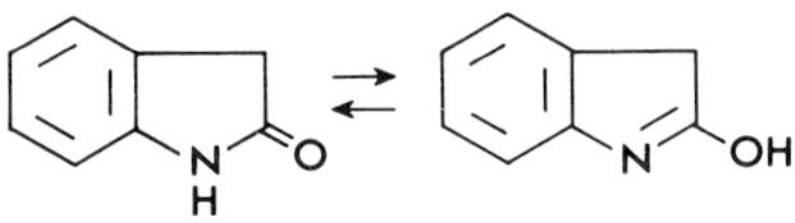

(45) C (46) D

Representatives of this class are found in the genus *Gelsemium* (Family: *Loganiaceae*), and in the genera *Mitragyna* and *Ourouparia* (Family: *Rubiaceae*). The structure of gelsemine (47), one of these alkaloids, has been elucidated through the aid of X-ray analysis.[38] The same structure has been simultaneously proposed on the basis of biogenetic considerations and nmr spectral data obtained on the alkaloid and its derivatives.[39] It is interesting to note that all conventional degradation methods failed in attempts to elucidate this structure.

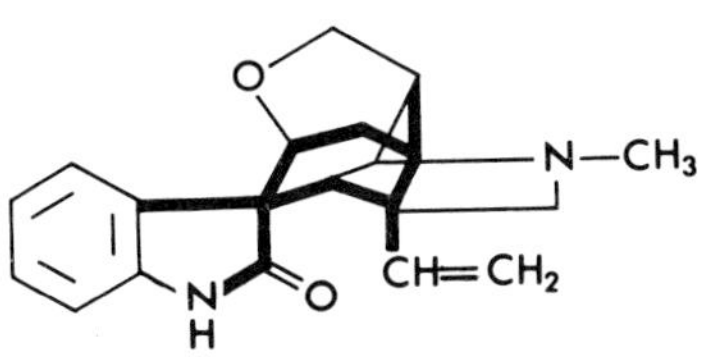

(47)

Alkaloids from *Mitragyna* and *Ourouparia* consist of an oxindole and alicyclic moieties present in ajmalicine or corynantheine type. Two compounds fall into this category: mitraphylline (48) and rhyncophylline (49).

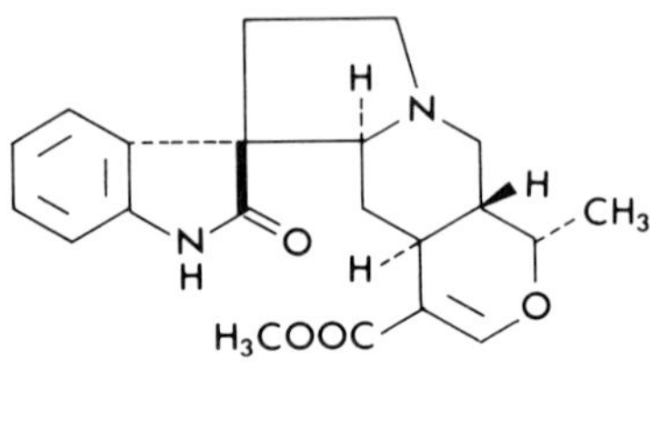

(48)

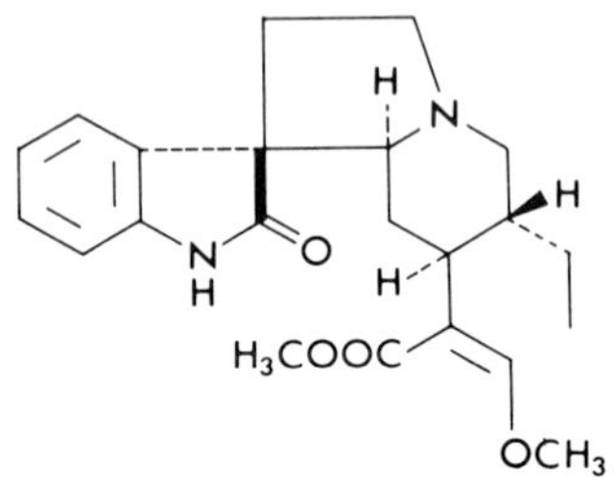

(49)

The presence of the oxindole moiety in mitraphylline (48) was shown from one of the products isolated from the mixture resulting from the zinc dust distillation, namely, 3-spirocyclopropano-oxindole (50).[40] This product is of greatest importance in the recognition of the presence of oxindole alkaloids.

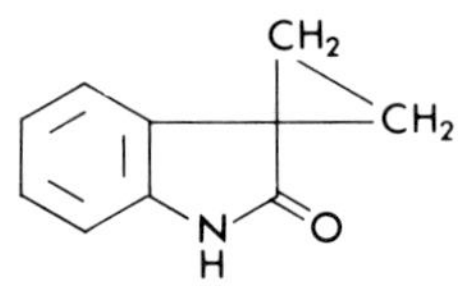

(50)

Oxindole alkaloids undergo isomerization upon treatment with a variety of acidic and basic reagents. For instance, refluxing mitraphylline (48) or its isomer isomitraphylline in pyridine gives a mixture of 20% mitraphylline and 80% isomitraphylline. A similar transformation occurs with rhyncophylline (49) and isorhyncophylline (the ratio being 30–70, respectively). The mechanism of these transformations is portrayed below:[41]

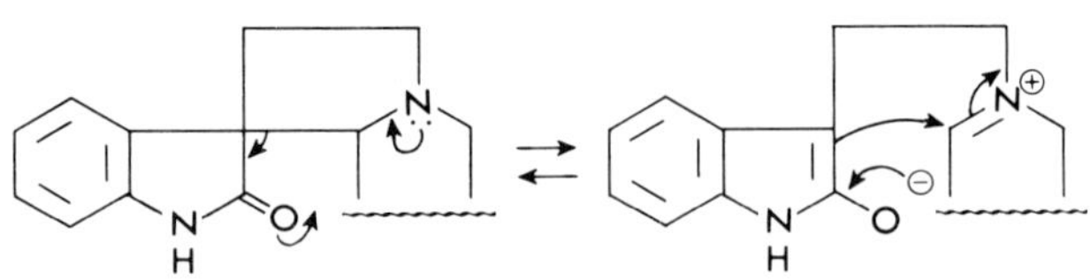

Mitraphylline and isomitraphylline have been synthesized from ajmalicine (21). The sequence of reactions is shown in Chart 9.1.[42] It was established by thin layer chromatography study that the primary product of acid hydrolysis is isomitraphylline.

CHART 9.1. Synthesis of Mitraphylline and Isomitraphylline from Ajmalicine

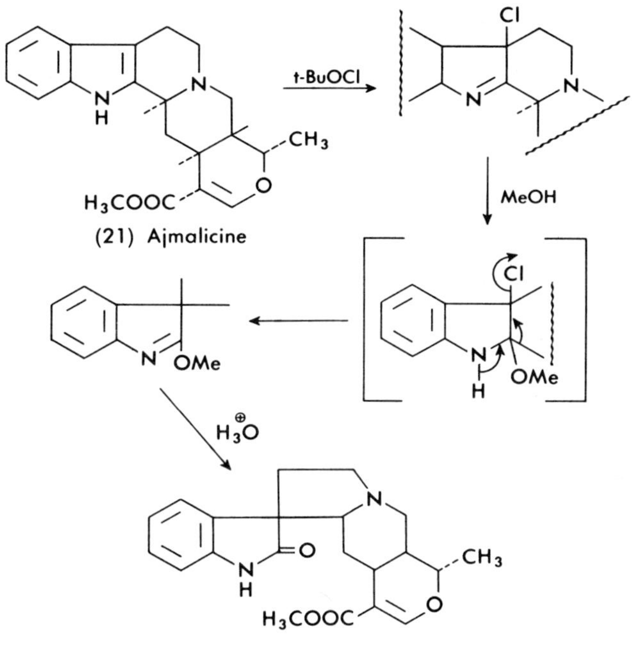

Mixture of Mitraphylline and Isomitraphylline

9. IBOGAMINE TYPE (51)

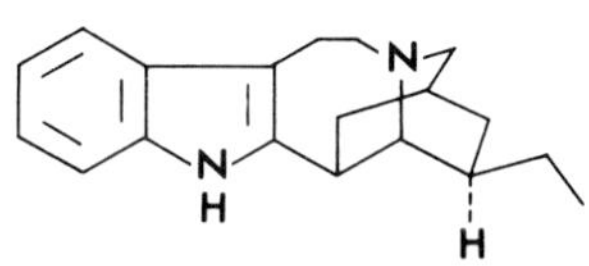

(51)

These alkaloids contain an isoquinuclidine ring system fused to an indole moiety. The investigations leading to their isolation and structure elucidation were stimulated by reports that the roots of *Tabernanthe iboga*, an *Apocynaceous* shrub from French Equatorial Africa, were used by natives to combat sleep and hunger.

Closely related genera such as *Voacanga, Ervatamia, Tabernaemontana, Conopharyngia,* and *Gabunia* were examined also, and a number of related alkaloids found. The minor constituents isolated from *Tabernanthe* were found to contain oxidation products of the indole moiety. It is possible, however, that the latter are artifacts produced in the plant during the extraction procedures.[43]

An unusual member of this group of compounds is catharanthine (52), one of the main alkaloids isolated from the leaves of *Vinca rosea* Linn. Its occurrence seems to be restricted so far only to this plant.[44]

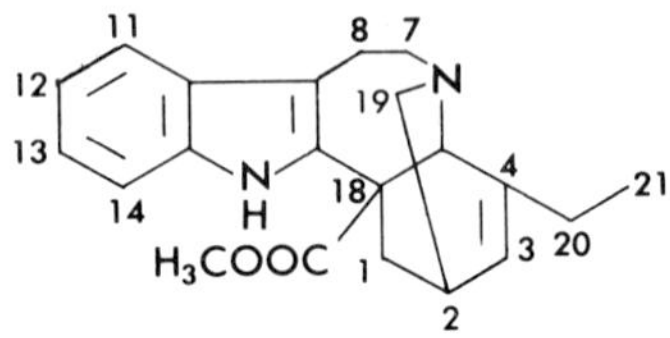

(52)

The structure elucidation of ibogaine (53), the main alkaloid of *Tabernanthe iboga*, as well as ibogamine (51) and tabernanthine (54), can be regarded as one of the last successful structural problems based on degradative chemical work without the aid of nuclear magnetic resonance and mass spectrometry.[45]

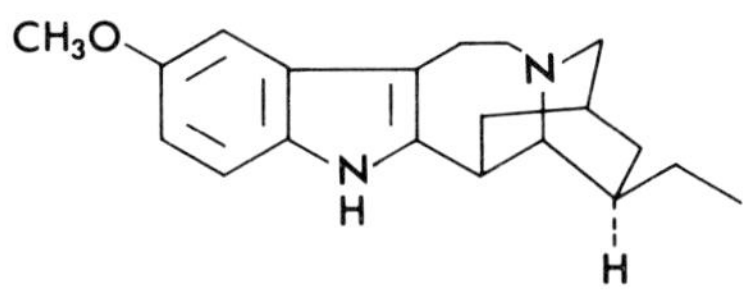

(53)

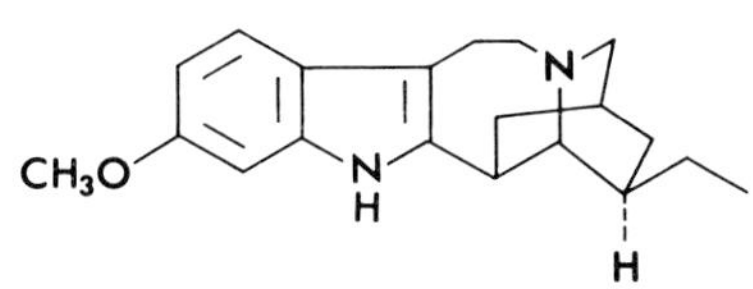

(54)

The most significant feature of this work was the isolation and correct assignment of the structure of two selenium dehydrogenation products of ibogaine (53), compounds (55) and (56). From the dehydrogenation of ibo-

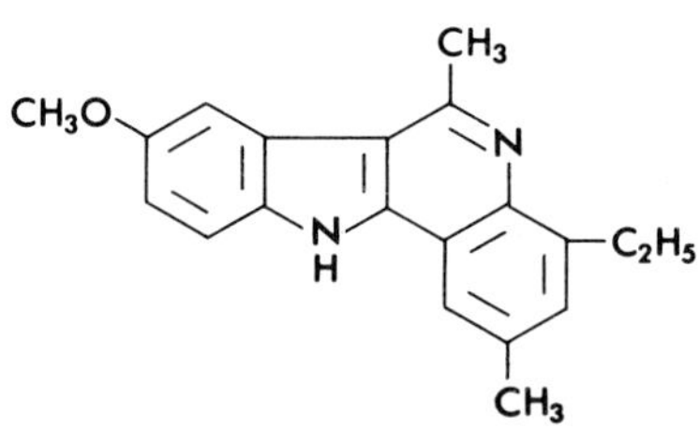

(55)

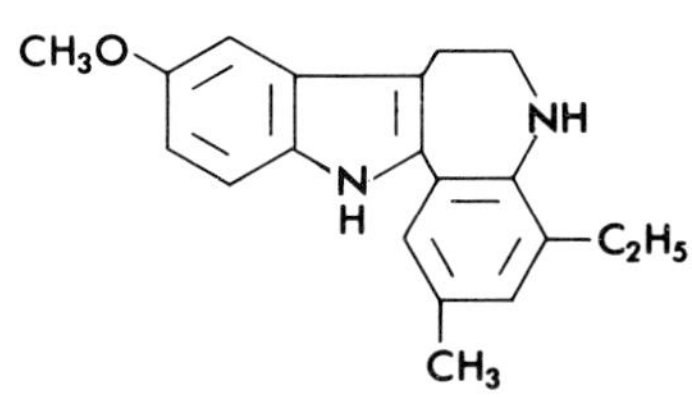

(56)

gamine, the corresponding demethoxy compounds resulted. The confirmation of the structure of indoloquinoline (55) was provided by its synthesis.[46] Later the X-ray analysis of ibogaine hydrobromide[47] confirmed the proposed structure of ibogaine and permitted the assignment of the C(4) ethyl group as shown (53). The absolute configuration of the alkaloid is based on chemical correlations with cleavamine[48] (57) and vincaleukoblastine[49] (vinblastine, VLB) (58) whose absolute configurations have been ascertained by the X-ray method.

Mass spectral studies made possible the correlation of methoxy congeners of ibogamine[50] such as ibogaline (59), a minor alkaloid of *Tabernanthe iboga*. Its structure was suggested on the basis of spectral data and biogenetic considerations.[51] The mass spectra of ibogamine (51) (A), ibogaine (53) (B), tabernanthine (54) (C), and ibogaline (59) (D) are shown on Figure 9.1. Examination of the spectra reveals immediately in the lower part of the spectra a group of peaks of similar intensity (e.g., m/e-122, 124, 135, 136, and 149). Obviously, these peaks are due to fragments

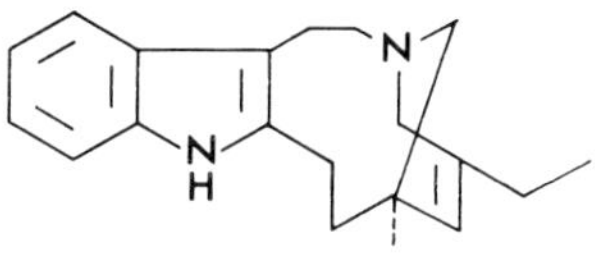

(57)

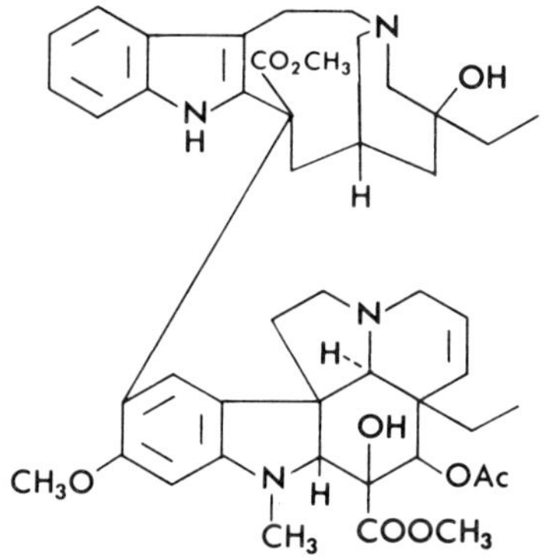

(58)

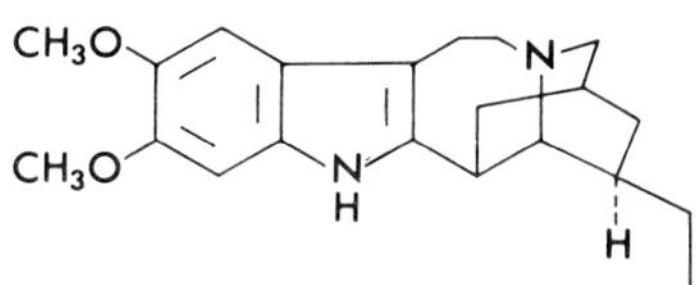

(59)

of the alicyclic portion of the molecule. The other group of peaks, which is found in the spectrum of ibogamine at m/e = 156, 195, 251, 265, and 280, is found with almost identical intensity but 30 amu higher in ibogaine (53) and tabernanthine (54) and 60 amu higher in ibogaline (59). It is apparent that these peaks represent the indole-containing moiety that is substituted differently in the benzene nucleus of these molecules. Since ultraviolet and infrared spectra indicated the presence of 5,6-dimethoxy indole moiety in ibogaline (59), its mass spectrum definitely proves it to be 12,13-dimethoxyibogamine (51).

Work towards a total synthesis of ibogamine type alkaloids has been carried out in several laboratories in the course of the last few years. These

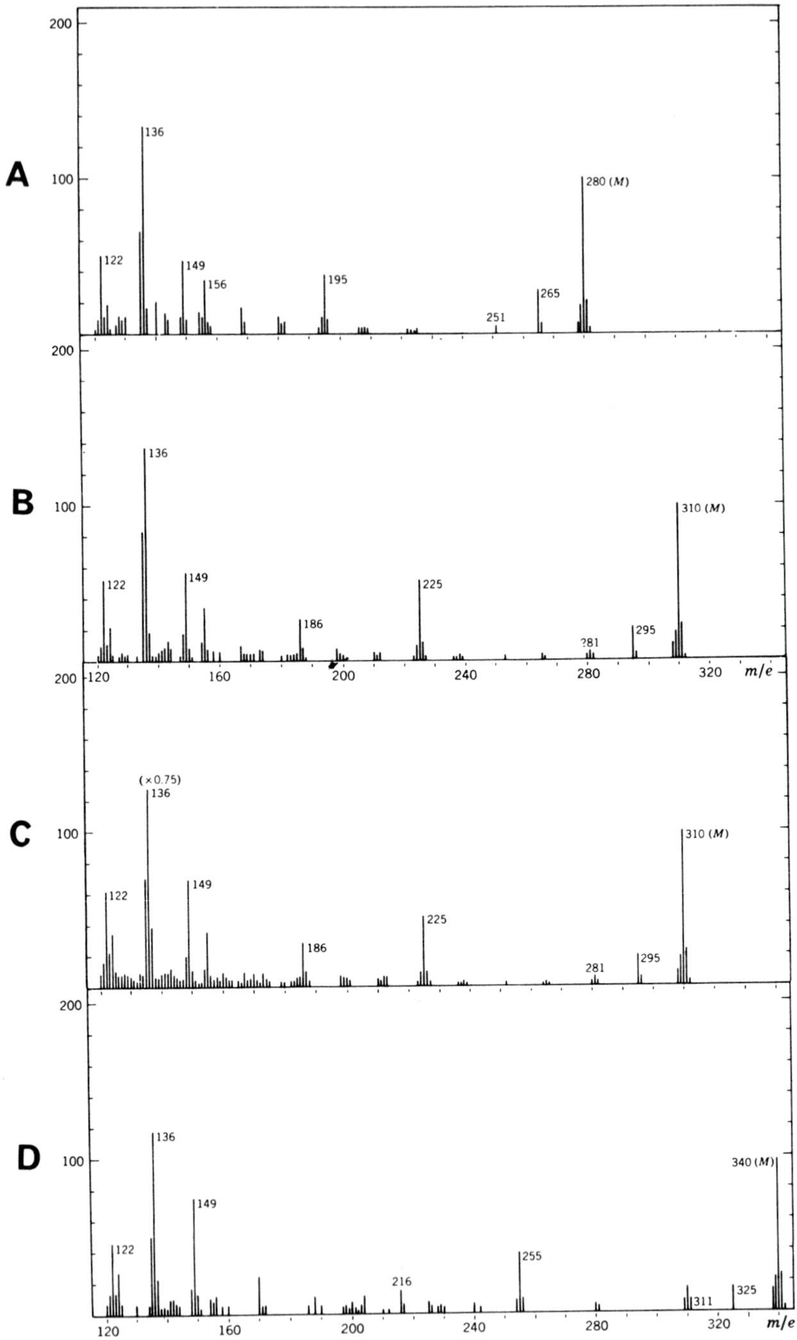

Figure 9.1 **Mass spectra of ibogamine and congeners. Relative intensity (M^+ = 100). A = Ibogamine; B = Ibogaine; C = Tabernanthine; D = Ibogaline.**

CHART 9.2 Synthesis of Ibogamine (51) and Epi-Ibogamine (62)

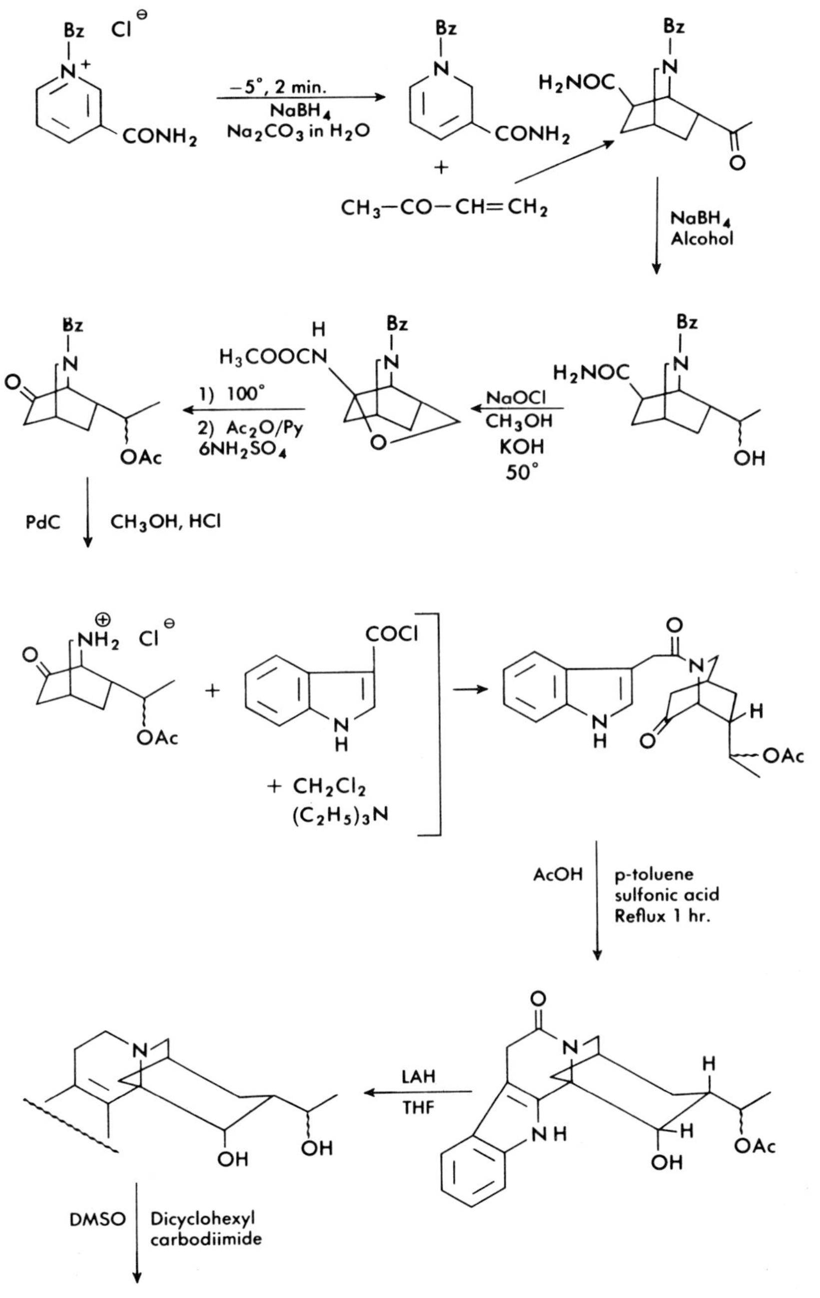

CHART 9.2 (Continued)

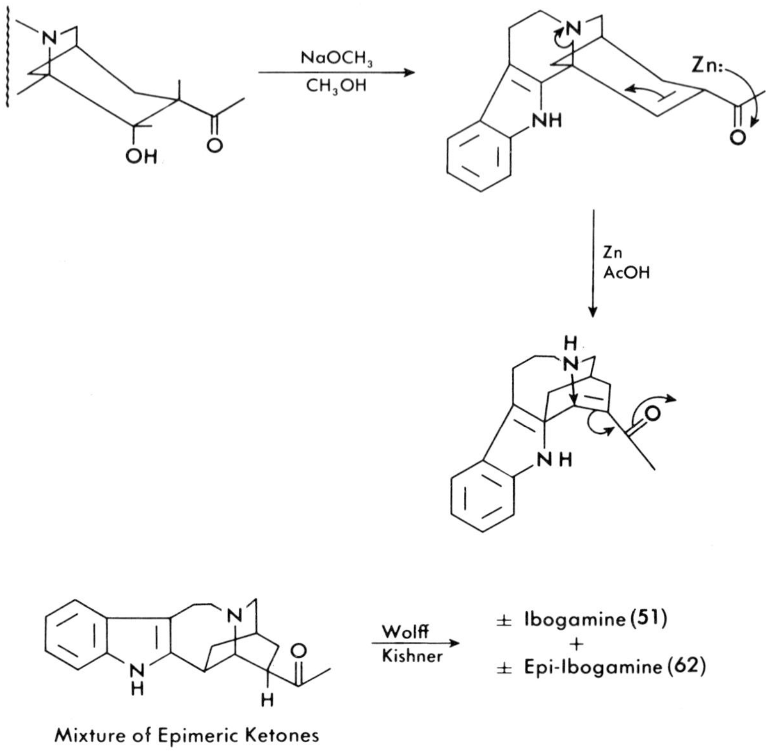

efforts resulted in the synthesis of ibogamine (51), using 3-carboxamido-N-benzylpyridinium chloride as the starting material.[52] This synthesis is shown in Chart 9.2. More recently, a synthesis representing a different approach has been published.[53]

9.1 Catharanthine (52)

Catharanthine (52) is one of the major constituents in the leaves of *Vinca rosea* Linn. The alkaloid is closely related to coronaridine (60), an alkaloid from *Tabernaemontana coronaria*. Of particular interest is its hydrogenation product dihydrocatharanthine (61) that, upon decarboxylation, yields a new base epi-ibogamine (62), differing from the known ibogamine (51) only in the orientation of the C(4) ethyl group. The configuration of the ring system was determined by the isolation of 2,6-dimethyl-4-ethyl-11-H-indolo [2,3-c] quinoline (55) from its selenium dehydrogenation product. An identical compound has been obtained by the same treatment of

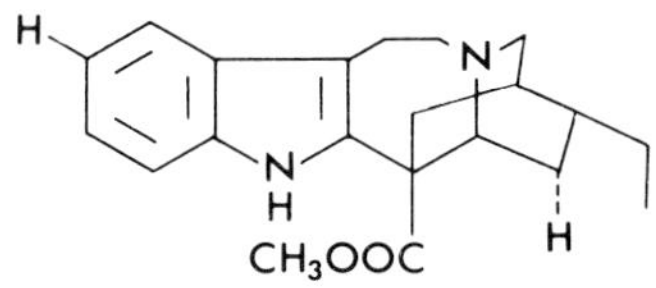

(60)

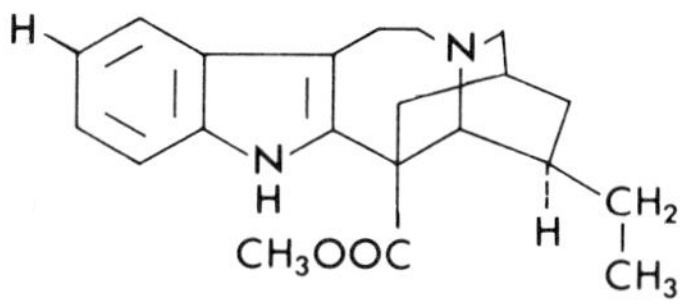

(61)

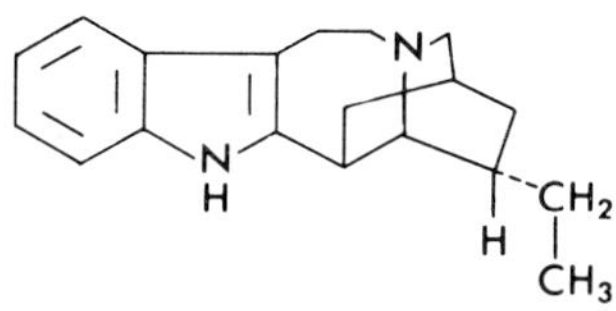

(62)

ibogamine (51). When catharanthine (52) is treated with concentrated hydrochloric acid at reflux, two products are obtained, decarbomethoxycatharanthine (63) and cleavamine (57). The latter is hydrogenated in

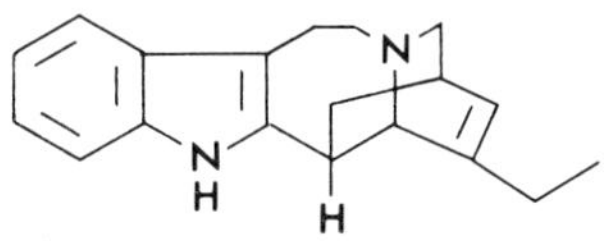

(63)

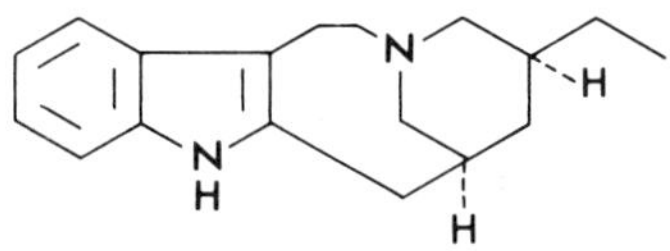

(64)

platinum oxide–ethanol to 4β-dihydrocleavamine (64) isomeric with quebrachamine (65). The structures of these compounds were suggested on the basis of the mass spectral evidence. The X-ray analysis of cleavamine methiodide confirmed the proposed structure of cleavamine (57) and established its absolute configuration.[54]

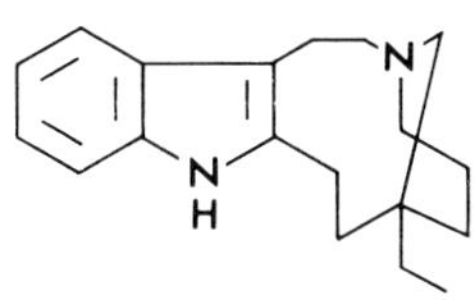

(65)

The formation of cleavamine (57) and decarbomethoxycatharanthine (63) from catharanthine (52) cannot be explained by the conventional mechanism for decarboxylation of isoquinuclidine alkaloids.

10. 2-ACYLINDOLE ALKALOIDS

The compounds included in this group occur in a number of genera of the *Apocynaceae* family: *Ervatamia, Tabernaemontana, Voacanga, Peschiera,*

Gabunia, *Ochrosia*, and *Vinca*. The first alkaloid recognized to be of this type was tabernaemontanine (66).[55] Its ultraviolet spectrum ($\lambda_{max}^{(EtOH)}$ 237, 312

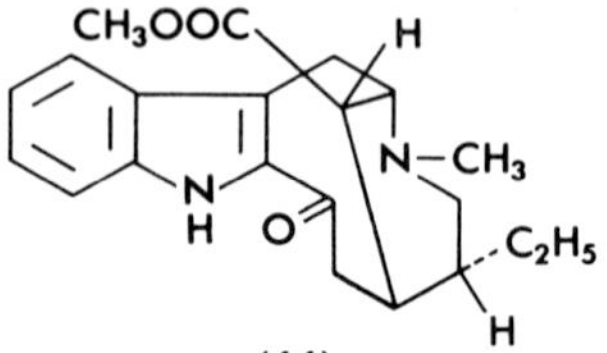

(66)

mμ; a_m 14,500, 17,500) is similar to that of 1-keto-1,2,3,4-tetrahydocarbazole (67) ($\lambda_{max}^{(EtOH)}$ 236, 309 mμ; a_m 15,900, 22,400, respectively). The in-

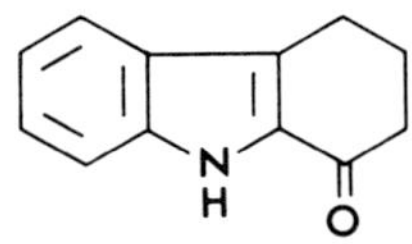

(67)

frared spectrum of tabernaemontanine (66) also shows similarities to that of compound (67) at the following wavelengths: $\nu_{max}^{CHCl_3}$ 3472, 3333 (N–H), 1650 (carbonyl) and 1333 cm.$^{-1}$. The presence of a methyl ester is indicated by bands at $\nu_{max}^{CHCl_3}$ 1727 and 1242 cm.$^{-1}$ Quite similar maxima are found in the spectra of other members of this family of alkaloids, such as vobasine (68), dregamine (69), and perivine (70).[56] The structure of vobasine (68) was elucidated by degradative chemical work combined with the use of uv, ir, and nmr data[57] and confirmed by mass spectra.[58]

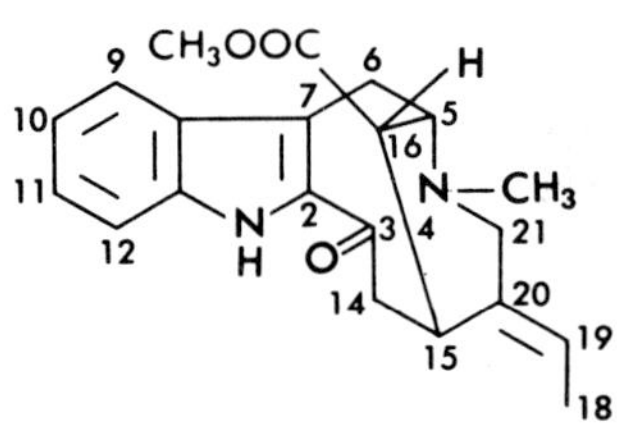

(68)

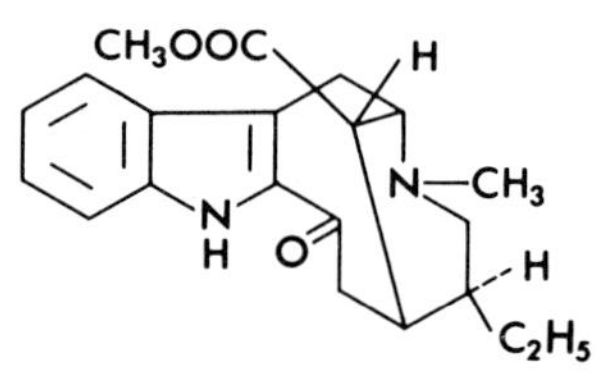

(69)

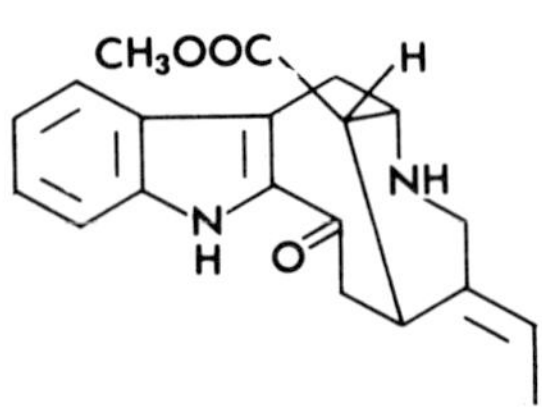

(70)

The alkaloid perivine (70) from *Vinca rosea* L. was early recognized as lacking the N_b methyl group.[59] The nmr spectrum of its alcohol, perivinol, shows one 3-proton singlet at $\delta=2.45$ assigned to a carbomethoxyl function magnetically shielded by the indole ring. This singlet is missing in the spectrum of the LAH reduction product of perivinol obtained by the reduction of perivine (70) using potassium borohydride in methanol. The major difference in the spectra of vobasinol and perivinol was the absence of an $N–CH_3$ signal at $\delta=2.54$ in the latter. As expected, reductive methylation using formaldehyde with palladium on carbon in dioxane afforded vobasine (68).

The alkaloid vobasine (68) represents one moiety of the dimeric alkaloid voacamine (71). This aspect of the chemistry of vobasine (68) is discussed in the section on dimeric alkaloids.

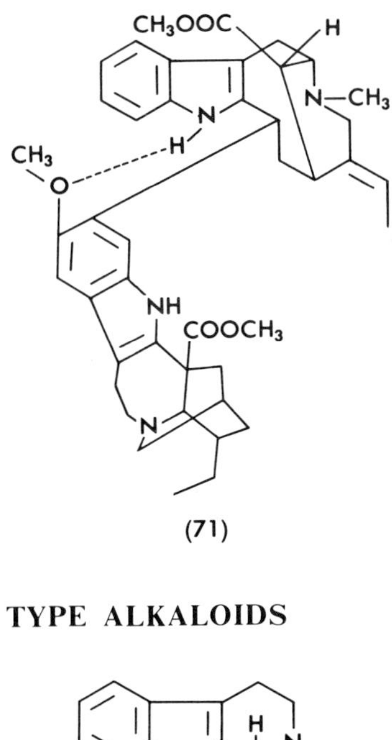

(71)

11. EBURNAMINE TYPE ALKALOIDS

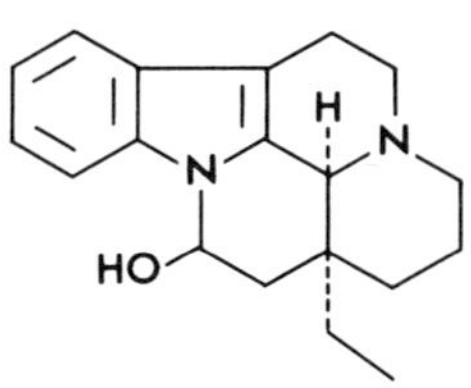

(72)

The alkaloids included in this group are compounds derived from canthinone (73), and are found in various members of the *Apocynaceae*. In the group we find such alkaloids as eburnamenine (74), eburnamonine (75), and eburnamine (72), to mention a few, all isolated from *Hunteria eburnea.*[60] Their structures have been elucidated and confirmed by the synthesis of eburnamonine (75). Although the structure elucidation and synthesis of eburnamonine[60] (75) secured the ring system of the alkaloid, it did not establish unequivocally the stereochemistry of the D/E ring junction. Canthinone (73), itself, occurs in *Pentaceras australis*.[61]

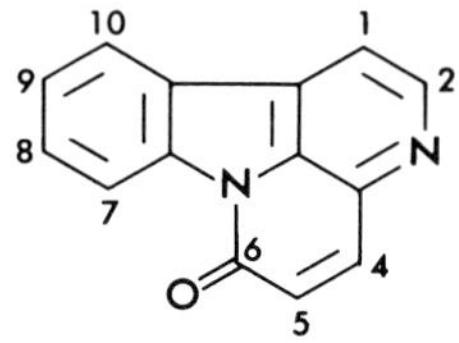

(73)

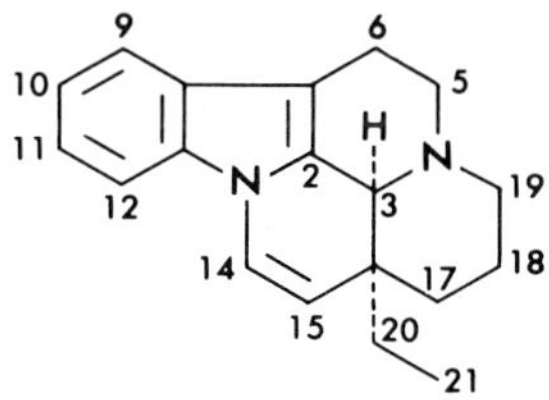

(74)

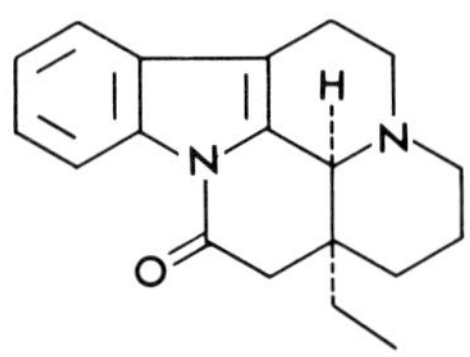

(75)

Other members of this group have been found in various species of *Vinca*, *Aspidosperma*, and *Pleiocarpa*. An interesting variant of this type of ring system is found in schizozygine (76), an alkaloid from *Schizozygia* caffaeoides.[62]

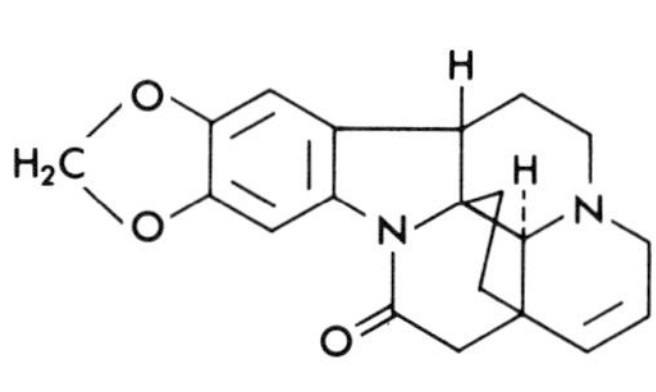

(76)

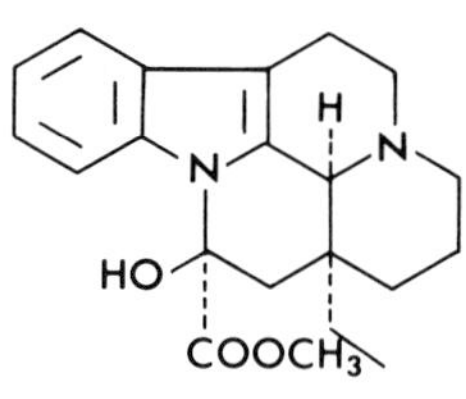

(77)

An alkaloid related to eburnamonine (75) is vincamine (77) from *Vinca minor* Linn.[63] Its structure has been corroborated by a synthesis involving the combination of the two major segments, tryptamine (78) and a C(10) unit in the form of dimethyl-3-ethyl-3-formylpimelate (79). The

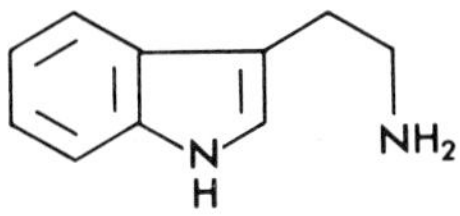

(78)

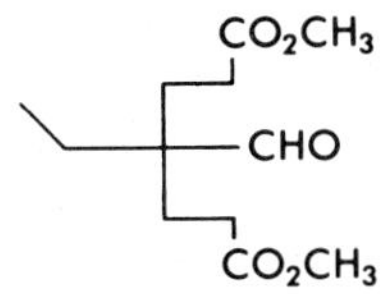

(79)

latter was obtained by the enamine alkylation of butyryl aldehyde with two equivalents of methyl acrylate in methanol. The diester was obtained by hydrolysis. Acid catalyzed condensation of tryptamine (78) and the aldehyde diester at room temperature led to a Schiff's base which cyclized on heating to form epimeric tetracyclic lactam esters. In the following sequence of reactions two epimeric amino esters were obtained. These were separated by chromatography, oxidized, and hydrolyzed to afford *d,l*-vincamine (77). The study of physical properties of two isomers and their comparison with related alkaloids of known stereochemistry led to the establishment of stereochemistry of vincamine (77) as shown (D/E ring junction *cis*).

12. ASPIDOSPERMA ALKALOIDS

Throughout this review we have been disregarding the conventional grouping of various classes according to their occurrence in certain plants. Therefore, in this section we include only those *Aspidosperma* alkaloids that can be directly correlated with the aspidospermine ring system. They include compounds isolated from various genera of *Apocynaceae* (*Aspidosperma, Vallesia, Kopsia, Vinca, Pleiocarpa*, and *Hunteria*) and are represented by seven main types: (a) aspidospermine (80); (b) quebrachamine (65); (c) aspidofractine (81)[64]; (d) vindolinine (82)[65]; (e) aspidoalbine (83)[66]; (f) obscurinervine (84)[67]; and (g) condylocarpine (85). The latter is of *Strychnos* type.

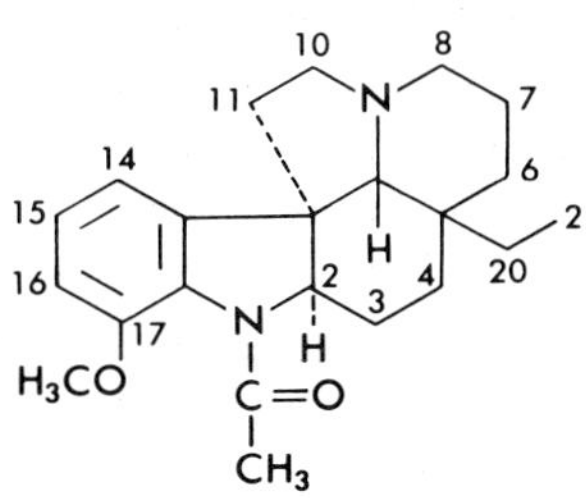

(80)

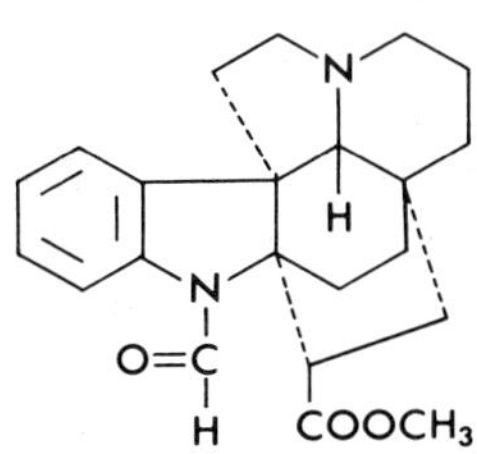

(81)

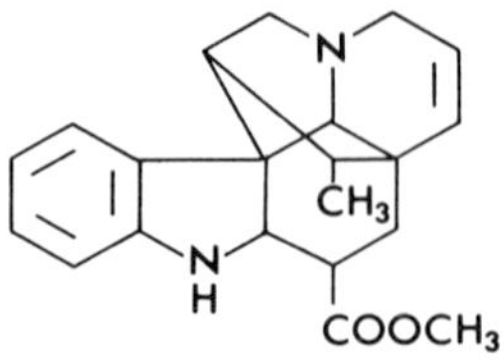

(82)

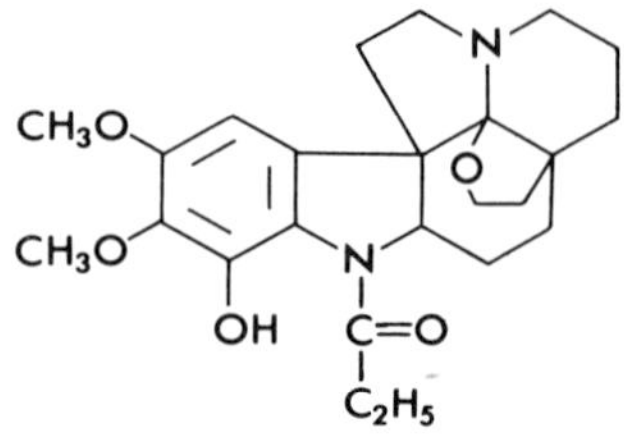

(83)

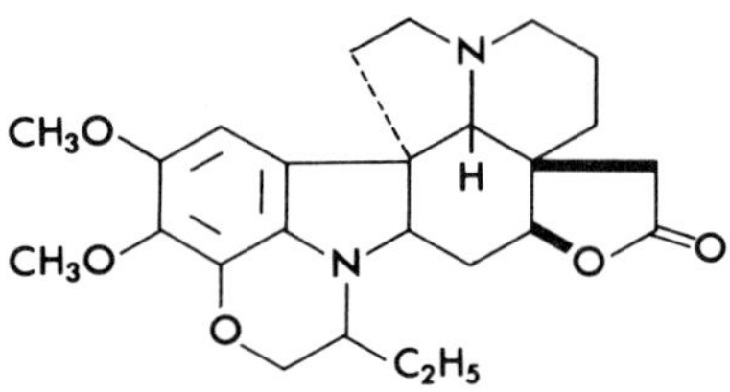

(84)

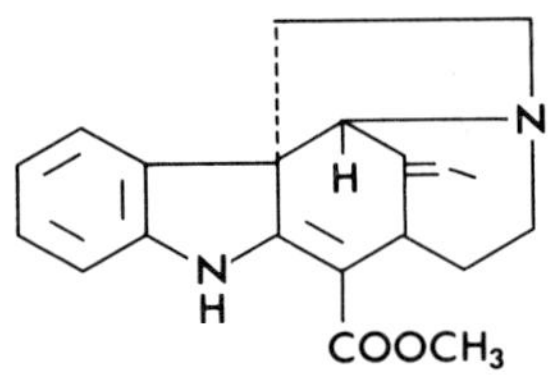

(85)

The greatest variety of these alkaloids has been found in *Aspidosperma* and *Vallesia* species.[68]

12.1 Aspidospermine (80)

The most important alkaloid in this group is aspidospermine (80). Its structure elucidation presented great difficulties and was resolved only in 1959 by a combination of degradative work[69] and X-ray analysis.[70] Four years later the alkaloid was synthesized.[71]

Once the structure of aspidospermine (80) was elucidated, the application of nmr spectroscopy and mass spectrometry led to the structure determination of numerous alkaloids of the seven types mentioned above. The importance of these techniques is well illustrated in a paper dealing with the isolation and structure determination of alkaloids from *Vallesia dichotoma*.[68]

In this study, twenty-eight indole and dihydroindole alkaloids have been isolated. All but seven of these have been characterized largely through the combined use of uv, ir, nmr spectroscopy and mass spectrometry. Several compounds were present in such small amounts that their purification could only be accomplished by preparative tlc.

The recent progress in the understanding of the chemistry of *Aspidosperma* alkaloids, some of which were available in only trace amounts, is due largely to the introduction of mass spectrometry for the study of these compounds. Therefore, it is appropriate to summarize at this point some pertinent data available from the mass spectra of aspidospermine (80) and its congeners. The two most significant features of the mass spectrum of deacetylaspidospermine (86) are the intense peak at m/e = 124

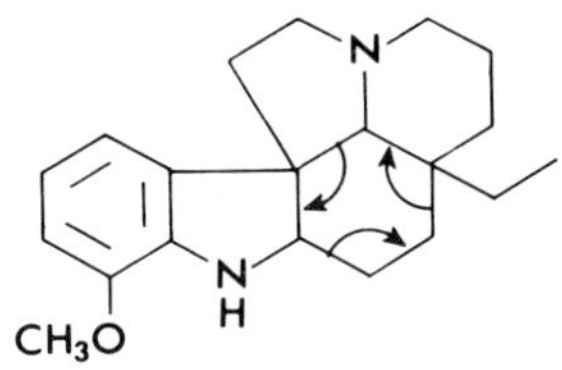

(86)

and a much smaller peak at m/e = 284, the latter arising from the loss of 28 amu from the molecular ion found at m/e = 312. (Chart 9.3) It should be pointed out that the molecular ion is of great value in determining the elemental composition of the alkaloid and allows one an unequivocal decision in this respect when the mass spectrum has been obtained on a high resolution instrument.[72] (See the section on dimeric alkaloids.)

CHART 9.3. Fragmentation of Deacetylaspidospermine (86)

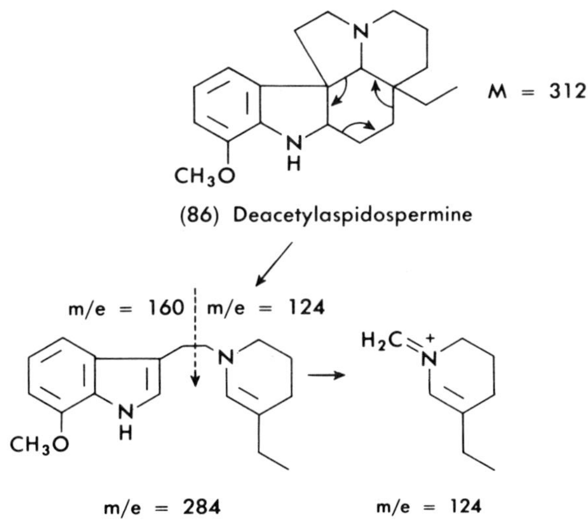

The proposed scheme explains satisfactorily the formation of these fragments and has been corroborated by fragmentation patterns of deacetylaspidospermine (86) containing deuterium at C(2).[73] It is apparent how various substitutions in the ring of aspidospermine (86) affect m/e values of various peaks of related alkaloids and thereby permit their structure assignment. This point will be discussed in the paragraph dealing with structure determination of a *Vinca* alkaloid, vindoline, a highly substituted dihydroindole alkaloid containing the aspidospermine ring system.

The absolute stereochemistry of several *Aspidosperma* alkaloids has been correlated with that of *Strychnos* alkaloids using comparisons of the molecular rotations and more recently optical rotatory dispersion (ord) curves.[74] These findings appear to have a bearing on the stereochemistry of co-occurring indole alkaloids in various plants. The group of *Aspidosperma* alkaloids do not appear to occur in nature in the same configuration. Thus, quebrachamine (45) and a few other related alkaloids occur in different plants in (+) or (−) configuration.

12.2 Vindoline (87)

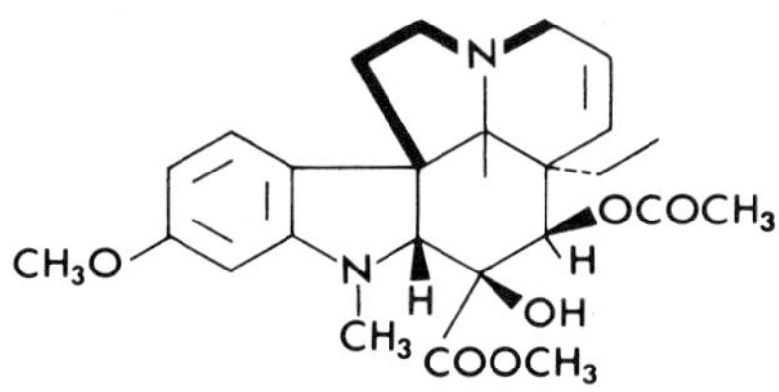

(87)

Vindoline (87) is the most abundant alkaloid in the leaves of *Vinca rosea* Linn. It is a very weak base (pK'_a 5.4) and represents a highly substituted variation of the aspidospermine ring system. Functional groups of the alkaloid were determined by preparation of appropriate derivatives. The presence of the aspidospermine ring system was established by the mass and nmr spectra of a keto derivative (88) obtained by pyrolysis of dihydrovindoline hydrochloride (89).

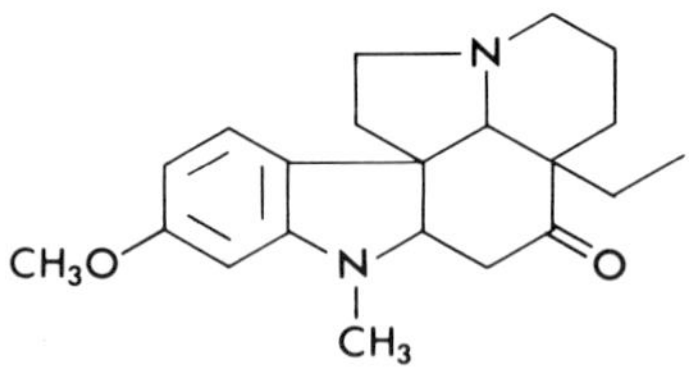

(88)

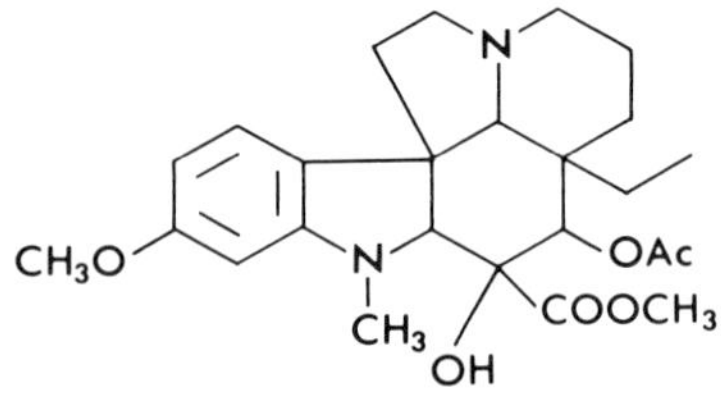

(89)

The nmr spectrum of vindoline is characterized by extremely well-defined frequencies that can be assigned to every proton[75]; this resulted in a very good understanding of the structure of this molecule (Figure 9.2

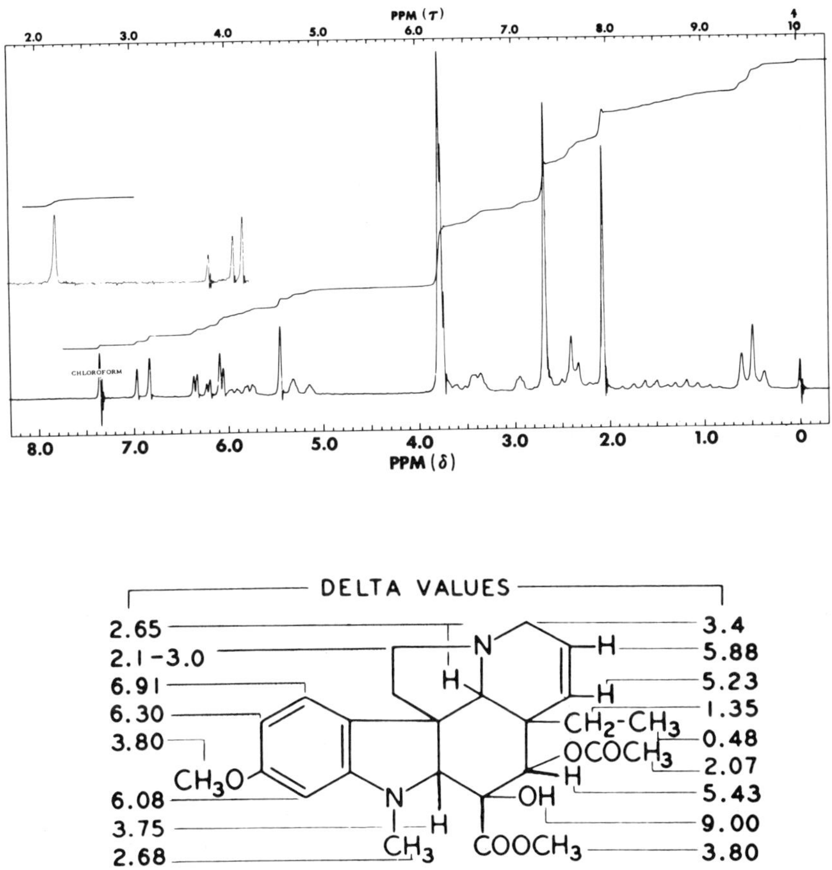

Figure 9.2 (a) Chemical shifts of vindoline in ppm (δ) values in $CDCl_3$ with Me_4Si as internal standard. (b) NMR spectrum of vindoline in $CDCl_3$.

a and b). Since vindoline represents the dihydroindole portion of dimeric oncolytic *Vinca* alkaloids, the X-ray analysis of vincristine methiodide also established the stereochemistry and absolute configuration of vindoline (87).

13. VINCA ALKALOIDS

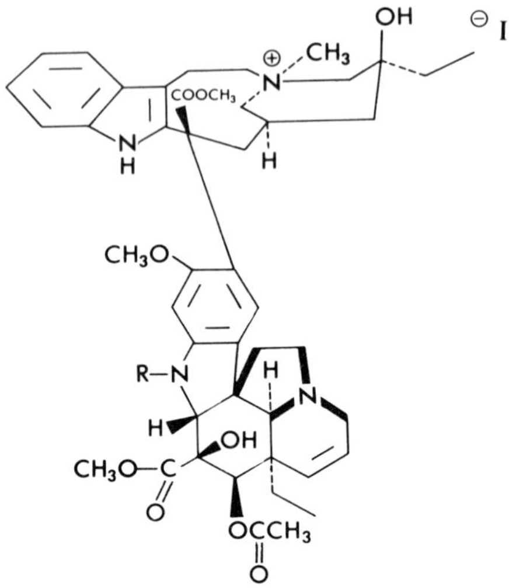

(58) R = CH_3
(90) R = CHO

The discovery of oncolytic properties of the dimeric alkaloids leurocristine (90) and vincaleukoblastine (58) from *Vinca rosea* Linn. (*Catharanthus roseus* G. Don) stimulated chemical investigations that resulted in the isolation of some sixty alkaloids of which fifty-five represent new compounds.[76] Since the two alkaloids mentioned above have found use in the chemotherapy of cancer, large amounts of alkaloidal residues from production lots were available for chemical work. Therefore, some of the alkaloids reported as minor constituents are truly present in only trace amounts. A few also could be artifacts formed during the work-up of extracts.[77]

Vinca rosea alkaloids can be divided into three groups. The dimeric indole-indoline-containing alkaloids such as vincaleukoblastine (58) represent a group derived by the condensation of a vindoline (87) derivative with a tetracyclic indole of velbanamine (91) type. The most important

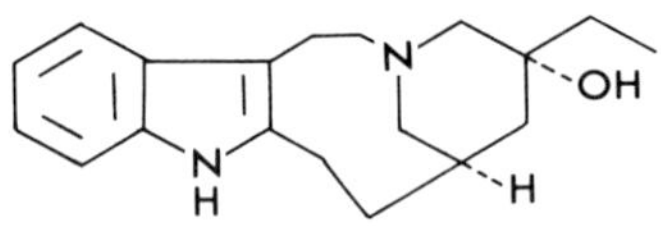

(91)

compounds in this group are the oncolytic alkaloids vincaleukoblastine[78] (58), leurosine[79,80] (90), leurosidine,[79,80] and leurocristine.[81,82] One of these substances, vincaleukoblastine (58), is described in detail in Section 14. A second group of compounds encompasses also dimeric alkaloids of miscellaneous structures which on the basis of the physical data are different from the group mentioned above. They are also mentioned in the section dealing with dimeric alkaloids. The remaining alkaloids represent a structurally heterogeneous selection of C_{18} to C_{25} substances containing two nitrogens.

The abundance of alkaloids of various types encountered during the studies of *Vinca rosea* alkaloids furnishes additional evidence of the variety of biogenetic pathways leading to the formation of various indole alkaloids. For a comprehensive treatment of the chemical and biological properties of *Vinca* alkaloids, the reader is referred to several review articles.[83]

14. DIMERIC ALKALOIDS

The search for new alkaloids has yielded a number of compounds which include in their molecules two indoles; two dihydroindoles; or one indole and one dihydroindole moiety attached to each other via C–C or C–N bonds. These compounds are often called dimeric alkaloids although the term is not strictly correct.

14.1 C-Dihydrotoxiferine (92)

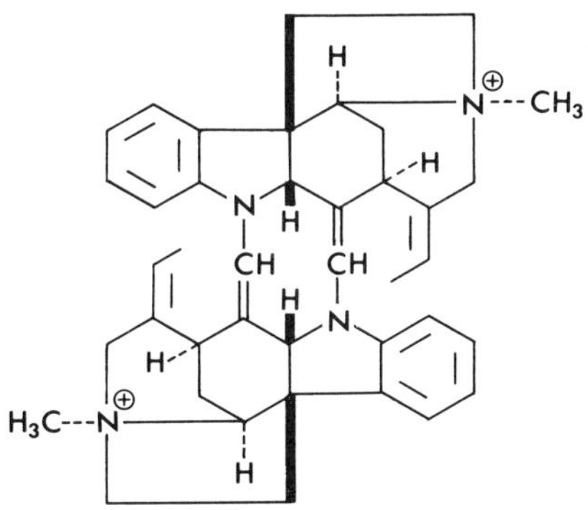

(92)

A truly dimeric alkaloid is C-dihydrotoxiferine (92), a member of Calabash-curare group. The compound represents a self-condensation product of hemidihydrotoxiferine (118). This is a truly dimeric alkaloid since in an aqueous acidic medium monomer and dimer exist in equilibrium.[84,85]

14.2 Folicanthine (93)

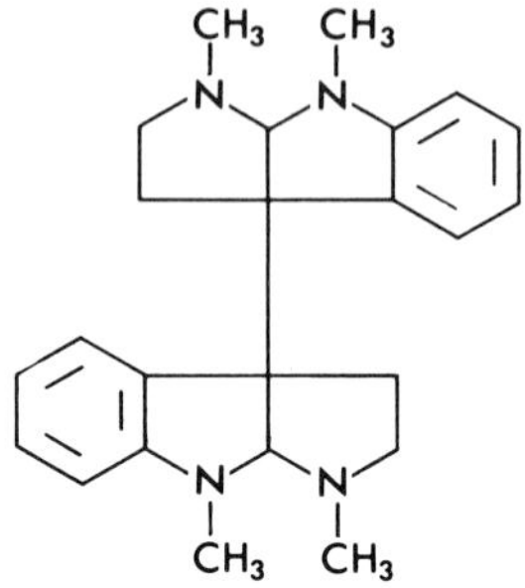

(93)

The dimers of this type are represented in *Calycanthus* alkaloids (*Calycanthaceae*). One of the major alkaloids occurring in the leaves of *Calycanthus floridus* L. is folicanthine[86] (93). Its structure was established by X-ray analysis[87] and confirmed by the total synthesis of the racemate.[88]

14.3 Geissospermine (94)

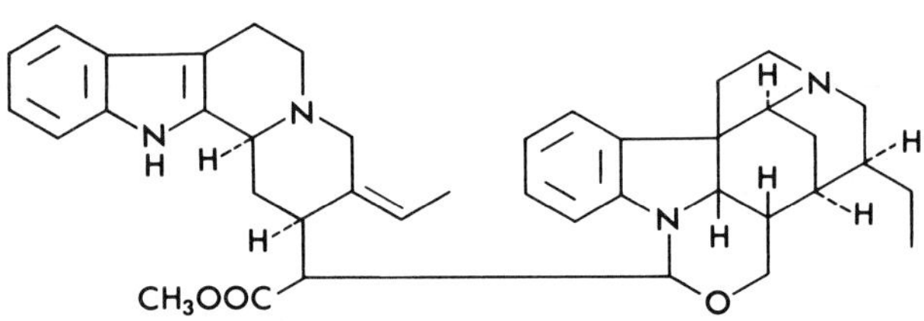

(94)

The third type of indole-dihydroindole alkaloid is represented by geissopermine (94) (*Geissospermum vellosii*), an amino acetal resulting from the union of the aldehyde geissoschizine[89] (97) with the amino alcohol geissoschizoline[90] (95).

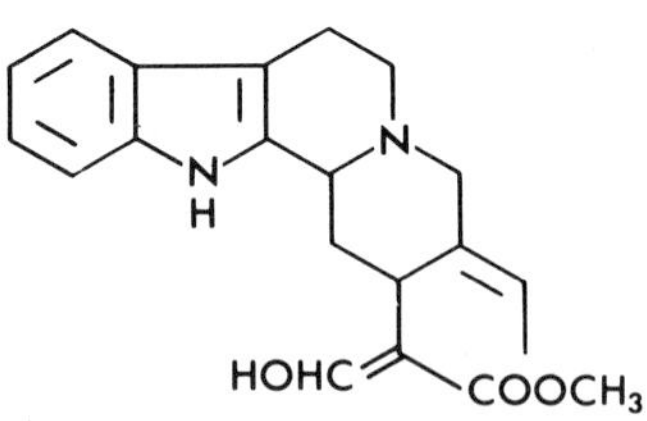

(97)

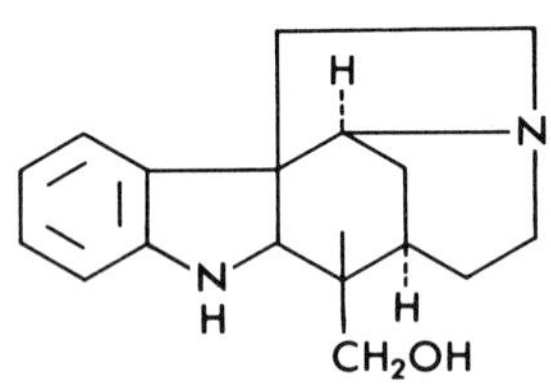

(95)

Cleavage and recombination reactions in this series of compounds are analogous to those observed with C-dihydrotoxiferine (92). The structure of geissospermine (94) was reconfirmed by partial synthesis, starting from geissoschizine (97) and geissoschizoline (95) at pH 5.[91]

14.4 Serpentinine (96)

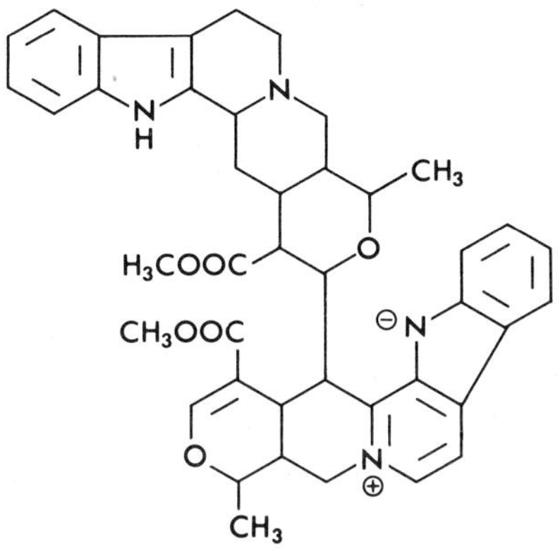

(96)

Serpentinine, one of the first alkaloids isolated from *Rauwolfia serpentina*,[92] was recognized only a few years ago as a "dimeric" alkaloid on the basis of new chemical evidence and nmr interpretation.[93]

The alkaloid incorporates serpentine (35) and 16,17-dihydro derivative of ajmalicine (21).

14.5 Vincaleukoblastine (Vinblastine, VLB) (58)

This indole-dihydroindole alkaloid is a typical representative of a large group of alkaloids from *Vinca rosea* Linn. Together with leurocristine[82] (90) [des N_a-methyl-N_a-formyl vincaleukoblastine (vincristine, VCR)]; it has found use in the treatment of various forms of malignant diseases in man. In addition to these two alkaloids, leurosine[79,80] and leurosidine[82] have also shown activity in experimental tumors in various animal species.

The structures of these oncolytic alkaloids have been elucidated by the combined use of conventional degradative methods in conjunction with ir, nmr, and high resolution mass spectrometry.[94]

Acid cleavage of the alkaloid afforded cleavamine (57), velbanamine (91), and deacetylvindoline (98). The structure of velbanamine (91) was established by mass spectrometry. The mode of attachment of the indole moiety to vindoline is apparent from the 1,4 resonance pattern in the nmr spectrum of VLB, which differs from the typical 1,2,4 pattern in the nmr spectrum of vindoline, and indicates C(15) of vindoline as the point of attachment.

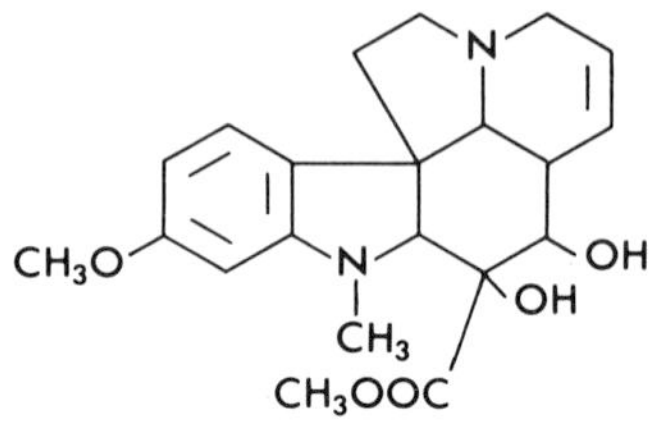

(98)

Elemental analyses of VLB and several derivatives, as well as molecular weight determinations from X-ray data, were in good agreement with the empirical formula $C_{46}H_{56-58}N_4O_9$ assigned to VLB; however, the question of the number of hydrogens (± 2H) and "*eo ipso*" the presence of tetracyclic or pentacyclic ring system in the indole portion remained unresolved. The answer to this problem was furnished by high resolution mass spectral analysis of decarbomethoxydeacetyl-VLB-monohydrazide (99), a reaction product of VLB with anhydrous hydrazine. This derivative

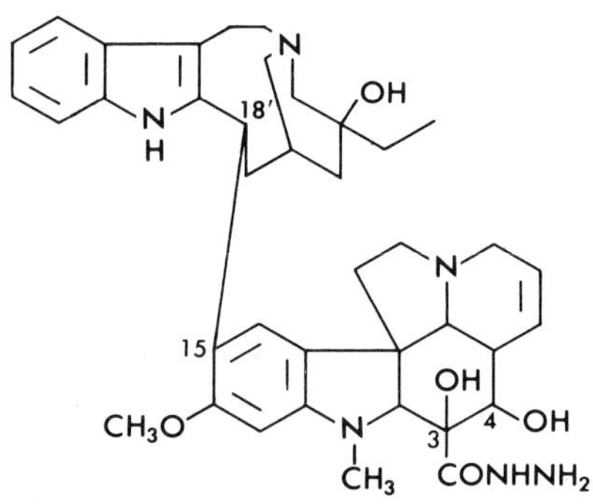

(99)

was shown to be a $C_{41}H_{54}N_6O_5$ compound, and its formation from VLB involved decarbomethoxylation at C(18), deacetylation at C(4), and acid hydrazide formation at C(3). As expected, the cleavage of the hydrazide (99) with aqueous hydrochloric acid in the presence of $SnCl_2$ and tin furnished velbanamine (91), obtained earlier from the acid cleavage of VLB, and deacetylvindoline hydrazide (100), identical with an authentic specimen prepared from vindoline itself.

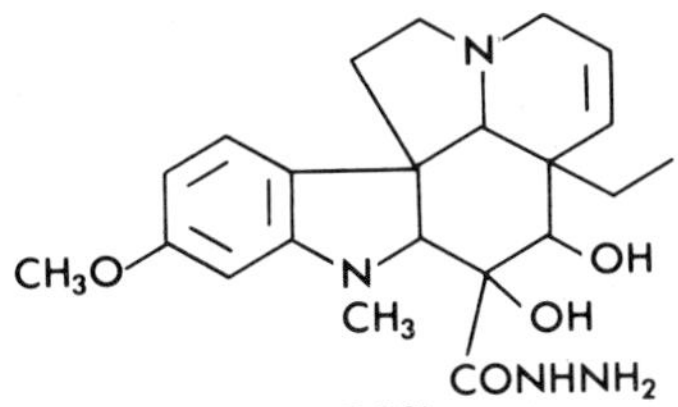

(100)

The molecular formula of the hydrazide could only be rationalized with an empirical composition of $C_{46}H_{58}N_4O_9$ for VLB. The latter was also substantiated by high resolution mass spectrometry (Table 9.1) and provided additional evidence for the tetracyclic nature of the indole moiety.[72] Although cleavage of VLB afforded a tetracyclic indole derivative, velbanamine (91), this finding did not prove the presence of a tetracyclic indole

TABLE 9.1. Fragmentation of VLB (58) and VLB-Hydrazide (99)

Mass Found	Error[a]	Composition of Ion	Elements Lost[b]
Vinblastine (58)			
838.4538	+2.4	$C_{48}H_{62}N_4O_9$[c]	
824.4372	+1.2	$C_{47}H_{60}N_4O_9$[d]	
811.4234	−0.3	$C_{45}C^{13}H_{58}N_4O_9$	
810.4219	+1.5	$C_{46}H_{58}N_4O_9$	
752.4153	+0.4	$C_{44}H_{56}N_4O_7$	$C_2H_2O_2$
751.4080	+1.0	$C_{44}H_{55}N_4O_7$	$C_2H_3O_2$
750.4014	+2.2	$C_{44}H_{54}N_4O_7$	$C_2H_4O_2$
651.3906	−0.5	$C_{40}H_{52}N_4O_4$	$C_6H_7O_5$
650.3858	+2.6	$C_{40}H_{50}N_4O_4$	$C_6H_8O_5$
591.3723	+2.6	$C_{38}H_{47}N_4O_2$	$C_8H_9O_7$
154.1219	−1.3	$C_9H_{16}NO$	$C_{37}H_{42}N_3O_8$
VLB-Hydrazide (99)			
711.4198	+1.1	$C_{40}C^{13}H_{54}N_6O_5$	
710.4126	−2.7	$C_{41}H_{54}N_6O_5$	
651.3841	−2.8	$C_{40}H_{52}N_4O_4$	CH_3N_2O
593.3841	−1.2	$C_{38}H_{49}N_4O_2$	$C_3H_5N_2O_3$
592.3775	−4.5	$C_{38}H_{48}N_4O_2$	$C_3H_6N_2O_3$
509.2711	+3.4	$C_{32}H_{35}N_3O_3$	$C_9H_{19}N_3O_2$
154.1232	+0.5	$C_9H_{16}NO$	$C_{32}H_{36}N_5O_4$

[a] Difference in millimass units between mass found and value calculated for the elemental composition in the third column (^{12}C = 12.000000).
[b] From mass 810 in case of (58).
[c] Dimethylation product.
[d] Monomethylation product.

moiety in the intact molecule of VLB since the pentacyclic isoquinuclidine alkaloid catharanthine was shown earlier to undergo acid-catalyzed transformation to cleavamine (57), a tetracyclic isoquinuclidine derivative, $C_{19}H_{24}N_2$.

The position of the tertiary hydroxyl group at C(4) in the indole portion of VLB was elucidated by mass spectrometric examination of velbanamine (91). This function was shown to occupy identical positions in both substances because upon cleavage in DCl and D_2O (*vide infra*) no deterium was found to be present in the hydroxyl-bearing piperidine ring of deuterio-velbanamine prepared from VLB.

The mass spectrum of velbanamine (91) has its most intense peak at m/e = 154 (101) requiring the hydroxyl group to be situated in the alicyclic portion of the molecule but not at C(8) and C(18). Furthermore, the accurate mass of a characteristic fragment due to the loss of 57 amu was determined to be 241.1709 (102). This value corresponds to $C_{16}H_{21}N_2$ and indicates the elimination of C_3H_5O which is in agreement with structure (102).[72] The shift of m/e = 241 in velbanamine (91) to m/e = 243 in dideuteriovelbanamine (mol. wt. 300) corroborates this suggestion.

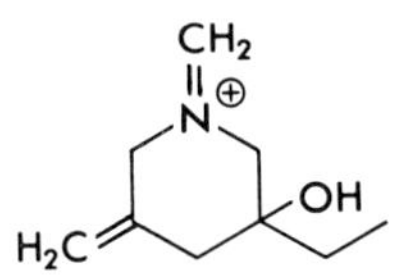

(101)

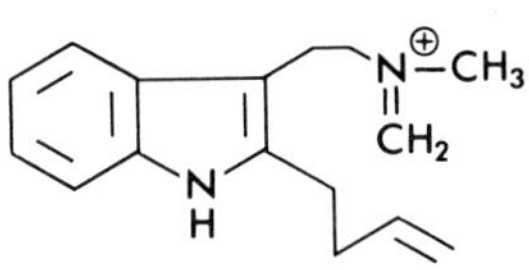

(102)

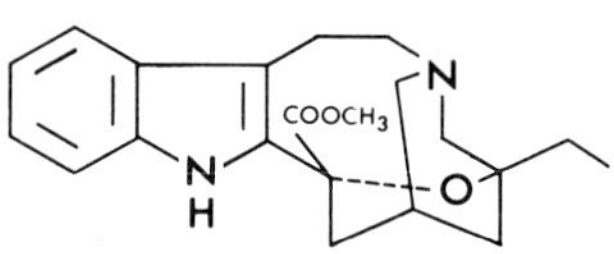

(103)

To establish finally the structure of VLB, the carbon-carbon bond linking C(15) of vindoline with the indole moiety had to be located. Cleavage of VLB in D_2O solution in the presence of DCl, $SnCl_2$, and tin gave deuteriovelbanamine. Its mass spectrum (mol wt 304) showed absence of deuterium in the piperidine ring because the prominent peak remained at m/e = 154 in undeuterated velbanamine (91). Six deuterium atoms were located in a fragment of m/e = 247 corresponding to m/e = 241 in velbanamine (91) itself. Four deuterium atoms are in the aromatic ring (nmr evidence) while the remaining two had to be at C(8) and/or C(18). Consequently, one of these two carbon atoms represents the vindoline terminus in VLB, and C(8) was eliminated as follows: cleavage of VLB in boiling 40% aqueous sulfuric acid (two hours) in the absence of reducing agents afforded an amino acid oxide that was esterified to the corresponding methyl ester (103), $C_{21}H_{26}N_2O_3$. The treatment of the ester with DCl in D_2O,

$SnCl_2$, and tin afforded deuteriovelbanamine whose mass spectrum was identical with that of a sample prepared analogously from VLB.[94] The stereochemistry and absolute configuration of VLB was deduced from the X-ray analysis of vincristine methiodide,[95] since the latter has been correlated chemically with VLB.

The structure elucidation of VLB furnishes an excellent example of the use of chemical and physical chemical methods in the understanding of structures of complex natural products.

14.6 Voacamine (71)

Another group of compounds consisting of two different indole moieties is represented by several alkaloids from plants belonging to the tribe of *Tabernaemontaneae*. One of the alkaloids of this type, voacamine (71) (*Voacanga africana Stapf.*), was shown to consist of vobasine (68) and voacangine (104) moieties. Its structure was established by the use of chemical and spectral methods and confirmed by partial synthesis from voacangine (104) and vobasinol[96] (105). Voacamidine (106) was obtained as a minor product in this reaction.

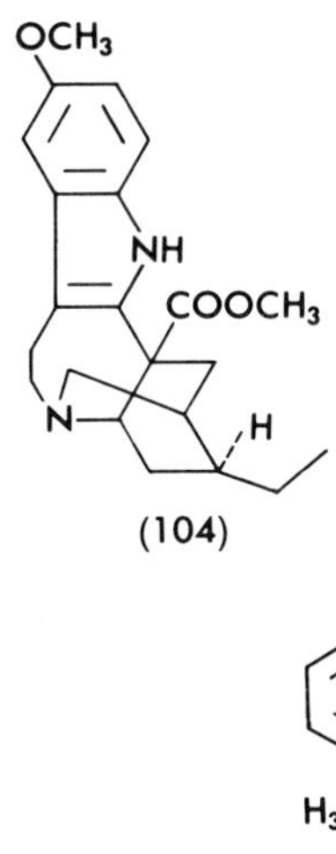

(104)

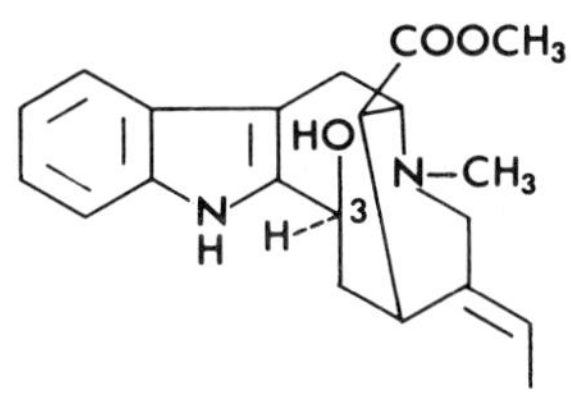

(105)

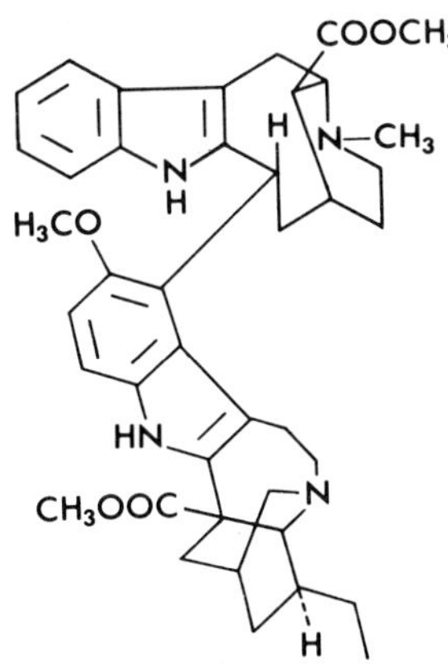

(106)

An interesting observation was made during the mass spectral analysis of voacamine (71). The correct molecular ion for the compound should be 704 ($C_{43}H_{52}N_4O_5$). Instead, the initial mass spectrum showed M+ = 718. When voacamine (71) was vaporized directly into the ion source, an intermolecular methylation on N_b took place. When voacamine acetate, which lacks the carbomethoxy group ($C_{42}H_{52}N_4O_3$), was used, the correct molecular ion m/e = 660 was obtained. A similar observation was made in the mass spectrum of VLB (Table 9.1).

15. STRYCHNOS ALKALOIDS

No other group of dihydroindole alkaloids has been subjected to such intensive chemical investigation as the group of *Strychnos* alkaloids, and particularly strychnine itself. This becomes apparent when one examines volumes of publications between the years elapsed since the discovery of strychnine and its structure elucidation. The reader is advised to examine some of these papers in order to appreciate the intellectual challenge offered by the seven ring system of strychnine in the days before the ir, nmr, and mass spectroscopy.

Strychnos represents a genus of small trees and climbing shrubs, members of the family of *Loganiaceae*, distributed throughout Africa, Asia, and South America. The South American plants are the source of curare, the notorious arrow and dart poison of Indians. The African and Asiatic plants serve as starting materials for the manufacturing of strychnine and closely related alkaloids. While strychnine still enjoys limited use in modern medicine as a respiratory stimulant, many question the therapeutic value of this alkaloid. Its greatest use is in the form of rat biscuits still employed for the extermination of rats and other rodents. Constituents of curare, such as C-toxiferine I and its derivatives, found use in the induction of muscular relaxation so important in the treatment of tetanus or surgery requiring light anesthesia.

15.1 Strychnine (107)

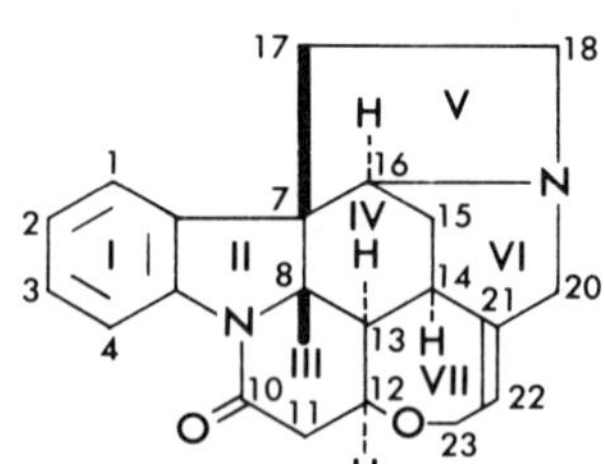

(107)

Strychnine (107) and brucine, its 2,3-dimethoxy congener, were among the first alkaloids discovered in the plant kingdom. Two pioneers of alkaloid chemistry, Pelletier and Caventou, isolated it in 1819 from the seeds and bark of *Strychnos nux vomica*.[97] Since that time strychnine (107) has been found in some thirteen species of the genus *Strychnos*.[98] The structure elucidation of this alkaloid remained for over sixty years one of the greatest challenges of natural products chemistry and culminated in the correct structure proposal.[99] Unequivocal proof was furnished by X-ray analysis of its salts[100] and by the total synthesis[101] of the alkaloid.

In the course of chemical investigations of strychnine, the alkaloid and its derivatives have been subjected to a great variety of chemical reactions with the purpose of opening various rings and stripping the molecule of its functional groups.[102,103]

To illustrate this approach, a few reactions have been selected and are depicted below. For example, ring III of strychnine can be opened leading to the formation of strychnic acid (108) that in turn can be reconverted to the starting material (107).

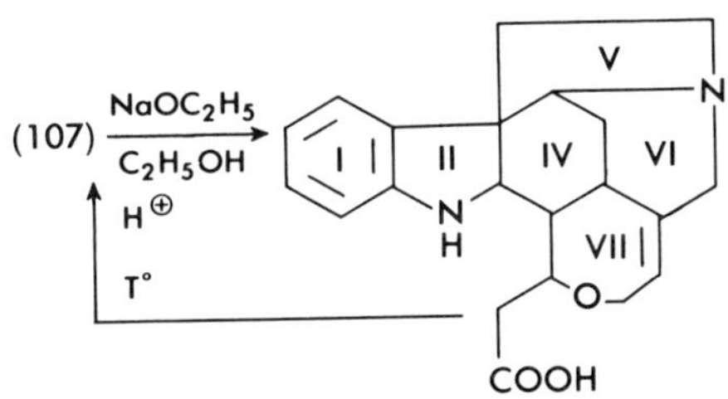

(108)

In another sequence ring VII was opened leading to the formation of isostrychnine[104] I (109).

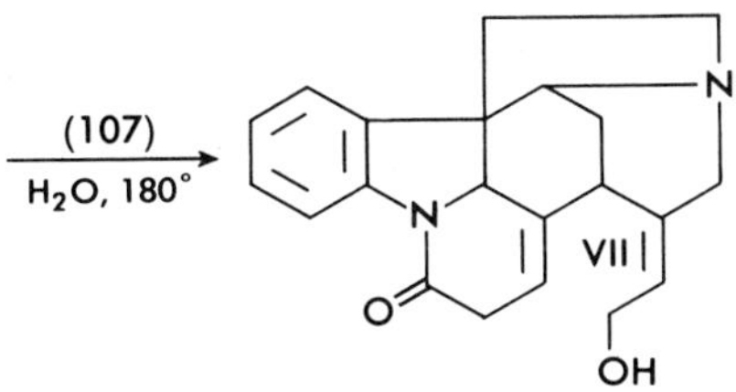

(109)

Reduction with sodium amalgam (Emde reduction) gave rise to a new derivative in which two rings were opened[105] (110).

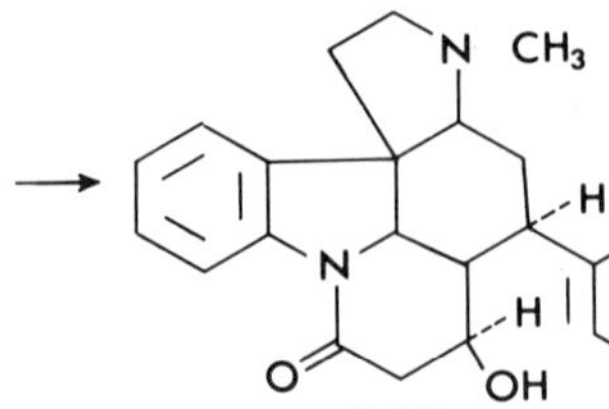

(110)

Oxidation of strychnine (107) with chromic acid under controlled conditions afforded a compound, Wieland acid (111), a derivative in which three rings, I, II, and III, have been degraded.[106]

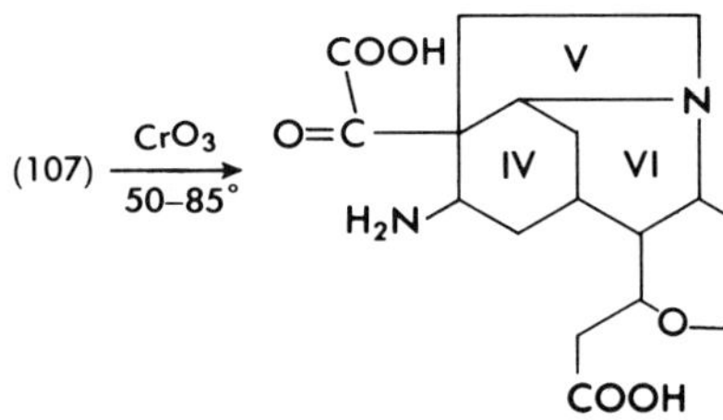

(111)

A critical appraisal of data of the type mentioned above together with the intuitive genius of Sir Robert Robinson[99] led to the proposal of the correct structure. The confirmation was furnished by X-ray data and by the brilliantly conceived total synthesis[101] of strychnine (107).

The sequence of steps in this synthesis contains several unusual and novel reactions, and stands out as a milestone in synthetic organic chemistry. Particularly unusual steps of the synthesis, shown in Chart 9.4, are emphasized by the use of heavy arrows.

CHART 9.4. Total Synthesis of Strychnine (107)

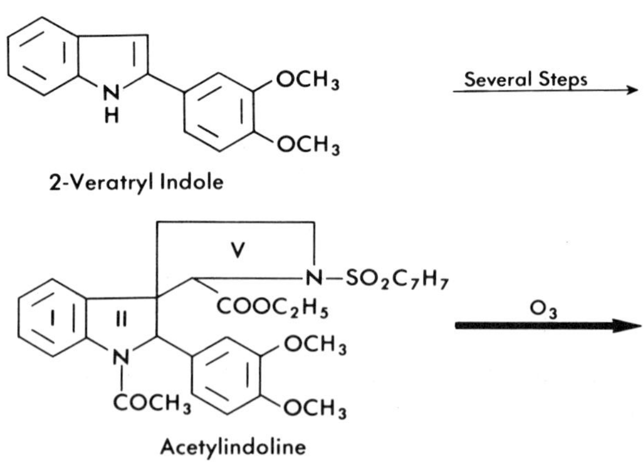

CHART 9.4. (Continued)

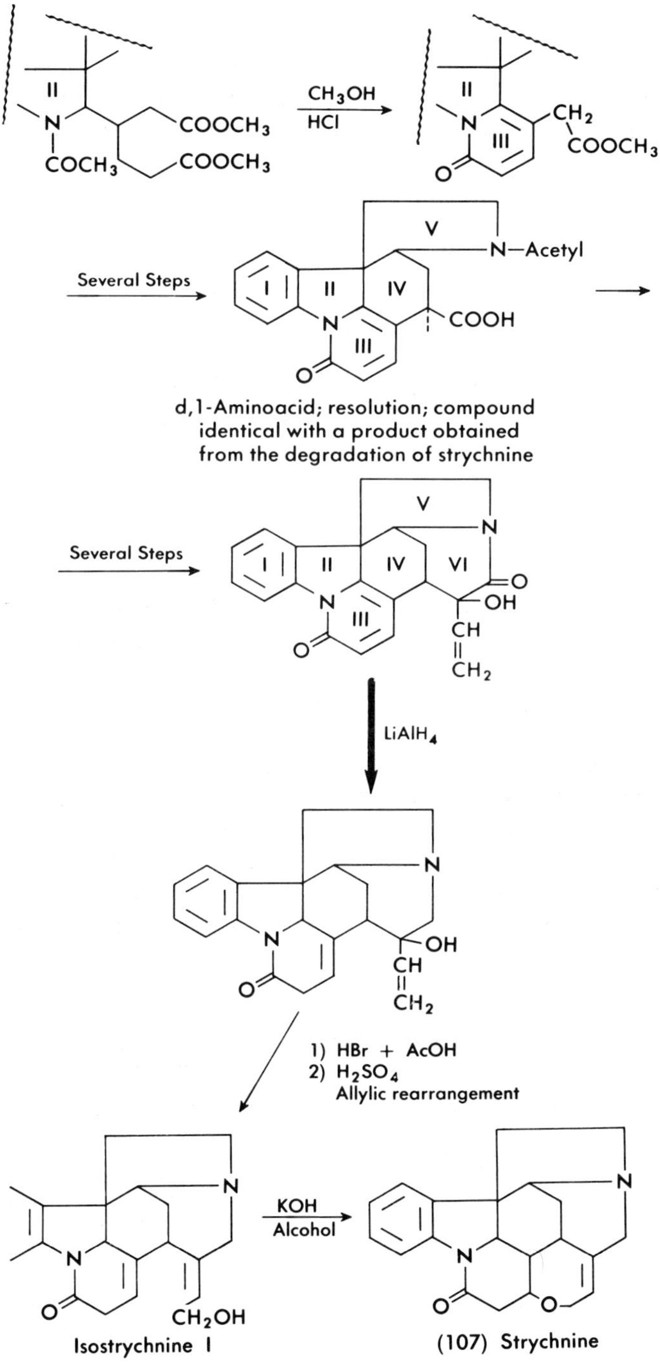

15.2 Calabash Curare

There are several types of arrow poisons (curare) made by South American Indians. Those packed in bamboo tubes are called *tube-curare* and consist of alkaloids from the genus *Chondrodendron* (Family: *Menispermaceae*). D-tubocurarine (112) is one of many bis-benzyl isoquinoline alkaloids of this type.

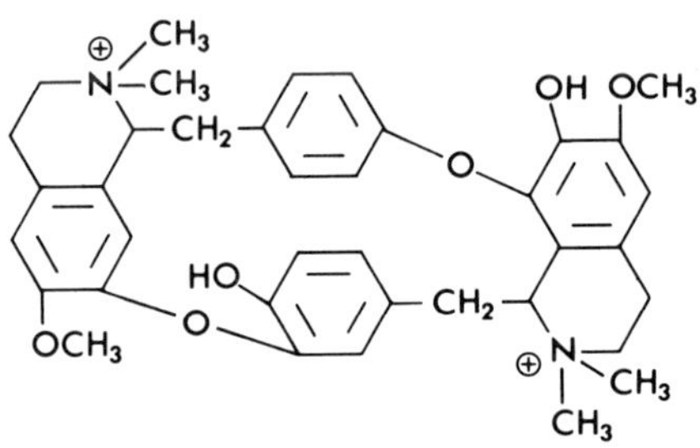

(112)

Calabash curare contain alkaloids prepared from *Strychnos* genus and are packed in calabashes or gourds, hence, the designation calabash-curare.

There are two main subfamilies of alkaloids in these poisons: monomeric C_{20} and dimeric, highly active C_{40} compounds. All occur as quaternary alkaloid salts. Thanks to the development of partition chromatography on cellulose, it was found[107,108] that even pure species of *Strychnos toxifera* yield some forty different quaternary alkaloids.[109]

One can distinguish here different types of alkaloids such as C-mavacurine (113) [modified cathinone (73) type], C-fluorocurine (114) (indoxyl type), macusine B (115) (sarpagine type), and C-toxiferine I (118) (a dimeric alkaloid).[110]

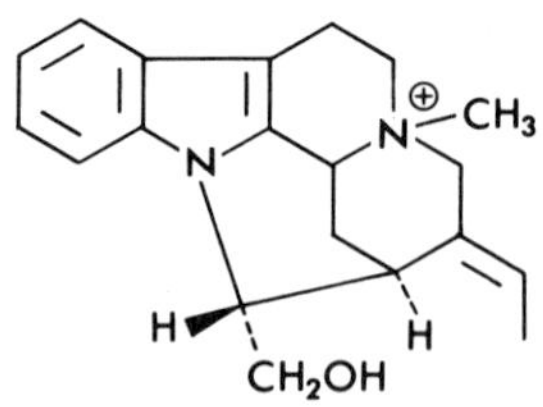

(113)

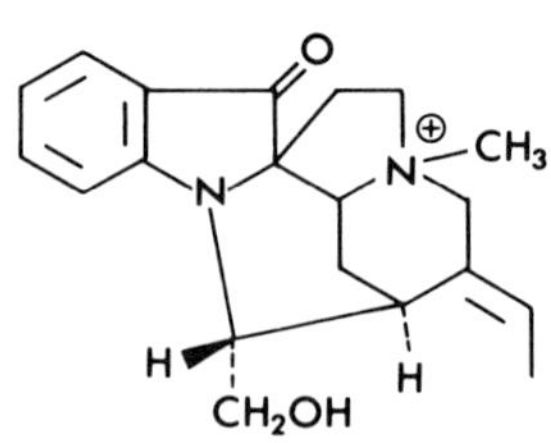

(114)

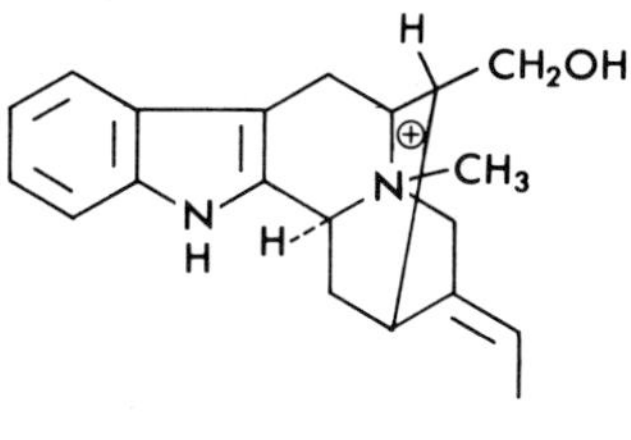

(115)

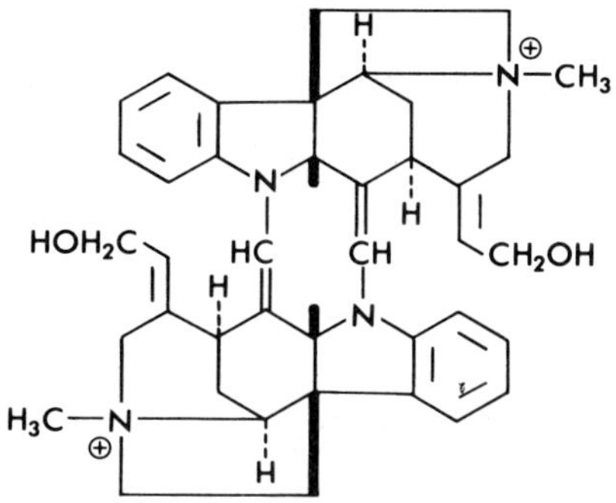

(116)

15.2.1 Dimeric Alkaloids of Calabash Curare. Dimeric C_{40} curare alkaloids are characterized by a high curarizing activity. They contain four nitrogens, two of which are weakly basic and two are quaternized.

The dimeric nature of these alkaloids was discovered[111] only in the late fifties and demonstrated in the following way. Pyrolysis of the quaternary alkaloid gives in the case of a monomeric alkaloid, species (A), and in the case of a dimeric alkaloid, species (B). The monomeric-norbase (A) is

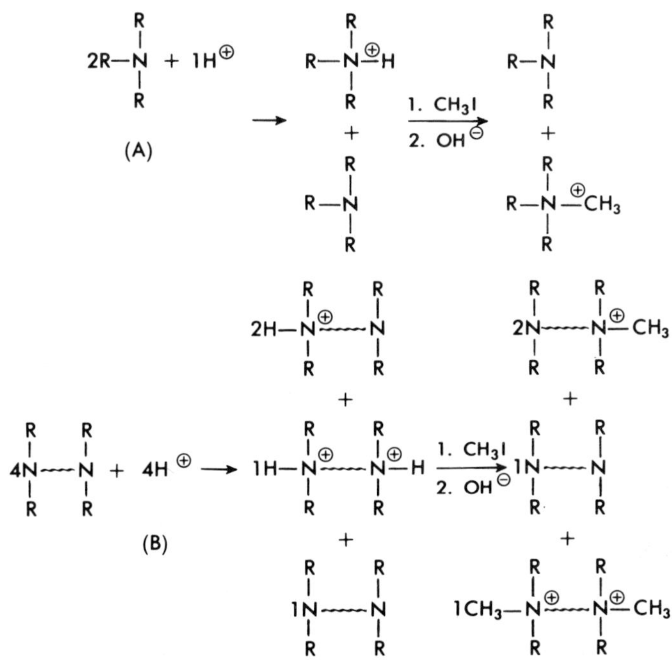

treated with one-half equivalent of acid, and an equilibrium mixture (1:1) of free base and protonated base obtained. By methylation with methyl iodide one obtains a methylated compound while the protonated base is unchanged. Products in the form of chlorides are placed on paper and subjected to chromatography using alkaline solvent mixtures. These transform chlorides into free bases that give two spots with different R_f values. The dimeric norbases (B) are treated analogously and afford three spots with different R_f values, namely, the bis-tertiary base, the mono-tertiary-monoquaternary base, and the bis-quaternary base.

Structures of dimeric alkaloids remained unknown until the alkaloid caracurine VII[112] was recognized as Wieland–Gumlich aldehyde (117) obtained many years earlier by the degradation of strychnine (107).[113]

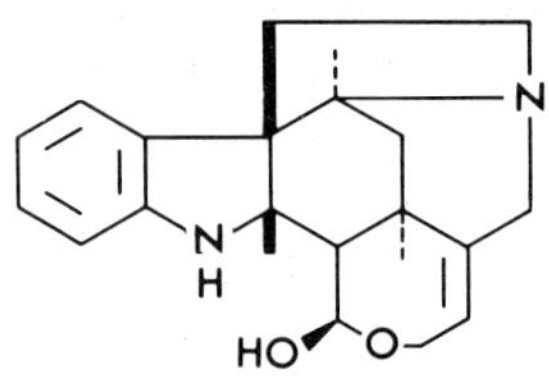

(117)

The aldehyde function in this compound is present in the cyclic hemi-acetal form. A free aldehyde occurs in nature as hemidihydrotoxiferine I (118).

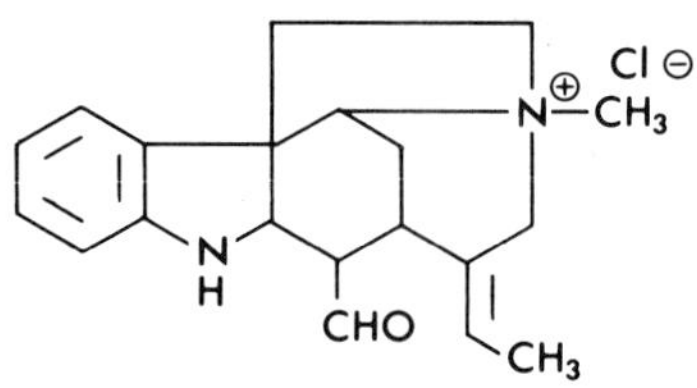

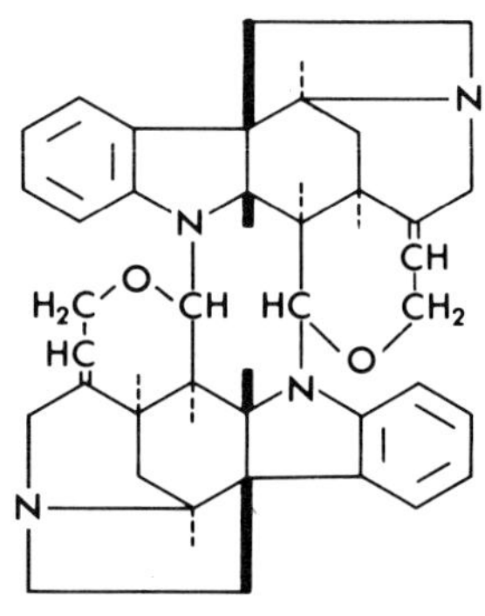

(118) (119)

Condensation of Wieland–Gumlich aldehyde (117) in acetic acid affords mainly caracurine V (119) besides some minor products.[114]

Further treatment of caracurine with hydrogen bromide followed by reduction of the bromide with zinc and acetic acid and quaternization gives C-dihydrotoxiferine (99). Interesting modifications of these structures are present in the alkaloid C-calebassine[115] (120) and C-curarine[116] (121).

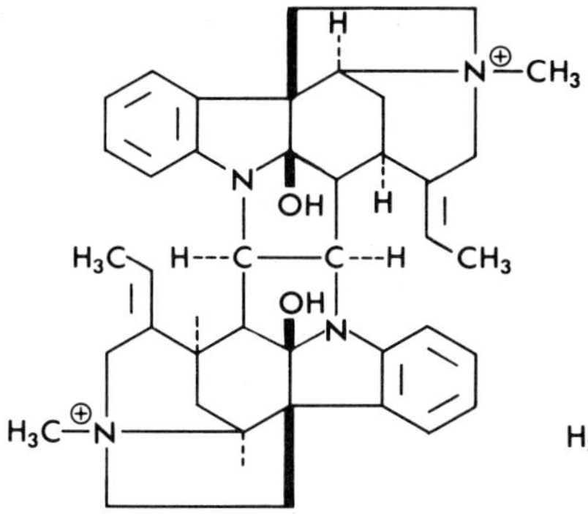

(120)

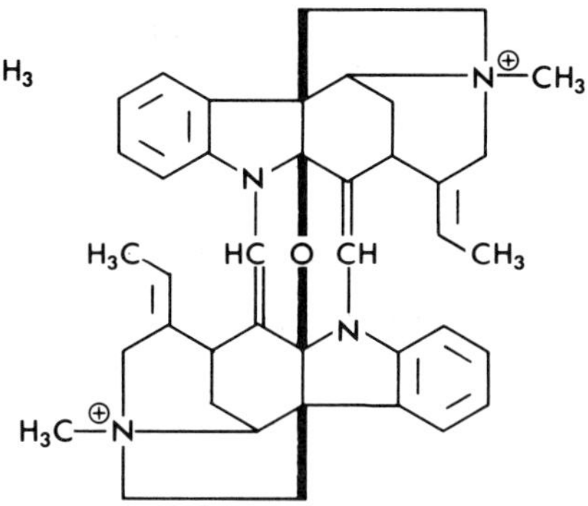

(121)

16. MISCELLANEOUS

To further illustrate structural varieties of indole-containing natural products, a few selected compounds deserve to be mentioned. These are either not representative of any of the foregoing classifications or possess interesting characteristics not usually associated with indole alkaloids.

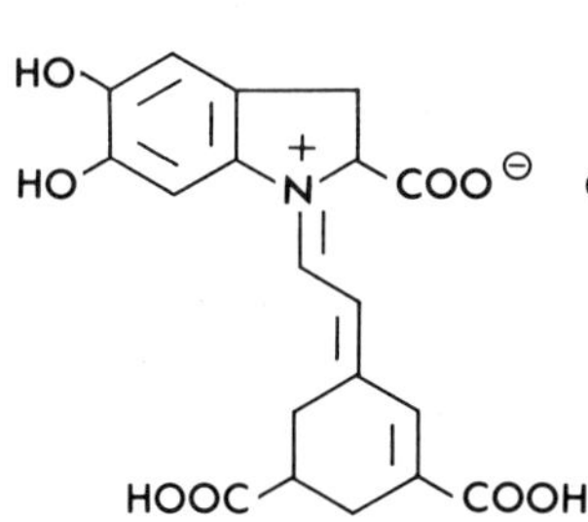

(122)

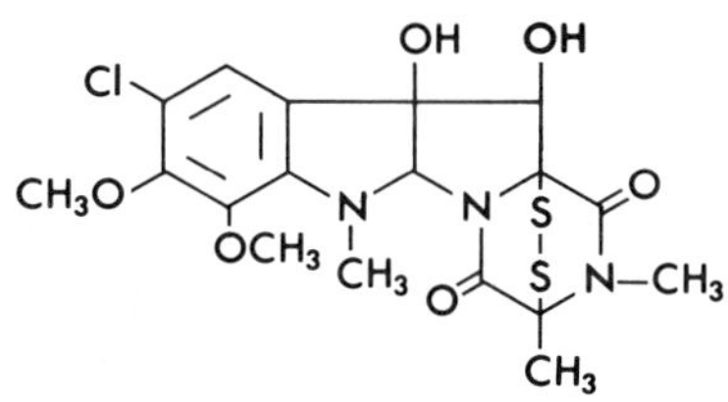

(123)

The nonalkaloidal substances betanidine (122) and sporidesmine (123), along with the alkaloids echitamine (124) and flavopereirine (125), illustrate the structural variety of indole-containing natural products.

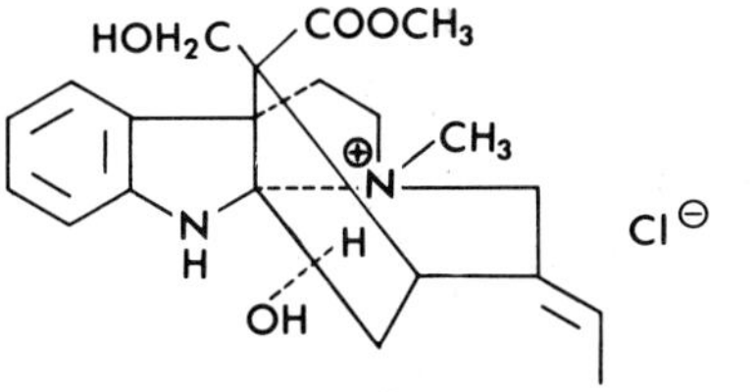

(124)

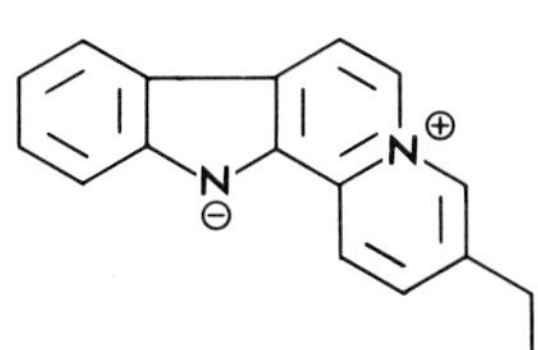

(125)

As representatives of indole alkaloids that are not apparently biogenetically derived from tryptophan, uleine (126) and apparicine (127) are included.

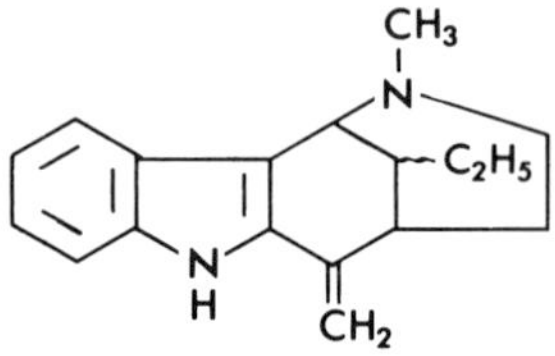

(126)

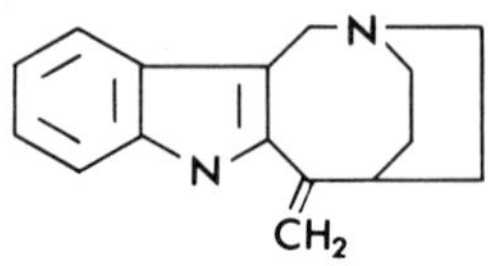

(127)

Since the quinazolinocarbolines rutaecarpine (128) and evodiamine (129) do not appear in the chapter devoted to this group of alkaloids, they

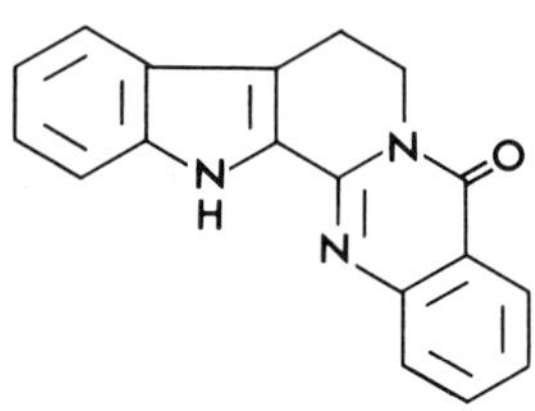

(128)

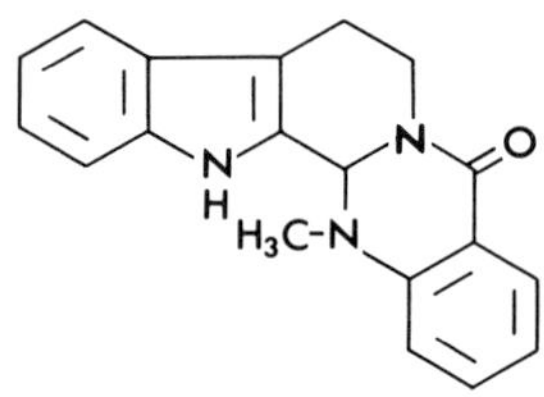

(129)

are mentioned here. These alkaloids and their congeners occur in the genera *Evodia*, *Hortia*, and *Zanthoxylum*, all members of the *Rutaceae* family. Several syntheses of rutaecarpine (128) have been reported. One synthesis was accomplished under so-called "physiological" conditions and involved reacting 3,4-dihydro-2-carboline (130) and *o*-aminobenzaldehyde (131) as hydrochlorides in aqueous solution at 25° for 10 hours. The intermediate compound was converted to rutaecarpine (128) in 70% yield by the use of potassium ferricyanide in a phosphate buffer at pH 6.9 and 25°.[117]

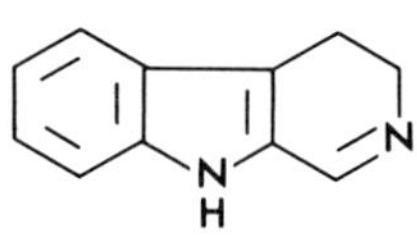

(130)

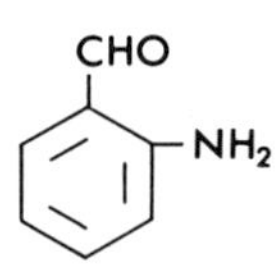

(131)

Betanidine (122) is an aglucon of betanine, a member of a large group of betacyanines, pigments of the beet *Beta vulgaris*.[118]

Echitamine (124) was isolated from *Alstonia scholaris*. The structure of the free base is not known. The structure of the chloride has been elucidated by the X-ray analysis.[119]

Sporidesmine (123) was found to be the toxic agent responsible for producing facial eczema in sheep. Its structure has been determined by X-ray analysis.[120]

Flavopereirine (125) occurs in *Geissospermum vellosii* and is readily available by several total syntheses.[121]

Apparicine (127) was found in various *Aspidosperma* species[122]; however, it has also been reported as tabernoschizine from *Conopharyngia durissima* and *Schizozygia caffaeoides*[123] and pericalline from *Vinca rosea* Linn.[124] The structure of this alkaloid can be regarded as a novel variant of uleine (126). Each of these bases contains the same carbon skeleton, but in apparicine (127) the N-methyl group of uleine (126) has become the methylene bridge. Both compounds do not appear to be derived from tryptophan since the two carbon bridge between N_b and the β-position of the indole nucleus is missing.

It is interesting to note that the alkaloid isolated from *Aspidosperma gomezianum*, *A. olivaceum*, *A. eburneum*, and *A. multiflorum* is present as (−)-whereas, *A. pasycarpon* affords (+)-apparicine[122] (127).

17. CONCLUDING REMARKS

Consideration of structural varieties of indole alkaloids prompts the basic question as to their role in nature. Despite numerous attempts to give a satisfactory explanation for their occurrence, no convincing theory is as yet available. The structure elucidation of many of these alkaloids contributed not only to the development of organic chemistry but also enriched the armamentarium of the modern medicine.

It is hoped that by illustrating the structural variations in indole alkaloids, the reader will feel compelled to pursue chemical investigations of various plant materials as sources of yet uncovered chemical entities of potential value in the understanding of biogenetic processes and leading towards the discovery of new biologically useful compounds.

ACKNOWLEDGMENTS

The invaluable help of Nancy De La Higuera in checking the bibliography is gratefully acknowledged.

The expert handling of the revised manuscript including drawing of over 200 structures by Mrs. Phylis Hager, Secretary of the Chemical Research Division, is gratefully acknowledged.

REFERENCES

1. F. Kroemeke, "F. L. Sertürner, der Entdecker des Morphiums," p. 61, Gustav Fischer, Jena, Germany, 1925.
2. B. Robinson, "The Fischer Indole Synthesis," *Chem. Rev.* **63,** 373 (1963).
3. G. M. Robinson and R. Robinson, *J. Chem. Soc.* **113,** 639 (1918); **125,** 827 (1924).
4. K. Clusius and H. R. Weisser, *Helv. Chim. Acta* **35,** 400 (1952).
5. C. Bulard and A. C. Leopold, *Compt. Rend. Acad. Sci.* **247,** 1382 (1958).
6. V. E. Tyler, Jr., *Science* **128,** 718 (1958).
7. A. Hofmann, R. Heim, A. Brack, H. Kobel, A. Frey, H. Ott, Th. Petrzilka, and F. Troxler, *Helv. Chim. Acta* **42,** 1557 (1959).
8. H. G. Boit, "Ergebnisse der Alkaloid Chemie bis 1960," p. 549, Academie-Verlag, Berlin, Germany, 1961.
9. T. L. Sourkes, "Biochemistry of Mental Disease," p. 382 Harper & Row, Publishers, New York, N.Y., 1962.
10. G. R. Clemo and R. J. Holt, *J. Chem. Soc.* 1313 (1953).
11. M. A. Ondetti and V. Deulofeu, *Tetrahedron Letters* **1,** 18 (1960).
12. R. B. Woodward, G. A. Iacobucci, and F. A. Hochstein, *J. Amer. Chem. Soc.* **81,** 4434 (1959).
13. J. M. Mueller, E. Schlittler, and H. J. Bein, *Experientia* **8,** 338 (1952).
14. M. Hesse, "Indolalkaloide in Tabellen," Springer-Verlag, Berlin, Germany, 1964.
15. R. E. Woodson, Jr., H. W. Youngken, E. Schlittler, and J. A. Schneider, "Rauwolfia: Botany, Pharmacognosy, Chemistry, and Pharmacology," Little, Brown and Company, Boston, Massachusetts, 1957.
16. R. B. Woodward and B. Witkop, *J. Amer. Chem. Soc.* **70,** 2409 (1948).
17. E. E. van Tamelen, M. Shamma, A. W. Burgstahler, J. Wolinsky, R. Tamm, and P. E. Aldrich, *J. Amer. Chem. Soc.* **80** 5006 (1958).
18. J. Poisson, *Ann. Chim* (Paris) **9,** 99 (1964).
19. E. Wenkert and N. V. Bringi, *J. Amer. Chem. Soc.* **81,** 1474 (1959).
20. N. Neuss, H. E. Boaz, and J. W. Forbes, *J. Amer. Chem. Soc.* **76,** 2463 (1954).
21. L. Dorfman, A. Furlenmeier, C. F. Huebner, R. Lucas, H. B. MacPhillamy, J. M. Mueller, E. Schlittler, R. Schwyzer, and A. P. St. Andre, *Helv. Chim. Acta* **37,** 59 (1954).
22. P. E. Aldrich, P. A. Diassi, D. F. Dickel, C. M. Dylion, P. D. Hance, C. F. Huebner, B. Korzun, M. E. Kuehne, L. H. Liu, H. B. MacPhillamy, E. W. Robb, D. K. Roychaudhuri, E. Schlittler, A. F. St. Andre, E. E. van Tamelen, F. L. Weisenborn, E. Wenkert, and O. Wintersteiner, *J. Amer. Chem. Soc.* **81,** 2481 (1959).
23. W. E. Rosen and J. N. Shoolery, *J. Amer. Chem. Soc.* **83,** 4816 (1961).
24. R. B. Woodward, F. E. Bader, H. Bickel, A. J. Frey, and R. W. Kierstead, *J. Amer. Chem. Soc.* **78,** 2023, 2657 (1956) and *Tetrahedron* **2,** 1 (1958).
25. E. E. van Tamelen and I. G. Wright, *Tetrahedron Letters* 295 (1964).
26. E. E. van Tamelen and J. B. Hester, Jr., *J. Amer. Chem. Soc.* **81,** 3805 (1959).
27. F. L. Weisenborn, M. Moore, and P. A. Diassi, *Chem. Ind. (London)* 375 (1954).
28. E. Wenkert and N. V. Bringi, *J. Amer. Chem. Soc.* **80,** 3484 (1958).
29. E. Wenkert, B. Wickberg, and C. L. Leicht, *J. Amer. Chem. Soc.* **83,** 5037 (1961).
30. M. Shamma and J. B. Moss, *J. Amer. Chem. Soc.* **84,** 1739 (1962).
31. E. E. van Tamelen and C. Placeway, *J. Amer. Chem. Soc.* **83,** 2594 (1961).
32. W. I. Taylor, *Alkaloids* R. H. F. Manske, Ed. Academic Press, New York, **8,** 759 (1965).
33. E. Leete, G. Ghosal, and P. N. Edwards, *J. Amer. Chem. Soc.* **84,** 1068 (1962).
34. Sir Robert Robinson, *Chem. Ind. (London)* 285 (1955).

35. R. B. Woodward, *Angew. Chem.* **68,** 13 (1956).
36. M. F. Bartlett, R. Sklar, W. I. Taylor, E. Schlittler, R. L. S. Amai, P. Beak, N. V. Bringi, and E. Wenkert, *J. Amer. Chem. Soc.* **84,** 622 (1962).
37. N. V. Sidgwick, "The Organic Chemistry of Nitrogen," 3rd ed., p. 654 Ian T. Millar and H. D. Springall, Eds.), Clarendon Press, Oxford, 1966.
38. F. M. Lovell, R. Pepinski, and A. J. C. Wilson, *Tetrahedron Letters* **4,** 1 (1959).
39. H. Conroy and J. K. Chakrabarti, *Tetrahedron Letters* **4,** 6 (1959).
40. J. W. Cook, R. M. Gailey, and J. D. Loudon, *Chem. Ind. (London)* 640 (1953).
41. J. C. Seaton, M. D. Nair, O. E. Edwards, and L. Marion, *Can. J. Chem.* **38,** 1035 (1960).
42. N. Finch and W. I. Taylor, *J. Amer. Chem. Soc.* **84,** 1318, 3871 (1962).
43. W. I. Taylor, *Lloydia* **27,** 368 (1964).
44. N. R. Farnsworth, R. N. Blomster, D. Damratoski, W. A. Meer, and L. V. Cammarato, *Lloydia* **27,** 302 (1964).
45. W. I. Taylor, *J. Amer. Chem. Soc.* **79,** 3298 (1957).
46. H. B. MacPhillamy, R. L. Dziemian, R. A. Lucas, and M. E. Kuehne, *J. Amer. Chem. Soc.* **80,** 2172 (1958).
47. G. A. Jeffrey, G. Arai, and J. Coppola, *Acta Cryst., 13,* 553 (1960).
48. N. Camerman and J. Trotter, *Acta Cryst.* **17,** 384 (1964).
49. J. W. Moncrief and W. N. Lipscomb, *J. Amer. Chem. Soc.,* **87,** 4964 (1965).
50. K. Biemann and M. Friedmann-Spiteller, *J. Amer. Chem. Soc.* **83,** 4805 (1961).
51. N. Neuss, *J. Org. Chem.* **24,** 2047 (1959).
52. G. Büchi, D. L. Coffen, K. Kocsis, P. E. Sonnet, and F. E. Ziegler, *J. Amer. Chem. Soc.* **88,** 3099 (1966).
53. S. I. Sallay, *J. Amer. Chem. Soc.* **89,** 6762 (1967).
54. J. P. Kutney, J. Trotter, T. Tabata, A. Kerigan, and N. Camerman, *Chem. Ind. (London)* 648 (1963).
55. M. Gorman, N. Neuss, N. J. Cone, and J. A. Deyrup, *J. Amer. Chem. Soc.* **82,** 1142 (1960).
56. J. A. Weisbach and B. Douglas, *Chem. Ind. (London)* 623 (1965).
57. U. Renner, *Experientia* **15,** 185 (1959).
58. U. Renner, D. A. Prins, A. L. Burlingame, and K. Biemann, *Helv. Chim. Acta* **46,** 2186 (1963).
59. M. Gorman and J. Sweeny, *Tetrahedron Letters* 3105 (1964).
60. W. I. Taylor and M. F. Bartlett, *J. Amer. Chem. Soc.* **82,** 5941 (1960).
61. H. F. Haynes, E. R. Nelson, and J. R. Rice, *Australian J. Sci. Res.* **A5,** 387 (1952).
62. U. Renner, *Lloydia* **27,** 406 (1964).
63. M. Kuehne, *Lloydia* **27,** 435 (1964).
64. C. Djerassi, H. W. Brewer, H. Budzikiewicz, O. O. Orazi, and R. A. Corrai, *J. Amer. Chem. Soc.* **84,** 3480 (1962).
65. C. Djerassi, S. E. Flores, H. Budzikiewicz, J. M. Wilson, L. J. Durham, J. Le Men, M. M. Janot, M. Plat, M. Gorman, and N. Neuss, *Proc. Natl. Acad. Sci. U. S.* **48,** 113 (1962).
66. C. Djerassi, L. D. Antonaccio, H. Budzikiewicz, J. M. Wilson, and B. Gilbert *Tetrahedron Letters* 1001 (1962).
67. K. S. Brown Jr. and C. Djerassi, *J. Amer. Chem. Soc.* **86,** 2451 (1964).
68. A. Walser and C. Djerassi, *Helv. Chim. Acta* **48,** 391 (1965).
69. H. Conroy, P. R. Brook, and Y. Amiel, *Tetrahedron Letters* **11,** 4 (1959).
70. J. F. D. Mills and S. C. Nyburg, *Tetrahedron Letters* **11,** 1 (1959).
71. G. Stork and J. E. Dolfini, *J. Amer. Chem. Soc.* **85,** 2872 (1963).

72. K. Biemann, *Lloydia* **27,** 397 (1964).
73. K. Biemann, "Mass Spectrometry, Organic Chemical Applications," pp. 297–333, McGraw-Hill Book Company, Inc., New York, N.Y., 1962.
74. W. Klyne, R. J. Swan, B. W. Bycroft, D. Schumann, and H. Schmid, *Helv. Chim. Acta* **48,** 443 (1965).
75. M. Gorman, N. Neuss, and K. Biemann, *J. Amer. Chem. Soc.* **84,** 1058 (1962).
76. G. H. Svoboda in "Antitumoral Effects of *Vinca rosea* Alkaloids," in Proceedings First Symposium G.E.C.A., Paris, June, 1965, p. 9, Excerpta Medica Foundation, Amsterdam, New York, London, Milan, Tokyo, Buenos Aires, 1966.
77. G. H. Svoboda, *Lloydia* **27,** 275 (1964).
78. R. L. Noble, C. T. Beer, and J. H. Cutts, *Ann. N. Y. Acad. Sci.* **76,** 882 (1958).
79. G. H. Svoboda, *J. Amer. Pharm. Assoc.* **47,** 834 (1958).
80. N. Neuss, L. L. Huckstep, and N. J. Cone, *Tetrahedron Letters* 811 (1967).
81. N. Neuss, M. Gorman, N. J. Cone, and L. L. Huckstep, *Tetrahedron Letters* 783 (1968).
82. G. H. Svoboda, *Lloydia* **24,** 173 (1961).
83. N. Neuss, M. Gorman, and I. S. Johnson, "Natural Products in Cancer Chemotherapy" in "Methods in Cancer Research," p. 633 (H. Busch, Ed.), Academic Press Inc., New York, N.Y., 1967.
84. W. Arnold, M. Hesse, H. Hiltebrand, A. Melera, W. von Philipsborn, H. Schmid, and P. Karrer, *Helv. Chim. Acta* **44,** 620 (1961).
85. V. Boekelheide, O. Ceder, T. Crabb, Y. Kawazoe, and R. N. Knowles, *Tetrahedron Letters* **26,** 1 (1960).
86. H. F. Hodson, G. F. Smith, and J. T. Wrobel, *Chem. Ind. (London)* No 47 1551 (1958).
87. I. J. Grant, T. A. Hamor, J. M. Robertson, and G. A. Sim, *Proc. Chem. Soc.* 148 (1962).
88. T. Hino and S. Yamada, *Tetrahedron Letters* No 25 1757 (1963).
89. H. Rapoport, R. J. Windgassen, N. A. Hughes, and T. P. Onak, *J. Amer. Chem. Soc.* **81,** 3166 (1959).
90. M. M. Janot, J. Le Men, A. Le Hir, J. Levy, and F. Puisieux, *Compt. Rend. Acad. Sci.* **250,** 4383 (1960).
91. F. Puisieux, R. Goutarel, M. M. Janot, J. LeMen, and A. Le Hir, *Compt. Rend. Acad. Sci.* **250,** 1285 (1960).
92. S. Siddiqui and R. H. Siddiqui, *J. Ind. Chem. Soc.* **8** 667 (1931).
93. H. Kaneko, *J. Pharm. Soc. Japan* **80** 1357, 1493 (1960).
94. N. Neuss, M. Gorman, W. Hargrove, N. Cone, K. Biemann, G. Büchi, and R. Manning, *J. Amer. Chem. Soc.* **86,** 1440 (1964).
95. J. W. Moncrief and W. N. Lipscomb, *J. Amer. Chem. Soc.* **87,** 4963 (1965).
96. G. Büchi, "I.U.P.A.C., The Chemistry of Natural Products," p. 21 (Special lectures presented at the Third International Symposium of Natural Products, Kyoto, Japan, April 17–18, 1964), Butterworths, London, 1964.
97. P. J. Pelletier and J. B. Caventou, *Ann. Chim. (Paris)* **10,** 144 (1819).
98. M. Hesse, "Indolalkaloide," p. 47, Springer-Verlag, Berlin, Germany, 1964.
99. L. H. Briggs, H. T. Openshaw, and Sir Robert Robinson, *J. Chem. Soc.* 903 (1946).
100. J. H. Robertson and C. A. Beevers, *Acta Cryst.* **4,** 270 (1951).
101. R. B. Woodward, M. P. Cava, W. D. Ollis, A. Hunger, H. V. Daniker, and K. Schenker, *J. Amer. Chem. Soc.* **76,** 4749 (1954).
102. J. Tafel, *Ann. Chem.,* **264,** 33 (1891).
103. W. F. Loebisch and H. Schopp, *Monatsh. Chemie* **7,** 83 (1886).
104. A. Bacovescu and A. Pictet, *Ber.* **38,** 2787 (1905).
105. W. H. Perkin, Jr., Sir Robert Robinson, and J. C. Smith, *J. Chem. Soc.* **135,** 1239 (1932).
106. F. Cortese, *Ann. Chem.,* **476,** 283 (1929).

107. H. Schmid, J. Kebrle, and P. Karrer, *Helv. Chim. Acta* **35,** 1864 (1952).
108. T. Wieland and H. Merz, *Chem. Ber.* **85,** 731 (1952).
109. A. R. Battersby, R. Binks, H. F. Hodson, and D. A. Yeowell, *J. Chem. Soc.,* 1848 (1960).
110. M. Hesse, W. von Philipsborn, D. Schumann, G. Spiteller, M. Spiteller-Friedmann, W. I. Taylor, H. Schmid, and P. Karrer, *Helv. Chim. Acta* **47,** 878 (1964).
111. W. von Philipsborn, H. Schmid, and P. Karrer, *Helv. Chim. Acta* **39,** 913 (1956).
112. K. Bernauer, S. K. Pavanaram, W. von Philipsborn, H. Schmid, and P. Karrer, *Helv. Chim. Acta* **41,** 1405 (1958).
113. H. Wieland and W. Gumlich, *Ann. Chem.* **494,** 191 (1932); H. Wieland and K. Kaziro, *Ann. Chem* **506,** 60 (1933).
114. K. Bernauer, F. Berlage, W. von Philipsborn, H. Schmid, and P. Karrer, *Helv. Chim. Acta* **42,** 201 (1959).
115. M. Hesse, H. Hiltebrand, C. Weissmann, W. von Philipsborn, K. Bernauer, H. Schmid, and P. Karrer, *Helv. Chim. Acta* **44,** 2211 (1961).
116. J. Nagyvary, W. Arnold, W. von Philipsborn, H. Schmid, and P. Karrer, *Tetrahedron* **14,** 138 (1961).
117. C. Schoepf and H. Steuer, *Ann. Chem.* **588,** 124 (1947).
118. T. J. Mabry, H. Wyler, G. Sassu, M. Mercier, I. Parikh, and A. S. Dreiding, *Helv. Chim. Acta* **45,** 640 (1962).
119. J. A. Hamilton, T. A. Hamor, J. M. Robertson, and G. A. Sim, *Proc. Chem. Soc.* 63 (1961).
120. J. Fridrichsons and A. McL. Mathieson, *Tetrahedron Letters*, 1265 (1962).
121. Y. Ban and M. Seo, *Tetrahedron* **16,** 5 (1961).
122. J. A. Joule, H. Monteiro, L. J. Durham, B. Gilbert, and C. Djerassi, *J. Chem. Soc.* 4773 (1965).
123. U. Renner and P. Kernweisz, *Experientia* **19,** 244 (1963).
124. G. H. Svoboda, *J. Pharm. Sci.* **52,** 407 (1963).

REVIEWS AND MONOGRAPHS 1960–1969

A. R. Battersby, Recent Researches on Indole Alkaloids, *Pure Appl. Chem.* **6**, 471–481 (1963).

K. Bernauer, Alkaloide aus Calebassen-Curare und Strychnos Arten, *Planta Medica* **9** 340 (1961).

K. Biemann, "Mass Spectrometry," McGraw-Hill Book Co., Inc., New York, N.Y., 1962.

H. D. Boit, "Ergebnisse der Alkaloide-Chemie bis 1960," Academie-Verlag, Berlin, Germany, 1961.

H. Budzikiewicz, C. Djerassi, and D. H. Williams, "Structure Elucidation of Natural Products by Mass Spectrometry," Vol. 1, Alkaloids, Holden-Day, Inc., San Francisco, Calif., 1964.

A. Cerletti, Uber Vorkommen und Bedeutung der Indolstruktur in der Medizin und Biologie, in "Progress in Drug Research," Vol. 2, pp. 227–249, Interscience Publishers, New York, N.Y., 1960.

A. Chatterjee and G. Ganguli, The Chemistry of Bis-indole Alkaloids, *J. Sci. Ind. Res. (India)* **23**, 178–204 (1964).

A. Chatterjee and A. B. Ray, Recent Developments in the Chemistry of Indole Alkaloids from *Rauwolfia serpentina* Benth., *J. Sci. Ind. Res. (India)* **21A,** 515–527 (1962).

T. R. Govindachari and S. Rajappa, Recent Developments in the Chemistry of Dihydroindole Alkaloids, *J. Sci. Ind. Res. (India)* **20A**, 267–693 (1961).

M. Hesse, "Indolalkaloide in Tabellen," Springer-Verlag, Berlin, Germany, 1964 and 1968.

R. H. F. Manske, *Alkaloids* **8** (1965) and **11** (1968) Academic Press, New York.

N. Neuss, "Physical Data of Indole and Dihydroindole Alkaloids," revised ed., Eli Lilly and Company, Indianapolis, Indiana, 1966 and 1968.

N. Neuss, I. S. Johnson, J. G. Armstrong, and C. J. Jansen, The Vinca Alkaloids, in "Advances in Chemotherapy," Vol. 1, pp. 133–174, Academic Press, Inc., New York, 1964.

N. Neuss, M. Gorman, and I. S. Johnson, Natural Products in Cancer Chemotherapy, in "Methods in Cancer Research," pp. 633–702, Academic Press, Inc., New York, 1967.

J. Poisson, Biosynthesis of Indole Alkaloids, *Bull. Soc. Franc. Physiol. Vegetale.* **7**, 36–41 (1961).

R. F. Raffauf and M. B. Flagler, Alkaloids of the Apocynaceae, *Econ. Botany* **14**, 37–55 (1960).

R. Robinson, Some Recent Correlations in the Structure of Indole Alkaloids and Their Biogenetic Significance, *Abhandl. Deut. Akad. Wiss. Berlin Kl. Chem. Geol. Biol.* 299–304 (1963).

A. W. Sangster and K. L. Stuart, Ultraviolet Spectra of Alkaloids, *Chem. Rev.* 65, 69–103 (1965).

J. E. Saxton, The Indole Alkaloids, *Alkaloids* R. H. F. Manske, Ed. **7**, 4–199 (1960) Academic Press, New York

G. H. Svoboda, I. S. Johnson, M. Gorman, and N. Neuss, Current Status of Research on the Alkaloids of *Vinca rosea* Linn., *J. Pharm. Sci.* **51**, 707–720 (1962).

G. H. Svoboda, The Current Status of Research on the Alkaloids of *Vinca rosea* Linn., in "Antitumoral Effects on *Vinca rosea* Alkaloids," p. 9 Excerpta Medica Foundation International Congress Series, No. 106, Amsterdam, New York, London, Tokyo, Buenos Aires, 1966.

Symposium on the Chemistry and Biological Activity of *Catharanthus*, *Vinca*, and Related Indole Alkaloids, *Lloydia* **27**, 275–485 (1964).

W. I. Taylor, Indole Alkaloids, "An Introduction to the Enamine Chemistry of Natural Products," Pergamon Press, Inc. N.Y., 1966.

J. A. Weisbach and B. Douglas, The Chemistry of 2-Acylindole Alkaloids, *Chem. Ind.* (London) **1965**, 623–633 (1965).

E. Wenkert, Biosynthesis of Indole Alkaloids, *J. Amer. Chem. Soc.* **84**, 98–102 (1962).

G. Zetler, Some Pharmacological Properties of 12 Natural and 11 Partially Synthetic Indole Alkaloids from Tropical Apocyanaceae of the Subtribe Tabernaemontaninae, *Arzneimittel-Forsch.* **14** 1277–1286 (1964).

10

The Chemistry of the Ergot Alkaloids

A. STOLL and A. HOFMANN
Pharmaceutical-Chemical Research Laboratories, Sandoz LTD, Basel, Switzerland

1. INTRODUCTION

The current interest in the chemical structure and behavior of the ergot alkaloids is enhanced by the vital position which these compounds hold in the field of pharmaceuticals. Aside from their classical application in obstetrics, the ergot alkaloids and their derivatives are being utilized to an ever increasing extent in internal medicine, neurology, and psychiatry.

2. SOURCES OF ERGOT ALKALOIDS

As the term ergot alkaloids indicates, ergot is the principal source of these alkaloids. Ergot or *Secale cornutum* is the product of a filamentous fungus *Claviceps purpurea* (Fries) Tulasne that grows parasitically on rye and other plants of the Gramineae family. The rye grains attacked by the fungus develop into dark brown, hornshaped pegs, projecting from the ripening ears. Histologically speaking, ergot consists of compactly interwoven hyphae; biologically speaking, these compact grains are sclerotia, the form in which the fungus passes the winter. The fungus *Claviceps purpurea* and related species also attack other genera of the Gramineae and form sclerotia, the shape and size of which vary with the species of the host plant. The official form of ergot, however, is the product that forms on the ears of rye.

Long before ergot was discovered to be a rich source of therapeutically useful alkaloids it played a tragic role being the cause of devastating poisonings, epidemics called in the middle ages "St. Antony's fire," *ignis sacer*, etc. The history of ergot and ergot poisonings has been exhaustively described by Barger in his monograph "Ergot and Ergotism."[1]

The extensive use of ergot alkaloids as medicaments has resulted in a substantial increase in the demand for ergot, a demand that has for many years exceeded the naturally occurring supply. The major portion of the naturally grown ergot originally came from Spain and Portugal where, in good summers, crops of over one hundred tons were obtained. Of late, the ergot fungus required for the industrial production of ergot alkaloids is to a large extent cultivated artificially, on rye and also *in vitro* in saprophytic culture.

In the case of the parasitic cultivation of ergot on rye, the rye ears, when in bloom, are infected with a conidia suspension, obtained by culture in vitro, either by spraying or, more effectively, by injection. For the purpose of industrial scale production, inoculation machines, which permit large fields to be infected economically by the injection method, are currently being used.

The cultivation of the ergot fungus on artificial nutrient media has also been developed recently. Originally, it was only possible to produce alkaloids *in vitro* with fungi of the *Claviceps* type that grow on wild grasses. Most of these alkaloids are of the Clavine type and hence, have found no therapeutic application. The first step of commercial significance was the discovery that the fungus strain *Claviceps paspali* (Stevens et Hall) is capable of producing considerable quantities of lysergic acid derivatives, especially lysergic acid amide and isolysergic acid amide by means of saprophytic submerged culture.[2] Lysergic acid, which is used as starting material for the production of pharmaceuticals, is not obtained by this process on an industrial scale. Another strain of *Claviceps paspali* obtained from the ergot of *Paspalum dilatatum* of Portuguese origin produces free isomeric lysergic acids in a very good yield by the submerged culture technique.[3]

3. STRUCTURAL TYPES OF ERGOT ALKALOIDS

The ergot alkaloids belong to the large and important class of indole alkaloids. In the ergot alkaloids, the indole group is built into a tetracyclic ring system that has been named ergoline (1)

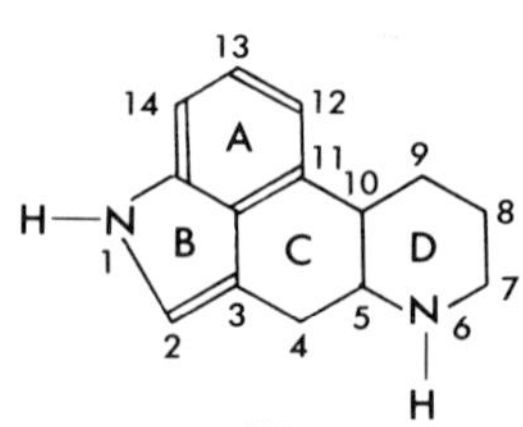

(1)

On the basis of their structural differences, the ergot alkaloids may be divided into two main classes:

Group A: Acid amide-like derivatives of lysergic acid with the general structure (2):

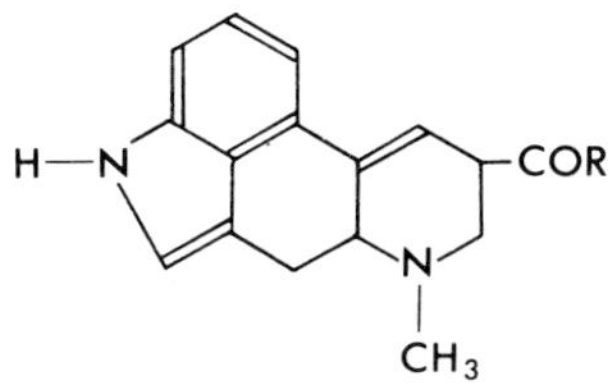

(2)

where R = the residue of a cyclic tripeptide: *Peptide type*

Ergotamine–Ergotaminine Ergosine–Ergosinine	Ergotamine group
Ergocristine–Ergocristinine Ergocryptine–Ergocryptinine Ergocornine–Ergocorninine	Ergotoxine group

Ergostine–Ergostinine
Ergosecaline-Ergosecalinine

R = L-2-aminopropanol: *Alkanolamide type*
Ergometrine–Ergometrinine

R = NH_2: *Amide type*
Ergine-Erginine

R = $NHCHOHCH_3$: *Carbinolamide type*
Lysergic acid methylcarbinolamide

Group B: Ergot alkaloids of the Clavine series with the general structure (3):

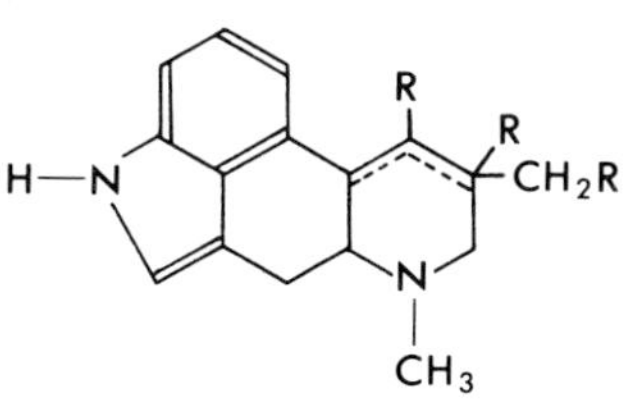

R = H or OH or $OCOCH_3$

(3)

All alkaloids belonging to this structural group were given names ending in *-clavine*, e.g., agroclavine, elymoclavine, etc. The alkaloids of the Clavine series occur primarily in ergot growing on wild grasses (e.g., *Agropyrum semicostatum* Ness and *Elymus mollis* Tri.).

4. LYSERGIC ACID

The structural element common to all the ergot alkaloids of group A is lysergic acid (4) or its stereoisomer, isolysergic acid (5). Fresh ergot, almost exclusively, contains the pharmacologically, highly active lysergic acid alkaloids. These are all levorotatory or only slightly dextrorotatory. By using old ergot or unsuitable methods of isolation, however, considerable quantities of the isomeric alkaloid forms are obtained. Isolysergic acid is the characteristic component of those alkaloids that have only weak activity and are strongly dextrorotatory. The names of all the members of this group end in *-inine*. Lysergic acid and isolysergic acid, as well as the alkaloids derived therefrom, are readily interconvertible, and an equilibrium of the isomeric forms is reached quickly in an alkaline medium.

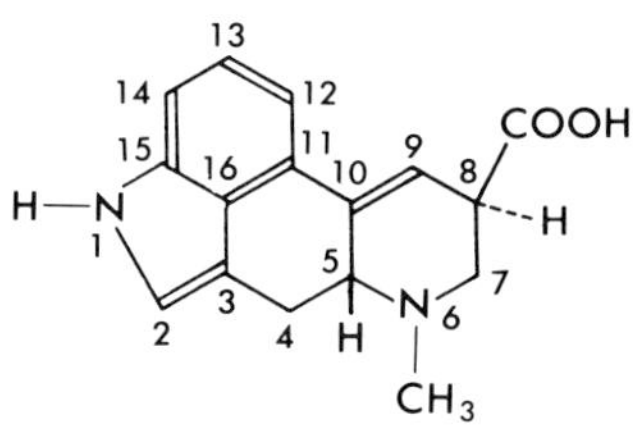

(4)
Lysergic Acid
(d-Lysergic Acid)

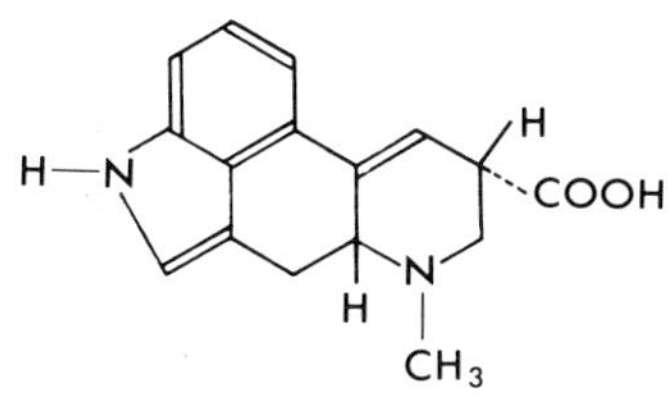

(5)
Isolysergic Acid
(d-Isolysergic Acid)

Lysergic acid is prepared by vigorous alkaline hydrolysis of ergot alkaloids of group A. The natural dextrorotatory lysergic acid was named d-lysergic acid and its isoform d-isolysergic acid. Their optical antipodes, l-lysergic and l-isolysergic acid, do not occur in nature but may be produced by racemization of d-lysergic acid by heating in a solution of Ba $(OH)_2$ or via the racemic lysergic and isolysergic hydrazides resulting from the splitting of ergot alkaloids with hydrazine.

Lysergic acid, isolysergic acid and their derivatives, the ergot alkaloids, give typical color reactions. The color reactions most commonly used are those of Keller with concentrated sulfuric acid and glacial acetic acid containing a trace of iron chloride, and of van Urk–Smith with p-dimethylaminobenzaldehyde in a sulfuric acid solution. A blue color results from both reactions and is used for the qualitative and quantitative determination of ergot alkaloids in galenical preparations and in ergot samples.

4.1 Elucidation of Structure

The structure of the ring skeleton of lysergic acid was elucidated by Jacobs and Craig of the Rockefeller Institute.[4–7] By oxydation of ergot alkaloids or lysergic acid they obtained the quinoline-betaine-tricarboxylic acid (6); and by fusion of dihydrolysergic acid with alkali, methylamine, and propionic acid, l-methyl-5-aminonaphthalene (7) and 3,4-dimethylindole (8).

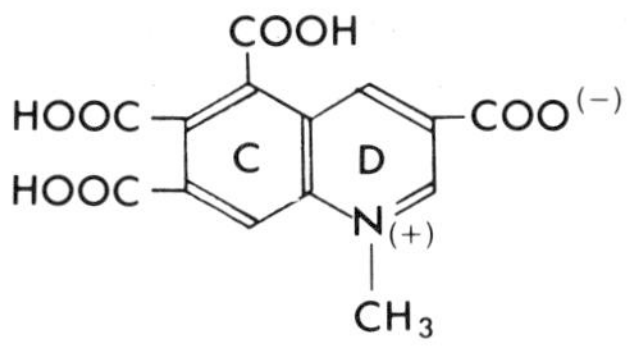

(6)

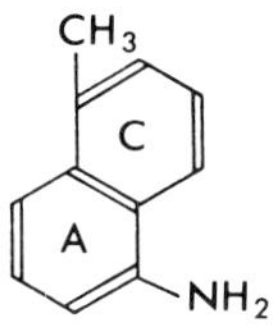

(7)

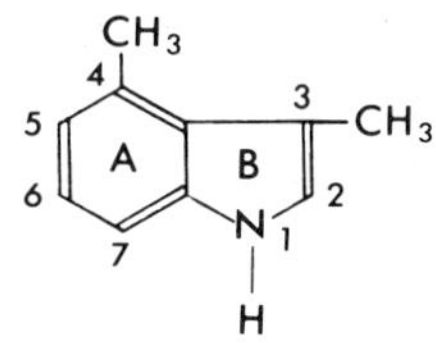

(8)

These cleavage products indicated that lysergic acid contained a new tetracyclic ring system, which was named ergoline (1). By comparing the

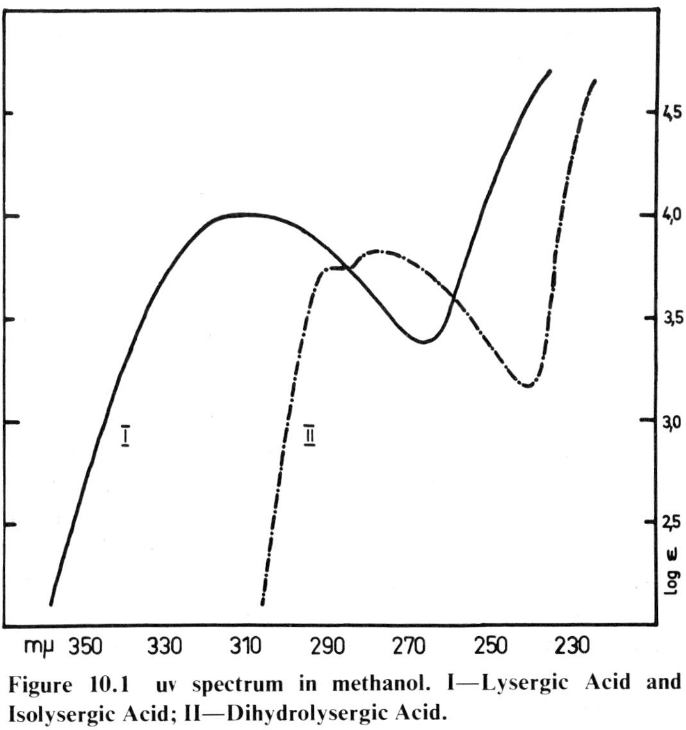

Figure 10.1 uv spectrum in methanol. I—Lysergic Acid and Isolysergic Acid; II—Dihydrolysergic Acid.

uv spectra of lysergic acid and isolysergic acid with that of dihydrolysergic acid, the latter being a typical indole spectrum (Figure 10.1), it was deduced that lysergic acid and isolysergic acid possessed a double bond in conjugation with the indole system. The position of the carboxyl group was ascertained from pK measurements and the finding that dihydrolysergic acid behaved like a β-aminocarboxylic acid when submitted to thermal scission. Finally Stoll, Hofmann, and Troxler, showed that removal of the asymmetry at C(8) at which the carboxyl group is situated, yielded identical products in the case of lysergic acid and isolysergic acid.[8] This proved lysergic acid and isolysergic acid to be diastereomers at C(8), as shown by structures (9) and (11).

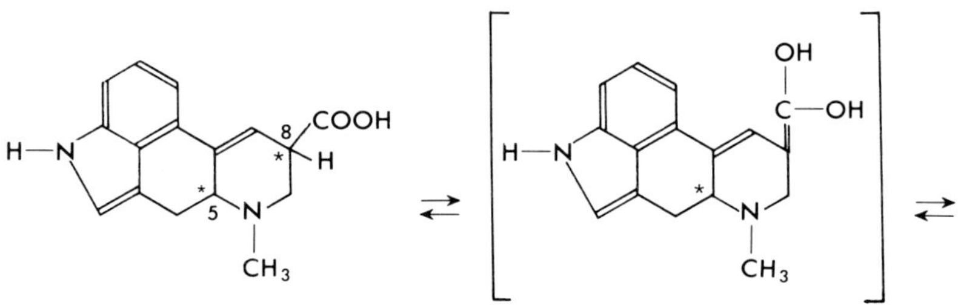

(9)
Lysergic Acid

(10)

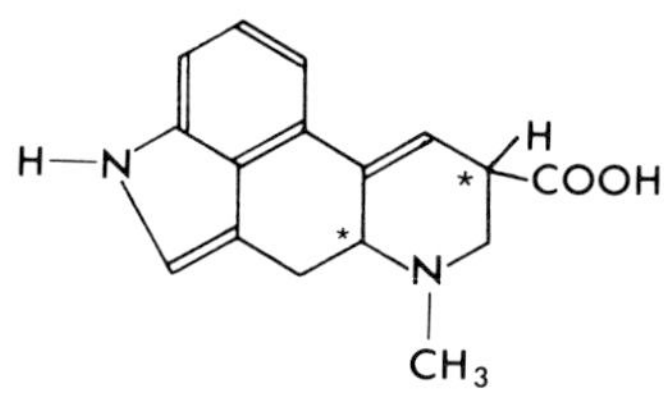

(11)
Isolysergic Acid

The isomerization proceeds via an intermediate compound that is symmetrical at C(8) and that exhibits a continuous conjugated system of double bonds from the enol double bond to the indole system. Thus, isomerization of dihydrolysergic acids, in which this continuous conjugation is interrupted, as well as that of ergolene derivatives, which have no carbonyl function that can be enolated, is more difficult or even impossible. The speed of the isomerization and the position of the equilibrium are both strongly influenced by the nature of the substituent at the carboxyl radical of lysergic acid.

4.2 Stereochemistry of Lysergic Acid and the Dihydrolysergic Acids

Investigations on the stereoisomeric dihydro acids and application of conformational analysis allowed the question of the configurational relationships between the two asymmetrical centers in lysergic acid and in isolysergic acid to be solved.

Catalytic hydrogenation of the $\Delta^{9,10}$-double bond of lysergic acid, isolysergic acid, and their derivatives afford products with structures (12) to (17). The absolute configurations shown anticipate results to be discussed later.

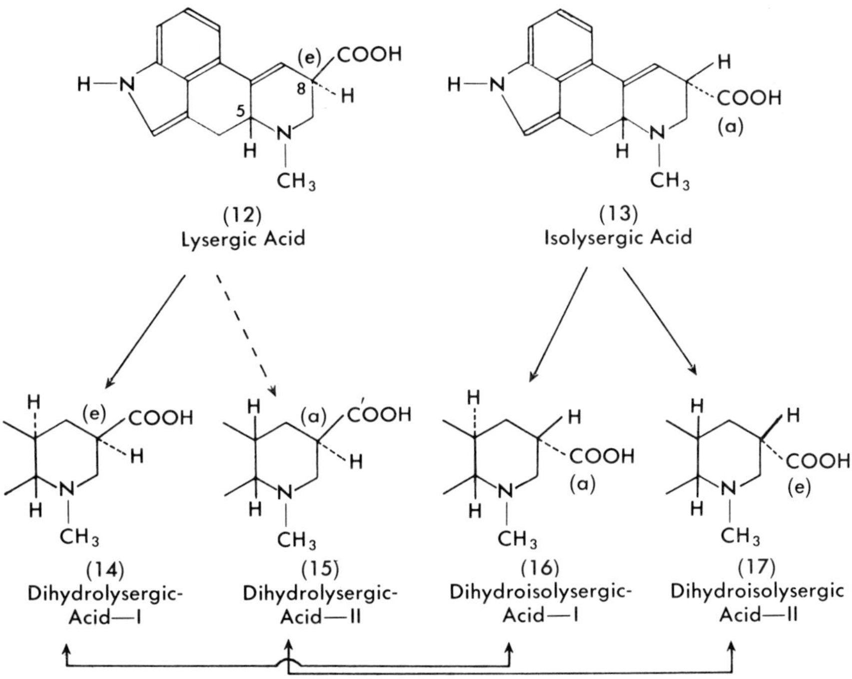

(12) Lysergic Acid; (13) Isolysergic Acid; (14) Dihydrolysergic-Acid—I; (15) Dihydrolysergic-Acid—II; (16) Dihydroisolysergic-Acid—I; (17) Dihydroisolysergic Acid—II

Hydrogenation of lysergic acid and its alkaloids yielded only one dihydrolysergic acid. In the iso series, however, both dihydroisolysergic acid-I and dihydroisolysergic acid-II were formed, the latter stereoisomer being predominant upon rapid hydrogenation. In an alkaline medium, dihydroisolysergic acid-I was irreversibly rearranged to form dihydrolysergic acid-I. Dihydrolysergic acid-II could only be produced in very small yield by alkaline saponification of dihydroisolysergic acid-II hydrazide. Conversely, saponification of dihydrolysergic acid-II hydrazide yields only dihydroisolysergic acid-II. In this case the equilibrium is on the side of the iso compound.

The question of *cis*- or *trans*-linkage of rings C and D was solved by noting the behavior upon hydrogenation. The fact that only one stereoisomer forms upon hydrogenation of lysergic acid, namely, dihydrolysergic acid-I, can be explained by assuming that the carboxyl group at C(8) and the hydrogen atom at C(5) are on the same side of the molecule in lysergic acid, thus shielding the double bond from the top side. The addition of hydrogen must then necessarily occur from the α-side, giving the *trans*-compound. On the other hand, the preponderant formation of dihydroisolygeric acid-II under conditions that favor the *cis*-compounds (rapid hydrogenation in glacial acetic acid with Pt catalyst) indicates a *cis*-configuration in the -II series.

As epimerization conditions favor the formation of dihydrolysergic acid-I (14) and dihydroisolysergic acid-II (17), it may be assumed that these compounds carry their carboxyl group in the equatorial position. In agreement is the fact that, in comparison with their epimers, they are less hindered in the C(8) position during saponification and condensation reactions. Furthermore, their derivatives exhibit a greater retention time on chromatography.

Dihydrolysergic acid-I thus has the C/D-trans-8-equatorial formula (14), whereas the epimeric dihydroisolysergic acid-I has the C/D-trans-8-axial arrangement (16). In the case of dihydrolysergic acid-II (structure 15) and dihydroisolysergic acid-II (structure 17), both of which have a *cis*-linkage of rings C and D, the carboxyl radical is in the axial and the equatorial position, respectively. Thus, the carboxyl group is in an equatorial position in lysergic acid (12) and in an axial position in isolysergic acid (13). Consideration of space models shows that such a conformation is only possible if the ring D has a pseudo-chair form in both epimers.

The foregoing deductions relate only to the *relative* spatial arrangement at the various asymmetric centers of isomeric lysergic and dihydrolysergic acids. The absolute configuration as depicted by structures (12)–(17) was elucidated by analysis of rotation-dispersion curves from which Leemann and Fabbri concluded that natural dextrorotatory lysergic acid has its hydrogen atom at C(5) in the β-position.[9]

This result was confirmed by oxidative degradation of the d-lysergic acid lactam (19), to a D-aspartic acid derivative.[10] d-Lysergic acid as such could not be used for the degradation, as it would yield an aspartic acid (23), dialkylated at the nitrogen atom that was known to racemize quickly. To obtain a sterically stable degradation product having a secondary nitrogen atom, N-norlysergic acid would have had to be used as starting material. However, as demethylation of lysergic acid was unsuccessful, the lactam (19), produced by heating lysergic acid in acetic acid anhydride for a short time, was employed. The lactam (19) was ozonized in a methylene chloride–methanol-water solution and then oxidized with hydrogen peroxide and anhydrous formic acid. As attempts to isolate the anticipated tricarboxylic acid (20) proved unsuccessful, the reaction mixture was treated with 1 N HCl to split off the oxalyl radical. After subsequent esterification of (21) with n-propanol, a yield of 4% of an optically active amino-acid ester was obtained in pure form. This proved to be identical with authentic D-(+)-N-methylaspartic acid di-n-propylester (22). Thus, the natural lysergic acid has the (5R:8R) configuration depicted by formula (18).

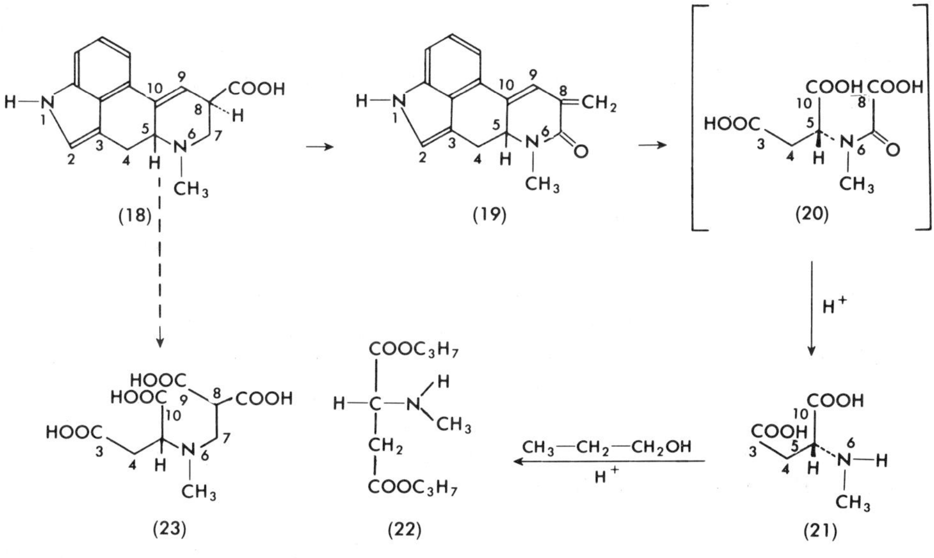

4.3 Syntheses of Dihydrolysergic Acid

Prior to the total synthesis of lysergic acid it was possible to synthesize its dihydro derivative. The latter lacked the double bond in the^{9,10}-position and was, therefore, being more stable. The introduction of this double bond proved to be very difficult, and it made a different synthetic approach necessary.

Racemic dihydrolysergic acid was synthesized by Uhle and Jacobs.[11] 4-Aminonaphthostyril (24), in which the rings A, B, and C of the ergoline system are already present, was used as starting material. Condensation of (24) with cyanomalonic dialdehyde, cyclizing with zinc chloride, and hydrolysis of the nitrile group with HCl, gave the ring system (25) of lysergic acid containing a carboxyl group at C(8). Catalytic hydrogenation of the chloromethylate of (25) yielded the tetrahydro-N-methylcarboxylic acid (26) which, upon reduction with sodium and butanol, yielded a small quantity of racemic dihydrolysergic acid (27).

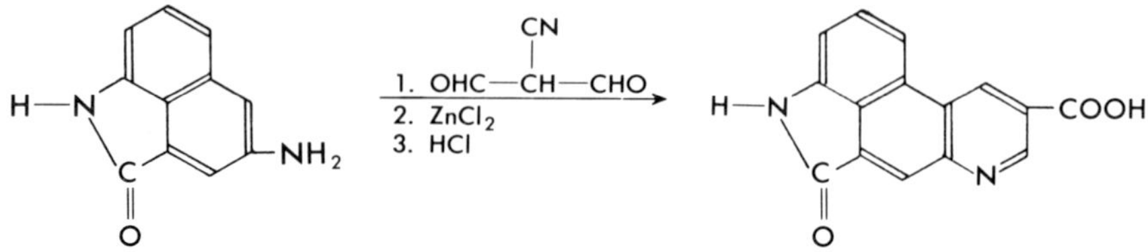

(24) (25)

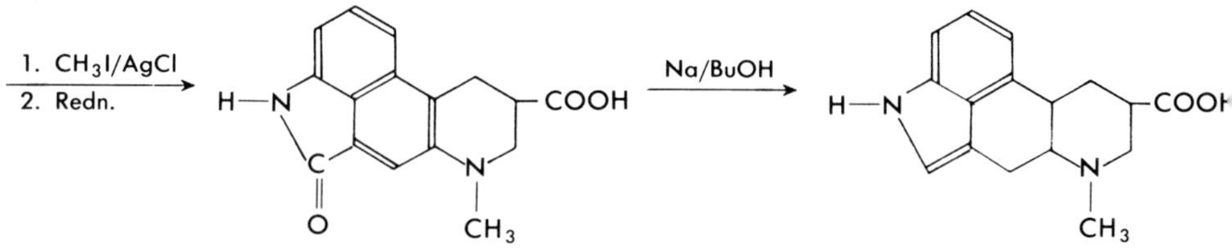

(26) (27)

(±)-Dihydrolysergic Acid

An improved synthesis of dihydrolysergic acid, which yielded homogeneous racemates and also led to the optically active dihydrolysergic acids, was effected via racemic dihydronorlysergic acids. As a starting material for this synthesis, Stoll and Rutschmann used the quinolone (28).[12] To ensure a successful reaction, the butanol used must contain a trace of water so that the methyl ester group of (29) is saponified before reduction to the corresponding alcohol occurs.

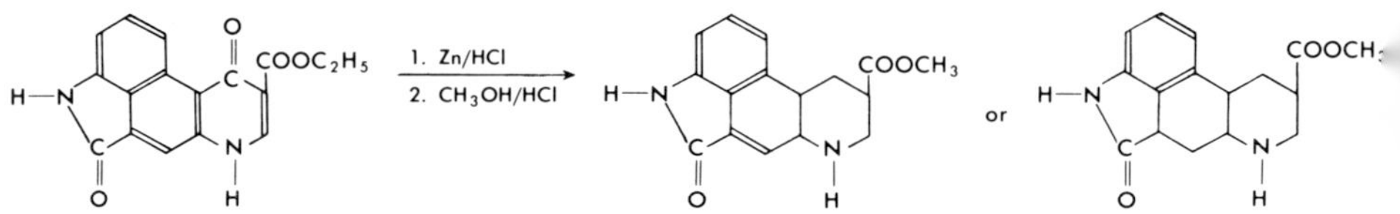

(28) A (29) B

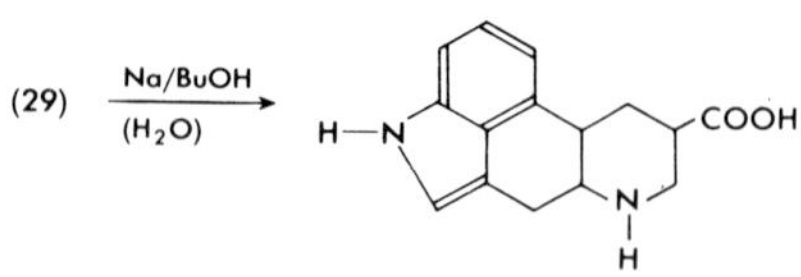

(30)
(±)-Dihydronorlysergic Acid

Compound (30) is a racemic mixture of three stereoisomers which, after esterification with methanol, may be resolved chromatographically to give racemic dihydronorlysergic acid-I methyl ester, racemic dihydronorisolysergic acid-I methyl ester, and racemic dihydronorisolysergic acid-II methyl ester.

The dihydronorlysergic acids were converted to the corresponding stereoisomer racemic dihydrolysergic acids by migration of the methyl radical. This migration was effected by heating the methyl esters of the dihydronorlysergic acids or by reduction in the presence of formaldehyde. Resolution of the resulting dihydrolysergic acid-I racemate in the form of the L-norephedride gave d-(−)-dihydrolysergic acid-I. This compound is the basic constituent of all the dihydrogenated natural ergot alkaloids.

4.4 Synthesis of Lysergic Acid

The first, and only, synthesis of lysergic acid was effected by the research group of Kornfeld at Eli Lilly.[13] In this particular synthesis a N-acyl-2,3-dihydroindole derivative was used as starting material, thus allowing the formation of rings C and D by classical methods. Dehydrogenation of the 2,3-dihydroindole system to the indole system was effected in the last stage.

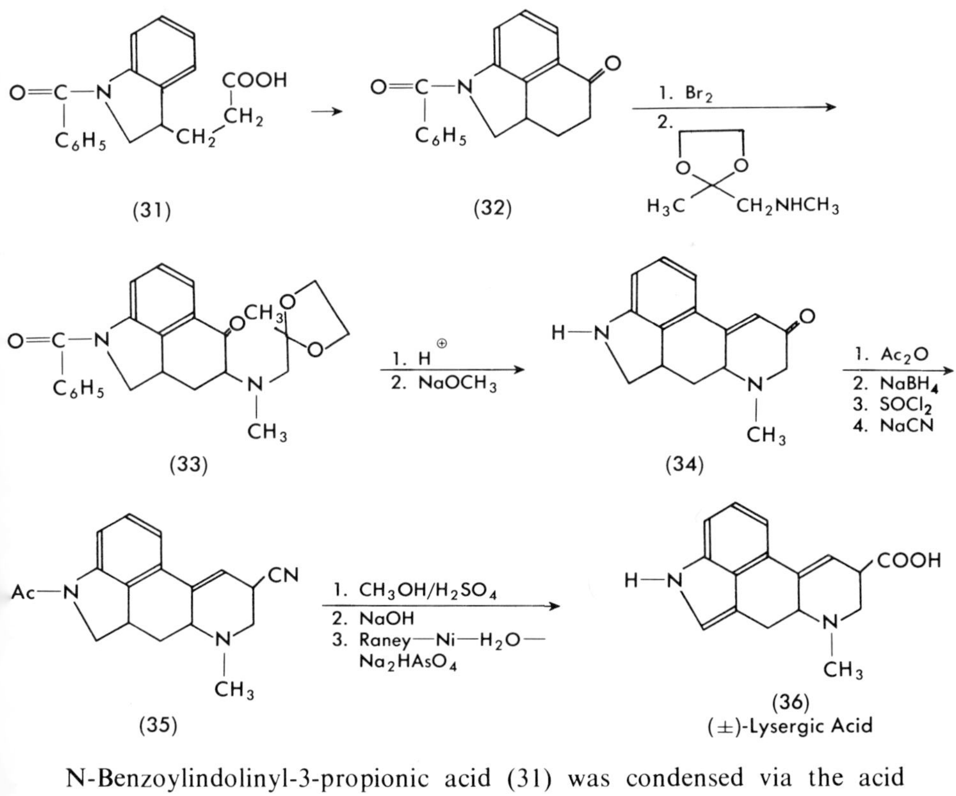

N-Benzoylindolinyl-3-propionic acid (31) was condensed via the acid chloride to form the tricyclic ketone (32). The bromination to the α-bromoketone was followed by the conversion with methylaminoacetone ethylene ketal to form compound (33). After acid hydrolysis to form the corresponding methylketone, this was cyclized with sodium methylate to give the tetracyclic, unsaturated ketone (34). The liberated secondary amino group was again protected by acetylation and the ketone reduced with $NaBH_4$ to a secondary alcohol which in turn was converted to the corresponding chloride with thionyl chloride in liquid sulfur dioxide. Treatment with sodium cyanide in liquid hydrocyanic acid yielded the nitrile (35), which was converted by methanolysis to the corresponding ester. This was then subjected to alkaline saponification to form 2,3-dihydrolysergic acid. Dehydrogenation with deactivated Raney nickel in an aqueous solution in the presence of sodium arsenate yielded racemic lysergic acid (36).

4.5 $\Delta^{8,9}$-Isomer of Lysergic Acid

From saprophytic cultures of a strain of *Claviceps paspali* (Stevens et Hall), a new isomer of lysergic acid (37) was isolated.[3] The structure of this

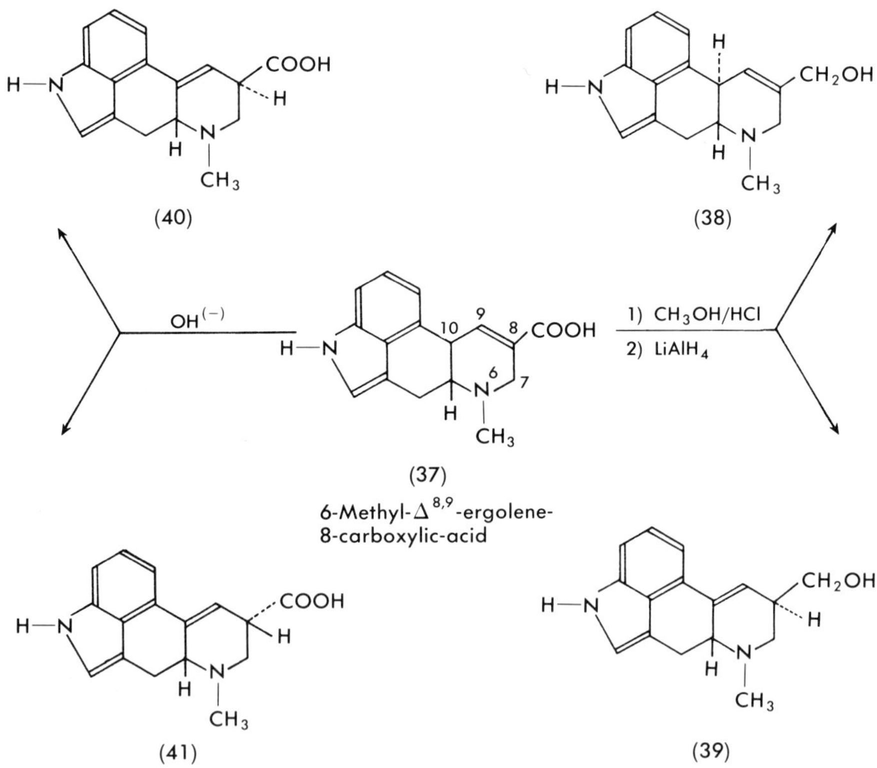

(40) (38)

(37)
6-Methyl-$\Delta^{8,9}$-ergolene-8-carboxylic-acid

(41) (39)

isomer was proved to be 6-methyl-$\Delta^{8,9}$-ergolene-8-carboxylic acid by hydrogenation of its methyl ester to elymoclavine (38) and lysergol (39). It differs from d-lysergic acid only by the position of the nonindolic double bond. The $\Delta^{8,9}$-isomer of lysergic acid is easily transformed into a mixture of lysergic acid (40) and isolysergic acid (41) in alkaline solution.

5. SIMPLE LYSERGIC ACID AMIDES

In this section only the naturally occurring amides of lysergic acid are considered. Besides these natural alkaloids, a great number of lysergic acid amides, discussed in section 8, have been synthesized.

5.1 Ergine–Erginine

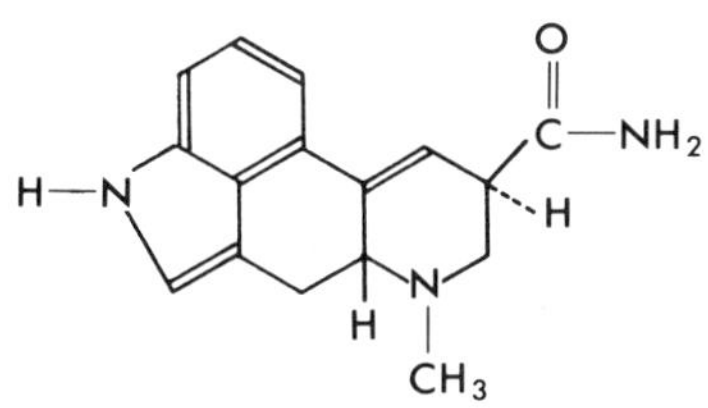

(42)

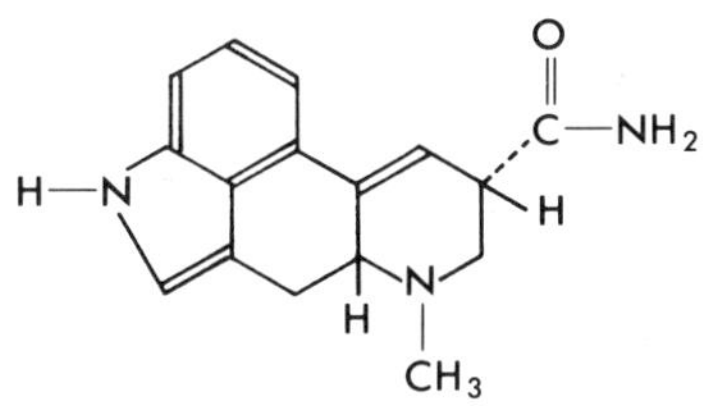

(43)

Lysergic acid amide (42) and isolysergic acid amide (43) which, for a long time, were only known as the hydrolysis products of ergot alkaloids, were recently isolated as the main constituents of the alkaloid mixture from ergot of *Paspalum distichum* L.[2] They are also to be regarded as genuine ergot alkaloids. The term *ergine*, which was originally applied to the first characteristic hydrolysis product of ergot alkaloids before it was identified as isolysergic acid amide, is now correctly used in connection with lysergic acid amide. In accordance with the conventional nomenclature used in the case of ergot alkaloids, the isolysergic acid amide has been named erginine.

Recently, ergine and erginine were also isolated as the main components, together with other alkaloids of the clavine group, from seeds of *Rivea corymbosa* (L.) Hall.f. and *Ipomoea tricolor* Cav.[14] These seeds were used centuries ago by Central American indians as a magic drug under the name of "Ololiuqui." The occurrence of lysergic acid alkaloids in the plant family of Convolvulaceae is a completely unexpected phytochemical discovery as these alkaloids had been found only in the lower fungi of the genus *Claviceps*.

5.2 Lysergic Acid Methylcarbinolamide

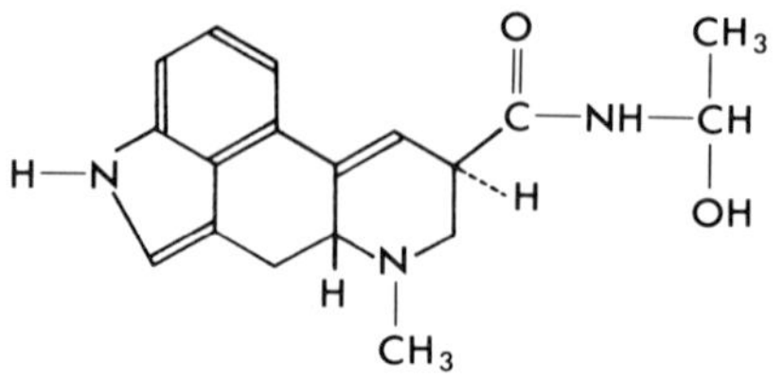

(44)

Lysergic acid methylcarbinolamide (44) was isolated together with ergine and erginine from saprophytic cultures of *Claviceps paspali*. It decomposes easily in a weak acid solution to form ergine and acetaldehyde.

5.3 Ergonovine–Ergonovinine (Ergometrine–Ergometrinine)

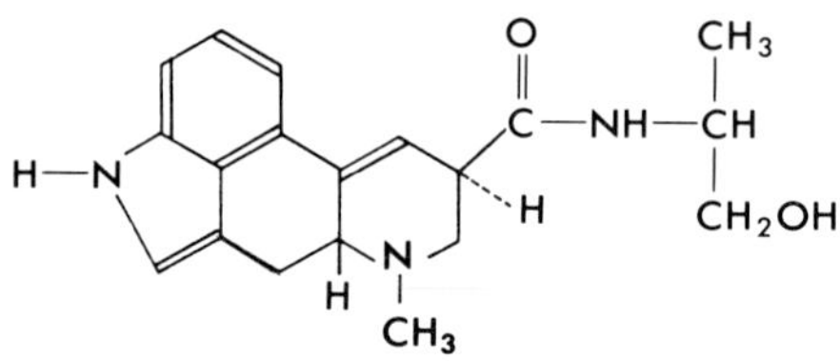

(45)

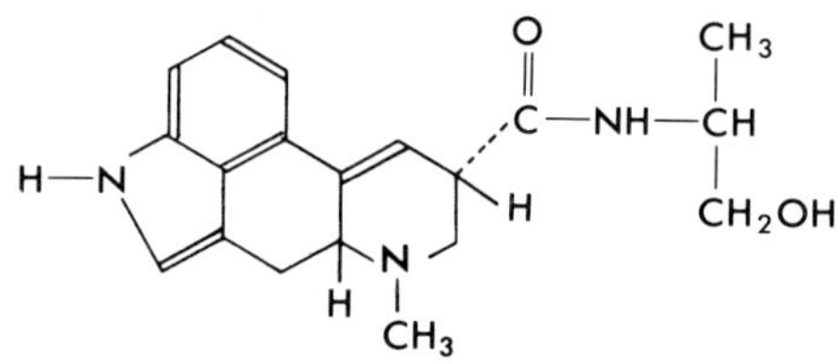

(46)

Ergonovine (45), the specific uterotonic ergot alkaloid, was discovered in 1935 in four different laboratories almost simultaneously and described under four different names. Although the names ergometrine and ergobasine have remained in use in Europe, ergonovine was adopted in the USA as the official name.

Ergonovinine (ergometrinine) (46) occurs only in small quantity in ergot together with ergonovine. However, it may easily be produced by rearrangement of ergonovine.

The relatively simple structure of ergonovine was elucidated by Jacobs and Craig[15] who showed that alkaline hydrolysis furnishes lysergic acid and L(+)-2-aminopropanol.

The synthesis of ergonovine from these two components was effected by Stoll and Hofmann.[16] Racemic isolysergic acid hydrazide, which is formed by heating lysergic acid methylester or an ergot alkaloid of the peptide type with hydrazine, served as the starting material for this synthesis. Racemization and isomerization occurs. The racemic isolysergic acid hydrazide was resolved into its optical antipodes with di-(p-toluyl) tartaric acid. d-Isolysergic acid hydrazide was transformed into the corresponding azide, which after condensation with L(+)-2-aminopropanol yielded d-isolysergic acid-L-propanolamide-(2), which proved to be identical with ergonovinine. This could be isomerized by alkaline treatment into ergonovine. By the same procudure, all eight theoretically possible stereoisomeric forms of ergonovine were synthesized.

6. THE PEPTIDE-TYPE ERGOT ALKALOIDS

The majority of alkaloids produced from ergot fungi are peptides of lysergic acid. The therapeutically most important ergot alkaloids belong to this type.

6.1 Elucidation of Structure

Upon hydrolysis the ergot alkaloids of the peptide type decompose to give lysergic acid, two amino acids (one of which is always proline), one α-keto acid, and one equivalent of ammonia. L-Phenylalanine, L-leucine, and L-valine were found as variable amino acids, and pyruvic acid, dimethylpyruvic acid or α-ketobutyric acid as α-keto acids. The amino acid proline, common to all peptide alkaloids, was obtained in the D-form upon acid hydrolysis or as L-proline under mild alkaline conditions. Table 10.1 gives a summary of the hydrolysis products.

The two alkaloid pairs yielding pyruvic acid have been named the ergotamine group after their most prominent representative. The three alkaloid pairs yielding dimethylpyruvic acid upon hydrolysis are known as the ergotoxine group. This nomenclature is due to the fact that, for numerous decades, mixtures of ergocristine, ergocryptine and ergocornine, which contained varying proportions of these constituents, were assumed to be single homogeneous individuals and were named "ergotoxine" by Barger and Carr in 1906[17]. It was only in 1943 that ergotoxine was separated into its three components with the aid of di-p-toluyl-L-tartaric acid.[18] In the "periodic system" of natural ergot alkaloids the alkaloid corresponding to ergocornine and having valine as the second amino acid is still missing from the ergotamine group. This alkaloid, which has not been found in nature, was, however, produced synthetically not long ago. (see Section 6.2).

Recently, a new alkaloid was discovered in the ergot of rye, which, upon hydrolysis, yields α-keto butyric acid as the α-keto acid, while the other cleavage products correspond to those of ergotamine and ergocristine. The new alkaloid, of which only a very small quantity is present together with ergotamine, was named ergostine whereas the double bond isomer of ergostine, was called ergostinine.

The α-keto acids are not present as such in the alkaloids as no free keto

TABLE 10.1. Cleavage Products Resulting from Hydrolytic Degradation of Ergot Alkaloids of the Peptide Type

Lysergic acid (Isolysergic acid)	NH_3 Proline (D- or L-)	Pyruvic acid	L-Phenylalanine	= Ergotamine (Ergotaminine)	Ergotamine Group
			L-Leucine	= Ergosine (Ergosinine)	Ergotamine Group
			L-Valine	= —	Ergotamine Group
		Dimethylpyruvic acid	L-Phenylalanine	= Ergocristine (Ergocristinine)	Ergotoxine Group
			L-Leucine	= Ergocryptine (Ergocryptinine)	Ergotoxine Group
			L-Valine	= Ergocornine (Ergocorninine)	Ergotoxine Group
		α-Ketobutyric acid	L-Phenylalanine	= Ergostine (Ergostinine)	

group can be detected. It is assumed that an α-hydroxy-α-amino acid group is present in the alkaloids that decompose to the corresponding α-keto acid and one equivalent of NH_3 upon hydrolysis.

As neither a free carboxyl group nor a basic amino group could be detected in the peptide portion, a cyclic configuration of the various constituents had to be assumed. Since mild alkaline hydrolysis yielded lysergic acid amide instead of lysergic acid, it was obvious that an acid–amide linkage is present between lysergic acid carboxyl and the α-hydroxy-α-amino acid moiety. As a result of these considerations, structural formulas (47) and (48) were postulated for the peptide alkaloids.

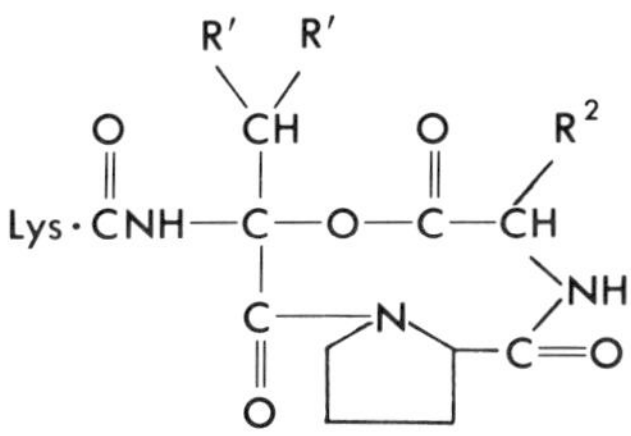

(47)

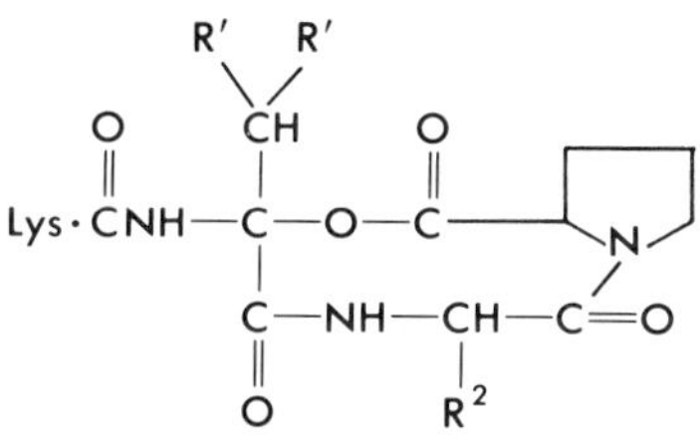

(48)

These hypothetical structures differ only in the sequence of the amino acids. This was determined as a result of large cleavage products obtained by a partial hydrolysis of the alkaloids or their dihydro derivatives. Thus, for example, the cleavage of dihydroergotamine and of dihydroergocristine with hydrazine yielded, aside from dihydrolysergic acid hydrazide, propionyl-L-phenylalanyl-L-proline, or isovaleryl-L-phenylalanyl-L-proline, respectively. The keto acid component was reduced by hydrazine to form the corresponding fatty acid. Upon mild hydrolysis with one equivalent of alkali in alcohol, however, the keto acid as such remained, and pyruvoyl-L-phenylalanyl-L-proline and dimethylpyruvoyl-L-phenylalanyl-L-proline, respectively was obtained from the above-mentioned alkaloids. Thus, the sequence of the amino acids was found to be that depicted by formula (48). This however, did not explain subsequent observations, all of which suggested that the peptide portion must necessarily contain a diketopiperazine ring. Cleavage with hydrazine and hydrolysis with one equivalent of alkali yielded, aside from the already mentioned acyl dipeptides, a considerable quantity of diketopiperazine, consisting of proline and the different amino acids, i.e., phenylalanylproline lactam in the case of ergotamine and ergocristine.

Even more informative were the results of the reduction with lithium aluminum hydride and thermal cleavage. These results are depicted by structures (49)–(55).

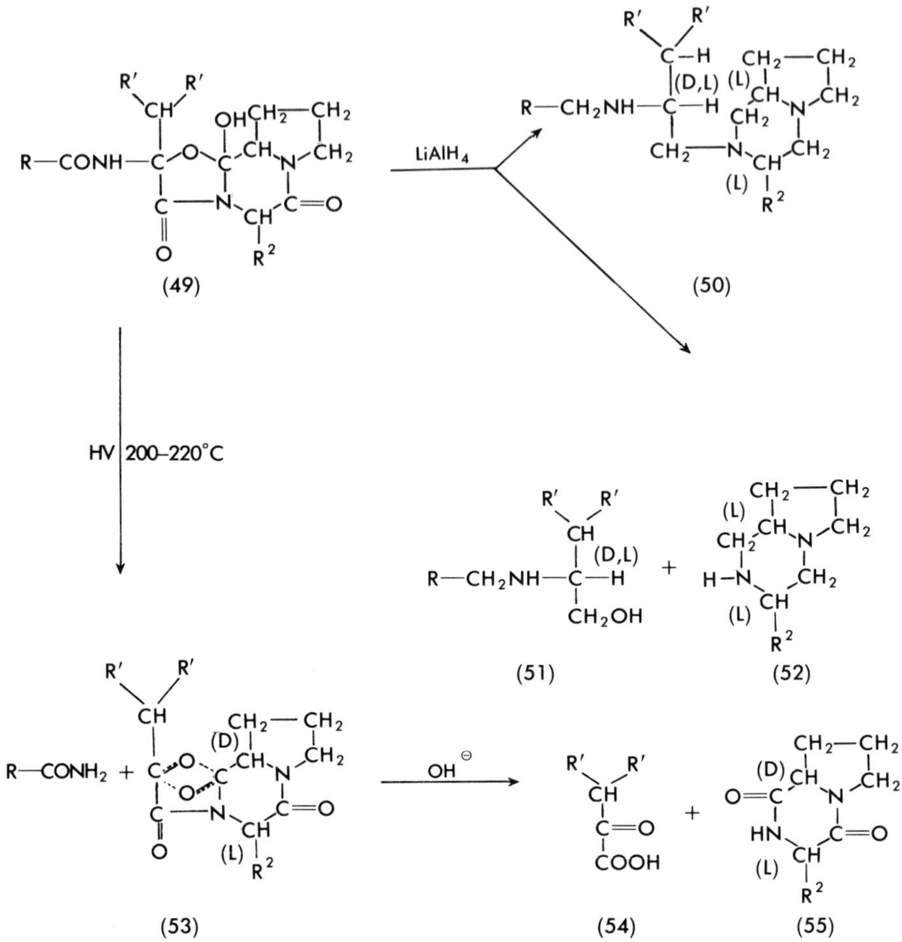

It was to be expected that a lactone group as in (48) would be split with $LiAlH_4$ to form a primary and a tertiary hydroxyl function. Instead, however, the reduction products (50), (51), and (52) were obtained.

The thermal cleavage of the dihydro alkaloids yielded, aside from the dihydrolysergic acid amide, the corresponding pyruvoyl- or dimethylpyruvoyl-diketopiperazines (53). Proline is present in the D form in compound (53) and (55). Inversion must have occurred during the thermal cleavage for, in the alkaloids, the proline radical has the L configuration, as may be seen from the isolation of (50) and (52). In (53) the carbonyl function of the α-keto acid radical cannot be detected chemically; obviously, the capacity for reaction is impaired by its proximity to the neighboring lactam carbonyl group.

The above findings may be satisfactorily explained by ascribing formula (49) to the peptide portion. In this formula the nine-membered lactam–lactone ring of (48) is divided into a five- and a six-membered ring. The structure resulting from cyclizing and migration of a hydrogen atom from the nitrogen atom of an amide group to the oxygen atom of a neighboring lactone group was named the cyclol grouping. Wrinch first used this term to describe cyclic peptide structures which were, at the time, still hypothetical.[19] The orthocarboxylic acid form of proline is the basis of the novel amino acid grouping of the ergot alkaloids. One of the three hydroxyls has a lactone-like linkage, the other a lactam-like linkage, and the third is free. The last-mentioned hydroxyl still exhibits very weak acidic properties and is responsible for the solubility of the peptide alkaloids in strong aqueous sodium hydroxide solution.

This structure (49) for the peptide alkaloids, which was proposed by Stoll, Hofmann, and Petrzilka, was confirmed ten years later by the total synthesis of ergotamine.[20]

6.2 Synthesis of Ergotamine and related Alkaloids

The major problem in the synthesis of the peptide portion of the ergot alkaloids was the assembly of the extremely labile α-hydroxy-α-amino acid grouping and the formation of the cyclol structure. The discovery that cyclol formation occurs spontaneously, if certain structural conditions are fulfilled, and that the system undergoes considerable stabilization upon cyclolization, combined with the idea of using this stabilization for the formation of the labile α-amino-α-hydroxy grouping by introducing the α-amino group only in the last step, provided a successful approach to the synthesis of the cyclol peptide.

Using this approach, Hofmann, Frey, and Ott, synthesized the peptide portion of ergotamine via steps (56)–(61) and by linking the resulting peptide with lysergic acid, synthetically produced this alkaloid.[21]

Methylbenzyloxymalonic acid ester, which may be produced from methylmalonic acid diethylester by bromination and treatment with sodium benzylate, was used as a starting material for the production of the α-hydroxy-α-amino acid moiety. By saponification of one of the ester functions and subsequent treatment with thionyl chloride, the semiester acid chloride (56) resulted. This was treated with L-phenylalanyl-L-proline lactam (57) in pyridine to produce the acylated diketopiperazine (58). As compound (56) was in its racemic form, compound (58) consisted of a mixture of two diastereoisomer forms. The acyl radical of compound (58), however, is, again, easily split off hydrolytically. For this reason, compound (58) was immediately treated with palladium-hydrogen in order to

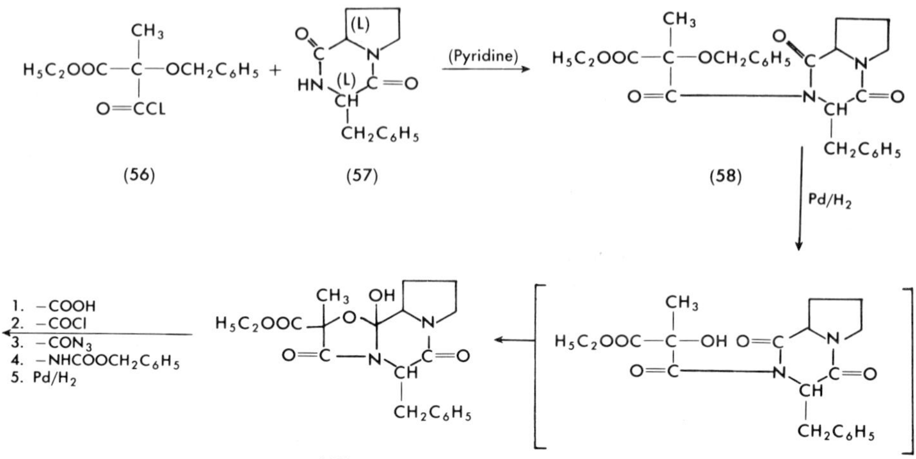

(56) (57) (58) (59) (60)

(60)
a): m.p. 135–136°
b): m.p. 202–204°

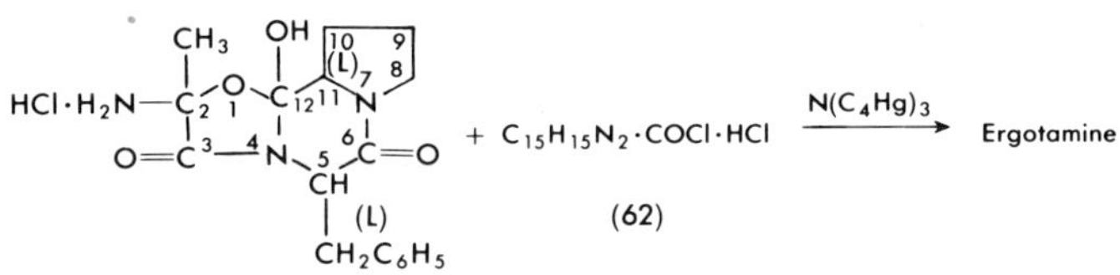

(61)
a): m.p. 180–183°
b): m.p. 131–133°

(62)

remove the benzyl group. The resulting compound (59), having a free hydroxyl group, cyclized spontaneously to form the stable cyclolcarboxylic acid ester (60). The cyclol structure in (60) was detected by the presence of an acid hydroxyl group and by analysis of the ir and nmr spectra. As compound (60) was stable it could be resolved into stereoisomeric forms by fractional crystallization. Two stereoisomers resulted having melting points of 135°–136° (60a) and 202°–204° (60b). These compounds differ in their configuration at the C(2′) atom [compare numbering in (61)]. The occurrence of only two stereoisomers of (60) shows that cyclol formation, as a result of which a new center of asymmetry is formed at C(12′), takes place stereospecifically. In both isomers the carbethoxy group was converted to the amino group by means of a Curtius degradation. Owing to the stability of the cyclol system, the ester could be saponified with sodium hydroxide solution to form the corresponding acid. This acid was treated, in the form of the sodium salt, with oxalyl chloride to yield the acid chloride which, upon reaction with sodium azide, gave the corresponding acid azide. A Curtius rearrangement of the azide, effected by heating with benzyl alcohol, yielded a benzyl urethane, which, upon hydrogenolysis, decomposed to form the amino cyclol (61). Although compound (61) is not stable as a free base, it was crystallized in the form of its hydrochloride.

Treatment of isomer (61a) with lysergic acid chloride hydrochloride (62) in chloroform in the presence of tributylamine, yielded a compound that was identical with the natural alkaloid ergotamine.

Acylation of compound (61b) with (62) yielded an alkaloid, which is an isomer of ergotamine, differing therefrom only by the spatial arrangement at the C(2′) atom of the peptide portion.

An improved process[22] for the production of ergotamine utilizes the optically active acid chloride of the methylbenzyloxymalonic acid semi-ester of known absolute configuration (66) as starting material. From this the absolute configuration at the C(2′) atom of the peptide portion of ergotamine can be deduced as depicted in (71a), i.e., corresponding to α-hydroxy-L-alanine.

Furthermore, with the aid of degradation reactions of intermediate products formed in the synthesis of the peptide portion, the configuration at the C(12′) atom could also be elucidated. It could be shown by pK-measurements and by the reaction sequence (63)–(65) that the cyclol hydroxyl is in the *cis*-position to the amino group at C(2′). A cyclic urethane (64) can be formed only if these two groups are situated on the same side of the molecule.

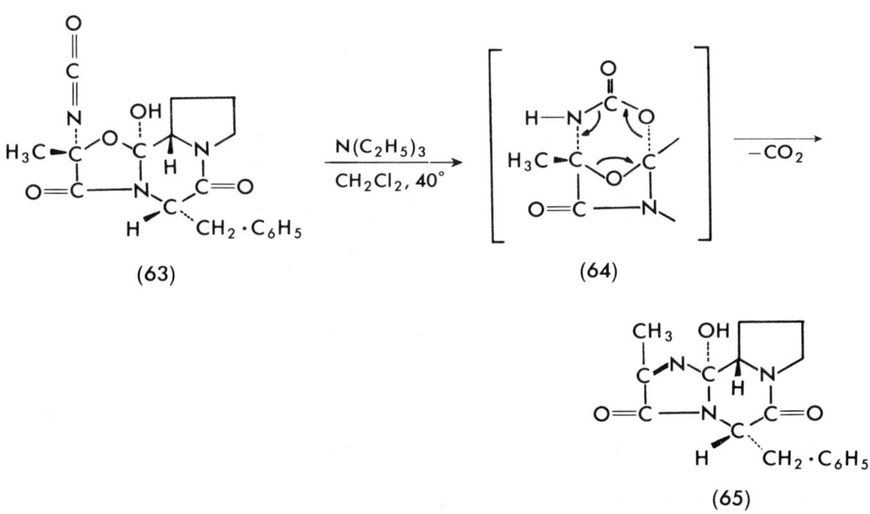

By the same improved process used for the synthesis of ergotamine (71a), ergosine (71b) as well as an alkaloid of the ergotamine group that had not been found in nature, [viz., that alkaloid having L-valine as variable amino acid and which could be referred to as "ergovaline" (71c)] were produced synthetically.[23] [See formulas (66)–(71).]

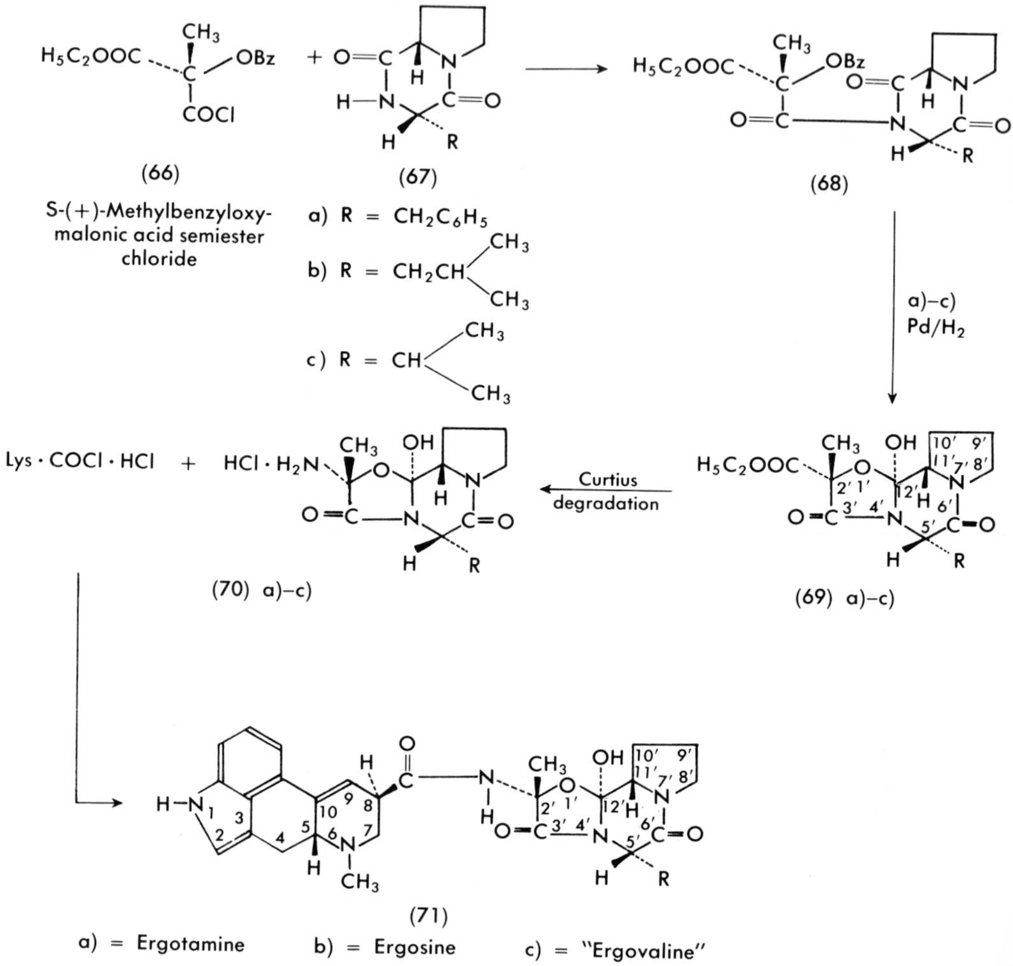

Ergostine was synthesized by the same process using (+)-ethylbenzyloxy malonic acid semiester chloride instead of the methyl analog (66).[24]

6.3 The individual Alkaloids of the Peptide Type

Up to the present time a total of six alkaloid pairs of the peptide type have been isolated from ergot. An additional alkaloid pair of uncertain structure, ergosecaline–ergosecalinine, also appears to have a peptide character.[25]

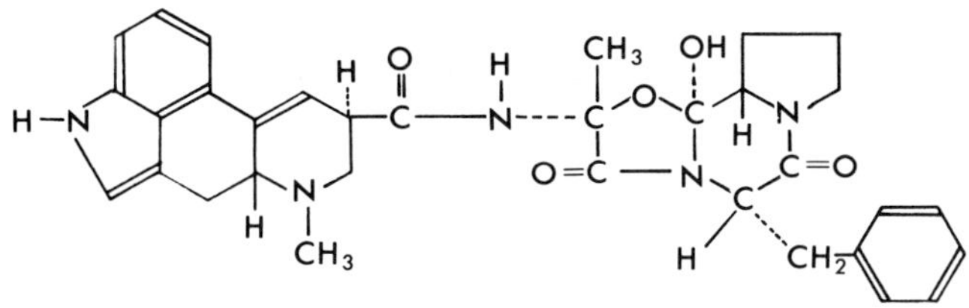

(72)

6.3.1 Ergotamine-Ergotaminine Ergotamine (72), isolated from Swiss ergot, was the first chemically homogeneous and fully active ergot alkaloid to find widespread medical application.[26] The major portion of ergotamine is currently obtained from ergot cultivated in the Swiss midlands by artificial mechanical infection of rye cultures.

Ergotaminine, the isolysergic acid isomer of ergotamine, is not easily dissolved in most solvents; and as a result, it crystallizes rapidly from the equilibrium set up between ergotamine and ergotaminine in hydroxyl-containing solvents. As is the case with most of the isolysergic acid derivatives, ergotaminine is pharmacologically ineffective.

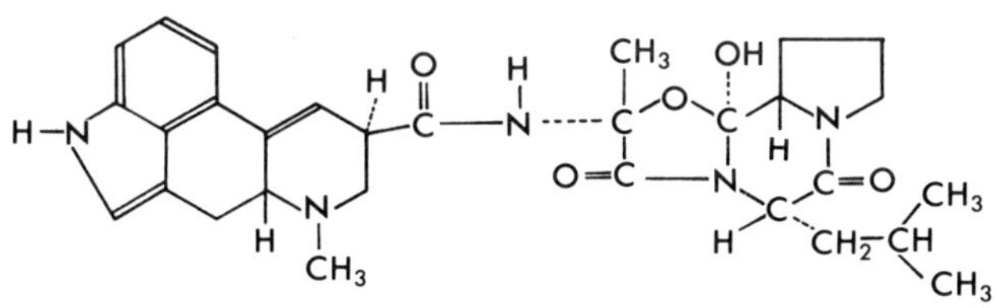

(73)

6.3.2 Ergosine-Ergosinine Ergosine (73) and ergosinine, which were isolated from Iberian ergot,[27] have so far found no medicinal application.

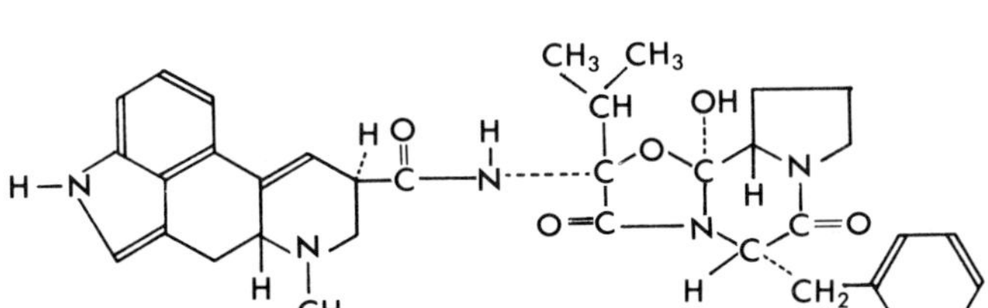

(74)

6.3.3 Ergocristine–Ergocristinine Ergocristine (74) which was isolated from Iberian ergot,[28] was found to be a constituent of the alkaloid mixture known as ergotoxine.[18] As a result of the strong sympathicolytic action of its dihydro derivative, it has found clinical application, e.g., as a component of Hydergin.®

Ergocristinine can be produced by rearrangement of ergocristine.

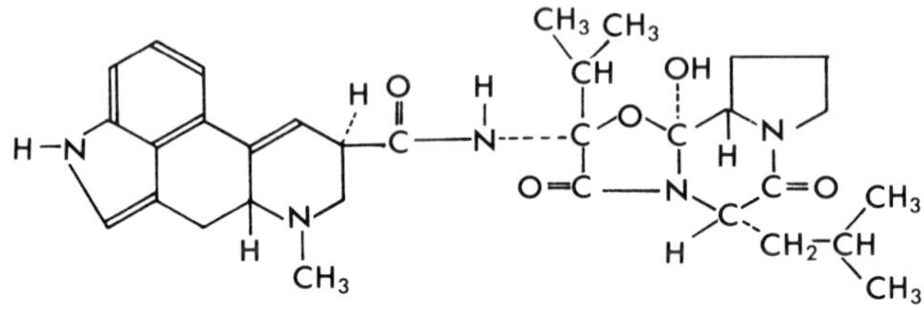

(75)

6.3.4 Ergocryptine-Ergocryptinine Ergocryptine (75), a component of the ergotoxine complex,[18] is the main alkaloid in ergot of Japanese and South American wild grasses. Its dihydro derivative is used as a constituent of Hydergin.®

Ergocryptinine is produced by rearrangement of ergocryptine in boiling methyl alcohol.

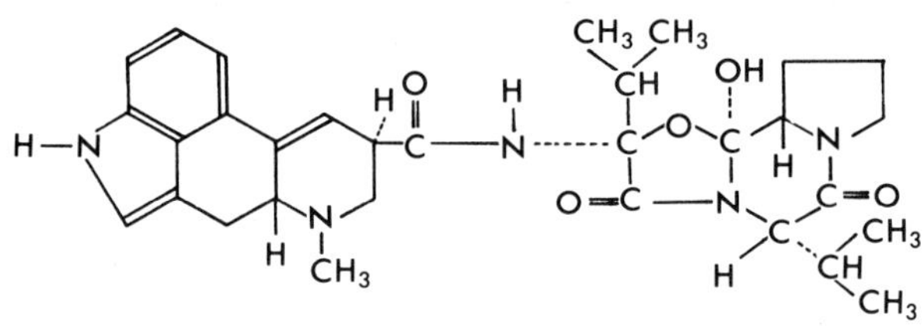

(76)

6.3.5 Ergocornine–Ergocorninine Ergocornine (76) was first discovered[18] during a separation of the ergotoxine complex. It is used medically in the form of its dihydro derivative as a constituent of Hydergin.®

Ergocorninine is obtained by isomerization of ergocornine.

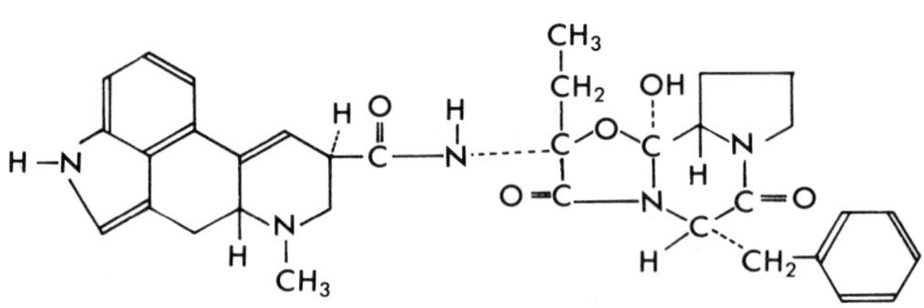

(77)

6.3.6 Ergostine–Ergostinine Ergostine (77) is found in very small quantities as a minor alkaloid in Swiss ergot.[24]

Ergostinine is prepared by rearrangement of ergostine.

7. THE ALKALOIDS OF THE CLAVINE SERIES

The first representatives of this second main class of ergot alkaloids were discovered in 1951 in the ergot of various species of grass growing in the Far East.[29] These alkaloids, given names ending in "*-clavine*," differ from the classical lysergic acid alkaloids in that the carboxyl group of the lysergic acid moiety has been reduced to the hydroxymethyl or methyl group.

Recently, with the aid of the new, efficient methods of detection, e.g., paper and thin-layer chromatography, clavine alkaloids have also been found in ergot obtained from rye. Furthermore, in 1960 Clavine-type alkaloids were found in higher plants, i.e., in genera of the family of twining plants (Convolvulaceae).[14]

TABLE 10.2. Clavine Alkaloids

Alkaloid	Formula	Structural Formula No.	Source
Agroclavine	$C_{16}H_{18}N_2$	(79)	*Agropyrum semicostatum* *A. ciliare* Fr. *Pennisetum typhoideum* Rich.
Elymoclavine	$C_{16}H_{18}ON_2$	(78)	*Elymus mollis* Tri. *Pennisetum* spp. Seeds of *Rivea corymbosa* (L.) Hall.f. Ipomoea tricolor Cav.
Lysergine	$C_{16}H_{18}N_2$	(80)	*Agropyrum* spp.
Lysergol	$C_{16}H_{18}ON_2$	(86)	Seeds of *Rivea corymbosa, Ipomoea tricolor* *Elymus* spp.
Lysergene	$C_{16}H_{16}N_2$	(87)	*Elymus* spp.
Setoclavine	$C_{16}H_{18}ON_2$	(88)	*Pennisetum typhoideum* *Elymus mollis*
Isosetoclavine	$C_{16}H_{18}ON_2$	(89)	*Pennisetum* spp. Japanese grasses
Penniclavine	$C_{16}H_{18}O_2N_2$	(90)	*Pennisetum typhoideum* Japanese grasses Rye
Isopenniclavine	$C_{16}H_{18}O_2N_2$	(91)	*Pennisetum* spp.
Festuclavine	$C_{16}H_{20}N_2$	(84)	*Agropyrum* and *Phalaris* spp. *Aspergillus fumigatus* Fres.
Pyroclavine	$C_{16}H_{20}N_2$	(83)	*Agropyrum* and *Phalaris* spp.
Costaclavine	$C_{16}H_{20}N_2$	(85)	*Agropyrum* spp.
Fumigaclavine A	$C_{18}H_{22}O_2N_2$	(92)	*Aspergillus fumigatus*
Fumigaclavine B	$C_{16}H_{20}ON_2$	(93)	*Aspergillus fumigatus*
Chanoclavine	$C_{16}H_{20}ON_2$	(94)	*Pennisetum typhoideum* Seeds of *Rivea corymbosa, Ipomoea tricolor*

Table 10.2 gives a survey of the isolated clavine alkaloids and indications on the source of these alkaloids.

7.1 Elucidation of Structure

The structural and configurational relationships between the alkaloids of the clavine group on the one hand, and the derivatives of lysergic acid alkaloids on the other, are shown by the structural scheme (78)–(85).

Catalytic hydrogenation of elymoclavine (78) yielded a mixture of d-dihydrolysergol-I (81) and d-dihydroisolysergol-I (82). Elymoçlavine has the same ring system and the same configuration at C(5) and C(10) as d-dihydrolysergic acid. Reduction of (78) with sodium in butanol yielded a mixture of agroclavine (79), lysergine (80), pyroclavine (83), festuclavine (84) and costaclavine (85) in addition to d-dihydrolysergol-I (81) and d-dihydroisolysergol-I (82).

The position of the isolated double bond in agroclavine, for which positions 7–8 or 8–9 were considered, was determined from the fact that agroclavine gave nearly the same pK value as its dihydro derivatives. Heating agroclavine in a sodium butylate solution produced only lysergine, whereas the same treatment of elymoclavine yielded a mixture of lysergol (86) and lysergene (87).

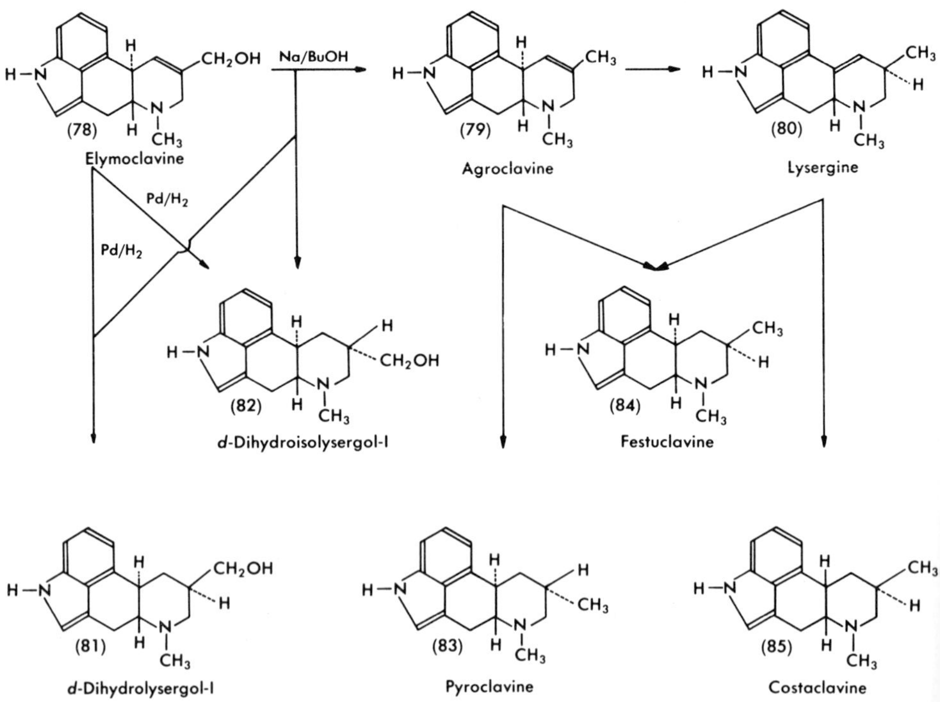

The structures of setoclavine (88), isosetoclavine (89), penniclavine (90), and isopenniclavine (91) were, for the major part, obtained from the oxidative formation of these alkaloids from agroclavine (79) and elymoclavine (78).

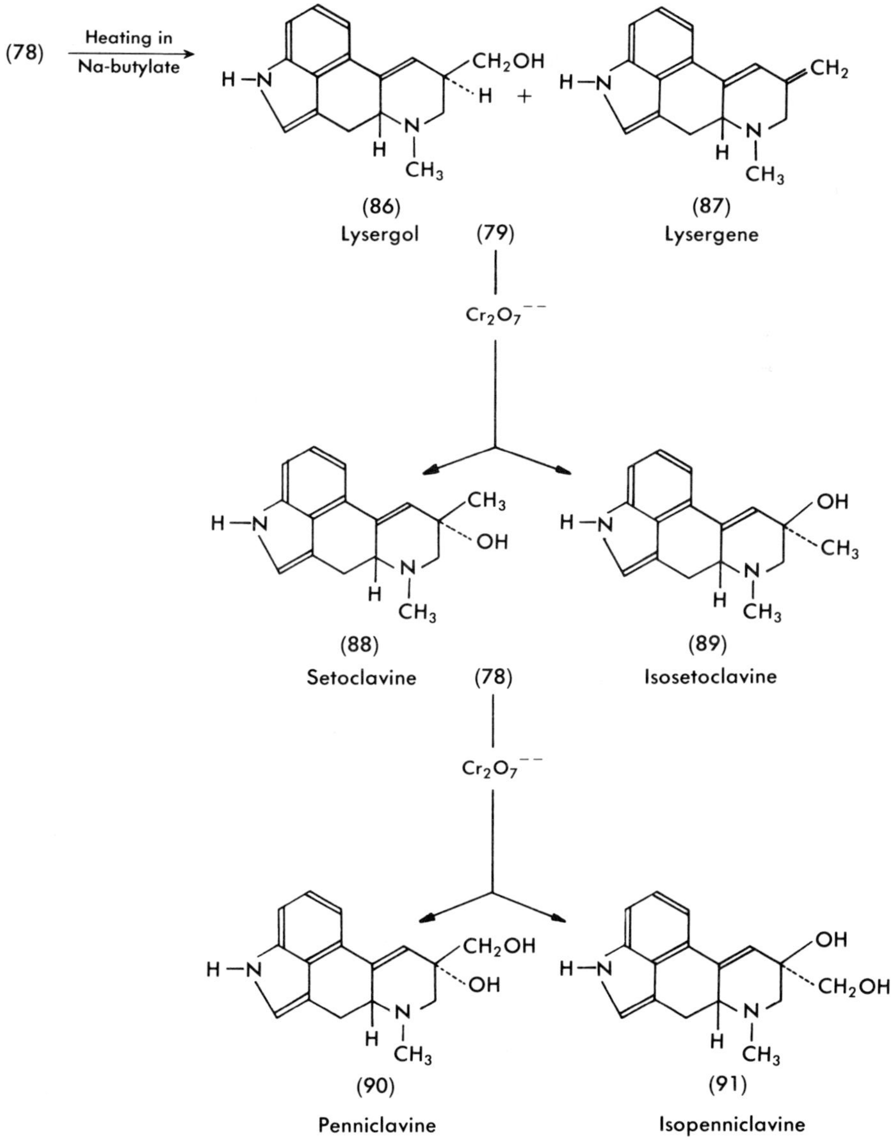

Deacetylation caused fumigaclavine-A (92) to be converted to fumigaclavine-B (93). The ease of deacetylation, as well as the ir bands at 1241

and 1725 cm^{-1}, indicated the presence of an ester group in compound (92). Heating (93) with NaOH caused water to be eliminated and resulted in the formation of lysergine (80). The configurations at C(9) and C(10) of compounds (92) and (93) have not, as yet, been elucidated.

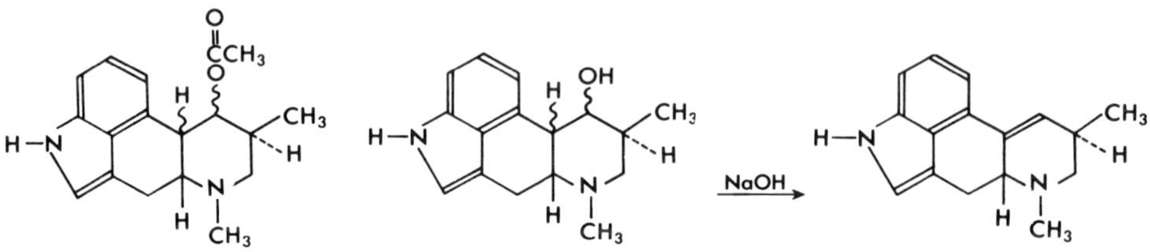

(92) Fumigaclavine-A (93) Fumigaclavine-B (80) Lysergine

Chanoclavine (94) is the only known ergot alkaloid in which the D ring of the ergoline system is open. (94) easily forms an O,N-diacetyl derivative in which the N atom at C(6) is acetylated. The configuration of (94) was deduced by comparison with festuclavine (84), which is obtained in small yield by catalytic hydrogenation of chanoclavine.

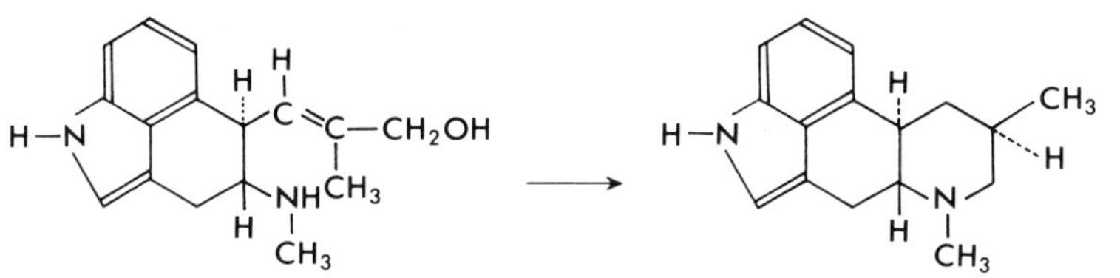

(94) Chanoclavine (84) Festuclavine

8. DERIVATIVES OF ERGOT ALKALOIDS

The potent and versatile pharmacological activity of natural ergot alkaloids prompted investigations into the relationship between their chemical structure and physiological action. The natural alkaloids were modified chemically and the resulting variations in pharmacological and therapeutic action were studied.

8.1 Lysergic Acid Derivatives of the Acid Amide Type

The azide process, employed in the first partial synthesis of ergonovine (v. Section 5.3) was used in the production of a great number of lysergic acid derivatives of the acid amide type.[16]

Other methods for the preparation of lysergic acid amides utilized mixed anhydrides of lysergic acid and sulfuric acid or trifluoroacetic acid. Lysergic acid chloride can also be employed.

Of the large number of synthetic lysergic acid amides two compounds have found medical application. d-Lysergic acid (+)-butanolamide, the next highest homolog of ergonovine, finds wide application in obstetrics, owing to its uterotonic action and ability to arrest post partum hemorrhage. d-Lysergic acid diethylamide, which is characterized by its exceptional hallucinogenic and psycholytic action, has become known in experimental psychiatry as LSD 25 and is used as a drug aid in psychotherapy.

8.2 Amino and Carbamic Acid Derivatives of 6-Methylergolene and 6-Methylergoline

The isomeric 6-methyl-8-amino-ergolenes (95a) and ergolines (96a) were produced by a modified Curtius degradation of d-lysergic acid azide, of d-isolysergic acid azide, or of the isomeric d-dihydrolysergic acid azides, respectively.

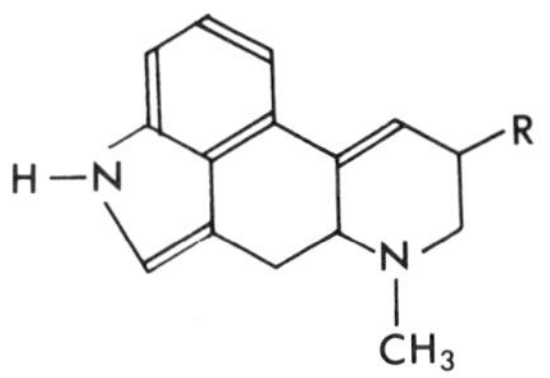

(95)

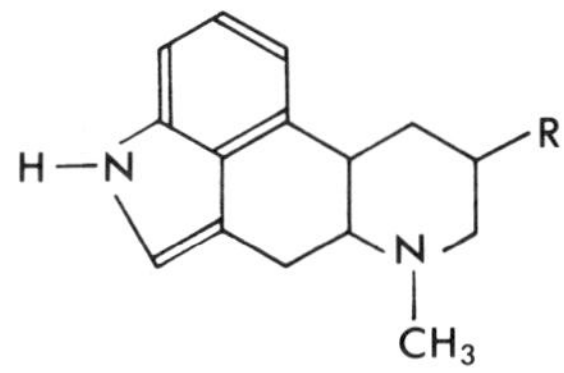

(96)

a: R = NH_2
b: R = NHCOOR′
c: R = $NHCON\langle^{R''}_{R''}$

Conversion of the corresponding isocyanates, resulting from the heating of the azides in benzene, with alcohols yielded the carbamic acid ester (95b), and with amines the urea derivatives (95c) and (96c).

8.3 Substitutions in the Ring System of Lysergic Acid

Positions 1 and 2 of the ring structure of lysergic acid are especially reactive and various substituents may be introduced.

In derivatives of lysergic acid and dihydrolysergic acid, the indole nitrogen atom can be acetylated with ketene to give (97a) and (97b) and substituted with the hydroxymethyl radical (97c) using formaldehyde, with a dialkylaminomethyl radical (97d) by means of a Mannich reaction, or with the cyanoethyl radical (97f) by the alkaline catalyzed addition of acrylonitrile.

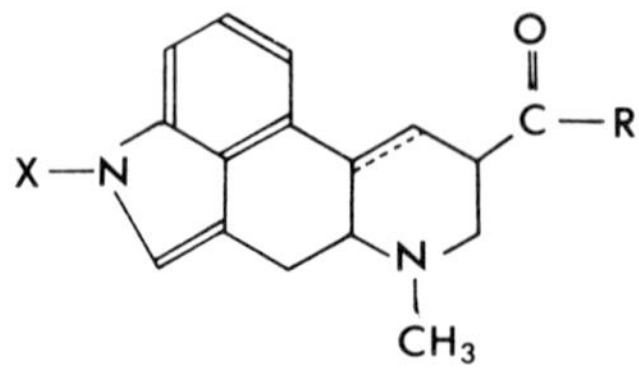

a: X = $COCH_3$
b: X = $COCH_2COCH_3$
c: X = CH_2OH
d: X = CH_2N (alkyl)$_2$
e: X = alkyl, benzyl
f: X = CH_2CH_2CN

(97)

The hydrogen atom at the indole nitrogen atom of lysergic acid derivatives has been replaced by a methyl group and by larger alkyl radicals, e.g., ethyl, propyl, allyl, and benzyl radicals (97e). The alkylation is effected by alkyl halides in liquid ammonia.

Derivatives of lysergic acid or dihydrolysergic acid may be halogenated at C(2) with bromo- or iodosuccinimide, N-2,6-trichloro-4-nitroacetanilide, and similar mild reagents to compound (98 a–c).

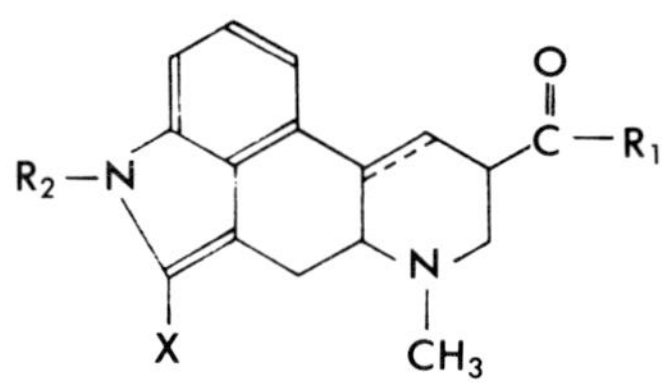

a: X = Cl
b: X = Br
c: X = I

(98)

One molecule of water is added at the $\Delta^{9,10}$ double bond in an acidic solution of the alkaloids under intense irradiation with uv light. The structures and configurations of the so-called lumi derivatives are depicted by formulas (99)–(102).

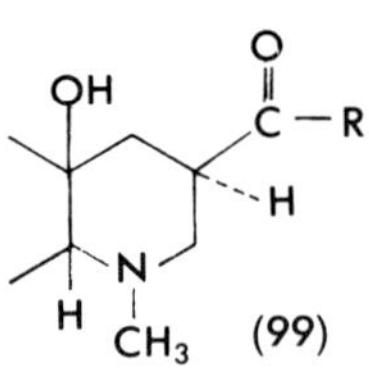

(99) Lumi-lysergic acid-I derivative

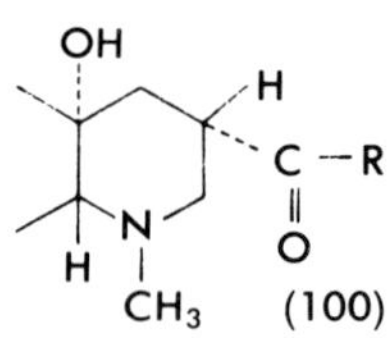

(100) Lumi-isolysergic acid-I derivative

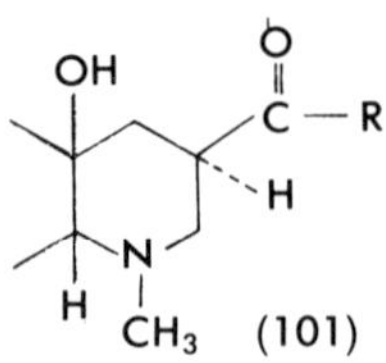

(101) Lumi-lysergic acid-II derivative

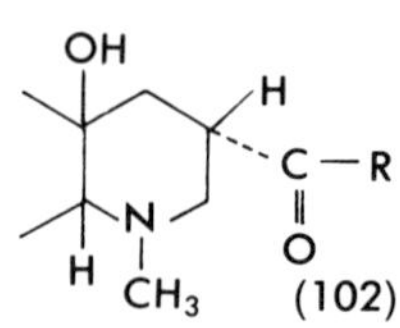

(102) Lumi-isolysergic acid-II derivative

9. PHARMACOLOGY AND THERAPEUTIC USE OF ERGOT ALKALOIDS AND THEIR DERIVATIVES

The pharmacodynamic properties of the ergot alkaloids cover an unusually broad spectrum of activity that can be classified as in the accompanying tabulation.[30,31]

Direct peripheral effects (on smooth muscles)	(1) Uterine contraction (2) Vasoconstriction
Indirect peripheral (humoral) effects	(3) Serotonin antagonism (4) Adrenergic blockade
Central nervous effects	(5) Bulbomedullary components: Vomiting, bradycardia, inhibition of the vasomotor center, and of the baroceptive reflexes (6) Mesodiencephalic components: Syndrome of ergotropic excitation with mydriasis, hyperglycemia, and hyperthermia

All natural ergot alkaloids possess these six principal effects to a greater or lesser degree. The vasoconstrictor and the contractile effect on the smooth muscle of the uterus are the most important peripheral effects. The latter accounts for the classical indication of the ergot alkaloids in obstetrics where they are used to treat post partum hemorrhage and to accelerate uterine involution in the puerperium. The widespread and successful use of ergotamine to mitigate migraine attacks is due to its tonifying effect on the smooth muscle of the blood vessels.

The neurohumoral effects of the ergot alkaloids are manifested in an antagonism to adrenaline and noradrenaline on the one hand and to 5-hydroxytryptamine (serotonin) on the other. The adrenolytic effect accounts for the use of the ergot alkaloids in internal medicine for the treatment of sympathetic overexcitation. The antagonism of the natural alkaloids to serotonin is not present to a marked degree. However, certain derivatives exhibit a marked and specific antagonism to serotonin.

The effects of the ergot alkaloids on the central nervous system are very diverse as sites of action are situated in the vasomotor center and the cardiac inhibitory center in the medulla oblongata as well as in the sympathetic structures of the diencephalon, particularly the hypothalamus. The inhibition of the vasomotor center and of the baroceptive reflexes and the stimulation of the vagal nuclei are responsible for the vasodilator, hypotensive, and bradycardic effects, especially in the case of the peptide type of alkaloid. Some also have a stimulating effect on the vomiting center.

Most ergot alkaloids stimulate the sympathetic structures of the mesencephalon and diencephalon, particularly the hypothalamus, leading to a syndrome of excitation with mydriasis, hyperglycemia, tachycardia, etc. This syndrome may be closely related to the psychotomimetic effects of certain alkaloids derivatives, e.g., lysergic acid diethylamide.

Ergotamine possesses the typical effect of the ergot alkaloids in a well-balanced manner. Ergotamine exerts a full-strength contractile effect on the uterus and on the smooth muscle of the vessels, reduces adrenergic activity, and elicits central effects by inhibiting the vasomotor centers. The stimulation of higher nervous structures is less pronounced and occurs only after the administration of toxic doses. Based on this spectrum of activity, ergotamine is an excellent hemostatic in obstetrics while it is also used in internal medicine and neurology as a sympathicolytic for the treatment of sympathicotonic conditions and as a vasoconstrictor in migraine and related vascular headaches.

The other natural polypeptide alkaloids, e.g., the alkaloids of the ergotoxine group (ergocristine, ergocryptine, and ergocornine) have a spectrum of activity similar to that of ergotamine but their toxic effects are more pronounced. For this reason they have not attained the clinical importance of ergotamine.

The activity spectrum of ergonovine is quite different from that of ergotamine. It exerts a marked uterotonic effect, whereas its adrenolytic action is practically insignificant and the central nervous effects are only manifest after high doses. Ergonovine is, therefore, used mainly in obstetrics.

Substitution of the propanolamine moiety in ergonovine by butanolamine yields methylergometrine, which exerts a greater effect than the natural alkaloid on smooth muscle and is, therefore, used on a large scale under the name Methergin® as a uterotonic and hemostatic in obstetrics.

Lysergic acid diethylamide (LSD 25) exhibits a quite unexpected type of effect. This product elicits marked excitation of central nervous structures. In minimal doses it elicits mydriasis, hyperthermia, and hyperglycemia. This syndrome of central excitation is closely related to the psychotomimetic effects of LSD 25, which has become of considerable importance in experimental psychiatry and has given a substantial impetus to modern psychopharmacology. The most striking effects of LSD are colored visions and hallucinations. LSD 25 is still regarded as an experimental tool in psychotherapy and its use as an adjuvant in psychoanalysis meets a growing interest. Active doses usually range from 30–150 μg.

Hydrogenation of the $\Delta^{9,10}$ double bond of the lysergic acid moiety results in fundamental changes in the pharmacodynamic action. The vasoconstrictor and uterotonic effects of the dihydrogenated derivatives have been markedly attenuated, as has the stimulation of central structures, so that these effects are hardly elicited by therapeutic doses. On the other hand the adrenolytic effect and the central inhibition of the vasomotor centers are markedly enhanced. This is manifested clinically in vasodilatation, hypotension, and a certain sedative action.

These properties are especially marked in the case of the hydrogenated derivatives of the ergotoxine alkaloids. A combination of dihydroergocristine, dihydroergocryptine, and dihydroergocornine in equal proportions has been widely used under the name Hydergin® for the treatment of peripheral and cerebral vascular disorders and of essential hypertension.

Methylation at the indole nitrogen group specifically enhances the serotonin antagonism exhibited by all ergot alkaloids. For example, 1-methyl-LSD and 1-methyl-2-bromo-LSD exhibit antiserotonin activity several times greater than that of their nonmethylated parent compounds. 1-Methyl-d-lysergic acid butanolamide (Deseril®, Sansert®) is, depending on the test employed, four to six times more powerful than LSD 25 as a serotonin antagonist. It is, so far, the most potent serotonin antagonist discovered.

REFERENCES

1. G. Barger, "Ergot and Ergotism," Gurney and Jackson, London, England, 1931.
2. F. Arcamone, C. Bonino, E. B. Chain, A. Ferretti, P. Pennella, A. Tonolo, and L. Vero, *Nature* **187,** 238–239 (1960).
3. H. Kobel, E. Schreier, and J. Rutschmann, *Helv. Chim. Acta* **47,** 1052–1064 (1964).
4. W. A. Jacobs and L. C. Craig, *J. Biol. Chem.* **106,** 393–399 (1934).
5. W. A. Jacobs and L. C. Craig, *J. Biol. Chem.* **111,** 455–465 (1935).
6. W. A. Jacobs and L. C. Craig, *J. Biol. Chem.* **113,** 767–778 (1936).
7. W. A. Jacobs and L. C. Craig, *J. Biol. Chem.* **128,** 715–719 (1939).
8. A. Stoll, A. Hofmann, and F. Troxler, *Helv. Chim Acta* **32,** 506–521 (1949).
9. H. G. Leemann and S. Fabbri, *Helv. Chim. Acta* **42,** 2696–2709 (1959).
10. P. A. Stadler and A. Hofmann, *Helv. Chim Acta* **45,** 2005–2011 (1962).
11. F. C. Uhle and W. A. Jacobs, *J. Org. Chem.* **10,** 76–86 (1945).
12. A. Stoll and J. Rutschmann, *Helv. Chim. Acta* **33,** 67–75 (1950).
13. E. D. Kornfeld, E. J. Fornefeld, G. B. Kline, M. J. Mann, R. G. Jones, and R. B. Woodward, *J. Amer. Chem. Soc.* **76,** 5256–5257 (1954).
14. A. Hofmann and H. Tscherter, *Experientia* **16,** 414 (1960).
15. W. A. Jacobs and L. C. Craig, *Science* **82,** 16–17 (1935).
16. A. Stoll and A. Hofmann, *Helv. Chim. Acta* **26,** 944–965 (1943).
17. G. Barger and F. H. Carr, *Chem. News* **94,** 89 (1906).
18. A. Stoll and A. Hofmann, *Helv. Chim. Acta* **26,** 1570–1601 (1943).
19. D. M. Wrinch, *Nature* **137,** 411–412 (1936); **138,** 241–242 (1936).
20. A. Stoll, A. Hofmann, and Th. Petrzilka, *Helv. Chim. Acta* **34,** 1544–1576 (1951).

21. A. Hofmann, A. J. Frey, and H. Ott, *Experientia* **17,** 206 (1961).
22. A. Hofmann, H. Ott, R. Griot, P. A. Stadler, and A. J. Frey, *Helv. Chim. Acta* **46,** 2306–2328 (1963).
23. P. A. Stadler, A. J. Frey, H. Ott, and A. Hofmann, *Helv. Chim. Acta* **47,** 1911–1921 (1964).
24. W. Schlientz, R. Brunner, P. A. Stadler, A. J. Frey, H. Ott, and A. Hofmann, *Helv. Chim. Acta* **47,** 1921–1933 (1964).
25. M. Abe, T. Yamano, S. Yamatodani, Y. Kozu, M. Kusumoto, H. Komatsu, and S. Yamada, *Bull. Agr. Chem. Soc. Japan* **23,** 246–248 (1959).
26. A. Stoll, Sandoz Ltd. Swiss Patent 79'879; *Helv. Chim. Acta* **28,** 1283–1308 (1945).
27. S. Smith and G. M. Timmis, *J. Chem. Soc.* 396–401 (1937).
28. A. Stoll and E. Burckhardt, *Z. Physiol. Chem.* 250, 1–6 (1937).
29. M. Abe, T. Yamano, Y. Kozu, and M. Kusumoto, *J. Agr. Chem. Soc. Japan* **25,** 458 (1952).
30. E. Rothlin, *Bull. Schweiz. Akad. Med. Wiss.* **2,** 249–273 (1946–1947).
31. A. Cerletti, *Proc. Intern. Congr. Neuro-Pharm. 1st, Rome* (1958); "Neuro-Psychopharmacology," p. 117–123 (P. B. Bradley, P. Deniker, and C. Radouco-Thomas, Eds.), Elsevier Publ. Co., New York, N.Y., 1959.

11

The Cinchona Alkaloids

W. SOLOMON
The Wellcome Research Laboratories
Beckenham, Kent, England

1. INTRODUCTION

1.1 Literature

The most informative account of the cinchona alkaloids is that of Turner and Woodward[1] in 1953. Further developments up to 1960 are dealt with by Boit.[2] For the early history of the chemistry of the alkaloids, the reader may be directed to Brühl,[3] Königs,[4] and Comanducci.[5] Detailed information about individual compounds, and the history, cultivation, and analysis of cinchona bark can be found in Henry[6] and in Henry and Solomon.[7] Authoritative accounts of the ancient history of the subject are given by Flückiger and Hanbury,[8] Haggis,[9] and Jaramillo-Arango[10]—the reader should consult these for botanical information and for the etymology of the words *cinchona* and *quinine*.

In the ensuing text, only the landmarks of the subject are highlighted by individual literature references.

1.2 Scope of the Chapter

This chapter is divided into six sections. The first section, designed to place the cinchona alkaloids in suitable perspective, is followed by an account of the way in which their plane structure was elucidated. Some characteristic reactions are reviewed in Section 3, but no attempt is made to discuss reaction mechanisms or to deal with biosynthetical theories. The stereochemistry of the cinchona molecule is given careful attention, and

Section 5 describes the synthesis of the alkaloids. The chapter concludes with a discussion of three nontypical ingredients of cinchona bark, but the numerous class of quinoline alkaloids botanically unrelated to the cinchona group is omitted.

1.3 Cinchona Bark

These alkaloids, of which quinine is perhaps the most important, occur in the bark of trees or shrubs of the various species of two Rubiaceous genera, *Cinchona* and *Remijia*, indigenous to the higher eastern slopes of the tropical Andes in South America. The botanical classification and the term *Cinchona* itself are due to Linnaeus, who established the genus in 1742. He derived the word from the (misspelt) name of one of the consorts of the Spanish Viceroy of Peru who reigned at Lima from 1628 to 1639. According to a widely celebrated but wholly discredited legend, the lady, the Countess of Chinchon, was dramatically cured of tertian fever by the powdered bark of the tree; the preparation came to be known as "Countess's Powder" and the bark itself was called Jesuit's or Peruvian bark. The word quinine, despite some slight phonetic kinship with the name of the Countess (the resemblance is more striking in German), is of different etymology. It stems from the Quichuan language (the ancient language of the Peruvian Indians) in which "quinua-quinua" was the word for a local balsam tree, *Myroxylon peruiferum*, Lin. fil., the nearly worthless bark of which was long confused, either ignorantly or fraudulently, with the genuine and much-prized febrifuge. The same etymology underlies the name of the simple heterocycle quinoline, C_9H_7N, first called "chinolein" by Gerhardt in 1842 because he obtained it from quinine. Hence, although the alkaloid is structurally derived from quinoline, etymologically, as so often happens in chemical nomenclature, it is the other way about, the word quinoline being later than and derivative from the name of the alkaloid.

When the medicinal properties of cinchona bark were first recognized in the 17th century, the harvesting of the material in its native habitat was undertaken with so much vigor that, in spite of the great difficulties attending such operations, the threat developed before the middle of the 19th century of a complete denudation of the natural cinchona forests and of a consequent cinchona famine. Determined attempts to transplant the shrub from South America to India and to the East Indies were made, and by 1880 the Dutch had ousted their various competitors in these enterprises, especially the British, and had begun the cultivation of extensive plantations on the island of Java where in due course a considerable industry was built up. Dutch cinchona became and until 1942 remained virtually the world's sole source of quinine, a drug which was for years the only existing specific antimalarial remedy and the only cinchona alkaloid of any considerable value. In recent years another alkaloid, quinidine, has gained a reputation in the treatment of certain heart conditions.

Although the efficacy and safety of quinine as an antimalarial agent may have come to be overrated, the drug achieved considerable strategic importance. The abrupt loss of supplies, which was occasioned by Japanese military action during the second World War, was regarded as an alarming disaster; it greatly stimulated the then 20-year-old research programs in the field of synthetic antimalarials. In due course, powerful drugs were developed some of which had important advantages over quinine.

Cinchona bark is currently produced in the East Indies as well as in various other parts of the world.

1.4 Isolation and Classification

Extracts of cinchona bark were first used in the 18th century and Fourcroy is said to have obtained crude quinine in 1792. The first crystalline preparation derived from the bark was that of the Lisbon physician Gomes who named it cinchonine. This alkaloid was isolated in a pure condition in 1820 by Pelletier and Caventou[11] who obtained it from extracts of "quinquina gris," which they identified as *Cinchona condaminea*. In a sequel,[12] the French chemists found another and distinct alkaloid in "quinquina jaune" (*C. cordifolia*), and to this they gave the name quinine.

In the course of time, many other alkaloids were found in various cinchona barks, and by 1884 upwards of twenty-five had been described. Quinidine was discovered in 1833, cinchonidine in 1847, dihydrocinchonine in 1870, and the other dihydrocinchona alkaloids were isolated about ten years later. *epi*Quinine and *epi*quinidine, stereoisomerides of quinine, were isolated in 1935. There exist in the literature thirty cinchona alkaloids in all, the latest to be announced, in 1941, being *hetero*quinine; 40 mg of this structural isomeride of the alkaloid were obtained from ¾ kg of commercial quinine.

These alkaloids rarely occur elsewhere in nature; the leaves of *Cinchona* contain only small amounts, but recently several new indole alkaloids were isolated from this part of the plant.

The principal cinchona alkaloids are white, crystalline solids that have a bitter taste and are sparingly soluble in water. They have high melting points and optical rotations. The alkaloids, some of which fluoresce in dilute sulphuric acid solution, are diacidic bases and form two series of salts of the type B·HCL and B·2HCL; the former, or "neutral" salts, tend to be sparingly soluble in water; the acid salts are more freely soluble.

The alkaloids are chemically reactive and are easily attacked by oxidizing agents and by acids. Even a mild agent such as boiling dilute acetic acid brings about the well-known rearrangement to the cinchonatoxines.

With four notable exceptions, the cinchona alkaloids of known structure are quinoline derivatives. Many other compounds belonging to this heterocyclic system occur in nature and are widely distributed among, but largely restricted to, the genera of the Rutaceae family; among these may be mentioned quinoline itself, cusparine and galipine (in Angostura bark), dictamnine, fagarine and, one of the most widely distributed alkaloids in nature, skimmianine. This newer group of alkaloids has excited a good deal of attention during the last ten or twenty years and they now far outnumber the original cinchona bases but, numerous though they be, they are of relatively small interest and significance and are not discussed in this Chapter.

The cinchona alkaloids can be classified in the following manner:

(a) Eight major alkaloids (Table 11.1), which are quinoline derivatives of known structure; they form the principal subject matter of this chapter. Read horizontally, they constitute four pairs of stereoisomerides.

(b) Nine minor alkaloids (Table 11.2); these also are of established structure. The five in the left-hand column are closely related to the major alkaloids and are quinoline derivatives. The occurrence of cupreine is not open to question, but the four quinine isomerides may be artefacts, i.e., have no independent preexistence in cinchona bark. The first three indole alkaloids in the right-hand column of Table 11.2 share with the cinchonas the feature of having a quinuclidine moiety, but aricine is a typical indole alkaloid which, apart from its botanical origin, bears no relation to the cinchona bases.

(c) Thirteen little-known alkaloids of unestablished structure—in certain of these cases their very existence as distinct chemical entities may be open to doubt. These rare cinchona alkaloids are neither discussed nor even listed in the present chapter and the interested reader must be referred to other sources[3,6,7] for further information.

According to some authorities cinchotoxine also occurs naturally in cinchona bark; it does not seem possible to substantiate this claim, but if allowed, the compound would belong in the left-hand column of Table 11.2 and there would be thirty-one recorded cinchona alkaloids.

The eight major alkaloids (Table 11.1) are quinolyl–quinuclidyl carbinols [(1); X = OH].

The name quinuclidine for the bicyclic system 1-azabicyclo-[2.2.2] octane (2) was proposed by Königs in 1904. The unsubstituted heterocycle

TABLE 11.1. Major Cinchona Alkaloids.

Dextrorotatory Series			Levorotatory Series		
Alkaloid	Formula (1); X = OH		Alkaloid	Formula (1); X = OH	
	R	R′		R	R′
Cinchonine	H	$—CH=CH_2$	Cinchonidine	H	$—CH=CH_2$
Quinidine	OMe	$—CH=CH_2$	Quinine	OMe	$—CH=CH_2$
Dihydrocinchonine	H	Et	Dihydrocinchonidine	H	Et
Dihydroquinidine	OMe	Et	Dihydroquinine	OMe	Et

TABLE 11.2. Minor Cinchona Alkaloids.

Quinoline Derivatives		Indole Derivatives	
Alkaloid	Formula	Alkaloid	Formula
Cupreine	(1); R = X = OH; $R' = —CH=CH_2$	Cinchonamine	(54)
*epi*Quinine	(1); R = OMe; X = OH; $R' = —CH=CH_2$	Quinamine	(53)
*epi*Quinidine	(1); R = OMe; X = OH; $R' = —CH=CH_2$	Conquinamine	(53)
Quinotoxine	(7); R = OMe; $R' = —CH=CH_2$	Aricine	*See* Chapter 9
*hetero*Quinine	(27); $R' = —CH=CH_2$; X = OH; Q = [(3); R = OMe]		

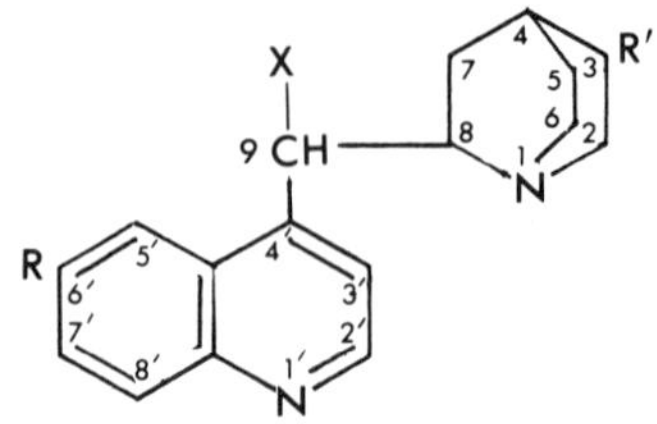

(1)

Major Cinchona Alkaloids

R = H or OCH_3

R′ = $\overset{10}{—CH}{=}\overset{11}{CH_2}$ or $\overset{10}{—CH_2}\overset{11}{CH_3}$

X = OH

(2)

Quinuclidine (Plane)

is, itself, a known compound; it will be noted that its numbering in (1) differs from that adopted in "The Ring Index" and in "Beilstein" for the heterocycle itself as shown in (2). Throughout this chapter the system of numbering shown in (1), which is due to Rabe, is used even for simple quinuclidine derivatives.

Quinuclidine is a highly symmetrical structure and an attempt should be made to see it in three dimensions (2a) and to realize that no two of the

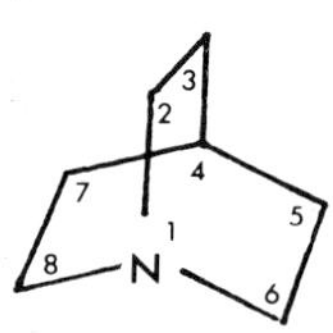

(2a)

Quinuclidine (Three-dimensional)
(Rabe Numbering)

three fused piperidine rings are co-planar; the three dimethylene chains are wholly equivalent and the one shown in (2) as occupying the central position [C(5) and C(6)] is no more (and no less) a "bridge" than are either of the two "sides."

Rabe proposed the (1) system of numbering in 1922 and, except for purposes of comparative biogenetics, this system has been in universal use ever since. As frequent identifications are needed in the ensuing text of individual positions and atoms, both in the cinchona molecule and in its simpler degradation products, these numbers are reproduced for the convenience of the reader in most of the formulas in this chapter; the same system of numbering is retained even in those cases [(9), (11), (12), (20), (26), (31)–(35), (39), (48)–(52), (55)] in which it has become systematically inappropriate or in which, as a consequence, the numbering in the formulas differs from the nomenclature used in the text. Wherever adequate, the symbol Q in these formulas stands for 4-quinolyl or 6-methoxy-4-quinolyl (3); the letter R is reserved throughout for the substituent at

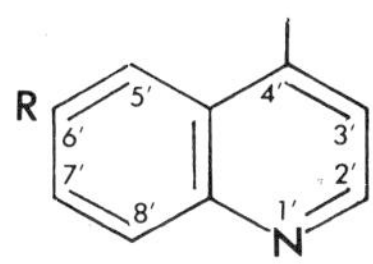

(3)

"Q" (R = H or OCH_3)

position 6′ and mostly stands for either H or OMe but occasionally for OH and NH_2; the symbol R′ is similarly retained for the substituent at position 3 and is either vinyl ($—CH{=}CH_2$) or ethyl. The letter X stands for OH or for a halogen atom.

In 1922, when Rabe numbered the cinchona skeleton, he also suggested the name ruban for the parent heterocycle [(1); R = R′ = X = H]. The term derives from the botanical family Rubiaceae to which the genera *Cinchona* and *Remijia* belong. The compound has one center of dissymmetry, that at C(8), and has been synthesized and resolved. According to the ruban nomenclature, quinine, for example, is one of the stereoisomeric 6′-methoxy-3-vinylruban-9-ols; quinidine is another, and their dihydro derivatives are the corresponding 3-ethyl-6′-methoxyrubanols. The system readily accommodates functional derivatives, ketones, and so forth, and also transformation products such as the cinchonatoxines and *hetero* isomerides.

The major alkaloids fall into two configurational and polarimetric groups, dextrorotatory and levorotatory. Cinchonine, quinidine, and their dihydro derivatives belong to the dextro group and all have the same configuration; cinchonidine and quinine for their part share a different configuration and are levorotatory. Although they are related in this way, the pairs of stereoisomerides, quinine and quinidine for example, are not mirror-image enantiomers. Dextrorotatory quinidine is 6′-methoxycinchonine; quinine is 6′-methoxycinchonidine. The rare alkaloid cupreine (Table 11.2) is levorotatory and is 6′-hydroxycinchonidine; it gives rise to quinine (and not quinidine) on methylation. The corresponding dextrorotatory cupreidine is not known to occur in nature. It is due to an unfortunate historical accident of nomenclature that the configurationally identical pairs are cinchonine–quin*id*ine and cinchon*id*ine–quinine, and not vice versa.

2. ELUCIDATION of PLANE STRUCTURE

The history of the elucidation of the plane structure of the cinchona alkaloids may be said to commence with the destructive potash distillations of Gerhardt in 1842 and to terminate in 1907 when Rabe's[13] decisive oxidation of cinchonine to cinchoninone [(4); R = H; R′ =—CH=CH$_2$] proved

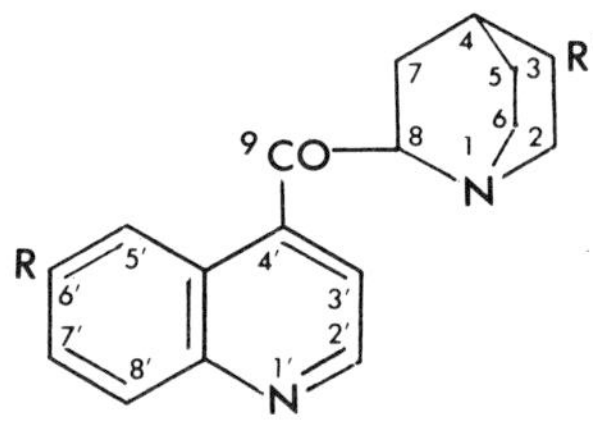

(4)

Cinchona Ketones

that the alkaloid was a secondary and not a tertiary carbinol; this enabled him to select the correct formula from among several otherwise equally tenable alternatives current at the time. It is instructive to see this sixty-five-year period, especially its first two or three decades, in proper perspective in relation to the history of chemistry. The early work on the cinchona alkaloids was done before structural formulas had come into general use and when oxygen was still counted as 8; quinoline was discovered long before Kekulé conceived the benzene ring, and Pasteur converted cinchonine to cinchonicine twenty years before van't Hoff and Le Bel laid the foundations of stereochemistry. It is important to realize that the first studies of pyridine and quinoline (and so of heterocycles generally) were incidental to or initiated by investigations into the nature of the cinchona alkaloids.

2.1 Empirical Composition—Functional Groups

Early elementary analyses of quinine and cinchonine had been carried out by Liebig, by Laurent, and by Strecker, but the final assurance that in modern terms cinchonine was $C_{19}H_{22}N_2O$ and quinine $C_{20}H_{24}N_2O_2$ was due to Skraup and dates from 1879.

The experiments that were to reveal the structure of the cinchona alkaloids were of three kinds:

(a) those which, leaving the molecular skeleton largely intact, yielded information about functional groups;
(b) studies of the important rearrangement of the alkaloids to "-icines" or "toxines" (discovered by Pasteur[14] in 1853);
(c) those in which the molecule underwent cleavage.

That both nitrogen atoms were basic followed from the existence of two series of salts, and that both were tertiary emerged from a study of the action of alkyl halides. These researches were carried out by Skraup, and by Claus and their collaborators between 1881 and 1894. Either the quinuclidine nitrogen atom [N(1)] or the quinoline nitrogen atom [N(1′)] could be quaternized, depending on whether the free base or the alkaloidal salt was employed. In the former case, the products were white; in the latter, colored. Colored dialkyl derivatives could be obtained by more vigorous treatment, the same dialkyl derivative resulting from both monoalkyl derivatives in any one case. On the other hand. two different alkylating agents gave two isomeric dialkyl derivatives; thus, quinine methiodide gave with ethyl iodide, a product distinct from that resulting from the action of methyl iodide on quinine ethiodide. The behavior with alkyl halides may be summarized:

(a) One alkylating agent

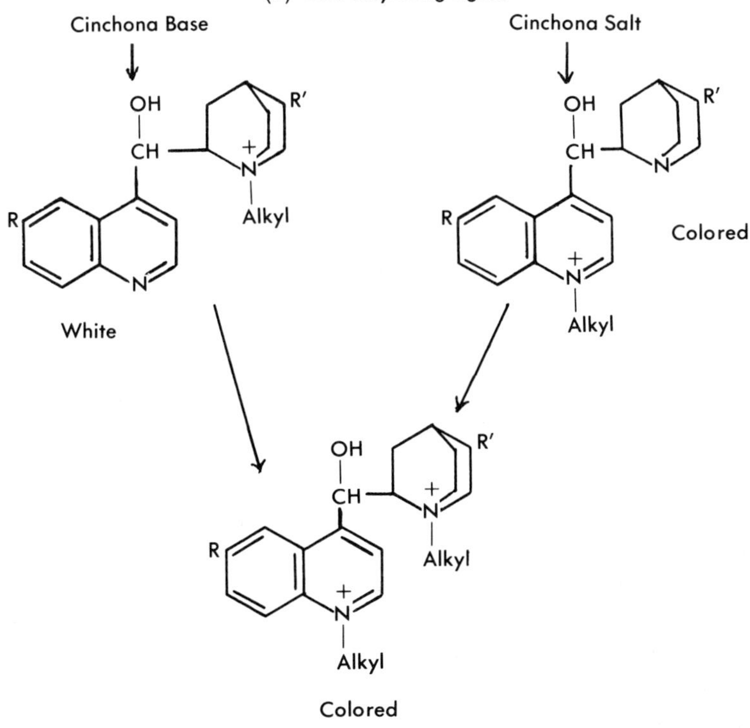

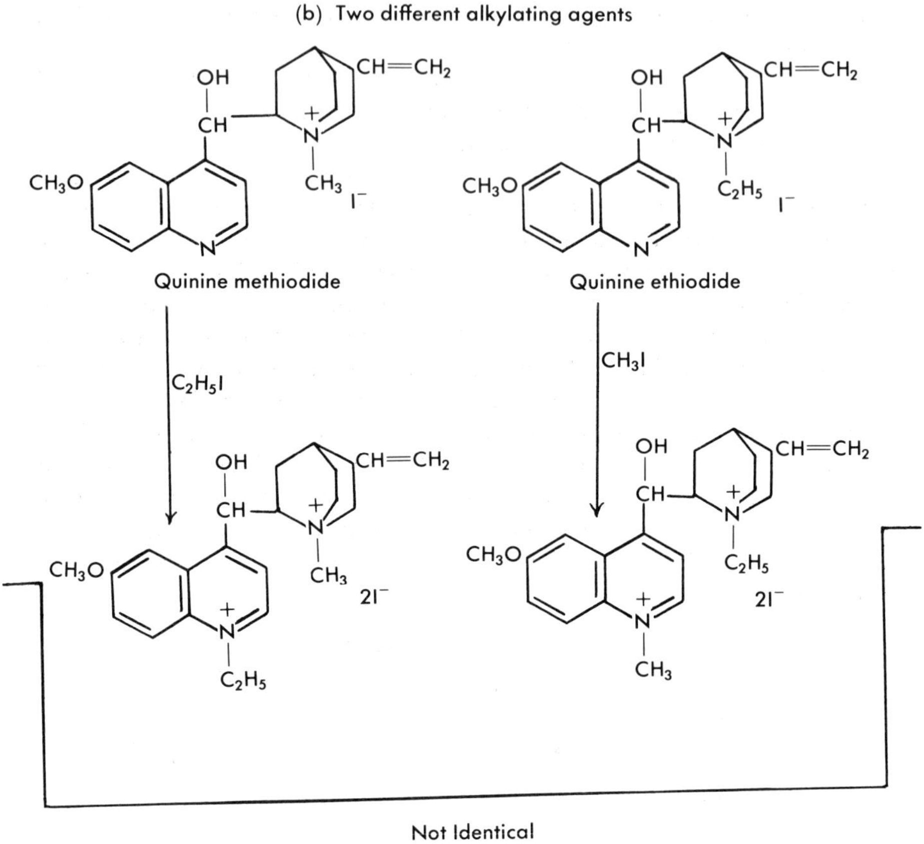

It had been shown earlier[15] that the alkaloids could be acetylated and benzoylated and that from the resulting acyl derivatives the original substances could be regenerated unchanged by a process of hydrolysis. From the vantage point of our present knowledge, these results must be regarded as remarkable; however that may be, such experiments suggested the presence of a hydroxyl group even though the evidence did not become unequivocal until later, when the nitrogen atoms had been shown to be tertiary, and when it was found that the putative hydroxyl group could be replaced by chlorine by the action of phosphorus pentachloride. This accounted for the sole oxygen atom of cinchonine and cinchonidine and for one of the two oxygen atoms in quinine and quinidine. That the second oxygen atom of the latter two alkaloids belonged to a methoxyl group was made evident when on heating with concentrated hydrochloric acid they furnished alkali-soluble phenolic products (R = OH) with evolution of methyl chloride. Furthermore, it was established that, whereas on degradation cinchonine and cinchonidine gave rise to fragments which were simple quinolines, the analogous fragments from quinine and quinidine were the corresponding 6-methoxyquinoline derivatives.

In 1845 it was known that cinchonine gave rise to a dichloride and a dibromide, and in 1874 it was shown that the alkaloid could add on the elements of hydrogen chloride. In 1892 Königs treated cinchonine dibromide with alcoholic potash and obtained dehydrocinchonine [(1); R = H; X = OH; R′ = —C≡CH]. The evidence of unsaturation so accumulated, however, gave no indication of its position in the molecule. The point was partly established by a study of the effect on the alkaloids of dilute, ice-cold potassium permanganate. This gave rise to a series of carboxylic acids [(1); X = OH; R′ = CO_2H], no longer unsaturated, and containing one carbon atom less than the bases which furnished them. The missing carbon atom turned up as formic acid and the products were distinguished by the generic suffixes "-tenines" and "-tenidines." The first of these acids, the one from cinchonine, was obtained in 1869 and was termed cinchotenine. Similarly, quinine, quinidine, and cinchonidine gave quitenine, quitenidine, and cinchotenidine, respectively. Königs's acetylenic dehydrocinchonine could be oxidized in the same way but the cinchotenine duly produced was in this instance attended by the evolution of carbon dioxide and not of formic acid.

These experiments gave clear evidence that the point of unsaturation was not present in the nucleus but belonged to a sidechain vinyl group; but still there was no indication of its position in the molecule.

2.2 The Dihydro Alkaloids

The saturated dihydro alkaloids [(1); X = OH; R′ = Et] often accompanied as impurities their commercially available and better-known vinyl congeners. The very existence of such compounds remained unsuspected until 1869, and it was then recognized mainly as a result of the permanganate oxidations of the kind just described. Provided the conditions were mild enough, the saturated alkaloids were largely unaffected by this reagent and, being strongly basic, were easily detected in and separated from the tenines and tenidines which were the major and acidic products of the reaction. In our own day, catalytic hydrogenation of the vinyl-bearing bases, rather than isolation of their dihydro derivatives from natural sources, has become the most advantageous method of obtaining these materials.

2.3 Anhydro and Desoxy Bases

In 1880, Königs found that the hydroxyl group in cinchonine could be replaced by chlorine when the alkaloid was treated with phosphorus pentachloride. The other cinchona alkaloids behaved in a similar manner and during the next decade Königs and Comstock made an intensive study of the important products of this reaction [(1); X = Cl]. About fifty years later, Rabe found that a corresponding reaction with phosphorus pentabromide ran a more complicated course. Cinchonine chloride as it was called (better names for it now would be 9-chlorodesoxycinchonine or 9-chloro-3-vinylruban) and analogous chlorides derived from the other alkaloids, lost the elements of hydrogen chloride on treatment with alcoholic potash and gave rise to a centrally situated double bond. The resulting products (5) were called "anhydro bases." In this way while

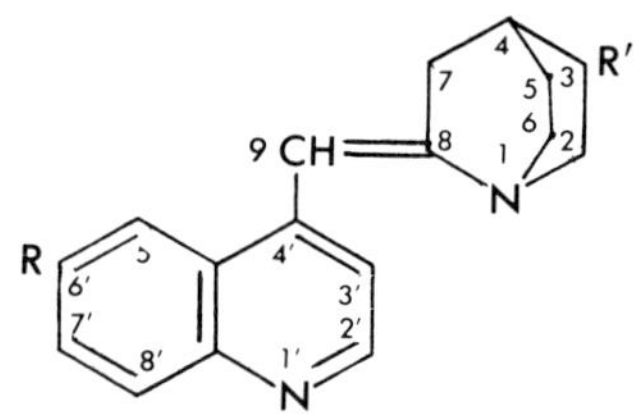

(5)

Anhydro Bases

cinchonine gave cinchene [(5); R = H; R′ = —CH=CH$_2$], quinine went to quinene [(5); R = OMe; R′ = —CH=CH$_2$]. Stereoisomeric pairs of alkaloids produced one and the same anhydro base; thus, cinchonine and cinchonidine furnished the same cinchene, and quinine and quinidine gave only one quinene. Analogous chloro and anhydro derivatives were obtained from the dihydro alkaloids (R′ = Et).

When the cinchona chlorides were reduced with iron and dilute sulphuris acid they yielded the desoxy bases (6) which are quinolyl-quinuclidyl

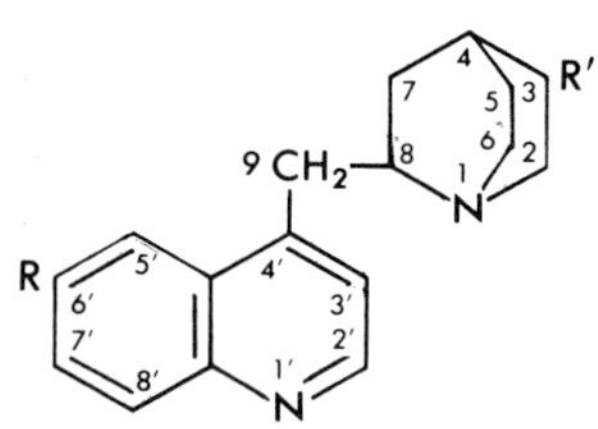

(6)

Desoxy Bases

methanes. Unlike cinchene and quinene, the desoxy bases are uniquely related to their parent alkaloids: cinchonine afforded a desoxycinchonine distinct from the desoxycinchonidine derived from cinchonidine; the eight major alkaloids of Table 11.1 eventually gave rise each to its own desoxy base. At the present time, better methods exist for the preparation of these substances, expecially the Wolff–Kishner reduction of the cinchona ketones (4).

2.4 The Toxines

The rearrangement of the cinchona alkaloids to toxines (7) is of central significance not only in the history of the elucidation of their structure

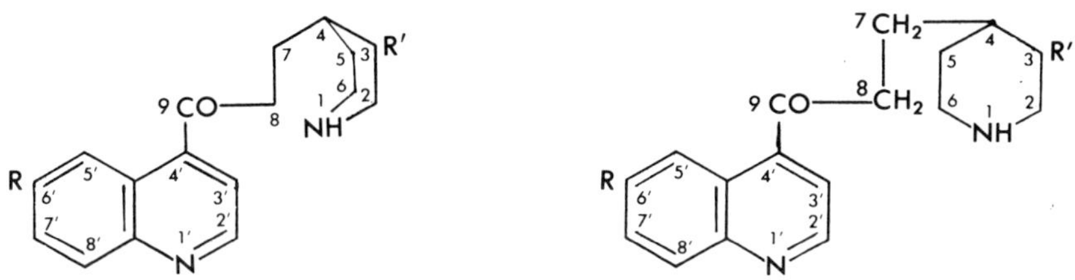

(7)

Toxines

but also in all synthetical approaches to these and to analogous bases. It is a form of intramolecular de-amination. The delicate grouping $—{}^{9}CH(OH)—{}^{8}CH(\overset{|}{\underset{|}{N}})$, which is involved, and which, as will be shown in the next section, is also the site of the reaction that gives rise to the *hetero* cinchona isomerides, is sensitive to various forms of attack. In the cinchona alkaloids, this group is especially labile because it is situated at the reactive γ-position of the quinoline ring: the alkaloids must be seen as derivatives of 4-methylquinoline, in which the reactive methyl group is the C(9) of the cinchona skeleton.

Pasteur,[14] who first observed the reaction in 1853, carried it out essentially by heating the alkaloids with a limited amount of sulphuric acid. He called the products cinchonicine and quinicine; the latter was subsequently isolated from cinchona bark. Pasteur noted with interest the greatly reduced optical power of these compounds as compared with the alkaloids that gave rise to them. Just as in the case of the anhydro bases, cinchonine and cinchonidine gave one and the same cinchonicine, and similarly quinicine could be prepared equally from either quinine or quinidine.

In 1878, Claus and his co-workers began a study of the action of alkali on the methiodides and similar quaternary salts of the cinchona alkaloids; instead of obtaining the expected characteristic water-soluble quaternary ammonium hydroxides, Claus isolated a series of insoluble bases, which he regarded as homologs of the original alkaloids. Thus, cinchonine methiodide gave him "methylcinchonine," etc. In 1894, while investigating these "homologs," von Miller and Rohde discovered[16] that the natural alkaloids, on prolonged boiling with dilute acetic acid, gave rise to what they called "toxines," and these were identified with Pasteur's "-icines." Contrary to the normal course of events, the German authors' later nomenclature entered into common use and Pasteur's "cinchonicine" and "quinicine" became obsolete; furthermore, their acetic acid procedure became the preferred method of effecting the toxine rearrangement. von Miller and Rohde showed that Claus's "homologs" were tertiary bases, ketonic in character, and identified them as the N(1)-alkyl-derivatives of their own toxines. They proved that his "methylcinchonine" and "methylcinchonidine" were one and the same substance, N(1)-methylcinchotoxine. Therefore, it was clear that, unlike the parent alkaloids which were known to be tertiary bases, the toxines were secondary and that in the course of the toxine rearrangement a C–N linkage must undergo rupture. Also, it seemed that the rearrangement might be favored by quaternization. This conclusion was in good agreement with the supposition (which, having regard to Pasteur's original conditions, seemed plausible at first sight) that the rearrangement was promoted by hydrogen ions, but seemed to be in conflict with the newly established fact that dilute acetic acid was more effective than the mineral acids. This point was later made the subject of detailed physical-chemical studies by H. C. Biddle. A simple way to resolve the seeming contradiction is to suppose that two factors are involved: (a) the reaction is favored by the tetracovalent state of the N(1) atom, obtained in acid solutions of the alkaloids or in their quaternary salts and hydroxides, and (b) the reaction is favored not by hydrogen ions but, on the contrary, by hydroxyl ions. In the case of the parent alkaloids, the two factors are in conflict, whatever the pH; but they reinforce each other in the case of the alkaline quaternary ammonium hydroxides. This account of the matter probably oversimplifies a complex reaction. Rabe showed that under certain conditions (depending inter alia on the nature of the quaternizing radical) the quaternized alkaloids gave rise on treatment with caustic alkali not to toxines but to oxiranes (ethylene oxides) of type (8). For this reason some of Claus's pairs of "homologs" (especially "benzylcinchonine" and "benzylcinchonidine"), being stereoisomeric oxiranes, were neither identical with one another nor with the corresponding *N*-substituted cinchotoxine.

The toxines can be brominated either at the imine nitrogen atom [N(1)] or at the C(8) methylene group adjoining the carbonyl function. Both bromo-derivatives so produced, when subjected to a variety of alkaline conditions, undergo intramolecular elimination of the elements of hy-

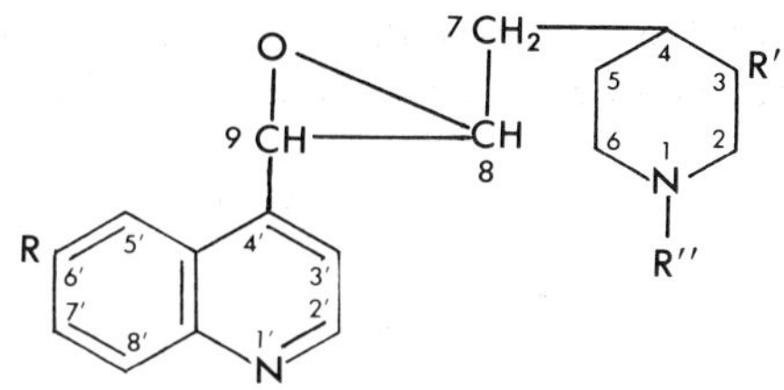

(8)

Oxiranes

drogen bromide to give the cinchona ketones (4). In these reactions the bicyclic quinuclidine system, first ruptured in the toxine rearrangement, is reconstituted. The ketones (4) were first obtained by oxidizing the alkaloids with chromic acid under very mild conditions; but the yields were exiguous, and the chain of operations, alkaloid → toxine → bromotoxine → ketone affords one of the only two practicable methods of preparing them in quantity. Stereochemical complications notwithstanding, the ketones can be reduced back to the carbinols [(1); X = OH] and this chain of operations then becomes the cycle: alkaloid → toxine → bromotoxine → ketone → alkaloid. The cinchonatoxines are intermediates in the synthesis of the cinchona alkaloids and the last three stages of this cycle constitute the final steps of the synthesis.

2.5 Cleavage

There are various reactions in which the cinchona alkaloids are apt to split into two more or less equal fragments traditionally known as "halves;" the quinoline part of the structure was known as the "first half," while the quinuclidine moiety, which in nearly all such experiments gave rise to simpler piperidine derivatives, was referred to as the "second half." Apart from some recent instances, such processes have been of four general types.

(a) The earliest cleavages (1840's–1850's) were effected by destructive distillation with potash. This indiscriminate attack on the molecule gave rise to quinoline and to several simple quinoline and pyridine derivatives. Seen in historical perspective, it would be difficult to overstate the interest and importance of the formation of these heterocycles, but in their own day these compounds were not very illuminating. They were new and of unknown structure and could not be architecturally assigned to specifiable areas of the cinchona skeleton. This method of cleavage and its products cannot be further considered here.

(b) The second method was by vigorous oxidation. The reactions and their products were studied for about twenty years (ca. 1874–1894) mainly by Königs and by Skraup. The substances oxidized were most of the alkaloids, the two unsaturated toxines, and the proximate products of oxidation. Cinchonine and cinchonidine gave cinchoninic acid [(9); R = H]; while from

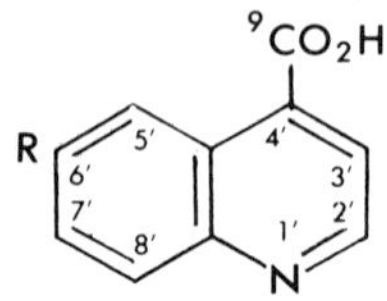

(9)

Cinchoninic Acid (R = H). Quininic Acid (R = OCH_3)

quinine and quinidine, quininic acid [(9); R = OMe] was obtained. These quinolinecarboxylic acids stemmed from the "first half." Oxidation products derived from the "second half" were piperidine derivatives; the unsaturated alkaloids furnished meroquinene [(10); R′ = —CH=CH_2],

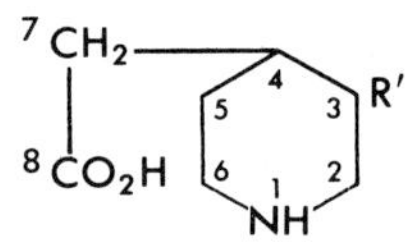

(10)

Meroquinene (R′ = $\overset{10}{\text{—CH}}=\overset{11}{\text{CH}_2}$)

Cincholoipon (R′ = $\overset{10}{\text{—CH}_2}\text{—}\overset{11}{\text{CH}_3}$)

Cincholoiponic Acid (R′ = $\overset{10}{\text{CO}_2\text{H}}$)

while in the corresponding saturated case, dihydrocinchonine gave the analogously constituted cincholoipon [(10); R′ = Et]. These piperidine acids could be further oxidized to cincholoiponic acid [(10); R′ = CO_2H] and finally to loiponic (piperidine-3,4-dicarboxylic) acid (11). All cinchona

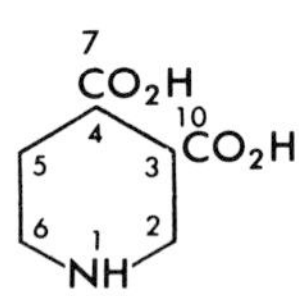

(11)

Loiponic Acid

alkaloids of known structure can potentially give rise to, or have in fact furnished, these last two dicarboxylic acids.

(c) In 1894, Königs[17] heated cinchene [(5); R = H; R′ = —CH=CH_2] and quinene [(5); R = OMe, R′ = —CH=CH_2] with aqueous phosphoric acid at 170°. Smooth hydrolytic cleavage occurred at the central double bond, and lepidine (4-methylquinoline) [(12); R = H] and 6-methoxy-

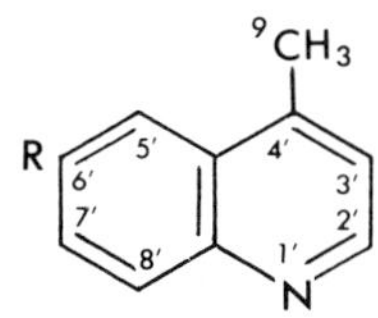

(12)

Lepidine (R = H)

lepidine [(12); R = OMe], respectively, were produced in high yield. In addition to these simple quinoline derivatives, meroquinene [(10); R = —CH=CH$_2$] was formed. It was during the course of these experiments, and not on oxidative cleavage, that meroquinene was first discovered. The analogous compound cincholoipon [(10); R′ = Et] was later obtained in a similar fashion from dihydrocinchene [(5); R = H; R′ = Et]. In due course, these products were correlated with those obtainable on oxidation. These results cast a clearer light on the structure of the cinchona alkaloids than any previous experiments had provided.

When the anhydro bases (5) were heated with hydrobromic instead of with phosphoric acid, the vinyl-bearing compounds (R′ = —CH=CH$_2$) underwent a remarkable reaction to which their saturated analogs (R′ = Et) were not subject. Cinchene gave *apo*cinchene [(13); R = H]

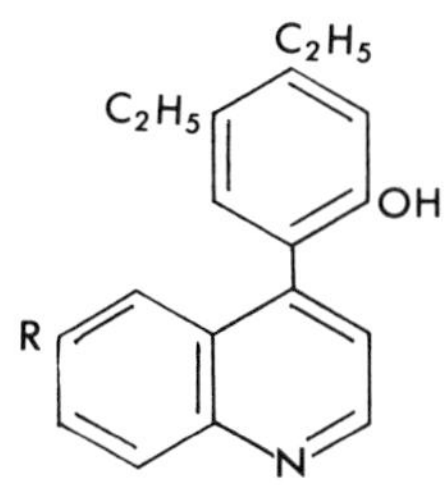

(13)

Apo Bases. *apo*Cinchene (R = H)
*apo*Quinene (R = OH)

and quinene gave *apo*quinene [(13); R = OH]. The structure of these compounds was not finally clarified until 1935; Königs devoted a great deal of attention to them but the *apo* bases must in retrospect be dismissed as having shed no light on the constitution of the cinchona alkaloids.

(d) The last group of cinchona fissions dates from the early years of this century. The method, which was first studied by Rabe[18] and by Königs,[19]

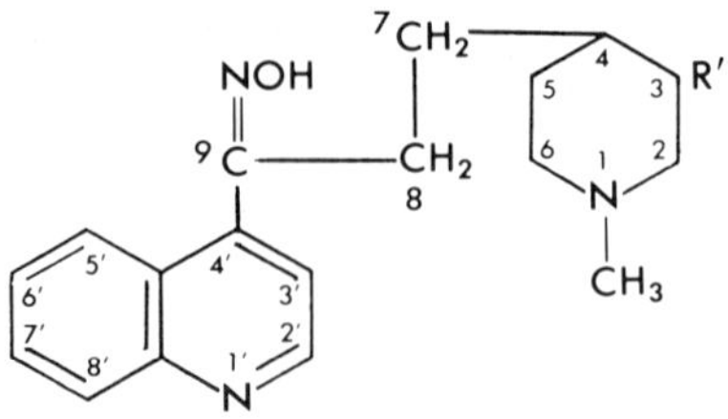

(14)

consisted in subjecting the oxime [(14); R′ = —CH=CH$_2$] and the isonitroso derivative [(15); R′ = —CH=CH$_2$] of N(1)-methylcinchotoxine to

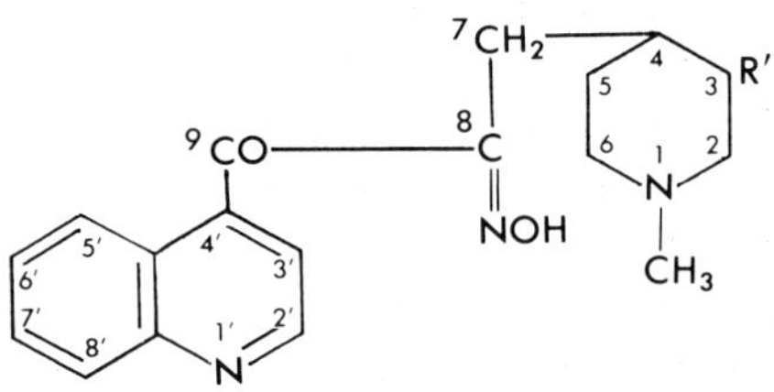

(15)

the Beckmann rearrangement. The corresponding dihydrocinchotoxine derivatives (R′ = Et) were also used but the experiments were not extended to the methoxylated (quinotoxine) analogs. The isonitroso derivatives (15) rearranged in the normal way as shown in the following scheme (Q = quinolyl; P = substituted piperidyl):

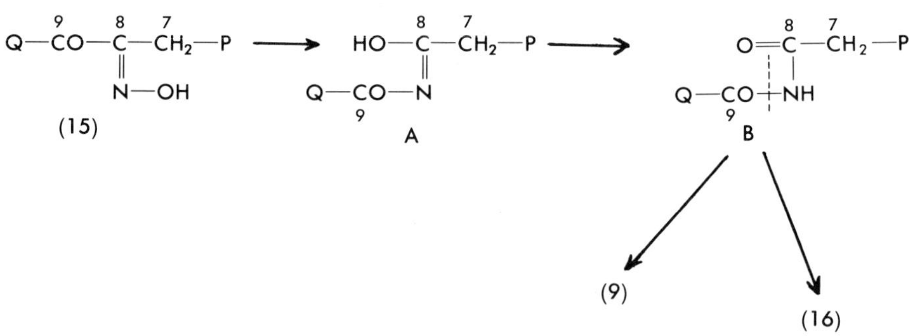

The interchange of the hydroxyl group and the Q—CO residue led to the hypothetical intermediate imidic acids (A) and acid amides (B) which on alkaline hydrolysis cleaved at the point indicated by the dotted line. The one part of the molecule then furnished cinchoninic acid (9), while from the remainder of the structure there resulted the nitriles (16).

The rearrangement of the oximes (14) ran a more complicated course, it being assumed by the authors that the complication was due to the oximes being mixtures of *syn-* and *anti-* forms:

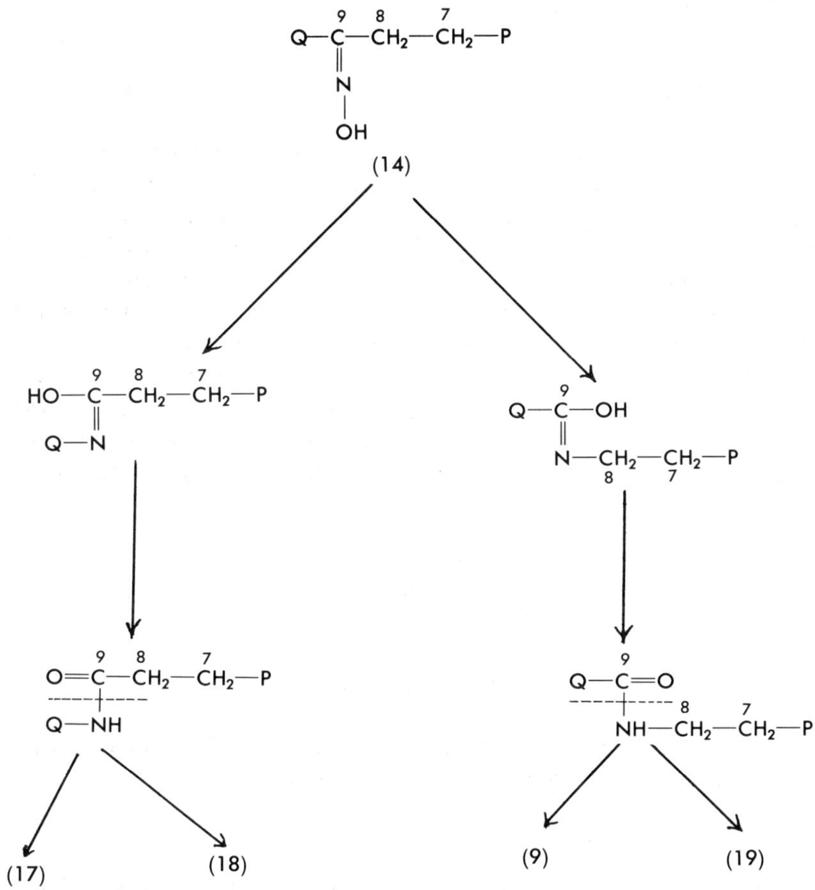

Cinchoninic acid (9) was at this time (1907) a well-known compound but the other Beckmann fragments [(16)–(19)] were new or had not previously been obtained by the direct degradation of the cinchona alkaloids.

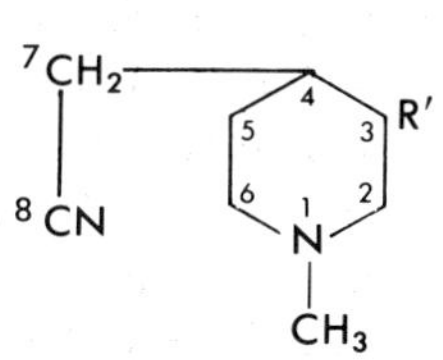

(16)

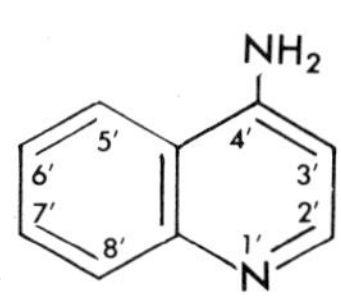

(17)

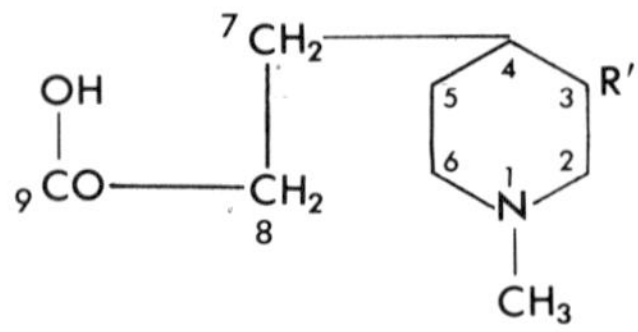

(18)

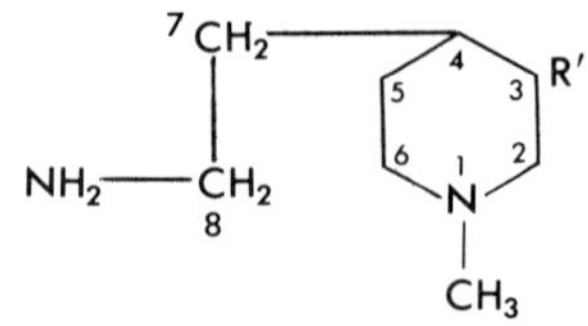

(19)

These Beckmann rearrangements were much improved several years later when it was found that the *N*-benzoyl analogs were more suitable than the *N*-methyl derivatives used by Rabe and by Königs. In particular, the rearrangement of *N*-benzoyl-cinchotoxineoxime acquired great importance because, as is shown in Section 5, the successful production from it of the homo-acid [(18); R′ = —CH═CH_2; with benzoyl in place of methyl] furnished the key to the partial synthesis of quinine.

Closely related to the Beckmann method of cleavage was Rabe's elegant one-step scission of cinchoninone [(4); R = H; R′ = —CH═CH_2] by means of amyl nitrite and ethanolic sodium ethoxide to give cinchoninic acid (9) and the oxime of 5-vinylquinuclid-2-one (20) in good yield; this was the

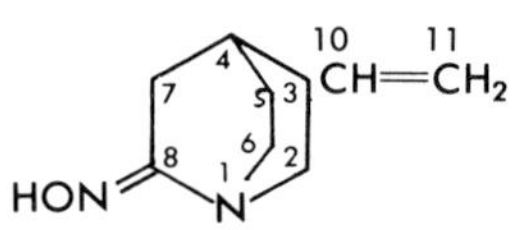

(20)

5-Vinylquinuclidin-2-one Oxime

first occasion on which the bicyclic quinuclidine skeleton emerged intact from a cleavage of a cinchona alkaloid into "halves."

2.6 Final Elucidation

The nature of the second-half fragments (10, 11) had long been understood but conclusive proof of their structure and stereochemical configuration did not become available until 1909 when cincholoiponic acid [(10); R′ = CO_2H], optically identical with that obtainable from the natural alkaloids, was synthesized. This was two years *after* the structure of cinchonine had been settled beyond dispute. However, attempts had been made previously to arrive at a formulation of the alkaloid on the available evidence. One of the key considerations was that of von Miller and Rohde[20] who argued correctly that when the tertiary nitrogen atom and carbinol function of cinchonine were converted in the toxine rearrangement to the secondary state and to a ketone group respectively, a rupture of a C–N linkage must have occurred, and, seeing that this happened without loss and that the resulting toxine still contained a piperidine ring, the second half in the original alkaloid must have a bicyclic structure. This led to a number of suggestions, but by 1907 only two formulas remained to be considered—the tertiary carbinol (21) and the secondary carbinol [(1); X = OH; R = H;

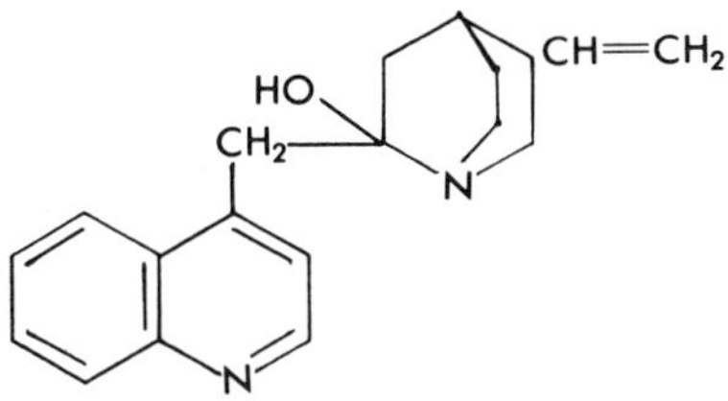

(21)

Cinchonine. One of the pre-1907 formulations

R′ = —CH=CH$_2$]. In that year, Rabe's oxidation of cinchonine to cinchoninone[13] decided in favor of the latter alternative and finally established the plane structure of the cinchona alkaloids.

3. CHEMICAL BEHAVIOR

In the preceding section, certain reactions were encountered but allusion to these was incidental to the tracing of a historical process. In the present section, five other types of reaction are discussed that have, so far, received only scant attention:

(a) action of mineral acids
(b) other reactions at the vinyl group
(c) oxidation to ketones
(d) formation of *hetero* cinchona alkaloids
(e) attacks on the quinoline ring

3.1 Action of Mineral Acids[21]

The two centers which are most vulnerable to attack by mineral acids are the methoxyl group in the 6′-position in quinine, quinidine, and their dihydro derivatives, and the vinyl group attached at C(3) in the four unsaturated alkaloids, cinchonine, cinchonidine, quinine, and quinidine. Of these two, the vinyl group is the more sensitive so that while it is possible to modify it in quinine and quinidine without demethylating these alkaloids, it is not possible to demethylate them directly without affecting the vinyl group. For this reason the phenolic bases cupreine and cupreidine [(1); R = X = OH; R′ = —CH=CH$_2$], which correspond to quinine and quinidine, respectively, cannot be obtained by the action of acids upon the latter.

The delicate C(8)–C(9) system, which succumbs to attack by weak acids in the toxine rearrangement, is relatively stable to strong mineral acids, and dihydrocinchonine and dihydrocinchonidine [(1); R = H; X = OH; R′ = Et], which have neither methoxyl nor vinyl weaknesses, are able to withstand conditions under which their unsaturated analogs (R′ = —CH=CH_2) give rise to various transformation products. Similarly, dihydroquinine and dihydroquinidine [(1); R = OMe; X = OH; R′ = Et] can resist the action of *cold* mineral acids which cannot demethylate them, under conditions in which quinine and quinidine cannot survive unchanged; even when *heated* with mineral acids they only undergo demethylation to the nonnaturally occurring bases dihydrocupreine and dihydrocupreidine [(1); R = X = OH; R′ = Et], respectively, as demonstrated by the fact that these phenols can be methylated and the starting alkaloids recovered unchanged. This relative stability of the dihydro alkaloids is often exploited in studies in which the incidence of transformation products would be irrelevant or otherwise undesirable.

The changes at the vinyl group, which result from the action of mineral acids, are of four types.

Type 1. Possibly a preliminary stage in all such reactions is the addition of HX to the double bond with the production of (22); when more or less

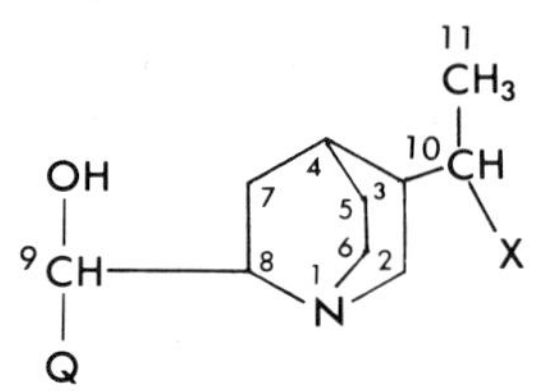

(22)

dilute sulphuric acid is used, X = OH; when the acid is a concentrated halogen acid, X = halogen. These products occur in two stereoisomeric forms owing to the creation of a new asymmetric center at C(10). When the elements of HX are withdrawn again by suitable reagents, the original vinyl group is not restored and the parent alkaloid not regenerated; instead, there result products belonging to one or other of the remaining three types mentioned below. This leads to the generalization that the vinyl alkaloids, once affected by mineral acids, cannot normally be reconstituted.

Type 2. In this case an isomeride (23) is produced resulting from a shift of the double bond from C(10)–C(11) to C(10)–C(3). There is some possibility of a further shift with involvement at C(2). The double bond, to start with symmetrical, is in its new position unsymmetrical, and compounds of this type are subject to *cis-trans* isomerism and apt, as in the

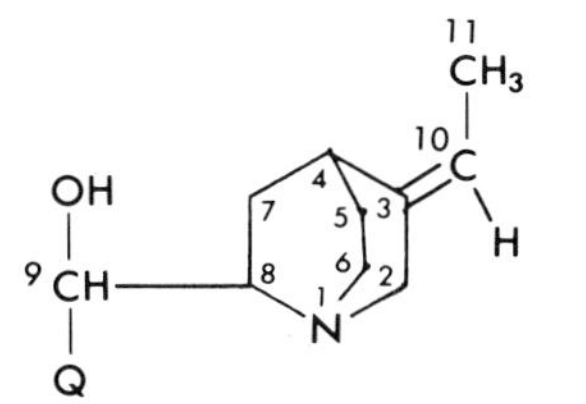

(23)

case of Type 1, to turn up in two forms. Carbon atom C(3) is one of the original points of dissymmetry: these ethylidene derivatives therefore are liable, especially when mixtures of both geometrical isomerides are used, to give rise on hydrogenation to two C(3) epimerides, only one of which is identical with a naturally-occuring dihydro alkaloid.

Type 3. In this reaction internal ethers (24) are produced. Only the dextrorotatory alkaloids, cinchonine and quinidine, can give rise to compounds of this structure and it is the facility with which they are formed, contrasted with the inability of cinchonidine and quinine to react in an analogous manner, that provides the accepted evidence for the stereochemical configuration at C(3)–C(8), namely, *cis-* in the *dextro* series, *trans-* in the *levo*. These cyclic ethers contain no point of unsaturation and no replaceable hydrogen, and, just as in the two preceding cases, they occur in two stereoisomeric forms. Thus, cinchonine gives rise to α-isocinchonine (cinchoniline) and β–isocinchonine (cinchonigine). Both dextrorotatory alkaloids, cinchonine and quinidine, have furnished a third isomeride of this type for which the structure (25) has been proposed.

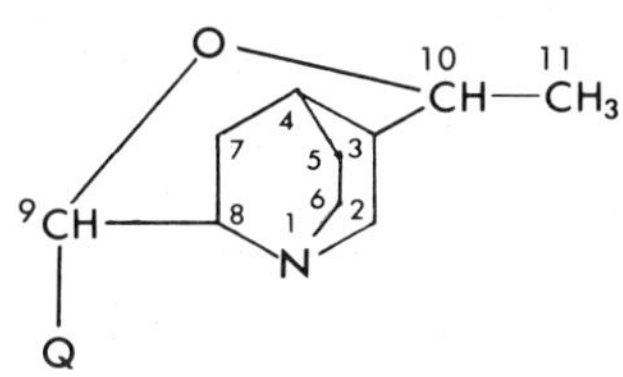

(24)

Cyclic Ethers

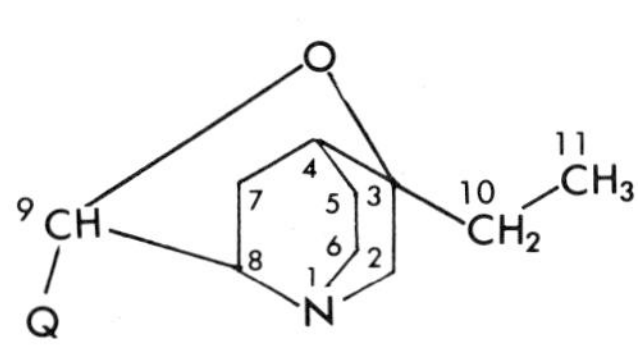

(25)

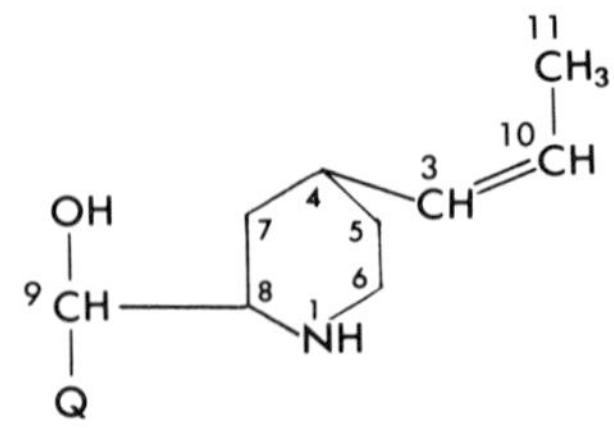

(26)

The Niquine-Type Fragmentation Products

Type 4. In this last case, a Grob[22]-type fragmentation occurs (26); C(2) is eliminated as formaldehyde, N(1) becomes secondary, and C(3) becomes the α-carbon atom of a propenyl sidechain attached to a piperidine ring. This remarkable reaction is best carried out by withdrawing by means of alkali or silver salts the elements of HX from the addition compounds of type 1 [(22); X = Halogen], and its products (26) once again allow of *cis-trans* isomerism. In this way cinchonine gives rise to a mixture of cinchonhydrines, first known as "δ-cinchonine"; not two but three of these isomerides have been isolated, a fact which still awaits explanation. Quinidine fragments to a mixture of niquidine and isoniquidine, but in the case of quinine, which gives niquine, the existence of a second isomeride is in doubt. The fourth unsaturated alkaloid, cinchonidine, has never been reported to give rise to fragmentation products of this kind. Unlike compounds of type 2 (23), which furnish on hydrogenation pairs of epimeric dihydro alkaloids, compounds of type 4 (26) cannot be expected to give more than one dihydro derivative. Thus, niquidine and isoniquidine both give one and the same dihydroniquidine (26a); it was synthesized in 1952.

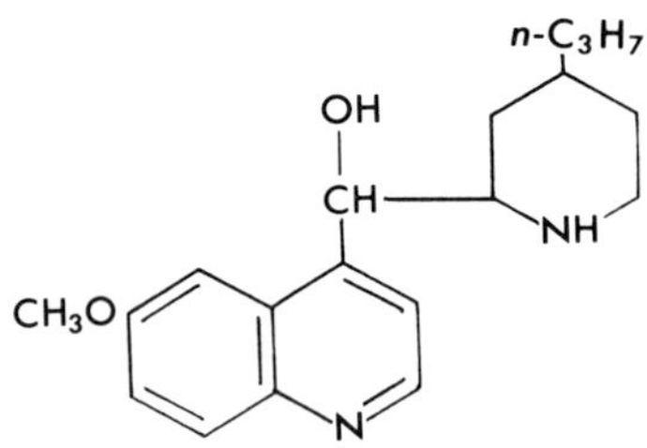

(26a)

Dihydroniquidine

3.2 Other Reactions at the Vinyl Group

Only two cases need be considered here and these constitute exceptions to the rule that any interference with the vinyl group is irreversible.

The first is the action of mercuric acetate; this gives rise to easily isolable mercury derivatives from which the original vinyl alkaloids can be regenerated unchanged by treatment with phosphoric acid. The dihydro alkaloids do not furnish derivatives of this kind so that the reaction affords an elegant, efficient, and now preferred method of separating the saturated cinchonas from the unsaturated; in particular it is superior to the older permanganate oxidation method in which the unsaturated alkaloids had to be sacrificed.

The other case to be mentioned here is the discovery that the vinyl alkaloids can be regenerated unchanged from their dibromide addition compounds by stripping the halogen atoms by means of sodium iodide in acetone or alcohol. As the dibromides are fairly stable to various reagents, this makes it possible to subject the vinyl alkaloids (or their precursors in a synthetical sequence) to processes which the original bases would be unable to withstand. After completing the operation, the protecting bromine atoms are removed and the vinyl group is restored. Long ago Rabe first showed how to convert dihydrocinchotoxine to the alkaloids, dihydrocinchonine and dihydrocinchonidine; similarly, dihydroquinotoxine was converted to a mixture of dihydroquinine and dihydroquinidine. These operations are the final steps in the synthesis of the saturated alkaloids and the dibromide device has made more feasible the corresponding stages in the unsaturated series. The same device also has made it possible to obtain the vinyl-phenolic alkaloid cupreine [(1); R = X = OH; R′ = $-CH=CH_2$] as well as its three nonnaturally occurring stereoisomerides, cupreidine, *epi*cupreine and *epi*cupreidine, which, as we have seen, are otherwise inaccessible. It is doubtful in this connection whether Cuprea bark (*Remijia pedunculata*), which was the sole natural source of cupreine, is now obtainable.

3.3 The Cinchona Ketones

These compounds (4) were first obtained[13] in minute yield by the gentle oxidation of cinchonine and its congeners with chromic acid. While historically the method is a landmark because it furnished the last link in the chain of evidence for the structure of the cinchona alkaloids, the process is in practice valueless, and the ketones can be better prepared from the bromo derivatives of the toxines. The best procedure[23] however is to oxidize the alkaloids by means of a modified Oppenauer reaction using potassium *t*-butoxide and benzophenone or fluorenone. The reaction can be reversed by the addition of isopropanol (Meerwein–Ponndorf) and in this way the cinchona ketones can be reduced back to the carbinols [(1); X = OH], a process needed in the last stage of the synthesis of the cinchona alkaloids. A reduction of this kind can potentially give rise to four stereoisomerides and to obtain any one of these a process of separation must be undertaken. By subjecting a carbinol in turn first to an Oppenauer oxidation and then to a Meerwein–Ponndorf reduction it is possible to stereoisomerise the cinchona alkaloids in fair yield. In this way, for example, quinine has been converted to quinidine. Interconversions of this kind were first achieved in 1896 by means of boiling amyl alcoholic potash but the Oppenauer–Meerwein–Ponndorf technique renders these earlier processes not only obsolete but crude and primitive.

At first sight, it may seem as though the cinchona ketones (4) corresponding formally to the desoxy bases (6) should, in common with these, each be uniquely related to its own parent alkaloid. Thus, cinchonine could be expected to furnish cinchoninone, whereas cinchonidine would give a distinct cinchonidinone; similarly there might be a quininone and a quinidinone, and so on. However, cinchonine and cinchonidine actually give rise to one and the same crystalline ketone, which Rabe called cinchoninone; quinine and quinidine likewise give only one crystalline quininone. This is due to the keto-enol equilibrium in the C(8)–C(9) system:

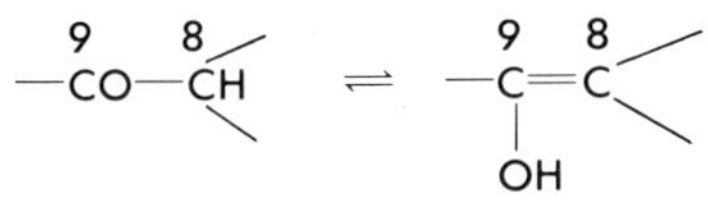

The dissymmetry at C(8) is lost in the enol form so that the keto form regenerated from it reappears in both possible epimeric modifications. Therefore, the same equilibrium mixture is produced from both cinchonine and cinchonidine, for, as is shown in the next section, the two alkaloids are distinguished solely by their stereochemical configuration at C(8) and C(9). Similarly, quinine and quinidine also give rise to one and the same keto-enol equilibrium mixture. However, because of the dissymmetry elsewhere in the molecule [C(3)–C(4)] and owing to other steric factors, equilibrium mixtures of this kind are not necessarily composed of equal parts of the two epimeric modifications. Moreover, when such a mixture crystallizes only one component is deposited to the exclusion of the other, its identity being determined by various factors, one of which can be presumed to be its relative solubility. When a crystalline ketone so obtained is redissolved, it slowly relapses to the same equilibrium mixture from which it first separated. For this reason, the cinchona ketones show mutarotation. While aged solutions, therefore, are mixtures of epimerides, each crystalline substance has a configuration at C(8) corresponding to only one of the two desoxy bases. Crystalline cinchoninone, for example, irrespective of whether it has been obtained from cinchonine or from cinchonidine, must have the configuration *either* of desoxycinchonine *or* of desoxycinchonidine; similarly, crystalline quininone must correspond at C(8) either to desoxyquinine or to desoxyquinidine.

It happens[1] that the two ketones are in fact misnomers; cinchoninone has at C(8) the configuration of the levorotatory desoxycinchonidine and for this reason would be better named "cinchonidinone." Similarly, crystalline quininone is related to dextrorotatory desoxyquinidine and ought to be called "quinidinone."

3.4 *Hetero* Cinchona Alkaloids

All too frequently Rabe chose to consign important results to the relative obscurity of his students' doctorate theses. His discovery of the *hetero* cinchona alkaloids was a case in point. It was first announced in 1932 and 1933 in University of Hamburg dissertations submitted by three of Rabe's pupils. In 1934, a polemic made the first accessible allusion to the new compounds, and in 1941 Rabe isolated *hetero*quinine from natural sources. But the first detailed report[24] did not appear in the ordinary literature until 1949.

The transition from the cinchona (1) to the *hetero* cinchona (27) series is a form of ring enlargement, the 1-azabicyclo[2.2.2]octane (2) (quinuclidine) group being converted to the 1-azabicyclo[3.2.2]nonane [(27); X = Q = R′ = H] system, which Rabe called *homo*quinuclidine. The central, hydroxyl-bearing carbon atom, C(9) is swallowed up in the larger ring, the bond at N(1) is switched from its original terminus at C(8) to its new terminus at C(9), and the hydroxyl group (X) is shifted from C(9) to C(8). Therefore, in the *hetero* alkaloids the two ring-systems, quinoline and *homo*quinuclidine, are directly linked like the two phenyl groups in biphenyl, whereas in the original cinchonas the two ring-systems, quinoline and quinuclidine, are separated by a carbinol function like the two phenyl groups in benzhydrol.

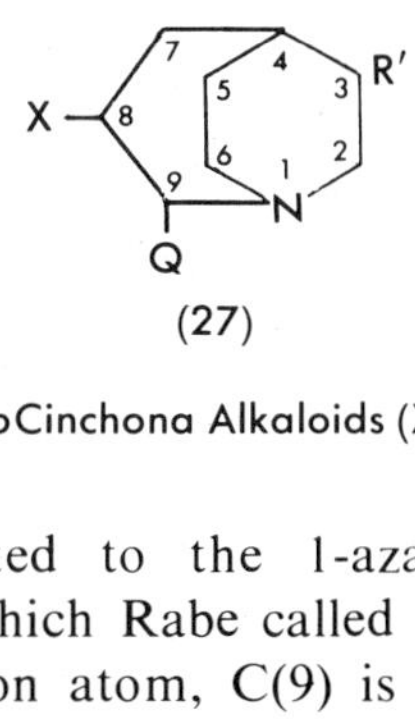

(27)

*hetero*Cinchona Alkaloids (X=OH)

This remarkable rearrangement, the mechanism of which has been discussed by Warnhoff[25], was first observed when dihydrocinchonine was treated with phosphorus pentabromide. Phosphorus pentachloride does not have this effect. The reaction, which under suitable conditions is reversible, is liable to occur when certain other attempts are made to modify or to replace the C(9) hydroxyl group. Thus, ring expansion occurs when the 9-acyloxy derivatives are treated with potassium or silver salts in dry organic solvents.

Although the *hetero* alkaloids resemble the ordinary cinchonas in some ways, there are two important differences. First, the *hetero* compounds are, surprisingly, almost devoid of all basic power; and secondly, they are stated not to give rise to toxines. A toxine in the *hetero* series would have to be formulated as (28) and it is of interest that in 1900 this

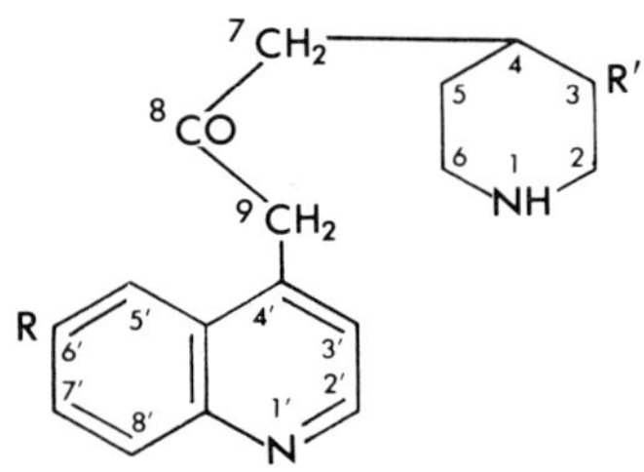

(28)

*hetero*Toxines

(R = H; R′ = —CH═CH$_2$) was one of the structures mooted for cinchotoxine itself and rejected on the ground that a compound so constituted would be expected to react with two molecules of amyl nitrite to give a diisonitroso derivative whereas only mono derivatives of this kind could actually be made. *N*-Benzoyl*hetero*dihydrocinchotoxine was described in 1962 but it was obtained by synthetic means.

All interference at C(9), even when it fails to result in a *hetero* rearrangement, is potentially capable of producing optical inversion at this center. In that event, the *epi* cinchona alkaloids are formed; in addition, there is the ever-present risk of a toxine transformation. It is for these reasons that one is entitled to some degree of scepticism concerning the ancient observations that the C(9) hydroxy group could be acylated and the original alkaloids then regenerated unchanged by hydrolyzing the resulting derivatives.

It is instructive to compare and to correlate the structure of these *hetero* isomerides with those of the other cinchona modifications that result from changes in the central C(8)–C(9)–N(1) system, especially with the toxines and oxiranes (see Section 2.4). The mechanism of the toxine transformation was for many years a constant preoccupation of Rabe's but has not been discussed since Turner and Woodward's treatment[1] of the subject in 1953. The relationship between these several cinchona forms is perhaps best highlighted by means of partial structures (see Figure 11.1). In the parent

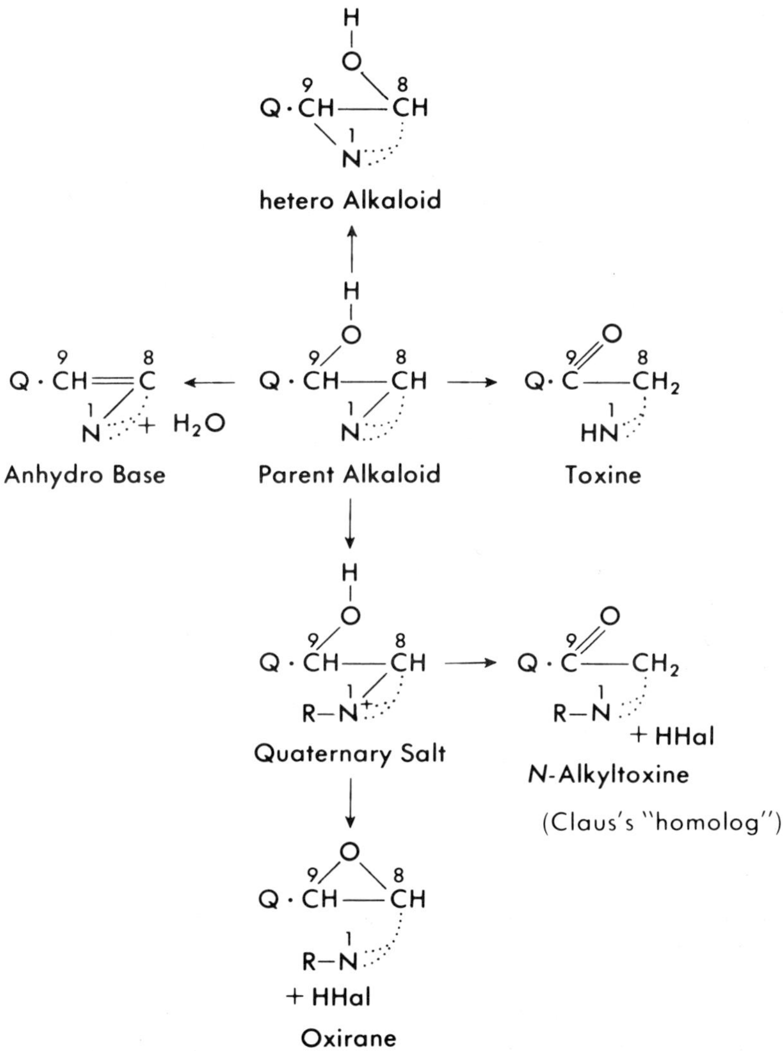

Figure 11.1. Changes at C(8)–C(9). The hetero bases, the toxines, and the oxiranes.

alkaloids the quinuclidine nitrogen atom [N(1)] is linked to C(8) and the central hydroxyl group to C(9). In their *hetero* isomerides these connections are, as we have seen, interchanged, the nitrogen atom now being joined to C(9) and the hydroxyl group to C(8). In the toxines and oxiranes the quinuclidine nitrogen atom is wholly severed from the system and retains its position in the molecule only by remote attachment through C(2)–C(3) and C(6)–C(5) via C(4)–C(7) (dotted lines in Figure 11.1) [for these numberings see the parent alkaloid formula (1)]. On severance, the nitrogen atom in the toxines claims one of the three hydrogen atoms of the system which, on being so deprived, rearranges with the formation of a ketone group. (An important difference between this rearrangement and the *hetero* transformation is that the latter is reversible whereas the toxines, once formed, cannot directly revert to the parent alkaloidal form.) In Claus's "homologs" (see Section 2.4—these tertiary bases are merely *N*-alkyl derivatives of the secondary toxines) and in the oxiranes, one of the three hydrogen atoms of the C(8)–C(9)–N(1) system is claimed not by the nitrogen atom but by the halide moiety of an extramolecular alkylating agent and comes away as H(Halogen); the alkyl group of the reagent is, in the meantime, retained by the newly-severed nitrogen atom; in the *N*-alkyltoxines (Claus's "homologs") the original system so impoverished then rearranges as before, but in the oxiranes a different type of rearrangement occurs in which a three-membered ring is formed, the oxygen atom being shared by the two carbon atoms of the system. It may be noted that while the dissymmetry not only of C(8) but also of C(9) is destroyed in the toxine rearrangement, in the oxiranes, on the contrary, both centers not only retain their unsymmetrical condition but the survival is held to be stereospecific. The last of the cinchona forms, which it may be instructive to mention here, are the anhydro bases; in this case it is the hydroxyl group (or a chlorine atom which has taken its place) rather than the nitrogen atom which undergoes severance. The group is wholly eliminated from the molecule together with the C(8) hydrogen atom to give rise to an ordinary double bond between C(8) and C(9).

3.5 The Quinoline Ring

When the cinchona alkaloids are treated with alkyl halides or attacked by per-acids the more strongly basic nitrogen atom in the quinuclidine ring is the first to be affected; it is quaternized by the first group of reagents and converted to an *N*-oxide by the second. Under more drastic conditions both nitrogen atoms are involved. Compounds (29) and (30), in which only the quinoline nitrogen atom is modified in this way, can be

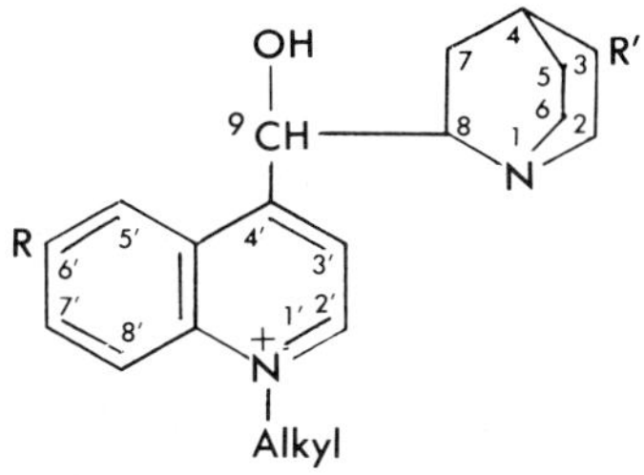

(29)

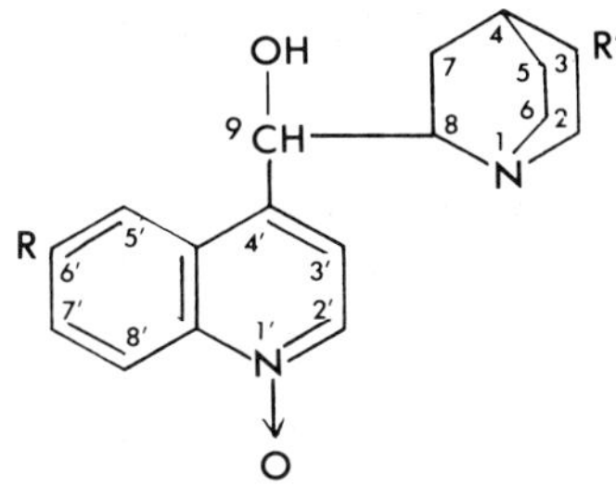

(30)

made accessible by quaternizing the neutral salts of the alkaloids instead of the free bases in the one case and, in the other, by treating the N(1),N(1′)-dioxides with sulphurous acid, a reagent which reduces the quinuclidine *N*-oxide grouping back to its original condition but is incapable of affecting the oxygen atom attached to the quinoline nitrogen.

It is well known that such changes profoundly modify the electronic mobility in pyridine and quinoline and that these substances, normally relatively inert, become susceptible to a variety of chemical reactions. Ochiai and his collaborators by using oxides of type (30) have gained access to a variety of 2′-substituted cinchona alkaloids, which in their turn have enabled the Japanese workers to effect ring-opening of the quinoline nucleus between N(1′) and C(2′). Recyclization between N(1′) and C(9) gave an indole ring, and it was ultimately possible to obtain products that could be identified with naturally-occurring alkaloids of the indole series. This remarkable interconversion[26] between two major groups of alkaloids is of great significance for theories regarding their stereochemistry and biogenesis.

The site of these reactions at N(1′)–C(2′) is in the pyridine part of the quinoline nucleus. The benzenoid moiety can also be attacked and on nitration, for example, the cinchona alkaloids give the 5′-nitro-derivatives from which the 5′-amino-analogs are obtainable. The isomeric 6′-amino-cinchona alkaloids, produced by applying the Bucherer reaction to 6′-hydroxy-compounds such as hydrocupreine and hydrocupreidine [(1); R = X = OH; R′ = Et], are also accessible. This last reaction is of some stereochemical importance; by diazotizing the 6′-amino radical and reducing the resulting diazonium grouping with hypophosphorous acid, it was possible to pass from the methoxylated (R = OMe) to the nonmethoxylated (R = H) alkaloids and so furnish the experimental evidence for the configurational identity, at each point of dissymmetry, of cinchonine with quinidine and of cinchonidine with quinine.

4. STEREOCHEMISTRY

This section is mainly phrased so as to refer only to cinchonine and its stereoisomerides but it must be understood to apply equally to the methoxylated analogs and to the dihydro derivatives, i.e., to all the major cinchona alkaloids.

The molecule of cinchonine [(1); R = H; X = OH; R′ = —CH=CH$_2$] contains four unsymmetrical carbon atoms, C(3), C(4), C(8), and C(9). There are, therefore, $2^4 = 16$ possible stereochemical modifications: cinchonine, cinchonidine, and fourteen other cases. The sixteen isomerides would constitute eight pairs of mirror-image enantiomers, the two members of any one pair giving rise on admixture to a racemic mixture. Cinchonine and cinchonidine are not such a pair; they do not form a racemate and they are substances of distinct properties having different melting points and solubilities; and although they are dextrorotatory and levorotatory, respectively, the two optical powers are different in arithmetical value.

4.1 Configuration at C(3)–C(4)

Cinchonine and cinchonidine give rise to one and the same cinchotoxine [(7); R = H; R′ = —CH=CH$_2$] in which C(8) and C(9) are no longer unsymmetrical. The two stereoisomerides therefore have identical configurations at C(3) and C(4) but differ from one another with respect to their configuration either at C(8), or at C(9), or at both C(8) and C(9). This is further borne out by the cleavage of both alkaloids to one and the same meroquinene [(10); R′ = —CH=CH$_2$], a substance in which C(3) and C(4) are stereochemically intact, and in which C(8) has become a symmetrical carboxyl group while C(9) has disappeared. Furthermore, seeing that the same meroquinine is obtained also from quinine and quinidine, all four vinyl alkaloids have the same configuration at C(3) and C(4). The dihydro alkaloids of course, obtainable by simple hydrogenation, also have the same configuration. *All cinchona alkaloids have the same configuration at C(3) and C(4).* This stereochemical pattern is widely prevalent in nature and reappears among the indole alkaloids; this fact is of great biogenetical importance.

The configuration in question, one of four geometrically possible, was identified[27] in 1944 by degrading cinchonine to two hydrocarbons, (31) and (32). The individual stages of the degradation are set forth in Figure 11.2, in which the numbering corresponds in all cases to that of the cinchona molecule (1). Cinchonine was first hydrogenated to dihydrocinchonine and this was cleaved into "halves" by means of an improved Beckmann rearrangement technique to which allusion was made at 2.5.

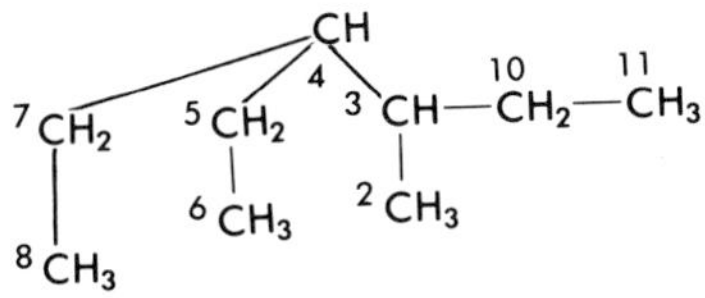

(31)

3-Ethyl-4-methylhexane

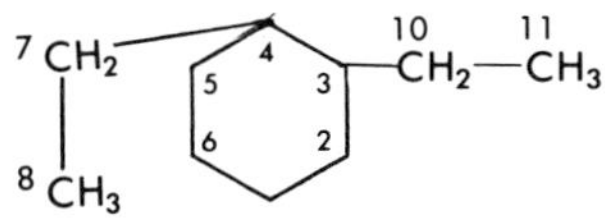

(32)

1,2-Diethylcyclohexane

The cincholoipon (10; $R' = C_2H_5$) resulting from this operation, in the form of its ethyl ester, gave, after a 5-stage process, a branched dibromononane which, on the one hand furnished on reduction the hydrocarbon (31) and on the other, when subjected to a further 6-stage sequence of operations, afforded the hydrocarbon (32). All the reactions were stereochemically unexceptionable and the two hydrocarbons must be configurationally identical with the corresponding parts of the alkaloid molecule.

The 3-ethyl-4-methylhexane (31) so obtained was levorotatory, and its configuration was correlated with that of the known (−)-2-methylbutyric acid, which was at that time (1944) arbitrarily represented by the Fischer projection formula (33). The correlation of hydrocarbon and

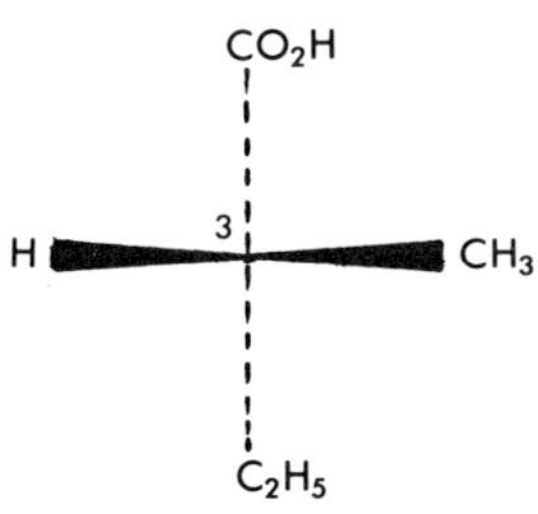

(33)

(−)-2-Methylbutyric Acid

acid was established by means of the sequence also shown in Figure 11.2. Actually, the optical rotation of (31) so obtained was lower than that of the endproduct of the degradation of cinchonine. This was due to the fact that in one of the two isomeric unsaturated predecessors of (31) the

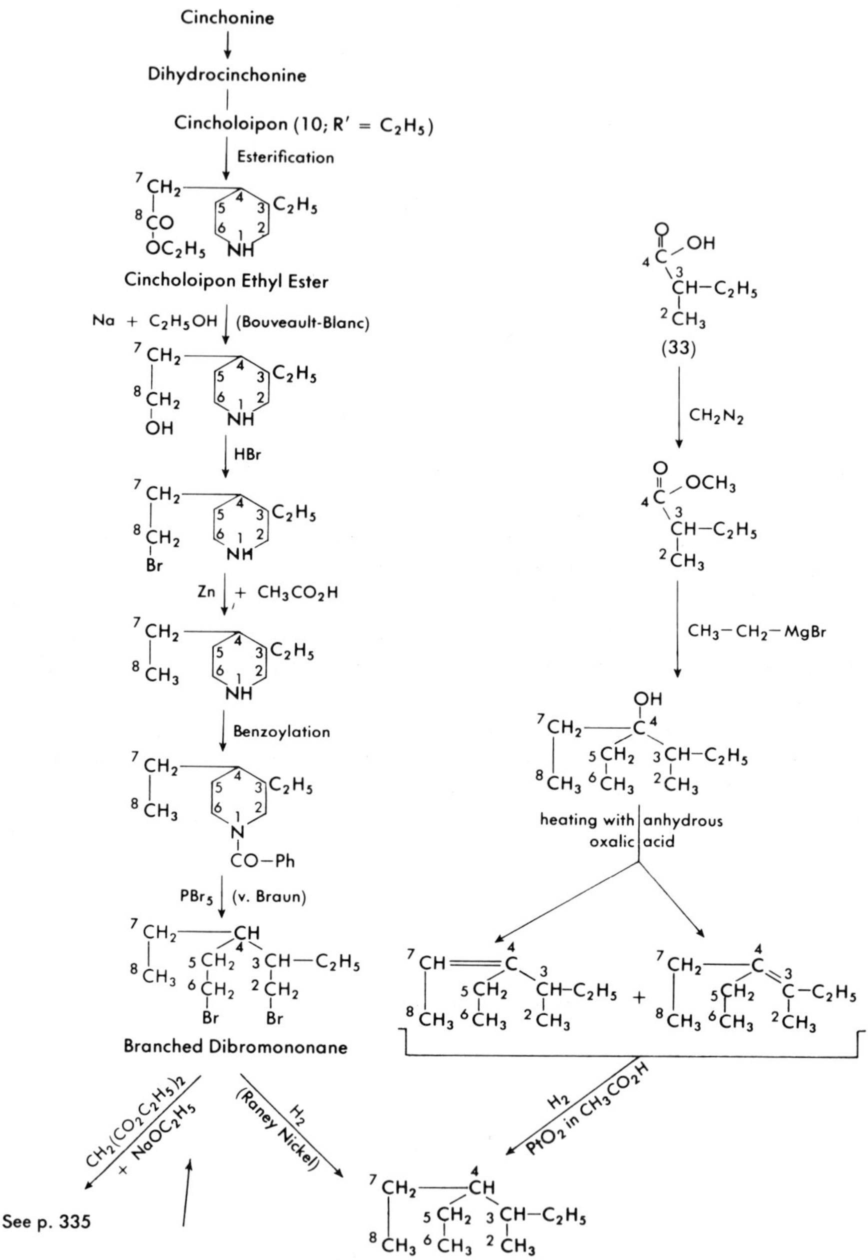

(31)

Figure 11.2. Configuration of cinchonine at C(3)–C(4). The Prelog–Zalán[27] degradation.

(continued from p. 334)

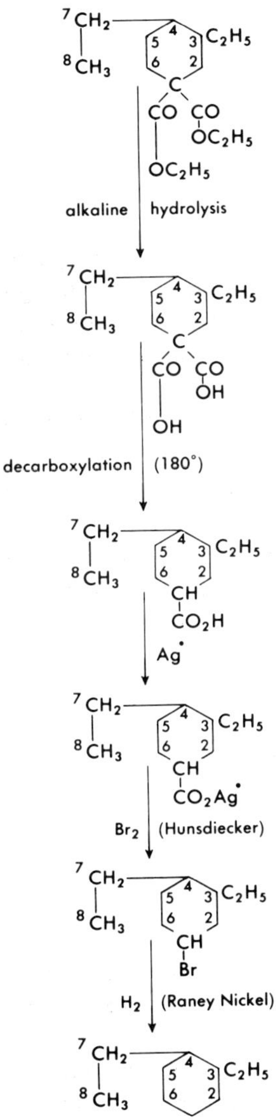

(32)

Figure 11.2. Continued.

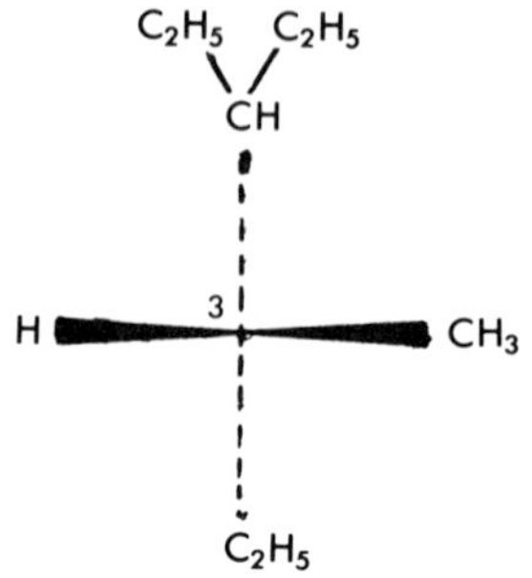

(34)

(−)-3-Ethyl-4-methylhexane

double linkage involved the (sole) asymmetric center [that at C(3)] and this served to "dilute" the effect of the other isomeride. It was as though the final product was partically racemized (not *inverted*) by the presence of the unwanted isomeride, and the argument, therefore, for the identity of configuration of hydrocarbon (31) as derived from the two sources was unimpaired. This compound was to be represented by the Fischer projection formula (34). In its molecule the sole center of dissymmetry is the cinchona C(3), and cinchonine has at that atom the configuration of (−)-2-methylbutyric acid.

The other hydrocarbon, 1,2-diethylcyclohexane (32), was optically inactive; it was, therefore, represented as the internally compensated *meso-* (or *cis-*) form (35). In this molecule, not only the cinchona C(3) but

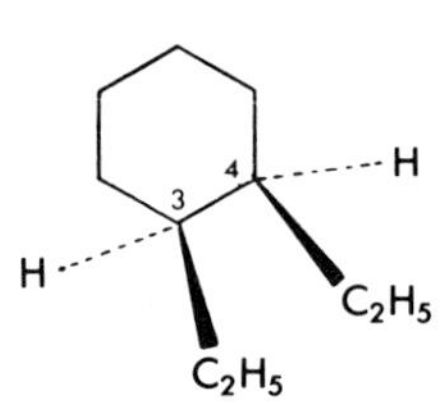

(35)

meso-1,2-Diethylhexane

also C(4) retain their asymmetry and differ from their original condition in the alkaloid in that they have become similar, like the two asymmetric centers in tartaric acid.

The stereochemical identification of these two hydrocarbons made it possible to deduce that cinchonine, and hence all the cinchona alkaloids, were to be represented in regard to their configuration at C(3) and C(4) by the space diagram (36) in which the configurations at C(8) and C(9) are uncommitted. These results were obtained in 1944. At that time the

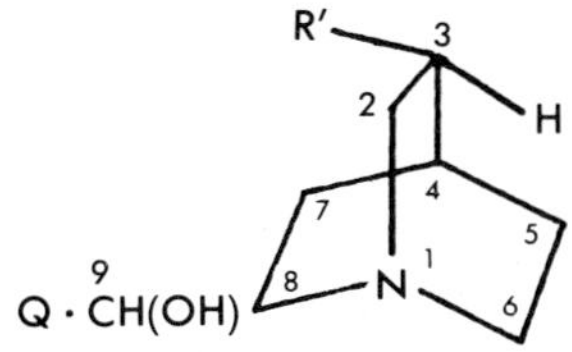

(36)

representation of (−)-2-methylbutyric acid by the Fischer projection formula (33) was relative and a matter, ultimately, of arbitrary convention. It has since turned out, by great good fortune, that the absolute configuration of this acid, as of other unsymmetrical molecules, is in accord with the original convention. Had this not been the case (and it was recognized by Prelog[27] himself as well as by Woodward[1] that there was an equal chance that the two configurations would prove to be contrary) the cinchona alkaloids would have had to be represented not by (36) but by its mirror image (36a). The complete space diagrams (40)–

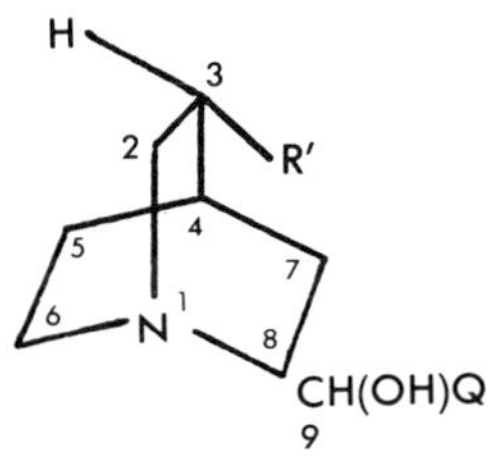

(36a)

(43) (v. inf.) would have had to be similarly inverted but would not have been affected in any other way.

4.2 Configuration at C(8)

We have seen that although cinchonine and cinchonidine furnish one and the same cinchotoxine, the alkaloids give rise to two different desoxy bases [(6); R = H; R′ =—CH=CH_2]; from cinchonine we get its own dextrorotatory desoxycinchonine, distinct in properties and identity from the levorotatory desoxycinchonidine obtainable from cinchonidine. In these compounds C(9) has become symmetrical and the difference between them [they are of course identical at C(3)–C(4)] must be ascribed to opposite configurations at C(8). The same difference must obtain in the parent alkaloids themselves: *cinchonine and cinchonidine have, therefore, opposite configurations at C(8).* The configuration of the dextrorotatory alkaloid at this point is deduced[28] from the fact that under suitable conditions it readily gives rise to the cyclic ethers cinchoniline and cinchonigine (24). Levorotatory cinchonidine cannot form ethers of this type. The facile production of such compounds, which owe their separate existence to the new center at C(10), is made possible by the close proximity of C(9) and C(10), and a brief examination of even a crude model will show that this in no way depends on the configuration at these latter centers but is solely determined by the relative configurations at C(3) and C(8). In dextrorotatory cinchonine, therefore, C(8) has with respect to C(3) the *cis* configuration shown in (37); in levorotatory cinchonidine the two centers are *trans* (38).

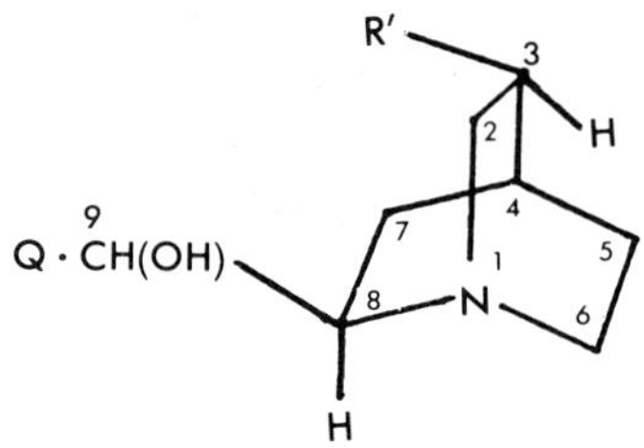

(37)

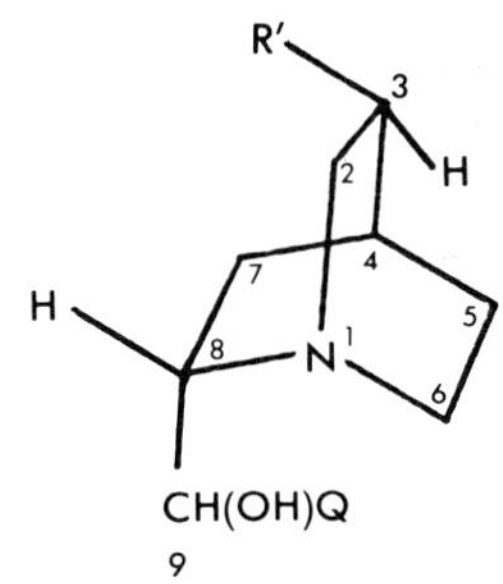

(38)

4.3 The *epi* Bases[29]

There are four possible arrangements at C(3)–C(4) and for each of these there are four possible arrangements at C(8)–C(9), making, as we know, sixteen cases in all. The four alkaloids which have the particular C(3)–C(4) configuration (36) that occurs in nature are cinchonine, cinchonidine, and two others; these are *epi*-C(9)-cinchonine and *epi*-C(9)-cinchonidine, commonly known as *epi*cinchonine and *epi*cinchonidine; they differ from their respective parent alkaloids *only* in regard to their configuration at C(9). Cinchonine and *epi*cinchonine for example have contrary configurations at C(9) but are identical at C(8).

All the major *epi* cinchona alkaloids of this type are known. Their existence was foreseen at an early stage and some of them were prepared in 1920. By 1932 all the major *epi* bases had been obtained and examined, and in 1935 two, *epi*quinine and *epi*quinidine, were isolated from cinchona bark. These substances can be formed by reducing the cinchona ketones (4) when all four stereoisomerides are liable to be produced; thus dihydroquininone (i.e. dihydroquinidinone) can give on reduction a mixture of dihydroquinine, dihydroquinidine, *epi*dihydroquinine and *epi*-dihydroquinidine, and ways have then to be devised for isolating the desired compounds. Alternatively the *epi* bases can be obtained by processes of optical inversion at C(9); among these are such methods as boiling with hydrochloric acid, treating the *O*-toluenesulphonates with aqueous tartaric acid, or treating the alkaloids with phosphorus halides. None of these procedures however is straightforward because, as we have seen, there is always the risk of bringing about a *hetero* or a toxine rearrangment instead of, or in addition to, the desired inversion.

4.4 Configuration at C(9)

Logically this problem falls into two separate parts but in practice these have not always been sufficiently clearly distinguished.

(a) Without any commitment as to its actual physical or geometrical significance, let us arbitrarily assign the symbol + to the configuration at C(8) in cinchonine. Then, the facts established so far [viz., cinchonine and cinchonidine are opposed at C(8); and cinchonine–*epi*cinchonine on the one hand, and cinchonidine–*epi*cinchonidine on the other, are in each case opposed at C(9) but identical at C(8)] lead to four, and only four, possibilities for the configuration at C(8) and C(9) in the four stereoisomerides. These are shown (a, b, c, and d) in Figure 11.3. It will be seen that in (a) and (c) cinchonine and cinchonidine are C(8)–C(9) antipodes as are *epi*cinchonine and *epi*cinchonidine. In (b) and (d) on the contrary, the C(8)–C(9) antipodal pairs are cinchonine–*epi*cinchonidine and cinchonidine–*epi*cinchonine. The crux of the first of our two problems is that at this stage of the argument it is impossible to decide between

(a)	Cinchonine	Epicinchonine	Cinchonidine	Epicinchonidine
Configuration at C(8)	+	+	−	−
Configuration at C(9)	+	−	−	+

(b)	Cinchonine	Epicinchonine	Cinchonidine	Epicinchonidine
Configuration at C(8)	+	+	−	−
Configuration at C(9)	+	−	+	−

(c)	Cinchonine	Epicinchonine	Cinchonidine	Epicinchonidine
Configuration at C(8)	+	+	−	−
Configuration at C(9)	−	+	+	−

(d)	Cinchonine	Epicinchonine	Cinchonidine	Epicinchonidine
Configuration at C(8)	+	+	−	−
Configuration at C(9)	−	+	−	+

Figure 11.3. The four possible configurations at C(8)–C(9).

these two alternatives; the C(8)–C(9) antipode of cinchonine *could* be cinchonidine [as in (a) or (c)], but it could equally well be *epi*cinchonidine [as in (b) or (d)]. It may be stated here at once that on this point there has never been any disagreement and all authorities, on whatsoever evidence, have opted for the former alternative (a–c) which makes cinchonine and cinchonidine C(8)–C(9) antipodes.

(b) The second part of the problem however is different in that it has long been the subject of controversy; however, this has probably been finally settled in a paper published in 1967.

The cinchona alkaloids are, at C(8)–C(9), 1,2-alkanolamines closely analogous to the four ephedrines (39). [The design and numbering in this

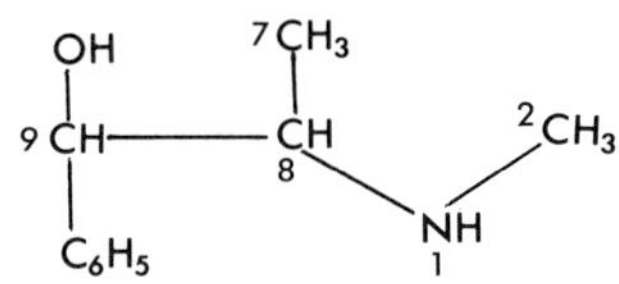

(39)

The Ephedrines and the Pseudoephedrines

formulation follow the cinchona (1) pattern.] One of the pairs of C(8)–C(9) antipodes in the cinchona series must therefore have the *erythro* configuration of (+)- and (−)-ephedrine, while the other and diastereoisomeric pair will have the *threo* configuration of the two enantiomeric *pseudo*ephedrines. The point is, of course, to decide which is which.

In summary, the twin problems of the orientation of groups at C(9) are first, to discover which of two possibilities, (a)–(c) or (b)–(d), is the correct way to assort the four stereoisomerides, cinchonine, *epi*cinchonine, *epi*cinchonidine, and cinchonidine, into two pairs of C(8)–C(9) antipodes, and then secondly, to establish to which pair to assign the *erythro* configuration of the two enantiomeric ephedrines and to which the *threo* disposition of the *pseudo*ephedrines.

It is now held that the C(8)–C(9) antipodes are cinchonine–cinchonidine and *epi*cinchonine–*epi*cinchonidine, respectively [as in (a) or (c)], and that while the former pair have the *erythro* configuration of the ephedrines, the *epi* bases are at C(8)–C(9) stereoanalogs of the *pseudo*ephedrines and have the *threo* structure of the latter.

We may refer to this as the Prelog configuration which, when translated into the terms of the *R–S* nomenclature,[30] is set forth in Table 11.3; the order of priority of groups according to the Sequence Rule of that system is given in Table 11.4

The main evidence[31] for the Prelog configuration was, until recently, drawn from a comparison of the relative basic strengths of the four stereoisomerides; it was corroborated, *inter alia*, by the facility with

TABLE 11.3. Rectus-Sinister[30] Configuration at C(8)–C(9).

Stereoisomeride	Configuration	
	at C(8)	at C(9)
Cinchonine	*R*	*S*
*epi*Cinchonine	*R*	*R*
Cinchonidine	*S*	*R*
*epi*Cinchonidine	*S*	*S*

TABLE 11.4. Order of Priority of Groups at C(8) and C(9) According to the Sequence Rule.[30]

Asymmetric Center	Sequence
C(8)	N(1) → C(9) → C(7) → H
C(9)	OH → C(8) → Q → H

which the *epi* bases and not the major alkaloids could form copper chelate derivatives. These results were long regarded in some quarters as inconclusive, and contrary verdicts, namely *threo* for cinchonine and cinchonidine, and *erythro* for the *epi* bases, were reached e.g. by means of O.R.D. studies. However, the Prelog configuration seems now to have been confirmed[32] by NMR measurements on the oxirane [(8); R = OMe; R′ = Et; R″ = CH_2Ph] derived from dihydroquinidine in which the fatal possibility of free rotation at C(9) is precluded by the stereospecifically created three-membered ring.

4.5 Final Elucidation

We are now in a position to represent the three-dimensional structure of the cinchona alkaloids by means of the space diagrams (40)–(43).

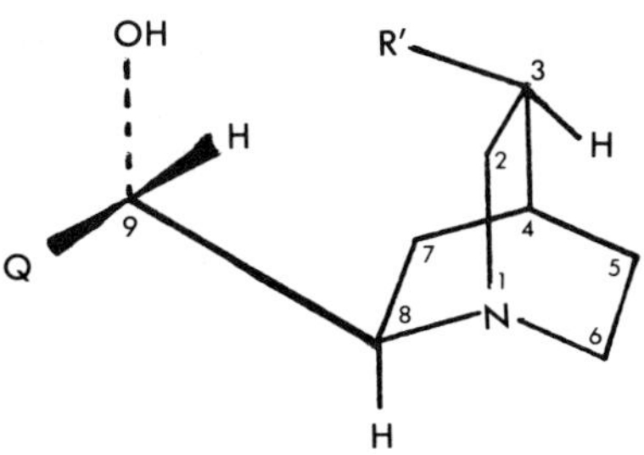

(40)

Cinchonine
Quinidine

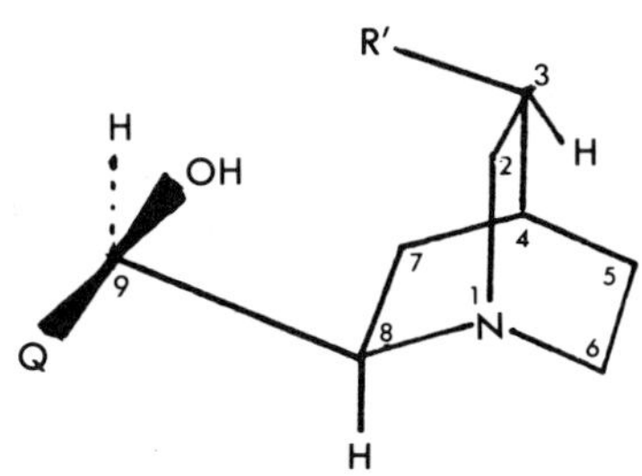

(41)

*epi*Cinchonine
*epi*Quinidine

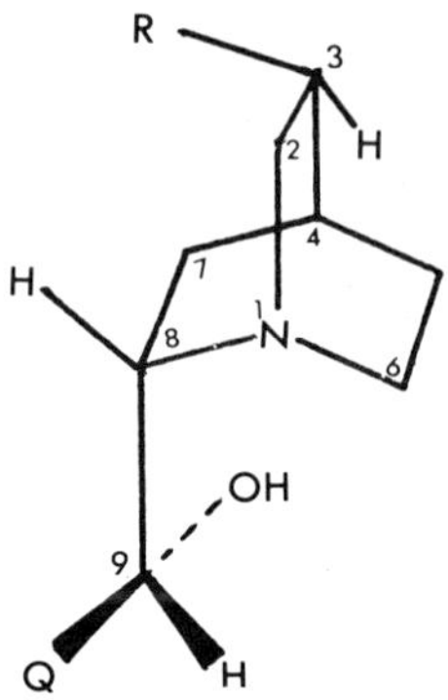

(42)

Cinchonidine
Quinine

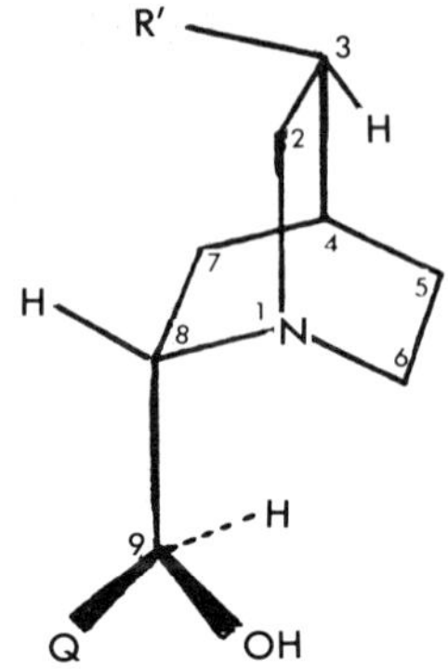

(43)

*epi*Cinchonidine
*epi*Quinine

4.6 Prelog and Rabe Configurations

It is of interest to compare this over-all picture with the results which were obtained long ago largely by applying the principle of optical superposition. Inferences of geometrical configuration from optical rotatory values are notoriously unsatisfactory; nevertheless, the attempt which was made to elucidate the stereochemistry of the cinchona alkaloids by such means merits some attention. The argument may be presented as follows.

Arithmetically, the major alkaloids have high rotatory powers; quinine and cinchonidine are strongly levorotatory (−158° and −111°, respectively), and quinidine and cinchonine are even more highly dextrorotatory +254° and +224°). Cinchotoxine and quinotoxine on the other hand, in which C(8) and C(9) are extinguished and only C(3) and C(4) are operative, are by contrast feebly active; both are dextrorotatory and their rotations fall fairly close together at around +40°. In the cinchona alkaloids, therefore, the "contribution" of C(3) and C(4) taken together is weakly dextrorotatory in total effect. It follows by algebraic subtraction that the contribution of C(8)–C(9) taken together is about −200° and −150° in the one case, and approximately +210° and +180°, respectively, in the other, i.e., of the order of magnitude of nearly ×200°. As to the individual contributions at these two centers, something can be deduced from the rotations of the desoxy bases in which C(9) is neutralized; of even greater significance is the rotatory power of the *epi* bases in which C(9) is not merely annulled but inverted and which are between 100° and 200° less active arithmetically than their parent alkaloids. It can be argued that the separate contributions of C(8) and C(9) in the highly active major alkaloids are both in the same sense, *dextro* in cinchonine and *levo* in cinchonidine, and that in the relatively feebly active *epi* bases, C(8) and C(9) are in mutual conflict and "rotate" in opposite directions. The position in the case of the four cinchonine stereoisomerides, of the two desoxy bases and of cinchotoxine, is set forth in Table 11.5. These results[29] were first announced in 1922,[33] and the C(8)–C(9) disposition here may be referred to as the Rabe configuration.

TABLE 11.5. Configuration at C(8) and C(9) from Optical Rotatory Values

Alkaloid	$[\alpha]_D$ in alcohol (approx.)	Configuration	
		at C(8)	at C(9)
Cinchonine	+ 230°	+	+
Desoxycinchonine	+ 180°	+	0
*epi*Cinchonine	+ 120°	+	–
Cinchotoxine	+ 50°	0	0
Cinchonidine	– 110°	–	–
Desoxycinchonidine	– 30°	–	0
*epi*Cinchonidine	+ 63°	–	+

These Prelog and Rabe systems are in accord in the answer they give to the first of our twin problems, which was concerned with the assortment of the four stereoisomerides. In both cases it is cinchonidine (and not *epi*cinchonidine) that is made the C(8)–C(9) antipode of cinchonine. As to the other problem however, which deals with the actual configuration of each pair, the two pictures are in conflict. This has never been acknowledged. The discrepancy is sufficiently manifest in Table 11.5 but is perhaps best brought out by applying the optical argument used in the cinchona case to the analogous circumstances obtaining in the four ephedrines. The two pseudoephedrines are much more highly active (about ±52°) than the two ephedrines (about ±6°, in alcohol). It would seem to follow (recapitulating the cinchona argument) that the separate contributions of the two centers of dissymmetry in the two highly active pseudoephedrines are both in the same sense (*dextro* or *levo*, as the case may be), while in the feebly active ephedrines the two centers are in mutual conflict and "rotate" in opposite directions. This would appear to be in conformity with the known configurations of the two pairs of bases, *erythro* for the ephedrines, *threo* for the pseudoephedrines. Clearly however the Rabe configuration at C(8)–C(9) would make the highly active cinchonine and cinchonidine the stereoanalogs not of the feebly rotating ephedrines but of the powerfully rotating pseudoephedrines and would give them the *threo* configuration of the latter. Similarly in the Rabe picture it is the *epi* cinchona alkaloids that have the *erythro* configuration of the ephedrines.

4.7 Conformation

Not a great deal of work has been done on the conformation of the cinchona alkaloids but it may be worthy of mention in this connection that carefully, though conventionally, purified quinine and quinidine have been separated chromatographically into almost equal fractions of distinct melting point and optical rotation; using the same technique cinchonine and cinchonidine appeared to be homogeneous.[34]

5. SYNTHESIS

5.1 The Perkin Experiment

The first attempt to synthesize quinine, $C_{20}H_{24}N_2O_2$, was made in 1856 by W. H. Perkin who forty years later gave an account of his experiments in a Hofmann Memorial Lecture. Perkin, working during a vacation in his "rough laboratory at home," treated allyltoluidine, $C_{10}H_{13}N$, with potassium dichromate in the hope of producing the reaction $2\ C_{10}H_{13}N + 3[O] = C_{20}H_{24}N_2O_2 + H_2O$. Having failed to "form" quinine in this way, he next tried the effect of replacing the allytoluidine by aniline. This experiment also failed but it gave the world a dyestuff.

5.2 Partial Syntheses

Half a century had to elapse before more rational attempts to build up the cinchona molecule could be envisaged and nearly as long again before this objective could be realized.

The dihydro alkaloids were synthesized by Rabe[35] in 1931, and their vinyl analogs by Woodward and Doering[36] in 1944; in both cases a four-stage process (Fig. 11.4) was used. Fairly good methods had been worked out mainly by Rabe and his collaborators between 1908 and 1919 for most of the individual steps of these sequences. Allusion has been made in earlier parts of this chapter, in discussing the toxines and the ketones and elsewhere, to the methods used in converting the toxines to the alkaloids (stages 3 and 4). By employing intermediates that had, themselves, been obtained from the natural bases, partial syntheses were achieved in the saturated series (R′ = Et). The vinyl-bearing alkaloids (R′ —CH=CH₂) were not so easily obtained; stages 3 and 4 were in this case much more difficult to carry out, while stage 1 was not merely difficult but impossible: the starting material [(44); R′ = —CH=CH₂] was not available.

The two compounds [(44); R′ = Et and —CH=CH₂] were the *N*-

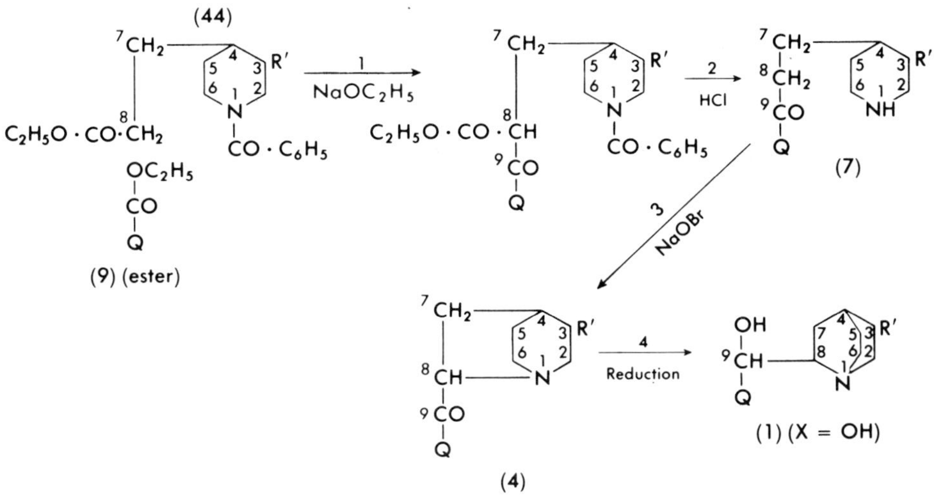

Figure 11.4. Synthesis of the cinchona alkaloids.

benzoyl derivatives of the esters of homocincholoipon [(45); R′ = Et] and homomeroquinene [(45); R′ = —CH═CH_2] respectively. Homocincholoipon was first obtained by Kaufmann[37] in 1916 by the permanganate oxidation of the N.1′-methosulphate of N.1-benzoyldihydrocinchotoxine. Homomeroquinene remained inaccessible until 1943 when Prelog[38] prepared it by a Beckmann rearrangement of N(1)-benzoylcinchotoxine oxime. But these preparations of the two piperidinepropionic acids were not syntheses.

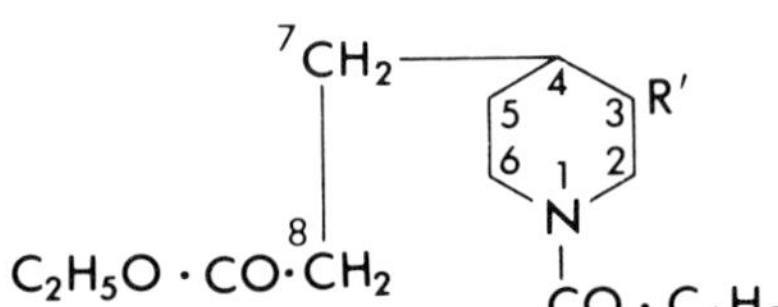

(44)

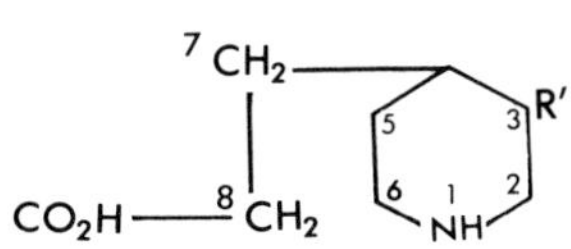

(45)

Homocincholoipon (R′ = $—\overset{10}{C}H_2—\overset{11}{C}H_3$)

Homomeroquinene (R′ = $—\overset{10}{C}H═\overset{11}{C}H_2$)

Of the two "first-half" starting materials (9), quinoline-4-carboxylic (cinchoninic) acid (R = H) and 6-methoxyquinoline-4-carboxylic (quininic) acid (R = OMe), the former had been synthetically accessible for many years; the latter was first synthesized in 1912. Both acids belong to the general chemical literature, and for further information the reader may be referred to standard texts.

The key linkup of the two cinchona halves (stage 1) was perfected in the saturated case by Rabe and Kindler[39] in 1918 and 1919. The two components were condensed in benzene solution by means of sodium ethoxide; it subsequently proved preferable to effect this condensation in the absence of a solvent. The ester and benzoyl groups were then removed and the resulting acid was decarboxylated simultaneously (stage 2) by boiling with 17% hydrochloric acid. This Rabe–Kindler process was later used in the synthesis of the vinyl alkaloids by Woodward and Doering and has been applied in one form or another in the preparation of a great number of cinchona analogs.

From 1919 onwards, (stages 3 and 4 having become feasible ten years earlier) the possibility of the total synthesis of the cinchona alkaloids turned on the problem of synthesizing homocincholoipon [(45); R′ = Et] and homomeroquinene [(45); R′ = —CH=CH_2]. The former task was accomplished by Rabe[35] in 1931. After a lifetime devoted to the study of the cinchona alkaloids, the synthesis of quinine itself eluded Rabe to the end and it was left to Woodward and Doering[36] in 1944 to describe the synthesis of homomeroquinene and with it the total synthesis of the principal alkaloid of cinchona bark.

Homocincholoipon and homomeroquinene contain, as their names suggest, one carbon atom more than the respective normal products of oxidation or cleavage of the natural alkaloids; they also contain two unsymmetrical carbon atoms, the cinchona C(3) and C(4). Their synthesis involved, therefore, the separation of mixtures of diastereoisomerides, the optical resolution of racemates at a suitable stage, and the judicious selection of the appropriate form in each case.

5.3 Homocincholoipon

Rabe's synthesis of this acid was based on a procedure of E. Koenigs and Ottmann[40] who in 1921 obtained the inactive compound from β-collidine (3-ethyl-4-methylpyridine) (46) which had itself been synthesized two years earlier. This pyridine homolog was condensed with chloral and the 3-ethyl-pyridylacrylic acid (47), obtainable from the product of this reaction by hydrolysis, was hydrogenated to the corresponding piperidylpropionic acid. Of the two possible racemic diastereoisomerides, the

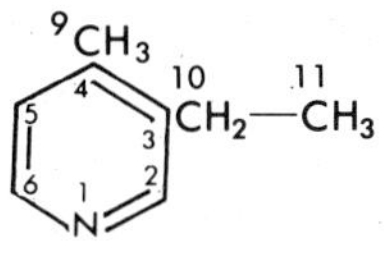

(46)

β-Collidine

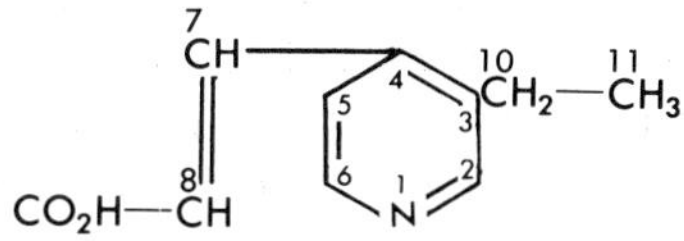

(47)

desired *cis*-form predominated under certain conditions, and the resulting *dl*-homocincholoipon, in the form of its ethyl ester, was resolved by means of the two tartaric acids, both enantiomers being isolated. The *dextro* form was identical with the corresponding compound obtainable from natural dihydrocinchotoxine. Its mirror-image isomeride enabled Rabe to obtain the interesting enantiomeric and racemic modifications of dihydroquinine and dihydroquinidine.

5.4 Homomeroquinene

The Woodward–Doering synthesis of *dl*-homomeroquinene was a twelve-stage process; here, it can be traced only in broad outline. 7-Hydroxyisoquinoline (48) was obtained from *m*-hydroxybenzylidineamino-acetal (49); a methyl group was introduced in the 8-position by a Mannich con-

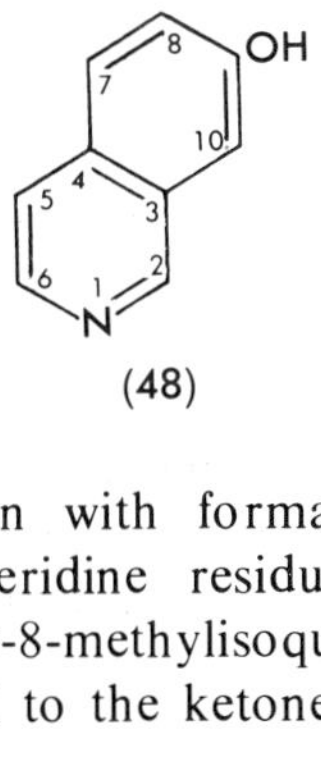

(48)

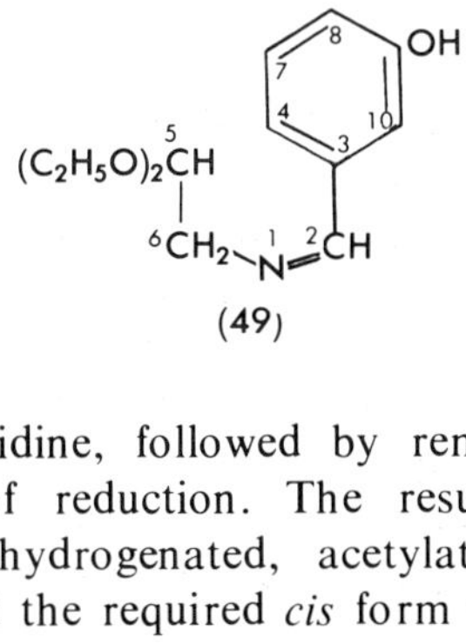

(49)

densation with formaldehyde and piperidine, followed by removal of the piperidine residue by a process of reduction. The resulting 7-hydroxy-8-methylisoquinoline was deca-hydrogenated, acetylated, and oxidized to the ketone (50). At this point the required *cis* form was sep-

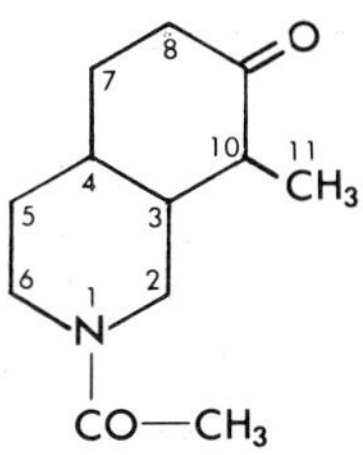

(50)

arated from the *trans* modification by taking advantage of the crystallizability of the *cis*-hydrate. The ketodecahydroisoquinoline ring was opened by ethyl nitrite and sodium ethoxide and the resulting oxime-ester (51) reduced to an unstable amine (52), which on exhaustive methy-

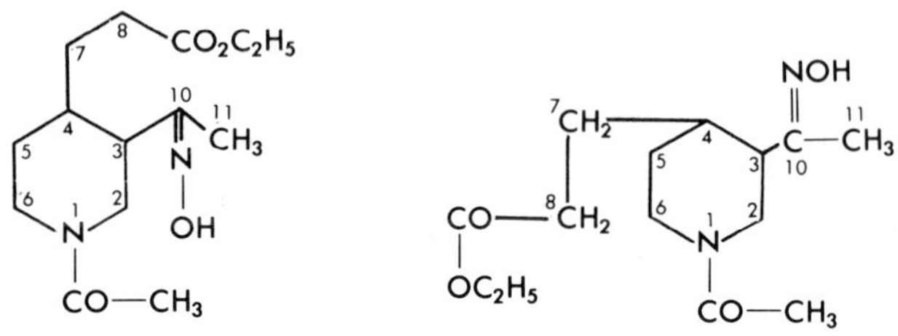

(51)

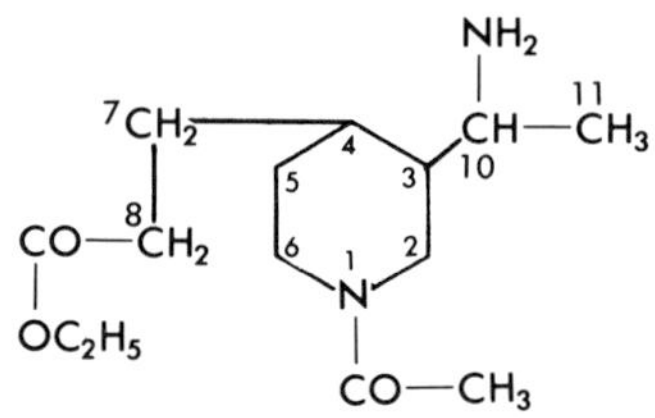

(52)

lation gave *dl-cis*-homomeroquinene [(45); R′ = —CH=CH_2], the acetamido and carbethoxy groups having undergone hydrolysis during the process. The *dl*-quinotoxine obtained from this by a Rabe-Kindler condensation with quininic ester gave the *dextro* form, identical with natural quinotoxine, by thrity-two crystallizations of its dibenzoyl-*d*-tartrate. As in the case of Rabe's dihydroquinine synthesis, the enantiomeric(−)-quinotoxine was also obtained, but the enantiomeric and racemic quinine and quinidine, which might have been prepared from this "unnatural" quinotoxine, have not been described.

6. QUINAMINE, CONQUINAMINE, CINCHONAMINE

These three minor alkaloids differ from the principal members of the cinchona group in being derived not from quinoline but from indole and indoline (2,3-dihydroindole). There is a further departure from the typical cinchona pattern in that we have here two heterocyclic systems joined directly and not by way of an intermediary carbinol function. On the other hand the new triad shares with quinine and its congeners not only their botanical origin but also the important structural feature that the "second half" consists of the same substituted quinuclidine ring system which is found in, e.g., cinchonine.

Quinamine and conquinamine are stereoisomerides [epimeric at C(8)] and have the structure (53). The "first half" is a tricyclic system

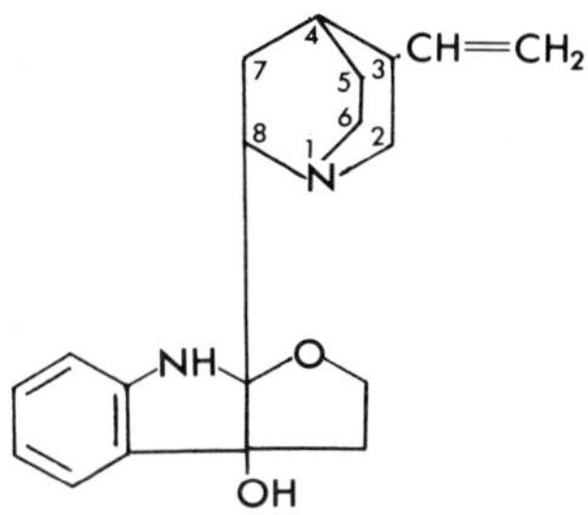

(53)

Quinamine and Conquinamine

in which indoline and oxolane rings are fused together. In the closely related cinchonamine (54), the oxolane ring is opened so that the "first half" is bicyclic; cinchonamine is a true indole derivative.

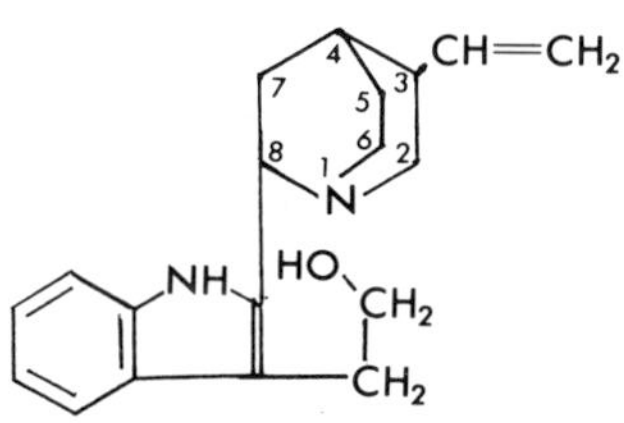

(54)

Cinchonamine

Quinamine and conquinamine which occur in small amounts in a number of *Cinchona* species, were first isolated in 1872 and 1877, respectively; cinchonamine was discovered in *Remijia purdieana* in 1881. The three alkaloids were fairly well characterized and their empirical composition was established by their discoverers and by several contemporaries. Thereafter, little interest was taken in these substances for over sixty years. Quinamine was reinvestigated in 1945 and several authors proposed formulas, none of which proved tenable; Witkop[41] deduced the correct structure (53) in 1950. The constitution of cinchonamine

(54) was worked out in the same year by Goutarel, Janot, Prelog, and Taylor.[42]

The natural alkaloid conquinamine is identical with the artefact *epi*-quinamine which can be obtained from quinamine by mild treatment with alcoholic potash. Quinamine can, in part, be regenerated from *epi*-quinamine by the same procedure because the action of the alcoholic alkali, which affects the configuration at C(8), leads to an equilibrium. *Epi*quinamine (conquinamine) is remarkable for its high rotatory power. The analogous epimeride of cinchonamine has been prepared but is not known to occur in nature.

Quinamine and cinchonamine are dextrorotatory, monacidic bases. The indole nitrogen atom in cinchonamine is almost devoid of basic power; the indoline nitrogen atom in quinamine is only slightly more basic. Quinamine is unstable and the action of acids under different conditions converts it to a variety of products.

There is evidence that, biogenetically, cinchonamine is a precursor of the major cinchona alkaloids. The reverse process, the conversion of cinchonine (and chinchonidine) to dihydrocinchonamine, has been achieved in the laboratory: this is one of the triumphs of recent alkaloid chemistry and is due to the skill, ingenuity, and patience of Ochiai, Ishikawa, and their collaborators in Japan.[26]

On oxidation with chromic acid both quinamine and cinchonamine give 5-vinylquinuclidine-2-carboxylic acid (55); these are the only two

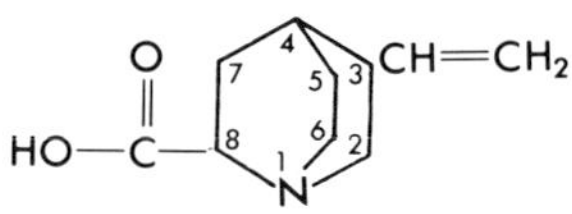

(55)

5-Vinylquinuclidine-2-carboxylic Acid

cases in which the quinuclidine skeleton survives intact when a cinchona alkaloid is subjected to oxidative cleavage. The acid exhibits the phenomenon of mutarotation—the carbon atom responsible being the same cinchona C(8) (in this case the carboxyl-bearing carbon atom) that gives rise to the epimerism of quinamine and conquinamine, and is involved in the alanogous mutarotation of the cinchona ketones.

As their empirical formulas show ($C_{19}H_{24}N_2O_2$ and $C_{19}H_{24}N_2O$, respectively), quinamine is in a higher state of oxidation than cinchonamine from which the former alkaloid can be obtained by the action of peracetic acid. The process can be reversed, and quinamine can be reduced back to cinchonamine by means of lithium aluminum hydride. This interconvertibility means that a Soviet synthesis[43] of cinchonamine described in 1958 constitutes also a synthesis of quinamine. However, the starting point of this synthesis was vinylquinuclidine carboxylic acid (55) which had not as yet become synthetically accessible. The situation was retrieved two years later when the same Russian team showed that the acid could be prepared by a Beckmann rearrangement of quinidinoneoxime, a substance ultimately obtainable from quinine (synthesised in 1944). The Russian synthesis of cinchonamine (and quinamine), therefore, partial when first announced in 1958, became total in 1960, *ex post facto*. Incomplete or unpublished syntheses of cinchonamine are known to have been under way in other laboratories at the same time, and the celebrated Japanese conversion[26] of the (synthetically accessible) cinchonine and cinchonidine to dihydrocinchonamine constitutes of course a synthesis of the latter substance.

The space diagram (56) has been proposed for the three-dimensional

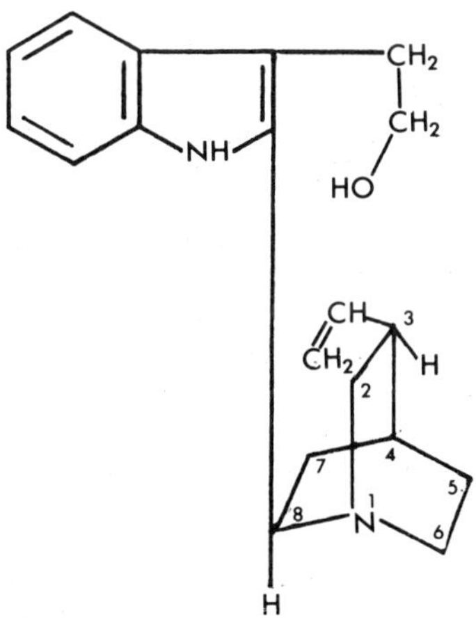

(56)

Cinchonamine in three dimensions

structure of cinchonamine but the argument on which this is based has been criticized and, except for the ever-present C(3)–C(4) pattern, the stereochemistry of cinchonamine and quinamine must remain in some doubt. Nothing is known of the configuration of the two indoline–oxolane carbon atoms in quinamine.

REFERENCES

1. R. B. Turner and R. B. Woodward, Alkaloids **3,** 1–63 (1953).
2. H. -G. Boit, "Ergebnisse der Alkaloid-Chemie bis 1960," pp. 572–582, Akademie-Verlag, Berlin, Germany, 1961.

3. J. W. Brühl, E. Hjelt, and O. Aschan, "Die Pflanzen-Alkaloide," pp. 161–240, F. Vieweg und Sohn, Brunswick, 1900 (Reprinted from "Roscoe-Schorlemmer's Lehrbuch der Organischen Chemie," Vol. 8).
4. W. Königs, *Ann.* **347,** 143–192 (1906).
5. E. Comanducci, "Die Konstitution der Chinaalkaloide," pp. 1–166, F. Enke, Stuttgart, 1911 [Reprinted from "Sammlung Chemischer und Chemisch-Technischer Vortrage," Vol. 16, (W. Herz, Ed.), Breslau].
6. T. A. Henry, "The Plant Alkaloids," pp. 418–483 J. and A. Churchill Ltd., London, England, 1949.
7. T. A. Henry and W. Solomon, "Thorpe's Dictionary of Applied Chemistry," Vol. 3, pp. 127–178 (J. F. Thorpe and M. A. Whiteley, Eds.), Longmans, Green and Co., London, England, New York, N. Y. 1943.
8. F. A. Flückiger and D. Hanbury, "Pharmacographia," pp. 338–370, Macmillan and Co., London, England, 1879.
9. A. W. Haggis, *Bull. Hist. Med.* **10,** 417 (1941).
10. J. Jaramillo-Arango, *J. Linnean Soc. London, Botany* **53,** 272 (1949).
11. P. J. Pelletier and J. B. Caventou, *Ann. Chim. et Phys.* **15,** 289 (1820).
12. P. J. Pelletier and J. B. Caventou, *Ann. Chim. et Phys.* **15,** 337 (1820).
13. P. Rabe, *Ber.* **40,** 3655 (1907).
14. L. Pasteur, *Compt. Rend.* **37,** 110 (1853).
15. P. Schützenberger, *Compt. Rend.* **47,** 233 (1858); G. H. Beckett and C. R. A. Wright, *J. Chem. Soc.* **29,** 655 (1876); O. Hesse, *Ann.* **205,** 314 (1880).
16. W. von Miller and G. Rohde, *Ber.* **27,** 1279 (1894).
17. W. Königs, *Ber.* **27,** 900 (1894).
18. P. Rabe and K. Ritter, *Ber.* **38,** 2770 (1905); *Ann.* **350,** 180 (1906); P. Rabe and E. Ackermann, *Ber.* **40,** 2013 (1907).
19. W. Königs, K. Bernhart, and J. Ibele, *Ber.* **40,** 648, 2873 (1907).
20. W. von Miller and G. Rohde, *Ber.* **27,** 1187 (1894).
21. E. Jungfleisch and E. Leger, *Ann. Chim.* **14,** 59–188 (1920).
22. C. A. Grob, *Experientia* **13,** 126 (1957).
23. R. B. Woodward, N. L. Wendler, and F. J. Brutschy, *J. Amer. Chem. Soc.* **67,** 1425 (1945).
24. P. Rabe, W. Hochstätter, F. Cohausz, K.-C. Scheel, K. Dussel, H. Haeuszler, H. Walther, R. Teske-Guttmann, and L. Cauer-Wilhelm, *Ann* **561,** 132 (1949).
25. E. W. Warnhoff, "Molecular Rearrangements," Vol. 2, pp. 877–879 P. de Mayo, Ed., Interscience Publishers, Inc., New York, N.Y., 1964.
26. E. Ochiai and M. Ishikawa, *Pharm. Bull. (Tokyo)* **5,** 498 (1957); E. Ochiai, M. Ishikawa, and Y. Oka, *Chem. Pharm. Bull. (Tokyo)* **7,** 744 (1959).
27. V. Prelog and E. Zalán, *Helv. Chim. Acta* **27,** 535, 545 (1944).
28. J. Kenner, *Ann. Rep. Progr. Chem. (Chem. Soc. London)* **19,** 157 (1922).
29. P. Rabe, A. Irschick, G. Suszka, E. Müller, A. Nielsen, F. Kolbe, W. von Riegen, and W. Hochstätter, *Ann.* **492,** 242 (1932).
30. R. S. Cahn, C. K. Ingold, and V. Prelog, *Angew. Chem. Intern. Ed. Eng.* **5,** 385 (1966).
31. V. Prelog and O. Häfliger, *Helv. Chim. Acta* **33,** 2021 (1950).
32. G. G. Lyle and L. K. Keefer, *Tetrahedron* **23,** 3253 (1967).
33. H. King and A. D. Palmer, *J. Chem. Soc.* **121,** 2577 (1922).
34. P. Baranger and R. Maréchal, *Compt. Rend.* **233,** 550 (1951).
35. P. Rabe, W. Huntenburg, A. Schultze, and G. Volger, *Ber.* **64,** 2487 (1931).
36. R. B. Woodward and W. E. Doering, *J. Amer. Chem. Soc.* **66,** 849 (1944); **67,** 860 (1945).

37. A. Kaufmann, E. Rothlin and P. Brunnschweiler, *Ber.* **49,** 2299 (1916).
38. M. Proštenik and V. Prelog, *Helv. Chim. Acta* **26,** 1965 (1943).
39. P. Rabe and K. Kindler, *Ber.* **51,** 1360 (1918); **52,** 1842 (1919).
40. E. Koenigs and W. Ottmann, *Ber.* **54,** 1343 (1921).
41. B. Witkop, *J. Amer. Chem. Soc.* **72,** 2311 (1950).
42. R. Goutarel, M.-M. Janot, V. Prelog, and W. I. Taylor, *Helv. Chim. Acta* **33,** 150 (1950).
43. Ch'en Ch'ang-pai, R. P. Evstigneeva, and N. Preobrazhenskii, *Proc. Acad. Sci. USSR. Chem. Sect. (English Transl.),* **123,** 927 (1958).

12

The Furoquinoline Alkaloids

PAUL J. SCHEUER
Department of Chemistry
University of Hawaii
Honolulu, Hawaii

1. INTRODUCTION

The furoquinoline alkaloids are characterized by their structural simplicity coupled with the remarkable fact that all naturally occurring representatives have been isolated from members of a single plant family, the Rutaceae. They are derivatives of the furo [2, 3-*b*] quinoline system (1). The

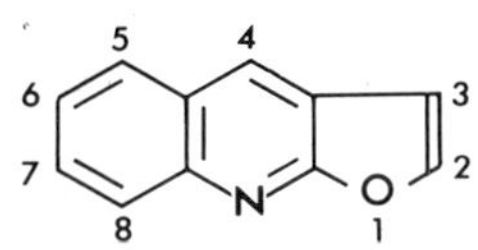

(1)

simplest representative of the group is dictamnine (4-methoxyfuro [2, 3-*b*] quinoline) and the alkaloid of widest occurrence is skimmianine (7.8-dimethoxydictamnine). Among the commonest variants of the parent compound dictamnine are those with methoxy or methylenedioxy substituents at positions 6, 7, and 8. An oxygen function at C(5) is rare. In addition to methyl ethers a few isopentyl ethers—occasionally hydroxylated—have been isolated. The most frequently encountered sidechain in the furan portion of the molecule is an isopropyl group at C(2), also at times hydroxylated.

This chapter presents some of the salient features of this group of compounds, their occurrence, their properties, and their synthesis—*in vitro* and *in vivo*. For more detailed information and data of an encyclopedic nature the reader should consult recent reviews by Openshaw,[1] by Boit,[2] and by Pakrashi and Bhattacharyya.[3] A synopsis of the furoquinoline group has recently become available in Swan's text.[4]

2. DISTRIBUTION

The rue family, or Rutaceae, comprises some 145 genera with about 1,400 species of tropical and subtropical trees, shrubs, and a few herbs, of which *Citrus* is the best known genus. Price[5] has studied the distribution pattern of alkaloids in the Rutaceae and has pointed out a number of interesting aspects. Occurrence of alkaloids has been reported in five of the seven subfamilies. Four of these five alkaloid-positive subfamilies contain furoquinolines, which at the time of Price's survey placed furoquinolines in 27 of the 36 genera in which alkaloids had been identified. This wide distribution of furoquinolines accompanied by their absence elsewhere in the plant kingdom renders these alkaloids typical of the Rutaceae. This phenomenon is even more surprising if one realizes that the Rutaceae is perhaps the plant group that is most prolific in the synthesis of varied alkaloid types and produces according to Price[5] no fewer than nine structural classes of alkaloids.

3. PHYSICAL AND CHEMICAL PROPERTIES

The structural variants that may be encountered in tricyclic and largely aromatic compounds are, of course, limited. As a consequence, general observations about furoquinoline alkaloids are nearly always unexceptional. This circumstance has greatly facilitated structural studies in this series, since it has allowed full use to be made of physical methods.

3.1. Ultraviolet Spectra

The furoquinoline alkaloids are colorless compounds and their electronic spectra are, therefore, confined to the ultraviolet part of the electronic spectral region. Sangster and Stuart in their comprehensive review[6] have pointed to the major features of these spectra. Two outstanding properties, which characterize the furoquinoline alkaloids, are reflected in their spectra. The spectrum of dictamnine [4-methoxyfuro [2,3-*b*] quinoline, (2)] may be considered the parent spectrum of the group. It consists of a maximum at ca. 245 nm and a broad band with a great deal of fine structure in the 290–330-nm region. The methoxylation pattern of the benzene ring exerts only a minor influence. However, ready isomerization of the dictamnines with methyl iodide leads to N-methyl-4-quinolones (3), which

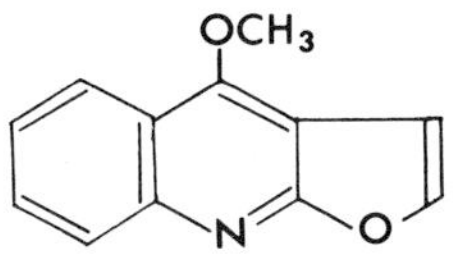

(2)

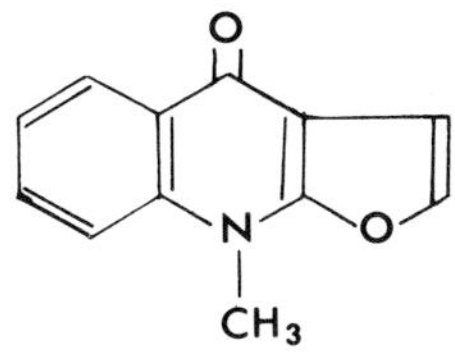

(3)

exhibit typical 4-quinolone spectra with a maximum at ca. 260 nm and a broad complex band at 300–340 nm. A second property, which is diagnostic of the group, is the hypsochromic shift of both ultraviolet bands accompanied by loss of fine structure when the spectrum is determined in acidic medium.[7] 2,3-Dihydrodictamnine derivatives, many of which occur in nature, exhibit parallel spectral behavior with the anticipated hypsochromic displacement (5–10 nm) of both absorption bands.

3.2 Infrared Spectra

Different aspects of the infrared spectra of furoquinolines have been examined by several groups of workers. Briggs and Colebrook[8] focused on the frequencies which may be ascribed to the furan portion of compounds bearing an aromatic system fused to the 2,3-positions of a furan. They noted some seven reasonably consistent band positions, but most of these vary in intensity and occur in regions which seriously overlap with other frequently encountered bands. Diagnostically the most useful band seems to be an invariably strong band at 1109–1088 cm^{-1}, which the authors however do not assign to a particular vibration mode.

Price and Willis[9] studied the use of infrared spectra for distinguishing between 2- and 4-quinolones. Since, as was mentioned earlier, 4-methoxyfuroquinolines are readily isomerized to N-methyl-4-quinolones, the spectra of 4-quinolones are pertinent to our discussion. The spectra of 2-quinolones are equally pertinent since hydrogenolysis of dictamnine (2) derivatives leads to derivatives of 3-ethyl-4-methoxy-2-quinolones (4).

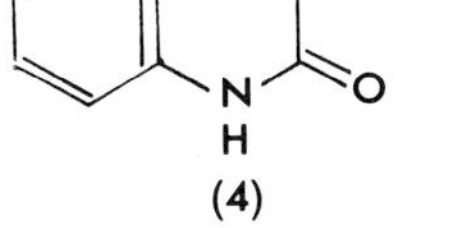

(4)

Price and Willis[9] concluded, however, that the carbonyl frequency is not a reliable parameter that allows discrimination between the two types. McCorkindale,[10] on the other hand, measured the intensities of the carbonyl bands of 2- and 4-quinolones and found that the two types can be differentiated by the much higher intensity of the carbonyl band of 2-quinolones as compared with 4-quinolones. McCorkindale[10] in fact applied this technique to confirm the structure of the alkaloid maculosidine (6,8-dimethoxydictamnine), where some doubt existed whether it was a linear [2,3-*b*] or an angular [3,2-*c*] furoquinoline. As the hydrogenolysis product was an unmistakable 2-quinolone, the alkaloid must be linear.

3.3 Nuclear Magnetic Resonance Spectra

Robertson's[11] comprehensive analysis of nmr spectra of furoquinolines has led to a number of valuable correlations. The furan protons form an AB quartet. The doublet at lower field ($\delta = 7.53$–7.58) is assigned to the proton at C(2) and the one at higher field ($\delta = 6.93$–7.02) to the proton at C(3). The 4-quinolone isomers retain the high field signal essentially intact and show a small (0.25 ppm) upfield shift for the proton at C(2).

When several methoxy groups are present in the molecule, the 4-methoxy signal is seen at lower field ($\delta = 4.4$) than the others (near $\delta = 4$).

An outstanding characteristic of the aromatic region is the signal of the C(5) proton. This is a characteristic signal of many alkaloids in the group since 5-methoxydictamnine is not known and of the known di- and trimethoxydictamnines only acronycidine (5,7,8-trimethoxy) bears a C(5) substituent. The C(5) proton invariably appears at lower field ($\delta = 7$–8) than other aromatic protons in the spectrum. Robertson[11] further showed that it is possible to differentiate between linear and angular furoquinolines on the basis of the chemical shift difference between the C(5) proton and the center of gravity of all other aromatic proton signals.

For derivatives of acronycidine which lack a C(5) but possess a C(6) proton Batterham and Lamberton[12] established that chemical shift values of the C(6) proton range from $\delta = 6.45$–6.63.

3.4 Mass Spectra

The mass spectra of the furoquinolines, which were examined by Clugston and MacLean,[13] offer an important clue for the presence or absence of an 8-methoxy group. All compounds, which were studied, are derivatives of dictamnine (2). All have intense molecular ion peaks, usually the base peak in the spectrum. Fragmentation is initiated by loss of methyl, followed by loss of carbon monoxide, which leads to M-15 and M-43 peaks of diminished intensity. In some of the compounds with an 8-methoxy group these two fragmentations are overshadowed by strong M-1 and M-29 peaks, which are completely absent in compounds without the 8-methoxy group. Not unreasonably, the proposed mechanism[13] postulates loss of hydrogen from the 8-methoxy group (M-1), followed by loss of formyl (M-29) to

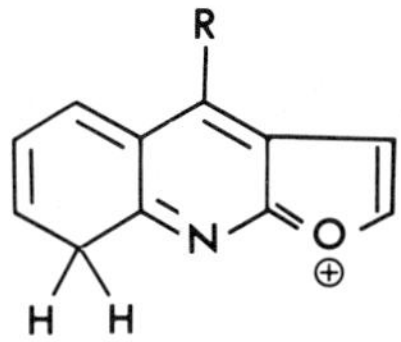

(5)

lead to a dihydrofuroquinoline ion (5). Interestingly enough, one of the 8-methoxy compounds, 7,8-dimethoxydictamnine (skimmianine), is exceptional in that its M-15 peak constitutes the base peak.

3.5 Chemical Properties

Perhaps the most characteristic reaction of this group of alkaloids is their facile isomerization to 4-quinolones. When dictamnine (2) is treated with methyl iodide at 80°, it is smoothly converted to isodictamnine (7), probably *via* the quaternary intermediate (6) by the indicated electronic shifts. The suggested intermediate (6) is supported by the reaction of O-methyl-

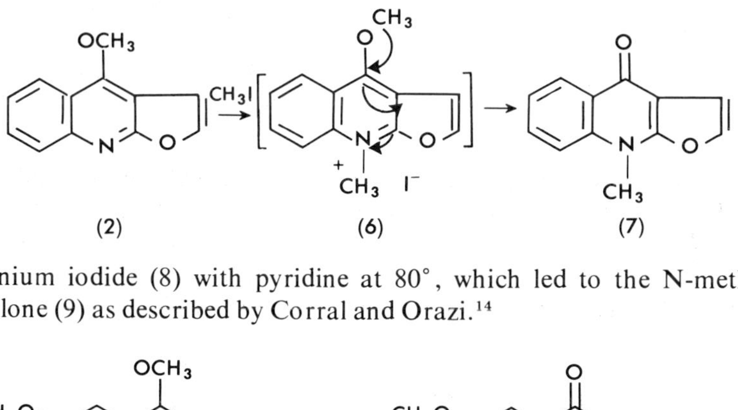

(2) (6) (7)

ribalinium iodide (8) with pyridine at 80°, which led to the N-methyl-4-quinolone (9) as described by Corral and Orazi.[14]

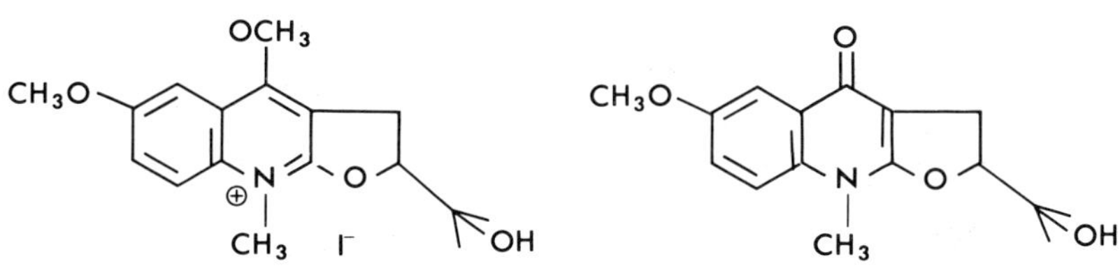

(8) (9)

A second characteristic reaction of the group is the hydrogenolysis with platinum in acetic acid to 3-ethyl-2-quinolones (4). These degradation products are vinylogous methyl esters and as such are readily demethylated. The ensuing 3-ethyl-4-hydroxy-2-quinolone (10) is a well-characterized compound.

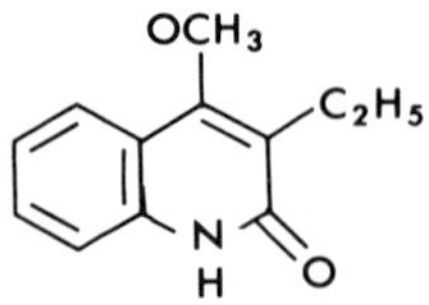

(4)

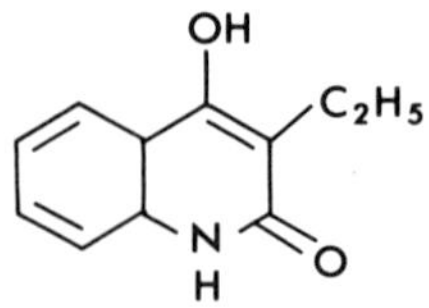

(10)

In contrast to this hydrogenolysis reaction, catalytic hydrogenation over palladium in ethanol leads to 2,3-dihydro compounds. Raney nickel reduction may lead to a mixture of compounds corresponding to the two modes of hydrogenation.[15]

Some interesting chemical transformations were carried out by Rapoport and Holden[16] in the course of their alkaloid studies of the South American shrub *Balfourodendron riedelianum*. Structure (10) had been assigned to the alkaloid balfourodine by the same workers,[17] but the isolation from the same plant of balfourolone (11)—subsequently recog-

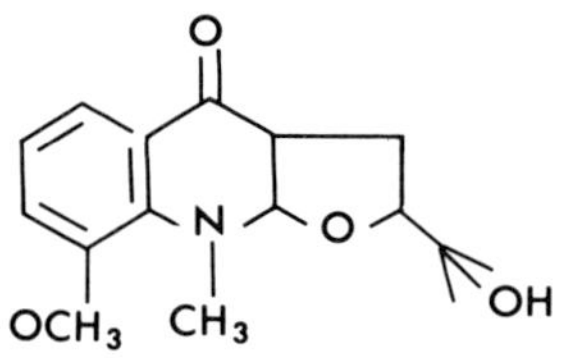

(10)

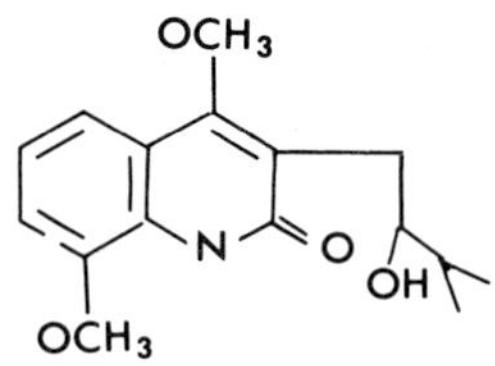

(11)

nized as an artifact—suggested the possibility that balfourodine might instead of structure (10) possess the dihydropyrano structure (12). Although dehydration of (10) with sulfuric acid led to a furan derivative, which on hydrogenation furnished the known (±)-lunacrine (13), acid-catalyzed rearrangement of (12) might have yielded the same result. In order to settle

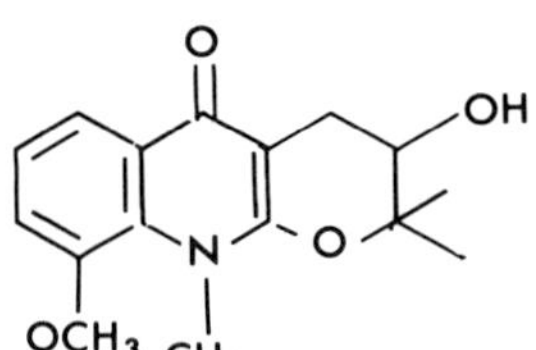

(12)

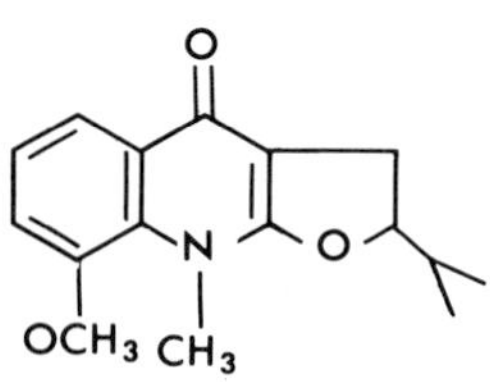

(13)

this point by comparative spectral studies, Rapoport and Holden[16] synthesized dihydrofuro- and dihydropyrano model compounds. The ultraviolet spectra of balfourodine (10) and of the model compound (14) were indeed identical, as were the ultraviolet spectra of the model compound (15) and of a minor alkaloid from the same plant, isobalfourodine. That

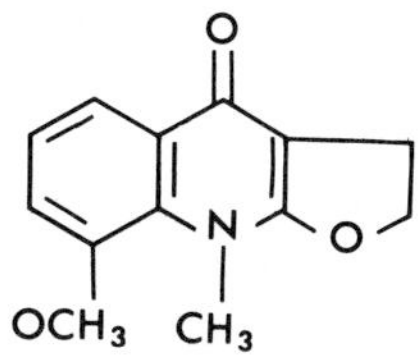

(14)

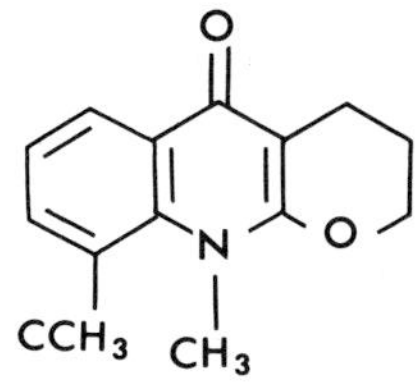

(15)

isobalfourodine indeed had structure (12), which had been considered for balfourodine, was shown by several interesting reactions. Acetylation with acetic anhydride and pyridine of (+)-isobalfourodine (12) at room temperature and of (−)-balfourodine (10) at boiling temperature led to the identical product, (+)-isobalfourodine acetate (16). Saponification of both

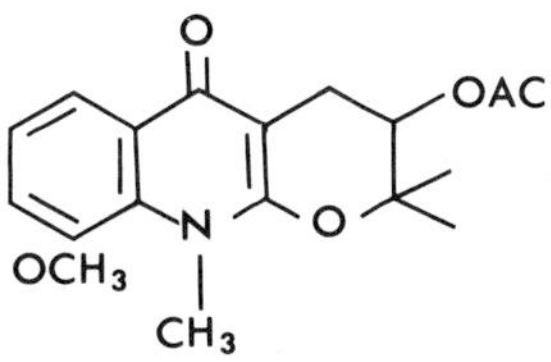

(16)

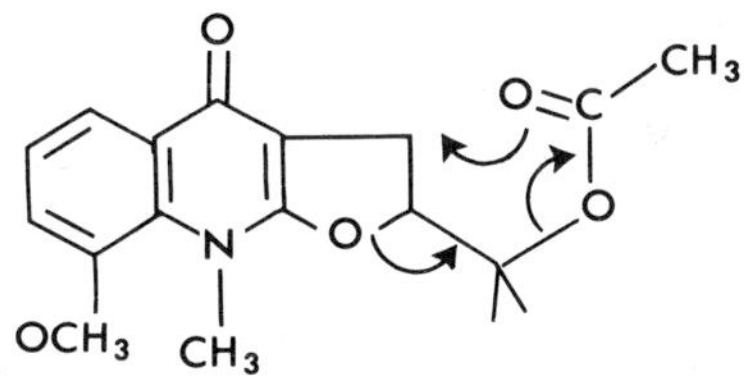

(17)

acetates furnisned (+)-isobalfourodıne. Optical rotation measurements indicated that isobalfourodine acetate and isobalfourodine derived from balfourodine were about 42% racemized; nevertheless, the ring expansion must have occurred by a stereospecific methanism as indicated in formula (17). Since both alkaloids have only one asymmetric center, an inversion took place during the ring expansion and the two alkaloids must have antipodal absolute configurations.

Ring expansion from the dihydrofuro to the dihydropyrano structure was also observed when the methosulfate of balfourodine (10) was prepared. Finally, treatment of balfourodine (10) with base led to a mixture of the two angular compounds of the ψ-series. (+)-ψ-balfourodine (18) and (−)-ψ-isobalfourodine (19).

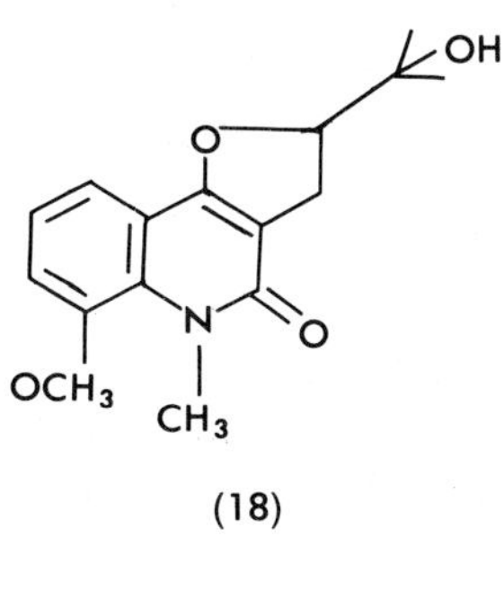

(18)

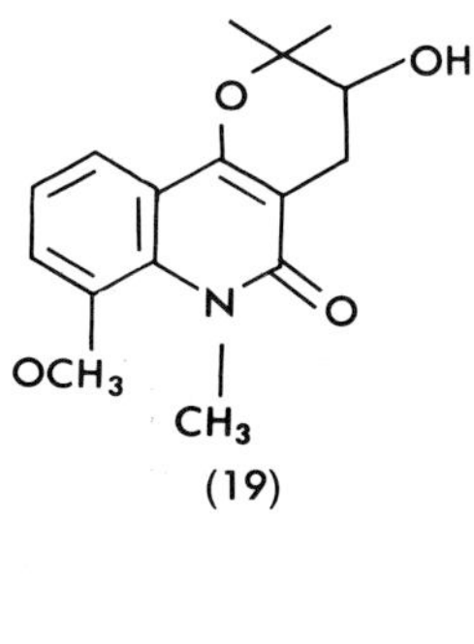

(19)

Another interesting feature of this group of furoquinoline alkaloids rests in the observation that, e.g., balfourodine (10) isolated from *Balfourodendron riedelianum* is dextrorotatory while hydroxylunacrine (10) derived from *Lunasia amara*[18] is its levorotatory antipode. In fact, several of the alkaloids from the two plant sources are enantiomerically related and because of their single asymmetric center at C(2) possess opposite, absolute configurations.

4. SYNTHESIS

Nearly all known naturally occurring furoquinoline alkaloids have been synthesized. Although formation of a linear or angular furoquinoline is a major problem in synthesis as in structure determination, several routes of general applicability have been worked out.

The method described by Tuppy and Böhm[19] is illustrated by a synthesis of dictamnine (2). Ethyl 2-anilino-4-hydroxyfuran-3-carboxylate (20) is cyclized in boiling diphenyl ether to 4-hydroxy-3-keto-2,3-dihydrofuro [2,3-*b*] quinoline (21). Treatment of (21) with diazomethane yields a mixture of O-methyl (22) and N-methyl derivatives. The desired (22) is transformed into the 3-chlorofuroquinoline with phosphoryl chloride containing a trace of water. The chloro compound furnishes dictamnine (2) by hydrogenolysis over palladized calcium carbonate. Completion of the sequence with the N-methyl derivative leads to the angular compound.

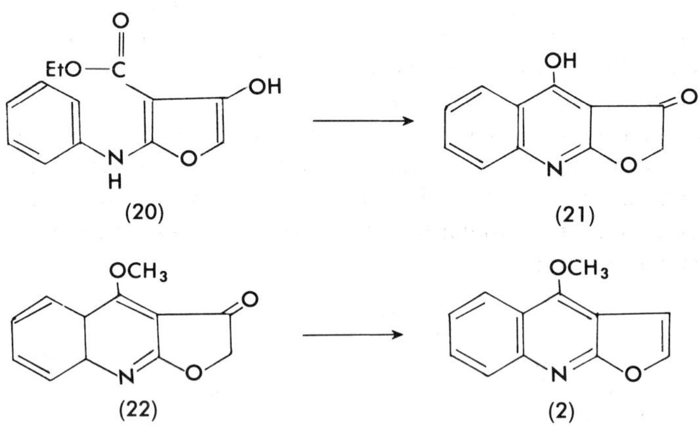

(20) (21) (22) (2)

A majority of furoquinoline alkaloids have been synthesized by Grundon and coworkers or by methods that were developed by this group. The synthesis of (±)-balfourodine (10) as described by Clarke and Grundon[20] will illustrate the method. Ethyl 3-methyl-2-butenylmalonate (23) was condensed with N-methyl-o-anisidine (24) in boiling diphenyl ether under nitrogen to yield 4-hydroxy-3-(3-methyl-2-butenyl)-2-quinolone (25) in 20-30% yields. This key intermediate may be hydroxylated in the β position of the sidechain and methylated to yield ($\pm$)-lunacridine (26); it may

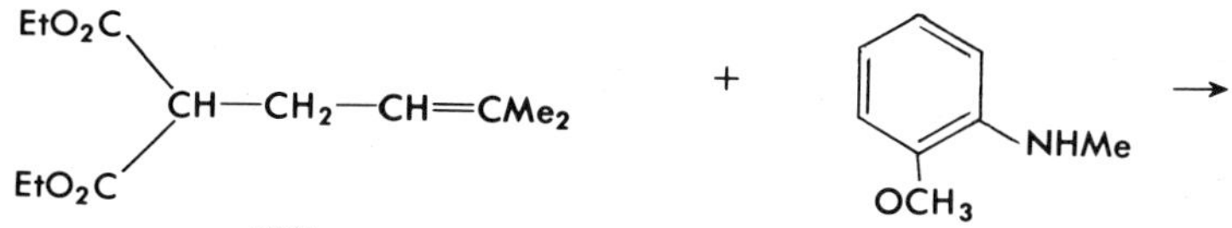

(23) (24)

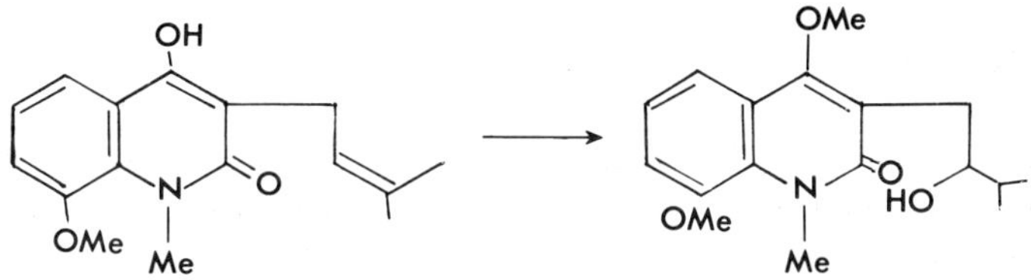

(25) (26)

be hydroxylated in the γ position and cyclized to a mixture of linear and angular dihydropyrano structures; or it may be treated with peroxylauric acid to yield ($\pm$)-balfourodine (10) without dihydropyrano byproducts.

The oxidative cyclization of quinolones without an N-methyl group by peroxylauric acid takes a slightly different course. Bowman and Grundon[21] showed that in this case a mixture of linear pyrano and furo compounds are obtained. By treatment of the quinolone (27), with 80% peroxylauric acid in chloroform these workers obtained the pyrano compound in 28% and the furo compound in 40% yield. Synthesis of the furo compound (28) constituted a structure proof of the alkaloid platydesmine, which Werny and Scheuer[22] had isolated from a Hawaiian rutaceous shrub. Structure (28), which had been suggested by Werny and Scheuer,[22] had been arrived at on the basis of spectral data and biogenetic analogies.

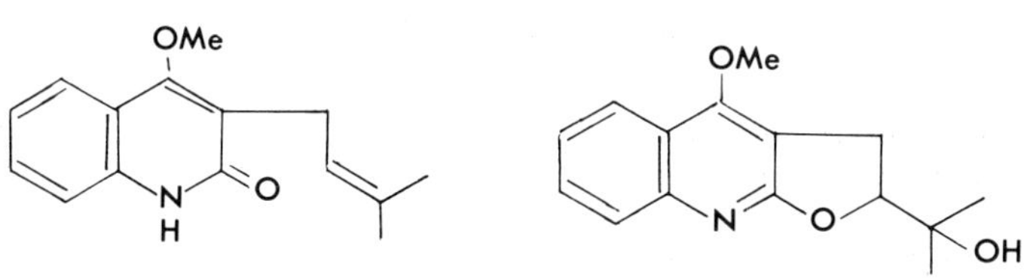

(27) (28)

In an interesting variation of the Grundon synthesis Reisch[23] condensed aniline and ethyl 2-propynylmalonate (29) in diphenyl ether with quantitative removal of ethanol and realized in a single step a 74% yield of 5-methyl [3,2-*c*] furo-2-quinolone (30) and a 16% yield of 2-methyl [2,3-*b*] furo-4-quinolone (31).

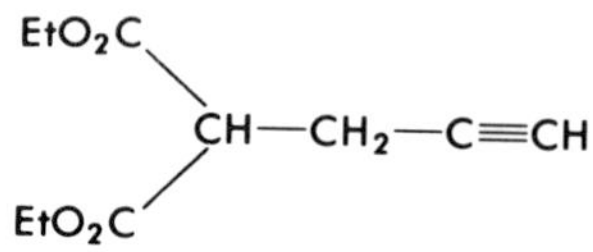

(29)

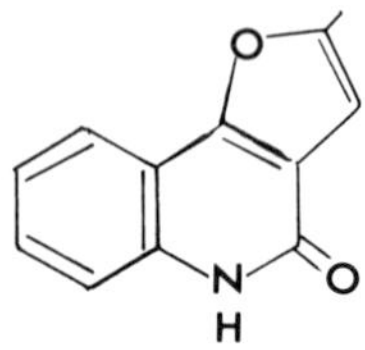

(30)

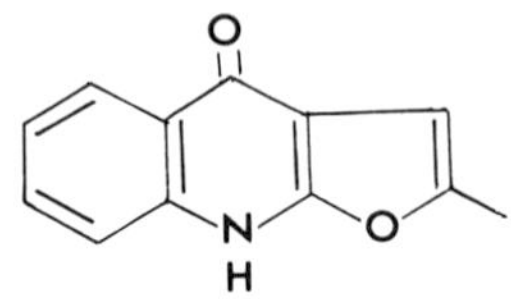

(31)

5. BIOGENETIC CONSIDERATIONS

The occurrence in the Rutaceae of simple quinoline, furoquinoline, acridine, and quinazoline alkaloids has long suggested an anthranilic acid (o-amino benzoic acid) precursor for these alkaloids. Price[24] in fact made this idea the basis for a discussion of these alkaloid types. That anthranilic acid is indeed the precursor of dictamnine (2) was recently shown by Monković, Spenser, and Plunkett.[25] In a series of tracer experiments on *Dictamnus albus* plants with doubly labelled (^{14}C, ^{15}N) anthranilic acid, carboxyl (^{14}C) labelled anthranilic acid, [1-^{14}C] and [2-^{14}C] acetic acid, and methyl (^{14}C) labelled methionine, these authors demonstrated that the quinoline nucleus of dictamnine is derived from anthranilic acid and acetic acid and that the methyl of the 4-methoxy group stems from methionine. Acetic acid was not significantly incorporated into the furan part of the molecule [C(2) and C(3)]. This finding coupled with the frequent occurrence of isopentyl moieties in the furoquinolines—not only attached to C(3) of quinoline but also as O-alkyl groups—strongly points to mevalonate as the missing building block. However, no biosynthetic experiments elucidating this point have been reported.

REFERENCES

1. H. T. Openshaw, *Alkaloids* **9,** 226–267 (1967); **7,** 233–246 (1960); **3,** 69–79 (1953).
2. H. G. Boit, "Ergebnisse der Alkaloid-Chemie bis 1960," pp. 707–729, Akademie-Verlag, Berlin, 1961.
3. S. C. Pakrashi and J. Bhattacharyya, *J. Sci. Ind. Res. (India)* **24,** 226 (1965).
4. G. A. Swan, "An Introduction to the Alkaloids," pp. 72–77, Blackwell Scientific Publications, Oxford, England 1967.
5. J. R. Price, "Chemical Plant Taxonomy," pp. 429–452 (T. Swain, Ed.), Academic Press, Inc., New York, N.Y., 1963.
6. A. W. Sangster and K. L. Stuart, *Chem. Rev.* **65,** 69 (1965).

7. F. Werny, Ph.D. Dissertation, University of Hawaii, 1962.
8. L. H. Briggs and L. D. Colebrook, *J. Chem. Soc.* 2458 (1960).
9. J. R. Price and J. B. Willis, *Australian J. Chem.* **12,** 589 (1959).
10. N. J. McCorkindale, *Tetrahedron* **14,** 223 (1961).
11. A. V. Robertson, *Australian J. Chem.* **16,** 451 (1963).
12. T. T. Batterham and J. A. Lamberton, *Australian J. Chem.* **18,** 859 (1965).
13. D. M. Clugston and D. B. MacLean, *Can. J. Chem.* **43,** 2516 (1965).
14. R. A. Corral and O. O. Orazi, *Tetrahedron* **21,** 909 (1965).
15. J. A. Lamberton and J. R. Price, *Australian J. Chem.* **6,** 66 (1953).
16. H. Rapoport and K. G. Holden, *J. Amer. Chem. Soc* **82,** 4395 (1960).
17. H. Rapoport and K. G. Holden, *J. Amer. Chem. Soc.* **81,** 3738 (1959).
18. S. Goodwin, A. F. Smith, A. A. Velasquez, and E. C. Horning, *J. Amer. Chem. Soc* **81,** 6209 (1959).
19. H. Tuppy and F. Böhm, *Angew. Chem.* **68,** 388 (1956).
20. E. A. Clarke and M. F. Grundon, *J. Chem. Soc.* 4190, 4196 (1964).
21. R. M. Bowman and M. F. Grundon, *J. Chem. Soc.* **C** 1504 (1966).
22. F. Werny and P. J. Scheuer, *Tetrahedron* **19,** 1293 (1963).
23. J. Reisch, *Arch. Pharm.* **300,** 533 (1967).
24. J. R. Price, *Fortschr. Chem. Org. Naturstoffe* **13,** 302 (1956).
25. I. Monkovic, I. D. Spenser, and A. O. Plunkett, *Can. J. Chem.* **45,** 1935 (1967).

13

Betalains, Red-Violet and Yellow Alkaloids of the Centrospermae

TOM J. MABRY
The Cell Research Institute and Department of Botany
The University of Texas at Austin

1. INTRODUCTION

One of the most striking features of flowering plants is their pigments, and it is, therefore, not surprising that these colored compounds were among the earliest natural products to be extensively investigated by organic chemists.

Thus, it is rather remarkable that the structure of the red-violet and yellow alkaloids which occur abundantly and uniquely in plants (for example, the red beet, cacti, cockscomb, pokeberry, and so on; see, for example, Figure 13.1) belonging to ten families presently allied in the Order Centrospermae[1,2] escaped elucidation until the 1960's. At the present time the structures of more than a dozen of the red-violet pigments, the betacyanins, and almost as many of the yellow types, the betaxanthins, are known. For reasons which will be evident in the course of this chapter, the expression "betalains" was recently introduced to describe the class of colored alkaloids to which both the betacyanins and betaxanthins belong.[3]

Like the more widespread plant pigments, the anthocyanins, which have not yet been found in any of the betalain-containing families, betalains occur in cell vacuoles of flowers, fruits and leaves. Moreover, betalains often accumulate in the stalk and, of course, in the beet they are found in high concentration in underground parts. It has also been noted that these alkaloids frequently accumulate at wound or injury sites in the plants that synthesize them normally. Betalains are water-soluble, occurring naturally as internal salts, i.e., zwitterions. They are readily recognized and charac-

Figure 13.1 Betalains account for red-violet and yellow flower colors in plants belonging to ten closely related plant families, one of which is the Cactaceae. Typical betalains are present in the yellow and red flowers of the two *Opuntia* cacti shown. (*Photographs by Dr. H. Rösler*) See Frontispiece

terized by their electrophoretic migration toward the anode in weakly acidic buffers, while anthocyanins migrate toward the cathode under similar conditions.

The present account emphasizes the structure determinations of the betalains as well as their chemistry and interconversions; in addition, the biogenesis and biological significance of these pigmented alkaloids will be discussed briefly.

2. THE BETACYANINS

2.1 Early Investigations

Interest in the red-violet betacyanins was stimulated as early as the 1860's when some European wine-makers enhanced the value of red wines by deepening their color with beet and pokeberry juices.[4] Thus, the need for recognizing these adulterants in the presence of the natural wine pigments, the anthocyanins, prompted the development of color and spectral tests for betacyanins; the techniques developed at that time remained in use as the primary detection methods for betacyanins until the application of paper chromatography and electrophoresis in the 1950's.

In the latter part of the nineteenth century a number of investigators[4] recognized that the betacyanins contained nitrogen (about 7%) and differed in chemical properties from the more common anthocyanins [a typical anthocyanin cation is cyanin (1), the red rose pigment]. The high nitrogen

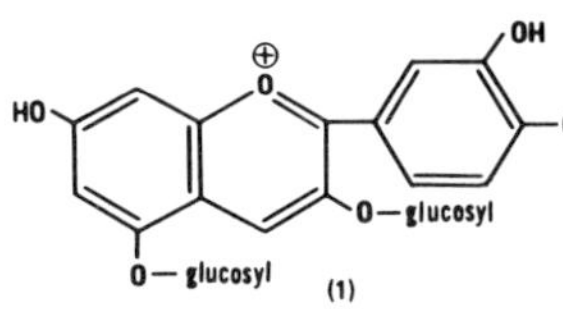

(1)

content of the betacyanins was an indication of the structural differences between betacyanins and anthocyanins that was not fully appreciated in later years.

Among the early workers, Schudel[5] (a student of Willstätter, the first major investigator of anthocyanins) probably made the most significant advances in understanding the nature of the betacyanins. In 1918 he obtained impure preparations of the beet (*Beta vulgaris*) pigment, which he called betanin. He also introduced the expression "betacyanins" to describe the class of pigments to which betanin belonged. Schudel showed that betanin was a nitrogen-containing glucoside that did not undergo the usual anthocyanin degradation with alkali to yield phloroglucinol (1,3,5-trihydroxybenzene).

In the 1930's investigators in Robinson's laboratory tackled the structure problems associated with the betacyanins, and, in addition, they designated the pigments as "nitrogenous anthocyanins," an expression which was to describe them for thirty years.[6–9] This name acknowledged the nitrogen content of the pigments, but erroneously presumed a direct structural relationship to the anthocyanins. Difficulty in obtaining crystalline preparations of the betacyanins finally led Robinson and co-workers to a structural approach that we now know could not have been successful. They synthesized several amine derivatives of the anthocyanin type, trying to prepare one related to the betacyanins. Much of the chemical data accumulated by Robinson's group on the betacyanins (e.g., the observation that at least two carboxyl groups could be methylated and that phloroglucinol was not obtained on alkaline fusion) was discounted because it was not compatible with the "nitrogenous anthocyanin" theory.[7] Robinson[9] commented on the structures of these pigments in the following manner: "When it became clear that the nitrogenous anthocyanins comprise a flavylium salt nucleus combined with an amino acid residue, synthetical work directed to the preparation of typical substances [e.g., (2)] was initiated."

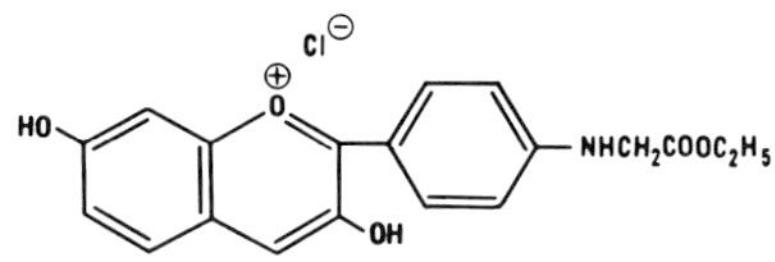

(2)

In 1956, more than fifteen years after Robinson's work, Schmidt and Schönleben[10] erroneously reported that pure betanin did not contain nitrogen, suggesting that the nitrogen content previously noted was attributed to associated amino acids.

Despite the strenuous efforts by workers in the Willstätter and Robinson laboratories, authentic crystalline preparations of betanin were not obtained until 1957, when both the Dreiding[11] and the Schmidt[12] groups succeeded almost simultaneously. Following this major advance, both groups were able to gain further insight into the structural features of this new class of alkaloids.

Most of the degradative work on the betacyanins was carried out with betanidin (3a) and its C(15) epimer, isobetanidin (4a); they are the two aglycones obtained by the acidic hydrolysis of betanin (3b).

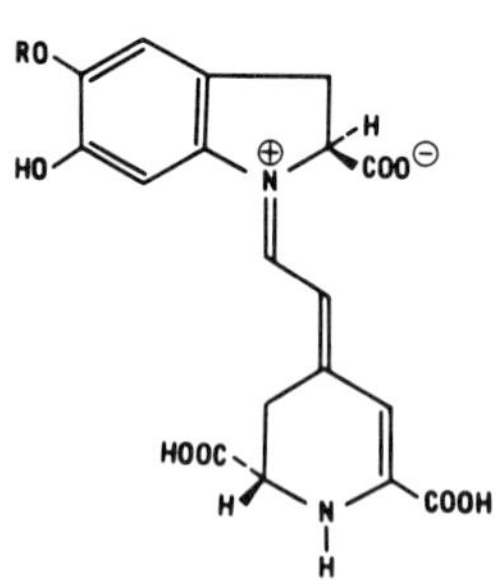

(3a), R = H; Betanidin

(3b), R = β-D-glucosyl; Betanin

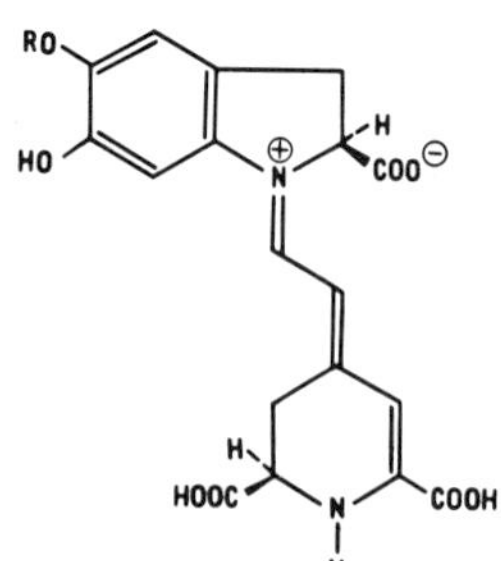

(4a), R = H; Isobetanidin

(4b), R = β-D-glucosyl; Isobetanin

In the period 1959–62, three important products were identified from the alkaline degradation of betanidin: 5,6-dihydroxy-2,3-dihydroindole 2-carboxylic acid (S-cyclodopa) (5), formic acid, and 4-methylpyridine 2,6-dicarboxylic acid (6).[13,14]

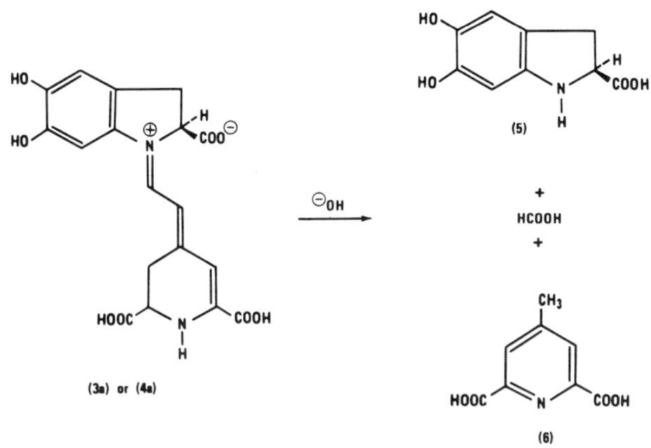

2.2 The Conversion of Betanidin and Isobetanidin to Neobetanidin Derivatives

The first indication of the novel ring system present in the betacyanins resulted from the structure analysis of a group of primary conversion products of betanidin, which were called neobetanidin derivatives.[15,16] The results from the neobetanidin studies suggested that betanidin and betanin could be formulated as (3a) and (3b), respectively, thus demonstrating that the betacyanins were not structurally related to the flavylium salts, as implied by the name "nitrogenous anthocyanins," but were a new class of pigmented alkaloids.

One of the neobetanidin derivatives important in the betanidin–betanin structure analyses was obtained by treatment of a deep red methanolic solution of betanidin hydrochloride (7) or isobetanidin hydrochloride (8) with diazomethane. From either aglycone, the reaction yielded a yellow, crystalline product, 5,6-di-O-methylneobetanidin trimethyl ester (9). The major absorption maximum of (9) (403 nm, $\epsilon = 32500$) was shifted in the presence of acid to 513 nm, $\epsilon = 44500$, producing a deep violet color. This suggested that the chromophore of protonated (9) was not significantly different from that of betanidin (λ_{max} 544 nm, $\epsilon = 51000$). That the diazomethane treatment of betanidin had introduced five methyl groups was evident from the nmr spectrum of (9), which displayed distinct signals for five methoxyl groups: one six-proton singlet at δ4.02 ppm and three three-proton singlets at δ3.97, 3.83, and 3.79.

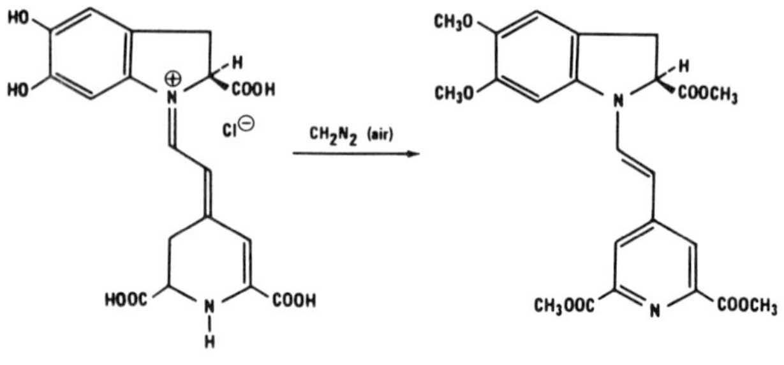

Another well-characterized neobetanidin derivative was obtained by acetylation of betanidin hydrochloride trimethyl ester (10). In acetic anhydride and pyridine, the trimethyl ester of betanidin yielded the yellow, crystalline 5,6-di-O-acetylneobetanidin trimethyl ester (11), which exhibited a bathochromic shift from an absorption maximum at 383 nm in methanol to 477 nm in acidic methanol. The nmr spectrum of (11) (Figure 13.2), which indicated two phenolic acetate groups (two three-proton

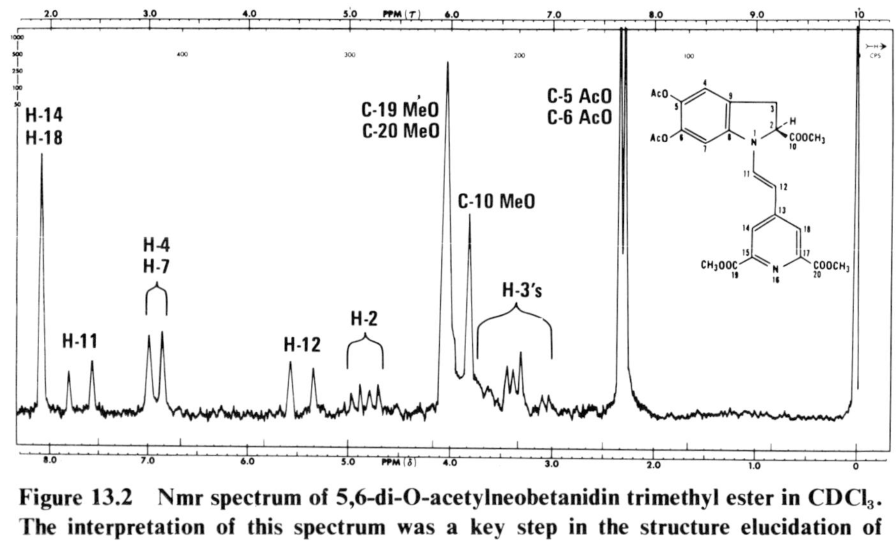

Figure 13.2 Nmr spectrum of 5,6-di-O-acetylneobetanidin trimethyl ester in $CDCl_3$. The interpretation of this spectrum was a key step in the structure elucidation of betanidin.

singlets at $\delta 2.28$ and 2.32 ppm) and three methoxyl groups (a six-proton singlet at $\delta 4.03$ and a three-proton singlet at $\delta 3.81$), confirmed that the eight oxygens of neobetanidin (and thus betanidin) corresponded to two phenolic hydroxyl groups and three carboxyl groups.

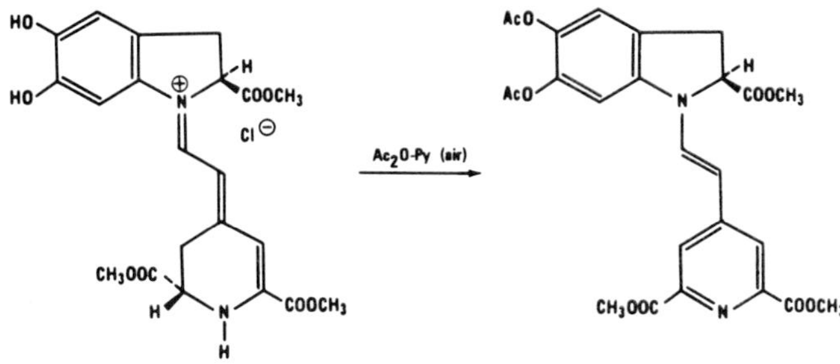

(10) (11)

Betanin was also converted to a neo-compound (15) during methylation with diazomethane (Figure 13.3).

The over-all reaction in the betanidin-to-neobetanidin conversions is a two-electron oxidation in which molecular oxygen is probably the oxidizing agent.

The presence of the connecting vinylene group [C(11) and C(12)] in the neobetanidin derivatives was confirmed by a palladium-catalyzed disproportionation reaction that converted the yellow 5,6-di-O-methylneobetanidin trimethyl ester (9) to the colorless 5,6-di-O-methyl-2,3-dehydro-11,12-dihydroneobetanidin trimethyl ester (12). The reaction consisted of an aromatization of the dihydroindole system at the expense of hydrogenation of the vinylene bridge.

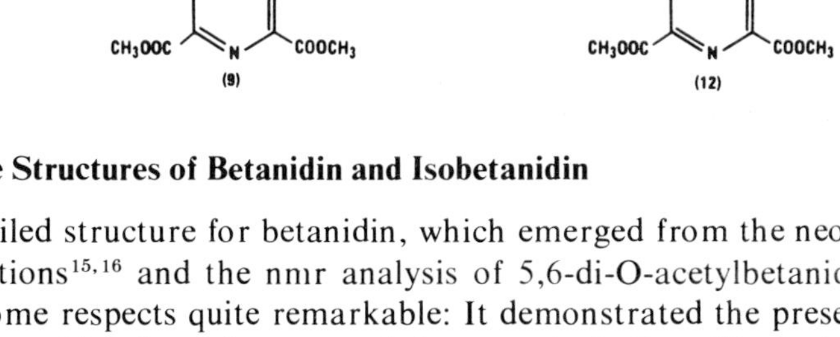

(9) (12)

2.3 The Structures of Betanidin and Isobetanidin

The detailed structure for betanidin, which emerged from the neobetanidin investigations[15,16] and the nmr analysis of 5,6-di-O-acetylbetanidin (13),[17] was in some respects quite remarkable: It demonstrated the presence of an unusual combination of structural features including dihydroindole and dihydropyridine rings and a polymethylene cyanine chromophore [in which a positive charge is partially localized on both nitrogen atoms, e.g., (14)].

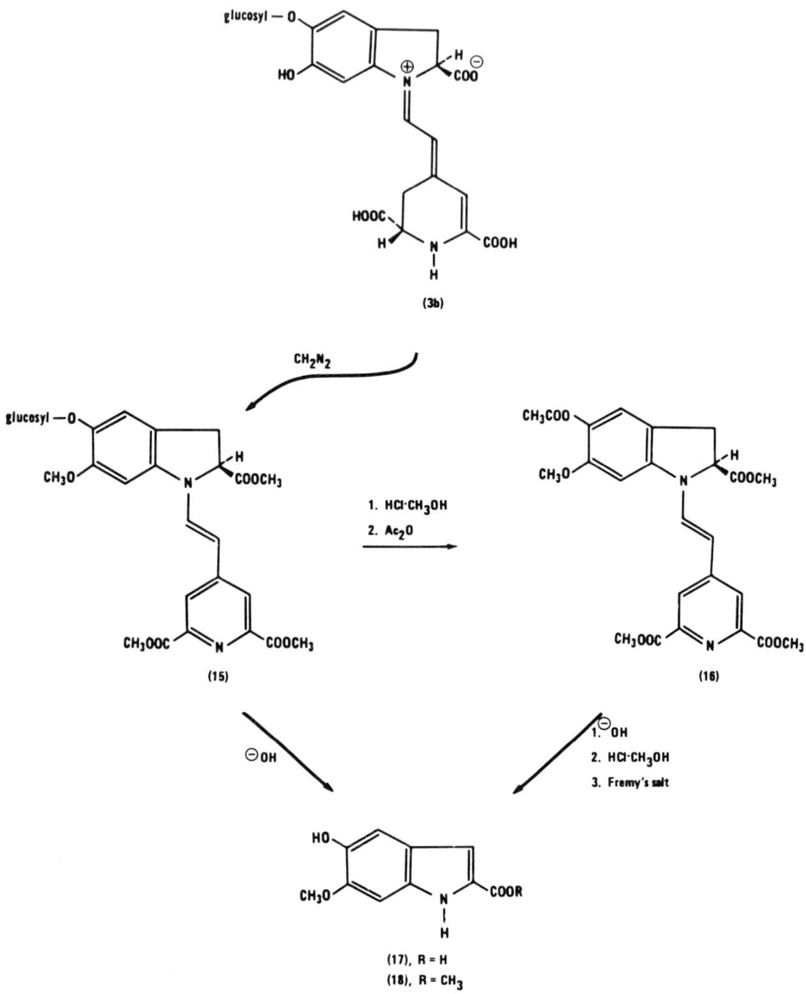

Figure 13.3 The isolation of (17) and (18) from betanin (3b) *via* neobetanin and neobetanidin derivatives provided proof that the glucose moiety is linked to C-(5) in betanin.

The absolute configuration at C(2) in both betanidin and isobetanidin was shown to be S by the isolation of 5,6-dihydroxy-2,3-dihydro-2S-indole 2-carboxylic acid (5) from their alkaline degradations.[14,17] Since both betanidin and isobetanidin yielded the same pentamethyl neobetanidin derivative (9) with diazomethane, they necessarily differed in configuration at C(15), the only asymmetric center other than C(2) present in the compounds. That the C(15) stereochemical assignments are those shown in

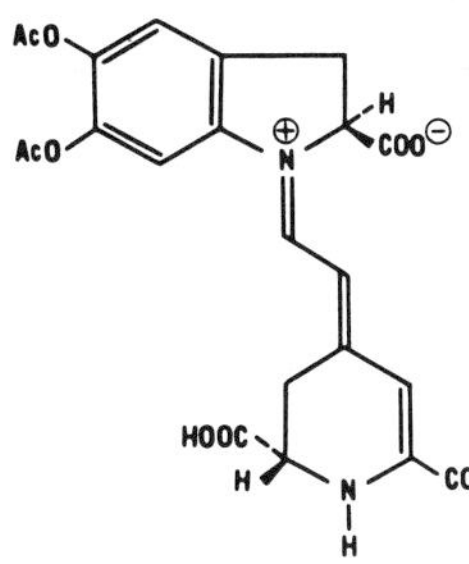

(13)

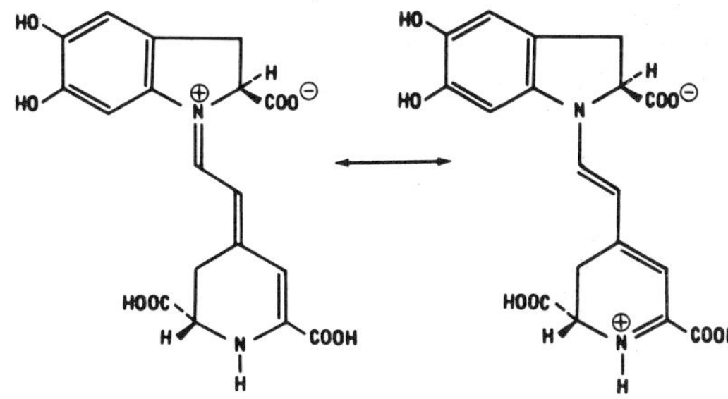

(14)

structures (3a) and (4a) will be clear from evidence presented in Section 4 of this chapter in connection with the discussion of the interconversions of betacyanins and betaxanthins.[18,19]

2.4 The Structures of Betanin and Isobetanin

Once structure (3a) had been established for betanidin, attention was focused on its monoglucoside, betanin. The only remaining problems in the structural analysis of betanin were the detailed structure of the glycosyl moiety and the manner and position of its attachment to betanidin. Hydrolytic studies with β-glucosidase[20,21] and the nmr spectral data for betanin in trifluoracetic acid[21] demonstrated that betanin was an O-β-D-glucopyranoside. The position of the linkage of the glucose in betanin was shown to be C(5) by the isolation of 5-hydroxy-6-methoxyindole 2-carboxylic acid (17) from the alkaline cleavage of the 6-methoxy-neobetanin trimethyl ester (15)[20] and by the isolation of the methyl ester (18) from the degradation of the 5-acetyl-6-methoxy-neobetanidin trimethyl ester (16) (Figure 13.3).[21] Therefore, betanin (3b) and isobetanin (4b) could be formulated as the 5-O-β-D-glucopyranosyl derivatives of betanidin and isobetanidin, respectively.

2.5 The Naturally Occurring Betacyanins

The hydrolysis of all betacyanins so far investigated leads to either betanidin (3a), isobetanidin (4a), or a mixture of the two C(15) epimers.[22,23] It is, therefore, likely that betacyanins differ only in their stereochemistry at C(15), the nature of the sugar moiety and/or in the esterification of the sugar and phenolic hydroxyl groups and the carboxyl groups.

The betacyanins, whose structures are known at this time, are summarized in Table 13.1; a number of other betacyanins, which are not recorded here, have been chromatographically, spectrally and/or electrophoretically detected[22,23] and, in some instances, partially structurally characterized.[26,27]

TABLE 13.1. The Structures of the Betacyanins.

Betanidin Series Isobetanidin Series

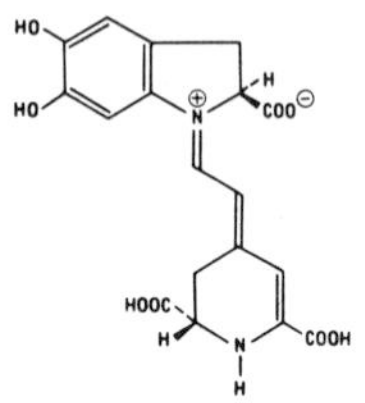

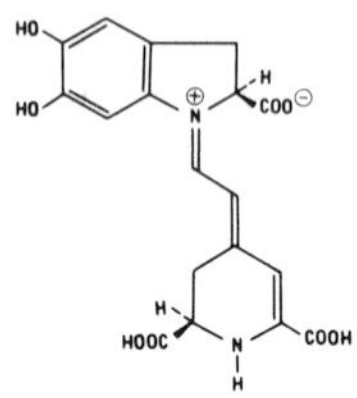

Compound	Compound	Substitution Pattern for Both Series	References
Betanidin	Isobetanidin	R = R′ = H	15–18
Betanin	Isobetanin	R = β-D-glucosyl; R′ = H	20, 21
Amarantin	Isoamarantin	R = 2′-0-(b-D-glucosyluronic acid)-β-D-glucosyl; R′ = H	24
Prebetanin	Isoprebetanin	R = 6′-0-(SO_3H)-β-D-glucosyl; R′ = H	25
Phyllocactin	Isophyllocactin	R = 6′-0-(malonyl)-β-D-glucosyl; R′ = H	26
Celosianin	Isocelosianin	R = 2′-0-[0-(trans-p-feruloyl)-β-D-glucosyluronic acid)]-0-(trans-p-coumaryl)-β-D-glucosyl; R′ = H	26
Iresinin-I	Isoiresinin-I	R = 2′-0-(β-D-glucosyluronic acid)-6′-0-(3-hydroxy-3-methyl-glutaryl)-β-D-glucosyl; R′ = H	26
Gomphrenin-I	Gomphrenin-II	R = H; R′ = β-D-glucosyl	27
Gomphrenin-III[a]		R = H; R′ = 6′-0-(cis-p-coumaryl)-β-D-glucosyl	27
Gomphrenin-V	Gomphrenin-VI	R = H; R′ = 6′-0-(trans-p-feruloyl)-β-D-glucosyl	27

[a]It was not established unambiguously that gomphrenin-III was a member of the betanidin series of betacyanins; the material used in the investigation may have been a mixture of the C(15) epimers.[27]

3. THE BETAXANTHINS

The yellow betalains are structurally related to the betacyanins and are called betaxanthins (*Gr., xanthos,* yellow); they have electrophoretic properties similar to betacyanins but show visible absorption maxima in the range 474–486 nm.

3.1 The Structure of Indicaxanthin

A long-suspected direct structural relationship between the red-violet betacyanins and the yellow betaxanthins was confirmed by the isolation and structure analysis of indicaxanthin (19), the betaxanthin from the fruits of the cactus *Opuntia ficus-indica.* The structure of indicaxanthin was established by (a) the acidic and alkaline degradations to L-proline (20) and 4-methylpyridine 2,6-dicarboxylic acid (6), and (b) the detection of L-aspartic acid (21) from the oxidative cleavage of indicaxanthin with hydrogen peroxide.[28] The latter result required an S-configuration at the asymmetric center in the dihydropyridine moiety of indicaxanthin.

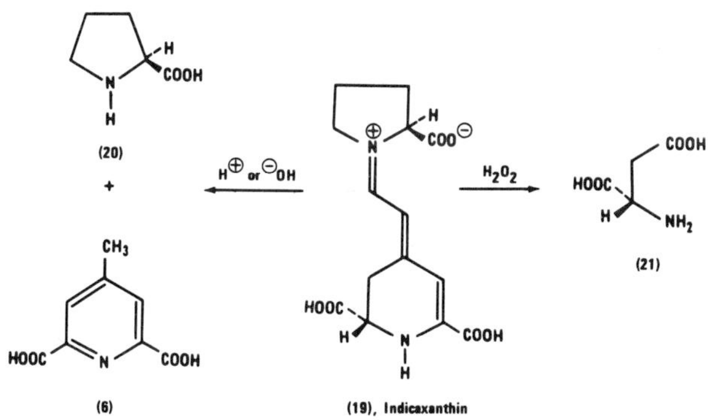

3.2 The Naturally Occurring Betaxanthins

Following the structure determination of indicaxanthin, the structures of six other betaxanthins were reported. Small quantities of two of the new betaxanthins, vulgaxanthin-I (λ_{max} 477 nm) and vulgaxanthin-II (λ_{max} 479 nm), were obtained from the red beet. Based on alkaline and acidic degradative reactions, structures (22) and (23) were proposed for the two new yellow alkaloids.[29]

In addition to indicaxanthin and vulgaxanthin-I, six other betaxanthins were isolated from the flowers of the common four o'clock, *Mirabilis jalapa*; the six were named miraxanthin-I, -II, -III, -IV, -V, and -VI. Structures were deduced for four of the new alkaloids [miraxanthin-I (24), -II

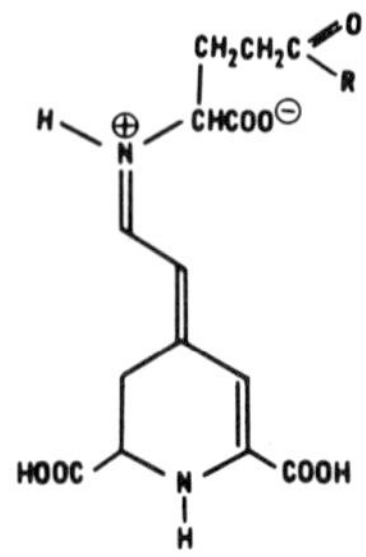

(22), R=OH; Vulgaxanthin-I

(23), R=NH_2; Vulgaxanthin-II

(25), -III (26) and -V (27)] on the basis of degradative reactions.[29] For example, all six of the alkaloids yielded 4-methylpyridine 2,6-dicarboxylic acid (6) on fusion with alkali, a result which indicated the presence of a dihydropyridine moiety in each of the pigments. On acid hydrolysis, miraxanthin-I afforded methionine sulfoxide, and miraxanthin-II gave aspartic acid. Miraxanthin-III and -V are noteworthy in that they are the first naturally occurring betalains that do not combine an α-amino acid with the dihydropyridine moiety.

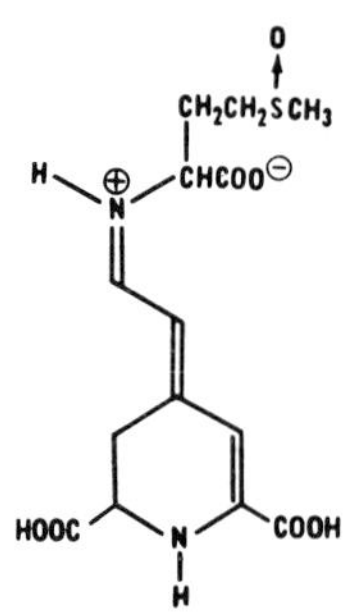

(24), Miraxanthin-I

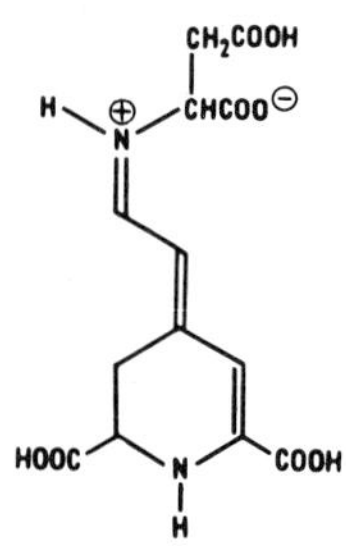

(25), Miraxanthin-II

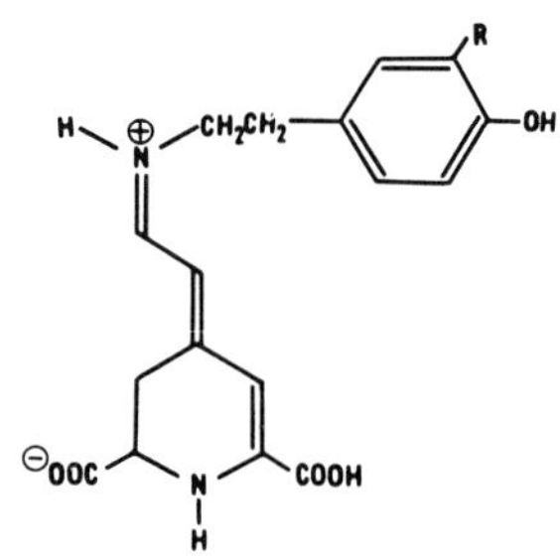

(26), R = H; Miraxanthin-III

(27), R = OH; Miraxanthin-V

Combinations of other amines or amino acids with the dihydropyridine group common to all betalains will probably account for the structures of additional betaxanthins.

4. THE INTERCONVERSION OF BETACYANINS AND BETAXANTHINS

In the presence of a weak base, the betacyanin and betaxanthin pigments undergo an exchange reaction in which the dihydropyridine portion of the pigments is transferred from one amino acid to a different amino acid. Similar *in vivo* conversions may account for the yellow-orange-red color changes exhibited by the flowers and fruits of some betalain-containing plants during different stages of their development.

4.1 Interconversion of Betanidin-Betanin with Indicaxanthin

In the presence of dilute ammonium hydroxide and excess L-proline (20), betanin (3b) yielded indicaxanthin (19).[19] Moreover, indicaxanthin was converted to betanidin (3a) in the presence of excess S-cyclodopa (5).[19] Although the mechanisms of these exchange reactions are not completely understood, it has been suggested[19] that they proceed by hydrolysis of the pigments to S-betalamic acid (28), which could then condense with available amino acids (or amines) to produce different betalains. Alternatively, the reactions outlined in Figure 13.4 might involve an intermediate substance such as (29).

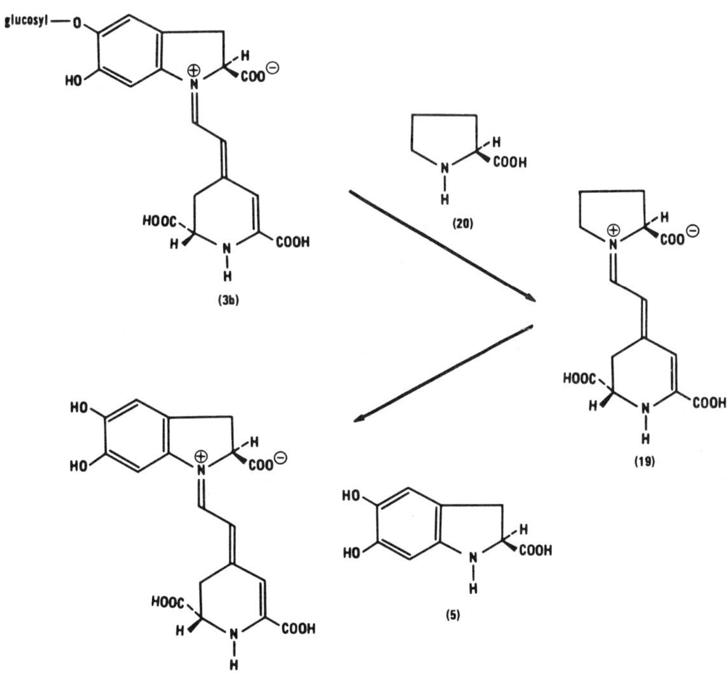

Figure 13.4 The interconversions of betanin-betanidin and indicaxanthin.

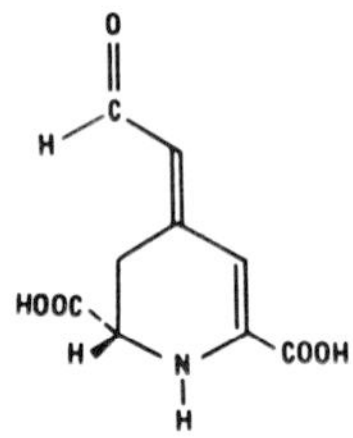

(28)

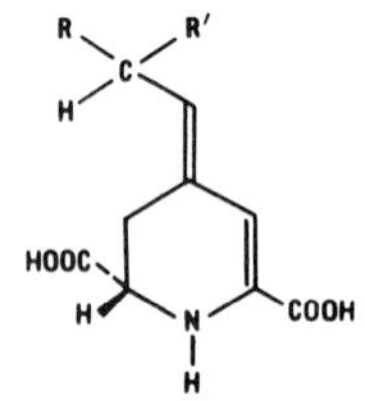

(29), R = cyclodopa moiety, R′ = proline group

4.2 Determination of the Stereochemistry at C(15) in Betanidin and Isobetanidin with the Imino-Acid Exchange Reaction

The imino-acid exchange reaction was used (a) to confirm that betanidin and isobetanidin differed only in their stereochemistry at C(15) and (b) to determine the absolute stereochemistry at C(15) in the two isomers.[18] The S-cyclodopa portions of both betanidin and isobetanidin were exchanged with 5,6-dihydroxy-2,3-dihydro-2R-indole 2-carboxylic acid (referred to as R-cyclodopa) (30).

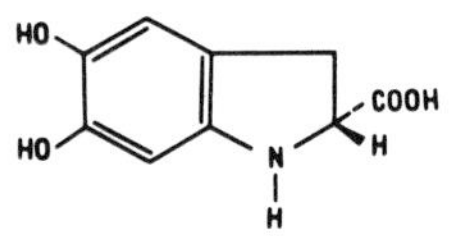

(30)

The new compound from betanidin was named enantio-isobetanidin because it exhibited a rotatory dispersion (o.r.d.) curve over the range of 300–700 nm in pH 4.5 potassium acetate-acetic acid buffer that was the mirror image of the curve obtained for isobetanidin. Analogously, the new compound from isobetanidin was named enantio-betanidin, as its o.r.d. curve was the mirror image of the one observed for betanidin. These results indicated that betanidin and isobetanidin differed in configuration at only one asymmetric center, namely at C(15).

That betanidin has the S configuration at C(15) (and thus isobetanidin the R) was demonstrated by converting indicaxanthin to betanidin. Indicaxanthin is known to have the S configuration at the chiral center equivalent to C(15) in betanidin. Therefore, the structure of betanidin can be formulated as (3a) and isobetanidin as the C(15) epimer (4a).

4.3 The Synthesis of Betaxanthins with an Imino-Acid Exchange Reaction

The imino-acid exchange reaction as carried out in the presence of dilute ammonium hydroxide alone[19] for the synthesis of indicaxanthin (19) from betanin does not appear to be suitable for the synthesis of all betaxanthins. Therefore, the discovery of conditions for a more general imino-acid exchange reaction was important in connection with the structure analysis of other betaxanthins, e.g., (24), (25), (26), and (27).

Miraxanthin-I (24), -II (25), -III (26) and -V (27) were synthesized from betanin using the following procedure.[30] Betanin was dissolved in water that was then saturated with sulfur dioxide; the resultant solution was concentrated *in vacuo*. The appropriate amino acid dissolved in water was added to the concentrate. The reaction mixture was made basic with ammonium hydroxide and allowed to remain at room temperature for an hour. After acidification, the newly formed betaxanthin was isolated by passing the solution over an ion exchange column. This procedure should be applicable to the synthesis of other betaxanthins.

5. THE CHROMOPHORE OF THE BETALAINS

The general formula (32) for the betalains represents formally the condensation product of a primary or secondary amine with betalamic acid (31). Betalamic acid has not been isolated from natural sources or prepared synthetically and may not be an intermediate in the biosynthesis of the betalains; however, it does represent a unifying structure for all known betacyanins and betaxanthins. Thus, the expression "betalains" has been adopted to describe the class of alkaloids to which both betacyanins and betaxanthins belong.[3] So far as is known, the structures of all betalains are based on the cation (32) in which R′ may be an H or be cyclized to R.

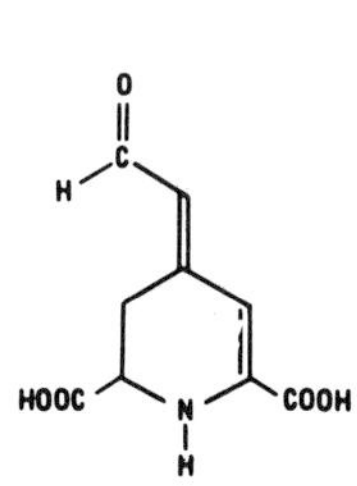

(31)

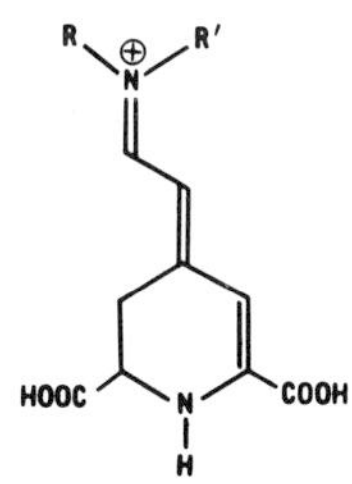

(32)

The chromophore responsible for the colors of all known betalains contains a cation (33) that has been described as a protonated 1,2,4,7,7-pentasubstituted-1,7-diazaheptamethin system.[16] When R and R′ in (32) do not extend the conjugation of the 1,7-diazaheptamethin system, the alkaloids are yellow (long wavelength absorption near 480 nm) and correspond to the betaxanthin type of betalain. In the betacyanins, R′ is a substituted aromatic group that is conjugated to the 1,7-diazaheptamethin cation; this extension of conjugation in the chromophore produces a bathochromic shift to around 540 nm, which accounts for the red-violet color of the betacyanins.

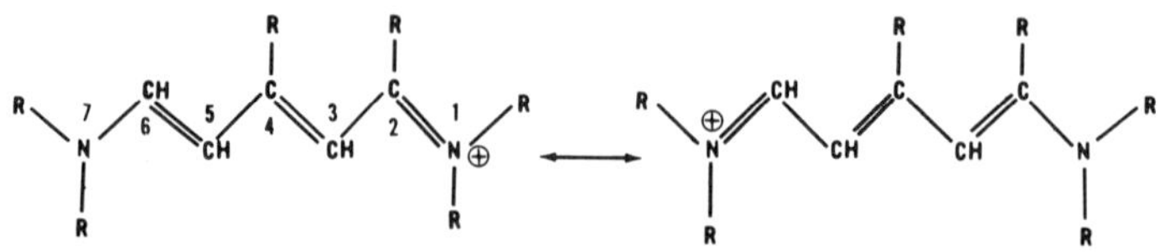

(33)

6. BIOGENESIS OF THE BETALAINS

A biosynthetic pathway for the betalains can be formulated on the basis of L-dopa (L-5,6-dihydroxyphenylalanine) (34) undergoing an oxidative cleavage and subsequent cyclization to betalamic acid (28) by one of two possible pathways (designated as A and B). The latter substance could then condense with available amines or amino acids to produce betalains. Condensation of (28) with cyclodopa or a derivative of cyclodopa (5) would yield betacyanins while other amino acids or amines would afford betaxanthins.

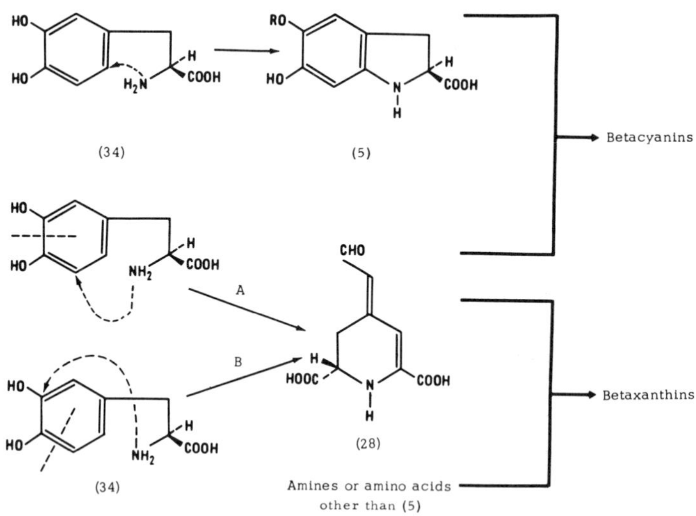

Despite preliminary studies by a number of workers[31–34] little firm evidence to support the biogenetic scheme for the betalains shown here was available until the recent investigations by the Dreiding and Mabry groups.[35] These workers established unambiguously that D,L-dopa-1-^{14}C and -2-^{14}C were incorporated in high yield (more than 10% in some experiments) into the dihydropyridine moiety of betanin in cactus fruits. Surprisingly, the same compounds were incorporated to a much lesser extent (less than 1%) into the dihydroindole portion of betanin. The data do not distinguish between the two possible pathways for cleaving dopa (A and B in the scheme presented here).

7. THE PHYLOGENETIC SIGNIFICANCE OF THE BETALAINS

Perhaps the most intriguing aspect of the betalain studies (see also Chapter 20) is that these highly pigmented alkaloids occur in only ten plant families, all of which are thought to be evolutionarily closely related. Moreover, so far as is known, the ten families do not contain the more common flower pigments, the anthocyanins. The ten families, which are now allied in the Order Centrospermae, are the following: Chenopodiaceae, Amaranthaceae, Portulacaceae, Nyctaginaceae, Phytolaccaceae, Stegnospermaceae, Aizoaceae, Basellaceae, Cactaceae, and Didiereaceae.[1,2]

REFERENCES

1. T. J. Mabry, A. Taylor, and B. L. Turner, *Phytochemistry* **2,** 61 (1963).
2. T. J. Mabry, "Comparative Phytochemistry," pp. 231–244 (T. Swain, Ed.), Academic Press, London, England, 1966; A. Wohlpart and T. J. Mabry, *Taxon* **17,** 148 (1968).
3. T. J. Mabry and A. S. Dreiding, "Recent Advances in Phytochemistry," pp. 145–160. (T. J. Mabry, R. E. Alston, and V. C. Runeckles, Eds.), Appleton-Century-Crofts, New York, N. Y., 1968.
4. See A. S. Dreiding, "Recent Developments in the Chemistry of Natural Phenolic Compounds," (W. D. Ollis, Ed.), Pergamon Press, Oxford, England, 1961, for a comprehensive review of the early literature pertaining to the betacyanins.
5. G. Schudel, Ph.D. Dissertation, Zurich-ETH, 1918.
6. A. M. Robinson and R. Robinson, *J. Chem. Soc.* 1439 (1932).
7. A. D. Ainley and R. Robinson, *J. Chem. Soc.* 446 (1937).
8. J. R. Price and R. Robinson, *J. Chem. Soc.* 449 (1937).
9. A. D. Ainley and R. Robinson, *J. Chem. Soc.* 453 (1937).
10. O. Th. Schmidt and W. Schönleben, *Naturwiss.* **43,** 139 (1956).
11. H. Wyler and A. S. Dreiding, *Helv. Chim. Acta* **40,** 191 (1957).
12. O. Th. Schmidt and W. Schönleben, *Z. Nuturforsch.* **126,** 262 (1957).
13. H. Wyler and A. S. Dreiding, *Helv. Chim. Acta* **42,** 1699 (1959).
14. H. Wyler and A. S. Dreiding, *Helv. Chim. Acta* **45,** 638 (1962).
15. T. J. Mabry, H. Wyler, G. Sassu, M. Mercier, I. Parikh, and A. S. Dreiding, *Helv. Chim. Acta* **45,** 640 (1962).
16. T. J. Mabry, H. Wyler, I. Parikh, and A. S. Dreiding, *Tetrahedron* **23,** 3111 (1967).
17. H. Wyler, T. J. Mabry, and A. S. Dreiding, *Helv. Chim. Acta* **46,** 1747 (1963).
18. M. E. Wilcox, H. Wyler, and A. S. Dreiding, *Helv. Chim. Acta* **48,** 1134 (1965).
19. H. Wyler, M. E. Wilcox, and A. S. Dreiding, *Helv. Chim. Acta* **48,** 361 (1965).

20. M. Piattelli, L. Minale, and G. Prota, *Ann. Chim.* **54,** 955 (1964).
21. M. E. Wilcox, H. Wyler, T. J. Mabry, and A. S. Dreiding, *Helv. Chim. Acta* **48,** 252 (1965).
22. H. Wyler and A. S. Dreiding, *Experientia* **17,** 23 (1961).
23. M. Piattelli and L. Minale, *Phytochemistry* **3,** 547 (1964).
24. M. Piattelli, L. Minale, and G. Prota, *Ann. Chim.* **54,** 963 (1964).
25. H. Wyler, H. Rösler, M. Mercier, and A. S. Dreiding, *Helv. Chim. Acta* **50,** 545 (1967).
26. L. Minale, M. Piattelli, S. De Stefano, and R. A. Nicolaus, *Phytochemistry* **5,** 1037 (1966).
27. L. Minale, M. Piattelli, and S. De Stefano, *Phytochemistry* **6,** 703 (1967).
28. M. Piattelli, L. Minale, and G. Prota, *Tetrahedron* **20,** 2325 (1964).
29. M. Piattelli, L. Minale, and G. Prota, *Phytochemistry* **4,** 121 (1965).
30. M. Piattelli, L. Minale, and R. A. Nicolaus, *Phytochemistry* **4,** 817 (1965).
31. L. Hörhammer, H. Wagner, and W. Fritzsche, *Biochem. Z.* **399,** 398 (1964).
32. L. Minale, M. Piattelli, and R. A. Nicolaus, *Phytochemistry* **4,** 593 (1965).
33. A. S. Garay and G. A. N. Towers, *Can. J. Bot.* **44,** 231 (1966).
34. A. Wohlpart, Ph.D. Dissertation, The University of Texas at Austin (1967).
35. H. E. Miller, H. Rösler, A. Wohlpart, H. Wyler, M. E. Wilcox, H. Frohfer, T. J. Mabry, and A. S. Dreiding, *Helv. Chim. Acta* **51,** 1470 (1968).

14

Pyrrolidine, Piperidine, Pyridine, and Imidazole Alkaloids

RICHARD K. HILL
Department of Chemistry
The University of Georgia
Athens, Georgia

This chapter presents an account of those alkaloids based on simple five- and six-membered nitrogen containing rings: pyrrolidine, pyridine, piperidine, and imidazole; the tropane group is treated in a separate chapter.

1. PYRROLIDINES

Pyrrolidine and its N-methyl derivative occur in tobacco, possibly as degradation products of nicotine. *Proline* is an important constituent of proteins, especially those of collagen, and its N-methyl betaine, *stachydrine*, occurs in alfalfa. The pyrrolidine ring appears to originate in nature generally from proline or its simpler relatives, glutamic acid and ornithine.

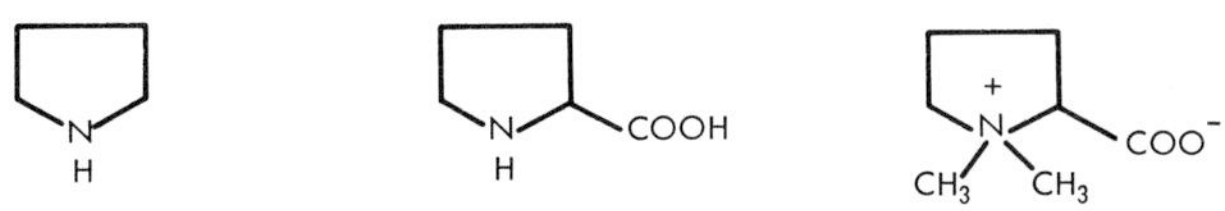

Pyrrolidine Proline Stachydrine

1.1 Simple Pyrrolidine Alkaloids[1]

The Peruvian coca shrub, source of cocaine, contains three simple pyrrolidine alkaloids as well: hygrine, hygroline, and cuskhygrine. *Hygrine*, $C_8H_{15}NO$, is a liquid tertiary aminoketone. Chromic acid oxidation afforded N-methyl proline, suggesting structure (1) for hygrine. This structure has been confirmed by a number of syntheses. The most efficient synthesis uses a Mannich condensation between acetoacetic acid and the immonium salt (2) prepared either by dehydrogenation of N-methylpyrrolidine or by partial reduction of N-methylpyrrolidone. Intermediate (2) has also been prepared by enzymatic oxidation of $CH_3NH(CH_2)_4NH_2$, a clue to the probable biosynthetic origin.

Freshly isolated hygrine has a small negative optical rotation but the free base rapidly racemizes, probably by β-elimination to (3) and reclosure. Oxidation of the levorotatory base showed the asymmetric carbon to have the same configuration as natural L-proline. *Hygroline* is the secondary alcohol related to hygrine.[2]

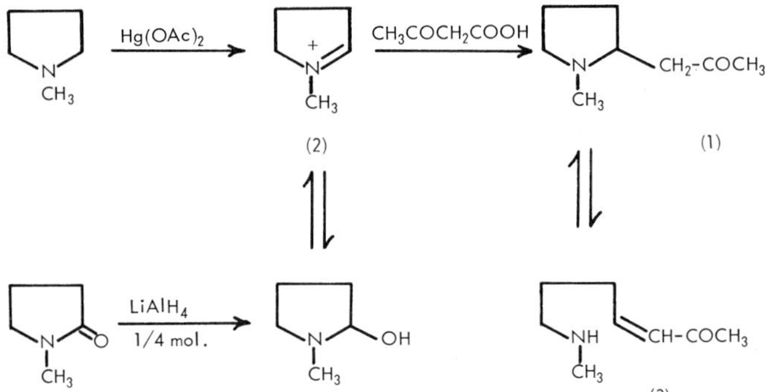

(1) (2) (3)

Cuskhygrine, $C_{13}H_{24}N_2O$, is a diaminoketone containing two N-methyl groups. Like hygrine, it is oxidized to N-methyl proline. The carbon skeleton was deduced by systematic Hofmann degradation to undecan-6-ol (5). Cuskhygrine has been synthesized by a route analogous to that used for hygrine, treating $HO_2C\text{-}CH_2\text{-}CO\text{-}CH_2\text{-}CO_2H$ with two equivalents of (2).

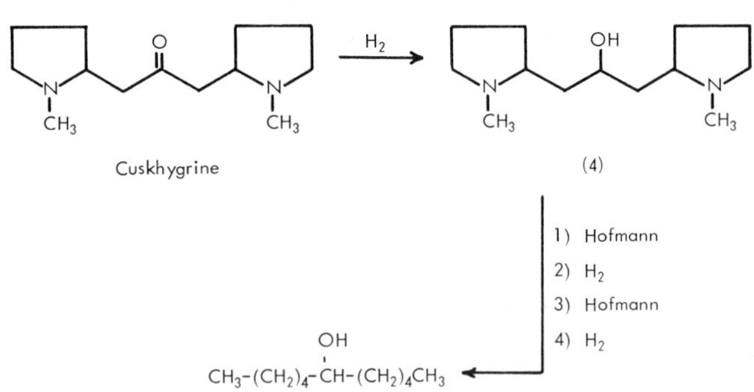

Cuskhygrine (4) (5)

Cuskhygrine is optically inactive; with two identical asymmetric centers, the question remains whether the alkaloid is the racemate or the *meso* compound. The fact that reduction gives a mixture of *two* epimeric alcohols (4) shows that it must be *meso*, since the racemate could afford only a single racemic alcohol.[3]

Shihunine, $C_{12}H_{13}NO_2$, has been isolated from a Chinese plant, *Dendrobium loliohense*.[4] Infrared absorption at 1731 cm^{-1} showed the presence of a γ-lactone, while permanganate oxidation to phthalic acid revealed the location of eight of the carbon atoms. Catalytic hydrogenolysis afforded an amino acid (7) that could be decarboxylated to 1-methyl-2-phenylpyrrolidine. These data, augmented by nmr evidence, led to structure (6).

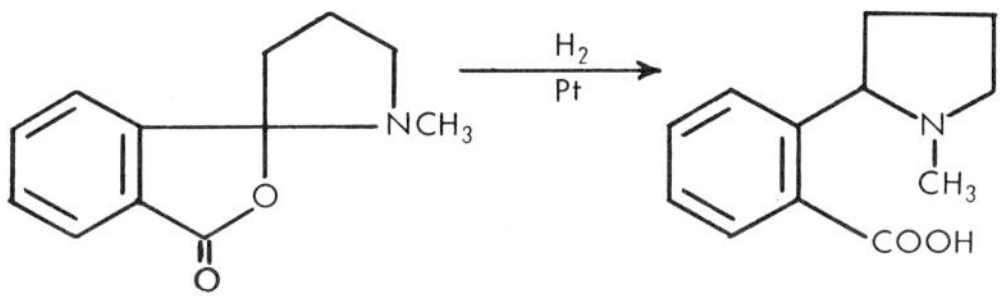

(6) (7)

1.2 Pyrrolizidine Alkaloids[5]

The pyrrolizidine alkaloids constitute a large family based on the pyrrolizidine nucleus. They occur in three main botanical families: *Senecio*, the largest genus of the *Compositae; Heliotropium, Trachelanthus,* and *Trichodesma,* three genera of the *Boraginaceae*; and the genus *Crotalaria* of the *Leguminosae*. These plant species are spread over the globe, and alkaloids have been isolated in the western U.S., Canada, New Zealand, Norway, and South Africa. *Senecio*, the most common in this country, comprises over 1000 species, mainly asters and ragworts. The name comes from the Latin, "old man," because of the resemblance of the flowers to the head of a bald or white-haired man.

Plants of these species, e.g., fireweed and poisonous ragwort, have been notorious for over a century as causing cattle diseases ("horse staggers" or "walking disease"). The alkaloids specifically attack the liver, causing tumors.

Chemically, the alkaloids are esters of aminoalcohols (called necines) and carboxylic acids (necic acids); approximately 100 are known. They may be grouped into three classes: simple monoesters, diesters with angelic acid plus another acid, and cyclic diesters with a dicarboxylic acid. In structure elucidation the alkaloid is generally subjected to alkaline hydrolysis to free the necine and necic acid. In many cases catalytic hydrogenation also liberates the necic acid, suggesting the presence of an allylic ester cleaved by hydrogenolysis. We will discuss here the elucidation of structure of several typical necines and necic acids.

1.2.1 Necines

1.2.1.1 *Structure* Retronecine (8), $C_8H_{13}NO_2$, is perhaps the most common necine. It is a tertiary amine, containing a carbon–carbon double bond and two hydroxyl groups, and so must be bicyclic. Catalytic hydrogenation of retronecine (or any of the monoester alkaloids based on it) gave successively (9) and (10), after the uptake of one or two equivalents of hydrogen. Kuhn–Roth oxidation showed the presence of a C-methyl group in (9) and (10) but not in (8), indicating the hydrogenolysis of a primary allylic alcohol ($C{=}C{-}CH_2OH \rightarrow C{=}C{-}CH_3$. Oppenauer oxidation converted (10) to ketone (11), showing that the remaining hydroxyl is secondary. To convert (10) to the parent skeleton (12) the hydroxyl was removed by reaction with $SOCl_2$ to give the chloride and hydrogenolysis. The structure of (12) was elucidated by a straightforward Hofmann exhaustive methylation; three steps were required to eliminate the nitrogen, revealing that the nitrogen is common to both rings. Synthesis of (13) and (14) established the basic skeleton of the necine as 1-methylpyrrolizidine, (12).

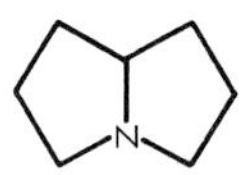

PYRROLIZIDINE

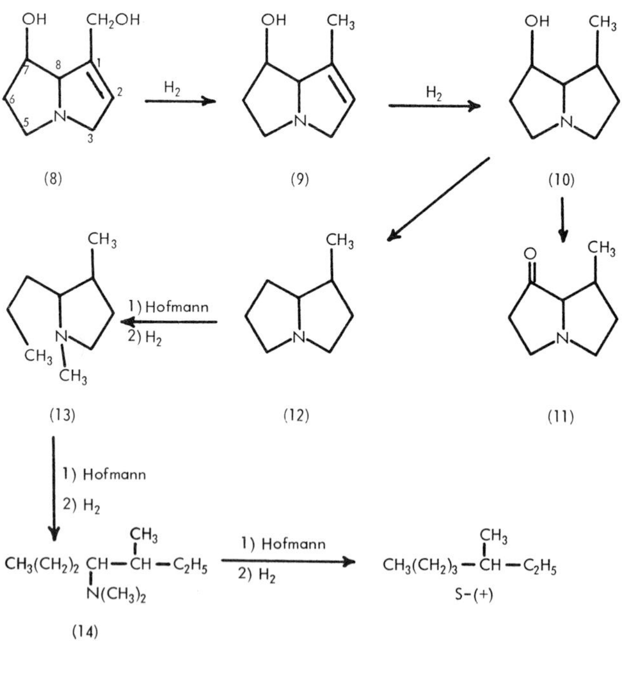

von Braun degradation of (10) with CNBr gave the bromide (15); refluxing with alkali converted it to a cyclic ether, (16). The ready formation of an ether implied that the ether ring was five-or six-membered, and suggested that the secondary hydroxyl was located at C(7). This was confirmed by synthesis of ketone (11). N-Benzoyl-4-methylpiperidine was oxidized to acid (17), which was resolved and then brominated to (18). Treatment with alkali effected ring closure to (19). Esterification and Michael addition to ethyl acrylate yielded diester (20), and Dieckmann cyclization and hydrolysis gave optically active (11).

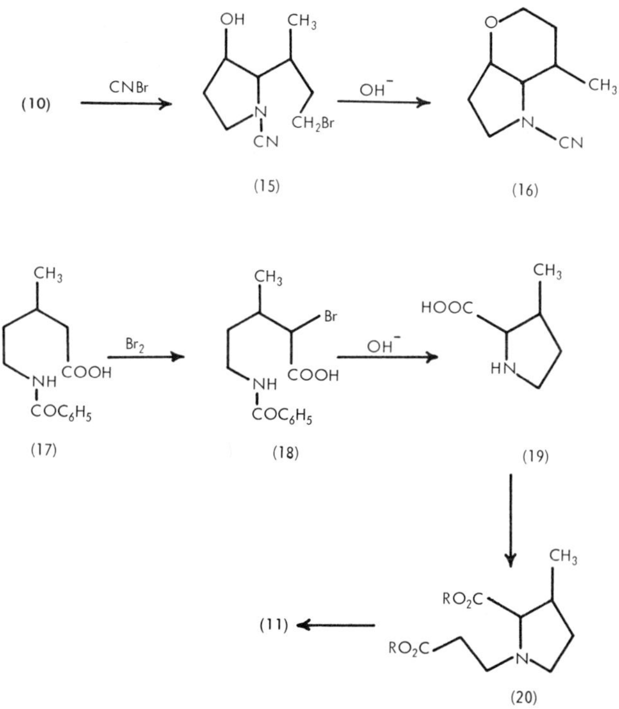

For the primary alcohol to be allylic the double bond must be in the 1,2- or 1,8-position. The latter would be an enamine, and retronecine does not exhibit the acid lability of an enamine. Confirmation of the 1,2-location for the double bond came from ozonolysis of (21) to a keto acid, (22).

These reactions established the structure of retronecine as (8). Other necines have been interrelated with (8), in particular *platynecine* (24), which can be isolated from the products of hydrogenation over Raney nickel, and *heliotridine*, epimeric with retronecine at C(7).

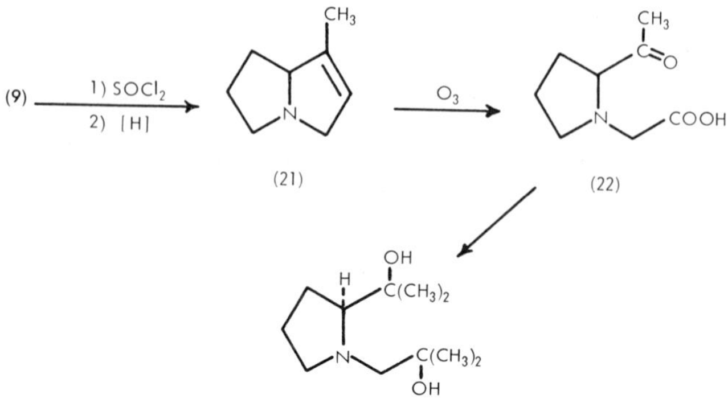

(21) (22) (23)

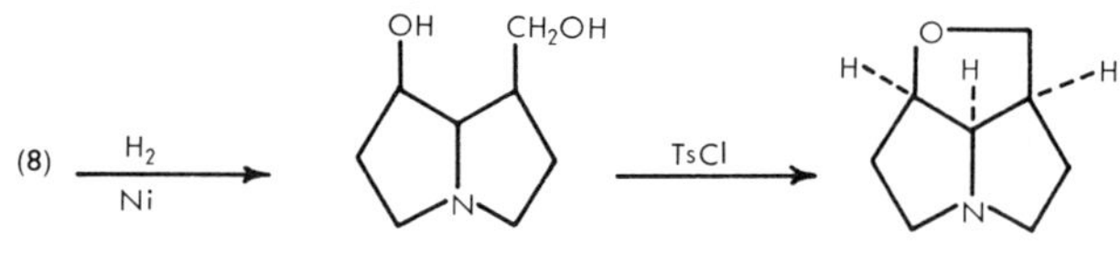

(24) (25)

1.2.1.2 ***Stereochemistry*** Retronecine has two asymmetric centers and consequently could be one of four possible stereoisomers. The relative configuration can be deduced unambiguously from the dehydration of platynecine, (24), to the cyclic ether, (25). This cyclization can be accomplished by a variety of acids or dehydrating agents, among them p-toluenesulfonyl chloride. The reaction is pictured as tosylation of the more reactive primary alcohol followed by internal displacement by the secondary hydroxyl. The fused tricyclic skeleton of (25) can be constructed only if the hydrogens at all three asymmetric centers are *cis*.

The absolute configuration has been assigned in several ways, e.g., (a) Hofmann degradation of (14) to (S)-(+)-3-methylheptane, (b) reaction of the methyl ester of (22) with methyl Grignard reagent to form diol (23), which was independently synthesized from L-proline.

1.2.1.3 ***Synthesis*** A number of syntheses have been recorded for the parent ring system, 1-methylpyrrolizidine. One of the most efficient used Michael addition of nitromethane, first to ethyl crotonate, then to ethyl acrylate to afford the nitro diester (26). Reductive cyclization with hydrogen over copper chromite catalyst gave racemate (12) and its diastereomer, separable by chromatography.[6]

Several of the necines themselves have been prepared by total synthesis. Substitution of $AcOCH_2—CH=CH—COOR$ for ethyl crotonate in the

Michael addition above led to dl-trachelanthamidine[6] (30). This base has also been prepared by other routes; a particularly noteworthy synthesis is patterned on the presumed biosynthetic pathway.[7] The amino bis-acetal (27) was hydrolyzed to the dialdehyde (28), which was allowed to stand for several days at pH 7. Under these conditions Mannich-type condensation occurred spontaneously, leading to aminoaldehyde (29). Borohydride reduction gave racemic trachelanthamidine. It is believed that dialdehyde (28) is closely related to the true biogenetic intermediate formed from two equivalents of ornithine followed by deamination, and that the Mannich cyclization is the most plausible pathway for the biosynthetic reaction.

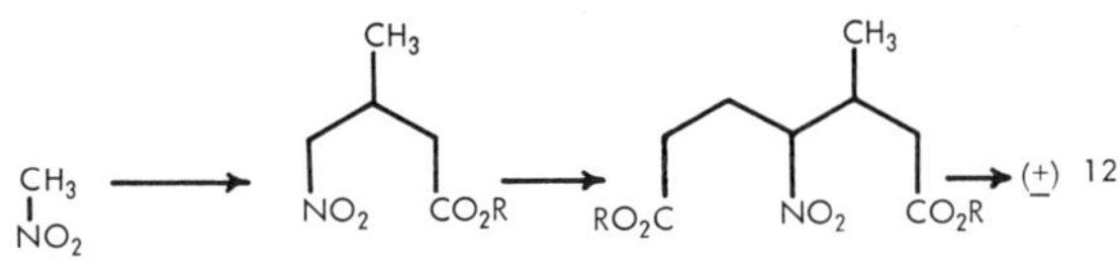

(26)

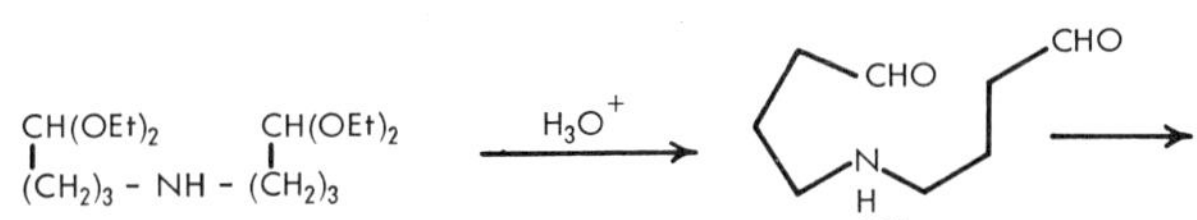

(27) (28)

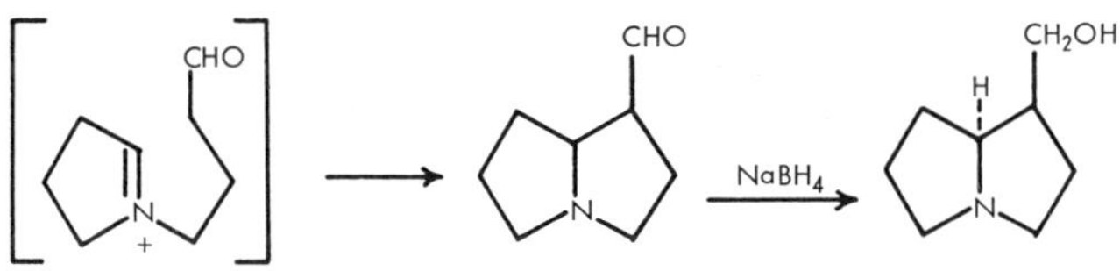

(29) (30)

Retronecine (8), a more difficult synthetic goal because of the double bond and additional hydroxyl, has been prepared by a stereospecific synthesis.[8] Michael addition of diester (31) to ethyl fumarate, followed by Dieckmann cyclization of the adduct, gave the ketodiester (32) after hydrolysis. Sodium borohydride reduction gave an alcohol which cyclized to the lactone (33). The carbamate was hydrolyzed and the amine alkylated with ethyl bromoacetate. Base-catalyzed cyclization of (34) then gave (35). Reduction led to a dihydroxyester, which underwent elimination during alkaline hydrolysis, affording (36). Lithium aluminum hydride reduction gave racemic retronecine, which was resolved to an optically active base identical with the natural necine.

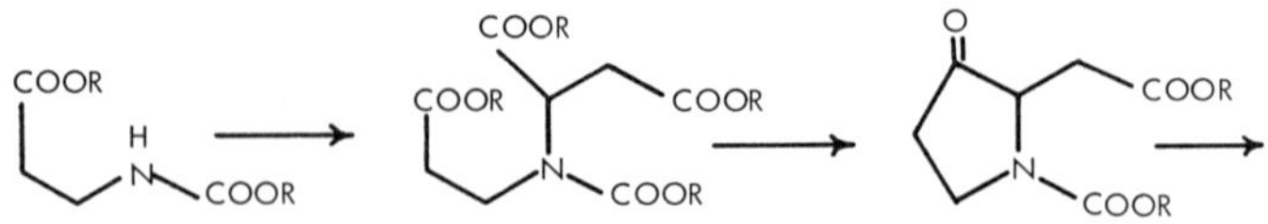

(31) (32)

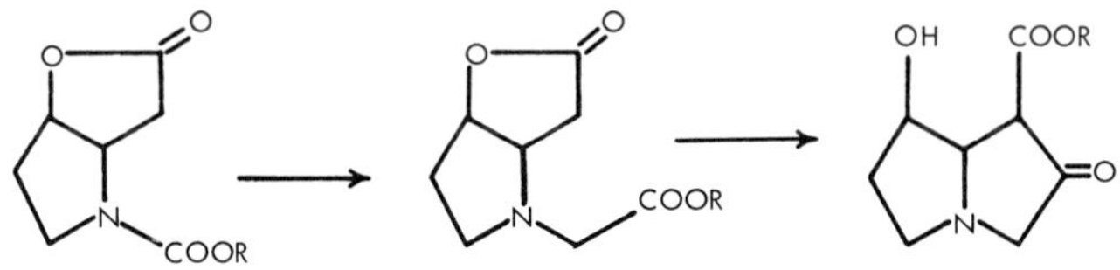

(33) (34) (35)

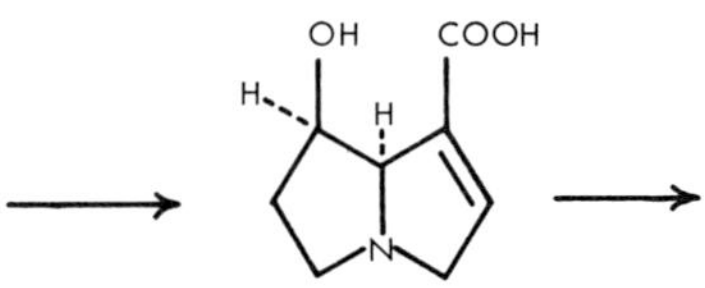

(8)

(36)

1.2.1.4. ***Other Pyrrolizidines*** A few pyrrolizidine bases occur free in nature, e.g., 1-methylenepyrrolizidine, (37), isolated from *Crotalaria* species. Another curious group of bases has been found in the *Gramineae*. These include *festucine* (38) from fescue hay and *loline* (39), as well as related compounds.

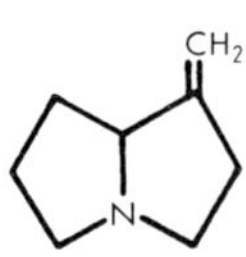

(37)

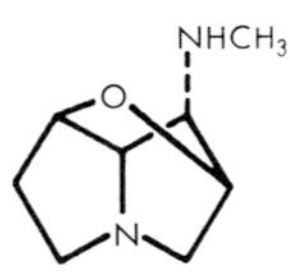

(38)

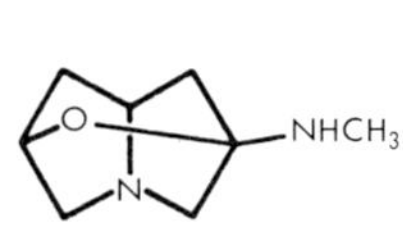

(39)

1.2.2 Necic Acids. (a) *Monocrotalic acid* (40), $C_8H_{12}O_5$, is obtained upon hydrogenolysis of monocrotaline. Hydrolysis of the alkaloid or of (40) gave CO_2 and monocrotic acid (41). The structure of the latter was quickly ascertained by the observation that hypobromite oxidation gave bromoform and α,β-dimethylsuccinic acid.

Monocrotalic acid behaves as a lactone of a dicarboxylic acid and contains a free hydroxyl group. Lithium aluminum hydride reduction of the methyl ester gave a tetrol, (42), which was cleaved by periodate to formaldehyde, acetic acid, and ketol (43). These cleavage products allow structure (40) to be assigned to monocrotalic acid.

(b) *Trachelanthic and viridifloric acids* both have the formula $C_7H_{14}O_4$, and are dihydroxyacids. Reduction with phosphorus and HI, followed by zinc and hydrochloric acid, converted both to ethylisopropylacetic acid, while mercuric oxide oxidation afforded diketone (45). Consequently the acids are diastereoisomers of structure (44). Both acids have been synthesized by hydroxylation of α-isopropylcrotonic acid (46). *cis*-Hydroxylation of *trans*- (46) with alkaline permanganate led to d,l-trachelanthic acid (47), while *trans*-hydroxylation (epoxidation with perbenzoic acid followed by acid hydrolysis) gave d,l-viridifloric acid (48).

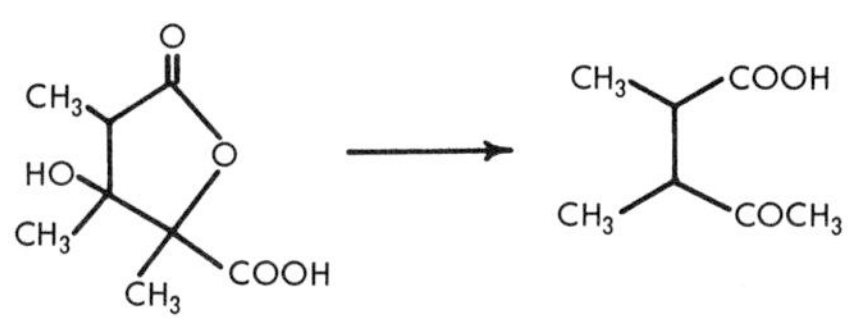

(40) (41)

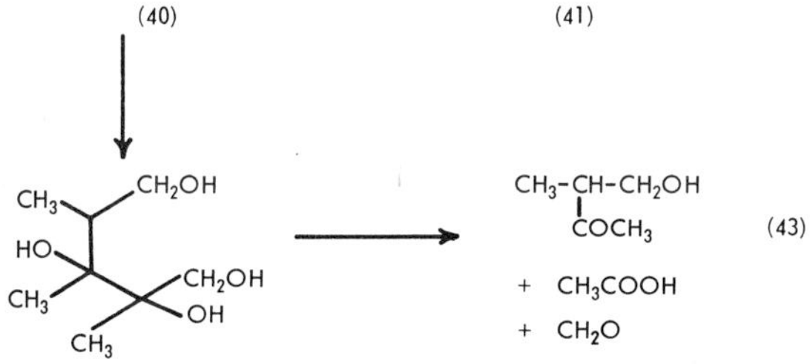

(42) (43)

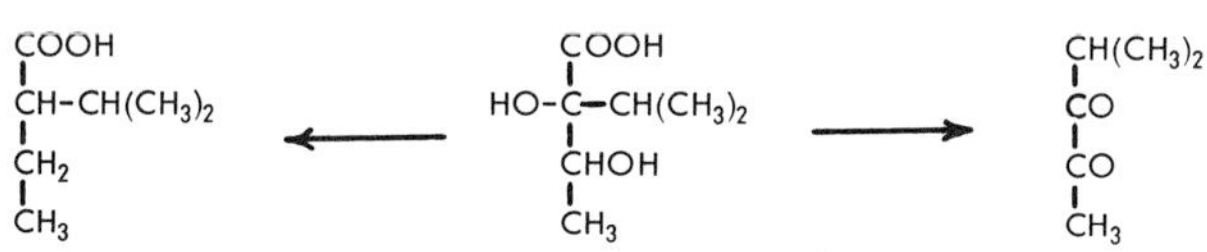

(44) (45)

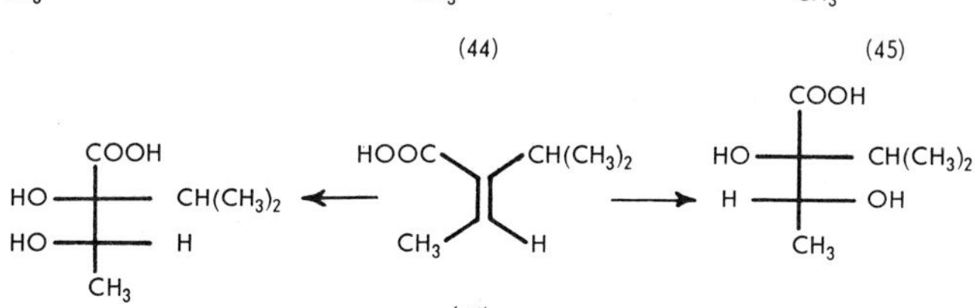

(47) (46) (48)

(c) *Integerrinecic and senecic acids* are isomers of the formula $C_{10}H_{16}O_5$, and both are hydroxydicarboxylic acids. Both contain one double bond, which the uv maximum at 215 nm shows to be conjugated with one carboxyl group. On catalytic hydrogenation, both yielded the same dihydroacid (50). Lead tetraacetate cleavage of (50) gave CO_2 (showing the presence of an α-hydroxy acid) along with a ketoacid (51), which was oxidized by hypobromite to bromoform and 2-ethyl-4-methyl glutaric acid (mixture of *cis* and *trans* isomers). Ozonolysis of both integerrinecic and senecic acids provided acetaldehyde and a diacid (52) that was not isolated, but cleaved by lead tetraacetate to ketoacid (53). Hypobromite oxidation of the methyl ketone grouping afforded methylsuccinic acid. These degradations led to the postulation of structure (49) for integerrinecic and senecic acids. It may be noted that the skeleton of these two acids contains two isopentane units, which suggests a relationship to terpenes. Integerrinecic acid has a uv extinction coefficient of about 9000 and senecic acid about 6000; in accord with other α,β-unsaturated acids, the isomer with the higher coefficient is assigned the *trans* geometry.

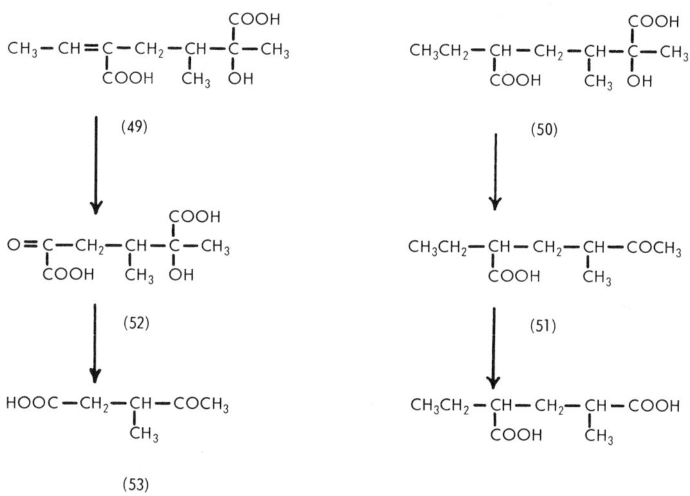

1.2.3. Structures of the Alkaloids Once the structures of the hydroxyamine and acid are known, the structures of the monoester alkaloids follow simply, as e.g., *viridiflorine*, which is hydrolyzed to (30) and (48). In the diester alkaloids it is still necessary to ascertain which acid, or which carboxyl of a diacid, is esterified with each hydroxyl. In the angelate group, e.g., *echiumine* (54), trachelanthic acid is liberated by hydrogenolysis, showing its attachment to the allylic hydroxyl. Similarly, hydrogenolysis of *riddelliine* (55) gives a monoester which liberates CO_2 on treatment with lead tetraacetate, whereas the alkaloid is inert to this reagent; consequently it must be the carboxyl that has an α-hydroxyl in ridellic acid that forms the allyl ester.

Monocrotaline cannot be represented as the monoester of retronecine and monocrotalic acid (40); the infrared spectrum does not show the five-membered lactone which is present in (40), and monocrotaline is a vicinal diol, as shown by its cleavage with lead tetraacetate. Monocrotaline must then be a cyclic diester, (56); once the allylic ester is cleaved by hydrogenolysis, the other ester function is broken by intramolecular attack of the γ-hydroxyl.

An appreciable number of the Senecio alkaloids exist naturally as N-oxides.

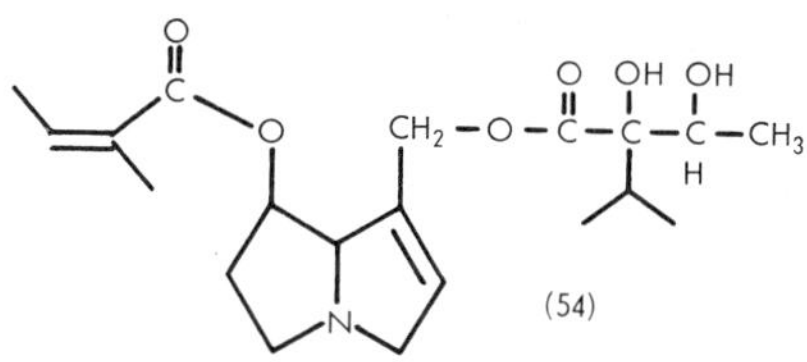

(54)

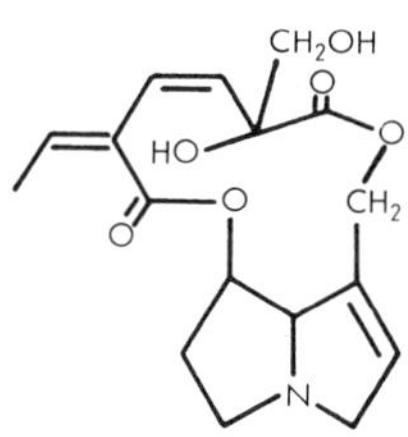

(55)

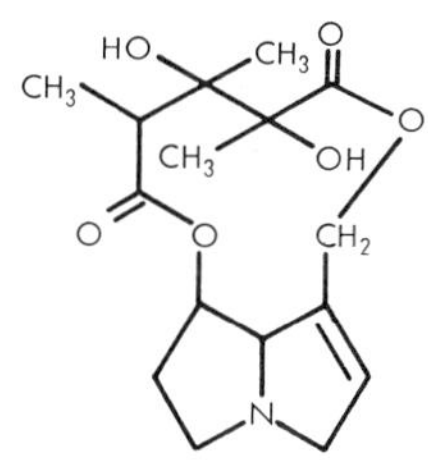

(56)

2. PIPERIDINE AND PYRIDINE ALKALOIDS[9]

Six-membered piperidine and pyridine rings are common among alkaloids, though biosynthetic studies in recent years have shown that they are formed from a variety of pathways.[10] In many (e.g., the lupin and lobelia alkaloids), the amino-acid, lysine, serves as the precursor, analogous to the production of pyrrolidine rings from ornithine. In others (hemlock group, himbacine, muscopyridine), the nitrogen appears to have been inserted into a straight carbon chain derived from acetate units. Still others (such as actinidine and gentianine) are clearly terpenoid skeletons incorporating nitrogen and are derived from mevalonate. Finally, the pyridine ring of nicotine and related tobacco alkaloids has been shown to originate from nicotinic acid, itself assembled from smaller units (glyceraldehyde and aspartic acid). Because of the diverse origins of piperidine and pyridine rings, we shall not attempt to organize them on a biogenetic basis, but by classes of related structure.

2.1 Piperidines and 2-Alkylpiperidines

2.1.1 Piperine and Chavicine, isomers of the formula $C_{17}H_{19}NO_3$, are weakly basic alkaloids from pepper. Alkaline hydrolysis affords piperidine and piperic acid from piperine, chavicinic acid from chavicine. The two acids have been shown to be geometric isomers of (58) and the alkaloids consequently have the amide structure (57).

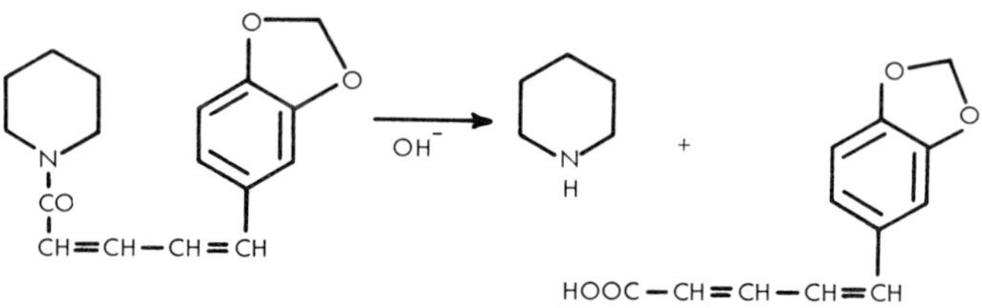

(57) (58)

2.1.2 Hemlock Alkaloids The toxic properties of the poison hemlock, *Conium maculatum,* have been known since antiquity, and extracts were used in ancient Greece for the execution of criminals. One of the earliest contributors to hemlock literature was Plato, who described the death of Socrates from hemlock poisoning.

Five alkaloids have been identified; the major one is *coniine*. It is an optically active secondary amine, $C_8H_{17}N$; both (+) and (−) forms occur in nature. Zinc dust distillation dehydrogenated it to 2-*n*-propylpyridine, while heating with hydriodic acid at 300° afforded n-octane. From this information coniine was assigned structure (59). This was confirmed by Ladenburg (1886), who carried out the first synthesis of an alkaloid. α-Picoline was condensed with acetaldehyde to yield (60), which was reduced to d,l-coniine and resolved. γ-*Coniceine* (61). the alkaloid present in largest amount in growing plants, is believed to be the precursor to the rest of the family.

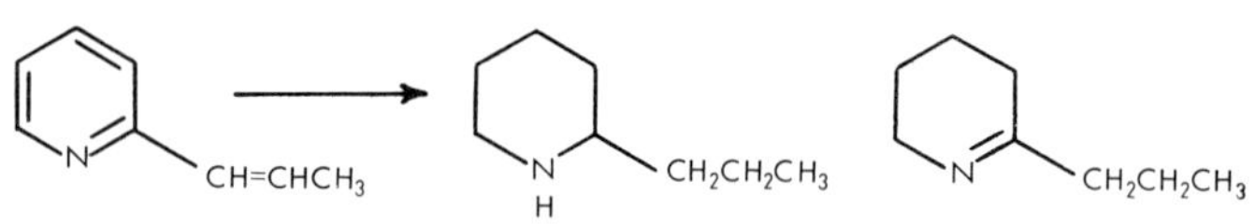

(60) (59) (61)

Two other hemlock alkaloids are hydroxylated derivatives of coniine. *Conhydrine* (62) and *pseudoconhydrine* (69) are both converted by hot HI to iodides which can be reduced to coniine. Chromic acid oxidation of conhydrine gave L-(−) pipecolic acid (63), so that the hydroxyl group must be located on the propyl sidechain. The location was ascertained by both degradation and synthesis. Hofmann degradation of conhydrine, taking the course typical of β-hydroxyamines, produced the epoxide (64); the second stage of the Hofmann sequence yielded the unsaturated epoxide (65). Hydrogenation and hydrolysis gave a crystalline diol, identified as octane-3,4-diol (66) from its oxidation products. Conhydrine thus must be the β-hydroxyamine (62). Synthesis was effected by Grignard synthesis of (67) and hydrogenation to both racemates of (62), which were separated by crystallization. Resolution of the higher-melting racemate gave (+)-conhydrine.

The relative configuration of conhydrine was assigned by determining the stereochemistry of the Hofmann product (66). The *erythro* isomer of (66), synthesized from *cis*-octene-3 by *cis*-hydroxylation, proved identical with the diol prepared by degradation of d,l-conhydrine.[11] Since epoxide formation (62)→(64) and hydrolysis (65)→(66) both occur with inversion of configuration, conhydrine must also be the *erythro* isomer (68).

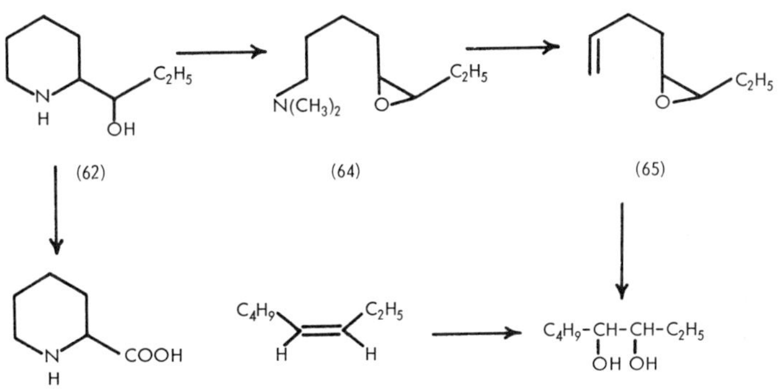

(62) (64) (65)

(63) (66)

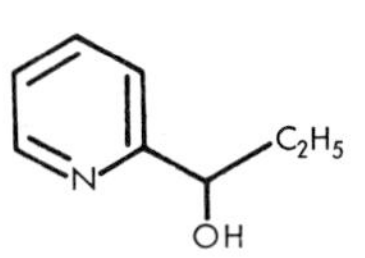

(67)

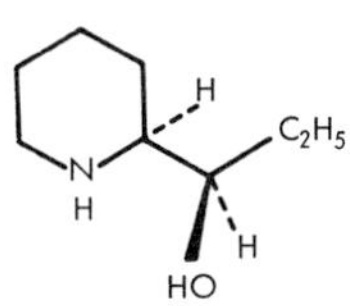

(68)

The location of the hydroxyl group in pseudoconhydrine (69) was shown by Hofmann degradation, which proceeded through (70) to (–) octene-1,-2-oxide (71), again with participation of the hydroxyl in epoxide formation. The structure thus deduced has been confirmed by several syntheses. One began with 5-hydroxy-2-picoline (72), prepared by KOH fusion of the corresponding sulfonic acid. After O-methylation, sodamide-catalyzed alkylation of the acidic methyl group with ethyl chloride, followed by ether cleavage, gave (73). Hydrogenation yielded d,l-(69), resolved to afford natural pseudoconhydrine.

X-ray analysis of pseudoconhydrine hydrobromide[12] has shown that the substituents on the ring are *trans*. The absolute configuration has been elucidated[11] by hydride reduction of oxide (71) to (R)-2-octanol.

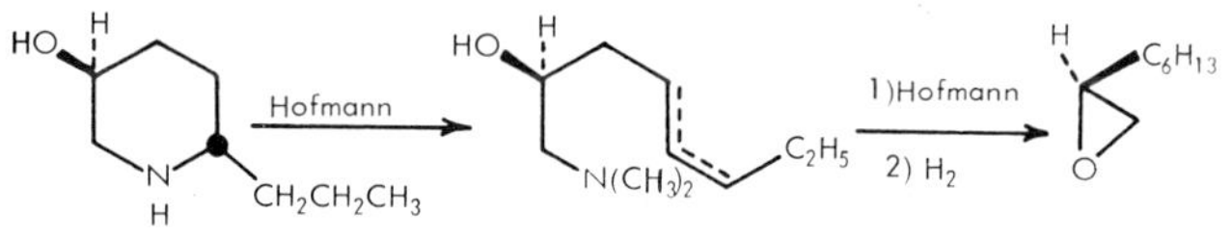

(69) (70) (71)

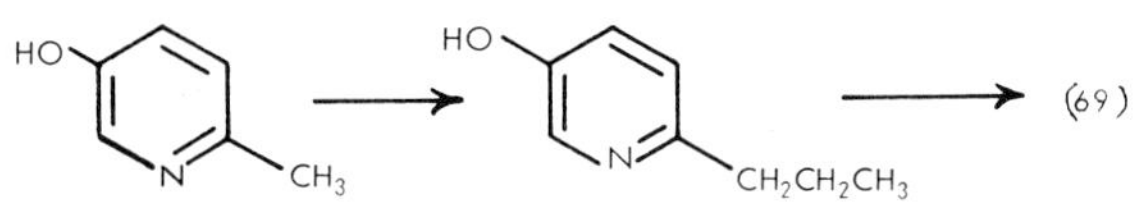

(72) (73)

2.1.3 Pomegranate Alkaloids The root bark of the pomegranate (*Punica granatum*) contains several alkaloids named in honor of the French chemist Pelletier. *Isopelletierine*, $C_8H_{15}NO$, is a secondary aminoketone; methylation yields N-methylisopelletierine (75), a congener. Wolff–Kishner reduction of (75) furnished N-methylconiine, and chromic acid oxidation gave N-methylpipecolic acid, so isopelletierine has the coniine skeleton with a carbonyl group in the sidechain. The structure was established by synthesis: acylation of α-picolyl lithium with acetic anhydride gave ketone (76), which was hydrogenated to d,l-isopelletierine (74). Isopelletierine, like hygrine, racemizes on treatment with bases and for this reason is usually isolated as the racemate, but careful extraction[13] avoiding alkali has afforded (R)-(−) (74) and (75).

Pelletierine has a curious history with a surprise ending. An optically active base isomeric with isopelletierine, it was regarded as an aldehyde because the oxime was dehydrated by PCl_5 to a substance $C_8H_{14}N_2$, thought to be a nitrile. Since pelletierine was reduced by the Wolff–Kishner method to coniine, it was assigned structure (77). Only after years of clever but fruitless efforts to synthesize this aminoaldehyde, attempts which led either to bicyclic compounds or polymers, was suspicion turned toward this formulation. Careful reinvestigation[13,14] showed that pelletierine was the (−) form of isopelletierine! The presumed nitrile $C_8H_{14}N_2$ was shown to lack cyano absorption in the infrared and proved to be the amidine (78), formed by dehydration of the normal Beckmann rearrangement product of isopelletierine oxime.[15] Thus, the aldehyde formula (77) does not represent a natural product.

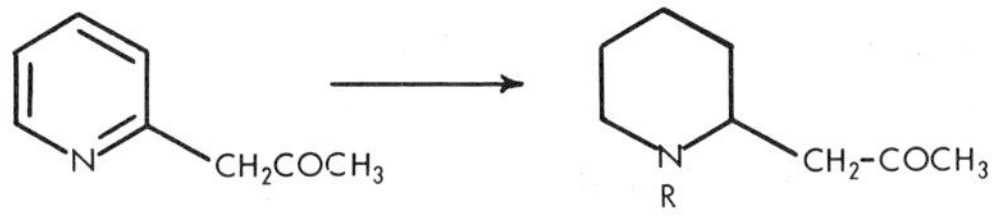

(76) (74) R = H (75) R = CH_3

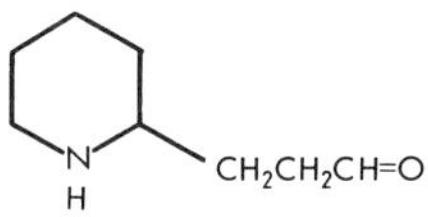

(77)

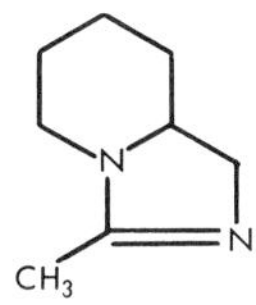

(78)

Pseudopelletierine (79), the remaining pomegranate alkaloid, is a tertiary amine of the formula $C_9H_{15}NO$. The oxygen is present as a carbonyl group, and formation of a dibenzylidene derivative upon condensation with benzaldehyde points to the moiety—CH_2COCH_2—. The carbon skeleton was elucidated by Willstäter in a classic application of the Hofmann exhaustive methylation. Electrolytic reduction of the keto group afforded the parent base (80). Two Hofmann eliminations yielded a nitrogen-free product, identified as 1,5-cyclooctadiene (82) by oxidation to succinic acid. Pseudopelletierine is optically inactive and cannot be resolved, so the nitrogen must be bonded to the cyclooctane skeleton in a symmetrical manner. Of the few possible *meso* structures, only (79) is consistent with the dehydrogenation of (80) to 2-n-propylpyridine. Pseudopelletierine is thus a higher homolog of tropinone. It has been synthesized by the same elegant Mannich-type condensation[16] used for tropinone: glutaraldehyde, methylamine, and acetone dicarboxylic acid, allowed to stand for several days at room temperature, afford a high yield of (79).

Having prepared a cyclooctadiene by Hofmann degradation of (80), Willstäter set out to synthesize the unknown cyclooctatetraene (86) to see if it had the same aromatic properties as its lower homolog benzene. Pseudopelletierine was reduced to the alcohol and dehydrated to (84), which was carried through two Hofmann sequences to the triene (85). Addition of bromine, replacement with dimethylamine, and a final Hofmann yielded cyclooctatetraene. It proved, of course, to be nonaromatic—it adds halogens, is easily oxidized and polymerized, and behaves completely as a polyene. A method of preparing cyclooctatetraene on an industrial scale was discovered during World War II, by tetramerizing acetylene in an autoclave with nickel (II) salts. Detailed study of its chemistry has led in recent years to a host of unexpected and intriguing findings[17]: the preparation of the "aromatic" dianion, cyclononatetraenyl anion, and homotropylium cation; valence tautomerism to bicyclo-(4.2.0)-isomers; oxidative rearrangements to tropylium and benzene derivatives; and synthesis of the fascinating bullvalene via a cyclooctatetraene dimer. Willstäter's early synthesis was indeed a seminal study.

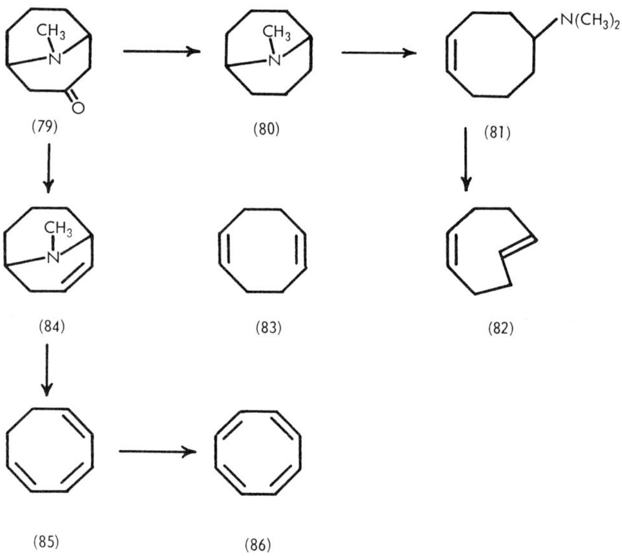

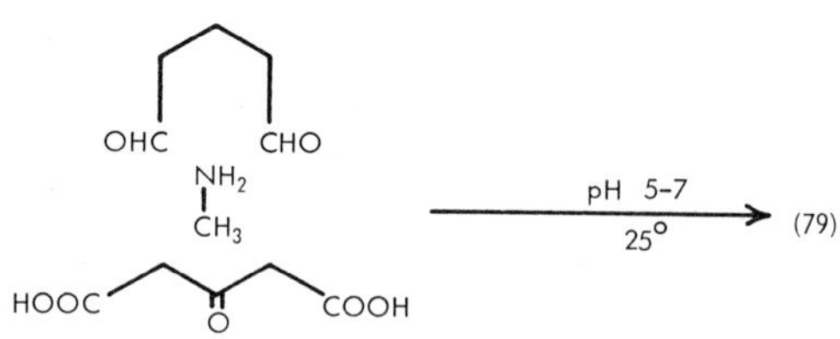

Still another interesting aspect of medium rings came from Willstäter's degradation of pseudopelletierine. The 1,5-cyclooctadiene obtained from Hofmann degradation was a labile, reactive diene, unlike the 1,5-cyclooctadiene prepared later by dimerization of butadiene. This has been explained on the basis of geometrical isomerism;[18] the butadiene dimer is the *cis*, *cis* isomer (83), while the Hofmann product is the *cis*, *trans* isomer (82). The eight-membered ring is the smallest in which geometrical isomers are possible, and both *cis* and *trans* cyclooctenes are known. The presence of a *trans* double bond in an eight-membered ring has a further consequence, that the molecule is held in a rigid, nonplanar, dissymmetric conformation and is accordingly resolvable. If amine (81) is resolved and then subjected to Hofmann elimination, the cyclooctadiene formed is optically active. Trans-cyclooctene has been resolved through the use of platinum complexes with optically active amines.[19]

2.1.4 Sedum Alkaloids[20] The stonecrop (*Sedum*) family contains several alkaloids structurally related to those of hemlock and pomegranate. *Sedridine*, $C_8H_{17}NO$, has secondary amino and hydroxyl groups. Chromic acid oxidation of (+)-sedridine gave (+)-isopelletierine, which established the structure as (87). von Braun degradation of the O,N-dibenzoate yielded dibromide (88), which was hydrogenolyzed to (S)-(+)-2-octanol benzoate, defining the configuration at the hydroxyl.[21]

Sedamine, $C_{14}H_{21}NO$, is a related secondary alcohol from *Sedum*, while a stereoisomer, *allosedamine*, has been found in *Lobelia inflata*. Both are oxidized to the same levorotatory ketone and further to a mixture of

(87) (88)

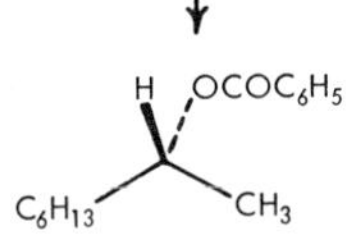

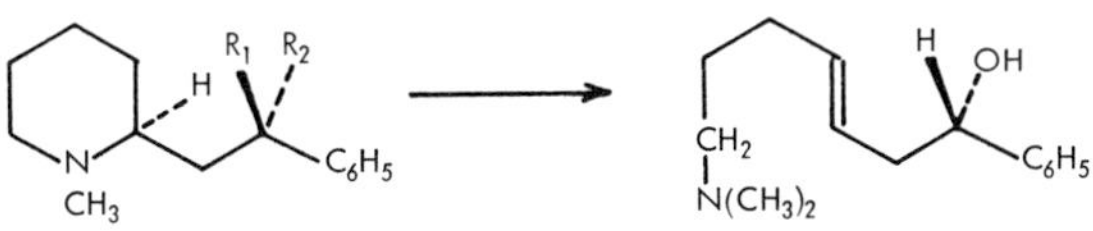

(89) R_1 = OH, R_2 = H

(90) R_1 = H, R_2 = OH

(92)

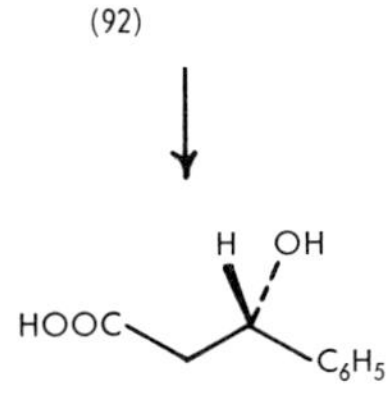

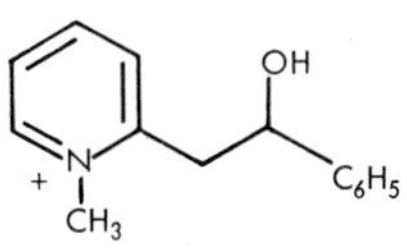

(93)

(91)

benzoic and (−) N-methylpipecolic acids, indicating structures (89) and (90) for the alkaloids. These have been confirmed by synthesis; base-catalyzed reaction of α-picoline methiodide with benzaldehyde gave (91), which was hydrogenated to a mixture of racemic sedamine and allosedamine. Hofmann degradation of allosedamine to (92), followed by oxidation to (S)-(−)-(93), showed that allosedamine is (90) and sedamine thus (89).

2.1.5 Other 2-Alkylpiperidines Another aminoalcohol related to the *Sedum* group is (94), which occurs in *Lobelia*. *Withania somnifera* contains cuskhygrine, isopelletierine, and two related ketones,[22] *anaferine* (95) and *anahygrine* (96). A general synthetic route to these, the *Sedum* alkaloids, and isopelletieriene is the addition of a β-ketoacid to piperideine (97) (cf. hygrine synthesis, Section 1.1) followed by reduction and N-methylation if necessary. Because of the generality of this scheme and the common occurrence of β-ketoacids in plants, it is believed that this represents the biosynthetic pathway as well.

Febrifugine (98), the alkaloid of the common hydrangea, is another hydroxypiperidine with a 2-propanone chain attached to quinazolone.[23] It has attracted interest because of its powerful antimalarial activity, and several syntheses have been recorded.

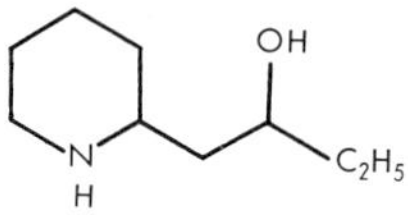

(94)

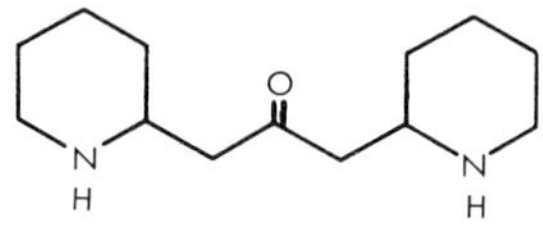

(95)

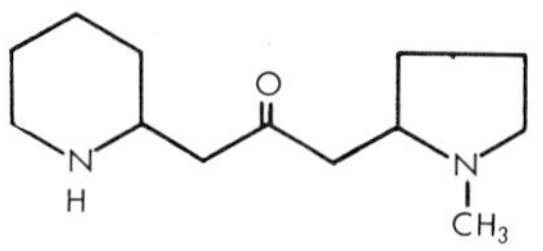

(96)

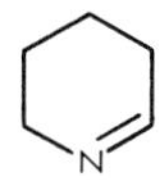

(97)

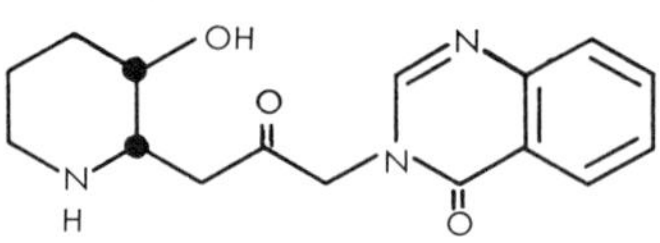

(98)

2.2 2,6-Dialkylpiperidines

2.2.1. Lobelia Alkaloids[24] *Lobelia inflata,* commonly known as Indian tobacco, is an emetic herb of the Great Smoky Mountains, rich in alkaloids. Formerly used to induce vomiting and to treat asthma and bronchitis, it has found a new application recently in antismoking preparations.

The main alkaloid is *lobeline* (99), $C_{22}H_{27}NO_2$, an optically active tertiary amine containing keto and secondary hydroxyl groups. The diol formed by reduction, *lobelanidine* (100), as well as the diketone *lobelanine* (101), accompany lobeline in the plant; both of these substances are optically inactive and therefore *meso* compounds. Hofmann elimination of (101) gave a diene (102), which was hydrogenated to 1,7-dibenzoylheptane, while Beckmann rearrangement of the dioxime and hydrolysis of the diamide formed led to aniline and diacid (103). This information was sufficient to assign structures to the alkaloids.

More than half a dozen different syntheses have been achieved. The most practical route, used on an industrial scale, involves a Mannich condensation of glutaraldehyde, methylamine, and benzoylacetic acid at pH 4 to give lobelanine in one step in 90% yield.[16]

Other alkaloids of *Lobelia* have one phenyl substituent replaced by ethyl; their structures have been deduced by similar methods. An example of this group is *lobinine* (104).

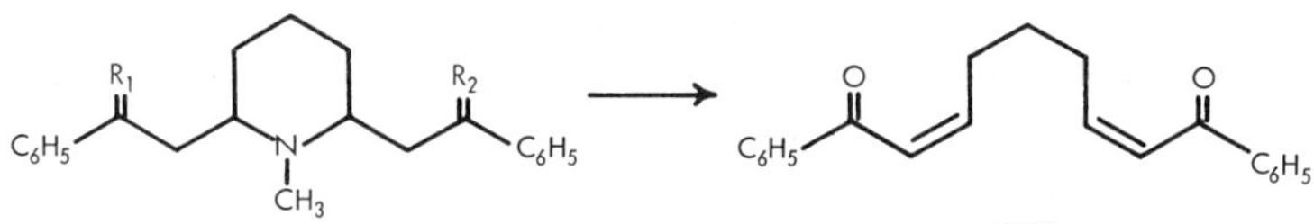

(99) R_1 = H, OH; R_2 = O

(100) $R_1 = R_2$ = H, OH

(101) $R_1 = R_2$ = O

(102)

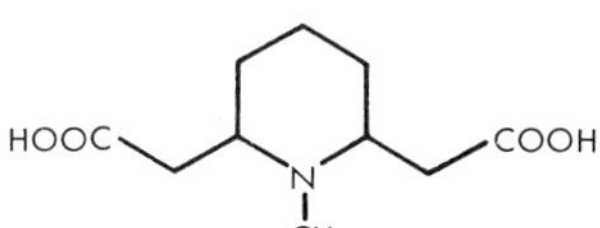

(103)

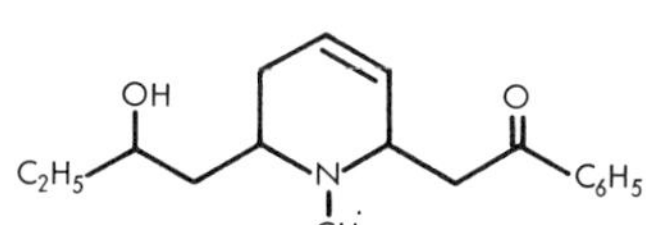

(104)

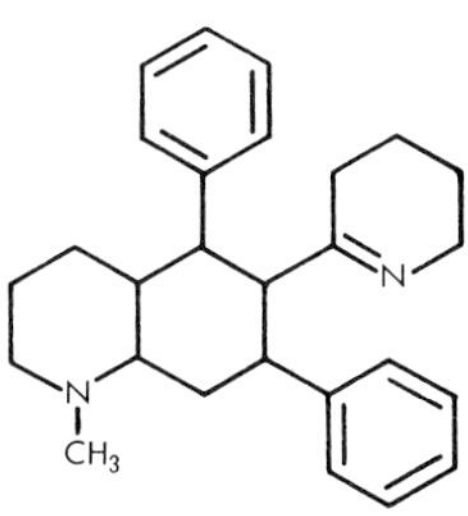

(105)

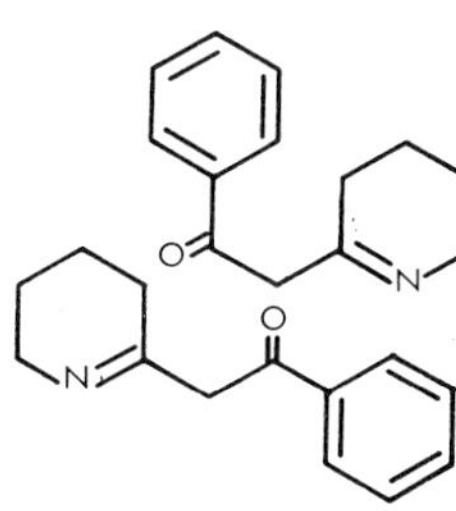

(106)

The major alkaloid of *Lobelia cardinalis, lobinaline,*[25] has a very different structure (105), which was deduced primarily by dehydrogenation to the derivative with all five rings aromatic. It seems probable that lobinaline is formed in nature by dimerization of a sedamine-like unit, e.g., (106), followed by reduction and methylation. This theory is strengthened by the occurrence of allosedamine in *Lobelia* species.

2.2.2 Pinidine[26] Though most pine species are devoid of alkaloids, certain of the western varieties, e.g., *Pinus sabiniana* Dougl., contain2-methylpiperidine and pinidine. The latter, $C_8H_{17}N$, was recognized as a piperidine by dehydrogenation to 6-methyl-2-*n*-propyl-pyridine. Ozonolysis afforded acetaldehyde, leading to structure (107) for pinidine. Confirmation came from oxidation of N-benzoylpinidine to acid (108), which was synthesized for comparison.

The relative configuration was established by converting (108) to a symmetrical derivative. Lithium aluminum hydride reduction gave alcohol (109). Further reduction of the corresponding chloride led to 2,6-dimethyl-N-benzyl-piperidine (110), identified as the *cis*-isomer by the lack of optical activity. An interesting outgrowth of this sequence was the observation that the *cis*-and *trans*-isomers of (110) could be differentiated by nmr spectroscopy; the benzyl methylene protons appear as a singlet in the *cis* amine, while in the *trans* isomer the attachment of the methylene group to a dissymmetric moiety results in magnetic nonequivalence of its protons, which then appear as a quartet.[27]

Hydrogenolysis-hydrogenation of N-methylpinidine methiodide gave (+)-2-dimethylaminononane (111), the enantiomer of which was prepared independently from (S)-(+)-2-nonanol, establishing the absolute configuration.

Simpler natural dialkylpiperidines are the dimethyl piperidines (112) (R = H and CH_3) found in *Nanophyton erinaceum*.[28] Both are optically active *trans* isomers.

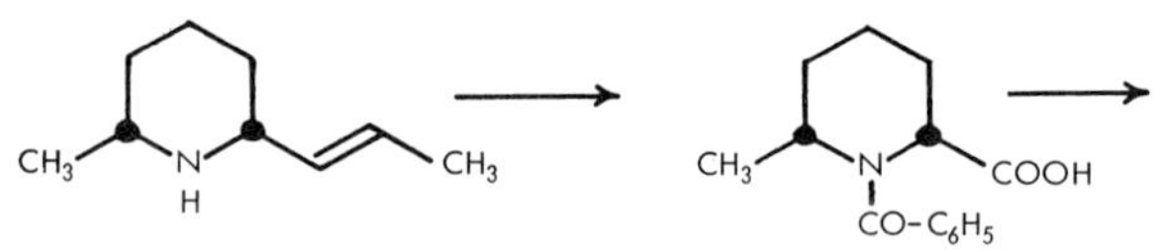

(107) (108)

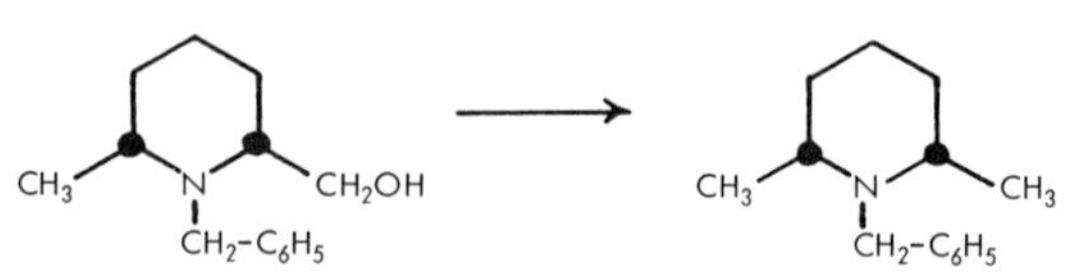

(109) (110)

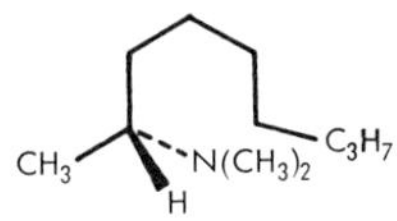

(111)

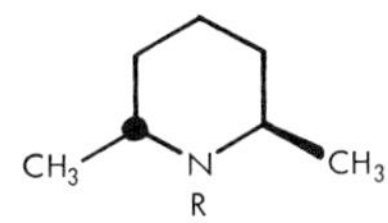

(112)

2.2.3 Carpaine and Cassine Carpaine ($C_{14}H_{25}NO_2$), a crystalline alkaloid found in the leaves of the papaya (*Carica papaya* L.), has the properties of a lactone and can be hydrolyzed to carpamic acid (113), a secondary amino acid. Oxidation of (113) with nitric acid afforded azelaic and suberic acids, indicating the presence of a straight chain of nine carbon atoms. When carpaine was subjected to Hofmann exhaustive methylation, hydrogenating after each elimination, two stages were required to eliminate the nitrogen, and the normal C_{14} acid, myristic acid, was formed. Apparently the lactone oxygen had also been eliminated at some step, but at any rate all fourteen carbons of carpaine were shown to lie in a straight chain.

Palladium dehydrogenation of carpaine gave a substituted pyridine (114) that was oxidized to 2,6-pyridine dicarboxylic acid. Similar dehydrogenation of the ethyl ester of (113) led to the related pyridine (115) in which the hydroxyl had been retained. Carpamic acid thus appeared to be a hydroxy derivative of 8-(6-methyl-2-piperidyl)-octanoic acid, and the remaining structural problem was to locate the hydroxyl on the piperidine ring.

To do this, the Hofmann elimination was repeated on carpamic acid; in contrast to the lactonic alkaloid, (113) retained the hydroxyl group throughout the sequence. After two Hofmann eliminations and subsequent hydrogenation, thc resulting hydroxyacid was oxidized by chromic acid to a C_{14} ketoacid along with the normal C_{12} dicarboxylic acid. Since the C_{14} keto acid gave a negative iodoform test it was presumed to be (116), a hypothesis quickly confirmed by synthesis. From this result structure (113) was deduced;[29] additional evidence for the relative position of the hydroxyl and amino groups was the ready cleavage by lead tetraacetate. Final proof of structure came from synthesis of the dehydrogenation products (114) and (115).

The stereochemistry of carpaine was deduced in a remarkably simple way.[30] The infrared spectrum of the methyl ester of (113), taken in dilute carbon tetrachloride solution, revealed a strongly hydrogen-bonded-OH whose relative intensity was undiminished on dilution, and which, therefore, must be due to intramolecular O—H · · · N bonding. For such an intramoleucular hydrogen bond to be possible the hydroxyl must be axial. But the only configuration in which the stable conformation would

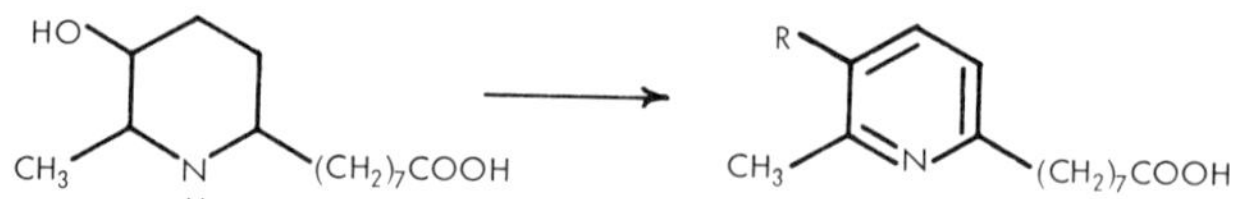

(113)

(114) R = H

(115) R = OH

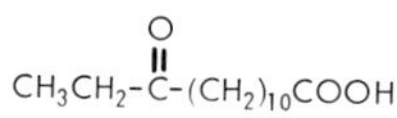

(116)

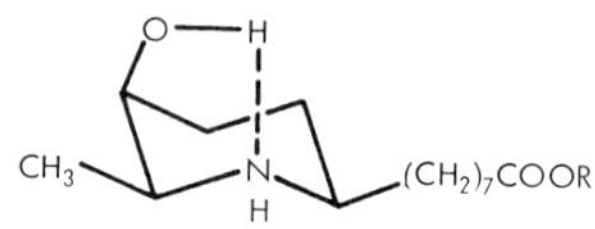

(117)

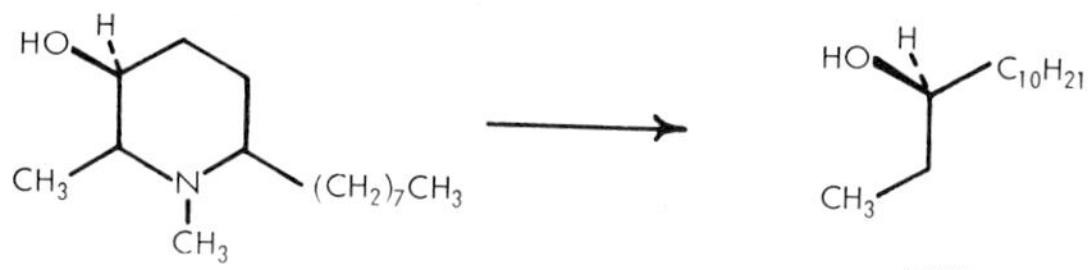

(118)

(119)

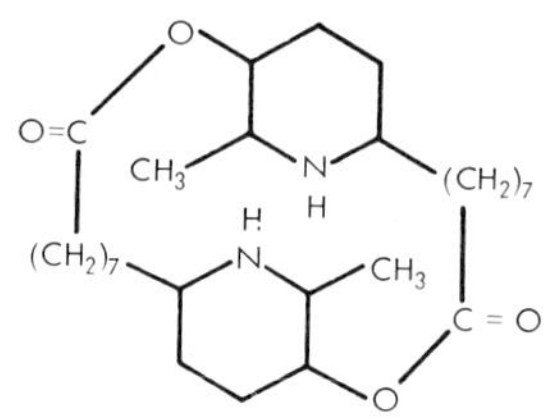

(120)

have an axial hydroxyl is that in which the two alkyl groups are equatorial, so carpamic acid must have the all *cis* configuration (117).

To determine the absolute configuration, (113) was first converted to (118) by a sequence of methylation, lithium aluminum hydride reduction, formation of the primary chloride, and another hydride reduction. Hofmann elimination (two stages) followed by hydrogenation gave (R)-(−)-3-tetradecanol (119), whose absolute configuration was established by synthesis from (R)-(+)-1,2-epoxybutane and n-decyl lithium.[31]

Once the structure of carpamic acid was settled, the formula of carpaine was accepted as the corresponding monomeric lactone for years. With the aid of mass spectrometers capable of handling high molecular weight materials it has been shown[32] that the parent peak of carpaine is twice that of a C_{14} formula, and that carpaine is a dimer of carpamic acid, the dilactone (120). Though carpaine consequently has a 26-membered ring, it can be prepared from carpamic acid by heating with thionyl chloride.

Cassine,[33] an alkaloid from a tropical American shrub, *Cassia excelsa* Shrad., has the formula $C_{18}H_{35}NO_2$, as shown by mass spectrometry. It is a secondary amine with a secondary hydroxyl and keto group. The nmr spectrum shows two methyl groups, one a doublet at $\tau 8.98$ ($-CHCH_3-$), the other a singlet at $\tau 7.95$, characteristic of a methyl ketone. This was confirmed by base-catalyzed exchange of five protons in deuteromethanol, indicating the grouping $-CH_2-CO-CH_3$. Dehydrogenation with palladium afforded (122), recognized by its uv spectrum as a 3-hydroxypyridine. In the nmr spectrum of (122) a singlet at $\tau 7.5$ (aromatic CH_3) had replaced the methyl doublet, showing that one methyl was located on the piperidine ring. Wolff–Kishner reduction of (122) and nitric acid oxidation yielded a small amount of lauric acid, revealing the nature of the long alkyl sidechain.

A novel application of the Beckmann fragmentation was used to locate the secondary methyl. The carbonyl group of N-methylcassine was re-

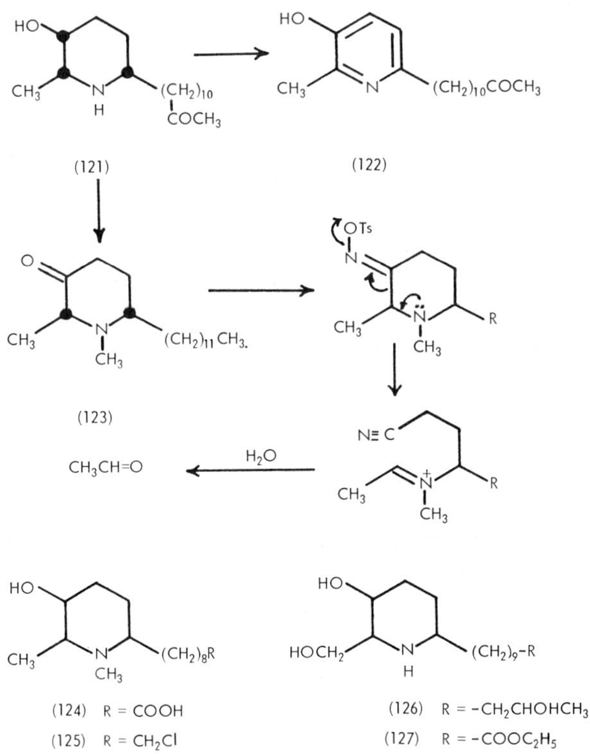

moved by Wolff–Kishner reduction and the secondary hydroxyl oxidized to ketone (123). Oximes of α-aminoketones are known to fragment under Beckmann conditions; the oxime of (123) liberated acetaldehyde on treatment with p-toluenesulfonyl chloride and pyridine, pinpointing the methyl between the nitrogen and carbonyl.

The structure thus derived for cassine (121) has been confirmed by partial synthesis from carpaine. N-Methylcarpamic acid was homologated to acid (124), which was reduced and converted to chloride (125). This chloride was then used to alkylate ethyl acetoacetate, leading after hydrolysis to a compound with the structure of N-methylcassine. The synthetic material was identical in every respect but the sign of optical rotation, showing that cassine and carpaine belong to enantiomeric series.

Two related alkaloids, *prosopine* (126) and *prosopinine* (127), have been found in *Prosopis africana*.

2.2.4 Galbulimima Alkaloids A curious group of alkaloids from giant trees of the Australian and New Zealand rain forests have somewhat related structures, in which one large substituent on a 2,6-dialkylpiperidine is a decalin derivative. A typical member is *himbacine* (128). The structures were derived by painstaking degradations and finally confirmed by x-ray crystallography.[34]

2.2.5 Muscopyridine Muscopyridine, $C_{16}H_{25}N$, is one of the few alkaloids from animal, rather than plant sources; it was isolated from the scent gland of the musk deer. It is an optically active pyridine, oxidizable to 2,6-pyridine dicarboxylic acid. The similarity in empirical formula to muscone (129), $C_{16}H_{30}O$, which accompanies it in the musk gland, led to the imaginative suggestion that muscopyridine might have the related structure (130). This structure was confirmed by an equally imaginative total synthesis.[35] Cyclododecanone was condensed with ethyl succinate in a Stobbe condensation to give (131), which was cyclized with acid, hydrolyzed, and decarboxylated to ketone (132). Wolff–Kishner reduction gave predominantly the olefin (133) in which the double bond had migrated. Application of the Schmidt reaction (HN_3) to (133), followed by immediate palladium dehydrogenation, gave a mixture of pyridines, mostly the desired (134). An acetoxy group was introduced into the α-position by rearrangement of the N-oxide with acetic anhydride, then saponified and oxidized to ketone (135). Monomethylation (methyl iodide and potassium t-butoxide), and reduction of the carbonyl group led to d,l-muscopyridine.

A final alkaloid which contains a long bridge connecting the 2- and 6-positions of a piperidine ring is *lythranidine* (135a), isolated from Japanese *Lythraceae* species.[36] The unusual skeleton has been confirmed by synthesis, involving double alkylation of 2,6-lutidine by the necessary biphenyl dihalide.

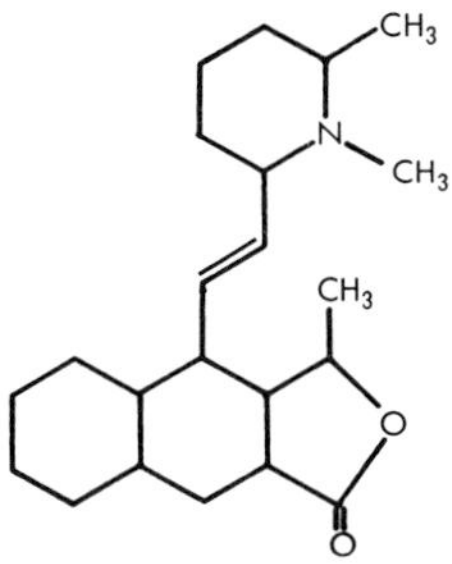

(128)

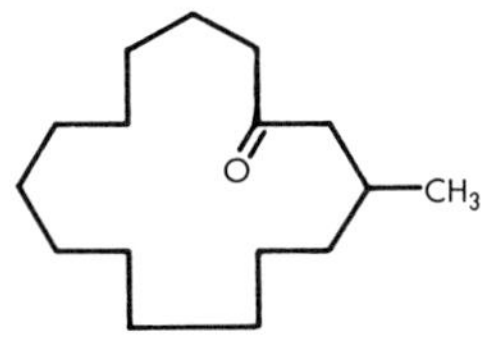

(129)

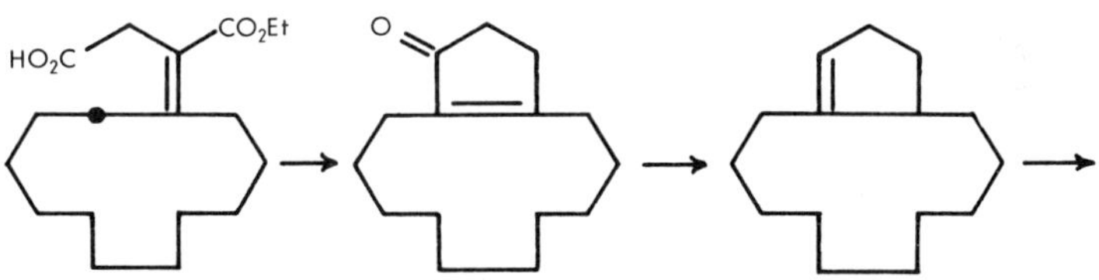

(131) (132) (133)

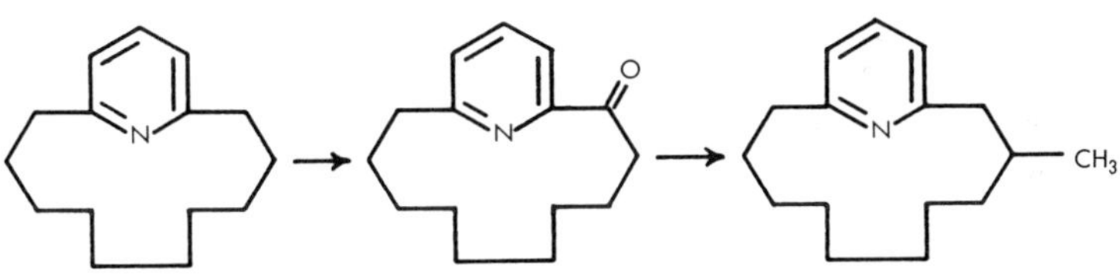

(134) (135) (130)

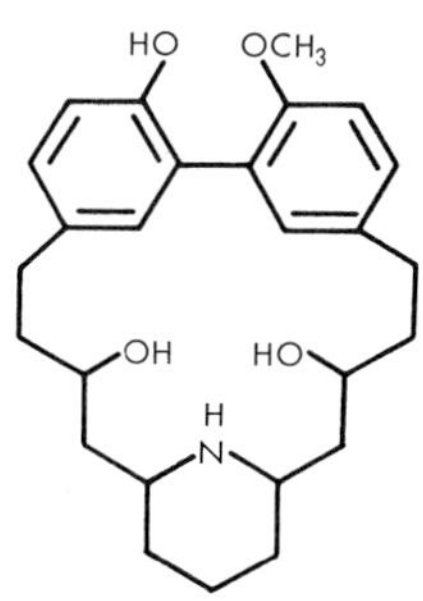

(135a)

2.3 Nicotine and Nicotinic Acid Derivatives

2.3.1 Nicotine Tobacco plants (genus *Nicotiana*) contain about a dozen pyridine alkaloids; the principal one, and indeed one of the longest known alkaloids, is *nicotine*, $C_{10}H_{14}N_2$. Both nitrogens are present as tertiary amines and one bears an N-methyl group. Chromic acid oxidation gave nicotinic acid (137), while distillation from lime afforded pyrrole and methylamine. Nicotine forms two different monomethiodides; when the one with the pyridine ring quaternized was oxidized with ferricyanide and then dichromate, (−) N-methylproline was formed. These reactions defined the structure of nicotine as (136).

Several total syntheses of nicotine have been accomplished. One began with Claisen condensation of ethyl nicotinate with N-methyl-2-pyrrolidone to form (138). Acid hydrolysis, accompanied by decarboxylation, gave (139), and after the carbonyl group was converted to an iodide by reduction and heating with HI, the amino iodide spontaneously cyclized to d,l-nicotine.[37]

Nicotine is a toxic alkaloid; 40 mg are fatal to man. At one time it was in wide use as an agricultural insecticide, but has now been largely supplanted by other chemicals.

Other tobacco alkaloids include the N-desmethyl compound, *nornicotine*; the pyrrole related to nicotine, *nicotyrine*; and *anabasine*. Anabasine is an isomer of nicotine; since it can be oxidized to nicotinic acid and dehydrogenated to 2,3′-bipyridyl, it has structure (140).

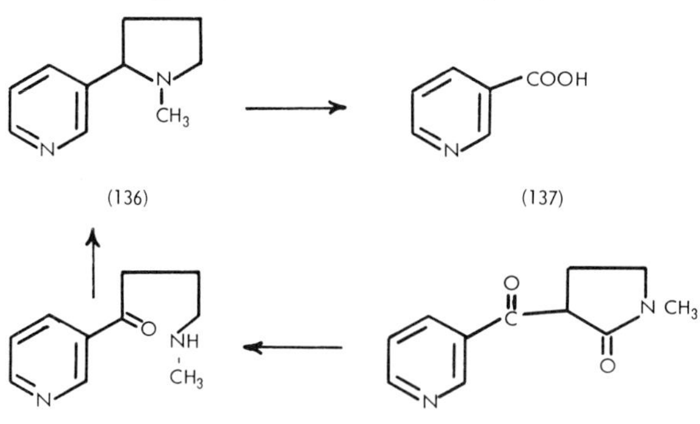

(136) (137)

(139) (138)

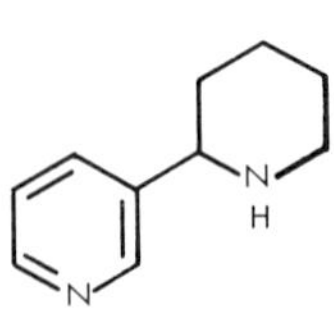

(140)

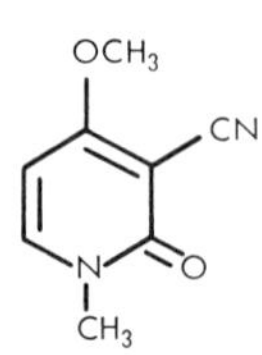

(141)

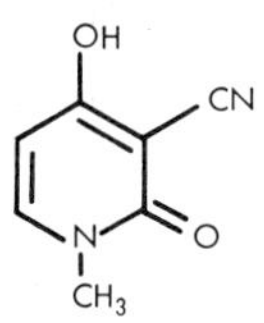

(142)

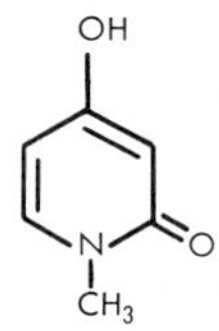

(143)

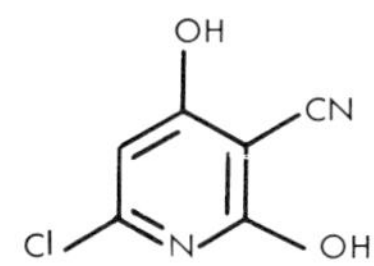

(144)

2.3.2 Ricinine (141) This alkaloid, $C_8H_8N_2O_2$, was isolated from castor bean seeds. It is one of the rare natural products containing a nitrile function. Alkaline hydrolysis brought about nucleophilic replacement of the methoxyl to form (142) and acid hydrolysis effected decarboxylation to pyridone (143). A simple synthesis of ricinine followed from the observation that cyanoacetyl chloride dimerizes to (144). Methyl sulfate alkylated both the nitrogen and the 4-hydroxyl of (144), and reduction of the chlorine with zinc and acid gave ricinine.

2.3.3 Areca Alkaloids Betel nuts from palm trees of the far East, when ground, mixed with slaked lime, and wrapped in betel leaves, are chewed in large quantity by natives. The nuts, which color the teeth and saliva red, contain alkaloids which have a stimulant and cathartic action. The parent member of this family is an amino acid *guvacine*, $C_6H_9O_2N$; its methyl ester is *guvacoline*. Guvacine is a secondary amine, is optically inactive and nonresolvable, and can be hydrogenated to hexahydronicotinic acid. Since it does not have the characteristics of an enamine, it was assigned structure (145). The corresponding N-methyl derivative, *arecaidine*, and its

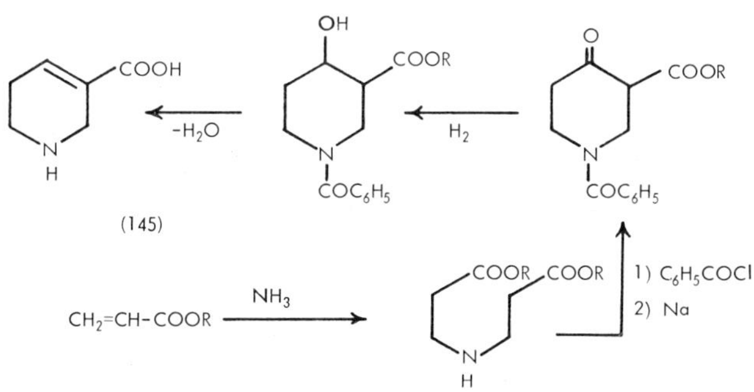

(145)

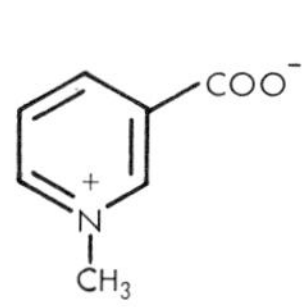

(146)

methyl ester *arecoline,* occur in the betel nuts, too; arecoline is a powerful stimulant of the central nervous system. A number of interesting syntheses have confirmed these structures, e.g., the synthesis of guvacine[38] shown below. The simplest preparation of arecoline is by partial reduction[39] of methyl nicotinate methiodide with sodium borohydride.

A final natural derivative of nicotinic acid is the N-methyl betaine *trigonelline* (146), found in a number of plants.

2.4 Monoterpenoid Pyridines and Piperidines

A handful of natural piperidines and pyridines incorporate nitrogen into one of the common monoterpenoid skeletons, the cyclopentane family represented by nepetalactone (147), the principle of catnip. One of these is *actinidine,*[40] which occurs in *Actinidia polygama,* another plant attractive to cats. Oxidation to 3,4,5-pyridinetricarboxylic acid, and the cooccurrence of dihydronepetalactone, led to the postulation of structure (148). This was confirmed by synthesis from an oxidation product of (147).

Skytanthine, $C_{11}H_{21}N$, was isolated from a Chilean plant, *Skytanthus acutus.*[41] It is a saturated, optically active, tertiary amine containing an N-methyl and two C-methyl groups. Hofmann elimination gave an olefin (150) containing a terminal vinyl group; ozonolysis yielded formaldehyde and a ketone (151) which gave a positive iodoform test. Baeyer-Villiger oxidation with trifluoroperacetic acid gave an acetate which was saponified and then oxidized to ketone (152); the infrared absorption at 1745cm^{-1} revealed this to be a cyclopentanone. Dehydrogenation of skytanthine gave a pyridine, $C_{10}H_{13}N$. Its nmr spectrum showed the presence of two aromatic protons located on the α-positions of a pyridine ring (τ1.99 and 1.89), an aromatic methyl at τ7.82, and a secondary methyl (τ8.73, d, J = 7). With this information structure (148) was deduced for the dehydrogenation product, which proved indeed to be identical with actinidine, so that skytanthine must be (149). Four of the possible stereoisomers have been synthesized from (147), and vpc analysis showed that natural skytanthine was a mixture of three of these.

Tecomine,[42] $C_{11}H_{17}NO$, a related base from *Tecoma stans* juss., is an α,β-unsaturated ketone, as revealed by strong uv absorption at 226 nm. Hydrogenation gave a dihydroketone whose ir absorption was characteristic of a cyclopentanone. Wolff–Kishner reduction of the dihydrocompound and dehydrogenation led to actinidine, revealing the carbon skeleton. Since the nmr spectrum showed one vinyl proton and two methyl doublets, tecomine must be (153).

A final member of this family to be mentioned here is *gentianine,* (154), which has been isolated from *Gentiana* species and other plants. The structure represents the closure into a pyridine ring of a C_{10} unit common among plant glycosides. In fact, careful work showed that gentianine was an artifact, not present in the plant but formed during isolation by reaction of the glycoside (e.g., *swertiamarin* (155)) with ammonium hydroxide.[43]

Two natural pyridines which have straight carbon chains attached and which may be derived from polyacetates are *anibine* (156), an alkaloid of a South American rosewood tree, and *fusaric acid* (157), a fungal metabolite.

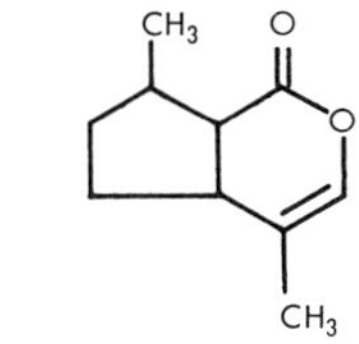
(147)

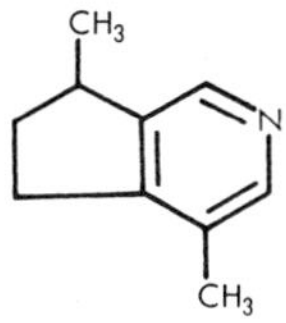
(148)

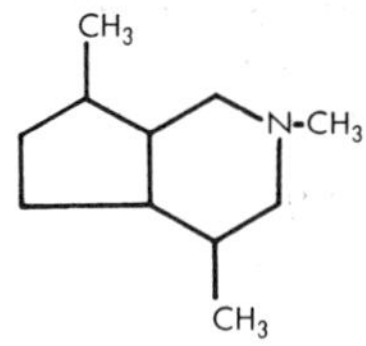
(149)

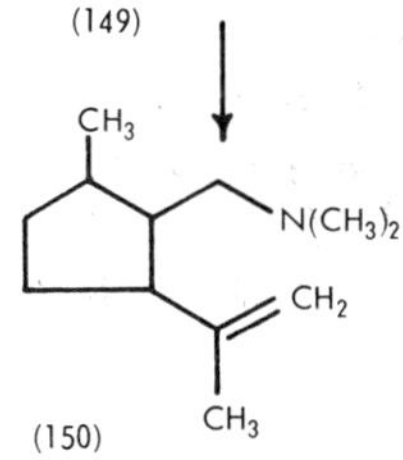
(150)

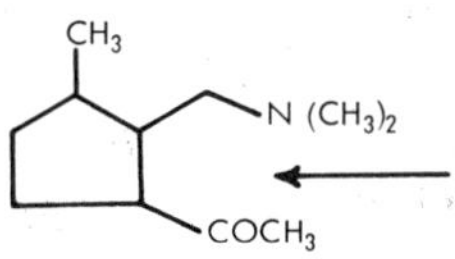
(151)

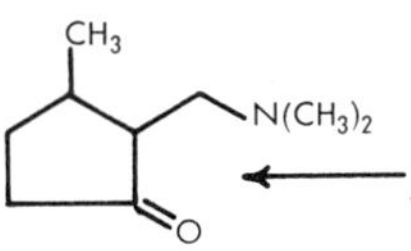
(152)

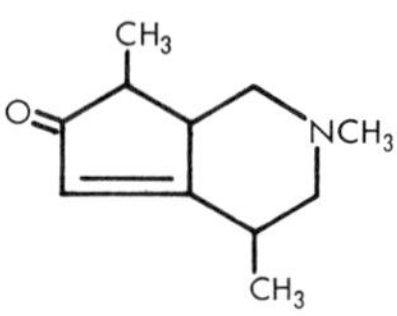
(153)

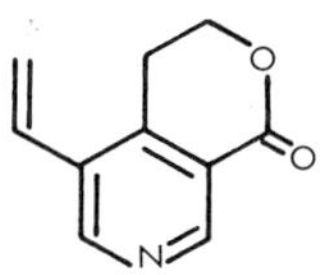
(154)

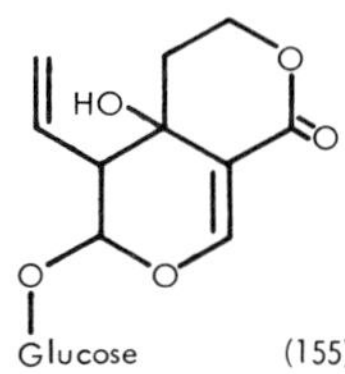
(155)

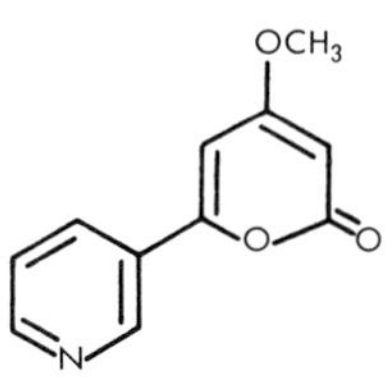
(156)

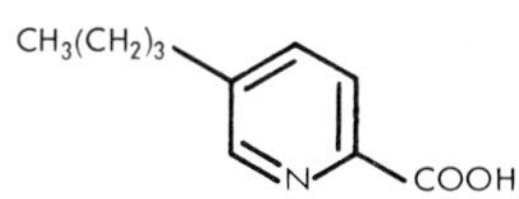
(157)

2.5 Piperideine Dimers

A small group of piperidine alkaloids has the skeleton of a reduced anabasine (140), but is considered separate from the tobacco alkaloids because of the probable origin from two lysine units rather than nicotinic acid. *Ammodendrine* (158) was found in *Ammodendron Conollyi* Bunge, a small tree of central Asia. It has the formula $C_{12}H_{20}N_2O$; one nitrogen is present as a secondary amine, the other as the acetyl derivative of a secondary amine. Hydrogenation gave a dihydro derivative, which was hydrolyzed and dehydrogenated to α,β-dipyridyl, showing the carbon skeleton. To discover which amino group was free, dihydroammodendrine was N-methylated and saponified, giving a compound identical with the hydrogenation product of N-methylanabasine.

A novel synthesis confirmed the structure and located the double bond.[44] C. Schöpf found that piperideine (97), prepared by dehydrohalogenation of N-chloropiperidine, spontaneously trimerized to a series of trimers, some merely acetal-type but one (160) involving the formation of a new carbon–carbon bond by aldol-type condensation. When (160) was acetylated and then heated, the ring joined by only acetal bonds was cleaved, leaving ammodendrine. This synthesis undoubtedly has some biogenetic significance, since (97) is the cyclic imine related to the aminoaldehyde (161) derivable from lysine by normal metabolic processes.

Adenocarpine is the corresponding cinnamoyl amide (159).

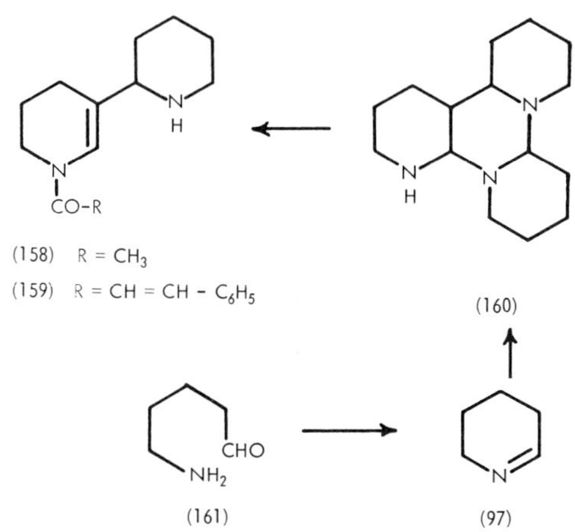

(158) R = CH_3

(159) R = CH = CH - C_6H_5

(160)

(161)

(97)

2.6 Quinolizidine Alkaloids

This section deals with those alkaloids that contain the quinolizidine (162) ring system. This is a common structural unit in alkaloids, though many compounds incorporating it belong really to the isoquinoline, indole, or lycopodium families and so are discussed in other sections. Here, we shall include simple quinolizidines and those more complex members derived from lysine. This system is found most frequently among the lupins, a large plant family of the *Leguminosae* containing gorse, broom, and laburnum among others.[45]

2.6.1 Lupinine The simplest member of the family, homologous with the pyrrolizidine skeleton, is *lupinine,* $C_{10}H_{19}NO$, isolated from yellow lupin seeds. The nitrogen is tertiary and the oxygen present as a primary alcohol, as shown by oxidation to an aldehyde and carboxylic acid. The carbon skeleton was deduced by Hofmann exhaustive methylation, hydrogenating after each elimination. Three stages were required to reach a nitrogen-free product (164), showing that the nitrogen is common to two rings. Dehydration of (164) to an olefin and ozonolysis afforded formaldehyde and 4-nonanone, from which the structure of (164) could be reconstructed and (163) postulated as the most likely for lupinine. This structure has been verified by a number of syntheses, of which two may be cited:

(a) Dehydrogenation of the substituted malonic ester (165) with mercuric acetate led to (166), which underwent base-catalyzed cyclization to (167). Hydrolysis and decarboxylation gave a mixture of monoacids, reduced by lithium aluminum hydride to d,l-lupinine and its diastereomer, epilupinine.[46]

(b) A synthesis along lines similar to the presumed biosynthesis was achieved[46] from (168), prepared by acyloin condensation of the requisite diester. Reduction to the diol, hydrogenolysis of the N-benzyl group, and cleavage with periodate gave dialdehyde (169), which underwent spontaneous Mannich condensation at pH 5 to aldehyde (170). Reduction furnished d,l-epilupinine.

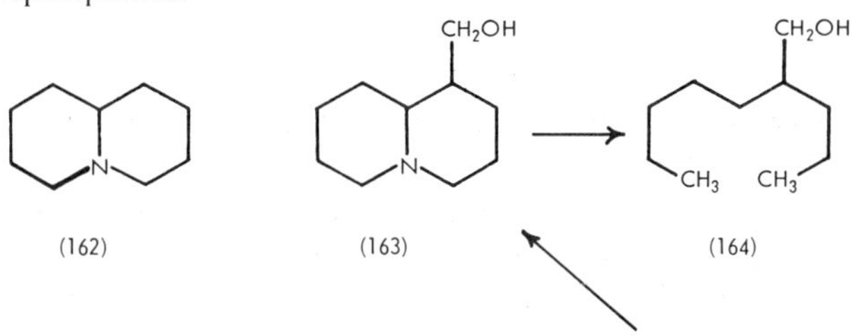

(162) (163) (164)

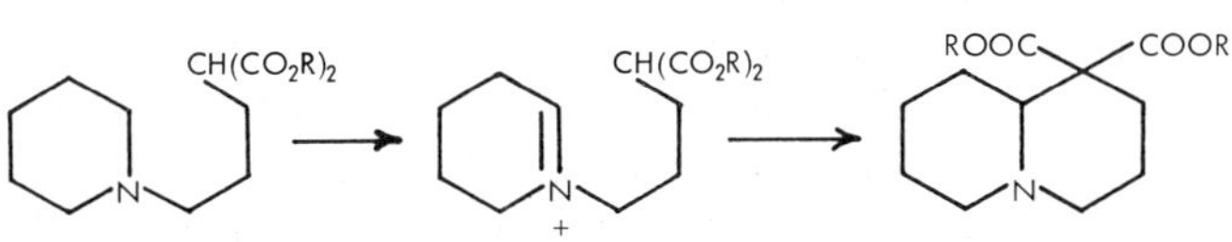

(165) (166) (167)

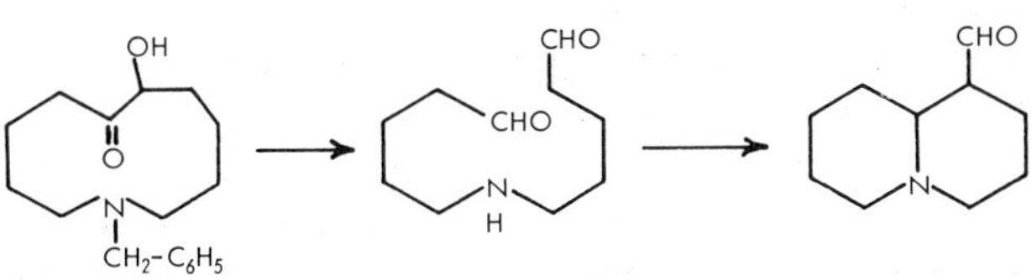

(168) (169) (170)

Refluxing a benzene solution of lupinine with sodium caused epimerization to epilupinine, undoubtedly by means of an oxidation-reduction system via the epimerizable aldehyde (170). This suggested that lupinine is the less stable isomer with the —CH_2OH axial (171). Proof of this configuration came from the observation that lupinine, but not epilupinine, shows an intramolecular O—H · · · N hydrogen bond in the infrared, and that lupinine tosylate cyclizes on standing to the quaternary salt (172).

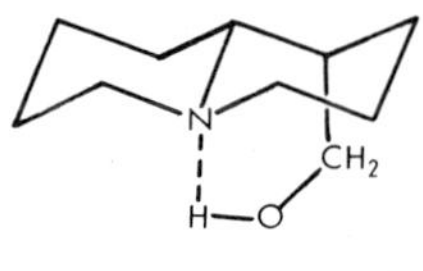

(171)

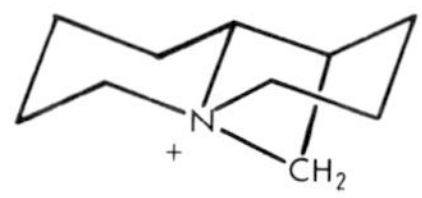

(172)

2.6.2 Cytisine This toxic alkaloid of a widely spread ornamental shrub, *Cytisus laburnam*, is responsible for occasional cases of poisoning, especially in children. It has the formula $C_{11}H_{14}N_2O$; one nitrogen is secondary, the other tertiary and weakly basic. An aromatic ring was apparent from substitution reactions and was eventually shown by color reactions to be an α-pyridone. Dehydrogenation experiments yielded a number of substituted quinolines, but these were later shown to have been formed by skeletal rearrangement.

The best evidence for the structure came from systematic Hofmann degradation. After two Hofmann eliminations, hydrogenating after each step, the secondary nitrogen was lost and pyridone (174) remained. This took up two moles of hydrogen to afford a saturated lactam which yielded glutaric acid on vigorous oxidation. Ozonolysis of (174), followed by hydrolysis, gave lactam (175), identified by oxidation to 2,4-dimethylglutaric acid. This sequence defined the structure of one end of the cytisine molecule; by reducing the pyridone ring a Hofmann series could then be carried out on the other nitrogen. Electrolytic reduction of cytisine gave a saturated diamine in which the secondary nitrogen was then protected by acetylation (176). Repeated Hofmann eliminations, again hydrogenating after each step, gave amide (177). Hydrolysis and dehydrogenation to a pyridine, which was oxidized to acid (178), completed the chain of evidence leading to structure (173) for cytisine.

The first synthesis of cytisine began with Michael addition of ethyl malonate to the vinyl pyridine (179), which was accompanied by elimination of acetic acid to furnish (180). The diacid (180) (R = H), treated with benzylamine and formaldehyde, took part in a combined Michael–Mannich reaction leading to (181). Reduction with lithium aluminum hydride

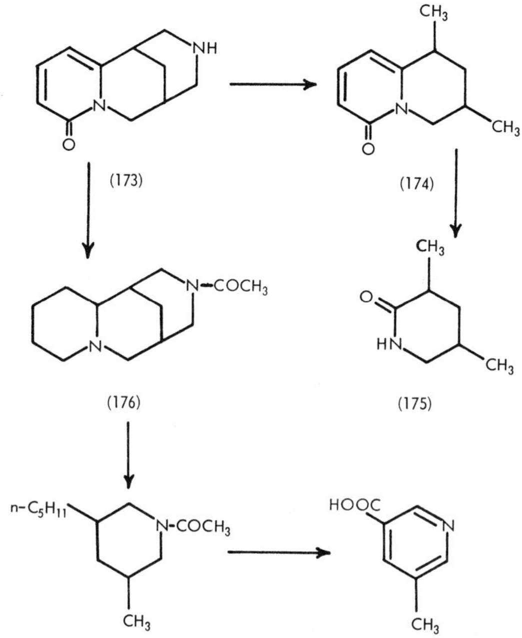

(173) (174) (175) (176) (177) (178)

gave carbinol (182), which was converted to the bromide with HBr; the bromide cyclized to the pyridinium salt (183) on neutralization. Alkaline ferricyanide oxidation yielded the pyridone, from which the benzyl group was cleaved by acid to afford d,l-cytisine.[47]

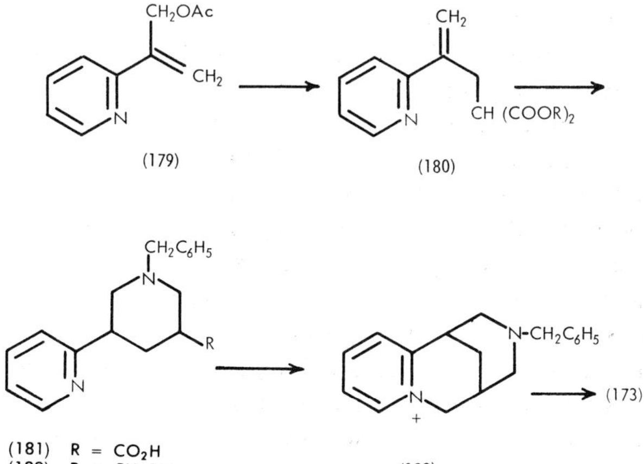

(179) (180)

(181) R = CO_2H
(182) R = CH_2OH

(183)

2.6.3 Tetracyclic Lupin Alkaloids A group of more than a dozen bridged tetracyclic quinolizidines occur in lupins and other plants; four of the most important members are shown below.

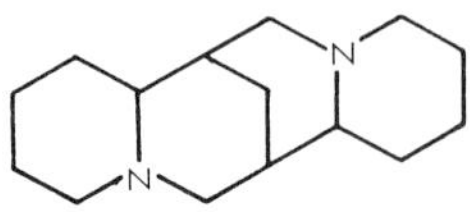

(184) Sparteine

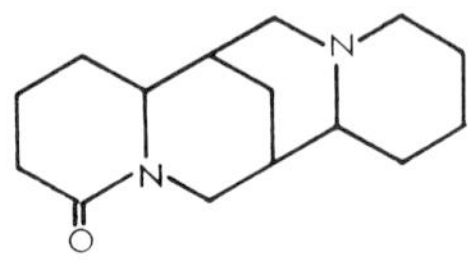

(185) Lupanine

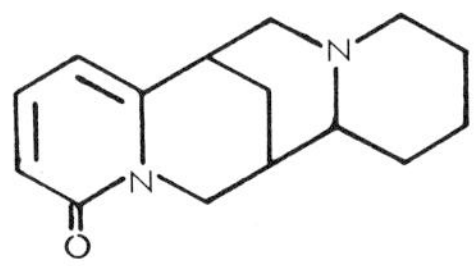

(186) Anagyrine

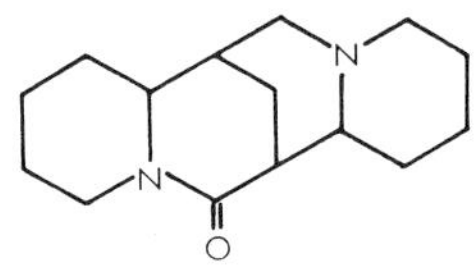

(187) Aphylline

Anagyrine (186), an alkaloid of composition $C_{15}H_{20}N_2O$ from *Anagyris foetida,* was the key to the structures of the entire group. Its similarity to cytisine (173) was noted early: the aromatic pyridone ring, the weakly basic nitrogen, etc. Systematic Hofmann degradation, carried out as had been done on cytisine, yielded pyridone (188). The aromatic ring was destroyed by ozonolysis and subsequent acid hydrolysis, leaving lactam (189), which was oxidized to 2-methyl-4-n-amyl-glutaric acid (190). This informative degradation plus other observations allowed structure (186) to be assigned to anagyrine.

An elegant synthesis[46] followed the pattern of the cytisine synthesis shown in Section 2.6.2. Intermediate (180) was again used in a Mannich reaction, this time with (97) to afford (191). Lithium aluminum hydride reduction and conversion to the bromide with HBr led again to a quaternary salt (192), which was oxidized by alkaline ferricyanide to d,l-anagyrine.

Most of the other tetracyclic lupin alkaloids have been interrelated with anagyrine. Saturation of the two double bonds of anagyrine afforded (–)-*lupanine* (185), while electrolytic reduction of anagyrine (or lithium aluminum hydride reduction of lupanine) gave (+)-*sparteine* (184).

Synthetic evidence in support of these structures came first from synthesis of oxosparteine (196), a crystalline lactam obtained by ferricyanide oxidation of sparteine. Condensation of ethyl 2-pyridylacetate (193) with ethyl orthoformate and acetic anhydride gave (194), which was reduced to (195) by consecutive catalytic hydrogenation and sodium-alcohol reduction. Conversion to the bromide and cyclization with potassium carbonate gave d,l-oxosparteine. Later syntheses have concentrated more

(186)

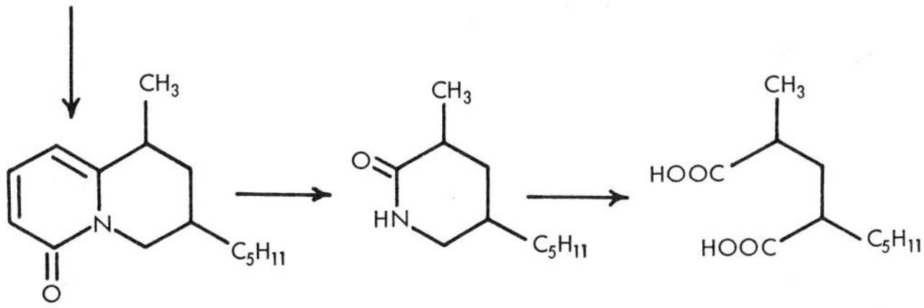

(188) (189) (190)

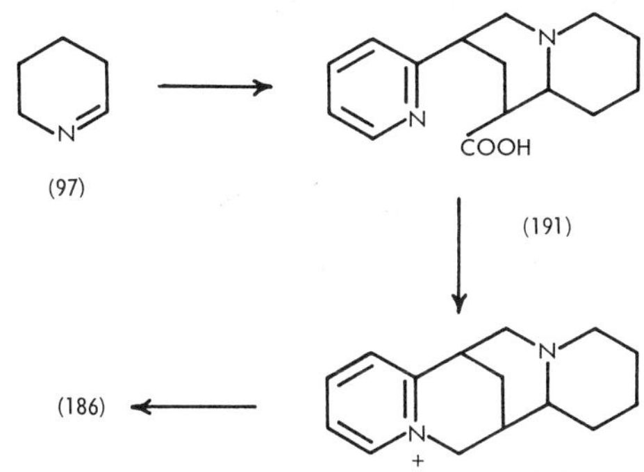

(180) + (97) (191) (186) (192)

on sparteine because of its symmetry. Condensation of (193) with formaldehyde gave (197), which was hydrogenated over copper chromite directly to a stereoisomeric mixture of sparteines (184). Another simple sparteine synthesis[48] involved mercuric acetate oxidation of (198), the Mannich condensation product of piperidine, formaldehyde, and acetone, to (199), which cyclized spontaneously to (200). Wolff–Kishner reduction completed the synthesis of d,l-sparteine.

Sparteine, with four asymmetric centers (two of them interdependent because of the bridge), may exist as three pairs of enantiomers, (184), (201), and (202). In fact, three different diastereomers of sparteine have been isolated from plants, so that all the configurational possibilities are represented in nature. All three racemates are formed in the synthesis from (197). A basis[49] for assigning configurations comes from the reactions of the synthetic intermediate, dioxosparteine (203). One, but not both, of the lactam carbonyls can be reduced by catalytic hydrogenation in acid. Similarly, the lactam group of *aphylline* (187), which belongs to the sparteine series, can be hydrolyzed though the isomeric lactam (204) is resistant to hydrolysis. The reactivity differences between the two halves of the molecule indicate that sparteine is the unsymmetrical isomer (184) rather than one of the symmetrical isomers.

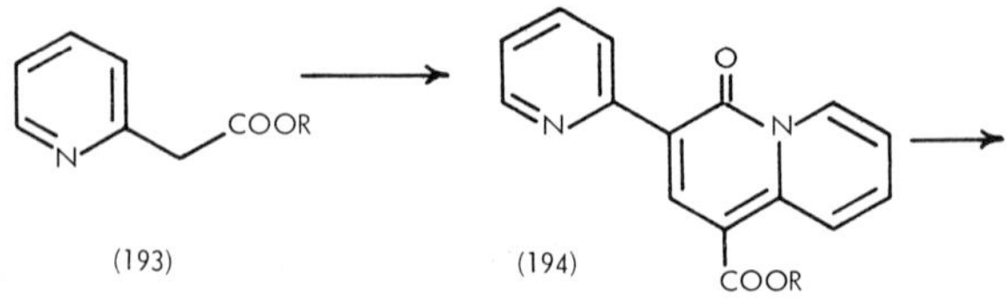

(193) (194)

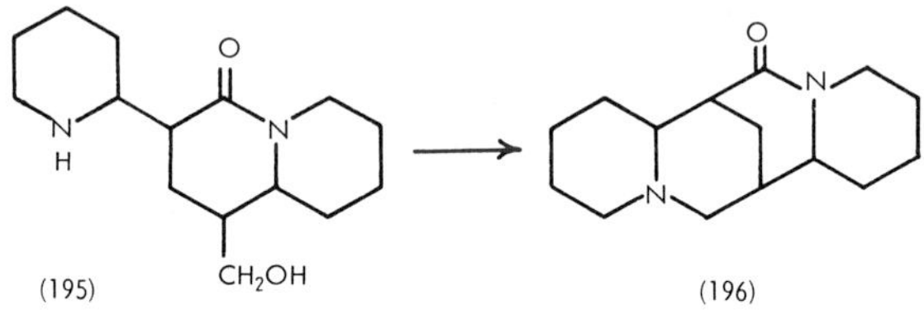

(195) (196)

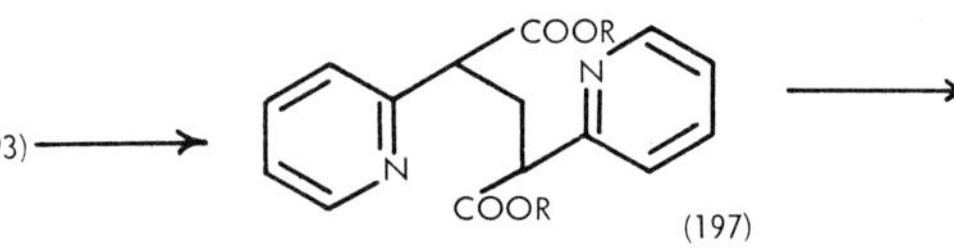

(193) → (197) → (184)

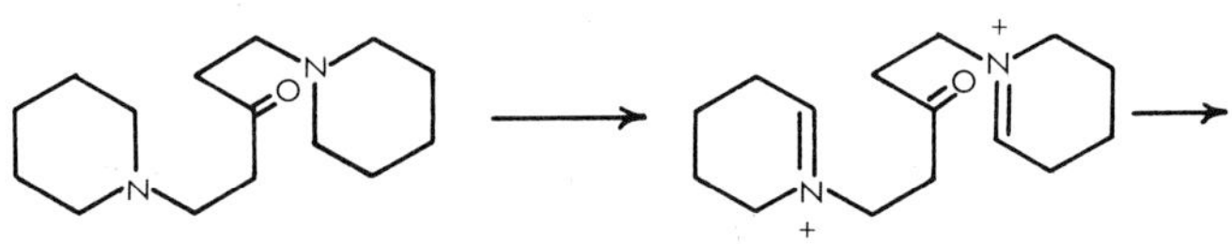

(198) (199)

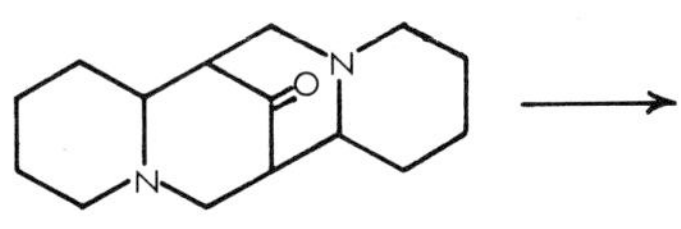

(200) → (184)

In the hydrogenation of anagyrine to lupanine, models show that the hydrogen at the newly created asymmetric center should be delivered *cis* to the methylene bridge, and since lupanine can be reduced to sparteine (184), anagyrine must have configuration (205). *Thermopsine,* a naturally occurring diastereomer of anagyrine, is reduced to α-isosparteine, which must consequently be (202). X-ray analysis of α-isosparteine has confirmed these arguments.

An isomeric tetracyclic skeleton is found in the matrine family. *Matrine* (206) occurs in many species of *Sophora,* along with the corresponding α-pyridone, sophoramine, and related compounds. Most of the features of

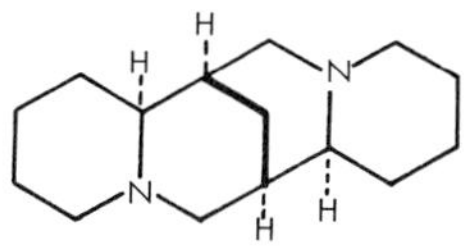

(201)

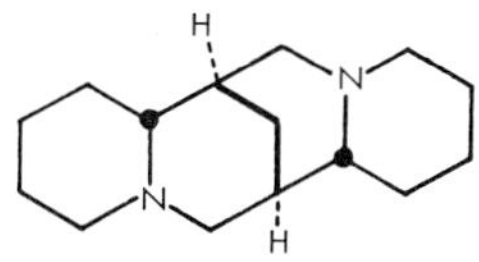

(202)

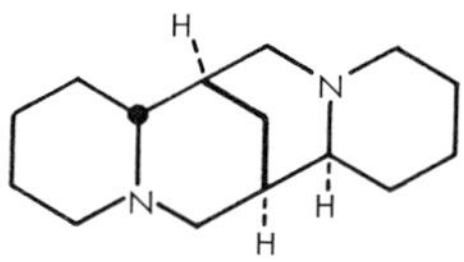

(184)

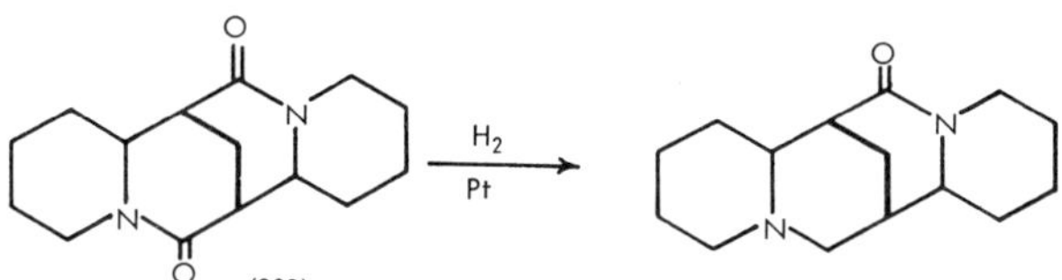

(203) (204)

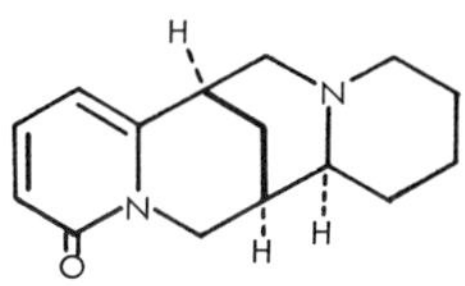

(205)

structure were recognized from degradation experiments, e.g., zinc dust distillation to simple quinolizidines and dehydrogenation to tricyclic pyridine derivatives. However, the real proof of structure came from a novel synthesis of the reduction product (207). Schöpf found that the piperideine trimers, e.g., (160), were converted by heating with ammonium chloride to a new isomer, (208), which has two carbon–carbon bonds formed by aldol-type condensation. Catalytic hydrogenolysis gave amine (209). The bromide formed by successive treatment of the primary amino group with nitrous acid and HBr was cyclized to (207), which after resolution was identical with the natural reduction product.[50]

A more straightforward synthesis of matrine itself began with (210), the Schiff base from β-alanine ethyl ester and ethyl β-ketopimelate. Hydrogenation of (210) led to lactam (211) which was converted by Dieckmann cyclization and hydrolysis to (212). Monocyanoethylation at both α-positions of the ketone was achieved via enamine alkylation to provide (213); hydrogenation then led directly to d,l-matrine.[51]

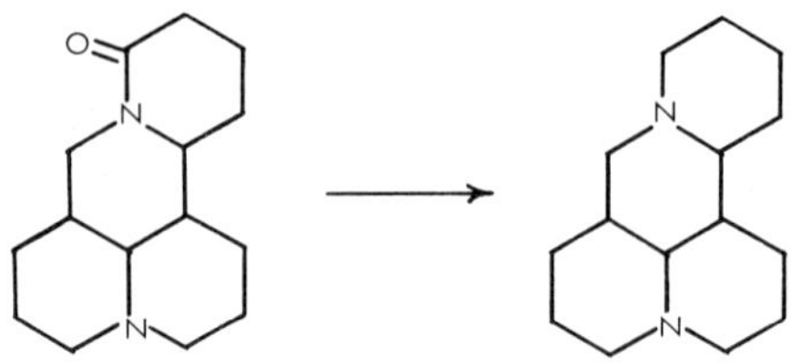

(206) (207)

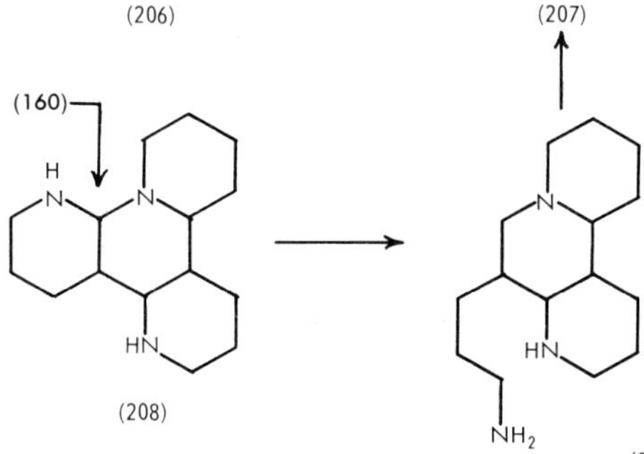

(208) (209)

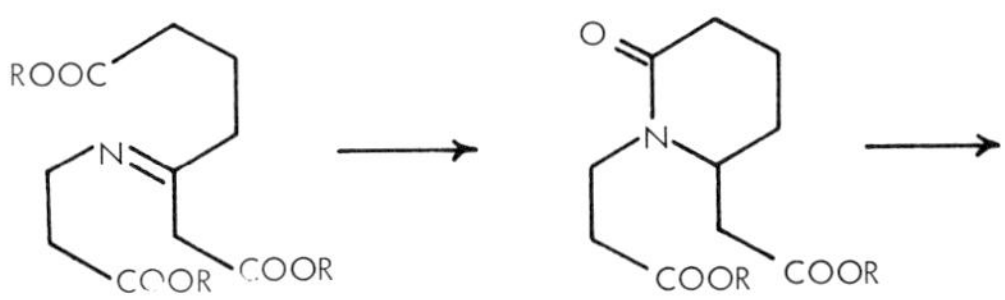

(210) (211)

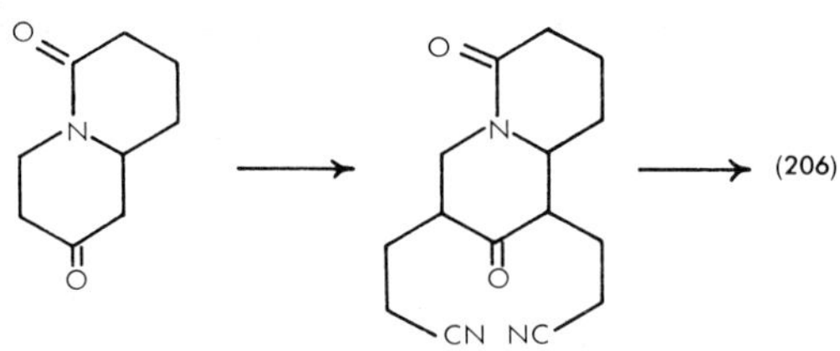

(212) (213)

2.6.4 Ormosia Alkaloids[52] Piperidine homologs of the tetracyclic lupin family are represented by a group of complex alkaloids from *Ormosia* species, two of which are *piptanthine* (214) and *panamine* (215).

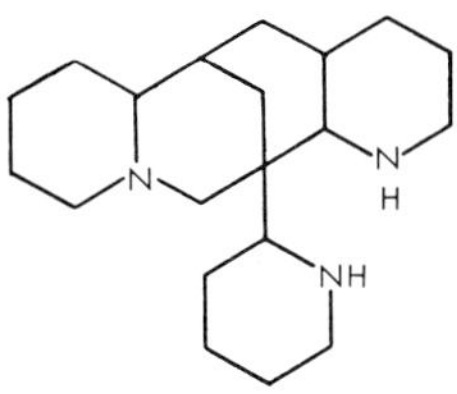

(214)

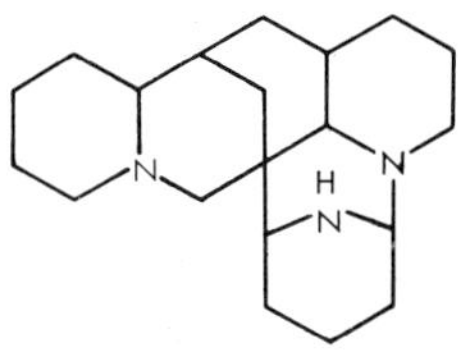

(215)

2.6.5 Other Quinolizidines Two final alkaloid familes containing quinolizidine rings are briefly mentioned here. One is a curous group of alkaloids from *Lythraceae* species[53] containing a 3,4-dimethoxyphenyl substituted quinolizidine ring with the aromatic ring joined in typical phenol oxidative patterns to a tyrosine unit. Two examples are the diphenyl ether *vertaline* (216) and the diphenyl alkaloid *lythridine* (217).

The genus *Nuphar* of the *Nymphaeaceae* (water-lilies) contains quinolizidines of a different sort,[54] exemplified by *nupharidine* (218). The carbon skeleton shows clearly that this family is sesquiterpenoid with oxygen and nitrogen forming heterocyclic rings based on the farnesol skeleton. A similar compound, *castoramine* (219), has been isolated from the scent gland of the North American beaver.

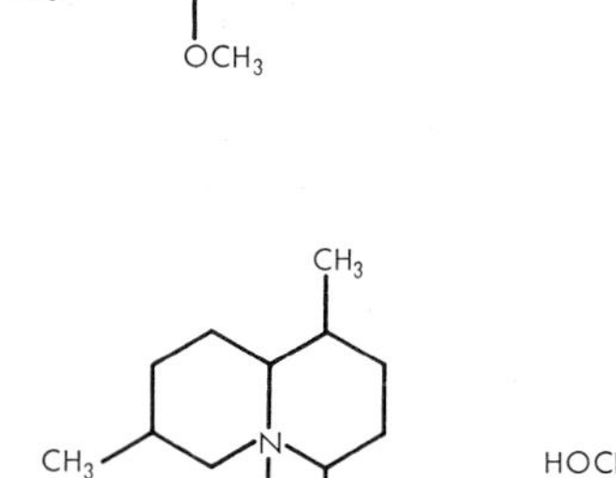

(216)

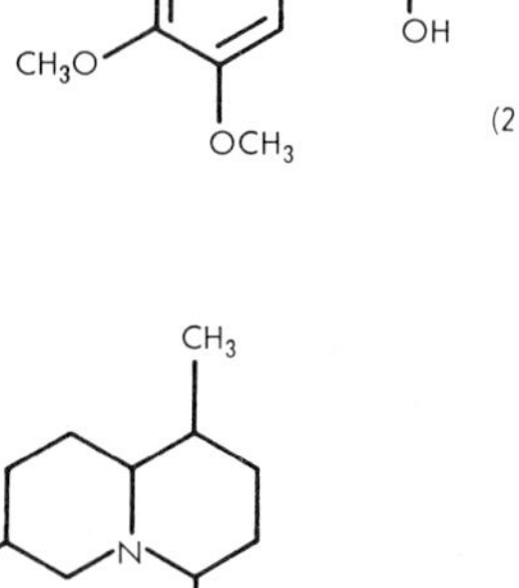

(217)

(218)

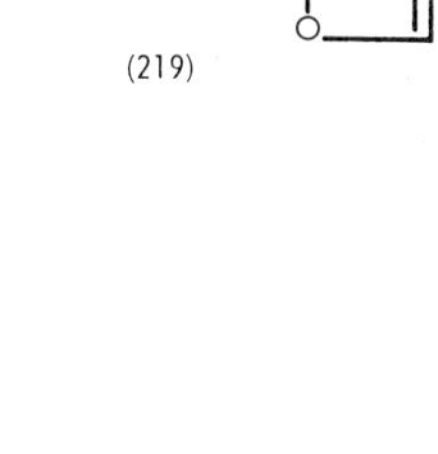

(219)

3. IMIDAZOLE ALKALOIDS

Relatively few simple derivatives of imidazole occur naturally. The amino acid histidine (220) is of course an important constitutent of proteins and the related amine histamine (221) plays a central role in allergic reactions in animal tissues. A few relatives of histidine are found in fungi, e.g., *ergothioneine* (222) in ergot.

The seeds of "Zapote blanco," the fruit of the Mexican tree *Casimiroa edulis*, contain the glycoside alkaloid *casimiroedine*, for which structure (223) has been shown by x-ray crystallography.[55] Though the fruit has a history of use in indigenous medicine the alkaloid is apparently without significant physiological activity.

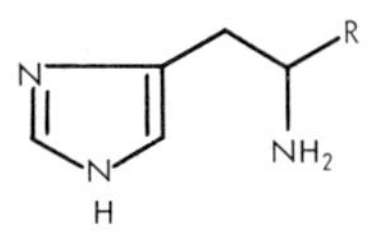

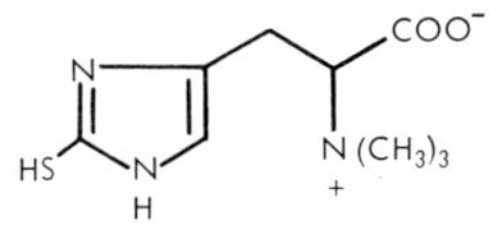

(222)

(220) R = COOH

(221) R = H

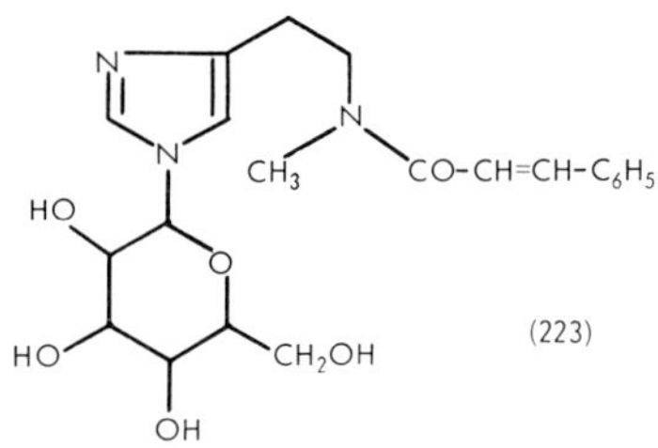

(223)

3.1 Pilocarpine

The most important imidazole alkaloid is *pilocarpine* (224), which occurs together with a steoisomer, isopilocarpine, in the leaves of South American *Pilocarpus* species, a family of shrubs of the order *Rutaceae* popularly known as jaborandi. Pilocarpine is a widely studied peripheral stimulant of the parasympathetic system, employed topically as a miotic and internally as a diaphoretic (i.e., to induce sweating) in the treatment of nephritis. The pure alkaloids were isolated before 1900 and their structures elucidated in the early part of this century.[56]

Pilocarpine is a low-melting, optically active, monoacidic base, of the formula $C_{11}H_{16}N_2O_2$. Both nitrogens are tertiary and an N-methyl group is present. Pilocarpine is insoluble in cold alkali but dissolves on heating, and acidification of the alkaline solution regenerates the alkaloid. This behavior typical of lactones reveals the nature of both oxygen atoms. Distillation of the alkaloid from soda lime gave a mixture of bases including 1,5-dimethylimidazole (225) and 1-methyl-5-n-amylimidazole (226), establishing both the identity of the heterocyclic base and the position of substitution. Partial structure (227) may be written for pilocarpine at this point.

The structure of the lactone ring was elucidated by oxidative degradation. Permanganate oxidation of pilocarpine or isopilocarpine gave a mixture of two homologous lactonic acids, (228) and (229). Acid (229) can be obtained more conveniently by an alternate method; ozonolysis of pilocarpine and isopilocarpine afforded respectively two optically active stereoisomers of amide (230). Acid hydrolysis of either yielded (229), epimerization accompanying hydrolysis of the amide from pilocarpine. Potassium hydroxide fusion of (229) gave the optically inactive triacid (231), showing the carbon skeleton of the C_8 acid. Since (228) and (229) are stable to heat, structures which represent derivatives of malonic acid can be ruled out. This information leaves (228) as the only reasonable structure for the lower acid, and the alkaloids must then be stereoisomers of structure (224).

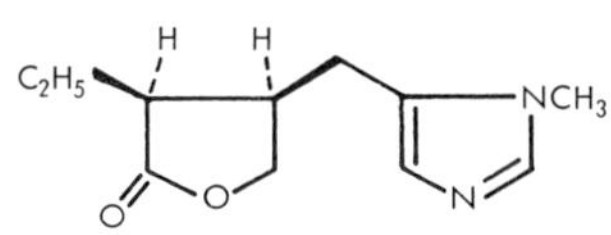

(224)

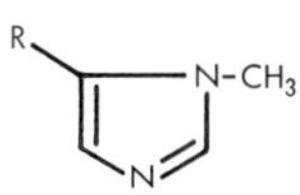

(225) R = CH_3
(226) R = n-C_5H_{11}
(227) R = $C_7H_{11}O_2$

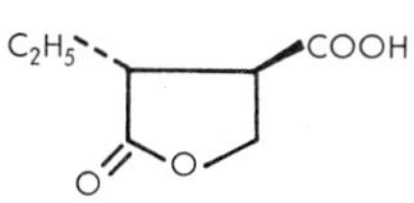

(228)

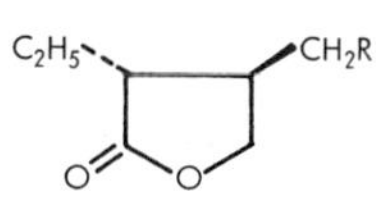

(229) R = COOH
(230) R = $CONHCH_3$

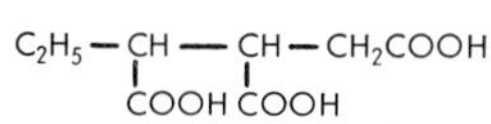

(231)

These structures have been confirmed by several syntheses. One began with Michael addition of ethyl malonate to (232), followed by ethylation, to afford (233). Acid hydrolysis and ether cleavage then yielded d, l-(229) and its diastereomer. Conversion to ketol (234) via the diazoketone was followed by oxidation to the α-ketaldehyde and standard construction of the imidazole ring with ammonia and formaldehyde; a final methylation gave pilocarpine.

The reactions of pilocarpine and isopilocarpine, particularly their ozonolysis to stereoisomeric amides, make it clear that they are stereo-

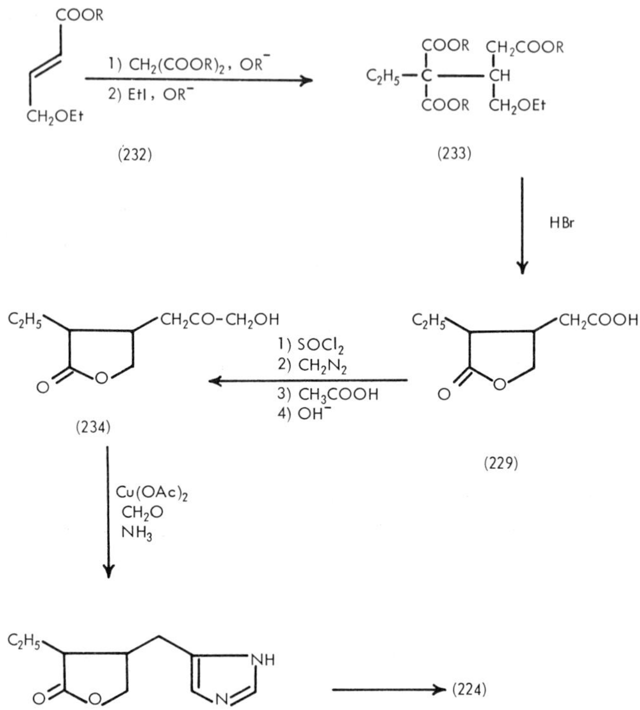

isomers, not structural isomers. Pilocarpine is isomerized to isopilocarpine by the action of acids, bases, or heat, and all its derivatives possessing the two asymmetric centers undergo the same easy isomerization. Since the *trans*-arrangement is more stable for adjacent substituents on a five-membered ring, pilocarpine is assigned the *cis*-configuration, with epimerization occurring at the labile asymmetric center contiguous to the lactone carbonyl to yield the more stable *trans*-isomers.

Unambiguous proof of the *trans*-configuration of the *iso* series came from conversion of (229) to the symmetrical diol (236). Mixed Kolbe electrolysis of (+)-(229) with acetic acid provided lactone (235), converted by lithium aluminum hydride reduction to the levorotatory, diol. The presence of optical activity in (236) is proof of the relative configuration of the two asymmetric centers in it and its precursors. Lactone (235)

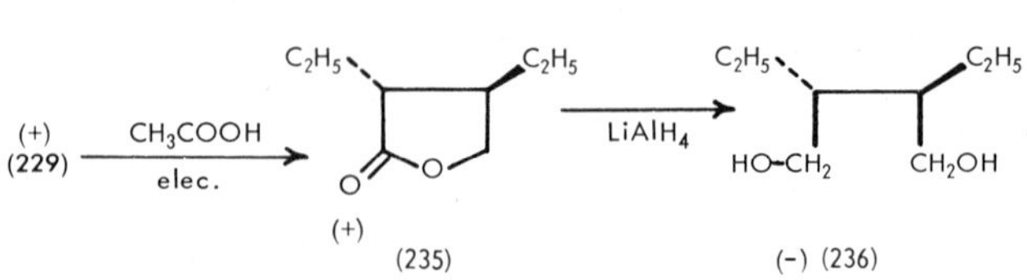

has been stereochemically correlated with strychnine,[57] enabling the assignment of absolute configuration; an independent determination of absolute configuration[58] of (228) has confirmed the configuration depicted in (224).

REFERENCES

1. L. Marion, *Alkaloids* **1,** 91 (1950); **6,** 31 (1960).
2. R. Lukeš, J. Kovář, J. Kloubek, and K. Blaha, *Coll. Czech. Chem. Commun.* **25,** 483 (1960).
3. F. Galinovsky and H. Zuber, *Monatsh. Chem.* **84**, 798 (1953).
4. Y. Inubushi, Y. Tsuda, T. Konita, and S. Matsumoto, *Chem. Pharm. Bull.* (Tokyo). **12,** 749 (1964).
5. N. J. Leonard, *Alkaloids* **1,** 107 (1950) **6,** 37 (1960); F. L. Warren, *Fortschr. Chem. Org. Naturstoffe* **12,** 198 (1955); **24,** 329 (1966).
6. N. J. Leonard and D. L. Felley, *J. Amer. Chem. Soc.* **71,** 1758 (1949); **72,** 2537 (1950).
7. N. J. Leonard and S. W. Blum, *J. Amer. Chem. Soc.* **82,** 503 (1960).
8. T. A. Geissman and A. C. Waiss, Jr., *J. Org. Chem.* **27,** 139 (1962).
9. L. Marion, *Alkaloids* **1,** 165 (1950); **6,** 123 (1960).
10. E. Leete, *Science* **147,** 1000 (1965); *J. Amer. Chem. Soc.* **86,** 2509 (1964).
11. R. K. Hill, *J. Amer. Chem. Soc.* **80,** 1609, 1611 (1958); J. Sicher and M. Tichy, *Coll. Czech. Chem. Commun.* **23,** 2081 (1958); K. Balenovic and N. Stımac, *Croat. Chem. Acta* **29,** 153 (1957).
12. H. S. Yanai and W. N. Lipscomb, *Tetrahedron* **6,** 103 (1959).
13. F. Galinovsky, G. Bianchetti, and O. Vogl, *Monatsh. Chem.* **84,** 1221 (1953); O. Vogl and G. Bianchetti, *Monatsh. Chem.* **86,** 1024 (1955).
14. F. Galinovsky and O. Vogl, *Monatsh. Chem.* **83,** 1055 (1952); F. Galinovsky and F. Höllinger, *Monatsh. Chem.* **85,** 1012 (1954); P. I. Mortimer, *Austral. J. Chem.* **2,** 82 (1958); R. E. Gilman and L. Marion, *Bull. Soc. Chim. France* 1993 (1961).
15. R. L. Augustine, *J. Amer. Chem. Soc.* **81,** 4664 (1959).
16. C. Schöpf and G. Lehmann, *Ann.* **518,** 1 (1935); A. C. Cope, H. L. Dryden, Jr., and C. F. Howell, *Org. Syn.* Coll. **4,** 816 (1963).
17. G. Schroeder, "Cyclooctatetraen," Verlag Chemie, Meinheim, 1965.
18. K. Ziegler and H. Wilms, *Ann.* **567,** 1 (1950).
19. A. C. Cope, C. R. Ganellin, H. W. Johnson, Jr., T.V. van Auken, and H. J. S. Winkler, *J. Amer. Chem. Soc.* **85,** 3276 (1963).
20. H. C. Beyermann, J. Eenshuistra, W. Eveleens, and A. Zweistra, *Rec. Trav. Chim.* **78,** 43 (1959) and earlier papers; H. C. Beyermann, L. Maat, A. van Veen, A. Zweistra, and W. von Philipsborn *Rec. Trav. Chim.* **84,** 1367 (1965); C. Schöpf, G. Dummer, and W. Wüst, Ann. **626,** 134 (1959).
21. G. Fodor and D. Butruille, Abstracts of Papers, 156th Meeting of the Amer. Chem. Soc., Sept. 1968, Org. Div., paper 137.
22. A. E. Schwarting, J. M. Bobbit, A. Rother, C. K. Atal, K. L. Khanna, J. D. Leary, and W. G. Walter, *Lloydia* **26,** 258 (1963).
23. B. R. Baker and F. J. McEvoy, *J. Org. Chem.* **20,** 136 (1955) and preceding papers; R. K. Hill and A. G. Edwards, *Chem. Ind., (London)* 858 (1962).
24. C. Schöpf, T. Kauffmann, P. Berth, W. Bundschuh, G. Dummer, H. Fett, G. Habermehl,

E. Wieters, and W. Wüst, *Ann.* **608,** 88 (1957); C. Schöpf and E. Schenkenberger, *Ann.* **682,** 206 (1965).
25. M. M. Robison, W. G. Pierson, L. Dorfman, B. F. Lambert, and R. A. Lucas, *J. Org. Chem.* **31,** 3206 (1966).
26. W. H. Tallent, V. L. Stromberg, and E. C. Horning, *J. Amer. Chem. Soc.* **77,** 6361 (1955); W. H. Tallent and E. C. Horning, *J. Amer. Chem. Soc.* **78,** 4467 (1956); R. K. Hill, T. H. Chan, and J. A. Joule, *Tetrahedron* **21,** 147 (1965).
27. R. K. Hill and T. H. Chan, *Tetrahedron* **21,** 2015 (1965).
28. A. D. Kuzovkov and G. P. Menshikov, *Zh. Obshch. Khim.* **20,** 1524 (1950); R. K. Hill and J. W. Morgan, *J. Org. Chem.* **31,** 3451 (1966).
29. H. Rapoport, H. D. Baldridge, Jr., and E. J. Volcheck, Jr., *J. Amer. Chem. Soc.* **75,** 5290 (1953).
30. M. Tichy and J. Sicher, *Tetrahedron Letters* 511 (1962).
31. J. L. Coke and W. Y. Rice, Jr., *J. Org. Chem.* **30,** 3420 (1965).
32. M. Spiteller-Friedmann and G. Spiteller, *Monatsh. Chem.* **95,** 1234 (1964).
33. R. J. Highet, *J. Org. Chem.* **29,** 471 (1964); R. J. Highet and P. F. Highet, *J. Org. Chem.* **31,** 1275 (1966); W. Y. Rice, Jr. and J. L. Coke, *J. Org. Chem.* **31,** 1010 (1966).
34. E. Ritchie and W. C. Taylor, *Alkaloids* **9,** 529 (1967).
35. K. Biemann, G. Büchi, and B. H. Walker, *J. Amer. Chem. Soc.* **79,** 5558 (1957).
36. E. Fujita, K. Fuji, K. Bessho, A. Sumi, and S. Nakamura, *Tetrahedron Letters* 4595 (1967); E. Fujita, K. Fuji, and K. Tanaka, *Tetrahedron Letters* 5905 (1968).
37. E. Späth and H. Bretschneider, *Ber.* **61,** 327 (1928).
38. S. M. McElvain and G. Stork, *J. Amer. Chem. Soc.* **68,** 1049 (1946).
39. N. Kinoshita, M. Hamana, and T. Kawasaki, *Chem. Pharm. Bull.* (Tokyo), **10,** 753 (1962).
40. T. Sakan, A. Fujino, F. Murai, Y. Butsugan, and A. Suzui, *Bull. Chem. Soc. Japan* **32,** 315, 1155 (1959).
41. C. Djerassi, J. P. Kutney, M. Shamma, J. N. Shoolery, and L. F. Johnson, *Chem. Ind. (London)* 210 (1961); E. J. Eisenbraun, A. Bright, and H. H. Appel, *Chem. Ind. (London)* 1242 (1962).
42. G. Jones, H. M. Fales, and W. C. Wildman, *Tetrahedron Letters*, 397 (1963).
43. T. R. Govindachari, S. S. Sathe, and N. Viswanathan, *Ind. J. Chem.* **4,** 201 (1966); H. G. Floss, U. Mothes, and A. Rettig, *Z. Naturforsch.* **19b,** 1106 (1964).
44. C. Schöpf and F. Braun, *Naturwiss.* **36,** 377 (1949).
45. N. J. Leonard, *Alkaloids* **3,** 119 (1953); **7,** 253 (1960); F. Bohlmann and D. Schumann, *Alkaloids* **9,** 176 (1967).
46. E. E. van Tamelen and R. L. Foltz, *J. Amer. Chem. Soc.* **82,** 502 (1960).
47. E. E. van Tamelen and J. S. Baran, *J. Amer. Chem. Soc.* **80,** 4659 (1958).
48. E. E. van Tamelen and R. L. Foltz, *J. Amer. Chem. Soc.* **82,** 2400 (1960).
49. F. Galinovsky, P. Knoth, and W. Fischer, *Monatsh. Chem.* **86,** 1014 (1955).
50. C. Schöpf, H. Arm, G. Benz, and H. Krimm, *Naturwiss.* **38,** 186 (1951).
51. L. Mandell, K. P. Singh, J. T. Gresham, and W. J. Freeman, *J. Amer. Chem. Soc.* **87,** 5234 (1965).
52. P. Deslongchamps, Z. Valenta, and J. S. Wilson, *Can. J. Chem.* **44,** 2539 (1966); Z. Valenta, P. Deslongchamps, M. H. Rashid, R. H. Wightman, and J. S. Wilson, *Can. J. Chem.* **44,** 2525 (1966); I. L. Karle and J. Karle, *Tetrahedron Letters* 1659 (1966); (1966); A. P. Davis and C. H. Hassall, *Tetrahedron Letters* 6291 (1966).
53. J. P. Ferris, C. B. Boyce, R. C. Briner, B. Douglas, J. L. Kirkpatrick, and J. A. Weisbach, *Tetrahedron Letters* 3641 (1966); S. C. Chu, G. A. Jeffrey, B. Douglas, J. L. Kirkpatrick, and J.A. Weisbach, *Chem. Ind. (London)* 1795 (1966).

54. J. T. Wróbel, *Alkaloids* **9,** 441 (1967).
55. S. Raman, J. Reddy, W. N. Lipscomb, A. L. Kapoor, and C. Djerassi, *Tetrahedron Letters* 357 (1962).
56. A. R. Battersby and H. T. Openshaw, *Alkaloids* **3,** (1953).
57. K. Nagarajan, C. Weissmann, H. Schmid, and P. Karrer, *Helv. Chim. Acta* **46,** 1212 (1963).
58. R. K. Hill and S. Barcza, *Tetrahedron* **22,** 1889 (1966).

15

Tropane Alkaloids

GABOR FODOR*
Departement de Chimie
Université Laval
Québec, Canada

1. INTRODUCTION

Although the tropane alkaloids were not the first ones detected in Nature they early attracted serious chemical and pharmaceutical interest. Prior to elucidation of the structures, the mydriatic action of many compounds was noted (*e.g.*, atropine from *Atropa belladonna*); others displayed strong anesthetic effects (e.g., cocaine from *Erithroxylon coca*). In addition to their pharmaceutical significance, the presence of an unusual ring system led to an intensive investigation of the chemistry of members of this series. In the last decade the stereochemistry of the key tropane alkaloids has been elucidated by using a number of stereospecific transformations while their syntheses have involved a number of new reactions and methods. Consequently, the initial primary interest in their pharmaceutical activity shifted towards a more thorough investigation of their structures. Since all structural problems have now been solved,[1–3] the main thrust of investigations concerns a correlation of structure with pharmacodynamic activities.[4–7] A number of nonnatural derivatives, mostly esters of the natural alcohols with a number of acids not occurring in the plant tissue, have been synthesized and screened. Quaternization of the same esters gives rise to a fundamental change in the pharmaceutical activities of the same esters leading to new ganglion blocking and central stimulating drugs. Both the chemical and pharmaceutical aspects of the tropanes secure them

*Present Address: Department of Chemistry, West Virginia University, Morgantown, West Virginia 26506.

a place among the most important alkaloids. In the following sections, a very condensed account of the constitution and configuration of the main natural tropanes is presented.

The approach to the structure of tropanes-which we describe, is based mainly on classical, chemical evidence corresponding to scientific reality and historical fidelity. Therefore, some of the structural problems may seem trivial or very simple when solved with modern tools. This does not diminish the importance of the achievements of the pioneers who were lacking nmr, mass spectroscopy, and other sophisticated methods. On the other hand, the constitution of phyllalbine, or the conformation of tropanes investigated recently, are discussed in up-to-date terms.

The basic skeleton present in the tropane alkaloids is nortropane, *aza*bicyclo-3,2,1^{1,5} octane. The tropane alkaloids occur as esters of amino alcohols. Hydrolysis of the alkaloids by baryta or aqueous mineral acids leads to the monohydroxylated alkamines: tropine, pseudotropine, nortropine; to the dihydroxylated valerine, or to the trihydroxylated teloidine. Scopolamine gives, depending on the conditions used, either scopine or oscine, i.e., oxydotropanols. The esterifying acids vary from aliphatic C_5-acids (2- and 3-methybutyric and tiglic) to aromatic (benzoic, veratric, vanillic, tropic, and cinnamic) acids; see Table 15.1. Except for oscine*, all the alkamines can be reesterified to afford the natural alkaloids.

Table 15.1 lists the most important esters and alkamines of known constitution and configuration in this series: α and β refer to stereochemical notations[8,42] (except for the truxillines) when they are merely a distinguishing symbol. The numbering of the carbon skeleton shown in the Table is followed throughout the text.

2. STRUCTURES OF THE MAIN TROPANE BASES

As with other natural products, the elucidation of the structures of the tropane bases involves two aspects: the investigation of gross structure and the determination of stereochemical configurations. Sometimes conformational analysis and biogenetic aspects attract interest too.

2.1 Tropeines

The generic name for esters of tropine is "Tropeines."

2.1.1 Atropine and Hyoscyamine Hydrolysis of atropine (3) affords tropine ($C_8H_{15}NO$) (1) and ($\pm$) tropic acid (2a)[9]; hyoscyamine (4) gives (−)

*Oscine is a product of an irreversible transformation of the product of hydrolysis, i.e., of scopine, see, Section 2.3.1.1.

TABLE 15.1.

General Formula

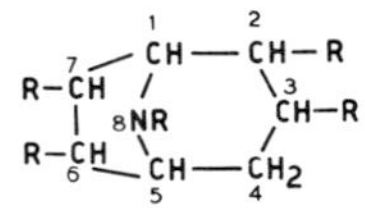

	Compound	Plant Source	Molecular Formula	m.p.	$[\alpha]_D$	N	C(2)	Substituents on C(3)	C(6)	C(7)
(1)	Tropine	*Atropa belladonna, Datura innoxia* etc.	$C_8H_{15}ON$	64°	—	CH_3	H	α-OH	H	H
(13)	Pseudotropine	*Datura innoxia (fastuosa)*	$C_8H_{15}ON$	108°	—	CH_3	H	β-OH	H	H
(3)	Atropine	*Atropa belladonna*	$C_{17}H_{23}O_3N$	116–7°	—	CH_3	H	α(±)-O-Tropoyl	H	H
(4)	(−) Hyoscyamine	*Atropa acuminata, Atropa belladonna* etc.	$C_{17}H_{23}O_3N$	108–11°	−21° (A)	CH_3	H	α(−)-O-Tropoyl	H	H
(30)	Tropacocaine	*Erythroxylum coca*	$C_{15}H_{19}O_2N$	49°	—	CH_3	H	β-O-Benzoyl	H	H
(31)	Phyllalbine	*Phyllantus discöides Muell arg.*	$C_{16}H_{21}O_4N$	209–210°	—	CH_3	H	α-O-Vanilloyl	H	H
(72a)	Dihydroxytropane	*Erythroxylum coca*	$C_8H_{15}O_2N$	212°	−25° (A)	CH_3	H	α-OH	β-OH	H
(73)	Valeroidine	*Datura sanguinea. Duboisia myoporoides*	$C_{13}H_{23}O_3N$	5°	− 9° (A)	CH_3	H	α-O-3-Methyl-Butyryl	β-OH	H
(85)	Meteloidine	*Datura ferox, Datura innoxia (Fastuosa)* etc.	$C_{13}H_{21}O_4N$	141–2°	—	CH_3	H	α-O-Tigloyl	β-OH	β-OH
(61)	Scopolamine	Presumably nonnatural, formed by racemization of hyoscine	$C_{17}H_{21}O_4N$	82–3°	—	CH_3	H	α-O-(±) Tropoyl	β-Oxide	
(62)	Hyoscine	*A. acuminata, Datura innoxia, A. belladonna, Duboisia myoporoides* etc.	$C_{17}H_{21}O_4N$	59°	−18° (A)	CH_3	H	α-O-(−) Tropoyl	β-Oxide	
(32)	(−) Cocaine	*Erythroxylum coca, E. monogynum*	$C_{17}H_{21}O_4$	98°	−16° (C)	CH_3	β-$COOCH_3$	β-O-Benzoyl	H	H

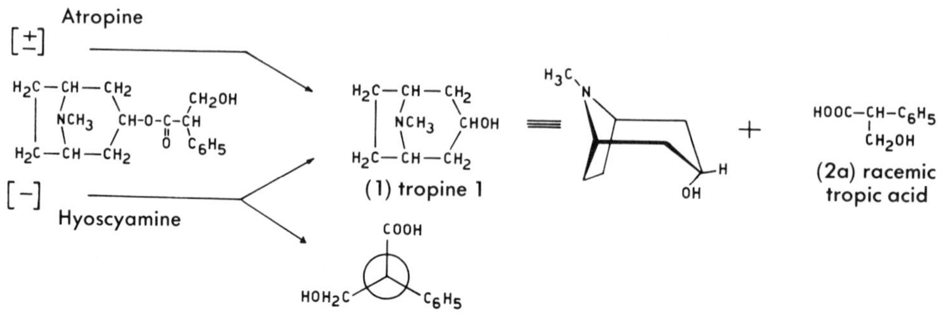

(2) Atropine hydrolysis.

tropic acid (2b). Alkali or heat converts[10,11] hyoscyamine into atropine. Racemic tropic acid has been synthesized[12] from acetophenone *via* atropic acid and resolved. It is, therefore, clear that both atropine and hyoscyamine are tropic acid esters of tropine.

2.1.1.1 ***The Constitution of Tropine*** Treatment of tropine with hydriodic acid followed by zinc and hydrochloric acid gave tropane (5),[13] which in turn was degraded[14] by zinc dust to 2-ethylpyridine (6).

Exhaustive methylation[15] of dehydrated tropine, 'tropidine' (2-tropene) (7) gave first β-methyl-tropidine (8) and this on hydrolysis provides 'tropylene' (9), a ketone that proved identical with $\Delta^{3,4}$ cycloheptenone. Further exhaustive methylation of β-methyl-tropidine led to 'tropylidene',[16]

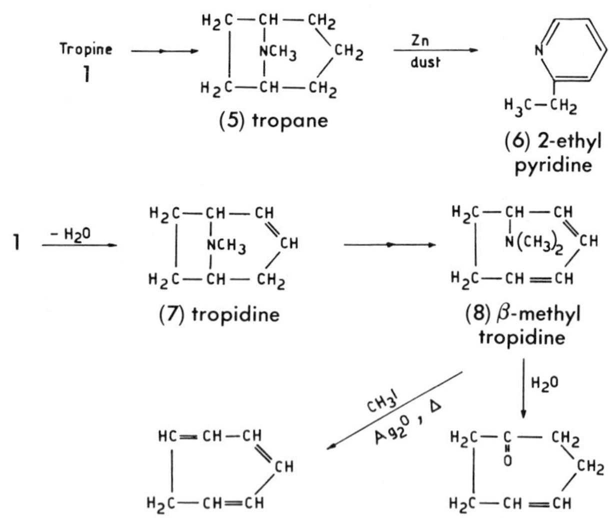

Degradation of tropine.

i.e., cycloheptatriene (10). These reactions demonstrate the presence of a tertiary ring-nitrogen and of a 7-membered carbocyclic system.

Oxidation of tropine (1) with chromic oxide gives a compound which, although it reduces ammonical silver oxide, proved to be ketone,[17] tropinone (11). Tropinone gives an oxime,[18] and it is nitrosated[19] to a dioximino ketone (12), thus proving the presence of 2 methylene groups adjacent to a ring-carbonyl group. Tropinone (11) was reduced by sodium and alcohol to pseudotropine (13)[20] and by magnesium and hydrochloric acid to tropine (1). These transformations demonstrate that tropine and pseudotropine are steroisomers at the hydroxyl bearing carbon.

Further oxidation of tropinone furnished[15] tropinic acid (14) that was degraded by repeated exhaustive methylation[21] to piperylenedicarboxylic acid (15). Hydrogenation[22] of the latter gave pimelic acid (16). Under more drastic conditions tropinone gave[23] with chromic oxide in sulfuric acid N-methyl succinimide (17), thus providing evidence for a pyrrolidine ring

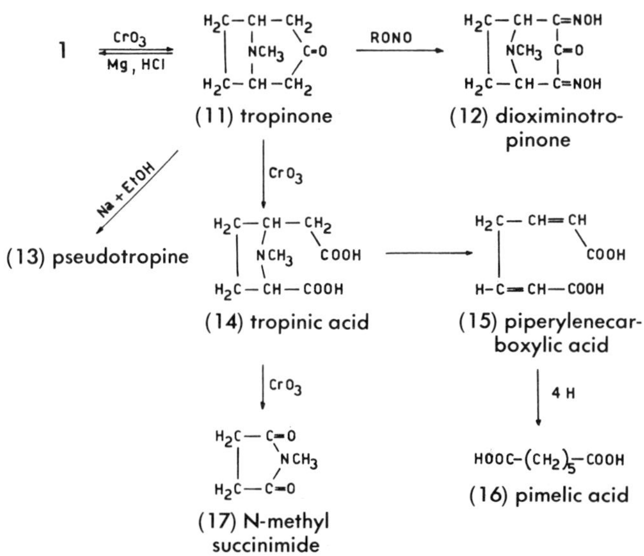

Oxidations and reductions of tropinone.

in the tropanes. Willstätter and Bode ascribed[24,25] to tropine and pseudotropine the structure of *cis* and *trans* tropan-3-ol. Definitive evidence was provided by the stepwise synthesis[26] of tropine from suberic acid, *via* tropidine (7). The more lucid and practical Robinson synthesis[27] started with succinic aldehyde (18a), methylamine and acetonedicarboxylic acid to give in one step tropinone (11). This synthesis has been extended to substituted succinic aldehydes; e.g., malic[28] (18b), *meso*-tartaric[29] (18c) and methoxy-succinic aldehydes,[30] (it failed[31] only with epoxy-succinic aldehyde) and amines (e.g., benzyl-amine). Use of the monoester of acetone dicarboxylic acid instead of the free dibasic acid enabled Willstätter and coworkers[32] to synthesize cocaines *via* ecgoninone.

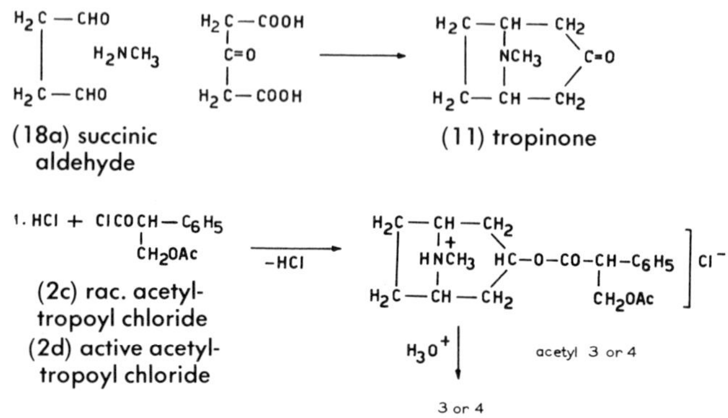

Synthesis of atropine.

Reduction of tropinone (11) (hydrogenation[33] over platinum oxide) gave tropine (1); reesterification[34] with acetyltropoyl chloride (2c) to acetylatropine and subsequent selective hydrolysis of the acetyl group afforded atropine (3). The latter has been resolved[35] with camphorsulfonic and certain other acids, e.g., benzoyltartaric,[36] into (+) and (−) hyoscyamine (4). The direct synthesis of the latter from tropine (1) hydrochloride and *S*(−)acetyltropoyl chloride (2d) *via* acetylhyoscyamine has been reported recently.[36]

2.1.1.2 ***The Configuration of Tropine and Pseudotropine*** The isomeric nature of tropine and pseudotropine at C(3) was recognized first by Willstätter and Iglauer[37] and differed from the belief[38] in N-diastereoisomerism. *Cis-trans* isomerization at the ring junction was considered impossible. This state of affairs did not change for five decades. Finally, stereospecific N → O acyl migration[39,40,44] was applied to N-acetyl-*nor*-pseudotropine hydrochloride (19), which rearranged[8,41,42] intramolecularly to O-acetyl-*nor*-pseudotropine hydrochloride (20). N-Acetyl-*nor*tropine[43] hydrochloride (21), under the same conditions, failed to react. This behavior is evidence for the *cis*-position of the C(3) hydroxyl in *nor*pseudotropine (13) and its derived esters, while tropine (1) and the tropeines are 3-trans-(with respect to the nitrogen bridge) tropanols.

Since N → O acyl migration takes places *via* 2-hydroxyoxazolidines[45] and in case of pseudotropine *via* a 2-hydroxy tetrahydro-1.3-oxazine (22), the ability of N-acyl-nortropanol to rearrange indicates that the N-acyl and hydroxyl groups are in steric proximity. This requires a transitory ring

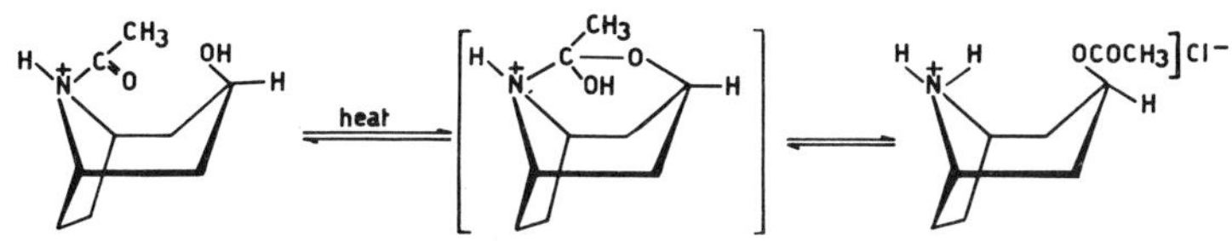

(19) N-ac-norpseudotropine HCl — (22) intermediate in acyl migration — (20) O-ac-norpseudotropine HCl

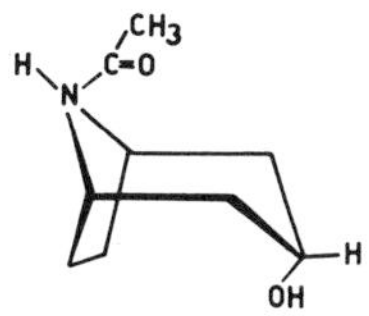

(21) N-ac-nortropine HCl

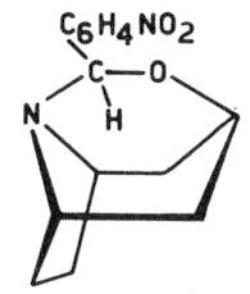

(23) p-nitrophenyloxazine from norpseudotropine

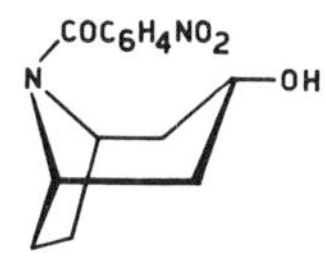

(24) N-p-nitrobenzoyl nortropine

Configuration of norpseudotropine.

closure involving the boat form of the tropane skeleton in the transition state.[46]

The same results have been achieved with the epimeric benzoates[47] and later by bridgeing the ring[48] from N to O in *nor*pseudotropine by *p*-nitrobenzaldehyde, giving a tetrahydrooxazine (23), closely related to the intermediate (22) in N → O acyl migration. Under the same conditions *nor*tropine afforded N-p-nitrobenzoyl-*nor*tropine (24) as a product.[49] Infrared data were interpreted in favor of an intramolecular H-bridge[50] in the case of pseudotropine; this was, however, later disproved.[51] Dipole moment measurements were complementary to the chemical results.[52]

The rates of hydrolysis of the epimeric benzoyl and p-nitrobenzoyl tropeines and pseudotropeines, and those of their methiodides have been determined.[53] Since tropine esters show smaller rates than pseudotropeines and since boat conformations (1B) and (13B) were assumed, these facts were explained in the case of the tropine esters by the postulation of a hydroxyl group hindered by the neighboring N–CH_3 group. Hence, configurations opposite to those established by chemical methods were assigned to the epimers. The error was one of interpretation, [54,55] since all these measurements agree with the true configurations *when a chair form is assumed* (1C) and (13C).

Actually, the chemical approach furnished definitive evidence for the configurations of the tropines.

Following a suggestion by Dr. Cahn,[42] the functional groups *cis* to the N-methyl bridge are designed as β, while the *trans* ones are designated as α, as in steroids. Accordingly, pseudotropine is tropan-3β-ol (13), while

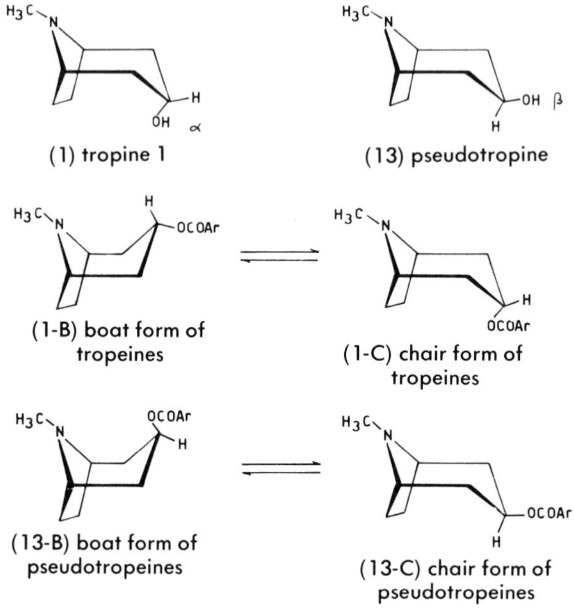

(1) tropine 1
(13) pseudotropine
(1-B) boat form of tropeines
(1-C) chair form of tropeines
(13-B) boat form of pseudotropeines
(13-C) chair form of pseudotropeines

Interpretation of kinetic data.

tropine is tropan-3α-ol (1). The same designations apply also to their esters, the tropeines.[1,2,56–61]

2.1.1.3. ***The Absolute Configuration of Tropic Acid*** The esterifying acid in hyoscyamine is *levorotatory* tropic acid (2b) to which the L-configuration was ascribed.[62] This assignment was based on the Freudenberg rule of optical shifts. The correlation of (2b) with alanine has been achieved recently.[63,64] (+) α-Methyl phenylacetic acid (25a) was converted

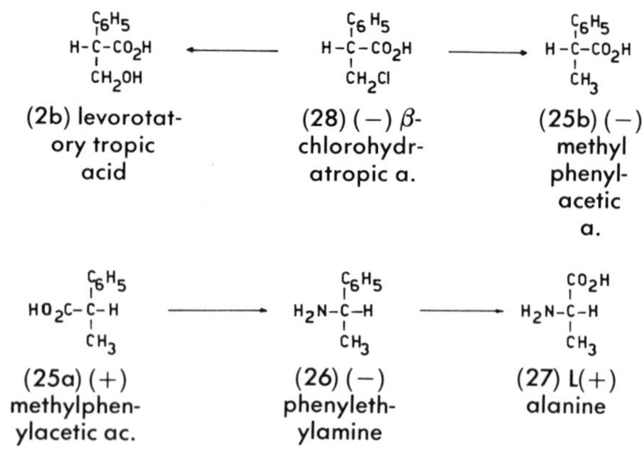

(2b) levorotatory tropic acid
(28) (−) β-chlorohydratropic a.
(25b) (−) methyl phenylacetic a.
(25a) (+) methylphenylacetic ac.
(26) (−) phenylethylamine
(27) L(+) alanine

Absolute configuration of (−) tropic acid.

earlier[65] by Curtius degradation to (−) α-phenyl ethylamine (26). This was followed by conversion to L(+) alanine (27). The absolute configuration[66] of (25a) is thus *S*. Furthermore, (−)β-chlorohydratropic acid (28) had been hydrolyzed[67] to (−)tropic acid. One link was missing. Reductive dechlorination of the (−)chloroacid (28) should provide either *S*(+) or *R*(−) methylphenylacetic acid (25a) or (25b). The latter (25b) was obtained by catalytic hydrogenolysis[63,64] of (−)β-chlorohydratropic acid (28). Consequently both this and (−) tropic acid (2b) have the *S*-configurations and hyoscyamine is *S*(−)tropoyl-tropan3α-ol.

2.1.2 Minor Alkaloids: Convolamine, Convolvine, Poroidine, Isoporoidine, Tropacocaine, Tigloidine *Convolamine*[56] gives tropine and veratric acid on hydrolysis, while *convolvine,*[57] *poroidine*[57] and *isoporoidine*[57] affords *nor*tropine (29) as well as veratric, 3-methyl butyric and 2-methyl butyric acids, respectively. Tropacocaine (30)[58] and tigloidine[59−61] furnish pseudotropine (13) together with benzoic and tiglic acids, respectively.

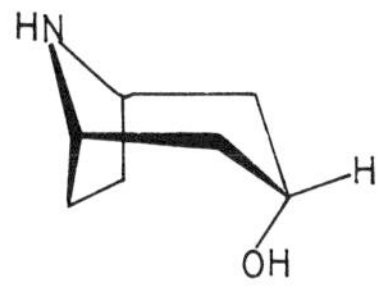

(29) nortropine

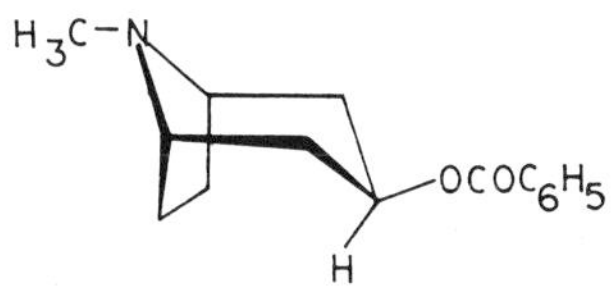

(30) tropacocaine

Nortropine and tropacocaine.

Since the corresponding esterifying acids are well known, the constitutions of these alkaloids are determined unequivocally.

2.1.3 Phyllalbine The molecular formula, $C_{16}H_{21}NO_4$, (291.3) was supported[68] by mass spectrometry showing a molecular ion M + 291. The infrared absorption at 1688 cm^{-1} indicated an aromatic ester grouping and the μ maximum at 221 mμ the presence of a phenolic hydroxyl. The nmr showed a multiplet at 3.22 ppm and a triplet at 5.25, width 12 cps, the latter being typical for the equatorial C(3) H in tropine and its esters. In contrast pseudotropine, has a quintuplet at 3.89 ppm indicating an *axial*

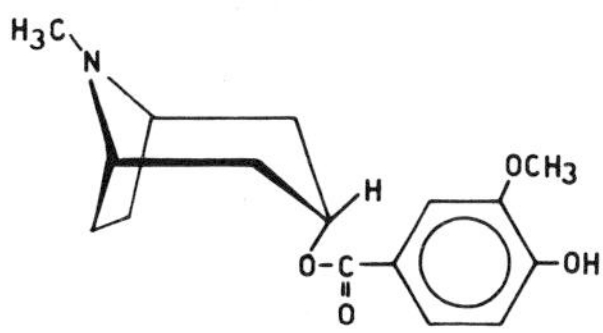

(31) phyllalbine

Phyllalbine.

C(3)H, which in tropacocaine (30) appears at 5.27 spread over 35 cps in both cases. This bandwidth indicates axial-axial and axial-equitorial couplings; in phyllalbine there is equatorial-axial and small equatorial-equatorial coupling. Hydrolysis of phyllalbine with 2N hydrochloric acid afforded tropine and vanillic acid. Thus phyllalbine is O-vanilloyl-3α-tropanol (31).

2.2 Ecgonines

2.2.1 Cocaine and Pseudococaine Cocaine (32) ($C_{17}H_{21}NO_4$) gives on acid hydrolysis[69] benzoic acid, methanol, and a carboxylic acid, $C_9H_{15}NO_3$, called ecgonine (33).[70] Chromic acid oxydation[71] of the

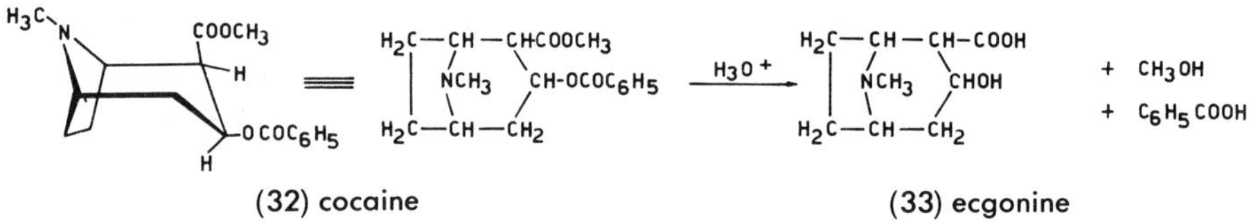

Hydrolysis of cocaine.

latter led *via* ecgoninone (34a) to tropinone (11) and then further to tropinic[72,73] (14) and ecgoninic acids[24] (35). Both acids were optically active. These data indicate that ecgonine is carboxy-tropine (or pseudo-tropine): the carboxyl group is located in the piperidine moiety. If this were not the case, a carboxy-substituted tropinic acid should be formed. Dehydration of ecgonine leads to anhydroecgonine[74] (36). This, in turn,

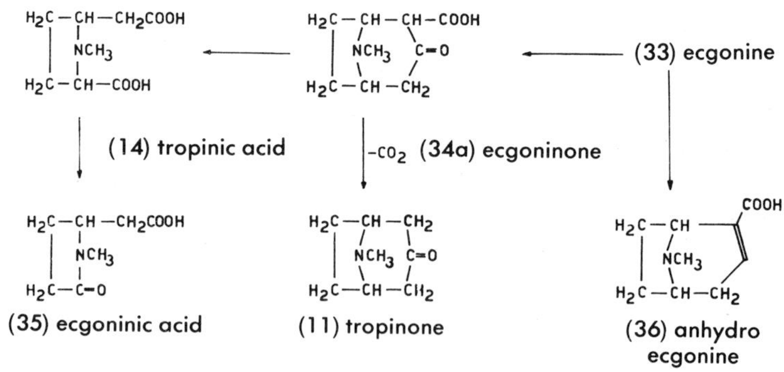

Degradation of cocaine.

by decarboxylation, should lead to tropidine[75,76] (2-tropene) (7). The final proof for the constitution of ecgonine as tropan-3-ol-2-carboxylic acid (33) was supplied by a total synthesis. Tropinone (11) sodium was carboxylated[25] to sodium tropinone-2-carboxylate (35b) (ecgoninone) and this was reduced to ecgonine (33) and pseudoecgonine.

The second synthesis[32] of ecgoninone methyl ester (34c) starts with succinic aldehyde (18a) and monomethyl acetonedicarboxylate by analogy with the Robinson route for tropinone (11). Reduction to ecgonine was followed by esterification of ecgonine with methanol and benzoylation of ecgonine methyl ester to (32).

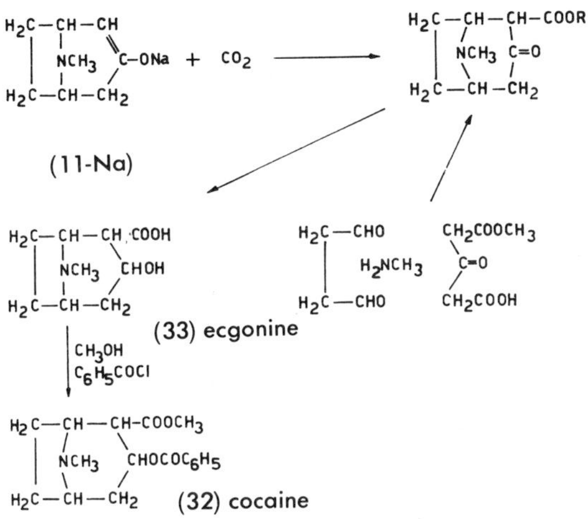

(34a) ecgoninone
R=H (34b) ecgoninone-Na
R=Na (34c) (±)ecgoninone methyl ester
R=CH_3 (34d) active in ecgonone methylester

Synthesis of ecgonine, etc.

(−)Ecgonine (33) was isomerized[77] to (+)-pseudoecgonine (37a) the alkamine of pseudococaine, by the action of alkali. This led to the presumption[32,82] of a C(3) epimerism in these bases similar to that encountered with the tropines. Later stereochemical work[78,79] disproved this theory and favored a C(2) epimerism. Pseudococaine (37b) was synthesized[32] from ecgoninone (34a) *via* pseudoecgonine, racemic (37a). Both cocaine epimers have a strong anesthetic effect; the same applies to cocaethylene, i.e., benzoylecgonine ethyl ester.

2.2.1.1 ***Cinnamylcocaine and α-and β-Truxilline*** α- and β-Truxilline,[80] as well as cinnamylcocaine,[81] were hydrolyzed to the same alkamine, ecgonine (33). Truxillic, isotruxillic, and cinnamic acid, respectively, were isolated and the constitution of the alkaloids is unequivocally known.

2.2.1.2 ***The Relative Configurations of the Ecgonines*** The configuration of ecgonine (33) and that of pseudoecgonine (37a) was established in three consecutive steps:

(a) Determination of whether the hydroxyl and nitrogen are in steric proximity or are well separated;

(b) Establishment of the relative steric position of the C(2)-COOH

and the C(3)-OH in the individual epimers; and
(c) Immediate proof for the *cis* or *trans* relationship of C(2)- carboxyl to the N-methyl bridge.

2.2.1.3 *Relative Configuration of the Nitrogen and C-(2) hydroxyl* Acyl migration N ⇌ O occurred reversibly[82] with N-acetyl-*nor*pseudoecgonine ethyl ester (38a) (38b) while N-acetyl-*nor*ecgonine ethyl ester gave an ill defined mixture. Therefore, the C(3)-hydroxyl is β in pseudoecgonine.

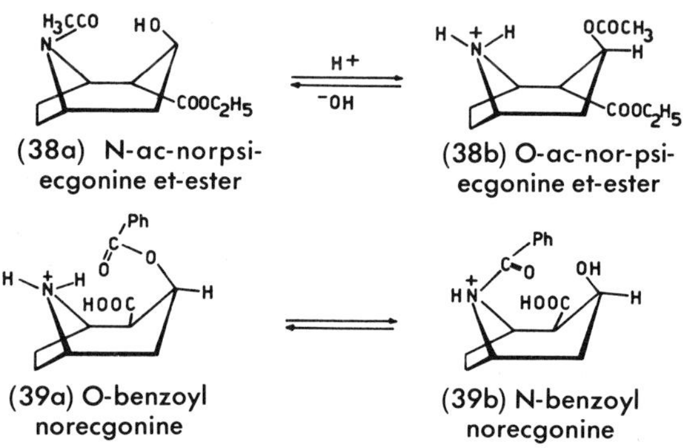

Acyl migration in norpseudoecgonine ester.

O-benzoyl-*nor*ecgonine (39a) rearranges easily to N-benzoyl-norecgonine[78] (39b); the reverse reaction, although it needs more drastic conditions, still proceeds.[79] Therefore, the C(3)-hydroxyl is β in ecgonine.

2.2.1.4 *C-(2)-COOH–C-(3)-hydroxyl Relationship* Curtius degradation[82] of O-benzoyl-ecgonine (40) and O-benzoyl-pseudoecgonine (41) gave two epimeric 2-benzamido-3-tropanols (42a) and (43), respectively. These isomers behave differently in N→O acyl migration. The ecgonine derivative (42a) showed an acyl shift to (42b) whereas the amide from pseudoecgonine (43) remained unchanged. Therefore, the functional groups at C(2) and C(3) are *cis*- in 2-amino-3-tropanol (and hence in ecgonine) and *trans* in pseudoecgonine and in the 2-amino-3β-tropanol into which it was degraded. Further support for this conclusion was provided by the ease with which ecgoninol (44a) (the product of LAH reduction of cocaine) gives a benzylidene ketal (45). Pseudoecgonine under the same conditions, failed to react with benzaldehyde. In fact, both epimeric ecgonines should be called 'pseudo' since each is derived from pseudo-(3β)-tropanol.

These stereochemical assignments were also demonstrated[78] by the conversion of (–) cocaine (32) to (37C) in the presence of catalytic amounts of sodium methoxide in methanol (37b).

Further systematic investigation involved a three-step conversion[83] of the carboxyl in the ecgonines into a methyl group and oxidation of the

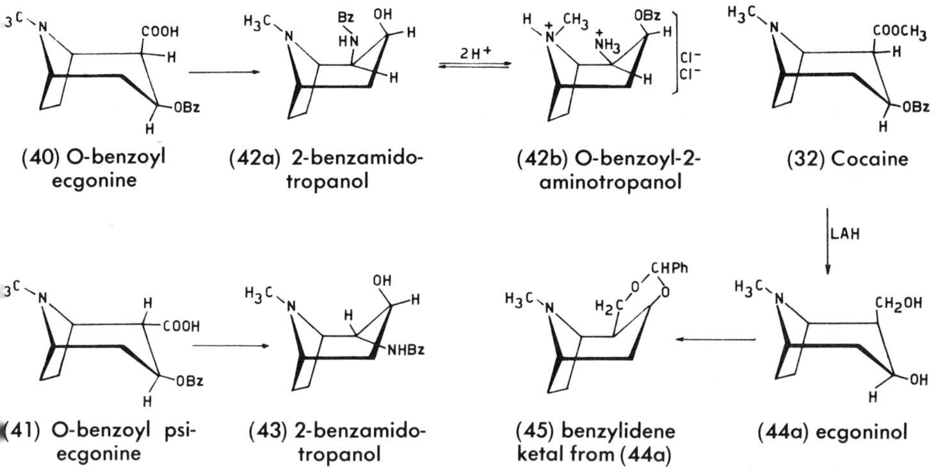

(40) O-benzoyl ecgonine; (42a) 2-benzamido-tropanol; (42b) O-benzoyl-2-aminotropanol; (32) Cocaine; (41) O-benzoyl psi-ecgonine; (43) 2-benzamido-tropanol; (45) benzylidene ketal from (44a); (44a) ecgoninol

Acyl migration in 2-benzamido-3-tropanols.

epimeric 2-methyl tropanols (46) and (47) into a pair of ketones (48a) and (48b). Compound (48b), derived from pseudoecgonine, gave an oxime (49) without inversion, while the ecgonine derivative (48a) irreversibly furnished the same oxime.[79,83] Thus, ecgonine and pseudoecgonine are epimeric at C(2).

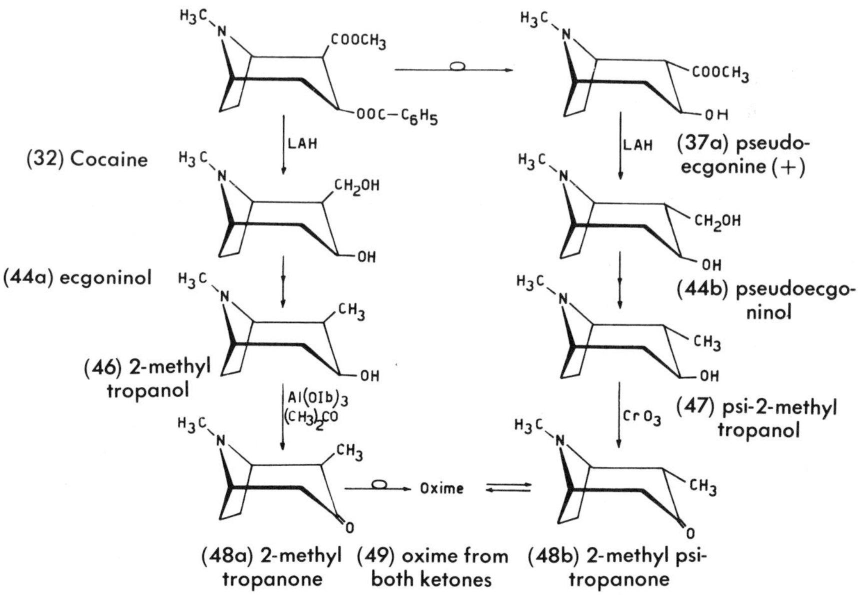

(32) Cocaine; (37a) pseudo-ecgonine (+); (44a) ecgoninol; (44b) pseudoecgo-ninol; (46) 2-methyl tropanol; (47) psi-2-methyl tropanol; (48a) 2-methyl tropanone; (49) oxime from both ketones; (48b) 2-methyl psi-tropanone

Correlation of cocaine and psicaine with 2-methyl-3-tropanone.

2.2.1.5 ***C-(2) and Nitrogen Relationship*** Cocaine (32) gave by von Braun degradation[84] with cyanogen bromide N-cyano-*nor*cocaine (50) which in turn, could be hydrated to N-carbamyl-*nor*cocaine (51). This compound was cyclized to the amide[79] (52). These transformations demonstrate the proximity of the carboxyl function and the ring nitrogen in cocaine.

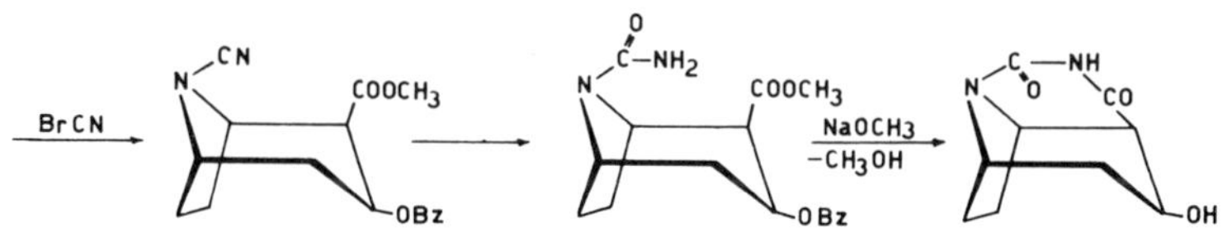

(50) N-cyano norcocaine (51) N-carbamyl norcocaine (52) lactam of (51)

N-cyano norcocaine; lactam.

2β-chloromethyl-3β-tropanol (53), obtained from ecgoninol (44a) by the action of thionyl chloride undergoes a spontaneous cyclization upon heating.[83] The salt-like compound m.p. 232° was supposed[83,85] to be an internal ether (54a). However, this salt does not move[86] from the origin in tlc, and its infrared spectrum showed clearly a hydroxyl at 3200 cm^{-1}, a fact incompatible with (54a). A more recent nmr study[87] showed explicitly a hydroyl at 4.82 ppm, as a singlet. There is no change in the signal of the C(3) proton during cyclization of (53) while the methylene protons at C(2′) were strongly deshielded. They appear at 4.35 ppm while the methine proton at C(1) is shifted to 5.25, and the proton at C(2) appears at 4.4 ppm. These assignments are based on a comparison with the *salt* prepared from C(2′)-d_2-ecgoninol, *via* C(2′)-d_2-2β-chloromethyl-3β-tropanol, by internal quaternization. Hence, (53) isomerizes to an azetidinium salt (55). This proves the cis, β-relationship of CH_2Cl to NCH_3 in (53) and by inference that of the methoxycarbonyl in cocaine. Cyclization of the methiodide of chloromethyl tropanol (53) gave the methiodide of the ring ether (54b), nmr data of which have been compared with those of (55). See Table 15.2.

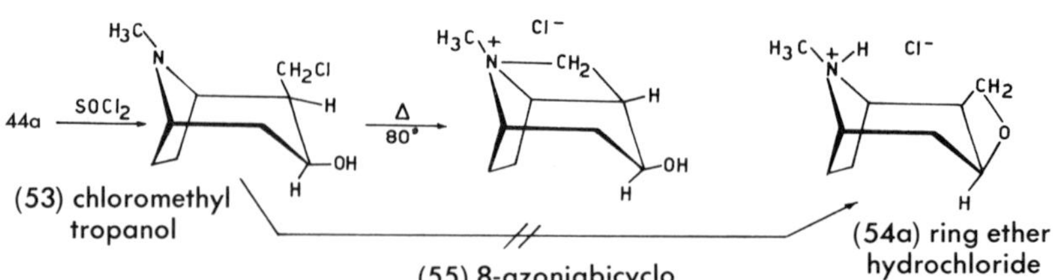

(53) chloromethyl tropanol

(55) 8-azoniabicyclo [2,2,1^{2,8}]-nonane chloride

(54a) ring ether hydrochloride

(54b) ring ether methohalide

Azetidinium salt from ecgoninol.

TABLE 15.2 Proton Chemical Shifts (ppm on δ scale) for Tropanes.

No.	Tropane	H-1 and H-5	H-6 and H-7	H-2 and H-4	N-Me	H-3	O-H	Miscellaneous
(11)	3-Oxo	3.35	1.4–2.0	2.2	2.41			—
(78a)	3-Oxo-6β-hydroxy	3.4	4.1–2.0	2.1–2.6	2.65		3.95	—
(11-d_4)	3-Oxo-2,4-d_4	3.38	1.4–2.0	–	2.42			–
(1)	3α-Hydroxy (ax)	3.0	2.05	1.6–1.9	2.16	3.90	4.3	–
(1-d_4)	3α-Hydroxy-2,4-d_4	2.97	2.00	–	2.18	3.88	3.56	–
(1-d_2)	3α-Hydroxy-6,7-d_2	2.98	2.05	1.6–2.0	2.18	3.89	3.4	–
(13)	3β-Hydroxy-(eq.)	2.07	1.5–1.8	1.5–1.8	2.22	3.70	3.51	–
(13-d_4)	3b-Hydroxy-2,4-d_4	3.09	1.8	–	2.20	3.73	4.79	–
(97)	3α-Cyano	3.17	3.8	2.2–2.8	2.51	3.95	–	–
(96)	3β-Cyano	3.10	1.5–1.9	1.5–1.9	2.2	3.4	–	–
(107a)	3α-Chloro	3.27	1.8–2.7	1.8–2.7	2.37	4.56	–	–
	3α-Methylsulfonyl	3.3	1.8–2.3	1.8–2.3	2.4	5.21	–	3.5 for Me protons
	3α-Benzoyloxy	3.08	1.52–2.2	1.5–2.2	2.25	5.25	–	8.08 and 7.45 for aromatic protons
	3β-Benzoyloxy	3.2	1.55–2.2	1.55–2.2	2.32	5.20	–	8.07 and 7.45 for aromatic protons
(44a)	3β-Hydroxy-2β-hydroxymethyl	3.28	1.48–2.15	1.48–2.15	2.2	4.0	5.45	4.1 for H-2
(53)	3β-Hydroxy-2β-chloromethyl	3.1 and 3.4	1.42–2.15	1.42–2.15	2.2	4.20	3.82	4.08 for H-2 (J = 2.5 cps)
(53) HCl	3β-Hydroxy-2β-chloromethyl (HCl)	3.95–4.7	1.9–2.9	1.9–2.9	2.92	4.0	4.72	3.95–4.78 for H-2
(55)	8-Methyl-8-azoniatricyclo [2.2.1.1^{2,7}] nonane chloride	5.15 and 4.28	2.0–2.8	3.0 and between 2.0–2.8	3.15	4.4	4.75	4.2 for H-9 (J = 4 cps)
(54b)	2′,3-Anhydro-2β-hydroxy methyl-3β-hydroxy methiodide	4.6	1.8–2.6	3.55 for H-2 and between 1.8–2.6	3.2 & 3.85	4.12	–	5.5 for H-9
(32)	(Cocaine) 2R:3S 2β-methoxycarbonyl 3β-benzoxy	3.55	1.58–2.18	2.95–2.6	2.22	5.28	–	8.1 and 7.5 for aromatic; 3.82 for ester methyl
(62)(−)	(Scopolamine) 3α-S-tropoyloxy 6,7β-epoxy	2.98	2.84–3.37	1.1–2.3	2.4	4.98	6.09	7.3 for aromatic; 3.5–4.3 for tropyl
(64a)	(Scopine) 3α-hydroxy-6,7β-epoxy	3.12	3.6	1.25–2.33	2.47	3.9	3.37	–
(63)	(Oscine) 3α, 6α-oxido	2.96–4.45	3.67–3.83	0.84–2.27	2.57	4.22	3.35	–
(3)	(Atropine) 3α-tropoyloxy	2.93	1.2–2.05	1.2–2.05	2.15	4.97	4.81	7.27 for aromatic; 3.53–4.3 for tropyl ester CH_2 and CH
(72a)	3α, 6β-Dihydroxy	3.02	3.3, 1.4–2.2	1.4–2.2	2.5	3.9	4.8	–
(87)	(Teloidine) 3α, 6β, 7β trihydroxy	2.85	4.45	1.15–2.2	2.4	3.8	4.65	–

2.2.1.6 ***The Absolute Configuration of (–) Cocaine*** In order to determine whether the structural formula used throughout this chapter for (–) cocaine (32), or its mirror image is correct, a correlation with L(+) glutamic acid (56) was made.[88] Ecgonine (33) was oxidized to ecgoninic acid (35), i.e., N-methyl-2-pyrrolidone-5-acetic acid. The same acid was synthesized from L (+) pyroglutamic acid (57) (2-pyrrolidone-5-carboxylic acid) via (58). An abbreviated scheme of this correlation is given below. It follows that (–)cocaine is (–)2*R*-methoxy-carbonyl-3S-benzoxyloxy-tropane.

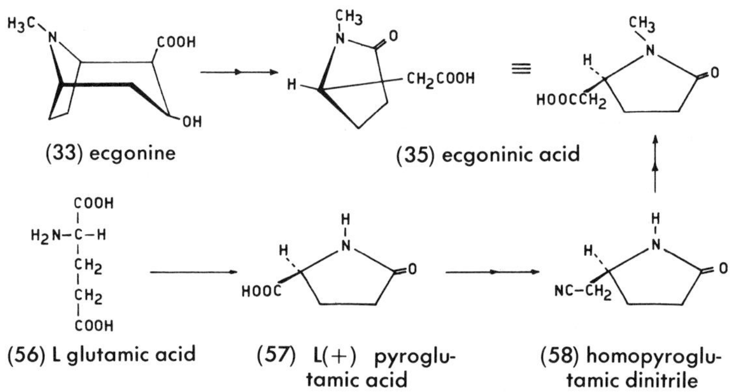

Absolute configuration of cocaine.

In addition to racemic ecgonine and pseudoecgonine a third racemate was synthesized.[32] The fourth racemate, was unknown until recently.

Unfortunately, there is a serious discrepancy between the data reported by different authors[89–91] for the individual racemates.

The most convincing approach to all possible diastereomers of ecgonine involves the oxidation of (+)pseudoecgonine methyl ester (37c) to *optically active* 2-methoxycarbonyl-tropinone (34d) followed by reduction. This leads to pseudoecgonine ester along with one of the epimers, presumably[90,92] the C(3) one, i.e. 2α-methoxycarbonyl-3α-hydroxy tropane (59).

On the other hand, one of the products of hydrogenation of (–)methoxycarbonyl tropinone (34d), nonidentical with either the 2β, 3β or with the 2α, 3β racemates, was epimerized during quaternization with methyl iodide, and this was interpreted as an axial→equatorial (β→α) shift of the C(2) group, similar to that observed in ecgonine. It is likely that this labile compound is (+)2β-methoxycarbonyl-3α-hydroxy tropane (60).[92]

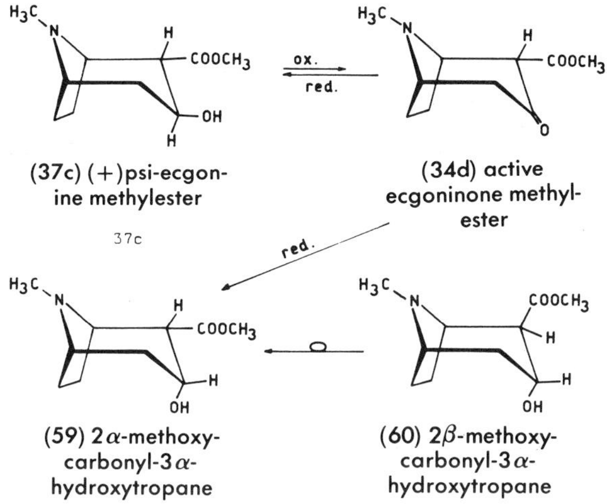

(37c) (+)psi-ecgonine methylester

(34d) active ecgoninone methylester

(59) 2α-methoxycarbonyl-3α-hydroxytropane

(60) 2β-methoxycarbonyl-3α-hydroxytropane

Ecgonine epimers.

2.3 Alkaloids Containing Oxygen Functions in the Pyrrolidine Moiety

2.3.1 Scopolamine (61) and Hyoscine (62)

2.3.1.1 ***Constitution*** The alkaloids[93] (61) and (62) are converted by both acidic[94] and alkaline hydrolysis into oscine[95] (63) (scopoline) ($C_8H_{13}NO_2$) along with racemic (2a) or (−)tropic acid (2b), respectively. Pancreatic lipase (or a Sorensen buffer solution) cleaves them into an isomeric alkamine,[96] scopine (64a). The optical activity of hyoscine is caused by the tropoyl group. Both oscine (63) and scopine (64a) are optically inactive, but the former can be resolved[94] by D_S(+)tartaric acid. Anhydrohyoscine is inactive, as is anhydroscopolamine. Neither is resolvable. Thus, scopine is the primary product of hydrolysis while oscine is a transformation product thereof. Since oscine (scopolamine) can be acetylated[97] and benzoylated,[98] it has a (secondary) hydroxyl group.

Treatment of oscine with hydrogen bromide affords: 'hydroscopoline bromide' (65) which was, in turn, debrominated by zinc and sulfuric acid to a diol,[99] 'hydroscopoline' (66). Oxidation of the latter with chromic acid led to scopolic acid (67), identified as 1-methyl-piperidine-*cis*-2,6-dicarboxylic acid.[100] It is clear that oscine has a piperidine skeleton and two oxygen functions elsewhere in the molecule. Hydroscopoline (66) is reduced[101] by phosphonium iodide to tropane (5); hence, oscine and scopolamine are tropane derivatives with oxygen functions in the pyrrolidine moiety.

Scopine (64a) is easily converted[96] into oscine (63). It has a hydroxyl group and another oxygen that is part of an ether group. Its epimer, pseu-

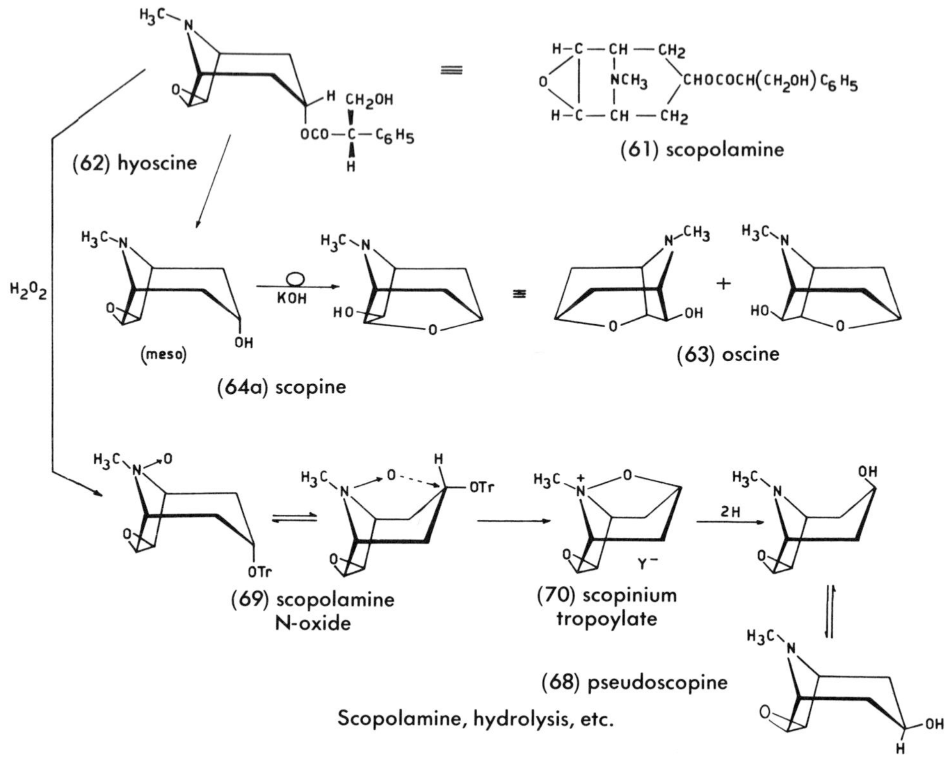

Scopolamine, hydrolysis, etc.

doscopine (68), was prepared[102] from scopolamine-N-oxide (69). Recent nmr work confirmed[140a] the position of the semipolar oxygen in (69) as axial. Heating of 69 produced scopinium tropoylate (70), the bromide of which gave on reductive cleavage pseudoscopine (68). This compound subsequently has been synthesized.[103]

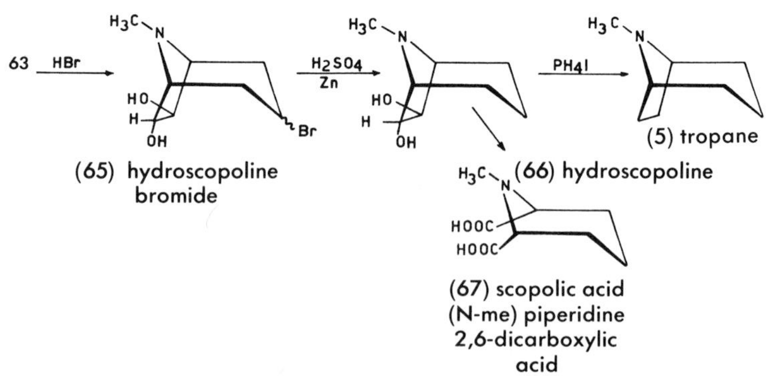

Degradation of oscine.

The transformation of scopine into oscine was assumed to involve an ethylene oxide→butylene oxide rearrangement. Scopine is 6,7-epoxy-3-hydroxy-tropane and scopolamine its 3-tropoyl ester. The final proof of both constitution and configuration was achieved much later (1956) by the stereospecific total synthesis of scopolamine and hyoscine.[104,105]

2.3.1.2 ***Configuration of Scopolamine and Hyoscine*** The conversion of the epoxide scopine (64a) into (±) oscine (63) a tetrahydrofuran derivative has been interpreted[82,106,107] as a backside nucleophilic attack of the hydroxyl oxygen upon either bridgehead of the epoxide ring at C(6), (7). Hence, the hydroxyl is α and the oxide function β. The N-oxide (69) can only undergo a transannular nucleophilic displacement[106] at C(3) (with inversion) to give (70) and (68) if the N-oxide function is axial and the hydroxyl on C(3) is α. This inference was supported experimentally by hydrogenolysis of (62) over Pd into diastereoisomeric tropandiol-S-tropoylates (71a,b), followed by hydrolysis to (±) tropandiol (72b) and resolution.[108,109] The (−)antipode[110] (72a) ('valerine') was identified[109] with the alkamine[110,112] of (−) valeroidine[111](73). Later it was revealed[113]

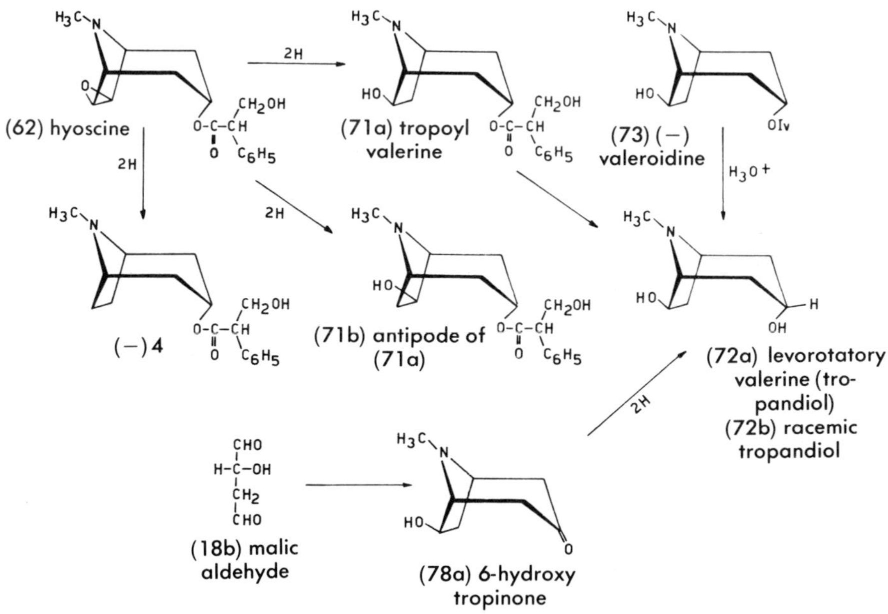

Configuration of scopolamine.

that hydrogenolysis of (−)hyoscine (62) over Raney-Ni gives a considerable amount of (−)hyoscyamine (4) as a product of complete deoxygenation. The same alkamine was synthesized[28] from malic aldehyde (18b) *via*

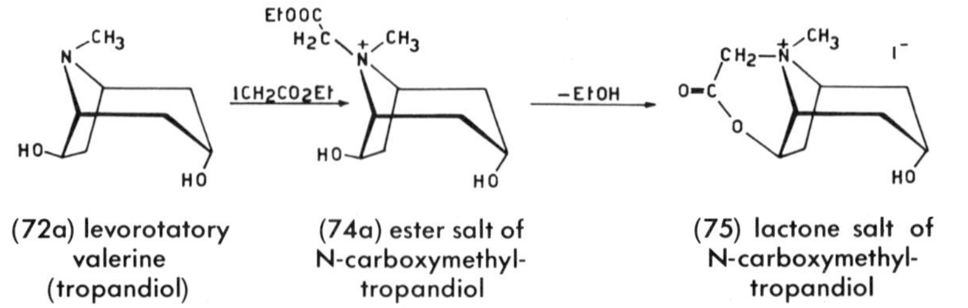

Lactone salt method.

(±) 6β-hydroxy-tropinone (78a) which was resolved, and the (+)antimer hydrogenated to (−)valerine (72a). Hence, valerine is a 3,6-tropandiol.

Now, the levorotatory alkamine (72a)[114] gave on quaternization with ethyl iodoacetate an ester-salt (−) (74a) that condensed spontaneously to a dextrorotatory lactone salt (75) during recrystallization. Thus, tropan 3,6-diol has a hydroxyl in β-position, otherwise the N-acetic ester group could not undergo transesterification. Scopolamine has an oxide ring at C(6) and C(7) with the same, β-configuration as the hydroxyl at C(6) in valerine.

2.3.1.3 ***Total Syntheses of Scopolamine and Hyoscine*** Direct evidence for the location of the C(3) hydroxyl in scopolamine (61) and valeroidine (73) was furnished by the *total* synthesis.[104,105] (±)-Tropan 3,6-diol (72b) was an important intermediate. This compound gave on acetylation, a diacetate, (83e). Selective hydrolysis of 83e affords a 3-monoacetyl-3,6-tropandiol (72c). The tosylate (76) cleaved by action of collidine (or better piperidine)[115] gave rise to 6-trop-3-enyl-acetate (77b), a key intermediate for scopolamine. This compound was hydrogenated to acetyl tropan-

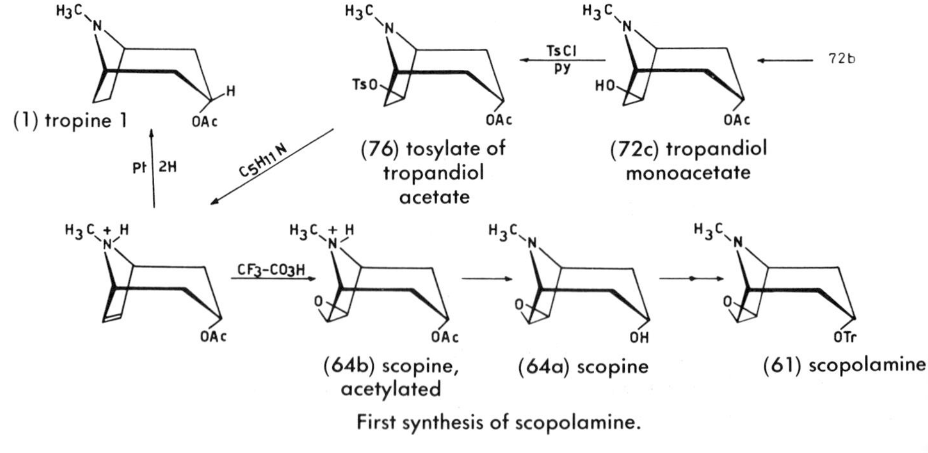

First synthesis of scopolamine.

3α-ol (1) of known configuration. Thus, valeroidine and scopolamine both have an α-hydroxyl at C(3) and valerine is 3α,6β-tropandiol.

6-tropenyl-3-acetate (77b) was oxidized[104,105] by trifluoroperacetic acid to acetyl scopine (64b). Since the oxygen addition to the double bond in that cage-like system was predicted[116] to take an exo-course, scopolamine is 6β,7β-epoxy-3α-S-tropoyloxytropane.

Acetylscopine was deacetylated to scopine (64a). Further acylation with (±) acetyltropoyl chloride gave acetylscopolamine which in turn, was deacetylated to give a mixture of (±) (61) as well as a number of byproducts.[104,105,116]

An improved and simplified total synthesis of (−) scopolamine, i.e., hyoscine (62), starting with trimethoxy tetrahydrofuran and proceeding *via* 6-methoxy-tropinone (78b), 3α,6α-oxido-tropane (79), 6β-bromo-3α-acetoxy-tropane (80) and 6-dehydro-hyoscyamine (81) has been outlined.[115,117] Some of the reagents used[115,117] are of interest: p-toluenesulfonic anhydride for intramolecular etherification, tungstic acid catalyzed epoxidation with aqueous hydrogen peroxide, and ether cleavage by acetyl bromide, etc.

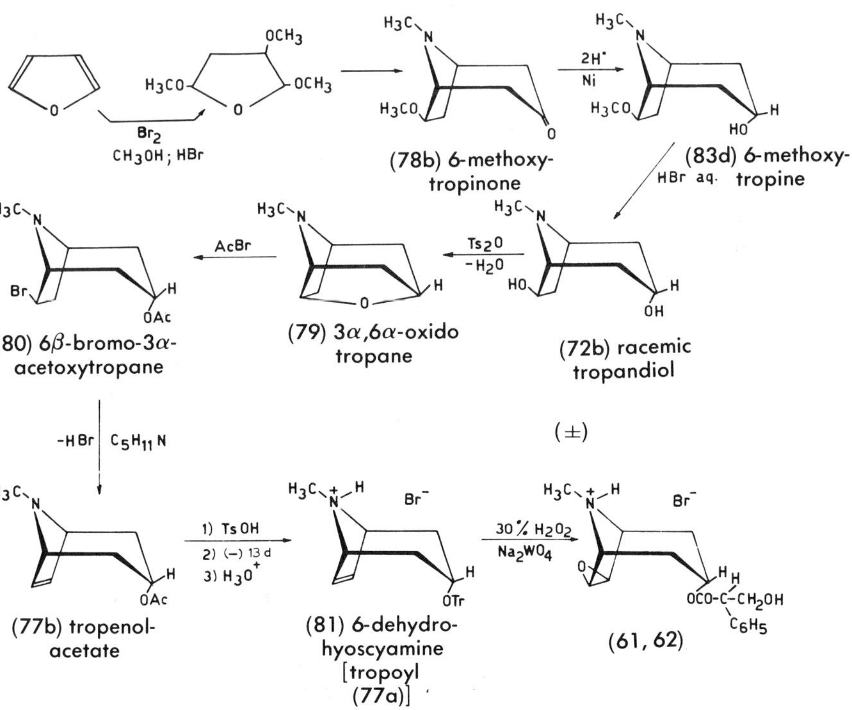

Second total synthesis of scopolamine.

2.3.2 Valeroidine and Mono- and ditigloyl-tropan-3.6-diol

2.3.2.1 ***Constitution*** Valeroidine[59,111] (73) gives 3-methyl butyric acid and a *levorotatory* base valerine (72a), $C_8H_{15}NO_2$, on hydrolysis. The antimer[118,119] of the latter arises from hydrolysis of the two other alkaloids. The (−)valerine base is also present in Javanese Coca leaves.[110]

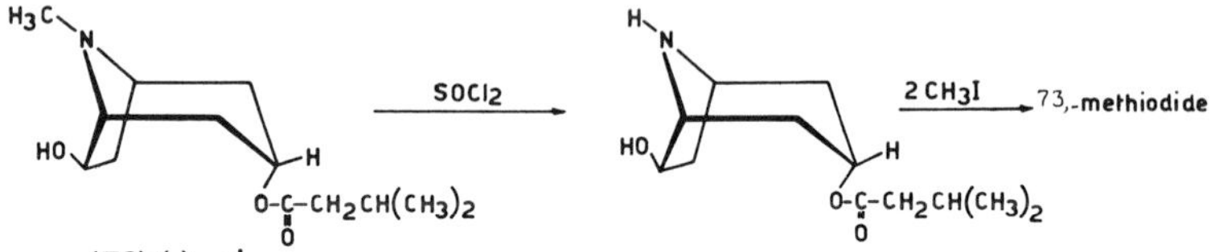

(73) (-) valeroidine

(82) norvaleroidine

Degradation of valeroidine.

Treatment with thionyl chloride resulted in the demethylation of valeroidine to form the nor-base (82) that was then reconverted[111] by methyl iodide into valeroidine methiodide.

The alkamine, $C_8H_{15}NO_2$, gives a dibenzoyl derivative and is dehydrated by phosphoryl chloride, or better with p-toluenesulfonic anhydride, into an internal ether (79), 'tropenoxide',[110] $C_8H_{13}NO$, presumably 6-deoxy-oscine. Since 'tropenoxide' gives tropane (5) when treated with hydriodic acid and phosphorus, it is a tropandiol, presumably the 3,6-diol.

The total synthesis of valeroidine was achieved in 1957.[120] Thermolysis of (+)6β-phenylcarbamyloxy-3α-tropanol isovalerate (83b) obtained from 6-hydroxytropinone (78a)[28] *via* 6-phenylcarbamyloxytropanone (84) and -tropandiol (83a) leaves no doubt as to the location of the

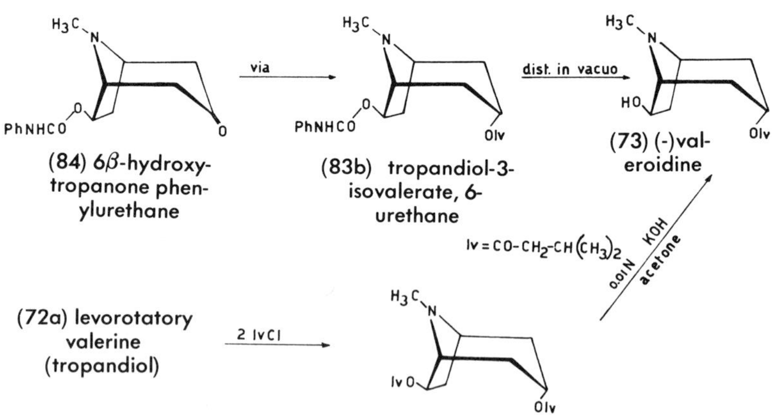

(84) 6β-hydroxytropanone phenylurethane

(83b) tropandiol-3-isovalerate, 6-urethane

(73) (-)valeroidine

(72a) levorotatory valerine (tropandiol)

(83c) diisovaleryl-(−)tropandiol

Syntheses of valeroidine.

valeroyl group. Selective hydrolysis[121] of the 3,6-diisovalerate (83c) gave better yields.

2.3.2.2 ***Relative Configuration of Tropan-3,6-diol*** The relative configuration of valeroidine as $3\alpha,6\beta$ was determined during the work on scopolamine (see Sections 2.3.1.2. through 2.3.1.3.)

2.3.2.3 ***Absolute Configuration*** Based upon the strong *dextro*-shift in optical rotation during lactone salt (75) formation from (−)N-ethoxycarbonylmethyl-$3\alpha,6\beta$-tropandiol (74a) following Hudson's lactone rule, the $3R$:6:R configuration was tentatively assigned[121] to (72a) and by inference, to (73).

Firm evidence was provided by the correlation of (−)valeroidine with S(−)methoxysuccinic acid in the following way.[122] First, (±)6-methoxytropinone—obtained from methoxysuccinic aldehyde by the Robinson-Stoll route[30]—was resolved and the (+)ketone (78b) hydrogenated to (−)6-methoxytropine (83d). Acetobromolysis of (83d) then gave (−)3α,-6β-diacetoxytropane (83e), i.e., the diacetate of the alkamine of (−)valeroidine of known *relative* configuration. This sequence clearly proves that the C(6)OH in natural valeroidine and the methoxyl group in the dextrorotatory ketone (78b) have the same relative configuration (β).

The only step where the C-O bond and, hence, the configuration of C(6) might have been affected is acetobromolysis. Therefore, selective methylation of the diol (72a) on N and O(6) was carried out and gave the

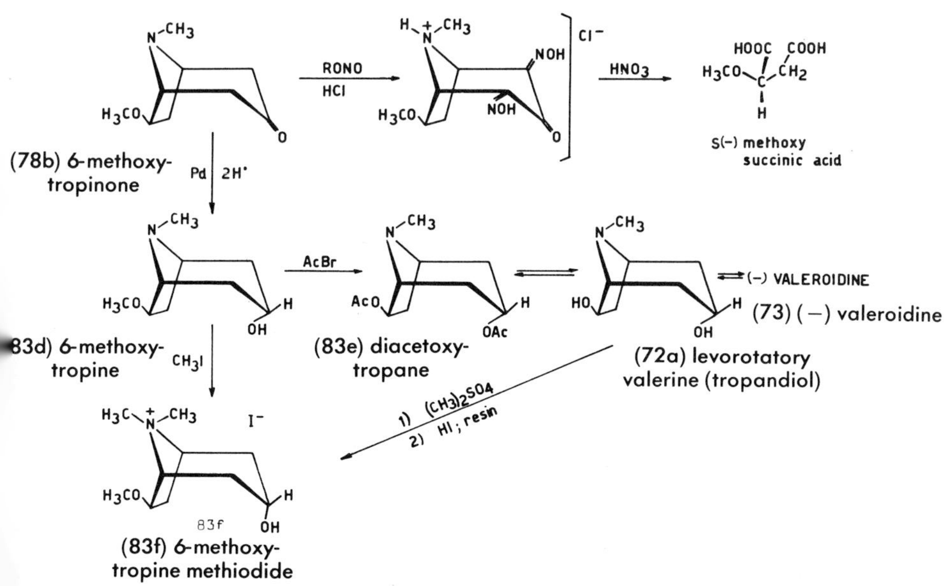

Correlation of valeroidine and δ-methoxysuccinic acid.

same methiodide (83f) as the one obtained from methoxytropine. In this way, all doubts were removed concerning the β-position of the C(6) methoxy group.

The last conclusive step involved nitrosation of the dextrorotatory ketone (78b) followed by oxidative degradation of the dioxime to S(−) methoxysuccinic acid. There are only two tropane compounds of the given constitution which could lead to S(−)methoxysuccinic acid: either 6β-methoxytropanone (78b) or 7α-methoxytropinone. However, correlation of the ketone with 6β,3α-tropanediol (72a) ruled out the second alternative. Thus, this sequence provides unequivocal evidence[122] for the 3*S*:6*S* configuration for valerine, the 3*R*:6*R* configuration corresponding to its dextrorotatory antimer.

Valeroidine is (−)3*S*-iso-valeroxy-6*S*-hydroxy tropane. 6β-Tigloyloxy and 3α,6β-ditigloyloxy tropane both have the 3*R*:6*R* absolute configuration.

2.3.3 Meteloidine (85) and Ditigloyl-teloidine.

2.3.1.1 ***Constitution*** *Meteloidine*[123] is not resolvable. On acetylation, it gives a diacetyl derivative. On hydrolysis with barium hydroxide, tiglic acid and the alkamine teloidine, $C_8H_{15}NO_3$, are formed; teloidine was thus assumed to have a tropane skeleton bearing three hydroxyl groups (87).

The final proof for the constitution was furnished by total synthesis[29] of (87) from meso-tartaric aldehyde (18c), methylamine and acetone-dicarboxylic acid. The resulting teloidinone (86a), on catalytic hydrogenation gave mainly teloidine (87) together with some pseudoteloidine (88). Since any tropane formed from meso-tartaric aldehyde must have *cis*-

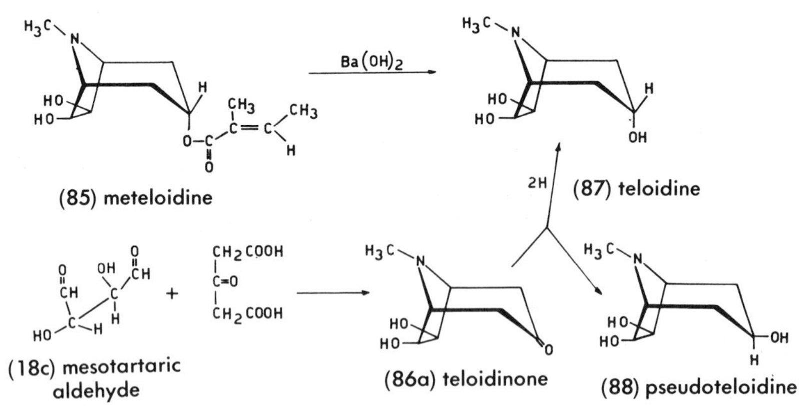

Degradation of meteloidine.

oriented hydroxyl groups, teloidine is a tropane 6,7-*cis*-diol, with a third hydroxyl at C(3) either *cis* or *trans* to the former one.

Tigloyl-meteloidine contains two moles of tiglic acid per mole of teloidine, thus one of the vicinal hydroxyls at C(6) or C(7) [in addition to C(3)] is certainly esterified in the latter. Since it is optically inactive, tigloyl meteloidine is the racemate of 3,6- (or 7) ditigloyl-tropane-3,6,7-triol.

2.3.3.2 *Configuration of Teloidine and Pseudoteloidine* Teloidine (87) gives a lactone salt with ethyl iodoacetate[114] the chloride of which (89) resisted periodate oxidation. Consequently, either the C(6) or the C(7) hydroxyl is involved in lactone salt formation and these groups are *cis* and β; teloidine is thus a 6β, 7β-tropandiol.

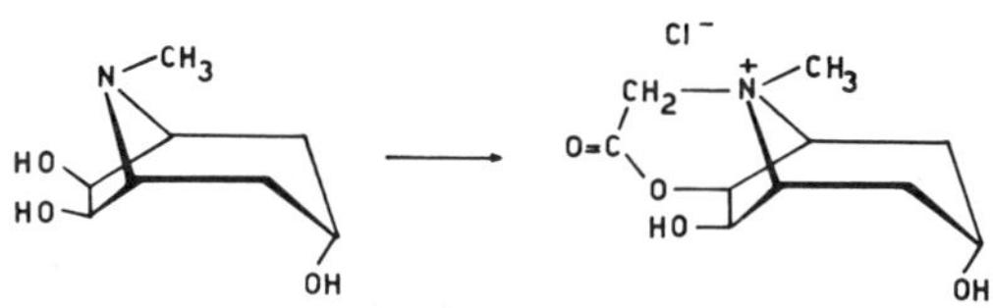

(87) teloidine

(89) lactone salt from teloidine

Configuration of teloidine.

Pseudonorteloidine 6,7-acetonide (90) gave a tetrahydrooxazine (91) with p-nitrobenzaldehyde with *nor*teloidine acetonide failed to do so.[124] Thus, the C(3) hydroxyl is *cis* in pseudoteloidine (88) and *trans* in teloidine. Teloidine is tropan 3α,6β,7β-triol.

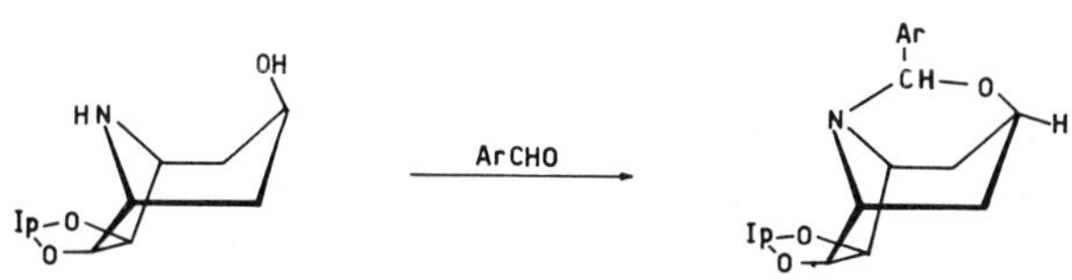

(90) isopropylidene-norteloidine

Ip = $(CH_3)_2C<$

(91) Ip-norteloidine-oxazine

Configuration of pseudoteloidine.

2.3.3.3 *Synthesis of Meteloidine and Dihydro-meteloidine* Dihydro-meteloidine (92C) has been obtained[125] by acylation of 6,7-benzal-teloidine (92a) with 2-methylbutyryl chloride and subsequent hydrogenolysis of the benzal group of (92b).

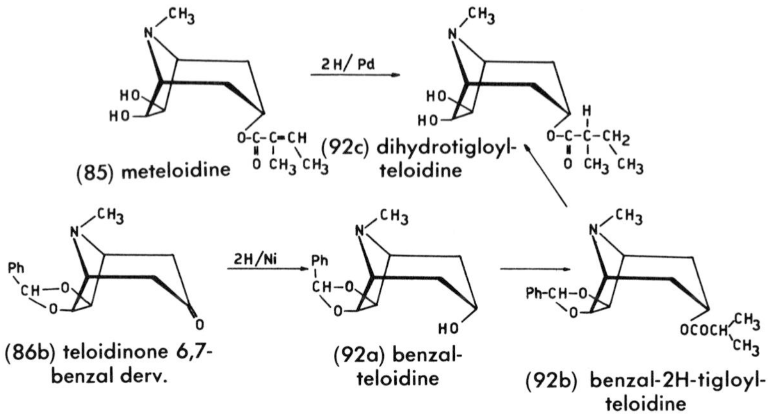

Synthesis of dihydrometeloidine.

Meteloidine (85) was synthesized[126] later from teloidine-6,7-carbonate (93) by acylation with tigloyl chloride and selective hydrolysis of the hydrochloride of 6β,7β-carbonyl-3α-tigloyl-teloidine (93)

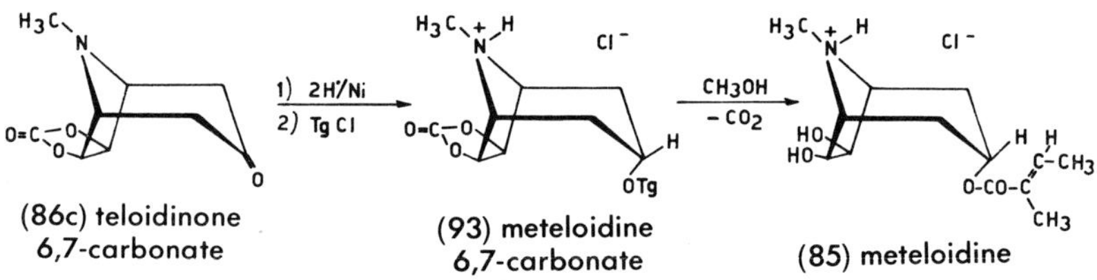

Synthesis of meteloidine.

2.3.4 Synthesis of Oscine Treatment of teloidine-6,7-ditosyl ester (94)[127,128] with bases afforded ($\pm$)7-tosyl-oscine (95), which either by hydrolysis or hydrogenolysis afforded oscine (63). A summary of the configurational assignments with the main alkaloids is given in a tabular survey.

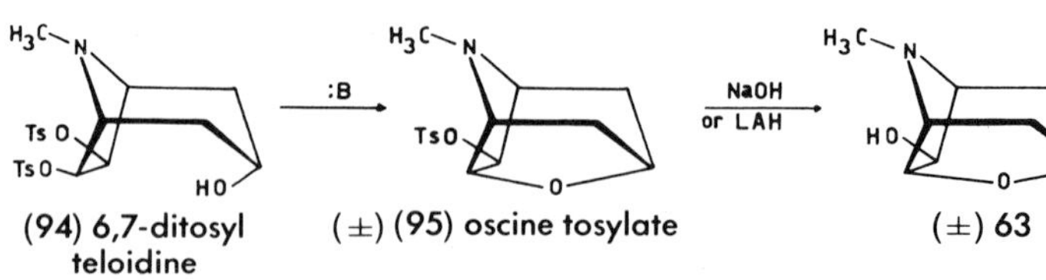

Synthesis of oscine.

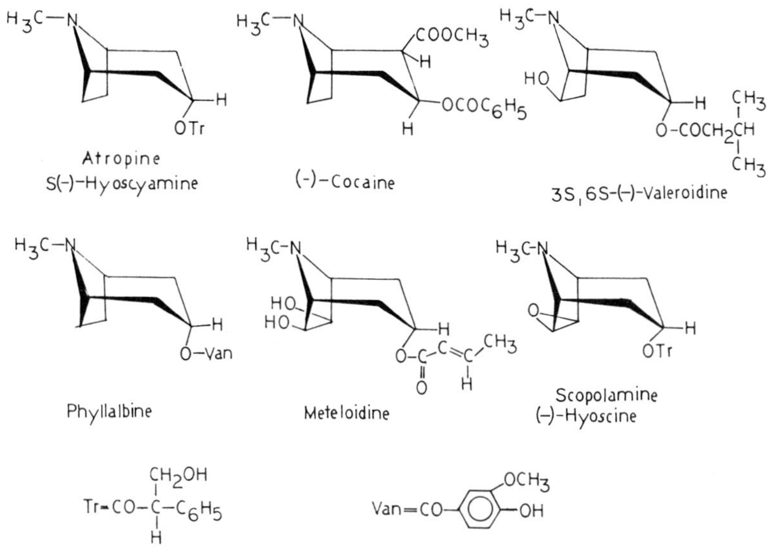

Table review on structures.

2.4 Conformation of the Piperidine Ring and of the N-Methyl Group in Tropanes[129–140]

X-Ray crystallographic investigations on tropanium salts, e.g., tropine (1) hydrobromide[133] showed a chair form of the piperidine ring with an equatorial N-methyl group. N_{ax}-Ethyl-N_{eq}-methyl nortropanium bromide (100) also assumes a chair conformation[134] with distortion of the angle between planes C(1)–N–C (5) and C(1)–C(7)–C(6)–C(5) to 142° [in tropine hydrobromide, this angle is considerably smaller, (126°)].

More recently the tertiary base pseudotropine (13) was also subject to x-ray analysis,[135] indicating equatorial orientation of the N-methyl group

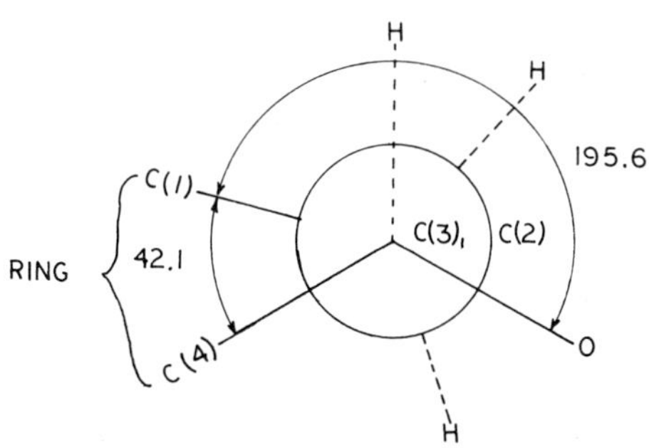

Newman diagram of pseudotropine.

with definite distortion of the angle between C(1)–C(2)–C(4)–C(5) and C(2)–C(3)–C(4), to 113°, also resulting in a decrease of the dihedral angle H-3ax/H-2 eq to 42° and in an increase of angle C(3)–O/C(2)–C(1) to 195.6° as visualized by the Newman diagram (see page 457).

Dipole moment measurements,[136] with 3β-cyano- and 3α-cyanotropanes, (96) and (97), gave the μ values of 4.04 ± 0.02 D for the β-form (calculated for the chair form 4.19 D) and 3.38 ± 0.02 D for the α-isomer. This discrepancy in case of the 3α-isomer was explained[136] in terms of considerable repulsion energy between α-hydrogens at C(6) and C(7), on the one hand, and the bulky α-substituent at C(3) on the other. Calculation of the repulsion energy curve as a function of angular distortion gives a μ value of 2.92 D for an angle of 35° between planes C(1)–C(2)–C(4)–C(5) and C(2)–C(3)–C(4). nmr studies with a number of tropine and

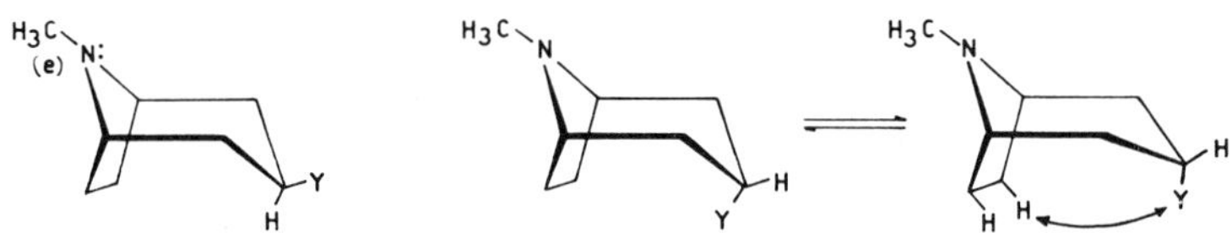

Y = OH; OAc ; CN (96) 3β-cyano-tropane (97) 3α-cyano-tropane

3α- and 3β-cyano tropanes.

pseudotropine derivatives allowed a different approach to conformation by determining the coupling constants of the H(3) and H(2) [or H(4)] protons. These could be correlated with dihedral angles, hence, with the actual shape of the ring. For instance, the axial H(3) in pseudotropine (13), which is coupled with the axial and the equatorial H(2) appears as a 9 line spectrum, with J = 10.2Hz for the ax-ax coupling. This corresponds, according to the Karplus curve, to a dihedral angle of 160° (instead of 180°), which again points to a distortion of the chair form of the piperidine ring. Similar results (J = 9Hz) were found[136] for (96).

Tropine (1), and 3α-cyanotropane (97) both have J values about 5Hz for the equatorial H(3)–axial H(2) coupling, interpreted in favor of a dihedral angle of 40°. However, the J value for scopine (64a) fits well with the one calculated for a less distorted chair form of the piperidine ring.[140]

The configuration of the N-methyl group in most of tertiary bases has been dealt with in the foregoing discussion. The configuration of the quaternary salts derived from them has also been investigated by the x-ray[134] nmr[136–140] and ir[138] techniques. In some cases, particularly with 3,6-tropandiol (172), and scopine (164a), the sequence of alkylating[129–132] a nor-base, e.g., (98) into a quaternary tropanium salt in two steps using two different groups seems to determine the preferential steric course of quaternization, hence the configuration of the N-alkyl groups. While (72b) gave preponderantly the lactone salt[114] (75) via (74a) 'reverse' quaternization gave the N-epimeric ester salt (74b), unable[130] to undergo transesterification. Thus, chemical evidence has been provided for their configuration.[130–132] Tropine ethiodide and bromide (99) have an equatorial N-ethyl group, as shown by nmr spectra, while N-ethyl nortropine methohalides have the opposite configuration (100), based on x-ray[134] and nmr[140b,151] evidence. In this way the absolute configuration of a number of quaternary salts on nitrogen became known.

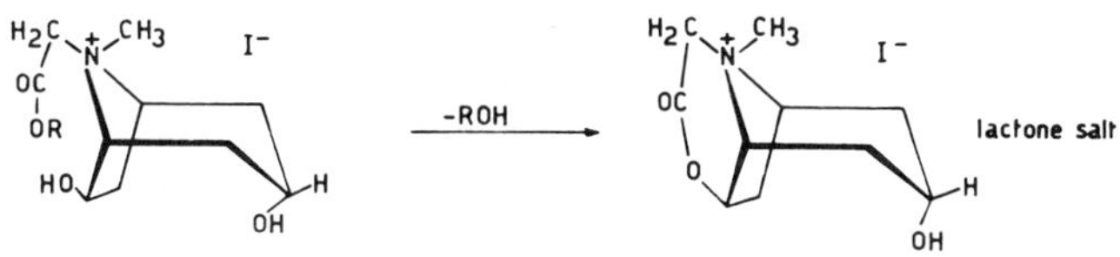

(74a) ester salt of N-carboxymethyl-tropandiol

(75) lactone salt of N-carboxymethyl-tropandiol

Lactone formation.

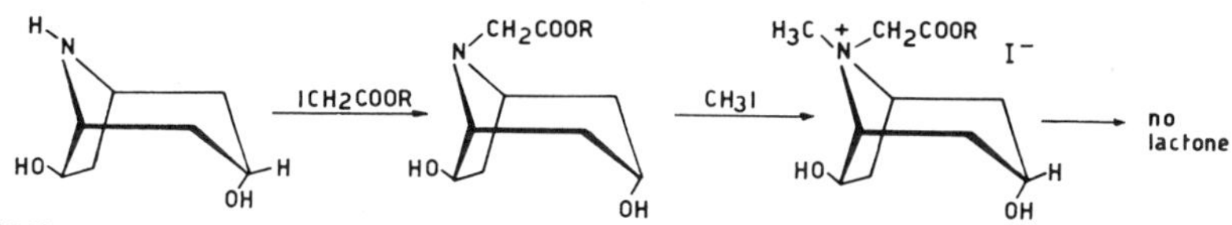

(98) nortropanediol

(74b) N-epimer of (74a)

No lactone formation.

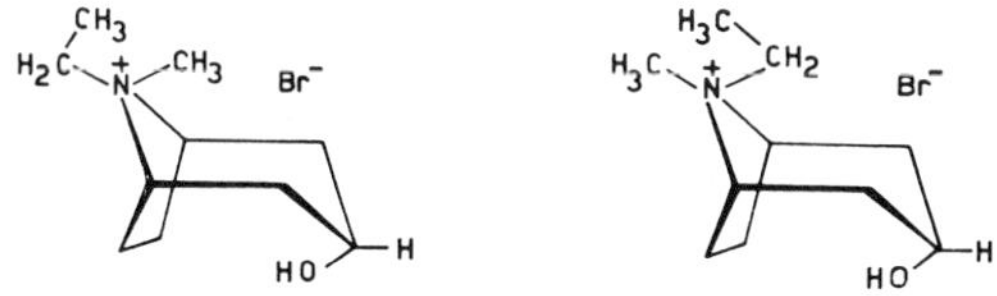

(99) tropine etho-bromide

(100) N-ethylnor-tropine metho-bromide

Me-et-nortropanium salts.

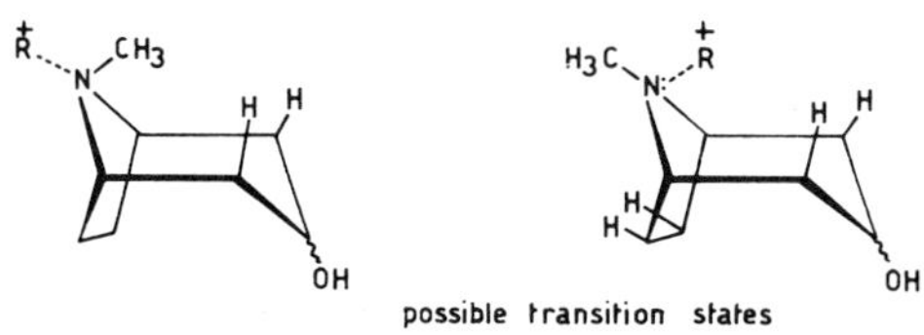

Transition states.

The ratio of the products of quaternization depends upon the difference in energy between the two possible transition states (see below) according to the Curtin–Hammett hypothesis.

3. CONVERSION OF TROPANES INTO TROPONES

Tropinone methiodide (101a) has been subjected to the action of sodium carbonate furnishing, besides dimethylamine, a nitrogen-free fragment believed[141] to be 'dihydrobenzaldehyde.' However, this constitution was discarded later when it proved to be a mixture of dihydrotropones[142] (102 a, b). Similarly, 2-bromotropinone (101b) and 6-hydroxy-tropinone methiodide (101c) could be cleaved[143] to tropone (103) itself in a poor yield. 6-tropen-3-one (104a) was prepared recently[144] from 6-tropen-3α-ol (77a) by chromic oxide in pyridine: its methiodide (104b) gave a quantitative yield of tropone (103) at pH 11 within a few minutes. The same applies to 6-methoxytropinone methiodide[145] (101d).

Polarographic analysis of that reaction at pH values below 9 detected 5-dimethylamino-cyclohepta-2,4-dienone (105) as an intermediate[145,146]. On quickly raising the pH to 11, the latter loses dimethylamine to give tropone. It is curious that this cleavage of tropanones occurs only

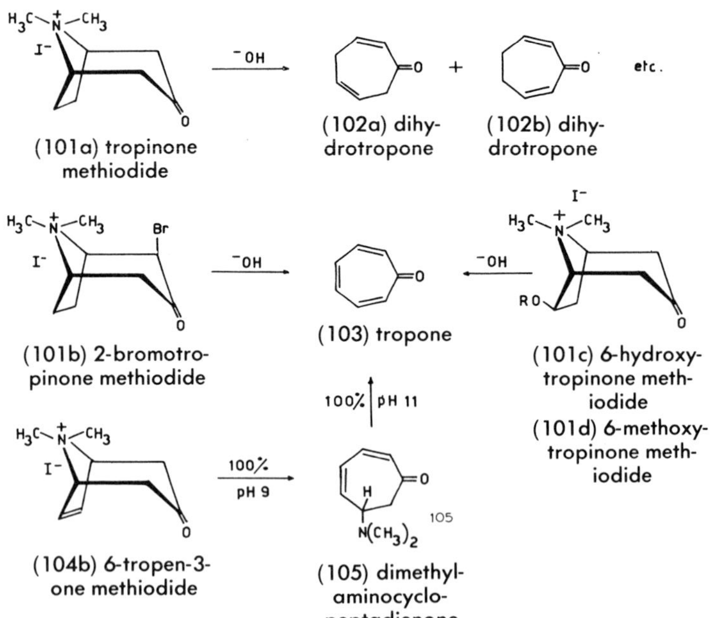

(101a) tropinone methiodide

(102a) dihydrotropone

(102b) dihydrotropone

(101b) 2-bromotropinone methiodide

(103) tropone

(101c) 6-hydroxytropinone methiodide

(101d) 6-methoxytropinone methiodide

(104b) 6-tropen-3-one methiodide

(105) dimethylaminocyclopeptadienone

Tropanes and tropone.

in the quaternary ammonium stage, i.e., when the nitrogen is positively charged.

The reversal of this elimination reaction could be achieved[147] too, by addition of methylamine to cycloheptatriene-carboxylic acid (106) leading to (−)anhydro-ecgonine, i.e., trop-2-ene-2-carboxylic acid (36).

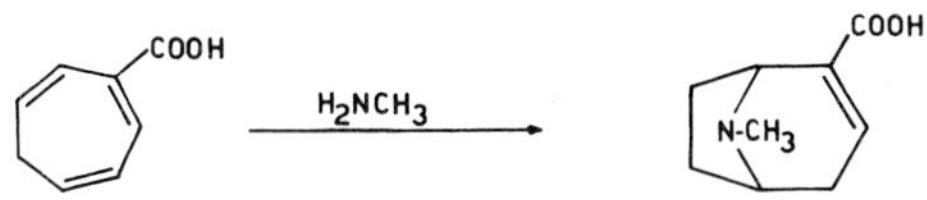

(106) cyclo-heptatriene carboxylic acid

(36) anhydro ecgonine

Incorporation of methylamine into cycloheptatriene.

4. MASS SPECTROMETRY AND OTHER FRAGMENTATIONS OF THE TROPANE BASES

A more profound fragmentation[148] of the tropanes takes place when potassium cyanide reacts with 3β-chloro-tropane (107b) forming N-methyl-2-allyl-5-cyano-pyrrolidine (108), involving a pyrrolinium salt as an intermediate. The epimeric 3α-chlorotropane (107a), gives on the contrary, 3α-cyano tropane (97) owing to transannular anchimeric assistance by the basic nitrogen.[148]

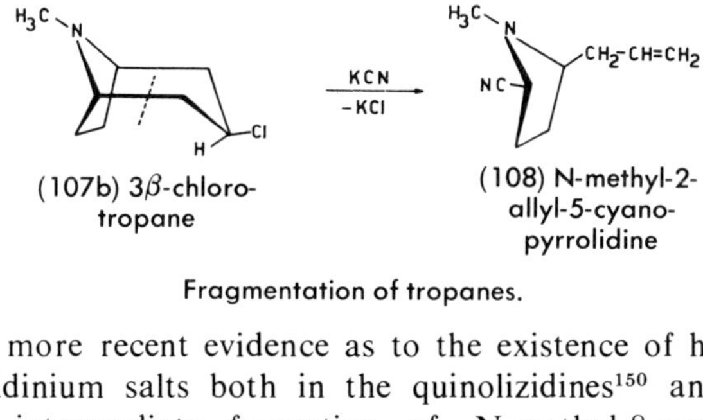

(107b) 3β-chloro-tropane

(108) N-methyl-2-allyl-5-cyano-pyrrolidine

Fragmentation of tropanes.

In view of more recent evidence as to the existence of highly strained tricyclic azetidinium salts both in the quinolizidines[150] and in the tropanes,[87] the intermediate formation of N-methyl-8-*azonia* tricyclo-

(107a) 3-α-chloro-tropane

(109) bridged tri-cyclic piperidine

(97) 3 α-cyano tropane

3α-Chlorotropane, cyanolysis *via* tricycle.

$[2,1,1^{2,8}]$octane chloride (109) by a rearward nucleophilic attack also has to be considered.

Very interesting information has been obtained[149] concerning the fragmentation of tropanes under electron impact. Investigations started with tropinone (11) and its 6,7-dideutero (11-d_2) and its 2,2,4,4-tetradeutero-derivatives (11-d_4). Shifts in the m/e ratio of certain species in the mass spectra or no change by going from one series to the other clearly indicates the original location of that fragment in the whole skeleton. For example, tropinone (11) and 6,7-dideutero-tropinone (11-d_2) both gave a fragment with m/e 83, while 2,2,4,4-tetradeutero-tropinone (11-d_4) afforded a peak at m/e 87. This species is very probably a part of the 6-membered ring: 5/6- and 1/7-fission with subsequent loss of CO are assumed to give the N-methyl-1(2) pyrrolenium ion (110) and (110-d_4), respectively.*

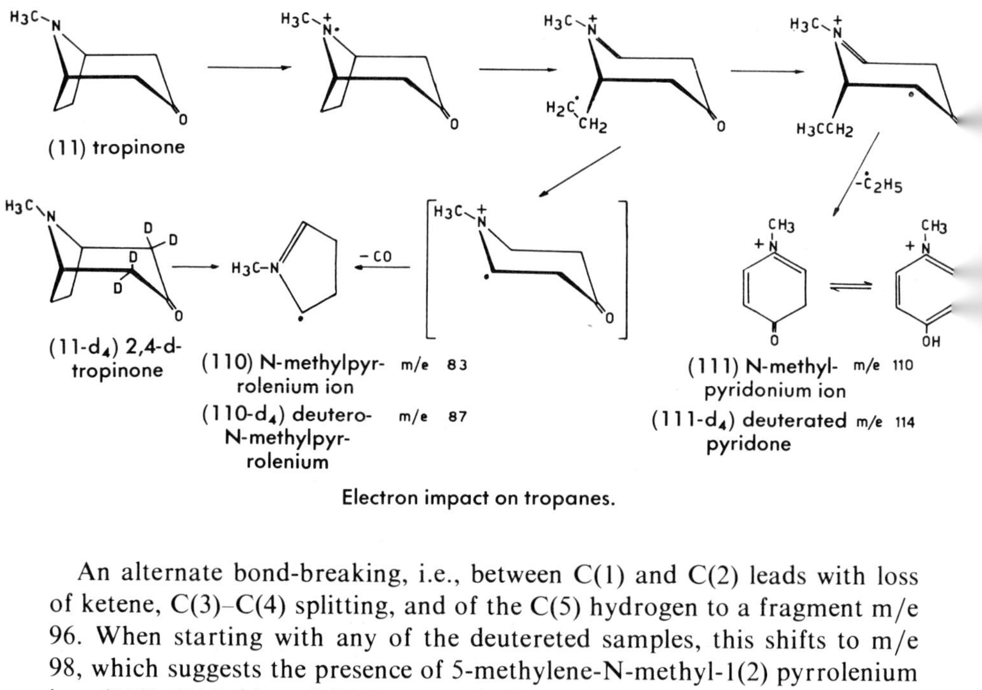

Electron impact on tropanes.

An alternate bond-breaking, i.e., between C(1) and C(2) leads with loss of ketene, C(3)–C(4) splitting, and of the C(5) hydrogen to a fragment m/e 96. When starting with any of the deutereted samples, this shifts to m/e 98, which suggests the presence of 5-methylene-N-methyl-1(2) pyrrolenium ions (112), (112-d_2), and (113), respectively.

The same technique was adapted in case of tropan-3α-ol (1) and 6,7-dideutero-tropan-3α-ol (1-d_2), where the main species had m/e 96: this was shifted, however, in the case of 2,2,3,4,4-pentadeutero-tropan-3α-ol

*Another fragment, m/e 110 is formed from (11), while (11-d_4) gave m/e 114, corresponding to species (111) and (111-d_4), respectively.

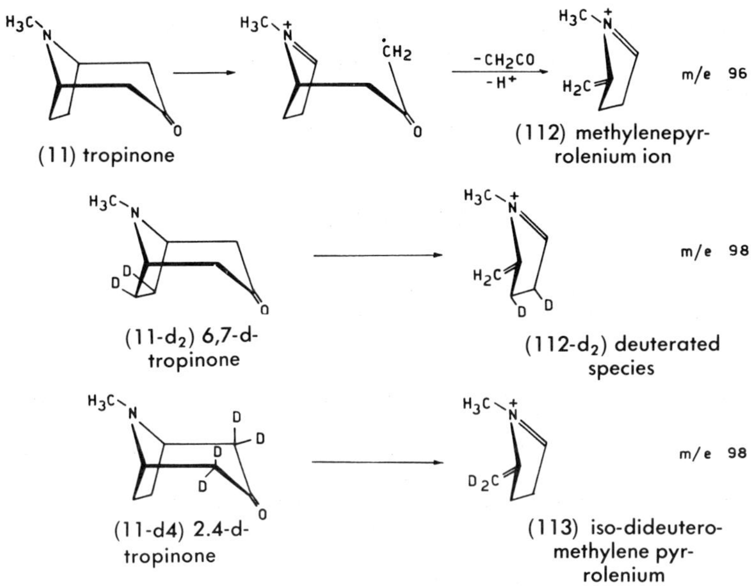

Further mass spectroscopic data.

($1\text{-}d_5$) to m/e 101. This shows that all five hydrogens have been maintained during impact; these data are best explained in items bridged N-methyl-/4H/-piperidinium ions (114) and ($114\text{-}d_5$), respectively.

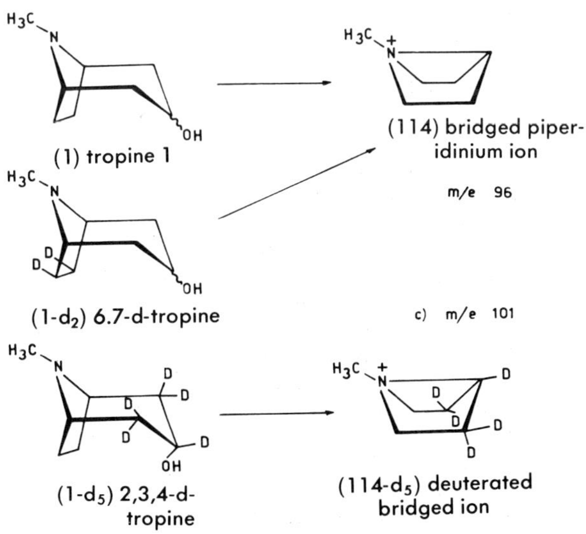

Fragmentation of tropine.

Ecgonines and some *thiadeaza*-tropanes have been submitted to mass spectrometric investigation, too. Using this technique the chemist can quickly gain information about the constitution of a hitherto unknown tropane base using the only a few micrograms of the substance.

REFERENCES

1. H. L. Holmes, *Alkaloids* **1,** 27–374 (1950).
2. G. Fodor, "Tropane Alkaloids," *Alkaloids* **6,** 145–177 (1960).
3. G. Fodor, "Tropane Alkaloids," *Alkaloids* **9,** 269–303 (1967).
4. L. Gyermek and K. Nádor, *J. Pharm. Pharmacol.* **9,** 209 (1957).
5. K. Nádor, in E. Jucker, "Fortschritte der Arzneimittelforschung," Vol. 2, Birkhäuser, Basel, Switzerland, 1968.
6. K. Nádor, M. Doda, and L. György, *Arch. Intern. Pharmacodyn.* **145,** 264 (1963).
7. K. Nádor, "Relationship between the Structure and Pharmacological Activity of Tropeines," in: "Recent Development in the Chemistry of Natural Carbon Compounds," Vol. 1, pp. 163–235, Publishing House of Academy of Science, Budapest, Hungary (1965).
8. G. Fodor and K. Nádor, *Nature* **169,** 462 (1952).
9. W. Lossen, *Ann.* **131,** 43 (1864).
10. J. Gadamer, *J. Prakt. Chem.* (ii) **64,** 565 (1901).
11. E. Schmidt, *Ber.* **21,** 2777 (1888).
12. A. McKenzie and K. K. Wood, *J. Chem. Soc.* **115,** 828 (1919).
13. A. Ladenburg, *Ann.* **217,** 74 (1883).
14. A. Ladenburg, *Ber.* **20,** 1647 (1887).
15. G. Merling, *Ber.* **24,** 3108 (1891).
16. R. Willstätter, *Ber.* **30,** 721 (1897).
17. R. Willstätter, *Ber.* **29,** 393 (1896).
18. R. Willstätter, *Ber.* **30,** 731 (1897).
19. R. Willstätter, *Ber.* **30,** 2679 (1897).
20. R. Willstätter, *Ber.* **29,** 936 (1896).
21. R. Willstätter, *Ber.* **28,** 3271 (1895).
22. R. Willstätter, *Ber.* **31,** 1534 (1898).
23. R. Willstätter, *Ber. Deutsch. Pharm. Ges.* **13,** 50 (1903).
24. R. Willstätter and A. Bode, *Ber.* **34,** 519 (1901).
25. R. Willstätter and A Bode, *Ber.* **33,** 416 (1900); **34,** 1457 (1901).
26. R. Willstätter, *Ann.* **317,** 204, 267, 307 (1901); *Ann.* **326,** 23 (1903).
27. R. Robinson, *J. Chem. Soc.* **111,** 762 (1917).
28. A. Stoll, B. Becker, and E. Jucker, *Helv. Chim. Acta* **35,** 1263 (1952).
29. Cl. Schöpf and H. Arnold, *Ann.* **358,** 109 (1947).
30. A. Stoll, E. Jucker, and A. Lindenmann, *Helv. Chim. Acta* **37,** 495 (1954).
31. Cl. Schöpf and A. Schmetterling, *Angew. Chem.* **64,** 691 (1952).
32. R. Willstätter, O. Wolfes, and O. Mäder, *Ann.* **434,** 11 (1923); R. Willstätter and M. Bommer, Ann. **422,** 18 (1922).
33. L. C. Keagle and W. H. Hartung, *J. Amer. Chem. Soc.* **68,** 1608 (1946).
34. R. Wolffenstein and L. Mamlock, *Ber.* **41,** 723 (1908).
35. M. Barowcliff and F. Tutin, *J. Chem. Soc.* **95,** 1966 (1909).

36. G. Fodor, J. Rákoczi, and Gy. Csepreghy, *Acta Chim. Hung.* **28,** 209 (1961); *Chem. Abstr.* **61,** 1903g.
37. R. Willstätter and F. Iglauer, *Ber.* **33,** 1170 (1900).
38. J. Gadamer, *Arch. Pharm.* **239,** 294 (1901).
39. G. Fodor and J. Kiss, Nature **163,** 287 (1949).
40. G. Fodor and J. Kiss, *J. Chem. Soc.* 1589 (1952).
41. G. Fodor, J. Kiss, K. Koczka and K. Nádor, *Acta Physiol. Hung.* **3,** 27 (1951).
42. G. Fodor and K. Nádor, *J. Chem. Soc.* 721 (1953).
43. M Polonovski and M. Polonovski, *Bull. Soc. Chim.* (IV) **41,** 1190 (1927).
44. G. Fodor and J. Kiss, *J. Amer. Chem. Soc.* **72,** 3495 (1950).
45. K. Koczka and G. Fodor, *Acta Chim. Acad. Sci. Hung.* **13,** 83 (1957); *Chem. Abstr.* **52,** 12845 f (1957).
46. G. Fodor, *Acta Chim. Acad. Sci. Hung.* **5,** 380 (1955); *Chem. Abstr.* **49,** 11673 b (1955).
47. A. Nickon and L. F. Fieser, *J. Amer. Chem. Soc.* **74,** 5566 (1952).
48. E. Hardegger and H. Ott, *Helv. Chim. Acta* **36,** 1186 (1953).
49. E. Hardegger and H. Ott, *Helv. Chim. Acta* **37,** 685 (1954).
50. B. L. Zenitz, C. M. Martini, M. Priznar, and F. C. Nachod, *J. Amer. Chem. Soc.* **74,** 5564 (1952).
51. H. S. Aaron and Ch. P. Rader, *J. Org. Chem.* **29,** 3426 (1964).
52. G. R. Clemo and H. K. Jack, *Chem. Ind. (London)* 195 (1953).
53. F. L. J. Sixma, C. M. Siegmann, and H. C. Beyermann, *Konikl. Ned. Akad. Wetenschap.* **54B,** 452 (1951); *Chem. Abstr.* **47,** 6957b (1951).
54. C. M. Siegmann and J. P. Wibaut, *Rec. Trav. Chim.* **73,** 203 (1954).
55. H. C. Beyermann, C. M. Siegmann, F. L. J. Sixma, and J. H. Wisse, *Rec. Trav. Chim.* **75,** 1445 (1956).
56. A. Orechoff and R. Konowalowa, *Ber.* **67,** 1153 (1934).
57. G. Barger, W. F. Martin, and W. Mitchell, *J. Chem. Soc.* 1685 (1938).
58. C. Liebermann, *Ber.* **24,** 2336 (1891).
59. G. Barger, W. F. Martin, and W. Mitchell, *J. Chem. Soc.* 1820 (1937).
60. F. L. Pyman, *J. Chem. Soc.* **111,** 1103 (1917).
61. A. Orechoff and R. A. Konowalowa, *Arch. Pharm.* **271,** 145 (1933).
62. K. Freudenberg, J. Todd, and R. Siedler, *Ann.* **501,** 206 (1933).
63. G. Fodor and Gy. Csepreghy, *Tetrahedron Letters* **7,** 16 (1959).
64. G. Fodor and Gy. Csepreghy, *J. Chem. Soc.* 3222 (1961).
65. H. I. Bernstein and F. C. Whitmore, *J. Amer. Chem. Soc.* **61,** 1324 (1939).
66. R. S. Cahn, C. K. Ingold, and V. Prelog, *Experientia* **12,** 81 (1956).
67. A. McKenzie and R. C. Strathern, *J. Chem. Soc.* **127,** 86 (1925).
68. J. Parello, P. Longevialle, W. Vetter, and J. A. McCloskey, *Bull Soc. Chim.* 2787 (1964).
69. F. Wöhler, *Ann.* **121,** 372 (1862).
70. W. Lossen, *Ann.* **133,** 351 (1865).
71. R. Willstätter and W. Müller, *Ber.* **31,** 1202 (1898).
72. C. Liebermann, *Ber.* **23,** 2518 (1890).
73. C. Liebermann, *Ber.* **24,** 606 (1891).
74. A. Einhorn, *Ber.* **20,** 1221 (1887).
75. A. Einhorn, *Ber.* **23,** 1338 (1890).
76. A. Einhorn and Y. Tahara, *Ber.* **23,** 234 (1893).
77. A. Einhorn and A. Marquardt, *Ber.* **23,** 468 (1890).
78. S. P. Findlay, *J. Amer. Chem. Soc.* **76,** 2855 (1954).
79. G. Fodor, O. Kovács, and I. Weisz, *Nature* **174,** 131 (1954).

80. C. Liebermann and W. Drory, *Ber.* **22,** 680 (1889).
81. C. Liebermann, *Ber.* **21,** 3372 (1888).
82. G. Fodor, *Nature* **170,** 278 (1952).
83. O. Kovács, G. Fodor, and I. Weisz *Helv. Chim. Acta* **37,** 892 (1954).
84. K. Hess, A. Eichel, and C. Uibrig, *Ber.* **50,** 351 (1917).
85. O. Kovács, I. Weisz, P. Zoller, and G. Fodor, *Helv. Chim. Acta* **39,** 99 (1956).
86. I. Weisz, unpublished results.
87. G. Fodor, N. Mandava, and I. Weisz, *Tetrahedron* **24,** 2357 (1968).
88. E. Hardegger and H. Ott, *Helv. Chim. Acta* **38,** 312 (1955).
89. K. Zeile and W. Schultz, *Chem. Ber.* **89,** 678 (1956).
90. S. P. Findlay, *J. Org. Chem.* **21,** 711 (1956).
91. M. S. Balinova, G. I. Bazilevskaya, and N. A. Preobrashenski, *Zur. Obsch. Khim.* **30,** 3258 (1960); **55,** 21155d (1960).
92. S. P. Findlay, *J. Org. Chem.* **24,** 1540 (1959); **22,** 1385 (1957).
93. E. Schmidt, *Arch. Pharm.* **230,** 207 (1892).
94. H. King, *J. Chem. Soc.* **115,** 476 (1919).
95. O. Hesse, *J. Prakt. Chem.* (ii) **64,** 353 (1901).
96. R. Willstätter and E. Berner, *Ber.* **56,** 1079 (1923).
97. K. Hess and O. Wahl, *Ber.* **55,** 1979 (1922).
98. F. Tutin, *J. Chem. Soc.* **97,** 1793 (1910).
99. E. Schmidt, *Arch. Pharm.* **243,** 559 (1905).
100. E. Schmidt, *Arch. Pharm.* **247,** 79 (1909).
101. K. Hess, *Ber.* **51,** 1007 (1918).
102. M. Polonovski and M. Polonovski, *Bull. Soc. Chim.* (IV) **43,** 79 (1928).
103. A. Heusner and K. Zeile, *Tetrahedron* **3,** 312 (1958).
104. G. Fodor, J. Toth, I. Koczor, and I. Vincze, *Chem. Ind.* (*London*), 1260 (1955); 764 (1956).
105. G. Fodor, J. Toth, I. Koczor, P. Dobó, and I. Vincze, *J. Chem. Soc.* 3461 (1959).
106. J. Meinwald, *J. Chem. Soc.* 712 (1953).
107. R. C. Cookson, *Chem. Ind. (London)* 337 (1953).
108. G. Fodor, O. Kovács, and L. Mészáros, *Research (London)* **5,** 534 (1952).
109. G. Fodor and O. Kovács, *J. Chem. Soc.* 2341 (1953).
110. O. Wolfes and O. Hromatka, *E. Merck's Jahresber.* **47,** 45 (1934).
111. W. F. Martin and W. Mitchell, *J. Chem. Soc.* 1155 (1940).
112. W. F. Mitchell and E. M. Trautner, *J. Chem. Soc.* 1330 (1947).
113. G. Fodor, I. Koczor, and G. Janzso, *Arch. Pharm.* **295,** 91 (1962).
114. G. Fodor, J. Toth, and I. Vincze, *Helv. Chim. Acta* **37,** 907 (1954).
115. G. Fodor, S. Kiss, and J. Rákoczi, *Chim. Ind.* **90,** 225 (1963).
116. G. Fodor, *Tetrahedron* **1,** 86 (1957).
117. G. Fodor et al., *Symposium on CNS Drugs*, Hyderabad, India, 1966, p. 270.
118. W. C. Evans and M. Wellendorf, *J. Chem. Soc.* 1991 (1958).
119. W. C. Evans and W. J. Griffin, *J. Chem. Soc.* 4348 (1963).
120. G. Fodor, J. Toth, and I. Vincze, *J. Chem. Soc.* 1349 (1957).
121. G. Fodor, I. Vincze, and J. Toth, *J. Chem. Soc.* 3219 (1961).
122. G. Fodor and F. Soti, *Tetrahedron Letters* **29,** 1917 (1964); *J. Chem. Soc.* 6830 (1965).
123. F. L. Pyman and W. C. Reynolds, *J. Chem. Soc.* **93,** 2077 (1908).
124. K. Zeile and A. Heusner, *Ber.* **90,** 2114 (1957).
125. J. C. Sheehan and E. R. Bissel, *J. Org. Chem.* **19,** 270 (1954).
126. K. Zeile and A. Heusner, *Z. Naturforsch.* **12b,** 661 (1957).
127. K. Zeile and A. Heusner, *Chem. Ber.* **90,** 2800 (1957).

128. K. Zeile and A. Heusner, *Chem. Ber.* **90,** 2809 (1957).
129. G. Fodor, K. Koczka, and J. Lestyân, *Magyar Kém.Folyóirat* **59,** 242 (1953); *Chem. Abstr.* **48,** 100029 (1954).
130. G. Fodor, J. Toth, and I. Vincze, *J. Chem. Soc.* 3504 (1955).
131. G. Fodor, *Bull. Soc. Chim.* 1332 (1956).
132. O. Kovács, G. Fodor, and M. Halmos, *J. Chem. Soc.* 873 (1956).
133. J. W. Visser, J. Manassen, and J. L. de Vries, *Acta Cryst.* **7,** 288 (1954).
134. C. H. MacGillavry and G. Fodor, *J. Chem. Soc.* 597 (1964).
135. H. Schenk, C. H. MacGillavry, S. Skolnik, and J. Laan, *Acta Cryst.* **23,** 423 (1967).
136. R. J. Bishop, G. Fodor, A. R. Katritzky, F. Soti, L. E. Sutton, and F. J. Swinbourne, *J. Chem. Soc.* (C) 74 (1966).
137. G. L. Closs, *J. Amer. Chem. Soc.* **81,** 5456 (1958).
138. J. McKenna, J. M. McKenna, A. Tulley, and J. White, *J. Chem. Soc.* 1711 (1965), and forthcoming papers.
139. D. R. Brown, R. Lygo, J. McKenna, J. M. McKenna, and B. G. Hutley, *J. Chem. Soc.* (B) 1184 (1967).
140. (a) N. Mandava and G. Fodor, *Can. J. Chem.* **46,** 2761 (1968); (b) G. Fodor, J. Medina, and N. Mandava, *Chem. Comm.* 581 (1968).
141. A. Eichengrün and A. Einhorn, *Ber.* **23,** 2870 (1890).
142. J. Meinwald, S. L. Emmermann, N. C. Yang, and G. Buchi, *J. Amer. Chem. Soc.* **77,** 4401 (1955).
143. E. E. vanTamelen, P. Barth, and F. Lornitzo, *J. Amer. Chem. Soc.* **78,** 5442 (1956).
144. G. Fodor, S. Kiss, and A. Heusner, *Chem. Ind. (London)* 372 (1963).
145. G. Fodor, *MTA Kém. Tud. Oszt. Közl.* **20,** 337 (1963).
146. P. Zuman, *Chem. Eng. News* **46,** 103 (1968).
147. Ch. Grundmann and G. Ottmann, *Ann.* **605,** 24 (1957).
148. S. Archer, T. R. Lewis, and B. Zenitz, *J. Amer. Chem. Soc.* **79,** 3603 (1957).
149. E. Blossey, H. Budzikiewicz, M. Ohashi, G. Fodor, and C. Djerassi, *Tetrahedron* **20,** 585 (1964).
150. D. E. Edwards, G. Fodor and L. Marion, *Can. J. Chem.* **44,** 13 (1966).
151. G. Fodor, N. Mandava and D. Fréhel, Lecture, Symposium on Conformational Analysis, Brussels, September 8–12, 1969; due to appear in the book with the same title, Butterworths, London.

16

The Lycopodium Alkaloids

D. B. MacLEAN
Department of Chemistry
McMaster University
Ontario, Canada

1. INTRODUCTION

The examination of a Lycopodium species for the presence of alkaloids was first reported in 1881 by Bödeker[1] who isolated an alkaloid from *L. complanatum* that corresponds in properties to the alkaloid now known as lycopodine. For almost sixty years there was little published work in this area and only a few reports of the presence of alkaloids in this family appeared. It was not until 1938 that a serious study appeared. At that time Achmatowicz and Uzieblo[2] examined Polish L. *clavatum* and isolated and characterized three alkaloids. Among them was lycopodine, the most widely distributed alkaloid of the group, to which they assigned the correct molecular formula. A few years later the first of an extensive series of papers[3–13] on this family of alkaloids was published by Manske and Marion. In their studies they isolated and characterized some thirty alkaloids and initiated structural studies on several of them. During the 1950's and 1960's many new alkaloids of this family were isolated until now nearly one hundred are known. Some of them were obtained by the examination of new species, others by reexamination, using modern methods of separation, of species investigated earlier.

The first alkaloid of the family to yield its structural secrets was annotinine, which, paradoxically, is unique in its ring system and more complex in structure than most members of the series. Its structure was finally clarified in 1957 by K. Wiesner[14] and his group at the University of New Brunswick. Wiesner's structure, arrived at in a series of elegant chemical studies, was confirmed shortly afterwards by Przybylska and Marion[15] in an x-ray analysis of annotinine bromohydrin.

In 1960 the structure of lycopodine[16] was elucidated and within a few months thereafter structures were assigned to α- and β-obscurine,[17] lycodine,[18,19] selagine,[20] and a number of alkaloids related in structure to lycopodine. New structural types have recently been found in lyconnotine,[21] cernuine,[22] annotine,[23] serratinine,[24] and serratinidine.[25]

The ring systems known in this family are shown in the accompanying structural formulas of a representative alkaloid of each system. The numbering system, which is applied in these formulas, was proposed by Wiesner.[26] It is based on the biogenetic proposal of Conroy[27] that the Lycopodium alkaloids may be derived by condensation of two 8-carbon polyacetate chains. Although it now appears that Conroy's proposal has no validity in fact (see Chapter 21), it has, nevertheless, played a useful role in the study of these alkaloids.

The material presented in this Chapter is not intended to constitute a comprehensive review of the subject, since such reviews are available.[26,28] An attempt has been made, however, to give an account of the work leading

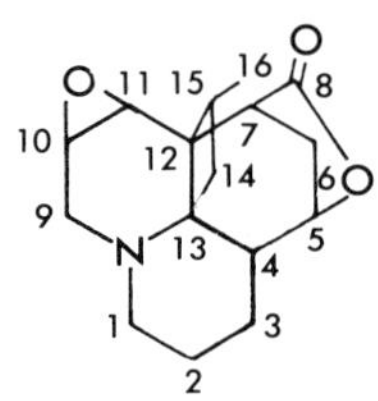

annotinine

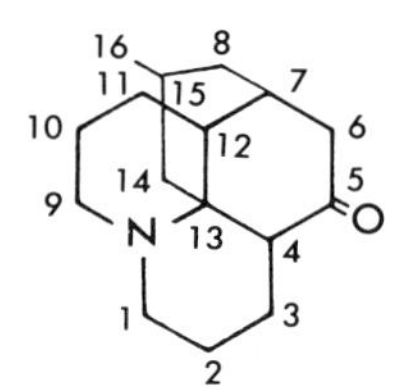

lycopodine

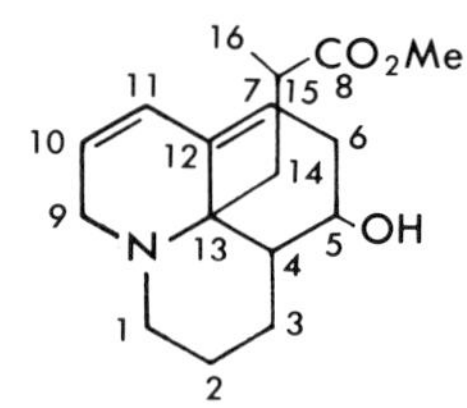

lyconnotine

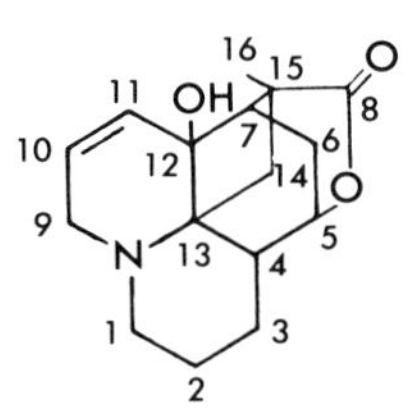

annotine

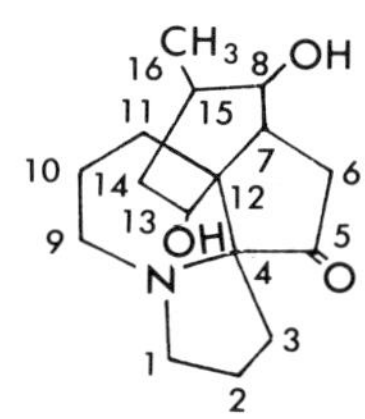

serratinine

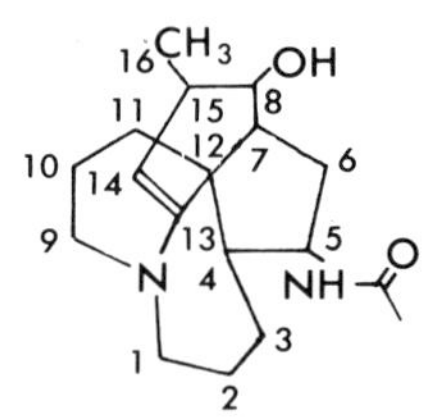

serratinidine

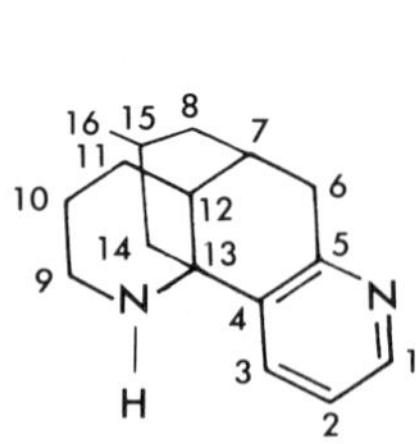

lycodine

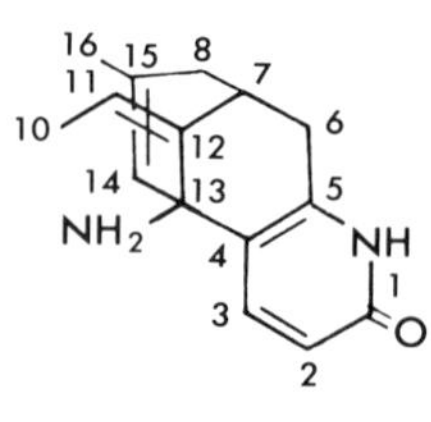

selagine

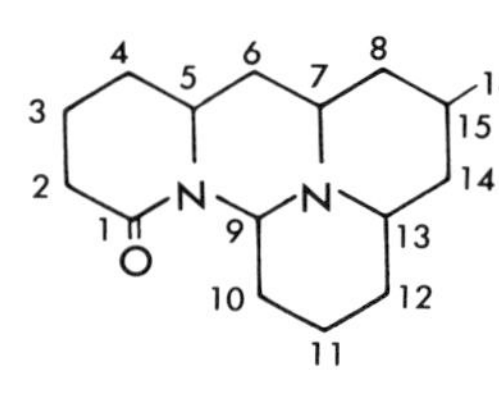

cernuine

to the elucidation of structure of a representative alkaloid of each structural type. In most cases, the work leading to the assignment of stereochemistry is also discussed. The review of synthesis in this series has been limited to the synthesis of annotinine and because of this, some excellent work, already reported or in progress, on the synthesis of the lycopodine and cernuine systems has not been included.

2. ALKALOIDS CONTAINING A HYDROJULOLIDINE RING SYSTEM

The majority of Lycopodium alkaloids of known structure fall into this category. Annotinine, annotine, lyconnotine and lycopodine represent the four structural types encountered. The first three are the sole representatives of their class but there are many alkaloids with a ring system similar to lycopodine. The structure of annotinine will be considered first, followed by lycopodine and alkaloids related to it, and finally lyconnotine and annotine.

2.1 Annotinine

Annotinine (1) $C_{16}H_{21}NO_3$, is of primary importance in any discussion of the Lycopodium alkaloids since the elucidation of its structure laid the foundation for much of the subsequent work. The alkaloid was isolated from *L. annotinum*[5] and from *L. annotinum* var. *acrifolium*[11] by Manske and Marion. In both species it is the major alkaloid. The chemical investigation of annotinine was initiated by its discoverers and carried on in a number of laboratories but the definitive chemical experiments, which resulted in the structural assignment, were conducted by K. Wiesner and his group.[29,30]

2.1.1 The Functional Groups of Annotinine The chemical investigation of annotinine was begun by Manske and Marion[11] who demonstrated the presence of a lactone function in the molecule and inferred from other experiments that the third oxygen was present in an ether linkage. The permanganate oxidation product (2),[11] $C_{16}H_{19}NO_4$, was a key compound in an investigation of MacLean and Prime.[31] In their study it was shown that the ether was an epoxide and that the structural units,

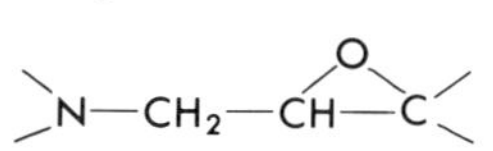 and 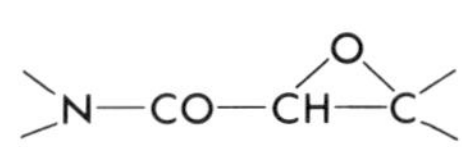

were present in 1 and 2, respectively. Compound 3 with the structural unit

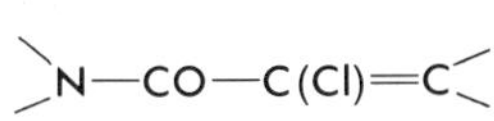

was used in a further elaboration of the structure.[32] It was found that permanganate oxidation of (3) yielded an amino acid (4), $C_{14}H_{19}NO_4$. The methyl ester of (4) was resistant to hydrolysis implying that the carboxyl group was attached to a quaternary center. Spectral measurements confirmed that (3) was a conjugated lactam (λ_{max} 230, log ϵ 4.2; ν_{max} 1657, 1616 cm^{-1}), that the lactam ring of 2 was six-membered or larger (ν_{max} 1645 cm^{-1}), and that annotinine was a γ-lactone. The partial structure of annotinine could then be further elaborated to,

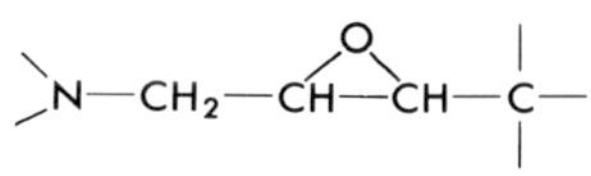

plus a γ-lactone.

The work discussed thus far is summarized in Scheme 16.1.

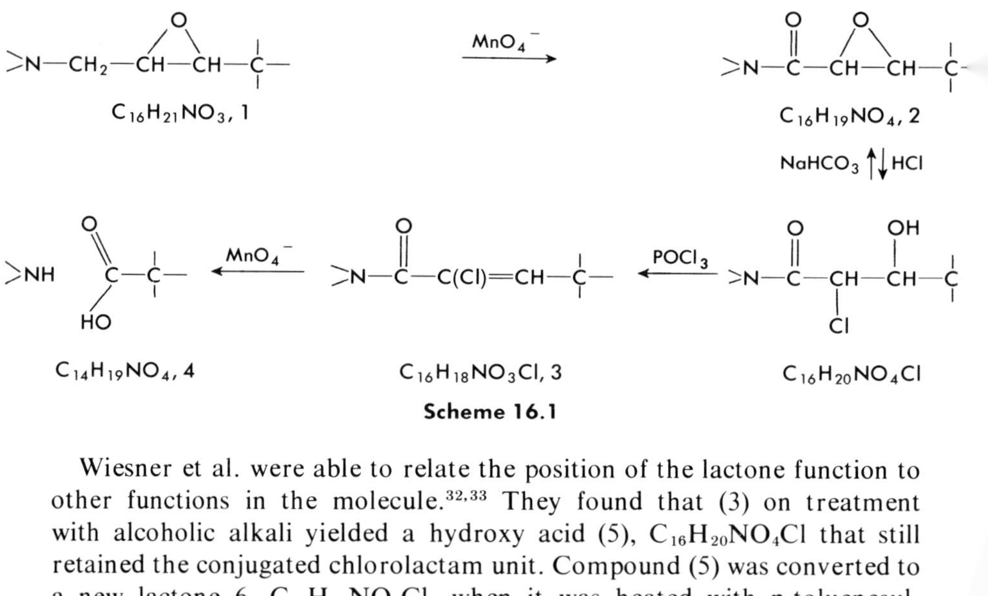

Scheme 16.1

Wiesner et al. were able to relate the position of the lactone function to other functions in the molecule.[32,33] They found that (3) on treatment with alcoholic alkali yielded a hydroxy acid (5), $C_{16}H_{20}NO_4Cl$ that still retained the conjugated chlorolactam unit. Compound (5) was converted to a new lactone 6, $C_{16}H_{20}NO_4Cl$, when it was heated with p-toluenesulfonic acid. The new lactone apparently formed by addition of the carboxyl group to the conjugated lactam. When (6) was heated in alkali a new hydroxy acid (7), formed in a β-elimination, was isolated which was apparently epimeric with (5) at the center carrying the carboxyl group. Later, they returned to this reaction and verified their original deductions.[29] The partial formulas of Scheme 16.2 summarize these reactions.

The recognition that lactonization at a new center in the annotinine system was possible permitted an interpretation of the formation of annotinine hydrate (8) $C_{16}H_{23}NO_4$. This compound, prepared[11] by reaction

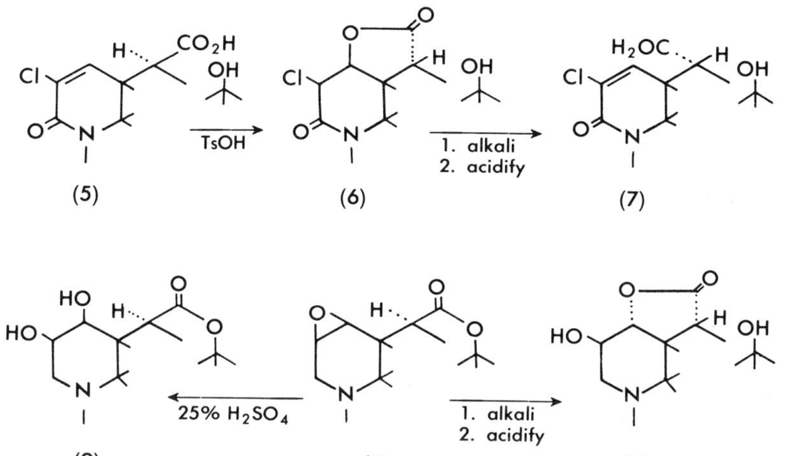

(5) (6) (7)

(9) (1) (8)

Scheme 16.2.

of annotinine with aqueous alkali followed by relactonization in an acidic medium, can be represented by partial structure (8).[32] In support of this structure, an isomer, to which partial structure (9) was assigned, was prepared by boiling annotinine in 25% aqueous sulfuric acid.[33]

The work outlined in Scheme 16.2 implies that the natural configuration of the carboxyl group as shown in (5) is unstable relative to its epimer as shown in (7). Verification of this point came from a study[29] of compound (10), $C_{16}H_{21}NO_3$, obtained originally[31] through simultaneous reduction and hydrogenolysis of (3). In methanolic potassium hydroxide, (10) was converted directly into a methyl ester (11), $C_{17}H_{25}NO_4$ while in ethanolic potassium hydroxide an acid (12), $C_{16}H_{23}NO_4$, was obtained. Esterification of (12), with diazomethane gave ester (13), epimeric with (11), but convertible to (11) upon treatment with methanolic KOH. The authors concluded that (11) was formed by attack of methoxide ion on the lactone ring yielding ester (13) that epimerized in the medium to (11). In the reaction in ethanolic potassium hydroxide it was proposed that the lactone ring was attacked by hydroxide ion yielding the carboxylate ion that failed to epimerize. Thus, acid (12) and ester (13) correspond to the natural configuration. Ester (13) was shown to epimerize readily to the more stable ester (11). In this study the authors also reported the preparation of the acid epimeric with (12) and showed that the hydroxy esters (11) and (13) were oxidized in chromic acid-pyridine to the corresponding keto esters, (14) and (15). The infrared spectra of the ketones suggested that the ketone group was present in a six-membered or larger ring. These reactions are summarized in Scheme 16.3.

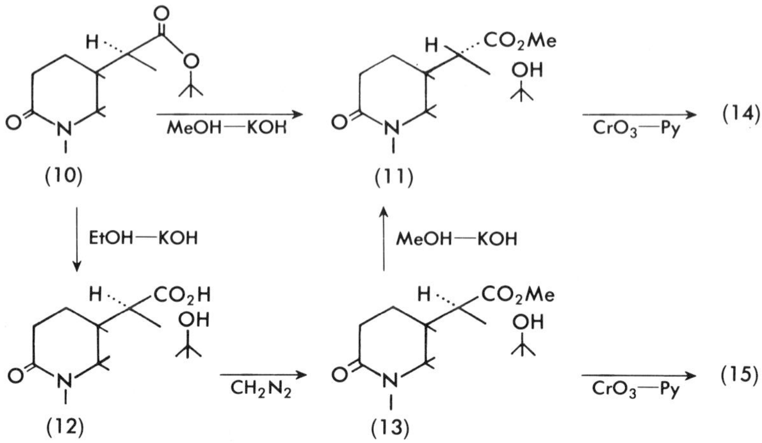

Scheme 16.3.

2.1.2 The Ring System Dehydrogenation studies on annotinine and several of its derivatives were used to establish the nature of its ring system. Selenium dehydrogenation of annotinine gave, among other products, 8-n-propylquinoline, thereby providing the first indication of a reduced quinoline system in the alkaloid.[34] Evidence for the presence of a bridged hydrojulolidine system in annotinine was gained from a study of the dehydration and dehydrogenation of the methyl ester (16) of hydroxy acid (7).[35–36] Dehydration of (16) over P_2O_5 yielded, along with other products, a simple dehydration product (17), $C_{17}H_{22}NO_4Cl$, which was dehydrogenated over Pd-C. The quinolone acid (18), which readily decarboxylated to the quinolone (19) of established structure, was isolated from the dehydrogenation reaction.[36] Reduction of (18) over sodium amalgam gave the dihydro compound (20), which had a uv spectrum superimposable on that of (21), a compound that had been synthesized earlier.[37] Thus, the position of the carboxyl group could be assigned unambiguously in (18) and (20). Since these dehydrogenation studies were of paramount importance in the structural assignment, considerable effort was expended to prove that (17) was a simple dehydration product and not a rearrangement product of (16).[30]

The dehydrogenation studies coupled with the earlier work on the interrelationship of the functional groups and the fact that annotinine had an analysis corresponding to a single C-methyl group led to the proposal of expression (1a) for annotinine.[36] There remained the problem of finding the terminus for the bridge, one end of which was affixed at C(12) since it seemed likely that this center was quaternary as pointed out before.

The dehydrogenation studies are summarized in Scheme 16.4.

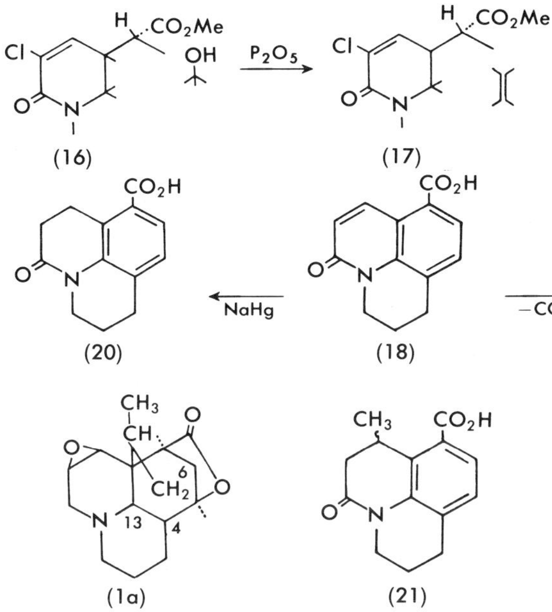

(16) (17) (20) (18) (19) (1a) (21)

Scheme 16.4.

Four compounds, isolated from the selenium dioxide oxidation of keto ester (14) (Scheme 16.3), provided the information necessary to define the periphery of the system.[14,30] A monounsaturated keto ester was assigned structure (22) (Scheme 16.5) on the basis of spectroscopic evidence and its conversion with sodium borohydride to an allylic alcohol that isomerized in base to the keto ester (14). Compounds (23) and (24) were shown to be epimeric hydroxy derivatives of (22). The fourth compound was assigned structure (25) on spectroscopic grounds and because of its conversion to hydroxy ester (13) on vigorous hydrogenation. The hydrogenation reaction was important in a stereochemical sense for it defined the relative configuration of three centers in the ring carrying the hydroxyl group. Arguments presented previously showed that the configuration of the three centers was not altered in the derivation of (13) from annotinine. The oxidation of (23) by periodate–permanganate gave formic acid and succinic acid. The former is apparently derived from C(6). The authors argued that succinic acid must originate from the chain, C(1) to C(4). These results, summarized in Scheme 16.5, allow only two sites, C(12) and C(13), for attachment of the bridge.

The results do not, however, define the structure of the bridge. Annotinine has a single C-methyl group (Kuhn–Roth) thus permitting two arrangements, $-CH_2-CH(CH_3)-$ and $>C(CH_3)_2$, of the three carbon

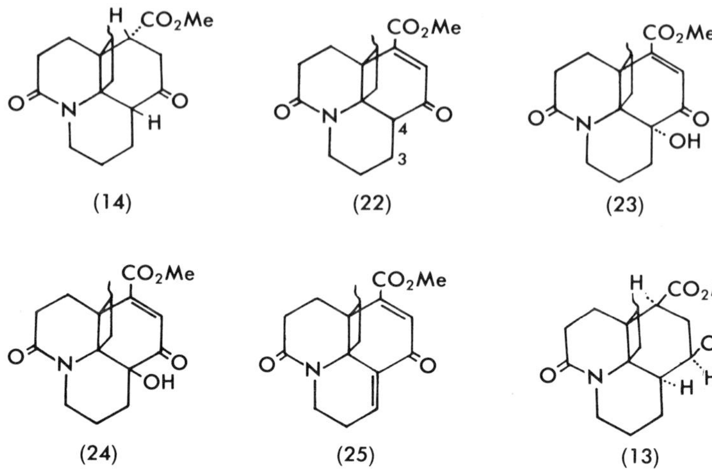

Scheme 16.5.

atoms in the bridge. Of the three possible structures for the bridge, that shown in Scheme 16.5 was selected because it accounted best for the structure of the dehydrogenation product, (21) (Scheme 16.4), obtained from amino acid (4).

The dehydrogenation product (21) was reported initially by Anet and Marion,[38–39] but it was Wiesner and his group who elucidated its structure.[37] In a later study the same group established the configuration of the asymmetric center in this optically active dehydrogenation product.[40] The assignment of configuration allowed them to delineate the bond cleavage and bond formation processes that must occur in the rearrangement of (4) to (21). Unfortunately, space does not permit a detailed examination of this remarkable rearrangement.

By collating the information at hand, it is now possible to assign structure (1) to annotinine. The relative stereochemistry at C(4), C(5), and C(7) follows from the conversion of (25) to (13) by catalytic hydrogenation. The *cis*-arrangement of the bridge and the lactone function is apparent from a consideration of the experiments outlined in Scheme 16.2 where lactonization at a new center is accompanied by epimerization of the carboxyl group. Wiesner's interpretation of the formation of annotinine hydrate provided the evidence for a *cis*-arrangement of the epoxide and the bridge. The relative configuration at C(15) however, could not be deduced from the chemical studies. The x-ray studies[15,41] reported soon afterward provided this information and also confirmed the structural and stereochemical assignment based on the chemical study. The absolute configuration of annotinine is shown in formula 1.[40]

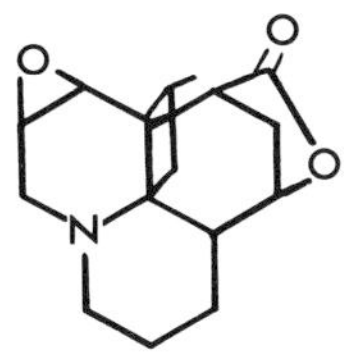

(1)

2.2 Lycopodine and Related Alkaloids

The structure of lycopodine was elucidated by Harrison and MacLean[16] in 1960. Within the year, structures were proposed for several other alkaloids with the same ring system but differing in the nature and number of their functional groups. Among these were acrifoline[42] and annofoline.[43] Anet[44] was able to interrelate the three alkaloids and, in doing so, confirmed their structures and established the stereochemistry of all three.

The majority of Lycopodium alkaloids of established structure fall into this group. A discussion of the studies leading to the structure of all of them is beyond the scope of this chapter. The reader, interested in pursuing the subject, will find the other alkaloids of this group listed in Table 16.1 (section 2.2.4) along with pertinent references to their structural elucidation.

2.2.1 Lycopodine The correct molecular formula, $C_{16}H_{25}NO$, was assigned to lycopodine by Achmatowicz and Uzieblo[2] who ascertained that the base did not contain O-methyl or N-methyl groups or groups with active hydrogen. Marion and Manske,[4] who carried out the first degradative study, isolated 7-methylquinoline and 5,7-dimethylquinoline from dehydrogenation of the alkaloid over palladium and selenium. The oxygen atom of lycopodine was shown to be present in a ketone function by MacLean, Manske, and Marion[45] who reported that the alkaloid was converted to an alcohol with $LiAlH_4$ and that it had an absorption band in its ir spectrum at 1693 $cm.^{-1}$ It was now apparent from a consideration of its molecular formula and of its functionality that lycopodine was a tetracyclic tertiary base. The structural study centered about two isomeric bromocyanamides obtained from the reaction of lycopodine with cyanogen bromide.[45] The periphery of the system was defined by a study of the reactions of these compounds and in conjunction with the dehydrogenation study permitted the assignment of structure (26) to the alkaloid.

The two isomeric bromocyanamides obtained in the reaction of lycopodine with cyanogen bromide may be assigned structures (27) and (28). The former compound is referred to in the literature as α-bromocyanolycopodine and was obtained in the higher yield. In a sequence of straightforward reactions,

$$(\text{—CH}_2\text{—Br} \rightarrow \text{—CH}_2\text{—OAc} \rightarrow \text{—CH}_2\text{—OH} \rightarrow \text{—C}\begin{smallmatrix}\nearrow O \\ \searrow OH\end{smallmatrix})$$

(27) was converted[45] to the carboxylic acid (29). Hydrolysis of (29) in aqueous mineral acid yielded an amino acid[46] that on treatment with diazomethane gave the lactam (30). The ir spectrum of 30, ν_{max} 1700 and 1635 cm^{-1}, attributed to carbonyl and lactam absorption, indicated that the lactam ring was six-membered or larger. A similar sequence of reactions carried out on the β-isomer (28) led to the carboxylic acid (31) that was hydrolyzed to an amino acid.[47] Treatment of the amino acid with diazomethane gave lactam (32), (ν_{max} 1700 and 1626 cm^{-1}), isomeric with (30). Again, it appears that the lactam is present in a six-membered or larger ring. Both (30) and (32) were converted to dihydrolycopodine (33) with lithium aluminum hydride. Since dihydrolycopodine is obtained by hydride reduction of lycopodine, no rearrangement of carbon skeleton has occurred in the formation of (30) and (32).

By hydrogenolysis, both (27) and (28) were converted to the bromine-free compounds (34) and (35), respectively.[47] Modified Kuhn-Roth oxidation of the alcohols derived from (34) and (35) led in both cases to the isolation of acetic, propionic and butyric acids. Since lycopodine gave only acetic acid on similar treatment it was obvious that a n-propyl group was present in each of (34) and (35). A relatively high yield of butyric acid was obtained in both oxidations indicating it was unlikely that the n-propyl groups were attached to quaternary carbon centers. These experiments indicated that the carbon skeleton of lycopodine is different from that of annotinine.

The borohydride reduction of keto acid (31) gave a hydroxy acid which spontaneously cyclized to the lactone (36) (ν_{max}^{nujol} 1743; $\nu_{max}^{CHCl_3}$ 1761 cm^{-1}).[46] Although the ir evidence was not conclusive it favoured a δ-lactone over a γ-lactone. Thus, the carbonyl group is separated from the nitrogen by three or more likely four carbon atoms about the periphery of the system. When the same reaction was applied to the acid (29), lactonization failed to occur.[46]

The environment about the carbonyl group was established through deuterium exchange studies and through a study of the benzylidene derivative[47,48] of compound (34). Deuterium exchange in basic medium on lycopodine and compound (34) showed that there were three hydrogens on the two carbons adjacent to the carbonyl group. The benzylidene derivative (37), obtained from (34) and benzaldehyde in basic medium, underwent a number of reactions that were valuable in the investigation of ring C.[47] It was converted by ozonolysis to the enolic diketone (38) and by oxidation with selenium dioxide to the two products, (39) and (40) that

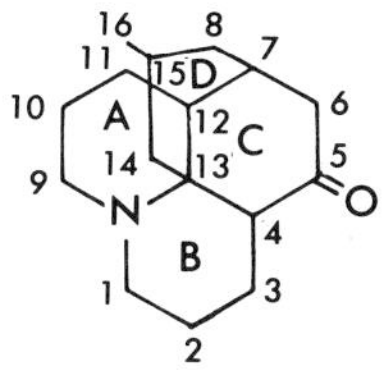

(26)

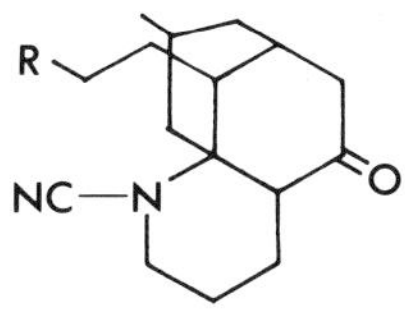

(27) R = CH_2Br
(29) R = CO_2H
(34) R = CH_3

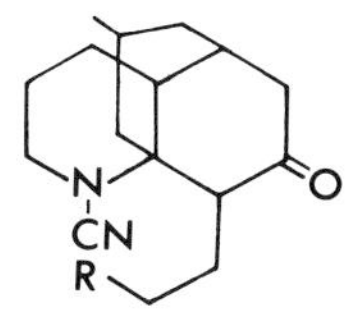

(28) R = CH_2Br
(31) R = CO_2H
(35) R = CH_3

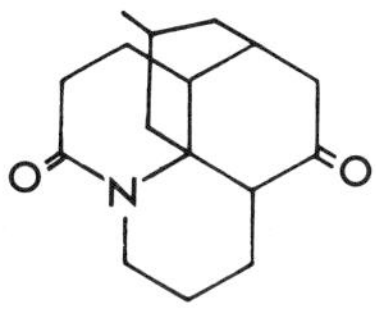

(30)

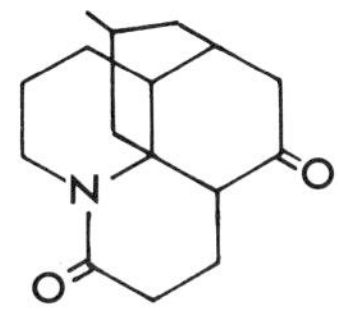

(32)

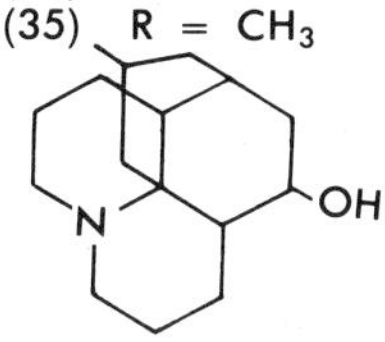

(33)

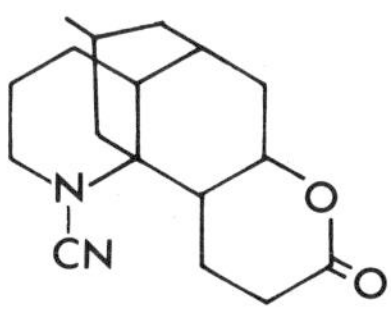

(36)

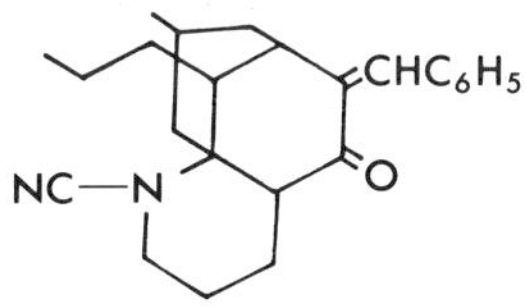

(37)

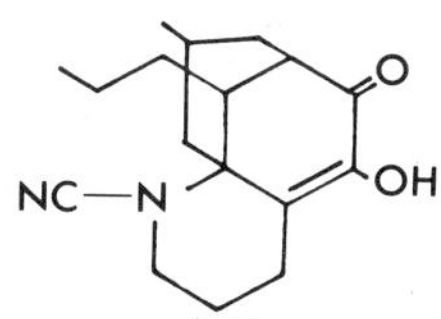

(38)

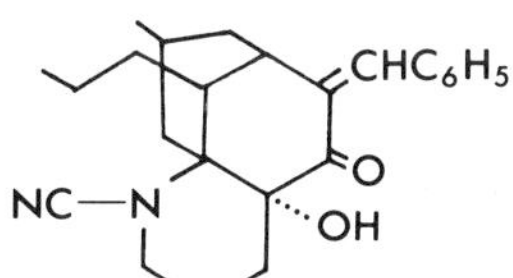

(39)

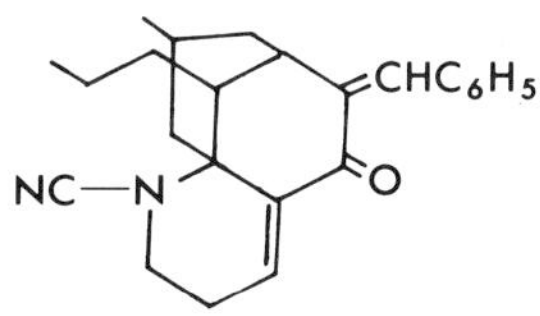

(40)

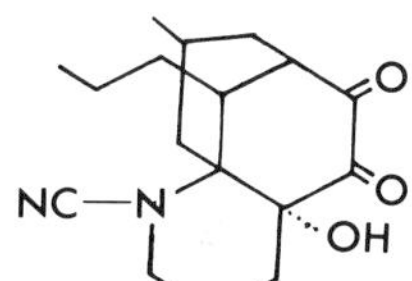

(41)

were assigned the structures shown on the basis of spectroscopic evidence and their reactions. Compound (39) on ozonolysis gave the yellow diketone (41) that failed to enolize even in basic medium. However, when (41) was treated with H_2 over Pt, it was converted to (38). The fact that (41) would not enolize indicated that the carbon center adjacent to the new carbonyl group in (41) was quaternary or at a bridgehead.

The reactions discussed thus far are in accord with the following partial structure for lycopodine,

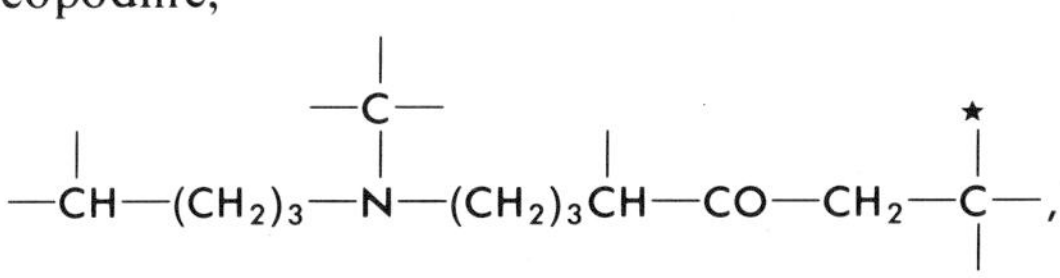

in which the starred carbon is quaternary or at a bridgehead. The partial structure accounts for twelve of the sixteen carbon atoms of lycopodine. Two more carbons are present in a >CH—CH_3 group since lycopodine has a single C-methyl group (Kuhn–Roth) and its nmr spectrum shows that the single methyl group is attached to a carbon carrying one hydrogen. nmr studies were also helpful in defining the character of the third carbon attached to nitrogen. In the cyanamides (34) and (35), the methylene protons α to nitrogen are shifted downfield relative to their position in lycopodine. Since in each case only two protons are shifted, it appears that the third carbon is quaternary.

The incorporation of the partial structures into a hydrojulolidine system led to the bridged structure (26) for lycopodine. This expression is compatible with all the reactions and properties of lycopodine and its derivatives discussed thus far. In particular, it accounts readily for the two dehydrogenation products, 7-methylquinoline and 5,7-dimethylquinoline. Examination of structure (26) shows that the quinoline nucleus found in the dehydrogenation products is derived from rings A and D.

The relative configuration of all centers except at C(15) was deduced by Harrison[49] and is illustrated in stereostructure (26a). The hydrogen at C(12) was known to be *cis* to the bridge because compound (27) yielded a cyclized product (42) on treatment with alcoholic KOH. The reaction of

(26a)

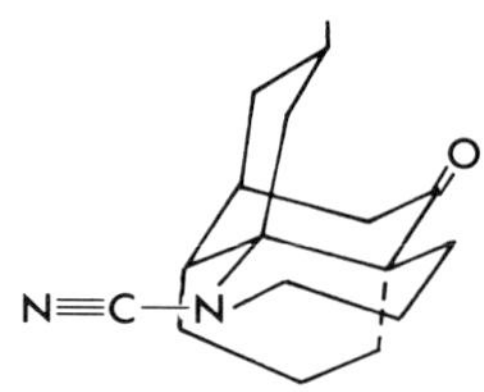

(42)

(27) with base was reported early in the structural study[45] but is was not until some years later[47] that it became apparent that cyclization had occurred at C(4). The configuration of C(4) in lycopodine is not rigorously defined but it seems likely that the C(3)—C(4) bond will be equatorial since lycopodine is stable to both acid and base. The configuration of C(4) in 5-epidihydrolycopodine has, however, been established by pyrolysis of its xanthate ester and conforms with that deduced for lycopodine.[65] The configuration at C(15) was established by Anet[44] whose work is discussed in Section 2.2.3.

2.2.2 Acrifoline and Annofoline Annofoline, $C_{16}H_{25}NO_2$ (43), is a simple dihydro derivative of acrifoline, $C_{16}H_{23}NO_2$ (44), but this relationship was not realized until structures had been assigned, independently, to each alkaloid.[44] The periphery of acrifoline was defined by the work of French and MacLean[42,50] who concluded that it must have the same carbon skeleton as that assigned to lycopodine. Anet and Khan[43] were able to cleave the bridge of annofoline and to dehydrogenate the cleavage product to julolidine. Once the alkaloids were interrelated the two studies complemented and strengthened one another.

Acrifoline,[11] possesses a double bond, a carbonyl group, and a hydroxyl function, and exists in the solid state as a hemiketal.[51] Earlier studies had shown that the base was tertiary and had no N—CH_3 or O—CH_3 groups.[52,53] nmr studies indicated that the double bond was trialkylated and that the alkaloid had a single >CH—CH_3 group.[50]

Degradation of acrifoline by the von Braun and Hofmann methods has been reported.[50] Acrifoline methiodide on treatment with strong base gave a conjugated diene (45) (λ_{max}^{EtOH} 240 mμ, $\epsilon = 24{,}000$), a reaction that correlated the position of nitrogen and the double bond. Catalytic reduction of (45) gave a mixture of dihydro and tetrahydro compounds from which acetic, propionic, and butyric acids were obtained on modified Kuhn–Roth oxidation. When acrifoline was treated with CN—Br two products, (46) and (47) were obtained. Compound (46) is related structurally to (45) since it was converted to a conjugated diene. Compound (47) was converted to a diene with isolated double bonds. Reduction of the diene derived from (47) gave a tetrahydro compound which gave acetic, propionic, and butyric acids on modified Kuhn–Roth oxidation. It was apparent then that the structural unit,

$$>C{=}CH{-}CH_2{-}CH_2{-}\overset{|}{N}{-}CH_2{-}CH_2{-}CH_2{-}\underset{\diagdown}{CH}{-}$$

was present in acrifoline.

Oxidation of acrifoline with selenium dioxide yielded a conjugated ketone (48). The uv, ir, and nmr spectra were consistent with the presence of the structural unit,

$$-\overset{\overset{\displaystyle O}{\|}}{C}-C(CH_3){=}CH-$$

in (48). (A similar oxidation study carried out on annofoline[43] led to a similar chromophore). The Oppenauer oxidation of dihydroacrifoline [an epimer of annofoline at C(12)] yielded a diketone (ν_{max} 1710, 1420 cm^{-1}). The ir data indicate that both carbonyl groups are in six-membered or larger rings and that the new ketone function generated from the hydroxyl group of acrifoline is apparently flanked by a methylene group. The partial structures derived from these studies are all found in expression (44) for acrifoline which also accounts for other reactions of the alkaloid.

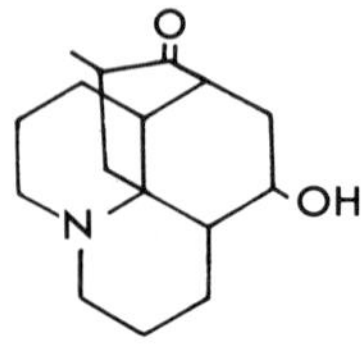

(43)

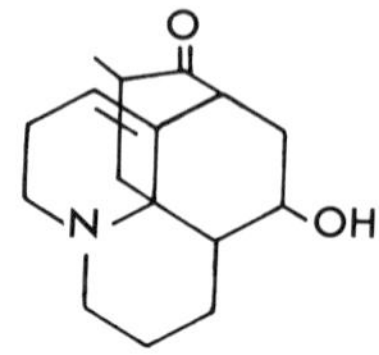

(44)

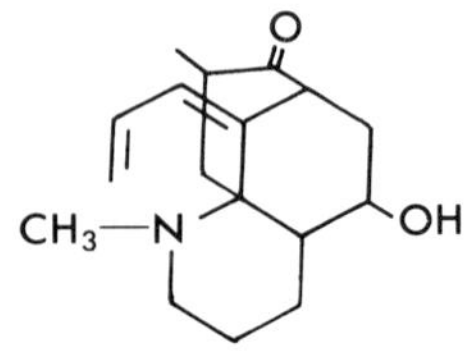

(45)

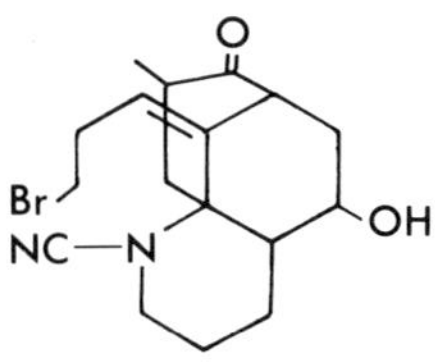

(46)

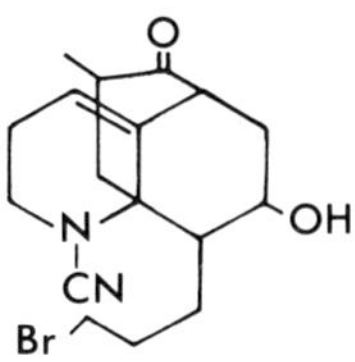

(47)

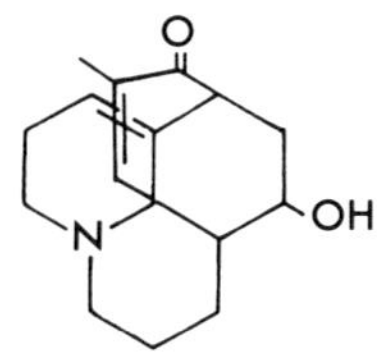

(48)

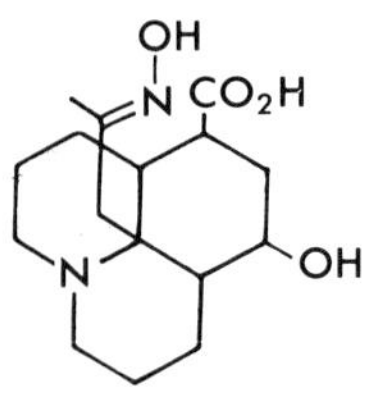

(49)

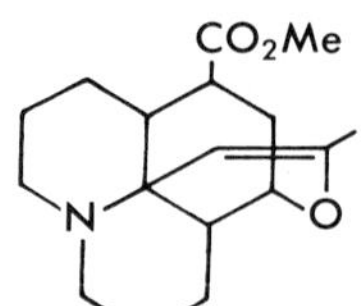

(50)

Annofoline was isolated and studied by Anet and Khan[43,54] who proposed structure (43) for the alkaloid. Of primary importance in their study was the reaction of annofoline with tert-butyl nitrite in alkaline medium from which an amphoteric substance (49) was isolated. Compound (49) was converted to (50) by acid hydrolysis and treatment of the hydrolysate with diazomethane. The dehydrogenation of (50) to julolidine occurred readily and provided the first direct evidence for the presence of this ring system in the alkaloids of the lycopodine group. On the basis of these and other studies Anet and Khan proposed structure (43) for annofoline. The subsequent conversion of acrifoline to annofoline confirmed the structure of both alkaloids.[44]

2.2.3 Interrelation of Annofoline and Lycopodine and Their Relative and Absolute Stereochemistry The relative configuration of annofoline was deduced in the following way.[44] The fact that annofoline is formed in much lower yield than dihyroacrifoline on catalytic reduction of acrifoline implies that it must have a configuration at C(12) similar to lycopodine with the H at C(12) *cis* to the bridge. The carbonyl absorption of annofoline, like that of acrifoline, has an abnormally low intensity and both bases in solution must exist as a mixture of hydroxyketone and hemiketal.[43] Hemiketal formation is sterically possible only if ring D assumes a boat conformation as in stereostructure (43a). Anet reasoned that the methyl group would assume an equatorial position in ring D. Wolff–Kishner reduction of annofoline proceeded normally yielding a product isomeric with dihydrolycopodine (33). Oxidation of the Wolff–Kishner product gave an isomer of lycopodine and it may be concluded that lycopodine and the new compound are epimeric at C(15).

A direct correlation with lycopodine was, however, achieved. Reduction of annofoline with $NaBH_4$ in the presence of base yielded two products that were not epimeric alcohols.[43] One of these was considered to be the normal reduction product (51), the other a reduction product of the C(15) epimer of annofoline, namely (52).[44] When (52) was dehydrated with $SOCl_2$ it gave the $\Delta^{4,5}$ anhydro compound (53) that on oxidation gave the ketone (54). In compound (54) ring D will maintain its chair conformation and the methyl group should retain its equatorial position. Wolff–Kishner reduction of (54) gave a compound (55) identical with the dehydration product of dihydrolycopodine. Thus, the two alkaloids were interrelated and the stereochemistry at C(15) of lycopodine established.

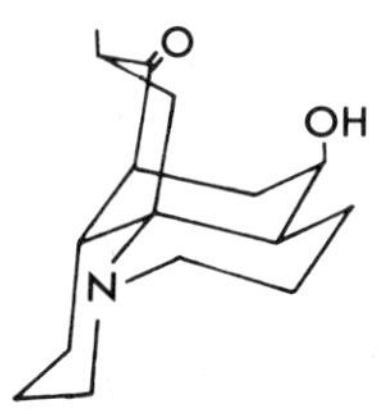

(43a)

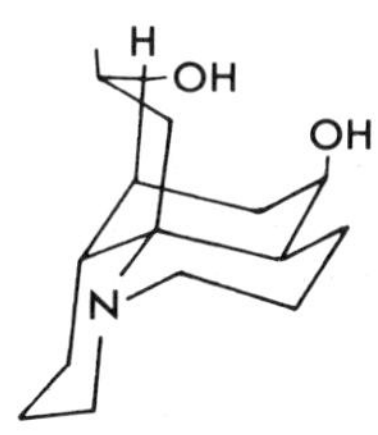

(51)

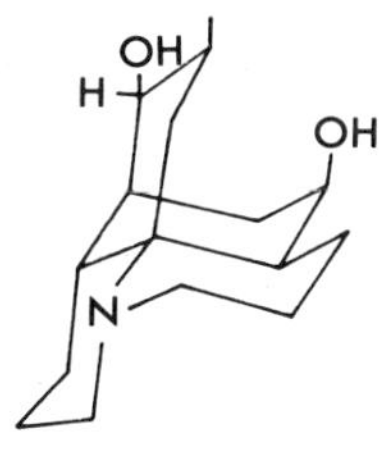

(52)

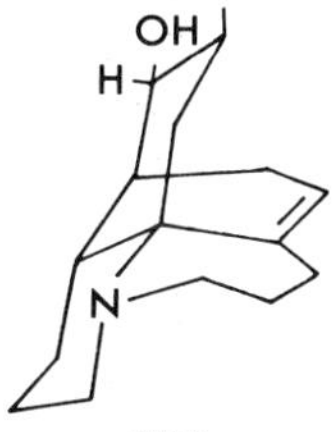

(53)

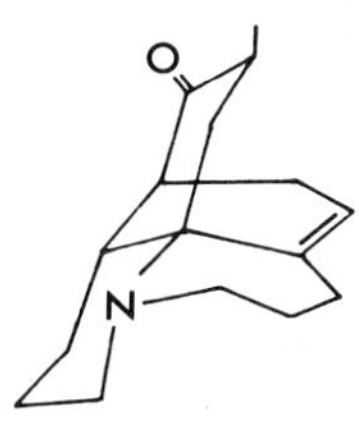

(54)

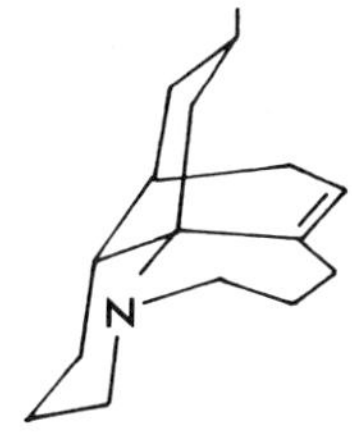

(55)

The absolute configuration of lycopodine shown in the formulas was deduced independently by Wiesner et al.[40] and by Burnell and Taylor.[55]

2.2.4 Other Alkaloids of the Lycopodine Group Table 16.1 lists other alkaloids of established structure which have the lycopodine skeleton. The position of the functional groups are designated by number and their configuration, where appropriate, by the designation α or β. The terms α and β have the same connotation as in steroid nomenclature. Unless otherwise stated the configuration at C(4), C(12), and C(15) is similar to lycopodine. References are given to the investigations that led to the structural elucidation of the alkaloids of Table 16.1. Compounds, which are simple esters or hydrolysis products of other alkaloids, are listed but their functional groups are not designated.

2.3 Lyconnotine

Lyconnotine, $C_{17}H_{25}NO_3$, a minor alkaloid of *L. annotinum*,[21] has been assigned structure (56). From spectroscopic studies it was evident that the alkaloid contained an ester function, a conjugated diene system, and a hydroxyl group. The formation of a methiodide on treatment of the base with one mole of methyl iodide showed that the nitrogen was tertiary. Hydrolysis, followed by acidification, gave a lactone (57), $C_{16}H_{21}NO_2$. It was then apparent that the base is tricyclic.

An nmr examination of lyconnotine and the derived lactone established the nature of the carbon skeleton from the nitrogen through the diene system to C(4). The substitution at C(15) was also derived from the nmr study. The presence of a hydrojulolidine system in the alkaloid was inferred from an examination of the mass spectrum of lyconnotine. The relative stereochemistry of the C(5)-hydroxyl and the C(13) sidechain follow from the formation of lactone (57). The configuration at C(4) was deduced from the ir spectrum of the tetrahydrolactone (58), which exhibited Bohlmann bands, that are possible only with the configuration and conformation shown. Neither the relative configuration at C(15) nor the absolute configuration of lyconnotine is known.

The skeletal structure, deduced from the chemical and spectroscopic studies, was confirmed by conversion of lyconnotine to compound (59) which was synthesized by an unambiguous route.[69]

2.4 Annotine

This alkaloid, $C_{16}H_{21}NO_3$, isolated from *L. annotinum*,[5] has been assigned structure (60) by Szarek et al.[23] In an earlier study Perry and MacLean[51] concluded that the alkaloid had a double bond, a hydroxyl group,

TABLE 16.1. Alkaloids with the Lycopodine Skeleton.

Formula	Name	Functional Groups	References
$C_{16}H_{25}N$	anhydrodihydrolycopodine	$\Delta^{4,5}$ double bond	44, 56
$C_{16}H_{23}NO$	anhydrolycodoline	5 keto; $\Delta^{11,12}$ double bond	57
$C_{16}H_{27}NO$	dihydrolycopodine (complanatine)	5β-OH	3, 47
$C_{16}H_{25}NO_2$	alkaloid L20	5 keto; 6α-OH	58
$C_{16}H_{25}NO_2$	clavalonine	5 keto; 8α-OH	55, 59
$C_{16}H_{25}NO_2$	flabelliformine	5 keto; 4α-OH	60
$C_{16}H_{25}NO_2$	lycodoline	5 keto; 12β-OH	61, 62
$C_{16}H_{25}NO_2$	lycofoline	5β-OH; 8α-OH; $\Delta^{11,12}$ double bond	63, 64
$C_{16}H_{27}NO_2$	deacetylfawcettiine		55, 59
$C_{18}H_{29}NO_2$	acetyldihydrolycopodine	5β-OAc	47, 56
$C_{18}H_{25}NO_3$	o-acetylacrifoline		51
$C_{18}H_{27}NO_3$	acetyllycofoline	5β-OAc; 8α-OH; $\Delta^{11,12}$ double bond	64
$C_{18}H_{29}NO_3$	fawcettiine	5β-OAc; 8α-OH	55, 59
$C_{18}H_{29}NO_3$	lofoline	5β-OAc; 8β-OH	43, 54
$C_{18}H_{29}NO_3$	lycoclavine	5β-OAc; 6αOH	65
$C_{18}H_{29}NO_4$	lycofawcine	5β-OAc; 8αOH, 12βOH	66
$C_{20}H_{29}NO_4$	diacetyllycofoline		64
$C_{20}H_{31}NO_4$	o-acetylfawcettiine		55, 59
$C_{20}H_{31}NO_4$	o-acetyllofoline		67
$C_{20}H_{31}NO_4$	o-acetyllycoclavine		65
$C_{20}H_{31}NO_4$	o-acetyllycofawcine		66
$C_{18}H_{28}N_2O$	flabelline	5-acetamido; $\Delta^{4,5}$ double bond	68

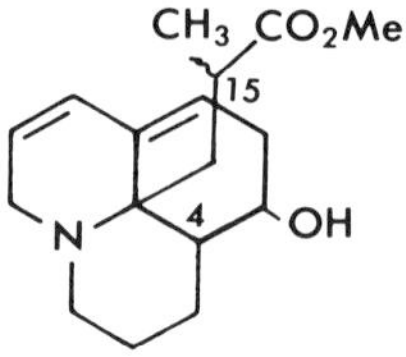

(56)

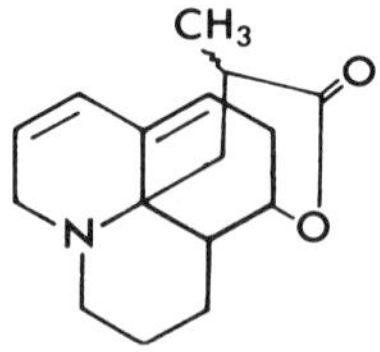

(57)

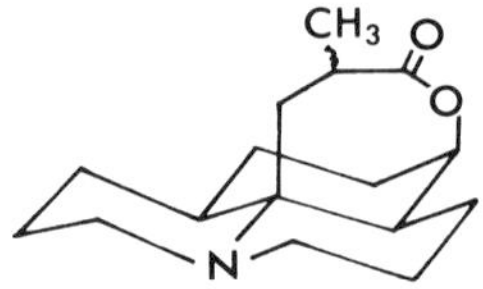

(58)

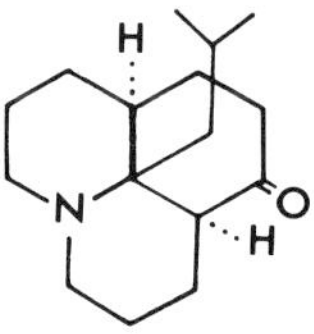

(59)

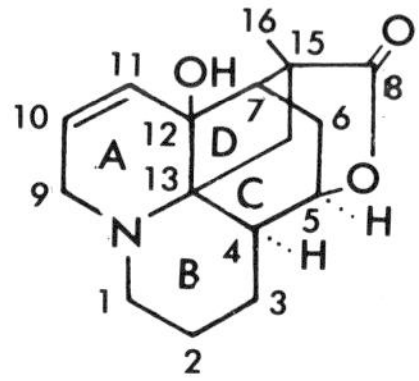

(60)

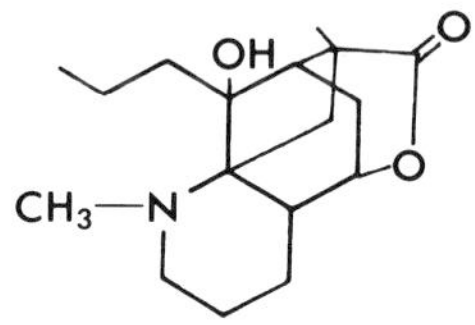

(61)

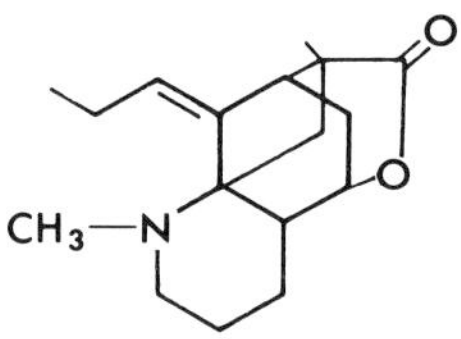

(62)

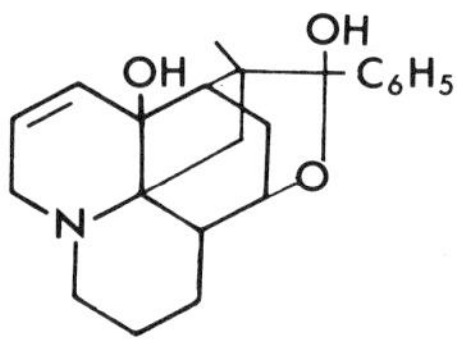

(63)

a carbonyl group, and a third oxygen that they considered to be an ether. The later work[23] showed that the carbonyl and third oxygen were present in a six-membered lactone function and that the base was tertiary. It was then known that annotine had a combination of four carbocyclic and heterocyclic rings aside from the lactone ring.

Ring A opened readily in an Emde degradation of annotine methiodide. The ring-opened product (61) was subjected to a series of degradation reactions akin to those used in the lycopodine study. The degradation studies, coupled with an nmr examination of annotine, the Emde product (61), and the dehydration product (62) established that the structural unit,

$$-CH_2-\underset{|}{N}-CH_2-CH{=}CH-\overset{OH}{\underset{|}{\overset{|}{C}}}-\overset{/}{\underset{\backslash}{C}}H,$$

was present in the base. The nmr study also showed that the single C-methyl group was uncoupled to neighboring protons and, because of its chemical shift, attached to the carbon carrying the carbonyl function. A single proton appearing at $\delta = 4.16$ was ascribed to a methine hydrogen associated with the lactone function. The signal width indicated coupling of this proton to 3 or 4 protons.

Reaction of annotine with borohydride led to a hemiacetal while reaction with phenyl lithium led to the hemiketal (63). From the zinc dust dehydrogenation of (63) α-methylacrylophenone was isolated indicating that the structural unit,

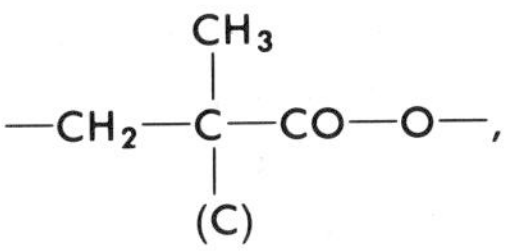

was present in the alkaloid.

The two structural units derived from the degradation and nmr studies could not be accommodated in the ring system of annotinine or lycopodine. Structure (60) was, therefore, proposed for the alkaloid. This structure, which accounts for the reactions and properties of annotine, has a bridged hydrojulolidine system like other alkaloids of the series. It is closely allied in structure to lyconnotine lactone but an interconversion of the two has not been realized. A mass spectrometric investigation of annotine and some of its derivatives has been reported.[70] Its fragmentation upon electron impact is very similar to that of other Lycopodium alkaloids[71] with a bridged hydrojulolidine system. The mass spectrometric study thus provided evidence, not obtained in the chemical examination, for the presence of the hydrojulolidine system in the alkaloid.

The relative configuration shown in formula (60) has been assigned to annotine on the following grounds. The asymmetric centers at C(5), C(7), C(13), and C(15) are fixed by the nature of the system. The hydroxyl group is considered to be *cis* to the bridge because it is intramolecularly bonded to the nitrogen atom and because there are no Bohlmann bonds in the ir spectrum of dihydroannotine. The assignment at C(4) was based on the argument used for assignment of configuration at C(4) in lyconnotinine. Thus, anhydrotetrahydroannotine, formed from dihydroannotine by dehydration and hydrogenation and expected to have the 12α-H configuration, shows Bohlmann bonds in its ir spectrum. The absolute configuration of annotine is not known.

3. MODIFIED HYDROJULOLIDINE ALKALOIDS

3.1 Serratinine and Serratine

Structure (64) has been assigned to serratinine, $C_{16}H_{25}NO_3$.[24] The presence of a five-membered cyclic ketone function flanked by a methylene group was established by ir examination of the base (ν_{max} 1724, 1427 cm^{-1}). The recovery of a basic diacetate (65) after acetylation of the base established the presence of two hydroxyl groups and a tertiary nitrogen in the system while nmr examination of (65) showed that both hydroxyl groups are secondary and that the alkaloid has a $>CH{-}CH_3$ group. Since the alkaloid had no N—Me groups and no double bond, it was inferred to be tetracyclic.

The reaction of serratinine with Zn-acetic anhydride gave compound (66) which is evidence for an α-aminoketone system in the alkaloid. The reduction of (66) with borohydride yielded an alcohol that dehydrated to an alkene showing a single vinyl proton in its nmr spectrum. This result established the quaternary character of the carbon between the nitrogen and the carbonyl function.

A conjugated ketone (67) was obtained from the selenium dioxide oxidation of (65). The nmr spectrum of (67) showed a single vinyl proton and also revealed that one of the two hydrogens geminal to the acetoxy groups was allylic to the new double bond and adjacent to a carbon carrying a single hydrogen. The carbon chain from nitrogen through the carbonyl to the first hydroxyl group was now known and is embodied in the partial structure,

$$-N-C-CO-CH_2-CH-CH(OH)-CH.$$

The extension of this chain to include the $>CH{-}CH_3$ group and the second hydroxyl function was accomplished in the following way. The monacetate (68), obtained by partial hydrolysis of (65), was dehydrated to the trisubstituted alkene (69) that showed a single vinyl hydrogen and a vinyl methyl group in its nmr spectrum. This unconjugated enone was converted with selenium dioxide to the conjugated dienone (70) which had two vinyl protons and a vinyl methyl group. Hydrolysis and oxidation of (69) gave an unconjugated diketone (ν_{max} 1740, 1700 cm^{-1}) that isomerized readily to the conjugated ketone (71). The ir and nmr data clearly indicated that the double bond of (71) was conjugated with the new ketone function. Thus, partial structure,

$$-N-C-CO-CH_2-CH-CH(OH)-CH(CH_3)-CH_2-CH(OH)-C-,$$

is present in serratinine, and the relative positions of the functional groups are known. An nmr study of the diketone (72) obtained from (68) by oxidation, bromination, and dehydrobromination provided evidence for the quaternary character of the carbon adjacent to the second hydroxyl function of serratinine.

The environment about the nitrogen atom was investigated by Hofmann degradation of desoxodihydroserratinine, obtained by Wolff–Kishner reduction of the base. The first methine, formed by fission of the N-C-4 bond was converted to a dihydro derivative and the degradation pursued through two more stages to a nitrogen free product. The latter compound has two vinyl groups and thus the structure about nitrogen can be represented in the partial structure,

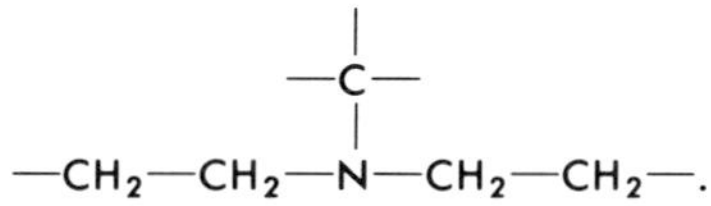

The studies discussed to this point account for 14 of the 16 carbon atoms of the system. When incorporated into structure (64), all the reactions and properties of serratinine are accounted for. In particular, this structure explains the carbonyl absorption of serratinine itself, of those derivatives with a ketone function in ring D, and of a lactam (ν_{max} 1640 cm^{-1}) formed by oxidation of (65). This structure is favored over an alternative structure, in which the size of rings A and B is interchanged, on mass spectrometric grounds and by the fact that structure (64) may be formally derived by a simple rearrangement of an alkaloid of the lycopodine group.

The stereochemistry of serratinine was deduced from other reactions and properties of the base and its derivatives.[72] The facile formation of ketal (73) from bromocyanamide (74) demanded that rings C and D be *cis*-fused and that the hydroxyl function at C(13) be axially oriented on ring D. The authors presented convincing evidence that it was the C(13) and not the C(8)-hydroxyl that was involved in this reaction. The facile formation of (73) suggested that the C(12)-C(13) bond and the N-C(4) bond are *cis* to ring C.* This assignment was strengthened by pKa measurements of serratinine and its diacetyl derivatives. The relative configuration at C(8) was deduced from an examination of the borohydride reduction product of (75), which, after hydrolysis, yielded a mixture of serratinine and 8-episerratinine separable by chromatography on alumina. Since serratinine was eluted first, they concluded that it had an axial hydroxyl group. Examination of the nmr spectrum of 8-acetyl-8-episerratinine provided the evidence necessary to assign the relative configuration of the methyl group. The hydrogen geminal to the 8-acetoxy group was coupled to adjacent hydrogens at C(7) and C(15) with coupling constants J_1 and J_2 of 4 cps and 11 cps, respectively. Since the C-D ring fusion has been shown to

*Recent X-ray work has shown that the N-C(4) bond is trans to the C-(12)–C-(13) bond in ring C but otherwise the structural and steroechemical assignments are correct. [K. Nishio, T. Fujivara, K. Tomita, H. Ishii, Y. Inubushi, and T. Harayama, *Tetrahedron Letters*, No. 11,861 (1969)].

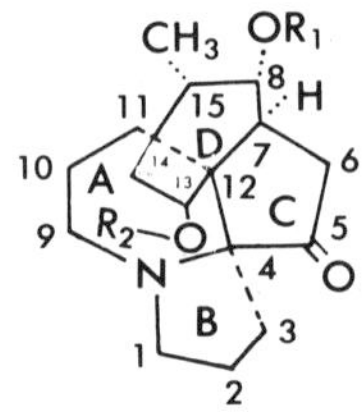

(64) $R_1 = R_2 = H$
(65) $R_1 = R_2 = Ac$
(68) $R_1 = H$; $R_2 = Ac$

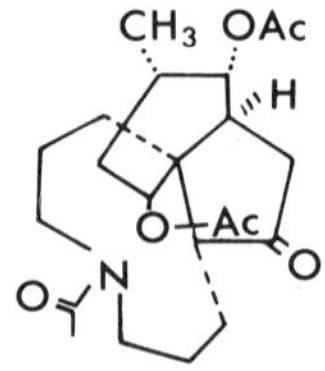

(66)

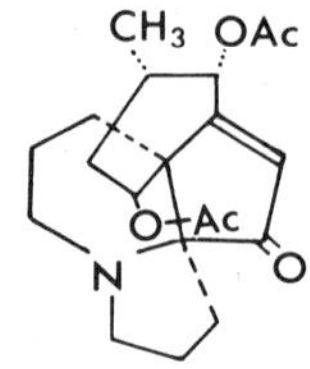

(67)

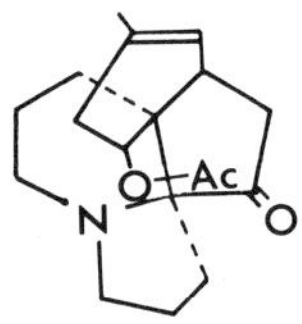

(69)

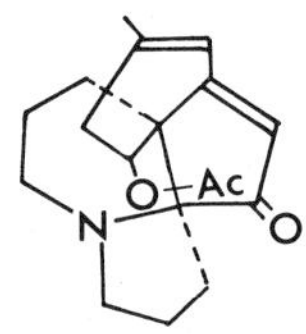

(70)

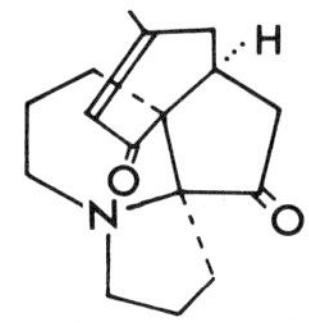

(71)

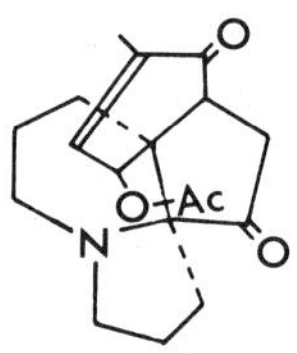

(72)

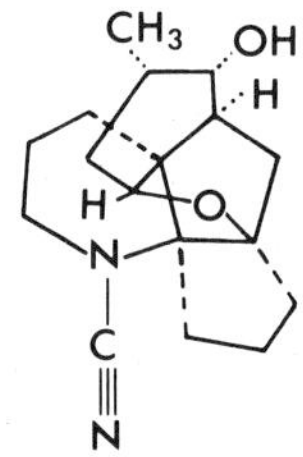

(73)

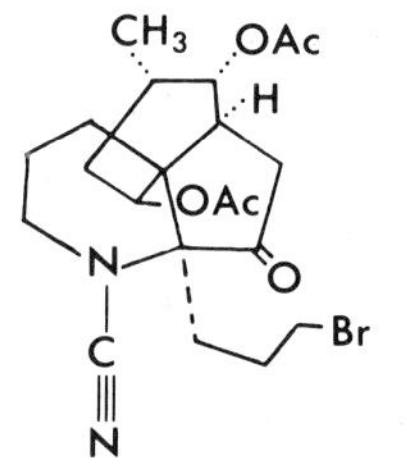

(74)

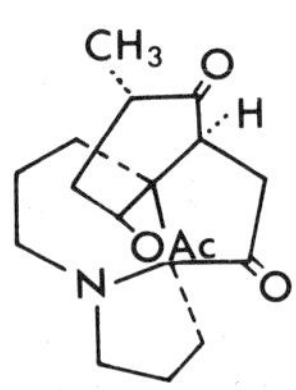

(75)

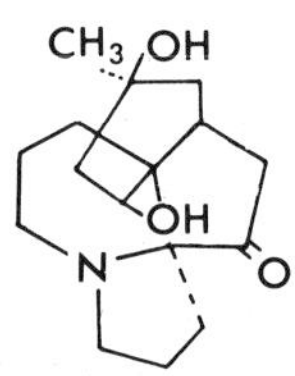

(76)

be *cis* and the C(8) hydrogen to be axial to ring D the coupling J_1 must be assigned to C(7) and the coupling J_2 to C-15. Thus, the hydrogens at C(8) and C(15) are in a *trans*-diaxial relationship and, therefore, the methyl group must be equatorial to ring D. Other experiments showed that stereostructure (64) represents the absolute configuration of the base.

Serratine, $C_{16}H_{25}NO_3$, an isomer of serratinine, has been assigned structure (76) on the following evidence.[73] Dehydration of monoacetylserratine gave a product identical with compound (69) previously derived from serratinine, thereby establishing the equivalence of the two ring systems. It was inferred from the nmr spectrum of diacetylserratine that it had one secondary and one tertiary hydroxyl group and that the methyl group was uncoupled to a neighboring hydrogen and geminal to an acetoxy group. The *cis*-relationship of the two hydroxyl groups of serratine was established by formation of a cyclic carbonate on treatment of the base with phosgene in pyridine and thus its structure and stereochemistry was known.

3.2 Serratinidine, Fawcettidine and Fawcettimine

The three alkaloids have been correlated with the known serratinine and their structures established in this fashion.

In the same communication in which they elucidated the stereochemistry of serratinine, Inubushi et al.[72] postulated that the serratinine system might be derived by rearrangement of an alkaloid of the lycopodine type, e.g., lycodoline (77). The steps, (77) → (78) → (79) → (64) which they postulated in this rearrangement, are summarized in Scheme 16.6.

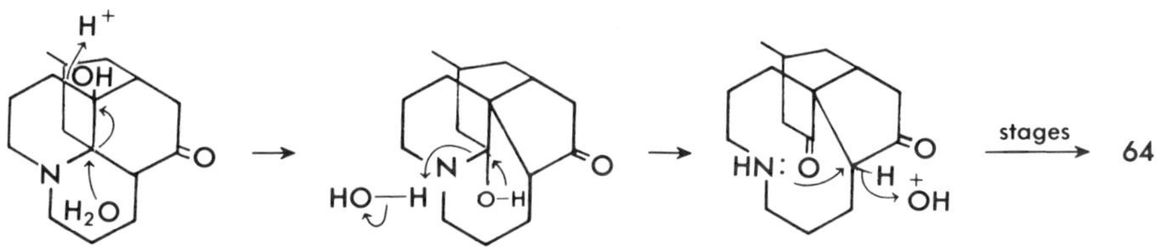

(77) (78) (79)

Scheme 16.6.

Inubushi et al.[25] state that in attempting to check this biogenetic proposal they converted serratinine to compound (80) which, with zinc in acetic acid, yielded the two products (81) and (82). The latter is a ring-chain tautomer of (83). This interconversion provided the basis for the derivation of the structures of the three alkaloids dealt with in this section.

The infrared spectrum of serratinidine (84) demonstrated the presence of a secondary amide function, a double bond, and an NH or OH group. The nmr spectrum demonstrated a single vinyl hydrogen, a $>CHCH_3$ group, a $-COCH_3$ group probably in an amide ($-NHCOCH_3$), a single H geminal to the acetamido group and a single hydrogen geminal to an OH group. The hydrogen geminal to the OH group was shifted downfield upon acetylation. A double irradiation experiment at $\tau = 7.5$ in the region of the tertiary hydrogen associated with the $>CHCH_3$ group resulted in the collapse of the doublet of the olefinic proton to a singlet and the doublet of the $>CHCH_3$ group to a singlet. This experiment suggested that the sequence,

$$>C{=}CHCH(CH_3){-}(C)$$

was present in the system.

At this point the authors observed that the nmr spectrum of acetylserratinidine was very similar to the nmr spectrum of compound (81). Conversion of (81) to acetylserratinidine was effected by straightforward steps and hydrolysis of the O-acetyl group gave a compound identical with natural serratinidine. These experiments did not, however, establish the stereochemistry at C(4) or C(5) in serratinidine.

Fawcettidine, $C_{16}H_{23}NO$, isolated from *L. fawcetii* [74,75] has been assigned structure 85.[76] Because of its weak basicity (pKa 6.2) and the presence of a carbonyl band at 1740 cm^{-1} in its ir spectrum, the Japanese workers suspected that fawcettidine might be related to alkaloids of the serratinine type. Accordingly, they converted serratinine to compound (86) that, with zinc in acetic acid, gave an oil, from which a crystalline picrate

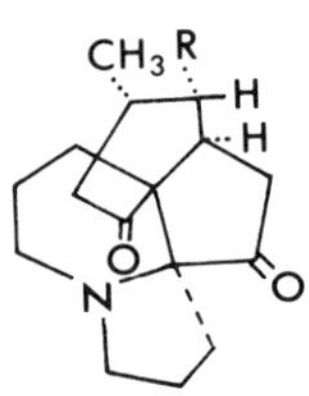

(80) R = OAc
(86) R = H

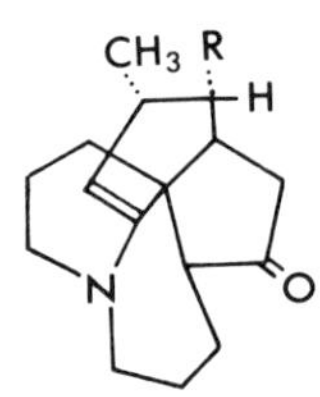

(81) R = OAc
(85) R = H

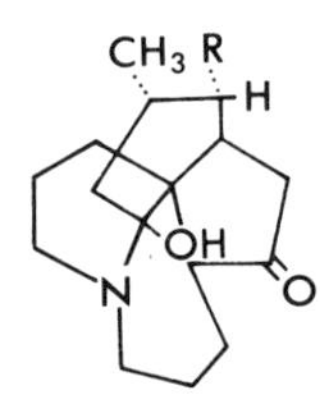

(82) R = OAc
(87) R = H

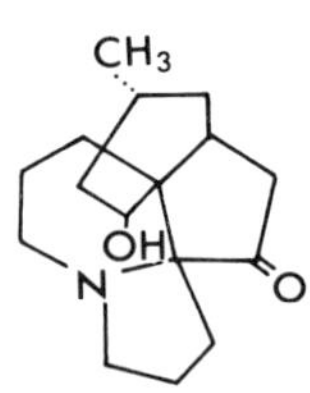

(90)

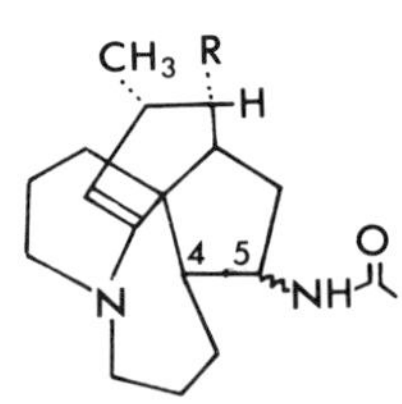

(84) R = OH

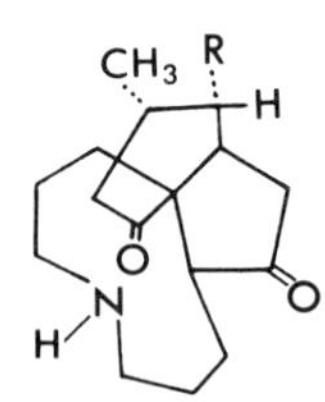

(83) R = OAc
(89) R = H

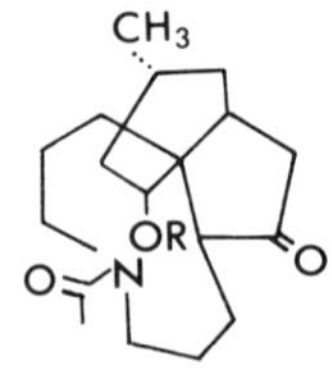

(91) R = Ac
(92) R = H

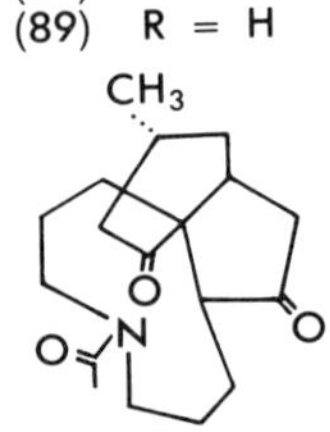

(88)

and methiodide were isolated. These salts were identical with those derived from natural fawcettidine and thus structure (85) is an adequate representation of this alkaloid.

Fawcettimine, $C_{16}H_{25}NO_2$, isolated from *L. fawcettii,* [74,77] has recently been assigned structure (87). Fawcettimine[78] shows carbonyl absorption in its ir spectrum in CCl_4 at 1732 and 3585 cm^{-1} indicative of a five-membered ketone and a hydroxyl group. Upon acetylation, however, fawcettimine yields an acetamide (88) ν_{max} 1730, 1690, and 1615. The absorption bands are indicative of a five-membered ketone, a six-membered ketone and an acetamide group but there were no bands attributable to -NH or -OH absorption. These results suggest that in fawcettimine there is a transannular reaction between a secondary amino group and a six-membered ketone, (87)$\rightleftarrows$(89), and that in solution fawcettimine exists predominantly in the form (87). The fact that a five-membered ketone was present in the base suggested that it, like fawcettidine, might be related to the serratinine alkaloids. A direct correlation was achieved by converting (86) to compound (90) that with Zn-Ac_2O gave the compound 91. Hydrolysis of the O-acetyl group of (91) gave compound (92) identical with the borohydride reduction product of N-acetyl fawcettimine (88). Fawcettimine (87) was also converted to fawcettidine (85) by dehydration with $POCl_3$ in pyridine.

4. ALKALOIDS WITH TWO NITROGEN ATOMS

Lycodine, the obscurines, and selagine which are closely allied in structure to the alkaloids of the lycopodine group are found in this group. A direct correlation of lycodine and the obscurines with lycopodine has been realized. Selagine, unlike the other alkaloids, which are tetracyclic, is a tricyclic base with a primary amino group but is, nonetheless, closely related in structure to the obscurines. Cernuine and lycocernuine, also discussed in this section, have a ring system entirely different from any of the alkaloids discussed thus far. They may be formally derived, however, by condensation of two 8-carbon polyacyl chains like other alkaloids of this family.

4.1 Lycodine and the Obscurines

The alkaloid, obscurine, found in several Lycopodium species,[3,5,7] was later found to be a mixture of two bases. The two compounds (93), $C_{17}H_{26}N_2O$ and (94), $C_{17}H_{24}N_2O$, were named α- and β-obscurine, respectively. Moore and Marion,[79] who carried out the separation of the bases, also examined them spectroscopically and reported the dehydrogenation of α-obscurine. They found that α-obscurine had $\nu_{max}^{CHCl_3}$ 3411 and 1667 cm^{-1}, attributed to an NH group and a carbonyl group in a lactam, respectively, and λ_{max}^{EtOH} 255 log $\epsilon = 3.75$. β-obscurine had spectral properties characteristic of an α-pyridone. Dehydrogenation of α-obscurine over Pd-charcoal gave 6-methyl-α-pyridone and 7-methylquinoline.

Ayer and Iverach took up the study of the obscurines and in 1960, shortly after the structure of lycopodine was elucidated, they proposed the structures (93) and (94) for the two alkaloids.[17,80] They found that the assigned molecular formulas were correct, that both bases were tertiary and that each had a single >N-CH_3 and >CH-CH_3 group. They were able to convert α-obscurine to β-obscurine by dehydrogenation with N-bromosuccinimide thereby establishing the identity of the ring systems in the two bases. An nmr study showed that β-obscurine had two vinyl protons in an AB pattern whereas α-obscurine had none. It was shown that the ir and uv data presented earlier for α-obscurine were expected for a 3,4-dihydro-2-pyridone. These data, coupled with dehydrogenation data, led Ayer and Iverach to propose structures (93) and (94) for the two bases. The two dehydrogenation products account for all the carbon atoms of α-obscurine except that of the N-CH_3 group. At the same time Ayer and Iverach[17] proposed that lycodine, $C_{16}H_{22}N_2$, had structure (95). Lycodine had been isolated from *L. annotinum* by Anet and Eves[81] who noted that it had a 2,3-disubstituted pyridine ring, a C-methyl group, and a secondary nitrogen atom. In a subsequent study[18] β-obscurine was converted via the corresponding chloropyridine derivative to N-methyllycodine.

The structures assigned to the three alkaloids were confirmed and their stereochemistry established by Anet and Rao who succeeded in converting β-cyanobromolycopodine (28) to lycodine.[19] The bromide was converted to the azide which on reduction gave the amine. The crude amine, which partially cyclized, was subjected to dehydrogenation. Lycodine was isolated in good yield after hydrolysis of the dehydrogenated product to remove the cyano group. This experiment showed that the asymmetric centers of lycodine were identical with the corresponding centers in lycopodine, and, therefore, of the obscurines. Thus, the stereochemistry of the alkaloids was known.

Three alkaloids, which are related to the obscurines and lycodine, other than N-methyl or des-methyl derivatives are known. A base, $C_{16}H_{24}N_2O_2$, isolated from *L. flabelliforme* has been assigned the structure 12-hydroxy-des-N-methyl-α-obscurine.[82] Sauroxine, $C_{17}H_{26}N_2O$, isolated from *L. saururus* is considered to be 12-epi-α-obscurine.[83,84] Flabellidine, $C_{18}H_{28}N_2O$, is a $\Delta^{4,5}$ tetrahydrolycodine acetylated at the pyridine nitrogen.[3,82]

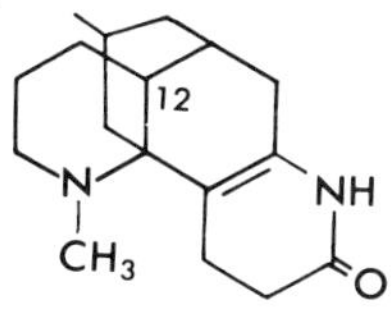

(93)

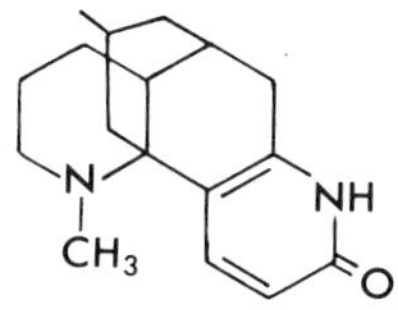

(94)

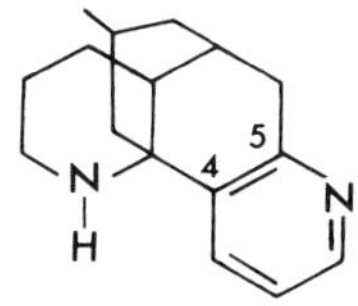

(95)

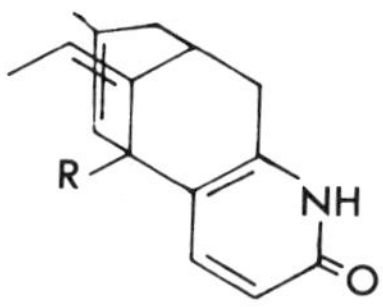

(96) R = NH_2
(97) R = OH

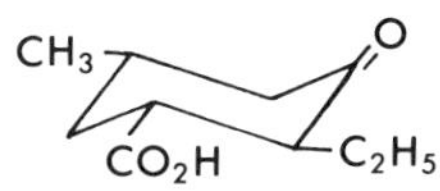

(98)

4.2 Selagine

Selagine (96), $C_{15}H_{18}N_2O$, found in *L. selago*, has been investigated by Valenta et al.[85] who were able to show by spectroscopic examination of the alkaloid that it contained an α-pyridone ring substituted as in β-obscurine, two vinylic methyl groups, and two vinylic protons other than those in the pyridone ring. The second nitrogen is present in a primary amino group because selagine was converted to an alcohol, selaginol, (97) by treatment with nitrous acid. Reduction of the base gave both a dihydro and a tetrahydro compound. The uv spectra of selagine and the two reduction products were identical showing that the double bonds were not conjugated to one another or to the pyridone ring. Modified Kuhn-Roth analysis of selagine gave only acetic acid whereas the dihydro and tetrahydro compound gave both acetic and propionic acid implying that an ethylidene group is present in selagine and an ethyl group in each of the reduction products. The nmr spectrum of the dihydro compound showed that the single vinyl proton remaining was uncoupled to neighboring protons. Dehydrogenation of the base yielded 6-methyl-α-pyridone as in the case of α-obscurine. From these and other studies it was concluded that selagine had structure (96). This structure was confirmed[20] when the oxidation of tetrahydroselaginol with permanganate-periodate yielded the keto acid (98). The structure of this acid was demonstrated by a synthesis of its racemate.

4.3 Cernuine and Lycocernuine

In 1948 Marion and Manske[12] examined *Lycopodium cernuum* L. from which they isolated two alkaloids, cernuine, $C_{16}H_{26}N_2O$, and alkaloid

L33, which melted at 218°, but which was not characterized further. Ayer et al.[22] renewed the study of the two alkaloids and in 1964 proposed structures for cernuine and L33 to which they assigned the molecular formula, $C_{16}H_{26}N_2O_2$, and the trivial name, lycocernuine. The two alkaloids are represented in structures (99) and (100), respectively.

Both alkaloids contain one basic and one nonbasic nitrogen. From the ir spectrum of cernuine ($\nu_{max}^{CCl_4}$ 1640, 1410 cm^{-1}), it was concluded that the nonbasic nitrogen was present in an amide or lactam function and that the amide carbonyl was flanked by a methylene. Lycocernuine had in addition to the bands found in cernuine, a band at 3620 cm^{-1} characteristic of a nonbonded hydroxyl group. Examination of the nmr spectrum of lycocernuine and its O-acetyl derivative established the secondary character of the hydroxyl group. By oxidation, lycocernuine was converted to a ketone 101 (ν_{max} 1705 cm^{-1}), which readily exchanged three hydrogens for deuterium under conditions that left lycocernuine unaffected, and which was reconverted to lycocernuine with borohydride.[86] Evidence was presented that compound 101 was an α-aminoketone. Wolff–Kishner reduction converted 101 to cernuine showing that lycocernuine was a simple hydroxyl derivative of cernuine.

The presence of a >CH—CH$_3$ group in each alkaloid and the environment about the two nitrogens was established by nmr examination of cernuine, lycocernuine, and their transformation products. Both alkaloids show a one proton quartet at $\tau = 4.54$ which is shifted to $\tau = 4.13$ in cernuine methiodide and to $\tau = 6.40$ in the spectrum of the product (102) formed from cernuine and lithium aluminum hydride. This proton is apparently located on a carbon between the two nitrogens and is adjacent to a methylene group. Lycocernuine was converted through its mesylate to anhydrolycocernuine 103, which had the properties of an enamine (ν_{max} 1650 cm^{-1}) and which had a single vinyl proton as a multiplet at $\tau = 5.22$. These data allow the development of the partial structure,

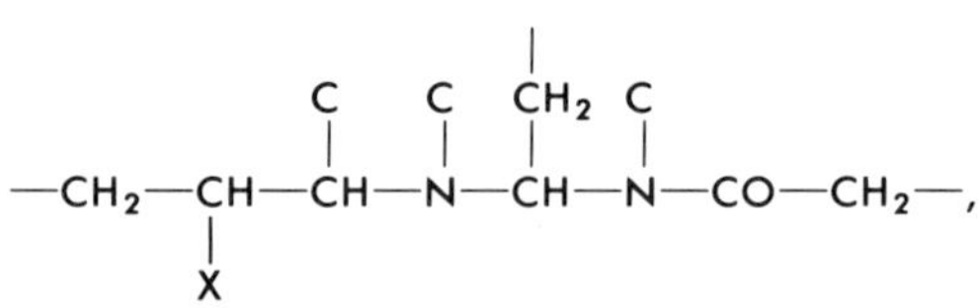

(for cernuine X = H; for lycocernuine X = OH), and demonstrate that the cernuine system is different from other lycopodium alkaloids.

Dehydrogenation studies led to the development of the gross structure of the alkaloids. From the dehydrogenation of cernuine over palladium-charcoal, 2-n-butyl-4-methyl-6-n-pentylpyridine was isolated and identi-

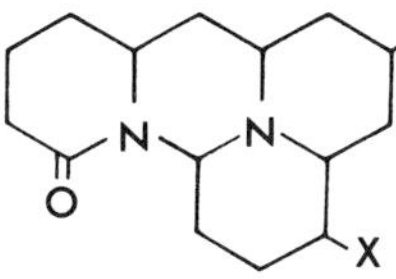

(99) X = H
(100) X = OH

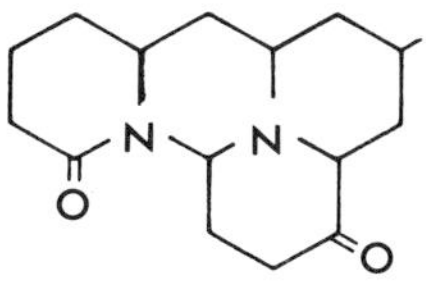

(101)

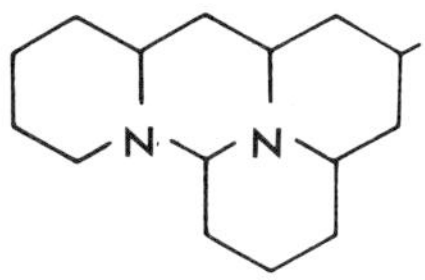

(102)

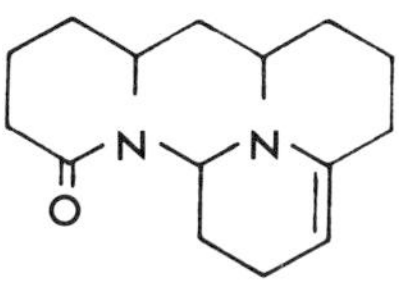

(103)

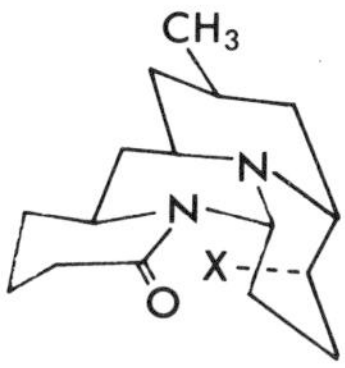

(99) A X = H
(100) A X = OH

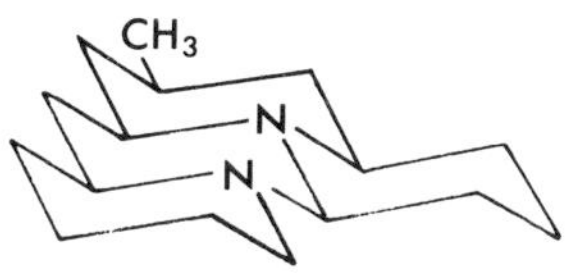

(104)

fied. The structure of this compound, which contains all but one of the carbon atoms of cernuine, was confirmed by synthesis. A similar dehydrogenation of 102 apparently led to a homologous pyridine in which a hexyl group replaced the pentyl group.[86] The authors reasoned that, if dehydrogenation proceeded without skeletal rearrangement, structures (99) and (100) adequately account for the reactions and properties of the two alkaloids. Moreover, they noted that the cernuine system, like other alkaloids of this family, may be formally derived by condensation of two triketooctanoic acid chains. Further degradative evidence in support of the structural assignment has been offered by Ayer et al.[86]

The relative and absolute stereochemistry of the two bases have been discussed.[87] The authors arrived at the assignments shown in structures (99a) and (100a) on the basis of further chemical and physical examination of the alkaloids and their degradation products. A synthesis of compound (104), a transformation product of lycocernuine, has recently been reported from the same laboratory.[88] The synthesis established beyond doubt the ring system of the alkaloids and at the same time supported the previous stereochemical assignment.

5. SYNTHETIC STUDIES

Prior to 1967, the interconversion of a number of alkaloids had been realized and a transformation product of lyconnotine had been synthesized (Section 2.3). Early in 1967, Ayer and Piers reported the synthesis of the cernuine ring system and were able to relate their synthetic product to a transformation product of lycocernuine.[88] However, the stereochemistry of the synthetic material differed from that of the natural bases. Later in the same year, the construction of the lycopodine ring system was realized but again the product, 12-epilycopodine, did not conform in configuration to the natural base.[89] A successful stereospecific synthesis of annotinine was, however, completed late in 1967.[90] It is the latter synthesis alone that will be dealt with in this section.

The successful outcome of the annotinine synthesis rested on the development of a method for the construction of the four-membered ring of annotinine. In a series of experiments on the model compound (105), conditions, which led to the photochemical addition of allene to the conjugated system and the conversion of the photoadduct (106) stereospecifically to the ketolactam (107), were found. The adduct (106) is favored over the alternative product of addition as the size of R increases, while the addition of hydrogen to the exomethylene group of (106) proceeds exclusively in the manner shown when the reduction is conducted after ketalization of the ketone function.[91] Application of the same methods to the tricyclic ketolactam (108), obtained as shown in the accompanying formulas, led to the tetracyclic compound (109) in high over-all yield.[92] None of the isomeric photoadduct was obtained. There remained now to introduce the lactone ring and the epoxide ring making use of the functionality already present in rings A and C.

The introduction of the lactone ring was accomplished in the following way.[92] The ketone function of (109) was reduced to an alcohol with sodium borohydride and the alcohol in turn converted to the mesylate. In an elimination reaction the mesylate yielded a mixture of the desired alkene (110) and a ring expanded isomer (111). The mixture was not separated at this stage but instead was subjected to oxidation by SeO_2 from which a mixture of allylic alcohols was obtained. Further oxidation by CrO_3-

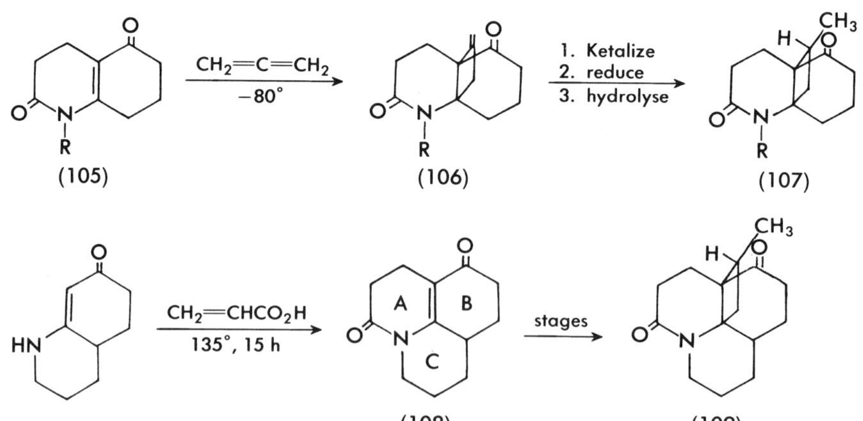

(105) (106) (107) (108) (109)

Pyridine yielded a separable mixture of the isomeric enone lactams (112) and (113). Compound (112) was the racemic form of a degradation product obtained in the earlier structural investigation.[30] A correlation with the natural base had thus been achieved.

The final carbon required to complete the carbon skeleton of annotinine[93] was introduced by conjugate addition of HCN to the enone system of (112). The resulting ketonitrile (114) was converted with $MeOH—H_2SO_4$ to the racemic form of (14) (Sections 2.1.1 and 2.1.2). Structural identity of the synthetic racemate with the naturally derived, optically-active compound was established by spectral and chromatographic comparison. To convert compound (14) to compound (10), the next objective of the synthetic sequence, it is necessary to epimerize centers 4 and 7, to reduce the ketone to hydroxyl and finally to form the lactone ring. Aside from the final step these transformations had been performed during the structural investigation. The conversion of (14) to the hydroxy ester (13) is outlined in Scheme 16.5 (Section 2.1.2) and the epimerization of (13) to (11) in Scheme 16.3 (Section 2.1.1). Hydrolysis of (11) gave the corresponding hydroxy acid (115). Its conversion to compound (10) was

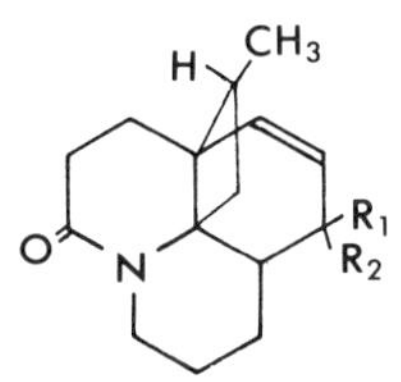

(110) $R_1 = R_2 = H$
(112) $R_1 + R_2 = O$

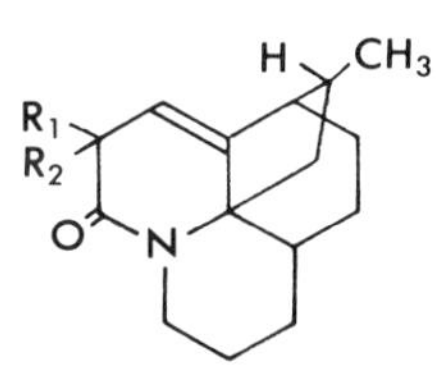

(111) $R_1R_2 = H$
(113) $R_1 + R_2 = O$

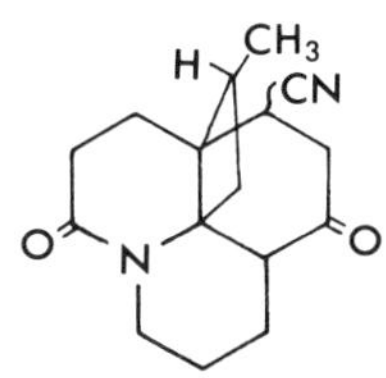

(114)

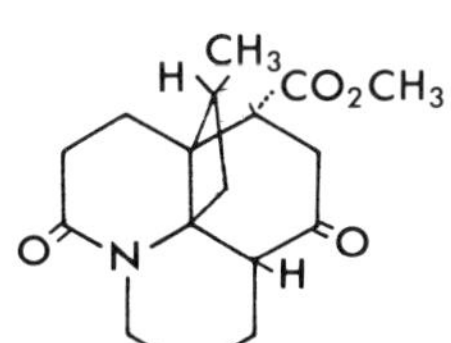

(14)

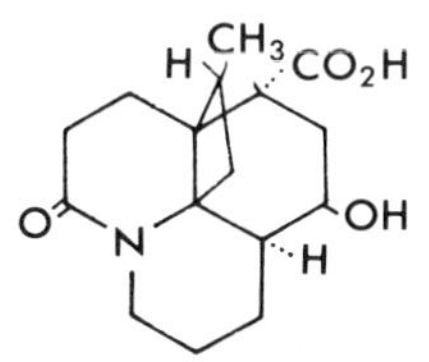

(115)

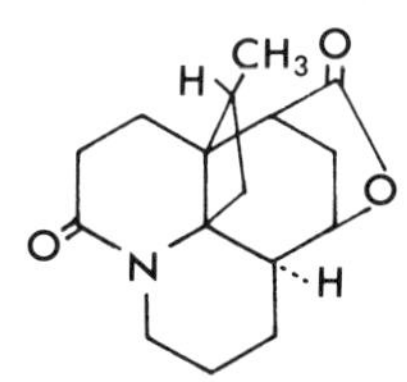

(10)

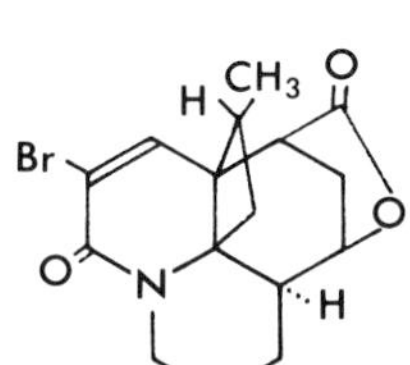

(116)

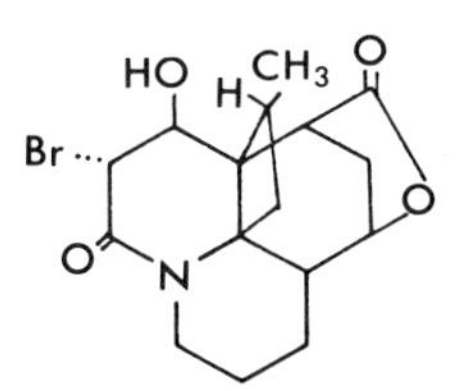

(117)

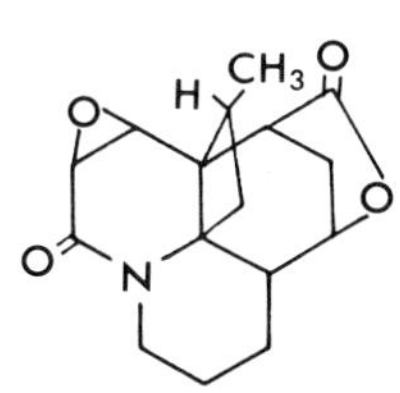

(2)

accomplished by prolonged heating with p-toluenesulfonic acid accompanied by azeotropic removal of water with refluxing benzene.[93]

The introduction of an epoxide into compound (10) was accomplished in three steps.[90] On treatment with N-bromosuccinimide compound (10) was converted to the brominated unsaturated lactam (116) and then with boiling 10% HBr to the bromohydrin (117). With sodium bicarbonate in acetone the bromohydrin yielded the lactam 2 identical with the product of permanganate oxidation of annotinine. Since the conversion of 2 to annotinine had been reported by Betts and MacLean[94] in 1957, the total synthesis of annotinine was complete.

REFERENCES

1. K. Bödeker, *Ann. Chem.* **208,** 363 (1881).
2. O. Achmatowicz and W. Uzieblo, *Roczniki Chem.* **18,** 88 (1938).
3. R. H. F. Manske and L. Marion, *Can. J. Res.* **20B,** 87 (1942).
4. L. Marion and R. H. F. Manske, *Can. J. Res.* **20B,** 153 (1942).
5. R. H. F. Manske and L. Marion, *Can. J. Res.* **21B,** 92 (1943).
6. L. Marion and R. H. F. Manske, *Can. J. Res.* **22B,** 1 (1944).
7. R. H. F. Manske and L. Marion, *Can. J. Res.* **22B,** 53 (1944).
8. L. Marion and R. H. F. Manske, *Can. J. Res.* **22B,** 137 (1944).
9. R. H. F. Manske and L. Marion, *Can. J. Res.* **24B,** 57 (1946).
10. L. Marion and R. H. F. Manske, *Can. J. Res.* **24B,** 63 (1946).
11. R. H. F. Manske and L. Marion, *J. Amer. Chem. Soc.* **69,** 2126 (1947).
12. L. Marion and R. H. F. Manske, *Can. J. Res*, **26B,** 1 (1948).
13. R. H. F. Manske, *Can. J. Chem.* **31,** 894 (1953).
14. K. Wiesner, W. A. Ayer, L. R. Fowler, and Z. Valenta, *Chem. Ind. (London)* 564 (1957).
15. M. Przybylska and L. Marion, *Can. J. Chem.* **35,** 1075 (1957).
16. W. A. Harrison and D. B. MacLean, *Chem. Ind. (London)* 261 (1960).
17. W. A. Ayer and G. G. Iverach, *Tetrahedron Letters* **10,** 19 (1960).
18. W. A. Ayer and G. G. Iverach, *Can. J. Chem.* **38,** 1823 (1960).
19. F. A. L. Anet and M. V. Rao, *Tetrahedron Letters* **20,** 9 (1960).
20. H. Yoshimura, Z. Valenta, and K. Wiesner, *Tetrahedron Letters* **12,** 14 (1960).
21. F. A. L. Anet, M. Z. Haq, N. H. Khan, W. A. Ayer, R. Hayatsu, S. Valverde-Lopez, P. Deslongchamps, W. Riess, M. Ternbah, Z. Valenta, and K. Wiesner, *Tetrahedron Letters* 751 (1964).
22. W. A. Ayer, J. K. Jenkins, and S. Valverde-Lopez, *Tetrahedron Letters* 2201 (1964).
23. W. A. Szarek, K. A. H. Adams, M. Curcumelli-Rodostamo, and D. B. MacLean, *Can. J. Chem.* **42,** 2584 (1964).
24. Y. Inubushi, H. Ishii, B. Yasui, M. Hashimoto, and T. Harayama, *Tetrahedron Letters* 1537 (1966).
25. B. Yasui, H. Ishii, T. Harayama, R. Nishino, and Y. Inubushi, *Tetrahedron Letters* 3967 (1966).
26. K. Wiesner, "Fortschritte der Chemie Organischer Naturstoffe," Vol. 20, p. 271 (L. Zechmeister, Ed.), Springer-Verlag, Vienna, Austria, 1962.

27. H. Conroy, *Tetrahedron Letters* **10,** 34 (1960).
28. D. B. MacLean, *Alkaloids* **8** (1968).
29. Z. Valenta, F. W. Stonner, C. Bankiewicz, and K. Wiesner, *J. Amer. Chem. Soc.* **78,** 2867 (1956).
30. K. Wiesner, Z. Valenta, W. A. Ayer, L. R. Fowler, and J. E. Francis, *Tetrahedron* **4,** 87 (1958).
31. D. B. MacLean and H. C. Prime, *Can. J. Chem.* **31,** 543 (1953).
32. D. R. Henderson, F. W. Stonner, Z. Valenta, and K. Wiesner, *Chem. Ind. (London)* 544 (1954).
33. D. R. Henderson, F. W. Stonner, Z. Valenta, and K. Wiesner, Chem. Ind. (London) 852 (1954).
34. C. Bankiewicz, D. R. Henderson, F. W. Stonner, Z. Valenta, and K. Wiesner, *Chem. Ind. (London)* 1068 (1954).
35. K. Wiesner, Z. Valenta, and C. Bankiewicz, *Chem. Ind. (London)* R41 (1956).
36. K. Wiesner, Z. Valenta, W. A. Ayer, and C. Bankiewicz, *Chem. Ind. (London),* 1019 (1956).
37. Z. Valenta, K. Wiesner, C. Bankiewicz, D. R. Henderson, and J. S. Little, *Chem. Ind. (London)* R40 (1956).
38. F. A. L. Anet and L. Marion, *Chem. Ind. (London)* 1232 (1954).
39. F. A. L. Anet and L. Marion, *Can. J. Chem.* **33,** 849 (1955).
40. K. Wiesner, J. E. Francis, J. A. Findlay, and Z. Valenta, *Tetrahedron Letters* 187 (1961).
41. M. Przybylska and F. R. Ahmed, *Acta Cryst.* **11,** 718 (1958).
42. W. N. French and D. B. MacLean, *Chem. Ind. (London)* 658 (1960).
43. F. A. L. Anet and N. H. Khan, *Chem. Ind. (London)* 1238 (1960).
44. F. A. L. Anet, *Tetrahedron Letters* **20,** 13 (1960).
45. D. B. MacLean, R. H. F. Manske, and L. Marion, *Can. J. Res.* **28B,** 460 (1950).
46. D. B. MacLean and W. A. Harrison, *Can. J. Chem.* **37,** 1757 (1959).
47. W. A. Harrison, M. Curcumelli-Rodostamo, D. F. Carson, L. R. C. Barclay, and D. B. MacLean, *Can. J. Chem.* **39,** 2086 (1961).
48. L. R. C. Barclay and D. B. MacLean, *Can. J. Chem.* **34,** 1519 (1956).
49. W. A. Harrison, Ph.D. Thesis, McMaster University, September 1960.
50. W. N. French and D. B. MacLean, *Can. J. Chem.* **39,** 2100 (1961).
51. G. S. Perry and D. B. MacLean, *Can. J. Chem.* **34,** 1189 (1956).
52. A. Bertho and A. Stoll, *Ber.* **85,** 663 (1952).
53. O. Achmatowicz and W. Rodewald, *Roczniki Chem.* **29,** 509 (1955).
54. F. A. L. Anet and N. H. Khan, *Can. J. Chem.* **37,** 1589 (1959).
55. R. H. Burnell and D. R. Taylor, *Tetrahedron* **15,** 173 (1961).
56. B. Douglas, D. G. Lewis, and L. Marion, *Can. J. Chem.* **31,** 272 (1953).
57. W. A. Ayer, B. Altenkirk, S. Valverde-Lopez, B. Douglas, R. F. Kaffauf, and J. A. Weisbach, *Can. J. Chem.* **46,** 15 (1968).
58. W. A. Ayer, J. A. Berezowsky, and D. A. Law, *Can. J. Chem.* **41,** 649 (1963).
59. R. H. Burnell and D. R. Taylor, *Chem. Ind. (London)* 1239 (1960).
60. M. Curcumelli-Rodostamo and D. B. MacLean, *Can. J. Chem.* **40,** 1068 (1962).
61. W. A. Ayer and G. G. Iverach, *Tetrahedron Letters* 87 (1962).
62. W. A. Ayer and G. G. Iverach, *Can. J. Chem.* **42** 2514 (1964).
63. F. A. L. Anet, M. Ahmad, and N. H. Khan, *Can. J. Chem.* **40,** 236 (1962).
64. R. H. Burnell and D. R. Taylor, *Chem. Ind. (London)* 1399 (1961).
65. W. A. Ayer and D. A. Law, *Can. J. Chem.* **40,** 2088 (1962).
66. W. A. Ayer, W. R. Bowman, P. Kebarle, and R. H. Burnell, *Can. J. Chem.* **43,** 328 (1965).

67. W. A. Ayer, A. N. Hogg, and A. C. Soper, *Can. J. Chem.* **42,** 949 (1964).
68. J. C. F. Young and D. B. MacLean, *Can. J. Chem.* **41,** 2731 (1963).
69. Z. Valenta, P. Deslongchamps, R. A. Ellison, and K. Wiesner, *J. Amer. Chem. Soc.* **86,** 2533 (1964).
70. D. B. MacLean and M. Curcumelli-Rodostamo, *Can. J. Chem.* **44,** 611 (1966).
71. D. B. MacLean, *Can. J. Chem.* **41,** 2654 (1963).
72. Y. Inubushi, H. Ishii, B. Yasui, and T. Harayama, *Tetrahedron Letters,* 1551 (1966).
73. Y. Inubushi, H. Ishii, and T. Harayama, *Chem. Pharm. Bull. (Tokyo)* **15,** 250 (1967).
74. R. H. Burnell, *J. Chem. Soc.* 3091 (1959).
75. R. H. Burnell, C. G. Chin, B. S. Mootoo, and D. R. Taylor, *Can. J. Chem.* **41,** 3091 (1963).
76. H. Ishii, B. Yasui, T. Harayama, and Y. Inubushi, *Tetrahedron Letters* 6215 (1966).
77. R. H. Burnell and B. S. Mootoo, *Can. J. Chem.* **39,** 1090 (1961).
78. Y. Inubushi, H. Ishii, T. Harayama, R. H. Burnell, W. A. Ayer, and B. Altenkirk, *Tetrahedron Letters* 1069 (1967).
79. B. P. Moore and L. Marion, *Can. J. Chem.* **31,** 952 (1953).
80. W. A. Ayer, J. A. Berezowsky, and G. G. Iverach, *Tetrahedron* **18,** 567 (1962).
81. F. A. L. Anet and C. R. Eves, *Can. J. Chem.* **36,** 902 (1958).
82. S. N. Alam, K. A. H. Adams, and D. B. MacLean, *Can. J. Chem.* **42,** 2456 (1964).
83. V. Deulofeu and J. DeLanghe, *J. Amer. Chem. Soc.* **64,** 968 (1942).
84. W. A. Ayer, T. E. Habgood, V. Deulofeu, and H. R. Juliani, *Tetrahedron* **21,** 2169 (1965).
85. Z. Valenta, H. Yoshimura, E. F. Rogers, M. Ternbah, and K. Wiesner, *Tetrahedron Letters* **10,** 26 (1960).
86. W. A. Ayer, J. K. Jenkins, S. Valverde-Lopez, and R. H. Burnell, *Can. J. Chem.* **45,** 433 (1967).
87. W. A. Ayer, J. K. Jenkins, K. Piers, and S. Valverde-Lopez, *Can. J. Chem.* **45,** 445 (1967).
88. W. A. Ayer and K. Piers, *Can. J. Chem.* **45,** 451 (1967).
89. H. Dugas, M. E. Hazenburg, Z. Valenta, and K. Wiesner, *Tetrahedron Letters* 4931 (1967).
90. K. Wiesner and L. Poon, *Tetrahedron Letters* 4937 (1967).
91. E. H. W. Böhme, Z. Valenta, and K. Wiesner, *Tetrahedron Letters* 2441 (1965).
92. K. Wiesner, I. Jirkovsky, M. Fishman and C. A. J. Williams, *Tetrahedron Letters* 1523 (1967).
93. K. Wiesner and I. Jirkovsky, *Tetrahedron Letters* 2077 (1967).
94. E. E. Betts and D. B. MacLean, *Can. J. Chem.* **35,** 211 (1957).

17

The Diterpene Alkaloids

S. W. PELLETIER AND L. H. KEITH
Department of Chemistry, University of Georgia
Athens, Georgia

1. GENERAL INTRODUCTION

Alkaloids modeled on diterpene skeleta[1] have been isolated from several genera of the families Ranunculaceae *(Aconitum, Delphinium)* and Garryaceae *(Garrya)* and from a single specie each of the Compositae *(Inula royleana)* and Rosaceae family *(Spiraea Japonica)*. These compounds are widely distributed throughout the plant world and have long been of interest because of their pharmacological properties and complex structures. Many of the compounds are intensely poisonous—the classical aconitine and pseudaconitine are among the most toxic materials of plant origin known to man. Extracts of the roots and leaves of *Aconitum* species were employed in ancient times as animal poisons, for treatment of such maladies as neuralgia, hypertension, gout and rheumatism, and in the medieval trials by ordeal.

The diterpene alkaloids may be divided into two broad catagories. The first group* includes a series of comparatively simple and relatively nontoxic amino alcohols (alkamines) that are modeled on a C_{20}-skeleton. These compounds, which are the subject of this chapter, are not extensively oxygenated and contain at most one methoxyl group. One of the distinguishing chemical features of this group is the formation of phenanthrenes

*This class is sometimes loosely referred to as the "Atisines," In this chapter a more restricted classification is used. The term "Atisines" refers only to those compounds modeled on an atisine skeleton.

when subjected to selenium or palladium dehydrogenation. A few of these compounds occur in the plant as monoesters of acetic or benzoic acid.

The second group (see Chapter 18) comprises the highly toxic ester bases called the *Aconitines*. These compounds are more heavily substituted than the C_{20}-bases, particularly by methoxyl and hydroxyl groups. Usually two of the alcoholic groups are esterified, one with acetic acid and the other with an aromatic acid such as benzoic or veratric acid. Hydrolysis of the aconitines furnishes the relatively nontoxic, parent alkamines known as aconines. The aconines are modeled on a hexacyclic C_{19}-skeleton rather than a C_{20}-skeleton. A number of alkaloids, which are intermediate in toxicity and substitution between the simple C_{20}-bases and the aconitines, but are normally included with the latter class because they have C_{19}-skeletons, are known. These are methoxyl-containing monoesters of the alkamine with benzoic or anthranilic acid derivatives. Examples of this intermediate class are benzoyl heteratisine and esters of lycoctonine.

The diterpene alkaloids are derived from tetra- or pentacyclic diterpenes in which carbon atoms 19 and 20 are linked with the nitrogen of a molecule of β-amino-ethanol, methyl- or ethylamine to form a heterocyclic ring. Thus far, four different types of skeleta* have been encountered among the diterpene alkaloids. They are the VEATCHINE (1), ATISINE (2), LYCOCTONINE (3), and HETERATISINE (4) types (see chart of skeletons). The VEATCHINE skeleton, which occurs in the Garrya alkaloids, e.g., veatchine, cuauchichicine, and songorine, incorporates a KAURANE (5) skeleton and obeys the isoprene rule. The ATISINE skeleton is modelled on an ATISANE (6) nucleus and differs from the veatchine type in that ring D is six- rather than five-membered; it does not obey the isoprene rule. The atisine skeleton appears in such alkaloids as atisine, atidine, hetisine, ignavine, and kobusine. The LYCOCTONINE skeleton, modelled on the ACONANE (7) framework, is found in one form or other in most of the aconitines and lycoctonines, e.g., aconitine, hypaconitine, delphinine, lycoctonine, and ajacine. It may be derived formally from the atisine skeleton by cleavage of the C(8)–C(9) bond, formation of new bonds between C(7)–C(20) and C(9)–C(15) and loss of the C(17) exocyclic methylene group. The HETERATISINE skeleton differs from the lycoctonine type in that expansion of ring-C by insertion of oxygen has occurred to give a lactone. All the diterpene alkaloids encountered

*The nomenclature of these alkaloids is based upon the standard skeletons ATISANE, KAURANE, and ACONANE with the numbering and stereochemistry illustrated. These skeletons have been incorporated into a proposal for the Common and Systematic Nomenclature of Cyclic Diterpenes which will be submitted to the IUPAC Commission on the Nomenclature of Organic Chemistry in 1969 and published in M. Fetizon and Le-Van-Thoi's forthcoming book on the cyclic diterpenes (Holden-Day).

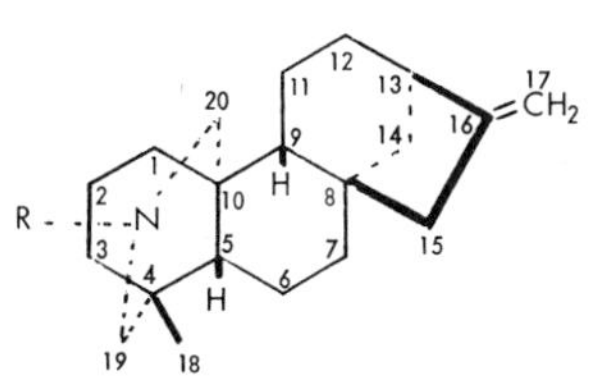

(1) VEATCHINE SKELETON

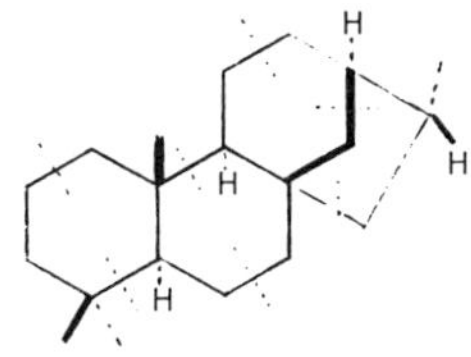

(5) KAURANE

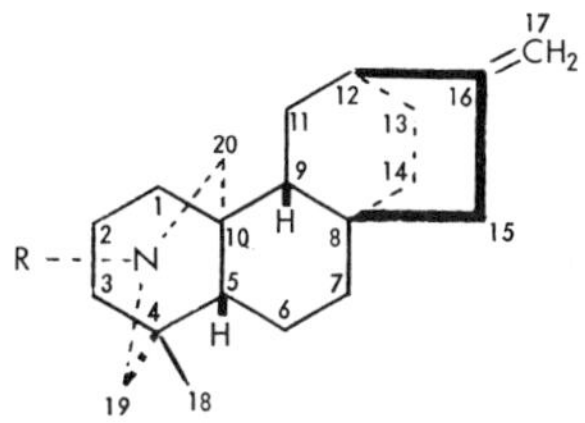

(2) ATISINE SKELETON

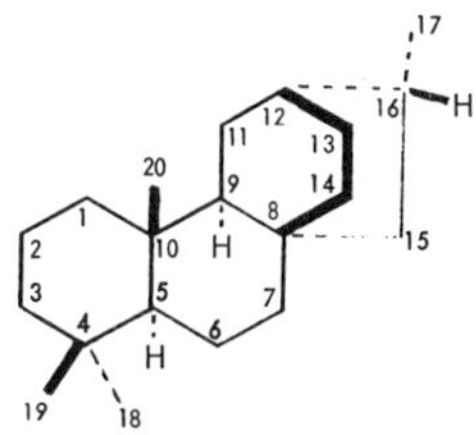

(6) ATISANE

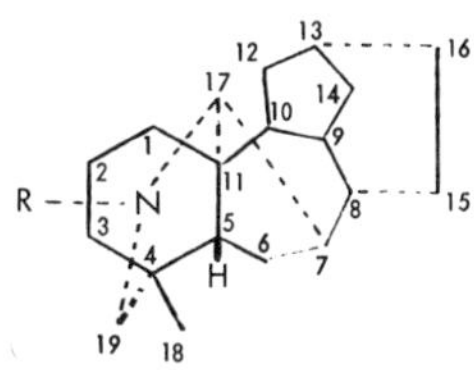

(3) LYCOCTONINE SKELETON

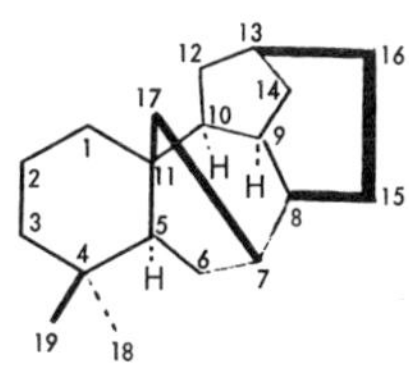

(7) ACONANE

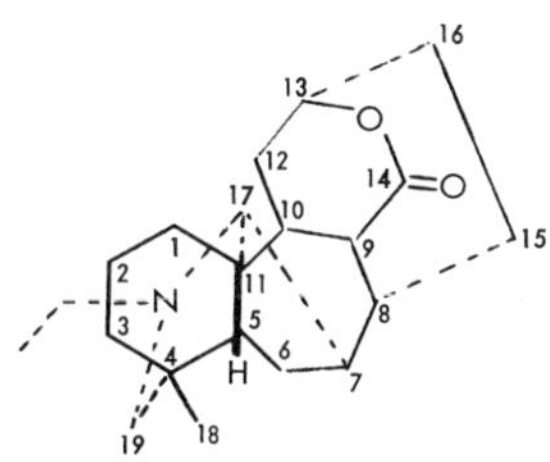

(4) HETERATISINE SKELETON

to date in nature are constructed on these four skeletal types. In certain alkaloids, however, one or more additional ring fusions are present. Thus, songorine has a bond between C(7) and C(20) of the veatchine skeleton and kobusine has bonds between C(14) and C(20) and between C(6) and the nitrogen of the atisine skeleton.

Careful chemical investigation of the alkaloids of species of *Aconitum* and *Delphinium* was initiated by Walter A. Jacobs and his collaborators at the Rockefeller Institute in 1942. These workers carried out a number of meticulous degradations that laid the foundation for subsequent contributions in this field. In later work on the Garrya alkaloids, K. Wiesner paralleled Jacobs' early studies on atisine and pointed out the striking similarity in the chemistry of the two groups of alkaloids. From then on advances in one field greatly assisted progress in the other. Since structures of the Garrya alkaloids were the first to be completed, these compounds are considered first.

2. THE GARRYA ALKALOIDS

The first report of crystalline bases from *Garrya* species is due to Oneto[2] who isolated the isomeric alkaloids veatchine and garryine from the bark of *G. Veatchii* Kellogg. Subsequent work has shown the presence of two other isomeric alkaloids, garryfoline and cuauchichicine, in the bark of the Mexican tree, *G. laurifolia* Hartiv.[3,4] Alkaloids possessing a veatchine skeleton also occur in certain *Aconitum* species, e.g., napelline and songorine.

2.1 Veatchine and Garryine

Veatchine (8), $C_{22}H_{33}NO_2$, is a strong tertiary base (pKa′ 11.5) containing one acylable hydroxyl, a C-CH_3 group and an exocyclic methylene group. On treatment with base veatchine is converted into the isomeric garryine (9) (pKa′8.7). Reduction of either alkaloid with $LiAlH_4$ results in hydrogenolysis of the ether bridge to give the same dihydroveatchine (10) (pKa′6.9), and due to hydrogenolysis of the ether ring and saturation of the exocyclic methylene group, catalytic hydrogenation of either alkaloid gives tetrahydroveatchine (11) (pKa′6.8). Both dihydro- and tetrahydroveatchine show two active hydrogens and give O,O-diacetate derivatives.[5–7]

Selenium dehydrogenation of veatchine or garryine yields 1-methyl-7-

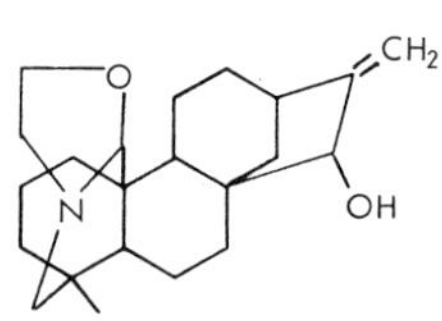

(8)

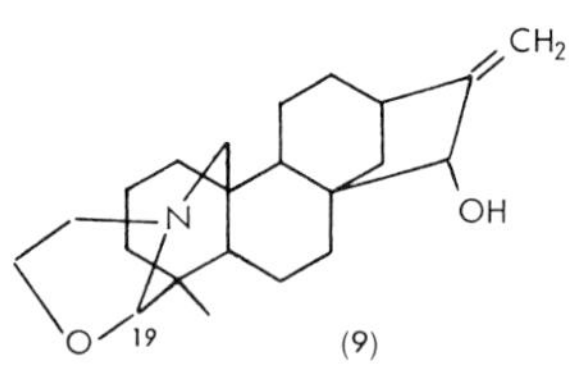

(9)

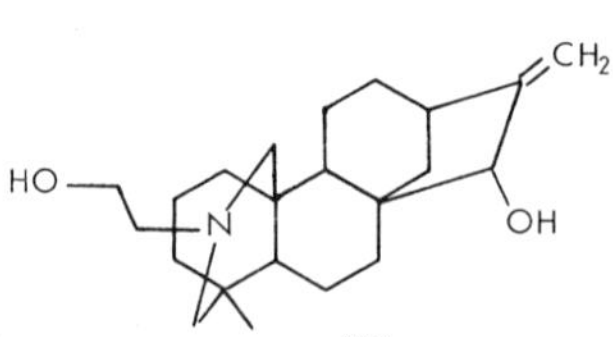

(10)

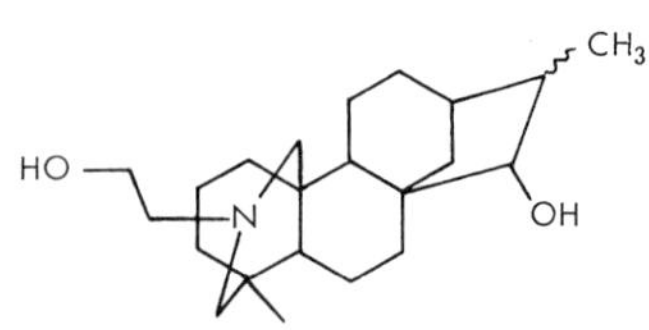

(11)

ethyl phenanthrene (12) and l-methyl-7-ethyl-3-azaphenanthrene (13); both products were identified by spectral studies and by synthesis.[5,6,8,9] On the basis of these products and other degradations to be described, together with biogenetic considerations, structure (10) was proposed for dihydroveatchine. Veatchine and garryine were then formulated as the isomeric cyclic ethers (8) and (9), respectively, each of which on reduction with $LiAlH_4$ would yield dihydroveatchine (10). These structural assignments were confirmed by pKa studies and by the results of extensive degradations that are briefly summarized.

Pyrolysis of veatchine or garryine with selenium at 290° gave the isomeric pyrolysis bases 14 and 15. Compound 14 shows absorption for the imine linkage at 1650 cm^{-1} and for a cyclopentanone ring at 1735 $cm.^{-1}$ The latter originates from isomerization of the allylic alcohol function. Reduction of 15 with $LiAlH_4$ gave the secondary base (16), which was alkylated with ethylene chlorohydrin to give dihydroveatchine (10). Oxidation of the latter with one equivalent of osmium tetroxide in ether afforded garryine (9), cyclization occurring at the less hindered 19-position. These reactions demonstrate the presence of an oxazolidine moiety in veatchine and garryine.[6,7,10]

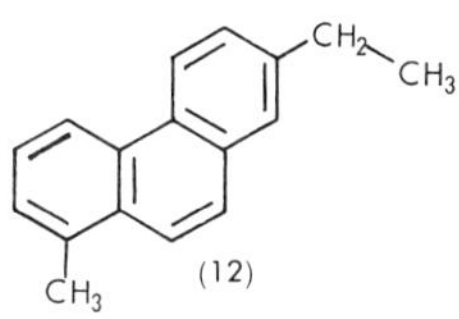

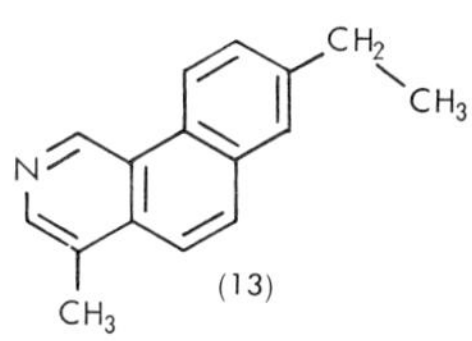

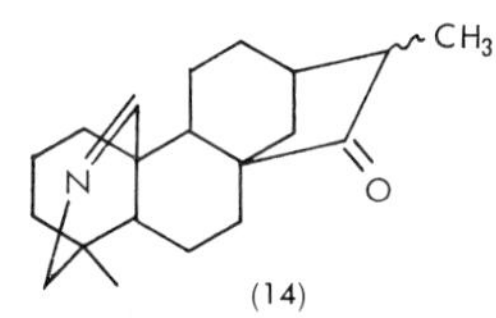

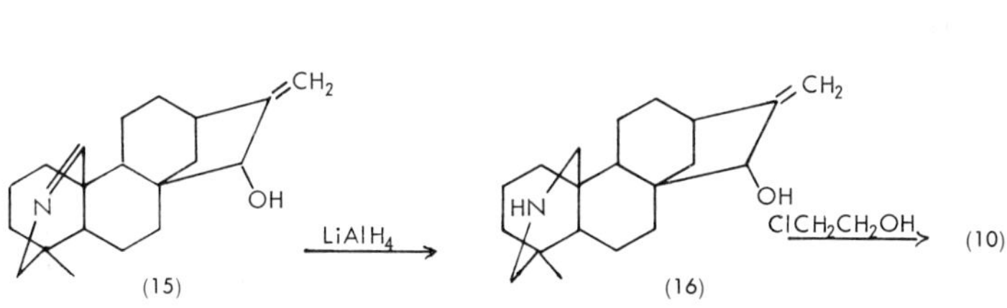

Oxidation of veatchine with potassium permanganate in acetone furnishes a γ-lactam (1700 cm^{-1}), oxoveatchine-A (17), and an isomeric δ-lactam (1630 cm^{-1}), oxoveatchine-B (18). Under the same conditions, garryine gives the δ-lactam (1618 cm^{-1}), oxogarryine (20), presumably via oxidation of the intermediate garryine hydrate (19). Reduction of oxygarryine with $LiAlH_4$ furnishes dihydroveatchine (10).[5,6]

Under more vigorous conditions of permanganate oxidation, veatchine furnishes the lactam dicarboxylic acids (21) and (22), formed by cleavage of the five-membered ring bearing the—CHOH group. These acids are

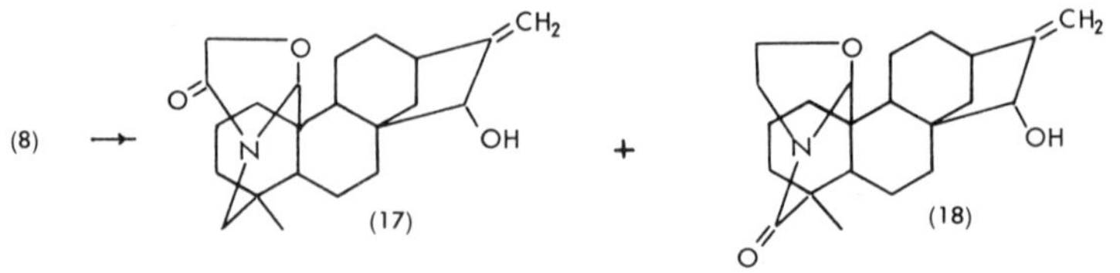

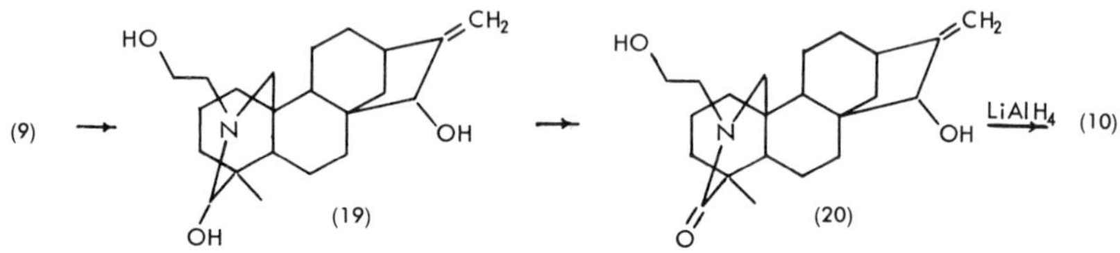

readily converted to the corresponding anhydrides (23) and (24), the infrared spectra of which are compatible with the presence of the γ-lactam (1700 cm^{-1}) and δ-lactam (1640 cm^{-1}), respectively, and of a glutaric anhydride grouping in each (1800, 1762, and 1800, 1775 cm^{-1}, respectively). The dimethyl esters (no hydroxyl absorption in infrared) or (21) and (22) on saponification give monomethyl esters that are resistant to further hydrolysis—an indication of the hindered nature of one of the carboxyl groups in each compound.[10]

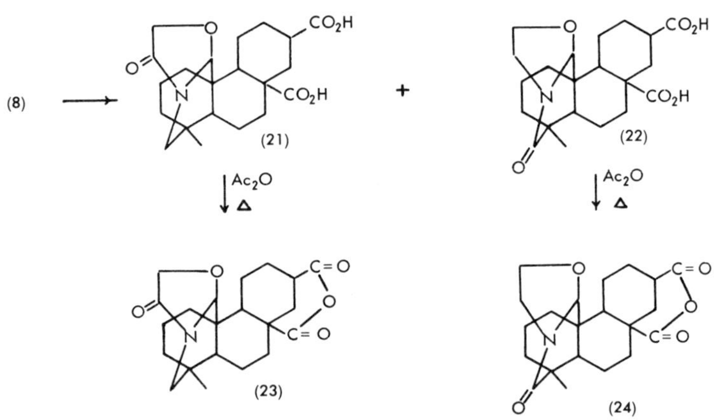

Dehydrogenation of acid (22) with selenium afforded pimanthrene (25) and 1-methylphenanthrene-7-carboxylic acid (26). The unusal conversion of the carboxyl group at C(13) to a methyl group during dehydrogenation may be explained by reduction of the anhydride carboxy group by the reducing medium of the dehydrogenation. Dehydrogenation of the triol (27), formed by lithium aluminum hydride reduction of the dimethyl ester of 22, also led to pimanthrene (25).[10]

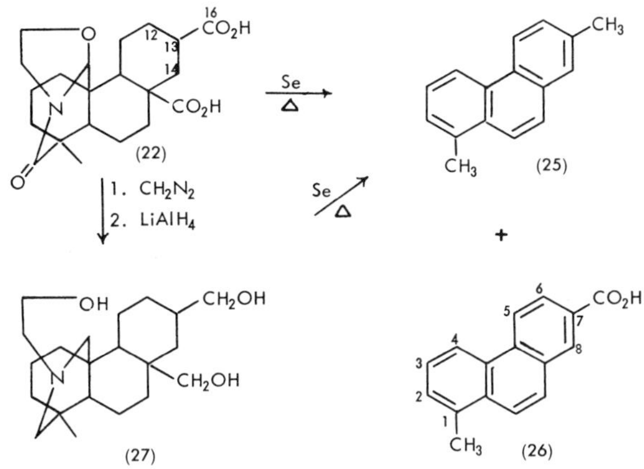

The reasons for the strong basicity of veatchine and the detailed stereochemistry of these compounds is considered in Section 5.

2.2 Cuauchichicine and Garryfoline[3,4]

These alkaloids of *Garrya laurifolia* Hartiv. are isomeric with veatchine and garryine. The infrared spectrum of cuauchichicine shows the absence of—OH or—NH absorption but a strong carbonyl band at 1730 cm^{-1} that can be attributed to a cyclopentanone ring. Kuhn–Roth determinations demonstrated the presence of two C-methyl groups and the lack of an exocyclic methylene group was confirmed by ozonization experiments. Selenium pyrolysis of cuauchichicine at 290° yielded a base (14), identical with that obtained earlier from veatchine by similar treatment. Moreover, reduction of the alkaloid with $LiAlH_4$ gives tetrahydroepiveatchine (29), a compound that has been synthesized from imine (14) by reduction to (30),

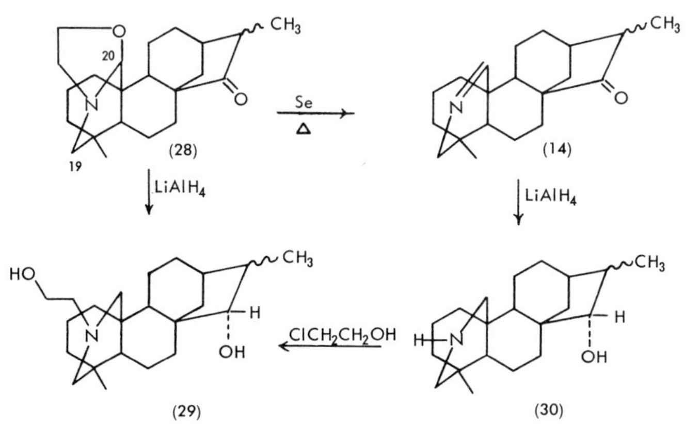

followed by alkylation with ethylene chlorohydrin. The termination point of the oxazolidine ring [C(19) or C(20)] was assigned on the basis of the pKa value for the alkaloid. Since the pKa′ value of cuauchichicine (11.15) is comparable with that of veatchine (11.5) rather than that of garryine (8.7), clearly in cuauchichicine the oxazolidine ring is fused at C(20) and the structure of the alkaloid may be formulated as (28).[3,4]

Garryfoline, $C_{22}H_{33}NO_2$, is similar to cuauchichicine in that it is a tertiary base containing a C-methyl group. However, it lacks a carbonyl group and contains instead a secondary hydroxyl and an exocyclic methylene group. That the skeleton of garryfoline is the same as that of cuauchichicine (28) has been demonstrated by its isomerization to the latter compound by acid. Garryfoline is reduced with $LiAlH_4$ to dihydrogarryfoline (32), a transformation analogous to the conversion of veatchine (8) to dihydroveatchine (10). Like garryfoline, the dihydroderivative (32) is isomerized with acid to a ketone, dihydrocuauchichicine (33) that is also obtained by catalytic reduction of cuauchichicine (28). On reduction with $LiAlH_4$, dihydrocuauchichicine (33) yields tetrahydroepiveatchine (34), also obtainable by catalytic hydrogenation of garryfoline itself. These transformations indicate that garryfoline must, therefore, have the structure of 15-epiveatchine (31). The attachment of the oxazolidine ring at

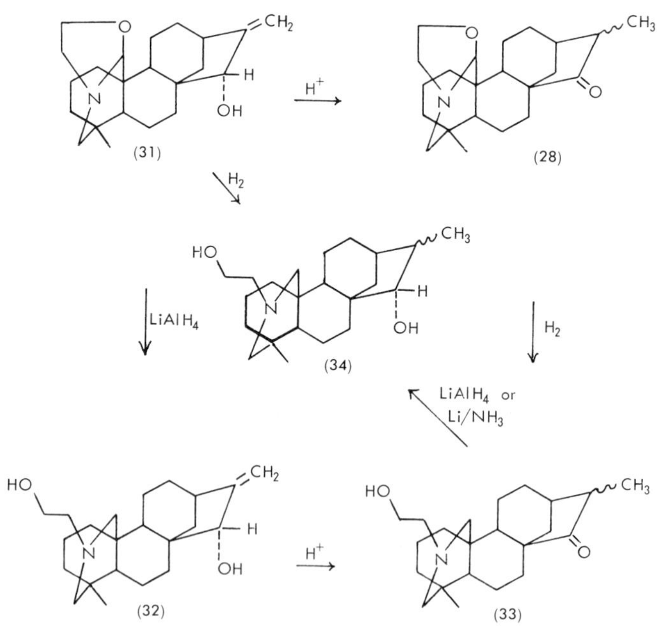

C(20) follows from a comparison of the pKa′ of garryfoline (11.8) with veatchine (11.5).[3,4]

Base catalyzed isomerization, comparable to the conversion of veatchine to garryine, also occurs with cuauchichicine and garryfoline. Thus, these alkaloids are isomerized readily by base to isocuauchichicine (35) and isogarryfoline (36), respectively—compounds in which the oxazolidine ring is attached at C(19). The products (35) and (36) are reduced by $LiAlH_4$ to tetrahydroepiveatchine (34) and dihydrogarryfoline (32), respectively.[3,4]

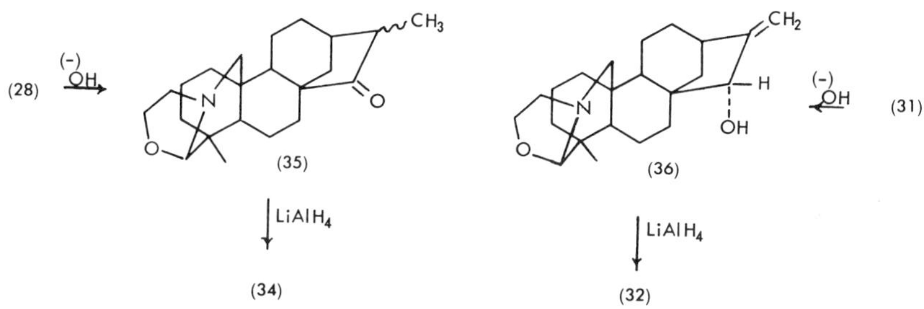

The acid catalyzed isomerization of garryfoline (31) to cuauchichicine (28) and of dihydrogarryfoline (32) to dihydrocuauchichicine (33) requires comment in view of the stability of veatchine to acidic conditions. This difference in behavior must be due to the difference in configuration of the C(15) hydroxyl group. The garryfoline → cuauchichicine isomerization may be formally represented as (37) → (39). The question as to which of the C(15) epimeric alcohols (veatchine or garryfoline) is thermodynamically more stable has been established by reduction of dihydrocuauchichicine (33) with lithium and alcohol in liquid ammonia,[3,4] conditions known to lead to the more stable epimeric alcohol.[11] The sole product is tetrahydroepiveatchine (34), previously obtained by hydride reduction of (33) or catalytic reduction of (31). Thus, garryfoline is thermodynamically more stable than veatchine. Yet garryfoline is the epimer that undergoes rearrangement of cuauchichicine. The isomerization can proceed only when the four centers involved are coplanar and the nonclassical carbonium ion (41) from garryfoline (40) has been invoked as an intermediate in this isomerization to the methyl ketone (42). In this cation, the quasiaxial (with respect to the 5-membered D-ring) C(15) hydrogen atom is coplanar with the bond joining C(12) and C(16). This leads to the complete relative stereochemistry for garryfoline as (37) ≡ (40); the hydroxyl groups at C(15) in garryfoline and veatchine are respectively quasi-equatorial and quasi-axial with reference to ring D.[3,4] The mechanism of

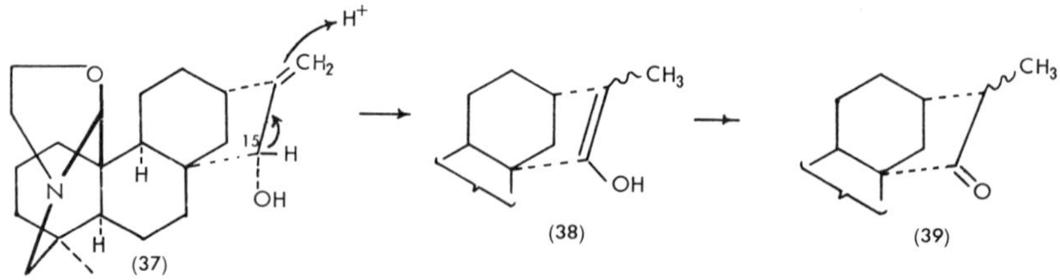

(37) (38) (39)

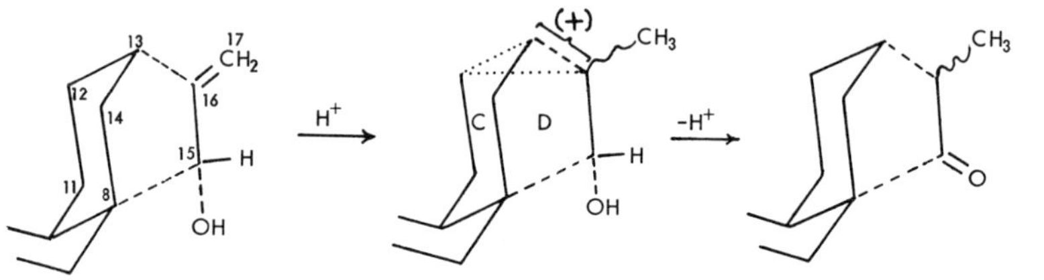

(40) (41) (42)

this reaction has been investigated recently using the epimeric (−)-kaur-16-en-15-ols as models.[12]

2.3 Napelline, Songorine*, and Lucidusculine

Napelline, $C_{22}H_{33}NO_3$, and songorine, $C_{22}H_{31}NO_3$, occur in *Aconitum napellus* L. **[13,14] The latter is a ketonic base which on reduction with $LiAlH_4$ yields napelline (luciculine), the corresponding secondary alcohol. Each base contains an N-ethyl, C-methyl, and exocyclic methylene group. The carbonyl group of songorine is part of a six-membered ring (1712 cm^{-1}). Napelline is readily isomerized to isonapelline, which contains a cyclopentanone system (1740 cm^{-1}) and two C-methyl groups, and in these respects resembles garryfoline. The presence of a veatchine-type skeleton in these bases is indicated by the formation 1-methyl-7-ethyl-3-azaphenanthrene (13) on dehydrogenation of isonapelline. Selenium dehydrogenation of songorine gives 1,9-dimethyl-7-ethyl-phenanthrene (46). The chemistry of napelline (43a), songorine (44) and isonapelline (45) will be discussed in terms of the structures† ultimately derived.[18–21]

Silver oxide converts dihydronapelline (47) to a weakly basic carbinolamine ether (48),‡ pKa′6.77, which forms a ternary iminium salt

*Also known as napellonine and shimoburo-base 1. Napelline is identical with luciculine.

**Songorine also occurs in *A. Soongaricum*[15] and various Japanese *Aconitum* species.[16,17]

†Wiesner assigned the hydroxyl in ring A to the 2-position. Later work by Sugasawa supports the 1-hydroxyl structure.

‡In order to be able to form a carbinolamine ether by oxidation of the C(1) hydroxyl, models show this hydroxyl must be α-equatorial and that ether formation is accompanied by an inversion of ring A from a chair to a boat form.

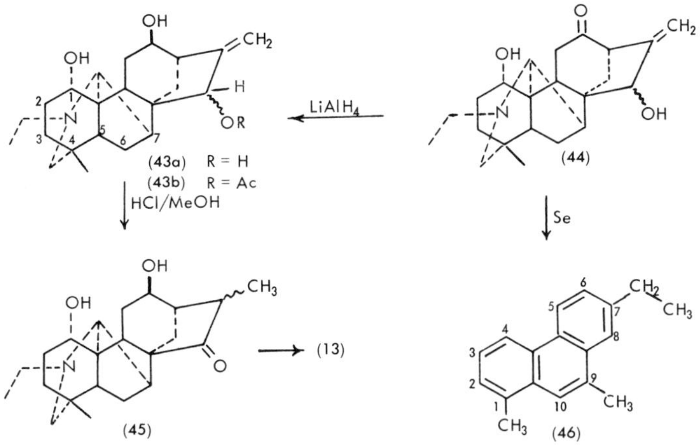

(49) (1670 cm^{-1}). Reduction of (48) with $LiAlH_4$ regenerates dihydronapelline. Oxidation of (47) with CrO_3–pyridine gives a triketolactam (50) [1642, 1740, 1712 cm^{-1}].[19]

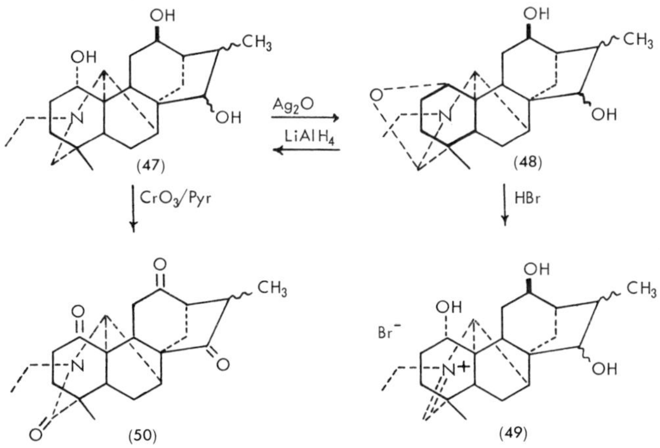

Isosongorine (51) [1736, 1695 cm^{-1}] available from the allylic rearrangement of songorine (44) with palladium-charcoal, gave on Huang–Minlon reduction the desoxy derivative (52). Oxidation of the latter with CrO_3–pyridine gave a keto δ-lactam (53) [1707, 1635 cm^{-1}]. The latter gave an amorphous monobenzylidine derivative and exchanged 1.82 atoms of deuterium when equilibrated in the presence of NaOD in MeOD and D_2O. These results indicate the presence of one methylene group adjacent to the keto function and therefore limit the site of the ring-A hydroxyl to either C(1) or C(3). To distinguish between these possibilities the diketolactam (54) was treated with isoamylnitrite to give the isonitroso-derivative (55). Cleavage of the latter with benzenesulfonyl chloride and alkali gave a nitrile (56) that was stable at its melting point and did not decarboxylate in boiling ethanolic hydrogen chloride. Had the keto function been at C(2), the expected product would be an easily decarboxylated malonamide (57). These results permit assignment of the hydroxyl in ring A to C(1); napelline and songorine are, therefore, represented by structures (43a) and (44), respectively.[20,21] The structure and absolute configuration of lucidusculine (43b) has been determined by x-ray analysis.[18,22]

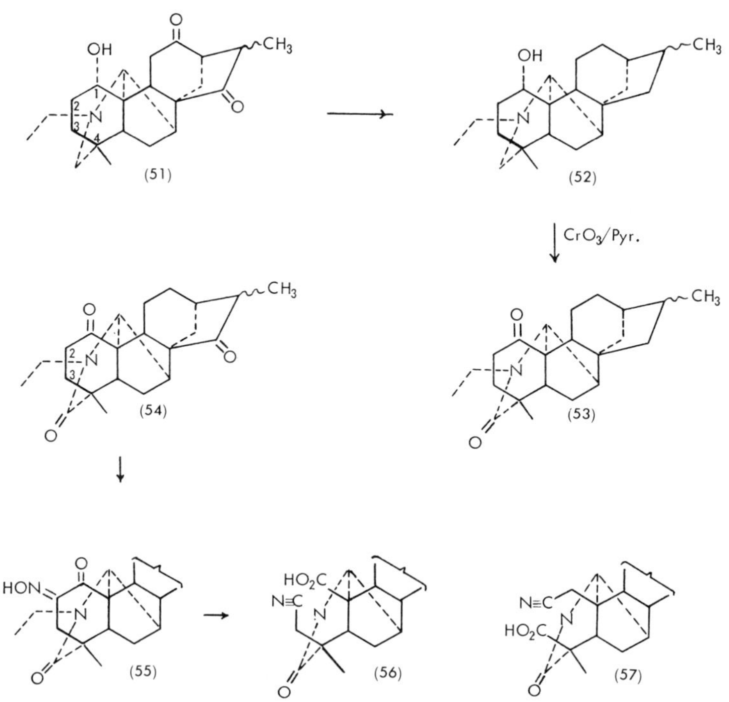

3. THE ATISINE ALKALOIDS

There is a close parallelism between the chemistry of the Garrya alkaloids and of the atisine bases. Recognition of this relationship greatly simplified the understanding of the structures of both groups of compounds. In this section, the chemistry of alkaloids possessing an atisine-type skeleton are discussed. Alkaloids possessing an atisine skeleton with extra ring fusions are discussed in Section 6.

3.1 The Structure of Atisine

3.1.1 General Structural Considerations Atisine, $C_{22}H_{33}NO_2$, the predominant alkaloid of *Aconitum heterophyllum*, is a tertiary base that contains an exocyclic methylene (1647, 984 cm^{-1}, $\tau = 5.0$, 2H-mult.), a secondary hydroxyl (3067 cm^{-1}), and a C-methyl group (1370 cm^{-1}, $\tau = 9.30$). N-alkyl determinations indicate the presence of an N-ethyl group or suitable precursor in the molecule.[23–25] Atisine is isomeric with the Garrya alkaloids and closely resembles them in its chemical behavior. Thus, atisine is a very strong base (pKa′12.8) that may be easily isomerized to isoatisine (pKa′10.3).[26,27] Reduction with hydride gives a dihydro derivative analogous to dihydroveatchine (10).[28] Studies on the tetrahydro derivative of atisine have shown the presence of a β-hydroxyethyl group attached to the nitrogen.[28] Oxidation of isoatisine with potassium permanganate gives a lactam, oxoisoatisine, $C_{22}H_{33}NO_3$, analogous to oxogarryine (20). Atisine contains one active hydrogen and on hydrogenation gives a mixture of tetrahydroderivatives.

3.1.2 Skeleton of Atisine One important difference noted between the two groups of alkaloids is their behavior on selenium dehydrogenation. While veatchine gave 1-methyl-7-ethyl-phenanthrene and 1-methyl-7-ethyl-3-azaphenanthrene, atisine gave 1-methyl phenanthrene, 1-methyl-6-ethyl-phenanthrene (62) and 6-ethyl-1-methyl-3-azaphenanthrene (61).[23,29,30] This behavior prompted the suggestion that the difference between atisine and the Garrya alkaloids lies in the position of attachment of ring D. In particular structure (58) was suggested for dihydroatisine and structures (59) and (60) for atisine and isoatisine, respectively.[8,31] Evidence for specific structural features as well as the stereochemistry of atisine are now considered.

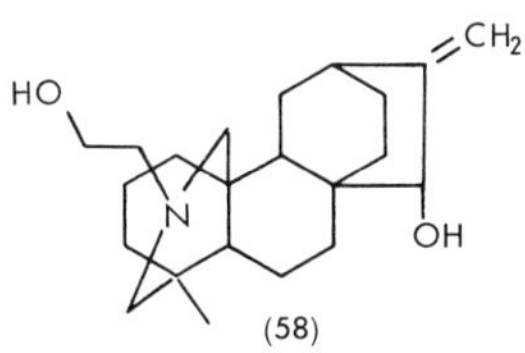

(58)

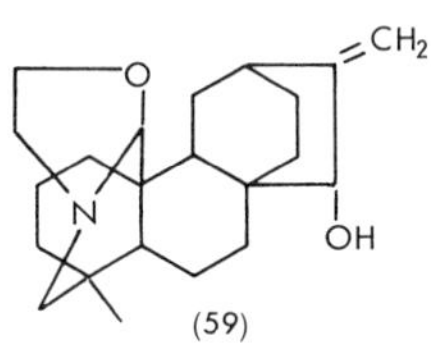

(59)

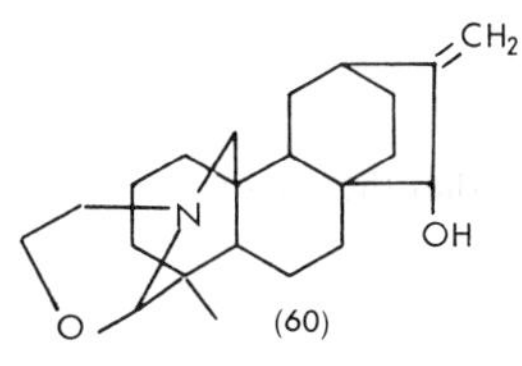

(60)

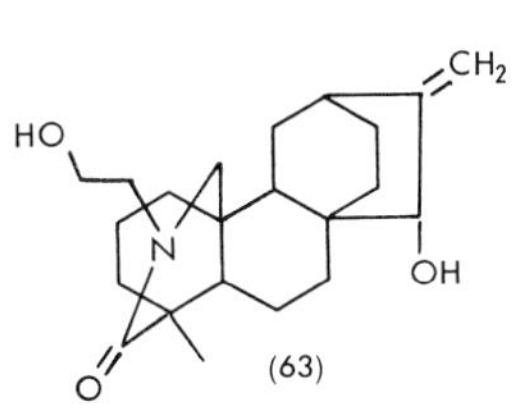

(63)

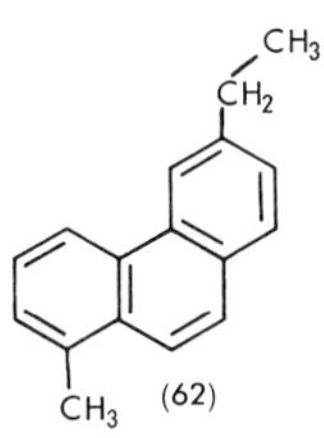

(62)

+

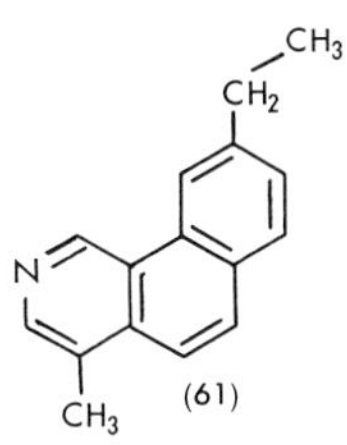

(61)

The relative positions of the secondary alcohol and exocyclic methylene group were originally assigned on the basis of the selenium dehydrogenation products from atisine and its derivatives. Thus atisine (59), oxoisoatisine (63) and tetrahydroatisine give 1-methyl-6-ethylphenanthrene (62), while the di- and tricarboxylic acids from isoatisine give 1,6-dimethylphenanthrene.[32] By analogy with veatchine (8), which gives 1,7-disubstituted phenanthrenes, the secondary hydroxyl and exocyclic methylene functions were located as in (59).

Definitive evidence on this point is available from two sources. Dvornik and Edwards[33,34] converted the axomethine alcohol (64) to the bisnor-keto acid (65), 1712 cm^{-1}, 1602 cm^{-1} Dibromination of the latter followed by dehydrohalogenation gave a crystalline phenol (66) and a keto γ-lactone (67) (1796, 1734, 1639 cm.$^{-1}$) [bridging of lactone ether oxygen either at C(11) or C(13)]. Formation of the latter, involving displacement of bromine α- to the ketone by carboxylate anion, demonstrates a 1,4-relationship of the ketone and carboxyl groups in (65). These results taken with dehydrogenation evidence for substitution at C(12) [C(6) on phenanthrene nucleus], show clearly that a bicyclo-[2,2,2] octane system is present and that the exocyclic methylene is located at C(16) [or C(13)]* and the secondary hydroxyl at C(15) [or C(14)],* respectively. Additional evidence on this point is available from the following sequence of reactions. Isomerization of isoatisine (60) with ethanolic hydrochloric acid gave a mix-

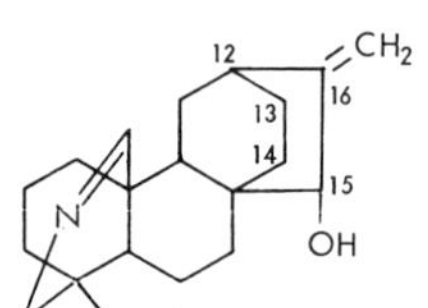

(64)

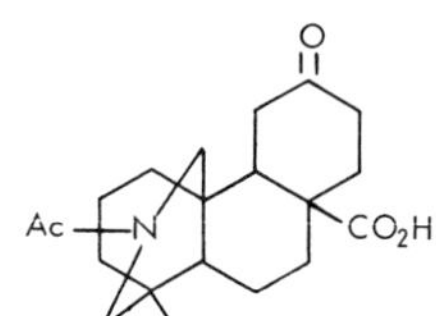

(65)

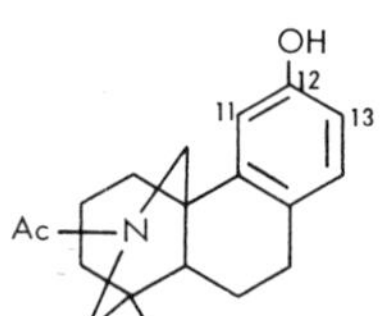

(66)

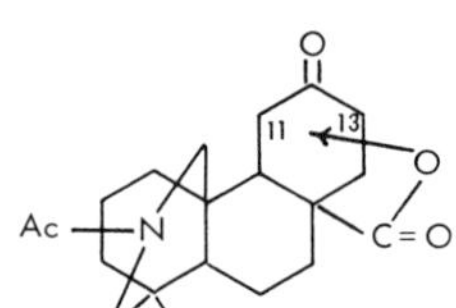

(67)

ture of epimeric 16-methyl ketones (68). (1715 and 1712 cm^{-1}). Wolff–Kishner reduction of (68) gave the desoxy derivative (69) and dehydrogenation of the latter with selenium afforded the epimeric azomethines (70), the lactam (71) and 1-methyl-6-isopropylphenanthrene (72). Had the

*Since the bicyclooctane system is symmetrical, the allylic alcohol system could be located on either the cis- or trans-branch.

methyl group in isoatisine been located at C(15) [or C(14)],* one would expect to obtain either 1,6-dimethylphenanthrene or 1-methyl-6-*n*-propylphenanthrene. The formation of the isopropyl derivative constitutes additional evidence for assigning the methylene to C(16) [or C(13)].*[35,36]

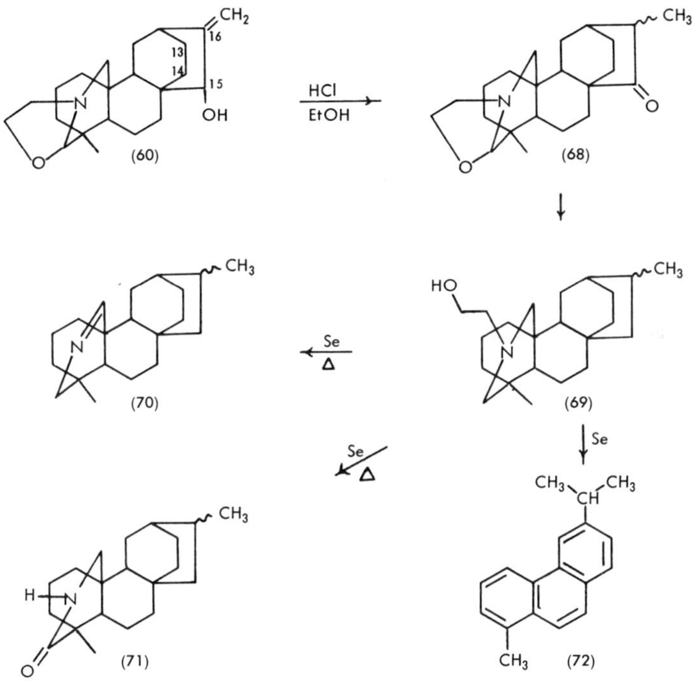

3.1.3 Oxazolidine Ring System Evidence for the existence of an oxazolidine ring in atisine and isoatisine derives from two experiments.[31,36] (a). Oxidation of atisine with permanganate in acetone gives lactam dicarboxylic acids (73) and (74)[32,37] whose methyl esters (75 and 76) show no band indicative of a hydroxyl group and give negative tests for active hydrogen. A similar oxidation of isoatisine furnishes oxoisoatisine dicarboxylic acid (77) [compare oxidation of garryine (9) to oxogarryine (20)] that has been related to (73) by catalytic reduction of the latter. (b). The second piece of evidence for the presence of an oxazolidine ring is the oxidation of both atisine and isoatisine to conjugated enones [(78) (λ_{max} 229 mμ, 9500) and (79) (λ_{max} 227 mμ, 8070), respectively] that show no hydroxyl absorption in the infrared and give negative tests for active hy-

*Since the bicyclooctane system is symmetrical, the allylic alcohol system could be located on either the cis- or trans-branch.

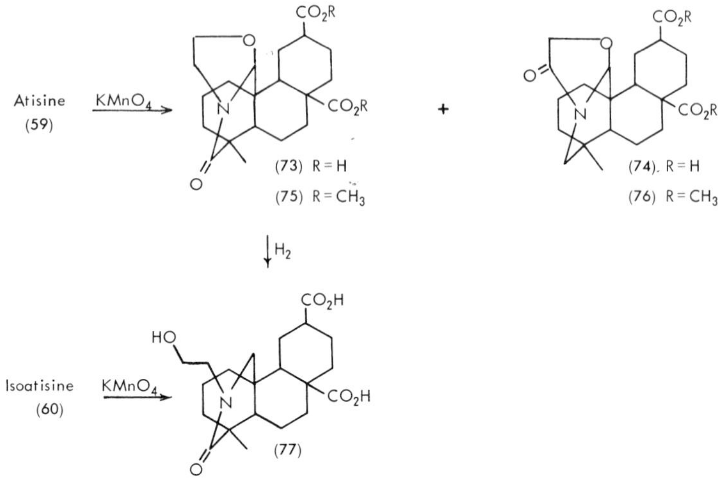

drogen. Refluxing compound (78) in methanol causes rearrangement of the oxazolidine ring to give a good yield of the iso-enone (79), a reaction which parallels the facile veatchine → garryine and atisine → isoatisine isomerizations.

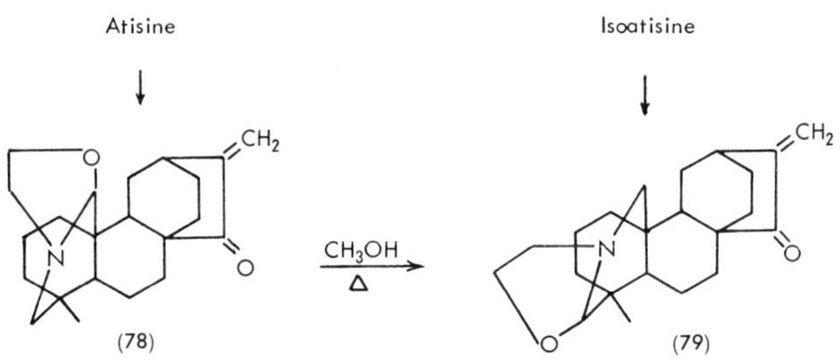

Final confirmation of the presence of an oxazolidine moiety in atisine was provided by reconstitution from the imino alcohol (80).[38] Reduction of (80) with sodium borohydride and alkylation of the resulting secondary amine with ethylene chlorohydrin gave the known dihydroatisine (58) in good yield. Cyclization of the β-hydroxyethyl group to give isoatisine (60) was effected with either one equivalent of osmium tetroxide in ether (cf. cyclization of dihydroveatchine to garryine) or one equivalent of mercuric acetate in aqueous acetic acid. The conversion of isoatisine to atisine has been effected by two different routes.[28,39,40] Atisine may also be obtained directly from (80) by treatment with ethylene chlorohydrin in dimethylformamide.[41] This reaction provides proof of the position of the imine linkage in 80.[41]

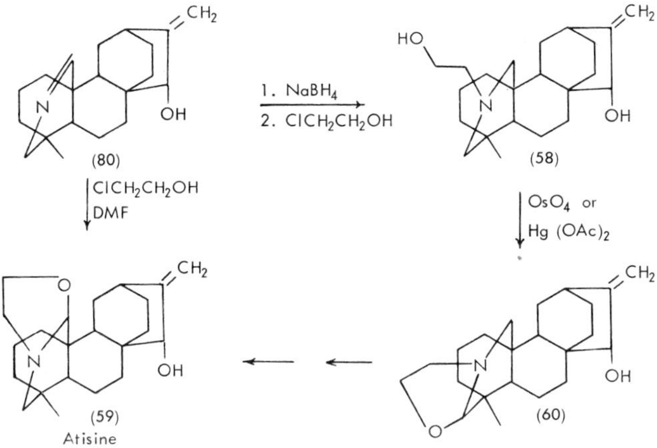

(80) (58) (59) Atisine (60)

3.1.4 Stereochemistry of Atisine At the time when the selenium dehydrogenation results discussed above were available it was assumed that atisine possessed the *trans*-anti skeleton that is common to most diterpenes.[33] Conformational arguments[42] relative to the structure of ajaconine have shown that A–B-*trans* stereochemistry is present in atisine. Reduction of ketones (81) and (83) with sodium borohydride gives in each case a pair of epimeric alcohols (82a, b) and (84a, b) each of which readily forms an O-acetate. This relatively unhindered character of the hydroxyl group supports location of the allylic alcohol group on the *trans* bridge of the bicycloöctane system, for if the group were on the *cis*-bridge only one epimer would be expected on reduction owing to the severe crowding at C(14). Moreover this view is confirmed by a study of the pKa's of the epimeric N-ethyl compounds (85a, b) and their acetates. Detailed experimen-

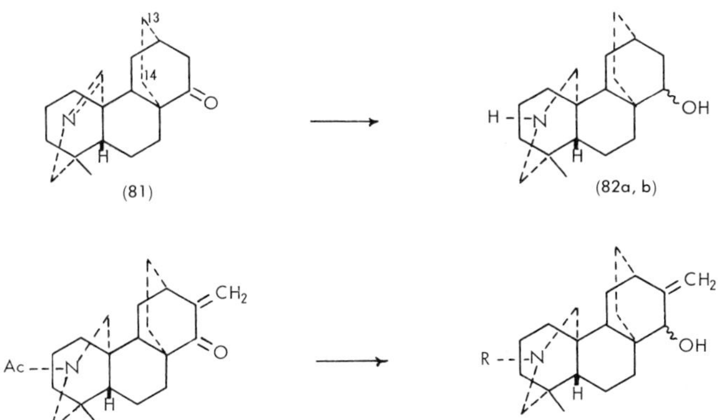

(81) (82a, b) (83) (84a, b) R = Ac (85a, b) R = Et

tal work indicates that the complete relative stereochemistry of atisine is expressed by structure (86).[33,35,36,42–44]

The assignment of configuration of the secondary hydroxyl in atisine is made difficult by the high degree of symmetry of the bicyclo[2,2,2] octane system. A tentative assignment of the β-configuration as in (86) has been made on the basis of the difference in the absorption of the epimeric alcohols (84a, b) on alumina.[44] The question of the absolute configuration of atisine is considered under Section 4.3.

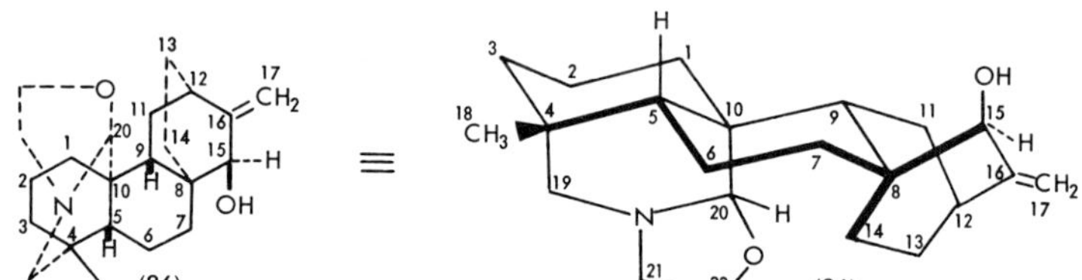

(86) ≡ (86)

3.2 The Structure of Atidine

Atidine, $C_{22}H_{33}NO_3$, another constituent of *Aconitum heterophyllum,* is a tertiary base possessing two hydroxyl groups (diacetate), a ketone function (oxime) in a six- or larger membered ring (1695 cm^{-1}), an exocyclic methylene group (3086, 1658, 900 cm^{-1}, τ = 4.9, 2H-double triplets) a C-methyl group (1376 cm^{-1}, $\tau = 9.23$) and an N—CH_2CH_2OH group. Atidine was shown to be a keto dihydroatisine (87) by Huang–Minlon reduction to dihydroatisine (58). A subsequent correlation of atidine and ajaconine via the dihydroderivative (see below) allows the keto function to be assigned to C(7) in atidine (87).[45–47]

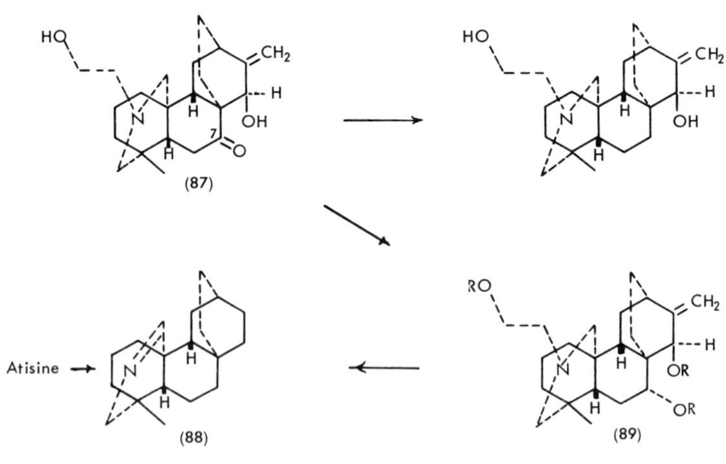

(87) (88) (89)

3.3 The Chemistry of Ajaconine

Ajaconine, $C_{22}H_{33}NO_3$, an alkaloid of *Delphinium ajacis* L., is of especial interest since it is the first reported example of an atisine derivative occurring in a *Delphinium* species. Ajaconine was shown to have the same carbocyclic skeleton as atisine by conversion to the oxygen-free axomethine base (88), obtained earlier from atisine.[48] That the allylic alcohol system of ajaconine has the same position and stereochemistry as in atisine was shown by reduction of atidine (87) (previously correlated with dihydroatisine) to a mixture of epimers, one of which is dihydroajaconine (89).[46,47]

The work of Dvornik and Edwards[43,49] has shown that ajaconine (90) contains a carbinol amine ether involving an oxygen atom at C(7). The N-methyl carbinolamine (91) derived from ajaconine was transformed by methanolic alkali into a mixture of the iso-compound (92) and a hydroxyl lactam (93) [3380, 1054, 1620 cm^{-1}]. The latter results from an intramolecular Cannizzaro-type reaction involving a transannular hydride transfer from C(20) to C(7). Oxidation of (93) followed by reduction gave the epimeric hydroxy lactam (94) with the hydroxyl (1082 cm^{-1}) in the original orientation of ajaconine. Bromination of the ketone (95), derived from ajaconine, followed by dehydrobromination gave an α,β-unsaturated ketone (96), 1664 cm^{-1}, λ_{max} 250 mμ (ϵ 9300) and 328 mμ (ϵ 96). This sequence locates the hydroxyl of (93) at C(7). That the carbinolamine function in ajaconine is derived from this 7-oxygen function is indicated by other transformations.

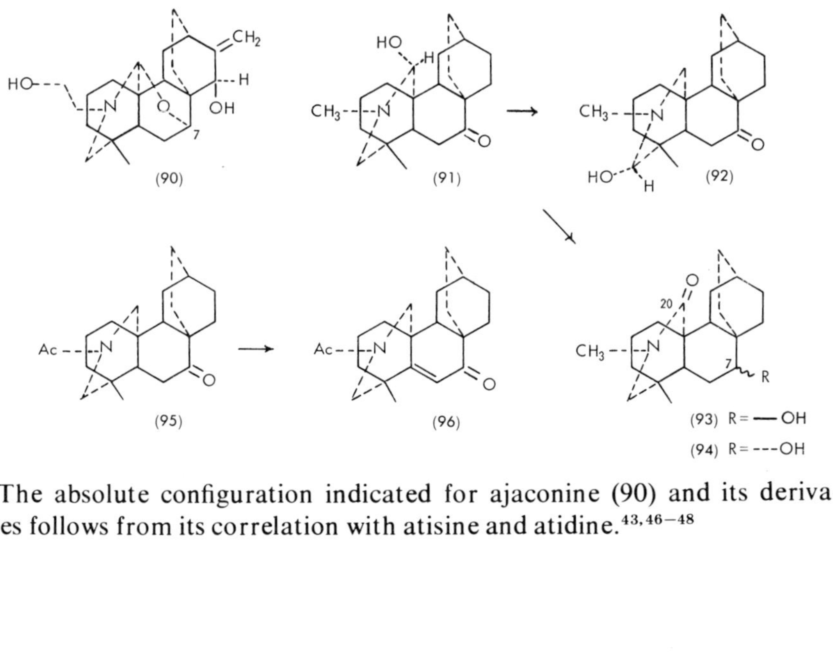

The absolute configuration indicated for ajaconine (90) and its derivatives follows from its correlation with atisine and atidine.[43,46–48]

4. CORRELATIONS AND ABSOLUTE STEREOCHEMISTRY OF ATISINE AND GARRYA ALKALOIDS

4.1 Correlation of Atisine and Garrya Alkaloids

The Atisine and Garrya alkaloids have been interrelated by converting both atisine and veatchine, by a parallel sequence of degradations to the same N-acetyl ester (103).[50,51] The respective azomethine acetates (97a and 97b) derived from atisine [12–16 bond] and veatchine [13–16 bond] were converted to the N-acetyl derivatives (98a and 98b) by reduction, acetylation and saponification. Oxidation of 98a and 98b with permanganate-periodate under controlled conditions gave the respective carboxylic acids (99a and 99b). Saponification of the dimethylesters (100a and 100b) gave 101a and 101b, respectively, which were transformed to the corresponding monobromides (102a and 102b) by the Hunsdiecker method. Reductive debromination of 102a and 102b with zinc dust in acetic acid gave the same acetyl ester (103). Compound 103 was also obtained from atisine by desulfurization of the thioketal (105) derived from 104. This correlation demonstrates that the Garrya alkaloids have the same stereochemistry of ring fusions as atisine.

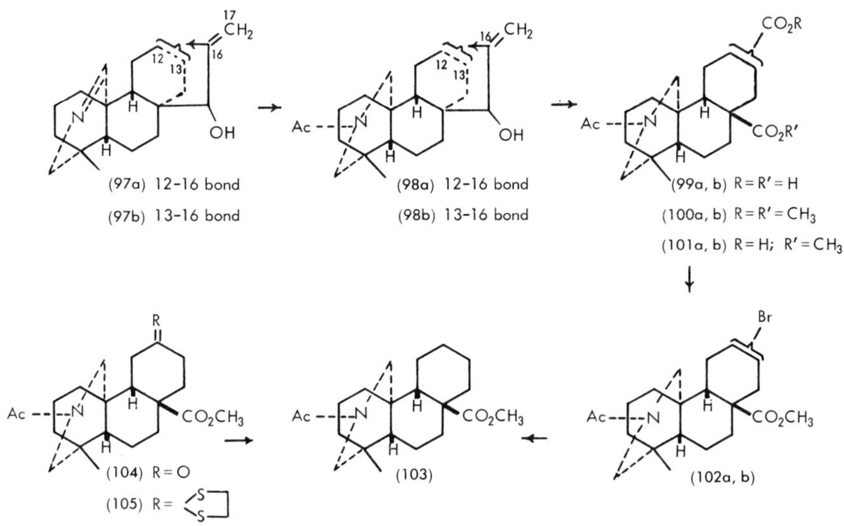

(97a) 12-16 bond
(97b) 13-16 bond
(98a) 12-16 bond
(98b) 13-16 bond
(99a, b) R = R′ = H
(100a, b) R = R′ = CH_3
(101a, b) R = H; R′ = CH_3
(104) R = O
(105) R = (dithiolane)
(103)
(102a, b)

4.2 Correlation of the Atisines with the Diterpenes

4.2.1 Removal of the Nitrogen from Diterpene Alkaloids Degradation of diterpene alkaloids to diterpenes of established configuration was delayed by the lack of a suitable method of removing the nitrogen atom. Since both positions β- to the nitrogen are quaternary, Hofmann-type degradations are ineffective. Success in removing the nitrogen from (106) has been achieved through a mild reaction with aqueous nitrous acid.[52,53] The major product is the hemiacetal (107). This hemiacetal has been converted successively to a primary alcohol (108), aldehyde (109) and carboxylic acid (130). The same procedure has been applied with success to the azomethines derived from the Garrya alkaloids (see below).

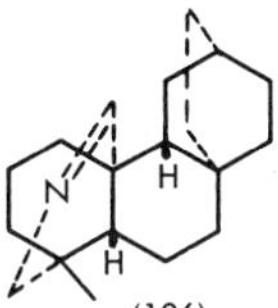
(106)

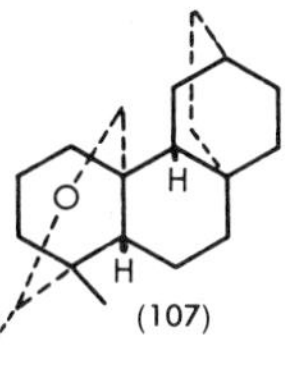
(107)

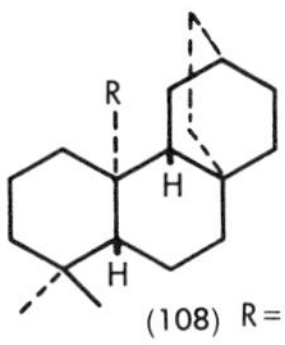
(108) R=CH₂OH

(109) R=CHO

(110) R=CO₂H

4.2.2 Correlation of Garryfoline with Kaurene and Stevane-B The acetate iminium chloride (111) derived from garyfoline was degraded by the internal Hofmann procedure to the azomethine (112). Elimination of the nitrogen was effected by treatment of the azomethine with nitrous acid. The resulting hemiacetal (113) was subjected to the Wolff–Kishner reduction, which effected simultaneous reduction of the 15-keto and masked 19-aldehydo functions. Oxidation of the resulting primary alcohol (114) with CrO_3–pyridine gave aldehyde (115) which was transformed by vigorous Wolff–Kishner conditions to the hydrocarbon (116). This hydrocarbon was identical with (–)-"β"-dihydrokaurene, the minor hydrogenation

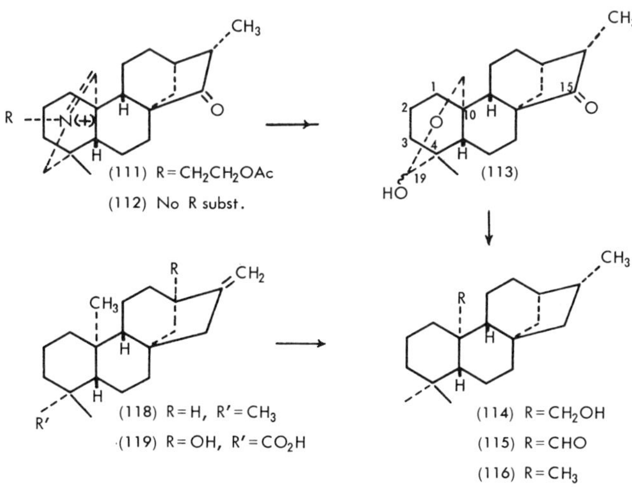
(111) R = CH_2CH_2OAc

(112) No R subst.

(113)

(114) R = CH_2OH

(115) R = CHO

(116) R = CH_3

(117) R = CO_2H

(118) R = H, R′ = CH_3

(119) R = OH, R′ = CO_2H

product of (−)-kaurene (118) and with "stevane-B," a degradation product of steviol (119). The evidence for structure (113) for the hemiacetal is based on the extremely hindered nature of the derived aldehyde (115) and acid (117) [pK^*_{MCS} 9.49].[54–57]

4.2.3 Correlation of Atisine and the Resin Acids The aldehyde (109) derived from atisine has been converted by Wolff–Kishner reduction to the hydrocarbon (120) that is enantiomorphic with a hydrocarbon (121) prepared by a long degradative sequence from abieta-6,8-diene (122).[58] This work represents the first correlation of the tetracarbocyclic ring system of atisine with the resin acids and confirms the mirror image relationship at C(5), C(9), and C(10). Hydrocarbon (133) has also been synthesized from abietic acid via maleopimaric acid.[59]

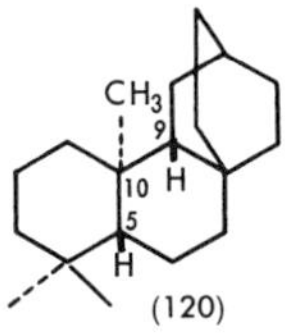

(120)

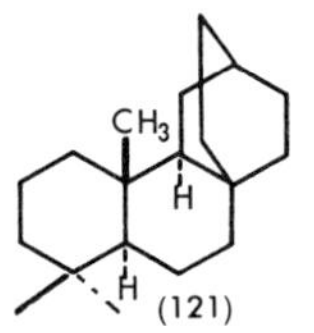

(121)

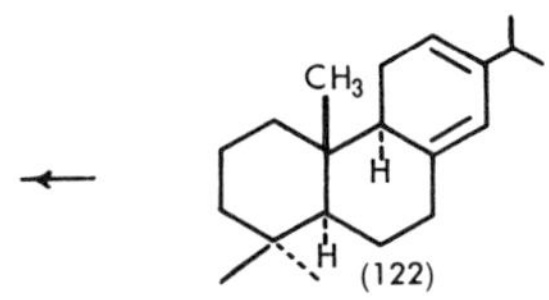

(122)

4.3 Absolute Configuration of Atisine and Garrya Alkaloids

The rotation of phenol (66) led Dvornik and Edwards to first suggest "that if the structure of atisine is based on an unrearranged diterpenoid skeleton, its A–B ring system bears a mirror image relation to that of the common diterpenes."[33] Furthermore, the rotational arguments based on ajaconine chemistry reinforced this perceptive suggestion.[43] Recently, the absolute configuration of the Atisine and Garrya alkaloids has been derived by an elegant series of experiments by Vorbrueggen and Djerassi.[55] Veatchine (123) and garryfoline (124) were converted into the 17-nor-16 ketones (125) and (126). These compounds exhibited a positive Cotton effect of very similar amplitude to that observed for the 17-nor-16-ketone (127) from phyllocladene (128). Since the absolute configuration of phyllocladene has been established and since the configuration at C(9) should not affect the sign of the Cotton effect, the absolute configuration of C(8) and C(13) of the Garrya alkaloids [and hence of C(8) and C(12) of atisine] is established. The correlation of garryfoline with (−)-"β"-dihydrokaurene (116) has been discussed previously. These results lead to the complete absolute configurational representations (123) and (124) for veatchine and garryfoline. In view of the correlation of atisine and veatchine,[50,51] this absolute configurational assignment also applies to atisine (86) and its relatives such as ajaconine and atidine.

ApSimon and Edwards[60] have provided independent evidence for the

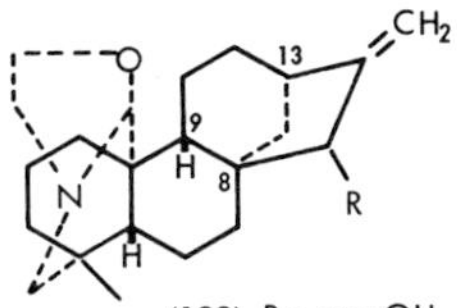

(123) R = H, OH

(124) R = OH, H

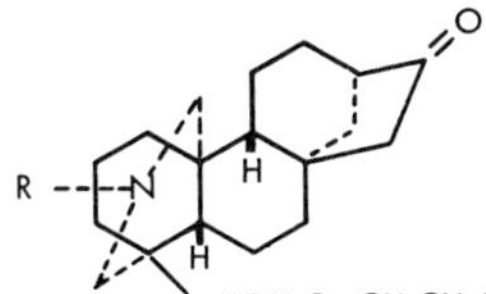

(125) R = CH_2CH_2OAc

(126) R = $COCH_3$

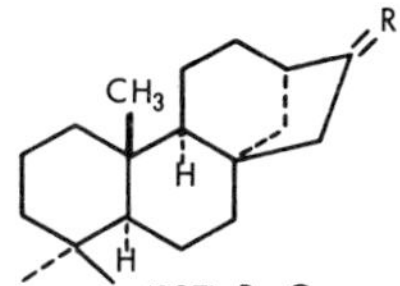

(127) R = O

(128) R = CH_2

absolute configuration of atisine by a synthesis from podocarpic acid (129) of the antipode of the phenol (66) originally obtained by degradation of atisine. The synthesis involved photolysis of the azide (130) derived from the methyl ether of podocarpic acid. One of the products of the photolysis reaction was the δ-lactam (131) which was converted by conventional procedures to the enantiomer of (66).

With the absolute configuration of the Atisine and Garrya alkaloids established, it is interesting to note that these compounds join the ranks of terpenes with antipodal stereochemistry of the A–B-ring fusion as compared with that of the steroids.

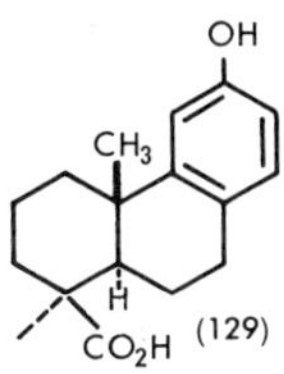

(129)

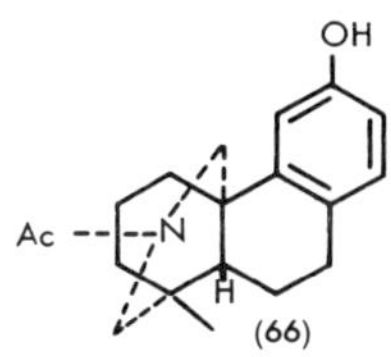

(66)

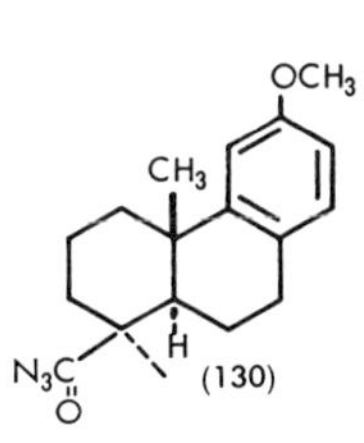

(130)

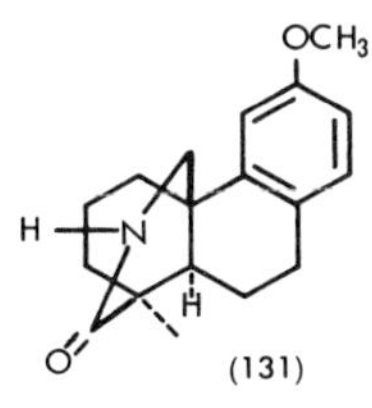

(131)

5. THE TERNARY IMINIUM SALTS OF THE ATISINE AND GARRYA ALKALOIDS

Early in their studies on the structure of atisine Jacobs and Craig[26] noted that treatment with dilute base isomerized atisine to isoatisine. Subsequent studies[4,5] have shown that veatchine, garryfoline, and cuauchichicine are similarly isomerized to garryine, isogarryfoline, and isocuauchichicine, respectively. The isomerization (132)→(135) proceeds even at room temperature in alcohol without external base. Another noteworthy feature is that members of these pairs of isomers manifest a remarkable difference in basic strengths. In 50% methanol, atisine has a pKa′ of 12.8 while isoatisine has a value of 10.35. Similar differences exist for the Garrya alkaloid pairs.

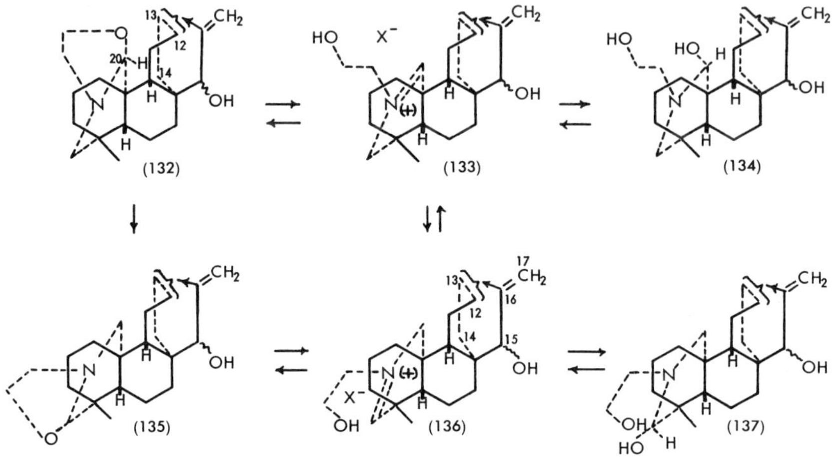

The salts of these alkaloids have been shown to exist in the ternary iminium form. In hydroxylic solvents the "normal" bases (132) exist almost completely as the ternary iminium hydroxides (133, X = OH). In the case of the "iso"-bases a high proportion of the oxazolidine (135) or pseudobase form (137) is present in the equilibrium. The reason for the preponderance of the ternary iminium hydroxide in the "normal" base equilibrium has been recognized as the steric interference with substituents on the tetrahedral C(20) in the oxazolidine form (132). Serious repulsive interactions occur between the hydrogens on C(20), C(13), and C(14) when C(20) is tetrahedral. However, the strain is relieved in the ternary iminium form since C(20) is trigonal. Consequently in solution the "normal" bases will exist almost completely as ternary iminium hydroxides (133, X = OH) and show high pKa's whereas the "iso"-bases, in which the steric factor does not operate to the same extent, will exist mostly in the oxazolidine form (135) and show lower pKa's. The same explanation accounts for the isomerization of the "normal" to the "iso"-bases. In solution the isomerization proceeds through the ternary iminium forms (133) and (136) by prototropy. Since steric factors are responsible for the "iso"-bases having a lower free energy than the "normal" bases, the equilibrium is shifted toward the sterically more favored "iso"-form.[1h,27,34,44,61–63]

In the case of the salts of the "normal" and "iso"-bases, the reverse of the situation described for the bases would be expected. The normal salts have a less bulky trigonal carbon in the hindered 20-position and, therefore, should be more stable than the iso-salts that have the more bulky tetrahedral carbon in this position. Experience has shown that this generalization is correct.[39] Thus isoatisinium chloride (136, X = Cl, 12–16 bond) may be converted to atisinium chloride (133, X = Cl, 12–16 bond) by refluxing in such solvents as DMSO, DMF, DEF, or high boiling alcohols. Garryinium chloride (136, X = Cl, 13–16 bond) under similar conditions is isomerized to veatchinium chloride (133, X = Cl, 13–16 bond). The yields for this conversion are in the order of 80–85%. Since the normal salts can be readily converted, without isomerization, to the corresponding bases by treatment with cold, aqueous alkali, this thermal isomerization of the salts provides a convenient practical method of reversing the facile "normal" base (132)→"iso"-base (135) isomerization. Detailed rate studies on this reaction in several organic solvents have been carried out.[40]

6. THE CHEMISTRY OF ALKALOIDS WITH A MODIFIED ATISINE SKELETON

In recent years a wide range of *Aconitum* species native to Japan and India have been examined for alkaloids. Among those encountered are several that are modeled on an atisine skeleton but possess additional ring fusions. This section surveys the chemistry of these interesting alkaloids.

6.1 Hetisine

Hetisine, $C_{20}H_{27}NO_3$, a minor constituent of *Aconitum heterophyllum*,[64,65] and *Delphinium cardinale* Hook,[66] represents an interesting variant of the atisine skeleton. The alkaloid has one hydrogenable double bond, three active hydrogens and a tertiary nitrogen. N-alkyl and methoxyl determinations are negative. The presence of an exocyclic methylene group is indicated by the infrared (3003, 1659, 899 cm^{-1}) and nmr spectra (τ-5.28, 5.46) and confirmed by the isolation of formaldehyde upon ozonolysis. C-methyl determinations and infrared absorption (1379 cm^{-1}) show the presence of one C-methyl in hetisine and two in dihydrohetisine. This is confirmed by the appropriate nmr spectra. Since dihydrohetisine shows no adsorption in the near uv, it is clear that hetisine must have a heptacyclic skeleton.[65,67,68]

The nature of the oxygen functions is indicated by formation of a crystalline diacetate (3247, 1742, 1252, 1225 cm^{-1}) and an amorphous triacetate, each of which regenerates hetisine on saponification. Furthermore, the alkaloid is inert to both periodate and lead tetraacetate and does not form an acetonide. It, therefore, possesses three acylatable hydroxyls that are nonvicinal and are not in a 1,3-*cis*-diaxial relationship.

Dehydrogenation of hetisine yields a complex mixture of hydrocarbons from which pimanthrene (25) has been isolated. The fact that hetisine lacks a free N-alkyl group and compares in basicity (pKa′ 9.85) with quinuclidine (10.3), suggests a quinuclidine-type structure with bonding from the nitrogen to either C(1), C(2), C(3), C(6), or C(7). One additional ring and three hydroxyls are necessary to complete the structure. An x-ray diffraction study has established the structure of hetisine as 138.[69]

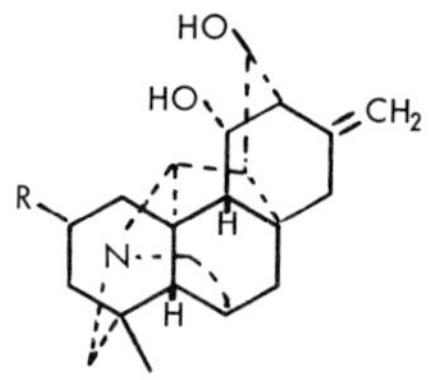

(138) R = α-OH

(142) R = O

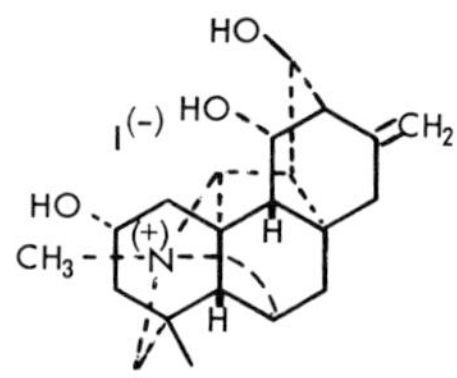

(139)

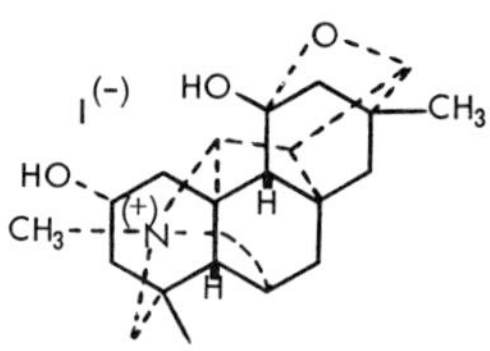

(141)

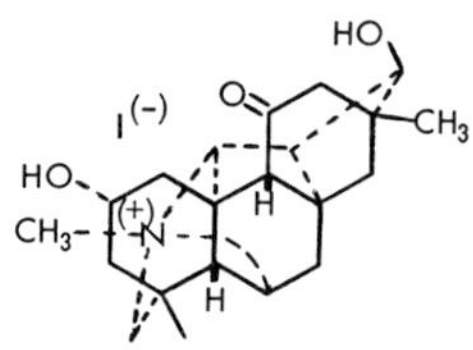

(140)

Oxidation of hetisine diacetate followed by hydrolysis, affords a dehydrohetisine, named hetisinone, that proved to be identical with a dehydrohetisine later isolated from *Delphinium cardinale* Hook[66] and from *Delphinium denudatum*.[71] The structure originally suggested for hetisinone (142)[1d] has been confirmed by chemical and spectroscopy data.[70]

Structure 138 allows the interpretation of an unusual rearrangement that hetisine methiodide undergoes. Jacobs and Huebner[65] had described a hetisine methiodide, mp 325°, formed in a sealed tube at 100°, which on Hofmann degradation gave a des-methylhetisine, $C_{21}H_{29}NO_3$. The latter showed the presence of only *one* double bond by hydrogenation and no carbonyl group. Wiesner et al[72] have shown that under milder conditions of quaternization (refluxing methanol) an isomeric methiodide (139), which undergoes a Hofmann degradation to give a normal des-methylhetisine, is obtained. The latter possesses an exocyclic methylene group and a new double bond, both of which can be hydrogenated. Clearly during quaternization Jacobs' methiodide rearranged with participation of the original exocyclic methylene group. This rearrangement is interpreted as proceding via the ketone (140) to the hemiketal structure (141).[72]

Dihydro-des-N-methylhetisine (143) is oxidized by permanganate to a carbinolamine ether (144). Reduction of the latter with sodium borohydride regenerates (143).[67] This oxidation is analogous to the cyclization of des-N-methyl hypognavinol effected by oxidation with silver oxide or alkaline ferricyanide (see Section 6.3).

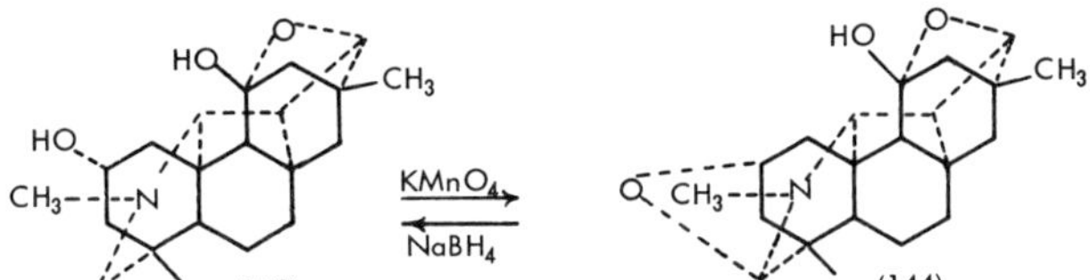

(143) (144)

6.2 Ignavine and Anhydroignavinol

These alkaloids occur in the roots of *Aconitum sanyoense* Nakai, *A. tasiromontanum* Nakai, and *A. japonicum*.[73–75] Early studies showed that ignavine, $C_{27}H_{31}NO_6$, lacks methoxyl, methylenedioxy, or N-methyl groups and is unreactive toward the usual carbonyl reagents. Saponification of ignavine gives one mole of benzoic acid and anhydroignavinol, $C_{20}H_{25}NO_4$. Methiodide formation is also accompanied by the loss of water.

Of the six oxygens in ignavine, two occur in the benzoyloxy group and four in hydroxyls. One of these hydroxyls is adjacent to the benzoyloxy group since the hydrolysis product is susceptible to periodate cleavage while ignavine is not. Diacyl derivatives are formed which contain a mole of water less than calculated on the basis of ignavine and are, therefore, anhydroignavinol derivatives. Saponification of these derivatives does indeed afford anhydroignavinol.[76]

That the exocyclic methylene group (1645, 892 cm^{-1}) is involved in a secondary allylic alcohol system as in atisine was demonstrated by catalytic isomerization to a methyl ketone (1692 cm^{-1}) and by oxidation to a conjugated enone (1615, 1687 cm^{-1}).[77]

Dehydrogenation of anhydroignavinol yields a complex mixture of hydrocarbons from which 1,7-dimethyl-6-*n*-propylphenanthrene, 1,8-dimethyl-3-ethylphenanthrene, and 1,8-dimethyl-3-isopropylphenanthrene have been isolated.[78–80] These products account for 19 of the 20 carbon atoms of anhydroignavinol. The assumption of a bicyclo[2,2,2]octane-allyl alcohol system in ignavine is given credence by oxidation of des-N-methyl-oxo-anhydroignavinol to a related dicarboxylic acid in which one of the carboxyls is tertiary. Furthermore, the three phenanthrene dehydrogenation products suggest a bond between C(14) and C(20) and a methyl group at C(1). Numerous degradations have established the position of the nitrogen and shown the presence of hydroxyls at C(2) and C(3) in anhydroignavinol.[77,81] Since ignavine has no N-alkyl group, a third bond must extend from nitrogen to one of the rings. The Japanese workers selected C(6) as the most likely site.

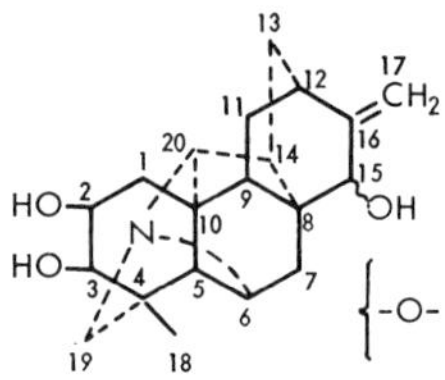

(145)

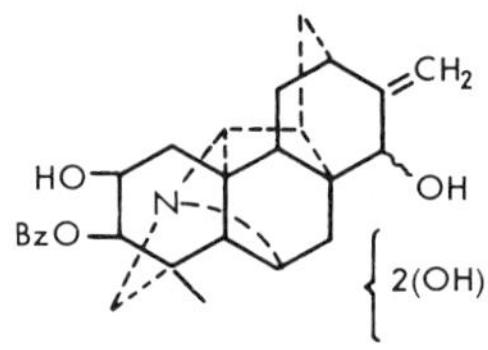

(146)

The published data clarify the nature of four of the six oxygens of ignavine and three of the four of anhydroignavinol. The fourth oxygen in anhydroignavinol is thought to exist as an ether since no hydroxyl band is observable in the infrared spectra of tribenzoyl anhydroignavinol and certain other anhydro-derivatives. Therefore, it is likely that loss of water accompanying many of the reactions of ignavine involves *ether* rather than double-bond formation. Possible positions for the hydroxyls involved in the elimination assuming a β-glycol moiety are C(11–13), C(7–14), C(7–9), and C(5–9). Since ignavine has a normal pKa′ value (7.7) for a tertiary amine, it is unlikely that a hydroxyl is at C(19) or C(20). In view of the above, anhydroignavinol and ignavine have been provisionally represented by partial structures (145) and (146), respectively.*

6.3 Hypognavine and Hypognavinol

Certain varieties of *A. sanyoense* contain an ester alkaloid that has one less oxygen atom than ignavine and is in many respects similar to it. This alkaloid, hypognavine, $C_{27}H_{31}NO_5$, is a benzoyl ester, has no methoxyl or N-methyl groups, fails to react with the usual carbonyl-test reagents and contains one hydrogenable double bond. Like ignavine, it has two acylatable hydroxyls and an exocyclic methylene group (1666, 895 cm,$^{-1}$ ozonization gives formaldehyde). In contrast to ignavine, hypognavine may be saponified to an alkamine (hypognavinol, $C_{20}H_{27}NO_4$) without

*The absolute configuration shown is based on analogy to that of the other diterpene alkaloids.

loss of a mole of water. Acylation reactions and methiodide formation are also straightforward in the case of hypognavinol.[82,83]

The exocyclic methylene in hypognavine undergoes rearrangement to a methyl ketone by hydrogenation over palladium-charcoal or by treatment with acid. Hypognavine may be oxidized to a conjugated enone [1705, 1630 cm.$^{-1}$; λ_{max} 231 (ϵ 22,400)], which on reduction over palladium-charcoal gives dihydrohypognavinone, the methyl ketone referred to earlier. These reactions afford clear evidence for the existence of a secondary allyl alcohol system such as occurs in atisine, ignavine, and songorine.

Selenium dehydrogenation of hypognavinol furnishes 1,8-dimethylphenanthrene, 1,7-dimethyl-6-*n*-propylphenanthrene and 1,8-dimethyl-3-ethylphenanthrene, the latter two being characteristic products of anhydroignavinol. This suggests that hypognavine has the same skeleton as derived from ignavine. Formula (147) will be used as a basis for interpreting the chemistry of hypognavinol.[79,83,84]

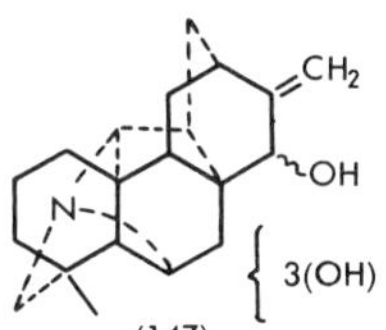

(147)

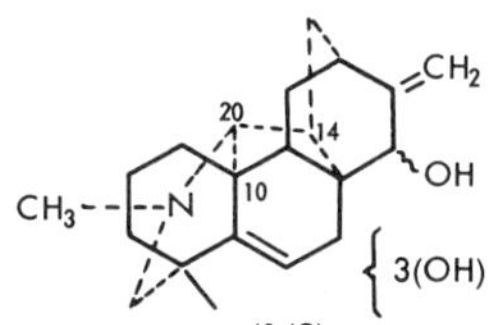

(148)

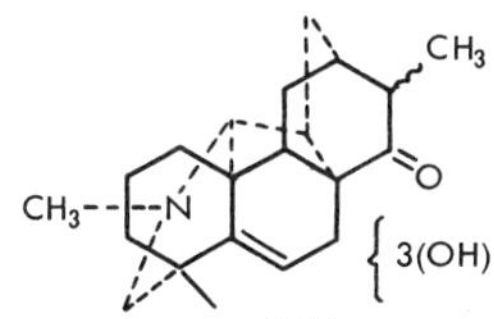

(149)

Hofmann degradation of hypognavinol methiodide gave the des-base (148) and the isomerization product (149). The resistance of des-N-methylhypognavinol toward a second Hofmann degradation is well accommodated by structure (148), for the β-positions to the nitrogen at C(4) and C(10) bear no hydrogen and formation of a double bond between C(14) and C(20) would violate Bredt's rule. The size of the heterocyclic ring in (147) was indicated by oxidation of three hypognavinol derivatives to δ-lactams.[84–86]

The three remaining oxygens of hypognavinol exist as hydroxyls of which two are acylatable. The nonacylatable hydroxyl is shown to be tertiary by oxidation experiments. Detailed transformations have shown that the α-glycol system in hypognavinol is masked by a benzoyloxy group in hypognavine. The location of the α-glycol system in ring-A of hypognavinol is fixed by oxidation of des-N-methylhypognavinol (150) with silver oxide or alkaline ferrocyanide to a carbinolamine ether (151) from which (150) may be regenerated by reduction with $NaBH_4$. Salts (152) of the carbinol-amine ether show infrared absorption typical of the $>C=N<$ group (1686–1679 cm^{-1}) and regenerate the parent base on treatment with alkali. Models show that an axial hydroxyl at C(2) is most favorable for ether formation. It should be noted that oxidation at C(20) is unlikely since the salt of the corresponding carbinolamine ether (153) would violate Bredt's rule. These transformations parallel the formation of a carbinolamine ether on oxidation of dihydro-des-N-methylhetisine, (143) → (144).

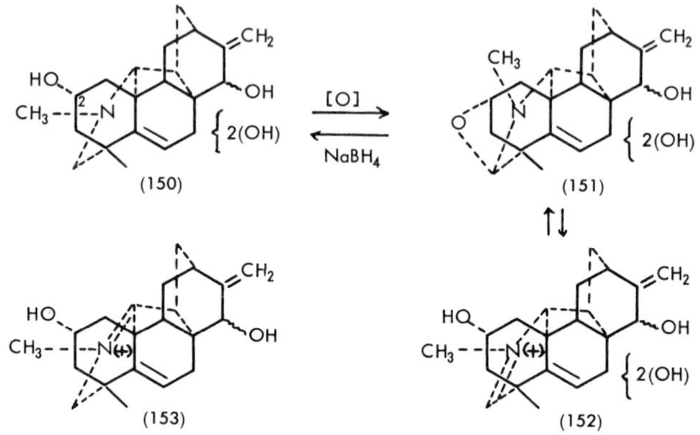

The assignment of the second hydroxyl of the glycol system at C(1) or C(3) is not yet settled. Periodate cleavage experiments indicate a *trans*-glycol system is present. The benzoyloxy group in hypognavine is accordingly assigned as 1β- or 3β-configuration. The site of the tertiary hydroxyl is unknown. Hypognavinol is thus represented by structure (154a) or (155a) and hypognavine by (154b) or (155b).[86]*

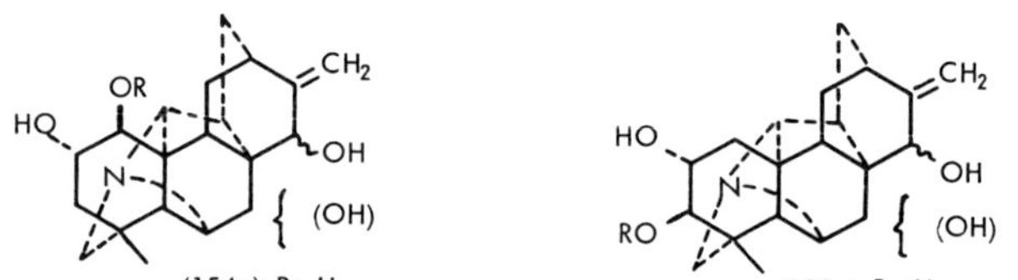

(154a) R = H
(154b) R = COC_6H_5

(155a) R = H
(155b) R = COC_6H_5

6.4 Kobusine*

This alkaloid, $C_{20}H_{27}NO_2$, has been isolated from *A. Kamtschaticum (fischeri), A. sachalinense* Fr. Schmidt, *A. lucidusculum* Nakai, and *A. yesoensis* Nakai.[87–90] It possesses two secondary hydroxyl groups located in six-membered rings, one of which is involved in an allyl alcohol group such as is common to atisine, ignavine, and hypognavine. Selenium dehydrogenation furnished 1,7-dimethyl-6-*n*-propylphenanthrene, a characteristic product obtained also from ignavine and hypognavine. These

*The absolute configuration shown is based on analogy to that of the other diterpene alkaloids.

results suggest that kobusine has the same skeleton as ignavine and hypognavine. The chemistry of kobusine is discussed in terms of the ultimately assigned structure (156).*[91–93]

When kobusine (156) is warmed with dilute HCl, compounds (157), (158), and (159) are formed. Compounds (158) and (159) are reducible to the same glycol (160). Oxidation of (160) with CrO_3–pyridine gave a γ-lactone (161) showing the proximity of the two hydroxyls. Treatment of kobusine with sodium in propanol gave the des-hydroxyl derivative (163) that was related to the methyl ketone (157) via (162). Oxidation of (163) with OsO_4–HIO_4 followed by CrO_3–pyridine gave a γ-lactone (164), thus establishing the relationship between the two hydroxyl groups.

The correlation of kobusine and pseudokobusine is discussed below.

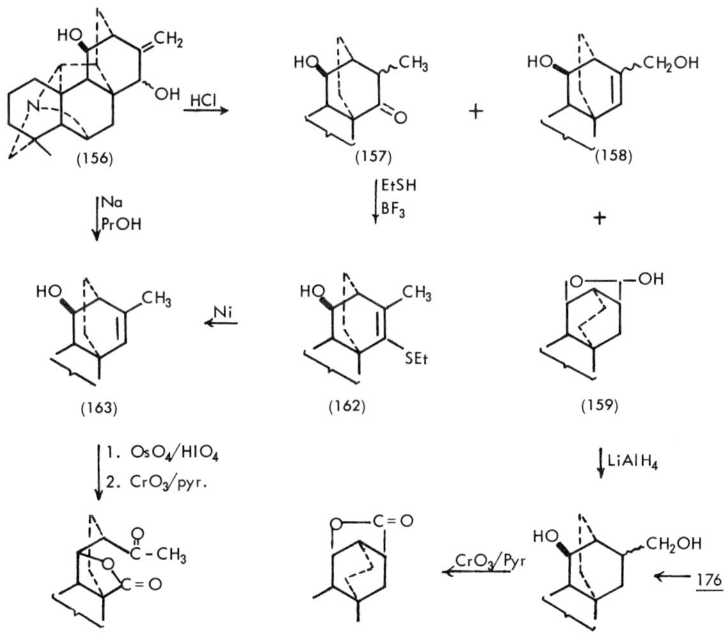

6.5 Pseudokobusine

Isolated from *A. Yesoensis* Nakai and *A. lucidusculum* Nakai,[89,94] this alkaloid, $C_{20}H_{27}NO_3$, is closely related to kobusine. It is a tertiary base

*Though a hydroxyl group is assigned to C(11), none of the previously published evidence precludes the location of this hydroxyl at C(13). A recent X-ray analysis has shown that the hydroxyl group is, indeed, at C(11). Also, the hydroxyl group at C(15) is shown to be β-oriented. (S. W. Pelletier, L. H. Wright, M. G. Newton and H. Wright, unpublished work.)

containing three acylatable hydroxyls (tribenzoate), an exocyclic methylene group (900 cm^{-1}), a C-methyl group and no methoxyl. Oxidation reactions and palladium-catalyzed rearrangement such as were used in the case of ignavine, demonstrate that one of the hydroxyls is involved in an allyl alcohol grouping. Oxidation experiments show the allylic hydroxyl and one other are on six-membered rings and the third hydroxyl is tertiary since it is inert to Kiliani reagent.[95]

Selenium dehydrogenation of pseudokobusine gave 1,7-dimethyl-6-*n*-propylphenanthrene, a result which suggests that pseudokobusine has fundamentally the same structure as kobusine, with the addition of one tertiary hydroxyl. The chemistry of pseudokobusine is discussed in terms of structure (165).*[93,96,97]

Acetylation of pseudokobusine gave besides the normal O-acetate, an N-acetyl-seco-derivative (166) [1596, 1700 cm,$^{-1}$ λ_{max} 299 (ϵ33), τ = 7.87s]. An analogous N-cyano-seco-derivative (167) (2218, 1697 cm^{-1}) was prepared by treatment of (165) with cyanogen bromide. Both seco-compounds (166) and (167) regenerated pseudokobusine when hydrolyzed with 20% potassium hydroxide. Pseudokobusine methiodide (168) gave with ammonium hydroxide the N-methylketone (169) (1675 cm^{-1}), which regenerated the methiodide (168) when treated with hydriodic acid. To establish that the tertiary hydroxyl of (165) is part of the masked aminoketone

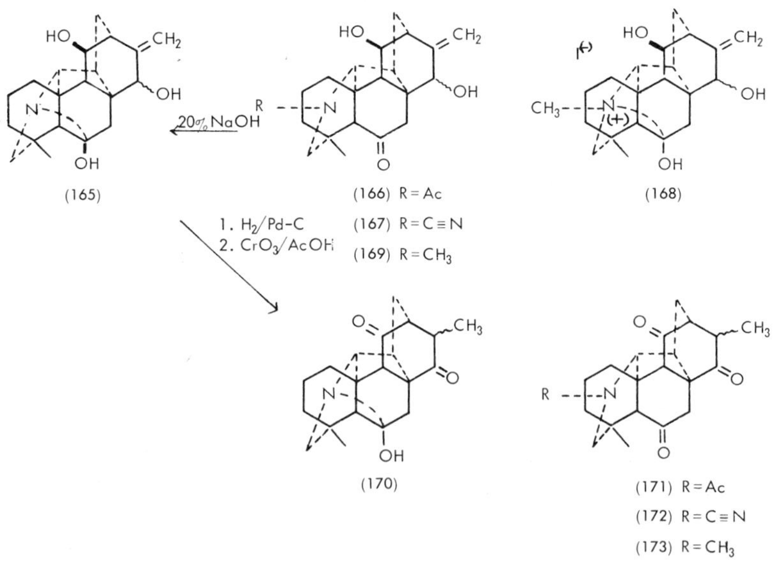

(165) (166) R = Ac (167) R = C≡N (169) R = CH_3 (168) (170) (171) R = Ac (172) R = C≡N (173) R = CH_3

*A recent x-ray analysis of kobusine has shown the configuration of the hydroxyl group at C(15) to be β. The configuration of this hydroxyl in pseudokobusine is, therefore, identical. (S. W. Pelletier, L. H. Wright, M. G. Newton and H. Wright, unpublished work).

group, ketodihydropseudokobusinone (170) was converted to the N-acetyl triketone (171; 1632 cm^{-1}), N-cyano triketone (172; 2221 cm^{-1}) and N-methyl triketone (173; 1714, 1708, 1683 cm^{-1}) by reactions parallel to those already described. Hydrolysis of (171) regenerated (170).[96]

The location of the third hydroxyl was determined by the following sequence. Oxidation of N-acetyl-seco-pseudokobusine (166) with OsO_4-$NaIO_4$ gave a hemiacetal monocarboxylic acid (174), (1720, 1650 cm^{-1}), the methyl ester of which was oxidized with CrO_3–pyridine to the γ-lactone 175 (1778, 1728, 1699, 1630 cm.$^{-1}$). The same γ-lactone was obtained by oxidation of (166) with manganese dioxide to the enone (176) (3420, 1699, 1692, 1630 cm^{-1}), and oxidation of the latter with OsO_4–$NaIO_4$, followed by esterification with diazomethane to give (175). These transformations limit the position of the third hydroxyl group to position C(11) or C(13). The Japanese workers assign the hydroxyl to C(11), though as in the case of kobusine, evidence for this assignment is not given.

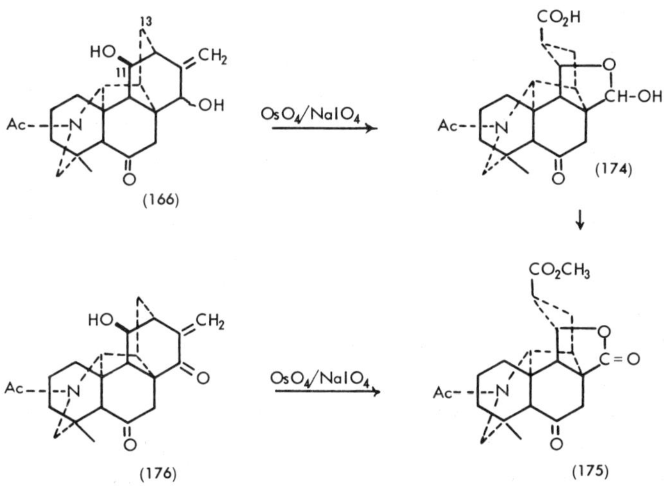

(166) (174) (176) (175)

6.6 Correlation of Kobusine and Pseudokobusine*

The correlation of kobusine and pseudokobusine has been effected by several routes. Two of these are illustrated: Dihydropseudokobusine (178) on reduction with sodium in propanol gave dihydrokobusine (179). Also reduction of ether kobusine or pseudokobusine with sodium in propanol gave the same alcohol (177).[93,97]

*A recent x-ray analysis of kobusine has shown the configuration of the hydroxyl group at C(15) to be β.The configuration of this hydroxyl in pseudokobusine is, therefore, identical. (S. W. Pelletier, L. H. Wright, M. G. Newton and H. Wright, unpublished work).

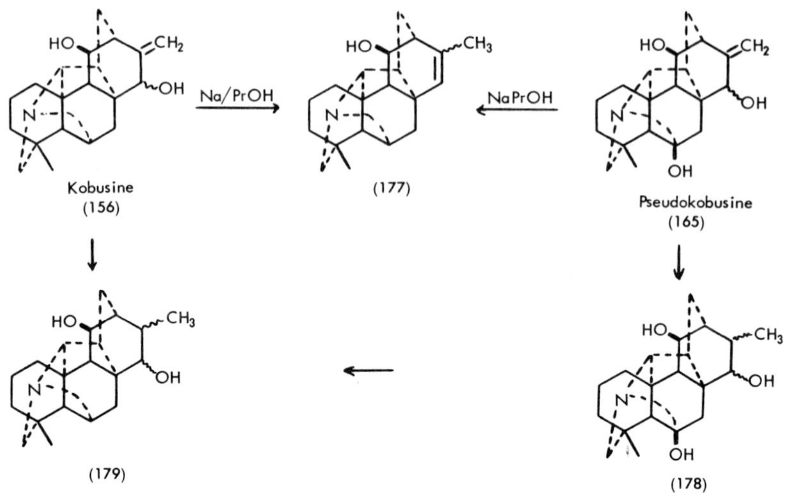

6.7 Isohypognavine*†

This alkaloid $C_{27}H_{31}NO_4$, (180b) occurs in the roots of *A. majimai Nakai* and *A. japonicum* Thumb[15,75] and is a benzoate of the alkamine, isohypognavinol (180a). The latter has three acylatable hydroxyls, one of which is involved in the typical allyl alcohol system as shown by oxidation with manganese dioxide to a conjugated enone, isohypognavinolone [181; 1630, 1703 cm^{-1}, λ_{max} 227 (ϵ6730)] and by isomerization of (180b) over palladium-charcoal to the epimeric methyl ketones (182a). Isohypognavine (180b) has been correlated with kobusine (156) by the following route. Oxidation of α-dihydroisohypognavinone (182a) with CrO_3–AcOH gave (182b), which was hydrolyzed to 182c. Conversion of the latter to the mesylate (182d) followed by refluxing in pyridine gave (183) (τ = 4.2–4.5, 2H mult.). Hydrogenation of (183) gave ketodihydrokobusinone (184) previously prepared from kobusine. Reduction of isohypognavinol (180a) with sodium in propanol gave a desoxy derivative (186), the ethiodide of which afforded on Hoffman degradation a tertiary base (185). Catalytic hydrogenation of (186) followed by oxidation gave the diketone (187) that afforded a monobromoderivative (188). Existence of a single proton adjacent to the bromine (τ = 5.99 s) restricts the position of the hydroxyl in (186) to C(2). The deshielding of the C-methyl in going from (187) to (188) also supports introduction of the bromine at C(3) in a position β-to the methyl.[98]

*The name "isohypognavine" is unfortunate, since the compound is not isomeric with hypognavine. A misinterpretation of early analytical data led to assignment of an incorrect molecular formula.

†The absolute configuration shown is based on analogy to that of the other diterpene alkaloids.

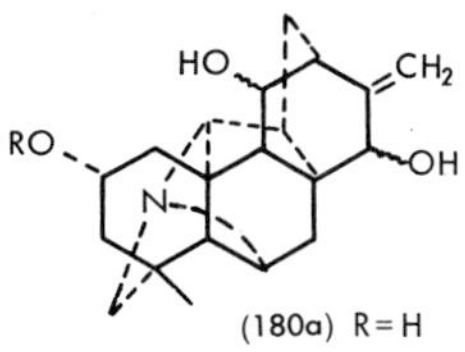

(180a) R=H

(180b) R=Bz

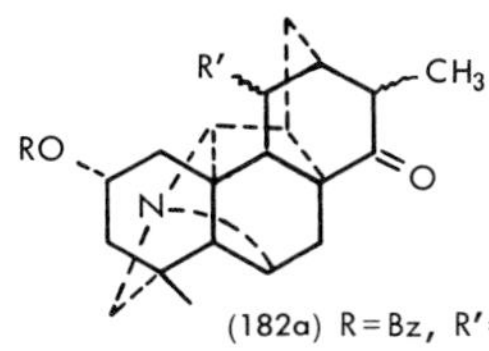

(182a) R=Bz, R'=OH

(182b) R=Bz, R'=O

(182c) R=H, R'=O

(182d) R=SO2CH3, R'=O

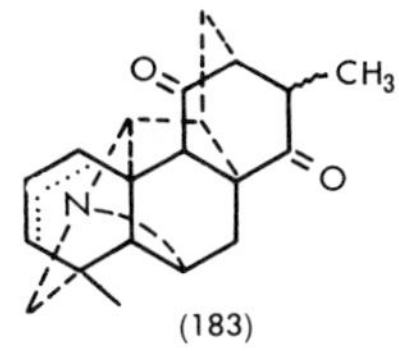

(183)

↓ MnO_2

↓ H_2/Pd

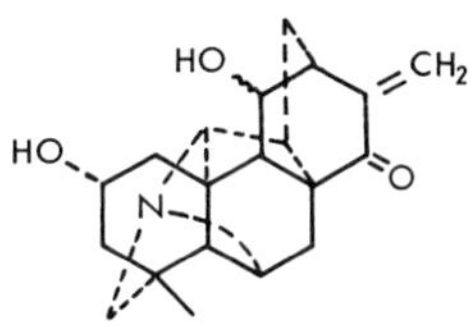

(181)

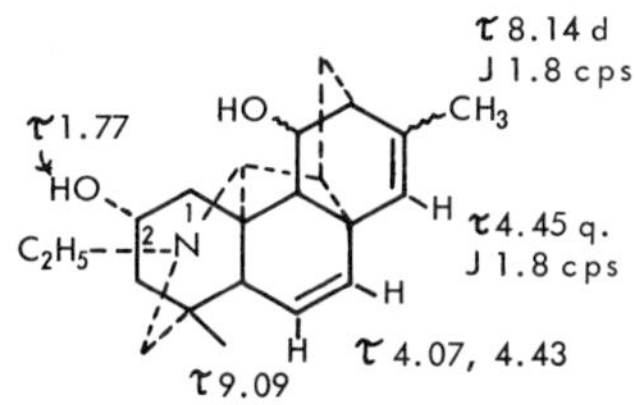

(185)

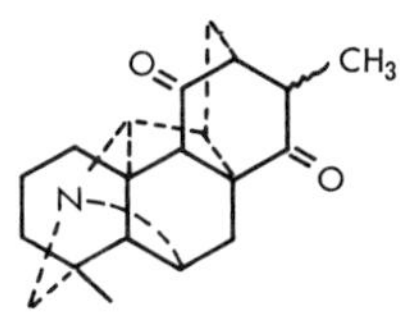

(184)

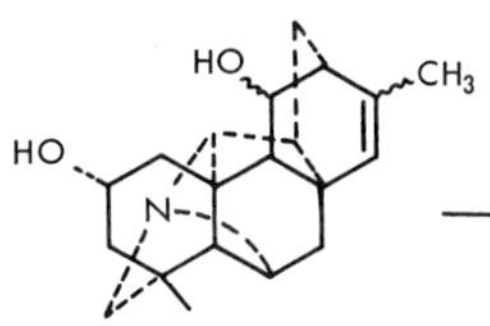

(186)

→

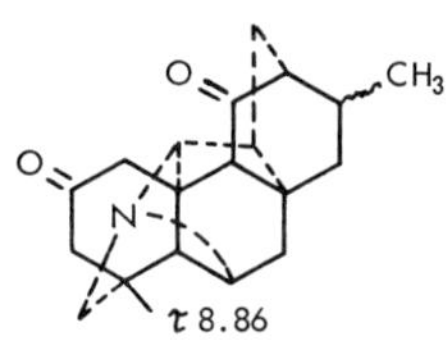

(187)

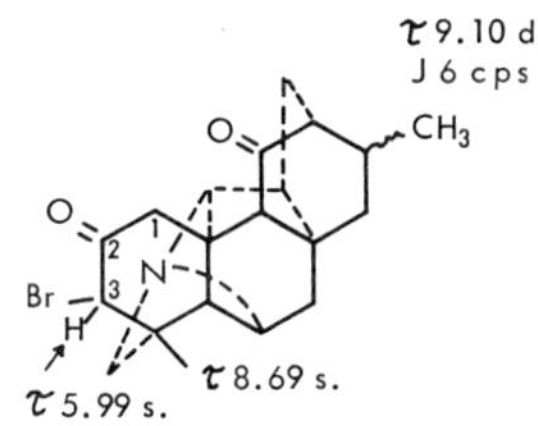

(188)

6.8 Alkaloids from *Spiraea Japonica* Rosaceae L

Ten new alkaloids have been isolated from this plant by Goto et al.[99] Structures for three of these, spiradine A (189), spiradine B (190) and spiradine C (191) have been reported. Reduction of spiradine A with sodium borohydride gave spiradine B and oxidation of the latter with CrO_3–pyridine regenerated spiradine A. Treatment of spiradine C with KOH in ethanol gave spiradine B. An x-ray analysis of the methiodide of spiradine A gave the structure and absolute configuration shown.[99]

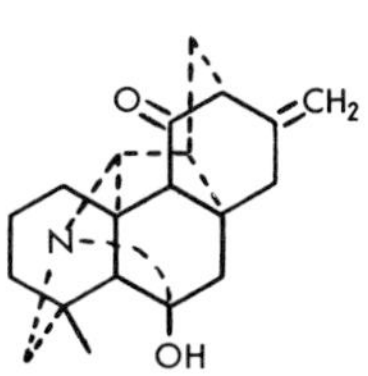

(189)

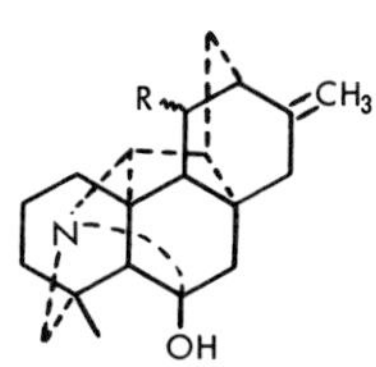

(190) R = OH

(191) R = OAc

7. SYNTHESES OF C_{20}-DITERPENE ALKALOIDS

7.1 Partial Synthesis of Atisine

The N-acetyl ester (192), a key intermediate in the correlation of the Atisine and Garrya alkaloids, has been converted to atisine by a twelve-step sequence.[100] The significant features of this sequence involved homologation of the carboxylic acid, Dieckman cyclization to a ketone (193) which on α-methylation, α-bromination, and dehydrobromination gave the enone (194). Reduction of (194) gave an alcohol which was hydrolyzed with KOH and a trace of hydrazine in diethylene glycol to the amine (195). The latter had already been converted to atisine.[38]

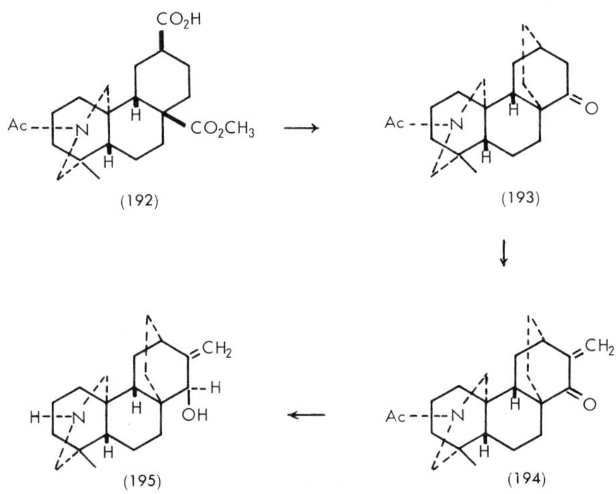

(192) (193) (194) (195)

7.2 Total Synthesis of Atisine

The first complete stereospecific synthesis of *dl*-atisine was reported by Nagata et al.[101] starting from ketone (196) which by a twenty-three step process was also converted into ketone (194). Hydrocyanation of 196 gave the *trans*-cyano ketone (197), which by a Wittig reaction followed by acid hydrolysis and methylation afforded (198). The latter was converted to the tetracyclic derivative (199) by a three step sequence via the lactamol. By Birch reduction, mesylation, and hydrolysis, 199 was converted to the enone (200). To construct the D-ring, (200) was hydrocyanated to the *trans*-cyano ketone (201) which was converted into the cyclic ketol (202) via the 15α-acetyl compound. The mesylate (203) derived from (202), on treatment with alkali gave (204) which yielded via (205) the pentacyclic ketone (206). After conversion of the latter into the exo-olefin, the protecting mesyl group was removed by Birch reduction and the desired acetyl group was then introduced (207). Finally, introduction of the β-hydroxyl at C-15 was effected by a three-step process: treatment of (207) with N-bromo-succinimide afforded a rearranged allylic bromide, which after epoxidation was treated with zinc to give the desired alcohol (208) and its epimer (209). Compounds (208), (209), and the enone (194) proved to be the racemic forms of the corresponding naturally derived materials. Since (194) and (208) had already been converted to atisine by Pelletier and Parthasarathy,[100] this sequence represents a stereospecific total synthesis of *dl*-atisine. The over-all yield from (196) to (207) is about 1.7%.

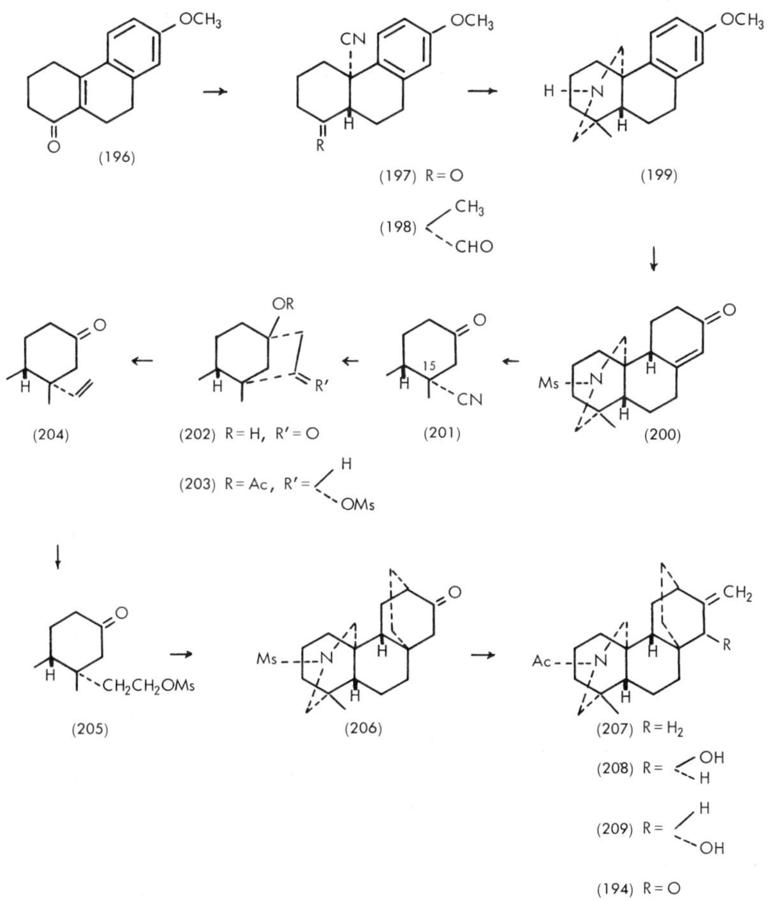

7.3 Total Synthesis of Garryine and Veatchine

A synthesis of *dl*-garryine and *dl*-veatchine has also been reported by Nagata et al.[102] The dimesyl derivative (203) was refluxed with collidine to give olefin (210). Sterically controlled hydroboration of (210) with bis-3-methyl-2-butylborane followed by oxidation and hydrolysis yielded the 1,2-diol (211). Rearrangement of the brosylate (212) to the bridged ketone 213 followed by Wittig condensation afforded the exomethylene derivative (214). The secondary amine from (214) was converted to the N-carbethoxy derivative (215). Introduction of the hydroxyl at C(15) was effected as in the atisine synthesis by successive bromination, epoxidation, and debromination. The products were (216) and its epimer. Removal of the carbethoxy group followed by alkylation of the resulting amine (217) gave *dl*-dihydroveatchine (218). Since dihydroveatchine has been converted to garryine[6] and garryine to veatchine[39,40], this work constitutes a total synthesis of the racemic forms of these alkaloids.

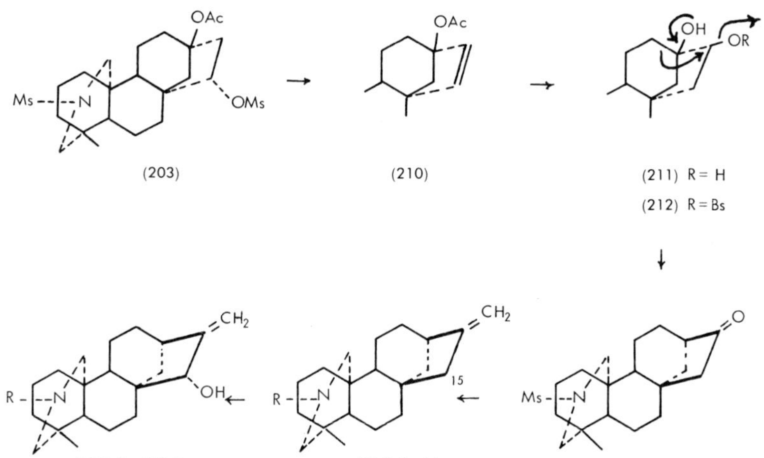

(203) (210) (211) R = H (212) R = Bs

(216) R = CO_2Et (217) R = H (218) R = CH_2CH_2OH

(214) R = Ms (215) R = CO_2Et

(213)

A totally different synthesis of garryine and veatchine has been reported by Masamune.[103–105] The carboxylic acid (219) was converted via the bromoketone (220) to the tetrahydropyranyl ethers (221). Hydrogenolysis of (221) gave the corresponding phenols (222). Base treatment of (222) effected cyclization of only one isomer to furnish a dienone ether (223). Catalytic hydrogenation of the benzoate (224) gave the tetrahydro *cis*-fused derivative (225). Carbomethoxylation of (225) with triphenylmethyl sodium and carbon dioxide followed by methylation afforded the β-keto ester (226). Annelation of (226) with ethyl vinyl ketone followed by exhaustive methylation, removal of the protective group and oxidation gave the diketone (227). Catalytic hydrogenation of the diketone gave a single saturated compound (228). Wolff–Kishner reduction of the monoketal of (228) gave an acid (229) that afforded the intermediate (230). Since veatchine azomethine acetate (15-OAc) has been converted to the levorotatory enantioners of (229) and (230), the stereochemistry indicated for all intermediates is correct.

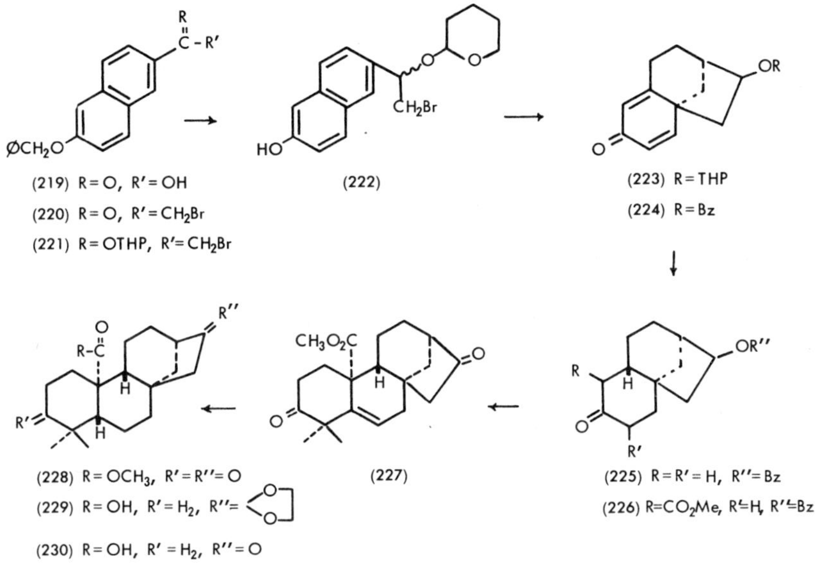

(219) R = O, R′ = OH
(220) R = O, R′ = CH_2Br
(221) R = OTHP, R′ = CH_2Br
(222)
(223) R = THP
(224) R = Bz
(225) R = R′ = H, R″ = Bz
(226) R = CO_2Me, R′ = H, R″ = Bz
(227)
(228) R = OCH_3, R′ = R″ = O
(229) R = OH, R′ = H_2, R″ = (ethylene ketal)
(230) R = OH, R′ = H_2, R″ = O

Intermediate (230) was converted by several steps via photolysis of the azide (231) to a ketoamide (232) that was identical with the compound previously prepared by Vorbrueggen and Djerassi[55] from veatchine azomethine acetate. Introduction of a methylene group (233) followed by

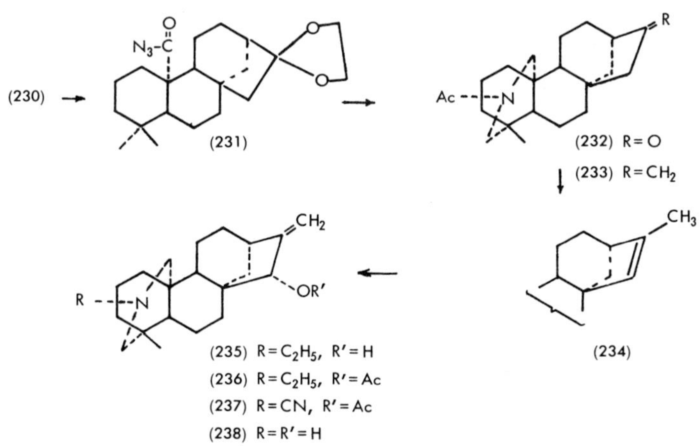

(231)
(232) R = O
(233) R = CH_2
(234)
(235) R = C_2H_5, R′ = H
(236) R = C_2H_5, R′ = Ac
(237) R = CN, R′ = Ac
(238) R = R′ = H

isomerization gave (234). Photosensitized oxygenation of (234) and hydride reduction of the resulting hydroperoxide gave the alcohol (235), which was converted to the secondary amine (238) via (236) and (237). Since Wiesner et al.[6] had previously transformed (238) to garryine, this work completes a total synthesis of garryine. Garryine has been converted to veatchine by Pelletier, Kawazu, and Gopinath.[39,40]

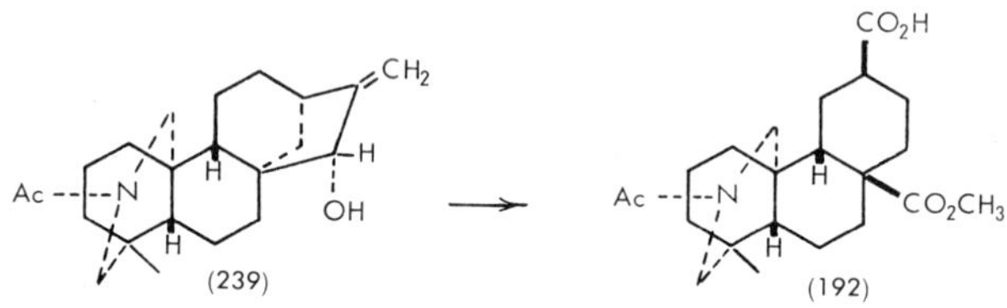

Masamune[105] has also converted compound (239), obtained from veatchine azomethine acetate, by a multistep procedure to the monoester carboxylic acid (192). Since the latter has already been converted to atisine,[100] this work completes in a formal sense the synthesis of atisine also.

Still a third synthesis of the Garrya alkaloids has been reported.[106,107] This synthesis starts with 5-methoxy-2-tetralone and proceeds through the key intermediates (240)–(245).[108] The *dl*-amine (245) was converted to (232) by a series of steps involving intermediates (246)–(248). The syn-

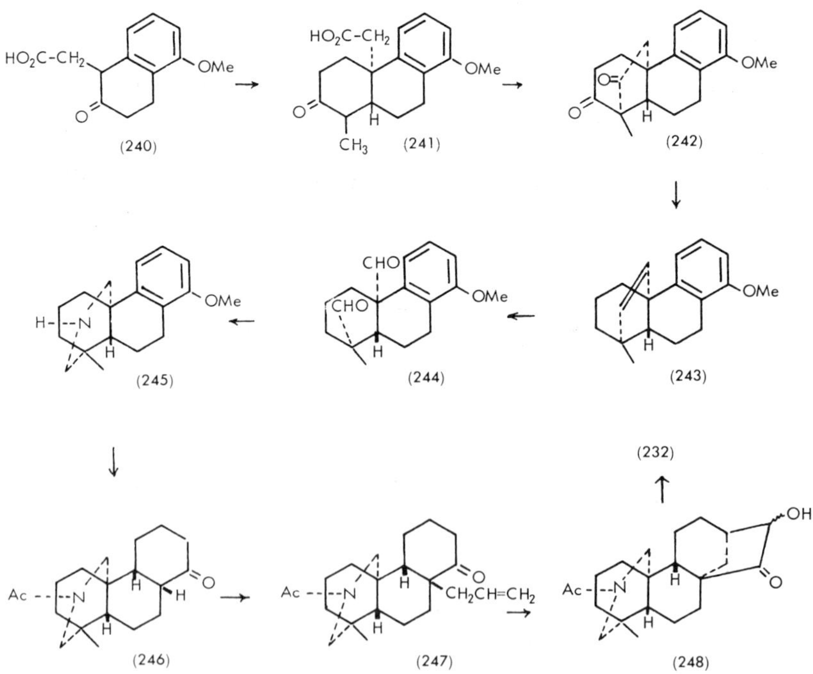

thesis of (232) constitutes a total synthesis of the Garrya alkaloids since the conversion of (232) to the amine (238) has been reported by Masamune[104] and the transformations of (238) to garryine[6] and of garryine to veatchine[39,40] have been reported.

In recent papers the New Brunswick group[109,110] has described two new synthetic sequences that can be used to elaborate both Atisine- and Garrya-type structures. An interesting conversion of the diterpene trichokaurin (249) to atisine, garryine and veatchine has been described.[111] A synthesis of these three alkaloids from (−)-abietic acid has also been recorded.[112]

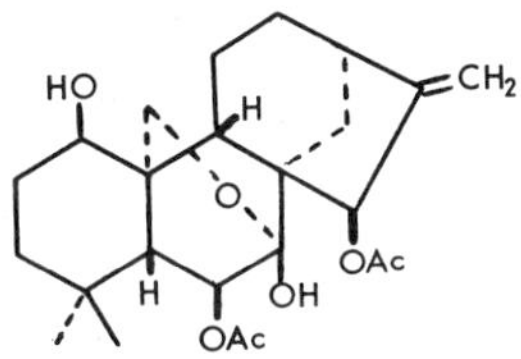

(249)

The diterpene songorine (44) despite its close relationship to veatchine (8), requires synthetic methods unlike those used for the synthesis of veatchine. A promising approach to the synthesis of songorine-type alkaloids has been reported.[113]

Several other interesting approaches to the C_{20}-diterpene alkaloids have been reported[59,114–138]—some involving total syntheses and others partial syntheses from natural products. Very brief surveys of diterpene alkaloid chemistry have also appeared.[139–143]

ACKNOWLEDGMENT

The author's work reviewed in this chapter was supported by grants from the National Institutes of Health, U.S. Public Health Service. Appreciation is expressed to the Chemical Society (London) and the editors of the Pergamon Press (London) and of Experientia (Basel), for permission to use in this chapter certain sections of my review articles published in *Quart. Rev. (London)* **21,** 525–548 (1967); *Tetrahedron* **14,** 76–112 (1961), and *Experientia* **20,** 1–10 (1964), respectively. We also gratefully acknowledge permission from the Academic Press to utilize material appearing in my Introduction and Chapter 2 of vol. 12 of R.H.F. Manske's "The Alkaloids."

REFERENCES

1. Detailed reviews on various aspects of diterpene alkaloid chemistry have been published by:
   (a) S. W. Pelletier and L. H. Keith, *Alkaloids* **12** (1969) (R. H. F. Manske, Ed.) in press.

(b) S. W. Pelletier, *Quart. Rev. London* **21,** 525–548 (1967).
(c) S. W. Pelletier, *Experientia* **20,** 1–10 (1964).
(d) S. W. Pelletier, *Tetrahedron* **14,** 76–112 (1961).
(e) H. G. Boit, "Ergebnisse der Alkaloid—Chemie bis 1960," pp. 851–905, 1009–1011, Academie-Verlag, Berlin, Germany, 1961.
(f) A. R. Pinder, in "Chemistry of Carbon Compounds," Vol. 4-C, pp. 2019–2033 (E. H. Rodd, Ed.), Elsevier Pub. Co., Amsterdam, 1960.
(g) E. S. Stern, *Alkaloids* **7,** 473–503 (1960).
(h) K. Wiesner and Z. Valenta, *Progr. Chem. Org. Nat. Prod.* **16,** 26–89 (1958).

2. J. F. Oneto, *J. Amer. Pharm. Assoc.* **35,** 204 (1946).
3. C. Djerassi, C. R. Smith, S. K. Figdor, J. Herran and J. Romo, *J. Amer. Chem. Soc.* **76,** 5889 (1954).
4. C. Djerassi, C. R. Smith, A. E. Lippman, S, K. Figdor, and J. Herran, *J. Amer. Chem. Soc.* **77,** 4801 (1955).
5. K. Wiesner, S. K. Figdor, M. F. Bartlett, and D. R. Henderson, *Can. J. Chem.* **30,** 608 (1952).
6. K. Wiesner, W. I. Taylor, M. F. Bartlett, J. R. Armstrong, and J. A. Edwards, *Chem. Ber.* **86,** 800 (1953).
7. M. F. Bartlett, W. I. Taylor, and K. Wiesner, *Chem. Ind. (London),* 173 (1953).
8. K. Wiesner, R. Armstrong, M. F. Bartlett, and J. A. Edwards, *Chem. Ind. (London)* 132 (1954).
9. M. F. Bartlett and K. Wiesner, *Chem. Ind. (London)* 542 (1954).
10. K. Wiesner, R. Armstrong, M. F. Bartlett, and J. A. Edwards, *J. Amer. Chem. Soc.* **76,** 6068 (1954); M. F. Bartlett, J. Edwards, W. I. Taylor, and K. Wiesner, *Chem. Industry* (London), 323 (1953).
11. D. H. R. Barton, *J. Chem. Soc.* 1027 (1953).
12. M. F. Barnes and J. MacMillan, *J. Chem. Soc.* (C), 351 (1967).
13. K. Wiesner, Z. Valenta, J. F. King, R. K. Maudgal, L. G. Humber, and S. Ito, *Chem. Ind. (London)* 173 (1957).
14. A. D. Kuzovkov, *J. Gen. Chem. (U.S.S.R.)* **23,** 521 (1953).
15. S. Junussow, *J. Gen. Chem. (U.S.S.R.)* **18,** 515 (1948).
16. E. Ochiai, T. Okamoto, S. Sakai, and S. Inoue, *J. Pharm. Soc. (Japan)* **75,** 638 (1955).
17. E. Ochiai and M. Kaneko, *J. Pharm. Soc. (Japan)* **76,** 469 (1956).
18. T. Okamoto, M. Natsume, Y. Iitake, A. Yoshino, and T. Amiya, *Chem. Pharm. Bull. (Tokyo)* **13,** 1270 (1965).
19. K. Wiesner, S. Ito, and Z. Valenta, *Experientia* **14,** 167 (1958).
20. T. Sugasawa, *Chem. Pharm. Bull. (Tokyo)* **9,** 889 (1961).
21. T. Sugasawa, *Chem. Pharm. Bull. (Tokyo)* **9,** 897 (1961).
22. A. Yoshino and Y. Iitake, *Acta Cryst.* **21,** 57 (1966).
23. W. A. Jacob and L. C. Craig, *J. Biol. Chem.* **143,** 589 (1942).
24. C. F. Huebner and W. A. Jacobs, *J. Biol. Chem.* **170,** 515 (1947).
25. C. F. Huebner and W. A. Jacobs, *J. Biol. Chem.* **174,** 1001 (1948).
26. W. A. Jacobs and L. C. Craig, *J. Biol. Chem.* **147,** 567 (1943).
27. S. W. Pelletier and W. A. Jacobs, *Chem. Ind. (London)* 1385 (1955).
28. O. E. Edwards and T. Singh, *Can. J. Chem.* **32,** 465 (1954); **33,** 448 (1955).
29. C. F. Huebner and W. A. Jacobs, *J. Biol. Chem.* **170,** 203 (1947).
30. D. M. Locke and S. W. Pelletier, *J. Amer. Chem. Soc.,* **80,** 2588 (1958); **81,** 2246 (1959).
31. S. W. Pelletier and W. A. Jacobs, *J. Amer. Chem. Soc.* **76,** 4496 (1954).

32. W. A. Jacobs, *J. Org. Chem.* **16,** 1593 (1951).
33. D. Dvornik and O. E. Edwards, *Chem. Ind. (London)* 623 (1958).
34. D. Dvornik and O. E. Edwards, *Can. J. Chem.* **42,** 137 (1964).
35. S. W. Pelletier, *Chem. Ind. (London)* 1116 (1958).
36. S. W. Pelletier and P. C. Parthasarathy, *J. Amer. Chem. Soc.* **87,** 777 (1965).
37. S. W. Pelletier and W. A. Jacobs, *J. Amer. Chem. Soc.* **78,** 4139 (1956).
38. S. W. Pelletier and W. A. Jacobs, *J. Amer. Chem. Soc.* **78,** 4144 (1956).
39. S. W. Pelletier and K. Kawazu, *Chem. Ind. (London)* 1879 (1963).
40. S. W. Pelletier, K. Kawazu, and K. W. Gopinath, *J. Amer. Chem. Soc.* **87,** 5229 (1965).
41. D. Dvornik and O. E. Edwards, *Can. J. Chem.* **35,** 860 (1957).
42. A. J. Solo and S. W. Pelletier, *Chem. Ind. (London),* 1108 (1960).
43. D. Dvornik and O. E. Edwards, *Tetrahedron* **14,** 54 (1961).
44. S. W. Pelletier, Tetrahedron **14,** 76 (1961).
45. S. W. Pelletier, *Chem. Ind. (London)* 1016 (1956).
46. S. W. Pelletier, *Chem. Ind. (London)* 1670 (1957).
47. S. W. Pelletier, *J. Amer. Chem. Soc.* **87,** 799 (1965).
48. D. Dvornik and O. E. Edwards, *Chem. Ind. (London)* 952 (1957).
49. D. Dvornik and O. E. Edwards, *Proc. Chem. Soc. (London)* 305 (1958).
50. S. W. Pelletier, *J. Amer. Chem. Soc.* **82,** 2398 (1960).
51. S. W. Pelletier and D. M. Locke, *J. Amer. Chem. Soc.* **87,** 761 (1965).
52. O. E. Edwards and R. Howe, *Proc. Chem. Soc. (London)* 62 (1959).
53. J. W. ApSimon, O. E. Edwards, and R. Howe, *Can. J. Chem.* **40,** 630 (1962).
54. H. Vorbrueggen and C. Djerassi, *Tetrahedron Letters* 119 (1961).
55. H. Vorbrueggen and C. Djerassi, *J. Amer. Chem. Soc.* **84,** 2990 (1962).
56. E. Mosettig, P. Quitt, U. Beglinger, J. A. Waters, H. Vorbrueggen, and C. Djerassi, *J. Amer. Chem. Soc.* **83,** 3163 (1961).
57. C. Djerassi, P. Quitt, E. Mosettig, R. C. Cambie, P. S. Rutledge, and L. H. Briggs, *J. Amer. Chem. Soc.* **83,** 3720 (1961).
58. W. A. Ayer, C. E. McDonald, and G. G. Iverach, *Tetrahedron Letters* **17,** 1095 (1963).
59. L. H. Zalkow and N. N. Girotra, *J. Org. Chem.* **28,** 2037 (1963); **29,** 1299 (1964).
60. J. W. ApSimon and O. E. Edwards, Can. J. Chem. **40,** 896 (1962).
61. K. Wiesner and J. A. Edwards, *Experientia* **11,** 255 (1955).
62. S. W. Pelletier, *Experientia* **20,** 1 (1964).
63. A. J. Solo and S. W. Pelletier, *Proc. Chem. Soc. (London)* 14 (1961).
64. W. A. Jacobs and L. C. Craig, *J. Biol. Chem.* **143,** 605 (1942).
65. W. A. Jacobs and C. F. Huebner, *J. Biol. Chem.* **170,** 189 (1947).
66. M. H. Benn, *Can. J. Chem.* **44,** 1 (1966).
67. A. J. Solo and S. W. Pelletier, *J. Amer. Chem. Soc.* **81,** 4439 (1959).
68. A. J. Solo and S. W. Pelletier, *J. Org. Chem.* **27,** 2702 (1962).
69. M Przybylska, *Can. J. Chem.* **40,** 566 (1962); *Acta Crystall., 16,* 871 (1962).
70. R. T. Aplin, M. H. Benn, S. W. Pelletier, J. Solo, S. A. Telang, and H. Wright, *Can. J. Chem.* 2635 (1968).
71. S. W. Pelletier, L. H. Keith, and P. C. Parthasarathy, *J. Amer. Chem. Soc.* **89,** 4146 (1967).
72. K. Wiesner, Z. Valenta, and L. G. Humber, *Tetrahedron Letters* **14,** 621 (1962).
73. E. Ochiai, T. Okamoto, T. Sugasawa, and H. Tani, *J. Pharm. Soc. (Japan),* **27,** 1605 (1952).

74. E. Ochiai, T. Okamoto, T. Sugasawa, H. Tani, and H. S. Hai, *J. Pharm. Soc. (Japan)* **27,** 816 (1952).
75. E. Ochiai, T. Okamoto, S. Sakai, M. Kaneko, K. Fujisawa, U. Nagai, and H. Tani, *J. Pharm. Soc. (Japan)* **76,** 550 (1956).
76. E. Ochiai, T. Okamoto, T. Sugasawa, H. Tani, S. Sakai, H. S. Hai, and H. Endo, *Pharm. Bull. (Japan)* **1,** 60 (1953).
77. E. Ochiai and T. Okamoto, *Chem. Pharm Bull. (Tokyo)* **7,** 550 (1959).
78. E. Ochiai, T. Okamoto, T. Sugasawa, and S. Sakai, *Pharm. Bull. (Japan)* **2,** 388 (1954).
79. E. Ochiai, T. Okamoto, S. Hara, S. Sakai, and M. Natsume, *Chem. Pharm. Bull. (Tokyo)* **6,** 327 (1958).
80. S. Hara and Y. Tokoushige, *Chem. Pharm. Bull. (Tokyo)* **8,** 976 (1960).
81. E. Ochiai and T. Okamoto, *Chem. Pharm. Bull. (Tokyo)* **7,** 556 (1959).
82. E. Ochiai, T. Okamoto, T. Sugasawa, H. Tani, and S. Sakai, *Pharm. Bull. (Japan)* **1,** 152 (1953).
83. S. Sakai, *J. Pharm. Soc. (Japan)* **76,** 1054 (1956).
84. S. Sakai, *Chem. Pharm. Bull. (Tokyo)* **6,** 448 (1958).
85. S. Sakai, *Chem. Pharm. Bull. (Tokyo)* **7,** 50 (1959).
86. S. Sakai, *Chem. Pharm. Bull. (Tokyo)* **7,** 55 (1959).
87. H. Suginome and F. Shimanouchi, *Ann.* **545,** 220 (1940).
88. H. Suginome and S. Umezawa, *J. Fac. Sci. Hokkaido Univ. Ser. III,* **4,** 14 (1950).
89. H. Suginome, S. Kakimoto, and J. Sonoda, *J. Fac. Sci. Hokkaido Univ. Ser. III,* **4,** 25 (1950).
90. H. Suginome and S. Imato, *J. Fac. Sci. Hokkaido Univ. Ser. III,* **4,** 33 (1950).
91. M. Natsume, *Chem. Pharm. Bull. (Tokyo)* **7,** 539 (1959).
92. T. Okamoto, *Chem. Pharm. Bull. (Tokyo)* **7,** 44 (1959).
93. T. Okamoto, M. Natsume, H. Zenda, S. Kamata, A. Yoshino, *Abstr. I.U.P.A.C. Sympos. Kyoto* 115 (1964).
94. H. Suginome, T. Koyama, and Y. Kunimatsu, *J. Fac. Sci. Hokkaido Univ. Ser. III,* **4,** 16 (1950).
95. M. Natsume, *Chem. Pharm. Bull. (Tokyo)* **8,** 374 (1960).
96. M. Natsume, *Chem. Pharm. Bull. (Tokyo)* **10,** 879 (1962).
97. M. Natsume, *Chem. Pharm. Bull. (Tokyo)* **10,** 883 (1962).
98. T. Okamoto, M. Natsume, and S. Kamata, *Chem. Pharm. Bull. (Tokyo)* **12,** 1124 (1964).
99. G. Goto, K. Sasaki, N. Sakabe, and Y. Hirata, *Tetrahedron Letters* **11,** 1369 (1968)
100. S. W. Pelletier and P. C. Parthasarathy, *Tetrahedron Letters* **4,** 205 (1963).
101. W. Nagata, T. Sugasawa, M. Narisuda, T. Wakabayashi, and Y. Hayase, *J. Amer. Chem. Soc.* **85,** 2342 (1963); *ibid* **89,** 1483 (1967).
102. W. Nagata, M. Narisuda, T. Wakabayashi and T. Sugasawa, *J. Amer. Chem. Soc.* **86,** 929 (1964); **89,** 1499 (1967).
103. S. Masamune, *J. Amer. Chem. Soc.* **86,** 288 (1964).
104. S. Masamune, *J. Amer. Chem. Soc.* **86,** 290 (1964).
105. S. Masamune, *J. Amer. Chem. Soc.* **86,** 291 (1964).
106. Z. Valenta, K. Wiesner, and C. M. Wong, *Tetrahedron Letters*, **36,** 2437 (1964).
107. R. W. Guthrie, W. A. Henry, H. Immer, C. M. Wong, Z. Valenta, and K. Wiesner, *Coll. Czech. Chem. Comm.* **31,** 602 (1966).
108. J. A. Findlay, W. A. Henry, T. C. Jain, Z. Valenta, K. Wiesner, and C. M. Wong, *Tetrahedron Letters* **19,** 869 (1962).
109. R. W. Guthrie, A. Philipp, Z. Valenta, and K. Wiesner, *Tetrahedron Letters* **34,** 2945 (1965).

110. R. W. Guthrie, Z. Valenta, and K. Wiesner, *Tetrahedron Letters* **38,** 4645 (1966).
111. E. Fujita, J. Fujita, and M. Shibuya, *Chem. Comm. (London)* 468 (1967).
112. A. Tahara and K. Hirao, *Tetrahedron Letters* **14,** 1453 (1966).
113. K. Wiesner and A. Philipp, *Tetrahedron Letters* **14,** 1467 (1966).
114. R. A. Bell and R. E. Ireland, *Tetrahedron Letters* **4,** 269 (1963).
115. I. Iwai, A. Ogiso and B. Shimizu, *Chem. Ind. (London)* 1288 (1962).
116. I. Iwai and A. Ogiso, *Chem. Ind. (London)* 1084 (1963).
117. B. Shimizu, A. Ogiso, and I. Iwai, *Chem. Pharm. Bull. (Tokyo)* **11,** 333 (1963).
118. B. Shimizu, A. Ogiso, and I. Iwai, *Chem. Pharm. Bull. (Tokyo)* **11,** 766 (1963).
119. A. Ogiso, B. Shimizu, and I. Iwai, *Chem. Pharm. Bull. (Tokyo)* **11,** 770 (1963).
120. A. Ogiso, B. Shimizu, and I. Iwai, *Chem. Pharm. Bull. (Tokyo)* **11,** 774 (1963).
121. A. Ogiso and I. Iwai, *Chem. Pharm. Bull. (Tokyo)* **12,** 820 (1964).
122. A. A. Othman and N. A. J. Rogers, *Tetrahedron Letters* **20,** 1339 (1963).
123. W. L. Meyer and A. S. Levinson, *Proc. Chem. Soc. (London)* **15** (1963).
124. W. L. Meyer and A. S. Levinson, *J. Org. Chem.* **28,** 2859 (1963).
125. T. Matsumoto and A. Suzuki, *Bull. Chem. Soc. (Japan)* **34,** 274 (1961).
126. T. Matsumoto, M. Yanagiya, E. Kawakami, T. Okuno, M. Kakizawa, S. Yasuda, Y. Gama, J. Omi, and M. Matsunaga, *Tetrahedron Letters* **9,** 1127 (1968).
127. L. H. Zalkow and N. N. Girotra, *Chem. Ind. (London)* 704 (1964).
128. N. N. Girotra and L. H. Zalkow, *Tetrahedron* **21,** 101 (1965).
129. L. H. Zalkow, N. N. Girotra, and V. B. Zalkow, *J. Org. Chem.* **32,** 806 (1967).
130. L. H. Zalkow, B. Kumar, D. H. Miles, and J. Nabors, *Tetrahedron Letters* **16,** 1965 (1968).
131. D. H. R. Barton and J. R. Hanson, *Chem. Comm. (London)* 117 (1965).
132. R. A. Finnegan and P. L. Bachman, *J. Org. Chem.* **30,** 4145 (1965).
133. A. Tahara, K. Hirao, and Y. Hamazaki, *Chem. Ind. (London)* 850 (1965).
134. A. Tahara, K. Hirao, and Y. Hamazaki, *Tetrahedron* **21,** 2133 (1965).
135. A. Tahara, K. Hirao, and Y. Hamazaki, *Chem. Pharm. Bull. (Tokyo)* **15,** 1785 (1967).
136. A. A. Othmann, M. A. Qasseem, and N. A. J. Rogers, *Tetrahedron* **23,** 87 (1967).
137. K. Wiesner, A. Philipp, and P. Ho, *Tetrahedron Letters* **10,** 1209 (1968).
138. J. R. Hanson, *Tetrahedron* 1701 (1966).
139. E. Fujita, *Bull. Inst. Chem. Res. Kyoto Univ.* **43,** 278 (1965).
140. E. Fujita, *Bull. Inst. Chem. Res. Kyoto Univ.* **44,** 239 (1966).
141. E. Fujita, *Bull. Inst. Chem. Res. Kyoto Univ.* **45,** 252 (1967).
142. K. Wiesner, S. Uyeo, A. Philipp, and Z. Valenta, *Tetrahedron Letters* **60,** 6279 (1968).
143. E. Fujita, T. Fujita, M. Schibuya and T. Shingu, *Tetrahedron,* **25,** 2517 (1969).

18

The C_{19}—Diterpene Alkaloids

L. H. KEITH and S. W. PELLETIER
Department of Chemistry
The University of Georgia
Athens, Georgia

1. INTRODUCTION AND GENERAL FUNCTIONAL GROUP CHARACTERISTICS

The diterpene alkaloids modeled on a C_{19}-skeleton comprises the highly toxic ester bases that are heavily substituted by methoxy and hydroxy groups. The lethal dose for man is about 2–5 mg and because of their great toxicity the alkaloids from the *Aconitum* and *Delphinium* species in India have, from ancient times, been used extensively as poisons, particularly as a dip for arrowheads. The drug attacks the autonomic nervous system causing a slowing of the heart rate and a lowering of the blood pressure. Interestingly, the hydrolyzed bases are relatively nontoxic.

The structures of the C_{19}-diterpene alkaloids may be subdivided into two general types based on their functional groups and a third type, a small group of diterpene alkaloids occurring in the mother liquors of *Aconitum heterophyllum*, which are modeled on a modified lycoctonine-type skeleton and contain a lactone ring. The first is the lycoctonine-type, which is based on the parent alkaloid lycoctonine, and the second is the aconitine-type based on the parent alkaloid of the same name. Derivatives of both lycoctonine and aconitine have been analyzed by x-ray crystallography, and their absolute configurations have been established from these studies. The other alkaloids of these two groups have either been directly correlated with one of these two alkaloids or with other alkaloids which, in turn, have been correlated with one of them whenever possible. There are several features common to both subgroups, the most obvious being the hexacyclic

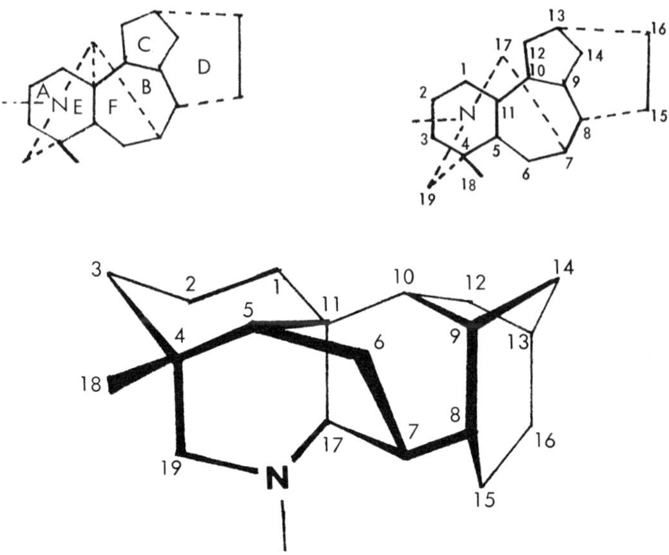

Figure 18.1 The lettering and numbering of the aconitine- and lycoctonine-type skeletal systems.

skeletal system comprised of one 7-membered, three 6-membered, and two 5-membered rings. Figure 18.1 illustrates the ring lettering and the numbering of the positions.

The lycoctonine- and aconitine-type skeletons possess a tertiary nitrogen substituted with either an ethyl or methyl group. In addition, they all possess a C(1) and C(8) oxygen functional group (OH, OCH_3, or ester) as well as a C(14) α-oxygen functional group and a methoxyl at C(16) in the β-configuration. Most possess a C(18) oxygen functional group as a methoxyl or an N-substituted ester of anthranilic acid. This substitution pattern may not prevail in all future alkaloids of this type,* but it certainly should occur in the majority of them.

The most important difference between an aconitine- and a lycoctonine-type skeleton is that the latter contains an oxygenated functional group at C(7). The presence or absence of a ditertiary α-glycol system—either as its free hydroxyl or in its methylenated form—is thus the determining factor in the classification, as well as much of the chemistry, of the lycoctonine- and aconitine-type alkaloids, respectively. Two further trends are to be noted. All of the known lycoctonine-type alkaloids have a β-methoxyl at C(6) while in all of the aconitine-type alkaloids that contain a C(6) meth-

*The chemistry of lappaconitine, while still anomolous in some respects, gives indication that a C(8) oxygen functional group may be absent.[1]

oxyl, it is in the α-configuration. Finally, whereas many of the aconitine-type alkaloids possess a bridgehead hydroxyl at C(13), none of the lycoctonine-type alkaloids have yet been isolated with a substituent at this position.

Recently, Achmatowicz et al.[2] demonstrated a chemical method of ascertaining the configuration of the C(6)-methoxyl, previously obtainable only by x-ray crystallography or correlations. Brief treatment of the diterpene alkaloids containing an α-C(6)-methoxyl with neutral permanganate gave the N-dealkylated derivatives as the predominant product. On the other hand, similar treatment of those alkaloids bearing a β-C(6)-methoxyl gave almost exclusively the lactam derivative. A suggested mechanism[2] is shown in Figure 18.2. Steric hindrance offered by a C(6)-α-

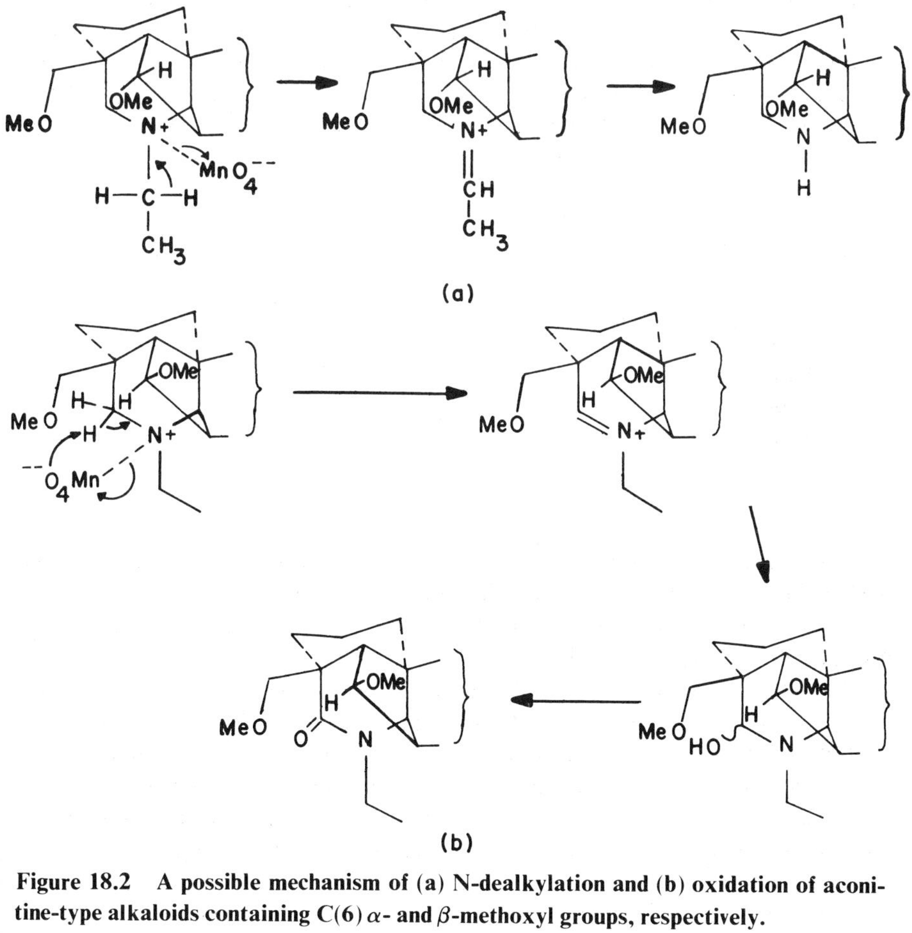

Figure 18.2 A possible mechanism of (a) N-dealkylation and (b) oxidation of aconitine-type alkaloids containing C(6) α- and β-methoxyl groups, respectively.

methoxyl to the vulnerable C(19) methylene prevents oxidation of the latter, thereby favoring attack on the imino-ethyl group instead. The β-oriented C(6)-methoxyl of the lycoctonine-type alkaloids, however, offers no interferences with the C(19) methylene so that normal oxidation occurs at this position.

2. THE ACONITUM-TYPE ALKALOIDS

2.1 Aconitine and Delphinine

The structural elucidations of aconitine and delphinine are so entwined that the chemistry of one cannot meaningfully be studied without the other. Aconitine (1) has been known since 1833[3] and is one of the most accessible and most complicated representatives of the aconitum alkaloids. The early work is characterized by confusion and, indeed, all but the most recent studies accomplished little more than clarification of the functional groups. Aconitine contains a tertiary nitrogen with an N-ethyl, one acetoxy (OAc) and one benzoyloxy (OBz) group, four methoxyls, and three hydroxyls; these studies have been reviewed by Stern.[4,5]

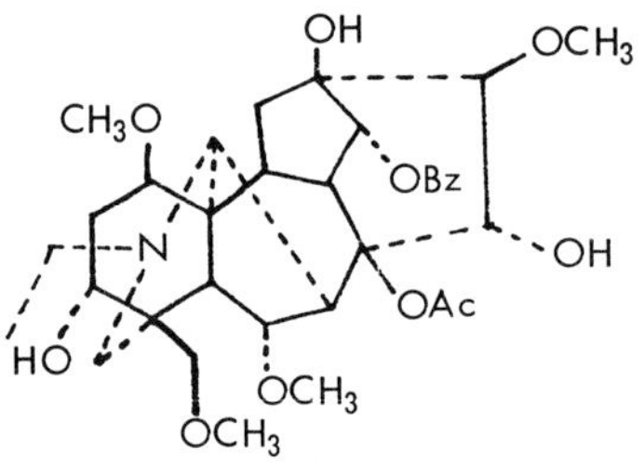

(1)

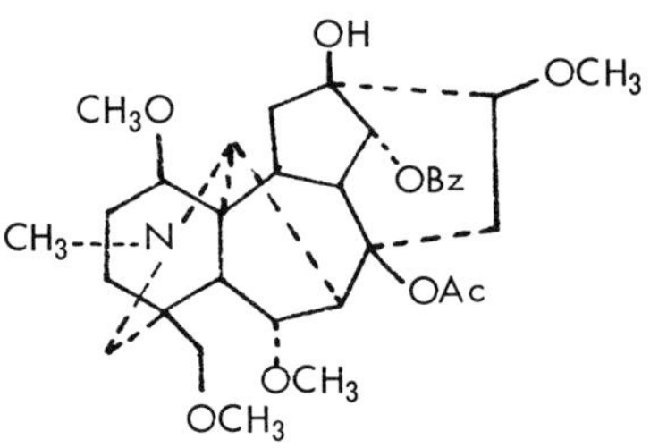

(2)

Delphinine (2) is the main alkaloid of *Delphinium staphisagria* and was first isolated in 1819. Early experimentation showed it to possess a hexacyclic skeleton with four methoxyls, one acetoxy and one benzoyloxy group, one tertiary hydroxyl and an N-methyl.[6]

2.1.1 The Chemistry of the C–D Ring Structures Analysis and repetition of previous studies with delphinine, plus additional experimentation, enabled Wiesner et al.[7–9] to deduce the C–D ring structure of delphinine as shown in partial structure (3); this deduction plus some conjecture as to the formation of the rings A, B, E, and F allowed structures for delphinine and aconitine to be proposed. Soon afterwards, the independent x-ray crystallographic studies of Przybylska and Marion[22,23] on desmethanol aconinone·HI confirmed the deductions of this classical degradative work.

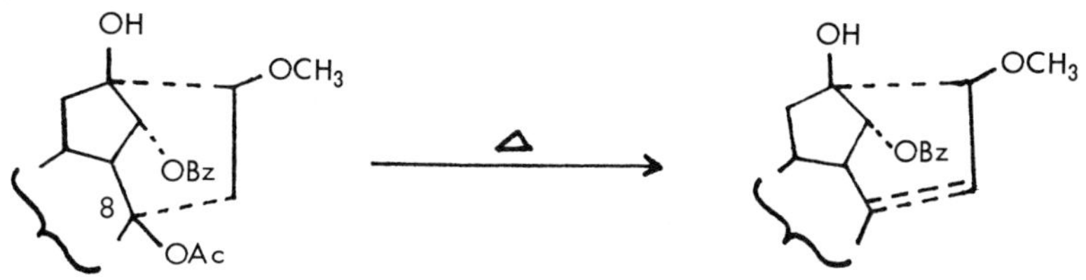

(3) (4)

2.1.1.1 ***Pyrolysis of Aconitine-type Alkaloids*** Pyrolysis of delphinine, like all aconitine-type alkaloids acetylated at C(8), proceeds with the loss of 1 mole of acetic acid. The product, pyrodelphinine (4) is an olefin with a bridgehead double bond.

Wiesner et al.[10] discovered an unusual spectroscopic property of these pyrolysis products. The uv spectra of these pyro-derivatives containing an N-alkyl group exhibit an absorption maximum between 235–245 mμ with an extinction coefficient of about 6,000 when the free base is used. This absorption disappears on acidification and reappears again on neutralization. However, if the nitrogen is substituted with an acetyl or formyl group, which removes the basic properties of the nitrogen, this characteristic absorption is not observed. Nor is it observed once the rearrangement to the corresponding isopyro-derivative is accomplished. This phenomenon appears to hold promise as a specific indication of an aconitine-type skeleton, particularly when correlation with an alkaloid of known absolute configuration is not readily amenable.

Since the bridgehead double bond in structure (5) is highly strained, the suggestion was made that the structure may be represented, to a small extent, by (6). If such a mesomerism is possible, it could give rise to the observed uv absorption and models show that there is an overlap of orbitals between the C(7)–C(8) or C(8)–C(15) double bond and the N–C(17) double bond.

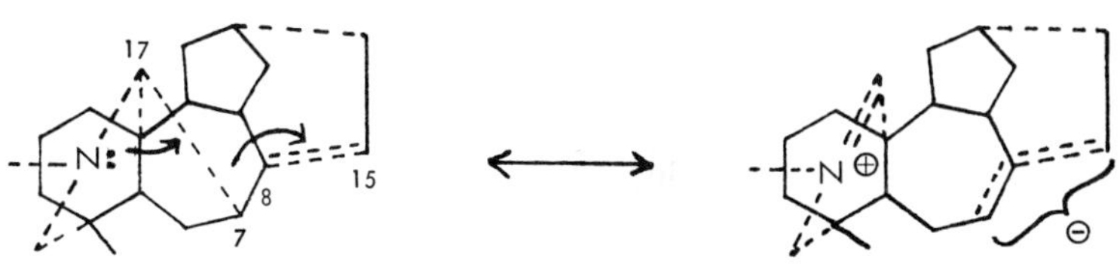

(5) (6)

On the other hand, pyrolysis of aconitine (and other aconitine-type alkaloids possessing a C(15) hydroxyl group) proceeds with the loss of acetic acid but the product is a ketone. The original degradative work employed oxonitine (7), an N-formyl derivative of aconitine produced by permanganate oxidation of the latter.[11] Pyrolysis of oxonitine (7) resulted in the formation of pyrooxonitine (8), an enol that tautomerized to the keto form and thus suggested that the C(8)-acetoxy group and the C(15)-hydroxyl of aconitine may be *trans*. The ketonic nature of pyrooxonitine was shown spectroscopically by carbonyl absorption characteristic of a cyclic 6-membered ketone (1718 cm^{-1}) in pyrooxonine (9).

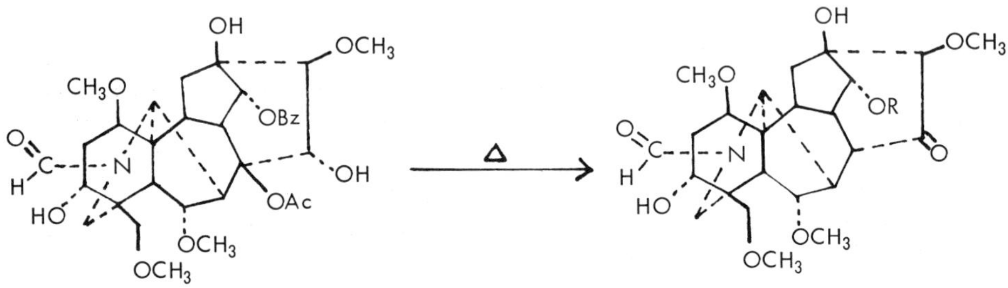

(7)

(8) R = Bz

(9) R = H

Chemical evidence for the ketonic nature of pyroaconitine (10), was demonstrated by subjecting it to Wolff–Kishner reduction.[12] The product contained no carbonyl absorption in its infrared spectrum, thus showing that not only had the ketone been reduced but also that the benzoyloxy group had been saponified; in addition, one of the methoxyl groups had been eliminated. Since it resisted catalytic hydrogenation, the Wolff–Kishner product was thus assumed to possess structure (11). A later correlation involving the conversion of another alkaloid, pseudaconitine (12), to the Wolff–Kishner reduction product of pyroaconitine showed that the latter was not (11) but the unsaturated derivative (13) instead.[13]

Most of the aconitine-type alkaloids contain an aromatic ester functional group. Tsuda and Marion[14] observed that a doublet of 1 proton intensity occurred at about $\tau = 5.1$ in the nmr spectra of all these compounds but that it was absent in the spectra of the corresponding alkamines. Since, in aconitine the acetoxy group is located on the quaternary C(8), this signal must be due to the proton geminal to the benzoyloxy group, and coupling with one neighboring proton. This condition is met if the aromatic ester is at C(14) but not if it is at the quaternary C(13). In addition, the acetoxy protons of these alkaloids all show signals at unusually high field (about $\tau = 8.7$) that cannot be attributed to their tertiary nature. The explanation is that the upfield shift is caused by the diamagnetic anisotropy of the aromatic ring which can easily come in close proximity to the C(8) acetoxy protons. This condition is satisfied if the benzoyloxy group is at C(14) but

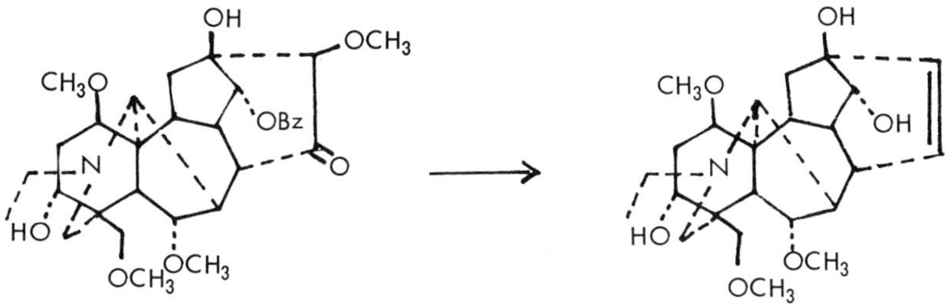

(10) (13)

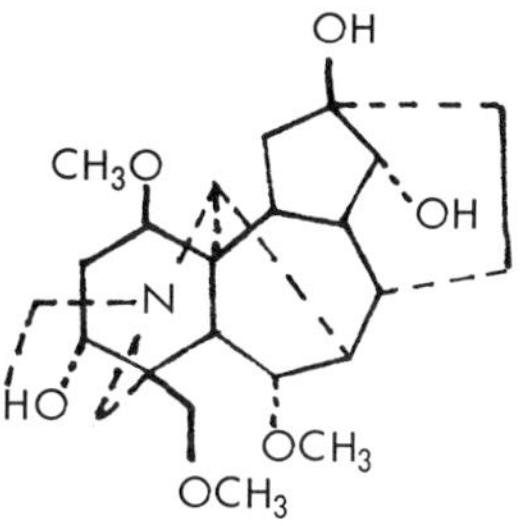

(11)

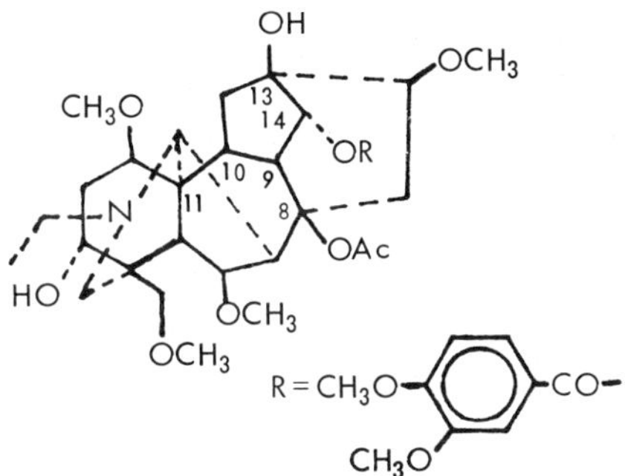

(12)

not if it is at C(12) or C(13). Dreiding models show that for a C(14)-α-benzoyloxy group the C(9)–C(14) H–H dihedral angle is about 40° while the dihedral angle for the β-configuration is about 80°; the Karplus values for dihedral angles of 40° and 80° are about 4 Hz and 0 Hz, respectively. The observed coupling constant of 4.5 Hz is thus in excellent agreement with the theoretical value.

2.1.1.2 *The Isopyro Rearrangement* Pyrodelphinine and the pyrolysis products of other aconitine-type alkaloids that lack a C(15) hydroxyl group undergo an acid catalyzed allylic rearrangement of the C(16) methoxyl and the bridgehead double bond. Thus, refluxing α-oxopyrodelphinine (14) in acidic methanol yields α-oxoisopyrodelphinine (15).

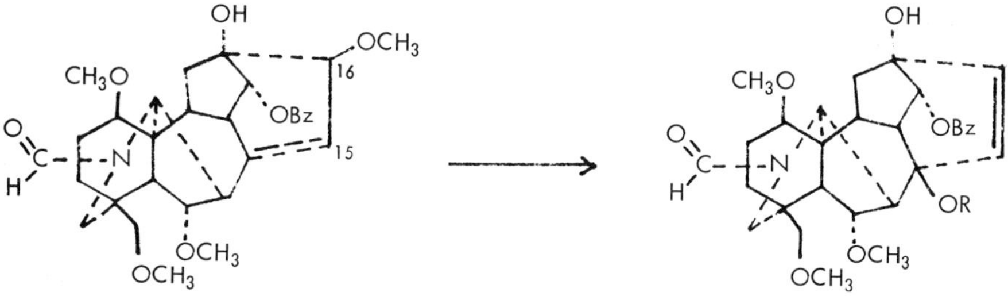

(14)

(15) R = CH_3

(16) R = Ac

Wiesner et al.[8] demonstrated that when the isomerization is carried out in ^{14}C-methanol, radioactivity corresponding to one mole of methanol was incorporated into α-oxoisopyrodelphinine. In addition, they found it was not possible to prepare α-oxoisopyrodelphinine in any medium other than methanol; thus, when α-oxopyrodelphinine is treated with *p*-toluenesulfonic acid in glacial acetic acid, the isomerization still occurred but the methoxyl was exchanged quantitatively for an acetoxy group (16).

2.1.1.3 ***The Acyloin Rearrangement*** Jacobs and Pelletier[15] noted the rather unusual isomerization of dextrorotatory ($[\alpha]_D = +33°$) α-oxodelphonone (17) to levorotatory ($[\alpha]_D = -80°$) iso- α-oxodelphonine (18), which occurred merely by passing (17) through alumina. The same product (18) could also be produced by treating (17) with base. Later, Wiesner et al.[9] formulated the isomerized C–D ring structure as it is shown in (18) and explained it as the product of a simple acyloin rearrangement. As evidence for the acyloin rearrangement, (18) was shown to take up one mole of periodic acid. The olefinic *seco*-keto acid isolated (20) must have been formed by spontaneous β-elimination of a hydroxyl from the intermediate *seco*-keto acid (19).[9]

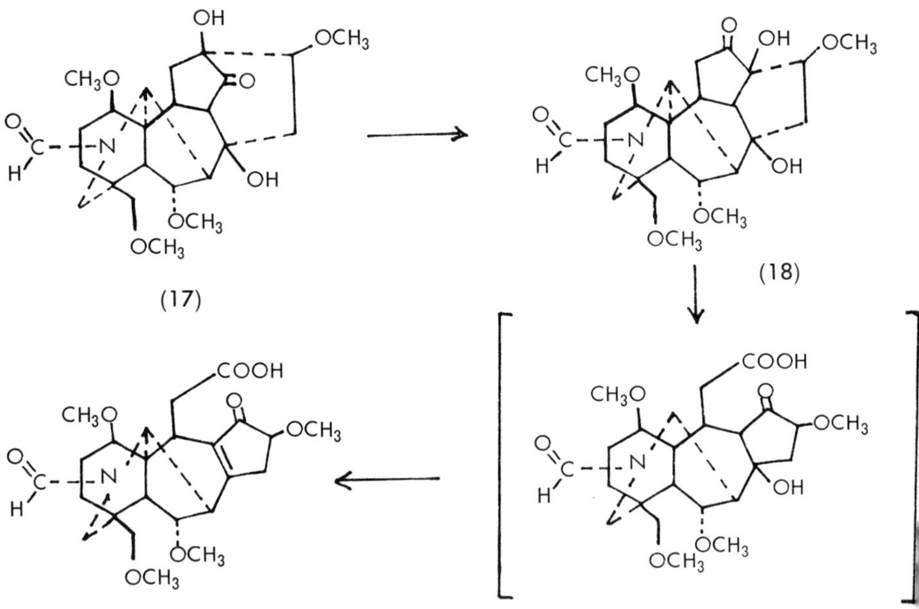

(17) (18) (20) (19)

The rearrangement with derivatives of aconitine is analogous. Oxidation of oxonine (21), the hydrolysis product of oxonitine (7), with chromium trioxide in pyridine (Sarett's reagent) gave the rearranged diketone (23). The latter is assumed to arise from a base catalyzed acyloin rearrangement of the intermediate oxidation product, (22), not isolated.[11] Cleavage of (23) with lead tetraacetate gave the *seco*-diketo acid (24) in an anologous manner with its delphinine counterpart. Both spontaneous β-elimination of a hydroxyl and β-elimination of the ring A methoxyl occurred from the original *seco*-diketo acid.

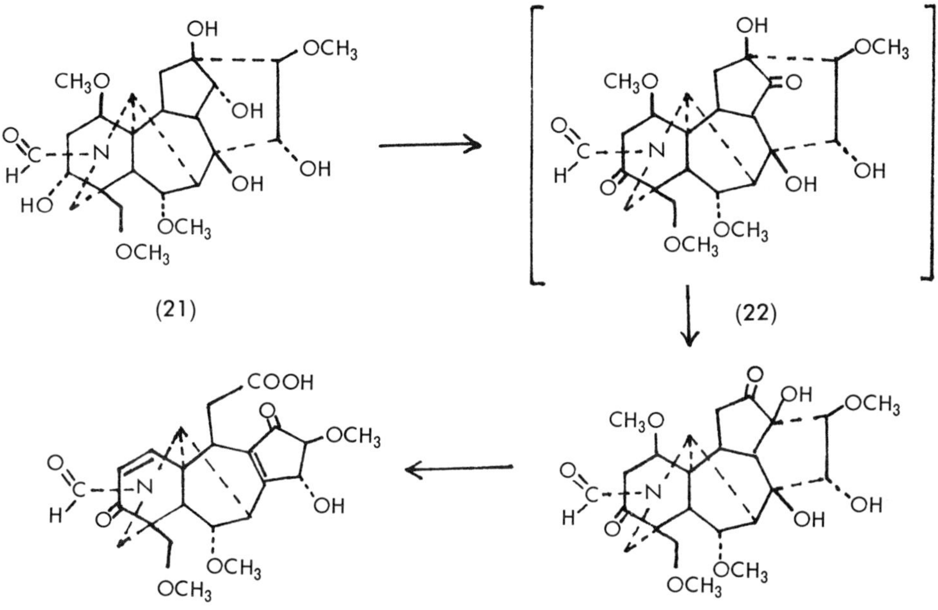

(21) (22) (24) (23)

If, on the other hand, oxonitine (7) is oxidized under the same conditions as oxonine, no rearrangement occurs and a high yield of the 3-keto derivative (25) is produced. Therefore, the C(14) position in aconitine is the site

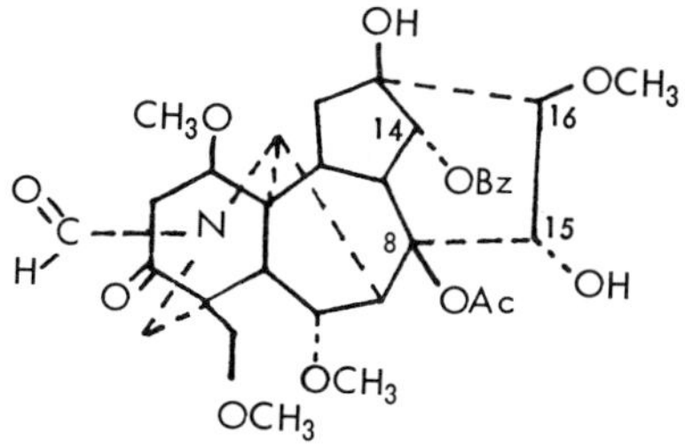

(25)

of one of the two ester groups, specifically the benzoyloxy group since pyrolysis had already demonstrated that the acetoxy group was at C(8).[16]

2.1.1.4 ***The Aromatization Rearrangement*** Oxonine (21) reacts with exactly 1 mole of periodic acid, and the *seco*-ketoaldehyde (26) produced undergoes an extensive and unusual rearrangement to a phenol (32) when heated in dilute aqueous base with air passing through the solution.[17] The

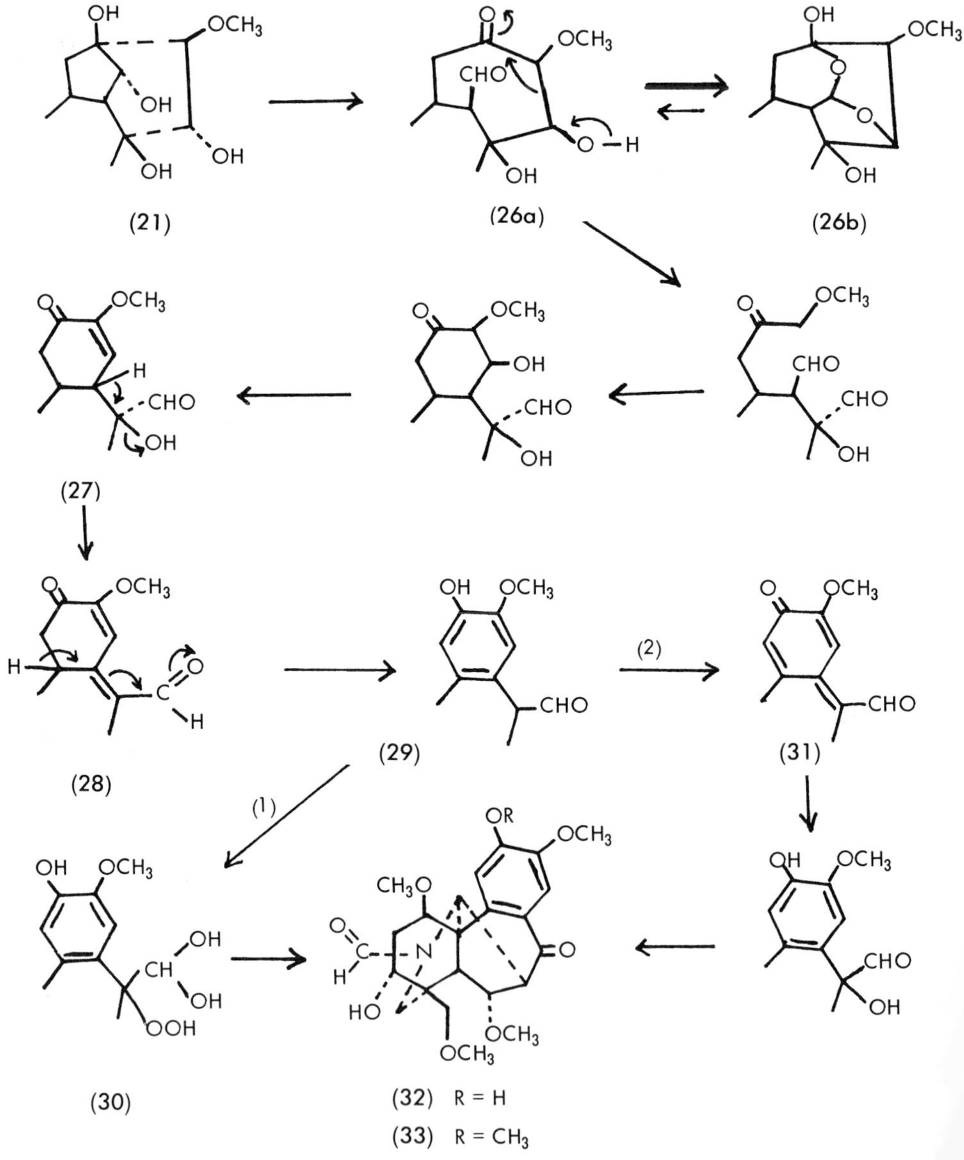

postulated mechanism[9,17,18] begins with a retroaldol cleavage of (26a). Then aldol condensation and dehydration with one aldehyde group occurs (26a) → (27) followed by dehydration to the dienonal (28) which, by a base-catalyzed double bond shift, could give the phenolic aldehyde (29). From (29) to (32), there appear to be at least two feasible paths, with the first seeming to be the most likely considering the catalytic effect of bubbling air through the solution. By path 1, air oxidation of (29) would give the peroxide (30) which, when cleaved by hydroxide, would give the phenol (32). Path 2 envisions an oxidation of (29) to give (31) followed by hydroxyl attack and spontaneous elimination of the aldehyde group. Compound (32) is an amorphous solid but the methyl ether (33) is crystalline. The existence of the *seco*-keto-aldehyde (26) in the cyclic acetal form (26b) is invoked to explain the otherwise anomolous facts that (a) sharp uptake of only one mole of periodate occurs and (b) the cleavage product shows no carbonyl group in its infrared spectrum.

2.1.2 The Hofmann Degradation The methohydroxide of delphonine (34) is postulated to undergo initial cleavage to the olefin, (35), not isolated.[9] The latter, *via* a reverse aldol, *β*-elimination of methoxyl and vinilogous *β*-keto aldehyde cleavage (although not necessarily in that order) gives rise to the isolated product, (36).

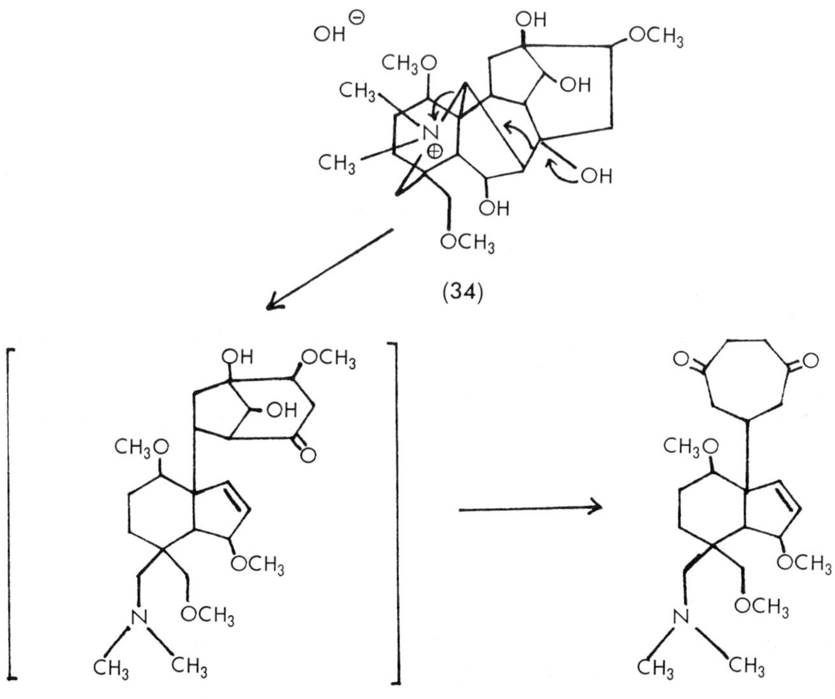

(34)

(35)

(36)

2.1.3 C(6)–C(18) γ-Lactone Formation Treatment of α-oxoisopyrodelphinine (37) with methanolic hydrochloric acid results in the replacement of two methoxyls by chlorine (38).[9] Substitution of the chlorine atoms by hydroxyl groups (39) followed by oxidation to (40) demonstrates that one of the replaced methoxyls was substituted on a primary carbon and the other on a tertiary carbon.[19] The same dichloride (38) is also produced from α-oxopyrodelphinine (41) making it clear that the tertiary methoxyl which was replaced is the same one involved in the pyro-isopyro rearrangement.

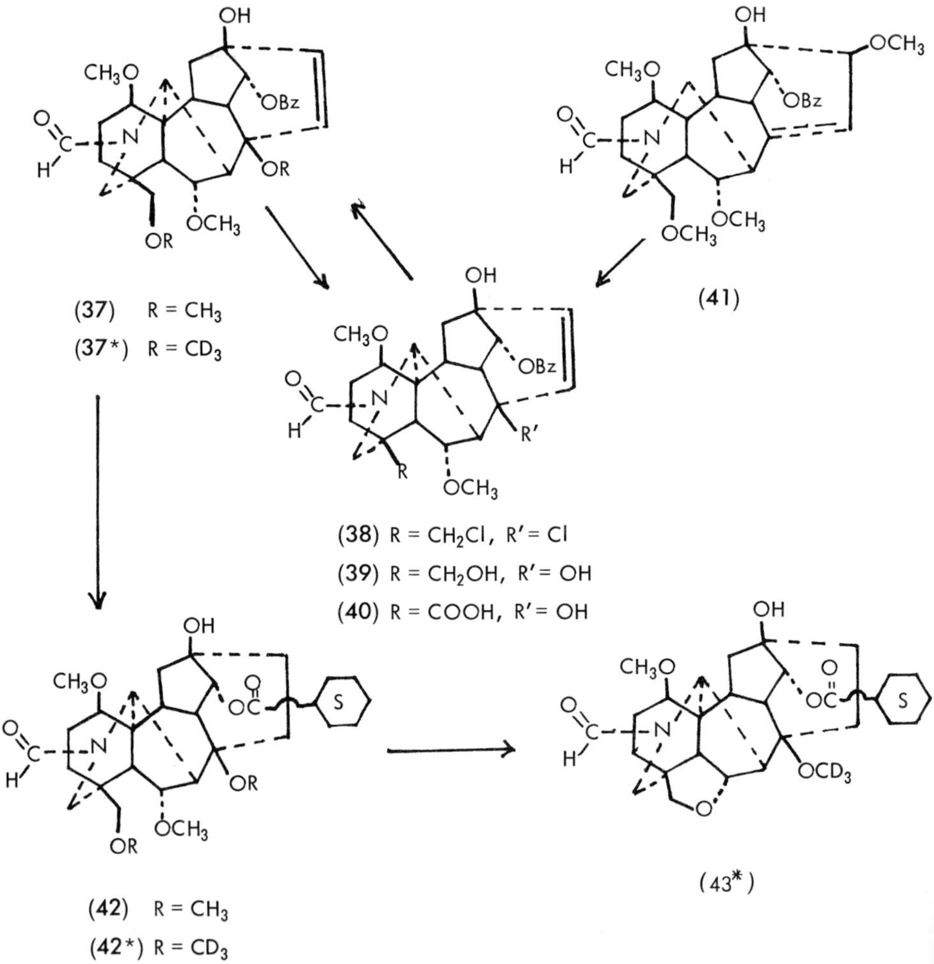

(37) R = CH_3
(37*) R = CD_3

(41)

(38) R = CH_2Cl, R′ = Cl
(39) R = CH_2OH, R′ = OH
(40) R = COOH, R′ = OH

(42) R = CH_3
(42*) R = CD_3

(43*)

Earlier experiments had shown that octahydro- α-oxoisopyrodelphinine (42) gives a dimethoxy ether with a corresponding loss of two methoxyls when treated with aqueous zinc chloride, and it had been reasonably assumed that internal ether formation took place between the same two methoxyls that were so easily displaced by chlorine.[20] A simple, yet elegant, labeling experiment showed, however, that only one of the two methoxyls displaced by chlorine is involved in the internal ether formation.[19] The dichloride (38) was converted into the deuterated α-oxoisopyrodelphinine (37*), and this labeled derivative was in turn converted into the cyclic ether (43*) *via* (42*). The ether was found to have retained one of its deuterated methoxyls.

2.1.4 Aconitine N-Oxide Pyrolysis and Oxidations Aconitine N-oxide (44), prepared by oxidation of aconitine (1) with peracetic acid and recon-

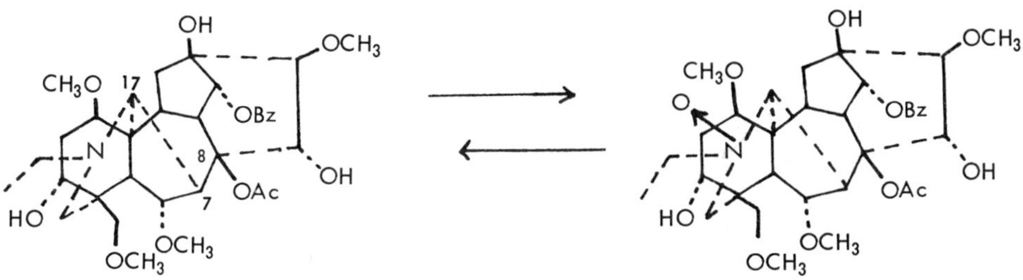

(1) (44)

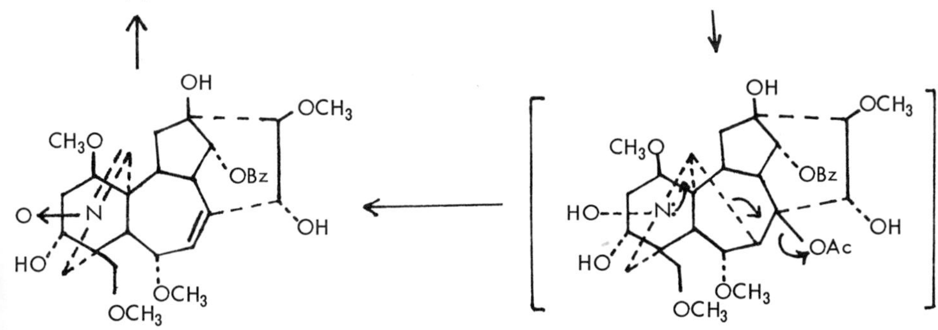

(46) (45)

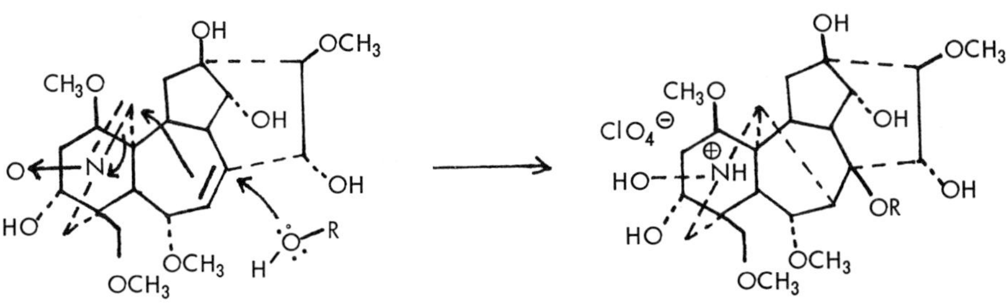

(47)

(48) R = CH_3

(49) R = C_2H_5

vertible to the latter by reduction, eliminates ethylene and acetic acid on pyrolysis to give, pentacyclic 46, presumably through the intermediate hydroxyamine (45).[17] Reduction of 46 followed by ethylation gave back aconitine and demonstrated that no rearrangements had occurred. The reformation of aconitine from (46) is explained by an attack at one end of the system by the solvent molecule with cyclization to the hydroxylamine which, in turn, is reduced to the secondary amine. This is further supported by the analogous cyclization of the desbenzoylnitrone (47) in methanolic or ethanolic perchloric acid. The formation of two different perchlorate salts, (48) and (49), depending on the choice of solvent, lends validity to the postulated mechanism. This reaction not only demonstrates the presence of the N-ethyl group but also the planar relationship described by the N—C(17)-C(7)-C(8) moiety.

A direct consequence of this postulation was the prediction that pyroaconitine N-oxide (50), lacking the crucial C(8)-acetoxy group, should produce a hexacyclic hydroxylamine instead of a pentacyclic one.[17] The pyrolytic product (51) was, indeed, a hexacyclic hydroxylamine and was further oxidized to (52) employing aerial oxidation in ammonium hydroxide in the presence of cupric ion. Hot aqueous periodic acid oxidation of (52) gave a neutral, blue compound (53) rather than the expected carboxylic acid.[21]

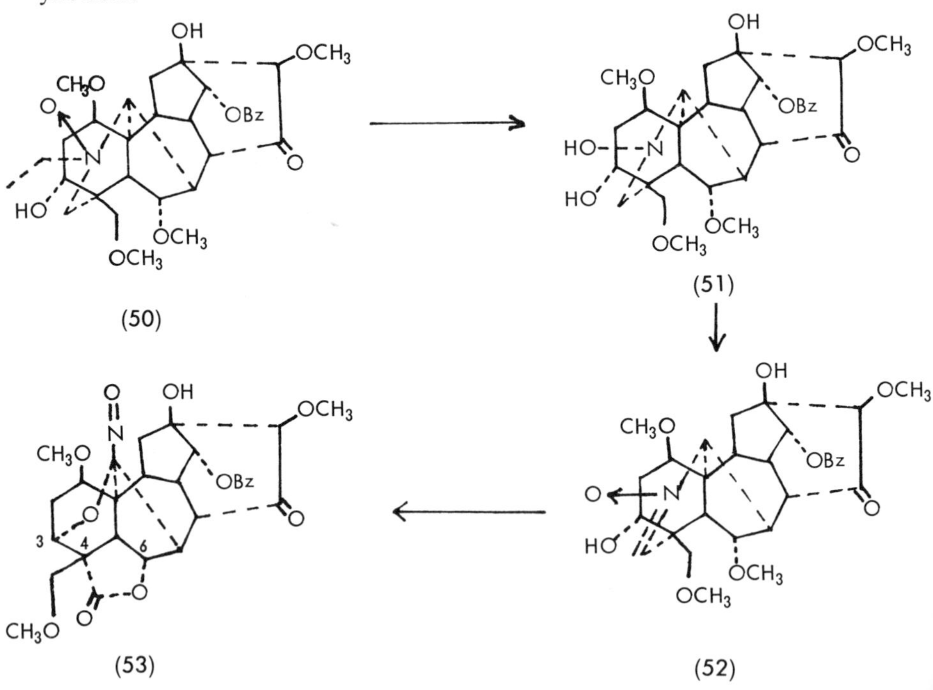

(50) (51) (52) (53)

Two significant facts are gained from this transformation: (a) the internal ether formation is sterically feasible only if the hydroxyl in ring A is at C(3) and α-oriented, and (b) the facile lactonization across the peri-positions between the generated C(4)-carboxylic acid group and the C(6)-methoxyl uniquely fixes both the location and configuration of the ring F methoxyl as designated.

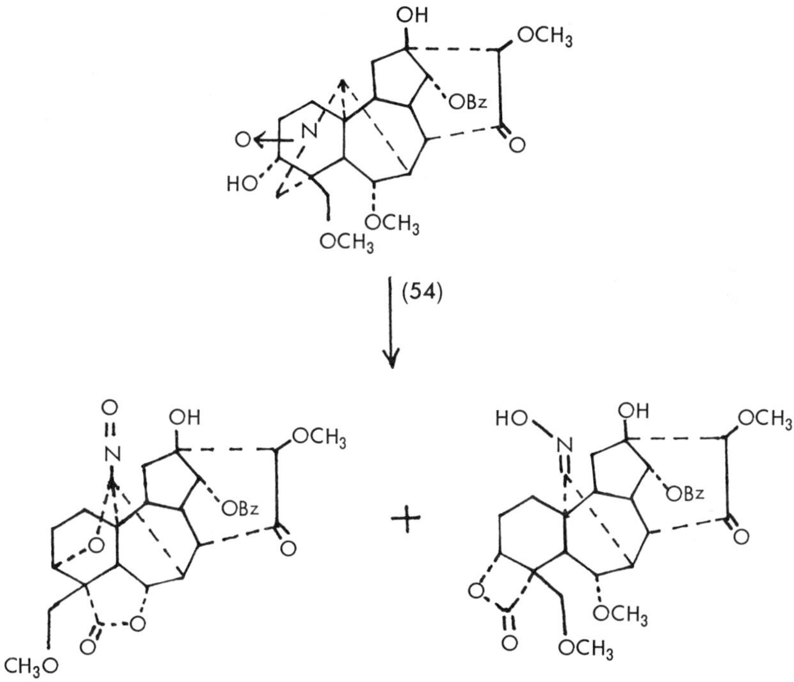

(55)

(56)

Similar oxidation of (54) yielded approximately equal amounts of the blue nitrosolactone (55) and the unusual oximino-β-lactone (56). It was then reasoned that since the oxidation of (52) gave essentially only γ-lactones while its desmethoxy counterpart produced both γ and β-lactones, that the C(1)-methoxyl opposed β-lactone formation.[21]

Then, if it is assumed that the fused β-lactone ring forces ring A into a pseudo-chair conformation, the bulky methoxyl-methyl substituent at C(4) is quasi-axial (57) and will destablize a β-lactone by nonbonded interaction with the methoxyl at C(1) if it is axially oriented.

This was the first evidence for the configuration of the C(1)-methoxyl group. The earlier x-ray crystallographic analysis by Przybylska and Marion[22,23] was an unquestionable milestone in the structural and skeletal

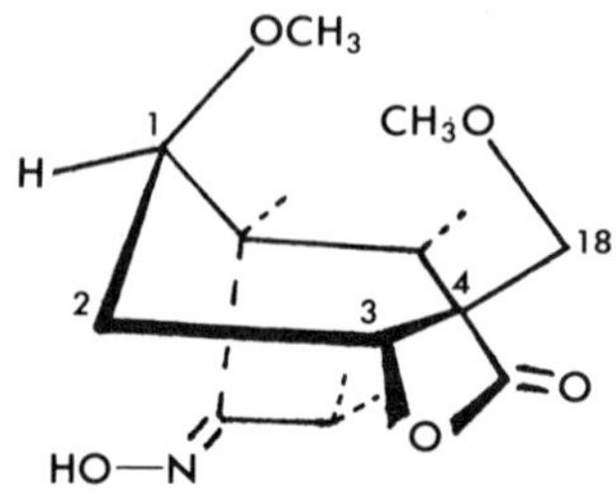

(57)

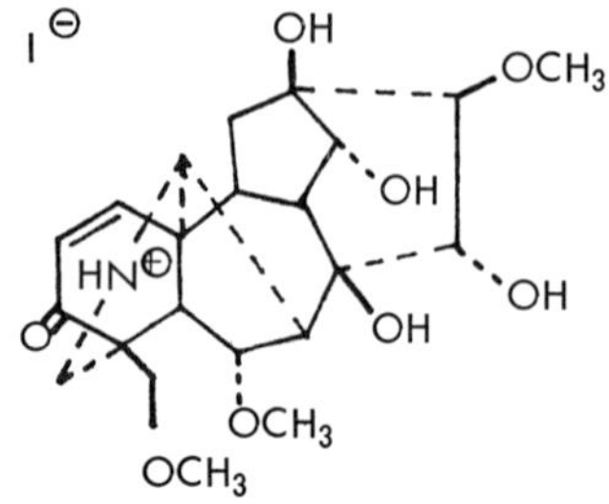

(58)

determination of aconitine. However, it had been done with demethanolaconinone hydriodide trihydrate (58) and although it substantiates the structural elucidations obtained by chemical means, as well as providing the absolute and relative configurations of thirteen out of the fifteen asymmetric centers present, it left the configuration of the C(1)-methoxyl undetermined.

2.1.5 The Correlation of Aconitine and Delphinine The correlation[24] of these two alkaloids involved the previously discussed aromatization reaction. α-Oxoisopyrodelphonine (59) was converted to its phenolic deriva-

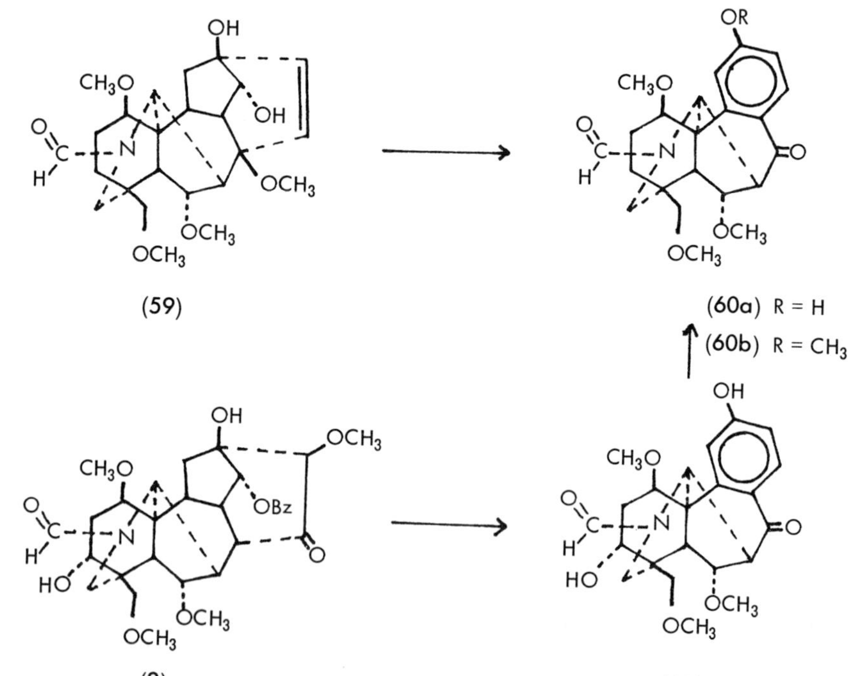

(59) (60a) R = H (60b) R = CH_3 (8) (61)

tive, (60a), and methylated (60b). Likewise, pryooxonitine (8) was converted to (61). Methylation of the phenolic hydroxyl (diazomethane) followed by chlorination of the C(3)-hydroxyl and then reduction with zinc gave a derivative identical with (60b).

2.2 Mesaconitine, Hypaconitine, and Deoxyaconitine

Both mesaconitine and hypaconitine are often found with aconitine and, therefore, it is not surprising that their structures are similar. The first report of these two alkaloids appeared in 1929.[25,26]

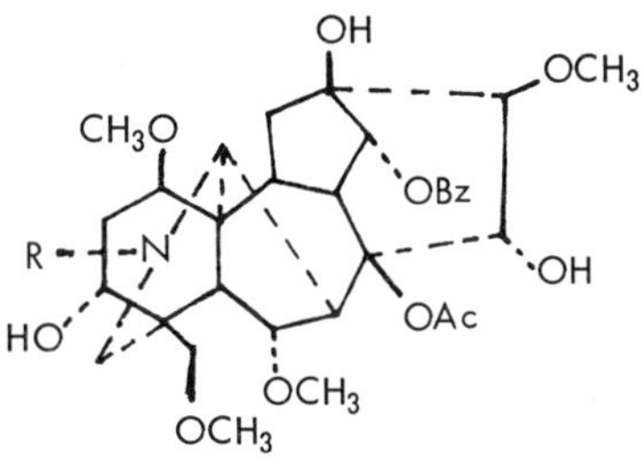

(1) R = Et

(62) R = CH_3

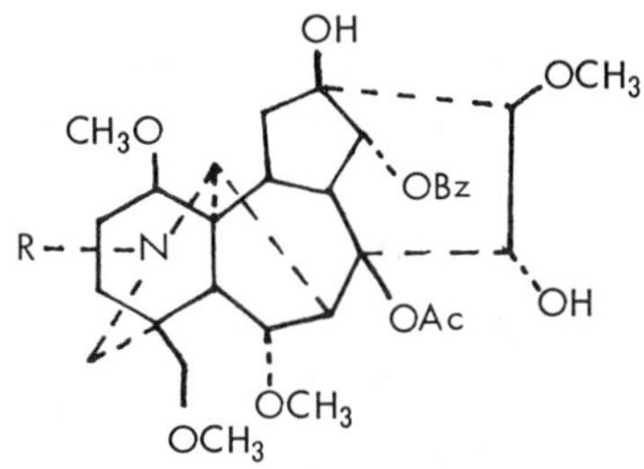

(64) R = CH_3

(65) R = Et

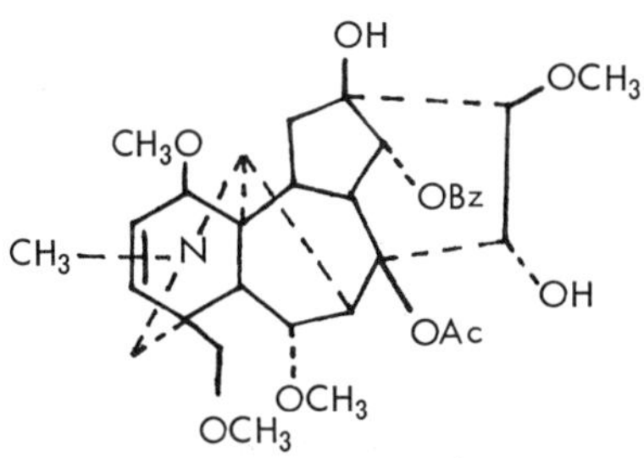

(63)

Dealkylation of aconitine (1) followed by methylation gave mesaconitine [(62), N-desethyl-N-methyl-aconitine)].[27] Vigorous treatment of (62) with thionyl chloride removed the hydroxyl (63) and hydrogenation gave a product identical with hypaconitine (64). The alternate sequence of reactions also produces hypaconitine.[28] An intermediate product of this sequence, deoxyaconitine, (65) is a recently reported alkaloid discovered as a contaminant in crude aconitine.[27]

2.3 Jesaconitine, Indaconitine, Pseudaconitine, and Bikhaconitine

Jesaconitine was first isolated in 1909[29] and differs from aconitine by only one ester functional group. Whereas, aconitine is hydrolyzed to acetic acid, benzoic acid, and the parent amino alcohol aconine (66), jesaconitine is hydrolyzed to acetic acid, anisic acid (4-methoxybenzoic acid), and aconine.[30] The nmr spectrum of jesaconitine, with its highly shielded acetoxy signal (τ = 8.64) and the doublet characteristic of the C(14) proton geminal to an ester (τ = 5.38, J = 4.5 Hz) left no doubt that positions C(8) and C(14) were the sites of esterification.[31,32] To distinguish between the two possible isomers, pyrolysis of the alkaloid was carried out in the nmr tube and the reaction was monitored by nmr. The shielded signal of the acetoxy protons was observed to slowly disappear while a new signal, due to acetic acid, appeared and grew to about the same intensity as the original shielded acetoxy signal. The signals due to the aromatic protons, however, remained completely unchanged. Thus, jesaconitine has structure (67) and the product of the pyrolysis, pyrojesaconitine, has structure (68).

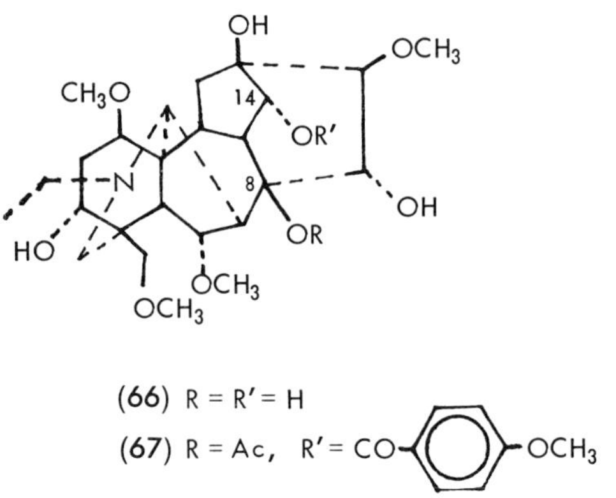

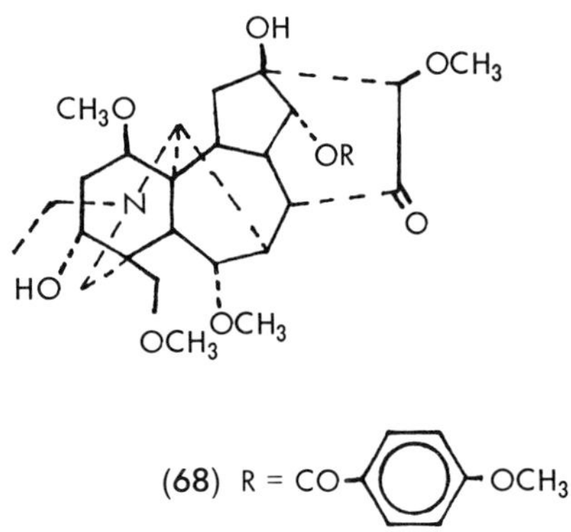

Indaconitine (69) and pseudaconitine (70) differ from one another by one ester functional group each giving the parent amino alcohol, pesudaconine (71), upon saponification. Conversion of indaconitine to delphinine (2) by removal of the C(3)-hydroxyl and replacement of the N-ethyl with an N-methyl was accomplished by Gilman and Marion.[33] Since the undisturbed rings A, B, E, and F of delphinine have been correlated with those of aconitine, this conversion proved the functional group placement and stereochimistry of this portion of the structures of both indaconitine and pseudaconitine with the exception of the ring A hydroxyl that had been removed.

Pyrolysis of pseudaconitine gave (72) and treatment of the latter with lithium aluminum hydride produced demethoxyisopyropseudaconine (73), a compound identical with the Wolff–Kishner reduction product of pyroaconitine.[13] The important points to note from this correlation are: (a) the position and configuration of the C(3)-hydroxyl of indaconitine

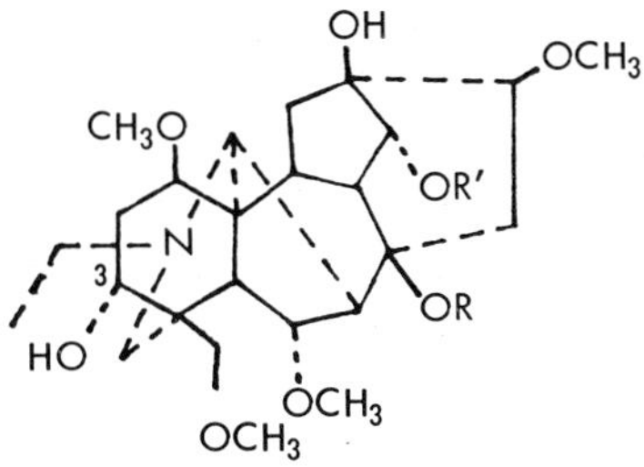

(69) R = Ac, R′ = Bz

(70) R = Ac, R′ = CO– 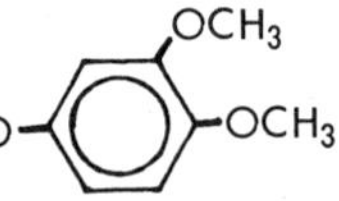

(71) R = R′ = H

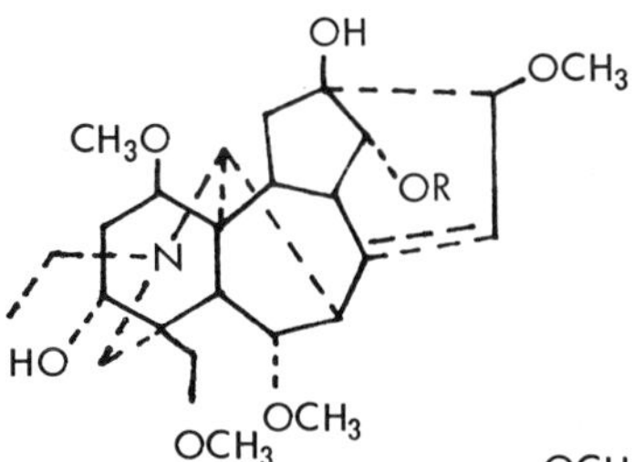

(72) R = CO– 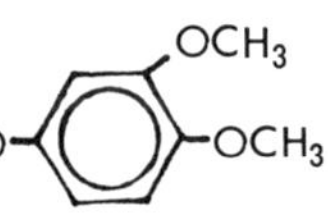

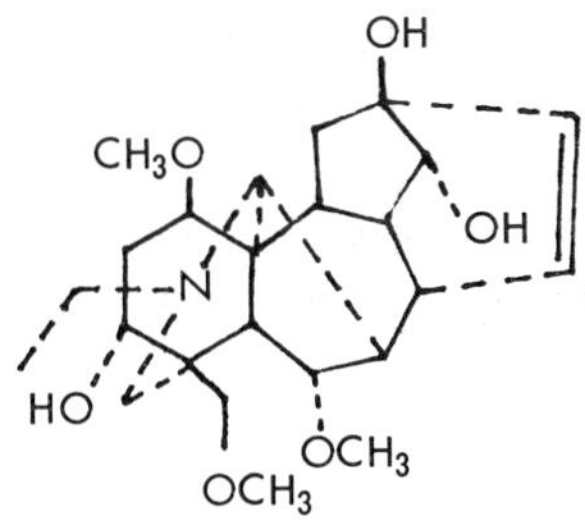

(73)

and pseudaconitine is rigorously proved, and (b) this correlation demonstrates that the stereochemistry of the C–D ring system of delphinine is the same as aconitine, a point reasonably assumed but unproven in previous discussions since the rearrangement involved in the correlation of aconitine and delphinine destroyed the existing C–D ring systems. The latter point follows from the previous conversion of indaconitine to delphinine.

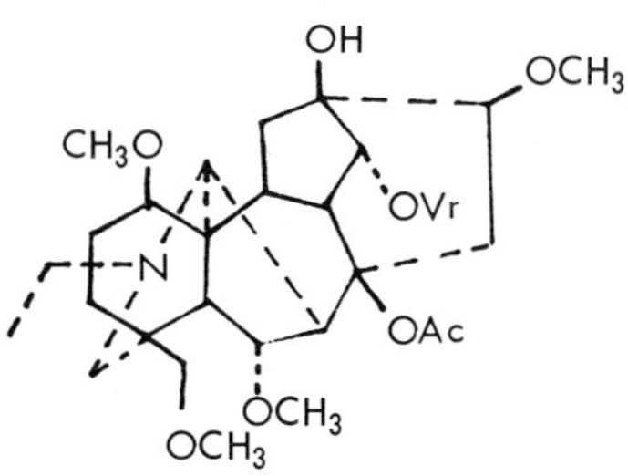

(74)

Vr = CO– 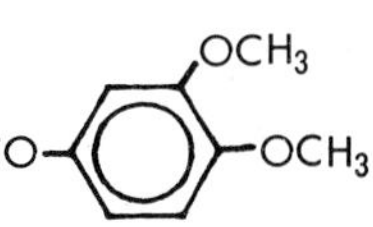

Bikhaconitine (74) was first isolated in 1905 from the roots of *Aconitum spictatum* Stapf, a plant used extensively as a poison by the Indian population and simply denoted by most of the languages of India as "bikh" or "bish" meaning *poison*.[34] Correlation with pseudaconitine (70) was accomplished by removal of the C(3)-hydroxyl from the latter followed by catalytic hydrogenation.[35] The deoxypseudaconitine produced was identical with bikhaconitine.

2.4 Chasmaconitine and Chasmanthinine

A reexamination[36–38] of the roots of *Aconitum chasmanthum* Stapf revealed the presence of four new alkaloids, chasmaconitine, chasmanthinine, chasmanine, and homochasmanine, in addition to indaconitine that was originally described in 1905.[39] Chasmanine and homochasmanine are discussed in a latter section.

Chasmaconitine (75) is identical with deoxyindaconitine; i.e. N-ethyl-N-desmethyldelphinine. Correlation with delphinine (2) was achieved merely by substituting the N-ethyl with an N-methyl group.[36]

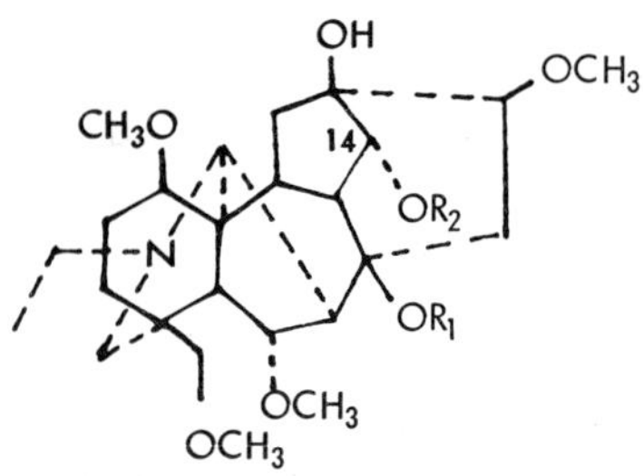

(75) R_1 = Ac, R_2 = Bz

(76) R_1 = Ac, R_2 = Cn

Chasmanthinine (76) contains C_2H_2 more than chasmaconitine in its empirical formula and differs from it only by one ester function group—a C(14) *trans*-cinnamyl group (Cn) in place of the benzoyloxy possessed by chasmaconitine.

2.5 Condelphine, Isotalatizidine, and Talatizidine

Condelphine (77) was first isolated in 1942 from *Delphinium confusum* and was found to be an O-acetate derivative of isotalatizidine (78).[40] The latter alkaloid, along with its C(1)-epimer talatizidine (79), had been isolated two years earlier from *Aconitum talassicum*, a rare species native to the mountainous regions of Talass-Alataou in Central Asia.[41] Further Soviet research allowed the postulation of partial structures for these three closely related compounds[42] and recent work by American researchers has established the complete structures of all three.[43,44] The latter work was done with condelphine and isotalatizidine isolated from *D. denudatum*, a plant native to the lower reaches of the Himalayas.

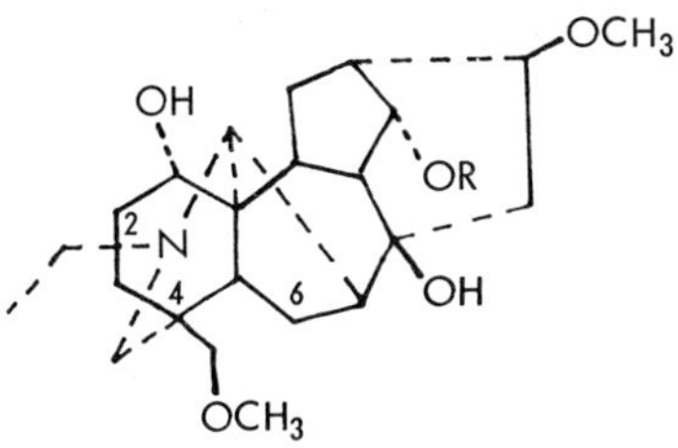

(77) R = Ac

(78) R = H

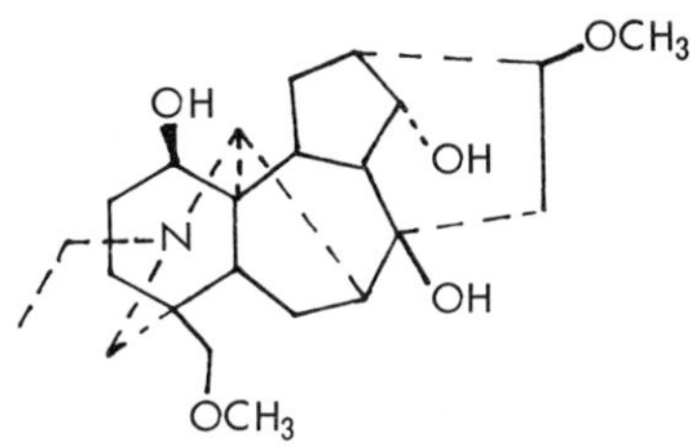

(79)

The derivations of structure for these three compounds are based on chemical and spectroscopic evidence, much of which has been covered earlier in this chapter. However, some insight into the conformations of ring A in condelphine and its C(1)-esters was gained from a study of the nmr spectra of some of these compounds.[44] The large coupling constants of the geminal proton of the C(1)-esters and (10 and 7 Hz) could only be explained by a chair conformation. Condelphine and isotalatizidine, however, must exist in the boat conformation an appreciable amount of their time as indicated by evidence in their IR spectra of intramoleuclar hydrogen bonding between the C(1)-hydroxyl and the nitrogen. Actually, they exist in a rapid equilibrium between the boat (80) and chair (81) conformations as evidenced by their nmr spectra, which always show the geminal C(1)-proton as a multiplet when the C(1)-hydroxyl is not esterified. Heating resolves this multiplet into a broad triplet whose coupling constant (J = 3 Hz) is in accord with that expected from this proton when ring A is

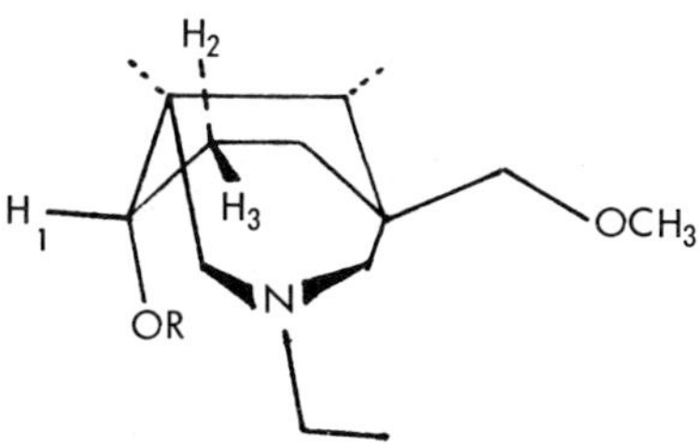

(80)

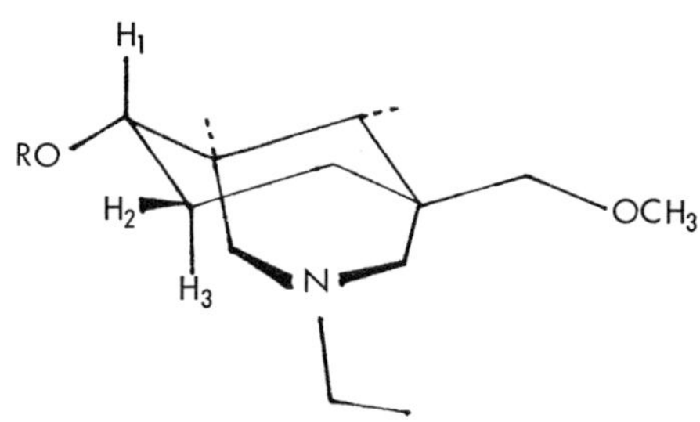

(81)

in the boat conformation, thereby making the dihedral angles involved nearly equal (50–60°). Thus, at higher temperatures the more highly energetic boat form predominates.

2.6 Chasmanine and Homochasmanine

These compounds are discussed after the section on browniine because, although classified as aconitine-type alkaloids, they have been correlated with browniine, a lycoctonine-type alkaloid.

3. THE LYCOCTONINE-TYPE ALKALOIDS

3.1 Lycoctonine

Lycoctonine is the parent amino alcohol of a number of ester derivatives found in plants of both the *Aconitum* and the *Delphinium* genera. Although great effort was put forth, the skeleton of lycoctonine eluded determination by chemical means. The structure of des-(oxymethylene)-lycoctonine hydroiodide monohydrate (82) was the first derivative of the C_{19}-diterpene alkaloids to be determined by x-ray crystallography[45,46] and the structure for lycoctonine (83) followed thereafter.[47]

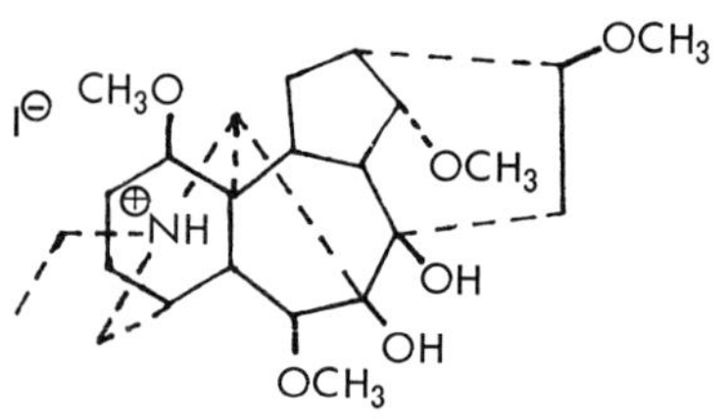

(82)

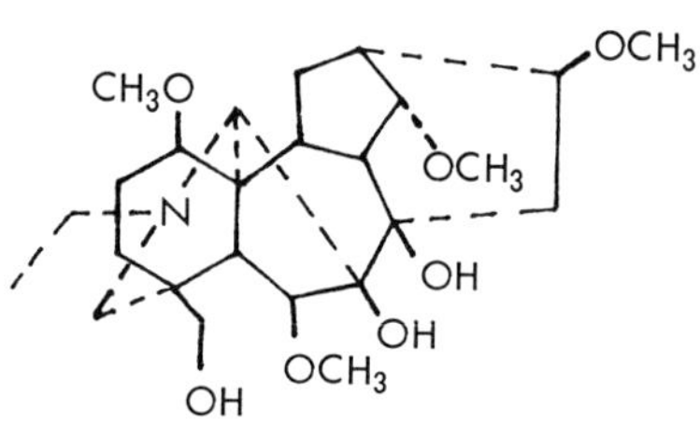

(83)

An unusual rearrangement is the reaction of either oxylycoctonine (lycoctonam) (84) or its pinacole dehydration product (85) to give lycoctamone (86), an α,β-unsaturated aldehyde, upon vigorous treatment with acid.[48]

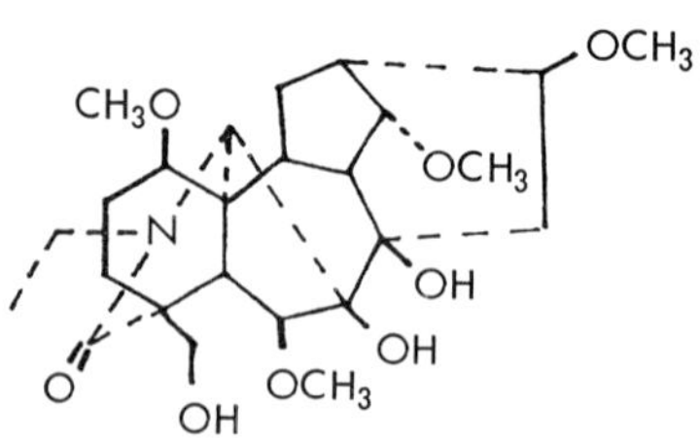

(84)

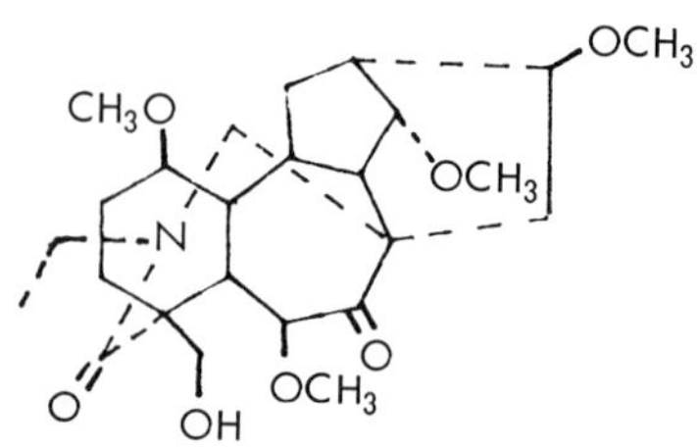

(85)

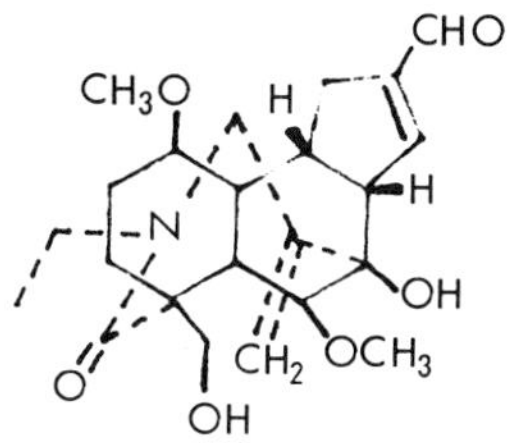

(86)

3.2 Elatine, Anthranoyllycoctonine, Ajacine, Methyllycaconitine, Delsemine, Avadharidine, and Lycaconitine

Although the structure of lycoctonine has been known since 1956, the structures of the above simple monoester derivatives were not deduced until three years later, primarily because of the difficulty in locating which of the three hydroxyls of lycoctonine bore the ester acid moiety. This problem was overcome by Kuzovkov and Platonova.[49]

Hydrolysis of elatine gives the amino alcohol elatidine (87) and the acid (88). Cleavage of the methylenedioxy group of elatidine produced lycoctonine (83). Thus, elatidine differs from lycoctonine only in the presence of a methylenedioxy group in place of the C(7)–C(8) glycol group and the ester elatine must have structure (89).

By heating elatine with phloroglucenol in acid, anthranoyllycoctonine (90) was produced. The anthranoylycoctonine obtained from this reaction was found to be identical with the cleavage products of ajacine, methyllycaconitine, delsemine, and lycaconitine. Therefore, each of these four alkaloids has the primary hydroxyl group of lycoctonine esterified with anthranilic acid and each differs only in the nature of the radical acetylating the amino group of the anthranilic acid.

It was known for many years that ajacine is N-acetylanthranoyllycoctonine. Hence, ajacine is represented by structure (91).

It had been known since 1954[50] that the esterifying acid of methyllycaconitine is (88). Thus, methyllycaconitine is represented by structure (92).

Delsemine is formed from methyllycaconitine by the cleavage of the methylsuccinimido group with ammonia, which converts it into a methylsuccinamido group. The structure of the esterifying acid (93) of delsemine had previously been established,[51–53] hence the structure of delsemine is represented by (94).

Avadharidine differs from delsemine only in that it is a succinic acid derivative whereas (94) is a methylsuccinic acid derivative. Acid cleavage of avadharidine produced anthranoylycoctonine (90); thus, the structure of avadharidine is (95).

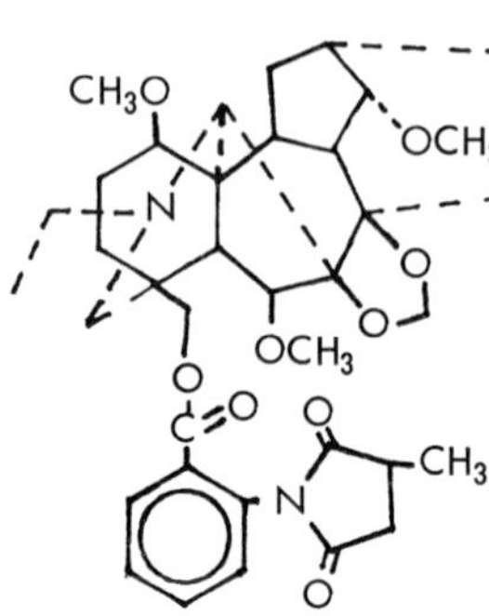

(89)

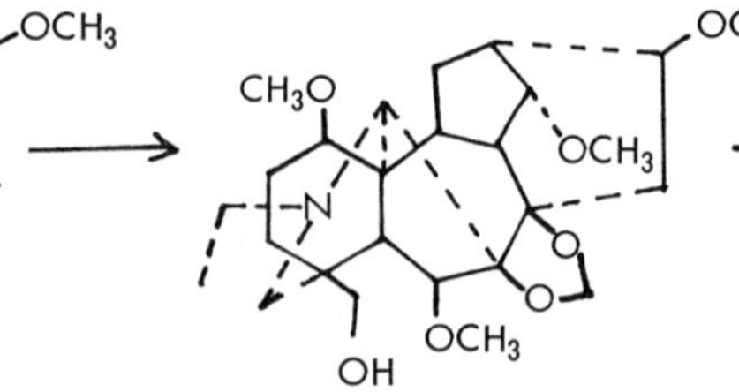

(87)

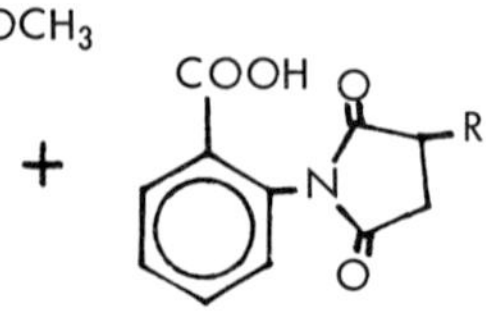

(88) R = CH_3

(96) R = H

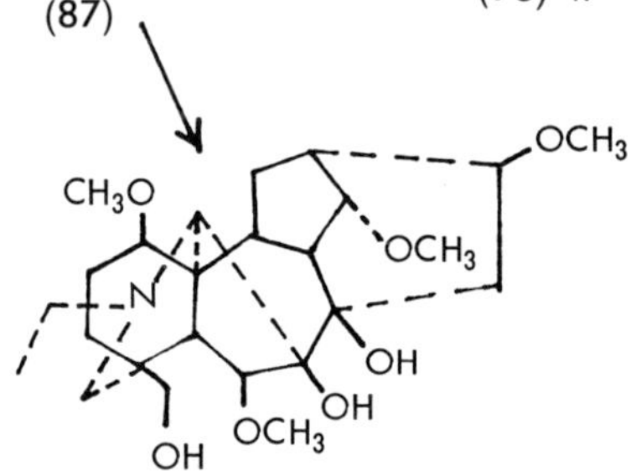

(83)

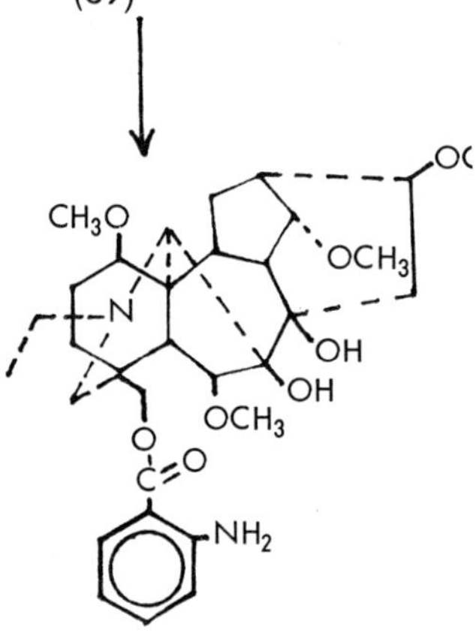

(90)

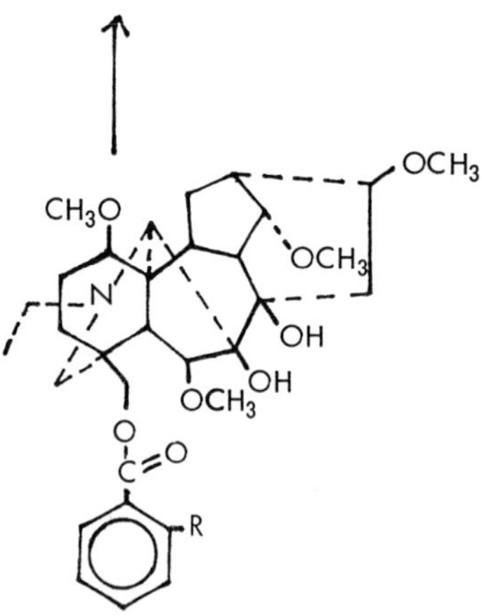

(91) R = $CH_3C(=O)$-NH-

(92) R = 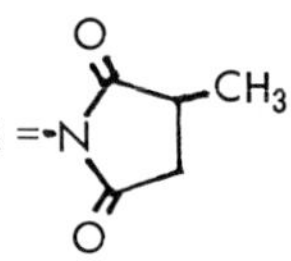

(95) R = 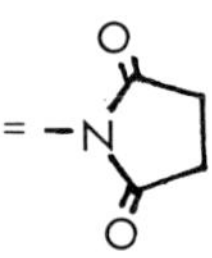

(94) R = -NH-C(=O)-CH(CH_3)-CH_2-C(=O)-NH_2

or -NH-C(=O)-CH_2-CH(CH_3)-C(=O)-NH_2

(97) R = -NH-C(=O)-CH_2CH_2-C(=O)-NH_2

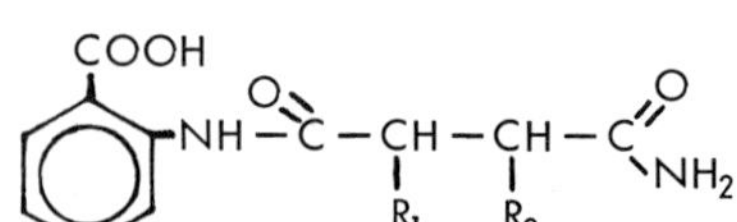

(93) R_1 = CH_3, R_2 = H

or R_1 = H, R_2 = CH_3

It is probable that avadharidine is formed from lycaconitine by reacting with ammonia during the extraction of the alkaloids from the plant in the same way that delsemine is formed from methyllycaconitine. The esterifying acid of lycaconitine is (96), and therefore the structure of lycaconitine is (97).

3.3 Delpheline, Deltaline, and Deltamine

Correlation of delpheline (98) with lycoctonine was achieved independently and simultaneously in two different laboratories; one method involved the conversion of lycoctonine to a delpheline derivative[54] and the other employed the reverse interrelation.[55] The conversion of lycoctonine (83) to a delpheline derivative[30] was accomplished by successive oxidation, reduction, oxidation steps, (99), (100), (101), followed by a facile pinacole rearrangement (102). Removal of the C(6)-methoxyl was accomplished with sodium amalgum in ethanol (103) and selenium dioxide oxidation of the latter gave the desired diketone (104), a compound identical with dehydrodemethyleneoxodelpheline pinacone.[56]

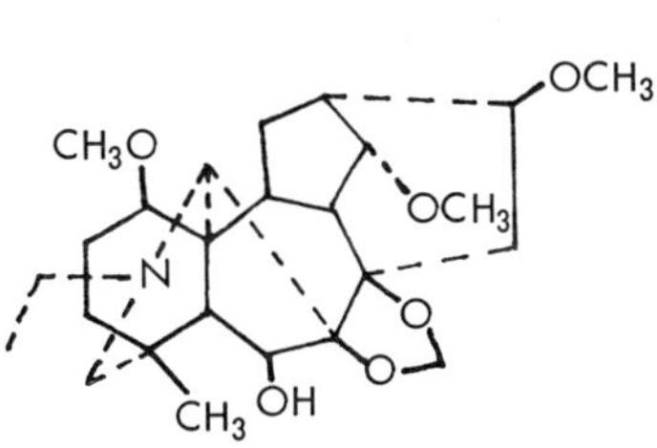

(98)

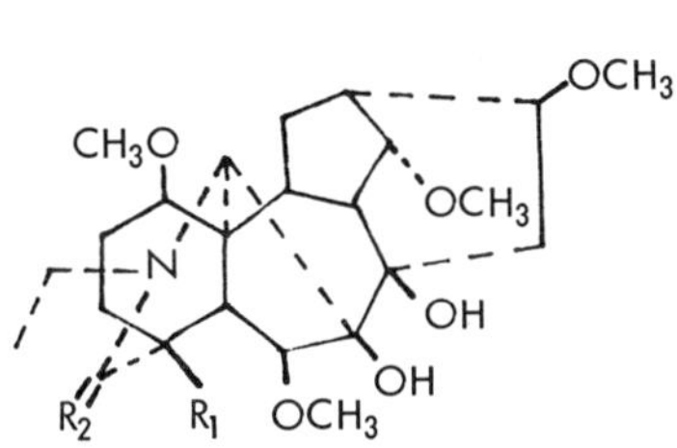

(99) R_1 = CHO, $R_2 = H_2$

(100) $R_1 = CH_3$, $R_2 = H_2$

(101) $R_1 = CH_3$, R_2 = O

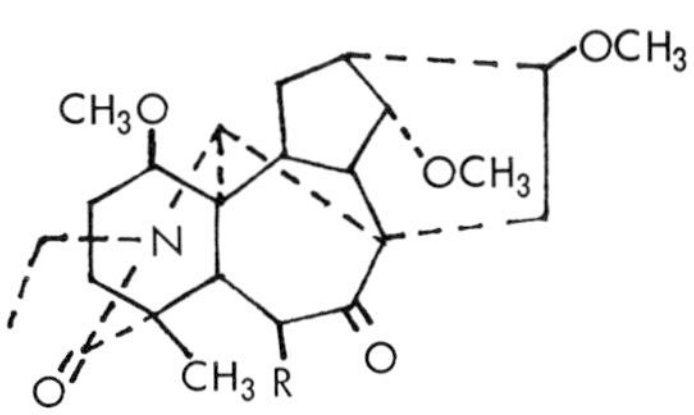

(102) R = OCH_3

(103) R = H

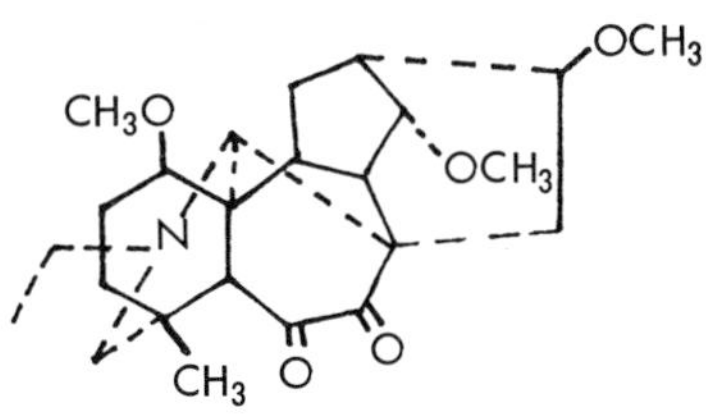

(104)

Deltaline (105) (also appearing as eldeline in past Soviet literature) is the acetyl ester of the amino alcohol deltamine (106). Treatment of deltaline with thionyl chloride followed by lithium aluminum hydride produced delphelene (98).[55] Methylation of the hydroxyl and hydrolysis of the methylenedioxy group gave a derivative (107) identical with desoxylycoctonine.

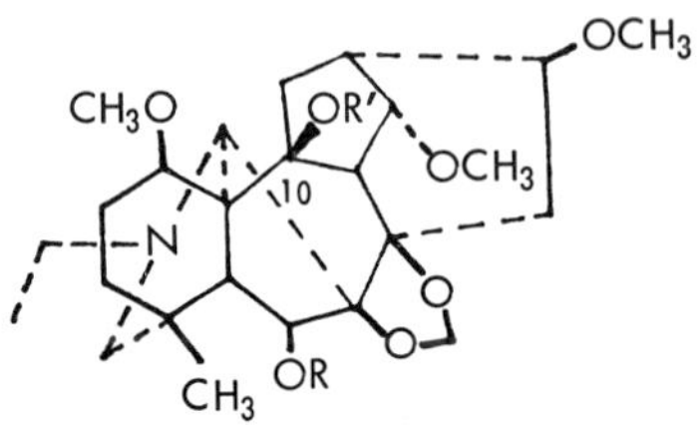

(105) R = Ac, R′ = H

(106) R = R′ = H

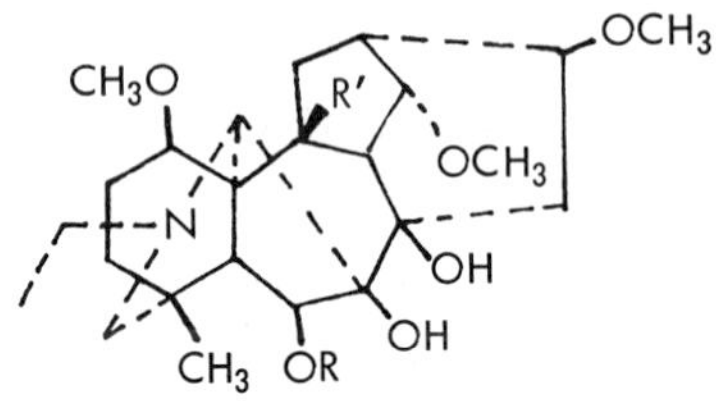

(107) R = CH_3, R′ = H

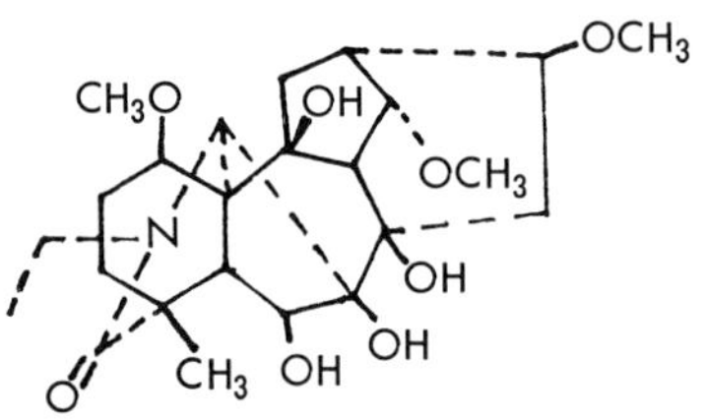

(108)

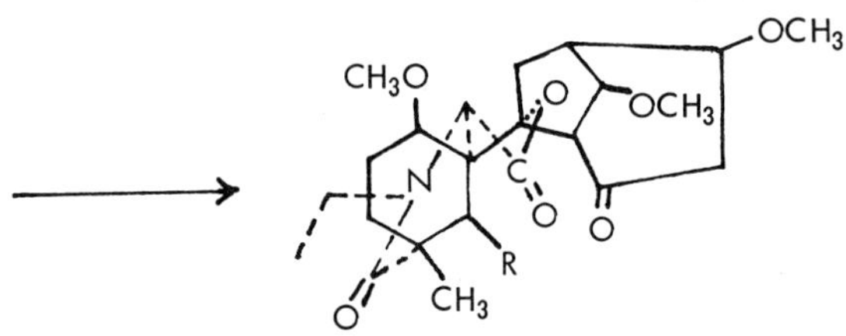

(109) R = CHO

(110) R = COOH

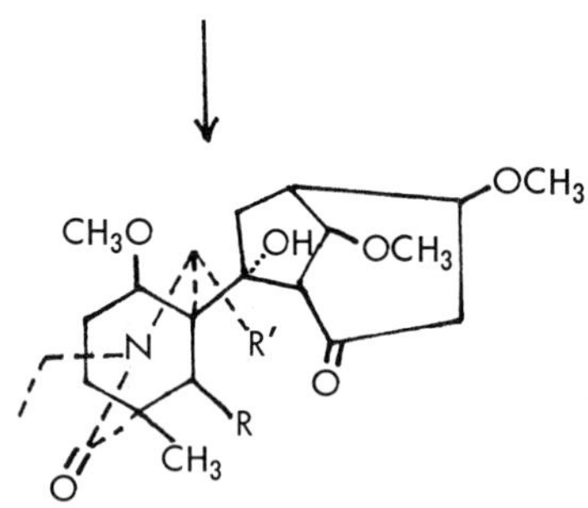

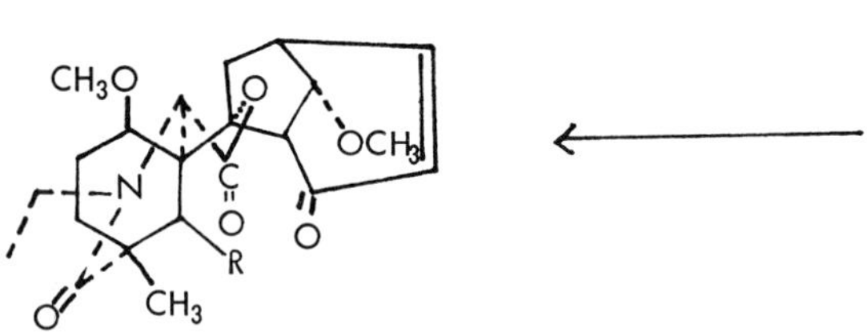

(112) R = CHO

(111) R = CHO, R′ = COOH

The determination of the position of the hydroxyl group that had been replaced by hydrogen in the conversion of deltaline to delpheline remained. Periodate oxidation of demethyleneoxodeltamine (108) gave the neutral γ-lactone (109) which was capable of being further oxidized to a carboxylic acid (110).[57] Treatment of (109) with alkali opened the lactone ring and, after acidification, (111) was obtained. The latter could be lactonized again by heating in sulfuric acid, but the product (112) lost one methoxyl group giving an α,β-unsaturated ketone in the process. Formation of a γ-lactone in the periodate oxidation of (108) and the fact that deltamine (106) is stable to periodate oxidation unequivocally determines the position of the hydroxyl at C(10).

3.4 Delcosine and Delsoline

The structural elucidation of delcosine (113) is an excellent example of the manifestation in the field of natural products chemistry of Murphy's well-known first law.* The first tentative structure suggested had the C(1)-hydroxyl and the C(6)-methoxyl switched.[58] Support for this assignment appeared to be the easy formation of carbinolamine ether linkages. How-

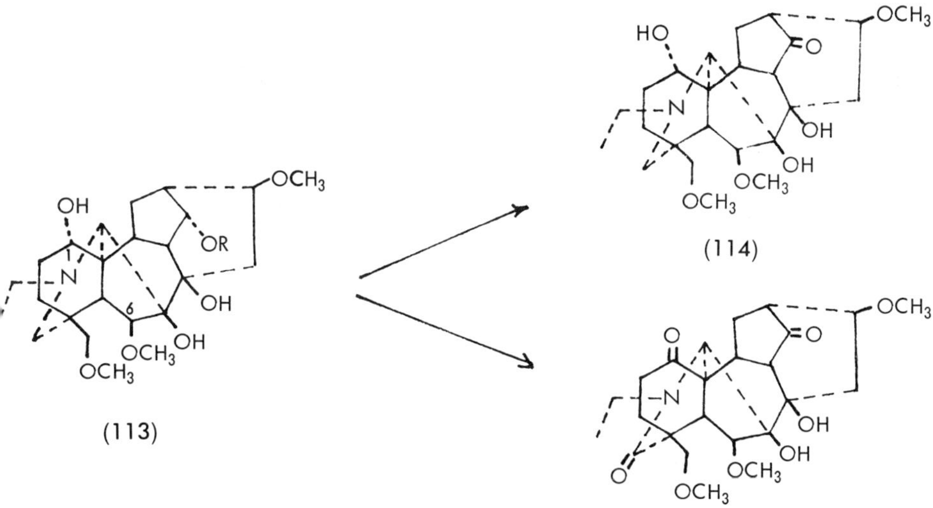

(113) (114) (115)

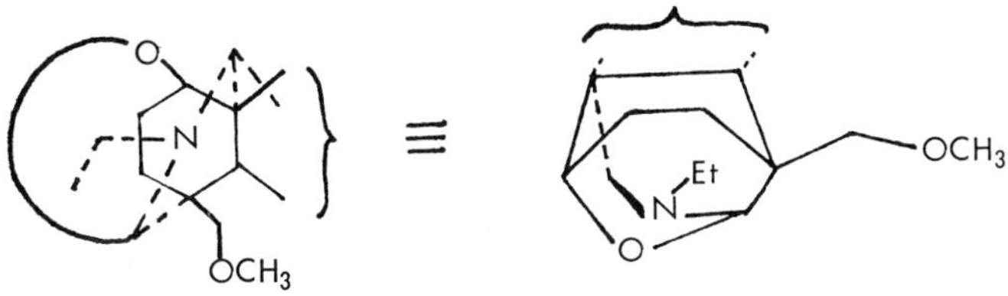

(116)

*In any worthy field of endeavor, if anything can possibly go wrong . . . it will!

ever, oxidation of delcosine with chromium trioxide in acetic acid was observed to produce a 5-membered ketone (114) while oxidation with chromium trioxide-pyridine complex (Sarett's reagent) gave one 5-membered and one 6-membered cyclic ketone (115). Had the C(1)-hydroxyl of (113) been switched with the C(6)-methoxyl, two 5-membered cyclic ketones would have been produced from the latter reaction.[59,60]

In retrospect then, formation of the facile carbinolamine ether linkages indicates that the secondary hydroxyl is at C(1) (where it can form a 6-membered ring as shown in partial structure 116) and defines the configuration of the C(1)-hydroxyl as *alpha*. Additional evidence for a C(1)-hydroxyl is the lead tetraacetate oxidation of oxodelcosine (117). The intermediate diketone (118) reacts further to give the hemiketal (119), thus limiting the secondary hydroxyl to C(1) since this appears to be the only position where formation of a cyclic hemiketal is geometrically possible.

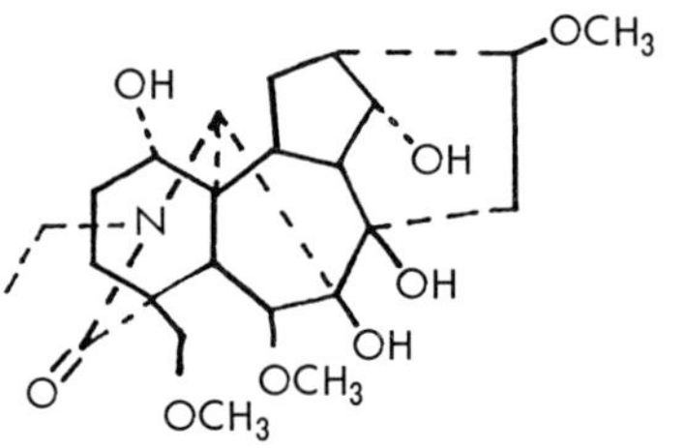

(117)

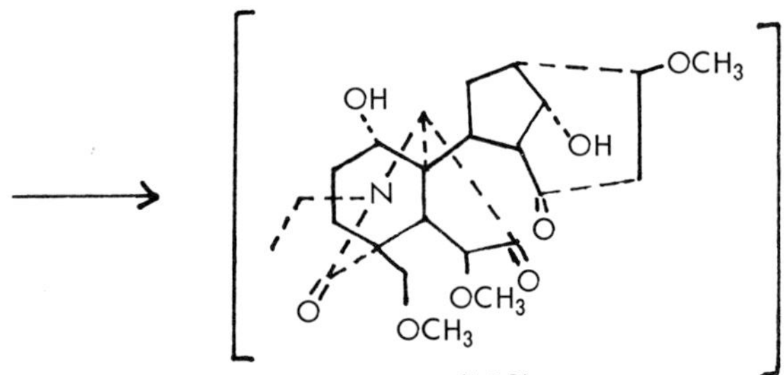

(118)

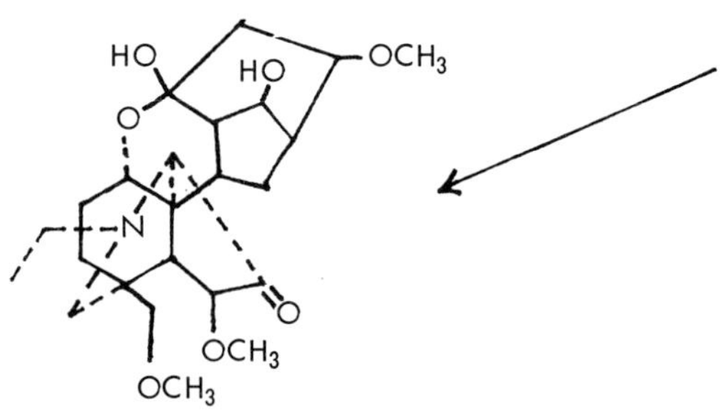

(119)

An attempt was made to correlate delcosine with lycoctonine.[61] The initial reactions involved an epimerization at C(1) since lycoctonine has a β-methoxyl at this position. Ethylation of the azomethine (120 → 121) was followed by epimerization with base to give what was believed to be 14-dehydrooxo-1-epidelcosine. Actually, unknown at the time, the epimerization reaction gave back the *original* C(1)-epimer,[62] 14-dehydrooxodelcosine (122). Sodium borohydride reduction of the latter surprisingly gave oxo-epi-14-delcosine (123). All other borohydride reductions of a C(14)-ketone have been almost completely stereospecific in producing the original epimer, the equitorial (α) alcohol. This stereospecificity has been at-

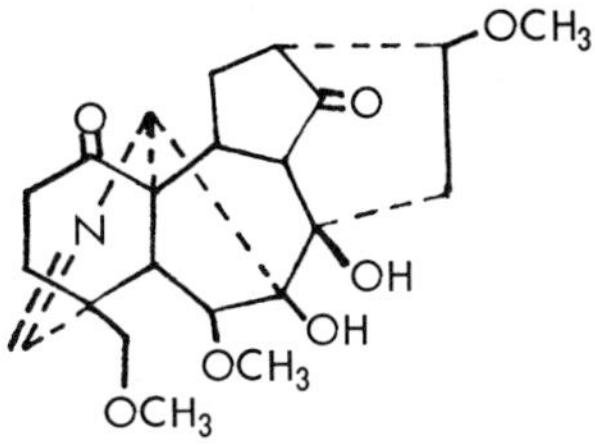

(120)

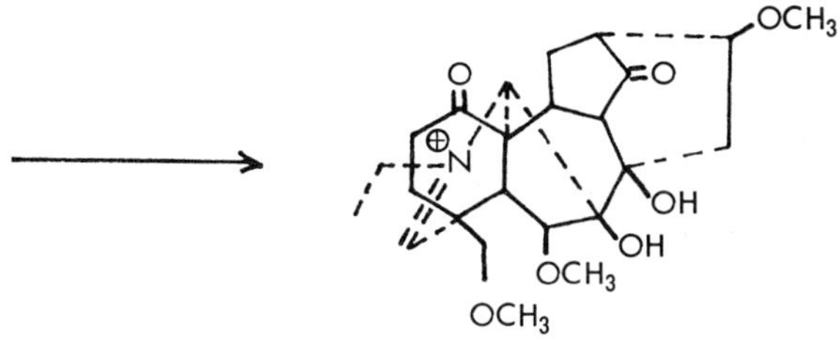

(121)

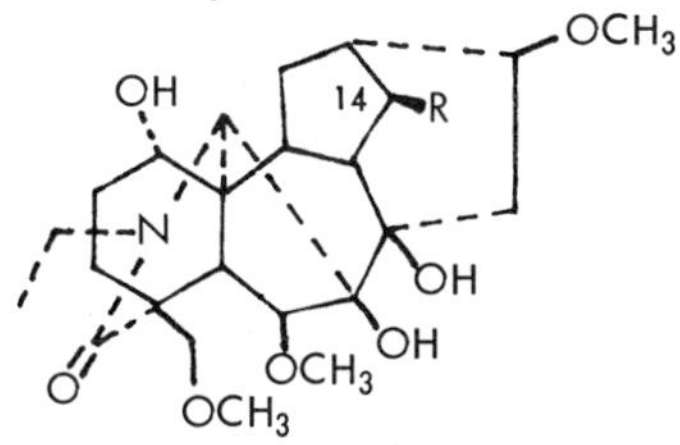

(122) R = O

(123) R = β OH

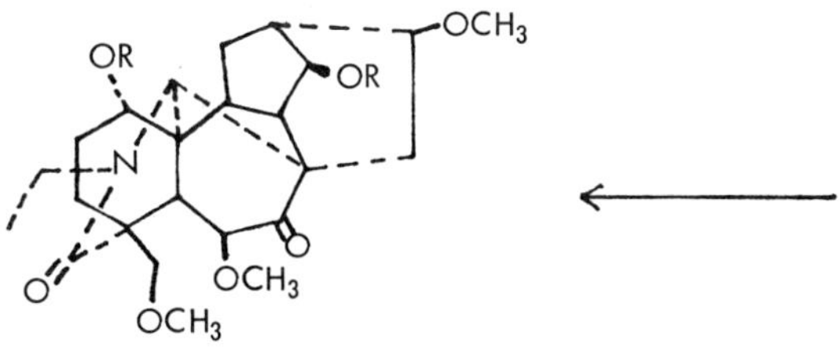

(124) R = H

(125) R = CH_3

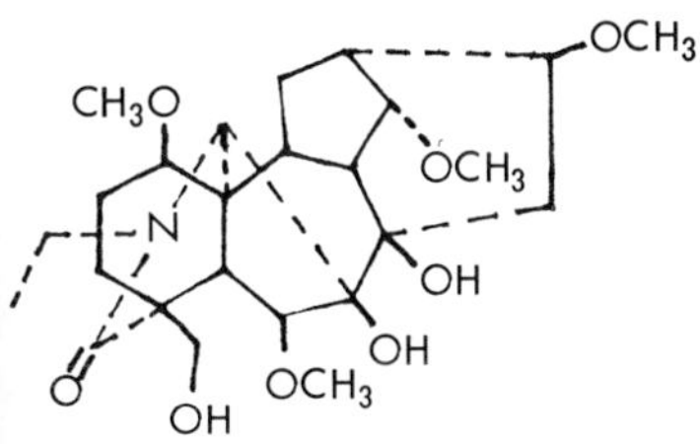

(84)

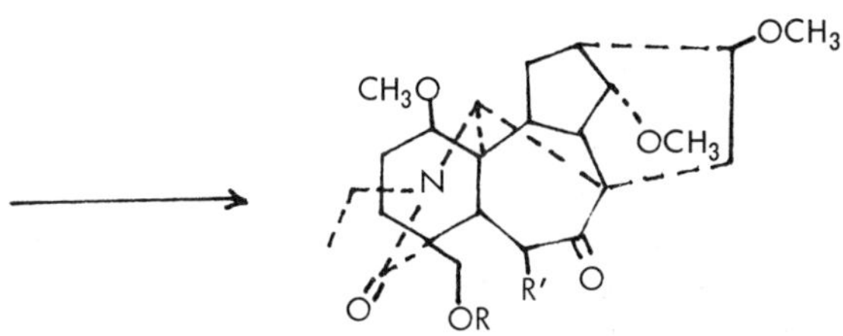

(126) R = H, R′ = OCH_3

(127) R = CH_3, R′ = OCH_3

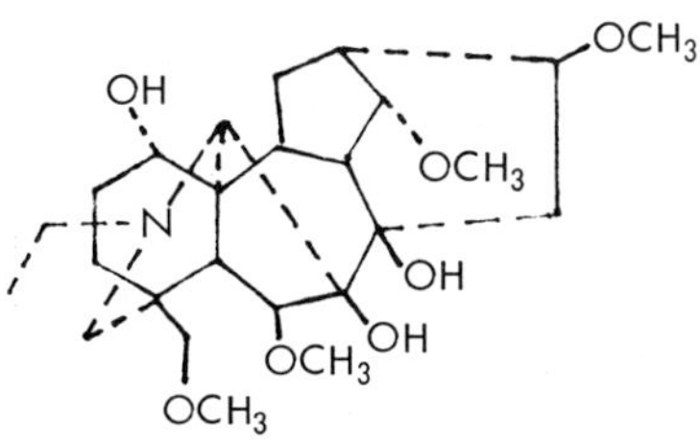

(128)

tributed to the presence of ring D, which blocks the approach of the attacking reagent from the side *cis* to the nitrogen bridge.

Proceeding on the erroneous assumption that epimerization had occurred at C(1) rather than at C(14), (123) was converted to the corresponding pinacone (124) and then methylated (125). By a similar acid-catalyzed pinacolic rearrangement, oxolycoctonine (lycoctonam) (84) was converted to anhydrooxolycoctonine (126) and methylated to 0-methylanhydrooxolycoctonine (127). The two derivatives, (125) and (127), were completely different. At this point, an x-ray crystallographic investigation of delcosine hydrobromide was initiated.[63] The preliminary study has shown that the skeleton of delcosine is the same as that of lycoctonine and that all of the substituents, with the exception of the C(1)-hydroxyl, are oriented as in lycoctonine.[62] In addition, ring A of delcosine exists in the boat conformation, in contrast to lycoctonine where it exists in the chair conformation. The difference is accounted for by the fact that intramolecular hydrogen bonding exists between the C(1)-hydroxyl and the nitrogen in delcosine.

Delsoline (128) is merely an 0-methyldelcosine. To prove this, delcosine was methylated with sodium hydride and methyl iodide in dioxane at room temperature.[64] This methylation is remarkedly stereospecific in that only the secondary hydroxyl of the 5-membered ring was methylated while the ring A hydroxyl was left untouched; when heated, the selectivity is lost and the dimethyl derivative is obtained. The 14-0-methyldelcosine thus produced was identical with delsoline.

3.5 Browniine

Browniine has recently been isolated from *D. brownii*, a plant native to Canada.[65] Oxidation of browniine (129) with Sarett's reagent gave a cyclic 5-membered monoketo lactam (130). Reaction with acetic anhydride–pyridine yielded the monoacetate (131) confirming the presence of only one secondary hydroxyl group. Permanganate oxidation of (131) produced (132) which, when hydrolyzed, yielded oxobrowniine (133). Reduction of dehydrooxobrowniine (130) with sodium borohydride gave predominantly oxobrowniine showing the usual selective C(14) ketone reduction. Both (130) and (133) rapidly consumed one mole of lead tetraacetate giving an uncharacterized *seco*diketone and a *seco*triketone (134) as the initial products, demonstrating the presence of two vicinal tertiary hydroxyl groups. Heating (134) caused elimination of methanol to form the α,β-unsaturated ketone (135), thus showing the presence of a C(16) methoxyl group. The two remaining oxygen functional groups, which occur at C(1) and C(18) in lycoctonine, must then be methoxyl groups in browniine and,

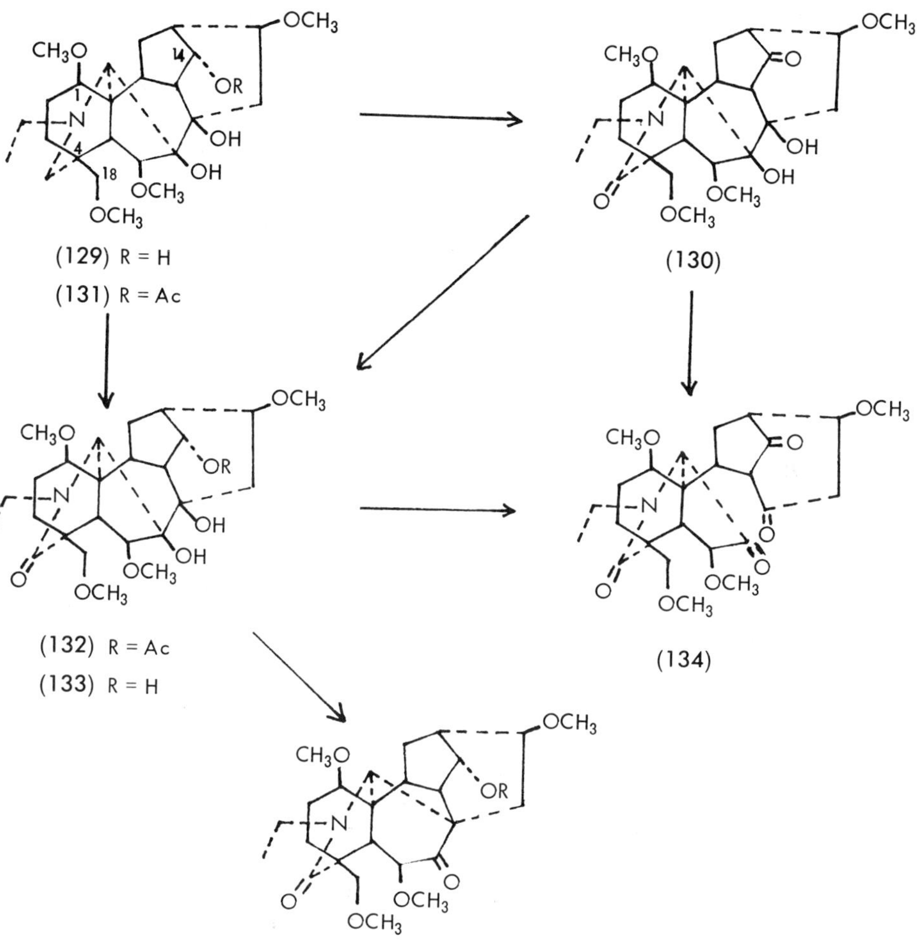

(129) R = H

(131) R = Ac

(130)

(132) R = Ac

(133) R = H

(134)

(136) R = H

(137) R = CH_3

as a working hypothesis, they were assumed to be at the same positions, reducing the probable structure of browniine to (129).

Correlation with a lycoctonine derivative would require 0-methylation of the C(18) hydroxyl of lycoctonine and 0-methylation of the C(14) hydroxyl of browniine. To remove all of the other hydroxyls, and eliminate the problem of tertiary hydroxyl methylation, oxobrowniine (133) was subjected to conditions under which the pinacol rearrangement takes place, and the resultant anhydrooxobrowniine (136) was then methylated. The 0-methyl derivative (137) was identical with the methyl ether of anhydro-lycoctonam. Thus, the position and configuration of the substituents of browniine were proved to be as shown in (129).

Treatment of (134) with hot sulfuric acid produces a mixture of acids. The structure of one, (138), containing only three methoxyl groups, was deduced from its spectra and empirical formula and, since (135) also produced (138), the eliminated methoxyl must be from C(16). Decarboxylation occurred upon melting to give the β,γ-unsaturated ketone 139 which seemed to rearrange to the corresponding α,β-unsaturated ketone (140) although none could be isolated.

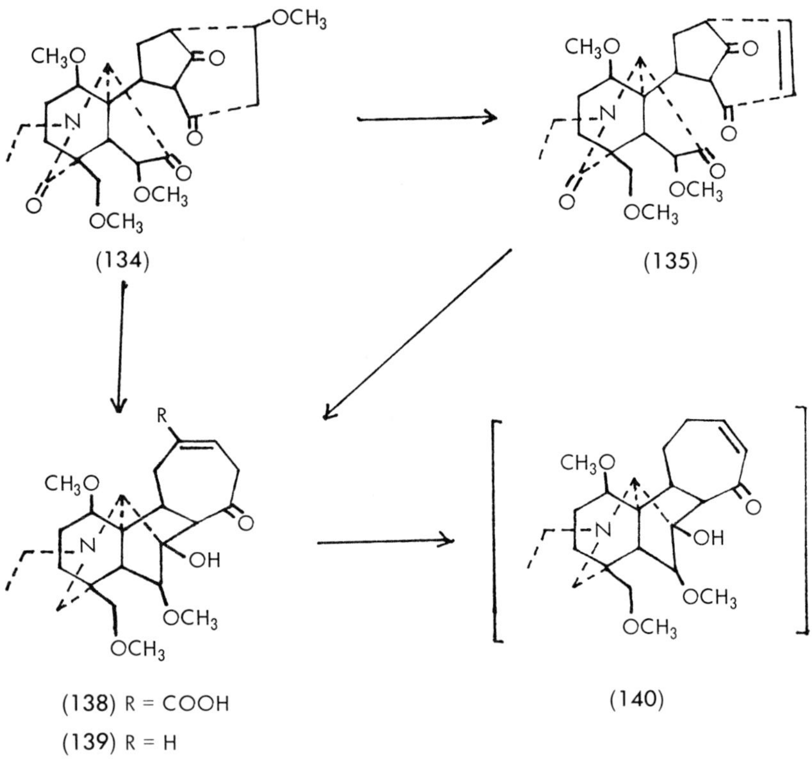

3.6 Chasmanine and Homochasmanine

The discussion of these two alkaloids has been delayed until now because, although they are classified as aconitine-type alkaloids, they have been correlated with browniine, a lycoctonine-type alkaloid. Chemical degradations and spectral characteristics similar to those previously discussed led to the structural formulation of chasmanine as (141).

After two abortive attempts to correlate chasmanine with aconitine-type alkaloids, a correlation with browniine was finally achieved.[66] The latter

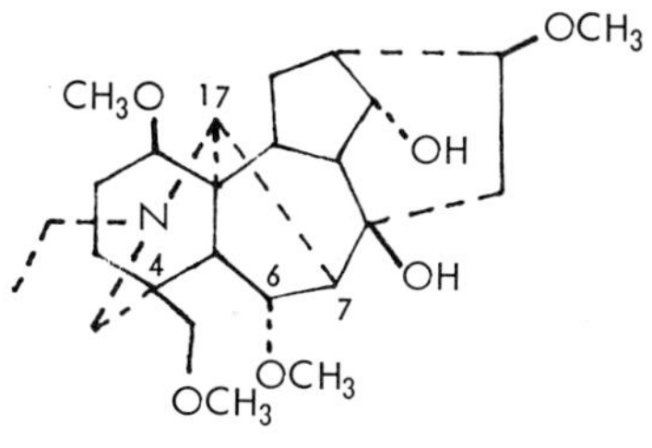

(141)

differs from chasmanine only in that it has a β-oriented C(6) methoxyl and a C(7) hydroxyl group. It was reasoned that the C(6) α-methoxyl of chasmanine, being crowded by the C(4) substituents, probably could be epimerized if a 7,17-*seco*-7-keto derivative of chasmanine could be prepared. Accordingly, diacetylchasmanine (142) was converted to the 7,17-*seco*-olefin (143) by treatment with lithium tri-*t*-butoxyaluminum hydride and hydration with diborane gave (144) which, in turn, was oxidized to (145).

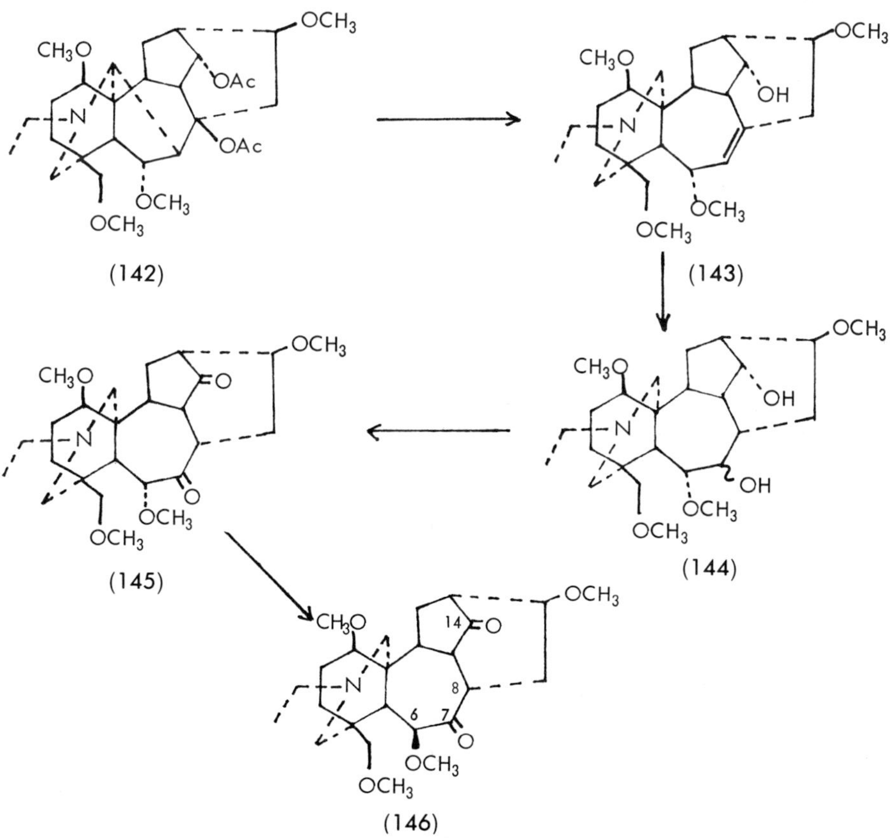

(142) (143) (144) (145) (146)

The desired epimer (146) was then obtained by refluxing (145) in basic solution.

Oxidation of browniine (129) with lead tetraacetate[66] gave hydroxybrowniine (147) and hydrogenation of the latter produced isobrowniine (148). A 3-minute oxidation of isobrowniine with permanganate yielded oxoisobrowniine (149) which, when reduced with zinc in acetic acid, gave a mixture of (150) and (151). Reduction of either (150) or (151) with lithium aluminum hydride gave (152) which, when oxidized with chromic oxide in acetone, produced 14-dehydro-8-deoxyisobrowniine (146); this diketone was identical with the corresponding diketone obtained from chasmanine.

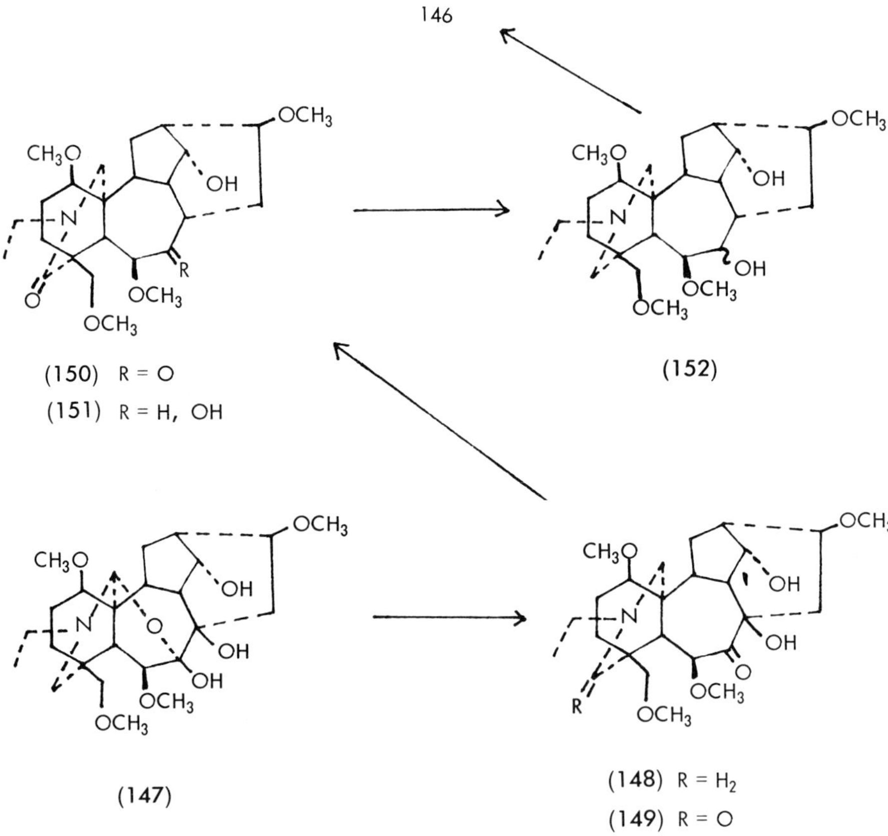

Homochasmanine (153) is the fifth alkaloid characterized from *Aconitum chasmanthum* Stapf and has the empirical formula $C_{26}H_{43}NO_6$ containing one CH_2 more than chasmanine.[38] The nmr spectrum showed the triplet characteristic of an N-ethyl group, plus five methoxyl signals.

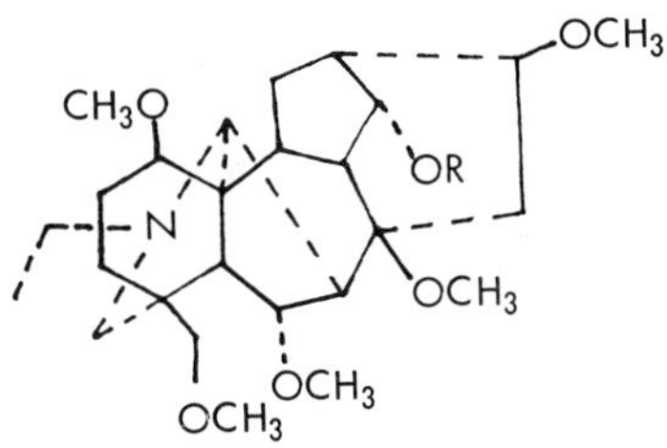

(153) R = H

(154) R = Ac

Since only an amorphous monoacetate (154) could be prepared, the compound can contain but one hydroxyl, thus accounting for all of the oxygens in the molecule. As a working hypothesis it was assumed that homochasmanine was an O-methylchasmanine. From the structure of chasmanine (141) it is seen that the additional methoxyl must be at either C(8) or C(14). The nmr spectrum of acetylhomochasmanine (154) contained a poorly resolved triplet at $\tau = 5.23$, which was attributed to the proton geminal to the acetoxy group and is consistent with a C(14) acetoxy group but not with a C(8) ester. Confirmation of this hypothesis was achieved using a reaction whereby the C(8) acetoxy group in aconitine-type alkaloids is replaced with a methoxyl group by heating under pressure in methanol. Heating diacetylchasmanine (142) in methanol under pressure gave an amorphous product (154) that was saponified affording a product identical to homochasmanine.

4. LACTONE-TYPE ALKALOIDS

4.1 Heteratisine

Heteratisine (155) occurs in the weak-base fraction of *A. heterophyllum* to the extent of .03%. It may be an artifact obtained during isolation by hydrolysis of its benzoyl ester. The molecular formula,[67] $C_{22}H_{33}NO_5$, has been confirmed by crystallographic studies[68] and the structure has been deduced independently by x-ray analysis of heteratisine hydrobromide monohydrate[69] and by chemical and spectral studies.[70,71]

Hot aqueous alkali with heteratisine or oxoheteratisine (156) caused only hydrolytic opening of the lactone ring. With the corresponding C(6)-ketones, however, profound degradation occurs under these conditions. The products lack lactone or carboxyl groups and may be formed by retro-aldol cleavage of the corresponding β-hydroxy ketone system (157, 158) followed by decarboxylation of the β-ketoacids (159) and (160), respectively.

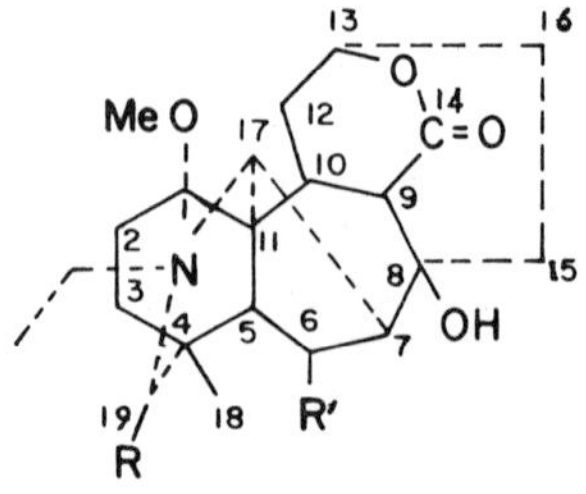

(155) R= H_2; R′= OH

(165) R= H_2; R′= OAc

(166) R= H_2; R′= OBz

(176) R= O; R′= OAc

(156) R= O; R′= OH

(161) R= R′=O

(175) R= H_2; R′= O

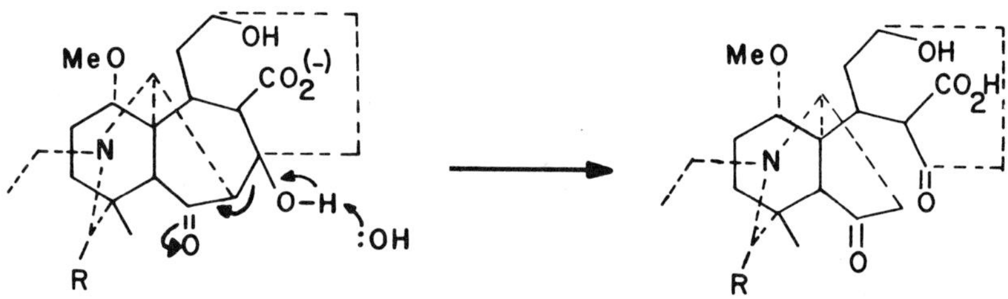

(157) R= H_2

(158) R = O

(159) R= H_2

(160) R=O

Treatment of dehydrooxoheteratisine (161) with potassium *t*-butoxide in *t*-butanol leads to the γ-lactone carboxylic acid characterized as its methyl ester (164). Its formation by cleavage of the initial retroaldol product (162) to a δ-lactone carboxylic acid (163) and subsequent isomerization to the γ-lactone provides decisive proof for location of the tertiary hydroxyl at a position β- to both the cyclopentanone and δ-lactone carbonyls and four carbons removed from the δ-lactone ether oxygen.[70]

Heteratisine esters undergo a facile thermally-induced ring cleavage which has unusual features.[72] Thus, pyrolysis of the acetate (165) or the benzoate (166) proceeds with elimination of one mole of the respective carboxylic acid to give pyroheteratisine (167) in about 90% yield and isopyroheteratisine (168) in about 1% yield. Brief treatment of pyroheteratisine with $NaOCH_3$ in CD_3OD effects exchange of *H*-C(9) by deuterium;

(161)

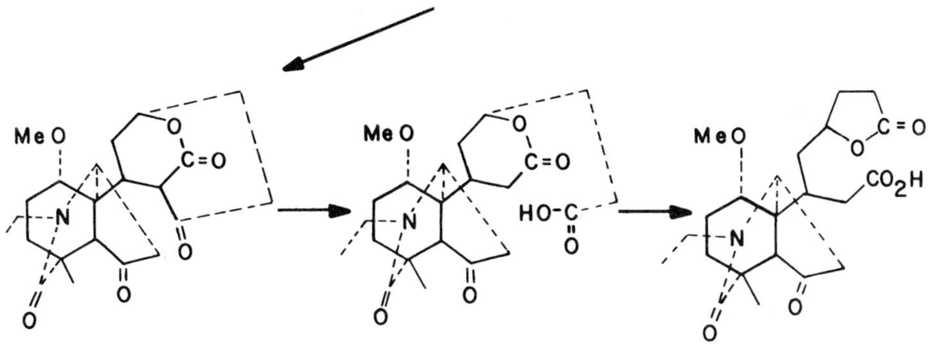

(162) (163) (164)

the H-C(9) signal vanishes and the vinylic proton doublet degenerates to a singlet. This facile deuterium exchange and the accompanying loss of allylic coupling supports the structure shown. Hydrogenation of pyroheteratisine gives a dihydroderivative and hydroxylation with osmium tetroxide gives a *cis*-diol (169) that forms a basic monoacetate (170). The H-C-OH signal at $\tau = 6.18$ in the diol and the H-C-OAc signal at $\tau = 5.30$ in the monoacetate both appear as sharp singlets.

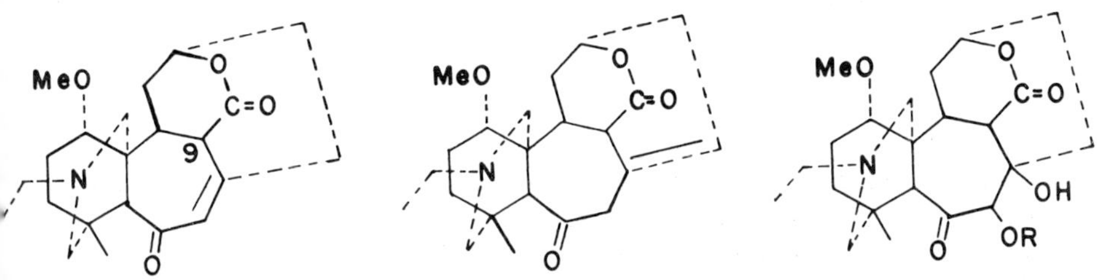

(167) (168) (169) R=H (170) R=Ac

Decisive proof for structure (167) by decarboxylation of the derived vinylogous β-ketocarboxylate looks facile, but it is not since generation of the Δ^8-double bond is sterically impossible. On extended boiling with aqueous alkali at pH 9, pyroheteratisine (167) suffers decarboxylative degradation and the reaction appears to follow the path (167) → (171) → · (172) → (173) → (174), since the spectral properties of the product permit its formulation as (174).

Additional facts relating to the pyrolysis reaction are the following: 8(OD)-des 8(OH)-benzoyl heteratisine yields benzoic acid-d_1. Dehydro-

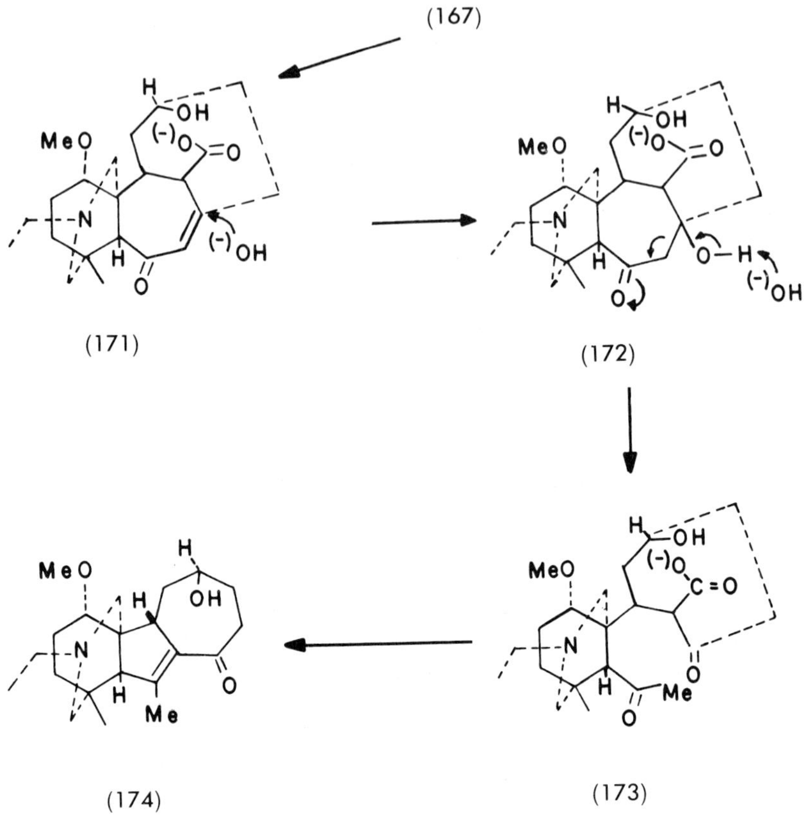

heteratisine (175) and oxoheteratisine acetate (176) fail to undergo a parallel reaction, the latter even on admixture with heteratisine acetate. The extreme facility and high yields of the pyrolysis reaction, in conjunction with the aforementioned facts suggest the reaction may proceed by a concerted process involving intramolecular participation of the basic nitrogen,[71] (177) → (178) → (169) → (167). An alternative mechanism suggests isomerization of heteratisine monoacetate (165) to the 8-acetate (180), fission to the ion pair (181), and hydride transfer from C(6) to C(17) to give pyroheteratisine (167)[73] (see p. 587).

4.2. Heterophyllidine, Heterophyllisine, and Heterophylline

These alkaloids were isolated in small quantities from the mother liquors of heteratisine and all contain a δ-lactone ring. The structures shown were derived mainly by comparison of the mass spectra of these alkaloids with that of heteratisine.[74]

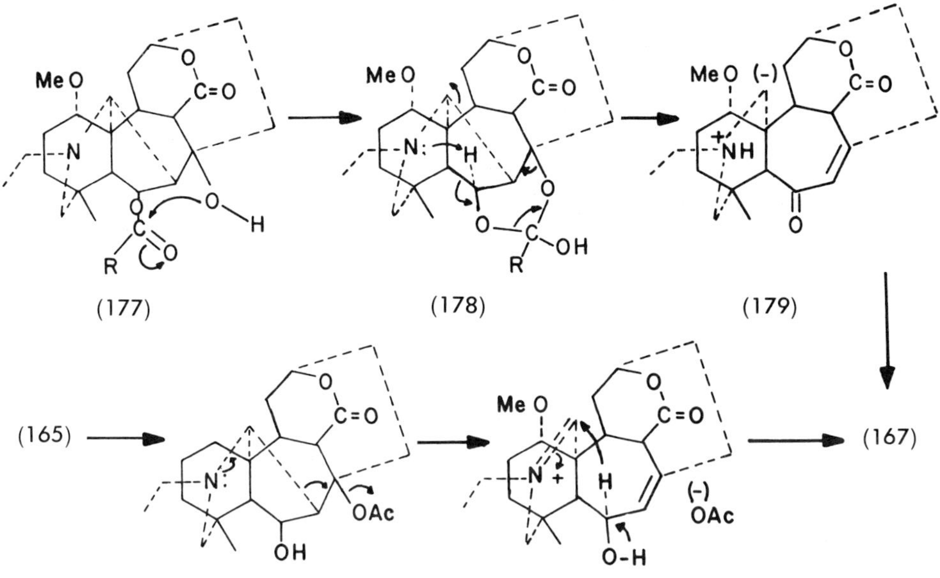

(177) (178) (179)

(180) (181)

The mass spectrum of heterophyllidine (182) in the m/e 98–300 region is identical with the spectrum of heteratisine (155). The molecular ion (M^+) at m/e 377 (14 less than the molecular ion of heteratisine), the absence of a peak at (M^+-31), and the presence of the most abundant peak at (M^+

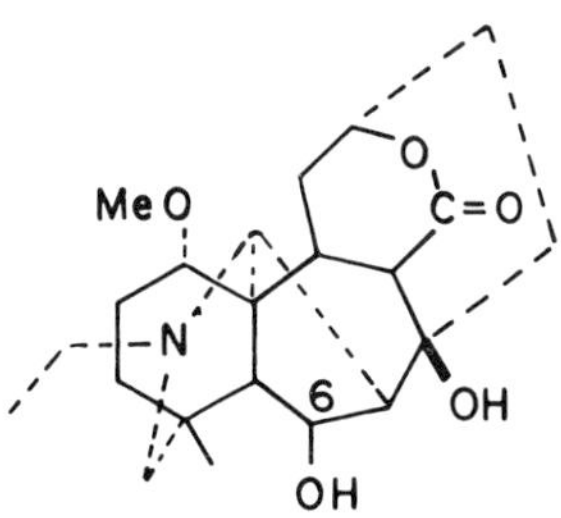

(155)

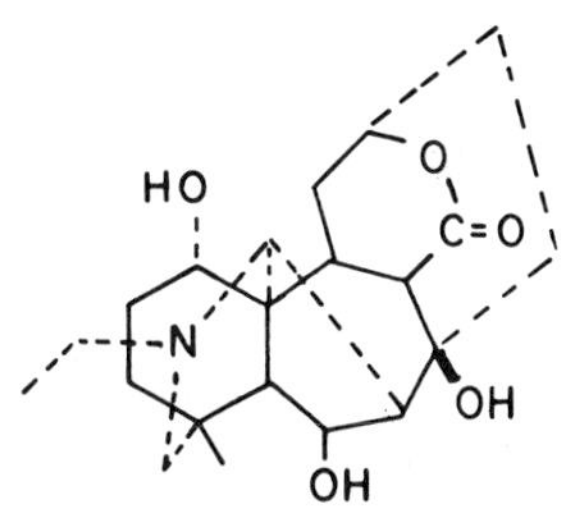

(183)

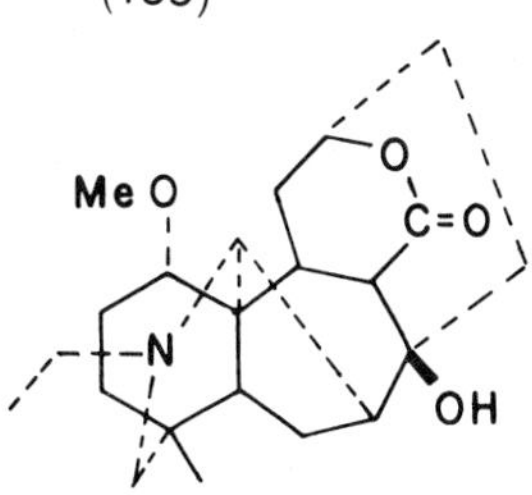

(183)

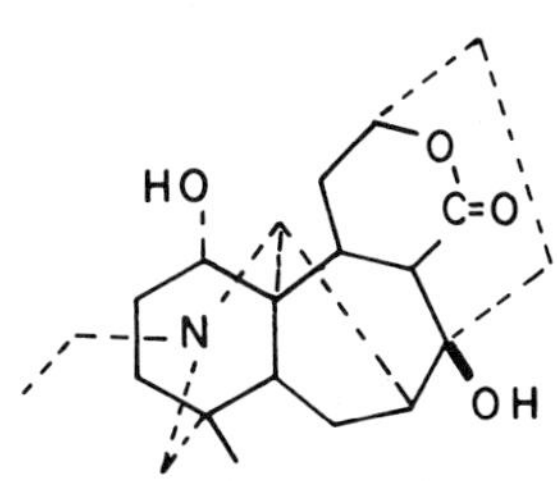

(184)

-17) suggest the 1-0-nor-heteratisine structure (182) for heterophyllidine. The nmr spectrum of heterophyllidine is also very similar to that of heteratisine, except that absorption due to —OCH_3 is absent.

Analysis of the mass spectra of heterophyllisine (183) and heterophylline (184) showed these alkaloids to be desoxyheteratisine and desoxyheterophyllidine, respectively. The nmr spectra of the pairs heterophylline–heterophyllidine and heterophyllisine–heteratisine are remarkably similar, but there are two notable differences. In the desoxy- compounds, signals assignable to *H*-C(6)—OH are absent and the 3H singlet for CH_3—C— occurs at $\tau = 0.16$ higher field, presumably because the deshielding effect of the neighboring hydroxyl at C(6) is absent.

REFERENCES

1. M. Khaimova, N. Mollov, P. Cerneva, A. Antonova and V. Ivanova, *Tetrahedron Letters* **38,** 2711 (1964).
2. O. Achmatowicz, Y. Tsuda, and L. Marionn *Can. J. Chem.* **43,** 2336 (1965).
3. P. L. Geiger, *Ann.* **7,** 269 (1833).
4. E. S. Stern, *Alkaloids* **4,** (1954).
5. E. S. Stern, *Alkaloids* **7,** (1960).
6. For a summarizing reference, see: W. A. Jacobs and S. W. Pelletier, *J. Amer. Chem. Soc.* **78,** 3542 (1956).
7. K. Wiesner, F. Bickelhaupt, and Z. Valenta, *Tetrahedron* **4,** 418 (1958).
8. K. Wiesner, F. Bickelhaupt, and D. R. Babin, *Experientia* **15,** 93 (1959).
9. For a summarizing reference, see: K. Wiesner, F. Bickelhaupt, D. R. Babin, and M. Gotz, *Tetrahedron* **9,** 254 (1960).
10. K. Wiesner, H. W. Brewer, D. L. Simmons, D. R. Babin, and F. Bickelhaupt, *Tetrahedron Letters* **3,** 17 (1960).
11. For a summarizing reference, see: K. Wiesner, M. Gotz, D. C. Simmons, and L. R. Fowler, *Coll. Czech. Chem. Comm.* **28,** 2462 (1963).
12. D. J. McCaldin and L. Marion, *Can J. Chem.* **37,** 1071 (1959).
13. Y. Tsuda and L. Marion, *Can. J. Chem.* **41,** 1485 (1963).
14. Y. Tsuda and L. Marion, *Can. J. Chem.* **41,** 1634 (1963).
15. W. A. Jacobs and S. W. Pelletier, *J. Org. Chem.* **22,** 1428 (1957).
16. O. Achmatowicz, Y. Tsuda, and L. Marion, *Can. J. Chem.* **43,** 825 (1965).
17. K. Wiesner, M. Gotz, D. C. Simmons, L. R. Fowler, F. W. Bachelor, R. F. C. Brown, and G. Buchi, *Tetrahedron Letters* **2,** 15 (1959).
18. P. De Mayo in "Molecular Rearrangements", Part 2, p. 912, Interscience Publishers, Inc., New York, N.Y., 1964.
19. K. Wiesner, F. Bickelhaupt, D. R. Babin, and M. Gotz, *Tetrahedron Letters* **3,** 11 (1959).
20. W. A. Jacobs and S. W. Pelletier, *J. Amer Chem. Soc.* **76,** 161 (1954).
21. F. W. Bachelor, R. F. C. Brown, and G. Buchi, *Tetrahedron Letters* **10,** 1 (1960)
22. M. Przybylska and L. Marion, *Can. J. Chem.* **37,** 1116 (1959).
23. Ref. 22, p. 1843.
24. K. Wiesner, D. L. Simmons, and L. R. Fowler, *Tetrahedron Letters* **18,** 1 (1959).
25. S. I. Morio, *Ann.* **476,** 181 (1929); *Chem. Abstr.* **24,** 619 (1930).
26. R. Majima and S. I. Morio, *Ann.* **476,** 171 (1929); *Chem. Abstr.* **24,** 619 (1930).

27. Y. Tsuda, O. Achmatowicz, and L. Marion, Ann. Chem. **680,** 88 (1964).
28. R. E. Gilman and L. Marion, *Can. J. Chem.* **40,** 1713 (1962).
29. K. Makoshi, *Arch. Pharm.* **247,** 243; *Chem. Abstr.* **3,** 2707 (1909); Chem. Abstr. **5,** 674 (1911).
30. R. Majima, H. Suginome and S. Morio, *Ber.* **57B,** 1486 (1924); *Chem. Abstr.* **19,** 291 (1925).
31. L. H. Keith and S. W. Pelletier, *Chem. Comm.* 993 (1967).
32. L. H. Keith and S. W. Pelletier, *J. Org. Chem.* **33,** 2497 (1968).
33. R. E. Gilman and L. Marion, *Tetrahedron Letters* **20,** 923 (1962).
34. W. R. Dunstan and A. E. Andrews, *J. Chem. Soc.* **87,** 1636 (1905).
35. Y. Tsuda and L. Marion, *Can. J. Chem.* **41,** 3055 (1963).
36. O. Achmatowicz and L. Marion, *Can. J. Chem.* **42,** 154 (1964).
37. O. Achmatowicz, Y. Tsuda, L. Marion, T. Okamoto, M. Natsume, H. H. Chang, and K. Kajima, *Can. J. Chem.* **43,** 825 (1965).
38. O. Achmatowicz and L. Marion, *Can. J. Chem.* **43,** 1093 (1965).
39. W. R. Dunstan and A. E. Andrews, *J. Chem. Soc.* **87,** 1620 (1905).
40. M. S. Rabinovich and R. A. Konovalova, *Zhur. Obshchei Khim.* **12,** 329 (1942).
41. R. A. Konovalova and A. P. Orekhov, *Bull. Soc. Chim.* **7,** 95 (1940); *Chem. Abstr.* **34,** 5450 (1940).
42. A. D. Kuzovkov and T. F. Platonova, *J. Gen. Chem. U.S.S.R.* **31,** 1286 (1961).
43. S. W. Pelletier, L. H. Keith, and P. C. Parthasarathy, *Tetrahedron Letters* **35,** 4217 (1966).
44. S. W. Pelletier, L. H. Keith, and P. C. Parasarathy, *J. Amer. Chem. Soc.* **89,** 4146 (1967).
45. M. Przybylska and L. Marion, *Can. J. Chem.* **34,** 185 (1956).
46. M. Przybylska and L. Marion, *Can. J. Chem.* **37,** 1843 (1959).
47. O. E. Edwards, L. Marion, and D. K. R. Stewart, *Can. J. Chem.* **34,** 1315 (1956).
48. O. E. Edwards, L. Marion, J. D. Connolly, M. Benn, and Z. Stojanac, *IUPAC Symposium of Natural Products,* Kyoto, Japan, Abstracts, p. 114, 1964.
49. A. D. Kuzovkov and T. F. Platonova, *J. Gen. Chem. U.S.S.R.* **29,** 2746 (1959).
50. R. C. Cookson, J. E. Page, and H. F. Trevett, *J. Chem. Soc.* 4028 (1954).
51. S. Yunusov and N. K. Abubakirov, *Zhur. Obshchei Khim.* **21,** 967 (1951)
52. S. Yunusov and N. K. Abubakirov, *Zhur. Obshchei Khim.* **22,** 1461 (1952).
53. S. Yunusov and N. K. Abubakirov, *Zhur. Obshchei Khim.* **24,** 733 (1954).
54. O. E. Edwards, L. Marion, and K. H. Palmer, *Can. J. Chem.* **36,** 1097 (1958).
55. M. Carmack, J. P. Ferris, J. Harvey, P. L. Magat, E. W. Martin, and D. W. Mayo, *J. Amer. Chem. Soc.* **80,** 497 (1958).
56. R. C. Cookson and M. E. Trevett, *J. Chem. Soc.* 3864 (1956).
57. A. D. Kuzovkov and T. F. Platonova, *J. Gen. Chem. U.S.S.R.* **29,** 3801 (1959).
58. W. I. Taylor, W. E. Walles, and L. Marion, *Can. J. Chem.* **32,** 780 (1954).
59. L. Marion and V. Skaric, *J. Amer. Chem. Soc.* **80,** 4434 (1958).
60. V. Skaric and L. Marion, *Can. J. Chem.* **38,** 2433 (1960).
61. V. Skaric and L. Marion, *Can. J. Chem.* **39,** 1579 (1961).
62. L. Marion, *Pure and Applied Chem.* **6,** 621 (1963).
63. G. A. Mair and M. Przybylska. Unpublished results; L. Marion, *Pure Appl. Chem.* **6,** 621 (1963).
64. F. Sparatore, R. Greenhalgh, and L. Marion, *Tetrahedron* **4,** 157 (1958).
65. M. H. Benn, M. A. M. Cameron, and O. E. Edwards, *Can. J. Chem.* **41,** 477 (1963).
66. O. E. Edwards, L. Fonzes, and L. Marion, *Can. J. Chem.* **44,** 583 (1966).
67. W. A. Jacobs and L. C. Craig, *J. Biol. Chem.* **143,** 605 (1942); *J. Amer. Chem. Soc.* **147,** 571 (1943).
68. R. Aneja and S. W. Pelletier, *Acta Cryst.* **17,** 457 (1964).

69. M. Przybylska, *Can. J. Chem.* **41,** 2911 (1963).
70. R. Aneja and S. W. Pelletier, *Tetrahedron Letters* **12,** 669 (1964).
71. R. Aneja and S. W. Pelletier, *Tetrahedron Letters* **3,** 215 (1965).
72. O. E. Edwards and C. Ferrari, *Can. J. Chem.* **42,** 172 (1964).
73. O. E. Edwards, *Chem. Comm.* 318 (1965).
74. S. W. Pelletier and R. Aneja, *Tetrahedron Letters* **6,** 557 (1967).

19

*Steroidal Alkaloids**

YOSHIO SATO
National Institute of Arthritis and Metabolic Diseases
National Institutes of Health
U.S. Public Health Service
Bethesda, Maryland

1. INTRODUCTION

Steroidal alkaloids are a class of compounds possessing the basic or modified steroidal skeleton with nitrogen incorporated as an integral part of the molecule either in the ring or in the side chain. Although there are a number of compounds of both animal and plant origin that fall into this category, owing to limitation in space, only those of botanical origin possessing the regular perhydrocyclopentenophenanthrene nucleus are discussed in this chapter.**

*A number of excellent publications have appeared on the chemistry of the steroidal alkaloids covering the subject till about 1963. For this the reader is referred to the following:

(a) H.-G. Boit, "Ergebnisse der Alkaloid-Chemie bis 1960," Akademie-Verlag, Berlin, 1961.
(b) L. F. Fieser and M. Fieser in "Steroids," Reinhold Publ. Corp., New York, N.Y., 1959.
(c) V. Prelog and O. Jeger, *Alkaloids* **3,** 247–312 (1953).
(d) V. Prelog and O. Jeger, *Alkaloids* **7,** 319–361 (1960).
(e) C. W. Shoppee in "Chemistry of the Steroids," pp. 433–463, Butterworth, Inc., Washington, 1964.
(f) R. Goutarel in "Les Alcaloides Steroidiques des Apocynacees," Hermann, Paris, 1964.

The author's intent in this short chapter has been to fill somewhat the void existing thereafter. However, an excellent and comprehensive review of the chemistry of the Solanum alkaloids written by K. Schreiber has appeared recently in *Alkaloids* **10,** 1–192 (1968).

**The veratrine alkaloids, which are mainly derivatives of C-nor-D-homosteroids, and the recently characterized Buxus alkaloids[1–4] of novel structure are considered in the next chapter.

For convenience of discussion the steroidal alkaloids can be divided into two general types: (1) those that are derived from pregnane with an amino substituent on either C(3), C(20) or at both, and (2) the other derived from a 20-piperidylpregnane moiety. To the former belong the Holarrhena and Funtumia alkaloids and to the latter the Solanum and Veratrum alkaloids. In the following, the alkaloids are arranged and discussed in the order of increasing complexity.

2. 3-AMINOPREGNANE BASES

2.1 Funtumine and Funtumidine

Reduction of funtumine (1) with sodium and alcohol gives funtumidine (2), and the former is regenerated by oxidation with chromium oxide. Deamination[5,6] by the Ruschig's[7] method yields 5α-pregnane-3,20-dione (3) from funtumine and 20α-hydroxy-5α-pregnan-3-one (4) from funtumidine. Since Wolff–Kishner reduction of funtumine yields 3α-amino-5α-pregnane[5] (5), the structure of funtumine is 3α-amino-5α-pregnan-20-one (1) and funtumidine (2) is 3α-amino-5α-pregnan-20α-ol. Funtumine has been suggested as the starting material for preparation of androstanolone derivatives since the side chain can be cleaved in good yields by peracids.[8]

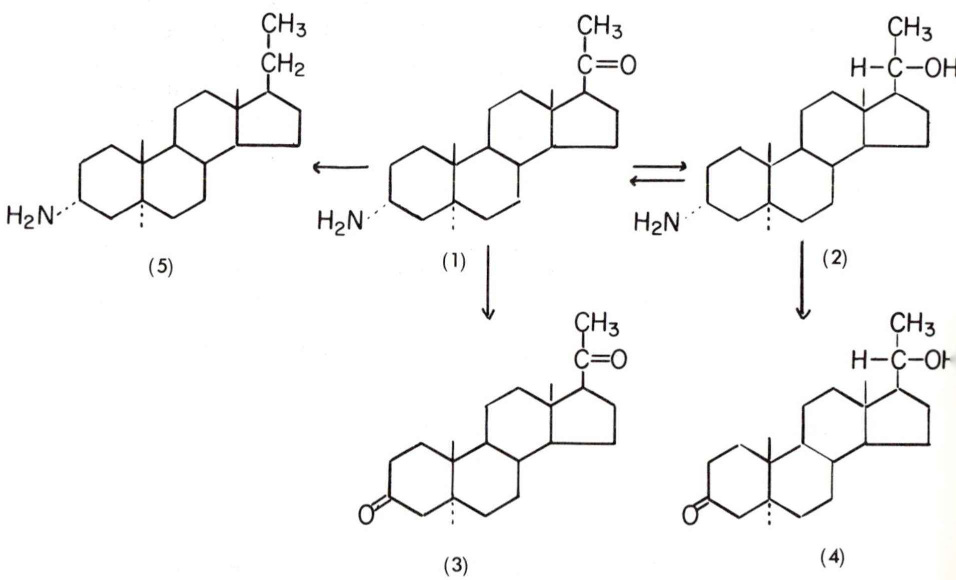

(5) (1) (2)

(3) (4)

Others of this group are holamine,[10] holaphyllamine,[10] and holaphylline[9] with the structures 3α-amino-, 3β-amino- and 3β-methylamino-5-pregnen-20-one, respectively.

2.2 Kurchiline

Kurchiline[11] isolated from *Holarrhena antidysenterica* is assigned the formulation 2α-hydroxy-3β-dimethylamino-5-pregnen-20-one (6) on the basis of its physical (nmr, mass spectroscopy) and chemical data. This was confirmed by the synthesis of dihydrodeoxokurchiline (7), the product resulting from the Wolff–Kishner reduction of 5,6-dihydrokurchiline. The synthesis consisted of converting 5α-pregnan-3-one (8) with lead tetraacetate to the 2α-acetoxy derivative (9) and to the oxime with hydroxylamine followed by reduction with lithium-aluminum hydride to the C(3) epimeric pair of amino compounds. Methylation with formic acid-formaldehyde and subsequent chromatography resolved the mixture to yield the desired 2α-hydroxy-3β-dimethylamino-5α-pregnane (7).

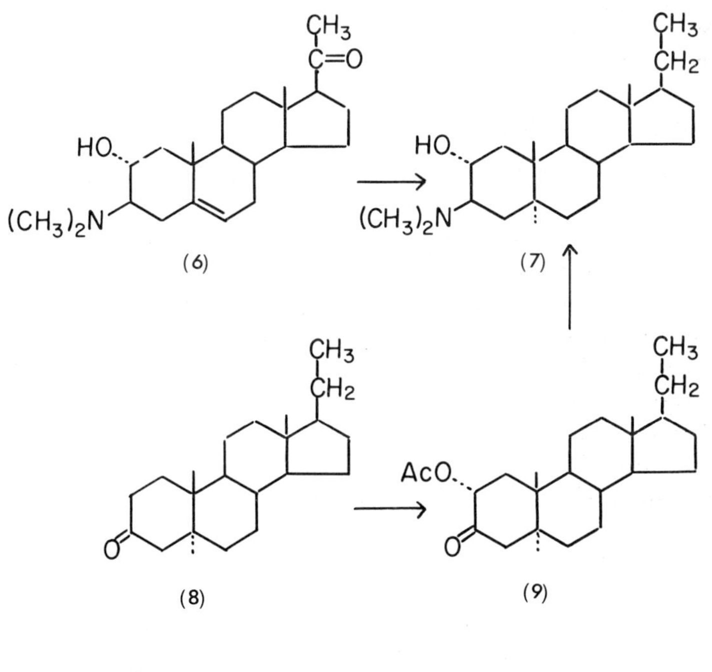

2.3 Other Alkaloids

Four other alkaloids, kurchiphylline (10), kurchiphyllamine (11), holadysine (12), and holadysamine were also isolated from the leaves of *H. antidysenterica*.[11] The structures were proposed mainly on the basis of spectral data (mass spectroscopy, ir, and nmr).[11] The occurrence of a 16-oxo function in ring D is of some interest. The formulation 3-methylamino-5,16-pregnadien-20-ol (13) has been proposed for holadysamine.[11]

TABLE 19.1. Other alkaloids.

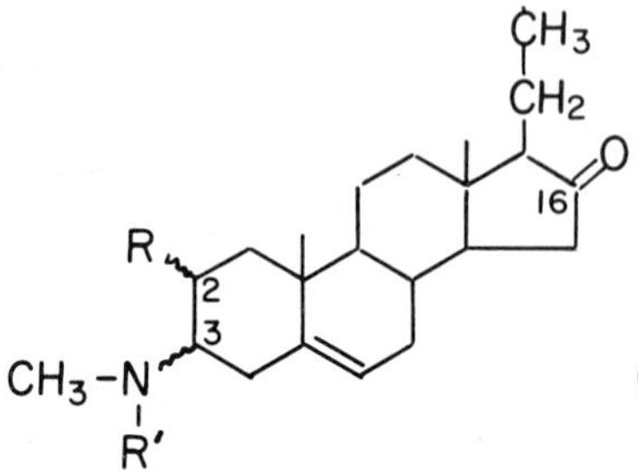

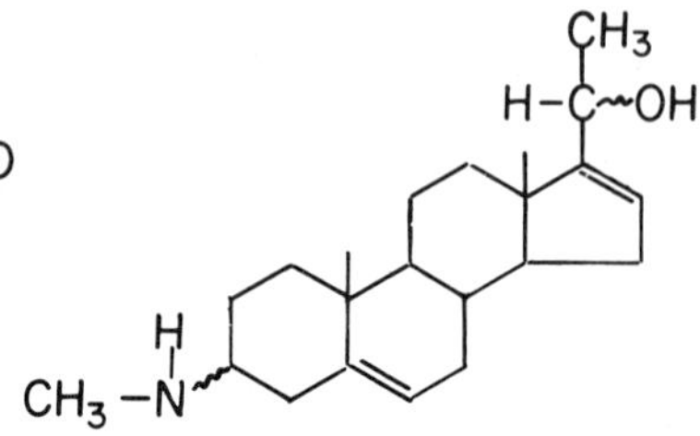

(13)

	R	R′
(10) Kurchiphylline	OH	CH_3
(11) Kurchiphyllamine	OH	H
(12) Holadysine	H	H

2.4 Paravallarine and Kibataline

2.4.1 Paravallarine The structure of paravallarine[12] (14) was deduced from the fact that Ruschig's deamination of 5,6-dihydroparavallarine afforded the known (20S)*-3-oxo-20-hydroxy-5α-pregnan-18-oic acid 18 → 20 lactone[13] (15). The β-orientation at C(3) was assigned on the basis of molecular rotation data. The structure was confirmed by partial synthesis of N-methyldihydroparavallarine by oximation and catalytic reduction of the oxolactone (15). The two C(3)-amino epimers were separated by chromatography and the 3β-epimer after methylation, furnished the desired N-methylidihydroparavallarine.[14] The other C(3α)-epimer upon methylation proved to be dihydrokibataline. Thus, kibataline[14] is (20S)-3α-dimethylamino-20-hydroxy-5-pregnen-18-oic acid 18 → 20 lactone. A 16α-hydroxy base of paravallarine named paravallaridine[15,16] has also been isolated from *Paravallaris microphylla.*[15]

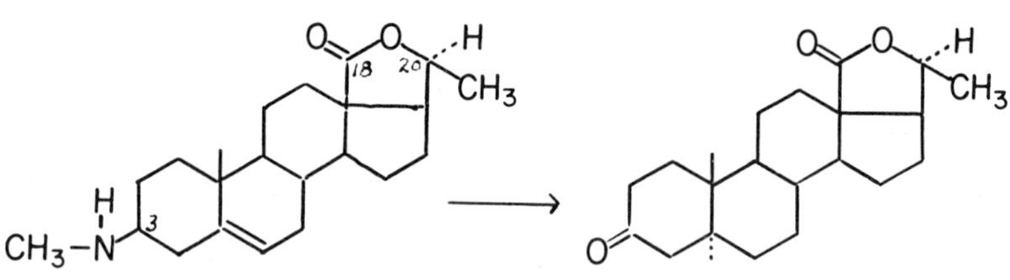

(14) (15)

*For nomenclature, see: R. S. Cahn, C. K. Ingold, and V. Prelog, *Experientia* **12,** 81 (1956); G. Bernath, *Nature* **189,** 304 (1961).

3. 20-AMINOPREGNANE BASES

3.1 Funtuphyllamines and Funtumafrines

Funtuphyllamine A has been synthesized by oximation and catalytic reduction of 3β-hydroxy-5α-pregnan-20-one and is 20α-amino-5α-pregnan-3β-ol. Formylation and lithium-aluminum hydride reduction of A give funtuphyllamine B,[17] the 20α-methylamino compound. Eschweiler–Clarke methylation of A or B yields funtuphyllamine C, 20α-dimethylamino-5α-pregnan-3β-ol,[17,18] a known, synthetically prepared compound.[19] Chromic acid oxidation of the funtuphyllamines B and C affords the corresponding 3-ketones, funtumafrines B and C.

3.2 Holafebrine, Irehamine, and Irehine

Holafebrine[20] is 20α-amino-5-pregnen-3β-ol and can be converted to funtuphyllamine A by catalytic reduction of the Δ^5 bond. The D(+) glucoside of holafebrine occurs in the roots of *Conopharyngia pachysiphon*[21,22] and shows hypotensive activity. Irehamine[23] is 20α-methylamino- and irehine[23] the 20α-dimethylamino-5-pregnen-3β-ol. Irehamine has been prepared from holafebrine by conversion with ethyl chloroformate to the N-carboethoxy compound and reduction with lithium-aluminum hydride. Methylation of holafebrine with formic acid-formaldehyde yields the 20α-dimethylamino derivative, irehine. Buxomegine[24] isolated recently from *Buxus sempervirens* has been shown to be identical with irehine.

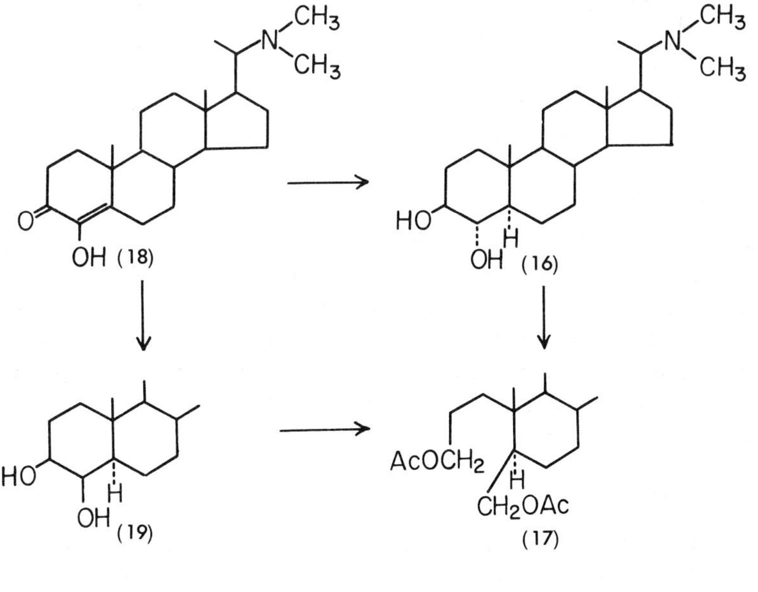

(18) (16) (19) (17)

Terminaline, another alkaloid isolated from *Pachysandra terminalis*, has been shown to possess the structure 20α-dimethylamino-5α-pregnane-3β,4α-diol (16).[25] The α-glycol group was shown by the oxidation of 16 with periodic acid, which yielded a dialdehyde convertible to the diol-diacetate (17) by sodium borohydride reduction and acetylation. The diol-diacetate (17) was also obtained from diosphenol (18) by sodium borohydride reduction to the 3β,4β-diol (19), followed by periodic acid oxidation, borohydride reduction and acetylation. The 3β,4α-diol orientation was assigned because the 19-methyl signal in the nmr spectrum occurred at $\tau = 9.18$, which is in the range for a compound unsubstituted at C(4) ($\tau = 9.1$–9.2),[26] whereas the axial hydroxyl at C(4) in the 3β,4β-diol (19) induced a downfield shift to $\tau = 8.98$. Furthermore, the base (16) failed to form an acetonide in contrast to the 3β,4β-diol. Finally, terminaline was prepared from the diosphenol (18) by reduction with sodium in boiling n-amyl alcohol.

4. 3,20-DIAMINOPREGNANE BASES

4.1 Irehdiamine A and B

Irehdiamine A and B, obtainable from *Funtumia elastica*,[27] are both converted to progesterone, by Ruschig deamination, which shows that the two amino functions are located at C(3) and C(20) and the double bond most probably at C(5,6). This was supported by the molecular rotation difference between A and its dihydro analog. Methylation of A with formic acid-formaldehyde gives the known 3β,20α-bisdimethylamino-5-pregnene.[28] Irehdiamine A is accordingly 3β,20α-diamino-5-pregnene. Conversion of irehdiamine B to dihydroirehdiamine B and correlation with 3β-methylamino-5α-pregnan-20-one derived from dihydroholaphylline[9] established the structure of irehdiamine B as 3β-methylamino-20α-amino-5-pregnene.

4.2 Chonemorphine

Chonemorphine, which occurs in the Chonemorpha species,[29] is converted by nitrous acid to deamino-oxychonemorphine. The latter has been shown to be 3β-hydroxy-20α-dimethylamino-5α-pregnane.[30] Chonemorphine is thus 3β-amino-20α-dimethylamino-5α-pregnane. Two isomeric bases, funtudiamine[31] (epipachysamine F[32]) and dictyodiamine[33] (epipachysamine C[25]), possessing the structures 3β-dimethylamino-20α-amino-5α-pregnane and 3β,20α-bismethylamino-5α-pregnane, have also been isolated.

4.3 Pachysamine A

Pachysamine A,[34] which shows a total of three N-methyl functions in the nmr, is converted to the N-acetate on acetylation. Methylation of the alkaloid affords the known 3α,20α-bisdimethylamino-5α-pregnane and Ruschig deamination yields 20α-dimethylamino-5α-pregnan-3-one (funtumafrine C). This accords well for the assignment of 3α-methylamino-20α-dimethylamino-5α-pregnane for pachysamine A. The 3β-isomer named dictyophlebine is also present in *Dictyophleba lucida.*[33]

4.4 Kurchessine, Dihydrokurchessine, and Holarrhena Alkaloid C

Kurchessine, isolated from the bark of *Holarrhena antidysenterica* and formerly ascribed the 3-dimethylamino-17(20)-conanene formulation,[58] has now been found to be 3β,20α-bisdimethylamino-5-pregnene.[35] It was identified by comparison with an authentic specimen.[36] The same alkaloid[37] was also isolated from *Pachysandra terminalis* of the family Buxaceae. It was the first pregnane type of alkaloid found outside of the family Apocynaceae. The 5,6-dihydro analog of kurchessine (Pachysandra alkaloid D), 3β,20α-bisdimethylamino-5α-pregnane, was also found in the plant extracts of *Pachysandra terminalis.*[37] Holarrhena alkaloid C isolated along with kurchessine is the 3α-epimer, 3α,20α-bisdimethylamino-5-pregnene.[35]

4.5 Malouetine

Malouetine[38] is unusual in that it exists as a diquaternary base in the plant. Its structure was deduced from the Hofmann degradation and subsequent hydrogenation of the quaternary ammonium hydroxide base to 3β-dimethylamino-5α-pregnane. It has been confirmed by its synthesis from the oxime (21)[19] of funtumafrine C by reduction with sodium and ethanol to the primary 3β-amine (22) followed by methylation to the 3β, 20α-bisdimethylamino compound and finally with methyliodide to the iodide of malouetine (20).[39]

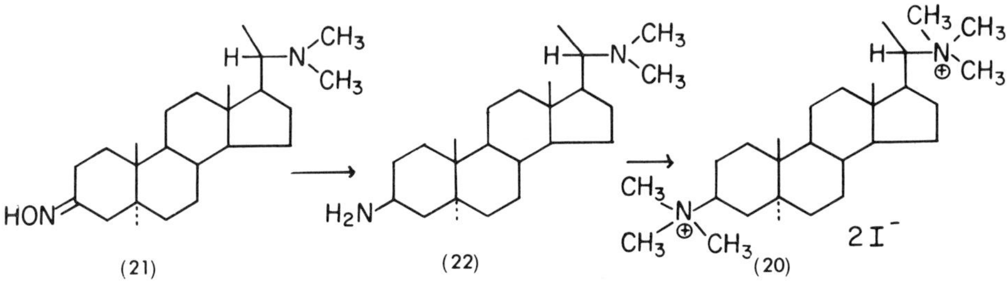

(21) (22) (20)

4.6 Pachysandra Alkaloids

4.6.1 Epipachysamine A and Saracocine Epipachysamine A exhibits a tertiary amide band in the ir spectrum. Treatment with phenyllithium[40] or by vigorous acid hydrolysis[41] leads to a desacyl compound that resembles 3β-dimethylamino-20α-methylamino-5α-pregnane[43] in its properties. Moreover, methylation with formic acid-formaldehyde furnishes the known 3β,20α-bisdimethylamino-5α-pregnane.[19] Thus, epipachysamine A[40] (saracodine)[41] is assigned the 3β-dimethylamino-20α-N-acetyl-N-methylamino-5α-pregnane structure. The Δ^5 analog of saracodine is known as saracocine.[41,42] The site of unsaturation in saracocine was ascertained by mass spectrometry.[41,44]

4.6.2 Pachysamine B Pachysamine B[34] is considered to be 3α-[N,N-(3′,3′-dimethylacrylyl, methyl)]amino-20α-dimethylamino-5α-pregnane on the basis of the nmr spectrum and by analogy with pachysandrine B (30). Its structure was confirmed by its synthesis from pachysamine A and β,β-dimethylacrylyl chloride.

4.6.3 Epipachysamine E The nmr spectrum of epipachysamine E exhibits signals for a β,β-dimethylacrylyl moiety and acid hydrolysis of the alkaloid yields chonemorphine. The structure, 3β-[N-(β,β-dimethylacrylyl)]amino-20α-dimethylamino-5α-pregnane, was proven by its synthesis from chonemorphine and β,β-dimethylacrylyl chloride.[32]

4.6.4 Epipachysamine B[25] The formulation 3β-(N-nicotinoyl)amino-20α-dimethylamino-5α-pregnane ascribed to the compound affords chonemorphine upon acid hydrolysis. The presence of a nicotinamide grouping in the molecule was deduced from spectral data and verified by the synthesis of the alkaloid from chonemorphine and nicotinic acid by the mixed anhydride procedure.[45]

4.6.5 Epipachysamine D The acid hydrolysis of the base followed by N-methylation yields 3β,20α-bisdimethylamino-5α-pregnane. The N-benzoyl moiety in the compound deduced from spectral data was confirmed by the synthesis of the base from chonemorphine and benzoyl chloride.[32] Epipachysamine D is, therefore, 3β-(N-benzoyl)amino-20α-dimethylamino-5α-pregnane.

4.7 Pachysandrine Alkaloids

4.7.1 Pachysandrine A Spectral data of pachysandrine A support the presence of an acetyl (—CH—CHOAc—CH—), conjugated N-methylamide (R—CO—N—CH_3), phenyl, and a dimethylamino moiety in the molecule. Mild hydrolysis of pachysandrine A (23)[46] with alkali gives an O-desacyl compound (24) that regenerates the parent compound upon acetylation. Vigorous treatment of the O-desacyl product (24) with strong alkali affords the O,N-desacyl compound (25) and benzoic acid. Oxidation of 25 with chromic acid yields the 4-oxo compound (26) that displays a negative Cotton ORD curve very similar to that of cholestan-4-one. The Wolff-Kishner (Huang-Minlon) reduction of 26 gives 20α-dimethyl-amino-5α-pregnane (27), while the treatment of the oxo compound (26) with base yields the diosphenol (18),[47] which is identical with a specimen prepared by the air oxidation[48] of 20α-dimethylamino-5β-pregnan-3-one (28). The diosphenol (18) is also converted to the base 27 by Huang-Minlon reduction. From these reactions the structure 3α-(N,N-benzoyl, methyl)amino-4β-acetoxy-20α-dimethylamino-5α-pregnane was proposed for pachysandrine A. The 3α,4β-configuration was assigned on the basis of the reaction of the O-desacyl compound (24) with phosphorous oxy-chloride in pyridine and subsequent alkaline hydrolysis to give the 4-epi

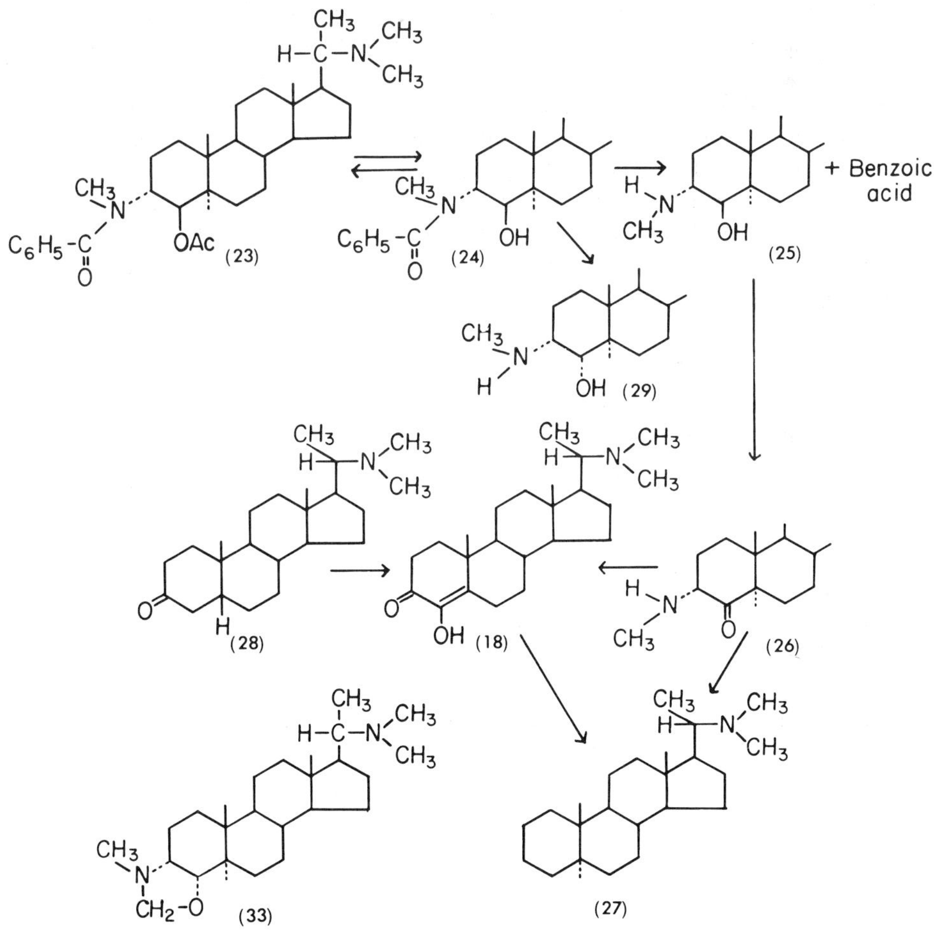

alcohol (29). The formation of the latter can be viewed as the result of an acyl migration[49] involving an oxazoline intermediate formed by the nucleophilic backside attack on the N-acyl group. This can occur only in a molecule which possesses a 1,2-diaxial system such as present in the 3α-acylamino-4β-hydroxy-5α-pregnane derivative. Nmr and ir data corroborated the above deduction.[46] The 3N-desmethyl-4O-desacyl compound of 23 is known as epipachysandrine-A.[178]

4.7.2 Pachysandrine B Spectral data indicate the presence of acetyl (—CH—CHOAc—CH—), conjugated N-methylamide (—C=C—CON CH_3), isobutenyl [$(CH_3)_2C{=}C$] and a dimethylamino moiety in addition to the two tertiary and one secondary methyl functions in pachysandrine B. On alkaline hydrolysis, the alkaloid (30) gives an O-desacetyl product (31) that reverts to the parent compound by acetylation. Acid hydrolysis of the alkaloid yields an ester (32) that appears to be formed by the migration of the N-acyl group with inversion, since mild alkaline hydrolysis furnishes the amino alcohol (29) identical with that derived from pachysandrine A. The amino alcohol (29) could also be obtained by the reaction of the acylamino alcohol (31) with phosphorus oxychloride followed by acid hydrolysis in a manner similar to the acylamino alcohol of pachysandrine A. This similarity in the behavior of pachysandrine B with A led to the assignment of 3α-[N,N-(3′,3′-dimethylacrylyl, methyl)]amino-4β-acetoxy-20α-dimethylamino-5α-pregnane for its structure (30).[46]

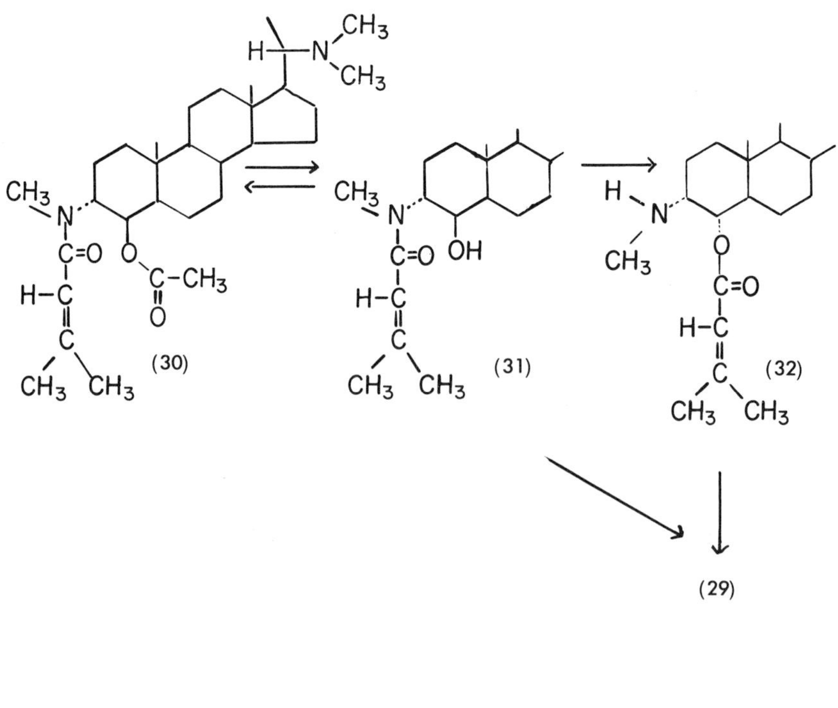

(30) (31) (32) (29)

4.7.3 Pachysandrine C and D Pachysandrine C(29)[40] has been identified as 3α-methylamino-4α-hydroxy-20α-dimethylamino-5α-pregnane by comparison with a specimen prepared from pachysandrine A.[46] It appears to be an artifact formed during the course of separation. It is interesting to note that an oxazolidine derivative (33) of pachysandrine C is also isolated from the same fraction. It is formed, presumably, during the N-methylation treatment (formaldehyde-formic acid) used to facilitate separation and is confined to the 4α-epimer (29). The 4β-isomer (25) forms the expected N-dimethyl derivative.

Pachysandrine D (32)[40] is hydrolyzed by mild alkali to yield pachysandrine C. This fact and the spectral data, together with the relationship of the alkaloid to pachysandrine B, lead to the formulation 3α-methylamino-4α-(3′,3′-dimethylacrylyloxy)-20α-dimethylamino-5α-pregnane for the alkaloid.

4.7.4 Pachystermine A and B Pachystermine A (34), one of the major alkaloids of *Pachysandra terminalis*,[50] can be converted to pachystermine B (35) by reduction with sodium borohydride and the latter in turn reconverted to A by oxidation with chromic acid. Pachystermine A is, therefore, the oxo derivative and B the alcohol. Treatment of A with ethanolic alkali yields the diosphenol (18). Furthermore, Huang-Minlon reduction of base A gives 20α-dimethylamino-5α-pregnan-4-one as well as the 4-deoxo compound.

The presence of the isopropyl group* in the sidechain ($C_6H_{10}O$) was detected from the nmr spectra. The remaining —C_2H_3—CO fragment appeared to form a ring system since no olefinic proton was observed in the nmr spectra. From the appearance of a β-lactam absorption in the ir spectra of the base and several derivatives, it was suspected that the —C_2H_3CO— moiety and the nitrogen atom formed a four-membered β-lactam ring.

The lithium-aluminum hydride reduction of pachystermine A (34) or B (35) leads to the same diol (36), which on N-methylation yields the N-methyl derivative (37). The alkaline hydrolysis of B affords a mixture of amino-acid diastereomers (38), which on conversion to the methyl ester, reduction to the alcohol, and N-methylation give rise to the diastereomeric mixtures of (37). By repeated chromatography, the mixture was separated into each diastereomer, one of which proved to be identical with the N-methyl diol product (37) of the lithium-aluminum hydride reduction of A or B.

The synthesis of the desoxy compound (39) derived from the N-methyl diol (37) by tosylation and lithium-aluminum hydride reduction provided

*The stereochemistry of the isopropyl moiety in 34 and 35 has now been shown to be α.[179]

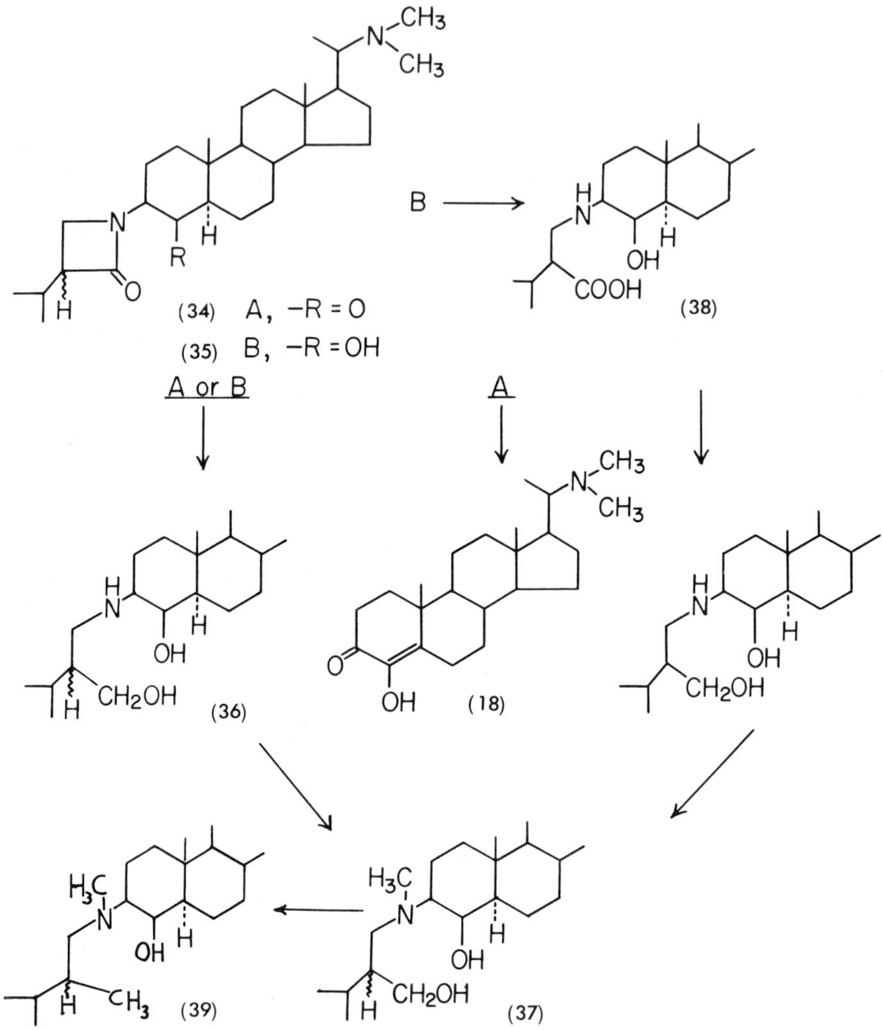

proof for the structures of A and B. This was achieved by condensation of 2,3-dimethylbutyraldehyde with the amino alcohol, 3β-amino-20α-dimethylamino-5α-pregnan-4β-ol, to the Schiff base, followed by borohydride reduction and N-methylation. The requisite aldehyde was obtained from the ozonolysis of Δ^{22}-ergosten-3-one.

4.7.5 Pachysantermine A Another alkaloid of novel type isolated from *Pachysandra terminalis* is pachysantermine A (40) bearing a 7-membered ring imino lactone structure on ring A.[51] Hydrogenation and N-methylation of 40 give the N-methyldihydro compound (41). The nmr spectra of 41 revealed the disappearance of two allylic protons and two allylic methyl signals [—N—CH_2—C=C—$(CH_3)_2$] and the appearance of protons of an isopropyl moiety [$(CH_3)_2CH$—]. Treatment of 41 with base affords an amino acid, which upon methylation and lithium-aluminum hydride reduction yields the diol (42). The properties of the diol (42) were in agreement with the diol obtained from pachystermine A.

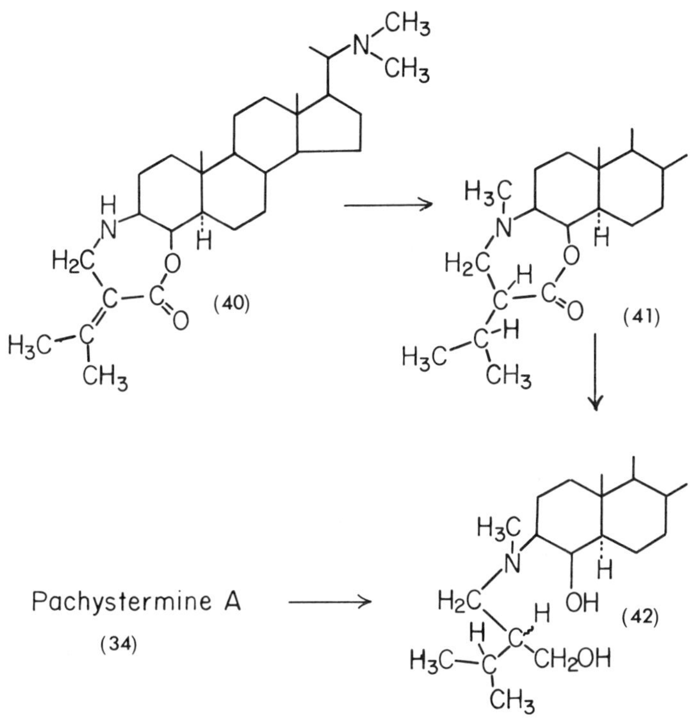

4.7.6 Holarrhimine Group

4.7.6.1 Holarrhimine, the Methylholarrhimes, and Holarrhidine Siddiqui and coworkers[52,53] demonstrated the presence of hydroxyl, two primary amino groups and a double bond in holarrhimine (43).[54] Methylation and catalytic reduction afforded dihydrotetramethylholarrhimine (44),[55] which on successive treatment with p-toluenesulfonyl chloride, sodium iodide, and methyl iodide resulted in the formation of dihydroconessine dimethiodide (45).[56] Oxidation of 44 with chromic acid to an aldehyde and to an acid, and the behavior of the acid ester toward hydrolysis and lithium-aluminum hydride reduction indicated the primary nature of the hydroxyl and its location at C(18).[57] The proof of the positions of the

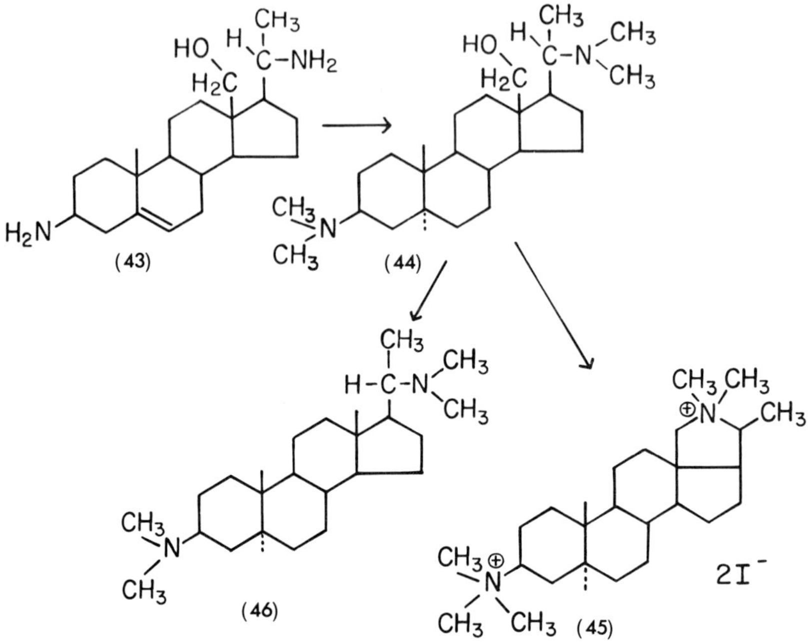

(43) (44) (46) (45)

amino group was obtained by Wolff–Kishner reduction of the aldehyde of 44 to 3β,20α-bisdimethylamino-5α-pregnane (46).[19] From these considerations holarrhimine (43) was assigned the 3β,20α-diamino-5-pregnen-18-ol structure. The N(3)methyl-,N(20)methyl- and the tetramethyl derivative of holarrhimine are also isolated from the same source.[58]

Holarrhidine[59] is the 3α-epimer of holarrhimine (44), present also in the kurchi bark.

5. 3-AMINOCONANINE BASES

5.1 Conessine

The structure of conessine (47) was established by Haworth et al.[55] Selenium dehydrogenation[60] of conessine yields 3′-ethylcyclopentenophenanthrene and Hofmann degradation gives apoconessine (48) that can be further degraded to pregna-3,5,20-triene (49) by Emde's procedure. The hexahydro derivative of apoconessine (50) is demethylated by von Braun's cyanogen bromide procedure and the resulting N-demethyl derivative (51) converted into 18-amino-5α-pregnane (52)[55] by treatment of the benzoyl derivative of the secondary amine with phosphorus pentachloride.

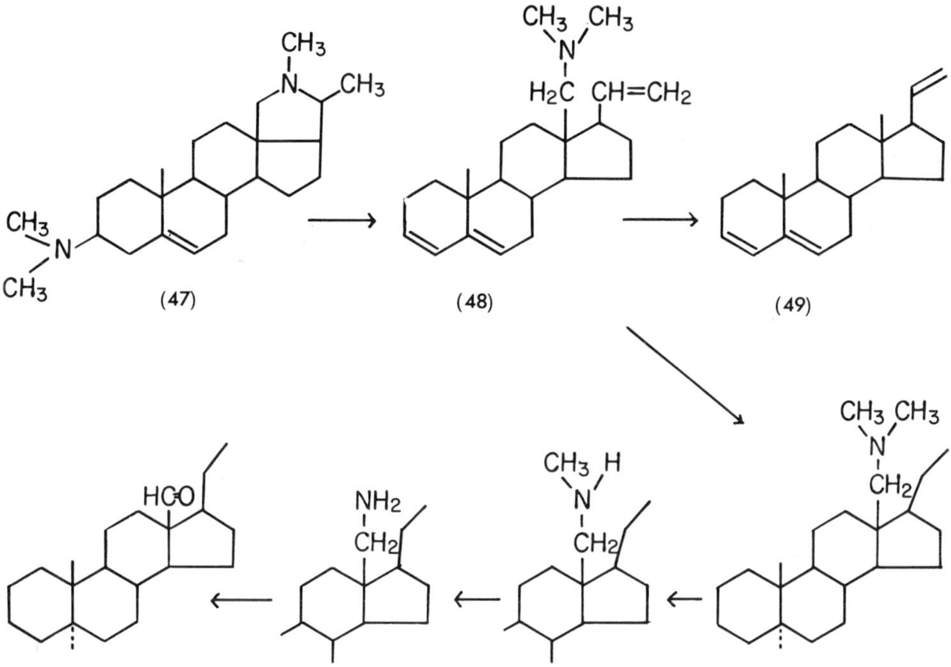

(47) (48) (49) (53) (52) (51) (50)

The Sommelet reaction readily transformed the 18-amino derivative into 18-oxo-5α-pregnane (53).[61] The demethylation of conessine with cyanogen bromide affords isoconessimine[66] that can be remethylated with formaldehyde-formic acid to conessine (47).[67] Likewise conessimine,[66] the N_3-dimethyl-, and conimine,[66] the N_3-methylconarrhimines are converted to conessine by methylation.[62] Conamine[65] is the N(20)-methylated base of the methyl free parent alkaloid, conarrhimine.[65] Besides the above bases, 3α-amino-5-conanene[63] (3-epiconamine), dihydroisoconessimine[63] and dihydroconessine (59)[63] have been isolated from *H. antidysenterica.*[63]

The partial synthesis of dihydroconessine (59) from 3β-acetoxy-20α-amino-5-pregnene (54) was achieved by Corey and Hertler[64] through the lithium-aluminum hydride reduction of the N-formyl derivative (55) to the N-methyl compound (56), conversion to the N-chloramine (57) by the Ruschig method followed by irradiation to the isomeric C(18)–chloride (58) and cyclization with base. In a similar manner, conanine, the des N-dimethylamino derivative of conessine, was synthesized by the Swiss group of investigators.[68] A number of elegant syntheses of conessine have since been perfected.[69–73] Conessine has also been transformed into 18-hydroxyprogesterone by several groups.[74–76]

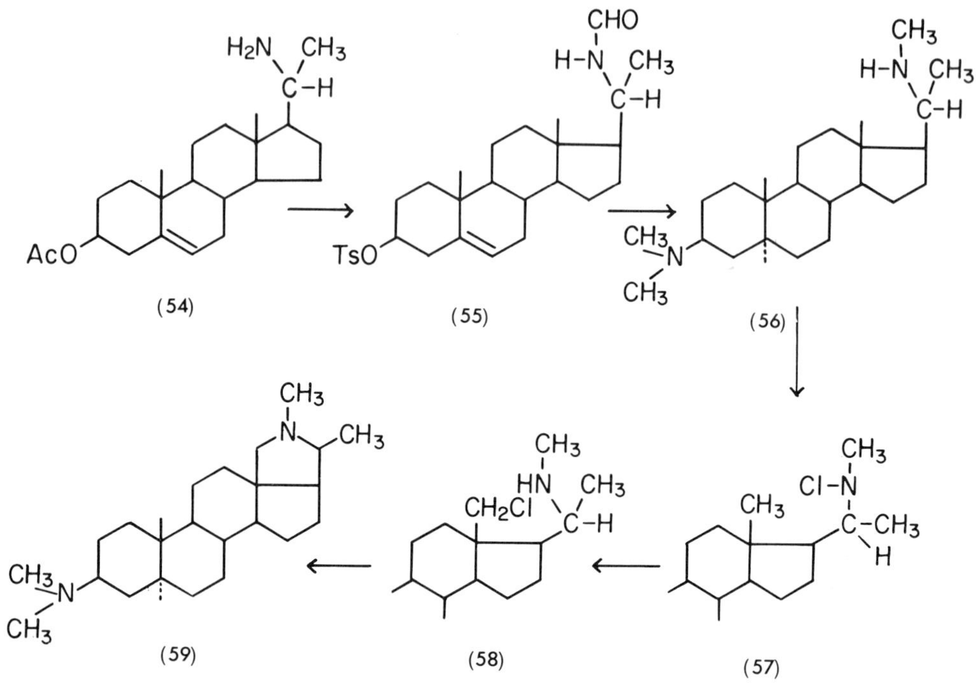

5.2 Oxygenated 3-Aminoconanenes

Holarrhenine (60)[79] has a hydroxyl function that can be oxidized to the oxo derivative, converted to the thioketal, and then be reduced to conessine (47) by Raney nickel.[77] The oxo compound possesses the characteristic ir absorption for a six-membered ketone. The exact location of the hydroxyl function was ascertained by the degradation of holarrhenine through the Hofmann and Emde reactions into the known 12β-hydroxy-5α-pregnane (61).[78] Holarrhenine (60) is, therefore, 12β-hydroxyconessine. The 12β-tosylate of holarrhenine has been found to undergo rearrangement to a C-nor-D-homosteroid[77] with lithium-aluminum hydride, a phenomenon previously observed to occur with the alkaline treatment of 12β-tosylates.[82] Holarrheline[80] is another related alkaloid that has been isolated from *Holarrhena floribunda*. It has been assigned the structure 3β-methylamino-12β-hydroxy-5-conanene by Janot et al.[80]

Holarrhetine and holafrine are two C(12)β-ester alkaloids isolated from the bark of *Holarrhena africana*.[81] Alkaline hydrolysis of holarrhetine yields holarrhenine (60) and pyroterebic acid, $(CH_3)_2C{=}CHCH_2COOH$, while holafrine gives 12β-hydroxyconessimine and the same acid.

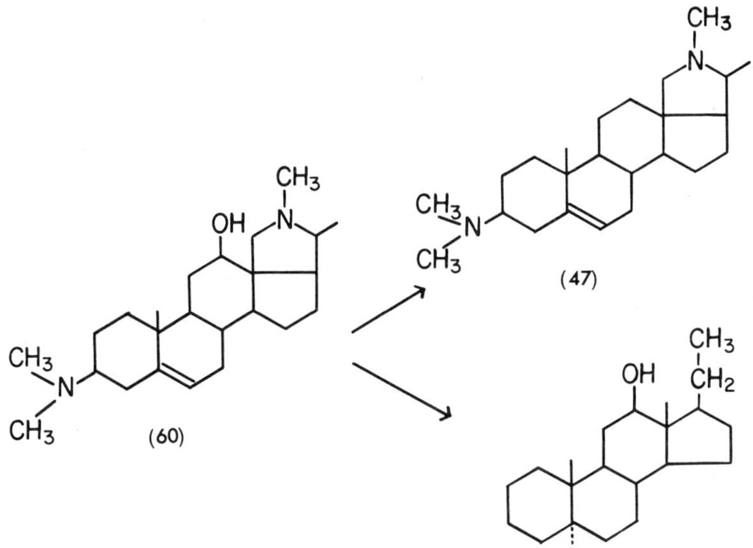

(60) (47) (61)

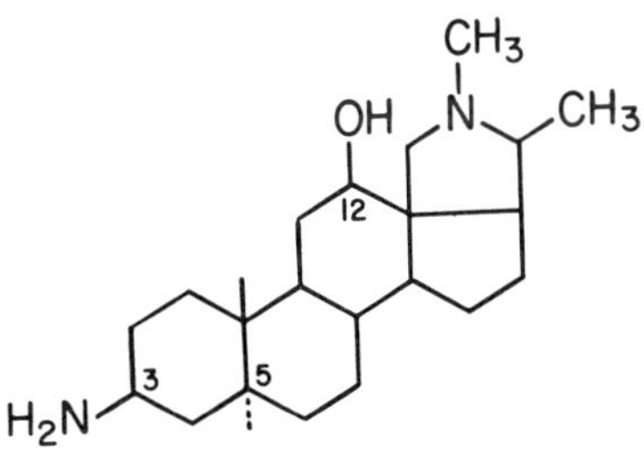

(62)

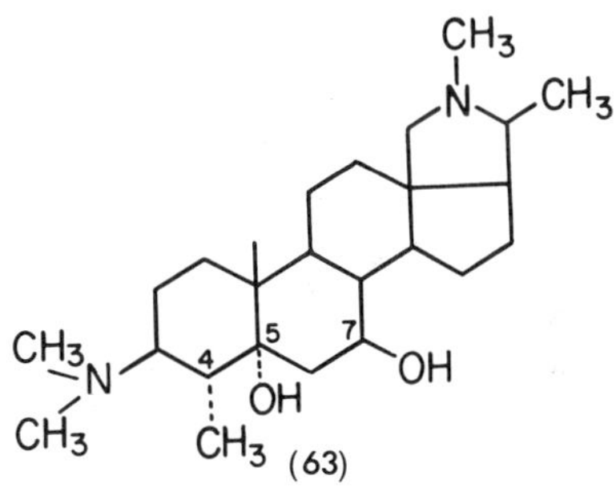

(63)

5.2.1 Other Oxygenated Derivatives 7α-Hydroxyconessine[83] has also been found to occur in the Kurchi bark *(Holarrhena antidysenterica)*. Its structure was confirmed through its synthesis by acetoxylation of conessine with t-butyl perbenzoate in acetic acid with a cuprous bromide catalyst,[84] saponification and chromatography of the epimeric mixture. Both 7α- and 7β-hydroxyconessine have also been prepared by microbial hydroxylation[85,86] of conessine. Holaline, which has been isolated from *Holarrhena floribunda*, is assigned the structure 5α-hydroxy-dihydroconessine.[80] The hydroxyl configuration is based on the fact that the hydroxyl is not acetylable under ordinary conditions and dehydrates readily to form conessine. The assignment was confirmed by its synthesis from 5α,6β-dihydroxy-dihydroconessine by mesylation to the 5α-hydroxy-6β-mesylate followed by alkaline treatment to the 5α,6α-epoxide and reduction with lithium-aluminum hydride. Funtessine (62),[31] which bears the structure 3β-amino-12β-hydroxy-5α-conanine, has recently been isolated from *Funtumia latifolia*. Its partial synthesis from holarrhenine (60) has been achieved by reduction to 5,6-dihydroholarrhenine, cyanogen bromide demethylation to the 3β-methylamino derivative, Ruschig deamination to the 3-oxo compound, oxime formation and reduction.[87] To kurcholessine[88,89] (63), another constituent of the kurchi bark, has been assigned the 3β-dimethylamino-5α,7β-dihydroxy-4α-methylconanine structure, derived mainly from spectral data of the alkaloid and its derivatives. The isolation of a substance retaining a methyl function at C(4) is of biogenetic interest.

5.3 3-Amino-N-Demethyl-5,18(N)-Conadienes

5.3.1 Irehline Irehline [(64), R = H], isolated from *Funtumia elastica*,[90] possesses an acetylable primary amino group and two double bonds. The possible site of unsaturation in rings B and E were indicated by the nmr spectrum that possessed signals for a C(21)-methyl (doublet) and two olefinic protons (Δ^5 and C(18) of R—CH═N—R). Confirmation of the 3β-amino-N-demethyl-5,18(N)-conadiene (64) structure for irehline was provided by the benzoylation of irehline monoacetate [(64), R = Ac] under Schotten-Baumann conditions that afforded the 18-oxo-20-benzoylamino derivative (65). Reduction of 65 with potassium borohydride followed by alkaline hydrolysis gave a product identical with holarrhimine (43).[90] Further proof was obtained by conversion of holarrhimine (43) into the

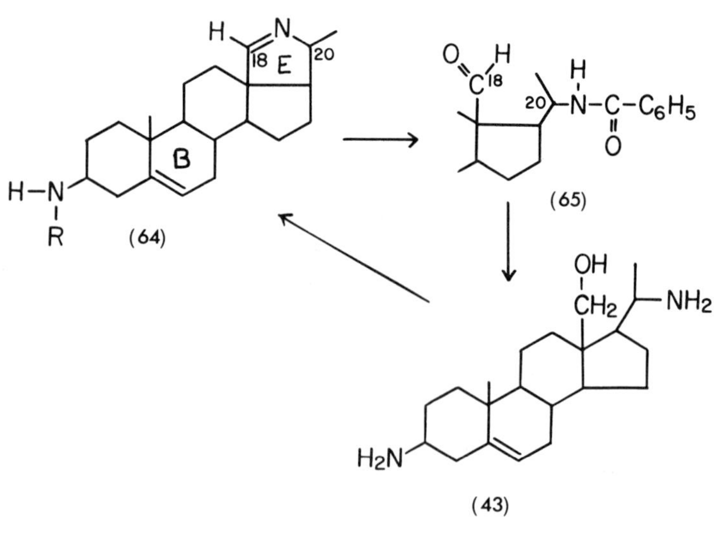

original alkaloid by oxidation with chromic acid.[91] Conkurchine, the alkaloid from *H. antidysenterica*[92] was formerly ascribed the structure of 3β-amino-N-demethyl-5,17(20)-conadiene.[93] The position of the double bond at $\Delta^{17(20)}$ was based mainly on the isolation of acetic acid from the ozonolysis of the pyrroline ring. The $\Delta^{18(N)}$ and $\Delta^{20(N)}$ pyrroline derivatives have now also been found to yield acetic acid.[94] Reexamination of conkurchine has shown it to be identical with irehline.[95]

5.3.2 Conessidine Conessidine is 3β-methylamino-N-demethyl-5,18(N)-conadiene,[95] [(64), R = CH_3]. The structure was confirmed by its partial synthesis by oxidation of N_3-methylholarrhimine with chromic acid.[91] The position of the second double bond in conessidine was formerly assigned at C(17,20).[93] Trimethylconkurchine,[93] which was obtained by Eschweiler–Clark methylation of either conkurchine or conessidine and assigned the structure 3β-dimethylamino-5,17(20)-conadiene has now been found to be conessine (47).[94] Apparently in the methylation of the alkaloids, the $\Delta^{18(N)}$ bond is reduced prior to methylation.

5.4 Oxygenated Conanenes

A number of bases have been isolated from *Funtumia latifolia* bearing a hydroxyl or oxo function on the conanene ring.

5.4.1 Latifoline Latifoline (66)[96] is 3β-hydroxy-5-conanene. It was identified by comparison with an authentic specimen prepared from cones-

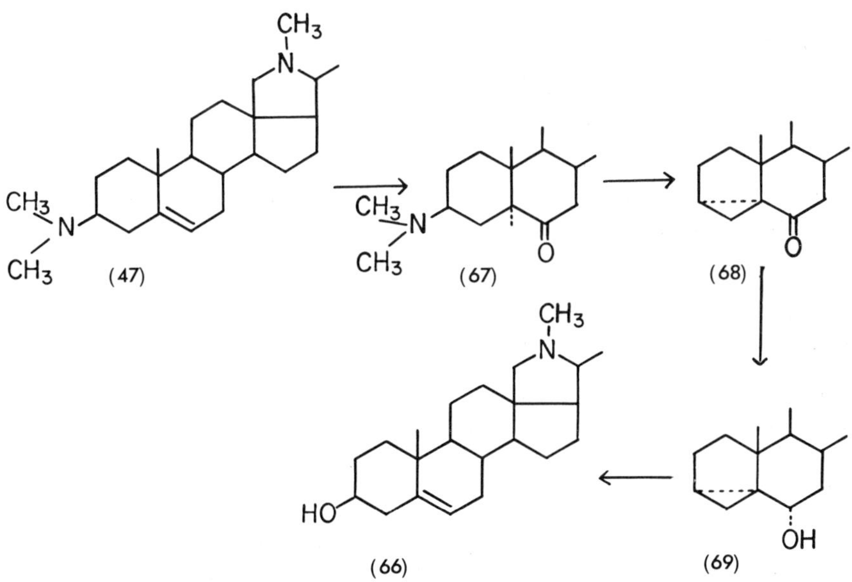

(47) (67) (68) (66) (69)

sine in the following manner. Conessine (47) was converted to the 6-ketone (67) by treatment with sodium borohydride and aluminum chloride and oxidation of the resulting 6-boron derivative with chromic acid.[75] Selective quaternization of the C(3)-dimethylamino group with methyl iodide to the C(3)-methiodide followed by treatment with potassium t-butoxide in t-butanol led to elimination of the basic moiety at C(3) and formation of the 3,5-cyclo-6-oxo base (68). Reduction with lithium-aluminum hydride to the 6α-ol (69) followed by acid catalyzed rearrangement gave latifoline (66).[76]

5.4.2 Latifolinine Latifolinine (71)[97] is 3-oxo-4-conanene. Its structure was proven by its partial synthesis from conessine (47) by von Braun demethylation to isoconessimine (70) followed by Ruschig deamination.

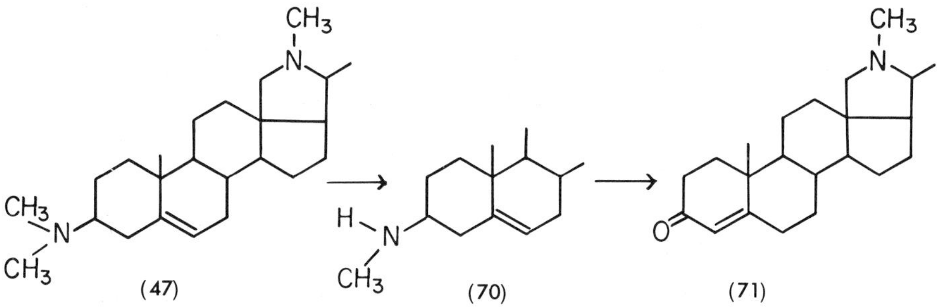

(47) (70) (71)

5.4.3 Funtuline Funtuline[98] possesses the 3β,12β-dihydroxy-5-conanene formulation. It was proven by partial synthesis of the 5,6-dihydro derivative from holarrhenine (60). Dihydroholarrhenine was demethylated to the 3β-methylamino base by von Braun's procedure, deaminated to the 3-one by Ruschig's method and reduced to the 3β-ol with potassium borohydride.

5.4.4 Funtudienine Funtudienine is 7-oxo-3,5-conadiene.[31] Its synthesis has been accomplished by chromic acid oxidation of O-acetyllatifoline (66) to the 7-one and elimination of the 3-acetoxy moiety with p-toluenesulfonic acid in benzene.

5.4.5 Holadienine Holadienine, the isomer of funtudienine, has been assigned the 3-oxo-1,4-conadiene structure.[80]

5.4.6 Nor-Latifoline Nor-latifoline is N-desmethyl-5-conanen-3β-ol.[97] The alkaloid is readily transformed into the O,N-diacetyl derivative by acetylation and into latifoline (66) by methylation with formic acid and formaldehyde.

5.4.7 Holonamine The uv data support the presence of a dienone structure that undergoes aromatization by treatment with acid. The nmr spectrum reveals doublets for the C(21)-methyl group and the C(18)-olefinic proton. The weak splitting of the latter is presumably due to long range coupling with the C(20) proton. The splitting pattern is exactly like that of irehline (64) which possesses a $\Delta^{18(N)}$ bond. The compound, moreover, reacts with o-aminobenzaldehyde to give a deep yellow dihydroquinazolinium derivative, typical of a Δ^1-pyrroline. The position of the hydroxyl group as 11α- was decided by comparison of the physical data (nmr, circular dichroism, uv) with holadysone[175] and 11α-hydroxy-1,4-pregnadien-3,20-dione which possesses a structure similar to holonamine in rings A, B, C, and D of the molecule. Holonamine[83,99] is thus thought to be 11α-hydroxy-N-demethyl-1,4,18(N)-conatrien-3-one.

6. ALKALOIDS DERIVED FROM 20-PIPERIDYLPREGNANES

The plants of the genus Solanaceae and Liliaceae yield a number of alkaloids that are frequently grouped under the headings solanum, veratrum or fritillaria alkaloids depending upon the species from which they are derived.

Although a few of these alkaloids have been isolated as free alkamines, most occur in the plant as glycosides with varying saccharide content. They are all characterized by possessing a pregnane or a modified pregnane skeleton with a methylpiperidine ring whose α-position is attached to C(20) of the steroid moiety. The various steroid alkaloids embodying the indolizidine (solanidine type) and oxa-azaspirodecane (solasodine type) system and the modified C-nor-D-homo steroid alkaloids with piperidine (veratramine type) and quinolizidine (cevine type) sidechains can all be visualized as arising from this prototype.

Veralkamine (72), isolable from *Veratrum album*[100] possesses a 20-piperidyl moiety with the unusual 17β-methyl-18-nor-17-isocholestane skeleton. It has been given the structure (22S,25S)-22,26-epimino-17β-methyl-18-norcholesta-5,12-diene-3β,16β-diol (72).[101,102] Its steroidal nature was indicated by selenium dehydrogenation, which afforded the hydrocarbon 1,2-cyclopentenophenanthrene and the base 2-ethyl-5-methylpyridine.[101] The absence of the normal C(18)-methyl function at C(13) was indicated by the isolation of the above hydrocarbon rather than the usual Diels hydrocarbon (3′-methyl-1,2-cyclopentenophenanthrene)[103] isolated from normal steroids. The presence of a Δ^5-3β-ol system was shown by Oppenauer oxidation, and oxidation of the 5,6-dihydro derivative afforded an unconjugated diketone one of which indicated a 5-membered ketone (1748 cm^{-1}). The spectroscopic data thus placed the second hydroxyl at C(16) and the second double bond at C(12,13). X-ray analysis of veralkamine hydroiodide confirmed these observations.[102]

Solacongestidine (73), the principal alkaloid isolated from *Solanum congestiflorum*[176] of Chilean origin, has been identified as (25R)-22,26-imino-5α-cholest-22(N)-en-3β-ol by its chemical degradation to 3β-acetoxybisnorallocholanic acid[104] and by comparison with a synthetic specimen.[105] The degradation was achieved by converting the alkaloid into the O,N-diacetyl-Δ^{22}-derivative (74) by refluxing with acetic anhydride and submission to acid hydrolysis to afford the 22-oxo-26-acetylamino derivative (75). Oxidation with chromic acid gave 3β-acetoxybisnorallocholanic acid (76). In the synthesis of the alkaloid,[105] 3β-acetoxy-5-pregnen-20-one (77) was reacted with 2-lithium-5-methylpyridine to give 20-(5′-methylpyridyl-2′)-5-pregnene-3β,20-diol (78), the 3-acetate of which was dehydrated by phosphoryl chloride-pyridine to the vinylpyridyl derivative (79). Catalytic reduction and hydrolysis followed by extensive chromatography afforded the appropriate (22S, 25R)-22,26-imino-5α-cholestan-3β-ol (80). Treatment of 80 with N-chlorosuccinimide gave the N-chloro derivative which was converted to the alkaloid (73) by methanolic alkali.

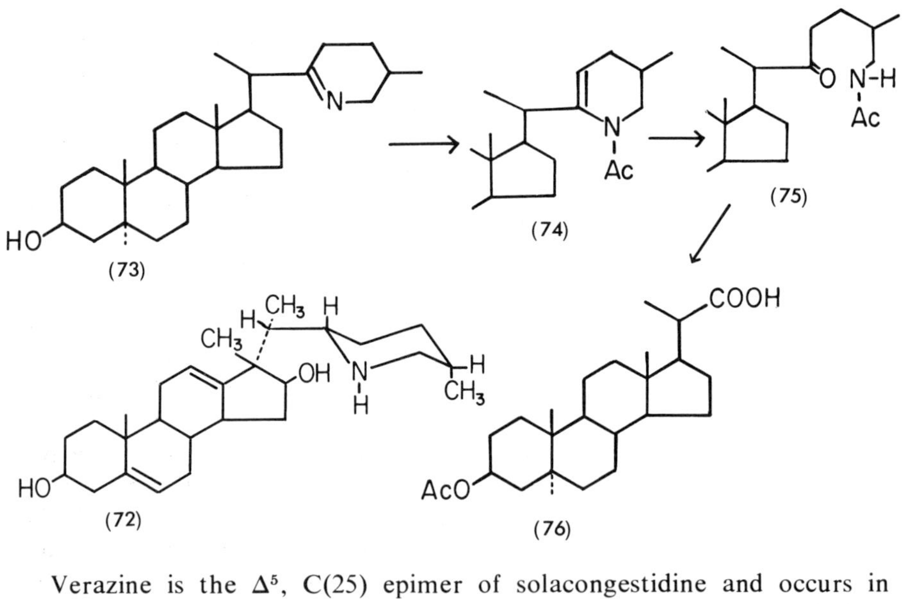

(73) (74) (75) (72) (76)

Verazine is the Δ^5, C(25) epimer of solacongestidine and occurs in *Veratrum album*.[106] It has been assigned the structure (25S)-22,26-imino-cholest-5,22(N)-dien-3β-ol[107] on the basis of its spectral measurements as well as its reduction to the known (22R,25S)- and (22S,25S)-22,26-imino-5α-cholestan-3β-ols.[108] Its synthesis[173] from Δ^5-tomatiden-3β-ol confirmed the assignment. These alkaloids bearing a piperideine moiety on the side-chain are of biogenetic interest since they can be regarded as attractive

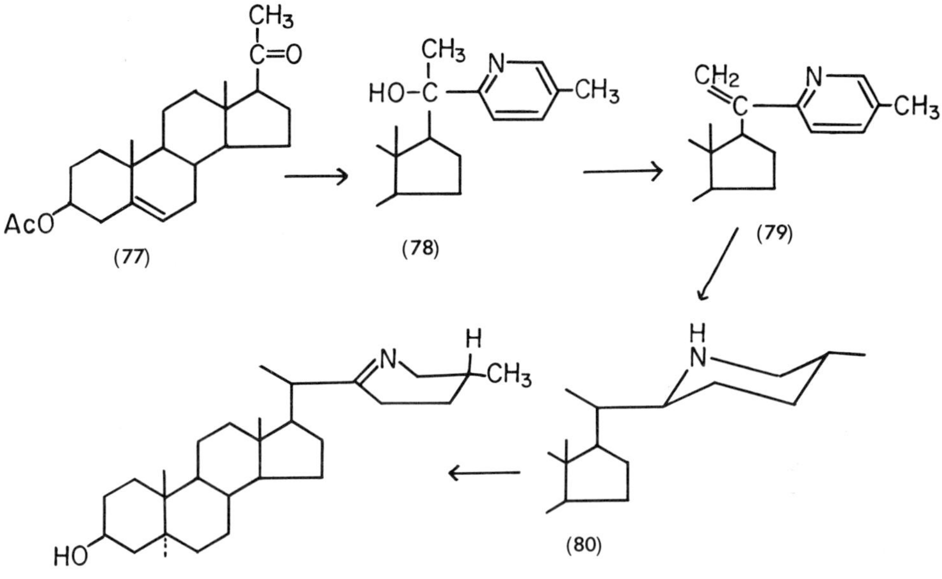

(77) (78) (79) (80) (73)

intermediates in the formation of a number of Solanum and Veratrum alkaloids.[109]

Tomatillidine occurs as a minor constituent along with solasodine in *Solanum tomatillo* (Philippi).[110] It has been assigned the structure (25R)-22,26-imino-5,22(N)-cholestadien-24-on-3β-ol since its 5,6-dihydrodeoxo derivative was found to be identical with the principal alkaloid isolated from *Solanum congestiflorum*. The site of the carbonyl function at C(24) was decided from spectral data.[104]

6.1 Spirosolanes

Spirosolanes[111] are characterized by possession of an oxaazaspirane side-chain structure and are exemplified by tomatidine (81), Δ^5-tomatidenol, soladulcidine, and solasodine (89).

6.1.1 Tomatidine Tomatine,[112] the glycoside of tomatidine, is present in a number of wild tomato plants. Upon acid hydrolysis it yields two moles of glucose, a mole each of galactose, and xylose and the aglycone, tomatidine. Tomatidine [(81), R = H] can be converted into the triacetylpseudo derivative (82) by refluxing in acetic anhydride and then to 3β-hydroxy-5α-pregn-16-en-20-one (83)[113] by oxidation with chromic acid and alkaline hydrolysis. When the O,N-diacetate of tomatidine (81, R = Ac) is oxidized with chromic acid. the product is tigogenin lactone acetate

(84).[114] These facts together with isolation of 5-methyl-2-ethylpyridine (85)[115] from the selenium dehydrogenation of tomatidine indicated the spiroaminoketal nature of the sidechain. The synthesis of tomatidine[116] by cyclization in phosphoric acid of the 27-aminopseudo derivative (86) obtained through the 27-tosylate, -iodide and -phthalimido pseudoneotigogenin derivatives ultimately proved the structure. Utilizing the method developed for the preparation of the azomethine (73),[105] Schreiber and Adam,[117] converted 3β,16β-diacetoxy-5α-pregnan-20-one (87) into (22S, 25S)-22,26-imino-5α-cholestane-3β,16β-diol (88), which was chlorinated to the N-chloro derivative with N-chlorosuccinimide and cyclized with sodium methylate to tomatidine.

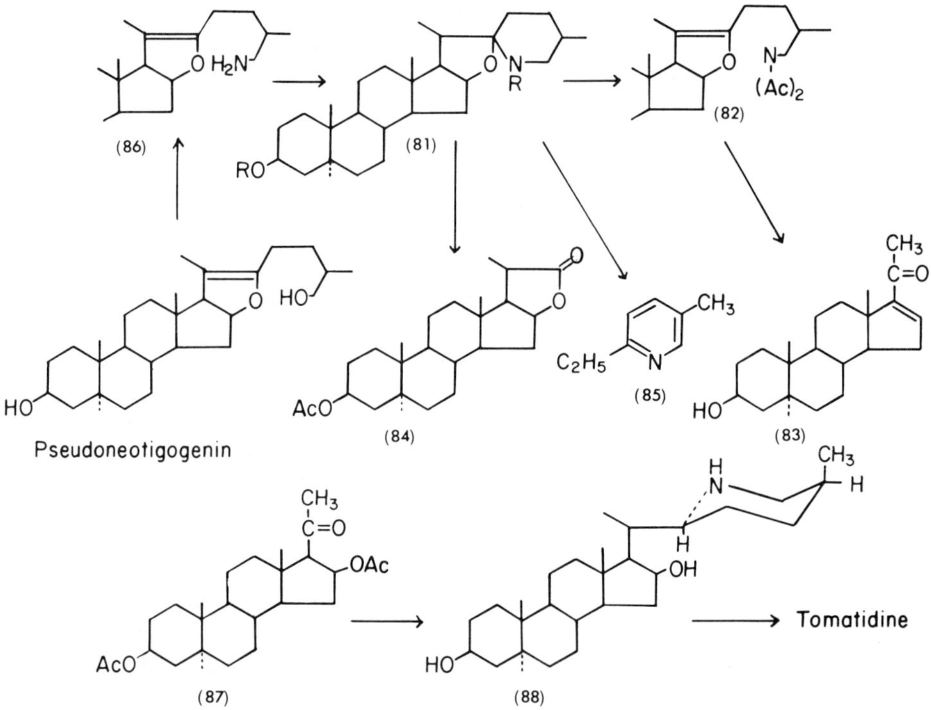

6.1.2 Tomatid-5-en-3β-ol Tomatidenol[118] is the C(5,6) unsaturated analog of tomatidine. Its catalytic reduction readily affords tomatidine. It has been synthesized[119] from the synthetically available (22S,25S)-22,26-imino-5α-cholestane-3β,16β-diol. In the plant it occurs as the glycoside (α,β,γ_1,γ_2,δ-solamarines).[120,121]

6.1.3 Solasodine Solasonine[122] and solamargine[125] are the two principal glycosides that afford solasodine. Upon hydrolysis, solasonine yields the alkamine solasodine and a molecule each of glucose, galactose, and rhamnose while solamargine gives solasodine, glucose, and two molecules of rhamnose. Acetylation of the alkamine (89) affords O,N-diacetylsolasodine (90) that isomerizes in boiling glacial acetic acid to 26-acetylamino-5,20(22)-furostadien-3β-ol acetate (91). The chromic acid oxidation of 91 and hydrolysis with acetic acid affords 3β-acetoxy-5,16-pregnadien-20-one (92) in good yields.[123] Since pregnadienolone is a key intermediate

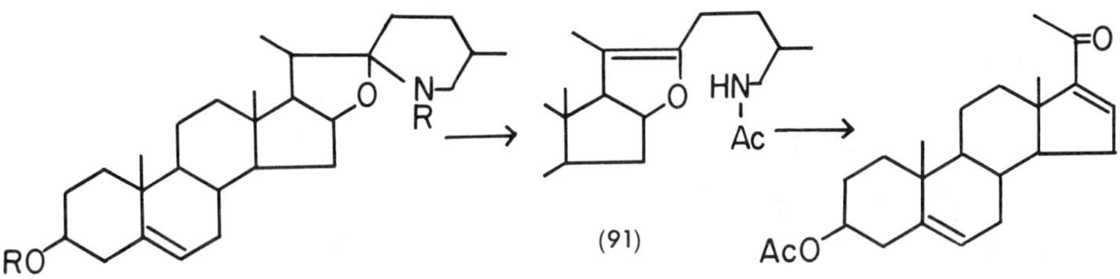

(91)

(92)

(89) R = H

(90) R = Ac

in the production of biologically active steroid hormones, considerable effort has been expended in the search for fruitful sources of solasodine, and up to now more than sixty species of the genus Solanum[122] have been shown to yield solasodine. The synthesis of solasodine has been achieved by several routes. The first synthesis developed by Uhle[124] consisted of converting the 27-tosylate of kryptogenin (93) through the iodide, phthalimide and epoxy-hydroxyphthalimide (94) to solasodine. The latter steps in this synthesis involved the reduction of the 16,22-diketo-27-phthalimido derivative with sodium borohydride to the 16β-hydroxy-22-oxo-27-phthalimide that cyclized to the hemiketal, the 16β,22-epoxy-22-hydroxy-27-phthalimido derivative (94). Cleavage with hydrazine to the 27-amine followed by cyclization with phosphoric acid afforded solasodine (89). In a similar manner the phthalimido derivative of pseudodiosgenin (95) was also converted into solasodine.[126]

Schreiber, Walther, and Rönsch[127] have converted the isomeric (22S,25R)-22,26-imino-5α-cholestane-3β,16β-diol[117] obtained from 3β,16β-diacetoxy-5α-pregnan-20-one into the Δ^5 analog (96), which was then transformed into solasodine through the alkaline treatment of the N-chloro derivative. The cyclization presumably proceeds through the azomethine (96a) formed from N-chloroamine and base.

6.1.4 Soladulcidine Soladulcidine[128] is 5,6-dihydrosolasodine, and can be obtained directly by the catalytic reduction of solasodine. It has been synthesized[117] from (22S,25R)-22,26-imino-5α-cholestane-3β,16β-diol by

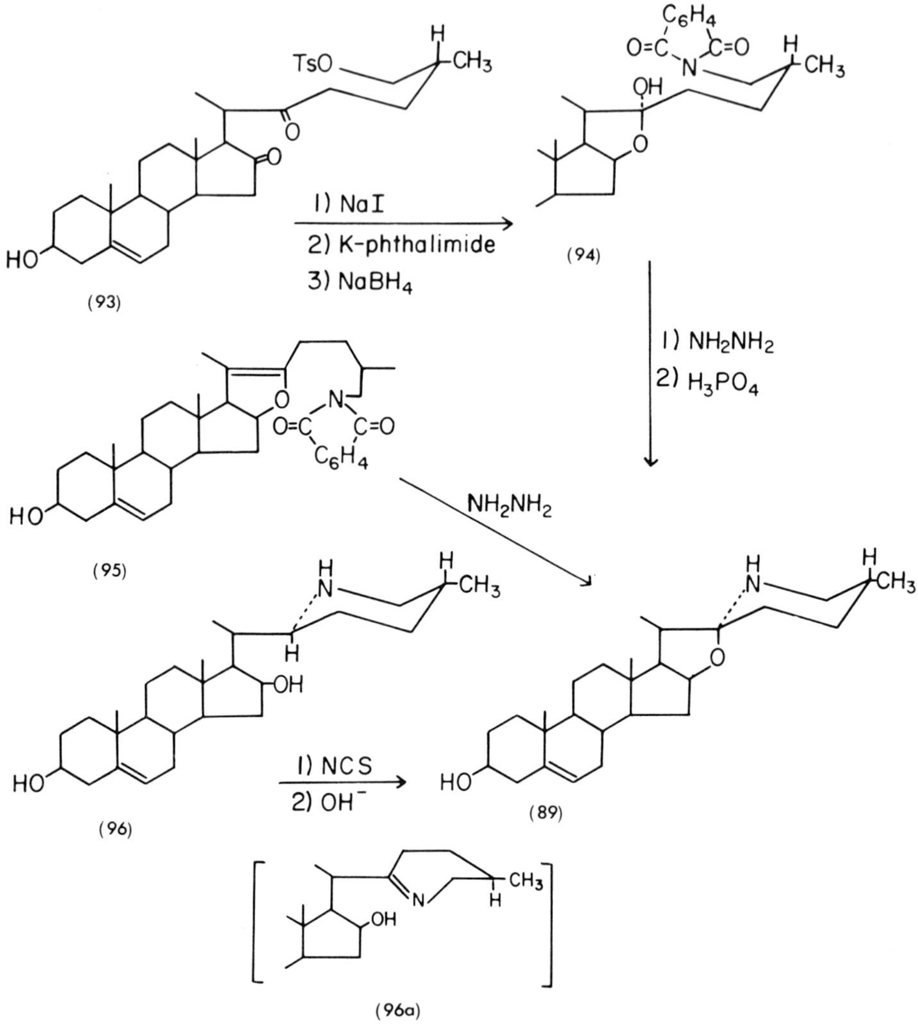

cyclization of the N-chloro derivative with base as in the manner of solasodine. Soladulcidine as well as tomatidine[123] can be converted in good yields to 3β-acetoxy-5α-pregn-16-en-20-one by isomerization of the O,N-diacetyl derivatives with glacial acetic acid to 3β-acetoxy-26-acetylamino-5α-furost-20(22)-ene followed by chromic acid oxidation and acetic acid hydrolysis.

6.1.4.1 Stereochemistry at C(22) and C(25) The stereochemistry at C(25) for solasodine and tomatidine has been indicated by a number of transformations[129, 130] and from the synthesis of the alkaloids from the appropriate sapogenins. The conversion in acetic acid of N-nitrososolasodine (97) to the steroidal sapogenin, diosgenin (98), and N-nitrosotomatidine (99) to the C(25) dihydro epimer of diosgenin, neotigogenin (100) is of interest. It is also significant that the main component from both reactions was the common entity 22,26-epoxyfurostan-3β-ol (101) in which asymmetry at C(25) had been destroyed.[131]

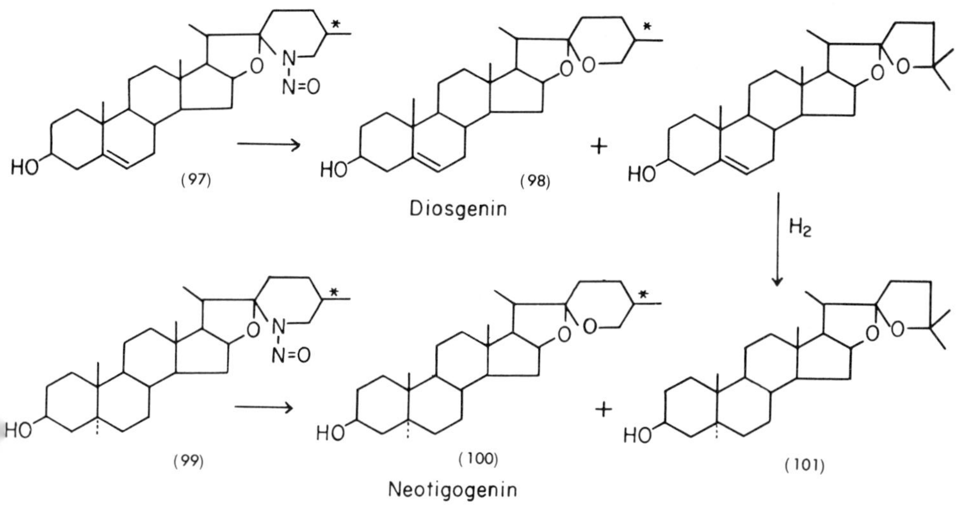

A direct proof for the existence of the C(25) isomers has been obtained by the degradation of soladulcidine and tomatidine into the enantiomeric 5-methyl-2-piperidones.[132] (25R)-3β-Acetoxy-26-acetylamino-5α-furost-20(22)-ene (102), which is obtainable through the acid catalyzed isomerization of O,N-diacetylsoladulcidine, is converted by chromic acid oxidation into allopregnenolone and (R)-(+)-5-methyl-2-piperidone (103), while the oxidation of the corresponding (25S)-derivative from tomatidine affords allopregnenolone and the enantiomeric (S)-(−)-5-methyl-2-piperidone (103).

In contrast to the sapogenins that differ only at C(25) between the iso and neo series,[133] accumulated evidence now points to the existence of an additional isomerism at C(22) in the case of the spirosolanes. Some of these observations are the difference in chemical behavior,[134,135] the large molecular rotation difference[132] between tomatidine and soladulcidine and their derivatives as compared to the difference between (25S) and (25R)-3β-acetoxy-26-acetylamino-5α-furost-20(22)-enes, and the dissimilarities in the ord curves[136] of the N-nitroso derivatives (positive Cotton effect in tomatidine and negative in solasodine).

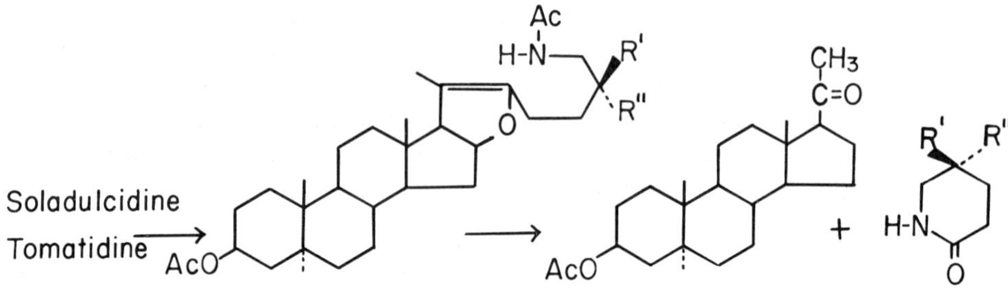

(102) (103)

R' = H, R'' = CH_3 (R series)

R' = CH_3, R'' = H (S series)

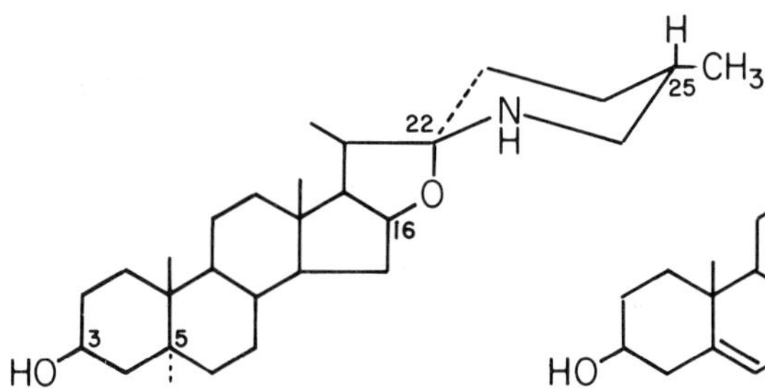

(81a)

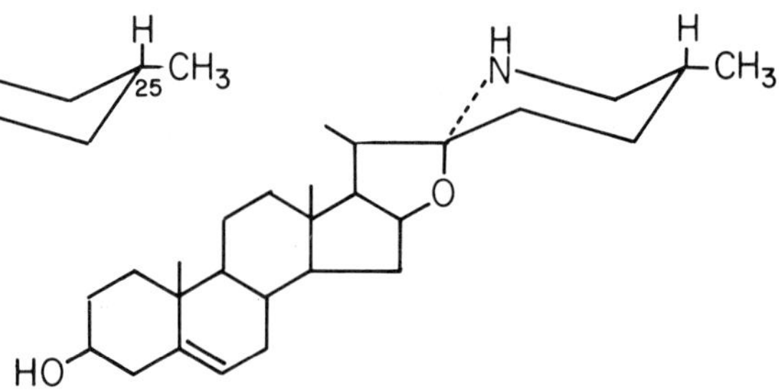

(89a)

The occurrence of isomerization at C(22) was confirmed by x-ray analysis of tomatidine hydroiodide,[137] which showed that it has the 22S configuration with the piperidine ring in the prone chair conformation and the C(25) methyl equatorially oriented. Tomatidine is, therefore, (25S)-5α, 22βN-spirosolan-3β-ol (81a), and solasodine (25R)-22αN-spirosol-5-en-3β-ol (89a).

Tomatidine has been hydroxylated[138] by the fungus *Helicostylum piriforme* to yield 7α,11α-dihydroxy-,7α-hydroxy- and 9α-hydroxytomatidine. Hydroxylation[139] of solasodine by the same fungus affords 9α-hydroxy-,11α-hydroxy- and 7β-hydroxysolasodine.

6.2 Solanidanes

6.2.1 Glycoalkaloids of Solanidane The principal glycoalkaloids[141,142] of solanidine[140] are the solanines and the chaconines. The β and γ-chaconines are partially hydrolyzed products of the trisaccharide α-chaconine, and β and γ-solanines, the partially hydrolyzed product of the trisaccharide α-solanine. Demissine[143] is the glycoalkaloid of demissidine (5α-solanidan-3β-ol).

6.2.2 Solanidine The selenium dehydrogenation of solanidine yields Diels hydrocarbon[144] and 5-methyl-2-ethylpyridine.[145] The structure inferred from the dehydrogenation experiment was confirmed by the synthesis of 5β-solanidan-3β-ol from sarsasapogenoic acid dioxime by Uhle and Jacobs[146] who converted the dioxime (104) to the amino derivative (105), pyrolyzed to the lactam (106) and catalytically reduced to the 5,6-dihydroalkaloid.

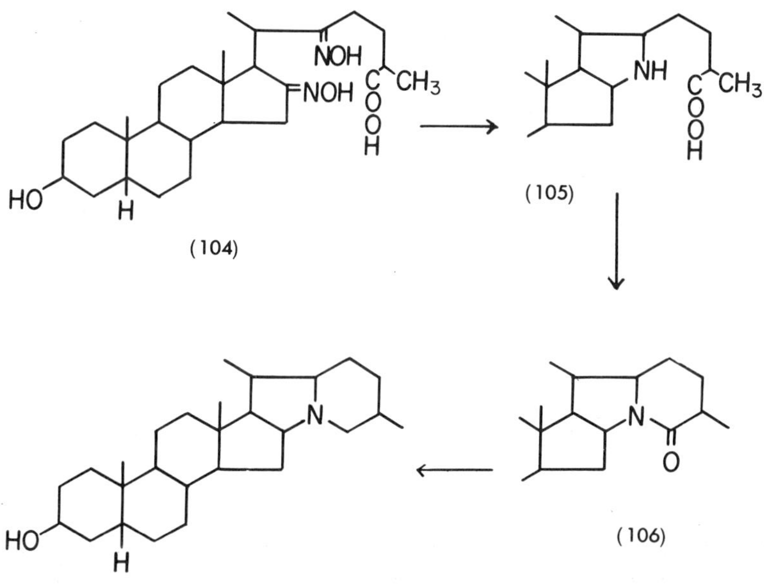

(104) (105) (106)

5β-Solanidan-3β-ol

A number of syntheses of solanidine have since been developed. Of these, the conversion of the C(22) isomeric dihydrotomatidines and tetrahydrosolasodines into the four C(22), C(25) epimeric solanidanones are of interest. The two C(22)-epimeric dihydrotomatidines (A = 107, B = 108), prepared from the reduction of tomatidine either catalytically or with lithium-aluminum hydride, were converted by oxidation and subsequent reduction to the C(22) epimeric (25S)-solanidan-3-ones[147] (109, 110). The isomer arising from dihydrotomatidine B has been correlated with natural solanidan-3-one. The C(22)-isomeric tetrahydrosolasodines (B = 111, A = 112) prepared by reduction of solasodine initially with lithium-aluminum hydride-aluminum chloride to the Δ^5-3β,16β-diols followed by catalytic reduction of the Δ^5 bond are likewise converted to the two C(22)-epimeric (25R)-solanidan-3-ones (113, 114). Thus, all four of the possible C(22)-and C(25)-isomers have been prepared.[148] X-ray analysis of demissidine hydroiodide[149] has settled the question of the configuration at C(22) that had been provisionally advanced for the solanidanones.[148,174]

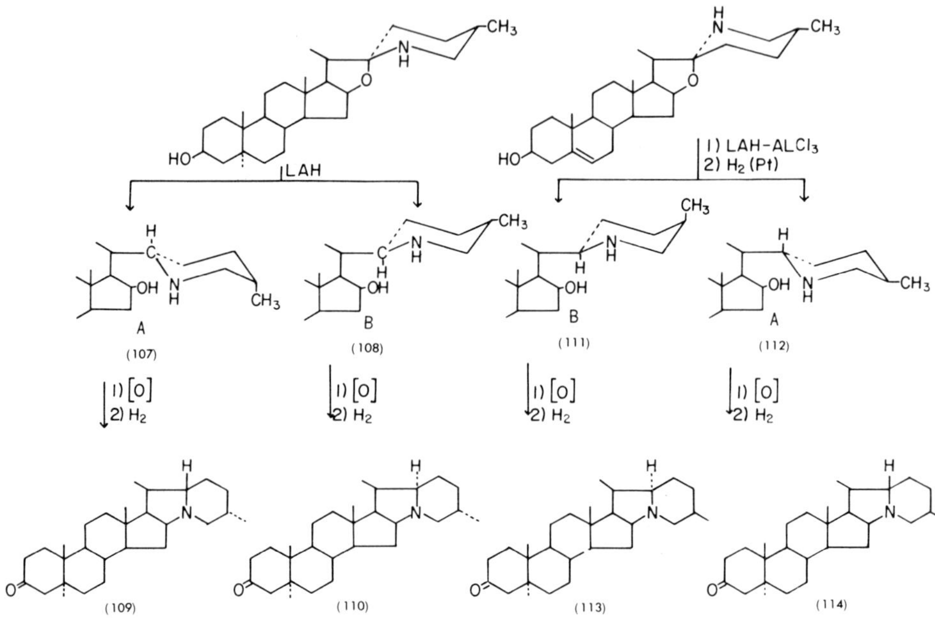

22-iso-(25S)-Solanidan-3-one (25S)-Solanidan-3-one (25R)-Solanidan-3-one 22-iso-(25R)-Solanidan-3-one

The naturally occurring solanidane alkaloids such as demissidine, solanidine, leptinidine, rubijervine, and isorubijervine are now shown to possess the (22R,NS)* (I) and the 22-isosolanidanes, the (22S, NR, 25S) (II)-configurations, (Fig. 19.1).

Schreiber, Horstmann, and Adam[150] have found that 22-iso-(25S)-solanidan-3-one (109) or the alcohol, 22-isodemissidine, can be isomerized into demissidine (116; 22αH,25βH) by treatment of the compound in a reducing atmosphere (H_2) in methanol over a platinum oxide catalyst.

Tomatiden-3β-ol, the Δ^5 analog of tomatidine, has also been transformed by way of the C(22) isomeric Δ^5-3β,16β-diols into solanidine and 22-isosolanidine[151] in a manner similar to the treatment of tomatidine.[147]

Another synthesis of demissidine (116) and its (22S,25R)-isomer has been achieved by Adam and Schreiber[152] through the Hofmann-Löffler-

*For nomenclature see R. S. Cahn, *J. Chem. Educ.* **41,** 119 (1964); G. Bernath, *Nature* **189,** 304 (1961).

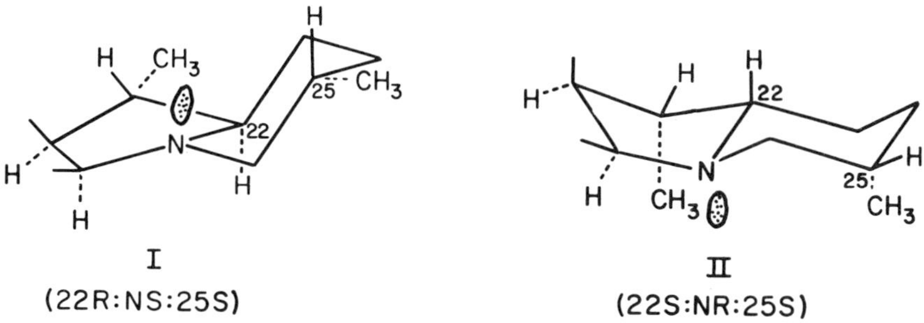

Figure 19.1

Freytag cyclization of the (22R,25S) and (22S,25R)isomeric N-chloro derivatives of 22,26-imino-5α-cholestan-3β-ol (115). In this process the N-chloro derivatives are submitted to uv irradiation in trifluoroacetic acid and the resulting 16-chloro compounds cyclized with alkali.

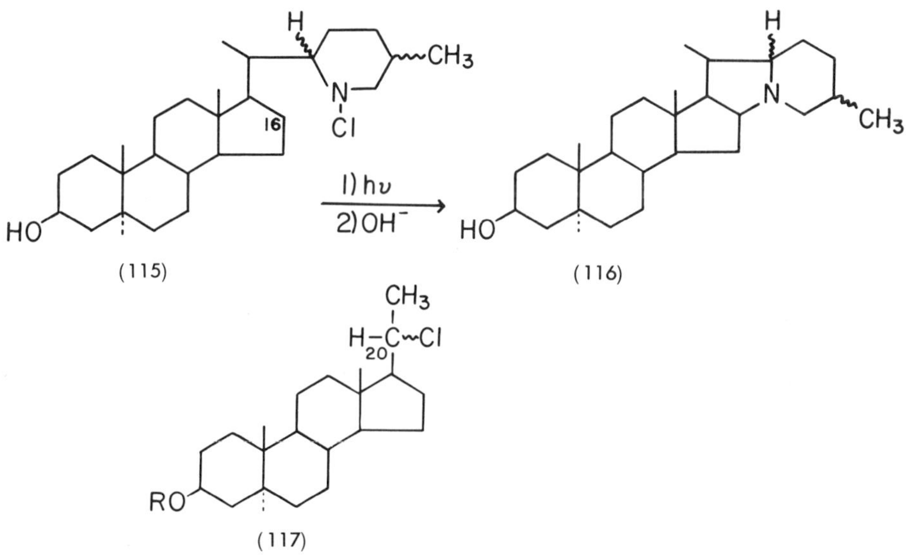

(115) (116)

(117)

It is interesting to observe that the irradiation of the N-chloro derivative of (22S,25S)-22,26-imino-5α-cholestan-3β-ol failed to yield isodemissidine but afforded instead a mixture of nitrogen free isomeric 20-chloropregnane derivatives(117).[153]

6.3 Hydroxysolanidines

6.3.1 Rubijervine[160] (118), Table 19.2 Selenium dehydrogenation of rubijervine yields 5-methyl-2-ethylpyridine and 1′-methyl-1,2-cyclopenteno-

TABLE 19.2. Hydroxysolanidines.

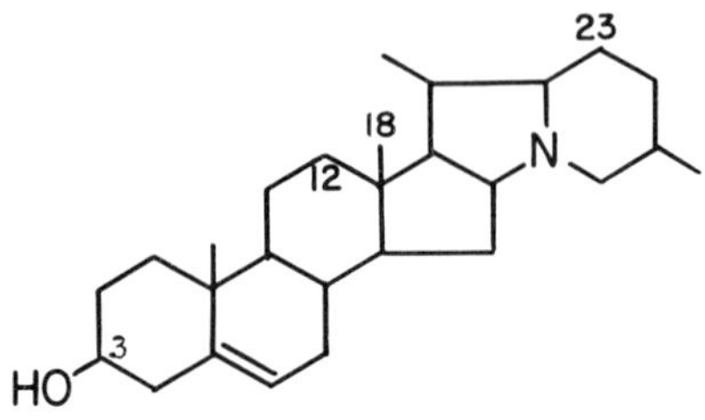

	Hydroxyl at	Source
(118) Rubijervine	12α	Veratrum album[160]
(119) Isorubijervine	18	Veratrum album[154]
(121) Leptinidine	23β	Solanum chacoense[165]

phenanthrene.[154] Rubijervine was shown to be a hydroxysolanidine from the fact that rubijervine 3-benzoate could be oxidized to the ketone and then reduced to solanidine by the Wolff-Kishner method.[155] Spectral studies[155,156] indicate the position of the carbonyl to be at C(12). The reduction of the 12-ketone by sodium borohydride affords a mixture of rubijervine (118) and 12-epirubijervine while reduction with sodium in propanol yields only the 12-epi alkaloid (12β-OH, equatorial).[156] Molecular rotation data also lend support to the 12α-configuration. Accordingly the 12α-hydroxysolanidine (118) structure is ascribed to rubijervine.

6.3.2 Isorubijervine[154] Selenium dehydrogenation of isorubijervine (119) yields 1,2-cyclopentenophenanthrene.[157] The alkaloid is a tertiary base which contains a primary alcoholic function along with the Δ^5-3β-ol system.[157,158] Thus, the oxidation of the 5,6-dihydro base affords the 3-oxo-18-carboxylic acid, the methyl ester of which is resistant to hydrolysis. The location of the primary alcoholic group at C(18) is confirmed by the formation of the neutral quaternary-type monotosylate (120). The reductive cleavage of 120 with sodium-ethanol results in the formation of solanidine.[159]

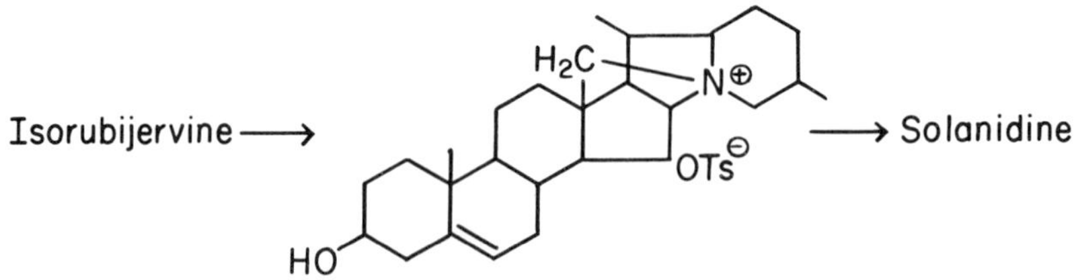

(120)

The 5,6-dihydro base is also converted into 5α-solanidan-3β-ol (116) by oxidation to the 18-aldehyde and Wolff-Kishner reduction.[161,162] The 18-hydroxysolanidine structure was thus assigned for isorubijervine (119) on these considerations. A Δ^4-3-oxo derivative of isorubijervine named veralobine has been isolated from *Veratrum album* subsp. *lobelianum* Bernh. Suessenguth.[163,164]

6.3.3 Leptinidine (121) Leptines isolated from the leaves of *Solanum chacoense*[165] are the glycosides of 23-acetoxysolanidine. The removal of the acetyl moiety gives the leptinines, the glycosides of 23-hydroxysolanidine. The hydrolysis of leptinine I yields the alkamine leptinidine and the sugars, glucose and rhamnose (2 moles) while leptinine II affords the sugars, glucose, rhamnose, and galactose:

Leptines = Leptinines + acetic acid

Leptinine-I = Leptinidine + glucose + 2 rhamnose

Leptinine-II = Leptinidine + glucose + rhamnose + galactose

Leptinidine possesses a double bond and two secondary hydroxyls. Precipitation of the alkaloid with digitonin, molecular rotation difference of the base with the dihydro base, dehydrogenation of the alkamine with copper-bronze to an α,β-unsaturated ketone, and the formation of a $\Delta^{3,5}$-diene by treatment of leptinine with acid, indicate that one of the hydroxyls is part of a Δ^5-3β-ol system. The identity of the parent ring system is revealed by the oxidation and subsequent Wolff-Kishner reduction of the dihydro base to solanidane.[166] The position of the second hydroxyl in leptinidine is provided by the isolation of 5-methyl-2-ethyl-3-hydroxypyridine from the selenium dehydrogenation of hydroxysolanthrene (3,5-diene of leptinidine).[167] The hydroxyl thus placed at C(23) was further defined by ir and nmr data as β-oriented.[168] Leptinidine is therefore 23β-hydroxysolanidine (121).

In the plant the leptines[165] are considered to be the factor responsible for resistance against the potato beetle.

6.4 Solanocapsine

Solanocapsine (122) isolable from several solanum species (*S. pseudocapsicum*,[169] *S. hendersonii*[170]) yields Diels hydrocarbon,[169] and 5-methyl-2-ethylpyridine[171] on selenium dehydrogenation. The molecule ($C_{27}H_{46}N_2O_2$) possesses a primary and a secondary amino group, a tertiary hydroxyl, and an oxidic function. The two amino functions are readily acetylated to the N,N′-diacetate. From considerations based mainly on the degradation of the alkaloid to tigogenin lactone acetate and (−)-4-amino-3-methylbutyric acid, Schreiber and Ripperger[172] deduced the structure of solanocapsine to be (22R,23S,25R)-3β-amino-

22,26-imino-16β,23-oxido-5α-cholestan-23-ol (122). The degradation is accomplished by conversion of the alkaloid to the N′-acetyl derivative (123) by acetylation after protection of the 3-amino function with salicylaldehyde. The blocking agent is removed by acid hydrolysis. Treatment of 123 with nitrous acid gives the 3β-ol, the acetate (124) of which is submitted to the action of boiling acetic acid. The resulting unsaturated derivative (125) is oxidized with chromic acid to the 22,23-diketone (126) and hydrolyzed with acid to 3β,16β-dihydroxybisnorallocholanic acid lactone (16→20) (tigogenin lactone) (127). In a similar manner, the N,N′-diacetate of solanocapsine (128) is converted by dehydration, oxidation and hydrolysis to 3β-amino-16β-hydroxy-5α-bisnorcholanic acid lactone (129) and D (–)-4-amino-3-methylbutyric acid (130). The latter has been correlated with D (+)-methylsuccinic acid to establish the orientation at C(25).

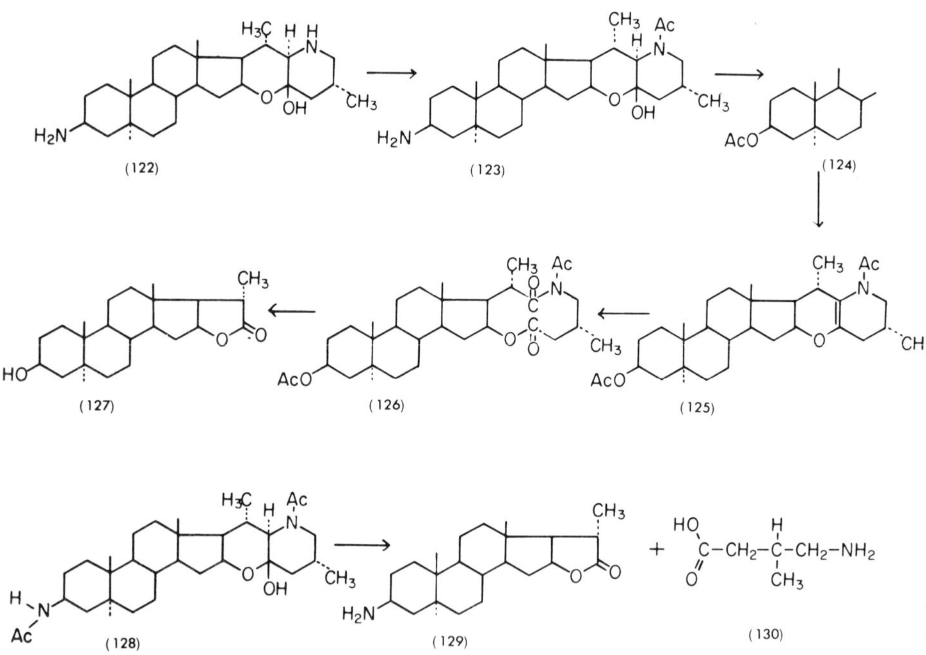

(122) (123) (124) (125) (126) (127) (128) (129) (130)

More recently Ripperger and Schreiber[177] have synthesized compound 122 and found it to be different from solanocapsine. Based on studies of the derivatives of the alkaloid and molecular rotation differences of the epimeric 16-hydroxy derivatives obtained from the alkaloid and 122, solanocapsine has now been reassigned the 16α-epoxy structure.

REFERENCES

1. K. S. Brown and S. M. Kupchan, *J. Amer. Chem. Soc.* **84,** 4590 (1962).
2. Ref. 1, p. 4592.
3. T. Nakano and S. Terao, *Tetrahedron Letters* 1035 (1964).
4. Ref. 3, p. 1045.
5. M.-M. Janot, Q. Khuong-Huu, and R. Goutarel, *Compt. Rend.* **246,** 3076 (1958).
6. M.-M. Janot, Q. Khuong-Huu, and R. Goutarel, *Compt. Rend.* **248,** 982 (1959).
7. H. Ruschig, W. Fritsch, J. Schmidt-Thome, and W. Haede, *Chem. Ber.* **88,** 883 (1955).
8. M.-M. Janot, Q. Khuong-Huu, and R. Goutarel, *Bull. Soc. Chim. France* 1640 (1960).
9. M.-M. Janot, A. Cavé, and R. Goutarel, *Bull. Soc. Chim. France* 896 (1959).
10. M.-M. Janot, A. Cavé, and R. Goutarel, *Compt. Rend.* **251,** 559 (1960).
11. M.-M. Janot, P. Longevialle, and R. Goutarel, *Bull. Soc. Chim. France* 2158 (1964).
12. J. LeMen, *Bull. Soc. Chim. France* 860 (1960).
13. B. Kamber, G. Cainelli, D. Arigoni, and O. Jeger, *Helv. Chem. Acta* **43,** 347 (1960).
14. A. Cavé, P. Potier, A. Cavé, and J. LeMen, *Bull. Soc. Chim. France* 2415 (1964).
15. J. LeMen, C. Kan, and R. Beugelmans, *Bull. Soc. Chim. France* 597 (1963).
16. R. Beugelmans, C. Kan, and J. LeMen, *Bull. Soc. Chim. France* 1306 (1963).
17. M.-M. Janot, Q. Khuong-Huu, and R. Goutarel, *Compt. Rend.* **250,** 2445 (1960).
18. M.-M. Janot, F. Lainé, Q. Khuong-Huu, and R. Goutarel, *Bull. Soc. Chim. France* 111 (1962).
19. V. Černý, L. Lâbler, and F. Šorm, *Coll. Czech. Chem. Commun.* **22,** 76 (1957).
20. M.-M. Janot, X. Monseur, C. Conreur, and R. Goutarel, *Bull. Soc. Chim. France* 285 (1962).
21. D. Dickel, R. Lucas, and H. B. MacPhillamy, *J. Amer. Chem. Soc.* **81,** 3154 (1959).
22. R. A. Lucas, D. F. Dickel, R. L. Dziemian, M. J. Ceglowski, B. L. Hensle, and H. B. MacPhillamy, *J. Amer. Chem. Soc.* **82,** 5688 (1960).
23. M. Truong-Ho, X. Monseur, Q. Khuong-Huu, and R. Goutarel, *Bull. Soc. Chim. France* 2332 (1963).
24. Z. Voticky and J. Tomko, *Coll. Czech. Chem. Commun.* **30,** 348 (1965).
25. T. Kikuchi, S. Uyeo, and T. Nishinaga, *Tetrahedron Letters* 1993 (1965).
26. Y. Kawazoe, Y. Sato, M. Natsume, H. Hasegawa, T. Okamoto, and K. Tsuda, *Chem. Pharm. Bull. (Tokyo)* **10,** 338 (1962).
27. M. Truong-Ho, Q. Khuong-Huu, and R. Goutarel, *Bull. Soc. Chim. France* 594 (1963).
28. D. H. R. Barton and L. R. Morgan, *J. Chem. Soc.* 622 (1962).
29. A. Chatterjee and B. Das, *Chem. Ind. (London)* 1445 (1959).
30. A. Chatterjee and B. Das, *Chem. Ind. (London)* 1247 (1960).
31. Q. Khuong-Huu, C. Monneret, J. Yassi, and R. Goutarel, *Bull. Soc. Chim. France* 2169 (1964).
32. T. Kikuchi, S. Uyeo, and T. Nishinaga, *Tetrahedron Letters* 3169 (1965).
33. M.-M. Janot, C. Monneret, X. Monseur, Q. Khuong-Huu, and R. Goutarel, *Compt. Rend.* **260,** 6118 (1965).
34. M. Tomita, S. Uyeo, and T. Kikuchi, *Tetrahedron Letters* 1641 (1964).
35. L. Lâbler and F. Šorm. *Coll. Czech. Chem. Commun.* **28,** 2345 (1963).
36. V. Černý, L. Lâbler, and F. Šorm, *Coll. Czech. Chem. Commun.* **24,** 378 (1959).
37. W. F. Knaack and T. A. Geissman, *Tetrahedron Letters* 1381 (1964).
38. M.-M. Janot, F. Lainé, and R. Goutarel, *Ann. Pharm. France* **18,** 673 (1960).
39. R. Goutarel, *Tetrahedron* **14,** 126 (1961).
40. T. Kikuchi, S. Uyeo, M. Ando, and A. Yamamoto, *Tetrahedron Letters* 1817 (1964).
41. A. Chatterjee, B. Das, C. P. Dutta, and K. S. Mukherjee, *Tetrahedron Letters* 67 (1965).

42. J. M. Kohli, A. Zaman, and A. R. Kidwai, *Tetrahedron Letters* 3309 (1964).
43. E. J. Corey and W. R. Hertler, *J. Amer. Chem. Soc.* **81,** 5209 (1959).
44. W. Vetter, P. Longevialle, F. Khuong-Huu-Lainé, Q. Khuong-Huu, and R. Goutarel, *Bull. Soc. Chim. France* 1324 (1963).
45. J. R. Vaughan, *J. Amer. Chem. Soc.* **73,** 3547 (1951).
46. M. Tomita, S. Uyeo, and T. Kikuchi, *Tetrahedron Letters* 1053 (1964).
47. K. S. Brown and S. M. Kupchan, *J. Amer. Chem. Soc.* **84,** 4590 (1962).
48. B. Camerino, B. Patelli, and R. Sciaky, *Tetrahedron Letters* 554 (1961).
49. J. Attenburrow, D. F. Elliott, and G. F. Penny, *J. Chem. Soc.* 310 (1948).
50. T. Kikuchi and S. Uyeo, *Tetrahedron Letters* 3473 (1965).
51. Ref. 50, p. 3487.
52. S. Siddiqui, *Proc. Indian Acad. Sci.* **3A,** 249 (1936).
53. S. Siddiqui and R. H. Siddiqui, *Proc. Indian Acad. Sci.* **3,** 257 (1936).
54. S. Siddiqui and P. P. Pillay, *J. Indian Chem. Soc.* **9,** 553 (1932).
55. H. Favre, R. D. Haworth, J. McKenna, R. G. Powell, and G. H. Whitfield, *J. Chem. Soc.* 1115 (1953).
56. L. Lábler, V. Černý, and F. Šorm, *Chem. Ind. (London)* 1119 (1955).
57. V. Černý and F. Šorm, *Coll. Czech. Chem. Commun.* **20,** 1473 (1955).
58. R. Tschesche and K. Wiensz, *Chem. Ber.* **91,** 1504 (1958).
59. L. Lábler and V. Černý, *Coll. Czech. Chem. Commun.* **24,** 370, 378 (1959).
60. R. D. Haworth, J. McKenna, and N. Singh, *J. Chem. Soc.* 831 (1949).
61. N. L. McNiven, *Chem. Ind. (London)* 1296 (1957).
62. S. Siddiqui, *J. Indian Chem. Soc.* **11,** 283 (1934).
63. V. Černý, L. Dolejs, and F. Šorm, *Coll. Czech. Chem. Commun.* **29,** 1591 (1964).
64. E. J. Corey and W. R. Hertler, *J. Amer. Chem. Soc.* **80,** 2903 (1958).
65. S. Siddiqui and R. H. Siddiqui, *Proc. Indian Acad. Sci.* **173,** 249 (1936).
66. S. Siddiqui and R. H. Siddiqui, *J. Indian Chem. Soc.* **11,** 787 (1934).
67. H. Warnecke, *Ber.* **19,** 60 (1886).
68. P. Buchschacher, J. Kalvoda, D. Arigoni, and O. Jeger, *J. Amer. Chem. Soc.* **80,** 2905 (1958).
69. G. Stork, S. D. Darling, I. T. Harrison, and P. S. Wharton, *J. Amer. Chem. Soc.* **84,** 2018 (1962).
70. J. A. Marshall and W. S. Johnson, *J. Amer. Chem. Soc.* **84,** 1485 (1962).
71. D. H. R. Barton and L. R. Morgan, *J. Chem. Soc.* 622 (1962).
72. W. Nagata, T. Terasawa, and T. Aoki, *Tetrahedron Letters* 869 (1963).
73. D. H. R. Barton and A. N. Starratt, *J. Chem. Soc.* 2444 (1965).
74. F. Buzetti, W. Wicki, J. Kalvoda, and O. Jeger, *Helv. Chim. Acta* **42,** 388 (1959).
75. R. Pappo, *J. Amer. Chem. Soc.* 81, 1010 (1959).
76. J. Hora and V. Černý, *Coll. Czech. Chem. Commun.* **26,** 2217 (1961).
77. A. Uffer, *Helv. Chim. Acta* **39,** 1834 (1956).
78. L. van Hove, *Tetrahedron* **7,** 104 (1959).
79. F. L. Pyman, *J. Chem. Soc.* **115,** 163 (1919).
80. M.-M. Janot, P. Devissaguet, Q. Khuong-Huu, and R. Goutarel, *Compt. Rend.* **260,** 6631 (1965).
81. H. Rostock and E. Seebeck, *Helv. Chim. Acta* **41,** 11 (1958).
82. C. F. Hiskey, R. Hirschmann, and N. L. Wendler, *J. Amer. Chem. Soc.* **75,** 5135 (1953).
83. R. Tschesche and H. Ockenfels, *Chem. Ber.* **97,** 2316 (1964).
84. L. Stárka, *Coll. Czech. Chem. Commun.* **26,** 2452 (1961).
85. S. M. Kupchan, C. J. Sih, S. Kubota, and A. M. Rahim, *Tetrahedron Letters* 1767 (1963).

86. E. L. Patterson, W. W. Andres, and R. E. Hartman, *Experientia* **20,** 256 (1964).
87. Q. Khuong-Huu, J. Yassi, C. Monneret, and R. Goutarel, *Bull. Soc. Chim. France* 1831 (1965).
88. R. Tschesche and P. Otto, *Chem. Ber.* **95,** 1144 (1962).
89. R. Tschesche, W. Meise, and G. Snatzke, *Tetrahedron Letters* 1659 (1964).
90. M.-M. Janot, M. Truong-Ho, Q. Khuong-Huu, and R. Goutarel, *Bull. Soc. Chim. France* 1977 (1963).
91. Q. Khuong-Huu, L. Lábler, M. Truong-Ho, and R. Goutarel, *Bull. Soc. Chim. France* 1564 (1964).
92. A. Bertho, G. von Schuckmann, and W. Schönberger, *Ber.* **66,** 786 (1933).
93. R. Tschesche and A. C. Roy, *Chem. Ber.* **89,** 1288 (1956).
94. Q. Khuong-Huu, M. Truong-Ho, L. Lábler, R. Goutarel, and F. Šorm, *Coll. Czech. Chem. Commun.* **30,** 1016 (1965).
95. M.-M. Janot, F.-X. Jarreau, M. Truong-Ho, Q. Khuong-Huu, and R. Goutarel, *Bull. Soc. Chim. France* 1555 (1964).
96. M.-M. Janot, Q. Khuong-Huu, and R. Goutarel, *Compt. Rend.* **254,** 1326 (1962).
97. Q. Khuong-Huu, J. Yassi, and R. Goutarel, *Bull. Soc. Chim. France* 2486 (1963).
98. M.-M. Janot, Q. Khuong-Huu, J. Yassi, and R. Goutarel, *Bull. Soc. Chim. France* 787 (1964).
99. R. Tschesche and H. Ockenfels, *Chem. Ber.* **97,** 2326 (1964).
100. J. Tomko, I. Bendik, S. Bauer, and I. Mokry, *Pharm. Zentralhalle* **99,** 373 (1960).
101. J. Tomko and I. Bendik, *Coll. Czech. Chem. Commun.* **27,** 1404 (1962).
102. J. Tomko, A. Vassova, G. Adam, K. Schreiber, and E. Höhne, *Tetrahedron Letters* 3907 (1967).
103. L. F. Fieser and M. Fieser, "Steroids," pp. 83–89, Reinhold Publ. Corp., New York, N.Y. (1959).
104. E. Bianchi, C. Djerassi, H. Budzikiewicz, and Y. Sato, *J. Org. Chem.* **30,** 754 (1965).
105. K. Schreiber and G. Adam, *Tetrahedron* **20,** 1707 (1964).
106. J. Tomko and A. Vassova, *Chem. Zvesti* **18,** 266 (1964).
107. G. Adam, K. Schreiber, J. Tomko, and A. Vassova, *Tetrahedron* **23,** 167 (1967).
108. G. Adam and K. Schreiber, *Tetrahedron Letters* 943 (1963).
109. K. Schreiber, *Abhandl. Deut. Akad. Wiss. Berlin, Kl. Chem. Geol. Biol.,* in press.
110. E. Bianchi, F. Diaz, and J. A. Garbarino, *Gazz. Chim. Ital.* **90,** 894 (1960).
111. K. Schreiber, *Z. Chem.* **3, 346** (1963).
112. T. D. Fontaine, G. W. Irving, R. Ma, J. B. Poole, and S. P. Doolittle, *Arch. Biochem.* **18,** 467 (1948).
113. Y. Sato, A. Katz, and E. Mosettig, *J. Amer. Chem. Soc.* **73,** 880 (1951).
114. R. Kuhn and I. Löw, *Chem. Ber.* **85,** 416 (1952).
115. R. Kuhn, I. Löw, and H. Trischmann, *Chem. Ber.* **86,** 372 (1953).
116. F. C. Uhle and J. A. Moore, *J. Amer. Chem. Soc.* **76,** 6412 (1954).
117. K. Schreiber and G. Adam, *Ann.* **666,** 155 (1963).
118. K. Schreiber, *Angew. Chem.* **69,** 483 (1957).
119. K. Schreiber and H. Rönsch, *Ann.* **681,** 196 (1965).
120. P. M. Boll, *Acta Chem. Scand.* **17,** 1852, 2126 (1963).
121. H. Rönsch and K. Schreiber, *Phytochemistry* **5,** 1227 (1966).
122. K. Schreiber, *Kulturpflanze* **11,** 451 (1963).
123. Y. Sato, N. Ikekawa, and E. Mosettig, *J. Org. Chem.* **25,** 783 (1960).
124. F. C. Uhle, *J. Amer. Chem. Soc.* **75,** 2280 (1953).
125. K. Schreiber, *Planta Med.* **6,** 435 (1958).
126. F. C. Uhle, *J. Amer. Chem. Soc.* **76,** 4245 (1954).
127. K. Schreiber, A. Walther, and H. Rönsch, *Tetrahedron* **20,** 1939 (1964).

128. P. Tuzson and Z. Kiss, *Acta Chim. Acad. Sci. Hung.* **12,** 31 (1957).
129. Y. Sato and H. G. Latham, *J. Amer. Chem. Soc.* **78,** 3150 (1956).
130. Y. Sato and H. G. Latham, *J. Org. Chem.* **22,** 981 (1957).
131. Y. Sato, H. G. Latham, L. H. Briggs, and R. N. Seelye, *J. Amer. Chem. Soc.* **79,** 6089 (1957).
132. K. Schreiber, *Ann.* **682,** 219 (1965).
133. R. K. Callow and P. N. Massey-Beresford, *J. Chem. Soc.* 4482 (1957).
134. L. Toldy, *Acta Chim. Acad. Sci. Hung.* **16,** 403 (1958).
135. P. M. Boll, *Acta Chem. Scand.* **14,** 783 (1960).
136. P. M. Boll, *Acta Chem. Scand.* **17,** 1176 (1963).
137. E. Höhne, H. Ripperger, and K. Schreiber, *Tetrahedron* **23,** 3705 (1967).
138. Y. Sato and S. Hayakawa, *J. Org. Chem.* **29,** 198 (1964).
139. Y. Sato and S. Hayakawa, *J. Org. Chem.* **28,** 2739 (1963).
140. K. Schreiber, *Kulturpflanze* **11,** 422 (1963).
141. R. Kuhn and I. Löw, *Angew. Chem.* **66,** 639 (1954).
142. K. Schreiber, *Chem. Ber.* **87,** 1007 (1954).
143. R. Kuhn and I. Löw, *Chem. Ber.* **80,** 406 (1947).
144. A. Soltys and K. Wallenfels, *Ber.* **69,** 811 (1936).
145. V. Prelog and S. Szpilfogel, *Helv. Chim. Acta* **25,** 1306 (1942).
146. F. C. Uhle and W. A. Jacobs, *J. Biol. Chem.* **160,** 243 (1945).
147. Y. Sato and H. G. Latham, *J. Amer. Chem. Soc.* **78,** 3146 (1956).
148. Y. Sato and N. Ikekawa, *J. Org. Chem.* **26,** 1945 (1961).
149. E. Höhne, K. Schreiber, H. Ripperger, and H.-H. Worch, *Tetrahedron* **22,** 673 (1966).
150. K. Schreiber, C. Horstmann, and G. Adam, *Chem. Ber.* **97,** 2368 (1964).
151. K. Schreiber and H. Rönsch, *Tetrahedron* **21,** 645 (1965).
152. G. Adam and K. Schreiber, *Tetrahedron* **20,** 1719 (1964).
153. G. Adam and K. Schreiber, *Tetrahedron Letters* 923 (1965).
154. W. A. Jacobs and L. C. Craig, *J. Biol. Chem.* **148,** 41 (1943).
155. Y. Sato and W. A. Jacobs, *J. Biol. Chem.* **179,** 623 (1949).
156. S. W. Pelletier and D. M. Locke, *J. Amer. Chem. Soc.* **79,** 4531 (1957).
157. W. A. Jacobs and L. C. Craig, *J. Biol. Chem.* **159,** 617 (1945).
158. Y. Sato and W. A. Jacobs, *J. Biol. Chem.* **191,** 63 (1951).
159. S. W. Pelletier and W. A. Jacobs, *J. Amer. Chem. Soc.* **75,** 4442 (1953).
160. C. R. A. Wright and A. P. Luff, *J. Chem. Soc.* **35,** 405 (1879).
161. D. Burn and W. Rigby, *J. Chem. Soc.* 963 (1953).
162. F. L. Weisenborn and D. Burn, *J. Amer. Chem. Soc.* **75,** 259 (1953).
163. J. Tomko and A. Vassova, *Pharmazie* **20,** 385 (1965).
164. J. Tomko, A. Voticky, G. Spiteller, and M. Spiteller-Friedmann, *Arch. Pharm.* **299,** 347 (1966).
165. R. Kuhn and I. Löw, *Chem. Ber.* **94,** 1088 (1961).
166. Ref. 165, p. 1096.
167. R. Kuhn and I. Löw, *Chem. Ber.* **95,** 1748 (1962).
168. K. Schreiber and H. Ripperger, *Chem. Ber.* **100,** 1381 (1967).
169. G. Barger and H. L. Fraenkel-Conrat, *J. Chem. Soc.* 1537 (1936).
170. K. Schreiber and H. Ripperger, *Z. Naturforsch.* **17b,** 217 (1962).
171. E. Schlittler and H. Uehlinger, *Helv. Chim. Acta* **35,** 2034 (1952).
172. K. Schreiber and H. Ripperger, *Ann.* **655,** 114 (1962).
173. G. Adam, K. Schreiber, and I. Tomko, *Ann.* **707,** 203 (1967).
174. K. Schreiber, Habilitationsschrift, University of Jena, 1961.
175. R. Tschesche, I. Mörner, and G. Snatzke, *Ann.* **670,** 103 (1963).

176. Y. Sato, Y. Sato, H. Kaneko, E. Bianchi, and H. Kataoka, *J. Org. Chem.* **34,** 1577 (1969).
177. H. Ripperger and K. Schreiber, *Ann.* **723,** 159 (1969).
178. T. Kikuchi, S. Uyeo, and T. Nishinaga, *Chem. Pharm. Bull. (Tokyo)* **15,** 577 (1967).
179. T. Kikuchi and S. Uyeo, *Chem. Pharm. Bull. (Tokyo)* **15,** 549 (1967).

20

Abnormal Steroidal Alkaloids

KEITH S. BROWN JR.
Centro de Pesquisas de Produtos Naturais
Faculdade de Farmácia, U.F.R.J.
Rio de Janeiro, Brazil

1. ALKALOIDS WITH THE C-NOR-D-HOMOSTEROID SKELETON

Members of the Liliaceous tribe Veratreae contain a complex insecticidal and hypotensive mixture of alkaloids. These compounds have been studied intensively for nearly thirty years, and several excellent reviews[1–5] have been published covering their chemistry and pharmacology.

The known *Veratrum* alkamines are all monoacidic and C_{27}†; they are divided into two groups[4] according to their oxygen content and normal natural conjugation pattern. The pharmacologically uninteresting *jerveratrum* alkaloids possess 1–4 oxygen functions and occur either as the free alkamines or as mono-D-glucosides; normal steroidal members of this group are discussed in the previous chapter. The six known *ceveratrum* alkamines have a high oxygen content (7–9) and usually occur in the plant as mono- or poly-esters with significant pharmacological activity.

The *Fritillaria* alkaloids (isolated mostly from Chinese drugs of the *pei-mu* class and the corresponding Liliaceous source-plants) show many similarities to the jerveratrum alkaloids ($C_{27}O_{1-4}$) and possess strong cardiotonic action.

1.1 Veratramine and Jervine

The two principal members of the aberrant jerveratrum alkamines were first studied separately, and not interrelated until 1952 when both struc-

†Alkaloids with as yet unelucidated structures have been assigned formulas with 19 to 28 carbon atoms.

tures were finally established. Jervine possesses a hindered conjugated ketone (no formation of an oxime; λ_{max} 252 mμ) with an unusually long wavelength n-π* band (λ_{max} 360 mμ), while veratramine possesses a benzene ring (λ_{max} 268 mμ).[6] Both alkaloids have a cholesterol-like homoallylic 3β-alcohol that gives a conjugated ketone upon Oppenauer oxidation.[7,8] The third oxygen in jervine is involved in an inert ether linkage; the second oxygen in veratramine is an acylable secondary hydroxyl group.

Selenium dehydrogenation of jervine gave 5-methyl-2-ethyl-pyridine [(1); R = H], already known from other steroidal alkaloids, and the new 3-hydroxy analog [(1); R = OH].[9,10] Similar dehydrogenation of veratramine gave 5-hydroxy-3-methylpyridine,[8,10] which implied cleavage of a benzylic bond and retention of the ethyl group on an aromatic homosteroid ring D. Both alkamines yielded N-acetates upon acetylation.

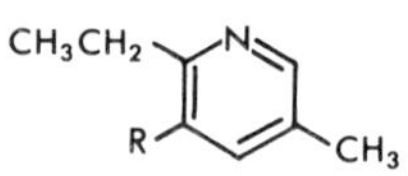

(1)

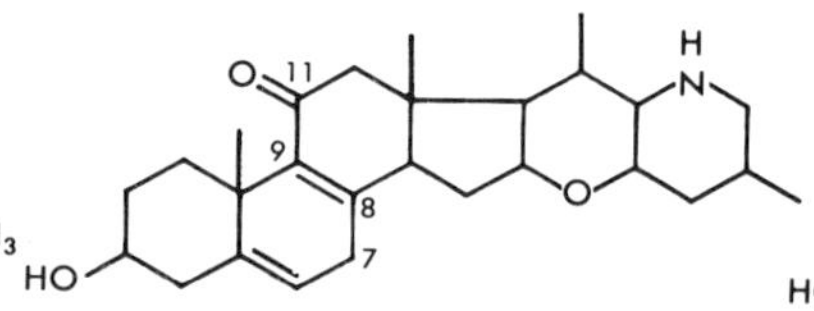

(2)

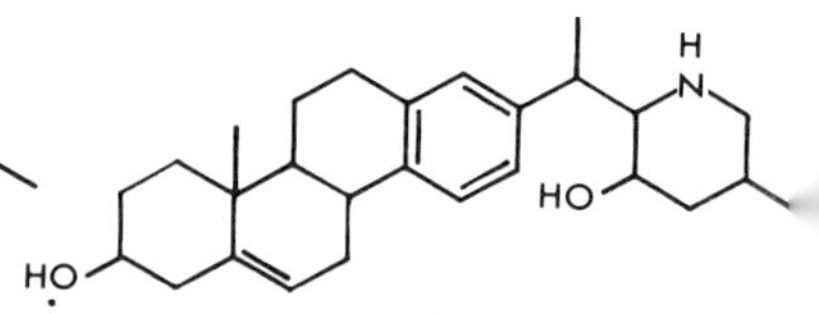

(3)

From these and other data, Jacobs and Sato published in 1949[10] and 1951[8] tentative structures for jervine (2) and veratramine (3), based on the assumption of a normal steroid nucleus and a D-homosteroid, respectively. These proposals were very reasonable for the data then available; the compounds, however, possessed a most unusual and unexpected feature that was not recognized in the early work.

1.1.1 Reinvestigation and Interrelation of Jervine with Veratramine Structure (2) would clearly be an interesting possible source material for the synthesis of the hormone cortisone. Beginning in 1950, jervine was, thus, reinvestigated by the group of Fried and Wintersteiner at Squibb. Evidence indicating that the alkaloid possessed an unprecedented modification in the basic steroid skeleton rapidly accumulated. Oxidation of O,N-diacetyljervine gave a 7-ketone that showed no signs of conjugation with the 11-ketone through the (proposed) 8,9-double bond.[11] Jervine was correlated with veratramine *via* energetic acid acetolysis followed by hydrogenation of the 5,6 double bond, to give an *indanone* (by uv)[12] synthesized from triacetyldihydroveratramine[13] by oxidation with chromic acid. Hot alkaline permanganate oxidation of veratramine gave benzene-1,2,3,4,-tetracarboxylic acid,[14] indicating that the aromatic ring was tetrasubstituted. Finally, acetolysis of O,N-diacetyljervine in the presence of zinc chloride[15] or boron trifluoride[16] gave fragmentation to a *nitrogen-free* dienone. Upon chromic acid oxidation, the dienone gave acetaldehyde and an enedione, reduced to a nonconjugated ketone with zinc and oxidatively rearranged to a phenol in alkali.[15] The above results could only be accommodated by the revolutionary assumption of a *five-membered C ring* and a *six-membered D ring* (C-nor-D-homosteroid ring system) in jervine (4) and veratramine (5).[13,15] The above indanone was formulated as (6)[12,13] and the dienone as (7).[15,16]

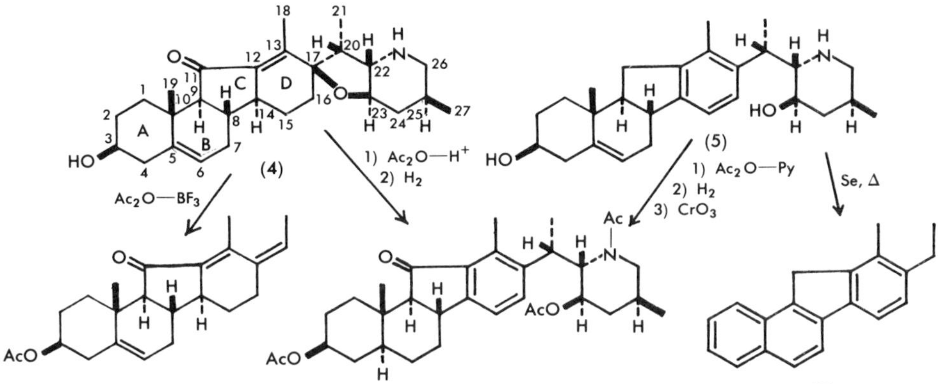

(7) (6) (8)

1.2 Unusual Chemical Reactions of the Alkaloids

During and after the structural investigations on jervine and veratramine, many truly new and interesting reactions were uncovered and explained. Selenium dehydrogenation of the bases gave, in addition to the pyridines (1), the Jacobs' hydrocarbon,[9,17] 8-methyl-7-ethyl-1,2-benzofluorene (8). Hydrogenation of jervine in the presence of alkali selectively reduced the conjugated 12,13-double bond[7,18]; with platinum in acetic acid, a tetrahydroderivative was obtained.[7] However, attempted hydrogenation with palladium in 10% aqueous acetic acid gave, in addition to known reduction products, a compound isomeric with jervine.[19] This product was much more stable to acid than jervine (implying that the oxide bridge was no longer allylic), but gave a conjugated dienone upon treatment with alkali. The latter could be oxidized to a 3-ketone of identical ultraviolet spectrum, indicating that the 5,6-double bond had migrated to a more distant position. On the basis of uv spectra, the isomer was formulated as $\Delta^{12(14)}$-jervine (9)* and the conjugated dienone as (10).[19]

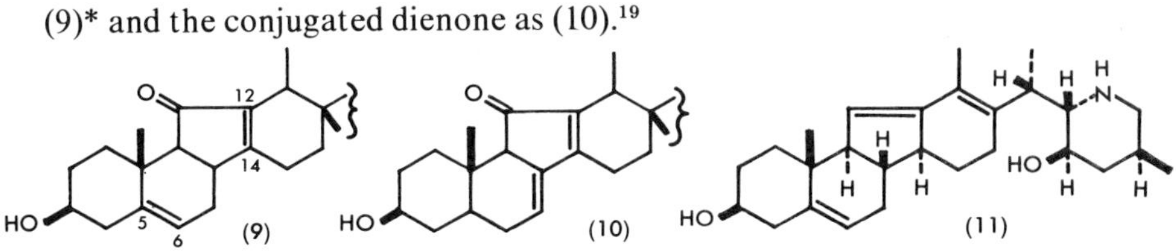

*Originally called "Δ^{13}-jervine" from a different numbering system still used in several of the chemical groups (12 and 13 become 13 and 17a).

Reduction of the relatively unreactive 11-ketone of jervine with sodium in *n*-butanol gave an 11α-alcohol, while with lithium aluminum hydride (LAH) an 11β-alcohol was obtained[20]; thus, the behavior paralleled that of normal steroidal 11-ketones. Wolff–Kishner reduction of jervine gave the natural alkamine 11-deoxojervine, along with the conjugated diene (11)[21] arising from oxide bridge cleavage.

1.2.1 Veratramine Ketoxide The oxidation of triacetyldihydroveratramine ($C_{33}H_{47}O_5N$) gave, besides the indanone (6) mentioned above,[13] an unusual and rather unstable major product $C_{33}H_{45}O_7N$, containing a hindered ketone. The reactions of this compound led to the proposal of structure (12) in which the 8,9-bond was cleaved (13) giving a nine-membered epoxyketone ring.[22] Treatment of this ketone with alkali, mineral acid, or boron trifluoride induced condensation of C(9) with C(7) to give an α-naphthol (by uv) with a free *p*-position (diazo coupling), formulated as [(14), R = OH]. Sodium borohydride reduction of the ketone (12) was accompanied by nonreductive epoxide ring opening to give the most exceptional glycol system of (15) that (presumably for steric reasons) reacted with neither acetic anhydride-pyridine, nor chromic acid, nor periodic acid or lead tetraacetate. Attempted energetic acetylation of (15) in the presence of boron trifluoride gave again the naphthalene system [(14); R = H] with loss of three molecules of water. Treatment of the ketone (12) with hydrobromic acid gave transannular cyclization to the benzofulvene (16), reducible with sodium in alcohol (in a 1,4-manner) to an indene (by uv).[22]

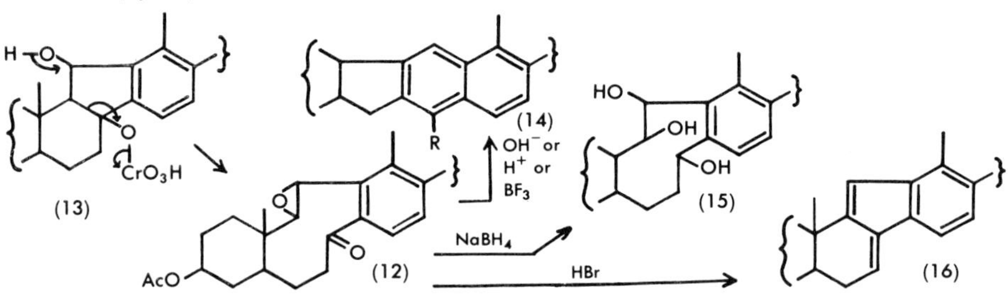

(13) (14) (12) (15) (16)

1.2.2 Ring D and the Sidechain; Jervisine An interrelation of veratramine and the steroid hecogenin was achieved by hydrogenation and Ruschig degradation of the former, which proceeded with rupture of the 22,23-bond (17) to the aldehyde (18). This was degraded with *n*-butylnitrite and sodium *n*-butoxide to the noroxime, which was hydrolyzed to ketone (19).[23] Sidechain degradation of hecogenin was followed by reduction to the corresponding 12β-alcohol-Δ^{16}-20-ketone. Tosylation and rearrangement[24] to the C-nor-D-homosteroid system was accompanied by aromatization to give (19).[25]

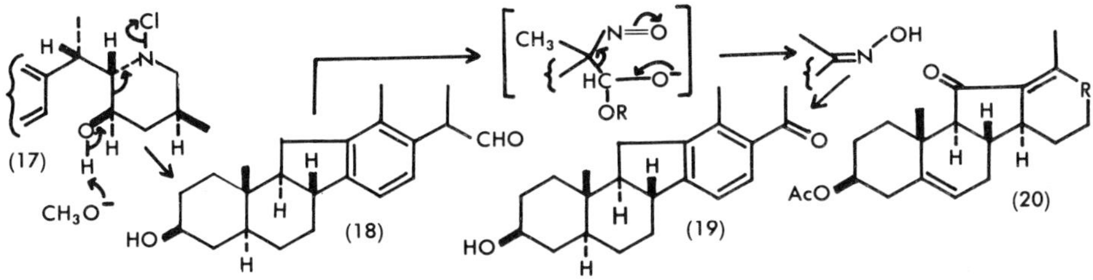

Photolysis of O-acetyljervine in dioxan at 253.7 mμ gave fragmentation of the sidechain with production of the aldehyde [(20); R = >CHCH(CH_3)CHO], the enedione [(20), R = >CO] [previously known from oxidation of (7)], and the vinylidene ether derivative [(20), R = —O—CH—$COCH_3$]. N-Acetyljervine was stable under similar photolysis conditions.[26] However, irradiation of the 23-mononitrite of O,N-diacetyljervine gave a fragmentation similar to that observed in the Ruschig degradation (17); the product was again the aldehyde [(20), R = >CHCH(CH_3)CHO].[27]

The three substituents on the piperidine ring have all been assigned as equatorial (22α, 23β, 25β) in recent work[28] which, along with many stereochemical investigations in ring D and the sidechain, is well summarized in Reference 1.

The acetolysis of O,N-diacetyltetrahydrojervine gave a 16,17 double bond and 23-acetate.[29] The double bond was hydroxylated normally with osmium tetroxide; but a major by-product was the hexacyclic tertiary amine-tertiary O-acetate (21), evidently formed by N-acetyl participation in the decomposition of the cyclic osmate ester.[29]

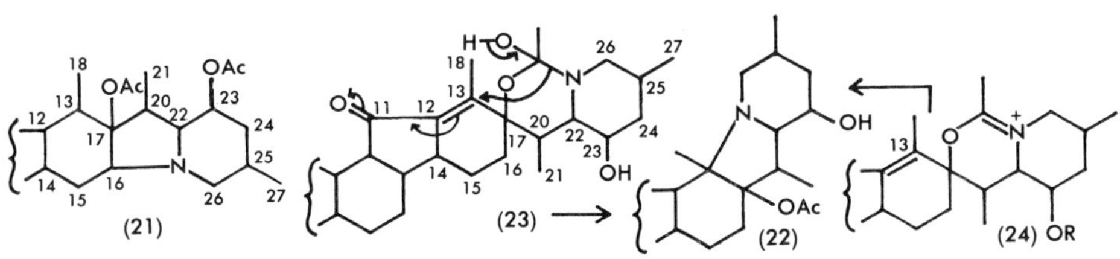

The acetolysis of diacetyljervine gave in addition to the 5,6-dehydro derivative of indanone (6)[12] the 17-alcohol 23-acetate, which upon hydrolysis in cold alkali underwent a most interesting rearrangement. The product, a very weakly basic tertiary amine-tertiary O-acetate different from (21), was formulated as (22),[12,30] arising from N→O acyl migration accompanying N addition to the electron-deficient terminal carbon of the conjugated ketone system (23). The dehydrated derivative of (23), quaternary iminium salt (24), was unexpectedly obtained by treatment of diacetyljervine with acetic anhydride-perchloric acid, or of N-acetyljervine with methanolic hydrogen chloride.[30] Alkali treatment of (24) also gave (22), named jervisine-17-acetate.

Mass spectral measurements have been shown to be useful in structure determination of jerveratrum alkaloids,[31] aiding especially in the confirmation of a 23-deoxyveratramine formulation for verarine.[31,32] The nmr spectra of a wide variety of jervine derivatives indicated predictability for the position of methyl group signals.[33] The small nmr interaction between an 11β-ol and the 18-methyl group was suggested as support for the twist conformation of ring D (in saturated derivatives),[33] also indicated by exclusive 19-oximation in the photolysis of 11β-nitrites.[34]

1.3 Isojervine

Perhaps the most unusual of all the derivatives of jervine is an isomer produced upon mineral acid treatment, known as isojervine. The N-methyl and N-acetyl derivatives of jervine, but not dihydrojervine, were similarly isomerized in acid. The product showed only a low-intensity uv shoulder at 252 mμ (ϵ 2900)[7]; however, the ir spectrum revealed the continued presence of a conjugated ketone (1640, 1690 cm^{-1}).[19] Failure of the ketone to react with carbonyl reagents indicated that it still occupied the hindered 11-position.

The structure elucidation of isojervine was accomplished competitively in three different groups in 1962.[35,36,37] Although the early work[7] had reported that the double bonds were all resistant to catalytic hydrogenation, in reinvestigations one mole of hydrogen uptake was observed; the product showed normal uv absorption (λ_{max} 238 mμ, ϵ 9500) for the conjugated ketone.[35–37] Oppenauer oxidation of isojervine gave a Δ^4-3-ketone (λ_{max} by subtraction 235 mμ, ϵ 16,500),[35,36] the N-acetyl derivative of which was also prepared by acid isomerization of N-acetyl-Δ^4-jervone.[37] Similar oxidation of dihydroisojervine, or reduction of isojervone, gave a saturated 3-ketone that also showed normal uv spectral behavior for the conjugated 11-ketone.[35] This "suppression of resonance interaction" relieved by monosaturation had also been observed in derivatives of 1-acetyl-1,4-cyclohexadiene.[37,38]

Isojervine possessed a new acylable secondary hydroxyl group. This was assumed to be in the 23-position, as selenium dehydrogenation of isojervine gave 3-hydroxy-5-methyl-2-ethylpyridine [(1) R = OH].[10] Proof for this assignment, and for the absence of skeletal rearrangement in the isomerization of jervine, came from interrelations of isojervine with jervine and veratramine. Lithium-in-ammonia reduction of N-acetyldihydroisojervine gave N-acetyl tetrahydroisojervinol,[37] prepared from jervine by hydrogenation, acetylation, and reduction with sodium in alcohol.[7] Similar reduction of dihydroisojervine gave a tetrahydroderivative[35,36] whose uv spectrum showed the three low-intensity absorptions (λ_{max} 311, 320, 330 mμ) ascribed to $\Delta^{13(17)}$-jervines. That the double bond was indeed in the β, γ-position to the ketone was shown by its epoxidation and acid treatment, leading to a quaternary base of the (24) type, with an hydroxyl at C(13). The triacetyl-epoxide rearranged on alumina to the 5,6-dihydro-17-epimer (25) of the tertiary alcohol obtained upon acetolysis of diacetyljervine (Section 1.2.2). Dehydration of the tertiary alcohol in (25) by prolonged heating with maleic acid, followed by aromatization with palladium, gave the known indanone (6).[35]

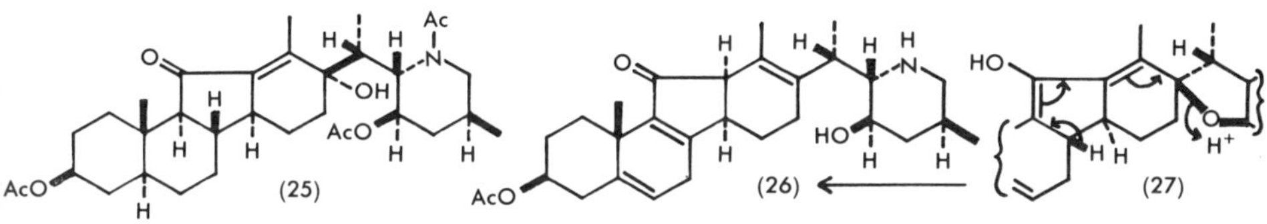
(25) (26) (27)

Thus, isojervine was formulated as (26)[35–37] with the abnormal uv absorption explained in terms of inhibition of resonance by the "cross-ring" double bond. The details of the stereochemistry, nmr spectra, and many additional reactions[35–37,39] leave no reasonable doubt as to the correctness of the structure proposed. The mechanism of formation was visualized as in (27).

1.4 Syntheses

In 1967, the total syntheses of jervine[40] and veratramine[41] were reported. The fundamental stepping-stone in both cases, 17-acetyl-$\Delta^{12,14,16}$-5α-etiojervatriene-3β-ol (19), was available from totally synthesized hecogenin (Section 1.2.2)[25] or by its tailored total synthesis.[42] Transformation of (19) to veratramine proceeded through the oxirane (28) to the known aldehyde (18)[23] that (as the bisulfite adduct) was treated with potassium cyanide and *l*-*t*-butyl-3-methyl-4-amino-butyrate; the product was benzoylated to the cyano ester (29).[41] Cyclization with methylsulfinylcarbanion gave an enamino-ester that was hydrolyzed and decarboxylated in acid to the ketone [(30); R = ϕCO] (mixture of stereoisomers). Transformation of one of the isomers of (30), identified as 20α-Me, 22β-H, to veratramine (5) was chemically straightforward.[41]

The transformation of (19) to jervine involved borohydride reduction of its 3-acetate and bromination of the resulting alcohol mixture. The higher-melting bromide (31) was treated with synthetic 1-acetyl-3(*S*)-methyl-5-piperidone enamine. There was isolated in 35% yield the 3O,N-diacetyl-

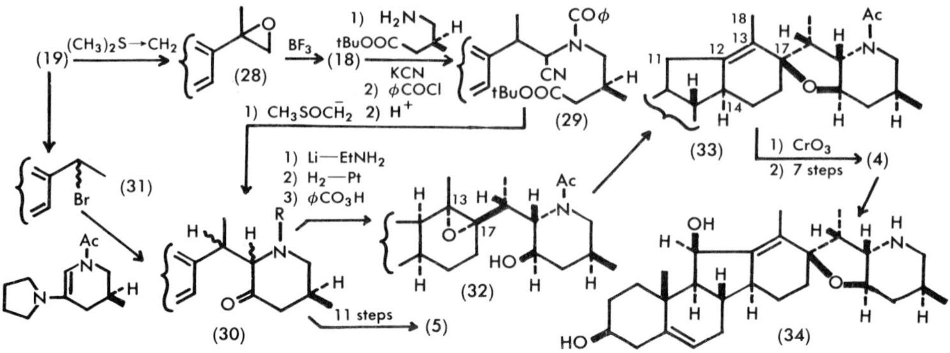

23-dehydro-5α,6-dihydroveratramine [(30); R = Ac] with the desired 20α-Me, 22β-H configuration, compared with material from degradation of veratramine.[40] Reduction of (30) with lithium in ethylamine and hydrogenation of the trisubstituted double bond gave the $\Delta^{13(17)}$-derivative, epoxidized to (32). Treatment of (32) with alkali to form the oxide bridge followed by dehydration gave the known diacetyl-dihydro-11-deoxojervine (33). Oxidation of (33) gave a very low yield of unsaturated 11-ketone which was converted to jervine (4) in a straightforward manner[40]; the major product of the oxidation was the 14-alcohol.

The synthesis of jervine was interpreted in terms of proof for the 17α-oxido bridge (previously proposed only on the basis of biogenetic precedent), since the epoxide (32), which was displaced with inversion at C(17) by the 23-oxygen, had a 19-methyl signal in the nmr at lower field than its isomer (also produced in the epoxidation), and thus should be β-oriented.[43] However, a recent x-ray determination of the structure of veratrobasine (34),[44] while confirming the 20α, 22α, 23β, and 25β stereochemistry proposed by Johnson and co-workers,[38,41] showed the presence of a 17β, 23β-oxido bridge in the molecule. Comparison of veratrobasine and jervine-11β-ol showed their identity by all criteria.[45] Thus, the correct complete structure and stereochemistry of jervine are as indicated in formula (4).

Very recently, a total synthesis of verarine, based upon a somewhat different conception from that of the jervine and veratramine syntheses, has been published.[45a]

1.5 Veramarine and the Fritillaria Alkaloids; Transition to the Ceveratrum Series

The highly oxygenated ceveratrum alkamines possess a tertiary nitrogen atom (no reaction with acetic anhydride or nitrous acid) and give upon selenium dehydrogenation a very large number of products. Among these are various pyridine (1) and 1,2-benzofluorene (8) derivatives and two important basic compounds containing almost all the skeleton, veranthridine (35) and cevanthridine (36).[46] The structures of (35) and (36) were suggested by a combination of uv spectral characteristics, reduction of the heterocyclic ring, and reversible oxidation to red fluorenone derivatives.[47] Thus, the *cevane* skeleton was assigned structure (37) (with 20β-H), formally derivable by cyclization of a $\Delta^{13(18),17(20)}$-jervadiene (38), which in turn could come from dehydrogenation of a normal steroid rearrangement product.[24] Some precedent for this sort of cyclization can be seen in the jervisine rearrangement. This derivation would also explain the frequent presence of a 20β-hydroxyl group (inverted configuration) in known cevane-type alkaloids.

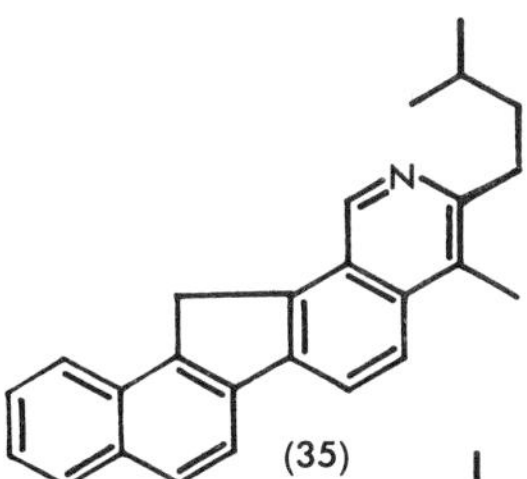

(35)

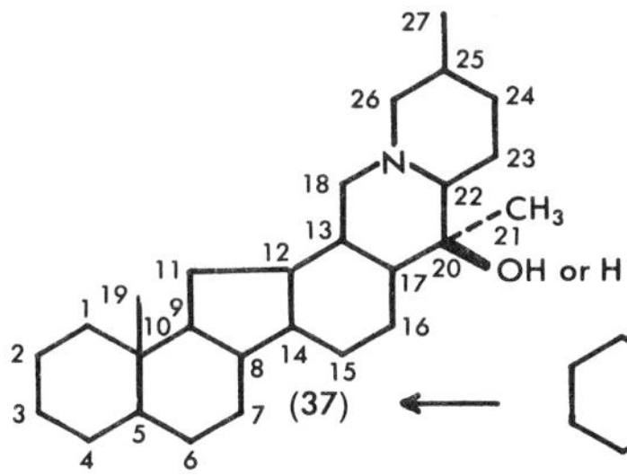

(37)

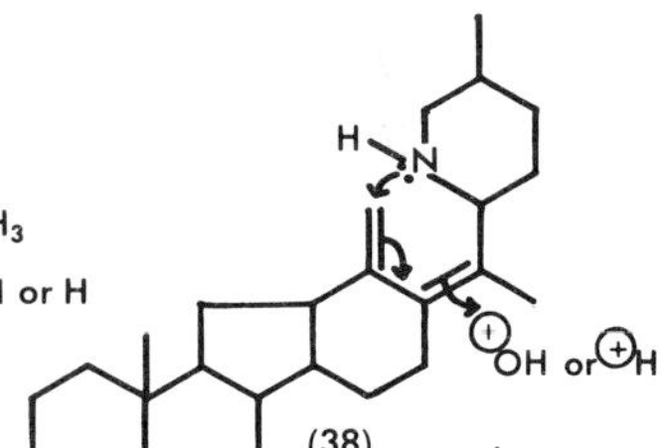

(38)

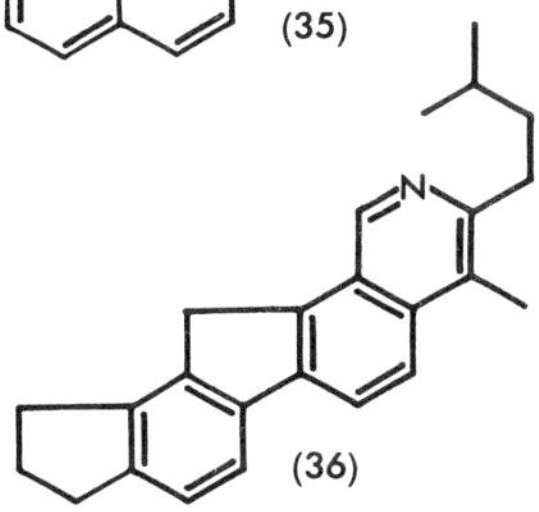

(36)

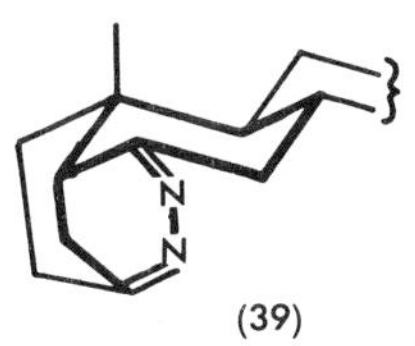

(39)

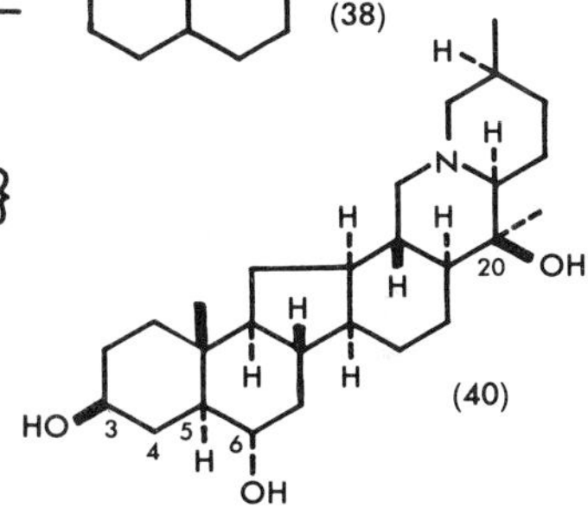

(40)

The *Fritillaria* alkaloids also give veranthridine upon dehydrogenation[48] and, therefore, should be derived from the cevane skeleton. However, they are not as highly oxygenated as the members of the ceveratrum group, and like the jerveratrum bases they occur conjugated only as β-D-glucosides. The most studied base, verticine (or peimine) from *F. verticillata*, is a tertiary amine with two secondary and one tertiary hydroxyl groups.[48–50] The derived diketone formed a dihydropyridazine system (39) with hydrazine hydrate, and was oxidized by air in alkaline medium to an enedione with uv spectral characteristics of Δ^4-cholestene-3,6-dione. Further oxidation experiments and spectral data led to the proposal of the 5α-cevanoid structure (40) for verticine.[50] Although this structure was recently supported by x-ray analysis of verticinone methobromide,[50a] conflicting conclusions have been published by the Russian workers.[50b] Similar structures have been proposed for several other *Fritillaria* alkaloids,[50b, 50c] including imperialine and some 20β-H and 3α-OH analogs, and also for the jerveratrum base veramarine.[51] These structures represent the transition between the typical jerveratrum alkaloids and the ceveratrum group discussed below.

1.6 The Structure and Configuration of Cevine

The above skeleton (37) for the alkaloids of the ceveratrum group was also indicated by the products obtained by chromic acid oxidation of the abundantly available hydrolysis product, cevine. Cevine ($C_{27}H_{43}O_8N$) was shown to contain a tertiary amine, no N-methyl or O-methyl group, and no C=C or C=O double bonds (therefore, heptacyclic). Destructive oxidation gave an acidic fraction containing seven components,[52] characterized during fifteen years of chemical work. The most significant of the products was a lactone-tricarboxylic acid $C_{14}H_{18}O_8$ that lost two moles of water upon heating or base treatment giving decevinic acid, $C_{14}H_{14}O_6$, λ_{max} 328 mμ. Upon mild sulfur dehydrogenation, decevinic acid gave 2-hydroxynaphthalic anhydride (41).[52] Base treatment followed by heat gave a six-membered-ketone-γ-lactone (42), with loss of two molecules of carbon dioxide. The methyl ester of (42) was allowed to react with methyl magnesium bromide and then dehydrogenated to give 2-methyl-5-isopropylnaphthalene[53]; and also rationally degraded to (+)-9-methyl-*cis*-1-decalone (43) with structure and absolute configuration proven by interrelation and synthesis.[54] As decevinic acid showed the same unusual ir and uv spectral characteristics as α-acetyl-α'-ethyl-β-methylglutaconic anhydride, its logical structure would be the analogous (44).[53] The two acidic groups 12 and 14, β and vinylogously β to the keto group, were lost upon heating; the 4-carboxyl then added to the 9,11-double bond to form the lactone. The lactone-tricarboxylic acid precursor (45) was derived from rings A, B, and C of the cevane skeleton as indicated in the numbering; decevinic acid was formed by condensation of C(3) with C(11).[54] Two of the remaining acidic oxidation products have been identified as the related tetracarboxylic acids (46) and (47).[55] Thus, cevine should be oxygenated

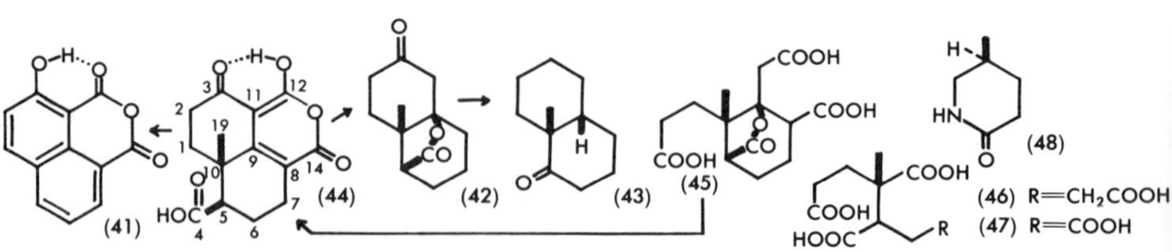

at positions 3,4,9,12, and 14, and C(10) possessed the normal steroid absolute configuration.

A further lactam product of the oxidation was (48) with the indicated absolute stereochemistry[56]; thus, the configuration at C(25) was the same in all the *Veratrum* alkaloids, and cevine possessed no oxygen functions in ring F.

1.6.1 Chemistry in Rings A, B, and C Cevine consumed two moles of periodate, and by cleavage of an appropriate derivative (see below) the product from one of the glycols could be identified as an aldehydo-γ-lactone. Since cevine also showed some characteristic reactions for an α-hydroxyketone (reaction with Tollens' and Fehling's reagents, reduction with sodium in alcohol), it was presumed to possess an α-ketol masked as a hemiketal.[57] That this function accounted for the oxygenation at C(3), C(4), and C(9) was suggested by the structure of the oxidation product (45) and confirmed by a variety of experiments. Although alkaline hydrolysis of cevadine and other natural monoesters usually produced cevine, milder hydrolysis gave a ketonic alkamine cevagenine,[58] convertible to cevine by base treatment or to a diosphenol by oxidation with bismuth oxide or 2,3,5-triphenyltetrazolium chloride.[59,60] Still milder hydrolysis gave the true nonketonic alkamine veracevine,[61] very similar in all reactions to cevine, esterifiable to the natural esters[62] and converted to cevagenine and thence to cevine by alkali. All three isomers gave with bismuth oxide an α-hydroxy-δ-lactone (49) arising from benzilic-acid type rearrangement of the intermediate α-diketone.[63] The three isomeric alkamines have been assigned structures (50)–(52) by consideration of the relative stabilities of axial and equatorial alcohols and the facile base-catalyzed epimerization of centers α to a ketone.[64]

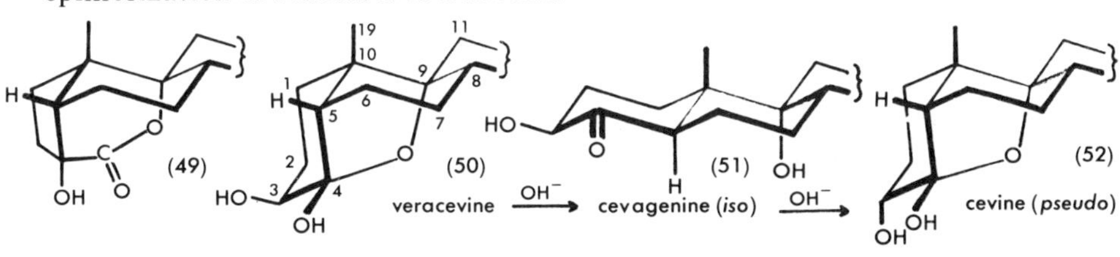

That the second glycol of cevine was present at C(12) and C(14) [suggested by oxidation product (45)] was supported by the consumption of one mole of lead tetraacetate by cevine triacetate, which contained no secondary hydroxyls. Furthermore, upon acid treatment cevagenine orthoacetate (see below) formed an isomeric C-orthoester (53) that was not transformed to a cevine derivative by strong base. Thus, the 9α-hydroxyl had been blocked,[65] and the ditertiary glycol could be assigned the α-configuration.

1.6.2 Chemistry in Rings D, E, and F. Five oxygen atoms of the eight present in cevine had now been accounted for. Only one of the remaining three hydroxyl groups was acylable. A second one was in ring D and α-oriented, for veracevine and cevine formed stable orthoacetates (54) upon acetylation in the presence of perchloric acid.[58,61] The orthoester structure for the "anhydro-tetraacetates" was suggested spectroscopically,[66] and confirmed by an unusual reduction and reoxidation: hydrogenation gave the 12O, 14O-acetaldehyde acetal,[67] which with chromic acid regenerated the orthoacetate.

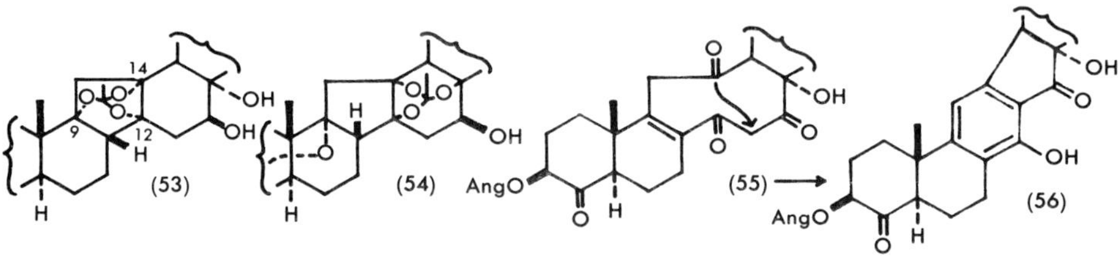

Of the possible positions for the additional two tertiary and one secondary hydroxyl groups[66] (carbons 13, 15–18, 20–27), the following were eliminated: C(21) (at least three C-methyl groups upon Kuhn–Roth oxidation of cevine): C(23)–C(27) [see the oxidation product (48)]; C(18) and C(22) (cevine showed no reactions of a carbinolamine); and C(13) and C(15) [adjacent to the ditertiary glycol at C(12) and C(14), thus would give more periodate consumption]. The only positions left were 16, 17, and 20. However, this vicinal triol would presumably consume lead tetraacetate or periodate. Indeed, a very slow additional uptake of these reagents by cevine after the initial rapid consumption of two moles could be seen. More significantly, epimerization[68] or oxidation[67] of the secondary hydroxyl at C(16) gave products showing much more rapid periodate uptake after two moles. Chromic acid oxidation of veracevine-3-angelate (cevadine) gave a transient monoene-tetraone (55) that cyclized and aromatized to a product (56)[69] showing characteristics of a nonenolizable 7-hydroxy-l-indanone; thus further indicated the presence of hydroxyl substituents at C(16) and C(17). The isolation of a hydroxyethyl-methylpyridine from selenium dehydrogenation of cevine[64] also indicated the presence of the 20-alcohol. Thus structure (57) could be written for cevine.[64]

The stereochemistry at carbons 3,4,5,8,9,10,12,14,17, and 25 had been suggested by the isomerizations, oxidations, and orthoester formations discussed above. The vicinal triol in 16,17, and 20 was presumably resistant to glycol-cleaving reagents for steric reasons (the 16-epi-alcohol was more rapidly cleaved; see also Section 1.2.1). Therefore, the 16β,

17α, 20β-triaxial triol conformation was probable. This conformational assignment was supported by the stereospecific catalytic hydrogenation of 16-keto-3,4-diacetylcevine to natural diacetylcevine[68]; the catalyst approach to rings DEF must be from the α-face, giving the 16β-ol as product. The 16-ketone was prepared after an unusual facilitated methanolysis of the 3,4,16-triacetate; simple heating in methanol selectively removed the 16-ester (see also Sections 2.3 and 2.3.2). Basic catalysis was provided by the tertiary nitrogen in the molecule,[68,70] and intramolecular facilitation by 1,3-diaxial interaction with the 20-hydroxyl group (58).[71] This methanolysis has been described as an intramolecular general acid-general base catalyzed ester solvolysis[72] of the sort that may operate in many biological esterase systems.[73]

Treatment of triacetylveracevine D-orthoacetate with N-bromosuccinimide, in the hopes of introducing an imine function, gave instead an oxide which no longer suffered facile methanolysis at C(16). Structure (59) was suggested for this oxide, for on oxidation with CrO_3-pyridine it gave a formamide (60) with an nmr signal for a methyl ketone.[68] Thus, the 22-hydrogen should have the α_F-configuration as shown.

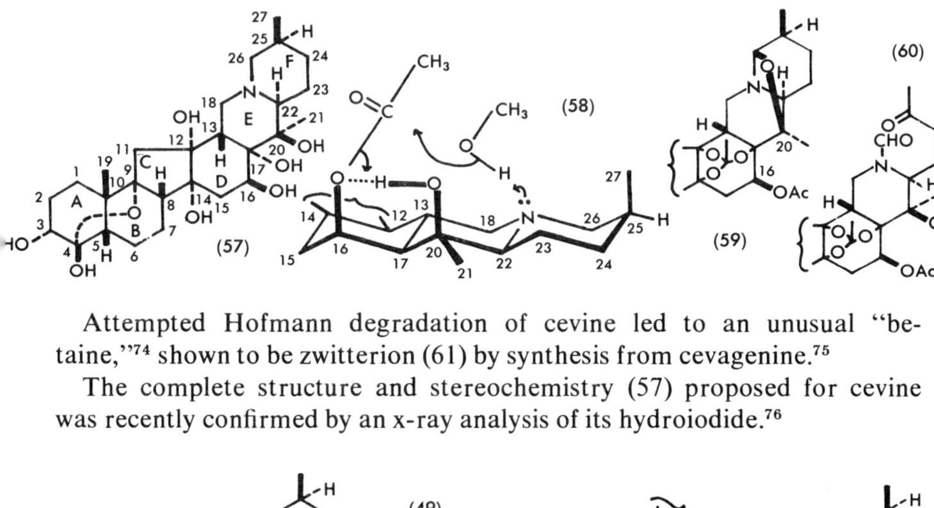
(57) (58) (59) (60)

Attempted Hofmann degradation of cevine led to an unusual "betaine,"[74] shown to be zwitterion (61) by synthesis from cevagenine.[75]

The complete structure and stereochemistry (57) proposed for cevine was recently confirmed by an x-ray analysis of its hydroiodide.[76]

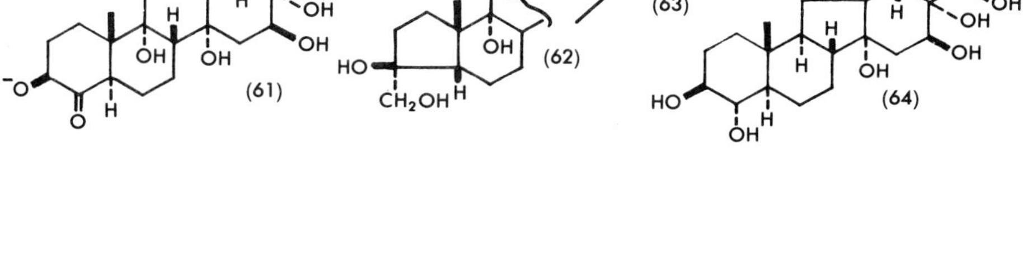
(61) (49) (62) (63) (64)

1.7 Additional Chemistry of the Ceveratrum Alkamines

The other alkamines obtained from *Veratrum* ester alkaloids show chemical reactions appreciably different from those of cevine. In germine and protoverine, the initial alkamine (type 50) is considerably more stable than veracevine, and is obtained upon normal alkaline hydrolysis of the natural esters. The ketonic *iso*-alkamine (type 51) and the nonketonic *pseudo*-alkamine (type 52) are obtained only with stronger alkali. Acetonide derivatives (see below) of germine and pseudogermine gave the same aldehydo-γ-lactone upon periodate treatment, indicating that the α-ketol-hemiketal function was present just as in veracevine and cevine. A third alkamine, zygadenine, showed the same stability as veracevine (pseudozygadenine $>$ neozygadenine $>>$ zygadenine), and (like veracevine), occurred in nature only as monoesters, but otherwise was more allied chemically to germine.

Another recently isolated alkamine proved to be identical to the bismuth oxide oxidation product (49) of pseudozygadenine, zygadenilic acid-δ-lactone[77]; it occurred in nature as its 16-angelate,[78] the only 16-ester known from *Veratrum*. Reduction of the alkamine acetonide with LAH followed by periodate cleavage gave the ring A cyclopentanone system (ir 1735 cm^{-1}) previously known in the cevanilic acid series [(49) → (62) → (63)].[63] The cevagenine diosphenol[60,79] and cevanilic acid-δ-lactone[79] have also been detected in natural alkamine mixtures; these three alkamines may be artifacts arising from labile natural 3-keto-4-esters during isolation.

The alkamine sabine[80] occurs in nature as the 3-acetate, sabadine.[79,81] It showed no evidences for the presence of the α-ketol hemiketal,[82] but gave D-orthoacetates, assisted hydrolysis of 16-acetates, and (with oxidation) diosphenol formation in ring A, identical to the reactions of cevine.[80,81,83] Failure to form a C-orthoacetate (53), rate constants observed for ring A glycol cleavage, and nmr measurements[84] were used to formulate sabine as 9-deoxy-5β-cevagenine-4α-ol (64).[85]

1.7.1 Chemistry in Rings B, C, and D; Germine and Protoverine. The most abundant and most hypotensive *Veratrum* alkaloids are polyesters derived from the alkamines germine and protoverine. Early investigations on germine indicated many similarities to cevine, but also important differences.[86] The relative stability of germine,[87] the isolation of only the hexane tetracarboxylic acid (47) upon oxidation,[86] and the formation of tetraacetates from both germine and isogermine, were explained in terms of a 7α-hydroxyl group strongly hydrogen-bonded to the 4α, 9α-oxide bridge and the 4β-hydroxyl group (65).[87,88] Removal of the 7-hydroxyl from germine by oxidation to a ketone and desulfurization of the propylenethioketal gave zygadenine (7-deoxogermine),[89] which did not show the unusual properties of the system (65).[90]

Germine, zygadenine, and protoverine contain a vicinal triol (giving formic acid upon periodate cleavage) that could be rendered inert to periodate by formation of an acetonide. Two of the triol hydroxyls, including the one not involved in the acetonide, were acylable. Isogermine, with a free 9-hydroxyl group, did not consume more periodate than its nonketonic isomers. However, 7-dehydrogermine diacetate consumed no less than five moles of periodate, indicating that (since periodate hydroxylates cyclic 1,3-diketones)[91] the part structure (66) could be present. Confirmation for this and for the placement of the vicinal triol at positions 14,15, and 16 was found in a wide variety of experiments. Germine 3,16-diacetate consumed one mole of periodate and gave a product with all the uv and ir characteristics of a *transoid*-cyclopentenone (67); its acetonide was very difficult to acetylate at C(7), suggesting steric hindrance from an α-oriented 14,15-acetonide. Methanolysis of 7-esters was unusually easy, indicating 1,3-diaxial participation by a 14α-hydroxyl group.

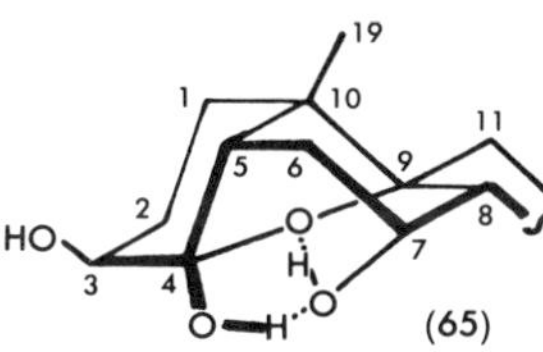
(65)

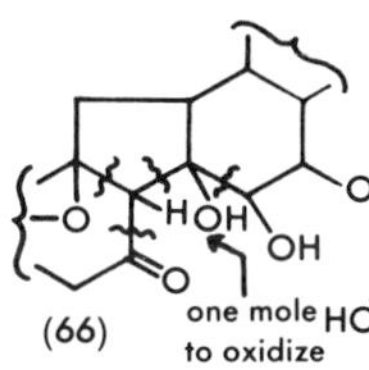

(66)

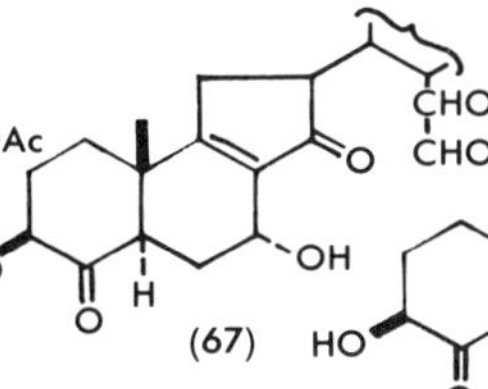
(67)

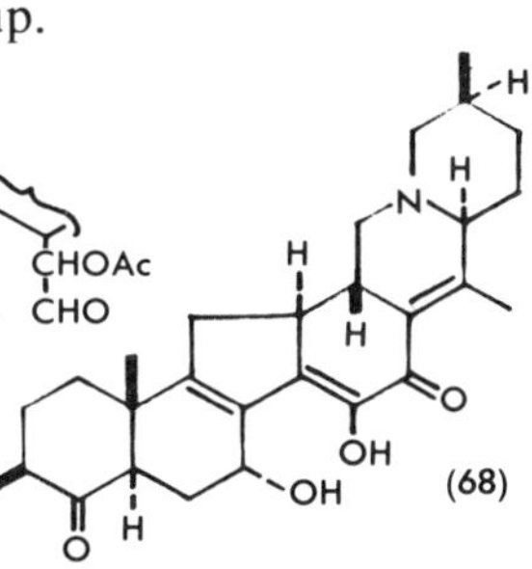
(68)

Selective transformation of the 16-hydroxyl group to the ketone (in a natural germine 3,7,15-triester) and subsequent base treatment gave an unusual cross-conjugated diosphenol (68) (λ_{max} 330 mμ), indicating the elimination of tertiary hydroxyls at 14 and 20. Facilitated methanolysis of 16-esters and formation of formamido-ketone (60) from germine tetraacetate paralleled perfectly the known chemistry of cevine, as did a large

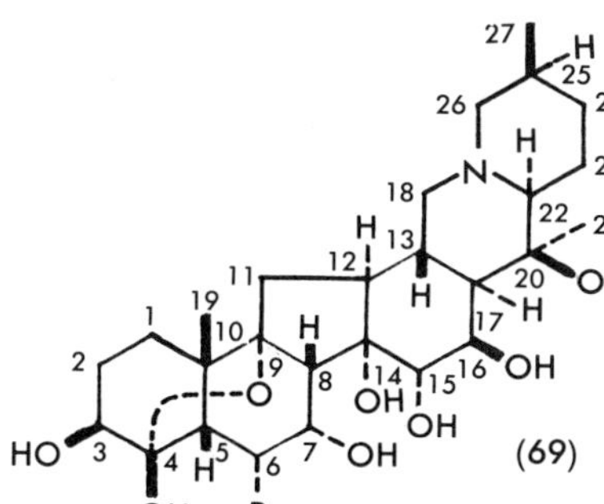
(69)

number of other comparisons. Thus, structure [(69), R = H] was assigned to germine.[88]

Protoverine contained one more hydroxyl group than germine, but otherwise was very similar in properties and reactions.[47,92,93] Under vigorous conditions (acetone-HI), isoprotoverine gave a diacetonide; this suggested the 6α-placement for the remaining hydroxyl group. The derived equatorial monotosylate [(69), R = OTs] was very resistant to S_N2 reduction by LAH. However, α-cleavage of 7-dehydro-3,6,16-triacetylprotoverine acetonide with calcium in liquid ammonia[94] gave the corresponding germine derivative, substantiating structure [(69), R = OH] for protoverine.[93]

Recently, nmr and mass spectral methods have been brought to bear on structure confirmation of ceveratrum alkaloids; some useful correlations have been discovered and applied[31,84] which should considerably facilitate structure elucidation in the future.

1.8 Ester Alkaloids

The natural veracevine and zygadenine 3-monoesters show principally insecticidal and sternutatory activity; they have received minimal pharmacological attention. However, the polyesters (especially tri- and tetraesters) of germine and protoverine are in current use in clinical treatment of hypertension. Thus, it was of great interest to elucidate the structures and determine the ester positions of these compounds, in order to develop a rationale for structure–activity relationships.[3]

Known esterifying acids include acetic, veratric, vanillic, angelic (often isomerized to tiglic upon hydrolysis),[95] *l*-2-methylbutyric, *d*-2-hydroxy-2-methylbutyric, and *erythro*-(*l*)- and *threo*-(*d*)-2,3-dihydroxy-2-methylbutyric. Acetic acid has been found esterifying all, and angelic four, of the six major alkamines. The aromatic acids are known only in the veracevine and zygadenine 3-monoesters, while 2-methylbutyric acid and hydroxylated derivatives occur only and invariably in polyesters of germine and protoverine. The relative and absolute stereochemistries of the various hydroxymethylbutyric acids were proven by appropriate syntheses.[96]

1.8.1 Methods of Determining Ester Positions In the germine and protoverine polyesters, ester position was determined by chemical methods related to those developed during the structural studies, and by selective esterifications and hydrolyses. For example, 7- and 16-esters were easily cleaved in neutral methanol; this, along with diosphenol (68) formation, helped to show that all natural esters contained a free 16β-hydroxyl group, while many were esterified at C(7), always by acetate.[97] Esters at C(15), almost always methylbutyrates, blocked glycol cleavage in ring D and acetonide formation, without affecting the assisted hydrolysis of 7-esters by the 14α-hydroxyl group. In protoverine, 6-esters (always acetates) blocked glycol cleavage in ring B, even in methanolized free 7-alcohols. Esters at C(3) blocked glycol cleavage in ring A and eliminated an oxidizable secondary hydroxyl group. Facilitated hydrolysis of esters at C(3), C(6), and C(15), while not so marked as for those at C(7) and C(16), was observable and attributed to 1,2-*cis*-interactions with hydroxyl groups at C(4), C(7), and C(14).[98] Weakly basic hydrolysis could remove esters at C(15) faster than at C(3), while 3-esters were cleaved rapidly by acid hydrolysis. Dihydroxymethylbutyryl esters were unusually reactive toward sodium borohydride and could be selectively removed by this reagent.[99] Most ester alkaloid structures were proven by interconversions with known esters, especially 15-monoesters from partial hydrolyses.

The structure proof of the active ester germitetrine provides an example of the application of these methods. Hydrolysis gave germine, two moles of acetic acid and one mole each of methyl-butyric and *erythro*-dihydroxymethylbutyric acid. Methanolysis produced first a mono-desacetyl triester, and then the natural ester neogermbudine by loss of a further acetate group; molecular rotation differences indicated that the first acetate was lost from the 7-position. As germitetrine was completely stable to sodium periodate whereas neogermbudine gave consumption of one mole, the second acetate was placed on the glycol of the dihydroxymethylbutyryl residue. Chromic acid titration of germitetrine gave one mole uptake; the ketonic product was transformed to the diosphenol (68) in base. Neogermbudine, acylated at 3 and 15 (as periodate was not consumed in the nucleus), was reduced with sodium borohydride to germine 15-(*l*)-2′-methylbutyrate, obtainable from at least five other natural germine esters by similar sequences and interrelations. Therefore, germitetrine was formulated as germine 3-*erythro*-2′-hydroxy-2′-methyl-3′-acetoxy-butyrate 7-acetate 15-(*l*)-2′-methylbutyrate.[99]

A full list of the known ester alkaloids along with the chemical elucidations of their structures may be found in Reference 1; a discussion of the relative pharmacological activity in relation to structure, in Reference 3.

1.9 Summary and Perspective

The structures of a very large number of isolated *Fritillaria* alkaloids and a smaller number of *Veratrum* bases still remain virtually unknown; other structures, while proposed, need confirmation and more intensive chemical investigation. Experiments on the biosynthesis of *Veratrum* alkaloids still await to be reported; the *in vitro* chemistry described in the preceding sections suggests many possible lines for fruitful experimentation in the living plant system.

As many of the recently isolated alkaloids[1] have been reported to possess a number of carbon atoms other than 27, it is probable that unusual structure modifications still await discovery in the C-nor-D-homo steroidal alkaloids.

The use of *Veratrum* alkaloids and their derivatives to rapidly control hypertension is well established. Efforts to synthesize and study the dihydroxy-perhydrobenzoquinolizine system (70) common to both the active *Veratrum* esters and the hypotensive alkaloid reserpine have given some results[100] which hopefully can be expanded and put to use in the future.

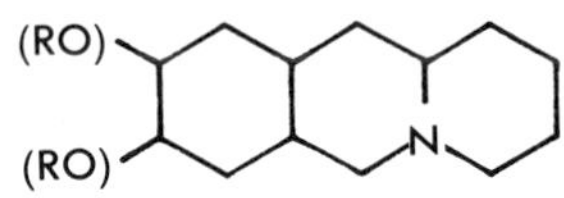

(70)

An approach to the total synthesis of various *Veratrum* alkaloids and their derivatives from hecogenin has been reported.[101] Hopefully, this could lead to readily available compounds of medicinal importance.

2. BUXUS ALKALOIDS

The boxwood (*Buxus sempervirens*) is a common ornamental shrub belonging to a small plant family (Buxaceae) of cosmopolitan distribution. Though the members of the family are best known for their fine-grained hardwoods, they have also found extensive use in folk medicine. Ancient (Old Testament, Homer, Aristotle) and modern references indicate that boxwood extracts possess activity against syphilis, rheumatism, dermatitis, rabies, malaria, cancer, and tuberculosis.[102,103] Although a British patent for curative action in the latter has been granted to Merck & Co.,[104] the drug has not been judged economically competitive with other common antitubercular agents such as isoniazid. The alkaloid mixture also produces powerful inhibition of the enzyme cholinesterase.[102,105] The pure alkaloid cyclobuxine-D has been used as a complexing agent to study the reversible helix-coil transitions of DNA,[106] suggesting that many interesting pharmacological activities remain to be discovered.

Chemical investigation of the exceedingly complex alkaloid mixture from boxwood began in 1830[102] but led to no conclusive results until Emil Schlittler, Karl Heusler, and Werner Friedrich (Basel) conquered the massive isolation problems in 1945–50.[102,107] Although data were published on seven individual alkaloids, the complexity and unusual nature of the principal bases prevented the suggestion of even partial structures. With the development of modern physical methods of structure analysis, interest in boxwood alkaloids revived independently in 1958–61 in no less than nine laboratories around the world: E.T.H. in Zurich, Sandoz A. G.

in Basel, Gif-sur-Yvette in France, Istituto Superior Sanita in Rome, Slovensk Akademie Vied in Bratislava, Czechoslovakia, Humboldt-Universität in Berlin, Kyoto University in Japan, and the Universities of Indiana and Wisconsin in the U.S.A. The last of these research groups, helped considerably by Dr. Schlittler (who arranged for large-scale isolation of the alkaloid mixture by CIBA) and by the development of an efficient partition-chromatographic separation of the closely related alkaloids,[108] eventually published the first complete structural and stereochemical representation of a boxwood alkaloid, cyclobuxine-D.[109] However, all of the groups have since contributed greatly to our present knowledge of *Buxus* alkaloids.

2.1 From Enigma to Breakthrough: Cyclobuxine-D

The alkaloid most intensively investigated by Schlittler and Heusler was that designated "A," isolated from the acetone-insoluble portion of the strong-base fraction.[102,107] In the resumed investigations (1960), an alkaloid with essentially identical properties, reactions, and derivatives proved to be the major component (35% upon partition chromatography[108]) of this same fraction; it was chosen for chemical investigation by several of the groups. The functional groups (two methylamino, one hydroxyl, one terminal methylene, a cyclopropane ring) were included in an unusual saturated C_{23} pentacyclic nucleus. Nuclear magnetic resonance spectroscopy, a very new tool in 1960, confirmed the above functions, showing two protons on the cyclopropane ring and suggesting the environment of the alcohol as —CH_2—CHOH—CH— in a rigid ring.[109] The spectrum also showed two methyl groups linked to quaternary carbons and one to a tertiary center, but gave no further information on the enigmatic ring system.

Selenium dehydrogenation of the alkaloid gave a complex hydrocarbon mixture from which both anthracenoid and phenanthrenoid components were isolated.[109] This unexpected finding was originally interpreted in terms of the presence of an unusual and nonsteroidal ring system. The true nature of the reaction was eventually understood (see below) only shortly

*In a laudable simplification of nomenclature, the principal research groups agreed to use trivial names for the various carbon-oxygen-olefin skeletons of known *Buxus* alkaloids and letter suffixes for the various N-methyl group substitutions possible (all of which are known from natural sources). Suffix *A* denotes the full N,N′-tetramethyl pattern; *D*, symmetrical and *E, F*, unsymmetrical di-substitution: *I* the completely unsubstituted di-primary amine. The full suffix system was published in 1964[110] and may also be found in an excellent and recent complete review on *Buxus* alkaloids.[111] The latter is recommended reading for those who wish more detailed information on the known structures and isolation work.

before final publication of the structure and configuration of alkaloid A, renamed cyclobuxine-D.[109*]

2.1.1 Hofmann Degradation, Oxidation, Formation of a Conjugated Diosphenol When the monomethiodide of the fully methylated cyclobuxine-A was pyrolyzed, a conjugated diene system appeared in the product with nmr and uv (229.5 mμ) spectral characteristics of the system —CH=CH—C(—)=CH₂. This suggested the proximity of one of the amino functions to the terminal methylene.

The secondary alcohol of N,N′-diacylated dihydrocyclobuxines could be easily oxidized with mild reagents (including manganese dioxide, which normally oxidizes only allylic alcohols) to a ketone in a five-membered ring (1730 cm^{-1}). When the parent ketone was prepared by hydrogenolytic cleavage of the oxidized di-*p*-nitrobenzylcarbamate, it decomposed rapidly in basic solution with loss of methylamine. The observation that the product was a pair of isomeric α,β-unsaturated cyclopentanones proved to be the turning point in the structure elucidation by the Wisconsin workers (June of 1962). A reconsideration of all the available facts resulted in the suggestion of reaction series (71) to (73) and complete elaboration of the structural hypothesis. The methylated-steroid proposal was supported by an additional unexpected and informative reaction. The 4-ketone prepared by ozonolysis and hydrogenolysis of cyclobuxine-D N,N′-di-*p*-nitrobenzylcarbamate was treated with base, in hopes of forming an unsaturated ketone in ring A. Instead, the amine was air-oxidized and hydrolyzed. The product and its acetate showed ir and uv characteristics of a *diosphenol*, of a type already known in the cholesterol[112] and cevagenine (see above Section 1.6.1)[59] cases; however, with *additional conjugation*

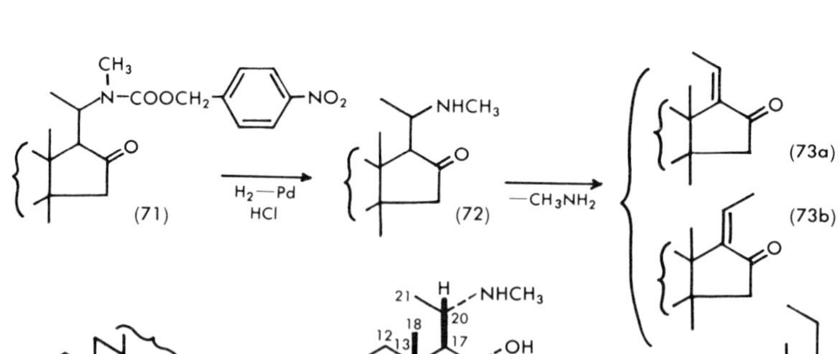

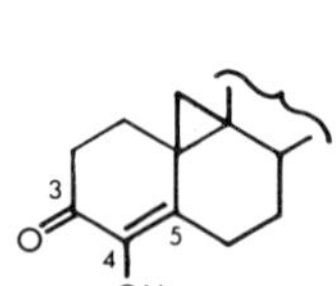

(74)

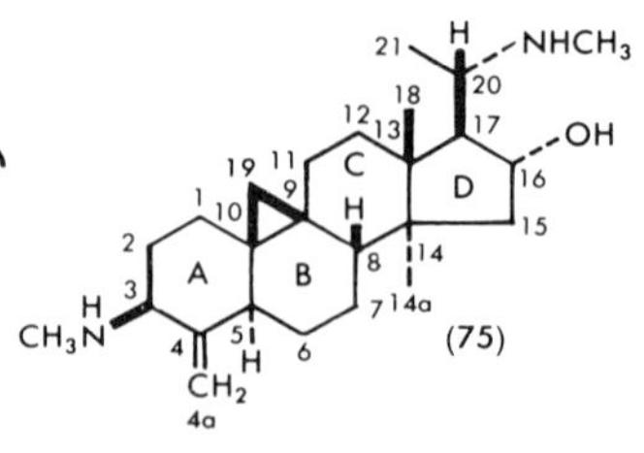

(75)

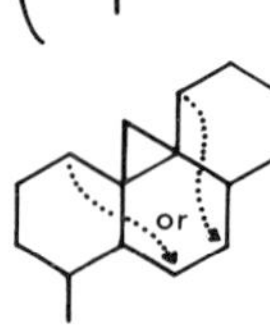

(76)

(bathochromic shift of 20–30 mμ) with the cyclopropane ring (74). The corresponding decyclized diosphenol and acetate showed the expected uv spectral positions.

The structure thus proposed for cyclobuxine-D [(75), without stereochemistry] was further confirmed by the failure of the double bond of decyclized compounds to migrate into conjugation with 16-ketones [indicating the intervention of a quaternary carbon, C(14)], and by a dehydrogenation of the acid-decyclized alkaloid. In this compound, the cyclopropane ring had been opened to give a new quaternary C-methyl group at C(10), visible in the nmr spectrum, and a double bond located primarily but not exclusively in the 9,11-position. Aromatization gave only phenanthrenoid compounds; thus, the cyclopropane ring was directly involved in the production of the anthracenes, possibly by one of the two rearrangements schematized in (76).[109]

2.1.2 Interrelation with Cycloeucalenol It now became necessary to fully prove the structure and determine the stereochemistry of cyclobuxine-D, through interrelation with a known steroid or triterpene. Cycloeucalenol (77) (from *Eucalyptus*) with a single methyl group at the 4-position, a cyclopropane ring in the desired location and a firmly established structure,[113] was a logical choice. Through the cooperation of T. J. King in Nottingham, a generous amount of cycloeucalenol was obtained for a side-chain degradation. Since Ruschig deamination[114] of dihydrocyclobuxine (78) to (79) proceeded smoothly (the reaction involved three eliminations and two hydrolyses in a single step), the saturated diketone (80) was chosen as a suitable interrelation compound. The first intermediate (81) from cycloeucalenol, already known,[113] was reproduced in 21% over-all

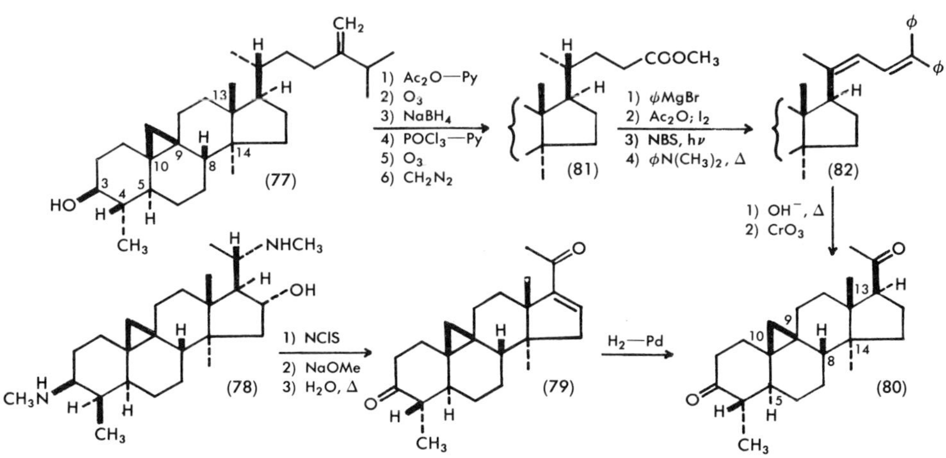

yield from cycloeucalenyl acetate. Meystre-Miescher degradation[115] of (81) nearly ended in disappointment due to the reaction of N-bromosuccinimide with the cyclopropane ring.[116] However, with carefully controlled conditions a moderate yield of crude diphenyl-butadiene (82) was obtained, which gave upon hydrolysis at C(3) and oxidation with chromium trioxide the diketone (80), identical with material from degradation of cyclobuxine-D; the total yield of the diketone from (81) was 18%. Thus, the structure of the first of the *Buxus* alkaloids was established, along with the stereochemistry at all of the ring junctions (5α, 8β, 9β, 10β, 13β, 14α). The stereochemistry at the remaining asymmetric centers (3, 16, 17, 20) was also ascertained and included in the initial publication[109]; it will be discussed in Section 2.3.

2.2 Further Structural Investigations: Cycloprotobuxine-C and Buxenine-G

Meanwhile, the acetone-soluble alkaloids were fractionated in several laboratories, yielding two principal components. One of these was evidently identical to the "Alkaloid L" originally isolated in impure form and studied by Friedrich.[107] It contained one methylamino and one dimethylamino group but no oxygen ($C_{27}H_{48}N_2$). Nmr spectroscopy confirmed that the cyclopropane ring was present, along with one tertiary and *four* quaternary C-methyl groups. Ruschig degradation of the chloramine gave a ketone showing a negative Cotton effect in the o.r.d. and no nmr signal for a methyl ketone; therefore, the methylamino group should be present at C(3) and not C(20). In independent interrelations, Calame and Arigoni[117] and Nakano and Hasegawa[118] showed that Alkaloid L (83) possessed the prototype triterpenoid skeleton; it was, therefore, renamed cycloproto-

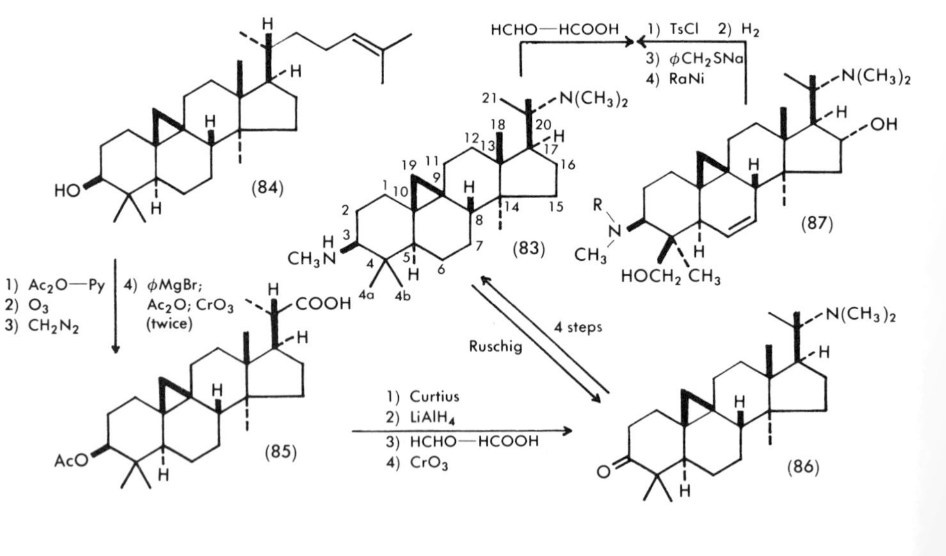

buxine-C. The Swiss workers synthesized (83) by a sidechain degradation of cycloartenol (84) (from *Artocarpus*, Jackfruit). Ozonolysis of the acetate and double Barbier-Wieland degradation gave the acid (85). Curtius degradation, LAH reduction of the resulting isocyanate, N-methylation and oxidation at C(3) gave the known Ruschig product (86), readily transformed to the natural alkaloid (83) through the oxime.[117] The Kyoto chemists removed the extra functions from cyclomicrophylline-A [(87), R = CH_3] (see Section 2.3) by tosylation, hydrogenation, displacement of the tosylates by sodium benzylmercaptide and desulfurization with Raney nickel, giving cycloprotobuxine-A, produced by N-methylation of (83) with formaldehyde-formic acid.[118]

2.2.1 Characterization of a Noncyclopropanoid Alkaloid The second principal component crystallized from acetone solution as an isopropylidene-imine, arising through spontaneous condensation of the solvent with a primary amino group of the alkaloid. This unusual behavior was observed in many *Buxus* alkaloids[119–122] and proved to be of considerable use in purification of individual bases that were noncrystalline in the free amino form. The isopropylidene-imine showed a mass spectral molecular ion at m/e 410 and had an analysis corresponding to $C_{28}H_{46}N_2$. One N-methyl group was visible in the nmr spectrum, along with one tertiary and four quaternary C-methyl groups, as in cycloprotobuxine-C, but no cyclopropyl protons could be detected. Hydrolysis of the imine with dilute HCl gave acetone (as the 2,4-DNP) and the pure parent amine, which resisted all attempts at crystallization. When the primary amino group was selectively chlorinated and the product degraded with alkali, a compound, which showed nmr signals for the N-methyl group and a methyl ketone, was obtained; therefore, the primary amino group was at position 20.

The most interesting feature of the alkaloid was the presence of a *heteroannular conjugated double bond system*, visible in the ultraviolet spectrum at 238, 247, and 256 mμ and carrying two protons (in the nmr, one a singlet and the other a broad quartet; six allylic protons were also present). The location of this system in the steroid nucleus proved to be very difficult. Both double bonds were easily reducible, and the uv spectrum implied that both were exocyclic to rings (calculated 245 mμ). A detailed study of the spectral data in Wisconsin led to the inspiration of structure (88) or (89), in which the cyclopropane had been opened to form not an angular methyl group, but rather a seven-membered ring and a double bond system satisfying all the physical data.[119] The same inspiration also visited Basel[120] and both groups proposed the structure, presently named buxenine-G,* almost simultaneously. A decision between the alternate

*We prefer buxenine to buxamine for the parent base, to avoid literature confusion with the antispasmodic buxamine = 2-hydroxy-3-aminobutyric acid.

structures was only possible after an x-ray crystallographic analysis of buxenine-G dihydroiodide,[103] which confirmed all elements of structure (88).

Buxenine-G is the prototype of a series of similar 9(10→ 19) *abeo*-pregnanes that include many variations in substitution at the 3, 4, 16 and 20-positions. One of these, buxpsiine (90) showed prominent peaks in the mass spectrum at m/e 71, 84, and 267 (91) consistent only with (and diagnostic for) the double-bond positions as shown.[123] The peak at m/e 84 was less intense than for normal 3-dimethylaminosteroids[124] owing to the more difficult cleavage of the vinylic 1-10 bond; the same phenomenon was observed in the spectra of cyclopropane-containing *Buxus* alkaloids.[125] Correspondingly, the intensity of the peak at m/e 71, arising from favored cleavage of the 1-2 bond, was generally increased to be the base peak in the spectra.

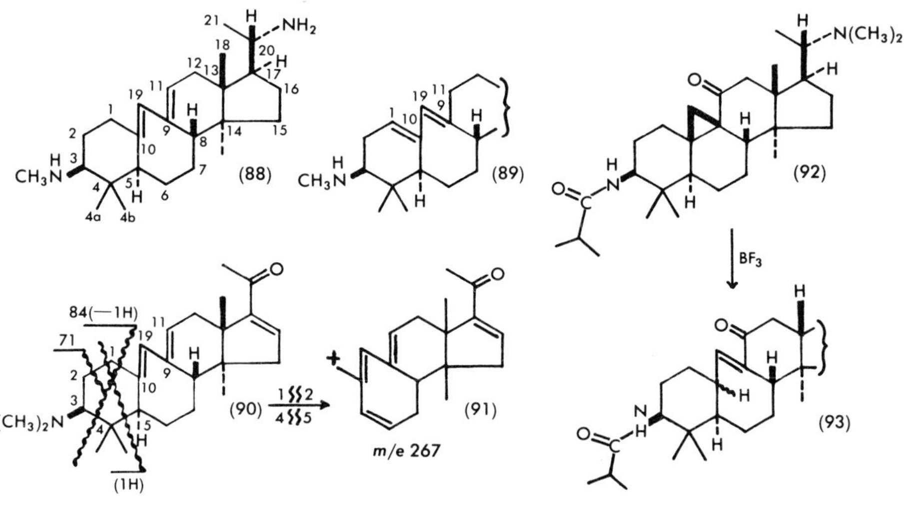

2.2.2 Syntheses of the Buxenine System; Cycloxobuxines and Buxidienines
Two unexpected entries into the 7-membered ring B system exemplified by buxenine-G have been reported. In the first of these,[126] N-isobutyryl-cycloxobuxine-F (92) (isolated from *Buxus balearica* and orginally called baleabuxine[127]) was treated with boron trifluoride; the product had a signal in the nmr spectrum for *one vinyl proton β to the ketone* (δ 6.51) and, therefore, was best represented as part structure (93).

The structure of N-isobutyrylcycloxobuxine-F had been proven[127] by reductive decyclization with sodium in propanol,[128] followed by cleavage of the exceptionally hydrolysis-resistant 3-amide with lithium in ethylamine, N-methylation, and oxidation. The major product was the fully

N-methylated 11-ketone (94) with an angular methyl group in C(10). Decyclization and chromic acid oxidation of cycloprotobuxine-A [N-methyl-(83)] gave a mixture of the corresponding 9-ene-12-one and the 8-ene-7,-11-dione. Selective Wolff–Kishner reduction of the latter, lithium-ammonia thermodynamic reduction of the conjugated 8-ene-11-one, and reoxidation at C(11) gave (94).

A further interrelation of cycloxobuxidine-F (95) with dihydrocyclomicrophylline-F [see Structure (87)] was achieved by energetic LAH reduction, giving hydrogenolytic removal of the cyclopropylcarbinylic 11-alcohol without opening of the three-membered ring.[126,127,131]

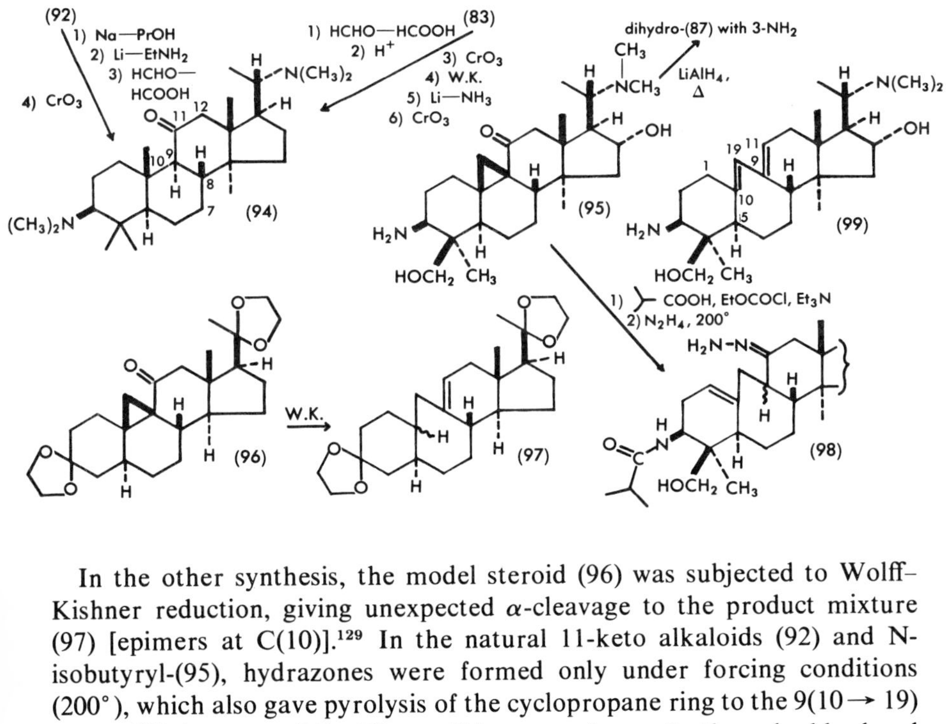

In the other synthesis, the model steroid (96) was subjected to Wolff–Kishner reduction, giving unexpected α-cleavage to the product mixture (97) [epimers at C(10)].[129] In the natural 11-keto alkaloids (92) and N-isobutyryl-(95), hydrazones were formed only under forcing conditions (200°), which also gave pyrolysis of the cyclopropane ring to the 9(10→ 19) *abeo*-$\Delta^{1(10)}$-isomers (98). These did not undergo further double bond migration upon completion of the Wolff-Kishner reduction.[130,131]

An elegant partial synthesis of the buxenine system itself has also been published by the French group: the 11β-alcohol resulting from mild LAH reduction of cycloxobuxidine-F (95) was treated with 30% sulfuric acid in dioxane; the product was identical with the natural base buxidienine-F (99).[131] Additional products were the nonconjugated $\Delta^{1(10),9(11)}$-and $\Delta^{5(10),9(11)}$-dienes.[130]

2.3 Chemistry and Stereochemistry in the Buxus Alkaloids: Ring A

The configuration of the amino-group at C(3) was initially assigned as β due to the weak negative Cotton effect (comparable to that of cholestane-4-one[132]) in the o.r.d. of ketones resulting from ozonolysis of a 4-methylene group.[109,133] A 3α(axial) substituent would surely modify the amplitude drastically. Furthermore, Hofmann degradation of 3-monomethiodides failed totally (except when the presence of the 4-methylene encouraged the formation of a conjugated system),[109,134] as had been observed for other 3β (equatorial) steroidal amines.[135] A 3α-amino group, similar to that known in many *Holarrhena* steroid alkaloids, has been suggested but not proven in cyclokoreanine-B (100).[136] Three 3-ketones have also been isolated from nature.[137,154]

The presence of the 4β-hydroxymethyl group in the cyclomicrophyllines (87) was indicated by its many unusual interactions with the 3β-amino group. The nmr signal positions for CH_2OH (δ 3.70) and CH_2OAc (δ 4.14)[134] indicated the axial orientation.[138] The hydroxyl and amino groups demonstrated strong intramolecular hydrogen-bonding[134]; the von Braun degradation of cyclomicrophylline-A [(87), R = CH_3] was only possible when the hydroxyl was acetylated.[134] An acetyl group on the 3-nitrogen migrated readily to the hydroxyl group.[134,139,140] Furthermore, 3-amides, excessively resistant to hydrolysis in normal alkaloids (see Section 2.2.2), were cleaved very easily in alkali when the 4β-hydroxymethyl group was present.[134] This phenomenon, indicating intramolecular facilitation of hydrolysis (101),[141] also operated in reverse, permitting the 3N-catalyzed cleavage of the O-acetate by simple refluxing in neutral methanol.[134]

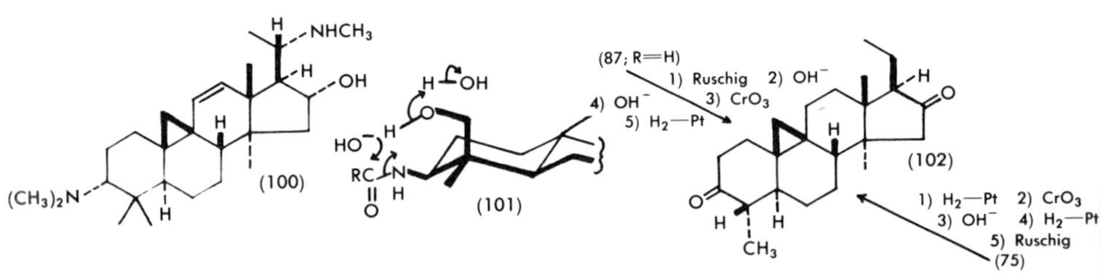

Transformation of the 3-amino group of cyclomicrophylline-C [(87), R = H] to a ketone by Ruschig deamination gave a product that readily lost formaldehyde in 1% alkali (retroaldol) to give the 4α-monomethyl derivative. This reaction permitted the original interrelation of the cyclomicrophyllines with cyclobuxine-D (75) as shown (102).[134]

The 3-amino group could be selectively methylated (before the 20N) with methyl iodide at low temperature.[142] Formaldehyde-formic acid treatment usually led to complete N-methylation, but in dihydrocyclomicrophylline-F or -C the product was the natural cyclobuxoxazine-A [(103), R = CH_3, R′ = H].[126,140,143] This type of alkaloid, arising from biological condensation of formaldehyde (R′ = H) or acetaldehyde (R′ = CH_3) with the 3β-nitrogen and the 4β-hydroxymethyl, had been isolated from two *Buxus* species.[140,143] Natural cyclobuxoxazine-C [(103), R = R′ = H] was synthesized by treatment of dihydrocyclomicrophylline-F with formaldehyde, giving a methoxymethylamine, which was transformed to the oxazine with activated alumina.[140] Cyclomethoxazine-A[(103); R = R′ = CH_3] (the natural alkaloid is B, with 20-$NHCH_3$) was produced by analogous treatment of dihydrocyclomicrophylline-C with acetaldehyde in acetic acid.[143] The natural and synthetic oxazines were reductively cleaved by LAH to the corresponding cyclomicrophyllines.[140,143]

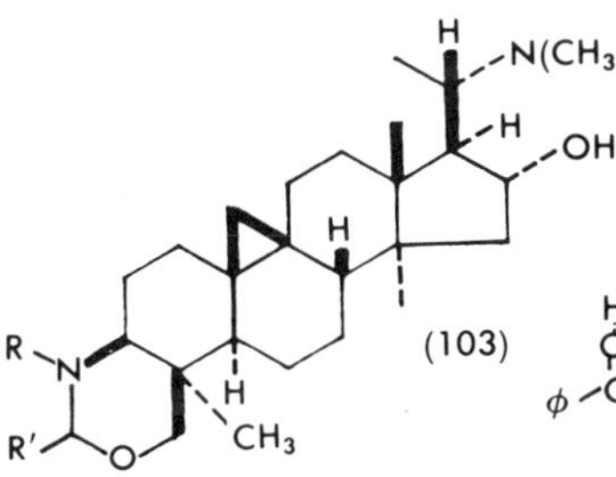
(103)

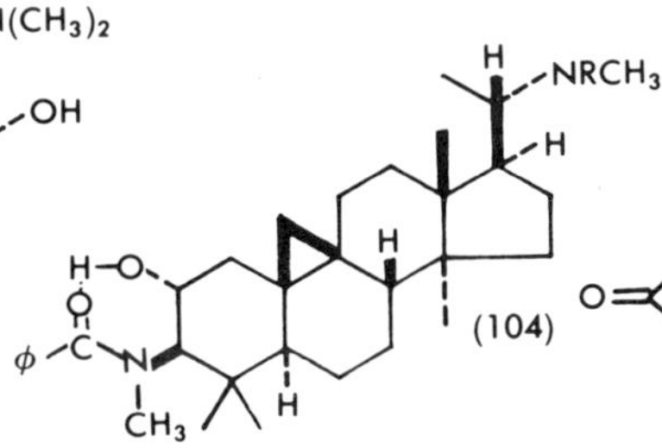
(104)

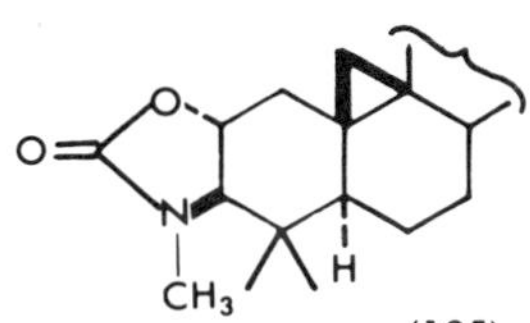
(105)

In the cycloprotobuxolines [(104), R = H or Me],[139] a secondary hydroxyl was present whose acetate showed an nmr signal for *CH*OAc of half-width 20 Hz, indicating that the alcohol was equatorial (the signal is spread due to the large vicinal *trans*-diaxial coupling constants). The benzamide present at C(3) was easily hydrolyzed in alkali (assisted by the neighboring hydroxyl); the product consumed one mole of periodate, and gave with phosgene an oxazolidone (105). Oxidation of (104) to a 2-ketone and reduction with LAH gave back (104) and also the 2β-epimer with a much narrower *CH*OAc signal in the nmr spectrum.[139]

Cyclobuxiviridine (106)[137] showed strong cyclopropyl conjugation with the unsaturated ketone (λ_{max}269 mμ; normal steroid Δ^1-3-ketone, 230 mμ) and two vinyl protons on a *cis*-double bond conjugated to the ketone (δ 5.96, 6.81); catalytic hydrogenation gave the known Ruschig product (86).[137]

The natural 4-monomethyl compounds were originally assigned as β[121] (and the products of methylene hydrogenation as α) through a combination of molecular rotation relationships, nmr spectra, Ruschig degradation results and hydrogenation behavior. It was shown that catalytic hydrogenation of acetylated derivatives of cyclobuxine-D gave an isomer at C(4) different from that produced by hydrogenation of the parent alkamine[121]; both products were stereochemically homogeneous. More recently, convincing nmr evidence has been advanced[137] for the reversal of the original assignment. The signals for the cyclopropyl protons appeared at appreciably lower field (average δ 0.7, 0.4) in the synthetic 4β-methyl and *all* 4,4-dimethyl compounds than in the natural and synthetic 4α-methyl isomers, including amines derived from cycloeucalenol (average δ 0.5, 0.2). The molecular rotation differences between many natural and synthetic 4α- and 4β-isomers included both positive and negative values.[137] The Ruschig degradation proceeded with isomerization at C(4), the same product being obtained from both isomers.[144] The exclusive β-face hydrogenation of the acetates was still not explained.

2.3.1 Rings B and C and the Cyclopropane Ring Much of the chemistry of the central part of the molecule has already been discussed (Sections 2.1.1 and 2.2.2). The cyclopropane ring showed interesting interactions with functions in rings A, B, and C, as indicated by its signals in the nmr spectra. In the unsaturated alkaloids (106) (Δ^1)[137] and (100) (Δ^{11}),[136] as well as in 11β-alcohols,[131,139] only half of the AB quartet was visible outside the "methylene envelope." In the 11-ketones,[126,127,139,145,131] no cyclopropyl protons were visible in the spectra. Slight interaction across the twist B ring between the cyclopropane and the 6,7 double bond of cyclomicrophyllines (87) not only shifted half of the cyclopropyl AB quartet to very high field (δ −0.2)[134,146,147] but also increased the intensity of the double bond absorption in the ultraviolet (λ_{max}204 mμ, ϵ 5800).[134]

Decyclization of the three-membered ring normally proceeded in hydrochloric acid to give an angular methyl group at C(10). In the conjugated olefin (100), the ring was opened upon attempted formic acid-formaldehyde methylation.[136] Vigorous decyclization (HCl, 100°) of cyclomicrophyllines gave the conjugated system (107). When deuterative decyclization was performed with DCl in D_2O, mild conditions led to the introduction of

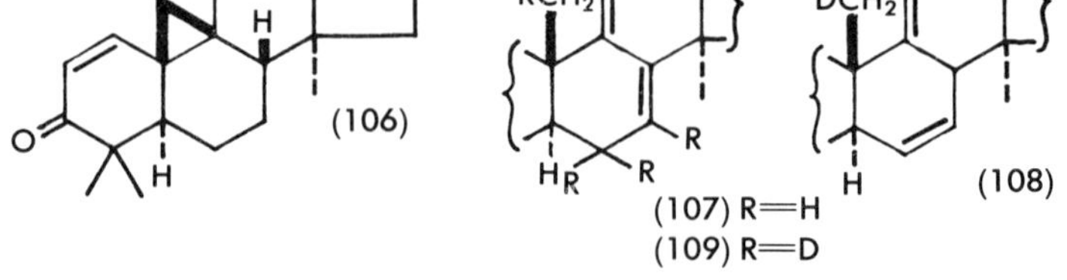

(106)

(107) R=H
(109) R=D

(108)

two atoms of deuterium (108) and vigorous conditions, seven atoms (109).[134]

Both the 11-ketones[126,127,139,131] and the 11-olefin (100)[136] showed conjugated systems in the uv spectra (λ_{max}218, ϵ 6350 and λ_{max}213, ϵ 14,800); the former also showed a characteristic ir absorption at 833 cm^{-1}.[139] Although the 11-ketone was smoothly hydrogenolyzed by LAH in boiling dioxan (Section 2.2.2), the product of milder reduction (an 11β-alcohol) did not undergo similar hydrogenolysis[139]; thus, the reaction did not proceed stepwise through the intermediate alcohol. A single 11-alcohol has been found in nature.[147a]

2.3.2 Ring D and the Sidechain The configuration of the 16α-hydroxyl was best indicated by its acylation, which invariably led to a negative molecular rotation increment.[109,134] Borohydride reduction of 16-ketones gave unstable alcohols giving a positive molecular rotation increment upon acetylation.[109] The saturated 16-ketones showed a very strong negative Cotton effect in the o.r.d., confirming the β-orientation of the sidechain at C(17). Although in 16-hydroxy-20-ketones the elimination of the hydroxyl occurred in both acidic and basic media, [109,142] it was not facile or quantitative[147]; a number of 20-ketones were isolated with the 16-hydroxyl group still intact, even after alumina chromatography.[133,142,143,148]

The nature of the sidechain at C(17) was easily determined by mass spectrometry; strong peaks appeared corresponding to the loss of the sidechain[125,149] and, if it were nitrogenated, for the resulting fragment.

All the known natural cyclopentenones (73) possess the *cis*-configuration (73a).[137,145,150] Whether this is due to biogenetic selectivity, or to specific elimination from natural 20-amino-16-ketones[151] upon alumina chromatography,[134,137] cannot be stated at present.

Selective methylation of the 20-nitrogen (before the 3N) of dihydrocyclobuxine-D was achieved by treatment with formaldehyde in methanolic acetic acid and platinum catalyst under hydrogen.[109] Thus, reactions are available for the selective methylation of either basic nitrogen (see Section 2.3). Whether the 16α-OH participates in the selective 20-methylation has not been established.

Transformation of N-isobutyryl-(95) to the 20N-oxide and pyrolysis gave the corresponding $\Delta^{17(20)}$-pregnene derivative without affecting the 16α-hydroxyl.[151a]

Three separate experiments have shown the α-configuration (well in accord with biogenetic precedent; see Chapter 19) for the 20-amino group. In the original work on cyclobuxine-D,[109] exceptional facilitation of hydrolysis of amides at C(20) was observed, which could best be explained in terms of participation by the 16α-hydroxyl. Repulsion between the 18- and 21-methyl groups forces the 20α-amino and 16α-hydroxyl functions to adopt approximately parallel conformations.[152] The corresponding 16-ketone did not show such facilitated hydrolysis.[109]

In buxenine-G, the free 20-amino group was condensed with salicylaldehyde in methanol. The resulting *o*-hydroxybenzylideneimine showed a strong positive Cotton effect in the o.r.d.,[119] indicating α-orientation for the amino group.[153] Treatment of 20-amino-16-ketones with alumina, or pyrolytic *cis*-elimination of the amino-group, gave exclusively *cis*-cyclopentenones (73a), requiring a 20α-configuration (110).[134]

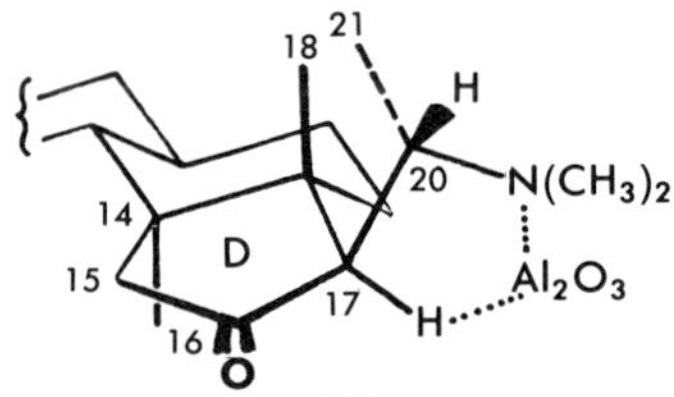

(110)

2.4 Summary and perspective

All *Buxus* alkaloids isolated to date, with the exception of buxandonine,[154] buxaltine,[147a] and irehine,[139,149] possess a 14α-methyl group and at least one methyl group at C(4), together with a degraded sidechain from the typical triterpene; the first two mentioned are among the very few known natural 14α-monomethyl steroids. The alkaloids thus represent a unique steroid-terpene class that should present a fruitful field for biogenetic investigations. The 9,19-cyclo-system may be a necessary and, as yet, little recognized intermediate in all tetracyclic triterpene biogenesis[155]; studies with *Buxus* may help to cast light on this possibility.

The problems of bacterial resistence and the complexity of disease mechanisms require the continued introduction of new drugs in the control of various common maladies. Thus, the broad spectrum of pharmacological activity known for *Buxus* alkaloids (see Section 2) suggests that they may yet find profitable application in pharmaceutical preparations.

ACKNOWLEDGMENTS

The author of this chapter is deeply grateful to Drs. S. W. Pelletier, S. M. Kupchan, T. Masamune, Y. Sato, T. Nakano, R. Goutarel, F. Khuong-Huu, and P. Longevialle for extensive technical revision, and to his wife Kay F. Brown for complete rhetorical revision of this manuscript on several separate occasions during its writing; to Mrs. Maria Luiza Conde Miranda for the complete typing of the several successive versions of the chapter; and to the National Science Foundation (U.S.A.), Grant No. GB 5389 X, and the Brazilian Conselho Nacional de Pesquisas and Banco Nacional de Desenvolvimento Econômico, for financial support during the period 1967–1969.

REFERENCES

1. S. M. Kupchan and A. W. By, *Alkaloids* **10,** 193–285 (1968).
2. S. M. Kupchan, J. H. Zimmerman, and A. Afonso, *Lloydia* **24,** 1 (1961).
3. S. M. Kupchan and W. E. Flacke, in "Antihypertensive Agents," pp. 429–458 (E. Schlittler, Ed.), Academic Press, Inc., New York, N.Y., 1967.
4. L. F. Fieser and M. Fieser, "Steroids," pp. 867–895, Reinhold Publ. Co., New York, N.Y., 1959.
5. H.-G. Boit, "Ergebnisse der Alkaloid-Chemie bis 1960," pp. 798–832, Akademie Verlag, Berlin, Germany, 1961.
6. W. A. Jacobs and L. C. Craig, *J. Biol. Chem.* **160,** 555 (1945).
7. W. A. Jacobs and C. F. Huebner, *J. Biol. Chem.* **170,** 635 (1947); W. A. Jacobs and Y. Sato, *J. Biol. Chem.* **175,** 57 (1948).
8. W. A. Jacobs and Y. Sato, *J. Biol. Chem.* **191,** 71 (1951).
9. W. A. Jacobs, L. C. Craig, and G. I. Lavin, *J. Biol. Chem.* **141,** 51 (1941).
10. W. A. Jacobs and Y. Sato, *J. Biol. Chem.* **181,** 55 (1949).
11. O. Wintersteiner, M. Moore, J. Fried, and B. M. Iselin, *Proc. Natl. Acad. Sci. U.S.* **37,** 333 (1951).
12. O. Wintersteiner and M. Moore, *J. Amer. Chem. Soc.* **75,** 4938 (1953).
13. Ch. Tamm and O. Wintersteiner, *J. Amer. Chem. Soc.* **74,** 3842 (1952); O. Wintersteiner and N. Hosansky, *J. Amer. Chem. Soc.* **74,** 4474 (1952).
14. O. Wintersteiner, M. Moore, and N. Hosansky, *J. Amer. Chem. Soc.* **75,** 2781 (1953).
15. J. Fried, O. Wintersteiner, M. Moore, B. M. Iselin, and A. Klingsberg, *J. Amer. Chem. Soc.* **73,** 2970 (1951); J. Fried and A. Klingsburg, *J. Amer. Chem. Soc.* **75,** 4929 (1953).
16. J. E. Herz and J. Fried, *J. Amer. Chem. Soc.* **76,** 5621 (1954).
17. L. Keller, Ch. Tamm, and T. Reichstein, *Helv. Chim. Acta* **41,** 1633 (1958).
18. R. Anliker, H. Heusser, and O. Jeger, *Helv. Chim. Acta* **35,** 838 (1952).
19. B. M. Iselin and O. Wintersteiner, *J. Amer. Chem. Soc.* **77,** 5318 (1955).
20. B. M. Iselin, M. Moore, and O. Wintersteiner, *J. Amer. Chem. Soc.* **78,** 403 (1956).
21. T. Masamune, Y. Mori, M. Takasugi, and A. Murai, *Tetrahedron Letters* 913 (1964); *Bull. Chem. Soc. Japan* **38,** 1374 (1965).
22. N. L. Hosansky and O. Wintersteiner, *J. Amer. Chem. Soc.* **78,** 3126 (1956).
23. R. W. Frank and W. S. Johnson, *Tetrahedron Letters* 545 (1963); R. W. Frank, G. P. Rizzi, and W. S. Johnson, *Steroids* **4,** 463 (1964); T. Masamune, M. Takasugi, and Y. Mori, *Tetrahedron Letters* 489 (1965); T Masamune, K. Kobayashi, M. Takasugi, Y. Mori, and A. Murai, *Tetrahedron* **24,** 3461 (1968).
24. R. Hirschmann, C. S. Snoddy, Jr., and N. L. Wendler, *J. Amer. Chem. Soc.* **74,** 2693 (1952); C. F. Hiskey, R. Hirschmann, and N. L. Wendler, *J. Amer. Chem. Soc.* **75,** 5135 (1953); R. Hirschmann, C. S. Snoddy, Jr., C. F. Hiskey, and N. L. Wendler, *J. Amer. Chem. Soc.* **76,** 4013 (1954); J. Elks, G. H. Phillipps, D. A. H. Taylor, and L. J. Wyman, *J. Chem. Soc.* 1739 (1954); R. Anliker, O. Rohr, and H. Heusser, *Helv. Chim. Acta* **38,** 1171 (1955); See also J. M. Coxon, M. P. Hartshorn, and D. M. Kirk, *Tetrahedron Letters* 119 (1965).

25. H. Mitsuhashi and K. Shibata, *Tetrahedron Letters* 2281 (1964); *Chem. Pharm. Bull. (Tokyo)* **16,** 814 (1968).
26. G. Bozzato, K. Schaffner, and O. Jeger, *Chimia (Aarau)* **20,** 114 (1966).
27. H. Suginome, M. Murakami, and T. Masamune, *Chem. Commun.* 343, (1966); *Bull. Chem. Soc. Japan* **41,** 468 (1968).
28. J. W. Scott, L. J. Durham, H. A. P. de Jongh, U. Burckhardt, and W. S. Johnson, *Tetrahedron Letters* 2381 (1967).
29. O. Wintersteiner, M. Moore, and B. M. Iselin, *J. Amer. Chem. Soc.* **76,** 5609 (1954).
30. O. Wintersteiner and M. Moore, *J. Amer. Chem. Soc.* **78,** 6193 (1956).
31. H. Budzikiewicz, *Tetrahedron* **20,** 2267 (1964).
32. J. Tomko and A. Vassová, *Chem. Zvesti* **18,** 266 (1964); J. Tomko and S. Bauer, *Coll. Czech. Chem. Commun.* **29,** 2570 (1964); T. Masamune, I. Yamazaki, and M. Takasugi, *Bull. Chem. Soc. Japan* **39,** 1090 (1966).
33. T. Masamune, N. Sato, K. Kobayashi, I. Yamazaki, and Y. Mori, *Tetrahedron* **23,** 1591 (1967).
34. H. Suginome, N. Sato, and T. Masamune, *Tetrahedron Letters* 1559 (1967); see, however, T. Masamune, A. Murai, H. Ono, K. Orito, and H. Suginome, *Tetrahedron Letters* 255 (1969), and T. Masamune, K. Orito, and A. Murai, *Tetrahedron Letters* 251 (1969).
35. O. Wintersteiner and M. Moore, *Tetrahedron Letters* 795 (1962); *J. Org. Chem.* **29,** 262 (1964).
36. T. Masamune, M. Takasugi, H. Suzuki, S. Kawahara, M. Gohda, and T. Irie, *Bull. Chem. Soc. Japan* **35,** 1749 (1962); *J. Org. Chem.* **29,** 2282 (1964).
37. W. G. Dauben, W. W. Epstein, M. Tanabe, and B. Weinstein, *J. Org. Chem.* **28,** 293 (1963).
38. E. A. Braude, E. R. H. Jones, F. Sondheimer, and J. B. Toogood, *J. Chem. Soc.* 607 (1947); E. R. H. Jones, G. H. Mansfield, and J. Meinwald, *J. Org. Chem.* **21,** 375 (1956).
39. O. Wintersteiner and M. Moore, *J. Org. Chem.* **29,** 270 (1964); **30,** 528 (1965).
40. T. Masamune, M. Takasugi, A. Murai, and K. Kobayashi, *J. Amer. Chem. Soc.* **89,** 4521 (1967).
41. W. S. Johnson, H. A. P. de Jongh, C. E. Coverdale, J. W. Scott, and U. Burckhardt, *J. Amer. Chem. Soc.* **89,** 4523 (1967).
42. W. S. Johnson, J. M. Cox, D. W. Graham, and H. W. Whitlock, Jr., *J. Amer. Chem. Soc.* **89,** 4524 (1967); W. S. Johnson, N. Cohen, E. R. Habicht, Jr., D. P. G. Hamon, G. P. Rizzi, and D. J. Faulkner, *Tetrahedron Letters* 2829 (1968).
43. K. Tori, K. Kitahonoki, Y. Takano, H. Tanida, and T. Tsuji, *Tetrahedron Letters* 559 (1964).
44. G. N. Reeke, Jr., R. L. Vincent and W. N. Lipscomb, *J. Amer. Chem. Soc.* **90,** 1663 (1968).
45. S. M. Kupchan and M. I. Suffness, *J. Amer. Chem. Soc.* **90,** 2730 (1968).
45a. J. P. Kutney, J. Cable, W. A. F. Gladstone, H. W. Hanssen, E. J. Torupka, and W. D. C. Warnock, *J. Amer. Chem. Soc.* **90,** 5332 (1968).
46. B. K. Blount, *J. Chem. Soc.* 122 (1935); 414 (1936); W. A. Jacobs and L. C. Craig, *J. Biol. Chem.* **129,** 79 (1939); L. C. Craig and W. A. Jacobs, *J. Biol. Chem.* **139,** 263 (1941); **139,** 293 (1941); L. C. Craig, W. A. Jacobs, and G. I. Lavin, *J. Biol. Chem.* **139,** 277 (1941).
47. W. A. Jacobs and S. W. Pelletier, *J. Org. Chem.* **18,** 765 (1953); S. W. Pelletier and W. A. Jacobs, *J. Amer. Chem. Soc.* **76,** 2028 (1954); **78,** 1914 (1956).
48. T.-T. Chu and J.-Y. Loh, *Hua Hsueh Hsueh Pao* **21,** 227 (1955) [*Chem. Abstr.*

51, 444 (1957)]; T.-T. Chu, W.-K. Hwang, and J.-Y. Loh, *Hua Hsueh Hsueh Pao* **21,** 232 (1955) [*Chem. Abstr.* **51,** 444 (1957)]; T.-C. Chu, J.-Y. Loh, W.-K. Huang, F.-C. Ho, and C.-C. Liu, *Hua Hsueh Hsueh Pao* **24,** 377 (1958) [*Chem. Abstr.* **53,** 20117 (1959)]; T.-C. Chu, J.-Y. Loh, W.-K. Huang, and F.-C. Ho, *K'o Hsueh T'ung Pao* 371 (1957) [Chem. *Abstr.* **55,** 27384 (1961)].

49. H. Morimoto and S. Kimata, *Chem. Pharm. Bull. (Tokyo)* **8,** 302, 871 (1960); F.-C. Ho, J.-Y. Loh, C.-C. Liu, and T.-C. Chu, *K'o Hsueh T'ung Pao* 372 (1957) [*Chem. Abstr.* **55,** 27384 (1961)].

50. S. Itô, M. Kato, K. Shibata, and T. Nozoe, *Chem. Pharm. Bull. (Tokyo)* **9,** 253 (1961); **11,** 1337 (1963).

50a. S. Itô, Y. Fukazawa, T. Okuda, and Y. Iitaka, *Tetrahedron Letters* 5373 (1968).

50b. R. N. Nuriddinov and S. Y. Yunusov, *Khim. Prir. Soedin.* **4,** 260 (1968) [*Chem. Abstr.* **70,** 68589 (1969)].

50c. R. N. Nuriddinov and S. Y. Yunsov, *Khim. Prir. Soedin.* **4,** 258 (1968) [*Chem. Abstr.* **70,** 68585 (1969)]; R. N. Nuriddinov, B. Babaev, and S. Y. Yunusov, *Khim. Prir. Soedin.* **4,** 261 (1968) [Chem. Abstr. **70,** 68588 (1969)].

51. J. Tomko and A. Vassová, *Pharmazie* **20,** 385 (1965); J. Tomko, Z. Votický, H. Budzikiewicz, and L. J. Durham, *Coll. Czech. Chem. Commun.* **30,** 3320 (1965); S. Itô, T. Ogino, and J. Tomko, *Coll. Czech. Chem. Commun.* **33,** 4429 (1968).

52. L. C. Craig and W. A. Jacobs, *J. Amer. Chem. Soc.* **61,** 2252 (1939); *J. Biol. Chem.* **134,** 123 (1940); **141,** 253 (1941).

53. K. J. Morgan and J. A. Barltrop, *Quart. Rev. London* **12,** 34 (1958) (summary of the unpublished work in Barton's group).

54. F. Gautschi, O. Jeger, V. Prelog, and R. B. Woodward, *Helv. Chim. Acta* **37,** 2280 (1954); **38,** 296 (1955).

55. O. Jeger, R. Mirza, V. Prelog, Ch. Vogel, and R. B. Woodward, *Helv. Chim. Acta* **37,** 2295 (1954).

56. O. Jeger, V. Prelog, E. Sundt, and R. B. Woodward, *Helv. Chim. Acta* **37,** 2302 (1954).

57. M. Freund and H. P. Schwartz, *Chem. Ber.* **32,** 800 (1899); D. H. R. Barton and J. F. Eastham, *J. Chem. Soc.* 424 (1953); D. H. R. Barton and C. J. W. Brooks, *Chem. Ind. (London)* **51,** 1366 (1953); W. A. Jacobs and L. C. Craig, *J. Biol. Chem.* **125,** 625 (1938).

58. A. Stoll and E. Seebeck, *Helv. Chim. Acta* **35,** 1270, 1942 (1952); H. Auterhoff, *Arch. Pharm.* **286,** 319 (1953).

59. E. Sundt, O. Jeger, and V. Prelog, *Chem. Ind. (London)* 1365 (1953).

60. A. Stoll, D. Stauffacher, and E. Seebeck, *Helv. Chim. Acta* **36,** 2027 (1953).

61. S. W. Pelletier and W. A. Jacobs, *J. Amer. Chem. Soc.* **75,** 3248 (1953); A. Stoll and E. Seebeck, *Helv. Chim. Acta* **36,** 189 (1953); S. M. Kupchan, D. Lavie, C. V. Deliwala, and B. Y. A. Andoh, *J. Amer. Chem. Soc.* **75,** 5519 (1953).

62. Z. J. Vejdělek, K. Macek, and B. Buděšinsky, *Chem. Listy* **50,** 603 (1956).

63. S. M. Kupchan and D. Lavie, *J. Amer. Chem. Soc.* **76,** 314 (1954); **77,** 683 (1955).

64. D. H. R. Barton, O. Jeger, V. Prelog, and R. B. Woodward, *Experientia* **10,** 81 (1954).

65. S. M. Kupchan, *J. Amer. Chem. Soc.* **77,** 686 (1955).

66. D. H. R. Barton, C. J. W. Brooks, and J. S. Fawcett, *J. Chem. Soc.* 2137 (1954).

67. D. H. R. Barton, C. J. W. Brooks, and P. de Mayo, *J. Chem. Soc.* 3950 (1954).

68. S. M. Kupchan and W. S. Johnson, *J. Amer. Chem. Soc.* **78,** 3864 (1956); S. M. Kupchan, W. S. Johnson, and S. Rajagopalan, *J. Amer. Chem. Soc.* **80,** 1769 (1958); *Tetrahedron* **7,** 47 (1959).

69. M. V. Mijovic, E. Sundt, E. Kyburz, O. Jeger, and V. Prelog, *Helv. Chim. Acta* **38,** 231 (1955).
70. W. J. Rosenfelder, *J. Chem. Soc.* 2638 (1954).
71. H. B. Henbest and B. J. Lovell, *J. Chem. Soc.* 1965 (1957); S. M. Kupchan, S. P. Eriksen, and M. Friedman, *J. Amer. Chem. Soc.* **84,** 4159 (1962); **88,** 343 (1966).
72. S. M. Kupchan, S. P. Eriksen, and Y.-T. Shen, *J. Amer. Chem. Soc.* **85,** 350 (1963); S. M. Kupchan, S. P. Eriksen, and Y.-T. S. Liang, *J. Amer. Chem. Soc.* **88,** 347 (1966).
73. M. L. Bender, *Chem. Rev.* **60,** 53 (1960); M. L. Bender, G. R. Schonbaum, G. A. Hamilton, and B. Zerner, *J. Amer. Chem. Soc.* **83,** 1255 (1961); M. L. Bender and F. J. Kezdy, *J. Amer. Chem. Soc.* **86,** 3704 (1964); R. M. Krupka and K. J. Lardler, *J. Amer. Chem. Soc.* **83,** 1458 (1961).
74. W. A. Jacobs and L. C. Craig, *J. Biol. Chem.* **125,** 625 (1938).
75. S. M. Kupchan and T. Masamune, *Chem. Ind. (London)* 632 (1959).
76. W. T. Eeles, *Tetrahedron Letters* 24 (1960).
77. B. Shimizu, *J. Pharm. Soc. Japan* **78,** 444 (1958); **79,** 993 (1959).
78. T. Tsukamoto and A. Yagi, *J. Pharm. Soc. Japan* **79,** 1102 (1959); A. Yagi and T. Kawasaki, *Chem. Pharm. Bull. (Tokyo)* **10,** 519 (1962).
79. H. Auterhoff, *Arch. Pharm.* **288,** 549 (1955).
80. H. Auterhoff and H. Möhrle, *Arch Pharm.* **291,** 288 (1958).
81. H. Möhrle and H. Auterhoff, *Arch. Pharm.* **292,** 337 (1959); S. M. Kupchan, A. D. J. Balon, and E. Fujita, *J. Org. Chem.* **27,** 3103 (1962).
82. H. Möhrle and H. Auterhoff, *Arch. Pharm.* **293,** 813 (1960).
83. H. Auterhoff, *Arch. Pharm.* **286,** 69 (1953); **287,** 380 (1954).
84. S. Itô, J. B. Stothers, and S. M. Kupchan, *Tetrahedron* **20,** 913 (1964).
85. S. M. Kupchan, N. Gruenfeld, and N. Katsui, *J. Med. Pharm. Chem.* **5,** 690 (1962).
86. L. C. Craig and W. A. Jacobs, *J. Biol. Chem.* **148,** 57 (1943); **149,** 451 (1943); H. Jaffe and W. A. Jacobs, *J. Biol. Chem.* **193,** 325 (1951); S. W. Pelletier and W. A. Jacobs, *J. Amer. Chem. Soc.* **75,** 3248 (1953); S. M. Kupchan, M. Fieser, C. R. Narayanan, L. F. Fieser, and J. Fried, *J. Amer. Chem. Soc.* **76,** 1200, 5269 (1954); **77,** 5896 (1955).
87. S. M. Kupchan and C. R. Narayanan, *Chem. Ind. (London)* **251,** 1092 (1956).
88. S. M. Kupchan and C. R. Narayanan, *J. Amer. Chem. Soc.* **81,** 1913 (1959).
89. S. M. Kupchan, *J. Amer. Chem. Soc.* **78,** 3346 (1956); **81,** 1925 (1959).
90. F. W. Heyl and M. E. Herr, *J. Amer. Chem. Soc.* **71,** 1751 (1949); S. M. Kupchan and C. V. Deliwala, *J. Amer. Chem. Soc.* **75,** 1025 (1953); S. M. Kupchan, D. Lavie, and R. D. Zonis, *J. Amer. Chem. Soc.* **77,** 689 (1955).
91. M. L. Wolfrom and J. M. Bobbitt, *J. Amer. Chem. Soc.* **78,** 2489 (1956).
92. L. C. Craig and W. A. Jacobs, *J. Biol. Chem.* **143,** 427 (1942); W. A. Jacobs and L. C. Craig, *J. Biol. Chem.* **149,** 271 (1943); H. Auterhoff and F. Günther, *Arch. Pharm.* **288,** 455 (1955).
93. S. M. Kupchan, M. Neeman, C. I. Ayres, R. H. Hensler, and S. Rajagopalan, *Chem. Ind. (London)* 1626 (1958); S. M. Kupchan, T. Masamune, and C. I. Ayres, *J. Amer. Chem. Soc.* **81,** 4753 (1959); S. M. Kupchan, C. I. Ayres, M. Neeman, R. H. Hensler, T. Masamune, and S. Rajagopalan, *J. Amer. Chem. Soc.* **82,** 2242 (1960).
94. J. H. Chapman, J. Elks, G. H. Phillipps, and L. H. Wyman, *J. Chem. Soc.* 4344 (1956).
95. S. M. Kupchan and A. Afonso, *J. Amer. Pharm. Assoc.* **48,** 731 (1959); **49,** 242 (1960).
96. H. A. Nash and R. M. Brooker, *J. Amer. Chem. Soc.* **75,** 1942 (1953); G. S. Myers, W. L. Glen, P. Morozovitch, R. Barber, G. Papineau-Couture, and G. A. Grant, *J.*

Amer. Chem. Soc. **78,** 1621 (1956); B. W. Christensen and A. Kjaer, *Proc. Chem. Soc.* 307 (1962); *Acta Chem. Scand.* **16,** 2466 (1962).
97. S. M. Kupchan, *J. Amer. Chem. Soc.* **81,** 1921 (1959).
98. J. Fried, H. L. White, and O. Wintersteiner, *J. Amer. Chem. Soc.* **72,** 4621 (1950); J. Fried, P. Numerof, and N. H. Coy, *J. Amer. Chem. Soc.* **74,** 3041 (1952); S. M. Kupchan, P. Slade, and R. J. Young, *Tetrahedron Letters* 22 (1960); S. M. Kupchan, P. Slade, R. J. Young, and G. W. A. Milne, *Tetrahedron* **18,** 499 (1962); H. G. Zachau and W. Karau, *Chem. Ber.* **93,** 1830 (1960); T. C. Bruice and T. H. Fife, *J. Amer. Chem. Soc.* **84,** 1973 (1962).
99. S. M. Kupchan and C. I. Ayres, *J. Amer. Pharm. Assoc.* **48,** 440 (1959).
100. S. M. Kupchan, G. R. Flouret, and C. A. Matuszak, *J. Org. Chem.* **31,** 1707 (1966).
101. J. W. Huffman, D. M. Alabran, and A. C. Ruggles, *J. Org. Chem.* **33,** 1060 (1968).
102. K. Heusler, doctoral dissertation, University of Basel, Switzerland, May 17, 1949, pp. 1–12, 47–49; R. Laurent, doctoral dissertation, University of Toulouse, France, 1947, pp. 17–34; E. Schlittler, K. Heusler and W. Friedrich, *Helv. Chim. Acta* **32,** 2209 (1949).
103. R. T. Puckett, G. A. Sim, E. Abushanab and S. M. Kupchan, *Tetrahedron Letters* 3815 (1966).
104. Merck & Co., Inc., British Patent 782,469 (1957).
105. D. Vincent and T. Mathou, *Compt. Rend.* **220,** 474 (1945); D. Vincent, I. Séro, and R. Laurent, *Thêrapie* **3,** 29 (1948); D. Vincent and M. Parant, *Compt. Rend. Soc. Biol.* **148,** 1878 (1954).
106. H. R. Mahler and G. Dutton, *J. Mol. Biol.* **10,** 157 (1964); H. R. Mahler, R. Goutarel, Qui Khuong-Huu and Minh Truong Ho, *Biochemistry* **5,** 2177 (1966).
107. K. Heusler and E. Schlittler, *Helv. Chim. Acta* **32,** 2226 (1949); W. Friedrich and E. Schlittler, *Helv. Chim. Acta* **33,** 873, 878 (1950).
108. K. S. Brown, Jr. and S. M. Kupchan, *J. Chromatography* **9,** 71 (1962).
109. K. S. Brown, Jr. and S. M. Kupchan, *J. Amer. Chem. Soc.* **84,** 4590, 4592 (1962); **86,** 4414, 4424 (1964).
110. K. S. Brown, Jr. and S. M. Kupchan, *Tetrahedron Letters* 2895 (1964).
111. V. Černy and F. Šorm, *Alkaloids* **9,** 375–426 (1967).
112. L. F. Fieser and R. Stevenson, *J. Amer. Chem. Soc.* **76,** 1728 (1954).
113. J. S. G. Cox, F. E. King, and T. J. King, *J. Chem. Soc.* 1384 (1956); 514 (1959); H. R. Bentley, J. A. Henry, D. S. Irvine and F. S. Spring, *J. Chem. Soc.* 3673 (1953); D. S. Irvine, J. A. Henry and F. S. Spring, *J. Chem. Soc.* 1316 (1955).
114. H. Ruschig, W. Fritsch, J. Schmidt-Thomé, and W. Haede, *Chem. Ber.* **88,** 883 (1955); L. Lábler and F. Šorm, *Coll. Czech. Chem. Commun.* **24,** 2975 (1960).
115. Ch. Meystre, H. Frey, A. Wettstein, and K. Miescher, *Helv. Chim. Acta* **27,** 1815 (1944); Ch. Meystre, H. Frey, R. Neher, A. Wettstein, and K. Miescher, *Helv. Chim. Acta* **29,** 627 (1946); Ch. Meystre, A. Wettstein, and K. Miescher, *Helv. Chim. Acta* **30,** 1022 (1947); C. S. Barnes, *Australian J. Chem.* **9,** 228 (1956).
116. M. Gaitonde, P. A. Vatakencherry, and Sukh Dev, *Tetrahedron Letters* 2007 (1964).
117. J. P. Calame and D. Arigoni, *Chimia (Aarau)* **18,** 185 (1964); J. P. Calame, doctoral dissertation, Zurich, Switzerland, 1965.
118. T. Nakano and M. Hasegawa, *Tetrahedron Letters* 3679 (1964).
119. S. M. Kupchan and W. L. Asbun, *Tetrahedron Letters* 3145 (1964).
120. D. Stauffacher, *Helv. Chim. Acta* **47,** 968 (1964).
121. K. S. Brown, Jr., and S. M. Kupchan, *J. Amer. Chem. Soc.* **86,** 4430 (1964).
122. S. A. Oletta, Belgian Patent 627,935 (1963); Sandoz A. G., Belgian Patent 637,016 (1964).

123. J. Tomko, O. Bauerová, Z. Votický, R. Goutarel and P. Longevialle, *Tetrahedron Letters* 915 (1966).
124. L. Dolejš, V. Hanuš, V. Černy, and F. Šorm, *Coll. Czech. Chem. Commun.* **28,** 1584 (1963); W. Vetter, P. Longevialle, F. Khuong-Huu-Lainé, Q. Khuong-Huu, and R. Goutarel, *Bull. Soc. Chim. France* 1324 (1963); Z. Pelah, D. H. Williams, H. Budzikiewicz, and C. Djerassi, *J. Amer. Chem. Soc.* **87,** 574 (1965).
125. D. Herlem-Gaulier, F. Khuong-Huu-Lainé, E. Stanislas, and R. Goutarel, *Bull. Soc. Chim. France* 657 (1965).
126. F. Khuong-Huu, D. Herlem-Gaulier, Q. Khuong-Huu, E. Stanislas, and R. Goutarel, *Tetrahedron* **22,** 3321 (1966).
127. F. Khuong-Huu-Lainé, D. Herlem-Gaulier and R. Goutarel, *Compt. Rend.* **261,** 4139 (1965); *Bull. Soc. Chim. France,* 3478 (1966).
128. H. Wehrli, M. S. Heller, K. Schaffner, and O. Jeger, *Helv. Chim. Acta* **44,** 2162 (1961).
129. S. M. Kupchan and E. Abushanab, *Tetrahedron Letters* 3075 (1965); S. M. Kupchan, E. Abushanab, K. T. Shamasundar, and A. W. By, *J. Amer. Chem. Soc.* **89,** 6327 (1967).
130. R. Goutarel and F. Khuong-Huu, personal communication, May 15, 1968.
131. D. Herlem-Gaulier, F. Khuong-Huu-Lainé, and R. Goutarel, *Compt. Rend.* **264,** 798 (1967); D. Herlem-Gaulier, F. Khuong-Huu-Lainé, R. Goutarel, and M.-J. Magdeleine, *Bull. Soc. Chim. France* 763 (1968).
132. C. Djerassi, "Optical Rotatory Dispersion," pp. 43, 50–51, McGraw-Hill Book Co., Inc., New York, N.Y., 1960.
133. Z. Votický and J. Tomko, *Tetrahedron Letters* 3579 (1965).
134. T. Nakano and S. Terao, *Tetrahedron Letters* 1035, 1045 (1964); *J. Chem. Soc.* 4512 (1965).
135. R. D. Haworth, J. McKenna, and R. G. Powell, *J. Chem. Soc.,* 1110 (1953); R. Ledger, J. McKenna and P. B. Smith, *Tetrahedron Letters* 1433 (1963).
136. T. Nakano, S. Terao, Y. Saeki, and K. D. Jin, *J. Chem. Soc.* 1805 (1966).
137. T. Nakano, S. Terao, and Y. Saeki, *J. Chem. Soc.* 1412 (1966).
138. E. Wenkert and P. Beak, *Tetrahedron Letters* 358 (1961); R. C. Cambie and L. N. Mander, *Tetrahedron* **18,** 465 (1962).
139. S. M. Kupchan, R. M. Kennedy, W. R. Schleigh, and G. Ohta, *Tetrahedron* **23,** 4563 (1967).
140. T. Nakano and S. Terao, *J. Chem. Soc.* 4537 (1965).
141. T. C. Bruice and S. J. Benkovic, "Bioorganic Mechanisms," Vol. I, pp. 146–166, Benjamin, New York, N. Y., 1966.
142. T. Nakano and M. Hasegawa, *J. Chem. Soc.* 6688 (1965).
143. F. Khuong-Huu-Lainé, A. Milliet, N. G. Bisset and R. Goutarel, *Bull. Soc. Chim. France* 1216 (1966).
144. K. S. Brown, Jr., T. Nakano, and S. Terao, unpublished experiments.
145. W. Döpke, B. Müller, and P. W. Jeffs, *Pharmazie* **21,** 643 (1967); *Naturwissenschaften* **54,** 249 (1967).
146. S. M. Kupchan and G. Ohta, *J. Org. Chem.* **31,** 608 (1966).
147. F. Khuong-Huu-Lainé, M. J. Magdeleine, N. G. Bisset, and R. Goutarel, *Bull. Soc. Chim. France* 758 (1966).
147a. W. Döpke, personal communication, August 1969 (*Pharmazie,* in press).
148. S. M. Kupchan and E. Abushanab, *J. Org. Chem.* **30,** 3931 (1965); Z. Votický, J. Tomko, L. Dolejš, and V. Hanuš, *Coll. Czech. Chem. Commun.* **30,** 3705 (1965).
149. J. Tomko, Z. Votický, V. Paulik, A. Vassová, and O. Bauerová, *Chem. Zvesti* **18,** 721 (1964).

150. W. Döpke and B. Müller, *Pharmazie* **21,** 769 (1966).
151. W. Dopke, B. Müller, G. Spiteller, and M. Spiteller-Friedmann, *Naturwissenschaften* **54,** 200 (1967).
151a. F. Khuong-Huu-Lainé, D. Herlem-Gaulier, and A. Milliet, *Bull. Soc. Chim. France* 256 (1969).
152. H. Hirschmann and F. B. Hirschmann, *J. Biol. Chem.* **184,** 259 (1950).
153. D. Bertin and M. Legrand, *Compt. Rend.* **256,** 960 (1963).
154. W. Döpke and B. Müller, *Pharmazie* **22,** 666 (1967).
155. G. Ponsinet and G. Ourisson, *Phytochemistry* **7,** 757 (1968) and references included therein.

21

Biosynthesis of Alkaloids

I. D. SPENSER
Department of Chemistry, McMaster University
Hamilton, Ontario, Canada

1. INTRODUCTION

1.1 Biogenesis and Biosynthesis

When the structure of an alkaloid has been established and its synthesis achieved, one question remains: How is the compound formed in its native habitat? The elucidation of the biosynthesis of a naturally occurring compound represents the last frontier of natural products chemistry. It is only since the advent of isotopic tracers, and particularly during the past fifteen years, that experimental progress has been made in the study of the biosynthetic formation of alkaloids. Yet the question of their origin has occupied chemists for many years and has provided a fertile field for speculation. Answers to the question were sought indirectly, on the basis of chemical reasoning. The ideas which emerged from this approach form the body of biogenetic theory. Many of the tenets of this theory were remarkably clairvoyant. Others proved to be wide off the mark.

Several lines of argument contributed to the development of biogenetic theory.

Firstly, it was assumed that complex natural products were assembled from simpler molecules which were, in turn, related to common cellular constituents. Recognition of such structural relationships provided the basis of biogenetic thinking. With the elucidation of the structures of the major groups of alkaloids, and the recognition of common features amongst the structural variants within each group, it became apparent that their skeletons could be visually dissected into fragments whose carbon–

nitrogen skeleton bore the features of one or the other of a small number of amino acids. Prominent among these were ornithine and lysine (Section 2), phenylalanine, tyrosine and dihydroxyphenylalanine (Section 3), and tryptophan (Section 4).

Secondly, it was assumed that if the fragments could indeed arise from common cellular constituents by chemically and biologically acceptable processes, their reassembly into the complex product should be explicable in terms of a rational chemical mechanism. Indeed, the derivation of the structures of a surprisingly large number of molecular types amongst the alkaloids could be rationalized, in chemical terms, by invoking a small number of simple organic reactions. Two of these, the Mannich reaction, i.e., the condensation of an amine, an aldehyde and a carbanion,

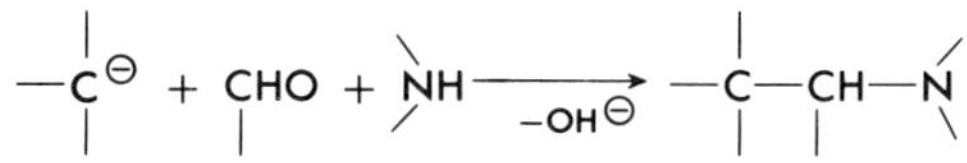

and the oxidative coupling of phenols, were of particular importance.

Finally, it was accepted that a successful chemical synthesis of the complex product, from suitably functionalized, "cell-possible" fragments, serves as confirmatory evidence for the validity of a biogenetic hypothesis. Mannich and phenol-coupling reactions have been employed in such "biogenetically modelled" syntheses of a large number of alkaloids. In many cases these syntheses were carried out in dilute aqueous solution in the pH range 4–8 at room temperature, conditions assumed to resemble those of the cellular environment.

Biogenetic hypotheses based on indirect evidence of this type are chemically satisfying models. Such models have served as stimuli for elegant syntheses of complex structures and have found important application as tools in structural elucidation.

But *biogenetic* hypotheses are not of direct biochemical relevance. Evidence bearing on the *biosynthesis* of an alkaloid cannot emerge from model reactions but must be sought within the biological material.

1.2 The Elucidation of Metabolic Pathways

The objective of the investigation of the biosynthesis of an alkaloid can be stated precisely. It is the elucidation of the normal and usually unique route, the metabolic pathway, whereby the complex product is generated *in vivo*. This objective is attained when every intermediate of the route has been recognized, when the enzymes which catalyze the individual steps have been characterized and the required co-factors are known, and when the stoichiometry, the kinetics, and the chemical mechanism of each enzyme-catalyzed step are understood.

Metabolic studies of this type have been carried out by biochemists for many years. The emphasis of their investigations has been on the origin and fate of primary cell constituents, that is of the substances which play an obligatory role in biological processes. As a result of these studies the major pathways of intermediary metabolism, mainly in mammalian tissues and in microorganisms, are well documented, and many of the enzymes which catalyze individual steps of these pathways have been isolated. Even so, present knowledge is far from complete. The study of the chemical mechanism of enzymic processes has gained momentum only in recent years. There is no justification for assuming *a priori* that primary metabolic processes follow the same pattern in different living forms. Each species presents a unique set of metabolic problems. Nevertheless, the metabolic routes of primary cell constituents in widely different organisms present a picture of biochemical unity.

By contrast, the occurrence of a wide spectrum of secondary metabolic products, compounds whose individual taxonomic distribution is severely restricted, emphasizes the diversity of biological material. The study of the metabolic origin and fate of these substances, of which the alkaloids are one segment, is in its infancy. Enzymic mechanisms are almost entirely unexplored. Experimental progress is largely confined to the recognition of precursor-product relationships. The progress which has been made in the study of the origin of alkaloids and of other secondary metabolites has resulted chiefly from the application of isotopic tracer methods and rests in no small measure on accumulated knowledge of primary metabolic pathways. It is from intermediates of these primary routes that secondary pathways of metabolism radiate.

The advent of isotopic tracers has greatly facilitated the investigation of metabolism. Yet a great deal of knowledge of primary metabolism, particularly in mammalian tissues and in microorganisms, was obtained by methods which predate the introduction of tracers.

The objective of the earliest of these methods was to define the nutritional requirements of heterotrophic organisms, i.e., those which depend for growth and survival on the supply of a series of more or less complex organic compounds. Such balance studies, which were based on the growth response of groups of closely matched individuals to defined diets, led to important findings. In animal studies, body weight was easily determined and served as an adequate experimental criterion. In studies with microorganisms, assay of population density served a similar purpose. A diet (or medium) which did not sustain growth lacked an essential component. Further experiments were aimed at finding substances which could replace the missing component and restore growth. Evidently, the missing component and its replacement were metabolically related. Studies of this type occasionally led to the recognition of metabolic intermediates. Under conditions of nutritional deprivation metabolites, which were not normally detectable, sometimes accumulated in blood and urine, or in the culture medium, and were isolated and characterized.

Such methods are unsuitable for an investigation of metabolism in green plants and other autotrophs, that is organisms whose survival and growth does not depend on a supply of preformed organic compounds, since they have the machinery to satisfy their needs by biosynthesis from carbon dioxide and inorganic nutrients. Balance studies with plants, aimed at gaining an insight into the origin of alkaloids were, nevertheless, carried out. The interpretation of the experimental observations derived from these experiments were, as we now know, highly misleading. The approach, which was used in these investigations, was to infuse individual plants with massive doses of putative precursors of a given endproduct that was known to occur normally. It was argued that if abnormally large amounts of a genuine precursor were made available to the plant, increased biosynthesis should follow, resulting in increased accumulation of product in the experimental, as compared to an untreated plant. If the administered material were not a precursor, on the other hand, only normal amounts of endproduct would be detectable in the experimental plant. As a result of studies of this type it was concluded that proline and ornithine served as precursors of trigonelline in dahlia and that lysine served as a precursor of ricinine in the castor plant and of morphine in the opium poppy. The actual precursors of trigonelline, of ricinine and of morphine, as established by isotopic tracer methods, are quite unrelated to ornithine and lysine, respectively. The increase in the amount of product, reported to have been observed in these experiments, was not due to a direct precursor-product relationship.

This experimental approach is questionable on several counts. Its rationale, that administration of large doses of a precursor will lead to increased accumulation of endproduct, is valid only if the rate of biosynthesis of a product is limited, because a precursor is in short supply. If precursor is normally available in excess, but biosynthesis is controlled by a rate-limiting step, enhanced supply of precursor will not lead to enhanced accumulation of endproduct.

Furthermore, the method departs significantly from the ideal experimental aim of metabolic studies, which is to examine pathways in the normal intact organism, under normal physiological conditions. It cannot be assumed that administration to a plant of a substance in massive doses does not lead to qualitative or quantitative physiological changes.

Another difficulty arises from the necessity of matching experimental with control plants. Individuals are rarely identical and the selection of plants is not a simple matter. Even under the most favorable conditions individuals are subject to natural variations which are interpretable only on the basis of statistical analysis of a large number of samples. Such analysis is not possible if experimentation is restricted to a small number of individuals.

The experimental methods for the investigation of metabolism, which have been discussed, make use of normal experimental material in an abnormal environment.

Significant biochemical findings in mammalian and microbiological systems result from studies of intact organisms with spontaneous or induced genetic abnormalities. The study of hereditary metabolic abnormalities in humans, the so-called "inborn errors of metabolism," is an early instance of the contribution of biochemical genetics to the understanding of metabolic processes. The examination of strains of microorganisms, whose genetic makeup had been altered by treatment with ionizing radiation or chemical mutagens, has proved a particularly fruitful source of biochemical knowledge. These studies, which led to fundamental discoveries in the field of biochemical genetics, yielded information on the identity of intermediates of numerous primary metabolic pathways. If the normal strain (the wild strain) of a microorganism grows on a medium containing a general carbon and nitrogen source (e.g., glucose and ammonia), but a mutant strain will not grow on this medium, but will grow if this basal medium is supplemented by an organic compound A or another compound B, then A and B may be metabolically related. The criteria for deducing normal biosynthetic relationships from experiments of this type have been critically examined by Adelberg.

Experimental biochemical genetics has not hitherto found application in the study of alkaloid biosynthesis. The genetic control of the distribution of plant pigments, however, has been studied. Correlation of the results of genetic analysis and chemical examination of pigments, in cultivated snapdragons, has led to a proposed pathway for the steps in the biosynthesis of flavonoids in this plant.

Much information on primary metabolism in mammalian tissues has emerged from studies using fragmented organisms of varying degree of complexity, ranging from surgically modified animals, excised organs, tissue slices, tissue homogenates, cellular fractions, to soluble cell extracts. Such techniques have found only limited application in investigations of metabolism in green plants, in general, and of alkaloid biosynthesis in particular.

Current knowledge on the biosynthesis of alkaloids has been obtained entirely from tracer experiments, mostly with intact plants, but in some instances also with excised shoots or with root tissue grown in sterile medium.

1.3 Isotopic Tracer Methods

The most common experimental procedure used in the study of alkaloid biosynthesis is to test a biogenetic hypothesis by allowing viable plant material to grow for a limited time in contact with a multicarbon substrate that is labelled with carbon-14 at a single, known, position. The desired product is then isolated from the plant, its carbon-14 content is assayed by quantitative measurement of radioactivity, and if labelled, the product is degraded by chemically unambiguous steps, such that the distribution of carbon-14 within its carbon skeleton may be deduced. The total activity within the intact product must be accounted for in terms of the activities at individual carbon atoms. If the activity within the product is confined to a single, or, in some cases, to several specific sites, incorporation of radioactivity is shown to have taken place non-randomly. The substrate is then said to have been incorporated specifically, and is regarded as a precursor.

Determination of the isotope distribution within a biosynthetic product by chemical manipulation, usually of very small quantities, is a time-consuming and often difficult operation. Recently a beginning has been made in the development of techniques which, when perfected, will yield the required information while obviating the need for chemical degradation. Direct mass spectrometry and nmr spectroscopy of biosynthetically labelled products are likely to find increasing application in the future. It cannot be stressed too strongly that incorporation studies in intact plants with specifically labelled substrates yield useful conclusions only if non-random incorporation of tracer into the products is demonstrated. Generally labelled substrates, i.e., multicarbon compounds containing carbon-14 at all carbon atoms, are not suitable for studies of this type.

Demonstration of specific entry of activity from a given singly labelled multicarbon precursor into a product does not constitute proof that the intact carbon skeleton of the precursor has been incorporated. It is only a first step. Further experiments are mandatory, which show specific incorporation of activity from other radiomers of the precursor, i.e., of precursor samples carrying carbon-14 at other single sites, into predictable positions of the product. The ultimate test of intact incorporation is the recovery, from an experiment with a multiply labelled precursor of defined isotope distribution, of a product whose quantitative isotope distribution matches that of the precursor. The rationale of this test is discussed in Section 2.1., in which one of its applications is illustrated.

Carbon-14 has found the widest application in investigations of alkaloid biosynthesis, but other radioactive isotopes, e.g., tritium, as well as stable isotopes, such as deuterium, carbon-13, nitrogen-15 and oxygen-18, have also been employed. The ideal objective of biosynthetic studies is to observe normal events in a normal, physiological environment. Since normal steady-state conditions of an intact plant are likely to be perturbed by administration of exogenous material, it is desirable to minimize such perturbation by keeping the quantity of administered precursor to a minimum. Since methods for the measurement of radioactivity are, in general, more sensitive than the usual mass-spectrometric methods for the quantitative assay of stable isotopes, larger quantities of precursor must be used when this is labelled with a stable rather than a radioactive isotope. Radioactive tracers are, therefore, to be preferred in biosynthetic investigations.

Even though a substrate has been elevated to the status of a *precursor* by demonstration of its specific incorporation into a product, this substrate is not necessarily an obligatory precursor, i.e., an intermediate of the normal, physiological, route of biosynthesis of the product. The tracer experiments to be discussed were successful in relating many of the structural types found among the alkaloids to intermediates of primary metabolism. Probable biosynthetic sequences have been proposed which are consistent with the results of these incorporation studies. But since tracer experiments involving the administration of labelled multicarbon precursors to intact systems lack the power to define obligatory routes of biosynthesis, or even the primary points of departure of the secondary pathways, these sequences must be regarded as tentative until further evidence accumulates from kinetic and enzymological studies. Even so, it is likely that in most instances incorporation studies provide a real insight into biosynthetic processes. It is from studies of this type that most of the knowledge of precursor-product relationships in alkaloid biosynthesis has emerged.

Radioactive tracers may be employed also to detect possible intermediates of biosynthetic pathways by two techniques which do not require degradation of the product. The first of these is the "carrier dilution" technique: If activity from a labelled precursor is specifically incorporated into a product, then this activity must have passed through every intermediate between precursor and product. Even though such intermediates may be present in the system in no more than minute amounts, and their direct isolation may, therefore, be technically impossible, their radioactivity may be high enough to be detectable. If the chemical identity of such an intermediate can be guessed at on biogenetic grounds or inferred from the chromatographic behavior of its radioactivity, and a sample of cold, i.e., nonradioactive, material of the inferred structure is added in quantities sufficient for recrystallization, reisolation and rigorous purification, and if this reisolated sample is found to be radioactive, its derivation in the experimental system from the precursor is proven. The compound may then be an intermediate.

Another approach to the tentative identification of possible intermediates in a biosynthetic pathway is the quantitative study of incorporation into a product of activity from a labelled general carbon source (e.g., carbon dioxide or glucose) in the presence of nonradioactive substrates. A compound whose addition causes a reduction in the efficiency of incorporation of radioactivity into the product may be an intermediate of the pathway of biosynthesis.

These methods are of limited usefulness, when applied to intact plants, and cannot elicit conclusive evidence. They can provide useful information when employed in conjunction with incorporation studies.

1.4 Origin of the Precursors

Recognition of a structural correspondence of the skeletons of the alkaloids and those of primary cell constituents was the cornerstone of biogenetic hypothesis. The first objective of biosynthetic investigations was to test these postulates by tracer methods. This first phase of the study of alkaloid biosynthesis is essentially complete. The tracer work to be described in later sections of this chapter was successful in relating most of the structural groups, found among the alkaloids, to intermediates of primary metabolism, and in establishing the biosynthetic anatomy of their origin. On this basis, probable reaction sequences have been proposed, which are consistent with the results of these incorporation studies. In the second phase of biosynthetic investigation specific entry into the alkaloids of intermediates of these postulated reaction sequences is being tested. These sequences, which in many instances refine and extend earlier biogenetic schemes, are of necessity tentative since the adopted experimental approach lacks the power to define obligatory routes of biosynthesis, or even the primary points of departure of the secondary pathways.

A discussion of the routes of primary metabolism is beyond the scope of this chapter. In the following discussion emphasis is placed on the stages of biosynthesis of each alkaloid from an arbitrarily chosen primary metabolite, which, in the absence of definite evidence, can be regarded as a convenient relay intermediate. This does not imply that the origin of this primary metabolite is not relevant to the problem. Indeed, important findings in the area of primary plant metabolism have emerged from investigations whose objective was the elucidation of the origin of alkaloids. This is illustrated by two notable instances. The discovery that in plants, unlike in mammalian systems, tyrosine is not derived by oxidation of phenylalanine, emerged from investigations of the biosynthesis of colchicine, of hydrastine and of the Amaryllidaceae alkaloids. Recognition that the biosynthesis of nicotinic acid, a primary metabolite, entailed an entirely different pathway in plants from that known to occur in microorganisms, was the outcome of studies of the biosynthesis of ricinine, and of the pyridine ring of nicotine and anabasine.

Specialized textbooks of biochemistry (e.g., "Metabolic Pathways," (D. M. Greenberg, Ed.), Academic Press, Inc., New York, N.Y., 1960; A. Meister, "Biochemistry of the Amino Acids," Academic Press, Inc., New York, N.Y., 1965) should be consulted for a detailed discussion of the evidence on which is based the current knowledge of the origin of the primary metabolites (glutamic acid, proline and ornithine, lysine, nicotinic acid, tyrosine and phenylalanine, tryptophan and anthranilic acid, mevalonic acid and acetate, methionine) which are here chosen as arbitrary points of departure. Most of this evidence has been gained from investigations of microbiological systems, and even though indications are that most of the relevant primary pathways are analogous in plants, it must be borne in mind that primary metabolism in plant systems is not extensively explored, and that as yet undiscovered differences may exist. The two examples, nicotinic acid and tyrosine, mentioned earlier, illustrate that arguments by analogy can be highly misleading, where metabolic pathways are concerned.

A further reservation must be made. The biosynthetic pathways which are proposed on the basis of incorporation studies freely invoke enzymatic reactions that are assumed to be of wide generality. Decarboxylation, transamination, and oxidative deamination of α-amino acids, methylation by transfer of the S-methyl group of methionine (transmethylation), and oxidative decarboxylation of α-keto acids, are examples of enzymatic reaction which are of widespread occurrence. Their mechanism is discussed in general textbooks of biochemistry (e.g., H. R. Mahler and E. H. Cordes, "Biological Chemistry," Harper & Row, New York, N.Y., 1966). It must be stressed, however, that assumption of unrestricted generality of such processes in biochemical systems is unjustified. Intracellular enzymes are of high substrate specificity. For example, an enzyme system which catalyzes the decarboxylation of one particular amino acid is unlikely to serve in the decarboxylation of another amino acid. Furthermore, it cannot be assumed that because a particular amino-acid decarboxylase has been detected in one plant species it is necessarily present in another plant species. It remains for the third phase of the investigation of alkaloid biosynthesis to establish the presence of the enzymes of each pathway in the plants in which particular alkaloids are generated and to study the kinetics of each step. Only when this has been done can obligatory routes of biosynthesis and points of departure of secondary pathways be rigorously defined.

2. THE BIOSYNTHESIS OF PYRROLIDINE AND PIPERIDINE ALKALOIDS

2.1 Ornithine and Lysine as Primary Precursors of the Ring Systems

Pyrrolidine and piperidine ring systems are found in nature in a variety of alkaloidal types. The structural analogy among representatives of the two series, summarized in Figure 21.1, is striking. It was one of the axioms of classical biogenetic hypothesis to regard the homologous amino acids ornithine (9) and lysine (12), respectively, as the progenitors of the two ring systems. 4-Aminobutanal and 5-aminopentanal were regarded as the biogenetically functionalized derivatives of these progenitors. The facile chemical syntheses of norhygrine (3) and of pelletierine (7), by condensation of acetoacetate with these aminoaldehydes, or rather with their cyclized derivatives, Δ^1-pyrroline and Δ^1-piperideine (Figure 21.4a), served as chemical confirmation of the ideas first proposed by Robinson in his "Theory of the Mechanism of the Phytochemical Synthesis of Certain Alkaloids." Tracer experiments with labelled ornithine and lysine confirm that these amino acids are indeed incorporated into the alkaloids. It is indicative of the decisive influence of Robinson's ideas that the first labelled compounds to be tested as precursors of the pyrrolidine system were ornithine (9) and putrescine (1,4-diaminobutane) (20) rather than proline (1), a common amino acid in which the pyrrolidine ring is preformed, or glutamic acid, the primary metabolite from which proline (1) and ornithine (9) are derived.

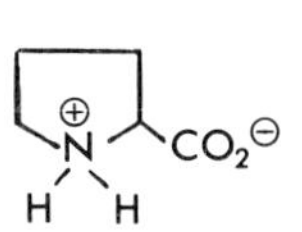

Proline
(1)

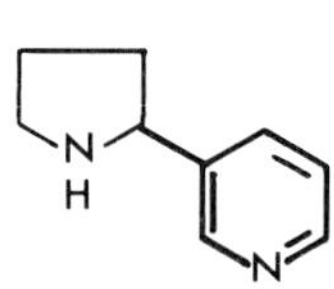

Nornicotine
(2)

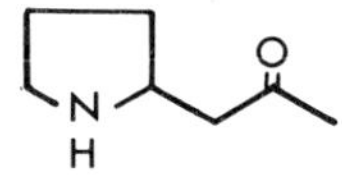

Norhygrine
(3)

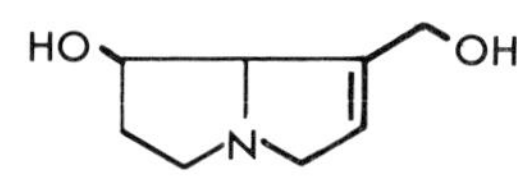

Retronecine
(4)

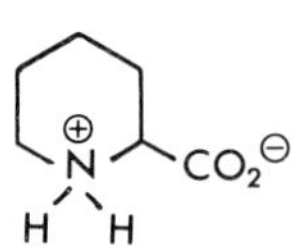

α-Pipecolic
Acid
(5)

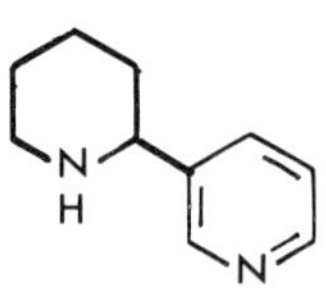

Anabasine
(6)

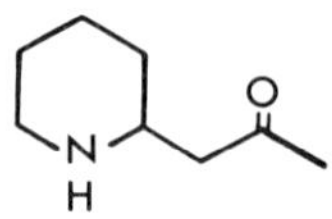

Pelletierine
(7)

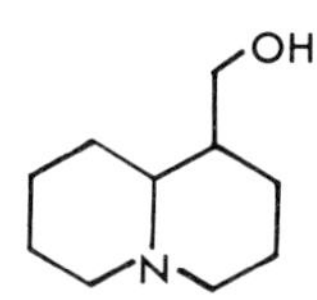

Lupinine
(8)

Figure 21.1 Pyrrolidine and piperidine alkaloids.

The labelled samples of the alkaloids, isolated from plants to which 2-^{14}C-ornithine or 2-^{14}C-lysine, respectively, had been administered, were degraded, to establish the position(s) of radioactive carbon. In several cases [nicotine (10), tropine (11), anabasine (6), sedamine (14)] degradation was not only unequivocal, but complete, and the *entire activity of the intact alkaloid was accounted for in terms of the activity at individual sites.* Definitive conclusions regarding the incorporation of precursor into product can be drawn on the basis of these degradations. Much of the value of a tracer experiment is lost if a complete balance sheet of activity distribution cannot be drawn up, since adequate internal control of the results is then lacking. If degradation is incomplete [e.g., retronecine* (4)], uncertainty remains. If degradation is only partial [e.g., lupinine (8), lobinaline (15), lycopodine (16)], conclusions cannot be drawn from a single experiment. Tentative conclusions are possible if parallel incorporation studies are carried out with several radiomers of a precursor.

Activity from 2-^{14}C-ornithine is indeed incorporated into nicotine (10), tropine (11) and retronecine (4) in a specific manner, indicating that four-carbon chains derived from ornithine (9) enter the alkaloids (Figure 21.2). But the mode of incorporation differs in each case. In nico-

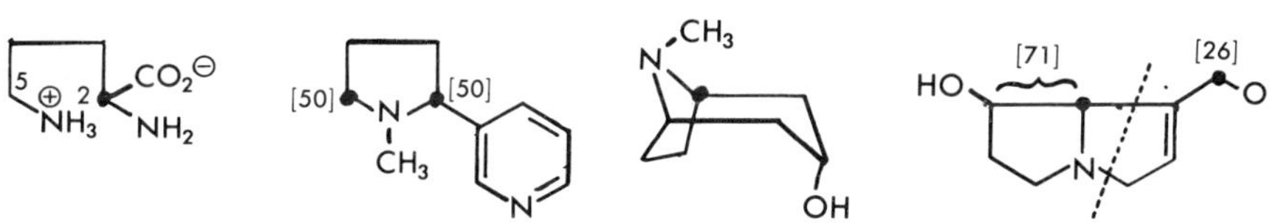

Ornithine (9) Nicotine (10) Tropine (11) Retronecine (4)

Figure 21.2 Incorporation of activity from 2-^{14}C-ornithine [•] into pyrrolidine alkaloids.

tine (10) two centers in the pyrrolidine nucleus are equally labelled. It must be concluded, therefore, that the pathway leads through a symmetrical intermediate, possibly putrescine (1,4-diaminobutane) (20). On the other hand, all activity in tropine (11) is confined to a single carbon atom, and a symmetrical intermediate cannot be involved. Two four-carbon chains from ornithine enter retronecine (4) nonsymmetrically. Since the level of activity at the two labelled centers differs,† the pathway involves two different intermediates, each derived by a separate route from ornithine.

*Degradation of retronecine established that 71% of the activity of the intact base was located in a 2-carbon fragment derived from C(7)(8) of the alkaloid, but these two C-atoms were not separated from one another.

†This conclusion is based on the assumption that activity of the 2-carbon fragment C(7)(8) is confined to C(8).

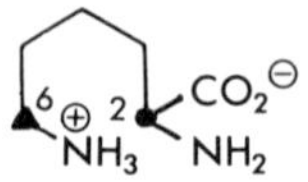

Lysine
(12)

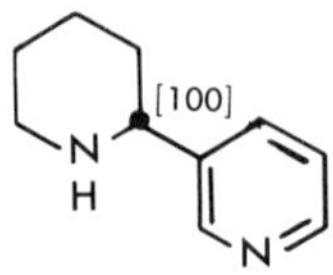

Anabasine
(6)

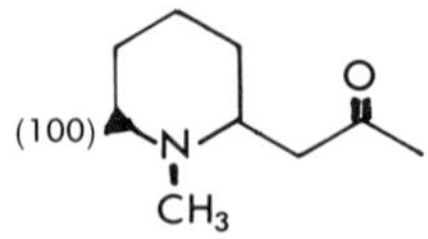

N-Methylpelletierine
(13)

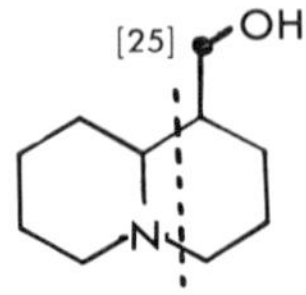

Lupinine
(8)

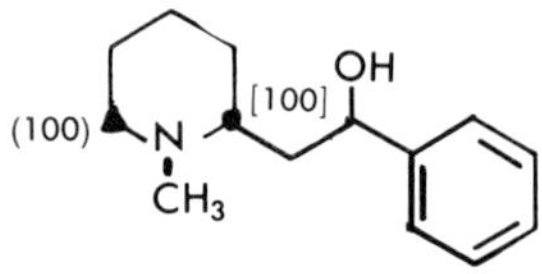

Sedamine
(14)

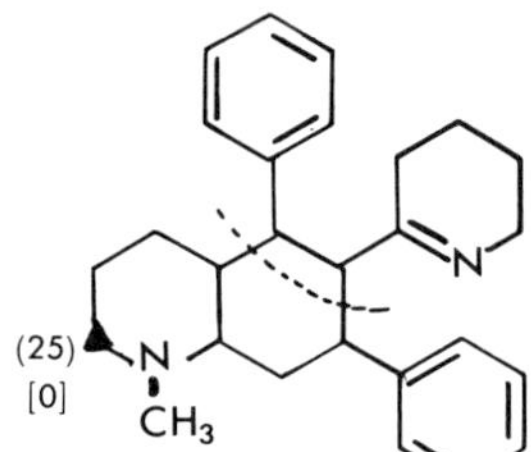

Lobinaline
(15)

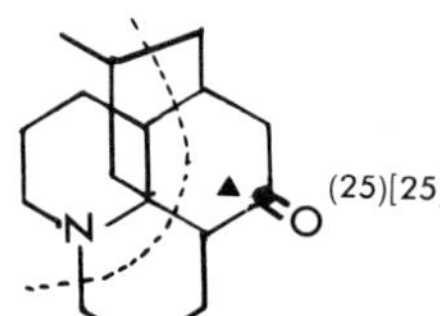

Lycopodine
(16)

Figure 21.3 Incorporation of activity from 2-^{14}C-lysine [•] and 6-^{14}C-lysine (▲) into piperidine alkaloids.

Activity from 2-^{14}C-lysine (12) is confined to a single carbon atom of the piperidine nucleus of anabasine (6) and of sedamine (14) (Figure 21.3). This result leads to the inference that a five-carbon chain, derived from lysine, is incorporated into the nucleus, and that the route excludes a symmetrical intermediate.

Incorporation of activity from a single radiomer of a multicarbon precursor, even though specific, does not conclusively demonstrate the intact incorporation of the multicarbon chain. The evidence is strengthened when specific entry of activity from other radiomers into corresponding positions is proven. Thus, activity from 5-^{14}C-ornithine was equally distributed over C(2) and C(5) of the pyrrolidine ring of nicotine (10) and label from 6-^{14}C-lysine entered C(6) of the piperidine ring of sedamine (14). These findings agree with and complement the results of the experiments with 2-^{14}C-ornithine and 2-^{14}C-lysine. On the other hand, 5-^{14}C-ornithine yielded retronecine (4) which contained one-fourth of its activity in the carbon atom of the hydroxymethyl sidechain. This result is not compatible with that obtained with 2-^{14}C-ornithine, and further data are required to resolve this apparent inconsistency.

The final proof for the incorporation into a product of an intact multicarbon chain is the demonstration that activity from a multiply labelled precursor of known isotope distribution enters the product specifically and without change in tracer content at the different positions, relative to one another. Such an experiment has not yet been performed with samples of ornithine or lysine, multiply labelled with ^{14}C at different carbon atoms. Lysine, multiply labelled at C(6) with ^{14}C, and at C(4) and C(5) with ^{3}H ($^{3}H/^{14}C$ ratio 10:1), yielded sedamine (14) and N-methylpelletierine (13) which showed $^{3}H/^{14}C$ ratios identical with that of the doubly labelled precursor. It is highly improbable that the isotope ratio would be maintained if, in the course of metabolism, the precursor molecule had been fragmented. Each fragment would then be expected to equilibrate with the metabolic reservoir (the "pool") of its own kind. Reassembly of such fragments after equilibration would be expected to lead to a product with an isotope distribution which is entirely different from that of the precursor, since the majority of the molecules in each metabolic reservoir are unlabelled, and the various reservoirs are most unlikely to be of equal size.

Quantitative retention of the isotope distribution of a precursor within a product is thus unlikely to be fortuitous and is regarded as diagnostic for intact incorporation.

The weight of evidence is compelling that an intact 4-carbon chain, derived from ornithine, is incorporated into the pyrrolidine rings of nicotine, tropine and retronecine (Figure 21.2), and an intact 5-carbon chain from lysine into the piperidine rings of anabasine, sedamine and N-methylpelletierine (Figure 21.3).

Complete degradation of lobinaline (15), lupinine (8) and lycopodine (16) has not yet been accomplished. Partial degradation indicated that one-fourth of the activity from 2-^{14}C-lysine (12) was confined to one center of lupinine (8) and lycopodine (16), and one-fourth of the activity from 6-^{14}C-lysine to one site in lobinaline (15) and lycopodine (16). Since each of these compounds contains more than four carbon atoms, incorporation is non-random, but since the location of three-quarters of the activity is unaccounted for, the results are indicative rather than conclusive, and inferences must be drawn with caution.

The hydroxymethyl carbon of lupinine (8) contained one-fourth of the activity of the alkaloid. Two five-carbon chains derived from lysine evidently generate its carbon skeleton. Lupinine, in turn, serves as an intermediate in the biosynthesis of sparteine and of the other $C_{15}N_2$ alkaloids of the lupine family (Chapter 14). Specific incorporation of activity from 2-^{14}C-lysine into several of these has been established by partial degradation. The degradation results are compatible with the interpretation that a symmetrical intermediate derived from lysine lies on the pathway. Other possibilities remain, however, and further work is required to establish beyond doubt the mode of incorporation of lysine.

Lobinaline (15) can be regarded, on a structural basis, as a modified dimer of sedamine (14). Specific incorporation of lysine was, therefore, not surprising. What was unexpected was the finding that only 25% of the activity from 6-^{14}C-lysine was located at C▲ (Figure 21.3). If lobinaline arose by dimerization of a monomer related to sedamine, and the monomer were generated, like sedamine, from lysine in a nonsymmetrical manner, 50% of the activity from 6-^{14}C-lysine might have been expected to reside at C▲ The result could be interpreted to show that incorporation of lysine into the piperidine ring of each monomer unit takes place symmetrically. Such an inference is based on the assumption that the unaccounted 75% of the activity of the intact lobinaline was equally distributed among the three other ring carbon atoms adjacent to nitrogen. A second tracer experiment showed this inference to be untenable: None of the activity of lobinaline, derived from 2-^{14}C-lysine was located at C▲. Incorporation of lysine was, therefore, nonsymmetrical.

On the basis of these incomplete results it must be presumed that the two monomer units, which combine to form lobinaline, are unequal, since they must be unequally labelled. Each must arise from a sedamine-like precursor by a different route, and the piperidine ring of at least one of them is shown to arise, like that of sedamine, in a nonsymmetrical fashion from lysine.

Label from 2-^{14}C- and from 6-^{14}C-lysine is specifically incorporated into lycopodine (16), the major alkaloid of the club moss family (Chapter 16). The carbonyl carbon accounts for one-fourth of the activity of the alkaloid derived from either precursor. These results do indeed point to a symmetrical incorporation of lysine. The hypothesis has been advanced that lycopodine and its relatives represent biosynthetic dimers of pelletierine (7). The structural relationship of pelletierine to the obscurine skeleton (19) (Chapter 16) is shown in Figure 21.4c. Lycopodine is thought to be derived from this skeleton by a rearrangement, after extrusion of one nitrogen atom.

An earlier hypothesis of biogenesis of the lycopodium alkaloids invoked acetate as their sole carbon source. The skeletons of lycopodine and its relatives can be formally derived from two tetraacyl chains (17). This hypothesis appeared to gain verisimilitude when it was shown that coniine (18), a hemlock alkaloid, was of polyketide origin. Activity from 1-^{14}C-acetate was incorporated into coniine. Degradation by unequivocal steps established the incorporation pattern shown in Figure 21.4b. Incorporation of activity from lysine (12) into coniine (18) was negligible. The structural relationship of coniine (18) and pelletierine (7) is self-evident, and a common biosynthetic origin might have been anticipated, but is ruled out by the tracer studies discussed in this section. These results indicate that struc-

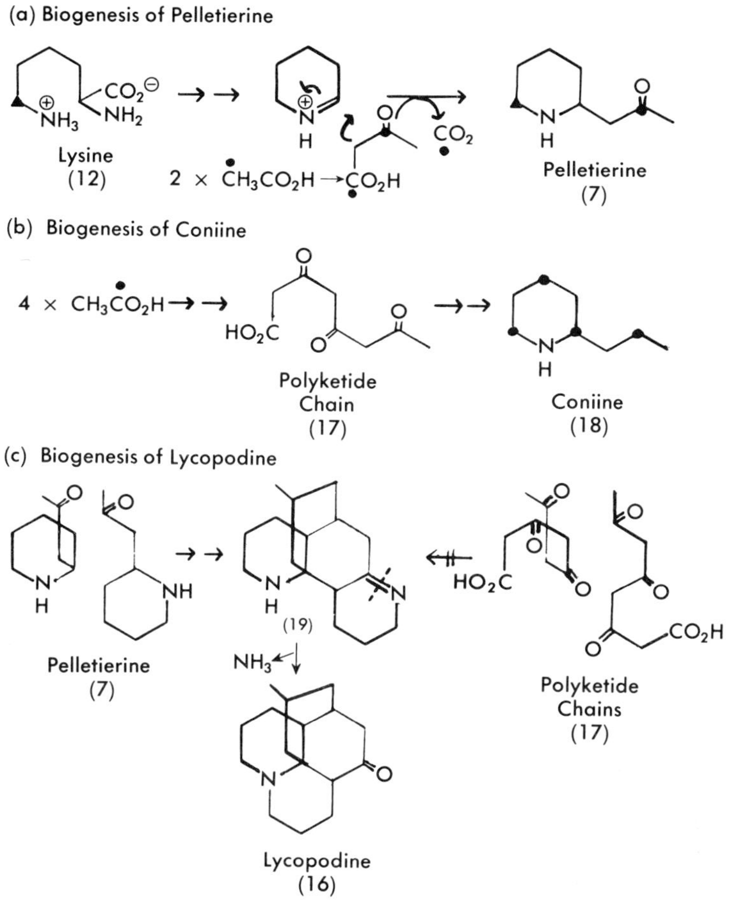

Figure 21.4 Classical and polyketide hypotheses of biogenesis of piperidine alkaloids.

tural relations [e.g., pelletierine (7)-coniine (18)] as well as structural analogies [e.g., nicotine (10)–anabasine (6)] can be biosynthetically misleading.

Incorporation of ornithine (9) and of lysine (12) into the alkaloids is invariably accompanied by loss of the carboxyl group as well as of one or [in the case of retronecine (4) and lupinine (8)] both of the nitrogens.

Attempts have been made to establish which one of the nitrogen atoms of the primary precursor is preserved in the product. Only few of these studies have yielded unequivocal results.

2-^{14}C,2-^{15}N-Lysine and 2-^{14}C,6-^{15}N-lysine were administered to isolated roots of *Nicotiana glauca* in separate experiments, and anabasine

(6) isolated in each case. Both anabasine samples were radioactive, and incorporation of activity was specific. But only the sample from the experiment with 2-^{14}C,6-^{15}N-lysine was substantially enriched with ^{15}N. The ^{14}C/^{15}N ratio of the isolated product (corrected for ^{14}N content of the pyridine nitrogen) was identical with that of the administered precursor. Clearly, it is the 6-amino nitrogen which is preserved in the piperidine ring of anabasine, and it is the 2-amino nitrogen which is lost in the course of biosynthesis.

In similar investigations, designed to elicit the origin of the nitrogen atom of tropine (11) and of several other alkaloids, anomalous or uninterpretable data were obtained. It is likely that these results reflect the limitations of the use of ^{15}N-enriched amino acids in studies with intact plant systems. Reversible cleavage of the C-N bond of metabolically active amino acids (transamination) is one of the fundamental biochemical reactions. Transaminase activities of a wide range of specificities have been detected in many plants. Reversible transamination of an ^{15}N-labelled substrate, which is supplied to an intact system in massive non-physiological doses (50-100 mg.) (doses necessitated by the relative insensitivity of the methods of detection of heavy isotopes), leads to swamping of the endogenous nitrogen pool. The observed ^{15}N-enrichment in a given product then represents the general level of isotopic nitrogen in the nitrogen pool, and the value of the ^{14}C/^{15}N ratio in the product is purely fortuitous and does not represent specific incorporation of the precursor into the product. Isolation of other nitrogenous components from the system and determination of their isotopic nitrogen enrichment would constitute a control. This was not done in the experiments under discussion.

2.2 Primary Precursors of Peripheral Groups

The carbon skeletons of only a few of the pyrrolidine and piperidine alkaloids are derived in their entirety from the primary amino-acid precursor. Stachydrine, the methyl betaine of proline, is generated from proline (1). Methionine serves as the methyl donor. The metabolic relationship of proline (1) and ornithine (9) has been referred to earlier. Pipecolic acid (5), found in plants as well as in microorganisms and in mammalian tissues, arises from lysine (12) by α-deamination and reduction of the intermediate Δ^1-piperideine-2-carboxylic acid. Two modified ornithine (9) chains account for the carbon skeleton of retronecine (4), two modified lysine (12) chains for that of lupinine (8).

All other alkaloids of the two series carry multicarbon substituents on one or both of the ring carbons adjacent to nitrogen, or can be regarded as dimers of such compounds. The origin of three such α-substituents has been investigated.

The pyridine ring of nicotine (10) and of anabasine (6) is derived from nicotinic acid (23), a primary cell constituent. The carboxyl group of nicotinic acid is lost in the course of the biosynthesis of these alkaloids. Decarboxylation takes place at a late stage of biosynthesis, since incorporation of the pyridine moiety into nicotine and anabasine takes place in a specific manner, the point of attachment of the pyridine ring corresponding to the carbon atom from which the carboxyl group has been shed. Most likely it is not free nicotinic acid which is involved in the biosynthetic process, but a nicotinic acid nucleotide. A nicotinamide nucleotide serves as the precursor, also, of ricinine (Chapter 14).

The C_6–C_2 unit which forms the sidechain of sedamine (14) and of its dimer, lobinaline (15), is derived from phenylalanine. Activity from 2-^{14}C-phenylalanine (24) is incorporated entirely into the corresponding position of sedamine. Label from 3-^{14}C-phenylalanine is recovered from the predicted positions of sedamine and lobinaline. That a C_6–C_2 unit derived from phenylalanine (24) is incorporated intact into the sedamine sidechain was established by an experiment with 2,3-^{14}C-phenylalanine of known ^{14}C-distribution. No change in isotope distribution occurred during biosynthesis. It is likely that the incorporation of the C_6–C_2 unit from phenylalanine into the sedamine sidechain takes place by way of cinnamic acid (25). Enzymic conversion of phenylalanine to cinnamic acid has been demonstrated in several plants. Activity from 3-^{14}C-cinnamic acid is incorporated specifically into sedamine and into lobinaline. It is as yet unknown whether cinnamic acid is further converted to benzoylacetic acid, as postulated by the classical hypothesis (cf. Figure 21.4a), or whether it reacts directly with the lysine-derived moiety, presumably Δ^1-piperideine, to yield the skeleton of sedamine.

The 3-carbon sidechain of pelletierine (7), and of its modified dimers of the lycopodium (16) series, as well as the non-ornithine-derived moiety of tropine (11) are derived from acetate. Activity from 1-^{14}C-acetate was recovered solely from the carbonyl carbon of N-methylpelletierine (13), and from the carbinol carbon of tropine (11). Activity from 2-^{14}C-acetate did not enter the carbinol carbon of tropine, but was present in the adjacent carbon atoms. Label from 2-^{14}C-acetate entered the C-methyl group of the lycopodium alkaloid annotinine, but not the adjacent ring carbon. It is likely, but has not been proven, that acetoacetate or its coenzyme A derivative serves as a component of the biosynthetic system, as predicted by Robinson's classical hypothesis.

2.3 Intermediates and Postulated Mechanism

Demonstration of specific incorporation of the primary precursors into the alkaloids delineates the gross anatomy of their biosynthetic origin. These

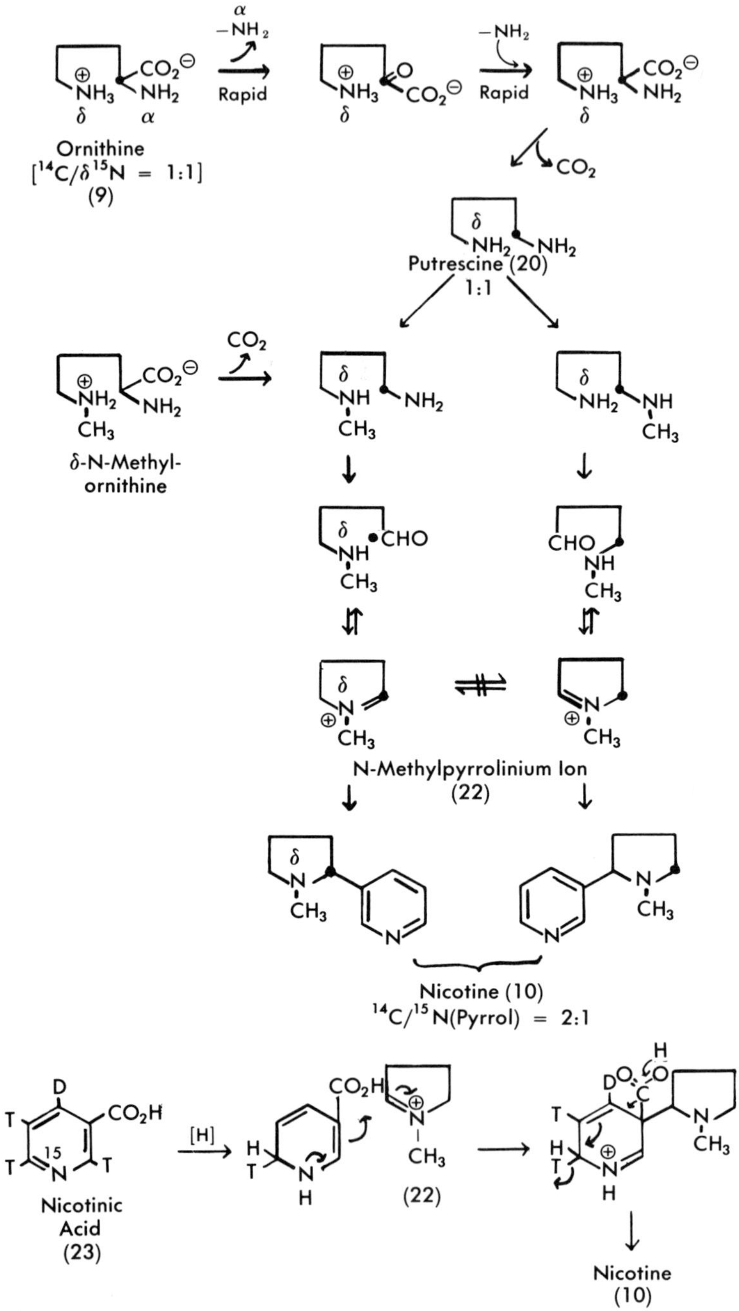

Figure 21.5 Biosynthesis of nicotine.

results do not clarify the pathway of biosynthesis. It is evident that conversion of primary precursor to each of the products requires a multistep sequence of reactions. In an attempt to throw some light on the steps of the sequences, studies have been carried out to test incorporation into the alkaloids of activity from compounds which were inferred to be possible intermediates.

Putrescine (1,4-diaminobutane) (20) and cadaverine (1,5-diaminopentane) are bases which can arise by decarboxylation of ornithine (9) and of lysine (12), respectively, in microorganisms. Enzymes catalyzing the decarboxylation of the two amino acids have not so far been detected in higher plants. Incorporation of radioactivity from 1,4-^{14}C-putrescine or from 1,5-^{14}C-cadaverine into several of the alkaloids has, nevertheless, been demonstrated. Label from 1,4-^{14}C-putrescine was located equally at C(2) and C(5) of the pyrrolidine ring of nicotine (10) and at the two bridgehead carbons of tropine (11). Retronecine (4), derived from 1,4-^{14}C-putrescine contained 25% of its activity at the hydroxymethyl carbon. Activity from 1,5-^{14}C-cadaverine entered the piperidine ring of anabasine (6) [50% at C(2)] and the quinolizidine system of lupinine (8) (25% at the hydroxymethyl carbon).

Several of these observations are at variance with the incorporation data obtained in the corresponding experiments with labelled ornithine (9) and lysine (12). Entry of 2-^{14}C-ornithine into tropine (11), and of 2-^{14}C-lysine into anabasine (6) led to a product labelled at a single carbon atom, a result which excludes intermediacy of a symmetrical molecule. Yet putrescine (20) and cadaverine, respectively, served as specific precursors. Specific incorporation of a substrate is not necessarily diagnostic of an obligatory intermediate of the pathway. If enzymes which normally catalyze obligatory reaction steps lack absolute specificity, it is possible that a substrate-analog, which is not normally present in the plant but is forced upon the system, may be converted to a normal intermediate and thence incorporated into the product. It is likely that the observed incorporation of putrescine into tropine and of cadaverine into anabasine represent abnormal pathways. These contradictory results expose the inherent limitations of an approach to biosynthesis that is limited to the study, in intact biological systems, of specific incorporation of putative substrates into selected endproducts, without reference to enzymic activities, rates of enzymic reactions, or even the normal presence of the substrate under investigation.

No unique pathway to nicotine emerges from experiments with singly and multiply labelled multicarbon tracers. If, as seems likely, it is the N-methyl-Δ^1-pyrrolinium ion (22) which, in the formation of the nicotine skeleton, combines with a moiety derived from nicotinic acid (23), several routes from ornithine to this ion are indicated by the tracer studies. These routes are summarized in Figure 21.5. It remains for a study of the kinetics of the relevant enzymes to determine which is the normal route of biosynthesis.

The steps of the routes leading from the primary precursors to the other alkaloids of the pyrrolidine and piperidine groups, and the mechanisms whereby the fragments derived from the different primary precursors combine to yield the skeleton of the intact alkaloids, are almost entirely unexplored.

3. ALKALOIDS CONTAINING PHENYLETHYLAMINE, ISOQUINOLINE AND RELATED STRUCTURES

3.1 Phenylalanine and Tyrosine as Primary Precursors

A phenylethylamine fragment is readily apparent in most of the alkaloids whose chemistry is discussed in the first seven chapters of this volume. This fragment is present in two ubiquitous amino acids, phenylalanine (24) and tyrosine (30), and a biogenetic relationship between these amino acids and the phenylethylamine and tetrahydroisoquinoline alkaloids was an early postulate of biogenetic thought. In one of the first contributions to biogenetic hypothesis, Winterstein and Trier suggested that the nucleus of the benzyltetrahydroisoquinoline, laudanosine, originated by interaction of 3,4-dihydroxyphenylethylamine (dopamine) (35) and 3,4-dihydroxyphenylacetaldehyde, two intermediates thought to be derivable from 3,4-dihydroxyphenylalanine (dopa) (34) by decarboxylation and oxidation, respectively. Mannich condensation of the amine and the aldehyde constituted a biogenetically modelled synthesis of the benzyltetrahydroisoquinoline skeleton.

This hypothesis was later refined and extended by Robert Robinson to encompass the biogenesis of several other alkaloidal types. He recognized structural relationships between the benzylisoquinoline skeleton [e.g., (54)] and the skeletons of the morphine (55) and aporphine [e.g., (78)] bases, and of the protoberberine [e.g., (59)], phthalideisoquinoline [e.g., (60)], protopine (67) and benzophenanthridine [e.g., (62)] alkaloids. More recently, the erythrina [e.g., (58), (85)] alkaloids have been added to this biogenetic family.

The evidence which has emerged from recent tracer work is entirely compatible with this classical picture of biogenesis.

A more recent contribution to biogenetic hypothesis, the concepts of intramolecular phenol coupling and rearrangement, first put forward by D. H. R. Barton, has been most fruitful in predicting further biogenetic relationships among the alkaloids related to the aromatic amino acids, and in providing the mechanistic rationale of postulated interconversions.

Biogenetic hypothesis has been remarkably successful in foreshadowing the biosynthetic events leading to the phenylethylamine-derived alkaloids, and has, in addition, provided a most rewarding tool in the structural elucidation of newly discovered bases and an important concept in the design of synthetic methods.

The alkaloids that are biogenetically related to the aromatic amino acids fall into a number of well-defined groups. The phenylethylamine bases originate from tyrosine (30) by modification of the sidechain and/or further hydroxylation of the nucleus. Phenylalanine (24) serves as the precursor of nitrogenous C_6–C_3 compounds [e.g., ephedrine (29)].

Simple tetrahydroisoquinoline alkaloids probably arise by interaction of a phenolic phenylethylamine moiety with an aliphatic fragment such as acetate [e.g., pellotine (38)] or mevalonate (e.g., lophocerine). Union of such a phenylethylamine moiety with a second aromatic fragment gives rise to three series of polycyclic bases. The first of these encompasses the alkaloids of the *Amaryllidaceae* (Chapter 6), whose parent, norbelladine (43), is derived by interaction of a tyramine (31) moiety with a C_6–C_1 unit. The second series is based on norlaudanosoline (64), whose benzylisoquinoline skeleton originates from the union of a dopamine (35) fragment with a C_6–C_2 unit. This group includes the alkaloid types (Chapters 3 and 5) whose biogenetic origin had been predicted by the Winterstein-Robinson hypothesis. Thirdly, combination of the dopamine (35) moiety with a C_6–C_3 unit leads to the 1-phenylethylisoquinoline (93) skeleton from which colchicine (92) (Chapter 8) is derived.

The C_6–C_1 and the C_6–C_3 unit are derived from phenylalanine (24) but not from tyrosine (30). The C_6–C_2 unit originates from tyrosine (30) and not from phenylalanine (24). It was only when the biochemical nonequivalence of phenylalanine and tyrosine in plants was recognized that apparently contradictory results of early tracer experiments with the two amino acids were reconciled. In mammalian tissues phenylalanine is converted to tyrosine in an enzyme-catalyzed aerobic hydroxylation reaction. In early biogenetic discussions phenylalanine and tyrosine were regarded as equivalent on the assumption that a similar process must occur in plants. Recent work shows that such an assumption was entirely unwarranted. In plants phenylalanine (24) suffers deamination which leads to *trans*-cinnamic acid (25). This cinnamic acid is the parent of a large number of phenolic C_6–C_3 compounds (phenylpropanoids) as well as of phenolic benzoic acids. Hydroxylation takes place at the cinnamic acid stage. The analogous deamination of tyrosine (30) to the corresponding p-hydroxycinnamic acid [p-coumaric acid (49)] is not of general occurrence in plants and has been detected only in grasses. Hydroxylation of tyrosine (30) to dihydroxyphenylalanine (34), a reaction which takes place in mammalian tissues, has been shown to occur in a number of plant species. These findings underline the danger of argument by analogy, where metabolic pathways are concerned.

3.2 Nitrogenous Phenylpropanoids, β-Phenylethylamines and Simple Isoquinolines

Phenylalanine (24) serves as the precursor of several nitrogenous phenylpropanoids. Label from β-^{14}C-phenylalanine was recovered quantitatively from the benzylic carbon atom of 3-dimethylamino-3-phenylpropionic acid (26), a component of the ester alkaloids of taxus. Activity from β-^{14}C-phenylalanine enters specifically into the benzylic carbon of pseudonorephedrine (28), and heavy nitrogen from ^{15}N-phenylalanine but not from ^{15}N-alanine was incroporated into ephedrine (29). The C-methyl group of these alkaloids corresponds in position to the carboxyl group of the amino-acid precursor and might be assumed to be derived from it by reduction. Since ^{14}C-formic acid serves as the precursor of the C-methyl group of ephedrine such a reduction process is unlikely. Incorporation of carboxyl-^{14}C-phenylalanine has not been tested, however. Radioactivity

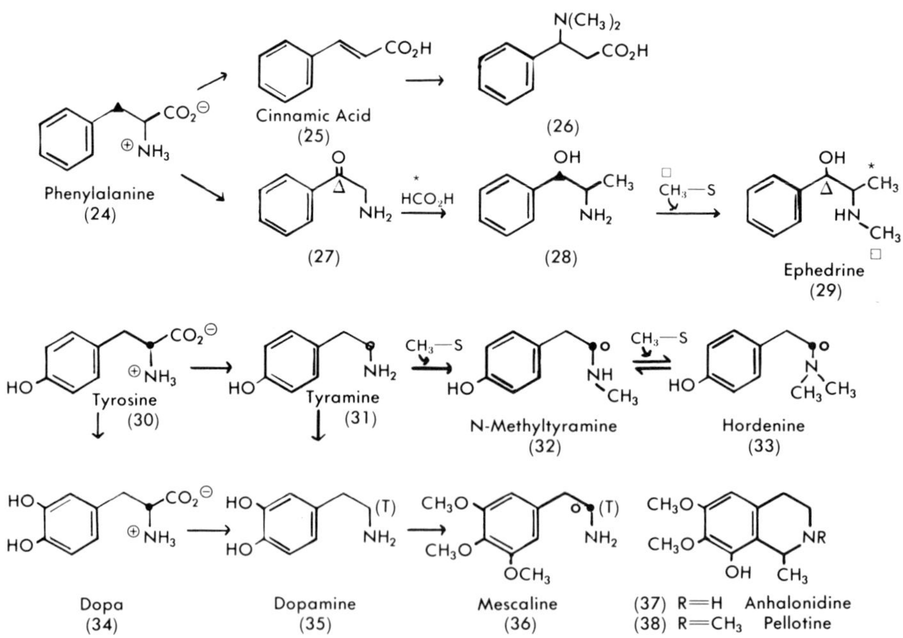

Figure 21.6 Biosynthesis of phenylethylamine alkaloids.

from carbonyl-^{14}C-ω-aminoacetophenone (27) enters the benzylic carbon of ephedrine (29), indicating an unusual pathway. Cinnamic acid (25) has not been tested as a precursor.

Several β-phenylethylamines and 1-methyltetrahydroisoquinolines have been shown to be derived from tyrosine. Label from α-^{14}C-tyrosine is incorporated specifically into the predicted positions of N-methyltyramine (32) and hordenine (33), found in barley roots, and into mescaline (36) and anhalonidine (37), in the peyote cactus. Activity from 1-^{14}C-tyramine (31) enters the corresponding position of both hordenine (33) and mescaline (36). The methyl groups, in each case, are supplied by methionine (91) (Figure 21.6).

Label from α-^{14}C-tyrosine (30) and from 1-^{14}C-dopamine (35) enters pellotine (38). A two-carbon fragment is required to complete its skeleton. Acetate is an obvious candidate. Yet activity from 1-^{14}C-acetate was not confined to C(1) of the isoquinoline nucleus but was scattered throughout the pellotine molecule. This finding is in direct contrast to the specific incorporation of acetate into the β-carboline alkaloid eleagnine (104), an indole analog of pellotine (Section 4.1.).

3.3 The Amaryllidaceae Alkaloids

In a major contribution to biogenetic theory Barton and Cohen suggested, in 1957, that the diverse skeletal structures found among the Amaryllidaceae alkaloids (Chapter 6) were all related to a hypothetical intermediate (43), now known as norbelladine. They outlined a mechanism whereby the three major structural groups among these alkaloids, the lycorine-, galanthamine- and crinine-types, were derivable from this parent compound by intramolecular oxidative coupling of its two phenol nuclei (Figure 21.7). Circumstantial support for the hypothesis came from the subsequent identification of belladine (44), the fully methylated derivative of the postulated intermediate, as a further structural type among the alkaloids.

The hypothesis is sustained by tracer evidence. Radioactivity from singly and multiply labelled samples of norbelladine (43) and p′-O-methylnorbelladine (39) is incorporated into alkaloids belonging to each of the three series. The distribution of label, found by unambiguous chemical degradation of the products, was exactly as predicted on the basis of Barton's hypothesis (Figure 21.8).

Specific entry of label from the O-methyl group of p′-O-methylnorbelladine (39) into the methylenedioxy function of haemanthamine (47) also provides experimental proof of the biosynthetic origin of this one-carbon function.

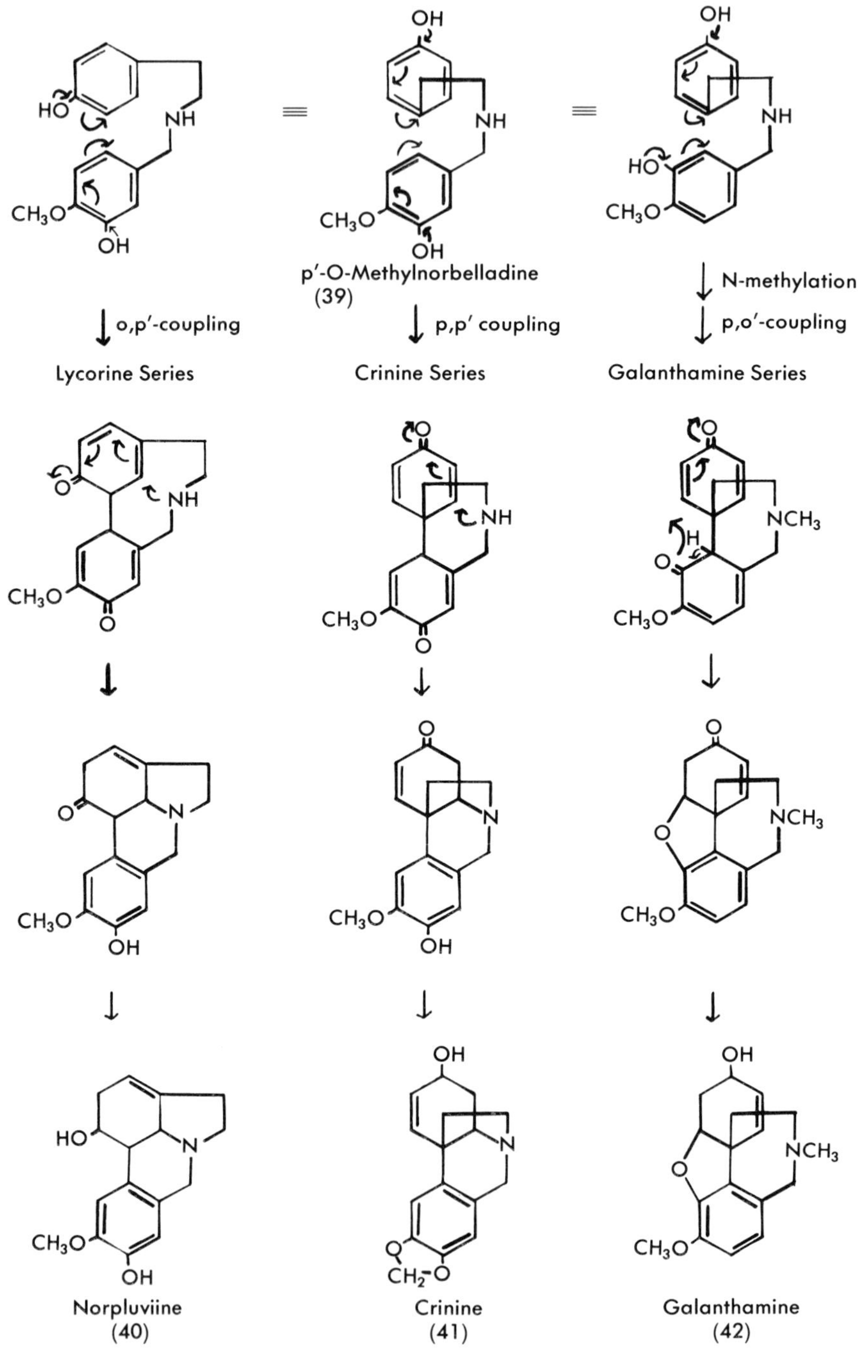

Figure 21.7 Biogenesis of the Amaryllidaceae alkaloids.

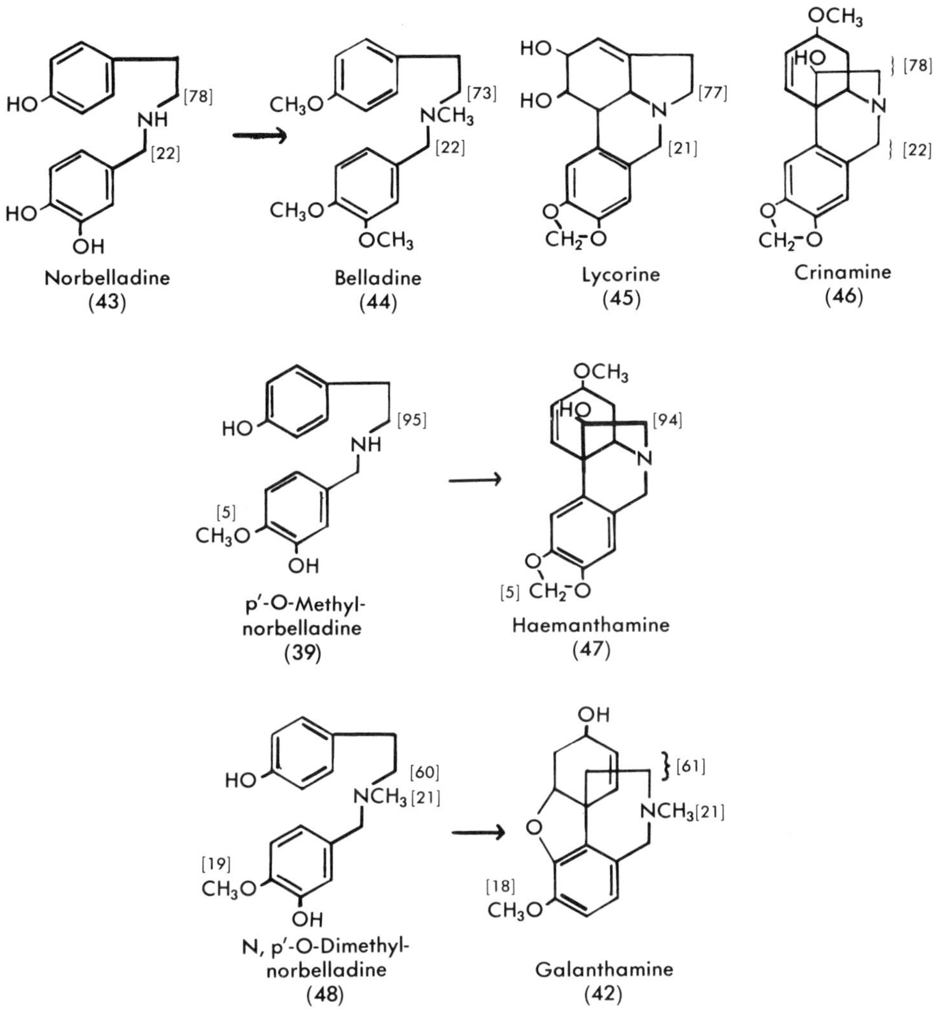

Figure 21.8 Incorporation of multiply labelled norbelladine into Amaryllidaceae alkaloids (Numbers are relative specific activities).

N-Methylation of norbelladine blocks entry of the precursor into alkaloids of the lycorine and crinine series (cf. Figure 21.7). It does not inhibit its incorporation into the galanthamine series. The methylation pattern of the precursor thus controls its entry into pathways leading to the different structural variants. Experiments with multiply labelled samples of p'-O-methyl- and N,p'-O-dimethylnorbelladine (39 and 48) show that, assuming these partially methylated derivatives to be indeed normal intermediates, the methylation processes are specific and irreversible.

Methylation at the m′-oxygen of norbelladine appears to inhibit the phenol coupling process: m′-O-Methylnorbelladine was not incorporated into haemanthamine (47). Its incorporation into belladine (44) has not been tested.

The pattern of evidence is consistent with the view that norbelladine (43) is the general precursor of the Amaryllidaceae alkaloids and that the

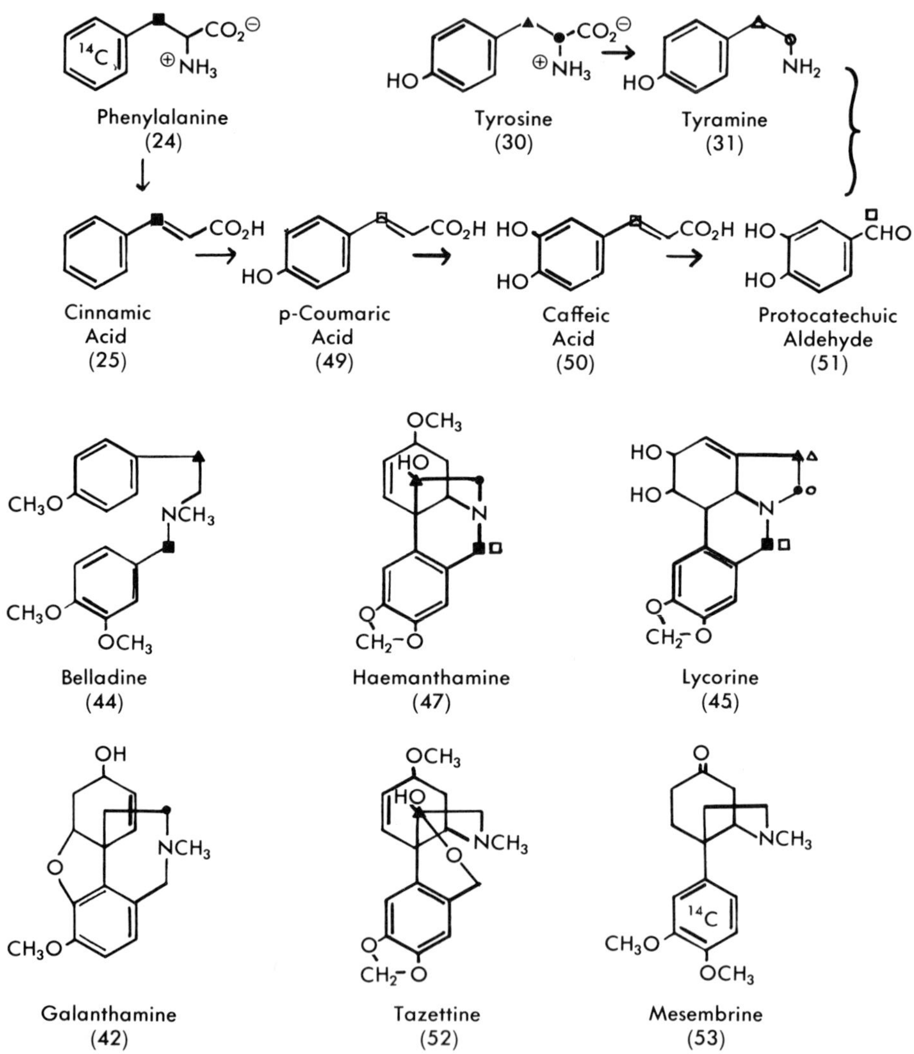

Figure 21.9 Incorporation of phenylalanine and tyrosine into Amaryllidaceae alkaloids.

biogenetic ideas first outlined by Barton and Cohen represent biosynthetic reality.

The origin of the C_6–C_2–N–C_1–C_6 system of norbelladine from primary metabolites remains to be discussed. Figure 21.9 summarizes the available data established by degradation studies on the incorporation of activity from labelled samples of tyrosine (30) and phenylalanine (24) into several of the alkaloids. The C_6–C_2 unit is derived from tyrosine. Activity from tyramine (31) is also incorporated into the same unit in a predictable manner. Even though this has not yet been confirmed by incorporation studies with ^{14}C,^{15}N-labelled tracers, it is likely that the C_6–C_2–N moiety of norbelladine (43), and hence of the alkaloids, is supplied by tyrosine (30) via tyramine (31). Dopa (34) did not replace tyrosine as a precursor, confirming that m-hydroxylation of the C_6–C_2 unit takes place at a late stage. Phenylalanine does not serve as a precursor of the C_6–C_2 unit (cf. Section 3.1.).

The C_6–C_1 unit is derived from phenylalanine (24), but not from tyrosine (30) or dopa (34). Label from β-^{14}C-phenylalanine is incorporated into this unit specifically. Activity from α-^{14}C- or carboxyl-^{14}C-phenylalanine does not enter the alkaloids. Since 3-^{14}C-cinnamic, -p-coumaric and -caffeic acid [(25), (49), (50)] serve as specific precursors, it is evident that the required hydroxylation of the C_6–C_1 precursor takes place at the level of cinnamic acid. This is borne out by the finding that protocatechuic aldehyde (51), but not p-hydroxybenzaldehyde, is incorporated. Dihydroxylation of cinnamic acid thus precedes sidechain cleavage. The mechanism of the shortening of the cinnamic sidechain is unknown.

It must be inferred from the available tracer evidence that norbelladine (43) originates by reduction of the condensation product of tyramine (31) with protocatechuic aldehyde (51), intermediates which originate from tyrosine (30) and phenylalanine (24), respectively, by the sequence outlined in Figure 21.9.

3.4 Benzylisoquinolines and Related Alkaloids

The classical biogenetic speculations of Robert Robinson, which placed a benzyltetrahydroisoquinoline, such as norlaudanosoline (64), at the hub of the postulated metabolic events leading to several families of alkaloids, not only served as a working hypothesis for biosynthetic investigations but correctly predicted many of the findings of recent tracer experiments.

The derivation of norlaudanosoline (64), from two C_6–C_2 units [(35) and (63)] related to dopa (34), has been suggested earlier by Winterstein and Trier. Dopa was assumed to arise in plants, as it does in animal tissues, by hydroxylation to tyrosine (30). Tracer experiments with α-^{14}C-tyrosine

(30) and 1-^{14}C-dopamine (35) established specificity of incorporation of these substrates into the benzylisoquinolines and their biogenetic relatives. In each case activity from α-^{14}C-tyrosine was located at two carbon atoms, activity from 1-^{14}C-dopamine at a single site within the products (Figure 21.10).

Tyrosine (30) Dopamine (35)

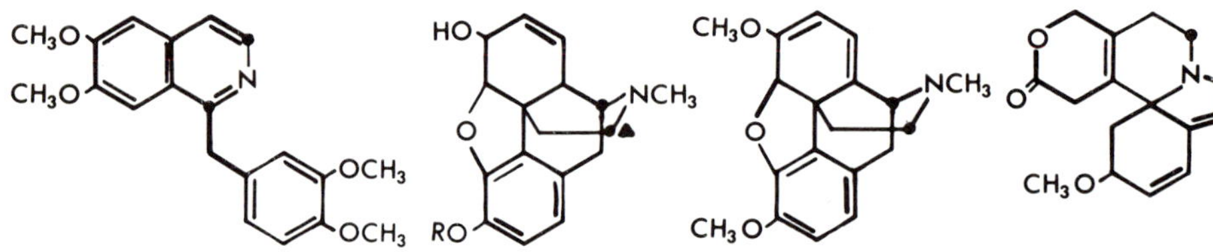

Papaverine (54) (55) R=H Morphine (56) R=CH_3 Codeine Thebaine (57) β-Erythroidine (58)

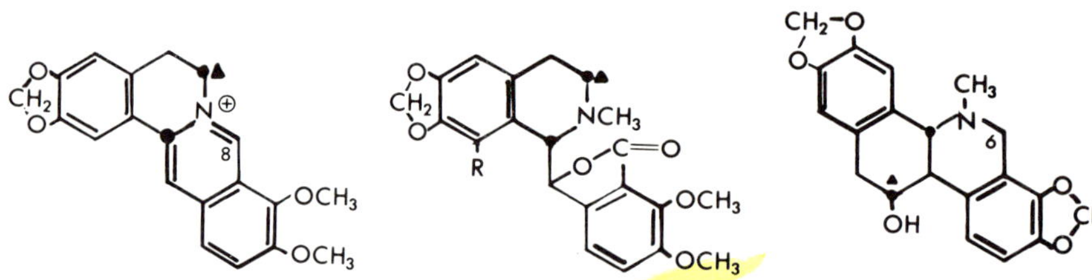

Berberine (59) (60) R=H Hydrastine (61) R=OCH_3 Narcotine Chelidonine (62)

Figure 21.10 Incorporation of label from α-^{14}C-tyrosine (•) and 1-^{14}C-dopamine (▲) into benzylisoquinoline and related alkaloids.

The incorporation data establish the derivation of the products from one precursor, tyrosine (30). These results suggest, but do not prove, a common route for the early steps of their biosynthesis. The data are consistent with the hypothesis that all the structural variants originate from a common intermediate, as envisaged by classical biogenetic concepts. Subsequent experiments confirm the validity of this view. Norlaudanosoline (64), the common intermediate of biogenetic hypothesis, as well as some of its O- and N-methylated derivatives, served as specific precursors of alkaloids belonging to the morphine, aporphine, erythrina, protoberberine, protopine, phthalideisoquinoline, and benzophenanthridine series (*vide infra*).

Since α-^{14}C-tyrosine (30) leads to products labelled at two centers, the common benzyltetrahydroisoquinoline intermediate is generated from two C_6–C_2 fragments derived from this amino acid. Dopa (34), which is also found to serve as a precursor, is presumably the next stage. Since activity from 1-^{14}C-dopamine (35) enters only a single carbon within the products, only one of the C_6–C_2 fragments is supplied by dopamine (35), obtainable from dopa (34) by decarboxylation. The other C_6–C_2 moiety cannot be 3,4-dihydroxyphenylacetaldehyde, a compound derivable from dopamine by amine-oxidase catalyzed oxidation, but is probably derived from 3,4-dihydroxyphenylpyruvic acid (63), the deamination product of dopa. It is not known whether the carboxyl group of this α-keto acid is detached before, during or following its condensation with dopamine. Figure 21.11 summarizes the steps from tyrosine to norlaudanosoline.

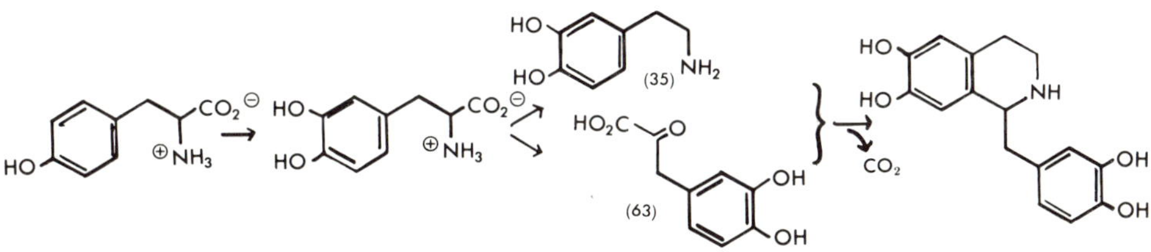

Tyrosine (30)   Dopa (34)   (35) Dopamine (63) 3,4-Dihydroxyphenyl-pyruvic Acid   Norlaudanosoline (64)

Figure 21.11 Origin of norlaudanosoline.

The norlaudanosoline-derived alkaloids fall into two major subgroups. The first of these includes the morphine (55), the aporphine [e.g., (78)] and the aromatic erythrina [e.g., (86)] bases, all of which contain a $C_{16}N$ nucleus, corresponding to that of the benzylisoquinoline precursor, which can thus supply the intact skeletons of these compounds. A further carbon atom is required to complete the skeletons of the second group, which includes the protoberberine [e.g., (59)], phthalideisoquinoline [e.g., (60)], protopine (67) and benzophenanthridine [e.g., (62)] bases. This additional one-carbon fragment, the "bridge" carbon [C(8)] of the protoberberines and protopine, the carbonyl carbon of the phthalideisoquinolines, and C(6) of the benzophenanthridines, is generated from a methyl group. Activity from methyl-^{14}C-methionine is incorporated specifically into this site in the nucleus of hydrastine (60), narcotine (61) and berberine (59).

The same source also supplies the peripheral one-carbon units, the N-methyl, O-methyl and methylenedioxy carbons, which are found in alkaloids of both subgroups. Methionine (91) serves as a one-carbon donor in many biochemical systems. Transfer of its S-methyl group to oxygen, nitrogen, and in some instances also to carbon, is catalyzed by specific enzymes. S-Adenosylmethionine is an intermediate in these reactions, in which the intact CH_3- group is transferred to the acceptor. It was likely that, in the present context, the acceptor was norlaudanosoline (64), and it was to be expected that N- and O-methylated derivatives of norlaudanosoline played a role as intermediates in the elaboration of the alkaloids.

Tracer experiments confirmed this inference. Several methylated derivatives of norlaudanosoline have been prepared with tagged atoms, and their incorporation into various alkaloids has been tested. These experiments, carried out by the schools of Barton and of Battersby, first with racemic compounds and later with optically pure precursors, have yielded important information on the final stages of the biosynthesis of the alkaloids. Since the fate of labelled methyl groups was to be investigated and the occurrence of partial or total demethylation of the tagged substrates and *de novo* incorporation of the label had to be excluded, these tracer experiments were carried out with multiply labelled compounds. If the distribution of multiple label in the product corresponds to that in the precursor, intact incorporation is demonstrated (Figure 21.12).

These experiments established the origin of the methylenedioxy function (cf. Section 3.3) and of the berberine bridge, by providing evidence that the methylated benzylisoquinoline precursor is converted into the products *in toto*. Methylenedioxy groups are formed by oxidative cyclization of an o-methoxyphenol function. The berberine bridge, and the corresponding carbon atom in the skeletons of protopine (67), narcotine (61) and chelidonine (62), are derived from the N-methyl group of the precursor. The interrelationship of the berberine-bridged alkaloids and their derivation from a protoberberine skeleton by oxidative steps had been an early contribution to biogenetic hypothesis. The biosynthetic conversion of a tetrahydroprotoberberine into a phthalideisoquinoline, a protopine, and a benzophenanthridine base has been demonstrated by tracer experiments.

Almost all the plants in which alkaloids related to norlaudanosoline are found contain compounds belonging to more than one skeletal type. A control mechanism must exist which directs norlaudanosoline into alternative routes leading to the different structural variants. Control of biochemical processes is, in general, exerted at the enzyme level. Metabolic homeostasis is often maintained by inhibition and re-activation of an enzyme, catalyzing an early step of a given metabolic sequence, by an end-product of the sequence. The enzymology of the biosynthetic processes leading to the norlaudanosoline alkaloids is entirely unexplored, and it is unknown whether the processes are controlled by such feedback inhibition. Evidence is accumulating, however, which can be interpreted in terms of chemical control. Two factors, methylation pattern and chirality,

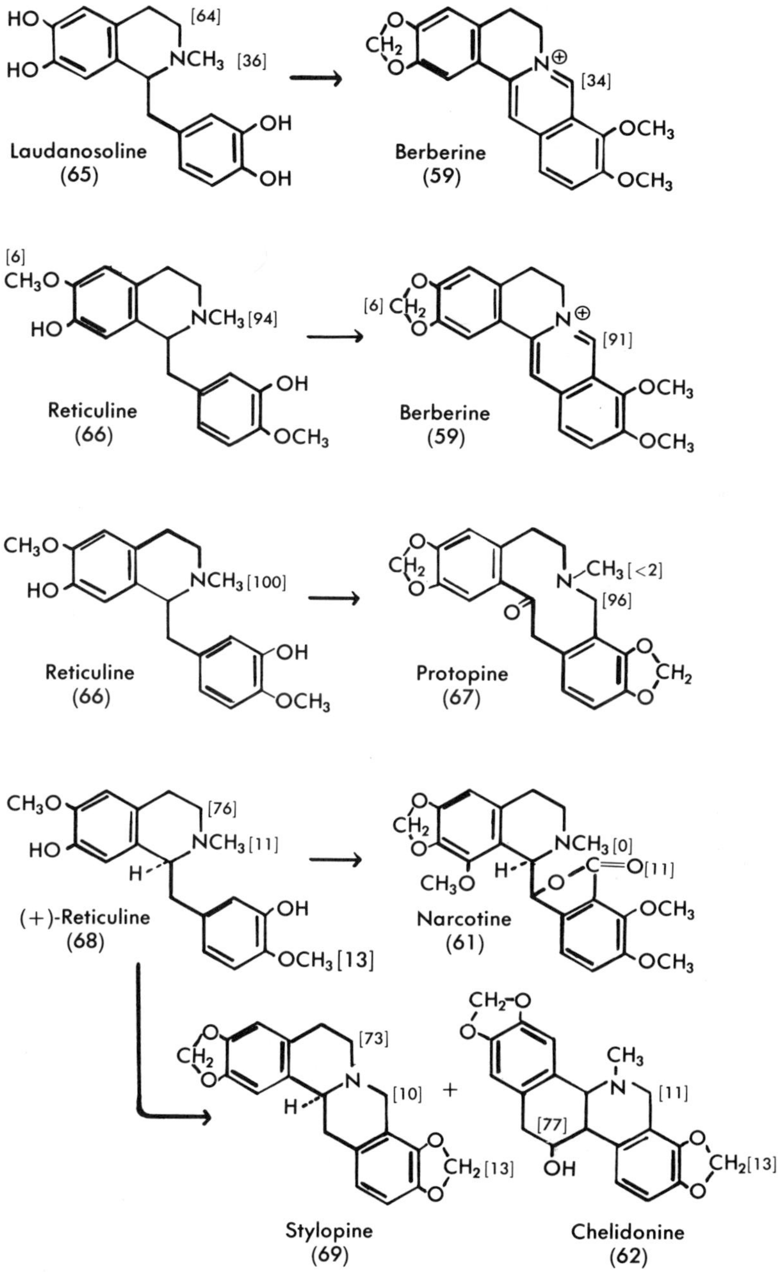

Figure 21.12 Incorporation of multiply labelled precursors into norlaudanosoline-derived alkaloids (Numbers are relative specific activities).

appear to direct the tetrahydroisoquinoline precursor into competing pathways.

The importance of the methylation pattern of the benzylisoquinoline precursor as a factor controlling the biosynthetic pathways is indicated by experiments with different O-methylated precursors.

Only one of these, reticuline (66), serves as a precursor of the morphine alkaloids in the opium poppy. Reticuline, a normal constituent of this poppy, has been established as an intermediate in the biosynthesis of thebaine (57) from carbon dioxide. It is the (−)-isomer (71 which corresponds stereochemically to thebaine (57), codeine (56) and morphine (55), and is their precursor. The (+)-isomer (68) of reticuline corresponds in chirality to narcotine (61), found in the opium poppy, and also to (−)-canadine (tetrahydroberberine) and (−)-stylopine (69), found in other plants. It is (+)-reticuline (68) which is preferentially incorporated into narcotine (61), into stylopine (69) and chelidonine (62), as well as into berberine (59) and protopine (67) (Figure 21.12). It appears that the berberine-bridged ($C_{17}N$) alkaloids are derived from one enantiomer of reticuline, and the non-berberine-bridged ($C_{16}N$) alkaloids from the other. An exception is sinomenine, a $C_{16}N$ alkaloid derived from (+)-reticuline (68). Protosinomenine, the structural isomer of reticuline whose O-methylation pattern corresponds to that of the product, is not incorporated into sinomenine. The O-methylation pattern of protosinomenine [cf. (79)] leads to the erythrina alkaloids [e.g., erythratine (85)]. The aporphine bases [e.g., isothebaine (78)] are derived from (+)-orientaline (75), yet another structural isomer of reticuline.

The mode of cyclization of the precursor is thus determined by the position of its free phenolic groups. It appears that the process of O-methylation directs cyclization by protection, and consequent inactivation, of phenolic groups in a specific manner. The lack of incorporation of the "wrongly" methylated benzylisoquinolines indicates that the enzymic methylation processes are highly specific and that they are irreversible. The fact that the fully O-methylated norlaudanosoline is not incorporated into any of the alkaloids of the opium poppy indicates that biosynthesis is blocked if free phenolic groups are not available.

The evidence is entirely in keeping with the predictions of Barton and Cohen, who rationalized the biogenesis of a large number of natural products, including the alkaloids here discussed, on the basis of oxidation of phenols by one-electron transfer processes, and the pairing of the radicals, so generated.

The postulated mechanism for the biogenesis of a number of alkaloids, derived from O-methylnorlaudanosolines by intramolecular phenol coupling processes and subsequent rearrangements, is shown in Figure 21.13. In each case evidence is available which provides experimental sup-

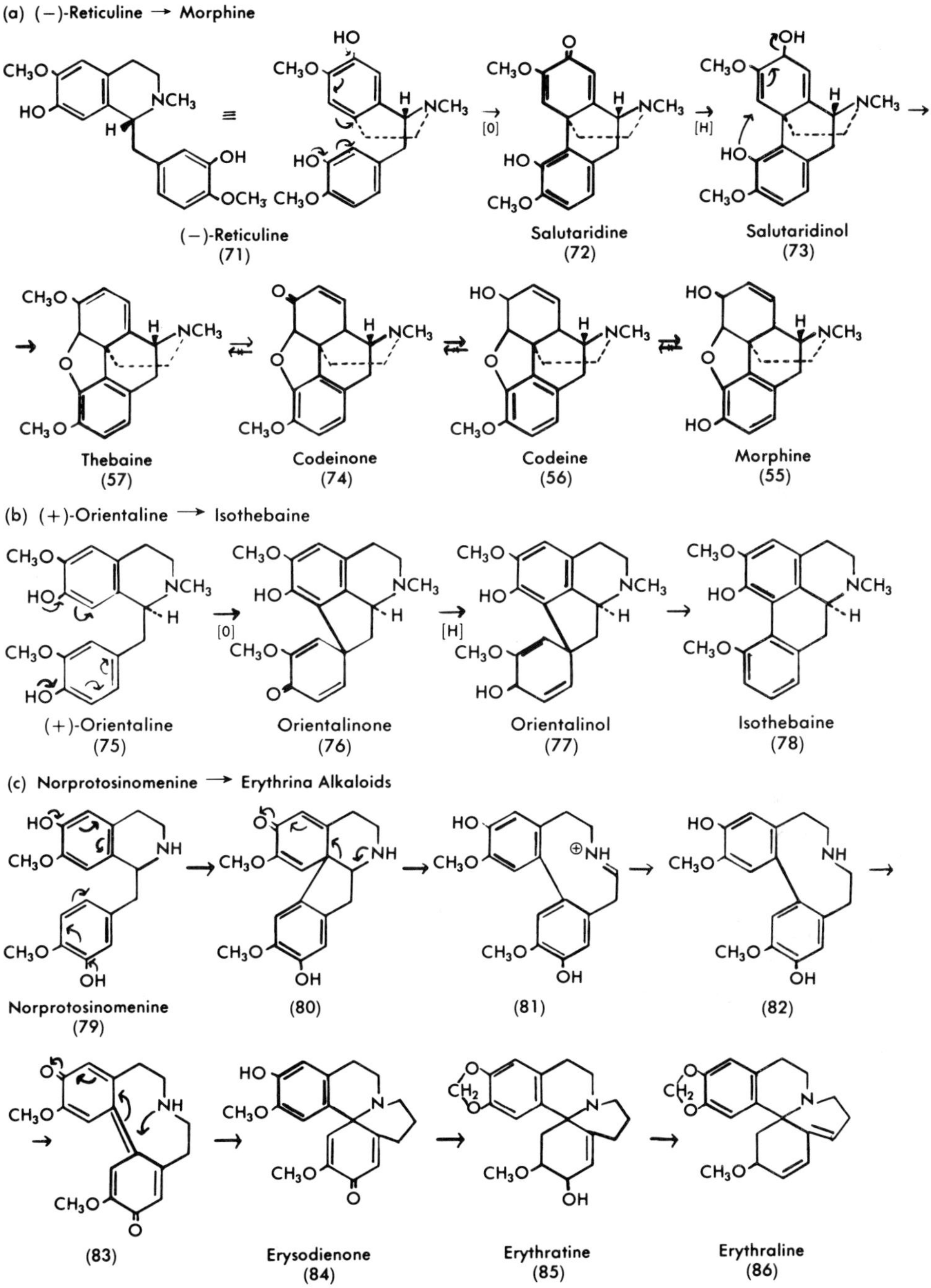

Figure 21.13 Phenol coupling processes in the biogenesis of norlaudanosoline-derived alkaloids.

port for the validity of the proposed sequences. Relevant tracer results with multiply labelled, sterochemically resolved substrates which had been predicted as precursors have been referred to earlier. In addition, the normal presence of these substrates [e.g., reticuline (66) and orientaline (75)] in the experimental plants has been demonstrated in most instances. The absolute stereochemistry of precursor and product has been found to correspond, and several of the postulated intermediates [e.g., salutaridine (72)] have been recognized as alkaloids in their own right.

One of the most dramatic confirmations of Barton's hypothesis came from investigations of the metabolic interrelationship of thebaine (57), codeine (56) and morphine (55) in the opium poppy. Since it had always been assumed that stepwise methylation of peripheral groups was a terminal process, a biogenetic sequence leading from a nonmethylated precursor, morphine (55), by way of a partially methylated intermediate, codeine (56), to a fully methylated endproduct, thebaine (57), was taken for granted. Contrary to this notion, Barton's hypothesis demanded the reverse, demethylation, sequence. Several lines of evidence confirm that thebaine is the first of these bases to be formed and that codeine and morphine are derived from it in successive irreversible demethylation steps. A rate study of the incorporation of ^{14}C-tyrosine into the alkaloids indicated that activity passed from thebaine through codeine to morphine. When opium poppies were exposed for a few hours to an atmosphere containing carbon-14 dioxide, the specific activities of the alkaloids showed the order thebaine > codeine > morphine. When labelled thebaine was administered to poppy plants, label entered both codeine and morphine. When labelled codeine was administered, morphine became active, but thebaine did not. Label from morphine failed to enter either codeine or thebaine. It was shown by partial degradation that the transfer of radioactivity was not due to nonspecific methyl exchange. By analogous studies codeinone (74) was shown to be an intermediate between thebaine and codeine.

In the area of the benzylisoquinoline and related alkaloids biogenetic hypothesis has been at its most successful in providing structural and mechanistic models which not only gave impetus and direction to biosynthetic investigations but were shown by these investigations to be essentially correct. In the same area biogenetic hypothesis suffered its most brilliant failure.

Structurally, the ipecacuanha alkaloids, of which emetine (89) is a prominent example, appear to be related to the norlaudanosoline series. The correct structure was deduced on the basis of biogenetic considerations. It was postulated that oxidative fission ("Woodward fission") of a norlaudanosoline-derived tetrahydroprotoberberine (87) system might yield an intermediate (90) which on condensation with a dopanine unit would yield the emetine skeleton (Figure 21.14). This biogenetic scheme acquired verisimilitude when the postulated intermediate (90) (protometine) was shown to accompany emetine in the plant. Notwithstanding the success of this model as a tool of structural chemistry, it does not reflect biosynthetic processes. Tracer work shows that the emetine skeleton is not

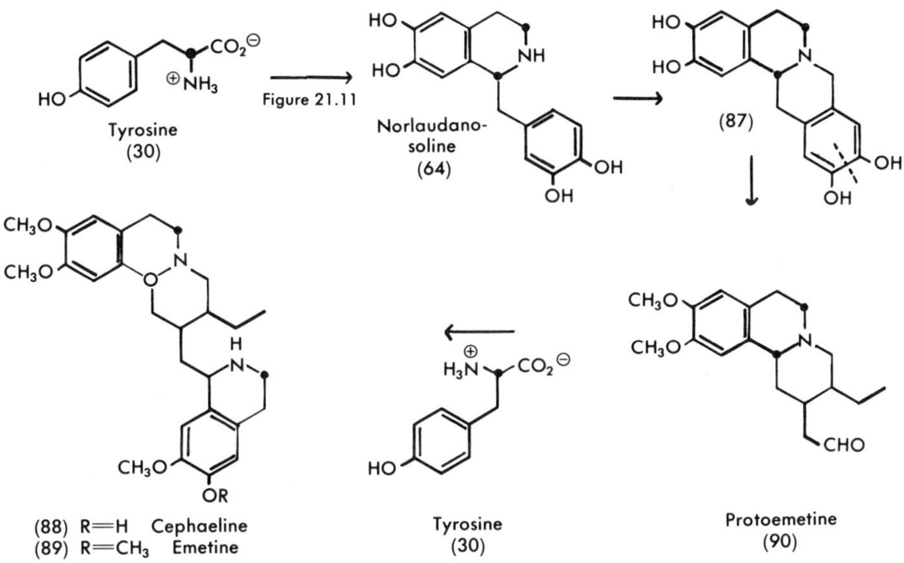

Figure 21.14 The Woodward fission.

related to norlaudanosoline. Degradation of radioactive cephaeline (88), a monophenolic analog of emetine (89), obtained from a feeding experiment with α-^{14}C-tyrosine (30), showed that activity was confined to two sites (● in 88), which were located within the phenylethylamine portions of each of the two isoquinoline nuclei. If norlaudanosoline (64) and a tetrahydroprotoberberine (87) had been intermediates in its formation, activity should have been present at a third carbon atom (○ in 88), corresponding to C(1) of norlaudanosoline. Furthermore, activity from one-carbon precursors entered the peripheral methyl groups, but not the center corresponding to the berberine bridge.

It is likely, but has not yet been proven, that the emetine skeleton (88) arises by reaction of two C_6–C_2 moieties, derived from tyrosine (30), with a C_{9-10} monoterpenoid unit of the corynanthe type (119) (Section 4.3.).

3.5 Colchicine, a Modified 1-Phenylethylisoquinoline Alkaloid

Biogenetic speculations proved singularly unsuccessful in predicting the mode of formation of colchicine. Structural relations failed to serve as a guide to its origin. Early incorporation studies established phenylalanine and tyrosine as primary precursors of the ring system but did little to advance the understanding of the mode of formation of the alkaloid from its progenitors. Discovery of a new class of alkaloids paved the way to a clearer understanding of colchicine biosynthesis.

Phenylalanine (24) and cinnamic acid (25) serve as precursors of the phenylpropanoid moiety of colchicine (92). The correspondence of the carboxyl group of the phenylalanine sidechain to C(7) of colchicine was demonstrated by tracer studies. The α- and the β-carbon atom of phenylalanine and the corresponding carbon atoms of cinnamic acid enter C(6) and C(5), respectively (Figure 21.15). Tyrosine (30) is not incorporated into this portion of the molecule.

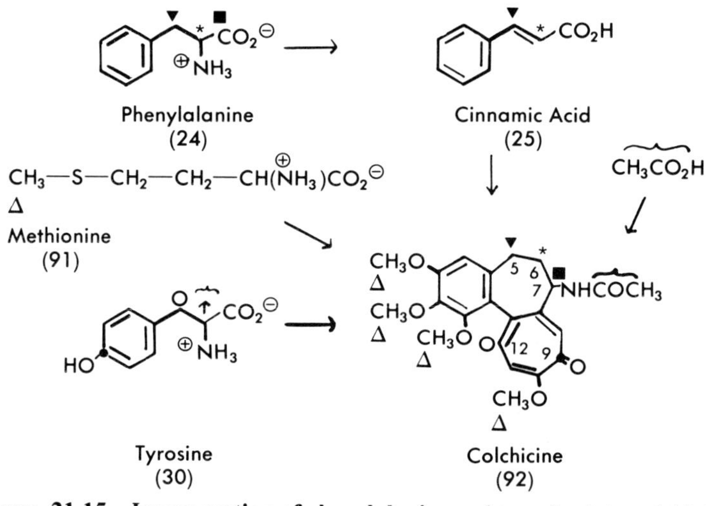

Figure 21.15 Incorporation of phenylalanine and tyrosine into colchicine.

The tropolone ring of colchicine, unlike tropolones of fungal origin which are derived from acetate, is generated from tyrosine (30). Activity from β-^{14}C-tyrosine enters C(12), whereas C(9) of colchicine originates specifically from C(4) of the aromatic nucleus of tyrosine (Figure 21.15). Thus, it appeared that the tropolone ring of the alkaloid represented the six nuclear carbons of tyrosine, expanded into a seven-membered ring by inclusion of the β-carbon of the sidechain. The α-carbon and the carboxyl group of the tyrosine sidechain are lost in the course of biosynthesis of the colchicine nucleus.

The tracer evidence fell into place with the elucidation of the structure of androcymbine (97 with H replacing CH_3 at *), a dienone alkaloid which accompanies colchicine and corresponds to it in stereochemistry. Reduction of O-methylandrocymbine (97) yields the phenylethylisoquinoline (95). Androcymbine may be derived from such a structure by phenol oxidation and coupling. The derivation of colchicine from phenylethylisoquinoline precursors, and its relationship to the androcymbine skeleton was demonstrated by a series of tracer experiments with doubly labelled precursors. Of particular significance was the finding that the $^{14}C/^{15}N$ ratio of colchicine (92), isolated from a feeding experiment with the phenylethylisoquinoline (94), doubly labelled as shown, matched that of the precursor. The results of these experiments provide strong evidence for several of the postulated steps of colchicine biosynthesis, shown in Figure 21.16. The sequence involves introduction into the dienone (97) of

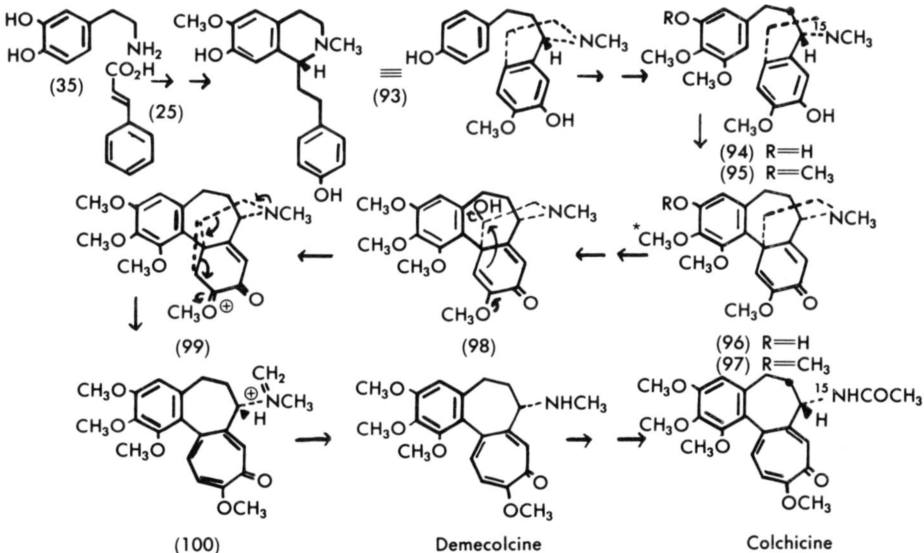

Figure 21.16 The biosynthesis of colchicine.

a hydroxyl or related group, whose elimination in a subsequent step provides the driving force for the ring expansion (98)→ (99)→ (100), whose occurrence was inferred from the tracer work with tyrosine.

4. THE BIOSYNTHESIS OF INDOLE AND QUINOLINE ALKALOIDS

4.1 Tryptophan as the Primary Precursor of Indole Alkaloids

The amino acid tryptophan (102), a constituent of most proteins, contains an indole nucleus. A biogenetic relationship between tryptophan and the indole alkaloids was postulated as early as 1919, when the structure of harmine, the first alkaloid to be recognized as a member of the β-carboline group, was elucidated. Biogenetically modelled syntheses of a variety of indole alkaloids have been carried out, starting from tryptamine (Chapter 9). Tracer experiments confirm the biosynthetic derivation of the indole alkaloids from tryptophan.

The pattern of incorporation of activity from α-^{14}C-or β-^{14}C-tryptophan into prototypes of the structural variants among these alkaloids was established by chemical degradation. The results, summarized in Figure 21.17, show that in each case the label is confined to a single site within the alkaloid. An indolylethylamine moiety is easily discernible in the

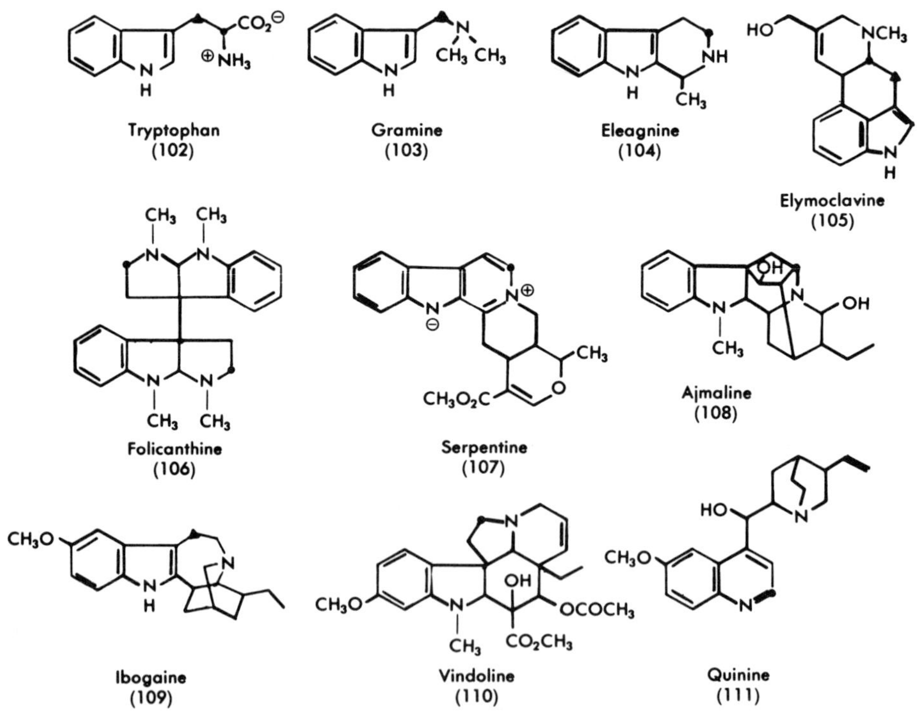

Figure 21.17 Incorporation of label from α-^{14}C-tryptophan (•) or β-^{14}C-tryptophan (▲) into a number of alkaloids.

skeletons of eleagnine (104), elymoclavine (105), serpentine (107), ajmaline (108), ibogaine (109) and vindoline (110). The skeleton of folicanthine (106) is accounted for in its entirety by two such moieties. Even though the incorporation of multiply labelled samples of tryptophan has been tested only in the case of elymoclavine (105), it is likely that tryptophan supplies the intact fragment. The carboxyl group of tryptophan is lost in the course of biosynthesis of the alkaloids. Since label from tryptamine is not incorporated into the clavine bases (e.g., elymoclavine), this loss occurs at a late stage in their biosynthesis. Incorporation of activity from labelled tryptamine into the other alkaloids has not been tested. It is unknown at what stage of their biosynthesis decarboxylation occurs. Nor is it known at what point methoxy functions [e.g., ibogaine (109), vindoline (110)] are introduced into the indole system. The chemical anatomy of the specific incorporation of tryptophan into the quinoline alkaloid quinine (111) is not immediately apparent from its structure. The origin of this compound will be referred to later (Section 4.5.).

The non-tryptophan-derived carbon atoms of the indole alkaloids arise from acetate. Acetate serves as the precursor of the two-carbon unit which completes the skeleton of eleagnine (104), activity from 1-^{14}C-acetate being confined to C(1) of its β-carboline nucleus. The five-carbon moiety which combines with the indolylethylamine fragment to yield the ergoline skeleton is derived from mevalonic acid (112), the biochemical isoprene precursor which, in turn, is generated from three acetate units. Two mevalonate-derived entities are required in the biosynthesis of the pentacyclic β-carboline alkaloids (107–110) and their biogenetic relatives.

4.2 Ergoline Alkaloids

L-Tryptophan is the precursor of the indolylethylamine unit of the ergoline skeleton of the clavine and ergot alkaloids (Chapter 10). Incorporation of D-tryptophan takes place only after conversion to the L-enantiomer, presumably via the corresponding keto acid. Incorporation of L-tryptophan must, in turn, occur with inversion of configuration, since the alkaloids possess a D-configuration at the asymmetric center derived from the α-carbon atom of L-tryptophan.

The tryptophan-derived fragment accounts for all but five carbon atoms of the ergoline sketeton. These five carbons have an isoprenoid arrangement and have been shown to be derived from mevalonic acid (112). Radioactivity from 2-^{14}C-mevalonic acid was incorporated almost exclusively into the hydroxymethyl group of elymoclavine (105), into the C-methyl group of agroclavine (113) and into the carboxyl group of lysergic acid.

Tracer evidence thus accounts for the origin of all atoms of the ergoline skeleton from tryptophan and mevalonic acid. The mode of combination of the two fragments has also been investigated. Two potential intermediates of the pathway have been considered. One of these, 4-dimethylallyltryptophan (114), might arise by electrophilic attack of isopentenyl pyrophosphate at C(4) of the indole nucleus of tryptophan. The other (115) might be formed by attack at the α-carbon of the tryptophan sidechain, with simultaneous decarboxylation and inversion of configuration. Labelled samples of both putative intermediates have been prepared and their incorporation into the ergoline skeleton tested. Label from both compounds entered elymoclavine (105), incorporation of activity from 4-dimethylallyltryptophan (114) being consistently higher than from (115). Since specificity of incorporation was not established, and since neither compound was incorporated as efficiently as tryptophan (102), the status of these intermediates in ergoline biosynthesis remains unsettled, even though a pathway via 4-dimethylallyltryptophan (114) is regarded as more likely. Conclusions regarding precursor sequences, based on a comparison of incorporation efficiencies of different substrates into a product are of doubtful validity. Such data are likely to reflect differences in rates of uptake of the various substrates by the cell rather than their relative positions in biosynthetic sequences. The mode and mechanism of combination of the tryptophan-derived and the mevalonate-derived

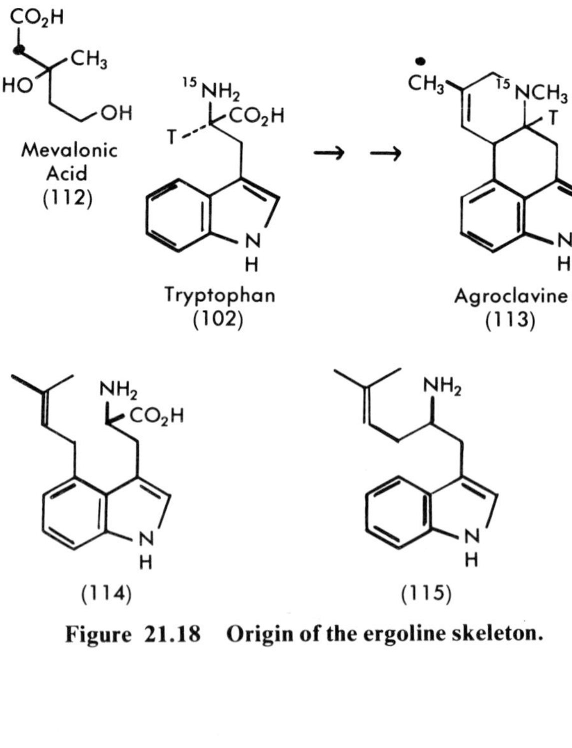

Figure 21.18 Origin of the ergoline skeleton.

fragments in the biosynthesis of the ergoline skeleton remains to be clarified.

4.3 The Non-tryptophan-derived Fragment of the Pentacyclic β-Carboline and Related Alkaloids

The skeletons of a large number of indole alkaloids contain, in addition to the tryptamine moiety, nine or ten other carbon atoms in a variety of arrangements. Several biogenetic schemes were proposed to account for the derivation of these multicarbon units. Tracer results support the hypothesis, first put forward by Wenkert, that this fragment is of monoterpenoid origin. According to this hypothesis the requisite nine or ten-carbon fragments may be formally derived by rearrangement of a cyclopentanoid monoterpene skeleton (117), which is itself generated from two mevalonic acid (112) units, by way of isopentenyl pyrophosphate and geraniol (116). Indeed, activity from 2-^{14}C- and 3-^{14}C-geraniol is incorporated entirely into the predicted site of catharanthine (121), vindoline (110) and quinine (111), alkaloids representing the different structural variants of the C_{9-10} unit. Activity from various radiomers of mevalonic acid also enters in a specific manner into the alkaloids. In accord with expectation, half of the label from 4-^{14}C-mevalonate is found at the carbon atom derived from 2-^{14}C-geraniol, half of the label from 3-^{14}C-mevalonate at the atom derived from 3-^{14}C-geraniol.

Contrary to expectations only one-quarter rather than one-half of the activity of the alkaloids, derived from 2-^{14}C-mevalonate, is recovered from the carboxyl carbon atom of catharanthine (121), ajmalicine (122), and vindoline (110). This must indicate that, at some stage in the biosynthetic process, C(2) and C(6) of one of the mevalonate precursor units become equivalent. A similar loss of identity of C(2) of mevalonic acid has been observed in investigations of the biosynthesis of skytanthine (124), a monoterpenoid alkaloid containing a carbon skeleton corresponding to that of the postulated cyclopentanoid C_{10} precursor of the indole alkaloids, and in studies of the origin of several other monoterpenes (e.g., plumieride). It is of interest that it is one of these C-atoms which is missing in the skeletons of ibogaine (109) (C_9 "iboga" type unit), quinine (111) (C_9 "corynanthe" type unit) and ajmaline (108) (C_9 "corynanthe" type unit).

The reason for this loss of identity of C(2) and C(6) of one mevalonate precursor, and the stage on the biosynthetic route at which it occurs, are unknown.

Current work is directed towards the identification of intermediates in the pathway beyond geraniol. Some evidence has been obtained that loganin (123) or a closely related monoterpene lies on the route of biosynthesis.

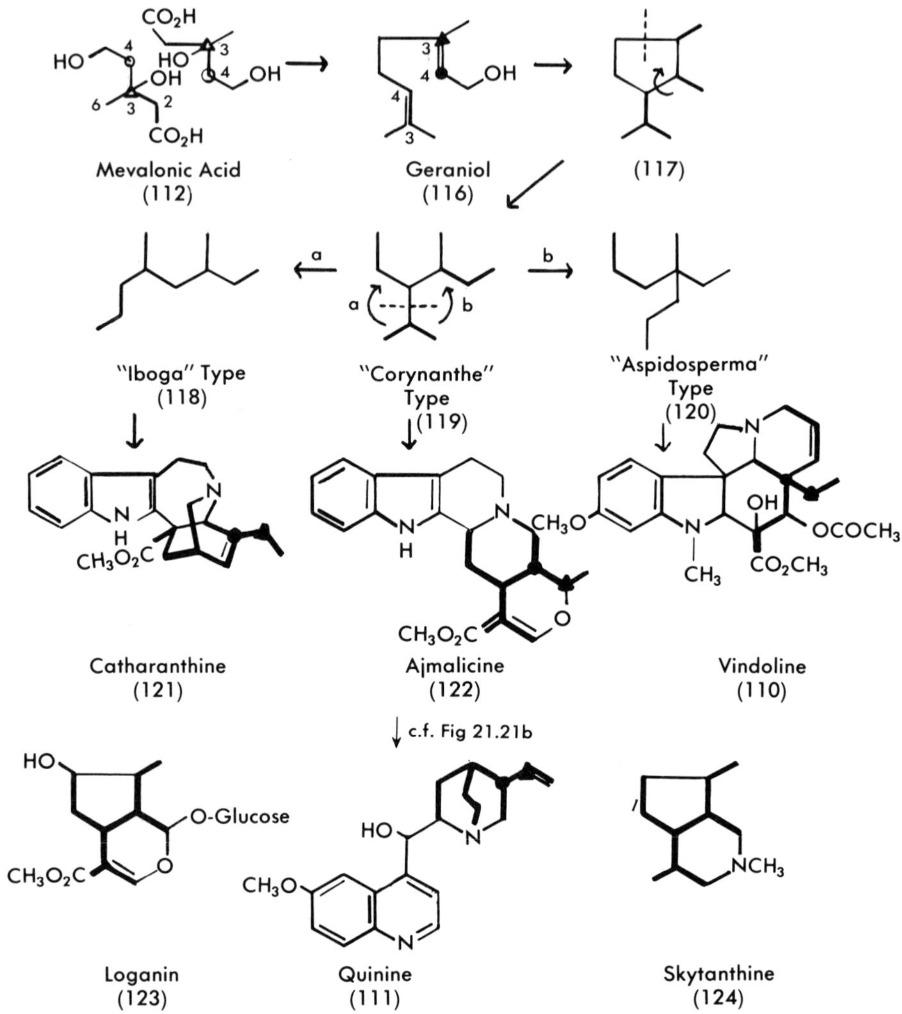

Figure 21.19 Incorporation of mevalonic acid and geraniol into the non-tryptophan-derived fragment of the indole alkaloids.

4.4 Gramine

Gramine (103) does not contain an indolylethylamine unit. One less carbon atom than in tryptophan (102) separates the sidechain nitrogen from the indole nucleus. In terms of biogenetic hypothesis specific incorporation of activity from β-^{14}C--tryptophan into the methylene carbon atom of the sidechain of gramine was unexpected. It had been assumed that tryptophan supplied only the indole nucleus and that gramine was produced in the barley plant in a manner analogous to its chemical synthesis from indole, formaldehyde, and dimethylamine. Even though incorporation of activity had been non-random, the experiment with β-^{14}C-tryptophan did not eliminate the possibility that incorporation of label had been indirect. The labelled precursor might have been cleaved, by an enzymic reaction analogous to one known to occur in microorganisms, into indole and a three-carbon unit, which on further degradation might have yielded a labelled one-carbon fragment, which could have entered the product in a specific manner. A further experiment eliminated this possibility. Tryptophan, doubly labelled at C(2) of the nucleus and at the β-carbon of the side-chain, yielded radioactive gramine whose activity was shown to be confined to the two predicted positions. The ratio of activities at these two labelled centers of gramine was identical with the ratio of activities of the two

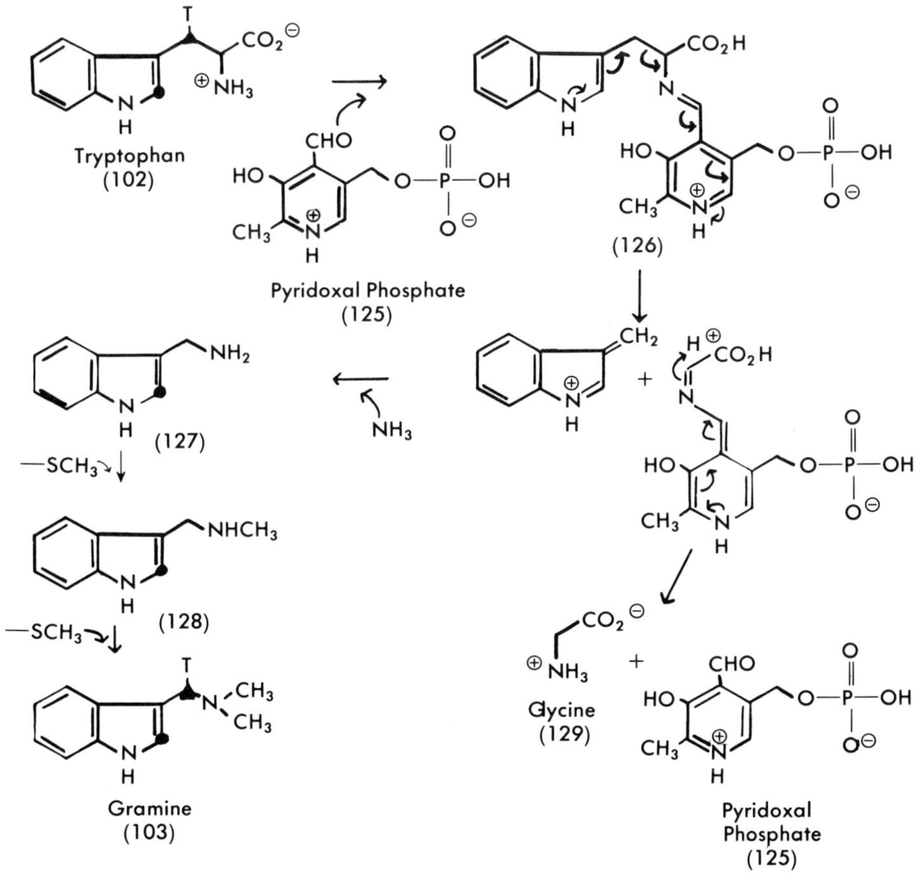

Figure 21.20 Biosynthesis of gramine.

labelled atoms in the precursor. This result shows that the β-carbon of the tryptophan sidechain remains attached to the indole nucleus throughout all stages of the pathway to gramine. Furthermore, an experiment with doubly labelled β-^{14}C, β-^{3}H-tryptophan demonstrates that the methylene group of the tryptophan sidechain remains intact. The resulting gramine showed a $^{14}C/^{3}H$ ratio identical with that of the precursor, and both tracers were found to reside exclusively at the methylene group of gramine. The evidence is entirely compatible with a hypothetical mechanism (Figure 21.20) involving α,β-cleavage of the tryptophan sidechain of an enzyme-bound tryptophan-pyridoxal phosphate adduct (126). 3-Aminomethylindole (127) and 3-methylaminomethylindole (128), the immediate precursors of gramine envisaged by this hypothesis, do in fact occur in barley and yield gramine when incubated with S-adenosylmethionine, the general biochemical methyl donor, in the presence of barley shoot extract. Both compounds serve as precursors of gramine, when administered to excised barley shoots. Tracer work on the precursors of gramine is complete. The final evidence for the proposed mechanism must be obtained by a search for the postulated pyridoxal-requiring enzyme. Indeed, a crude enzyme preparation has been obtained which catalyzes the conversion of tryptophan (102) into gramine (103) and glycine (129).

4.5 Tryptophan and Anthranilic Acid as Precursors of the Quinoline Alkaloids

The quinoline nucleus is found in a variety of naturally occurring compounds. Studies of the biosynthesis of a number of these indicate the existence of several pathways leading to this ring system.

Only one of these pathways is fully understood. Kynurenic acid (130) and its relatives, which are found in mammalian tissues, in insects, and in certain microorganisms, have been shown, by incorporation studies, by recognition of intermediates, and by elucidation of the reaction sequence at the enzyme level, to be metabolic products of tryptophan (102). The correspondence of the carbon skeleton of kynurenic acid to that of tryptophan is shown in Figure 21.21a.

The fragmentary knowledge of the biosynthesis of other quinoline derivatives rests entirely on tracer studies. Tryptophan serves as a precursor of the quinoline nucleus of quinine (111) (cf. Section 4.1.). The skeletal correspondence of quinine and tryptophan is illustrated in Figure 21.21b. This mode of conversion, in which C(2) of the indole nucleus of the precursor is preserved in the product thus differs from that of the kynurenic acid pathway, in which C(2) of the indole nucleus is extruded as formic acid.

The vinylquinuclidine system of quinine is of monoterpenoid origin,

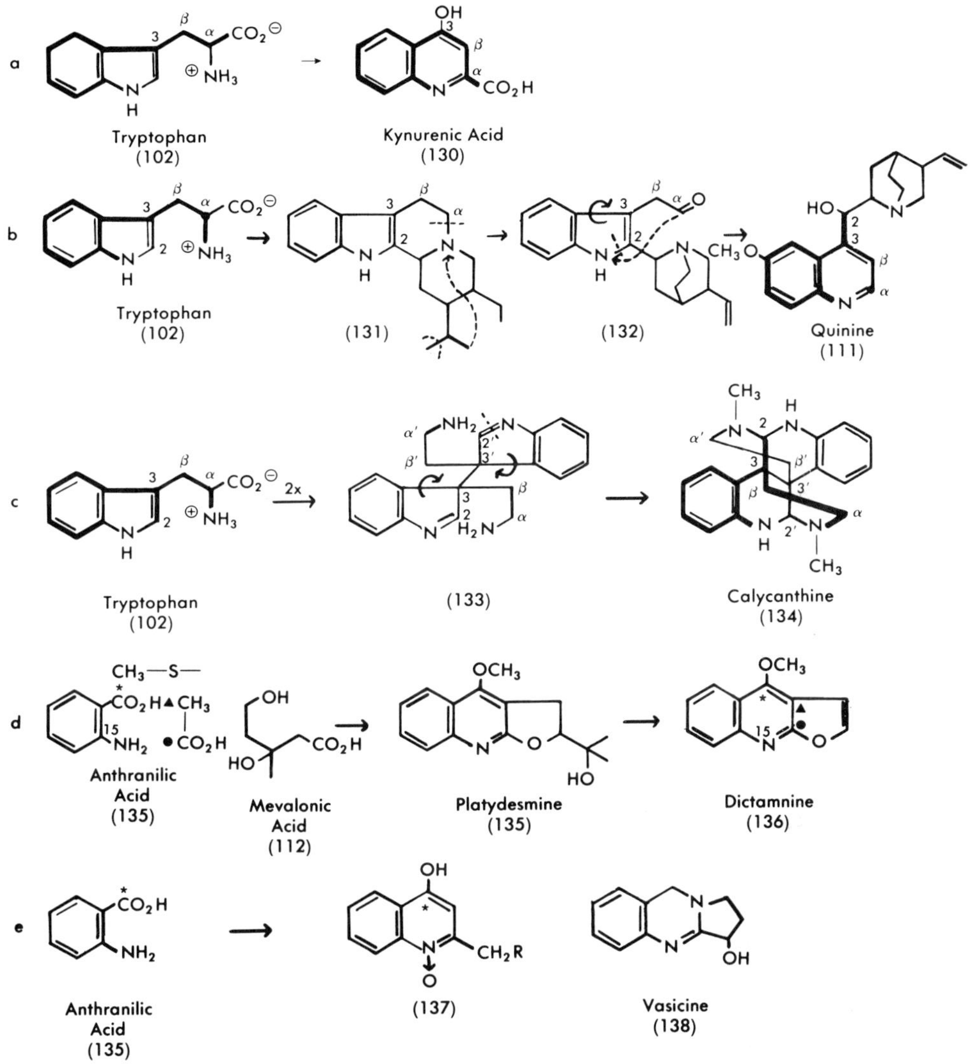

Figure 21.21 Biosynthesis of quinoline alkaloids.

activity from various radiomers of mevalonic acid and of geraniol being incorporated specifically (cf. Section 4.3.). The tracer evidence is consistent with the hypothesis (Figure 21.21b) that the quinine ring system arises by rearrangement of an indole alkaloid of the "corynanthe" type (cf. Figure 21.19). Compounds whose skeletal structure corresponds to (132), a postulated intermediate in this process, accompany quinine in cinchona bark.

Activity from labelled tryptophan enters calycanthine (134), an alkaloid containing two fused quinoline nuclei. Even though specificity of incorporation has not yet been demonstrated, it is likely that the alkaloid arises by rearrangement of a tryptamine dimer (133) (Figure 21.21c). It is from the same dimer that folicanthine (106) (cf. Section 4.1.) is very probably derived.

Anthranilic acid (135), rather than tryptophan, has been shown to be the nitrogenous precursor of several other quinoline alkaloids. Anthranilic acid can arise by biochemical degradation of tryptophan. Its benzene nucleus, carboxyl group and amino group arise, in this degradative process, from the phenyl ring, C(3) and the nitrogen atom of the indole nucleus of tryptophan, respectively. Anthranilic acid also serves as the precursor of tryptophan. This biosynthetic process is accompanied by decarboxylation, and only the aniline moiety of anthranilic acid is incorporated into the amino acid. If activity from carboxyl-^{14}C-anthranilic acid is incorporated specifically into C(4) of the quinoline nucleus of a product, the inference can be drawn that anthranilic acid is incorporated directly and that tryptophan is not an obligatory intermediate. If, in a feeding experiment with doubly labelled, carboxyl-^{14}C,^{15}N-anthranilic acid, the product shows a ^{14}C/^{15}N ratio identical with that of the precursor, it can be inferred that tryptophan does not lie on the pathway.

Radioactivity from carboxyl-^{14}C,^{15}N-anthranilic acid was confined to the predicted site of dictamnine (136), a furoquinoline alkaloid. Since the ^{14}C/^{15}N ratio of the product was identical with that of the anthranilic acid, decarboxylation had not taken place in the course of biosynthesis and anthranilic acid was thus incorporated directly, and not by way of tryptophan. The non-anthranilic carbon atoms of the pyridine ring of dictamnine were found, by unambiguous degradation of the product, to be derived from an acetate unit. No significant incorporation of activity from acetate into the remaining two-carbon unit of the furan nucleus was observed.

Since furoquinoline alkaloids, such as dictamnine, are accompanied by bases of the 2-isopropyl-2,3-dihydrofuroquinoline series [e.g., (135)], it is likely that such a structure is an intermediate in the biosynthesis of the furoquinoline system (Figure 21.21d). The two-carbon unit of its furan nucleus would represent a vestige of an isoprene unit, and should be derived from C(4) and C(5) of mevalonic acid (112). Attempts to demonstrate incorporation of activity from 5-^{14}C-mevalonic acid into dictamnine have not yet met with success, however.

Activity from carboxyl-^{14}C-anthranilic acid has been shown to enter a number of other quinoline and quinazoline alkaloids (Figure 21.21e). The origin of the non-anthranilic fragment of these bases is not firmly established.

5. CONCLUSION

The initial phase of the investigation of the biosynthesis of alkaloids is almost complete. Primary metabolites which account for the carbon skeletons of most of the structural groups of the alkaloids have been identified by incorporation studies in intact plants. In several groups, notably among the alkaloids derived from the aromatic amino acids, phenylalanine, tyrosine and tryptophan, a further stage has been reached. Significant advances have been made in the delineation of intermediates of the biosynthetic pathways, and in the clarification, on a purely chemical basis, of the mechanism of individual steps of the biosynthetic sequences.

The groundwork has been laid for an attack on the biochemical and plant physiological aspects of alkaloid biosynthesis. The enzymology of alkaloid formation and the significance of alkaloid biosynthesis as an aspect of the intermediary metabolism of plants provide challenging topics for future studies.

REFERENCES

SECTION 1

Alkaloid Biogenesis

E. Winterstein and G. Trier, "Die Alkaloide," p. 263, Borntraeger, Berlin, 1910.

R. Robinson, *J. Chem. Soc.* **111,** 876 (1917).

G. K. Hughes and E. Ritchie, *Rev. Pure Appl. Chem.* **2,** 125 (1952).

R. Robinson, "The Structural Relations of Natural Products," Clarendon Press, Oxford, 1955.

R. B. Woodward, *Angew. Chem.* **68,** 13 (1956).

E. Wenkert, *Experientia* **15,** 165 (1959).

E. Leete, "Biogenesis of Natural Compounds," 2nd edition. p. 953 (P. Bernfeld, Ed.), Pergamon Press, Macmillan Co., New York, 1967.

Alkaloid Biosynthesis

A. R. Battersby, *Quart. Rev.* **15,** 259 (1961).

K. Mothes and H. R. Schütte, *Angew. Chem. Intern. Ed. Engl.* **2,** 341, 441 (1963).

E. Leete, *Ann. Rev. Plant, Physiol.* **18,** 179 (1967).

I. D. Spenser, "Comprehensive Biochemistry," Vol. 20, p. 231 (M. Florkin and E. H. Stotz, Eds.), Elsevier Pub. Co., Amsterdam, 1968.

Metabolic Pathways

E. A. Adelberg, *Bacteriol. Rev.* **17,** 253 (1953).

B. D. Davis, *Advan. Enzymol.* **16,** 247 (1955).

Tracer Methods

D. P. Zilversmit, C. Entenman, and M. C. Fishler, *J. Gen. Physiol.* **26,** 325 (1943).

SECTION 2

Pyrrolidine alkaloids

Stachydrine

J. M. Essery, D. J. McCaldin, and L. Marion, *Phytochemistry* **1,** 209 (1962).

Nicotine

E. Leete and K. J. Siegfried, *J. Amer. Chem. Soc.* **79,** 4529 (1957).

R. F. Dawson, D. R. Christman, A. D'Adamo, M. L. Solt, and A. P. Wolf, *J. Amer. Chem. Soc.* **82,** 2628 (1960).

E. Leete, E. G. Gros, and T. J. Gilbertson, *Tetrahedron Letters* 587 (1964).

A. A. Liebman, B. P. Mundy, and H. Rapoport, *J. Amer. Chem. Soc.* **89,** 664 (1967).

T. J. Gilbertson and E. Leete, *J. Amer. Chem. Soc.* **89,** 7085 (1967).

Tropine

E. Leete, *J. Amer. Chem. Soc.* **84,** 55 (1962).

E. Leete, *Tetrahedron Letters* 1619 (1964).

J. Kaczkowski and L. Marion, *Can. J. Chem.* **41,** 2651 (1963).

Retronecine

C. A. Hughes, R. Letcher, and F. L. Warren, *J. Chem. Soc.* 4974 (1964).

W. Bottomley and T. A. Geissman, *Phytochemistry* **3,** 357 (1964).

Piperidine Alkaloids

Anabasine

E. Leete, *J. Amer. Chem. Soc.* **78,** 3520 (1956).

E. Leete, *J. Amer. Chem. Soc.* **80,** 4393 (1958).

E. Leete, E. G. Gros and T. J. Gilbertson, *J. Amer. Chem. Soc.* **86,** 3907 (1964).

Coniine

E. Leete, *J. Amer. Chem. Soc.* **86,** 2509 (1964).

Sedum and Lobelia Alkaloids

R. N. Gupta and I. D. Spenser, *Can. J. Chem.* **45,** 1275 (1967).

R. N. Gupta and I. D. Spenser, *Chem. Commun.* 84 (1968).

Lupinine and other Lupine Alkaloids

K. Mothes and H. R. Schütte, *Angew. Chem. Intern. Ed. Engl.* **2,** 341 (1963).

Lycopodine

R. N. Gupta, M. Castillo, D. B. MacLean, I. D. Spenser, and J. T. Wrobel, *J. Amer. Chem. Soc.* **90,** 1360 (1968).

SECTION 3

Phenylpropanoids

E. Leete and G. B. Bodem, *Tetrahedron Letters* 3925 (1966).

E. Leete, *Chem. Ind. (London)* 1088 (1958).

I. Imaseki, S. Shibata, and M. Yamazaki, *Chem. Ind. (London)* 1625 (1958).

Hordenine

E. Leete and L. Marion, *Can. J. Chem.* **31,** 126 (1953).

A. W. Frank and L. Marion, *Can. J. Chem.* **34,** 1641 (1956).

Cactus alkaloids

E. Leete, *J. Amer. Chem. Soc.* **88,** 4218 (1966).

A. R. Battersby, R. Binks, and R. Huxtable, *Tetrahedron Letters* 563 (1967).

S. Agurell, J. Lundström, and F. Sandberg, *Tetrahedron Letters* 2433 (1967).

Amaryllidaceae alkaloids

D. H. R. Barton, G. W. Kirby, J. B. Taylor, and G. M. Thomas, *J. Chem. Soc.* 4545 (1963).

A. R. Battersby, R. Binks, S. W. Breuer, H. M. Fales, W. C. Wildman, and R. J. Highet, *J. Chem. Soc.* 1595 (1964).

H. M. Fales and W. C. Wildman, *J. Amer. Chem. Soc.* **86,** 294 (1964).

J. Zulalian and R. J. Suhadolnik, *Proc. Chem. Soc.* 422 (1964).

D. H. R. Barton, *Proc. Chem. Soc.* 293 (1963).
P. W. Jeffs, W. C. Archie, and D. S. Farrier, *J. Amer. Chem. Soc.* **89**, 2509 (1967).

Phenol coupling

D. H. R. Barton and T. Cohen, "Festschrift A. Stoll," p. 117, Birkhäuser, Basel, 1957.
D. H. R. Barton, *Proc. Chem. Soc.* 293 (1963).
A. I. Scott, *Quart. Rev.* **19**, 1 (1965).
D. H. R. Barton, *Chemistry in Britain*, **3**, 330 (1967).

Benzylisoquinoline and related alkaloids

F. R. Stermitz and H. Rapoport, *J. Amer. Chem. Soc.* **83**, 4045 (1961).
J. R. Gear and I. D. Spenser, *Can. J. Chem.* **41**, 783 (1963).
A. R. Battersby, *Proc. Chem. Soc.* 189 (1963).
D. H. R. Barton, *Pure Appl. Chem.* **9**, 35 (1964).
A. R. Battersby, R. Binks, R. J. Francis, D. J. McCaldin, and H. Ramuz, *J. Chem. Soc.* 3600 (1964).
A. R. Battersby, D. M. Foulkes and R. Binks, *J. Chem. Soc.* 3323 (1965).
D. H. R. Barton, G. W. Kirby, W. Steglich, G. M. Thomas, A. R. Battersby, T. A. Dobson, and H. Ramuz, *J. Chem. Soc.* 2423 (1965).
A. R. Battersby, R. Binks, W. Lawrie, G. V. Parry, and B. R. Webster, *J. Chem. Soc.* 7459 (1965).
D. H. R. Barton, R. H. Hesse, and G. W. Kirby, *J. Chem. Soc.* 6379 (1965).
I. Monkovic and I. D. Spenser, *Can. J. Chem.* **43**, 2017 (1965).
G. W. Kirby, *Science* **155**, 170 (1967).
E. Leete and J. B. Murrill, *Phytochemistry* **6**, 231 (1967).
G. Blaschke, H. I. Parker and H. Rapoport, *J. Amer. Chem. Soc.* **89**, 1540 (1967).
A. R. Battersby, R. J. Francis, M. Hirst, R. Southgate, and J. Staunton, *Chem. Commun.* 602 (1967).

Colchicine

A. R. Battersby, R. Binks, J. J. Reynolds, and D. A. Yeowell, *J. Chem. Soc.* 4257 (1964).
E. Leete, *Tetrahedron Letters* 333 (1965).
A. R. Battersby, *Pure Appl. Chem.* **14**, 117 (1967).

SECTION 4

Incorporation of tryptophan

E. Leete, *Tetrahedron* **14**, 35 (1961).
N. Kowanko and E. Leete, *J. Amer. Chem. Soc.* **84**, 4919 (1962).
D. G. O'Donovan and M. P. Kenneally, *J. Chem. Soc.* (C), 1109 (1967).
D. G. O'Donovan and M. F. Keogh, *J. Chem. Soc.* (C), 1570 (1966).

Ergot alkaloids

F. Weygand and H. G. Floss, *Angew. Chem. Intern. Ed. Engl.* **2**, 243 (1963).
F. Weygand, H. G. Floss, U. Mothes, D. Gröger, and K. Mothes, *Z. Naturforsch.* **19b**, 202 (1964).
H. G. Floss, U. Hornemann, N. Schilling, D. Gröger, and D. Erge, *Chem. Commun.* **105** (1967).

Non-tryptophan moiety

E. Wenkert, *J. Amer. Chem. Soc.* **84**, 98 (1962).
A. R. Battersby, *Pure Appl. Chem.* **14**, 117 (1967).

Gramine

K. Bowden and L. Marion, *Can. J. Chem.* **29**, 1037 (1951).
E. Leete and L. Marion, *Can. J. Chem.* **31**, 1195 (1953).

D. O'Donovan and E. Leete, *J. Amer. Chem. Soc.* **85,** 461 (1963).
B. G. Gower and E. Leete, *J. Amer. Chem. Soc.* **85,** 3683 (1963).

Quinoline alkaloids

I. Monkovic, I. D. Spenser, and A. O. Plunkett, *Can. J. Chem.* **45,** 1935 (1967).

22

Alkaloids and Plant Systematics

TOM J. MABRY AND JAMES A. MEARS
The Cell Research Institute and Department of Botany
The University of Texas, Austin, Texas

1. INTRODUCTION

Systematics is concerned with evolutionary relationships among organisms. Historically, most classification schemes for plants were based on obvious morphological features such as leaf, stem, and flower structures. Today, the morphological data are often supplemented (but not replaced) by cytological, ultrastructural and, notably, biochemical information. We need not comment in detail on the broad chemical and taxonomic concepts involved in biochemical systematic investigations, since a number of excellent reviews and books have recently dealt with this topic.[1] Furthermore, this account of the application of alkaloid data for plant systematics will not cover the detailed distribution of the major classes of alkaloids; this information has been described in other chapters of this text and in even greater detail in Hegnauer's multivolume "*Chemotaxonomie der Pflanzen*,"[2] Boit's "*Ergebnisse der Alkaloid-Chemie*"[3] and the ten volumes edited by Manske (with Holmes, Volumes I-IV) entitled "*The Alkaloids*."[4] The main purpose of this chapter is to give the budding chemist not only some insights into the goals and aspirations of the systematist but also an awareness of both the potential and limitations of chemical data as an aid in the study of botanical systematics (Figure 22.1).

It is important to recognize that although alkaloid data are useful for some systematic studies, other kinds of chemical information may be employed and, in many instances, required to resolve other taxonomic problems. For example, flavonoids and terpenes often are valuable for analyz-

Figure 22.1. Perhaps the most important element in systematics is common sense.

ing species,[5] and comparative protein and nucleic acid studies[1] are beginning to provide insights about the evolutionary relationships of families and higher taxonomic categories.* These latter topics, however interesting, do not fall in the domain of this chapter; here we must restrict ourselves to an account of the application of the study of alkaloids to plant systematics.

Past investigations clearly demonstrated that certain classes of alkaloids are restricted to a few groups of plants which are considered for other reasons to be closely related. However, most of these early studies were not concerned with systematic problems and usually did not yield the sort of data required by the systematist today. It is, therefore, important for the

*The categories or taxa (singular, taxon) used in this chapter are orders, families, genera and species; orders are composed of families, which are composed of genera (singular, genus), which in turn are composed of species (singular, also species). The members of a species have many similarities but often show considerable variation in a number of features. Furthermore, the members of a species tend to interbreed freely.

chemist interested in contributing to systematics to be aware of some of the general problems encountered in any biochemical systematic study; thus, in the following two sections we discuss the necessity of documenting the plant population under investigation and the importance of establishing the variation of the chemical data with respect to such matters as the stage of development of the plant.

The fourth section of this chapter presents a few examples of alkaloid studies selected to illustrate the application of alkaloid data to different kinds of systematic problems, while the final section presents a tabulation of some alkaloids which appear to be characteristic for particular groups of plants.

2. DOCUMENTATION OF THE PLANT MATERIAL

One of the most important aspects of any chemotaxonomic investigation is the documentation of the plant population being studied. Thus, it is remarkable that many chemists who practice "chemotaxonomy" either completely ignore or pay only cursory attention to this facet of their natural products work. In our laboratory, we deal with this problem in the following manner: Before the chemical analysis of a given collection of plant material is undertaken, we give the curator of the university herbarium a dried, pressed plant specimen from the plant population under investigation. The selected plant should be typical for the population and, when possible, should have mature reproductive organs. Once the plant has been glued onto a special paper by the herbarium personnel, a permanent acquisition number is stamped onto the paper. The resident taxonomists usually can determine the binomial nomenclature* for a plant by careful examination of the voucher specimen. If any question remains regarding the identity of the plant material, we may send the herbarium specimen to a taxonomist who specializes in that group of plants.

In the chemical publication[6] describing the chemical results, we record the locations and dates of the plant collections, the parts of the plant used in the study, the name of the herbarium where the voucher specimens are deposited, the acquisition number of each voucher and, finally, the name of the taxonomist who identified the plants. With this information and with the possibility of comparing specimens collected at other times with the original vouchers, later investigators can usually determine the relationship of the plants, which were previously investigated, to the plant they are

*By convention the name of a plant consists of three terms: names of genus, and species and the author citation. For example, *Ambrosia psilostachya* DC. is the name of plants which belong to the species *psilostachya* of the genus *Ambrosia*. The author citation DC. refers to Auguste-Pyrame DeCandolle, the nineteenth century taxonomist who first described the species.

analyzing. With the plant well documented, the chemical data is of immediate value to the systematist.

Ettlinger and Kjaer[7] have commented succinctly on the matter of chemists documenting their plant material:

> "The necessity of the mechanical features of systematic work, the collection and preservation of herbarium specimens and the assignment of correct names, tends to be disregarded by most chemists. They instinctively rebel, in the first place, against what is dimly imagined to be the arbitrary and prescriptive character of systematics. More particularly, they are used to going quickly beyond seemingly passive and simple observation, and find it difficult to believe that the information gained by looking carefully at a plant is not just as superficial as when they describe a compound as colorless needles. Chemists rely on structures as primary symbols, seldom pondering names; they object to botanical nomenclature both because it shifts with increasing data, which in hope lead to more enlightened judgment, and because the Latin language and rules of priority make it seem antiquarian. Few chemists are at peace with a sense of history. We remark only that the binomial of genus and species often cannot be traced without the author citation that completes the name (omitted, we might add, in some botanic garden seed lists) and that the problem of identification is separable to some extent from that of nomenclature, in that a complete name, even if invalid or out of fashion, can frequently be recognized—indeed, the system of intergeneric transfers is designed for this end. The name chosen, however, should convey meaning with a minimum of effort and possible confusion, and ordinarily a person competent to identify a plant will use valid, consistent nomenclature. The pitfalls, gross and subtle, in use of material from botanic gardens and the necessity for scrupulous identification of all specimens have been emphasized in other discussions.[8] We may point out that search in keys is only a first step in accurate identification, that comparison with authentic herbarium specimens is generally indispensable, and that many sheets are wrongly designated even in first-class herbaria. Plants, therefore, should be identified by someone experienced and critically interested in their particular group. The herbarium specimens are the permanent, versatile and reusable, experimental record."

3. THE RELIABILITY OF THE ALKALOID DATA

It is well established that variation of both the morphological and chemical features of plants belonging to the same species, or even to the same population of a species, can occur. It has been our experience that alkaloids exhibit greater variation than do some other groups of natural products. We wish to emphasize that it is impossible to carry out meaningful chemotaxonomic studies without considering the extent and nature of the variation; once understood, however, chemical variation is predictable and is often an asset.

In this section, we comment on some of the possible sources of variation in the alkaloid content of plants and illustrate the phenomena with some of the studies from our laboratory on the variation of lupine alkaloids in the genus *Baptisia.* The genus, which is a member of the family Leguminosae, consists of sixteen species that occur in the eastern half of North America. *Baptisia* species have been the subject of a broad biochemical systematic study in which the flavonoids, phenolic acids, alkaloids, volatile constituents, amino acids and, to some extent, proteins have been examined. Hundreds of populations of each of the sixteen species have often been analyzed. Here, we restrict ourselves to a discussion of the variation in the alkaloid chemistry of *Baptisia*; the systematic significance of the alkaloid data is considered in Section 4.2.

Some of the lupine alkaloids that have been detected in *Baptisia*[9] are cytisine (1), methylcytisine (2), anagyrine (3) and its steric isomer thermopsine (4), baptifoline (5), sparteine (6), hydroxysparteine (7), and lupanine (8).

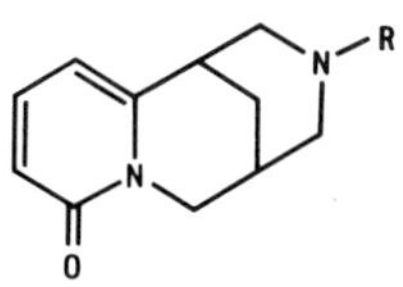

(1) R = H
(2) R = CH_3

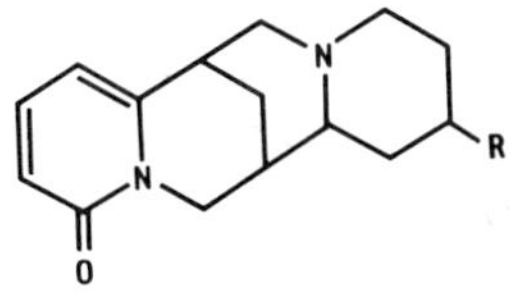

(3), (4) R = H
(5) R = OH

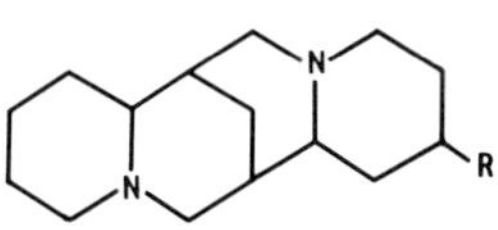

(6) R = H
(7) R = OH

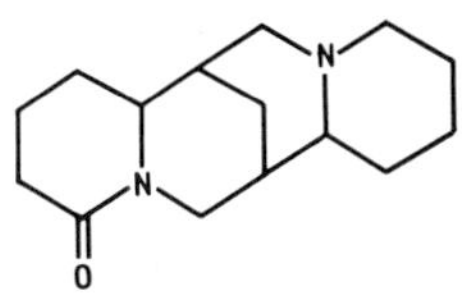

(8)

3.1 The Effect of Developmental and Environmental Changes on Alkaloid Content

Generally, it is accepted that developmental changes and environmental factors are among the more important causes of variation in the alkaloid content of plants. In order to study the variation of lupine alkaloids during development on a plant-to-plant and population-to-population basis for one *Baptisia* species, Cranmer analyzed the alkaloids of ten separate plants from two populations of *Baptisia leucophaea* Nutt. at three different stages of development.[10,11] The results are recorded in Table 22.1. The plants from both the Cat Spring and Bastrop, Texas populations (which are 120 miles apart) were similar in their alkaloid chemistry at the same stage of development, although individual plants at a given developmental stage exhibited considerable quantitative variation. Thus, in this instance, the environmental differences between the two populations did not produce any significant differences in the alkaloid content of the plants. In contrast, striking alkaloid variation occurred during development: as the plants matured, the relative amount of cytisine (1) decreased markedly while the relative amount of anagyrine (3) increased, and in most instances, became the dominant alkaloid. However, this information must not be interpreted as demonstrating that a net increase in anagyrine occurred during development since the percentage of total alkaloid per gram of dry plant material decreased from 1.02% to 0.65% to 0.08% during the three developmental stages.

Nowacki[12] encountered a similar variation of lupine alkaloids in species of another legume genus, *Lupinus*, and suggested that some of the variation, particularly with respect to a number of minor alkaloids, might be the result of nonenzymatic transformations. Whether or not this is true, the chemotaxonomist must take into account the fact that considerable variation in the alkaloid content of plants does occur during development.

3.2 Gene-Caused Difference in the Alkaloid Content of Plants

With few exceptions, the individual members of any population of plants have some genetic differences that can be expected to produce chemical and morphological differences. A number of workers[12,13] have investigated the problem of gene control of alkaloid synthesis by examining the alkaloid content of hybrid plants. The results, which suggest that the genetics of alkaloid formation are complex, clearly indicate that hybridization* and the related biological phenomenon of introgression† can produce considerable variation in the alkaloid content of plants within a

*Hybridization refers to the crossing of members of two genetically distinct groups of plants (often two species).

†The term introgression denotes the influx of genetic material from one genetically distinct group of plants into a population of a closely related group of plants. Introgression, which begins when hybrid plants cross back with plants belonging to one of the parental types, leads to populations in which eventually most of the individual plants contain various mixtures of the genetic material of both of the parent taxa.

TABLE 22.1. Changes in the Lupine Alkaloids for Leaves of Individual Plants of *Baptisia leucophaea* Nutt. from Two Populations at Different Developmental Stages.[a]

Cat Spring Population							
Young, Pre-Flowering Plants							
Plant No.	Cytisine	Methylcytisine	Anagyrine	Hydroxyspartiene	Baptifoline	Thermopsine	Sparteine
1	77	trace	23	trace			
2	70	2	28			trace	
3	78		22				
4	78		22			trace	
5	81	3	15	1			
6	82		18				
7	80		20				
8	70	3	27			trace	
9	83	2	15				
10	80	1	19				
Flowering Plants							
Plant No.	Cytisine	Methylcytisine	Anagyrine	Hydroxyspartiene	Baptifoline	Thermopsine	Sparteine
1	69	3	28	trace	trace	trace	
2	73	3	24	trace		trace	
3	68	3	29	trace		trace	
4	74	5	21	trace	trace	trace	
5	65	3	32		trace	trace	
6	67	2	31				
7	69	3	28	trace		trace	
8	79	3	23	trace	trace	trace	
9	72	3	25				
10	70	4	26	trace		trace	

Table 22.1 *(continued)*

Old Plants

Plant No.	Cytisine	Methylcytisine	Anagyrine	Hydroxyspartiene	Baptifoline	Thermopsine	Sparteine
1	63	3	33	trace	trace	trace	1
2	26	12	61	trace	trace	1	
3	25	17	55	1	1	1	
4	40	13	47	trace	trace	trace	
5	32	17	50		1	trace	
6	37	14	48		1	trace	
7	19	17	64		trace	trace	
8	29	10	59	trace	1	1	
9	30	19	51		trace	trace	
10	30	13	57		trace	trace	

Bastrop Population

Young, Pre-Flowering Plants

Plant No.	Cytisine	Methylcytisine	Anagyrine	Hydroxyspartiene	Baptifoline	Thermopsine	Sparteine
1	86	1	13				
2	88	trace	12				
3	81	2	17				
4	89	trace	11	trace		trace	
5	90	trace	10				
6	91	trace	9	trace			
7	85	1	14			trace	
8	92	trace	8				
9	92	trace	8				
10	89	1	10			trace	

Table 22.1 *(continued)*

Flowering Plants

Plant No.	Cytisine	Methylcytisine	Anagyrine	Hydroxyspartiene	Baptifoline	Thermopsine	Sparteine
1	60	6	31	1	1	1	trace
2	71	1	25	1	1	1	
3	69	3	27	trace		1	
4	69	3	27	trace		1	
5	73	1	24	1		1	
6	59	5	32	2	1	1	trace
7	64	4	29	2		1	
8	18	16	64	trace	trace	1	1
9	61	6	30	2	1		trace
10	56	7	33	2	1	1	

Old Plants

Plant No.	Cytisine	Methylcytisine	Anagyrine	Hydroxyspartiene	Baptifoline	Thermopsine	Sparteine
1	19	12	67	1	trace	1	trace
2	28	17	52	trace	1	1	1
3	41	8	47	1	1	1	1
4	23	16	57	1	1	1	1
5	23	18	58	1	trace	trace	trace
6	26	19	55	trace	trace	trace	
7	20	21	58		trace	1	trace
8	17	17	64	trace	trace	1	1
9	30	17	51	1	1	trace	
10	21	21	55	1	1	1	

[a] Each number represents the percentage of a given alkaloid relative to the total alkaloid material isolated from the plant.

TABLE 22.2. Relative Percent of Lupine Alkaloids in Seven Hybrid Plants and in the Two Parent Species.[a]

Parent *Baptisia* Species	Cytisine	Methylcytisine	Anagyrine	Hydroxysparteine	Baptifoline	Thermopsine	Unknown
B. sphaerocarpa Nutt.	9	74	6	–	1	11	–
B. leucophaea Nutt.	77	trace	23	–	–	trace	–
B. sphaerocarpa × *B. leucophaea* Hybrid Plants							
1	3	13	38	5	–	41	trace
2	3	21	39	3	–	34	trace
3	2	31	42	3	–	22	trace
4	1	6	90	2	–	1	–
5	2	96	2	trace	–	trace	–
6	2	36	36	4	–	20	2
7	2	15	70	1	–	12	–

[a] All the analyses with the exception of that for *B. sphaerocarpa* are for leaves and stems of young plants. The *B. sphaerocarpa* plants were flowering.

population. Chemists, many of whom are unaware of the widespread occurrence of natural hybridization, must recognize that populations which contain two or more closely related species often contain hybrid plants. Although such hybrid plants are occasionally indistinguishable in appearance from plants belonging to one or the other parent species, hybrids may exhibit unique alkaloid patterns. In addition, it is important to recognize that in some populations of a given species introgression may have occurred; as a result, many plants in such populations will exhibit considerable chemical differences.

Both hybridization and introgression occur in *Baptisia*. For most *Baptisia* species, there are several populations in which two or more species are present. These populations usually contain all of the possible hybrid combinations and innumerable plants derived by backcrossing of the hybrids with the parent species. The original work involving the chemical investigation of natural hybridization involved *Baptisia* species that occur mainly in southeast Texas. Initially, flavonoids were studied, and it was subsequently established that the hybrid plants combined the flavonoid chemistry of the parent species and occasionally produced new flavonoids that were unique to the hybrid plants. Later, the investigation was extended to include an analysis of the alkaloid content of seven authentic hybrid plants (*B. leucophaea* × *B. sphaerocarpa*) (Table 22.2). The data indicated that the hybrid plants not only failed to exhibit the alkaloid chemistry of the parent species either singly or combined but also showed some striking quantitative variation among the individual hybrid plants. With one exception (plant No. 5, Table 22.2), a tetracyclic alkaloid [either anagyrine (3) or its steric isomer, thermopsine (4)] was dominant in the hybrid plants whereas the major alkaloid in each of the parent species was a tricyclic type [cytisine (1) or its N-methyl derivative (2)].

The careful investigator can use to advantage the fact that hybridization and introgression alter the genetic makeup and thus the chemistry of certain populations of species. For example, populational analysis of some natural products (notably flavonoids and terpenes) represents one of the best available techniques for detecting and documenting natural hybridization and introgression.

3.3 Variation in Alkaloid Content for Different Organs of a Plant

Extensive chemical variation can occur among the various organs of an individual plant. For this reason, in comparative chemical studies of different plants, it is essential that the same organs at the same stage of development be analyzed. Perhaps the most striking chemical changes are associated with the reproductive parts of the plant—buds, flowers, fruits

TABLE 22.3. Relative Percent of Lupine Alkaloids in Organs of *Baptisia bracteata* Muhl.

Organ	Cytisine	Methylcytisine	Anagyrine	Hydroxysparteine	Baptifoline	Thermopsine	Percent Yield of Total Alkaloid Material on a Dry Weight Basis
Seed hulls	79	10	10	–	–	–	0.25
Leaves	22	7	45	9	15	2	0.31
Stems	60	10	28	–	1	1	0.15

and seeds—organs that also exhibit the greatest morphological changes with aging.

Cranmer[9–11] observed in the course of our *Baptisia* studies that alkaloids often showed greater variation between the organs of plants belonging to a single species than between the same organs for different species. The results for leaves, stems and seed hulls of *Baptisia bracteata* Muhl. showed that in leaves the major alkaloid was the tetracyclic type anagyrine (3), while the tricyclic alkaloid cytisine (1) predominated in both the stems and seed hulls (Table 22.3). Moreover, the total yield of alkaloid material from the different organs varied considerably, with the stems containing the least amount on a dry weight basis.

3.4 Summary

Among the more important factors that may produce both qualitative and quantitative variation in the alkaloid content of a species are (a) developmental changes, (b) genetic differences, and (c) the part of the plant being analyzed. Therefore, to obtain chemical results that can be used for systematics, it is absolutely essential for the chemist to determine the extent and nature of the variation in the alkaloid content of the plants under investigation.

4. THE SYSTEMATIC SIGNIFICANCE OF ALKALOIDS

Nitrogen-containing compounds occur abundantly in and are vital for all living organisms. However, the several thousand nitrogen-containing natural products, which are defined as alkaloids, are mainly restricted to the flowering plants (the angiosperms; see Figure 22.2). Evolutionary relationships among the higher taxonomic categories shown in Figure 22.2 cannot be deduced from the fragmentary alkaloid data, particularly since many different classes of alkaloids are synthesized by different biogenetic pathways,[14] and since in some instances the same class of alkaloids has arisen independently in different taxa. Therefore, at this time the systematic value of alkaloids is, with few exceptions, limited to problems at the family, generic, and species levels. In this section, we present a few examples that illustrate the application of the study of alkaloids to botanical systematics.

4.1 Betalains in the Order Centrospermae

Perhaps, the most remarkable example of the use of alkaloids for systematics concerns the occurrence of betalains in plants belonging to the or-

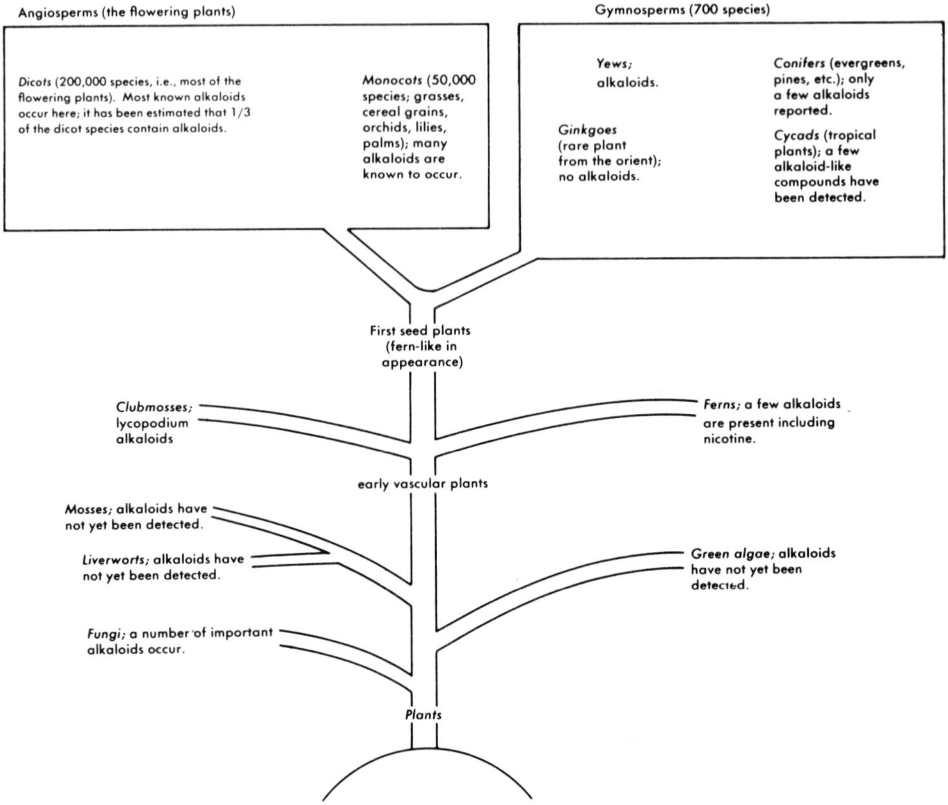

Figure 22.2. The general distribution of alkaloids in the plant kingdom is shown superimposed upon a hypothetical scheme for the evolutionary relationships of the plants.

der Centrospermae. These unique red and yellow alkaloids are restricted to ten plant families which, at one time or another, were positioned in the order Centrospermae for morphological and anatomical reasons. However the first time[15] that all of the ten betalain-containing families, and only these ten families, were placed in the Centrospermae was in 1963, shortly after it was recognized[16] that the betalains were structurally unrelated to the more common anthocyanins. Some systematists still object to the concept that the betalain-containing families constitute an evolutionary (phyletic) line distinct from the phyletic lines that produced the other flowering plants. Nevertheless, most present-day systematists now acknowledge that the betalain-containing families are evolutionarily related; however, they usually retain one or more anthocyanin-containing families in the order (see Figure 22.3).[17]

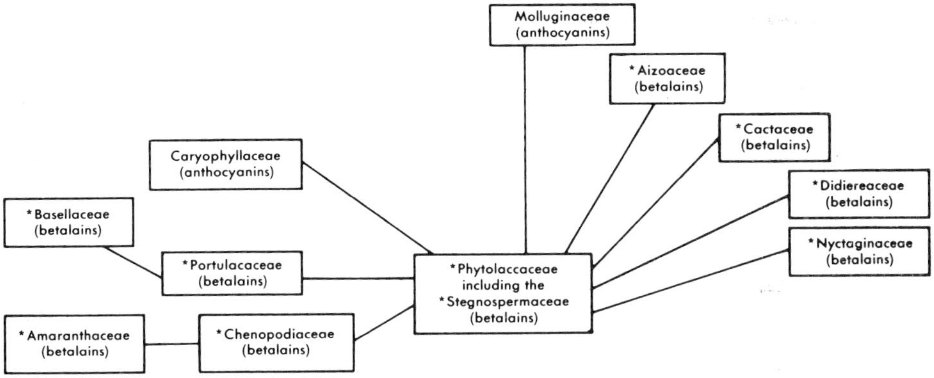

Figure 22.3. Possible relationships among families of the Centrospermae (or Caryophyllales) according to Cronquist.[17] The arrangement is based mainly on embryological and anatomical features. The phyletic arrangement of Mabry, Taylor, and Turner,[15] who placed greater emphasis on the chemical data, would include only the betalain-containing families in the Centrospermae, separating the anthocyanin-containing families into one or more additional orders.

The view[15] that the order Centrospermae should be reserved for the betalain-containing families rests, in part, on the following observations:

(1) The ten families each contain both types of betalains [general structure (9)]—the red-violet betacyanins and yellow betaxanthins.

(2) The betalains do not occur elsewhere.

(3) The ten betalain-containing families do not synthesize the common flower pigments, the anthocyanins [e.g., (10)].

(4) Each of the ten betalain-containing families has been considered to be a member of the Centrospermae on purely morphological grounds.

(5) Recent evidence indicates the the betalains are synthesized in these ten families by a unique biogenetic pathway.[18]

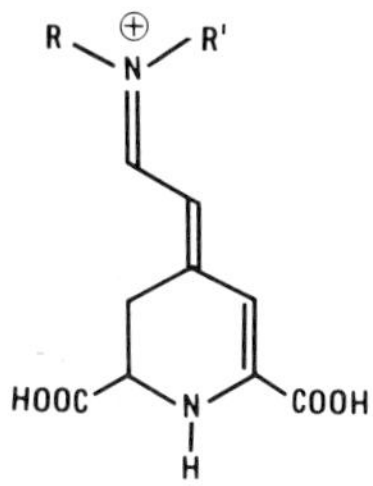

(9)

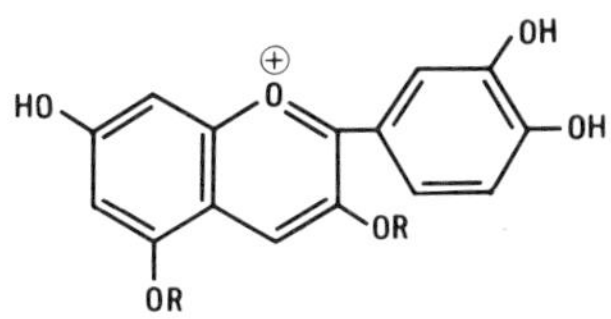

(10)

(6) Some systematists do not consider the morphological evidence sufficient to include such anthocyanin-containing families as the Caro-phyllaceae in the Centrospermae.

These observations suggest a common ancestor for the ten betalain-containing families. On the other hand it is not yet possible to exclude conclusively from the Centrospermae those families that contain anthocyanins and lack betalains but appear for other reasons to belong in this order.

4.2 Lupine Alkaloids in the Leguminosae

Except for isolated occurrences in the families Chenopodiaceae, Berberidaceae and Solanaceae, the lupine alkaloids [e.g., (1)–(8)] are found only in the three tribes Sophoreae, Podalyrieae, and Genisteae of the subfamily Papilionoideae, family Leguminosae. Several workers have presented arguments that the presence of lupine alkaloids in the three Papilionoideae tribes suggests that they originated from a common ancestral stock which contained a gene pool for synthesizing these compounds (Figure 24.4). Certainly, the striking similarities in the alkaloid

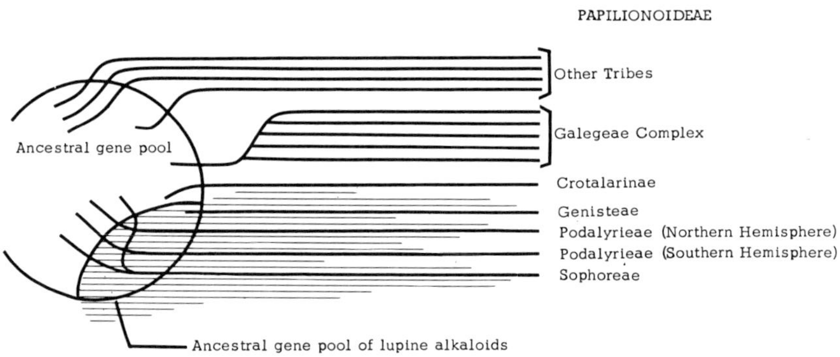

Figure 22.4. Tribes of the Leguminosae subfamily Papilionoideae, some of which contain lupine alkaloids, are arranged according to hypothetical phyletic affinities.[11]

chemistry of genera from each of the three tribes supports this view (Table 24.4).

The widespread occurrence of the same alkaloids within a genus or tribe generally reduces the value of the alkaloid data for understanding the relationships among the species of a single genus. For example, the lupine alkaloid data for all sixteen species of the genus *Baptisia*[9,11] (Table 22.5), does not aid in separating the species into phyletic groups. In all but one species either cytisine (1), methylcytisine (2) or their combination was the major alkaloid material detected. In addition, anagyrine (3) was present in all but three species and, when present, was usually the

TABLE 22.4. Distribution of lupine alkaloids[a] among legume genera.[11]

Genus	Cytisine	Methylcytisine	Anagyrine	Sparteine	Lupanine	Baptifoline	Hydroxysparteine	Thermopsine	No. of other alkaloids reported
Podalyrieae									
Baptisia	++	++	++	+	+	+	+	+	1–2
Thermopsis	++	++	++	++	+			+	5
Anagyris	++	+	++	++					0
Piptanthus	++			+					2
Podalyria					++				0
Genisteae									
Genista	++	++	++	++	+		++		2–4
Retama	+	+	++	++	+	+	++		3–4
Cytisus	+	++	+	++	++				4
Ulex	++		++	+					0
Spartium	++	+	+	++					0
Sarothamnus				++	+				1
Lupinus			+	++	++			+	9
Hovea				++					0
Templetonia	+			++					0
Adenocarpus				++					0
Laburnum	++								0
Sophoreae									
Sophora	+	+	+	+		+			6
Ammodendron			+	++					1
Ammothamnus				++					2
Virgilia					+				2
Ormosia				+					3–4

[a] If the alkaloid is a major component for the genus as a whole, it is indcated by ++ notations; minor alkaloids are denoted by +.

TABLE 22.5. Distribution of lupine alkaloids in *Baptisia*.[a] Species are grouped into four series, believed to be phyletic groupings.

Species	Alkaloids[b]									
	Cytisine	Methylcytisine	Anagyrine	Hydroxysparteine	Baptifoline	Thermopsine	Sparteine	Lupanine	Unknown	% dry weight
Group 1										
B. perfoliata R. Br.	84	5	12				trace			0.19
B. sphaerocarpa Nutt.	9	74	6			11				0.64
Group 2										
B. leucantha T. & G.	90	10								0.31
B. alba Hook.	31	16	26	5		trace			23	2.00
B. megacarpa Chapm.	19	45	36							0.94
Group 3										
B. cinerea Fern. & Schubert	1	2	45	1		51				2.20
B. bracteata Muhl.	41	9	37	4	8	1				0.23
B. leucophaea Nutt.	34	13	50	2	1	1	1			0.44
B. lanceolata Ell.	95	5								0.47
B. elliptica Small	7	93								0.11
B. nuttalliana Small	76	3	3		17					0.66
B. australis Lehm.	99	1								0.20
Group 4										
B. arachnifera Duncan	57	23	11	trace		9				2.40
B. simplicifolia Croom	26	21	32	12			trace	9		0.61
B. tinctoria R. Br.	72	9	10	9						0.08
B. lecontei T. & G.	54	46								0.46
B. calycosa Engelm.	97	1	2							0.21
B. hirsuta Small	60	5	28						7	0.42

[a] Adapted from Cranmer and Mabry.[9]
[b] Column figures refer to % of total alkaloids.

third major alkaloid. Thus, the lupine alkaloid data for *Baptisia* are best used to relate *Baptisia* to *Thermopsis* (Table 22.4) as well as to other members of the tribes Podalyrieae, Genisteae and Sophoreae.

4.3 Berberine, Pavine, and Protopine Alkaloids in *Argemone*

Stermitz[19] has summarized the systematic significance of the alkaloids produced by species belonging to two poppy genera, *Papaver* and *Arge-*

mone (family Papaveraceae). The alkaloid data appear to be especially valuable as an aid for subdividing *Argemone*, a genus which is very difficult to subdivide on other grounds. Eleven species of *Argemone* can be placed into three groups on the basis of their alkaloid content (Table 22.6): group I consists of those species which contain only protopine and berberine alkaloids, while group II species contain the pavine types. The group II species can be further separated into groups IIa and IIb depending on whether or not they contain only the pavine alkaloids.

The obvious question, of course, is whether or not the three groups shown in Table 22.6 each contain evolutionarily related species. The view that the alkaloids are valid evolutionary markers for the *Argemone* species is strengthened by considering all the information available for

TABLE 22.6. The distribution[19] of three types of alkaloids in eleven species of *Argemone*.[a]

Species	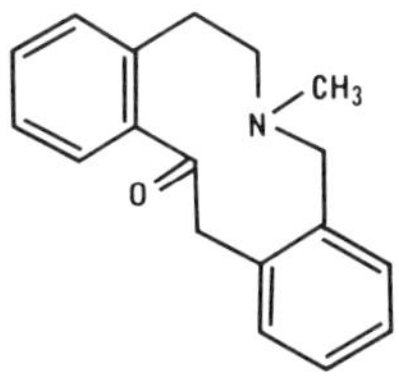 protopine type	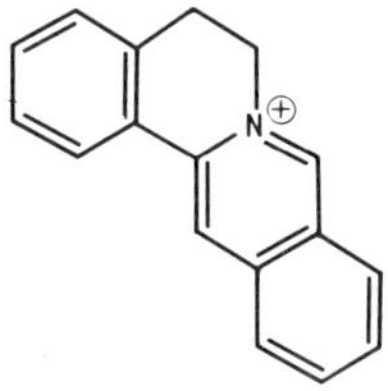 berberine type	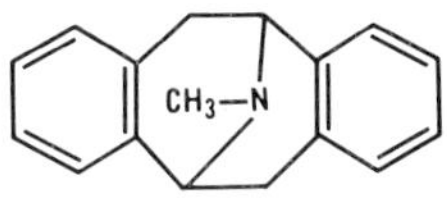 pavine type
Group I			
A. mexicana L.	+	+	–
A. aenea Ownbey	+	+	–
A. albiflora Horn.	+	+	–
A. polyanthemos Ownbey	+	+	–
A. corymbosa Greene	+	+	–
A. ochroleuca Sweet	+	?	–
Group IIa			
A. pleiacantha Greene	+	+	+
A. platyceras Link & Otto	+	+	+
Group IIb			
A. hispida Aibray	–	–	+
A. gracilenta Greene	–	–	+
A. munita Dur. & Hilg.	(trace)	–	+

[a] The species are arranged in what are considered to be three phyletic groups.

the eleven species. For example, using morphological data, Ownbey[20] had previously suggested similar groups but did not consider the evidence conclusive.

5. DIAGNOSTIC ALKALOIDS

In this final section, a list of alkaloids, useful for characterizing certain taxa, is presented (Table 22.7).

TABLE 22.7. The distribution of diagnostic[a] alkaloids[b]

Nonflowering Plant Families	Class of Alkaloids[c]	Typical Structure
Clavicipitaceae (fungi) *Claviceps purpurea* strains	Ergot	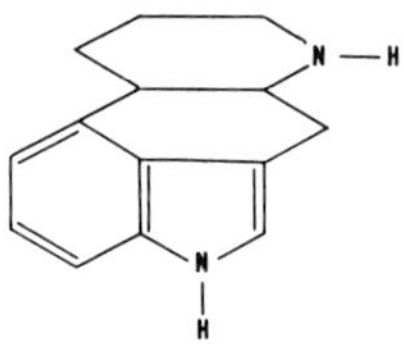
Lycopodiaceae (club mosses) *Lycopodium* species	Lycopodium	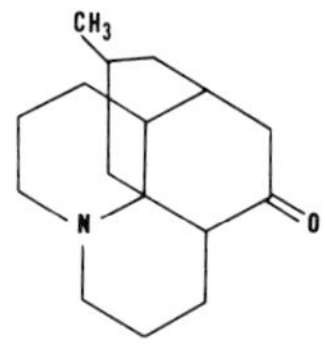
Taxaceae (yews) *Taxus* species	Taxine	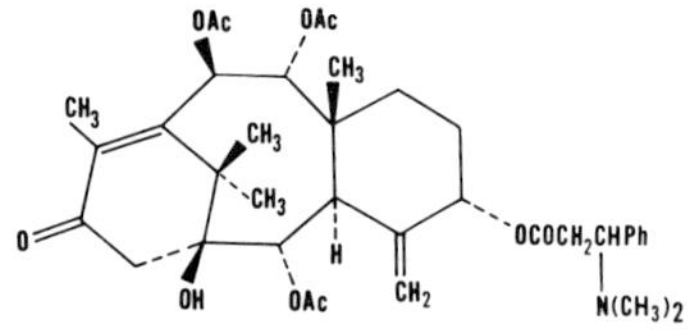
Flowering plant families	**Class of Alkaloids[c]**	**Typical Structure**
Aizoaceae *Mesembryanthemum* and related genera	Mesembrine	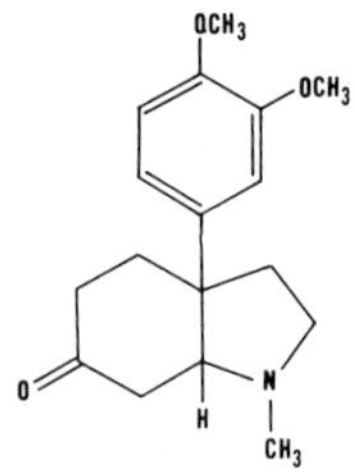

Table 22.7 *(continued)*

Flowering plant families	Class of Alkaloids[c]	Typical Structure
Amaryllidaceae *Veratrum* species	Steroid	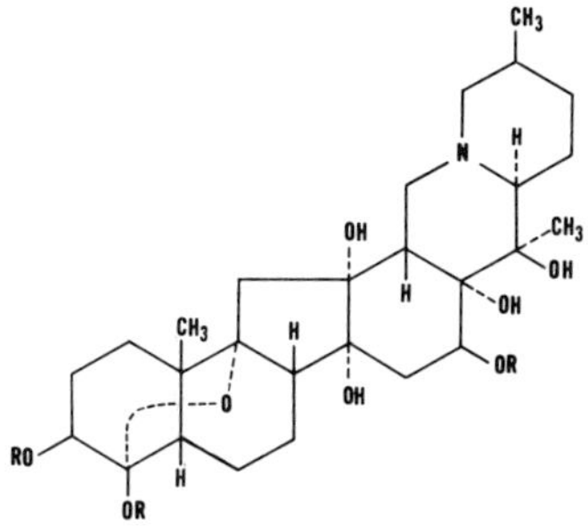
Species of several genera, particularly *Crinum*, *Narcissus*, *Amaryllis*, *Leucojum*, *Hymenocallis*, *Hippeastrum*	Amaryllidaceae	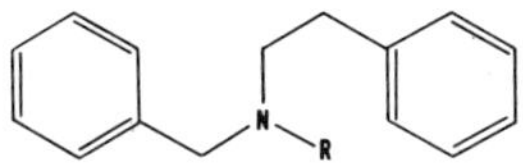 + various tetracyclic forms
Apocynaceae species of several genera	Ibogaine	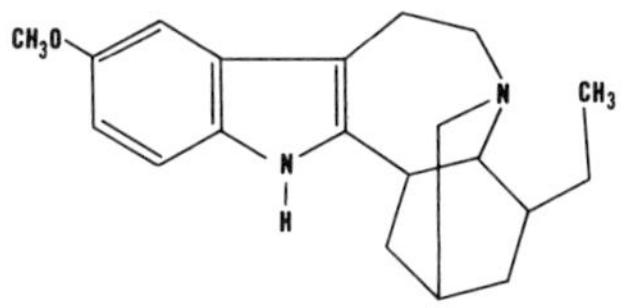
several genera, particularly *Rauwolfia*	Yohimbine	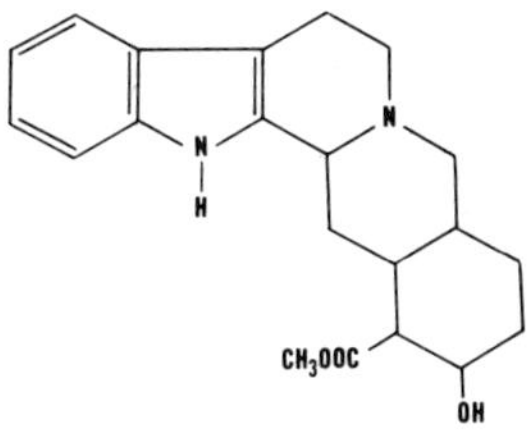
Rauwolfia species	Sarpagine	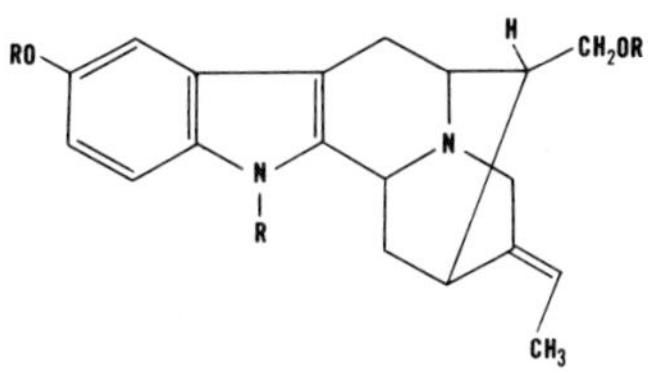
Holarrhena species	*Holarrhena*	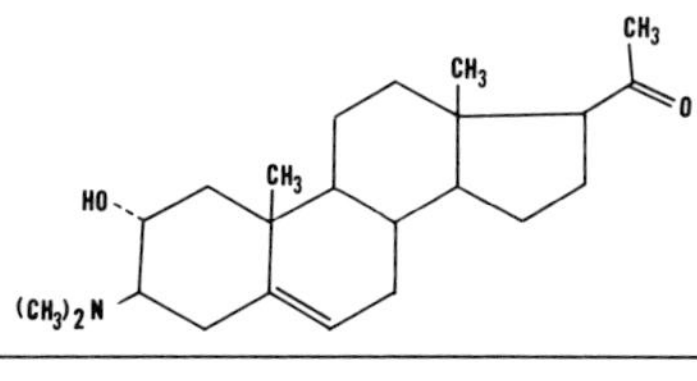

Table 22.7 *(continued)*

Flowering plant families	Class of Alkaloids[c]	Typical Structure
Berberidaceae most species examined	Protoberberine	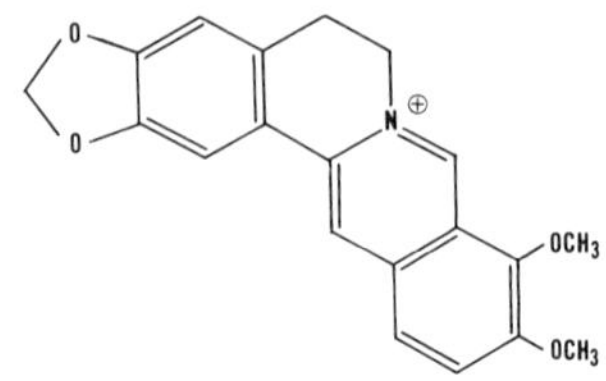
many species	Bisbenzylisoquinoline	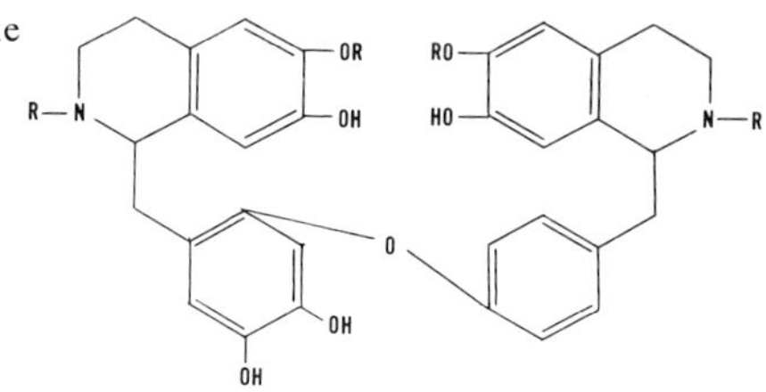
Buxaceae *Buxus* species	Buxamine	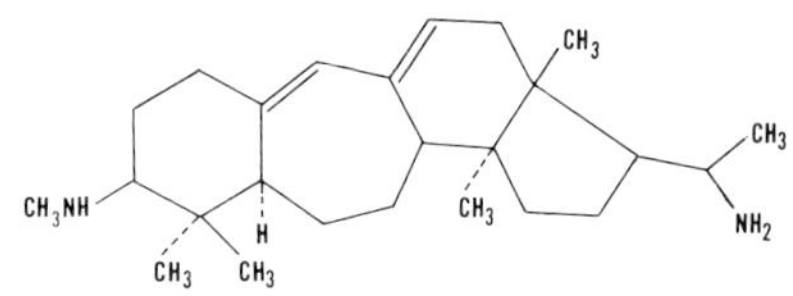
Cactaceae *Anhalonium* and *Lophophora* species	Tetrahydroiso- quinoline	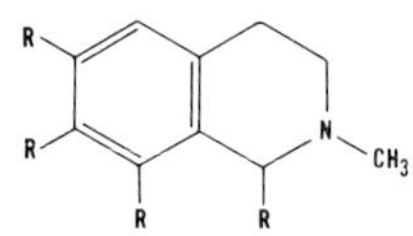
Calycanthaceae all species examined	Calycanthine	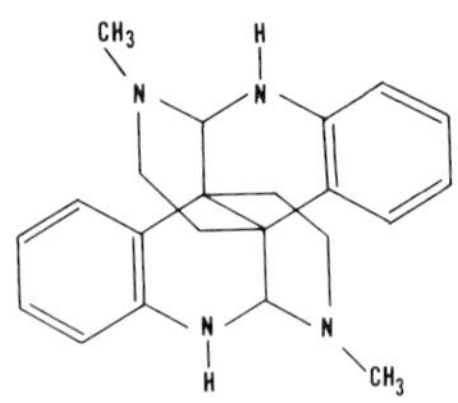
Campanulaceae *Lobelia* species	Piperidines (α, α'-substituted)	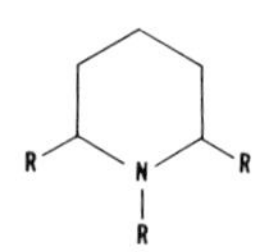

Table 22.7 *(continued)*

Flowering plant families	Class of Alkaloids[c]	Typical Structure
Centrospermae, families of this order (see Section 4.1)	Betalains	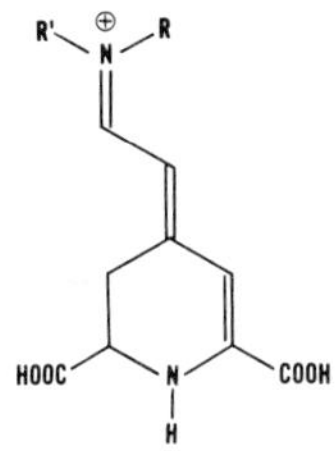
Compositae *Senecio* species	Pyrrolizidine	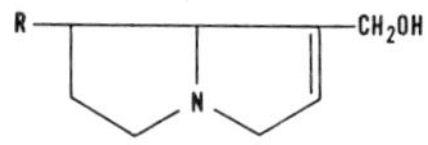
Daphniphyllaceae *Daphiphyllum* species[d]	Daphniphylline	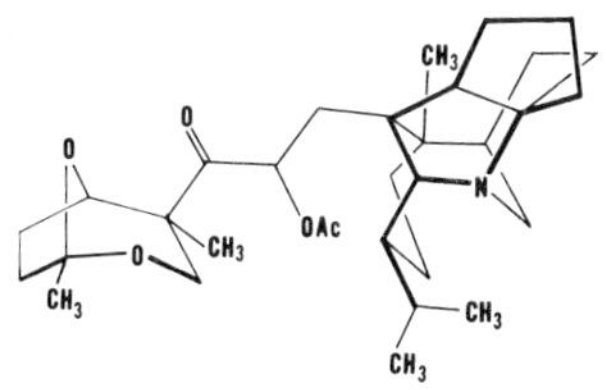
Garryaceae many species	Diterpenes	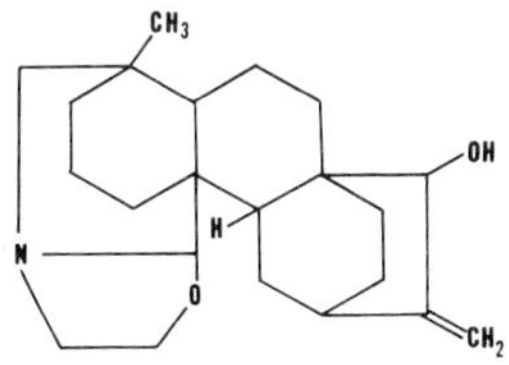
Himantandraceae in 1 of 3 species	Himbacine	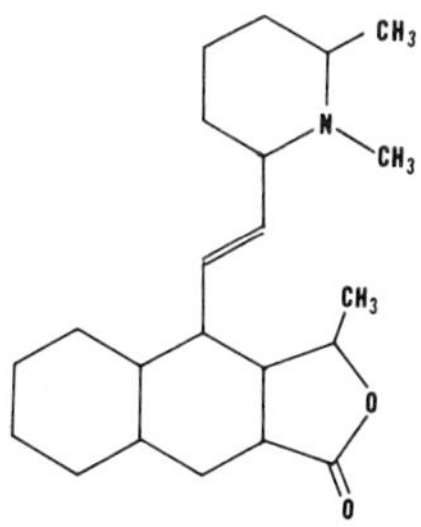
Leguminosae genera of the subtribes Sophoreae, Podalyrieae and Genisteae	Quinolizidine	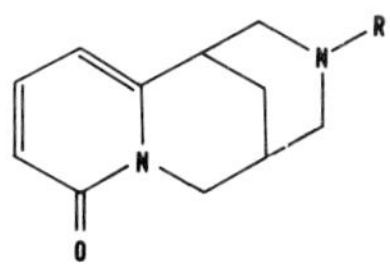 + various other cyclic forms

Table 22.7 *(continued)*

Flowering plant families	Class of Alkaloids[c]	Typical Structure
Leguminosae (*continued*)		
Crotalaria species	Pyrrolizidine	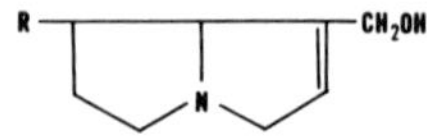
Erythrina species	*Erythrina*	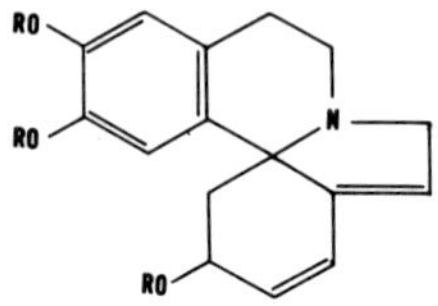
Liliaceae		
Colchicum and related genera	Colchicine	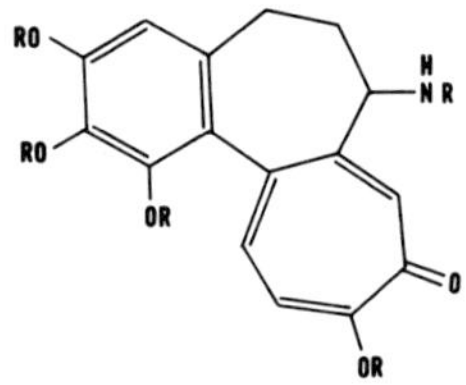
Stemona species	Tuberostemonine	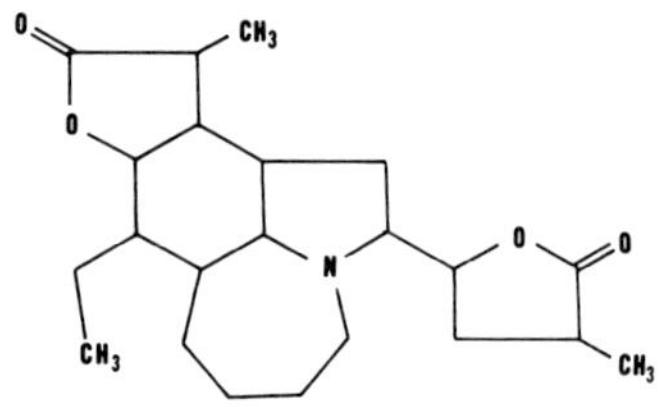
Loganiaceae		
Strychnos species	Strychnine	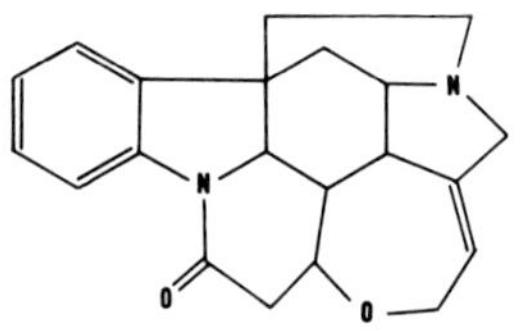
	Curare	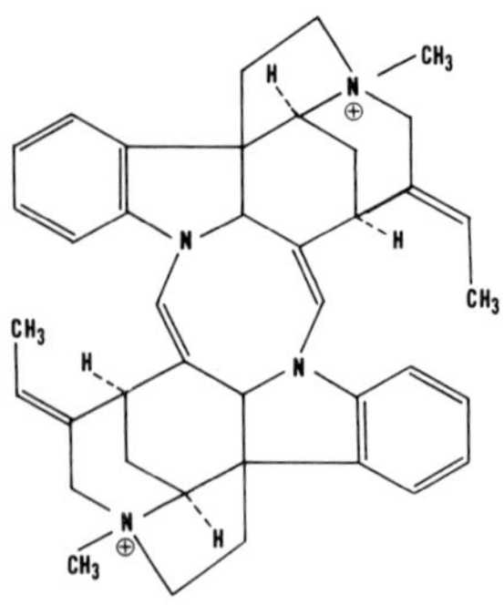

Table 22.7 *(continued)*

Flowering plant families	Class of Alkaloids[c]	Typical Structure
Menispermaceae		
most species examined	Protoberberine	(see above)
many species	Bisbenzyliso-quinoline	(see above)
Nymphaeaceae		
Nuphar species	*Nuphar*	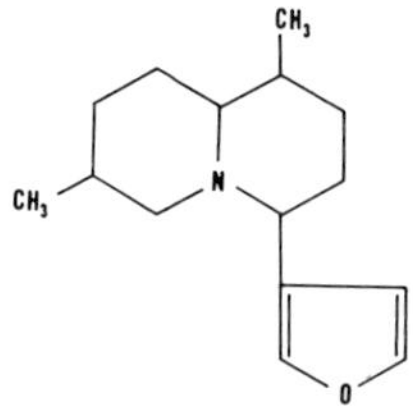
Papaveraceae		
many species	Aporphine	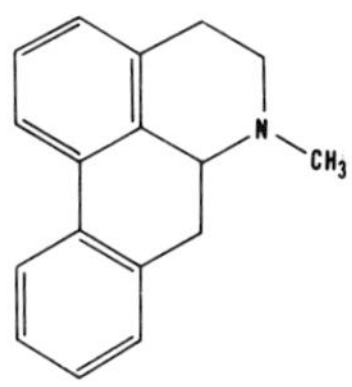
	Protopine	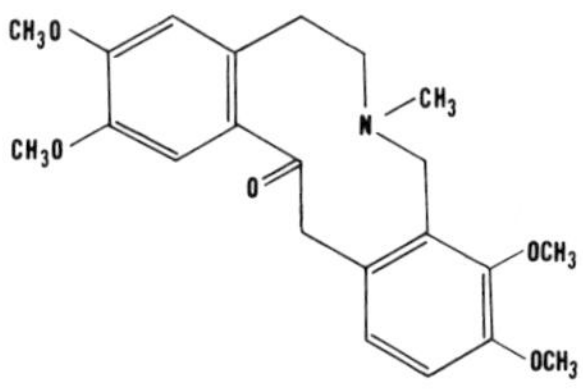
	Phthalideiso-quinoline	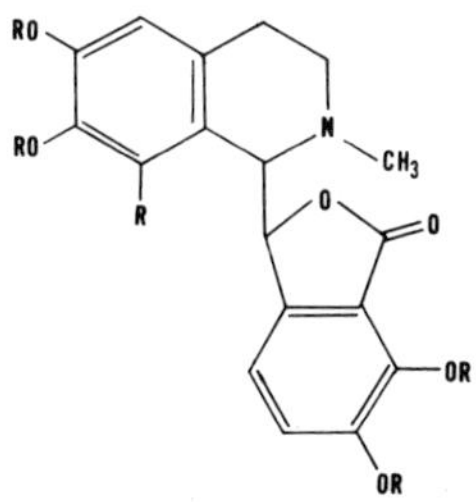
	Naphthaphenanthri-dine	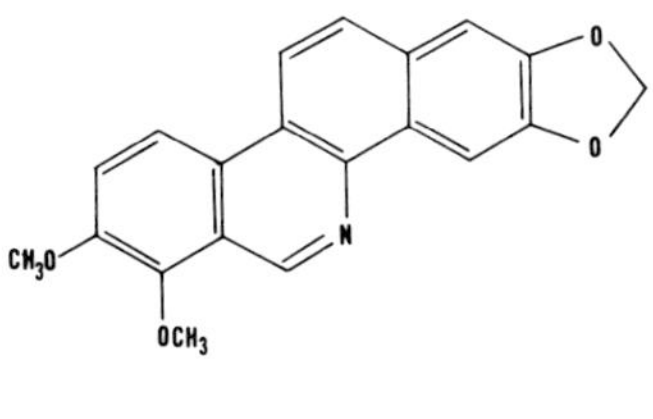

Table 22.7 *(continued)*

Flowering plant families	Class of Alkaloids[c]	Typical Structure
Papaveraceae (*continued*)		
Dicentra and *Corydalis* species	Cularine	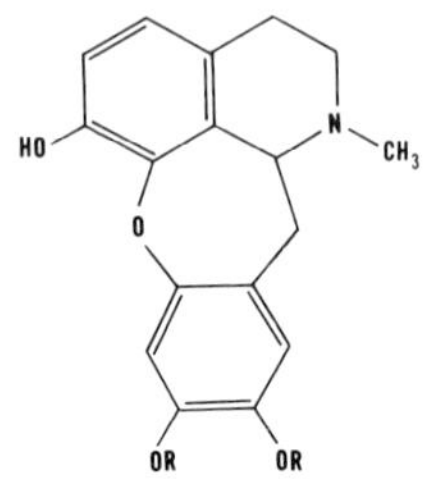
Corydalis species	Protoberberine	(see above)
Ranales, most families of this order Magnoliaceae, Berberidaceae, Menispermaceae, Ranunculaceae, Annonaceae etc.	Protoberberine Aporphine Bisbenzylisoquinoline	(see above) (see above) (see above)
Ranunculaceae *Aconitum* and *Delphinium* species	Diterpene	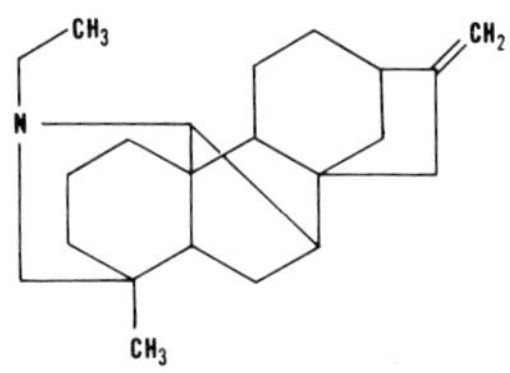 + various other forms
Rubiaceae several genera	Ipecac	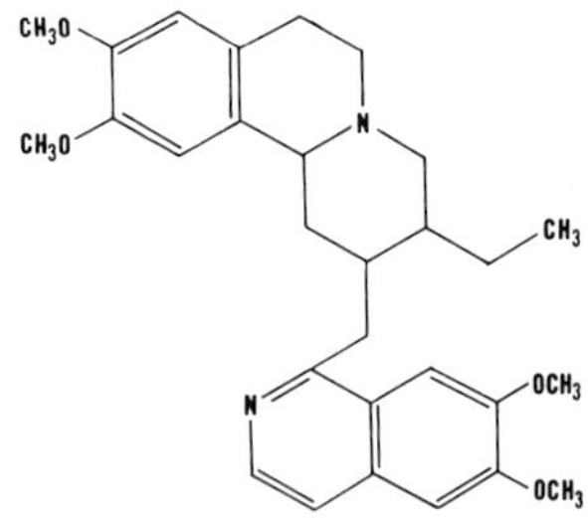
Mitragyne and *Ourouparia* species	Oxindole	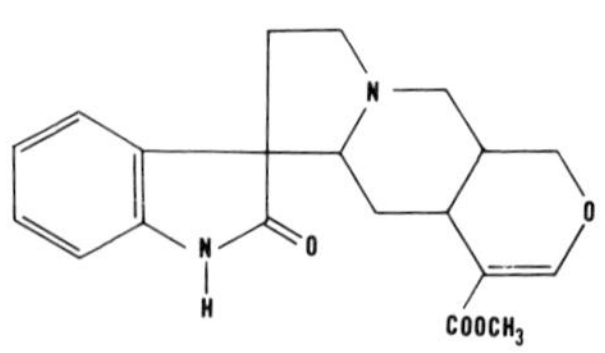

Table 22.7 *(continued)*

Flowering plant families	Class of Alkaloids[c]	Typical Structure
Mitragyne and *Ourouparia* (*continued*)		
Cinchona species	Cinchonamine	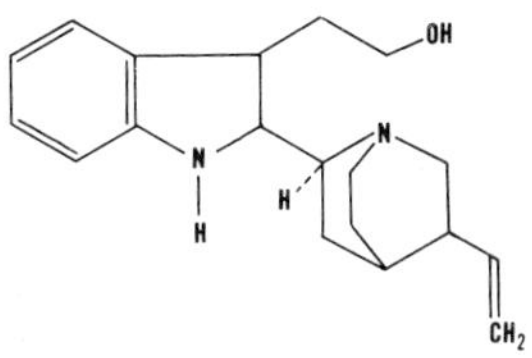
Rutaceae		
a few genera	Furoquinoline	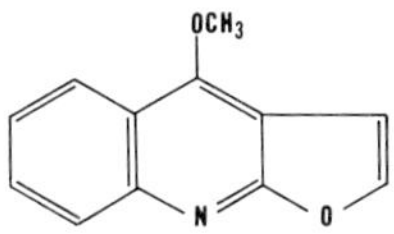
species of a few related Australian genera	Acridine	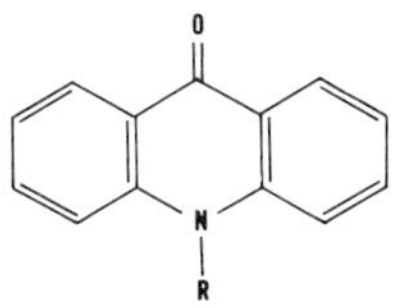
Solanaceae		
Solanum and *Lysopersicum* species	Steroid	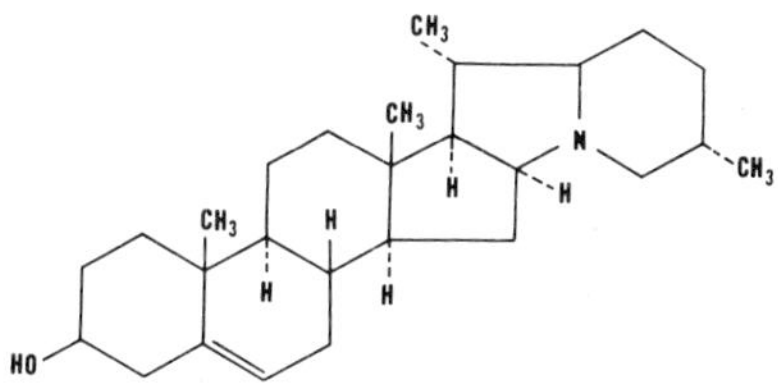
species of several genera, particularly *Datura*	Tropane	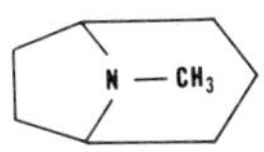

[a] The use of the expression "diagnostic" does not imply that these alkaloids are completely restricted to the taxa indicated, although they often are.

[b] The information presented here was compiled for the most part from references 1–4.

[c] The classes of alkaloids are usually denoted either by the name of the taxon for which the alkaloids are characteristic or by the name of a typical structural type.

[d] This family contains only a few species which are the only source of the unique alkaloids indicated.

REFERENCES

1. General reviews and books on chemotaxonomy: (a) R. E. Alston "Evolutionary Biology," pp. 197–305 (Th. Dobzhansky, M. K. Hecht and Wm. C. Steere, Eds.), Appleton-Century-Crofts, New York, N.Y., 1967; (b) R. E. Alston, T. J. Mabry, and B. L. Turner, *Science* **142,** 545 (1963); (c) B. L. Turner, *Pure & Appl. Chem.* **14,** 189 (1967); (d) R. E. Alston and B. L. Turner, "Biochemical Systematics," Prentice-Hall, Inc., Englewood Cliffs, New Jersey, 1963; (e) T. Swain, Ed., "Comparative Phytochemistry," Academic Press, Ltd., London, England 1966; (f) T. J. Mabry, R. E. Alston, and V. C. Runeckles, Eds., "Recent Advances in Phytochemistry," Appleton-Century-Crofts, New York, N.Y., 1968; (g) T. Swain, Ed., "Chemical Plant Taxonomy," Academic Press, Ltd., London, England, 1963.
2. R. Hegnauer, "Chemotaxonomie der Pflanzen," Vol. 1–5, Birkhauser Verlag, 1962–69.
3. H.-G. Boit, "Ergebnisse der Alkaloid-Chemie," Akademie Verlag, Berlin, Germany, 1961.
4. R. H. F. Manske and, in part **(1–4)**, H. L. Holmes, *The Alkaloids* (1950–1968).
5. See Reference 1(a) for a general review of much of this work.
6. See, for example, T. J. Mabry, H. E. Miller, H. B. Kagan, and W. Renold, *Tetrahedron* **22,** 1139 (1966).
7. M. G. Ettlinger and A. Kjaer in "Recent Advances in Phytochemistry," pp. 61–62 (T. J. Mabry, R. E. Alston, and V. C. Runeckles, Eds.), Appleton-Century-Crofts, New York, N.Y., 1968.
8. (a) A. Kjaer, *Fortschr. Chem. Org. Naturstoffe* **18,** 122 (1960); (b) V. H. Heywood, "Comparative Phytochemistry," p. 18 (T. Swain, Ed.), Academic Press, Ltd., London, England, 1966.
9. M. F. Cranmer and T. J. Mabry *Phytochemistry* **5,** 1133 (1966).
10. M. F. Cranmer, "Biochemical and Biosystematic Investigations of the Lupine Alkaloids of the Legume Genus *Baptisia*," Ph.D. thesis, The University of Texas at Austin, 1965.
11. M. F. Cranmer and B. L. Turner, *Evolution* **21,** 508 (1967).
12. E. Nowacki, *Genet. Polonica* **4,** 161 (1963).
13. (a) K. Mothes, *The Alkaloids* **6** (1960); (b) H. H. Smith, *Amer. Natur.* **99,** 73 (1965).
14. See, for example, E. Leete, "Biogenesis of Natural Compounds," 2nd ed. (P. Bernfeld, Ed.), Macmillan, New York, N.Y. 1967, pp. 953–1023.
15. T. J. Mabry, A. Taylor, and B. L. Turner, *Phytochemistry* **2,** 61 (1968).
16. T. J. Mabry, H. Wyler, G. Sassu, M. Mercier, I. Parikh, and A. S. Dreiding, *Helv. Chim. Acta* **45,** 640 (1962).
17. See for example, A. Cronquist, "The Evolution and Classification of Flowering Plants," Houghton Mifflin, Boston, Mass., 1968, pp. 172–184.
18. H. E. Miller, H. Rösler, A. Wohlpart, H. Wyler, M. E. Wilcox, H. Frohofer, T. J. Mabry, and A. S. Dreiding, *Helv. Chim Acta* **51,** 1470 (1968).
19. F. R. Stermitz, "Recent Advances in Phytochemistry," (T. J. Mabry, R. E. Alston and V. C. Runeckles, Eds.), Appleton-Century-Crofts, New York, N.Y. 1968, pp. 161–183.
20. G. B. Ownbey, *Mem. Torrey Botan. Club* **21,** 1 (1958).

Author Index

Reaction and Reagent Index

Subject Index